MOLECULAR BASIS OF OXIDATIVE STRESS

Chemistry, Mechanisms, and Disease Pathogenesis

Edited by

FREDERICK A. VILLAMENA
Department of Pharmacology and Davis Heart and Lung Institute
The Ohio State University
Columbus, Ohio

WILEY

Copyright © 2013 by John Wiley & Sons, Inc. All rights reserved

Published by John Wiley & Sons, Inc., Hoboken, New Jersey
Published simultaneously in Canada

No part of this publication may be reproduced, stored in a retrieval system, or transmitted in any form or by any means, electronic, mechanical, photocopying, recording, scanning, or otherwise, except as permitted under Section 107 or 108 of the 1976 United States Copyright Act, without either the prior written permission of the Publisher, or authorization through payment of the appropriate per-copy fee to the Copyright Clearance Center, Inc., 222 Rosewood Drive, Danvers, MA 01923, (978) 750-8400, fax (978) 750-4470, or on the web at www.copyright.com. Requests to the Publisher for permission should be addressed to the Permissions Department, John Wiley & Sons, Inc., 111 River Street, Hoboken, NJ 07030, (201) 748-6011, fax (201) 748-6008, or online at http://www.wiley.com/go/permissions.

Limit of Liability/Disclaimer of Warranty: While the publisher and author have used their best efforts in preparing this book, they make no representations or warranties with respect to the accuracy or completeness of the contents of this book and specifically disclaim any implied warranties of merchantability or fitness for a particular purpose. No warranty may be created or extended by sales representatives or written sales materials. The advice and strategies contained herein may not be suitable for your situation. You should consult with a professional where appropriate. Neither the publisher nor author shall be liable for any loss of profit or any other commercial damages, including but not limited to special, incidental, consequential, or other damages.

For general information on our other products and services or for technical support, please contact our Customer Care Department within the United States at (800) 762-2974, outside the United States at (317) 572-3993 or fax (317) 572-4002.

Wiley also publishes its books in a variety of electronic formats. Some content that appears in print may not be available in electronic formats. For more information about Wiley products, visit our web site at www.wiley.com.

Library of Congress Cataloging-in-Publication Data

Molecular basis of oxidative stress : chemistry, mechanisms, and disease pathogenesis / edited by Frederick A. Villamena, Department of Pharmacology and Davis Heart and Lung Institute, The Ohio State University, Columbus, Ohio, USA.
  pages cm
  Includes bibliographical references and index.
  ISBN 978-0-470-57218-4 (cloth)
  1. Oxidative stress. 2. Oxidation. I. Villamena, Frederick A.
  QD281.O9M65 2013
  571.9'453–dc23

2013002853

Printed in the United States of America

10 9 8 7 6 5 4 3 2 1

CONTENTS

Preface	xvii
About the Contributors	xix
Contributors	xxv

1 Chemistry of Reactive Species   1
Frederick A. Villamena

    1.1   Redox Chemistry, 1
    1.2   Classification of Reactive Species, 2
         1.2.1   Type of Orbitals, 3
         1.2.2   Stability of Radicals, 3
         1.2.3   ROS, 4
             1.2.3.1   Oxygen Molecule (O_2, Triplet Oxygen, Dioxygen), 4
             1.2.3.2   Superoxide Radical Anion ($O_2^{\bullet -}$), 5
             1.2.3.3   Hydroperoxyl Radical ($HO_2^{\bullet}$), 9
             1.2.3.4   Hydrogen Peroxide (H_2O_2), 10
             1.2.3.5   Hydroxyl Radical ($HO^{\bullet}$), 11
             1.2.3.6   Singlet Oxygen ($^1O_2{}^1\Delta_g$ or $^1O_2^*$), 13
         1.2.4   Reactive Nitrogen Species, 14
             1.2.4.1   Nitric Oxide (NO or $^{\bullet}$NO), 14
             1.2.4.2   Nitrogen Dioxide ($^{\bullet}NO_2$), 16
             1.2.4.3   Peroxynitrite ($ONOO^-$), 17
         1.2.5   Reactive Sulfur and Chlorine Species, 18
             1.2.5.1   Thiyl or Sulfhydryl Radical ($RS^{\bullet}$), 18
             1.2.5.2   Disulfide (RSSR), 19
             1.2.5.3   Hypochlorous Acid (HOCl), 20
    1.3   Reactivity, 22
         1.3.1   Thermodynamic Considerations, 22
         1.3.2   Kinetic Considerations, 24
             1.3.2.1   Unimolecular or First-Order Reactions, 25
             1.3.2.2   Bimolecular or Second-Order Reactions, 25
             1.3.2.3   Transition State Theory, Reaction Coordinates and Activation Energies, 26

1.4 Origins of Reactive Species, 26
    1.4.1 Biological Sources, 26
        1.4.1.1 NADPH Oxidase, 26
        1.4.1.2 Xanthine Oxidoreductase or Oxidase, 27
        1.4.1.3 Mitochondrial Electron Transport Chain (METC), 27
        1.4.1.4 Hemoglobin (Hb), 28
        1.4.1.5 Nitric Oxide Synthases, 28
        1.4.1.6 Cytochrome P450 (CYP), 29
        1.4.1.7 Cyclooxygenase (COX) and Lipoxygenase (LPO), 29
        1.4.1.8 Endoplasmic Reticulum (ER), 29
    1.4.2 Nonbiochemical Sources, 29
        1.4.2.1 Photolysis, 29
        1.4.2.2 Sonochemical, 30
        1.4.2.3 Photochemical, 30
        1.4.2.4 Electrochemical, 30
        1.4.2.5 Chemical, 30
1.5 Methods of Detection, 31
    1.5.1 *In Vitro*, 32
        1.5.1.1 Flourescence and Chemiluminescence, 32
        1.5.1.2 UV-Vis Spectrophotometry and HPLC, 33
        1.5.1.3 Immunochemical, 34
        1.5.1.4 Electron Paramagnetic Resonance (EPR) Spectroscopy, 34
    1.5.2 *In Vivo*, 38
        1.5.2.1 Histochemical, 38
        1.5.2.2 Immunocytochemical Methods, 38
        1.5.2.3 Low Frequency EPR Imaging, 38
        1.5.2.4 *In Vivo* EPR Spin Tapping-Ex Vivo Measurement, 38
References, 38

2 Lipid Peroxidation and Nitration   **49**
Sean S. Davies and Lilu Guo

Overview, 49
2.1 Peroxidation of PUFAs, 49
    2.1.1 Hydroperoxy Fatty Acid Isomers (HpETEs and HpODEs), 50
    2.1.2 Hydroxy Fatty Acids (HETEs and HODEs), 51
    2.1.3 Isoleukotrienes, 51
    2.1.4 Epoxy Alcohols, 52
2.2 Cyclic Endoperoxides and Their Products, 52
    2.2.1 Isoprostanes, 52
        2.2.1.1 Isoprostane Regio- and Stereoisomers, 54
        2.2.1.2 F_2-Isoprostanes, 54
        2.2.1.3 Other Major Isoprostane Products, 54
        2.2.1.4 Minor Isoprostane Products, 56
    2.2.2 Diepoxide Pathway Products, 57
        2.2.2.1 Isofurans and Related Compounds, 57
    2.2.3 Serial Cyclic Endoperoxides, 57
2.3 Fragmented Products of Lipid Peroxidation, 58
    2.3.1 Short-Chain Alkanes, Aldehydes, and Acids, 58
    2.3.2 Oxidatively Fragmented Phospholipids, 58
    2.3.3 PAF Acetylhydrolase, 59
    2.3.4 Hydroxyalkenals, 59

   2.3.5 Malondialdehyde, 61
   2.3.6 Acrolein, 61
 2.4 Epoxy Fatty Acids, 62
 2.5 Lipid Nitrosylation, 62
   2.5.1 Formation of Reactive Nitrogen Species, 63
   2.5.2 Lipid Nitration Reactions, 63
   2.5.3 Detection of Lipid Nitration *In Vivo*, 64
   2.5.4 Bioactivities of Nitrated Lipids, 64
Summary, 65
References, 65

3 Protein Posttranslational Modification 71
James L. Hougland, Joseph Darling, and Susan Flynn

Overview, 71
 3.1 Oxidative Stress-Related PTMs: Oxidation Reactions, 71
   3.1.1 Cysteine, 71
     3.1.1.1 Formation of Sulfur–Oxygen Adducts: Sulfenic, Sulfinic, and Sulfonic Acids, 72
     3.1.1.2 Formation of Sulfur–Nitrogen Adducts: *S*-Nitrosothiols and Sulfonamides, 73
     3.1.1.3 Formation of Sulfur–Sulfur Adducts: Disulfides and S-Glutathionylation, 74
     3.1.1.4 Redoxins: Enzymes Catalyzing Cysteine Reduction, 75
   3.1.2 Methionine, 76
   3.1.3 Oxidation of Aromatic Amino Acids, 78
     3.1.3.1 Tyrosine, 78
     3.1.3.2 Tryptophan, 79
     3.1.3.3 Histidine, 79
     3.1.3.4 Phenylalanine, 79
   3.1.4 Oxidation of Aliphatic Amino Acids, 79
 3.2 Amino Acid Modification by Oxidation-Produced Electrophiles, 80
   3.2.1 Electrophiles Formed by Oxidative Stress, 80
   3.2.2 Carbonylation Reactions with Amino Acids, 80
 3.3 Detection of Oxidative-Stress Related PTMs, 81
   3.3.1 Mass Spectrometry, 81
   3.3.2 Chemoselective Functionalization, 82
   3.3.3 Cysteine Modifications, 82
     3.3.3.1 Sulfenic Acids, 82
     3.3.3.2 Cysteine-Nitrosothiols, 82
     3.3.3.3 Cysteine-Glutathionylation, 82
   3.3.4 Protein Carbonylation, 83
 3.4 Role of PTMs in Cellular Redox Signaling, 84
Summary, 85
References, 85

4 DNA Oxidation 93
Dessalegn B. Nemera, Amy R. Jones, and Edward J. Merino

Overview, 93
 4.1 The Context of Cellular DNA Oxidation, 93
 4.2 Oxidation of Oligonucleotides, 94
 4.3 Examination of Specific Oxidative Lesions, 96
   4.3.1 8-Oxo-7,8-Dihydro-2′-Deoxyguanosine, 96

  4.3.2 Lesions on Ribose Bases Including Apurinic or Apyrimidinic Sites, 99
  4.3.3 Novel Types of Ribose and Guanine Oxidative Lesions and Future Outlook, 101
    4.3.3.1 Tandem Lesions, 101
    4.3.3.2 Hyperoxidized Guanine, 102
    4.3.3.3 Oxidative Cross-Links, 103
Future Outlook of DNA Oxidative Lesions, 103
References, 103

5 Downregulation of Antioxidants and Phase 2 Proteins 113
Hong Zhu, Jianmin Wang, Arben Santo, and Yunbo Li

Overview, 113
5.1 Definitions of Antioxidants and Phase 2 Proteins, 113
  5.1.1 Antioxidants, 113
  5.1.2 Phase 2 Proteins, 113
5.2 Roles in Oxidative Stress, 114
  5.2.1 Superoxide Dismutase, 114
  5.2.2 Catalase, 114
  5.2.3 GSH and GSH-Related Enzymes, 114
    5.2.3.1 GSH, 114
    5.2.3.2 Glutathione Peroxidase, 115
    5.2.3.3 Glutathione Reductase, 115
    5.2.3.4 GST, 115
  5.2.4 NAD(P)H:Quinone Oxidoreductase, 116
  5.2.5 Heme Oxygenase, 116
  5.2.6 Ferritin, 116
  5.2.7 UDP-Glucuronosyltransferase, 116
5.3 Molecular Regulation, 116
  5.3.1 General Consideration, 116
  5.3.2 Nrf2 Signaling, 116
  5.3.3 Other Regulators, 117
5.4 Induction in Chemoprevention, 117
  5.4.1 Chemical Inducers, 117
  5.4.2 Chemoprotection, 117
5.5 Downregulation, 117
  5.5.1 Selective Chemical Inhibitors, 117
    5.5.1.1 *N,N*-Diethyldithiocarbamate, 118
    5.5.1.2 3-Amino-1,2,4-Triazole, 118
    5.5.1.3 BSO, 118
    5.5.1.4 Sulfasalazine, 118
    5.5.1.5 Dicumarol, 118
  5.5.2 Drugs and Environmental Toxic Agents, 118
Conclusions and Perspectives, 119
References, 119

6 Mitochondrial Dysfunction 123
Yeong-Renn Chen

Overview, 123
6.1 Mitochondria and Submitochondrial Particles, 123
6.2 Energy Transduction, 125
6.3 Mitochondrial Stress, 125

6.4 Superoxide Radical Anion Generation as Mediated by ETC and Disease Pathogenesis, 126
    6.4.1 Mediation of $O_2^{\bullet-}$ Generation by Complex I, 126
        6.4.1.1 The Role of FMN Moiety, 126
        6.4.1.2 The Role of Ubiquinone-Binding Domain, 126
        6.4.1.3 The Role of Iron–Sulfur Clusters, 127
        6.4.1.4 The Role of Cysteinyl Redox Domains, 127
        6.4.1.5 Complex I, Free Radicals, and Parkinsonism, 129
    6.4.2 Mediation of $O_2^{\bullet-}$ Generation by Complex II, 129
        6.4.2.1 The Role of FAD Moiety, 129
        6.4.2.2 The Role of Ubiquinone-Binding Site, 129
        6.4.2.3 Mutations of Complex II Are Related with Mitochondrial Diseases, 129
        6.4.2.4 Mitochondrial Complex II in Myocardial Infarction, 130
    6.4.3 Mediation of $O_2^{\bullet-}$ Generation by Complex III, 130
        6.4.3.1 The Q-Cycle Mediated by Complex III, 130
        6.4.3.2 Role of Q Cycle in $O_2^{\bullet-}$ Generation, 131
        6.4.3.3 The Role of Cytochrome b_L in $O_2^{\bullet-}$ Generation, 132
        6.4.3.4 Bidirectionality of Superoxide Release as Mediated by Complex III, 132
    6.4.4 Complex IV, 132
Summary, 133
References, 134

7 NADPH Oxidases: Structure and Function   137
Mark T. Quinn

Overview, 137
7.1 Introduction, 137
7.2 Phagocyte NADPH Oxidase Structure, 137
    7.2.1 Flavocytochrome b, 138
    7.2.2 $p47^{phox}$, 139
    7.2.3 $p67^{phox}$, 140
    7.2.4 $p40^{phox}$, 141
    7.2.5 Rac1/2, 141
    7.2.6 Rap1A, 142
7.3 Phagocyte ROS Production, 142
    7.3.1 Superoxide Anion ($O_2^{\bullet-}$), 142
    7.3.2 Hydrogen Peroxide (H_2O_2), 143
    7.3.3 Hypochlorous Acid (HOCl), 143
    7.3.4 Hydroxyl Radical ($HO^{\bullet}$), 143
    7.3.5 Singlet Oxygen ($^1O_2^*$), 144
    7.3.6 Nitric Oxide ($^{\bullet}NO$) and Peroxynitrite ($OONO^-$), 144
7.4 Phagocyte NADPH Oxidase Function, 145
7.5 Nonphagocyte NADPH Oxidase Structure, 146
    7.5.1 NOX1, 147
    7.5.2 NOX3, 149
    7.5.3 NOX4, 149
    7.5.4 NOX5, 150
    7.5.5 DUOX1 and DUOX2, 150
    7.5.6 NOXO1, 150
    7.5.7 NOXA1, 151

x CONTENTS

    7.6 Nonphagocyte ROS Production, 151
    7.7 Functions of Nonphagocyte NADPH Oxidases, 152
        7.7.1 Cardiovascular System, 152
        7.7.2 Renal System, 154
        7.7.3 Pulmonary System, 155
        7.7.4 Central Nervous System, 156
        7.7.5 Gastrointestinal System, 157
        7.7.6 Hepatic System, 158
        7.7.7 Thyroid Gland, 159
Summary, 159
Acknowledgments, 159
References, 160

8 Cell Signaling and Transcription   **179**
Imran Rehmani, Fange Liu, and Aimin Liu

    Overview, 179
    8.1 Common Mechanisms of Redox Signaling, 179
    8.2 Redox and Oxygen-Sensitive Transcription Factors in Prokaryotes, 181
        8.2.1 Fe–S Cluster Proteins, 181
        8.2.2 Prokaryotic Hydrogen Peroxide Sensors: Proteins Utilizing Reactive Thiols, 182
        8.2.3 PerR: A Unique Metalloprotein Sensor of Hydrogen Peroxide, 182
        8.2.4 Summary, 184
    8.3 Redox Signaling in Metazoans, 185
        8.3.1 Primary Sources of ROS in Eukaryotic Redox Signaling, 185
        8.3.2 The Floodgate Hypothesis, 186
        8.3.3 Redox Regulation of Kinase and Phosphatase Activity, 187
        8.3.4 Communication between ROS and Calcium Signaling, 188
        8.3.5 Redox Modulation of Transcription Factors, 188
        8.3.6 Summary, 189
    8.4 Oxygen Sensing in Metazoans, 190
        8.4.1 HIF, 190
        8.4.2 PHD Enzymes, 190
        8.4.3 FIH, 191
        8.4.4 Factors Influencing Fe(II)/α-KG Dependent Enzymes, 192
        8.4.5 ROS and Oxygen Sensing, 193
        8.4.6 Summary, 193
    8.5 Medical Significance of Redox and Oxygen-Sensing Pathways, 194
        8.5.1 Cancer, 194
        8.5.2 Vascular Pathophysiology, 194
Concluding Remarks, 195
References, 195

9 Oxidative Stress and Redox Signaling in Carcinogenesis   **203**
Rodrigo Franco, Aracely Garcia-Garcia, Thomas B. Kryston,
Alexandros G. Georgakilas, Mihalis I. Panayiotidis, and Aglaia Pappa

    Overview, 203
    9.1 Redox Environment and Cancer, 203
        9.1.1 Pro-Oxidant Environment and Endogenous Sources of RS in Cancer, 203
            9.1.1.1 Reactive Oxygen Species (ROS)-Generating NADPH Oxidases and Cancer, 203

9.1.1.2 Mitochondria Mutations, Oxidative Stress, and Cancer, 205
9.1.1.3 Nitric Oxide Synthases (NOS) and Cancer, 205
9.1.1.4 Other Sources of RS in Cancer, 205
9.1.2 Alterations in Antioxidant Systems and Cancer, 205
9.1.2.1 Glutathione (GSH) and Glutathione-Dependent Enzymes in Cancer, 205
9.1.2.2 Catalase, 206
9.1.2.3 Superoxide Dismutases (SODs), 206
9.1.2.4 Peroxiredoxins (PRDXs), 207
9.1.2.5 Heme Oxygenase (HO), 207
9.1.2.6 TRX/TRX Reductase System, 207
9.2 Oxidative Modifications to Biomolecules and Carcinogenesis, 207
9.2.1 Oxidative Posttranslational Protein Modifications in Cancer, 208
9.2.1.1 Protein Carbonylation, 208
9.2.1.2 Protein Nitration, 208
9.2.1.3 Protein Nitrosylation or Nitrosation, 208
9.2.1.4 Protein Glutathionylation, 208
9.2.1.5 Methionine Sulfoxide, 208
9.2.2 Lipid Peroxidation (LPO) and Cancer, 209
9.2.3 Oxidative DNA Damage and Carcinogenesis, 209
9.2.3.1 Types of Oxidatively Induced DNA Damage, 209
9.2.3.2 Base and Nucleotide Excision Repair in Oxidative DNA Damage Processing, 211
9.3 Measurement of Oxidative DNA Damage in Human Cancer, 213
9.4 Epigenetic Involvement in Oxidative Stress-Induced Carcinogenesis, 213
9.5 Deregulation of Cell Death Pathways by Oxidative Stress in Cancer Progression, 216
9.5.1 Apoptosis, 216
9.5.2 Autophagy, 219
9.5.3 Redox Regulation of Drug Resistance in Cancer Cells, 219
Conclusions and Perspective, 220
Acknowledgments, 221
References, 221

10 Neurodegeneration from Drugs and Aging-Derived Free Radicals   **237**
Annmarie Ramkissoon, Aaron M. Shapiro, Margaret M. Loniewska, and Peter G. Wells

Overview, 237
10.1 ROS Formation, 237
10.1.1 Introduction to ROS, 237
10.1.2 CNS Sources of ROS, 238
10.1.2.1 Mitochondria, 238
10.1.2.2 Nicotinamide Adenine Dinucleotide Phosphate Hydrogen (NADPH) Oxidase (NOX), 239
10.1.2.3 Phospholipase A2 (PLA2), 239
10.1.2.4 Nitric Oxide Synthases (NOSs), 240
10.1.2.5 Monoamine Oxidase (MAO), 240
10.1.2.6 Cytochromes P450 (CYPs), 240
10.1.2.7 Xanthine Oxidoreductase, 240

CONTENTS

- 10.1.2.8 Excitotoxicity, 241
- 10.1.2.9 Immune Response Microglia, 241
- 10.1.3 Prostaglandin H Synthases (PHSs), 241
  - 10.1.3.1 Role of Prostaglandin Synthesis and Their Receptors, 241
  - 10.1.3.2 Genetics of PHS, 243
  - 10.1.3.3 Primary Protein Structures of PHSs, 246
  - 10.1.3.4 PHS Enzymology, 247
  - 10.1.3.5 Inhibition of PHSs, 248
  - 10.1.3.6 Cellular Localization and CNS Expression of PHSs, 249
  - 10.1.3.7 PHS in ROS Generation, Aging, and Neurotoxicity, 250
  - 10.1.3.8 PHS in Neurodegenerative Diseases, 253
- 10.1.4 Amphetamines, 255
  - 10.1.4.1 History and Uses, 255
  - 10.1.4.2 Pharmacokinetics, 256
  - 10.1.4.3 Distribution, 257
  - 10.1.4.4 Metabolism by Cytochromes P450 (CYPs) and Elimination, 257
  - 10.1.4.5 Receptor-Mediated Pharmacological Actions of METH, 259
  - 10.1.4.6 Effects of METH Abuse, 260
  - 10.1.4.7 Evidence from Animal and Human Studies for Neurotoxicity, 261
- 10.2 Protection against ROS, 263
  - 10.2.1 Blood Brain Barrier (BBB), 263
  - 10.2.2 Antioxidative Enzymes and Antioxidants, 263
    - 10.2.2.1 Glucose-6-Phosphate Dehydrogenase (G6PD), 263
    - 10.2.2.2 SOD, 266
    - 10.2.2.3 H_2O_2 Detoxifying Enzymes, 266
    - 10.2.2.4 Heat Shock Proteins, 267
    - 10.2.2.5 NAD(P)H: Quinone Oxidoreductase, 267
    - 10.2.2.6 GSH, 267
    - 10.2.2.7 Dietary Antioxidants in the Brain, 268
  - 10.2.3 DNA Repair, 268
    - 10.2.3.1 Ataxia Telangiectasia Mutated (ATM), 268
    - 10.2.3.2 Oxoguanine Glycosylase 1 (Ogg1), 268
    - 10.2.3.3 Cockayne Syndrome B (CSB), 269
    - 10.2.3.4 Breast Cancer 1 (Brca1), 269
- 10.3 Nrf2 Regulation of Protective Responses, 269
  - 10.3.1 Overview, 269
  - 10.3.2 Mechanism of Action of Nrf2, 269
  - 10.3.3 Genetics of Nrf2, 270
  - 10.3.4 Protein Structure of Nrf2, 271
  - 10.3.5 Regulators of Nrf2, 272
    - 10.3.5.1 Negative Regulation by Kelch-Like ECH-Associated Protein 1 (Keap1), 272
    - 10.3.5.2 Negative Regulation by Proteasomal Degradation, 272
    - 10.3.5.3 Regulation of Transcriptional Complex in Nucleus, 274
  - 10.3.6 ARE, 274

10.3.7 Activators of Nrf2, 276
10.3.8 Nrf2 in Neurotoxicity and CNS Diseases, 277
  10.3.8.1 Nrf2 Expression, 277
  10.3.8.2 Nrf2 in Neurodegenerative Diseases, 277
  10.3.8.3 Nrf2 in Chemically Initiated Neurotoxicities, 278
  10.3.8.4 Nrf2 in Fetal Neurodevelopmental Deficits, 279
10.3.9 Nrf KO Mouse Models, 280
10.3.10 Evidence for Polymorphisms in the Keap1–Nrf2–ARE Pathway, 280
Summary and Conclusions, 281
Acknowledgments, 281
References, 281

11 Cardiac Ischemia and Reperfusion   311
Murugesan Velayutham and Jay L. Zweier

Overview, 311
11.1 Oxygen in the Heart, 311
  11.1.1 Beneficial and Deleterious Effects of Oxygen in the Heart, 311
  11.1.2 Ischemia and Reperfusion, 311
  11.1.3 Oxidative Stress and Injury, 312
11.2 Sources of ROS during Ischemia and Reperfusion, 312
  11.2.1 Cellular Organelles, 312
    11.2.1.1 Mitochondria, 312
    11.2.1.2 Endoplasmic Reticulum (ER), 312
    11.2.1.3 Peroxisomes, 313
  11.2.2 Cellular Enzymes, 313
    11.2.2.1 Xanthine Oxidoreductase (XOR), 313
    11.2.2.2 Aldehyde Oxidase (AO), 314
    11.2.2.3 NADPH Oxidase (Nox), 314
    11.2.2.4 NADH Oxidase(s), 314
    11.2.2.5 Cyt c, 315
    11.2.2.6 NOSs, 315
    11.2.2.7 Nitrate/Nitrite Reductase(s), 316
11.3 Modulation of Substrates, Metabolites, and Cofactors during I-R, 316
  11.3.1 ROS, 316
  11.3.2 Hypoxanthine and Xanthine, 316
  11.3.3 NADH, 316
  11.3.4 BH_4, 317
  11.3.5 NO, 317
  11.3.6 Peroxynitrite ($ONOO^-$), 318
  11.3.7 Free Amino Acids, 318
11.4 ROS-Mediated Cellular Communication during I-R, 318
11.5 ROS and Cell Death during Ischemia and Reperfusion, 319
  11.5.1 Apoptosis, 319
  11.5.2 Necrosis, 319
  11.5.3 Autophagy, 319
11.6 Potential Therapeutic Strategies, 320
  11.6.1 Inhibitors of XDH/XO (Allopurinol/Febuxostat), 320
  11.6.2 BH_4 Supplementation, 320
  11.6.3 Nitrate/Nitrite Supplementation, 320

11.6.4 Ischemic Preconditioning (IPC), 321
11.6.5 Pharmacological Preconditioning, 321
Summary and Conclusion, 321
References, 321

12 Atherosclerosis: Oxidation Hypothesis 329
Chandrakala Aluganti Narasimhulu, Dmitry Litvinov, Xueting Jiang, Zhaohui Yang, and Sampath Parthasarathy

Overview, 329
12.1 Lipid Peroxidation, 329
12.2 Oxidation Hypothesis of Atherosclerosis, 330
    12.2.1 The Oxidized LDL (Ox-LDL), 330
12.3 Animal Models of Atherosclerosis, 331
    12.3.1 Human Atherosclerosis and Animal Models, 332
    12.3.2 Progression of Human Disease Calcification, 332
    12.3.3 Inflammation and Atherosclerosis, 333
12.4 Aldehyde Generation from Peroxidized Lipids, 333
    12.4.1 The Oxidation of Aldehydes to Carboxylic Acids, 333
    12.4.2 Proatherogenic Effects of Aldehydes, 334
    12.4.3 AZA: A Lipid Peroxidation-Derived Lipophilic Dicarboxylic Acid, 334
    12.4.4 Could Antioxidants Inhibit the Conversion of Aldehydes to Carboxylic Acids?, 334
Summary, 334
Acknowledgments, 335
References, 335

13 Cystic Fibrosis 345
Neal S. Gould and Brian J. Day

Overview, 345
13.1 Lung Disease Characteristics in CF, 345
    13.1.1 Lung Epithelial Lining Fluid (ELF), Host Defense, and CFTR, 346
    13.1.2 Lung Infection and Reactive Oxygen Species (ROS) in CF, 346
    13.1.3 Inflammation in CF, 347
    13.1.4 Airway Antioxidants in CF, 347
13.2 Role of CFTR in the Lung, 348
    13.2.1 Chloride Transport, 348
    13.2.2 GSH Transport, 348
    13.2.3 SCN Transport, 348
13.3 Oxidative Stress in the CFTR-Deficient Lung, 348
    13.3.1 The Importance of ELF Redox Status, 349
    13.3.2 Cellular Oxidative Stress, 349
    13.3.3 NO and CF, 349
    13.3.4 Oxidative Stress Due to Persistent Lung Infection, 349
13.4 Antioxidant Therapies for CF, 351
    13.4.1 Hypertonic Saline Inhalation, 351
    13.4.2 Pharmacologic Intervention, 351
    13.4.3 Oral Antioxidants, 352
    13.4.4 Inhaled Antioxidants, 352
Summary, 353
References, 353

14 Biomarkers of Oxidative Stress in Neurodegenerative Diseases   359
Rukhsana Sultana, Giovanna Cenini, and D. Allan Butterfield

    Overview, 359
    14.1 Introduction, 359
    14.2 Biomarkers of Protein Oxidation/Nitration, 361
           14.2.1 Protein Carbonyls, 361
           14.2.2 Protein Nitration, 362
    14.3 Biomarkers of Lipid Peroxidation, 363
    14.4 Biomarkers of Carbohydrate Oxidation, 366
    14.5 Biomarkers of Nucleic Acid Oxidation, 367
    Acknowledgments, 368
    References, 368

15 Synthetic Antioxidants   377
Grégory Durand

    Overview, 377
    15.1 Endogenous Enzymatic System of Defense, 377
    15.2 Metal-Based Synthetic Antioxidants, 378
           15.2.1 Mn^{III} Complexes (Salens), 379
           15.2.2 Mn^{III} (Porphyrinato) Complexes (Also Called Metalloporhyrins), 380
           15.2.3 Other Metal Complexes, 382
    15.3 Nonmetal-Based Antioxidants, 382
           15.3.1 Ebselen, 382
           15.3.2 Edaravone, 385
           15.3.3 Lazaroids, 388
    15.4 Nitrones, 389
           15.4.1 Protective Effects of Nitrones (with Particular Attention to PBN), 392
                  15.4.1.1 Protection against Endotoxic Shock, 392
                  15.4.1.2 Protection against Diabetes-Induced Damages, 392
                  15.4.1.3 Protection against Xenobiotic-Induced Damages, 392
                  15.4.1.4 Protection against Noise-Induced Hearing Loss, 392
                  15.4.1.5 Protection against Light-Induced Retinal Degeneration, 392
                  15.4.1.6 Protection against Fulminant Hepatitis, 393
                  15.4.1.7 Cardioprotective Effects, 394
           15.4.2 Antiaging Effects of Nitrones, 394
           15.4.3 Neuroprotective Effects of Nitrones, 394
           15.4.4 Clinical Development of the Disulfonyl Nitrone, NXY-059, 395
           15.4.5 The Controversial Mode of Action of Nitrones, 396
                  15.4.5.1 Antioxidant Property of PBN against Lipid Peroxydation, 396
                  15.4.5.2 Anti-Inflammatory and Anti-Apoptotic Properties of Nitrones, 396
                  15.4.5.3 Action on Membrane Enzymes, 397
                  15.4.5.4 Interaction with the Mitochondrial Metabolism, 397
    References, 398

Index   407

PREFACE

That life as we know it is built from but a handful of elements suggests that despite the necessary complexity of biomolecules to store and relay information, it is still highly regulated by one simple molecule—oxygen. More simply, if one theme can be reduced from the vastly circuitous biochemistry of the living cell, it is that of oxygen regulation. At the heart of this highly regulated system is the relatively predictable behavior of the key biological oxido-reductants. Most typical oxido-reductants are the reactive species of oxygen, nitrogen, sulfur, and halogens. Due to their highly reactive nature, these species can be difficult to observe; however, they are increasingly understood to play a key role in the regulation of vital cellular processes such as in proliferation, intracellular transport, cellular motility, membrane integrity, immune responses, and programmed cell death. Formed as by-products of the metabolism of oxygen, reactive species are regulated by powerful antioxidant defense systems within the cell to minimize their damaging effects. However, the imbalance between the pro-oxidant and antioxidant defense mechanisms of the cell or organism in favor of the former can result in oxidative stress. Prolonged oxidative stress conditions lead to the pathogenesis of various diseases such as cancer, neurodegeneration, cardiovascular, and pulmonary diseases to name a few.

In a most abstract sense, life itself is a cascade of events originating from the very fundamental nature of the electron, to the reactivity of molecules on which electrons reside, to the chemical modifications that these reactions cause to biomolecular systems that can lead to a variety of intracellular signaling pathways. Such communication signals the survival or death of the cell, and ultimately that of the whole organism. Thus, it follows that the most fundamental causes of disease are reactive species.

The goal of this book is to provide comprehensive coverage of the fundamental basis of reactivity of reactive species (Chapter 1) as well as new mechanistic insights on the initiation of oxidative damage to biomolecules (Chapters 2–4) and how these oxidative events can impact cellular metabolism (Chapters 5–8) translating into the pathogenesis of some disease states (Chapters 9–13). This field of study could hopefully provide opportunities to improve disease diagnosis and the design of new therapeutic agents (Chapters 14–15).

Frederick A. Villamena

Columbus, Ohio

ABOUT THE CONTRIBUTORS

D. Allan Butterfield was born in Maine. He obtained his PhD in Physical Chemistry from Duke University, followed by an NIH Postdoctoral Fellowship in Neurosciences at the Duke University School of Medicine. He then joined the Department of Chemistry at the University of Kentucky in 1975, rising to Full Professor in eight years. He is now the UK Alumni Association Endowed Professor of Biological Chemistry, Director of the Center of Membrane Sciences, Director of the Free Radical Biology in Cancer Core of the UK Markey Cancer Center, and Faculty of the Sanders-Brown Center on Aging at the University of Kentucky. He has published more than 550 refereed papers on his principal NIH-supported research areas of oxidative stress and redox proteomics in all phases of Alzheimer disease and in mechanisms of chemotherapy-induced cognitive dysfunction (referred to by patients as "chemobrain"). His chapter contribution was coauthored by Rukhsana Sultana and Giovanna Cenini.

Giovanna Cenini received her PhD in Pharmacology from the University of Brescia in Italy. After spending two years in the Butterfield laboratory as a predoctoral fellow and two years as a postdoctoral scholar, Dr. Cenini is now a postdoctoral scholar in Biochemistry at the University of Bonn. She has published approximately 15 papers from her time in the Butterfield laboratory mostly on oxidative stress and p53 in Alzheimer disease and Down syndrome.

Yeong-Renn Chen was born in Taipei, Taiwan, and received his PhD in Biochemistry from Oklahoma State University. Following as NIH-NIEHS IRTA postdoctoral fellow (under the mentorship of Dr. Ronald P. Mason), he joined the Internal Medicine Department of the Ohio State University, where he was promoted to the rank of Associate Professor. He is currently an Associate Professor of Physiology and Biochemistry at the Department of Integrative Medical Sciences of Northeast Ohio Medical University. His research focuses on mitochondrial redox, the mechanism of mitochondria-derived oxygen free radical production, and their role in the disease mechanisms of myocardial ischemia and reperfusion injury.

Joseph Darling received his BS in Chemistry from Lake Superior State University, and his doctoral research focuses on the role and specificity of posttranslational modifications involved in peptide hormone signaling.

Sean S. Davies was born in Honolulu, Hawaii. He obtained his PhD in Experimental Pathology from the University of Utah, followed by a postdoctoral fellowship in Clinical Pharmacology at Vanderbilt University, where he is now an Assistant Professor of Pharmacology. His research centers on the role of lipid mediators in chronic diseases including atherosclerosis and diabetes with an emphasis on mediators derived nonenzymatically by lipid peroxidation. His goal is to develop pharmacological strategies to modulate levels of these mediators and thereby treat disease. His chapter contribution was coauthored with Lilu Guo.

Brian J. Day was born in Montana. He obtained his PhD in Pharmacology and Toxicology from Purdue University, followed by an NIH Postdoctoral Fellowship in Pulmonary and Toxicology at Duke University. He then joined the Department of Medicine at National Jewish Health, Denver, Colorado in 1997 and is currently a Full Professor and Vice Chair of Research. He has published more than 120 refereed papers on his principal

NIH-supported research areas of oxidative stress and lung disease. He is also a founder of Aeolus Pharmaceuticals and inventor on its product pipeline. He currently serves as Chief Scientific Officer for Aeolus Pharmaceuticals that is developing metalloporphyrins as therapeutic agents. His chapter contribution was coauthored by Neal Gould.

Grégory Durand was born in Avignon, France. He obtained his PhD in Organic Chemistry from the Université d'Avignon in 2002. In 2003 he was appointed "Maître de Conférences" at the Université d'Avignon where he obtained his Habilitation Thesis in 2009. In 2007 and 2009 he spent one semester at the Davis Heart & Lung Research Institute (The Ohio State University) as a visiting scholar. He is currently the Director of the Chemistry Department of the Université d'Avignon. His research focuses on the synthesis of novel nitrone compounds as probes and therapeutics. He is also involved in the development of surfactant-like molecules for handling membrane proteins.

Susan Flynn received her BS in Medicinal Chemistry and B.A. in Chemistry and from SUNY-University at Buffalo, and her doctoral research focuses on determining the substrate reactivity requirements for *in vivo* posttranslational modification and activation of associated cellular pathways.

Rodrigo Franco was born in Mexico, City, Mexico, and received his BS in Science and his PhD in Biomedical Sciences from the National Autonomous University of Mexico, Mexico City. His postdoctoral training was done at the National Institute of Environmental Health Sciences-NIH in NC. Then, he joined the Redox Biology Center and the School of Veterinary and Biomedical Sciences at the University of Nebraska-Lincoln, where he is currently an Assistant Professor. His research is focused on the role of oxidative stress and thiol-redox signaling in neuronal cell death.

Aracely Garcia-Garcia coauthored the chapter by Rodrigo Franco. Born in Monterrey, Mexico, she received her PhD in Morphology from Autonomous University of Nuevo Leon. Following as Research Scholar at University of Louisville, KY, she joined the School of Veterinary Medicine and Biomedical Sciences of the University of Nebraska-Lincoln, where she is currently Postdoctoral Fellow Associate. Her research encompasses the understanding of the mechanisms of oxidative stress and autophagy in experimental Parkinson's disease models.

Alexandros G. Georgakilas is an Associate Professor of Biology at East Carolina University (ECU) in Greenville, NC and recently elected Assistant Professor at the Physics Department, National Technical University of Athens (NTUA), Greece. At ECU, he has been responsible for the DNA Damage and Repair laboratory and having trained several graduate (1 PhD and 8 MSc) and undergraduate students. His work has been funded by various sources like East Carolina University, NC Biotechnology Center, European Union and International Cancer Control (UICC), which is the largest cancer fighting organization of its kind, with more than 400 member organizations across 120 countries. He holds several editorial positions in scientific journals. His research work has been published in more than 50 peer-reviewed high-profile journals like *Cancer Research, Journal of Cell Biology*, and *Proceedings of National Academy of Sciences USA* and more 1000 citations. Ultimately, he hopes to translate his work of basic research into clinical applications using DNA damage clusters as cancer or radiation biomarkers for oxidative stress. Prof. Georgakila coauthored his chapter with Thomas Kryston.

Neal S. Gould received his PhD in Toxicology from the University of Colorado at Denver in 2011, and he is currently a Postdoctoral Fellow at the University of Pennsylvania in Dr. Ischiropoulos' research group. He has published seven refereed papers in the area of oxidative stress and lung disease.

Lilu Guo received her PhD in Chemistry from the University of Montana, and she is currently a postdoctoral research fellow in the Davies lab. Her research utilizes mass spectrometry and other biochemical techniques to characterize biologically active phosphatidylethanolamines modified by lipid peroxidation products.

James L. Hougland was born in Rock Island, Illinois. He obtained his PhD in Chemistry from the University of Chicago, followed by an NIH Postdoctoral Fellowship in Chemistry and Biological Chemistry at the University of Michigan, Ann Arbor. He then joined the Department of Chemistry at Syracuse University in 2010 as an assistant professor. His research focuses on protein posttranslational modification, in particular the specificity of enzymes that catalyze protein modification and the impact of those modifications on biological function. His chapter contribution was coauthored by Joseph Darling and Susan Flynn.

Xueting Jiang is currently a doctoral student at the Department of Human Nutrition, Ohio State University, and focusing on dietary oxidized lipids and oxidative stress. She is the recipient of the AHA predoctoral fellowship, and is pursuing her PhD in Dr. Sampath Parthasarathy's research group.

Amy R. Jones was born in Cincinnati, OH. She received a BA degree majoring in Chemistry from the University of Cincinnati. She is currently pursuing an MS degree in Biochemistry at the University of Cincinnati. Her research, under the direction of Dr. Edward J. Merino and Dr. Stephanie M. Rollmann, involves exploring the biochemisty of cytotoxic antioxidants.

Thomas B. Kryston, was born in Saint Petersburg, Florida, and received his MS in Molecular Biology and Biotechnology at East Carolina University. His graduate work focused on Oxidative Clustered DNA Lesions as potential biomarkers for cancer. Following his graduate studies, he was employed by The Mayo Clinic where his research interests were with Hexanucleotide expansions in ALS patients.

Yunbo Li is a professor and chair of the Department of Pharmacology and assistant dean for biomedical research at Campbell University School of Osteopathic Medicine. He is an adjunct professor at the Department of Biomedical Sciences and Pathobiology at Virginia Polytechnic Institute and State University, and an affiliate professor at Virginia Tech-Wake Forest University School of Biomedical Engineering and Sciences. He currently serves as Co-Editor-in-Chief for *Toxicology Letters* and on the editorial boards of *Cardiovascular Toxicology, Experimental Biology and Medicine, Molecular and Cellular Biochemistry, Neurochemical Research*, and *Spinal Cord*. Dr. Li is an active researcher in the areas of free radicals, antioxidants, and drug discovery, and the author of over 100 peer-reviewed publications and two recent monographs: Antioxidants in Biology and Medicine: Essentials, Advances, and Clinical Applications; and Free radical Biomedicine: Principles, Clinical Correlations, and Methodologies. The research in his laboratories has been funded by the United States National Cancer Institute (NCI), National Heart, Lung and Blood Institute (NHLBI), National Institute of Diabetes and Digestive and Kidney Diseases (NIDDK), American Institute for Cancer Research (AICR), and Harvey W. Peters Research Center Foundation. Dr. Li was joined by Hong Zhu, Jianmin Wang, and Aben Santo in his chapter.

Dmitry Litvinov received his PhD in Engelhardt Institute of Molecular Biology, Russia. He is currently working as a postdoctoral fellow at the University of Central Florida in Dr. Sampath Parthasarathy's research group.

Aimin Liu was born in China. He obtained his PhD from Lanzhou Institute of Chemical Physics, Chinese Academy of Sciences and from Stockholm University. He did postdoctoral research at Xiamen University, University of Newcastle upon Tyne, and University of Minnesota. He started his independent research career at University of Mississippi Medical Center in October 2002, rising to Associate Professor in 2008 with tenure. He joined the chemistry faculty of Georgia State University in 2008 and was promoted to tenured Full Professor in 2012. He has published more than 60 refereed papers reporting mechanisms of oxygen activation by metalloproteins and metal-mediated signal transduction. His chapter is coauthored by Imran Rehmani and Fange Liu.

Fange Liu was born in Beijing, China. After obtaining her Bachelors degree with honors, she joined Georgia State University in 2008 to pursue her PhD degree in the area of redox regulation by metalloproteins in cell signaling.

Margaret M. Loniewska is currently a doctoral student in toxicology in the Department of Pharmaceutical Sciences at the University of Toronto, focusing upon the role of glucose-6-phosphate dehydrogenase in neurodegeneration.

Edward J. Merino was born San Diego, CA and received his PhD in Bio-organic Chemistry from the University of North Carolina at Chapel Hill. Following as postdoctoral fellow at the California Institute of Technology, he joined the Chemistry Department of the University of Cincinnati, where he is currently an Assistant Professor. His research encompasses DNA damage, specifically DNA-protein cross-links and evaluation of DNA repair signaling, induced from reactive oxygen species and the design of novel cytotoxic antioxidants. His chapter contribution was coauthored by Dessalegn B. Nemera and Amy R. Jones.

Chandrakala Aluganti Narasimhulu received her PhD in Immunology from Sri Krishnadevarya University, India; and she is currently a postdoctoral fellow at the University of Central Florida in Dr. Sampath Parthasarathy's research group. She has published 13 peer-reviewed publications, 5 of which are in the area of oxidative stress and cardiovascular disease.

Dessalegn B. Nemera, is a predoctoral fellow in the lab of EJM. He immigrated to the United States from Ethiopia eight years ago. Dessalegn completed both an associate degree, from Cincinnati State Community College, and a Bachelor of Science, from the University of Cincinnati, with honors. He is studying the propensity of oxidative DNA-protein cross-links to form.

Mihalis I. Panayiotidis was born in Athens, Greece and received his PhD in Toxicology from the School of

Pharmacy at the University of Colorado, USA. After completion of an NIEHS-IRTA postdoctoral fellowship, he followed with Assistant Professor positions at the Department of Nutrition and the School of Community Health Sciences at the University of North Carolina-Chapel Hill, USA and the University of Nevada-Reno, USA, respectively. Currently, he has joined the Laboratory of Pathological Anatomy, University of Ioannina, Greece where he is an Assistant Professor of Molecular Pathology. His research encompasses the role of oxidative stress and natural products in cancer formation and prevention, respectively.

Aglaia Pappa was born in Ioannina, Greece and received her PhD in Biological Chemistry & Pharmacology from the University of Ioannina, Greece. After completion of a postdoctoral training at the School of Pharmacy, University of Colorado, USA, she has joined the Department of Molecular Biology & Genetics, Democritus University of Thrace, Greece as an Assistant Professor of Molecular Physiology & Pharmacology. Her research encompasses the role of oxidative stress in human disease, including carcinogenesis.

Sampath Parthasarathy obtained his PhD degree from the Indian Institute of Science, Bangalore, India in 1974. He spent one year at the Kyoto University, Japan as a postdoctoral fellow and subsequently joined the Duke University at Durham, NC. He then joined the Hormel Institute, University of Minnesota and became an Assistant Professor. From 1983–1993 Dr. Parthasarathy was a member of the faculty and reached the rank of professor at the University of California at San Diego. He developed the concept of oxidized LDL with his colleagues. In 1993, he was invited to become the Director of Research Division in the Department of Gynecology and Obstetrics at Emory University as the McCord-Cross professor. After serving 10 years at Emory, he joined Louisiana State University Health Science Center at New Orleans in November 2003 as Frank Lowe Professor of Graduate Studies and as Professor of Pathology. During 2006–2011, he served as the Klassen Chair in Cardiothoracic Surgery at the Ohio State University and was instrumental in developing a large animal model of heart failure. Currently, he is the Florida Hospital Chair in Cardiovascular Sciences and serves as Associate Director of Research at the Burnett School of Biomedical Sciences at the University of Central Florida in Orlando. Dr. Parthasarathy has published over 240 articles and has also written a book *Modified Lipoproteins in the Pathogenesis of Atherosclerosis*.

Mark T. Quinn was born in San Jose, CA and received a PhD in Physiology and Pharmacology from the University of California at San Diego. Following postdoctoral training at The Scripps Research Institute, he joined the Department of Chemistry and Biochemistry at Montana State University. Subsequently, he moved to the Department of Microbiology and then to the Department of Immunology of Infectious Diseases, where he is currently a Professor and Department Head. His research is focused on understanding innate immunity, with specific focus on neutrophil NADPH oxidase structure and function and regulation of phagocytic leukocyte activation during inflammation.

Annmarie Ramkissoon obtained her PhD in toxicology in 2011 from the University of Toronto, where she focused upon drug bioactivation and antioxidative responses in neurodegeneration. Dr. Ramkissoon received several honors including a national graduate student scholarship from the Canadian Institutes of Health Research (CIHR) and the Rx&D Health Research Foundation. She is currently a postdoctoral fellow in the Division of Oncology in the Cancer and Blood Diseases Institute at the Cincinnati Children's Hospital Medical Center.

Imran Rehmani was born in St. Louis, Missouri. He obtained his Bachelors degree at the University of Mississippi in 2007. He researched at Georgia Tech and Georgia Health Sciences University before entering Georgia State University in 2010 under the advisement of Aimin Liu. He recently graduated with an MS in Chemistry. He will be joining Centers for Disease Control and Prevention as an ORISE research fellow.

Arben Santo is a professor and chair of the Department of Pathology at VCOM of Virginia Tech Corporate Research Center. His research is centered on pathology of cardiovascular diseases and inflammatory disorders.

Aaron M. Shapiro received his MSc degree in interdisciplinary studies and toxicology from the University of Northern British Columbia in 2008, and is currently a doctoral student in toxicology in the Department of Pharmaceutical Sciences at the University of Toronto, focusing upon the role of oxidative stress and DNA repair in neurodevelopmental deficits. Aaron has won several awards for his research, including a national Frederick Banting and Charles Best Graduate Scholarship from the CIHR.

Rukhsana Sultana received her PhD in Life Sciences from the University of Hyderabad. After spending time as a postdoctoral scholar and research associate in the Butterfield laboratory, Dr. Sultana is now Research Assistant Professor of Biological Chemistry at the University of Kentucky. She has coauthored more than 100 refereed scientific papers, mostly on oxidative stress in Alzheimer disease.

Murugesan Velayutham was born in Tamil Nadu, India, and received his PhD in Physical Chemistry (Magnetic Resonance Spectroscopy) from the Indian Institute of Technology Madras, Chennai, India. He did his postdoctoral training at North Carolina State University and Johns Hopkins University. Currently, he is a research scientist at the Davis Heart Lung Research Institute, The Ohio State University College of Medicine. His research interests have been focused on understanding the roles of free radicals/reactive oxygen species and nitric oxide in biological systems as well as measuring and mapping molecular oxygen levels and redox state in *in vitro* and *in vivo* systems using EPR spectroscopy/oximetry/imaging techniques. He is a cofounding member of the Asia-Pacific EPR/ESR Society and a member of The International EPR Society.

Frederick A. Villamena was born in Manila, Philippines, and received his PhD in Physical Organic Chemistry from Georgetown University. Following as ORISE, CNRS, and NIH-NRSA postdoctoral fellow, he joined the Pharmacology Department of the Ohio State University, where he is currently an Associate Professor. His research encompasses design and synthesis of nitrone-based antioxidants and their application toward understanding the mechanisms of oxidative stress and cardiovascular therapeutics.

Jianmin Wang is the president of Beijing Lab Solutions Pharmaceutical Inc. His research interest focuses on drug discovery and development.

Peter G. Wells obtained his PharmD degree from the University of Minnesota in 1977, received postdoctoral research training in toxicology and clinical pharmacology in the Department of Pharmacology at Vanderbilt University from 1977 to 1980, and joined the University of Toronto Faculty of Pharmacy in 1980, where he is currently a professor in the Division of Biomolecular Sciences in the Faculty of Pharmacy, and cross-appointed to the Department of Pharmacology and Toxicology in the Faculty of Medicine. Dr. Wells' research has focused upon the toxicology of drugs that are bioactivated to a reactive intermediate, more recently in the areas of developmental toxicity, cancer, and neurodegeneration. He has received several honors for the research of his laboratory, most recently a Pfizer Research Career Award from the Association of Faculties of Pharmacy of Canada in 2011.

Zhaohui Yang is currently an associate professor in Wuhan University with a doctoral degree in Medical Science from Wuhan University. He worked as a postdoctoral fellow in Dr. Sampath Parthasarthy's research group from 2010 to 2012.

Hong Zhu is an assistant professor of physiology and pharmacology at VCOM of Virginia Tech Corporate Research Center. Dr. Zhu has authored over 50 peer-reviewed publications in the general areas of biochemistry, physiology, pharmacology, and toxicology. Her research currently funded by NIH is related to the inflammatory and oxidative basis of degenerative disorders and mechanistically based intervention.

Jay L. Zweier was born in Baltimore, Maryland, and received his baccalaureate degrees in Physics and Mathematics from Brandeis University. After PhD training in Biophysics at the Albert Einstein College of Medicine, he pursued medical training at the University of Maryland, School of Medicine and received his MD in 1980. Subsequently, he completed his residency in internal medicine followed by his cardiology fellowship at Johns Hopkins University. In 1987, he joined the faculty of The Johns Hopkins University School of Medicine. In 1998, he was promoted to the rank of Professor and in 2000 was appointed as Chief of Cardiology Research, at the Johns Hopkins Bayview Campus. He was elected as a fellow in the American College of Cardiology in 1995 and the American Society of Clinical Investigation in 1994. In July of 2002, Dr. Zweier joined The Ohio State University College of Medicine as Director of the Davis Heart & Lung Research Institute and the John H. and Mildred C. Lumley Chair in Medicine. Dr. Zweier is currently Professor of Internal Medicine, Physiology, and Biochemistry, Director of the Center for Environmental and Smoking Induced Disease and the Ischemia and Metabolism Program of the Davis Heart & Lung Research Institute. He has published over 400 peer-reviewed manuscripts in the fields of cardiovascular research, free radical biology, and magnetic resonance.

CONTRIBUTORS

D. Allan Butterfield, University of Kentucky, Lexington, Kentucky

Giovanna Cenini, University of Kentucky, Lexington, Kentucky

Yeong-Renn Chen, Northeast Ohio Medical University, Rootstown, Ohio

Joseph Darling, Syracuse University, Syracuse, New York

Sean S. Davies, Vanderbilt University, Nashville, Tennessee

Brian J. Day, National Jewish Health, Denver, Colorado

Grégory Durand, Université d'Avignon et des Pays de Vaucluse, Avignon, France

Susan Flynn, Syracuse University, Syracuse, New York

Rodrigo Franco, University of Nebraska-Lincoln, Lincoln, Nebraska

Aracely Garcia-Garcia, University of Nebraska-Lincoln, Lincoln, Nebraska

Alexandros G. Georgakilas, East Carolina University, Greenville, North Carolina

Neal S. Gould, Children's Hospital of Philadelphia, Philadelphia, Pennsylvania

Lilu Guo, Vanderbilt University, Nashville, Tennessee

James L. Hougland, Syracuse University, Syracuse, New York

Xueting Jiang, University of Central Florida, Orlando, Florida

Amy R. Jones, University of Cincinnati, Cincinnati, Ohio

Thomas B. Kryston, East Carolina University, Greenville, North Carolina

Yunbo Li, Edward Via Virginia College of Osteopathic Medicine, Blacksburg, Virginia

Dmitry Litvinov, University of Central Florida, Orlando, Florida

Aimin Liu, Georgia State University, Atlanta, Georgia

Fange Liu, Georgia State University, Atlanta, Georgia

Margaret M. Loniewska, University of Toronto, Toronto, Ontario, Canada

Edward J. Merino, University of Cincinnati, Cincinnati, Ohio

Chandrakala Aluganti Narasimhulu, University of Central Florida, Orlando, Florida

Dessalegn B. Nemera, University of Cincinnati, Cincinnati, Ohio

Mihalis I. Panayiotidis, University of Ioannina, Ioannina, Greece

Aglaia Pappa, Democritus University of Thrace, Alexandroupolis, Greece

Sampath Parthasarathy, University of Central Florida, Orlando, Florida

Mark T. Quinn, Montana State University, Bozeman, Montana

Annmarie Ramkissoon, Cincinnati Children's Hospital Medical Center, Cincinnati, Ohio

Imran Rehmani, Centers for Disease Control and Prevention, Atlanta, Georgia

Arben Santo, Edward Via Virginia College of Osteopathic Medicine, Blacksburg, Virginia

Aaron M. Shapiro, University of Toronto, Toronto, Ontario, Canada

Rukhsana Sultana, University of Kentucky, Lexington, Kentucky

Murugesan Velayutham, The Ohio State University Columbus, Ohio

Frederick A. Villamena, The Ohio State University, Columbus, Ohio

Jianmin Wang, Beijing Labsolutions Pharmaceuticals, Beijing, China

Peter G. Wells, University of Toronto, Toronto, Ontario, Canada

Zhaohui Yang, Wuhan University, Hubei Province, China

Hong Zhu, Edward Via Virginia College of Osteopathic Medicine, Blacksburg, Virginia

Jay L. Zweier, The Ohio State University, Columbus, Ohio

1

CHEMISTRY OF REACTIVE SPECIES

Frederick A. Villamena

1.1 REDOX CHEMISTRY

Electron is an elementary subatomic particle that carries a negative charge. The ease of electron flow to and from atoms, ions or molecules defines the reactivity of a species. As a consequence, an atom, or in the case of molecules, a particular atom of a reactive species undergoes a change in its oxidation state or oxidation number. During reaction, oxidation and reduction can be broadly defined as decrease or increase in electron density on a particular atom, respectively. A more direct form of oxidation and reduction processes is the loss or gain of electrons on a particular atom, respectively, which is often referred to as electron transfer. Electron transfer can be a one- or two-electron process. One common example of a one-electron reduction process is the transfer of one electron to a molecule of oxygen (O_2) resulting in the formation of a superoxide radical anion ($O_2^{\bullet-}$) (Eq. 1.1). Further one-electron reduction of $O_2^{\bullet-}$ yields the peroxide anion (O_2^{2-}) (Eq. 1.2):

$$O_2 + e_{aq}^- \rightarrow O_2^{\bullet-} \qquad (1.1)$$

$$O_2^{\bullet-} + e_{aq}^- \rightarrow O_2^{2-} \qquad (1.2)$$

Conversely, two-electron oxidation of metallic iron (Fe^0) leads to the formation of Fe^{2+} (Eq. 1.3) and further one-electron oxidation of Fe^{2+} leads to the formation of Fe^{3+} (Eq. 1.4). Electrons in this case can be introduced electrochemically or through reaction with reducing or oxidizing agents:

$$Fe^0 \rightarrow Fe^{2+} + 2e_{aq}^- \qquad (1.3)$$

$$Fe^{2+} \rightarrow Fe^{3+} + e_{aq}^- \qquad (1.4)$$

Another method by which oxidation state on a particular atom can be altered is through change in bond polarity. Electronegative atoms have the capability of attracting electrons (or electron density) toward itself. Listed below are the biologically relevant atoms according to their decreasing electronegativities (revised Pauling): F (3.98) > O (3.44) > Cl (3.16) > N (3.04) > Br (2.96) > S > (2.58) > C = Se (2.55) > H (2.20) > P (2.19). Therefore, changing the electronegativity (or electropositivity) of an atom attached to an atomic center of interest can result in the reversal of the polarization of the bond. By applying the "whose-got-the-electron-rule" will be beneficial in identifying atomic centers that underwent changes in their oxidation states. For example, based on the electronegativity listed above, one can examine the relative oxidation states of a carbon atom in a molecule (Fig. 1.1). Since carbon belongs to group 14 of the periodic table, the carbon atom has 4 valence electrons. When carbon is bonded to an atom that is less electronegative to it (e.g., hydrogen atom), the carbon atom tend to pull the electron density toward itself, making it electron-rich. The two electrons that it shares with each hydrogen atom are counted toward the number of electrons the carbon atom can claim. In the first example, methane has four hydrogen atoms attached to it. Since hydrogen is less electronegative than carbon, all eight shared electrons can be claimed by carbon, but since carbon is only entitled to four electrons by virtue of its valence electron, it has an excess of four electrons, making its oxidation state −4. However, when a carbon atom is covalently bound to a more electronegative atom (e.g., oxygen and chlorine), the spin density distribution around the

---

Molecular Basis of Oxidative Stress: Chemistry, Mechanisms, and Disease Pathogenesis, First Edition. Edited by Frederick A. Villamena.
© 2013 John Wiley & Sons, Inc. Published 2013 by John Wiley & Sons, Inc.

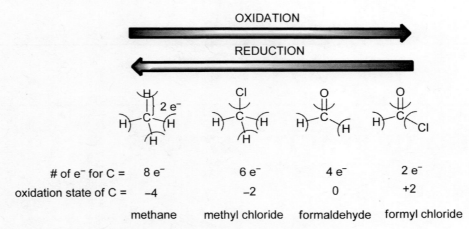

Figure 1.1 Oxidation states of the carbon atom calculated as number of valence electrons for the carbon atom (i.e., 4 e⁻) minus the number of electrons that carbon can claim in a molecule. Order of increasing electronegativity: H < C < O < Cl.

carbon atom decreases and are polarized toward the more electronegative atoms. In this case, the electrons shared by carbon with a more electronegative atom are counted toward the more electronegative atom. In the case of formyl chloride, only the two electrons it shares with hydrogen can be counted toward the total number electrons the carbon atom can claim since the four electrons it shares with oxygen and the two electrons it shares with chlorine cannot be counted toward the carbon because these electrons are polarized toward the more electronegative atoms. Hence, the carbon becomes deficient in electron density, and by virtue of its four valence electrons, it can only claim two electrons from the hydrogen atom, therefore, the net oxidation state can be calculated to be +2. The increasing positivity of the carbon from methane to formyl chloride indicates oxidation of carbon and therefore, oxidation can now be broadly defined as (1) loss of electron; (2) loss of hydrogen atom; and (3) gain of oxygen or halogen atoms, while reduction can be defined as (1) gain of electron; (2) gain of hydrogen atom; and (3) loss of oxygen or halogen atoms.

1.2 CLASSIFICATION OF REACTIVE SPECIES

Definition. Free radicals are integral part of many chemical and biological processes. They play a major role in determining the lifetime of air pollution in our atmosphere[1] and are widely exploited in the design of polymeric, conductive, or magnetic materials.[2] In biological systems, free radicals have been implicated in the development of various diseases.[3] So what are free radicals? The word "radical" came from the Latin word *radix* meaning "root. In the mid-1800s, chemists began to use the word radical to refer to a group of atoms. How the word "radical" had become a chemical terminology is not clear, but one could only speculate that these groups of atoms that make up a molecule was figuratively referred to as "roots" or basic foundation of an entity. In the early 1900s, early literature referred to metallic atoms as basic radicals and nonmetallic ones as acid radicals, for example, in $Mg(OH)_2$ or H_2S, respectively. During this time, radicals are still referred to as group entities that are part of a compound but not until Gomberg had demonstrated during this same time that radicals can indeed exist by themselves as exemplified by his synthesis of the stable triphenylmethyl radical **2** from the reduction of triphenylchloromethane **1** by Zn (Eq. 1.5):[4]

In the late 1950s, the electron paramagnetic resonance spectrum of **2** had been obtained, further confirming the radical nature of trityl which can indeed be stable enough to exist by itself and be spectroscopically detected.[5] Radical is defined in modern times as a finite chemical entity by its own that is capable of undergoing chemical reaction. Radicals carry an odd number of electrons in the form of an atom, neutral or ionic molecule. By virtue of Pauli's exclusion principle, the number of electrons occupying an atomic or molecular orbital is limited to two provided that they have different spin quantum number. This pairing of electron results in the formation of a chemical bond between atoms, existence of lone pair of electron or completion of the inner core nonbonding electrons. For radicals, electrons are typically on an open shell configuration in which the atomic or molecular orbitals are not completely filled with electrons, making them thermodynamically more energetic species than atoms or molecules with closed shell configuration or with filled orbitals. For example, the noble gases He, Ne, or Ar, with filled atomic orbitals, $1s^2$ (He), $1s^2 2s^2 2px^2 2py^2 2pz^2$ (Ne), $1s^2 2s^2 2p^6 3s^2 3px^2 3py^2 3pz^2$ (Ar), are known to be inert, while the atomic H, N, or Cl with electron configurations of $1s^1$ (H), $1s^2 2s^2 2px^2 2py^1 2pz^0$ (N), and $1s^2 2s^2 2p^6 3s^2 3px^2 3py^2 3pz^1$ (Cl) are known to be highly reactive and hence exist as diatomic molecules. Similarly, molecules with open shell molecular orbital configurations are more reactive than molecules with closed shell configuration. For example, hydroxyl radical has an open shell configuration of $\sigma_{pz}^2 p_x^2 p_y^1$ while the hydroxide anion has a closed shell configuration of $\sigma_{pz}^2 p_x^2 p_y^2$, making the former more reactive than the latter.

1.2.1 Type of Orbitals

Radicals can be classified according to the type of orbital (SOMO) that bears the unpaired electron as σ– or π–radicals. Radical stability is governed by the extent of electron delocalization within the atomic orbitals. In general, due to the restricted spin delocalization in the σ–radicals, these radicals are more reactive than the π–radicals. Examples of σ–radicals are H•, formyl-, vinyl-, or phenyl-radicals (Fig. 1.2).

Almost all of the radical-based reactive oxygen species (ROS) that will be discussed in this chapter fall under the π–type category but each will differ only on the extent of spin delocalization within the molecule. Examples of π–radicals with restricted spin delocalization are •CH$_3$, •SH, and HO• and are relatively less stable than π–radicals with extended spin delocalization (e.g., HOO•, O$_2$•−, and NO) (Fig. 1.3).

1.2.2 Stability of Radicals

Radicals can also be categorized according to their stability as stable, persistent, and unstable (or transient). Although the terms stable and persistent are often used interchangeably, free radical chemists agree that persistent radicals refer to the thermodynamic favorability of being monomeric as opposed to being dimeric as formed via radical–radical reaction in solution. Radical-based ROS are not persistent (or stable) making their detection in solution very difficult. ROS detection is commonly accomplished by detecting secondary products arising from their redox or addition reaction with a reagent as will be discussed in Section 1.5. Figure 1.4 shows examples of dimer formation from HO•, HO$_2$•, TEMPO, and trityl, and their respective approximate dissociation enthalpies. Rates of ROS decomposition in solution, of course, depend on the type of substrates that are present in solution but lifetimes of these radicals vary in solution since even one of the most stable radicals such as the trityl radical for example is not stable in the presence of some oxido-reductants.

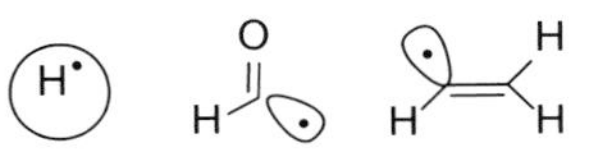

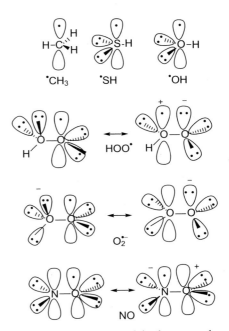

Figure 1.3 Methyl, thiyl, hydroxyl, hydroperoxyl, superoxide, and nitric oxide as examples of π–radicals.

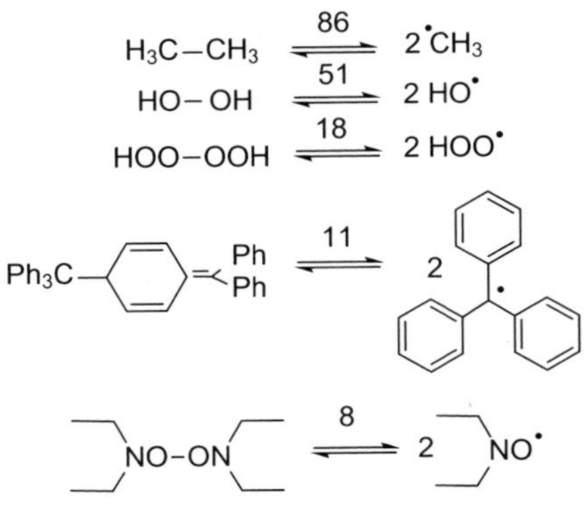

Figure 1.2 Hydrogen, formyl, and vinyl σ-radicals.

Figure 1.4 Dissociation enthalpies (ΔH^0 in kcal/mol) of various dimers showing nitroxide to be the most stable radical and the methyl radical being the least stable.

4 CHEMISTRY OF REACTIVE SPECIES

Classification of reactive species is sometimes cumbersome since, for example, a number of molecules contain more than one atom whose oxidation states are altered during reaction. Nitric oxide (NO), for example, can react with hydroxyl radical (HO•) to form nitrous acid (HNO_2), but in order to classify whether NO is a reactive nitrogen or oxygen species, one has to carefully examine the oxidation states of the relevant atoms of the reactants and the product (Fig. 1.5).

Using the "whose-got-the-electron-rule" mentioned earlier, one can assign the oxidation states for each of the species involved in the transformation. The nitrogen atom of NO underwent an oxidation since its oxidation state has increased from +2 to +3 in HNO_2, while the oxygen of HO• (not of NO) underwent reduction (from −1 to −2). We can therefore classify NO as reactive nitrogen species (RNS) while HO• as ROS since it was the nitrogen atom of NO and the oxygen atom of HO• that underwent oxidation state modification after reaction. Figure 1.6 shows the various reactive oxygen, nitrogen, and sulfur species with their respective oxidation states.

1.2.3 ROS

1.2.3.1 Oxygen Molecule (O_2, Triplet Oxygen, Dioxygen)
The electronic ground state of molecular oxygen is the triplet state, $O_2(X^3\Sigma_g^-)$. Dioxygen's molecular orbital $O_2(X^3\Sigma_g^-)$ has the two unpaired electrons occupying each of the two degenerate antibonding π_g-orbitals and whose spin states are the same or are parallel with each other (Fig. 1.7).

Owing to dioxygen's biradical (open-shell) property, it exhibits a radical-type behavior in many chemical reactions. Elevated physiological concentrations of O_2 (hyperoxia) have been shown to be toxic to cultured epithelial cells due to necrosis, while lethal concentrations of H_2O_2 and $O_2^{•-}$ cause apoptosis, suggesting that the mechanism of O_2 toxicity is distinct from other oxidants. However, in *in vivo* systems, apoptosis is predominantly the main mechanism of cell death in the lung upon breathing 99.9% O_2.[6]

Chlorinated aromatics have been widely used as biocides and as industrial raw materials, and they are ubiquitous as environmental pollutants. The toxicology of polychlorinated biphenyls (PCBs) have been shown to be due to the formation H_2O_2 and $O_2^{•-}$ from one-electron oxidation or reduction by molecular oxygen of reactive hydroquinone and quinone products, respectively, via formation of semiquinone radicals (Eq. 1.6).[7] Oxygenation of pentachlorophenol[8] (PCP) also leads to the formation of superoxide via the same mechanisms (Eq. 1.7):

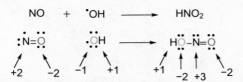

Figure 1.5 Reaction of nitric oxide with hydroxyl radical to produce nitrous acid showing pertinent oxidation states of the atoms undergoing redox transformation.

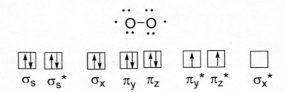

Figure 1.7 Molecular orbital diagram of dioxygen showing its biradical nature.

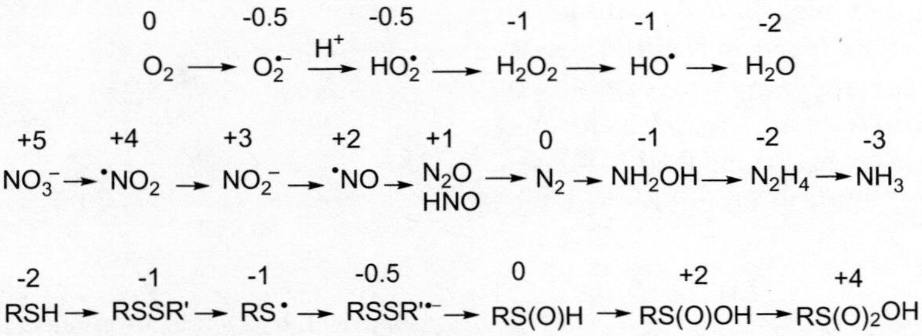

$$\overset{0}{O_2} \longrightarrow \overset{-0.5}{O_2^{•-}} \overset{H^+}{\longrightarrow} \overset{-0.5}{HO_2^{•}} \longrightarrow \overset{-1}{H_2O_2} \longrightarrow \overset{-1}{HO^{•}} \longrightarrow \overset{-2}{H_2O}$$

$$\overset{+5}{NO_3^-} \longrightarrow \overset{+4}{{}^{•}NO_2} \longrightarrow \overset{+3}{NO_2^-} \longrightarrow \overset{+2}{{}^{•}NO} \longrightarrow \overset{+1}{\underset{HNO}{N_2O}} \longrightarrow \overset{0}{N_2} \longrightarrow \overset{-1}{NH_2OH} \longrightarrow \overset{-2}{N_2H_4} \longrightarrow \overset{-3}{NH_3}$$

$$\overset{-2}{RSH} \longrightarrow \overset{-1}{RSSR'} \longrightarrow \overset{-1}{RS^{•}} \longrightarrow \overset{-0.5}{RSSR'^{•-}} \longrightarrow \overset{0}{RS(O)H} \longrightarrow \overset{+2}{RS(O)OH} \longrightarrow \overset{+4}{RS(O)_2OH}$$

Figure 1.6 Reaction of nitric oxide with hydroxyl radical to produce nitrous acid showing pertinent oxidation states of the atoms undergoing redox transformation.

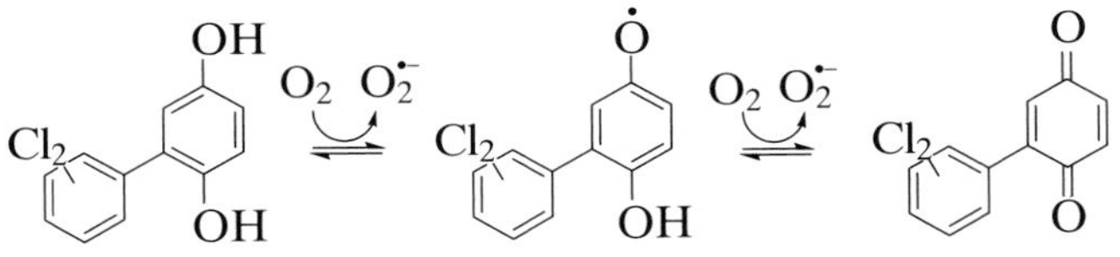

PCBs (1.6)

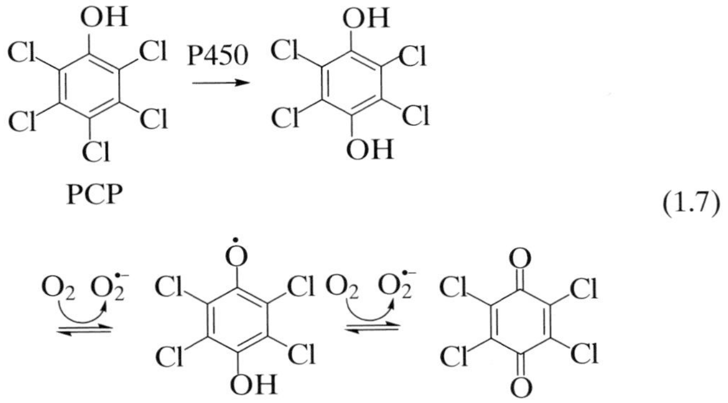

PCP (1.7)

Oxygen addition to 1,4-semiquinone radicals was observed to be more facile than their addition to 1,2-semiquinones with free energies of reaction of 7.4 and 10.3 kcal/mol, respectively (Eq. 1.8 and Eq. 1.9).[9] The experimental rate constants for the reaction of O_2 with 2,5-di-*tert*-butyl-1,4-semiquinone radicals were $2.4 \times 10^5\ M^{-1}\ s^{-1}$ and $2.0 \times 10^6\ M^{-1}\ s^{-1}$ in acetonitrile and chlorobenzene, respectively, similar to that observed in aqueous media at pH 7. The formation of quinones was suggested to occur via a two-step mechanism in which O_2 adds to the aromatic ring followed by an intramolecular H-atom transfer to the peroxyl moiety and concomitant release of $HO_2^{\bullet}$. This reactivity of O_2 to semiquinone to yield $HO_2^{\bullet}$ underlies the pro-oxidant activity of hydroquinones:[10]

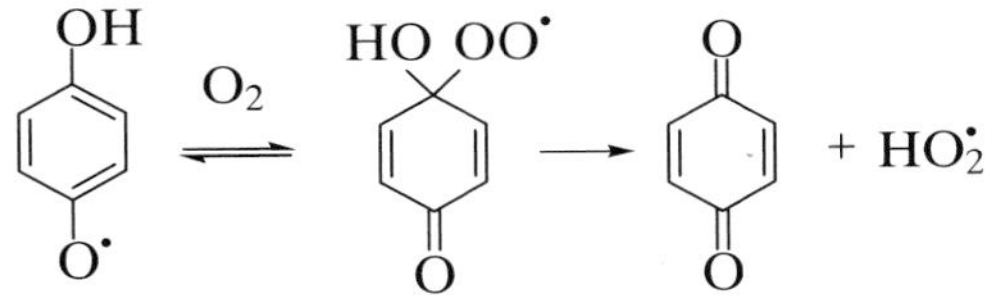

$$\quad (1.8)$$

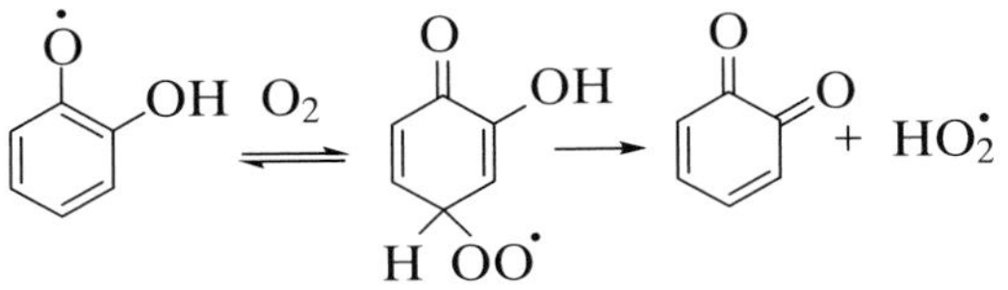

$$\quad (1.9)$$

Perhaps one of the most important reactions of O_2, although reversible in most cases, is its addition to carbon- or sulfur-centered radicals which is relevant in the propagation steps in lipid peroxidation processes or thiol oxidation, respectively. The reaction of dioxygen

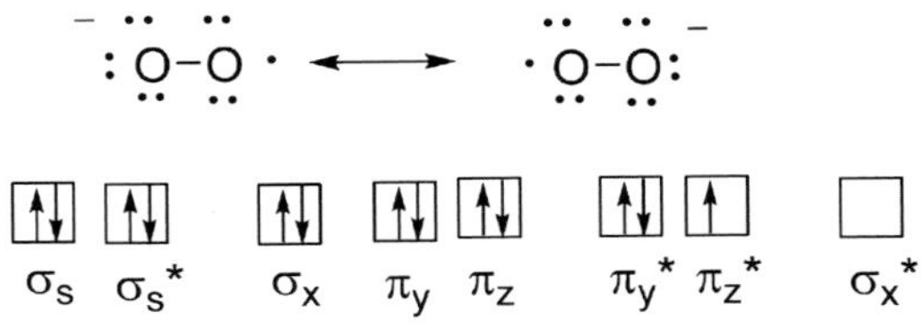

Figure 1.8 Molecular orbital diagram of $O_2^{\bullet-}$.

with lipid and thiyl radicals form peroxyl ($LOO^{\bullet}$) and thiol peroxyl ($RSOO^{\bullet}$) radicals, respectively, (Eq. 1.10 and Eq. 1.11):

$$L^{\bullet} \underset{}{\overset{O_2}{\rightleftharpoons}} LOO^{\bullet} \quad (1.10)$$

$$RS^{\bullet} \underset{}{\overset{O_2}{\rightleftharpoons}} RSOO^{\bullet} \quad (1.11)$$

1.2.3.2 Superoxide Radical Anion ($O_2^{\bullet-}$)
Superoxide is the main precursor of the most highly oxidizing or reducing species in biological system. The one-electron reduction of triplet dioxygen forms $O_2^{\bullet-}$ and initiates oxidative cascade. The molecular orbital of $O_2^{\bullet-}$ shows one unpaired electron in the antibonding π_g-orbital (Fig. 1.8) and is delocalized between the π^* orbitals of the two oxygen atoms.

Dismutation Reaction By virtue of superoxide's oxidation state, $O_2^{\bullet-}$ can either undergo oxidation or reduction to form dioxygen or hydrogen peroxide, respectively (Eq. 1.12),

$$\begin{aligned} O_2^{\bullet-} &\rightarrow O_2 + e^- \quad \text{(oxidation)} \\ O_2^{\bullet-} + e^- + 2H^+ &\rightarrow H_2O_2 \quad \text{(reduction)} \end{aligned} \quad (1.12)$$

thereby allowing $O_2^{\bullet-}$ to dismutate to H_2O_2 and O_2 according to Equation 1.13:

$$2O_2^{\bullet-} + H^+ \rightleftharpoons H_2O_2 + O_2 \quad K_{pH7} = 4 \times 10^{20} \quad (1.13)$$

The dismutation of two $O_2^{\bullet-}$ in the absence of proton is slow with $k < 0.3\ M^{-1}\ s^{-1}$ due to repulsive effects between the negative charges. However, in acidic medium, the rate $O_2^{\bullet-}$ dismutation significantly increases due to the formation of the neutral $HO_2^{\bullet}$ (Eq. 1.14 and Eq. 1.15) in which electron transfer between the radicals becomes more facile:

$$O_2^{\bullet-} + HO_2^{\bullet} \rightarrow O_2 + HO_2^- \quad k = 1 \times 10^8\ M^{-1}\ s^{-1} \quad (1.14)$$

$$HO_2^{\bullet} + HO_2^{\bullet} \rightarrow O_2 + H_2O_2 \quad k = 1 \times 10^5\ M^{-1}\ s^{-1} \quad (1.15)$$

The pK_a of the conjugate acid of $O_2^{\bullet-}$ was determined to be 4.69, which indicates that $O_2^{\bullet-}$ is a poor base but $O_2^{\bullet-}$ has strong propensity to abstract proton from protic substrates. For example, $O_2^{\bullet-}$ addition to water results in the formation of HO_2^- and HO^-, with an equilibrium constant equivalent to 0.9×10^9.[10] This indicates that $O_2^{\bullet-}$ can undergo proton abstraction from substrates to an extent equivalent to a conjugate base of an acid with a pK_a of 24 (Eq. 1.16):[10]

$$2O_2^{\bullet-} + H_2O \rightleftharpoons HO_2^- + O_2 + HO^- \quad K_{pH7} = 0.9 \times 10^9 \quad (1.16)$$

This ability of $O_2^{\bullet-}$ to act as "strong base" is due to its slow initial self-dismutation to O_2 and peroxide (O_2^{2-}) that can drive the equilibrium further right to form the hydroperoxide, HO_2^-. Since the pK_a of H_2O_2 is ~11.75,[11] the basicity of HO_2^- can approach those of RS^-.

Dismutation has also been reported to be catalyzed by SOD mimetics, fullerene derivatives, nitroxides, and metal complexes. Superoxide dismutation should meet the following criteria: (1) no structural or chemical modification of the mimetic upon reaction with $O_2^{\bullet-}$; (2) regeneration of O_2; (3) production of H_2O_2; and (4) absence of paramagnetic primary by-products. Tris-malonyl-derivatives of fullerene (C_{60}) have been shown to exhibit SOD mimetic properties with rate constants in the order of $10^6\ M^{-1}\ s^{-1}$ compared to dismutation rates imparted by SODs (i.e., ~$10^9\ M^{-1}\ s^{-1}$).[12] In vivo studies using SOD2–/– knockout mice indicate increased life span by 300% and show localization in the mitochondria functioning as MnSOD.[13] Computational studies show that electron density around the malonyl groups is low, thereby making this region more susceptible to nucleophilic attack by $O_2^{\bullet-}$ via electrostatic effects.[13] Osuna et al.[14] suggested a dismutation mechanism by which $O_2^{\bullet-}$ interacts with the fullerene surface and is stabilized by a counter-cation and water molecules. An electron is transferred from $O_2^{\bullet-}$ to the fullerene-producing O_2 and fullerene radical anion. Subsequent electron transfer from fullerene radical anion to another molecule of $O_2^{\bullet-}$ gives the fullerene–O_2^{2-} complex, and protonation of the peroxide by the malonic acid groups gives fullerene–H_2O_2, where H_2O_2 is released along with the regenerated fullerene (Fig. 1.9).

SOD exists in two major forms: as a Cu,ZnSOD that is primarily present in cytosol while MnSOD is located in the mitochondria. There is also an FeSOD that has chemical similarities with MnSOD such as being suscep-

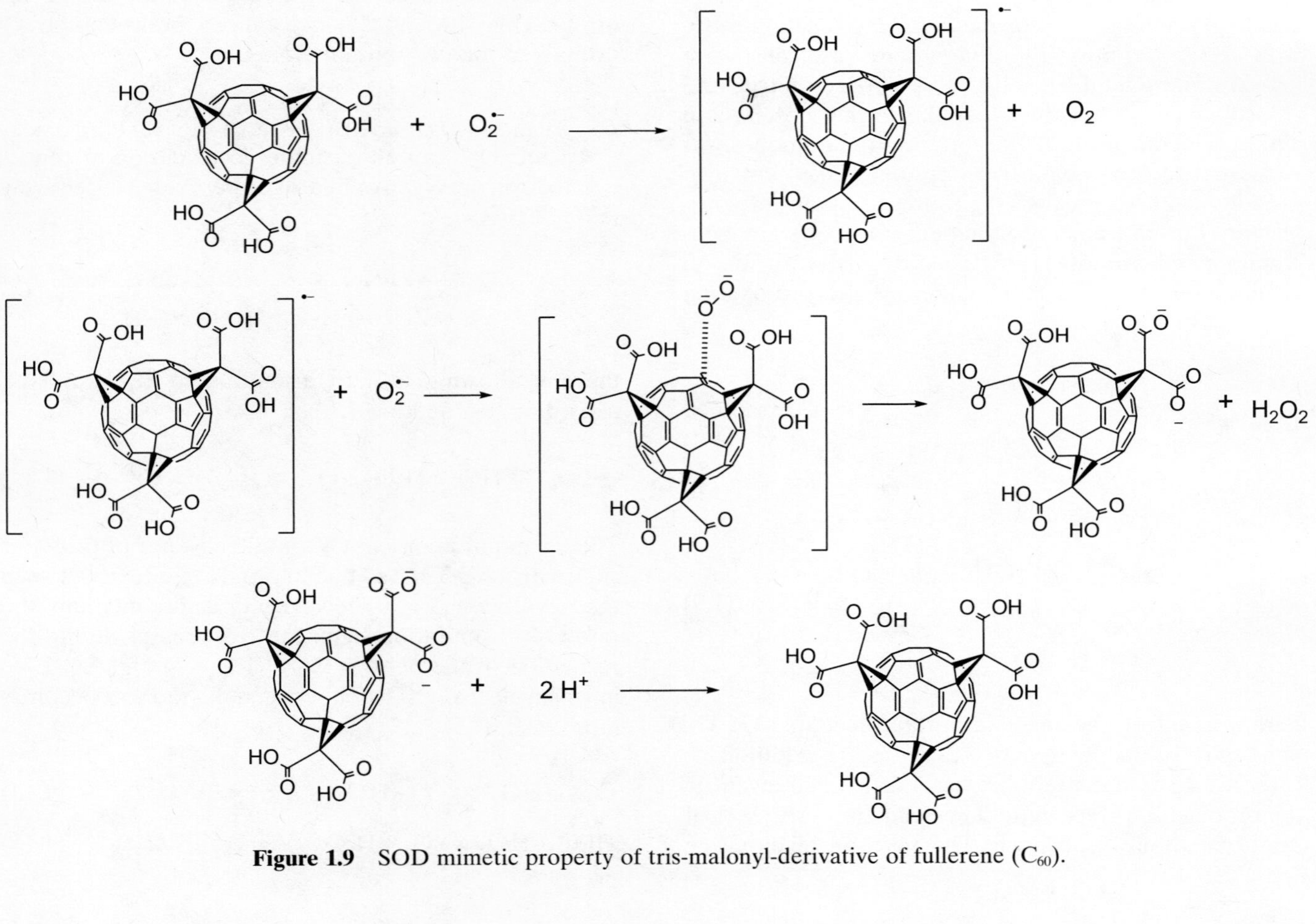

Figure 1.9 SOD mimetic property of tris-malonyl-derivative of fullerene (C_{60}).

$$M_a^n\text{-L} + O_2^{\bullet-} \longrightarrow M_b^{(n-1)}\text{-L} + O_2$$
$$M_b^{(n-1)}\text{-L} + O_2^{\bullet-} + 2H^+ \longrightarrow M_a^n\text{-L} + H_2O_2$$
$$\text{net reaction: } 2O_2^{\bullet-} + 2H^+ \longrightarrow H_2O_2 + O_2$$
where M_a/M_b = Cu(II)/Cu(I); Mn(III)/Mn(II); Fe(III)/Fe(II); Ni(II)/Ni(I); Mn(II)/Mn(I)

Figure 1.10 SOD mimetic property of metal-complexes.

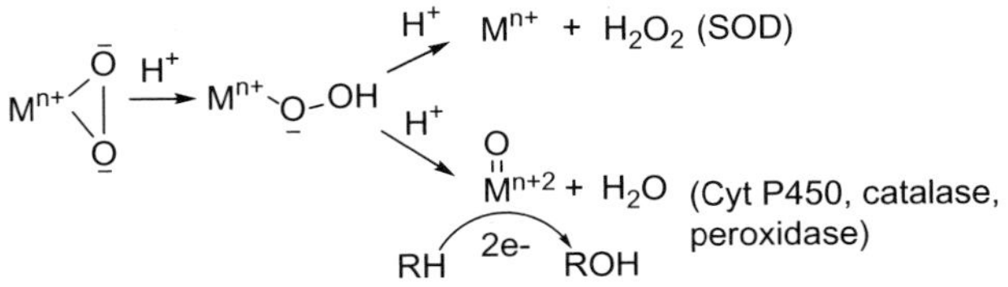

Figure 1.11 Activation of $O_2^{\bullet-}$ by metal ions.

tible to deactivation at high pH and resistance to CN^- inactivation. Over the past years, the synthesis of metal-complexes-based SOD mimetics involved the use of Ni(II),[15] Cu(II),[8] Mn(III),[16] Mn(II),[17] Fe(II), and Fe(III).[18] The overall dismutation reaction of metal-SOD/SOD mimetic involves the following redox reaction (Fig. 1.10):

Activation of $O_2^{\bullet-}$ by metal ions via the formation of metal-peroxo adduct ($M^{(n+1)}-O_2^{2-}$):

$$Fe(II) + O_2^{\bullet-} \rightarrow Fe(III)\text{-}O_2^{2-} \quad (1.17)$$

Formation of $M^{(n+1)}-O_2^{2-}$ can also be achieved through several pathways such as combination of $M^{(n-1)}$ and O_2, $M^{(n+1)}$ and O_2^{2-}, or $M^{(n)}$, O_2, and e-.[19] Protonation of metal-peroxo adducts can proceed via two different pathways, depending on the metabolizing enzyme involved. For example with SOD, release of H_2O_2 occurs with the metal oxidation state unchanged, while in the case of catalase, peroxidases, and cytochrome P450, O–O bond cleavage occurs with the formation of a high valent metal oxo-species (Fig. 1.11).[19]

Electrostatic effect plays an important role in enhancing SOD mimetic activity by introducing positively charged moieties.[13] For example, studies show that the presence of guanidinium derivative of an imidazolate-bridged dinuclear copper moiety enhances SOD activity by 30% compared to when the guanidinium is lacking.[8] Also, increasing the number of positive charge on the ligand and its proximity around the metal center give higher SOD mimetic activity by several-fold compared to the singly-charged analogue.[20]

Nitroxide or aminoxyl-type compounds have also been shown to impart SOD-mimetic properties with catalytic rates that are in the order of 10^5 M^{-1} s^{-1} at pH 7.[21,22] The mechanism was suggested to be catalyzed by formation of an oxoammonium intermediate which in turn converts $O_2^{\bullet-}$ to molecular O_2 according to the following reactions shown in Equation 1.18:

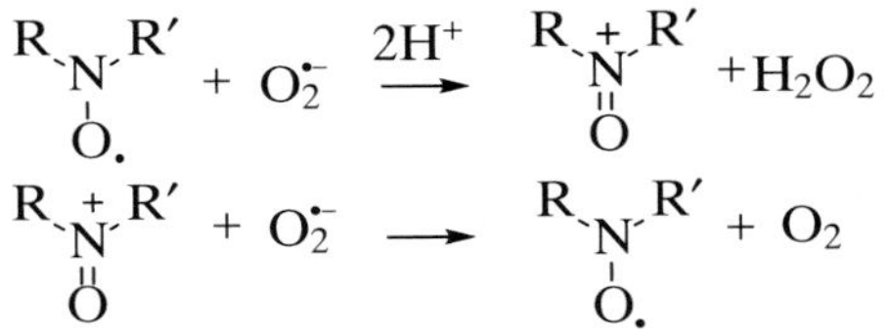

$$(1.18)$$

Nucleophilic Substitution Reaction Nucleophilic substitution reaction has also been observed for $O_2^{\bullet-}$ with alkyl halides and tosylates in DMSO leading to the formation of alkylperoxy radicals then to peroxy anions via one-electron reduction (Eq. 1.19):[23,24]

$$RX + O_2^{\bullet-} \longrightarrow ROO^{\bullet} + X^-$$
$$ROO^{\bullet} \xrightarrow[H^+]{e-} ROOH \quad (1.19)$$

Addition Reactions Reaction of $O_2^{\bullet-}$ with tyrosyl radical generated from sperm whale myoglobin was investigated, and results show that $O_2^{\bullet-}$ prevented myoglobin dimer formation as a mechanism for repairing protein tyrosyl radical.[25] Moroever, an addition product with $O_2^{\bullet-}$ at Tyr151 was identified using mass spectrometry as a more preferred reaction compared to dimer formation, and this addition reaction was enhanced in the presence of exogenously added lysine.[25] This study further supports previous observations on the formation of tyrosyl hydroperoxide generated from $O_2^{\bullet-}$ and tyrosyl radical as enhanced by the presence of H-bond donors.[26,27] Addition of $O_2^{\bullet-}$ and tyrosyl radical at the ortho-position is the most thermodynamically preferred addition product (Eq. 1.20).[27] In aprotic solvents, reaction of $O_2^{\bullet-}$ with α-dicarbonyl carbon involves nucleophilic addition to the carbonyl carbon followed by dioxetane formation via addition of the terminal O to the other carbonyl carbon. Reductive cleavage by the second $O_2^{\bullet-}$ yields benzoate and oxygen:[28]

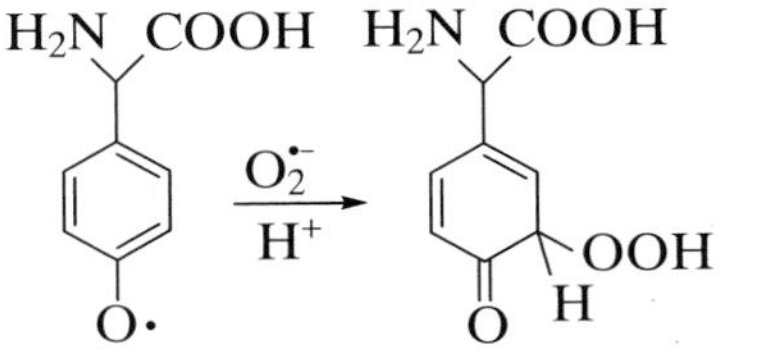

$$(1.20)$$

Proton-Radical Transfer By virtue of the pK_a of the conjugate acid of $O_2^{\bullet-}$ of 4.8, $O_2^{\bullet-}$ is considered a weak base. However, proton and radical transfer pathways have been proposed for the antioxidant property of monophenols and polyphenols, respectively, against $O_2^{\bullet-}$.[29]

For monophenols, electrogenerated $O_2^{•-}$ acts as weak base and the phenolic compound (PhOH) acting as Bronsted acid according to Equation 1.21 in which the formation of phenoxide PhO^- and $HO_2^•$ though thermodynamically unfavorable, can be driven to completion by the subsequent electron transfer reaction between $HO_2^•$ and $O_2^{•-}$ to form HO_2^- (a very strong base) and O_2 in which the former can further abstract proton from phenol to form the phenoxide (PhO^-) according to Equation 1.21:

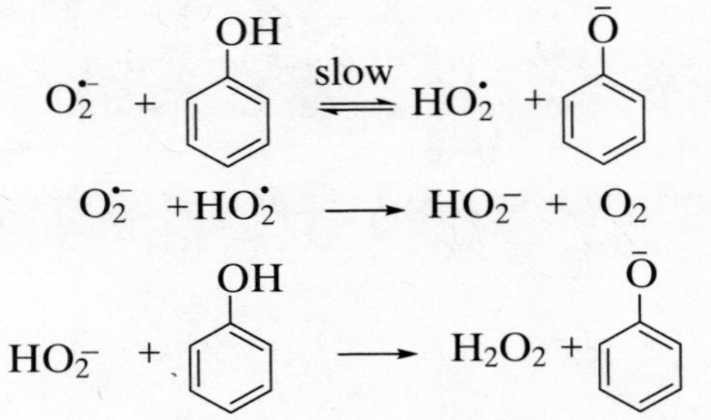

$$O_2^{•-} + PhOH \underset{}{\overset{slow}{\rightleftharpoons}} HO_2^• + PhO^-$$

$$O_2^{•-} + HO_2^• \longrightarrow HO_2^- + O_2 \quad (1.21)$$

$$HO_2^- + PhOH \longrightarrow H_2O_2 + PhO^-$$

Polyphenols, however, undergo radical (or H-atom) transfer reaction with $O_2^{•-}$ to form the phenoxyl radical ($PhO^•$) and HO_2^-; similarly with monophenols, HO_2^- can also abstract proton from PhOH to form phenoxide (PhO^-). The fate of $PhO^•$ was shown to form nonradical products via dimerization or oligomerization, or semiquinone formation. This difference in the pathway between monophenols and polyphenol decomposition with $O_2^{•-}$ can be due to the stabilization of the radical in polyphenols via resonance as evidenced by the higher reactivity of polyphenols containing o-diphenol rings with $O_2^{•-}$ according to Equation 1.22:

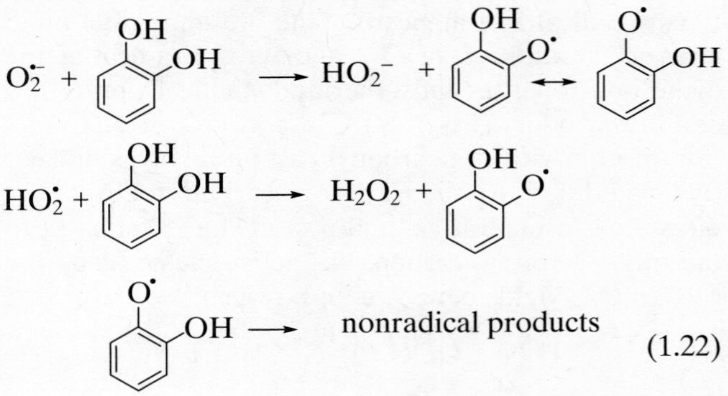

$$O_2^{•-} + \text{catechol} \longrightarrow HO_2^- + \text{semiquinone radical} \leftrightarrow \text{resonance form}$$

$$HO_2^• + \text{catechol} \longrightarrow H_2O_2 + \text{semiquinone radical}$$

$$\text{semiquinone radical} \longrightarrow \text{nonradical products} \quad (1.22)$$

Reactivity of $O_2^{•-}$ was also reported with cardiovascular drugs such as 1,4-dihydropyridine analogues of nifedipine to form pyridine (Eq. 1.23).[30] The proposed mechanism involves a two-electron oxidation of DHP to form the pyridine and hydrogen peroxide:

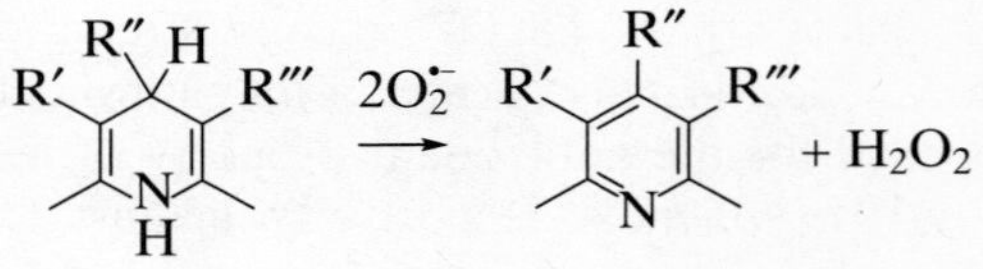

$$\text{DHP} \xrightarrow{2 O_2^{•-}} \text{pyridine} + H_2O_2 \quad (1.23)$$

Pathway 1

$$O_2^{•-} + GSH \longrightarrow GSO^• + OH^-$$
$$GSO^• + GS^- \longrightarrow GS^• + GSO^-$$
$$GSO^- + H^+ + GSH \longrightarrow GSSG + H_2O$$
$$GS^• + GS^- \rightleftharpoons GSSG^{•-}$$
$$GSSG^{•-} + O_2 \longrightarrow GSSG + O_2^{•-}$$

Pathway 2

$$O_2^{•-} + RSH + H^+ \longrightarrow RS^• + H_2O_2$$
$$RS^• + RS^• \longrightarrow RSSR$$

Net Reactions

$$4\,GSH + 2\,O_2 \longrightarrow GSSG + 4\,H_2O$$
$$2\,RSH + O_2 \longrightarrow RSSR + H_2O_2$$

Figure 1.12 Various pathways for the reaction of $O_2^{•-}$ with thiols.

Reaction of $O_2^{•-}$ with thiols were found to be highest for acidic thiols with approximated rate constants in the orders of $10–10^3$ M^{-1} s^{-1}.[31] Oxygen uptake shows concomitant formation of H_2O_2 in some thiols such as peniciallamine and cysteine via a complex radical chain reaction with the formation of oxidized thiols (Fig. 1.12), but this mechanism was not observed for GSH, DTT, cysteamine, and N-acetylcysteine. This difference in mechanisms among thiols for H_2O_2 formation is not clear but was proposed to be due to the nature of the thiol oxidation products formed during the propagation step and of the termination products; thus, stoichiometry could play an important factor in product formation.

Computational studies show that reaction of $O_2^{•-}$ with MeSH to give $MeSO^•$ and HO^- (Pathway 1) as the most favorable mechanism with ΔG_{aq} of -170.5 kcal/mol compared to the formation of $MeS^•$ and HO_2^- (Pathway 2) with endoergic ΔG_{aq} of 68.2 kcal/mol.[32] However, the free energies for the formation of $MeSO^- + HO^•$ and $MeS^- + HO_2^•$ are $\Delta G_{aq} = -52.5$ and 32.2 kcal/mol, respectively. Therefore, the proposed Pathway 2 is unfavorable unless the reacting species is $HO_2^•$ to give $MeS^•$ and H_2O_2 with $\Delta G_{aq} = -11.3$ kcal/mol but formation of $MeSO^•$ and H_2O from HO_2^- and MeSH is far more favorable with $\Delta G_{aq} = -278.7$ kcal/mol. As previously suggested,[32] the reactivity of other oxidants such as H_2O_2 and $HO^•$ to thiols should also be considered and may involve a more complex mechanistic pathway.

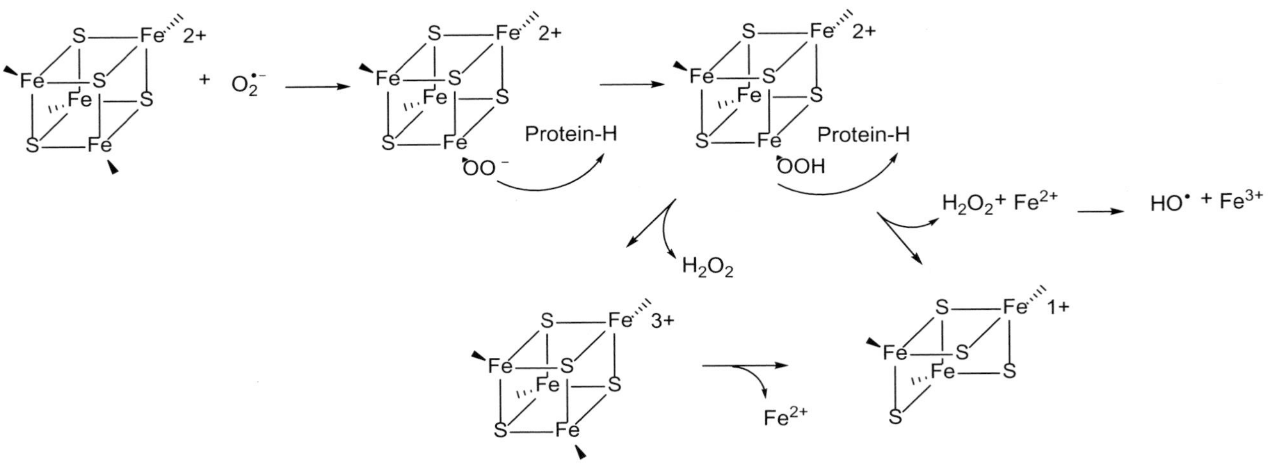

$$[4Fe\text{-}4S]^{2+} + O_2^{\bullet-} + 2H^+ \xrightarrow{-10.1} [3Fe\text{-}4S]^{1+} + H_2O_2 + Fe^{2+}$$

$$[4Fe\text{-}4S]^{2+} + 2/3\,O_2^{\bullet-} + 8/3\,H^+ \xrightarrow{-27.1} [3Fe\text{-}4S]^{1+} + 4/3\,H_2O + Fe^{3+}$$

$$[4Fe\text{-}4S]^{2+} + O_2 \xrightarrow{17.6} [3Fe\text{-}4S]^{1+} + O_2^{\bullet-} + Fe^{2+}$$

$$[4Fe\text{-}4S]^{2+} + 1/2\,O_2 + 2H^+ \xrightarrow{-23.5} [3Fe\text{-}4S]^{1+} + H_2O + Fe^{3+}$$

Figure 1.13 Free energies (in kcal/mol) of the reaction of $O_2^{\bullet-}$ and O_2 with $[4Fe\text{-}4S]^{2+}$ cluster.

Reaction with Iron–Sulfur [Fe–S] Cluster Iron–sulfur clusters are important cofactors in biological system. They serve as active sites in various metalloproteins catalyzing electron-transfer reactions and plays a role in other biological functions such as O_2 sensing ability (e.g., by the transcription factor FNR).[33] The ubiquitousness of [Fe–S] clusters in enzymatic systems such as in Complex II and III of the mitochondrial electron transport chain, ferredoxins, NADH dehydrogenase, nitrogenase, or hydro-lyases underlies their susceptibility for inactivation by ROS specifically by $O_2^{\bullet-}$ through formation of unstable oxidation state of the [Fe–S] cluster and their subsequent degradation (Fig. 1.13). For example, hydro-lyase enzymes such as dihydroxy-acid dehydratase, fumarase A and B and aconitase can be inactivated by $O_2^{\bullet-}$ with a second-order rate constant of 10^6–10^7 $M^{-1}\,s^{-1}$ while the rate of their inactivation by O_2 is orders of magnitude lower ($10^2\,M^{-1}\,s^{-1}$).[34] This difference in the rates of inactivation of $O_2^{\bullet-}$ versus O_2 can be accounted to the favorability of the initial steps in the oxidation of a $[4Fe\text{-}4S]^{2+}$ by $O_2^{\bullet-}$ and O_2 with ΔG of -10.1 kcal/mol and 17.6 kcal/mol, respectively.[34] However, these initial steps only represent formation of Fe^{2+}, H_2O_2, or $O_2^{\bullet-}$ and can further undergo redox reactions to form H_2O as end product. The overall free energies of oxidation of $[4Fe\text{-}4S]^{2+}$ by $O_2^{\bullet-}$ and O_2 leading to the formation of the most stable product (H_2O) and Fe^{3+} are comparable with ΔG of -27.1 kcal/mol and -23.5 kcal/mol, respectively.

1.2.3.3 Hydroperoxyl Radical ($HO_2^\bullet$)

Protonation of $O_2^{\bullet-}$ leads to the formation of $HO_2^\bullet$ whose concentration in biological pH exists a hundred times smaller than that of $O_2^{\bullet-}$; however, the presence of small equilibrium concentration of $HO_2^\bullet$ ($pK_a = 4.8$) can contribute to the $O_2^{\bullet-}$ instability in neutral pH due to dismutation reaction shown in Equation 1.14. In acidosis condition, the reactivity of $HO_2^\bullet$ is expected to be more relevant than $O_2^{\bullet-}$. Electrochemical reduction of O_2 in the presence of strong or weak acids such as $HClO_4$ or phenol, respectively, generates $HO_2^\bullet$.[35] Hydroperoxyl radical is a stronger oxidizer than $O_2^{\bullet-}$ with $E^{o\prime} = 1.06$ and 0.94 V, respectively, and due to its neutral charge, it is capable of penetrating the lipid bilayer and hence, it has been suggested that $HO_2^\bullet$ is capable of H-atom abstraction from PUFAs or from the lipids present in low-density lipoproteins. Cheng and Li[36] argued against the role of $HO_2^\bullet$ in LPO initiation since the concentration of $HO_2^\bullet$ at physiological pH is less than 1% of the generated $O_2^{\bullet-}$ and that SOD have little effect on peroxidation in liposomal or microsomal systems. However, it has been demonstrated that LOOH is more likely the preferred species for $HO_2^\bullet$ attack and not the LPO initiation process. H-atom abstraction from peroxyl-OOH and not from the alkyl C–H backbone is the preferred mechanism of $HO_2^\bullet$ reactivity, and therefore, $HO_2^\bullet$ is more important than $O_2^{\bullet-}$ in initiating LOOH-dependent LPO, but not as the H-abstraction initiator in LPO.[36]

Relevant to the antioxidant activity of catechols or hydroquinones (QH_2), the reactivity of $HO_2^\bullet$ with QH_2 involves H-atom transfer reaction to form semiquinone radical and H_2O_2 with a rate constant of 4.7×10^4 $M^{-1} s^{-1}$ for 1,2-dihydroquinone (Eq. 1.24):[37]

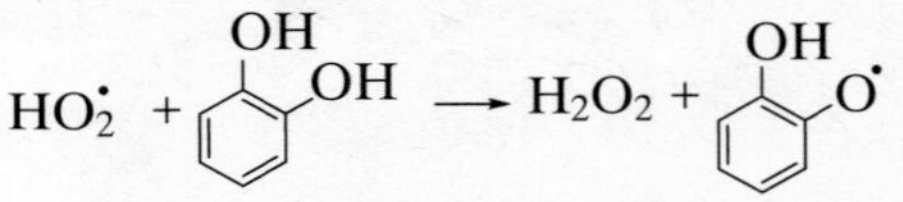

$$HO_2^\bullet + \text{catechol-OH} \longrightarrow H_2O_2 + \text{semiquinone-O}^\bullet \quad (1.24)$$

1.2.3.4 Hydrogen Peroxide (H_2O_2) Hydrogen peroxide is perhaps one of the most ubiquitous ROS present in biological systems due to its relative stability with an oxidation potential of 1.8 V compared to other ROS such as $O_2^{\bullet-}$, $HO_2^\bullet$, or $HO^\bullet$. Hydrogen peroxide is the protonated form of the two-electron reduction product of molecular oxygen and is a nonradical ROS with all the antibonding orbitals occupied by paired electrons (Fig. 1.14). Hydrogen peroxide undergoes highly exoergic disproportionation reaction to form two equivalents of water and one equivalent of oxygen where the rate of disproportionation is temperature dependent.

Perhaps the most common reaction of H_2O_2 is its metal-catalyzed reaction to produce $HO^\bullet$ and $HO_2^\bullet$ (the Fenton chemistry) as proposed by Haber and Weiss (Eq. 1.25, Eq. 1.26, Eq. 1.27, Eq.1.28, Eq.1.29, Eq.1.30, Eq. 1.31, and 1.32).[38] Perez-Benito[39] proposed that this reaction can undergo propagation in which the $HO^\bullet$ can further react with H_2O_2 to produce $HO_2^\bullet$ according to Equation 1.26. Depending on the pH, the equilibrium concentrations of $HO_2^\bullet$ and $O_2^{\bullet-}$ can vary (Eq. 1.27), and it has been suggested[39] that $HO_2^\bullet$ and $O_2^{\bullet-}$ are involved in the reduction and oxidation of Fe^{3+} (Eq. 1.28) and Fe^{2+} (Eq. 1.29), respectively. Iron (III) reaction with H_2O_2 can also lead to $HO^\bullet$ production in acidic pH via formation of $FeOOH^{2+}$ complex and its subsequent decomposition to Fe^{2+} and $HO_2^\bullet$ (Eq. 1.30 and Eq. 1.31) in which the formed Fe^{2+} can propagate the cycle to produce $HO^\bullet$ as shown in Equation 1.25, Equation 1.26, Equation 1.27, Equation 1.28, and Equation 1.29:

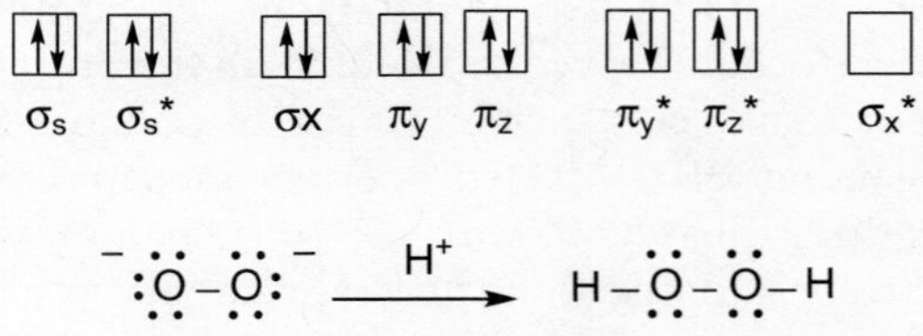

$$Fe^{2+} + H_2O_2 \rightarrow Fe^{3+} + HO^\bullet + HO^- \quad (1.25)$$

$$HO^\bullet + H_2O_2 \rightarrow H_2O + HO_2^\bullet \quad (1.26)$$

$$HO_2^\bullet \rightleftharpoons H^+ + O_2^{\bullet-} \quad (1.27)$$

$$O_2^{\bullet-} + Fe^{3+} \rightarrow O_2 + Fe^{2+} \quad (1.28)$$

$$2HO_2^\bullet \rightarrow H_2O_2 + O_2$$
$$Fe^{2+} + HO_2^\bullet \rightarrow Fe^{3+} + HO^- \quad (1.29)$$

$$Fe^{3+} + H_2O_2 \rightleftharpoons FeOOH^{2+} + H^+ \quad (1.30)$$

$$FeOOH^{2+} \rightarrow Fe^{2+} + HOO^\bullet \quad (1.31)$$

Shown in Figure 1.15 is the metal-independent generation of $HO^\bullet$ from H_2O_2, which was proposed to be formed from tetrachlo-bezoquinones (TCBQ)[8] through nucleophilic substitution reaction forming the hydroperoxyl-TCNQ and O–O homolytic cleavage to yield $HO^\bullet$ and TCBQ-O$^\bullet$. Subsequent disproportionation TCBQ-O$^\bullet$ yields TCBQ-O$^-$, which can further react with excess H_2O_2 to produce $HO^\bullet$.

Hydrogen peroxide oxidation of anions is not favorable. For example, oxidation of Cl^- to HOCl by H_2O_2 is highly endoergic with ~30 kcal/mol. However, myeloperoxidase-mediated oxidation of Cl^- in the presence of H_2O_2 gave rate constants that are dependent on the Cl^- concentration. It was proposed that Cl^- reacts with MPO-I (an active intermediate formed from the reaction of MPO with excess H_2O_2) to form the chlorinating intermediate MPO-I–Cl^-. The rate-limiting step is [Cl^-] dependent; that is, at low [Cl^-], k_2 is the rate-limiting step with $k_2 = 2.2 \times 10^6$ $M^{-1} s^{-1}$ and $k_3 = 5.2 \times 10^4$ s^{-1} (Eq. 1.32):[40]

Figure 1.14 Molecular orbital diagram of H_2O_2.

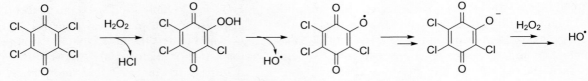

Figure 1.15 Metal-independent generation of $HO^\bullet$ from H_2O_2.

$$MPO + H_2O_2 \underset{k_{-1}}{\overset{k_1}{\rightleftharpoons}} MPO\text{-}I + H_2O$$

$$MPO\text{-}I + Cl^- \underset{k_{-2}}{\overset{k_2}{\rightleftharpoons}} MPO\text{-}I\text{-}Cl^- \qquad (1.32)$$

$$MPO\text{-}I\text{-}Cl^- \xrightarrow{k_3} MPO + HOCl$$

In the absence of ionic substrates, myeloperoxidase has been reported to degrade H_2O_2 to oxygen and water thereby imparting a catalase activity.[41] Kinetic analysis show that there is 1 mol of oxygen produced per 2 mol of H_2O_2 consumed with a rate constant of $\sim 2 \times 10^6$ M^{-1} s^{-1} which is an order of magnitude slower than the rate constant observed for catalase of 3.5×10^7 M^{-1} s^{-1}. Oxidation of nitrite to nitrate by H_2O_2 in the presence of catalase has been reported.[42] In the absence of catalase, nitrite reacts with H_2O_2 to form peroxynitrite.[43] Hydroxylation and nitration of tyrosine and salicylic acid by H_2O_2 in the presence of nitrite occur between the pHs of 2–4 and 5–6, respectively, as mediated by peroxynitrite formation.[44]

Four major detoxification pathways for H_2O_2 operate intracellularly: (1) catalase; (2) gluthathione peroxidase; (3) peroxiredoxin enzymes; and (4) nonenzymatic mean via oxidation of protein thiol residues.[45] These pathways will be discussed in detail in the succeeding chapters. Probably one of the most important reactions in biological systems is the reaction of H_2O_2 with thiols. The cellular signaling property H_2O_2 is mainly dependent on the oxidation of intracellular protein thiols in which majority of these reactions form protein disulfides as opposed to *S*-glutathiolation.[45] The H_2O_2 reaction with thiols is free radical mediated and the rate is dependent on the pK_a of the thiol in which the thiolate (RS^-) is the reacting species to form the sulfenic acid (RSOH) intermediate according to Equation 1.33.[31] The reported rate constant for the reaction of H_2O_2 with thiolates range from 18–26 M^{-1} s^{-1} which is relatively slow compared to the reaction of $O_2^{\bullet-}$ with thiols ($>10^5$ M^{-1} s^{-1}).[31] Catalysis of RSSR formation with Cu(II) from peroxides has also been reported:[46]

$$\begin{aligned} RS^- + H_2O_2 &\rightarrow RSOH + HO^- \\ RSOH + RSH &\rightarrow RSSR + H_2O \end{aligned} \qquad (1.33)$$

1.2.3.5 Hydroxyl Radical (HO•)

Hydroxyl radical originates from the three-electron reduction of oxygen. Among all the ROS, $HO^\bullet$ perhaps is the most reactive and short-lived. Aside from the $HO^\bullet$'s significant role in controlling atmospheric chemistry, it plays a direct role in the initiation of oxidative damage to macromolecules in biological systems. Unlike $O_2^{\bullet-}$ and H_2O_2 whose reactions are limited due to their lower oxidizing ability, $HO^\bullet$ can practically react with almost every organic molecules via H-atom abstraction, electrophilic addition, or radical–radical reactions, to name a few. The standard reduction potential for $HO^\bullet_{aq}/HO^-_{aq}$ couple was determined to be 1.77 V in neutral solution.[47] The half-life of $HO^\bullet$ is $\sim 10^{-9}$ s compared to $\sim 10^{-5}$ s and ~ 60 s for $O_2^{\bullet-}$ and H_2O_2, respectively.

Reactivity with ROS/RNS. Radical–radical reaction of $HO^\bullet$ proceeds at diffusion-controlled rate. For example, at neutral pH, reaction of $HO^\bullet$ with various ROS and non-ROS radicals ranges between $\sim 10^9$ and 10^{10} M^{-1} s^{-1} (Eq. 1.34). The reactions are characteristic of addition of the hydroxyl-O to the heteroatoms. In the case of $HO^\bullet$ reaction to $O_2^{\bullet-}$ and $HO_2^\bullet$, their oxidation via electron transfer reactions to form O_2 was observed (Eq. 1.35):

$$\begin{aligned} HO^\bullet + HO^\bullet &\rightarrow H_2O_2 & k &= 5.2 \times 10^9 \\ HO^\bullet + H^\bullet &\rightarrow H_2O & k &= 7 \times 10^9 \\ HO^\bullet + ClO_2^\bullet &\rightarrow H^+ + ClO_3^- & k &= 4 \times 10^9 \\ HO^\bullet + NO &\rightarrow H^+ + NO_2^- & k &= 1 \times 10^{10} \\ HO^\bullet + NO_2 &\rightarrow HO_2NO & k &= 1 \times 10^{10} \end{aligned} \qquad (1.34)$$

$$\begin{aligned} HO^\bullet + O_2^{\bullet-} &\rightarrow HO^- + O_2 & k &= 7 \times 10^9 \\ HO^\bullet + HO_2^\bullet &\rightarrow H_2O + O_2 & k &= 6.6 \times 10^9 \end{aligned} \qquad (1.35)$$

Theoretical studies show that hydrogen bonding between $HO^\bullet$ and H_2O_2 forms a five-membered ring structure with two distorted hydrogen bonds with a binding energy of ~ 4 kcal/mol.[48] This $HO^\bullet$–H_2O_2 interaction leads to H-atom abstraction to yield $O_2^{\bullet-}$. In pyridine, H_2O_2 reaction with $HO^\bullet$ has a relatively slower rate of 3×10^7 M^{-1} s^{-1} compared to most of $HO^\bullet$ reactions.[49]

Reactivity with ions. Reaction of $HO^\bullet$ to anions leads to a one-electron oxidation of the anion. It has been suggested that simple electron transfer mechanism from the anion to the $HO^\bullet$ is not likely the mechanism due to the large energy associated with the formation of the hydrated hydroxide ion.[50] Instead, an intermediate $HOX^{\bullet-}$ adduct is initially formed (Eq. 1.36). Reaction of $HO^\bullet$ to cations can also result in an increase in the oxidation state of the ion, but unlike its reaction with anions, the reaction occurs at a much slower rate constants that is no more than $\sim 3 \times 10^8$ M^{-1} s^{-1}/s via H-atom abstraction from the metal-coordinated water (Eq. 1.37)[50]:

$$\begin{aligned} HO^\bullet + Cl^- &\rightarrow ClOH^- & k &= 4.3 \times 10^9 \\ HO^\bullet + CO_3^{2-} &\rightarrow HO^- + CO_3^{\bullet-} & k &= 3.7 \times 10^8 \text{ (pH 11)} \end{aligned} \qquad (1.36)$$

$$\begin{aligned} HO^\bullet + Fe^{2+} &\rightarrow FeOH^{2+} & k &= 3.2 \times 10^8 \\ HO^\bullet + Cu^{2+} &\rightarrow CuOH^{2+} & k &= 3.5 \times 10^8 \end{aligned} \qquad (1.37)$$

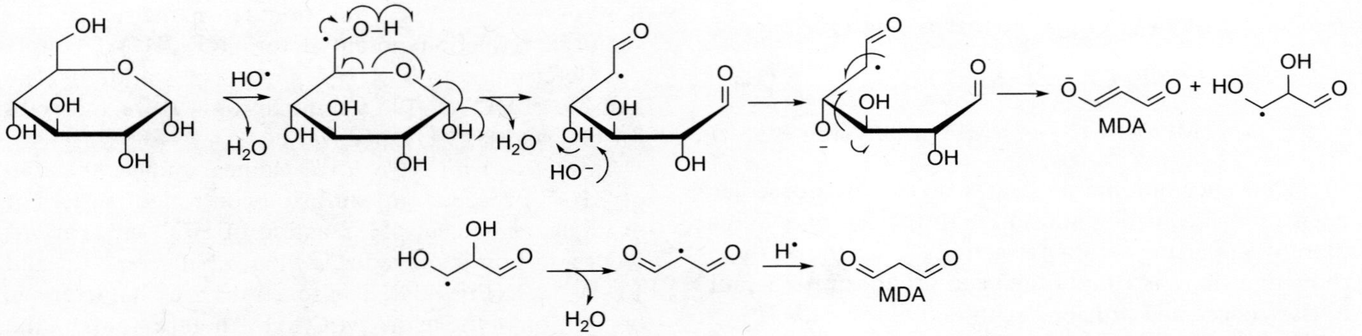

Figure 1.16 Malonaldehdye (MDA) formation from the reaction of hydroxyl radical to deoxyribose.

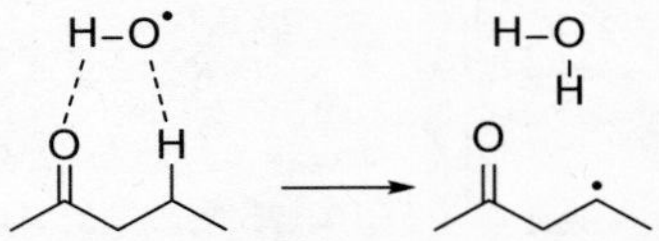

Figure 1.17 Transition state H-bonding interaction of hydroxyl radical to carbonyl leading to H-atom abstraction at the beta position.

Modes of reaction with organic molecules. There are two main mechanisms of HO• reaction with organic compounds, that is, H-atom abstraction and addition reaction. With protic compounds such as alcohols, reaction of HO• proceeds via H-atom abstraction from C–H bond and not from the O–H to form water and the radical species. The general reaction for HO• with alcohol is HO• + RH → R• + H_2O, and not HO• + ROH → RO• + H_2O. For example, ascorbate/ascorbic acid (AH-/AH_2) react with HO• to form ascorbate radical anion ($A^{•-}$) and ascorbyl radical (HA•) with rate constants of 1.1×10^{10} M^{-1} s^{-1} (pH = 7) and 1.2×10^{10} M^{-1} s^{-1} (pH = 1), respectively.[50] EPR studies revealed formation of a C-centered radical.[51] Reaction of HO• with aliphatic alcohols such as methanol and ethanol gave rate constants of 9.0×10^8 M^{-1} s^{-1} and 2.2×10^9 M^{-1} s^{-1}, respectively, using pulse radiolysis.[52] Preference to abstract H atom at the alpha position (i.e., the H attached to the C atom that is also attached to the OH group) was theoretically demonstrated and was found to be both kinetically and thermodynamically favorable. For example, the relative energies of H-atom abstraction as calculated at the CCSD(T) level of theory are as follows: α-H = −25.79 kcal/mol > β-H = −16.26 kcal/mol > OH = −15.67 kcal/mol.[53]

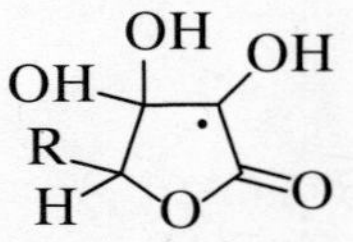

Ascorbyl Radical

Reaction of HO• with deoxyribose forms a C-centered radical which further decomposes to form malonaldehyde (MDA) (Fig. 1.16).[54] MDA is a toxic by-product of polyunsaturated lipid degradation.[55,56] Increase dose of HO• results in increase MDA-like products,[54] therefore, production of MDA in biological systems has become a popular biomarker of oxidative stress using thiobarbuturic acid (TBARS) via MDA electrophilic addition reaction to form an UV detectable adduct, TBARS-MDA. Radiolysis of D-glucose undergoes H-atom abstraction at the C-6 position and rearrangement leads to the initial elimination of two water molecules. Fragmentation yields MDA upon protonation and a dihydroxyaldehyde radical species which can further undergo dehydration to form another molecule of MDA.[57]

Reaction of HO• to ketones and aldehydes also gave preference to H-atom abstraction. Rate constants for H-atom abstraction in aqueous phase were faster 2.4–2.8×10^9 M^{-1} s^{-1} for acetaldehyde and propionaldehyde, compared to acetone with $k = 3.5 \times 10^7$ M^{-1} s^{-1}.[58] Computational studies show that for ketones with at least an ethyl group attached to the carbonyl carbon, the preference for H-atom abstraction is at the beta-position rather than the alpha position due to the presence of strong H-bond interaction forming 7-member ring transition state structure (Fig. 1.17)[59] In aldehydes, abstraction of the aldehydic-H was shown to be the most favored according to the equation, RHC = O + HO• → $[RC = O]^{•} + H_2O$.[60]

Reaction of HO• to carboxylic acids is also that of H-atom abstraction of the acidic-H and alpha-H. There are two possible reactions in acetic acid/acetate system. One that involves H-atom abstraction from C–H and the other from OH according to Equation 1.38 and Equation 1.39, respectively:

$$CH_3COO^- + HO^{•} \rightarrow {}^{•}CH_2COO^- + H_2O$$
$$k = 7.0 \times 10^7 \ M^{-1} \ s^{-1}$$
(1.38)

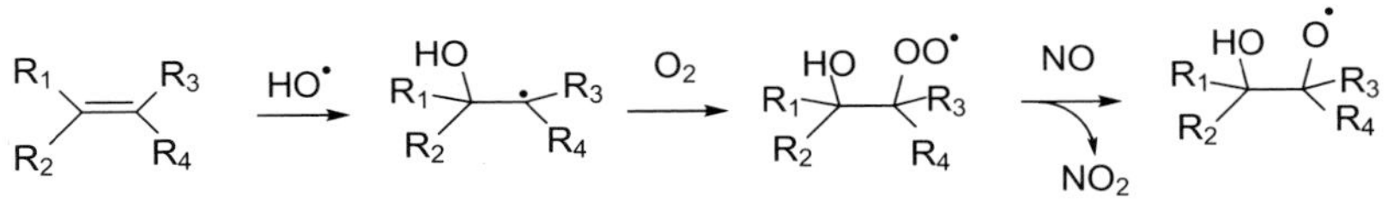

Figure 1.18 Addition reaction of hydroxyl radical to alkenes and subsequent reaction of O_2 and NO with the formed HO-alkene adduct.

$$CH_3COOH + HO^\bullet \rightarrow CH_3COO^\bullet + H_2O$$
$$k = 1.7 \times 10^7 \, M^{-1} \, s^{-1} \quad (1.39)$$

Rate constants for these reactions show that H-atom abstraction from C–H bond is 4× faster than abstraction from O–H in aqueous solution.[61] The same trend in the relative reactivities of $HO^\bullet$ with various acids and their respective conjugate base had been observed.[61]

The reaction of $HO^\bullet$ with alkenes is relevant in the initiation of lipid peroxidation processes and will be discussed in detail in the succeeding chapter. It has been demonstrated that increasing alkyl substitution on the C=C bond enhances its reaction rate with $HO^\bullet$ by two orders of magnitude.[62] In the gas phase, initial reaction of $HO^\bullet$ to alkenes forms the HO-alkene adduct which in the presence of O_2 gives the (β-hydroxylalkyl)peroxy radical. Further reaction with NO yields the β-hydroxyalkoxy radical and NO_2 according to Fig. 1.18.[63]

Reaction of $HO^\bullet$ with aromatic hydrocarbons mainly proceeds via addition reaction. Laser flash photolytic study in acetonitrile gave rate constants ranging from $1.2-7.9 \times 10^8 \, M^{-1} \, s^{-1}$ for one-ringed aromatic hydrocarbons compared to $1.8-5.2 \times 10^9 \, M^{-1} \, s^{-1}$ for naphthalenic systems.[64] Experimental and computational studies indicate that the electrophilic nature of $HO^\bullet$ addition was supported by the higher rate of $HO^\bullet$ addition reaction in aqueous solution compare to acetonitrile by a factor of 65. The stabilized aromatic ring-OH complex in the transition state has the aromatic unit and assumes a radical cation-like form and that the HO* like a hydroxide anion. This can have implication in the $HO^\bullet$ reactivity with DNA bases in which the stabilization of the radical cation form can increase $HO^\bullet$ reactivity to bases.[65] The same addition mechanism was proposed for benzaldehyde and its methoxy-, chloro- and nitro-analogues.[66]

Thiols, such as GSH or thiol-based synthetic antioxidants such as N-acetyl cysteine, are important biological species. H-atom abstraction is the main mechanism of $HO^\bullet$ reaction with thiols ($RSH + HO^\bullet \rightarrow RS^\bullet + H_2O$) with rate constants that range from $8.8 \times 10^9 \, M^{-1} \, s^{-1}$ to $2 \times 10^{10} \, M^{-1} \, s^{-1}$.[50] Computational studies also show that H-atom abstraction of the thiyl-H is the main reaction channel[67] via formation of a short-lived, weakly bonded

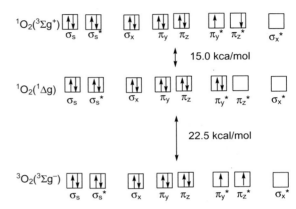

Figure 1.19 Bonding orbitals of singlet oxygens, $^1\Delta_g$ and $^3\Sigma_g^+$, in comparison to the triplet ground state, $^3\Sigma_g^-$.

adduct prior to the abstraction process.[68] Using peroxynitrite, formation of $RS^\bullet$ species as source of $HO^\bullet$ was demonstrated by spin trapping.[69]

1.2.3.6 Singlet Oxygen ($^1O_2{}^1\Delta_g$ or $^1O_2{}^*$) Singlet oxygen is the diamagnetic and less stable form of molecular oxygen. The energy separation between $^1O_2(^1\Delta_g)$ and the triplet ground state oxygen $^3O_2(^3\Sigma_g^-)$ was estimated to be 22.5 kcal/mol (94.3 kJ/mol), corresponding to a near-infrared transition of 1270 nm, while the energy separation between the $^1O_2{}^1\Delta_g$ and the singlet $^1O_2(^3\Sigma_g^+)$ is 15.0 kcal/mol.[70] Electronic configuration of the various spin states of oxygen show only variations in the electronic distribution at the pi-antibonding (π^*) orbitals. As shown in Figure 1.19, unlike the ground state oxygen ($^3\Sigma_g^-$), the electron distribution in $^1\Delta_g$ and $^3\Sigma_g^+$ have antiparallel spins where in the former, the two electrons occupy the same orbital while in the latter, each electron occupies two separate orbitals. Spin-forbidden transition from $^1\Delta_g$ and $^3\Sigma_g^-$ makes $^1O_2{}^*$ a relatively longer-lived species compared to the short-lived $^3\Sigma_g^+$ due to the spin-allowed transition. In solution, lifetimes of $^1O_2{}^*$ is solvent dependent and range from 10^{-3} to 10^{-6} s, with the shortest lifetime observed in water.[71]

Due to the high energy state of $^1O_2{}^*$, its generation in biological system usually involves photo-excitation via direct absorption through vibrationally excited water at 600 nm, or indirectly through photosensitization. Certain organic molecules absorb photons of a particular wavelength causing transition from singlet ground state (S_0) to one of the higher energy 1st or 2nd excited states, that is, S_1 and S_2, respectively. Through vibrational relaxation (VR) or internal conversion (IC) (a nonradiative transition), $S_2 \rightarrow S_1$ ($\tau_{S1} = 10^{-8}$ s) transition occurs which can further undergo conversion to $S_1 \rightarrow S_0$ via IC, or through emission of fluorescence which is a radiative transition between spin states of the same multiplicity. One has to note that these processes do not involve change in multiplicity ($S = 1$) where the lowest energy orbitals still have the two electrons of opposite spins and are usually referred to as "spin allowed" transitions. Transition from S_0 to excited triplet states (T_1), whereby two electrons with the same spins occupy different orbitals is "spin forbidden". However, the energy difference between S_1 and the lower lying T_1 is about ~12 kcal which can facilitate $S_1 \rightarrow T_1$ transition via inter-system crossing (ISC), another nonradiative process, for molecules with large spin-orbit coupling. Higher excited states transition ($S_2 \rightarrow T_2$) can also occur and through VR and IC, $T_2 \rightarrow T_1$ is possible. Photosensitizers typically have longer T_1 half-life than S_1 with $\tau_{S1} = 10^{-4}$–10^{-3} s and has a quantum yield of 0.7–0.9. Conversion of $T_1 \rightarrow S_0$ emits phosphorescence as a spin forbidden radiative transition.

The high quantum yield and longer half-life for T_1 state of photosensitizers have significant ramification in the initiation of a variety of chemical reactions. There are two major types of reaction resulting from T_1 quenching (i.e., Type I and II). Type I processes are typically characterized by H-atom abstraction or electron transfer between the excited sensitizer (A) to a substrate (X) (triplet oxygen for example to yield $O_2^{\bullet-}$) and sensitizer (A)$^{\bullet+}$ according to Equation 1.40:

$$A(T_1) + O_2 \rightarrow A^{\bullet+} + O_2^{\bullet-}$$
$$A(T_1) + X \rightarrow X^{\bullet+} + A^{\bullet-} \qquad (1.40)$$

where $O_2^{\bullet-}$ can further dismutate to H_2O_2 and to form $HO^{\bullet}$. Alternatively, $O_2^{\bullet-}$ can also be produced from $A^{\bullet-}$ as a secondary product depending on the direction of the electron transfer reaction (Eq. 1.41).

$$A^{\bullet-} + {}^3O_2 \rightarrow A + O_2^{\bullet-} \qquad (1.41)$$

Formation of ROS from $O_2^{\bullet-}$ can have implications in the initiation of oxidative damage to key biomolecular systems. Type II processes involve photosensitization of biological or synthetic compounds through energy-transfer mechanism (in contrast to electron-transfer mechanism for Type I) from a sensitizer triplet state molecule T_1 to the ground state triplet O_2, a spin-allowed process (Eq. 1.42).[71]

$$A(T_1) + {}^3O_2 \rightarrow A(S_0) + {}^1O_2^* \qquad (1.42)$$

Oxidative modification via Type I or Type II processes may depend on the O_2 concentration in which the former is more likely to occur at low O_2 concentration.

The generation of singlet oxygen through photosensitization has been widely exploited in photodynamic therapy, environmental remediation and synthesis.[70] In general, the reactivity of ${}^1O_2^*$ was found to be lower than that of $HO^{\bullet}$ but higher than $O_2^{\bullet-}$, and is ca. 1 V more oxidizing than ${}^3O_2$.[70] There are two major quenching mechanisms for singlet ${}^1O_2^*$, that is, through physical means where interaction of ${}^1O_2^*$ with substance A forms ${}^3O_2$; or chemical where ${}^1O_2^*$ reacts with A to form product B or a combination of both. Physical quenching of ${}^1O_2^*$ occurs mainly through its interaction with solvents, or other substrates such as, azide, carotene, or lycopene, but its most common reaction is chemical which accounts for its main mode of action in photodynamic therapy. For example, reaction of ${}^1O_2^*$ with double bonds results in the formation hydroperoxides via "ene"-reactions, or endoperoxides through Diels-Alder-type addition to unsaturated lipids (PUFA or cholesterol), amino acids (e.g., His, Trp, and Met), or nucleic acids (e.g., guanosine).[72] Singlet oxygen has also been shown to be chemically produced from H_2O_2 and hypochlorite, KO_2 reaction with water, and thermal decomposition of aryl peroxides.[71] In biological systems, ${}^1O_2^*$ can be endogenously produced from the decomposition of alpha-linolenic acid hydroperoxide by cytochrome c and lactoperoxidase,[73] metabolism of indole-3-acetic acid by horseradish peroxidase and neutrophils,[74] oxidation of NADPH by liver microsomes,[75] from myeloperoxidase-H_2O_2-chloride system,[76] or from horseradish peroxidase-H_2O_2-GSH system.[77]

1.2.4 Reactive Nitrogen Species

1.2.4.1 Nitric Oxide (NO or $^{\bullet}$NO) Nitric oxide is a paramagnetic molecule with a bond order of 2.5 where the unpaired electron occupies an antibonding orbital (Fig. 1.20). Nitric oxide is nonpolar and with solubility in aqueous solution of 1.94×10^{-6} mol/cm/atm at 298K.[78] The diffusivity (D) at 298 K of NO is similar to that O_2 with D_{NO} in water of 2.21×10^{-5} cm^2/s and 2.13×10^{-5} cm^2/s for O_2.[78]

Nitric oxide functions as an intracellular signaling molecule and is the main precursor of highly oxidizing RNS's in biological system. Nitric oxide's toxicity is generally limited to its reaction or oxidation to form the more highly reactive species such as $ONOO^-$ and $^{\bullet}NO_2$.[43]

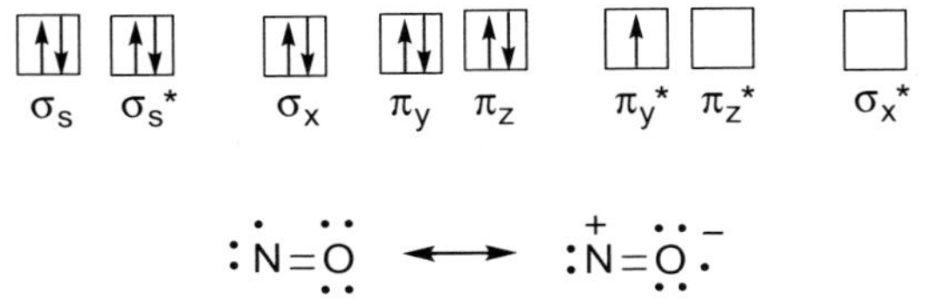

CLASSIFICATION OF REACTIVE SPECIES   15

Figure 1.20 Bonding orbitals of nitric oxide.

NO Radical Reaction Due to NO's radical nature, it exhibits rich chemistry and is capable of reacting with radicals or transition metals to form complexes. NO is relatively stable and unreactive to nonradical species. Theoretical evidence show dimerization of NO to $(NO)_2$ is only slightly favorable with $\Delta H = -2.3$ kcal/mol.[79] The facile reaction of NO with O_2 gave $k_1 = 2.1$–2.9×10^6/M^2/s at 22°C (based on the rate law: $4k_1 [NO]^2[O_2]$)[80,81] in aqueous solution and yields a variety of NO_x products such as NO_2, N_2O_3 and N_2O_4, as well as NO_2^- via a complex mechanism. Kinetic model for NO reaction with O_2 is shown in Equation 1.43.

$$2NO + O_2 \rightarrow 2NO_2 \quad (k = 2.9 \times 10^6 \, M^{-1} s^{-1})$$
$$2NO_2 + 2NO \rightleftharpoons 2N_2O_3 \quad (k_{forward} = 1.1 \times 10^9 \, M^{-1} s^{-1})$$
$$2N_2O_3 + 2H_2O \rightarrow 4NO_2^- + 4H^+ \quad (k = 530 \, s^{-1})$$
$$\text{Net:} \quad 4NO + 2H_2O + O_2 \rightarrow 4NO_2^- + 4H^+ \tag{1.43}$$

The reaction of NO with O_2 results in the formation of NO—O_2 weak complex, a nitrosyldioxyl radical intermediate, and further reaction to NO yields the dimerized NO_2 (ONOONO) which along with NO_2 and N_2O_3 are potent oxidants. Subsequent reaction of ONOONO with two equivalents of NO yields equimolar amounts of N_2O_3. Since N_2O_3 is not formed in the presence of NO_2 scavengers, it was assumed that ONOONO acts as a weak oxidant and its formation from NO/O_2 is the rate limiting step.[81]

Reactions of NO with short-lived radicals such as $(SCN)_2^{\bullet-}$, $CO_2^{\bullet-}$, $CO_3^{\bullet-}$, and hydroxyethyl radicals in aqueous solution have been reported with rate constant approaching diffusion controlled limit.[82] Reaction of NO with $(SCN)_2^{\bullet-}$ forms the NOSCN intermediate that upon hydrolysis yields NO_2^- (Eq. 1.44).

$$NO + (SCN)_2^{\bullet-} \rightarrow NO(SCN)_2^{\bullet-} \rightarrow NOSCN + SCN^-$$
$$(k = 4.3 \times 10^9 \, M^{-1} s^{-1})$$
$$NOSCN + H_2O \rightleftharpoons (HO\text{-}NOSCN)^- + H^+$$
$$\rightarrow HNO_2 + SCN^- \tag{1.44}$$

For $CO_3^{\bullet-}$, O-transfer to NO yields NO_2^- and CO_2 as the most preferred mechanism according to Equation 1.45:

$$NO + CO_3^{\bullet-} \rightarrow NO_2^- + CO_2 \quad (k = 3.5 \times 10^9 \, M^{-1} s^{-1}) \tag{1.45}$$

Reaction of NO with $CO_2^{\bullet-}$ forms the transient $NOCO_2^-$ and its subsequent decomposition yields hyponitrite radical anion and CO_2 (Eq. 1.46):

$$NO + CO_2^{\bullet-} \rightarrow NOCO_2^- \xrightarrow{NO} N_2O_2^{\bullet-} + CO_2 \tag{1.46}$$
$$(k = 2.9 \times 10^9 \, M^{-1} s^{-1})$$

With hydroxyethyl radical (radical derived from ethanol), its reaction with NO gives oximes/hydroxamic acids as the main products (Eq. 1.47).

$$CH_3\overset{\bullet}{C}HOH + NO \longrightarrow CH_3CHOH\text{—}$$
$$\phantom{CH_3\overset{\bullet}{C}HOH + NO \longrightarrow CH_3CH} \overset{|}{NO}$$
$$CH_3\underset{\|}{C}OH + CH_3\underset{|}{C}=O \quad (k = 3.0 \times 10^9 \, M^{-1} s^{-1}) \tag{1.47}$$
$$NOH NHOH$$

Reactions of NO with lipid alkoxyl (LO$^\bullet$) and peroxyl (LOO$^\bullet$) radicals are relevant in the termination of lipid peroxidation processes since NO, being more soluble in nonpolar solvents, can concentrate in lipid bi-layers, and therefore, can play a role in the regulation of lipid peroxidation. Reaction of NO with alkoxyl (RO$^\bullet$) and peroxyl (ROO$^\bullet$) radicals approaches that of diffusion controlled rate. Reaction of MeO$^\bullet$ with NO yields MeONO[83] in aqueous solution while reaction of MeOO$^\bullet$ with NO proceeds at a rate of constant of $3.7 \times 10^9 \, M^{-1} s^{-1}$ to yield ROONO and decomposes to free RO$^\bullet$ and $^\bullet NO_2$ via O–O homolysis (Eq. 1.48 and Eq. 1.49).[84]

$$RO^\bullet + NO \rightarrow RONO \tag{1.48}$$
$$ROO^\bullet + NO \rightarrow ROONO \rightarrow RO^\bullet + {}^\bullet NO_2 \tag{1.49}$$

NO-Metal Reaction Metal complexation of NO is important in the regulation of protein function. Aside from radicals, metals (mostly heme iron) are NO's principal target. A classic example is the activation of the enzyme guanylyl cyclase (sGC) via NO complexation with the ferro-heme. The formation of nitrosyl-Fe(II) complex results in changes in the electronic property of the heme iron such that the histidine ligand that was initially bound became labile and leads to change in the protein conformation. This change allows for the catalytic formation of the secondary messenger, cGMP from GTP which then causes relaxation of the smooth muscle tissue. Other metalloenzymes where NO plays a crucial role in their regulations are, *cyt* P450, cytochrome oxidases, catalase or peroxidases.[85] There are two major NO binding modes to metalloporphyrin depending on

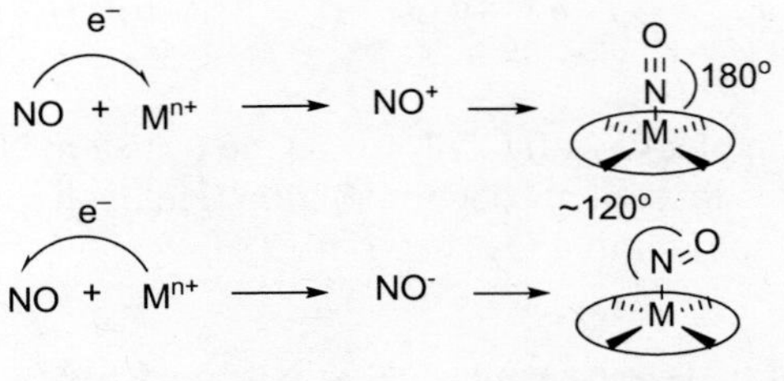

Figure 1.21 Binding modes of nitric oxide to metal ions.

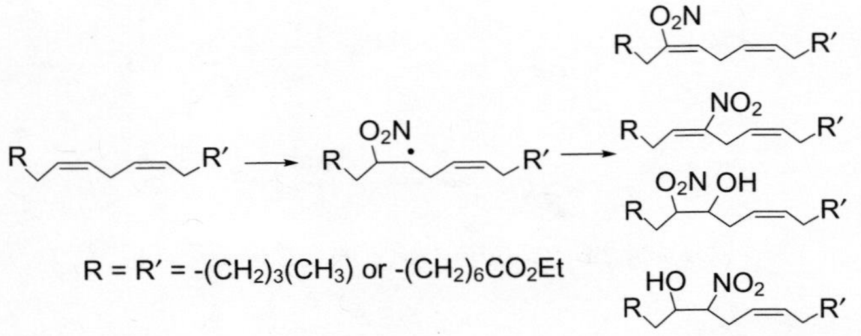

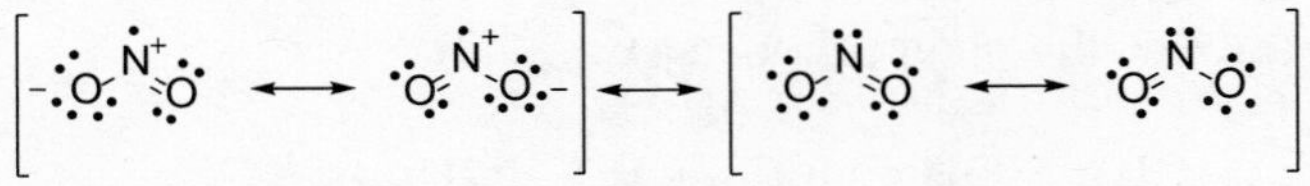

Figure 1.22 Mesomeric structures of nitrogen dioxide.

Figure 1.23 Nitration and hydroxylation of PUFA by $^\bullet NO_2$.

the direction of the charge transfer between the metal and NO: (1) NO assumed the NO^+ upon complexation and is characterized by a 180° M–N–O angle; and (2) NO assumed the NO^- upon complexation and is characterized by a 120° M–N–O angle (Fig. 1.21).

The direction of charge transfer is dependent on several factors such as the oxidation state and type of the metal ion, as well as, the coordination number and type of co-ligands bound to the metal other than the NO. The nature of the M-NO bonding mode determines protein conformation and NO-M reactivity (e.g., in the activation of O_2 to yield nitrate). For example, M–N–O bond angles for $Fe^{II}(OEP)(NO)$ and $[Fe^{III}(OEP)(NO)]^+$ (OEP = octaethylporphinato) are 143.6° and 176.9°, respectively.[86] Dependence of NO binding mode on the type of metal ions can also be seen with tetraphenylporphyrin complexes of Mn^{II}, Fe^{II} and Co^{II}, where the M–N–O bond angles are as follows, 176.1°, 142.1°, and 128.5°, respectively.[85]

1.2.4.2 Nitrogen Dioxide ($^\bullet NO_2$)

Nitrogen dioxide is one of the nitrogen oxides (NO_x) and is a paramagnetic π molecule where the unpaired electron is delocalized between the three atoms (Fig. 1.22). The polar property of $^\bullet NO_2$ is due to its bent structure with an O–N–O bond angle of 136° having negative charges on the O atoms and a positive charge on the N.

Nitrogen dioxide is sparingly soluble in water. Surface chemistry of adsorbed $^\bullet NO_2$ in aqueous system leads to its decomposition to H^+, nitrate and nitrous acid (HONO), and the presence of antioxidants such as ascorbate, urate and glutathione catalyzes this hydrolytic disproportionation.[87] In the gas and aqueous phases, $^\bullet NO_2$ dimerizes with a rate constant of $\sim 5 \times 10^8\ M^{-1}\ s^{-1}$ where it reacts rapidly with water to form nitrites and nitrates.[88] Nitrogen dioxide is a powerful oxidizer with a $E^0(^\bullet NO_2/NO_2^-) = 0.89–1.13\ V$.[89] Among the most important reactions of $^\bullet NO_2$ are: (1) H-atom abstraction from C–H bond; (2) addition reaction to C=C bonds; (3) O-transfer reactions; (4) radical–radical addition; (5) electron transfer. The H-atom abstraction involving $^\bullet NO_2$ is a slow process due to the weak H–O bond in HONO with a dissociation energy of ~79 kcal/mol compared to C–H bond of ~100 kcal/mol.[87]

Addition to Double Bonds Nitration of PUFA occurs via $^\bullet NO_2$ addition to C=C bond (Fig. 1.23). Nitrated PUFA are important biomarker of nitrosative stress due to their abundance in biological systems. Reaction of $^\bullet NO_2$ with 1,4-pentadiene moiety of ethyl linoleate[90] for example proceed via competition between H-atom abstraction and free radical combination. The formation of vicinal –OH along with –NO_2 results from the O-centered addition of another $^\bullet NO_2$ to the alkyl radical and the subsequent hydrolysis of the nitrite to form the hydroxyl group. Addition of $^\bullet NO_2$ to double bonds have also been observed in xenobiotics, food additives, retinoic acid, 17β-estradiol, or cinnamic acids.[91]

Radical–Radical Addition The major product in the reaction of $^\bullet NO_2$ with $MeO^\bullet$ is methyl nitrate ($MeONO_2$) through O–N bond formation.[83] However, reaction of $^\bullet NO_2$ with tyrosyl radical ($Tyr^\bullet$) forms the 3-nitrotyrosine via C–N bond formation which is one of the most studied biomarker of oxidative damage to protein systems due to their abundance in biological systems. One could initial assume that $HO^\bullet$ can abstract H-atom from tyrosine but its preferred mode of reaction is the ortho-directed addition to the aromatic ring to form the ortho-dihydroxytyrosine with $k = 7.0 \times 10^9\ M^{-1}\ s^{-1}$.[92] While $^\bullet NO_2$ is able to abstract H-atom from Tyr to form $Tyr^\bullet$, this reaction is relatively slower ($k = 3.2 \times 10^5\ M^{-1}\ s^{-1}$)[93] than the H-atom abstraction by $CO_3^{\bullet -}$ with $k = 4.5 \times 10^7\ M^{-1}\ s^{-1}$ (Fig. 1.24) Resonance structure of $Tyr^\bullet$ shows localization of the unpaired electron on the

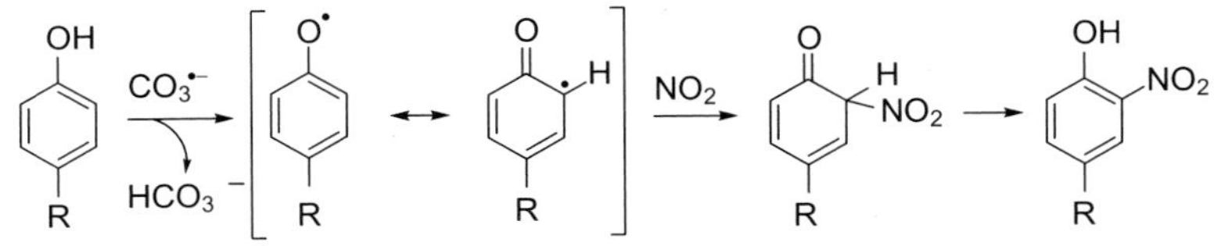

Figure 1.24 Radical–radical addition of $^\bullet NO_2$ to tyrosyl radical.

phenoxyl-O and the carbon atom ortho to the phenoxyl-O. Subsequent addition of $^\bullet NO_2$ to Tyr$^\bullet$ yields the 3-nitrotyrosine with $k = 3 \times 10^9\ M^{-1}\ s^{-1}$.[93] Lost of enzyme function has been correlated with the degree of tyrosyl nitration and has been observed in prostaglandin H2 synthase (PGSH-1),[94] MnSOD[95] and mitochondrial cytochrome c.[96]

Reaction with Thiols Nitrosation of thiol-containing biomolecules is one of the most important processes in posttranslational protein modification. Production of nitrosothiols (RSNO) is an important regulatory function of NO in cell signaling and pathology. Examination of RSNO formation at low micromolar concentrations of NO indicate N_2O_3 and $^\bullet NO_2$ as the nitrosating agents via a one-electron oxidation of thiols to RS$^\bullet$ (Eq. 1.50) and its subsequent radical–radical addition to NO to form S-nitrosothiols (RSNO).[97] Using laser flash photolysis, the rate of glutathiyl radical (GS$^\bullet$) reaction with NO to form GSNO was reported to be $2.8 \times 10^7\ M^{-1}\ s^{-1}$, which is lower than that expected for GSSG formation through radical–radical reaction further demonstrating that GS$^\bullet$ does react slowly with NO to form GSNO.[98]

$$RSH + NO_2 \rightarrow RS^\bullet + NO_2^- + H^+ \quad (k = 10^7\ M^{-1}\ s^{-1}) \quad (1.50)$$

1.2.4.3 Peroxynitrite (ONOO$^-$) Peroxynitrite is formed from the addition reaction of NO with superoxide ($O_2^{\bullet-}$) at a diffusion-controlled rate.[43,81] Peroxynitrite is known to exist in the relatively stable cis-conformation, or gain a proton to form peroxynitrous acid (ONOOH, pK_a 6.8). One relevant mechanism for ONOO$^-$/ONOOH decay is its homolytic cleavage through $^\bullet$ONO$\cdots$O$^{\bullet-}$ and $^\bullet$ONO$\cdots^\bullet$OH intermediates.[99] The higher membrane permeability of ONOOH compared to its unprotonated form can result in its homolysis to form HO$^\bullet$ and $^\bullet NO_2$ leading to the initiation of oxidation of key biomolecular systems. For ONOO$^-$, the rate of radical cleavage has been reported at $\approx 10^{-6}/s$, with negligible $^\bullet NO_2$ and O$^{\bullet-}$ release,[100] while for ONOOH, the rate of radical cleavage has been reported to be $0.35 \pm 0.03/s$, with about 30% of HO$^\bullet$ and $^\bullet NO_2$ being released at pH < 5 via escape from the solvent cage. Like $O_2^{\bullet-}$, ONOO$^-$ is capable of reacting with protein active sites containing Cu, Zn, sulfhydryl and Fe–S clusters to cause nitration and protein cleavage resulting in enzyme deactivation.[59,101–103] The rate constant for ONOOH isomerization to nitric acid (HNO$_3$) was found to be $1.1 \pm 0.1/s$.[104] While the low rate of homolytic cleavage of ONOO$^-$ makes the reaction trivial, ONOO$^-$ is known to react with dissolved CO$_2$ to form nitrosoperoxycarbonate (ONOOCO$_2^-$) at a rate constant of $3.0 \times 10^4\ M^{-1}\ s^{-1}$.[105] ONOOCO$_2^-$ is a two-electron oxidant that undergoes homolytic cleavage to form 30% CO$_3^{\bullet-}$ and $^\bullet NO_2$ (Eq. 1.51).[95,106]

$$ONOO^- + CO_2 \rightarrow ONOOCO_2^- \rightarrow CO_3^{\bullet-} + NO_2 \quad (1.51)$$

The modes of decay of ONOOCO$_2^-$ and ONOOH has been shown to vary depending on the ability of the solvent to hold the intermediate species in the solvent cage, and is therefore, dependent on the viscosity of a solvent.[107] Peroxynitrite is a strong nucleophile, and has been shown to cause β-scission of carbonyl groups,[108,109] where acyl- and H-spin adducts have been observed using EPR spin trapping.[110,111] Peroxynitrite has recently been shown to form peroxynitrate (O$_2$NOO$^-$) at neutral pH through combination of ONOO$^-$ and ONOOH to form O$_2$NOOH and nitrite (NO$_2^-$).[112]

Reaction of ONOO$^-$ with inorganic radicals such as CO$_3^{\bullet-}$, $^\bullet N_3$, ClO$_2^\bullet$ and HO$^\bullet$ involves a one-electron transfer process. For example, ONOO$^-$ oxidation by the inorganic radicals yields ONOO$^\bullet$ and the corresponding anions with varying rate constants (Eq. 1.52–1.55).[113] With NO, ONOO$^-$ forms $^\bullet NO_2$ and NO$_2^-$ with highly exoergic free energy of −113 kJ.[114]

$$ONOO^- + CO_3^{\bullet-} \rightarrow ONOO^\bullet + CO_3^{2-}$$
$$(k = 7.7 \times 10^6\ M^{-1}\ s^{-1}) \quad (1.52)$$

$$ONOO^- + ^\bullet N_3 \rightarrow ONOO^\bullet + N_3^-$$
$$(k = 7.2 \times 10^8\ M^{-1}\ s^{-1}) \quad (1.53)$$

$$ONOO^- + HO^\bullet \rightarrow ONOO^\bullet + HO^-$$
$$(k = 4.8 \times 10^9\ M^{-1}\ s^{-1}) \quad (1.54)$$

$$ONOO^- + ClO_2^\bullet \rightarrow ONOO^\bullet + ClO_2^-$$
$$(k = 3.2 \times 10^4\ M^{-1}\ s^{-1}) \quad (1.55)$$

Peroxynitrite acts as two-electron oxidant with thiols. Thiolates from low molecular weight thiols or protein thiols react with ONOOH to form sulfenic acid (RSOH). With low molecular weight thiols, the rates of thiol oxidation increases with decreasing thiol pK_a,[115] consistent with the mechanism of nucleophilic attack of the thiolate-O to the peroxyl-O of ONOOH with nitrite as the leaving group according to the mechanism shown in Equation 1.56:

$$RS^- \quad O=N-O-OH \rightarrow RSOH + NO_2^- \quad (1.56)$$

Rate constants for the reaction of $ONOO^-$ with free cysteine and the single thiol of albumin was reported to be in the order of $10^3 \ M^{-1} \ s^{-1}$.[116] The formation of RSSR' from RSOH in the presence of RS^- is fundamental to the regulation of protein function. With peroxidoxin thiol ($PrxS^-$), the reaction with ONOO to yield NO_2^- and PrxSOH is faster ($10^7 \ M^{-1} \ s^{-1}$)[117] than $ONOO^-$ reaction with small molecular weight thiols. Decomposition of $ONOO^-$ via one-electron or two-electron reduction processes can be catalyzed by metalloporphyrins of iron and manganese which can have protective effects against $ONOO^-$ induced damage. One electron reduction leads to the formation of $^\bullet NO_2$ while its two-electron reduction forms NO_2^-.[101] The formation of $^\bullet NO_2$ from $ONOO^-$ is shown to cause tyrosine nitration to form 3-nitrotyrosine.[118]

1.2.5 Reactive Sulfur and Chlorine Species

1.2.5.1 Thiyl or Sulfhydryl Radical (RS$^\bullet$)
Thiyl radicals are analogous to alkoxyl radicals (RO$^\bullet$) but there are important differences between the nature of the bonds involving S and O atoms. For example, the S–H bond in thiols is weaker than the O–H bond in alcohols with experimental bond dissociation energies of 88.0 kcal/mol and 104.4 kcal/mol for CH_3S–H and CH_3O–H, respectively.[119] The bond length for S–H is 1.33 Å compared to O–H of 0.96 Å. These differences in the structural and physical properties of thiols compared to alcohols play an important role in the reactivity of thiols compared to alcohols in which the S is more accessible. Since S is less electronegative than O, therefore, thiyls are more electrophilic than alkoxyl radicals with a longer C–S bond length of 1.81 Å compared to C–O of 1.42 Å.

Generation of RS$^\bullet$ in biological systems occurs via one-electron oxidation of thiols (RSH) by metal ions such as Cu^{2+} or Fe^{3+}, HO$^\bullet$, peroxynitrite, DNA or protein radicals. Disulfide formation (GSSG) from GS$^\bullet$ via radical–radical addition is fast with rate constant of $1.5 \times 10^9 \ M^{-1} \ s^{-1}$.[120] The susceptibility of RSH to oxidation is the basis of thiol antioxidant or repair mechanisms. GSH for example is the predominant intracellular antioxidant with cytosolic concentrations of up to 10 mM. Due to the high GSH concentration, the formation of disulfide is regulated. Gluthatione reacts with tyrosyl radical Tyr$^\bullet$ to yield GS$^\bullet$ and TyrOH ($k = 2 \times 10^6 \ M^{-1} \ s^{-1}$) as a repair mechanism but at a 220× slower rate than Tyr$^\bullet$ reaction with ascorbate. Ascorbate being more abundant in tissues makes GSH a minor player in this type of repair mechanism.[121]

Thiyl radicals can catalyze conversion of cis to trans isomerism in unsaturated systems. In lipid systems, the conversion of the naturally occurring *cis* unstaturated fatty acids to *trans* can cause morphological changes in the lipid bi-layers.[122] Reaction of thiyl with unstaturated compounds can also result to addition reaction where the preference for radical attack is the one with the highest electron density such as double bonds demonstrating the electrophilic nature of thiyl radicals which is due to the ability of the d-orbitals of sulfur to accommodate the negative charge. The rate constant for thiyl radical addition to monounsaturated fatty acid esters such as methyl oleate, methyl palmitoleate, methyl Z-vaccenate, and oleic acid in *tert*-butyl alcohol is in the order of k_a^Z and $k_a^E \sim 10^5 \ M^{-1} \ s^{-1}$, while the rate constant for the β-elimination to Z or E configurations are higher with k_f^Z and k_f^E of $\sim 10^7$/s and 10^8/s, respectively.[123] Thiyl radical induced isomerization for linoleic acid, linolenic acid and arachidonic acid gave k_a^Z and k_a^E of $\sim 10^6 \ M^{-1} \ s^{-1}$ and k_f^Z and k_f^E of $\sim 10^5$/s, respectively (Fig. 1.25).[51]

Relevant to the oxidation PUFA, thiyl can also undergo H-atom abstraction in bisallylic systems and, like HO$^\bullet$ (Eq. 1.57), demonstrates their pro-oxidative role in the initiation of lipid peroxidation. The rate constant for H-atom abstraction by thiyl radicals with PUFAs was in the order of $10^7 \ M^{-1} \ s^{-1}$.[124]

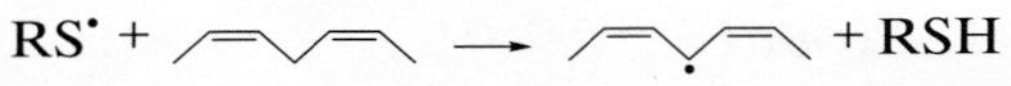

$$RS^\bullet + \diagup\!\!=\!\!\diagdown\!\!=\!\! \longrightarrow \diagup\!\!=\!\!\diagdown\!\!\cdot + RSH \quad (1.57)$$

H-atom abstraction from aliphatic alcohols and ethers has been shown to occur at a rate constant of 10^3–$10^4 \ M^{-1} \ s^{-1}$.[125] In peptidic systems, intramolecular H-atom transfer between cysteine thiyl radical and the $^\alpha$C-H bond occurs with rate constants that are in the order of 10^3–$10^5 \ M^{-1} \ s^{-1}$.[126] The favorability of this reaction was shown to be dependent on peptide and protein sequence as well as structure and can have implications in the

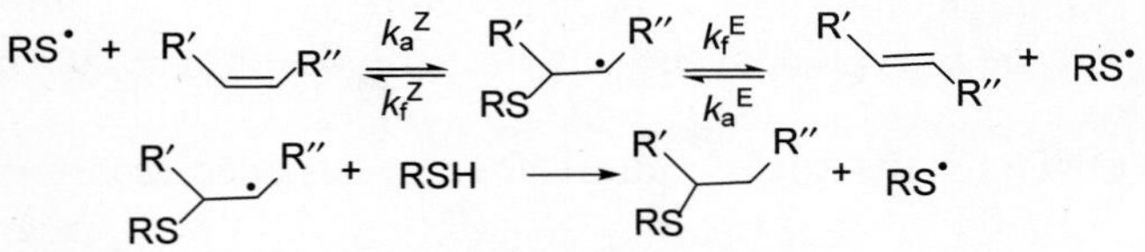

Figure 1.25 Thiyl radical mediated E and Z isomerization of monosaturated fatty acid.

catalysis of protein damage due to its potential irreversibility resulting in protein fragmentation and/or epimerization.[127] Interconversion between $^\alpha$C-, $^\beta$C-, and S-centered radicals in GS$^\bullet$ (Eq. 1.58) has been shown to proceed favorably and is pH dependent with an overall rate constants of $k_{forward} = 3.0 \times 10^5$/s, $k_{reverse} = 7.0 \times 10^5$/s and $K = 0.4$, with an equilibrium ratio at pH 7 of 8:3:1 for S:$^\beta$C-:$^\alpha$C-, centered radicals.[128]

$$RHN-\underset{\underset{\overset{|}{S^\bullet}}{CH_2}}{CHC}-\overset{O}{\overset{\|}{C}}-\overset{H}{\overset{|}{N}}-\underset{\overset{|}{CH_3}}{CHC}-\overset{O}{\overset{\|}{C}}-R'' \longrightarrow RHN-\underset{\underset{\overset{|}{SH}}{CH_2}}{CHC}-\overset{O}{\overset{\|}{C}}-\overset{H}{\overset{|}{N}}-\overset{\bullet}{C}-\underset{\overset{|}{CH_3}}{C}-R'' \tag{1.58}$$

H-atom abstraction from carbohydrates by thiyl radical have been reported.[129] H-atom transfer of C^1-H of 2-deoxy-D-ribose, 2-deoxy-D-glucose, α-D-glucose and inositol by cysteine-derived thiyl radical gave rate constants that are in the order of 10^4 M^{-1} s^{-1}.

Quenching of thiyl radicals by ascorbate results in the formation of ascorbyl radical and RSH while thiyl reaction with radicals such as NO, O$_2$, and R$^\bullet$ showed varying reactivity. GSNO formation from the addition of GS$^\bullet$ to NO was estimated to be much faster than the previously reported rate constant of 2.8×10^9 M^{-1} s^{-1} using laser flash photolysis.[98] Using pulse radiolysis, the rate constant for the reaction of NO with thiyl radicals of glutathione (Eq. 1.59), cysteine and penicillamine were reported to be in the range of $2-3 \times 10^9$ M^{-1} s^{-1}.[130]

$$GS^\bullet + NO \rightarrow GSNO \quad k = 2.7 \times 10^9 \ M^{-1} \ s^{-1} \tag{1.59}$$

Reaction of thiyl radicals with O$_2$ yields RSOO$^\bullet$ but the presence of excess RSH leads to the formation RSO$^\bullet$ and RSOH under normal conditions.[131] The reported rate constants for the reaction of GS$^\bullet$ with O$_2$ vary from 3.0×10^7 M^{-1} s^{-1} to 2.0×10^9 M^{-1} s^{-1} indicating a more complex mechanism resulting from this addition reaction (Eq. 1.60).[132,133]

$$GS^\bullet + O_2 \underset{k_{-1}}{\overset{k_1}{\rightleftharpoons}} GSOO^\bullet \tag{1.60}$$
$$k_1 = 2.0 \times 10^9 \ M^{-1} \ s^{-1}; k_{-1} = 6.2 \times 10^5 \ s^{-1}$$

Reaction of GS$^-$ with GS$^\bullet$ forms GSSG$^{\bullet-}$ with a rate constant of 4.5×10^8 M^{-1} s^{-1} with an equilibrium constant of 2.25×10^3/M.[134] Decay of RSSR$^{\bullet-}$ forms RS$^-$ and RS$^\bullet$, with RS$^\bullet$ further undergoing intramolecular H-atom abstraction mechanism to form the α-amino carbon-centered radical with rate constants ranging in the order of 10^4-10^5/s for cysteine, homocysteine and gluthathione at pH 10.5.[134] Protonation of RSSR$^{\bullet-}$ leads to its decomposition to RS$^\bullet$ and RSH and ultimately to RSSR with rate constants in the order of 10^5-10^6/s.[135] Reaction of GSSG$^{\bullet-}$ with O$_2$ has a rate constant of 1.6×10^8 M^{-1} s^{-1} (Eq. 1.61).[136]

$$GS^\bullet + GS^- \rightleftharpoons GSSG^{\bullet-} \xrightarrow{O_2} GSSG + O_2^{\bullet-} \tag{1.61}$$

1.2.5.2 Disulfide (RSSR)
Unlike the S–H bond dissociation energy being lower than the O–H, the S–S bond dissociation energy is higher compared to O–O. Reported BDE for MeS-SMe is 74 kcal/mol compared to MeO-OMe of 37.6 kcal/mol.[119,137] Thiol-disulfide interchange as described by Equation 1.62 and Equation 1.63 shows formation of a mixed disulfide intermediate RSSR' from the oxidation of RSH and reduction of RSSR'.[138] Thiol-disulfide interchange is an important biochemical process and occurs in many metabolic reactions of thiols either endogenously or from thiol-based drugs such as penicillamine. The rate constants for the symmetrical thiol-disulfide exchange reaction have been determined for several thiols such as GSH, Cys, or homocysteine in aqueous basic medium (pH > 10) with k in the range of 12–60 M^{-1} s^{-1}.[138]

$$RSH + R'SSR' \rightleftharpoons RSSR' + R'SH \tag{1.62}$$

$$\frac{RSH + RSSR' \rightleftharpoons RSSR + R'SH}{2RSH + R'SSR' \rightleftharpoons RSSR + 2R'SH} \tag{1.63}$$

Disulfide bonds play a major role in protein thermal stability but through chemical means, disulfide bonds can be broken down via several mechanisms. Under basic or neutral conditions, hydroxide (HO$^-$) is shown to attack the sulfur atom to form sulfenic acid and thiolate anion and can ultimately result in post-translational protein modification to form complex disulfides (Eq. 1.64) or mixed sulfenic acid/disulfides with another protein/s.

$$HO^- \overset{\frown}{S-S}\text{Protein} \longrightarrow HO-\underset{}{S} \ \underset{}{S}\text{-Protein} \xrightarrow{RSSR'} \underset{\underset{R}{\overset{|}{S}}}{\overset{|}{S}} \ \underset{\underset{R'}{\overset{|}{S}}}{\overset{|}{S}}\text{-Protein} \tag{1.64}$$

Hydroxide can also abstract the α- or β-protons of the Cys residue leading to C–S or S–S bond breakage, respectively, followed by β- or α-elimination according to Figure 1.26.[139]

Disulfide can be further oxidized to disulfide-S-monoxide and disulfide-S-dioxide. Oxidation of one of the sulfur atoms leads to the weakening of the S–S bond and is therefore more susceptible to reaction with RSH to form sulfenic (RSOH) and sulfenic acids (RSO$_2$H) to generate the mixed disulfide (Fig. 1.27).[140]

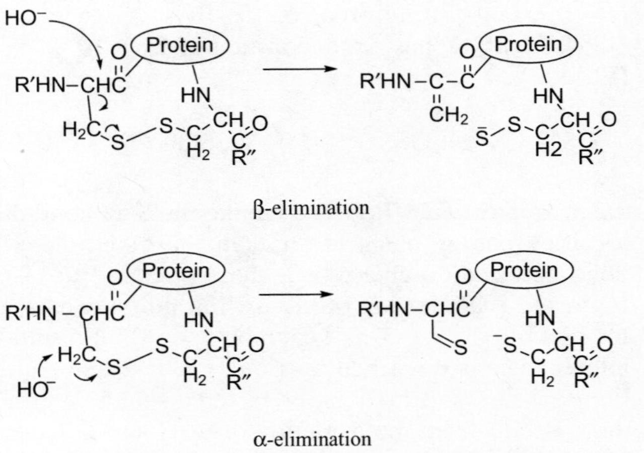

Figure 1.26 β- or α-elimination reactions of hydroxide on protein with cysteine residues.

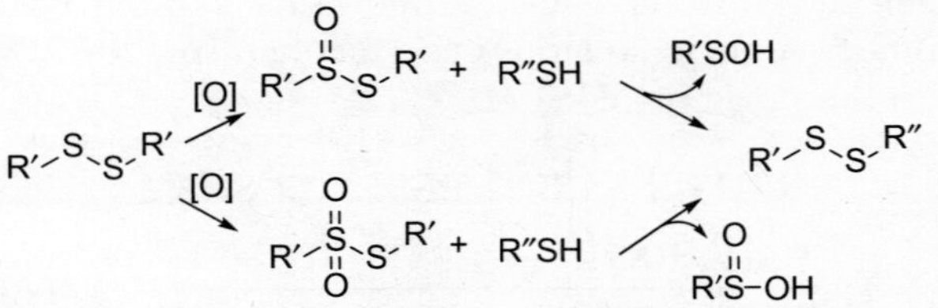

Figure 1.27 Formation of mixed disulfides through oxidation processes.

Disulfides can also be enzymatically reduced to RSH by glutathione reductase[141] or thioredoxin reductases[142] in the presence of NADPH, or chemically, by small molecules such as dithiothreitol, hydrazine or sulfones.[143]

1.2.5.3 Hypochlorous Acid (HOCl)

Hypochlorous acid is usually formed from the reaction of Cl_2 gas with water, however in biological systems, their formation have been mediated by a secreted heme protein, myeloperoxidase (MPO), which can convert H_2O_2 to HOCl in the presence of chloride ion (Cl^-) according to Equation 1.65.[144]

$$H_2O_2 + Cl^- + H^+ + MPO \rightarrow HOCl + H_2O \quad (1.65)$$

HOCl has a pK_a of 7.5, therefore, it co-exists with the ionized hypochlorite (^-OCl) in solution at physiological pH. The HOCl produced has been shown to be a potent 2-electron oxidant capable of chlorinating electron rich substrates and oxidation of heme, tyrosine or cysteine residues in proteins, DNA and lipids.

Hypochlorous acid reacts with various ROS such as H_2O_2 to generate stoichiometric amounts of $[O_2(^1\Delta_g)]$,[145]

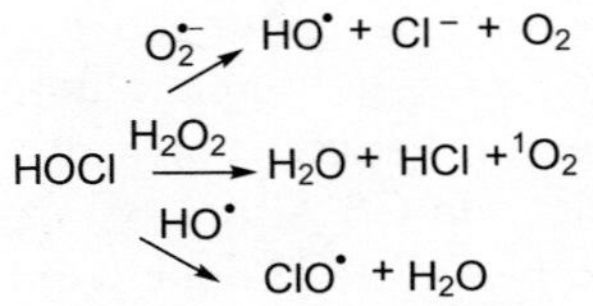

Figure 1.28 Reactions of hypochlorous acid with various reactive oxygen species.

with $O_2^{\bullet-}$ to generate $HO^{\bullet}$,[146] and with $HO^{\bullet}$ to form $ClO^{\bullet}$ (Fig. 1.28).[147]

Reaction of HOCl with hydroperoxide such as linoleic acid hydroperoxide (LA-OOH) mimics that of its reaction with H_2O_2 producing $[O_2(^1\Delta_g)]$ (13% yield) at physiological pH (Eq. 1.66).[148]

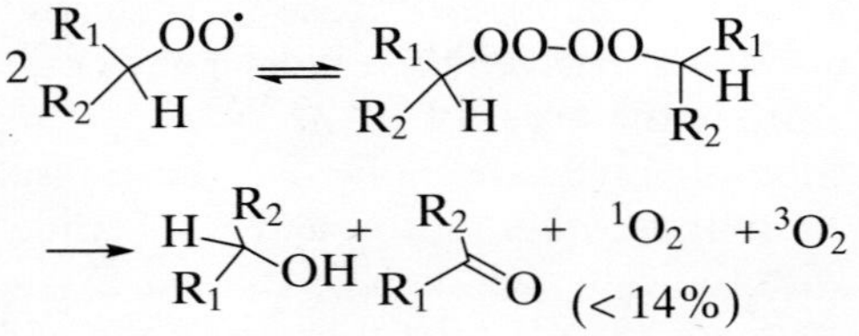

$$2 \begin{array}{c} R_1 \\ R_2 \end{array}\!\!\!\!\!\begin{array}{c} OO^{\bullet} \\ H \end{array} \rightleftharpoons \begin{array}{c} R_1 \\ R_2 \end{array}\!\!\!\!\!\begin{array}{c} OO\text{-}OO \\ H \end{array}\!\!\!\!\!\begin{array}{c} R_1 \\ H \\ R_2 \end{array}$$

$$\rightarrow H\!\!\begin{array}{c} R_2 \\ R_1 \end{array}\!\!\!\!\!\!OH + \begin{array}{c} R_2 \\ R_1 \end{array}\!\!\!\!\!\!O + {}^1O_2 + {}^3O_2 \quad (<14\%) \quad (1.66)$$

With anions such as NO_2^-, HOCl is capable of forming a reactive intermediate that can nitrate phenolic substrates such as tyrosine and 4-hydroxyphenyl acetic acid with high yield at physiological pH.[149,150] The nitrating intermediates were identified to be $^{\bullet}NO_2$ and nitryl chloride (NO_2-Cl) based on Equation 1.67.

$$HOCl + NO_2 \rightarrow HO + ClNO_2(Cl\text{-}ONO)$$
$$\rightarrow Cl^{\cdot+} \cdot NO_2^- + Cl^{\bullet} + {}^{\bullet}NO_2 \quad (1.67)$$

Sulfite reaction with HOCl gives the intermediate, Cl–SO_3^- and its subsequent hydrolysis forms Cl^- and SO_4^{2-}.[151] Reaction rate of HOCl with low molecular weight antioxidant such as ascorbate (AH^-) is 6×10^6 $M^{-1} s^{-1}$.[152]

$$AH^- + HOCl \rightarrow A + Cl^- + H_2O \quad (1.68)$$

Electron transfer reaction between Fe^{2+} and HOCl occurs with the generation of $HO^{\bullet}$ and $Cl^{\bullet}$ according to Equation 1.69 and Equation 1.70,

$$HOCl + Fe^{2+} \rightarrow Fe^{3+} + HO^{\bullet} + Cl^- \quad (1.69)$$

$$HOCl + Fe^{2+} \rightarrow Fe^{3+} + HO^- + Cl^{\bullet} \quad (1.70)$$

where the formation of $HO^{\bullet}$ predominates due to the electron transfer reaction between $Cl^{\bullet}$ and H_2O to further form $HO^{\bullet}$.[153]

Reaction of HOCl with free amino acid backbone generates chloramine species at the free amino moiety.

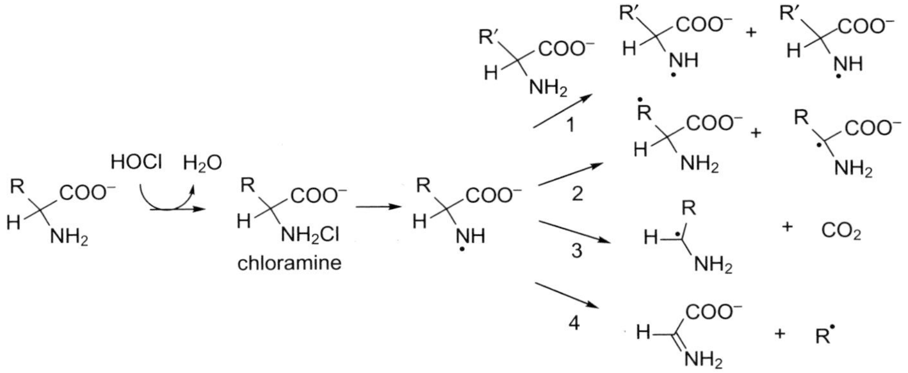

Figure 1.29 Reaction of hypochlorous acid with amino acids.

Chloramine undergoes further decomposition to nitrogen-centered radicals which subsequently undergo further decomposition pathways such as (1) intra- and intermolecular H-atom abstraction; (2) decarboxylation; (3) β-scission according to Figure 1.29.[154]

Analogous to the reaction of amines with HOCl, GSH forms S-chloro derivative with HOCl which can hydrolyse to yield the corresponding sulfenic acid (GSOH) (via formation of thiyl radical)[154] with an estimated rate constant of >10^7 M^{-1} s^{-1} (Eq. 1.71, Eq.1.72, and Eq. 1.73).[153] With amino acids containing thiols, methionine, or cysteine, the rates were estimated to be in the order of ~10^{4-5} M^{-1} s^{-1}).[155]

$$HOCl + GSH \rightarrow GSCl + H_2O \rightarrow GS^\bullet + Cl^\bullet \quad (1.71)$$

$$Cl^\bullet + H_2O \rightarrow HO^\bullet + Cl^- + H^+ \quad (1.72)$$

$$GS^\bullet + HO^\bullet \rightarrow GSOH \quad (1.73)$$

The formation of sulphonamide (RSO_2NHR) but not the formation of GSSG from HOCl and GSH via intramolecular cyclization reaction has also been observed.[156] Methionine oxidation by HOCl forms methionine sulfoxide and dehydromethionine according to Equation 1.74[157]:

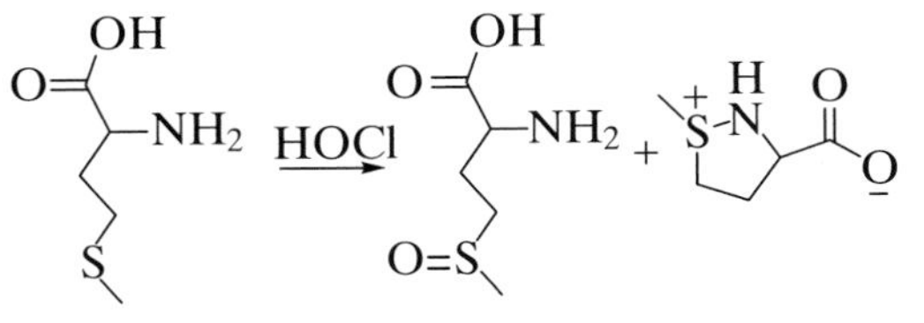

(1.74)

Reaction of HOCl with tyrosine and peptidyl-tyrosyl residues yielded 3,5-dichlorotryosine (diCl-Tyr) in addition to Cl-Tyr. Further reaction of the mono- and dichlorinated tyrosines gave the corresponding mono- and dichlorinated 4-hydroxyphenylacetaldehydes, Cl-HPAA and diCl-HPAA, respectively, according to Figure 1.30.[158]

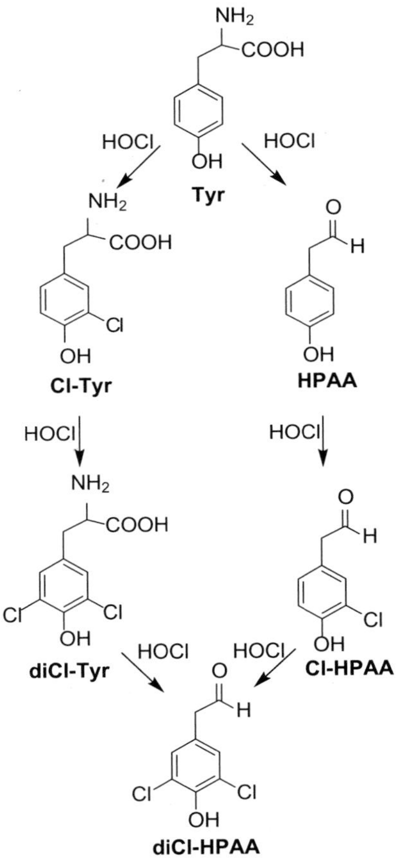

Figure 1.30 Reactions of hypochlorous acid with tyrosine.

Oxidation of cytochrome c by HOCl has rate constant of >3×10^5 M^{-1} s^{-1}. This reaction is not only selective toward the heme iron but also involves N-halogenation of the side chain amino groups and with concomitant generation of $HO^\bullet$ (Eq. 1.75).[159]

$$Fe(II)cyt\ c + HOCl \rightarrow Fe(III)cyt\ c + HO^\bullet + Cl^- + \text{other products} \quad (1.75)$$

HOCl reaction with lipids occurs at either the lipid head group or the unsaturated portion of the fatty acid side-chain. For example, reaction of HOCl with phosphoryl-serine and phosphoryl-ethanolamine are rapid with $k \sim 10^5$ M^{-1} s^{-1} yielding chloroamines as the major products.[44] Reaction with unsaturated fatty acid chains involves initial formation of chlorohydrins[160] followed by secondary dehydrohalogenation reactions to yield the epoxide (Eq. 1.76). The formed epoxide can further react with HOCl to form ROS and lipid peroxidation products.

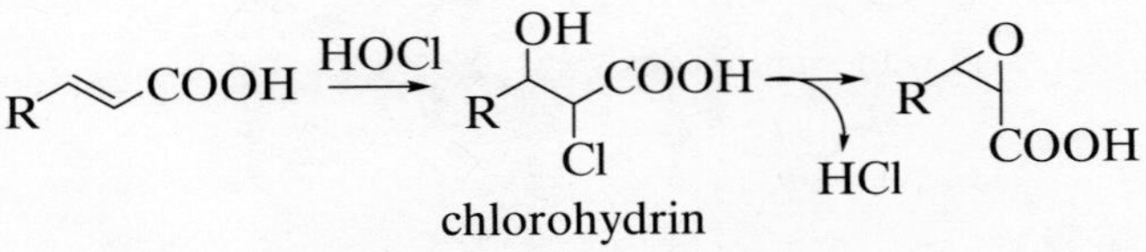

$$R-CH=CH-COOH \xrightarrow{HOCl} R-CH(OH)-CHCl-COOH \xrightarrow{-HCl} R-\text{(epoxide)}-COOH$$

chlorohydrin

(1.76)

Reaction of HOCl with nucleotide bases occur primarily on the exocyclic free amino group (e.g., of cytosine, adenosine and guanosine) or nitrogen atoms of the heterocyclic ring (e.g., of thymidine, uridine and guanosine) which contain lone pairs to form N–Cl bond. These adducts can result in miscoding and have been identified in tissues under inflammatory conditions. The rate constants for reactions within the heterocylic ring is in the order of 10^3–10^4 M^{-1} s^{-1}. With uridine for example, N–Cl formation leads to the formation of N-centered radical (Eq. 1.77).[161]

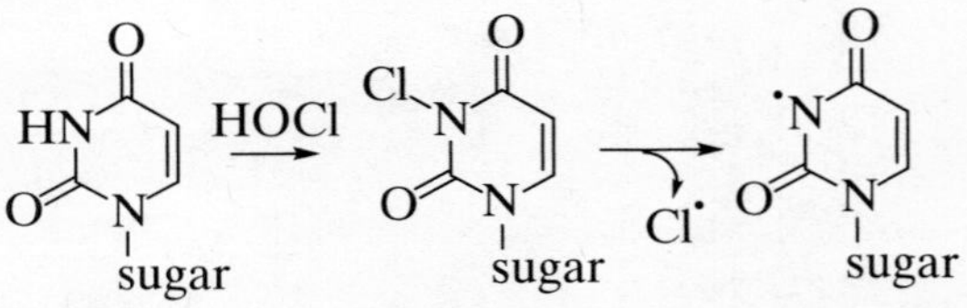

(1.77)

Direct chlorination on the carbon atom by HOCl of the heterocylic ring was also observed to give chlorinated products such as 5-chloro-2′-deoxycytidine, 5-chloro-uracil, 8-chloro-2′-deoxyguanosine, and 5-chloro-2′-deoxyadenosine[162] as well as hydroxylation of the pyrimidine moiety to give thymine glycol (cis/trans), 5-hydroxycytosine, 5-hydroxyuracil, 5-hydroxyhydantoin (Fig. 1.31).[163]

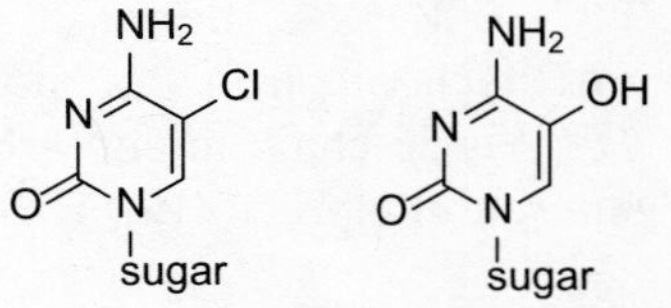

Figure 1.31 Chlorination and hydroxylation of pyrimidine by hypochlorous acid.

Reaction of related compound such as NADPH with HOCl is characterized by an initial fast reaction with $k = 4.2 \times 10^5$ M^{-1} s^{-1} leading to the formation of a stable pyridine product (Py/Cl). Subsequent reaction with HOCl ($k = 3 \times 10^3$ M^{-1} s^{-1}) leads to the total loss of the aromatic pyridine ring absorbance.[164]

1.3 REACTIVITY

As in all chemical reactions, reactions involving reactive species are governed thermodynamically and kinetically, and these two inter-related forces can offer insights into the favorability and rate of a reaction, respectively.

1.3.1 Thermodynamic Considerations

The favorability of redox reaction involving reactive species is governed by the overall change in the potential energy whereby the energy is released (in this case of an exothermic reaction) or addition of energy (endothermic reaction) to the system for the reaction to proceed. The thermodynamic favorability is defined by an entity called free energy (ΔG) which is either introduced or given off in a reaction. One can envision that reactants and products have stored energy in them. Calculation of ΔG can be theoretically and experimentally performed. As an example for the formation $ONOO^-$ from $O_2^{\bullet-}$ and $^{\bullet}NO$, one can calculate the favorability of this reaction by taking into account the potential energies of the individual species. One important theoretical consideration in determining the free energy of reaction (ΔG_{rxn}) is that the type and number of atoms in the product and reactant sides should be conserved as shown in the equation: $O_2^{\bullet-} + {^{\bullet}NO} \rightarrow ONOO^-$. Each of these species carries a potential energy originating from the separation of the individual nuclei and electrons from the molecule. For example, the following are the total electronic energies (ε_o) (with thermal free energies, G_{corr}) for $O_2^{\bullet-}$, $^{\bullet}NO$, and $ONOO^-$ formed from nuclei and electrons (Fig. 1.32).

The $\Delta G^o_{rxn,298K}$ for the reaction: $O_2^{\bullet-} + {^{\bullet}NO} \rightarrow ONOO^-$ can then be calculated using the Equation 1.78:

$$\Delta G^o_{rxn,298K} = \Sigma (\varepsilon_o + G_{corr})_{products} - \Sigma (\varepsilon_o + G_{corr})_{reactants}$$

$$\Delta G^o_{rxn,298K} = ((-280.402251) - (-150.482170 + -129.907204)) * 627.5095$$

$$\Delta G^o_{rxn,298K} = -8.08 \text{ kcal/mol} \quad (1.78)$$

The ΔG for the formation of $ONOO^-$ from $O_2^{\bullet-}$ and $^{\bullet}NO$ is therefore exothermic since the total energy of the reactant is greater than the reactants, and therefore,

$$O_2^{\bullet-} \longrightarrow 2\,O^{8+} + 17\,e^- \qquad -150.482170 \text{ hartrees}$$
$$NO \longrightarrow O^{8+} + N^{7+} + 15\,e^- \qquad -129.907204 \text{ hartrees}$$
$$ONOO^- \longrightarrow 3\,O^{8+} + N^{7+} + 32\,e^- \qquad -280.402251 \text{ hartrees}$$

Figure 1.32 Total electronic energies for $O_2^{\bullet-}$, $^{\bullet}NO$, and $ONOO^-$ formed from nuclei and electrons.

TABLE 1.1 Gibbs Energies of Formation for Various ROS/RNS[114,165]

Compounds	$\Delta_f G^o$ (kcal/mol)
HO$^{\bullet}$	15.7 (12.7)[166]
H$_2$O	−56.7
H$_2$O$_2$	−14.1
HO$_2^{\bullet}$	10.7 (1.7)[165]
HO$_2^-$	−7.6
HO$^-$	−28.1
NO$^{\bullet}$	24.4
NO$^+$	52.3
NO$^-$ (singlet)	32.5
NO$^-$ (triplet)	15.3
NO$_2^{\bullet}$	15.1
NO$_2^+$	52.1
NO$_2^-$	−7.7
NO$_3^{\bullet}$	31.3
NO$_3^-$	−26.6
N$_2$	4.2
N$_2$O	27.2
N$_2$O$_2^{\bullet-}$	33.7
N$_2$O$_3$	35.1
ONOO$^{\bullet}$	20.1
ONOO$^-$	10.1 (16.6)[167]
O$_2$	3.9
O$_2^{\bullet-}$	7.6
ONOOH	7.5

Adapted from Reference 114.

excess energy is given off, hence, the reaction is said to proceed spontaneously. In contrast, the dismutation reaction of two $O_2^{\bullet-}$ to form O_2^{2-} and O_2 according to the equation: $2O_2^{\bullet-} \rightarrow O_2^{2-} + O_2$, gave $\Delta G^o_{rxn,298K} = 35.7$ kcal/mol, which is endoergic and does not proceed spontaneously due to repulsion between the two $O_2^{\bullet-}$. The two contrasting equations demonstrate the relative thermodynamic stability of the two reactions in which the formation of $ONOO^-$ is preferred due to the less repulsion between reactants and the radical–radical nature of the reaction.

However, ΔG of formation for ROS/RNS can also be obtained experimentally. Koppenol had compiled a series free energies as shown in Table 1.1.[114]

The ΔG is defined by Equation 1.79,

$$\Delta G = \Delta H - T\Delta S \qquad (1.79)$$

where ΔH is the change in enthalpy, T is the absolute temperature and ΔS is the change in entropy. Although the exoergicity or endoergicity of a reaction is determined by the minimization of the total enthalpy (i.e., net heat change), the minimization of the total free energy of the system at constant temperature and pressure is the driving force for all reactions. Therefore, the sign of ΔG indicates favorability of a reaction, that is,

$\Delta G < 0$ (favored or spontaneous)
$\Delta G = 0$ (equilibrium, neither forward or backward reactions are favored)
$\Delta G > 0$ (not favorable, nonspontaneous)

The concept presented above assumes that the reaction is unidirectional, meaning that the products are perfectly thermodynamically stable and does not revert back toward the formation of the reactant. However, there are reactions involving reactive species that are not unidirectional. These reactions contain significant quantities of reactants and products at equilibrium (Eq. 1.80), a state in which the composition of the reactant and products remains unchanged.

$$A \underset{k_2}{\overset{k_1}{\rightleftharpoons}} B \qquad (1.80)$$

The relationship between free energy and thermodynamic equilibrium (K_{eq}) constant is described by Equation 1.81:

$$\Delta G^o = -RT \ln K_{eq} \qquad (1.81)$$

where R is the universal gas constant and T is the absolute temperature. Since K_{eq} represents the ratio of the molar concentrations of A relative to B, and of k_1 and k_2 at equilibrium, that is, $K_{eq} = [B]/[A] = k_1/k_2$, it is expected that ΔG^o will obviously be dependent on temperature as temperature affect the direction of the equilibrium. Examples of temperature-dependent reversible reaction is the transnitrosation reaction between thiol and S-nitrosothiol (Eq. 1.82):

$$RSH + R'SNO \rightleftharpoons R'SH + RSNO \qquad (1.82)$$

With R'SNO as S-nitroso-N-acetyl-penicillamine (SNAP), and with gluthathione or L-cysteine as RSH, the K_{eq}'s were determined to be 3.69 and 3.66, at 25°C. Using Equation 1.79, ΔG^o can be calculated to be −0.77 kcal/mol. With ΔG being negative, it is exoergic hence the equilibrium is shifted to the product side of the equation. At higher temperature (i.e., 33°C) for gluthathione or L-cysteine, the K_{eq} is lower with 3.0 and 2.58, which correspond to ΔG^o of −0.66 and −0.58 kcal/mol, respectively, indicating the equilibrium is shifted to the right.

Conversely, K_{eq} can be determined based on ΔG^o of formations. For example, in the ionization of ONOOH to ONOO$^-$ (Eq. 1.83),

$$ONOOH \rightarrow H^+ + ONOO^- \quad (1.83)$$

using Table 1.1, the ΔG^o for the formation of ONOOH and ONOO$^-$ is 7.5 and 16.6 kcal/mol, respectively. The free energy of ionization is then equal to ΔG^o(ONOO$^-$) − ΔG^o(ONOOH) = (16.6 kcal/mol) − (7.5 kcal/mol) = 9.1 kcal/mol using ΔG^o = 0 kcal/mol for H$^+$. Using Equation 1.79 and RT = 0.593 kcal/mol at 25°C, one can calculate the pK_a to be 6.7 which is consistent to that observed experimentally of 6.5 by absorption spectroscopy measurements.[168]

Free energy can also be described as a function of the cell potential (E^o_{cell}) which is characterized by electron transfer or redox reaction. Using Equation 1.84,

$$\Delta G^o = -nFE^o_{cell} \quad (1.84)$$

where n = is the number of electrons transferred in a half-reaction and F = Faraday's constant (23.06 kcal/mol/V), one can predict the spontaneity of a reaction based on the standard electrode potential of a half cell reaction. Buettner had complied an extensive list of one electron reduction potential for a variety of half-cell reactions at pH 7.[169] Table 1.2 lists some of the reduction potentials of half reaction couples. Half-cell reactions are presented such that the species on the right side is the reduced form of the species in the left side. For example, the half-cell reaction, HO$^\bullet$, e$^-$, H$^+$/H$_2$O, can be written as HO$^\bullet$ + e$^-$ + H$^+$ $\rightarrow$ H$_2$O with a reduction potential of E^o = 2.31 V at standard conditions. Oxidation of H$_2$O can be written in reverse, that is, H$_2$O $\rightarrow$ HO$^\bullet$ + e$^-$ + H$^+$ but the sign has to be reversed, that is, E^o = −2.31 V. It should be noted that half-cell reaction potentials involving H$^+$ or HO$^-$ can be pH dependent. Table 1.2 generally shows that the species with the most positive reduction potential (in this case HO$^\bullet$) is the most reducing and is therefore the easiest to oxidize.

To predict the spontaneity of a reaction based on reduction potentials, one can write two half-cell reactions where one is a reduction and the other is an oxidation process. For example, in Fenton chemistry, the reaction of Fe(II) with H$_2$O$_2$ is represented below. Note that the sign for the reduction potential of Fe(II) is negative (Eq. 1.85) since Fe(II) is oxidized to Fe(III) in this reaction.

$$Fe(II) \rightarrow Fe(III) + e^- \quad \Delta E^o = -0.11\ V \quad (1.85)$$
$$H_2O_2 + e^- + H^+ \rightarrow H_2O + HO^\bullet \quad \Delta E^o = +0.39\ V \quad (1.86)$$
$$\overline{H_2O_2 + Fe(II) + H^+ \rightarrow H_2O + HO^\bullet + Fe(III)} \quad \Delta E^o = +0.28\ V \quad (1.87)$$

TABLE 1.2 Reduction Potentials for Various Half-Cell Reactions Showing One-Electron and Two-Electron Oxidants[114]

Half-cell reactions	ΔE^o (vs NHE) At pH 7 in V at 25°C
HO$^\bullet$, e$^-$, H$^+$/H$_2$O	2.31
CO$_3^{\bullet-}$, e$^-$, H$^+$/HCO$_3^-$	2.10
RO$^\bullet$, e$^-$, H$^+$/ROH	1.60
2NO, 2e$^-$, 2H$^+$/N$_2$O, H$_2$O	1.59
H$_2$O$_2$, 2e$^-$, 2H$^+$/2H$_2$O	1.35 (1.78)[89]
HOO$^\bullet$, e$^-$, H$^+$/H$_2$O$_2$	1.05
ROO$^\bullet$, e$^-$, H$^+$/ROOH	1.00
NO$_3^-$, e$^-$, 4H$^+$/NO, 2H$_2$O	0.96
NO$_3^-$, 2e$^-$, 3H$^+$/HNO$_2$, H$_2$O	0.93
RS$^\bullet$, e$^-$/RS$^-$	0.92
O$_2^{\bullet-}$, e$^-$, 2H$^+$/H$_2$O$_2$	0.91
N$_2$O$_4$, 2e$^-$/2NO$_2^-$	0.87
O$_2$, 4e$^-$, 4H$^+$/2H$_2$O	0.85 (1.23)[89]
^{1}O$_2$, e$^-$/O$_2^{\bullet-}$	0.81
PUFA$^\bullet$, e$^-$, H$^+$/PUFA-H	0.60
$^\bullet$NO$_2$, e$^-$/NO$_2^-$	0.60
α-Tocopheroxyl$^\bullet$, e$^-$, H$^+$/α-Tocopherol	0.50
H$_2$O$_2$, e$^-$, H$^+$/H$_2$O, HO$^\bullet$	0.39
O$_2$, 2e$^-$, 2H$^+$/H$_2$O$_2$	0.36
ascorbate$^\bullet$, e$^-$, H$^+$/ascorbic acid	0.28
Fe(III), e$^-$/Fe(II)	0.11
NO$_3^-$, 2e$^-$, H$_2$O/NO$_2^-$ + 2HO$^-$	0.01
O$_2$, e$^-$/O$_2^{\bullet-}$	−0.18
FAD, 2e$^-$, 2H$^+$/FADH$_2$	−0.22
NADP$^+$, 2e$^-$, H$^+$/NADPH	−0.32
NAD$^+$, 2e$^-$, H$^+$/NADH	−0.32
O$_2$, e$^-$, H$^+$/HOO$^\bullet$	−0.46
NO, e$^-$/^{3}NO$^-$	−0.81
2NO$_3^-$, 2e$^-$, 2H$_2$O/N$_2$O$_4$ + 4HO$^-$	−0.85
GSSG, e$^-$/GSSG$^{\bullet-}$	−1.5

Adapted from References 89 and 114.

The net equation gave a positive ΔE^o value of +0.28 V (Eq. 1.87). For a reaction to occur spontaneously, the ΔG^o must be negative. However, according to Equation 1.84, E^o must be positive to meet the requirement for spontaneity, and therefore, reaction of H$_2$O$_2$ with Fe(II) is considered highly favorable.

1.3.2 Kinetic Considerations

Although free energies are useful entities to predict if a reaction will take place, it does not address the rate by which the process will occur. Thermodynamics only describes the relative stability of the reactants versus products. The rate of reaction is proportional to the molar concentration of a component (Eq. 1.88).

$$-d[\text{reactant}]/dt \text{ or } +d[\text{product}]/dt \quad (1.88)$$

at isothermal and constant volume. As the reaction proceeds, the reactant/s concentrations decrease and this is accompanied by a decrease in the rate of the reaction as they usually tend to slow down overtime. Since rates have variability, a way to quantify the rate of a chemical reaction is through the use of an experimental measure of a reaction rate which is usually referred to as rate constants (k). (Note that by convention, small letter k is referred to as the rate constant and the capitalized K as equilibrium constant). Rate constant is independent of how far the reaction proceeded and its scale. Reactive species in biological systems could exhibit unimolecular, bimolecular or higher order reactions and each of these types of reaction are described by a rate constant.

1.3.2.1 Unimolecular or First-Order Reactions
Only one reactant in which the rate of its reaction is solely proportional to its concentration at constant volume where the reaction is described in Equation 1.89,

$$A \rightarrow \text{products} \quad (1.89)$$

and where the rate law is described in Equation 1.90:

$$\text{Rate of reaction} = -d[A]/dt = k_1[A] \quad (1.90)$$

Experimentally, one can determine the first-order rate constant (k_1) by monitoring the formation or decay of A as a function of time (Eq. 1.91).

$$\ln\left(\frac{[A]_t}{[A]_0}\right) = -k_1 t \text{ or } \log\left(\frac{[A]_t}{[A]_0}\right) = \frac{-k_1 t}{2.303} \quad (1.91)$$

where $[A]_0$ and $[A]_t$ are concentrations at time = 0 and time = t, respectively. The first-order rate constant has a dimension of time^{-1} and is usually expressed in s^{-1} unit. The half-life ($t_{1/2}$) of a first-order reaction which is the time required for the [A] to decrease by 50% is described as

$$k_1 t_{1/2} = 0.693$$

Therefore, based on this equation, by knowing $t_{1/2}$, one will be able to determine k_1. Examples of this reaction is the decomposition of GSSG$^{\bullet-}$ to form GS$^-$ and GS$^{\bullet}$, or ONOOH to form NO$_2^{\bullet}$ and HO$^{\bullet}$.

1.3.2.2 Bimolecular or Second-Order Reactions
This reaction occurs from two reactants that are the same species (Eq. 1.92). The rate law is described in Equation 1.93.

$$A + A \rightarrow \text{products} \quad (1.92)$$

$$\text{Rate of reaction} = -d[A]/dt = k_2[A]^2 \quad (1.93)$$

where the rate is proportional to the instantaneous concentration of A. The second-order rate constant is usually expressed in M^{-1}/s unit.

Experimentally, one can determine the second-order rate constant (k_2) by monitoring the formation or decay of A as a function of time (Eq. 1.94).

$$\frac{1}{[A]_t} - \frac{1}{[A]_0} = k_2 t \quad (1.94)$$

The half-life for second-order reaction is described by Equation 1.95,

$$k_2 t_{1/2} = 1/[A] \quad (1.95)$$

which indicates that the $t_{1/2}$ of the second-order rate constant is inversely proportional to [A]. Examples of this reaction are the bimolecular reaction between two HO$^{\bullet}$ to form H$_2$O$_2$, or the dismutation of HOO$^{\bullet}$ to form H$_2$O$_2$ and O$_2$.

Majority of reactions, however, are between two different species (Eq. 1.96) as described by the rate law (Eq. 1.97):

$$A + B \rightarrow \text{products} \quad (1.96)$$

$$\text{Rate of reaction} = -d[A]/dt = -d[B]/dt = k_2[A][B] \quad (1.97)$$

The integrated rate law for k_2 determination is described by Equation 1.98:

$$\ln\left(\frac{[A]_t}{[B]_t}\right) = ([A]_0 - [B]_0) k_2 t + \ln\left(\frac{[A]_0}{[B]_0}\right) \quad (1.98)$$

To simplify the kinetic measurements, second-order kinetics can be investigated using first-order rate law by making one of the reagents in large excess. For example, if A is in large excess over B, that is, $[A]_0 >>> [B]_0$, then $[A]_t \sim [A]_0$, therefore, Equation 1.98 can be rewritten as $k_2[A]_0 = k_1'$ where k_1' is the pseudo-first-order rate constant that is related to the concentrations of B according to Equation 1.99,

$$\ln\left(\frac{[B]_t}{[B]_0}\right) = -k_1' t \quad (1.99)$$

Using the known initial concentration of the reactant that is in excess, that is $[A]_0$, the second-order rate constant k_2 can be calculated from k_1'.

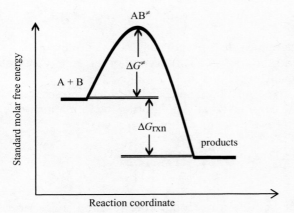

Figure 1.33 Classical reaction coordinate for an exothermic reaction showing the free energies of activation ($\Delta G^{\neq}$) and reaction (ΔG_{rxn}).

1.3.2.3 Transition State Theory, Reaction Coordinates and Activation Energies

Transition state theory is the current model used to describe a chemical reaction in terms of physical processes. It assumes that reactions are in equilibrium between the reactants and an activated transition state structure. By determining the reaction rate constants (k_c), the standard Gibbs free energy of activation ($\Delta G^{\neq}$) can be calculated using Equation 1.100,

$$k_c = \frac{k_B T}{h} e^{-\Delta G^{\neq}/RT} \quad (1.100)$$

where $k_B T/h$ is the universal factor composed of Boltzman (k_B) and Planck (h) constants and the absolute temperature (T).

In a simple reaction coordinate composed of reactants (A + B), activated complex (AB$^{\neq}$) and products, the potential energy diagram for an exothermic reaction is shown in Figure 1.33.

The activated complex lie at the saddle point (highest energy of a potential energy surface) and is in "quasi-equilibrium" with the reactant molecules which is later converted into products.

The magnitude of $\Delta G^{\neq}$ therefore determines the rate of the reaction; that is, the higher the activation barrier the slower the reaction rate will be. One also has to consider that free energy is temperature dependent and hence the kinetics of a reaction. Several external factors can affect the magnitude of $\Delta G^{\neq}$ and the rate of reactions. For example, increased temperature, concentration, and pressure can increase the probability of collision between two particles and therefore, the rate of reaction increases. Catalysts such as enzymes provide lower activation barrier by increasing the collision rate between reactants by arranging the orientation of the reactants for optimal reactivity; by changing the electronic property of the reactants though increased electrophilicity or nucleophilicity; through changes in intramolecular forces of attraction that can hinder reactants reactivity; or by simply providing alternative pathways for the reaction mechanism.

The range of rates by which reactions in biological system occurs is wide from very slow (<1) to diffusion controlled rate (10^9–10^{10}). Table 1.3 shows the various biologically relevant reaction and their experimental rate constants.

Based on Table 1.3, in general, the fastest reactions (10^9–10^{10}) involve either addition reaction or electron transfer reaction between two radicals. Intermediate rate reactions (10^5–10^8) are mostly characterized by H-atom abstraction, reaction between radical anions or electron transfer between the pi-radicals such as in the case of NO and O_2. Slow reactions (10^{-2}–10^4), are mostly unimolecular decomposition that involves bond breaking of N–O, O–O or N–N bonds and electron transfer between anions and neutral molecules.

1.4 ORIGINS OF REACTIVE SPECIES

1.4.1 Biological Sources

Among the numerous reactive species formed in biological systems, $O_2^{\bullet-}$ and NO are the two major precursors. The enzymatic generation of $O_2^{\bullet-}$ and $^{\bullet}$NO has been shown to originate from O_2 and arginine, respectively as substrates. These radicals are formed in various subcellular compartments such as membrane, mitochondria, endoplasmic reticulum[172] or golgi apparatus.[173] Below are the common sources of $O_2^{\bullet-}$ and NO but the mechanistic details will be left in the succeeding chapters and the list below only offers a general overview of the different enzymes responsible for their generation.

1.4.1.1 NADPH Oxidase

Superoxide radical anion are generated through stimulated professional phagocytes (e.g., neutrophils, macrophages monocytes, dendritic cells and mast cells).[174] Pentose phosphate pathway generates NADPH during the oxidative phase in which two molecules of NADP$^+$ are reduced to NADPH though the utilization of glucose-6-phosphate into ribulose 5-phosphate according to Equation 1.101,

$$\text{glucose-6-phosphate} \xrightarrow{\text{2 NADP}^+ \quad \text{2NADPH/H}^+} \text{ribulose 5-phosphate} \quad (1.101)$$

where NADPH subsequently reduce O_2 to $O_2^{\bullet-}$ via the NADPH oxidase pathway (Eq. 1.102). The details of which are discussed in Chapter 2.

TABLE 1.3 Various Reactions of Reactive Species and their Respective Rate Constants at Normal Conditions

Reaction	Rate Constants
$O_2^{\bullet-} + {}^{\bullet}NO \to ONOO^-$	$1.9 \times 10^{10}\ M^{-1}\ s^{-1}$
$O_2^{\bullet-} + HO^{\bullet} \to O_2 + HO^-$	$1.0 \times 10^{10}\ M^{-1}\ s^{-1}$
${}^{\bullet}NO + HO^{\bullet} \to HNO_2$	$1.0 \times 10^{10}\ M^{-1}\ s^{-1}$
${}^{\bullet}NO_2 + HO^{\bullet} \to {}^{\bullet}NO_2 + HO^-$	$1.0 \times 10^{10}\ M^{-1}\ s^{-1}$
${}^{\bullet}NO + R^{\bullet} \to RNO$	$1.0 \times 10^{10}\ M^{-1}\ s^{-1}$
$2O_2^{\bullet-} + 2H^+ \to O_2 + H_2O_2$ (SOD catalyzed)	$1.0 \times 10^{9}\ M^{-1}\ s^{-1}$
$O_2^{\bullet-} + {}^{\bullet}NO_2 \to O_2NOO^-$	$4.5 \times 10^{9}\ M^{-1}\ s^{-1}$
${}^{\bullet}NO_2 + {}^{\bullet}NO_2 \to N_2O_3$	$1.1 \times 10^{9}\ M^{-1}\ s^{-1}$
${}^{\bullet}NO + TyrO^{\bullet} \to Tyr\text{-}ONO$	$1.0 \times 10^{9}\ M^{-1}\ s^{-1}$
${}^{\bullet}NO_2 + TyrO^{\bullet} \to Tyr\text{-}NO_2$	$1.3 \times 10^{9}\ M^{-1}\ s^{-1}$
$NO_2^- + HO^{\bullet} \to {}^{\bullet}NO_2 + HO^-$	$5.3 \times 10^{9}\ M^{-1}\ s^{-1}$
$GSSG^{\bullet-} + O_2 \to GSSG + O_2^{\bullet-}$	$5 \times 10^{9}\ M^{-1}\ s^{-1}$
$2GS^{\bullet} \to GSSG$	$1.5 \times 10^{9}\ M^{-1}\ s^{-1}$
${}^{\bullet}NO + GS^{\bullet} \to GSNO$	$3 \times 10^{9}\ M^{-1}\ s^{-1}$
${}^{\bullet}NO_2 + GS^{\bullet} \to GSNO_2$	$3 \times 10^{9}\ M^{-1}\ s^{-1}$
$GS^{\bullet} + GSNO \to GSSG + {}^{\bullet}NO$	$1.7 \times 10^{9}\ M^{-1}\ s^{-1}$
$GS^{\bullet} + O_2 \to GSOO^{\bullet}$	$2 \times 10^{9}\ M^{-1}\ s^{-1}$
$GSOO^{\bullet} + {}^{\bullet}NO_2 \to GSOONO_2$	$1 \times 10^{9}\ M^{-1}\ s^{-1}$
$GSOO^{\bullet} + {}^{\bullet}NO \to GSOONO$	$3 \times 10^{9}\ M^{-1}\ s^{-1}$
$CO_3^{\bullet-} + {}^{\bullet}NO + HO^- \to HCO_3^- + NO_2^-$	$3.9 \times 10^{9}\ M^{-1}\ s^{-1}$
$GSOO^{\bullet} + GSNO \to GSSG + O_2 + {}^{\bullet}NO$	$3.8 \times 10^{8}\ M^{-1}\ s^{-1}$
$CO_3^{\bullet-} + O_2^{\bullet-} + H^+ \to HCO_3^- + O_2$	$4.0 \times 10^{8}\ M^{-1}\ s^{-1}$
$N_2O_3 + RH \to RNO + H^+ + NO_2^-$	$1.8 \times 10^{8}\ M^{-1}\ s^{-1}$
$CO_3^{\bullet-} + Tyr \to HCO_3^- + TyrO^{\bullet}$	$4.5 \times 10^{7}\ M^{-1}\ s^{-1}$
$2TyrO^{\bullet} \to diTyr$	$8.05 \times 10^{7}\ M^{-1}\ s^{-1}$
$N_2O_3 + GSH \to GSNO + H^+ + NO_2^-$	$6.6 \times 10^{7}\ M^{-1}\ s^{-1}$
${}^{\bullet}NO_2 + GSH \to GS^{\bullet} + H^+ + NO_2^-$	$2 \times 10^{7}\ M^{-1}\ s^{-1}$
$H_2O_2 + \text{Catalase-Fe(III)} \to \text{Compound 1}$	$k_1 = 1.7 \times 10^{7}$;
$\text{Compound 1} + H_2O_2 \to \text{Cat Fe(III)} + 2H_2O + O_2$	$k_2 = 2.6 \times 10^{7}\ M^{-1}\ s^{-1}$
$UH_2^- + {}^{\bullet}NO_2 \to NO_2^- + UH^{\bullet-} + H^+$	$1.8 \times 10^{7}\ M^{-1}\ s^{-1}$
$2{}^{\bullet}NO + O_2 \to 2{}^{\bullet}NO_2$	$2 \times 10^{6}\ M^{-1}\ s^{-1}$
$4{}^{\bullet}NO + O_2 + 2H_2O \to 4HNO_2$	$8.0 \times 10^{6}\ M^{-1}\ s^{-1}$
$GS^{\bullet} + GS^- \to GSSG^{\bullet-}$	$9.6 \times 10^{6}\ M^{-1}\ s^{-1}$
$CO_3^{\bullet-} + GSH \to HCO_3^- + GS^{\bullet}$	$5.3 \times 10^{6}\ M^{-1}\ s^{-1}$
$LOO^{\bullet} + TOH \to LOOH + TO^{\bullet}$	$2.5 \times 10^{6}\ M^{-1}\ s^{-1}$
$UH^{\bullet-} + Asc^- \to UH_2^- + A^{\bullet-}$	$1 \times 10^{6}\ M^{-1}\ s^{-1}$
$ROO^{\bullet} + UH_2^- \to ROO^- + UH^{\bullet-} + H^+$	$3 \times 10^{6}\ M^{-1}\ s^{-1}$
$CO_3^{\bullet-} + RH \to HCO_3^- + R^{\bullet}$	$4 \times 10^{5}\ M^{-1}\ s^{-1}$
${}^{\bullet}NO_2 + Tyr \to NO_2^- + TyrO^{\bullet}$	$3.2 \times 10^{5}\ M^{-1}\ s^{-1}$
$GSSG^{\bullet-} \to GS^{\bullet} + GS^-$	$1.6 \times 10^{5}\ s^{-1}$
$GSOO^{\bullet} \to GS^{\bullet} + O_2$	$6 \times 10^{5}\ s^{-1}$
${}^{\bullet}NO_2 + RH \to NO_2^- + R^{\bullet} + H^+$	$3.2 \times 10^{5}\ M^{-1}\ s^{-1}$
$GSH + TyrO^{\bullet} \to GS^{\bullet} + Tyr$	$3.5 \times 10^{5}\ M^{-1}\ s^{-1}$
$GS^{\bullet} + Tyr \to GSH + TyrO^{\bullet}$	$3.5 \times 10^{5}\ M^{-1}\ s^{-1}$
$2O_2^{\bullet-} + 2H^+ \to O_2 + H_2O_2$	$2.54 \times 10^{5}\ M^{-1}\ s^{-1}$
$N_2O_3 \to {}^{\bullet}NO + {}^{\bullet}NO_2$	$8.1 \times 10^{4}\ s^{-1}$
$ONOO^- + CO_2 \to NO_3^- + CO_2$	$2 \times 10^{4}\ M^{-1}\ s^{-1}$
$ONOO^- + CO_2 \to {}^{\bullet}NO_2 + CO_3^{\bullet-}$	$1 \times 10^{4}\ M^{-1}\ s^{-1}$
$Tyr\text{-}ONO \to {}^{\bullet}NO + TyrO^{\bullet}$	$1 \times 10^{3}\ s^{-1}$
$Urate + ONOO^- \to products$	$4.8 \times 10^{2}\ M^{-1}\ s^{-1}$
$ONOO^- + GSH \to NO_2^- + GSOH$	$6.6 \times 10^{2}\ M^{-1}\ s^{-1}$
$LOO^{\bullet} + LH \to LOOH + L^{\bullet}$	$10\text{–}50\ M^{-1}\ s^{-1}$
$GSOONO_2 \to GSOO^{\bullet} + {}^{\bullet}NO_2$	$0.75\ s^{-1}$
$ONOO^- + H^+ \to HNO_3$	0.568
$ONOO^- + H^+ \to {}^{\bullet}NO_2 + HO^{\bullet}$	$0.232\ s^{-1}$
$UH^{\bullet-} + O_2 \to$ no measurable reaction	$<10^{-2}\ M^{-1}\ s^{-1}$

Adapted from References 170, 171 and 277.

$$\text{NADPH} \longrightarrow \text{NADP}^+ \qquad (1.102)$$
$$O_2 \quad O_2^{\bullet-}$$

1.4.1.2 Xanthine Oxidoreductase or Oxidase

During ischemia, ATP is metabolized to adenosine and through adenosine deaminase, adenosine is converted to inosine which further decomposes to hypoxanthine (Eq. 1.103).[175] Although hypoxanthine can be converted to xanthine by xanthine oxidase (XO) via a reductive half-reaction, xanthine can be independently formed from GMP through purine metabolism. This catalytic purine degradation is also associated with the formation of H_2O_2 and $O_2^{\bullet-}$.

$$\text{ATP} \to \text{ADP} \to \text{AMP} \to \text{adenosine} \to \text{inosine} \to \text{hypoxanthine} \qquad (1.103)$$

XO belongs to a family of molybdoflavoenzymes and is released by a calcium-triggered protease during hypoxia (Eq. 1.104).

$$\xrightarrow[\text{Ca}^{2+}/\text{protease}]{\text{xanthine dehydrogenase}} \text{xanthine oxidase (XO)} \qquad (1.104)$$

Hypoxanthine or xanthine can undergo reductive half-reaction with XO at the Mo–Co centers. Two electrons are transferred to XO from xanthine, thereby reducing Mo(VI) to Mo(IV). The oxidative half-reaction then takes place at FAD where electron transfer between the reduced Mo–Co occurs with FAD as mediated by $Fe_2\text{-}S_2$ centers, thus maintaining Mo to be as Mo(VI) and FAD as $FADH_2$. Transfer of electrons from $FADH_2$ to NAD^+ or O_2 occurs during the reoxidation of fully six electron-reduced XO. The first two processes involve 2-electron reduction of O_2 to form H_2O_2, then the remaining two electrons are each used to reduced O_2 to $O_2^{\bullet-}$. The total ROS produced is, therefore, two molecules of each H_2O_2 and $O_2^{\bullet-}$ (Eq. 1.105).

$$(1.105)$$

1.4.1.3 Mitochondrial Electron Transport Chain (METC)

Metabolism of O_2 involves a series of electron transfer between an electron donor (NADH) and

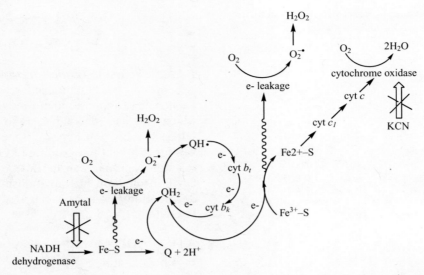

Figure 1.34 The ubiquinone cycle of the mitochondrial electron transport chain showing the formation of reactive oxygen species.

an electron acceptor, O_2, via the METC, with concomitant transfer of protons from the inner mitochondrial membrane.[176] This process involves transfer of four electrons from cytochrome c oxidase to O_2 to form two molecules of water and four molecules of H^+ according to Equation 1.106:

$$4\,\text{cyt}\,c_{\text{red}} + O_2 + 8H^+ \rightarrow 4\,\text{cyt}\,c_{\text{ox}} + 2H_2O + H^+ \quad (1.106)$$

However, partial metabolism of O_2 occurs prior to its full reduction to water by cytochrome c oxidase. Although it is estimated that under normal conditions, 1–2% of O_2 consumed by mitochondria are converted to ROS. This phenomenon called *electron leakage* maybe more prevalent in pathophysiological conditions.

The major sources of radical generation within the mitochondria have been identified to be the NADH dehydrogenase and ubiquinone. Figure 1.34 shows the ubiquinone cycle in which ubiquinone (Q) reduces cytochrome b through multiple processes that also leads to the oxidation of NADH dehydrogenase. The cycle is coupled to the electron transfer process that occurs between ubiquinol (QH2) and cytochrome c1 via proteins containing Fe–S clusters. At the site of this electron transfer process, an electron is "leaked" to the O_2 molecule to give $O_2^{\bullet-}$ then subsequently forming H_2O_2. Studies on heart and nonsynaptic brain mitochondria of mammals and birds show that oxygen radicals are generated at complex I in heart and brain mitochondria in States 4 and 3, while complex III (ubiquinone cytochrome c reductase) generates radicals only in heart mitochondria and only in State 4.[177] Other sources of ROS in the mitochondria are the dehydrogenases, quinone oxidoreductase and monoamine oxidase B.

1.4.1.4 Hemoglobin (Hb) Oxygen binds to the heme Fe(II) on a reversible and stable manner and is the basis of Hb function. However, the Fe(II) heme can undergo auto-oxidation (~3 within 24-hour period) to form Fe(III) and $O_2^{\bullet-}$ (Eq. 1.107) and is a common mechanism of oxidative stress in red blood cells.[178]

$$Hb(II)O_2 \rightarrow Hb^+ + O_2^{\bullet-} \quad (1.107)$$

1.4.1.5 Nitric Oxide Synthases Nitric oxide synthase catalyzes the production of nitric oxide from L-arginine via an electron flow from NADPH→ FAD → FMN → heme → oxygen based on Equation 1.108.[179]

$$\text{L-arginine} + \frac{3}{2}\text{NADPH} + H^+ + 2O_2$$
$$\rightarrow \text{citrulline} + NO + \frac{3}{2}\text{NADP}^+ \quad (1.108)$$

The Fe(III) heme upon reduction by $FMNH_2$ to Fe(II) enables binding to O_2 to form the ferrous-dioxy complex or $Fe(III)O_2^-$ (species 1). Species 1 can presumably further undergo a one-electron reduction by tetrahydrobiopeterin (H_4B) to form the iron-peroxo species (species II) and O–O bond cleavage yields water (Fig. 1.35) and iron-oxo species which is thought to hydroxylate the guanindino nitrogen of the L-arginine and ultimately leading to the generation of NO (Fig. 1.36).

Under oxidative conditions such as in the presence of ONOO$^-$, the oxidation state for H_4B is altered such that conversion of species 1 to 2 is hampered. The peroxo group of species 1 then decomposes to $O_2^{\bullet-}$ and Fe(III).

1.4.1.6 Cytochrome P450 (CYP)
CYP is one of the most important class of enzymes responsible for the oxidation of organic substances using lipids and steroids as well as xenobiotics as substrates.[180,181] The catalytic action of CYP mirrors that of NOS enzymes where the formation of oxo-ferryl ($Fe^{IV} = O$), species II (shown in Fig. 1.35) is the oxidizing form of the heme. Like in NOS, non-reduction of $Fe(III)O_2^-$ results in the production of $O_2^{\bullet-}$.

1.4.1.7 Cyclooxygenase (COX) and Lipoxygenase (LPO)
Arachidonic metabolism can mediate several important cellular events such as inflammation, chemotaxis, and regulation of muscle tone. However, the formation of metabolites such as prostaglandins, thromboxane and leukotriene generates ROS.[182] The formation of PGG_2 and HpETEs hydroperoxides has been shown to be mediated by COX and LPO. These unstable peroxides can yield $HO^\bullet$ and $RO^\bullet$ via O–O bond cleavage.

1.4.1.8 Endoplasmic Reticulum (ER)
ER is an organelle responsible for protein folding and maturation. Along with Golgi complex, it is involved in the transport of new proteins, lipids and other small molecules to their proper destination. Recently, ER has been implicated in hypoxia- and diabetes-mediated oxidative stress.[172] During accumulation of newly synthesized unfolded proteins, the unfolded pretein response (UPR) is activated and causes a variety of inflammatory and stress signaling responses. The mechanism of radical production from ER was proposed to originate from an enzyme Ero1p, a flavin-containing oxidase, due to its ability to reduce molecular O_2 to yield H_2O_2 when acting on thiol substrates according to Equation 1.109 and Equation 1.110.[183]

$$E\text{-}FAD + 2\ RSH \rightarrow EFADH_2 + RSSR \quad (1.109)$$

$$EFADH_2 + O_2 \rightarrow EFAD + H_2O_2 \quad (1.110)$$

Ero1p is an enzyme responsible for the disulfide bond formation in eukaryotic cells under aerobic and anaerobic conditions. The ability of Ero1p to transfer electron to other small molecules and macromolecular electron acceptor has also been demonstrated.

1.4.2 Nonbiochemical Sources

1.4.2.1 Photolysis
Shown in Equation 1.111 and Equation 1.112 is the generation of $O_2^{\bullet-}$ during ionizing radiation of air-saturated sodium formate using stopped-flow radiolysis apparatus on line with a Van de Graaff electron generator at 2-MeV.[184]

$$H_2O \rightsquigarrow H_2O_2 + H_3O^+ + H^\bullet + HO^\bullet + H_2 + e_{aq}^-$$

$$H^\bullet + O_2 \rightarrow HO_2^\bullet$$

$$e_{aq}^- + O_2 \rightarrow O_2^{\bullet-} \quad 2.3 \times 10^{10}\ M^{-1}\ s^{-1} \quad (1.111)$$

$$HO_2^\bullet \rightleftharpoons O_2^{\bullet-} + H^+$$

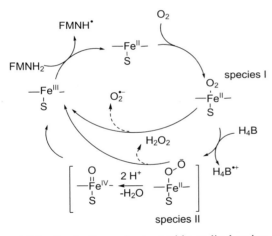

Figure 1.35 Production of superoxide radical anion from nitric oxide synthase. (Adapted with permission from *Chem. Rev.*, **2003**, *103*(6), 2365–2384. Copyright 2003 American Chemical Society.)

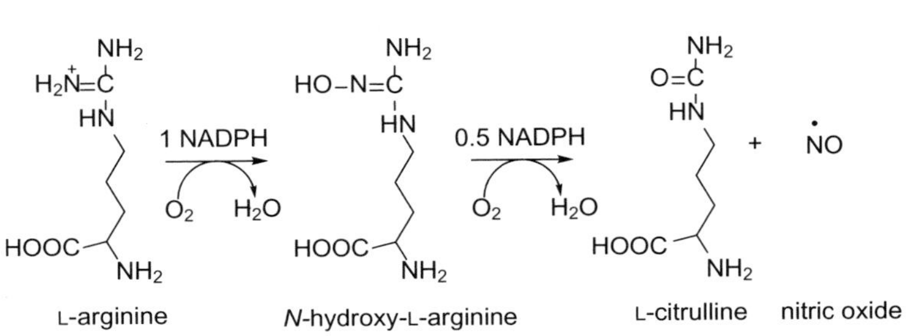

Figure 1.36 Production of nitric oxide from L-arginine, NADPH, and O_2.

$$HCOO^- + HO^\bullet \rightarrow CO_2^{\bullet -} + H_2O \quad 2.5 \times 10^9 \ M^{-1}\ s^{-1}$$
$$CO_2^{\bullet -} + O_2 \rightarrow O_2^{\bullet -} + CO_2 \quad 4.6 \times 10^9 \ M^{-1}\ s^{-1}$$
$$(1.112)$$

Superoxide can also be formed in O_2-saturated aqueous formate solution upon short UV irradiation by Xe or Ar lamp. The main process involves photochemical decomposition of water through its dissociation into $HO^\bullet$ and $H^\bullet$ (Eq. 1.113). In the presence of O_2, electron transfer occurs to produce $O_2^{\bullet -}$ (Eq. 1.114). Moreover, formate ($HCOO^-$) can react with $H^\bullet$ or $HO^\bullet$ to form a common product $CO_2^{\bullet -}$, where $CO_2^{\bullet -}$ can further reduce O_2 to $O_2^{\bullet -}$ (Eq. 1.115).[185]

$$H_2O + h\nu \rightarrow HO^\bullet + H^\bullet \quad (1.113)$$

$$H^\bullet + O_2 \rightarrow HO_2^\bullet = H^+ + O_2^{\bullet -} \quad 1.6 \times 10^{10} \ M^{-1}\ s^{-1} \quad (1.114)$$

$$H^\bullet + HCO_2^- \rightarrow H_2 + CO_2^{\bullet -} \quad 1.5 \times 10^8 \ M^{-1}\ s^{-1}$$
$$HO^\bullet + HCO_2^- \rightarrow H_2O + CO_2^{\bullet -} \quad 2.5 \times 10^9 \ M^{-1}\ s^{-1}$$
$$CO_2^{\bullet -} + O_2 \rightarrow CO_2 + O_2^{\bullet -} \quad 1.0 \times 10^{10} \ M^{-1}\ s^{-1}$$
$$(1.115)$$

Photolysis of H_2O_2 solution can also convert $HO^\bullet$ and $H^\bullet$ to $HO_2^\bullet$ via reactions shown in Equation 1.116 and Equation 1.117[185]:

$$H^\bullet + H_2O_2 \rightarrow HO^\bullet + H_2O \quad (1.116)$$

$$HO^\bullet + H_2O_2 \rightarrow HO_2^\bullet + H_2O \quad (1.117)$$

UV-photolysis of H_2O_2 yields $HO^\bullet$ via O–O bond homolytic cleavage,[186] while photolysis of alkyl disulfides results in C–S and S–S homolytic bond cleavage.[187]

1.4.2.2 Sonochemical Acoustic cavitation involves the nucleation, growth and violent collapse of gas-filled microbubbles in a liquid. The bubble collapse is associated with the creation of a transient region with very high temperature (>1000 K) and pressure (>100 atm) in pressure. It is known that the collapse of the bubbles is accompanied by the emission of light, a process known as sonoluminescence. Ultrasound-induced pyrolysis of argon-purged water showed formation of $H^\bullet$ and $HO^\bullet$ using spin trapping technique.[188,189] Sonolysis in the presence of drugs also known as sonosensitizers can yield $ROO^\bullet$ and $RO^\bullet$ exhibiting enhanced therapeutic action against cancer cells for example.[190]

1.4.2.3 Photochemical There are three major pathways for the photochemical generation of $O_2^{\bullet -}$ as shown in Figure 1.37: (1) via photoionization of a sensitizer molecule which generates hydrated electron (e_{aq}^-) which in turn directly reduces O_2 to $O_2^{\bullet -}$; (2) use of excited state acceptor (Sen) that can accept electron from a ground state electron donor such as an amine or other electron-rich substrates to form ($Sen^{\bullet -}$) which then leads to the reduction of O_2 by $Sen^{\bullet -}$ to form $O_2^{\bullet -}$; (3) and through electron transfer with O_2 or $^1O_2$ by a sensitized excited or ground state donor, respectively.[191]

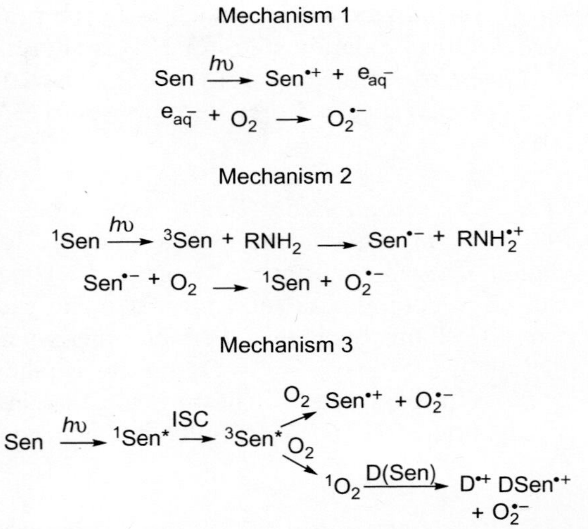

Mechanisms 1, 2, and 3 require O_2 for the production of $O_2^{\bullet -}$. For Mechanism 1, tryptophan and other amino acids as well as other aromatic compounds (such as amines, phenols, methoxybenzenes and indoles) are capable of generating $O_2^{\bullet -}$ under photoionizing conditions in near-UV light. Mechanism 2 involves charge-transfer mechanism which is very common among flavin and its analogues which usually occurs in the presence of an electron donor such as EDTA. Triplet state methylene blue in the presence of alkylamines results in electron transfer to generate $O_2^{\bullet -}$. Mechanism 3 shows the formation of $O_2^{\bullet -}$ from $^1O_2$ ($^1O_2$ is generated from O_2 sensitization by rose bengal) using furfuryl alcohol,[192] fullerenes,[193] or quinones[194] as specific $^1O_2$ quenchers.

Figure 1.37 Various photochemical mechanisms for the formation of reactive oxygen species.

1.4.2.4 Electrochemical The standard potential of $O_2/O_2^{\bullet -}$, $E^0 = -0.284$ V (vs. NHE). Using this potential, $O_2^{\bullet -}$ can be electrochemically generated from one-electron reduction of O_2 in alkaline aqueous solution,[195] DMSO[196,197] or ionic liquids.[198]

1.4.2.5 Chemical Tetramethylmmonium salt of $O_2^{\bullet -}$ (Me_4NO_2) can be prepared from solid state metathesis combination of KO_2 and Me_4NOH with high purity.[199] Also, Me_4NO_2 can be prepared from NH_3 treatment of KO_2 with Me_4NF or reaction of $Me_4NOH \cdot 5H_2O$ with excess KO_2.[200]

Quinones are active sites of mitochondrial *bc* complex (III) or as active moieties of xenobiotics.[201] One electron reduction of quinone leads to the formation of semiquinone. In general, reduction of quinone to hydroquinone can be accomplished nonenzymatically via two-electron reduction with the reducing equivalents of NADPH, or by one-electron reduction to semiquinone with microsomal or mitochondrial enzymes (Fig. 1.38). Semiquinone can reduce O_2 to form $O_2^{\bullet-}$ and the original quinone. The process is repeated until ROS production is at its maximum and semiquinone begins to accumulate, at which time the system becomes depleted with O_2. This process is usually referred to as redox cycling. The diagram below shows the redox cycling of ROS by quinone as mediated by an electron donor, NADPH. Redox cycling has also been observed in ortho-bezoquinones.

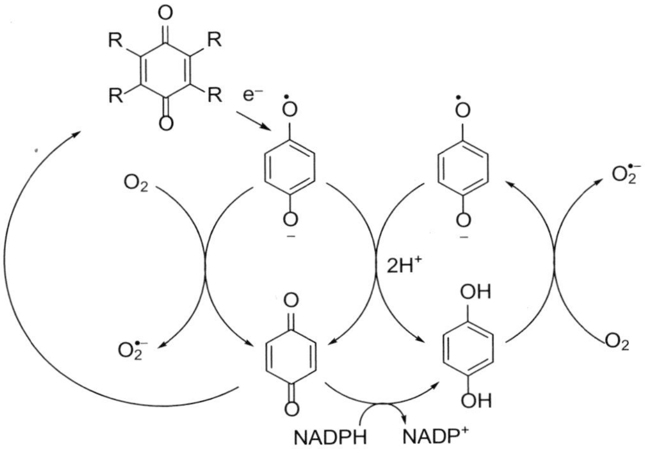

Nitric oxide can be generated from chemical sources directly or indirectly by enzymatic or nonenzymatic systems. Nitric oxide can be photochemically or thermally generated from metal-NO complexes, *N*-nitrosamines, *N*-hydroxyl nitrosamines, nitrosoimines, nitrosothiols, C-nitrosothiols, and diazetine dioxides. NO can also be generated indirectly through enzymatic metabolism of organic nitrates/nitrites, guanidines, hydroxyureas, oximes, oxatriazole-5-imines, or furoxans.[202]

Stable $ONOO^-$ solution can be generated directly from ozone and sodium azide,[203] nitrite and H_2O_2,[43] or organic nitrite and H_2O_2.[204] Since 3-*N*-morpholinosydnonimine (SIN-1) comes in solid form, the use of SIN-1 is the most common form of $ONOO^-$ delivery due to its ease of handling. Figure 1.39 shows the proposed decomposition pathway for SIN-1, which involves electron transfer reaction with O_2 to form $O_2^{\bullet-}$. The oxidized SIN-1 intermediate decomposes to form NO. Combination of the generated NO and $O_2^{\bullet-}$ in solution then yields $ONOO^-$.[205]

Figure 1.38 NADPH-mediated redox cycling of ROS by quinones.

1.5 METHODS OF DETECTION

Reactive species in *in vitro* and in *in vivo* systems can be directly or indirectly detected. Due to the instability and short half-lives of the common radicals, reagents are needed for their detection. By exploiting the chemistry of radical addition or electron transfer reactions to some reagents, one can use this process as an analytical tool to detect ROS production. In particular, the detection

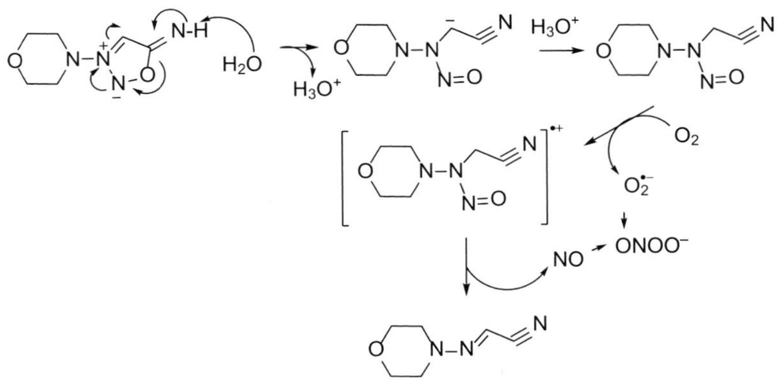

Figure 1.39 Proposed mechanism for the generation of peroxynitrite from SIN-1.[205]

of $O_2^{\bullet-}$ is the most relevant since it is the major precursor of most reactive species and it signals the first sign of oxidative burst in biological systems. However, in *in vitro* and *in vivo* systems where the flux of $O_2^{\bullet-}$ can be below the detection limit of the commonly used analytical techniques, secondary products such as the formation of other ROS, RNS, RSS as well as the formation of biomolecular radicals such as protein or nucleotide radicals can be directly or indirectly detected. Analysis of the primary or secondary addition products of radicals to substrates can be analyzed using various methods such as by chromatography, electrochemistry, mass spectrometry, spectrophotometry or by magnetic resonance spectroscopy.

1.5.1 In Vitro

1.5.1.1 Flourescence and Chemiluminescence
Fluorescence (FL) occurs when light is absorbed by a fluorophore (excitation) with subsequent emission of light, while chemiluminescence (CL) occurs with the emission of light as a result of a chemical reaction; the latter is more sensitive than the former by 2 orders of magnitude.[206] Although FL and CL are among the most sensitive techniques for radical detection *in vitro* (i.e., >1 nM), caution is needed for their application. FL and CL probes are capable of detecting various ROS/RNS via two-electron oxidation, and therefore, suffer from selectivity and may compete with endogenous intracellular antioxidants such as ascorbate, urate and thiols. Moreover, these probes can generate $O_2^{\bullet-}$ via formation of an active intermediate after reaction with ROS/RNS. Due to the lack of selectivity to a particular reactive species, FL and CL are more appropriately called redox probes to indicate their general reactivity to various reactive species. Some of the most common FL probes are dichlorodihydrofluorescein (DCFH$_2$), rhodamine (RhH$_2$) and ethidine (DHE) (Fig. 1.40). DCFH$_2$ and RhH$_2$ react with $O_2^{\bullet-}$ or H_2O_2 poorly, and fluorescence arising from this reaction could be catalyzed by metal ion impurities, and therefore are not suitable probes for ROS. Carbonate radical anion and $^\bullet NO_2$ are better detected using DCFH$_2$ due to their higher reactivity and higher fluorescence yield, however, this is not true for HO$^\bullet$ and HOCl. RhH$_2$ gives fluorescence with all of the reactive species. Unlike DCFH$_2$ and RhH$_2$, DHE is highly reactive to $O_2^{\bullet-}$ but yields two fluorescent products: (1) specific to $O_2^{\bullet-}$ (i.e., 2-hydroxyethidium, 2-OH-E$^+$); and (2) nonspecific to $O_2^{\bullet-}$ that can be formed photochemically (E$^+$). To differentiate between 2-OH-E$^+$ and E$^+$, the use of HPLC/FL assay has been suggested and provides unequivocal differentiation of the two products.[207] In spite of this complication, DHE is currently perhaps the most specific FL probe for $O_2^{\bullet-}$.

Lucigenin (LC) is the most commonly used CL probe for $O_2^{\bullet-}$ but like DCFH$_2$, it can also generate $O_2^{\bullet-}$ and is not specific for $O_2^{\bullet-}$ because it also gives luminescence in the presence of nucleophiles and reducing agents to form LC$^{\bullet+}$. Addition of $O_2^{\bullet-}$ to LC$^{\bullet+}$ (formed from its enzymatic reduction) forms a dioxetane intermediate that cleaves to form the excited state *N*-methylacridone which later can emit light (Fig. 1.41). Other reductants that can generate LC$^{\bullet+}$ are H_2O_2, flavoproteins, eNOS, NADPH reductases and cyt P450.[206]

Boronates have been shown to react with ONOO$^-$, HOCl, and H_2O_2 imparting fluorescence but with varying rates of reaction.[208,209] The second-order rate constants show ONOO$^-$ to be the most reactive (~10^5–10^6 M^{-1} s^{-1}) followed by HOCl (~10^3–10^4 M^{-1} s^{-1}) and by H_2O_2 (~2 M^{-1} s^{-1}).[209] The mechanism of oxidant reaction to boronates involves nucleophilic addition of the oxidant to the boron atom followed by the heterolytic

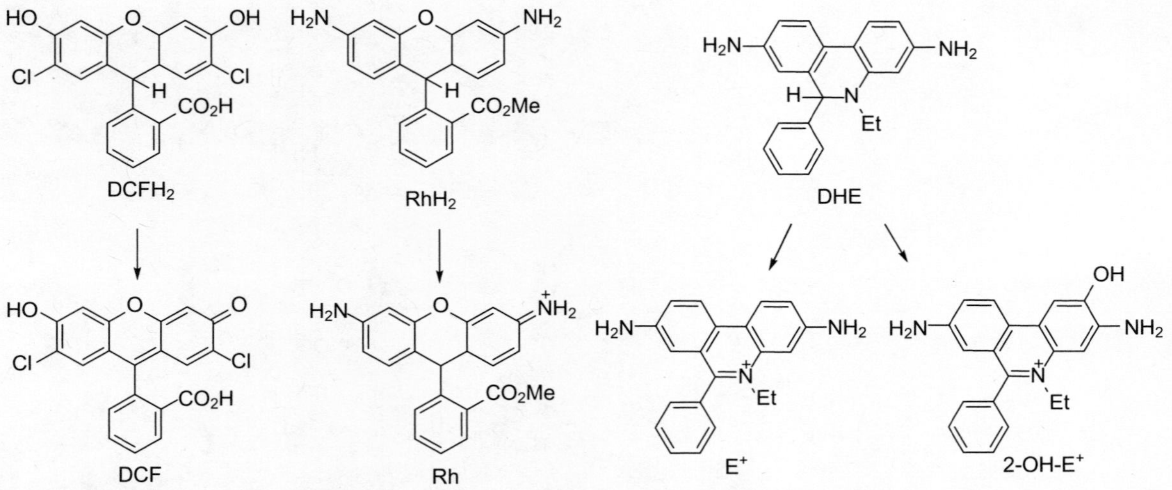

Figure 1.40 Fluorescence probes for ROS and their various products.

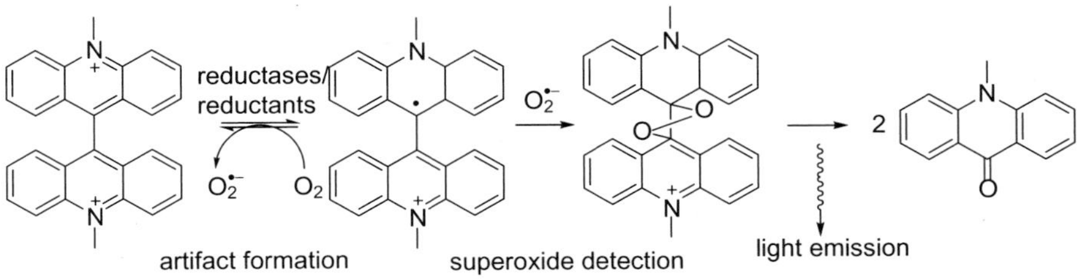

Figure 1.41 Formation of superoxide radical anion from lucigenin$^{\bullet+}$ and its reaction with superoxide resulting in chemiluminescence.

cleavage of the X–O bond to form the respective ion. Intramolecular rearrangement of the B–O yields the final fluorescent phenolic products (Eq. 1.118).

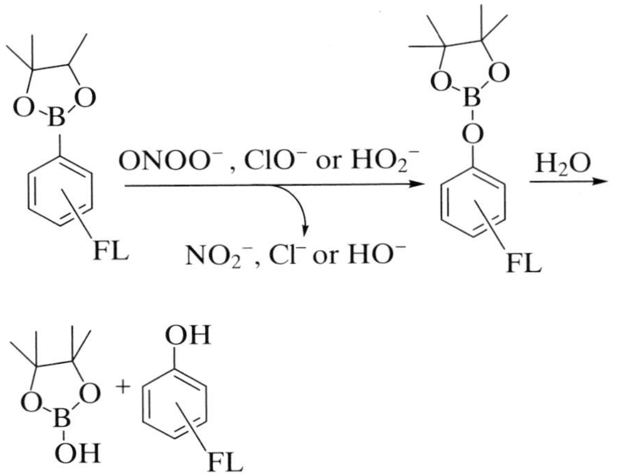

$$NO + O_3 \rightarrow NO_2^* + O_2 \quad (1.120)$$

$$NO_2^* \rightarrow NO_2 + h\nu \quad (1.121)$$

1.5.1.2 UV-Vis Spectrophotometry and HPLC

Several assays for $O_2^{\bullet-}$ based on 1-electron transfer reaction have been employed due to the high rate constants observed for this type of reaction. Cytochrome (cyt) c^{3+} can be reduced to cyt c^{2+} by $O_2^{\bullet-}$ and can be detected spectrophotometrically. Due to the relatively low rate constant of this reaction (~10^5 M^{-1} s^{-1}), the amount of $O_2^{\bullet-}$ generated can be underestimated. Another popular spectrophotometric technique for $O_2^{\bullet-}$ detection is through the use of p-nitrotetrazolium blue (NBT) which forms a colored monoformazan anion. However, the use of cyt c and NBT have limitations such that their reduction is not specific to $O_2^{\bullet-}$ and cannot be applied in *in vivo* systems.[212]

Nitric oxide can be measured by using reduced hemoglobin according to Equation 1.122. Oxidation of hemoglobin to methemoglobin can be detected spectrophotometrically with a detection threshold of 1 nmol.

$$NO + Hb(Fe^{2+})O_2 \rightarrow Hb(Fe^{3+}) + NO_3^- \quad (1.122)$$

Also, by using thioproline, NO can be trapped and the adduct formed can be detected using mass spectroscopy.[213] Nitrite as an oxidation end product of NO can also be detected spectrophotometrically using Griess assay. Nitrite is detected as red pink coloration produced from the reaction of sulphanilic acid with NO_2^- where the product formed reacts further with an azo dye (alpha-naphthilamine) giving a colored product. In systems where NO_3^- are present, prior reduction of NO_3^- to NO_2^- is required to obtain the total NO_2^-/NO_3^- content by treatment of the sample with sodium formate and nitrate reductase.[214]

For NO detection, fluorescence probes have been employed such as those containing the vicinal diamines (e.g., flourescein based, DAF-2; rhodamine-based, DAR-4M; BODIPY-based, DAMBO; and cyanine-based, DACs). Reaction of NO with diamine proceeds in the presence of oxygen to form the highly fluorescent *N*-nitrosated product (Eq. 1.119).[210]

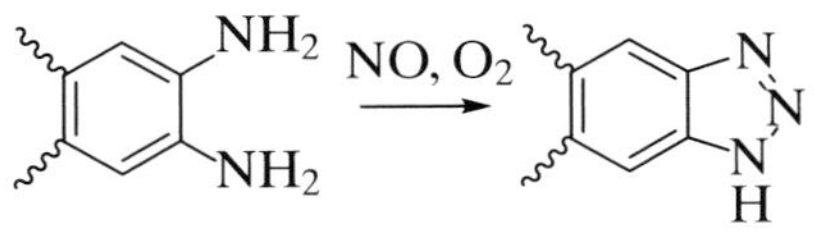

(1.119)

The reaction of NO with ozone imparts chemiluminescence and has been exploited to detect NO formation. This ozone-based detection of NO in the gas phase involves light emission along with the formation of $^\bullet NO_2$ (Eq. 1.120 and Eq. 1.121).[211] This technique, although very sensitive, requires the use of an NO analyzer equipped with ozone generator and sensitive photomultiplier tube and purging of NO from the sample is required.

Hypochlorous acid can be trapped by taurine forming taurine chloramine. Taurine chloramine can then be spectrophotometrically assayed using 5-thio-2-nitrobenzoic acid (TNB)[215] but has some limitations such as the need to predetermine the chloramine concentration for accurate measurements, poor selectivity as other oxidants can bleach TNB, and the light sensitivity of TNB. Iodide was proposed to be an alternative to TNB and by using 3,3′,5,5′-tetramethylbenzidine (TMB) as chromophore due to its ability to be oxidized by hypoiodous acid with a sensitivity of 1 μM of taurine chloramine. Dihydrorhodamine was also used as chromophore giving 10-fold greater sensitivity than TMB.[216]

Other radicals such as $HO^{\bullet}$, $^{\bullet}NO_2$ or $HO_2^{\bullet}$ have been shown to form adducts with various substrates such as amino acids, DNA bases or lipids via hydroxylation, nitration and hydroperoxide formation, respectively, which can be detected using a variety of analytical methods such as HPLC, electrochemical, spectrophotometric or by MS. Hydroxyl radical for example adds to 8-hydroxyguanine of the DNA to yield the 8-hydroxy-2-deoxygianosine (8-OHdG) which can be isolated and analyzed using HPLC/electrochemical methods.[217] Recently, more improved methods using HPLC/electrochemical detection for hydroxylation was proposed using 4-hydroxybenzoic acid and terephthalate assays which do not have the drawbacks associated with the use of salicylate or phenylalanine.[218] Using tandem mass spectroscopic techniques, nitration of amino acid residues in peptides[219] have been demonstrated while proteomic approach have been successful in identifying nitration in proteins.[220,221]

Ferrous oxidation-xylenol orange (FOX) assay has been used as a conventional technique for hydroperoxide formation in lipids,[222,223] and peptide/protein systems.[27,224,225] FOX assay technique uses the Fenton chemistry to generate the $HO^{\bullet}$ from the O–O homolytic cleavage from the ROOH which decolorizes the xylenol orange dye. RNS adduct of lipids have been characterized using HPLC coupled with UV and MS detection.[226] MDA, being one of the end products of lipid hydroperoxide formation, can be measured using TBARS assay with thiobarbituric acid (TBA) as a reagent. Caution is required in interpreting TBARS data since MDA participates in other reactions other than TBA and is not exclusively formed from lipid peroxidation. Moreover, MDA is only formed from a particular lipid peroxidation process out of a myriad of several lipid peroxidation decomposition reactions.[227] Analysis of F_2-isoprostanes (F_2–IPs) are more reliable marker of lipid peroxidation which possesses a 1,3-dihydroxycyclopentane ring with the OH groups in the *syn* position and are mainly formed from the arachidonic acid oxidation. Analysis of F_2–IPs can be carried *ex vivo* using LC/MS/MS or GC/MS.[48,228]

1.5.1.3 Immunochemical
Formation of macromolecular radical systems such as protein and DNA radi-

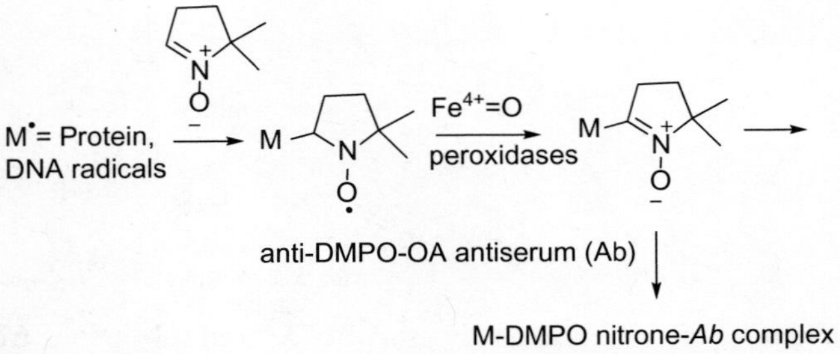

Figure 1.42 Immuno-spin trapping of macromolecular radicals using DMPO and anti-DMPO octanoic acid antiserum.

cals are relevant intermediates for the initiation of oxidative stress in biological systems. Nitrone spin traps, for example 5,5,dimethyl pyrroline N-oxide (DMPO) adds to protein radicals to form a longer-lived protein spin adduct. This radical adduct can later form the diamagnetic analogue, nitrone-protein adduct (formed via a variety of oxidative pathways mostly mediated by heme iron centers) which can be detected at 1 μM sensitivity using polyclonal antibodies against DMPO coupled to octanoic acid antiserum. This immune-spin trapping (ISP)[229] approach combines the specificity of spin trapping to free radical formation and antigen–antibody interactions (Fig. 1.42). Coupled with MS/MS technique, one would be able to also identify the specific site/s of radical formation. Immunochemical detection using ISP has been employed in various radical systems formed from hemoglobin-tyrosyl,[230] myoglobin-tyrosyl,[231] Cu,Zn-SOD,[232] thyroid peroxidase,[233] catalase-peroxidase,[234] and DNA.[235–237]

1.5.1.4 Electron Paramagnetic Resonance (EPR) Spectroscopy
EPR spectroscopy exploits the magnetic moment of an electron through absorption of microwave radiation in the presence of external magnetic field. As shown in Figure 1.43, there are three major approaches for the detection of $O_2^{\bullet-}$ using EPR and various probes, that is, (1) spin-quenching (or spin-loss) using trityl; (2) spin-formation using hydroxylamine; (3) spin trapping using nitrones where the former involves loss of signal and the latter two involve signal formation. Due to the inherent stability of trityl radicals, they have been employed as probes for the detection of $O_2^{\bullet-}$. The most common are the triarylmethyl (trityl)-based radicals[238–240] which can undergo electron transfer reaction with $O_2^{\bullet-}$ to yield O_2 and the EPR silent trityl anion. The synthetic trityl radicals, TAM OX063 and perchlorotriphenylmethyl (PCM-TC), have been shown to give high reactivity to $O_2^{\bullet-}$ with second-order rate constants of 3.1×10^3 M^{-1} s^{-1} and 8.3×10^8 M^{-1} s^{-1}, respectively.[238,239] Trityl radicals show inertness toward a majority of the

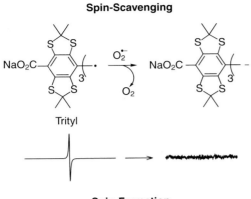

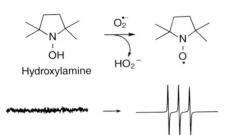

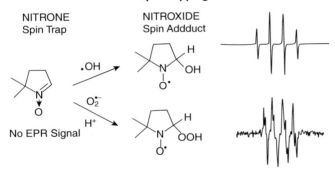

Figure 1.43 Radical detection using EPR spectroscopy and various radical probes.

common oxido-reductant species, however, they exhibit reactivity with other radical species, such as $HO_2^\bullet$, $RO_2^\bullet$ and $HO^\bullet$.[239]

Although trityls offer some degree of specificity to $O_2^{\bullet-}$, the loss of signal also brings concern about other unknown factors that can result in the loss of signal, and therefore, other non-EPR technique is recommended to confirm the results from using trityl as reagent. A spin-generating system uses diamagnetic probes such as hydroxylmines and nitrone spin traps which involves electron transfer or addition reactions with $O_2^{\bullet-}$, respectively, forming an EPR-detectable paramagnetic aminoxyl (or commonly called nitroxides) species. Hydroxylamines are oxidized by $O_2^{\bullet-}$ via a

Figure 1.44 Various N-hydroxy-pyrrole or piperidine derivatives used as probes for superoxide.

simple electron transfer mechanism to yield a paramagnetic aminoxyl species and H_2O_2. Figure 1.44 shows the commonly used hydroxylamines, TPO-H, TPL-H, TEMPONE-H,[241] CP–H,[242] and PP-H[243] which are N-hydroxy-pyrrole or piperidine derivatives able to react with $O_2^{\bullet-}$. Rate constants for hydroxylamine probe reaction with $O_2^{\bullet-}$ are dependent on the structure of the probe. In the case of the negatively charged probe, PP–H, its rate of reaction to $O_2^{\bullet-}$ was found to be slower with $k = 840 \pm 60$ M^{-1} s^{-1} due mostly to repulsive effect, while the neutral probes, TPO-H and TPL-H have higher rate constants in the range of $1–2 \times 10^3$ M^{-1} s^{-1}.[107] The redox reaction of $O_2^{\bullet-}$ with the hydroxylamine produces H_2O_2 and can be considered an artifactual source of other ROS. Since other ROS/RNS species as well as metal ions and O_2 can also give the exact same EPR triplet signal, caution should be practiced in data interpretation using hydroxylamine probes.

Detection of $^1O_2$ can be accomplished using the amine 2,2,6,6-tetramethyl-4-piperidone (TEMP). Reaction of $^1O_2$ with TEMP leads to the formation of the nitroxide 2,2,6,6-tetramethyl-4-piperidone-N-oxyl (TEMPO) which can be detected using EPR as shown in Equation 1.123.[244]

$$\text{TEMP} \xrightarrow{^1O_2} \text{TEMPO} \quad (1.123)$$

Spin traps are nitrone-based molecules. Although hydroxylamines exhibit 10- to 1000-fold higher reactivity to $O_2^{\bullet-}$ than the nitrones, the paramagnetic species generated from hydroxylamine does not allow discrimination between the different radicals generated.[243] Spin traps, however, add to a free radical at its α-carbon (C-2) position to form an aminoxyl adduct (or spin adduct), except that the signal is more complex than the ones observed from hydroxylamines (Fig. 1.43).[245] The complex spectrum of the spin adduct is due to the presence of a β-H and the nature of the radical moiety, and is the basis for their popularity not only in free radical detection but also in their identification. Shown in Figure 1.45 are the commonly used spin traps, and are divided into two major classes, the cyclic nitrones, 5,5-dimethyl-1-pyroline *N*-oxide (DMPO), 5-(ethoxycarbonyl)-5-methyl-1-pyrroline *N*-oxide (EMPO), and 5-(diethoxyphosphoryl)-5-methyl-1-pyrroline *N*-oxide (DEPMPO), and the linear, *N*-tert-butyl-α-phenylnitrone (PBN).

Aside from $O_2^{\bullet-}$, other radicals that can also be identified using spin trapping are $HO^{\bullet}$, $RO^{\bullet}$, $RS^{\bullet}$, $^{\bullet}NO_2$, $CO_3^{\bullet-}$, $CO_2^{\bullet-}$, $N_3^{\bullet-}$ and so on. The half-lives of the spin adducts vary significantly which range from seconds to hours depending on the type of the radical and nitrone used. The least stable are the $O_2^{\bullet-}$ adducts of DMPO and PBN with a half-life of <1 minute in aqueous solution. However, C-5 derivatized spin traps such as EMPO and DEPMPO exhibit longer $O_2^{\bullet-}$ adduct half-lives of ~8 and ~14 minutes, respectively. One major disadvantage of this technique, in spite of the improved $O_2^{\bullet-}$ adduct half-lives, is the slow reactivity of these spin traps with $O_2^{\bullet-}$ with rate constants ranging from <1–10 M^{-1} s^{-1} which requires the use of high concentrations (typically 10–100 mM) of these reagents in solution for $O_2^{\bullet-}$ detection. However, other radicals exhibit significantly fast reactivity and long adduct half-lives with spin traps.

Spin trapping has been employed to detect $O_2^{\bullet-}$ from xanthine oxidase,[238,246] the mitochondrial ETC,[247,248] and NADPH oxidase.[249] Nitrones have also been successfully used to detect $O_2^{\bullet-}$ generation in human epithelial cells,[250] human neutrophils,[251] reperfused cardiac tissue,[252] and small animals using *ex vivo* techniques.[253,254] *Ex vivo* spin trapping was also demonstrated in ischemia-reperfusion studies where the spin trap was administered to the animals before the onset of ischemia. Reperfusates were then collected and radical adduct generation was detected by EPR spectroscopy.[252,255]

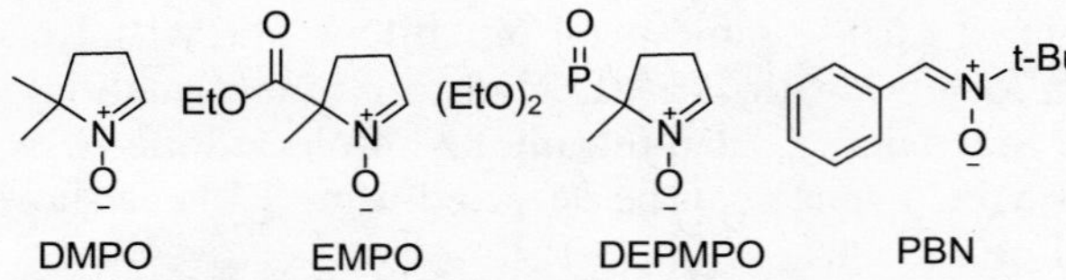

Nitric oxide does not add to nitrone spin traps but they undergo redox reaction with nitronyl nitroxides (NN) and addition reaction with iron-thiocarbamate complexes with fast rates of reaction whose products are detectable by EPR. Oxidation of NO by NN as opposed to reduction of O_2 by hydroxylamine occurs. There are two commonly used NN's, the 2-phenyl-4,4,5,5-tetramethylimidazoline-l-oxyl 3-oxide (PTIO) and its water soluble analogue 2-(4-carboxyphenyl)-4,4,5,5-tetramethylimidazoline-1-oxyl 3-oxide (C-PTIO) (Fig. 1.46).

Nitric oxide react with NN via addition reaction to the nitroxyl-O and subsequent liberation of $^{\bullet}NO_2$ to form the imino nitroxide (IN) as shown in Figure 1.46. The detection of NO through the use of NN shows clear distinction between the spectral profile imparted by the NN versus the IN product formed (Fig. 1.47). By virtue of symmetry, the spectral profile of NN is characterized by two equivalent N hyperfine splitting constants with $a_{N1,3}$ = 8.2 G, while the IN gives asymmetrical product with the two nitrogens giving two different hfsc's of a_{N1} = 9.8 G and a_{N3} = 4.4 G. The rate constant for thereaction of NO with NN is in the order of ~10^3 M^{-1} s^{-1} which is fast enough to compete with O_2 but not with $O_2^{\bullet-}$. Moreover, nitroxyl (HNO) also reacts with NN to form the similar IN product and that the $^{\bullet}NO_2$ formed can participate with other reactions involving NN and therefore requires careful consideration in the interpretation of the signal formed.[256,257] One main disadvantage of NN as probes for NO is their ability to be reduced to

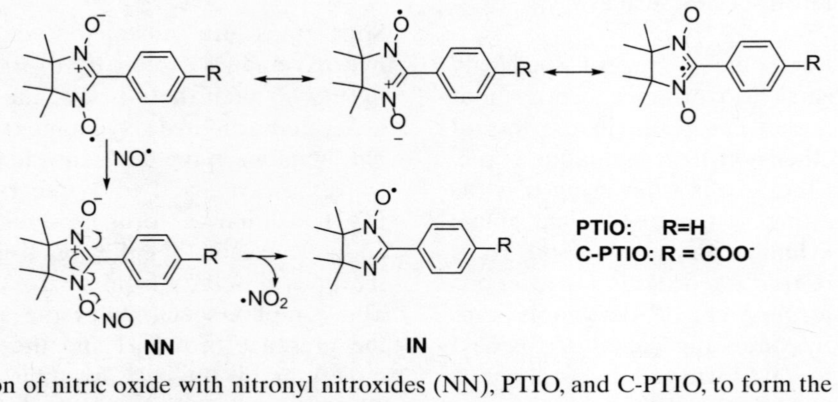

Figure 1.45 Commonly used spin traps for radical detection and identification.

Figure 1.46 Reaction of nitric oxide with nitronyl nitroxides (NN), PTIO, and C-PTIO, to form the imino nitroxide (IN).

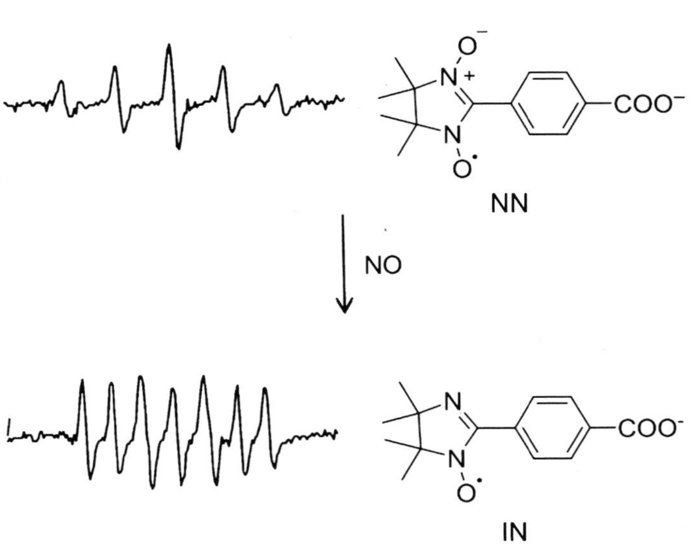

Figure 1.47 EPR spectra of nitronyl nitroxides (NN) and imino nitroxide (IN) after reaction with NO. (Adapted with permission from *J. Am. Chem. Soc.* **2010**, *132*(24), 8428–8432. Copyright 2010 American Chemical Society.)

EPR-silent hydroxylamine by reductants such as ascorbate or metal ions.

Several NO traps such as Fe^{2+}-dithiocarbamate complexes have been developed that allows NO detection using EPR (Fig. 1.48). This technique was first introduced by Vanin et al.[258–260] *In vivo* EPR experiments using mice showed that NO trapping by the hydrophobic Fe^{2+}-DETC is more efficient than by the hydrophilic Fe^{2+}-MGD due to the higher stability of the latter in animal tissues.[261] The redox state of Fe-dithiocarbamates plays a critical role in the detection of NO. Under aerobic condition, Fe^{2+} complex can be readily oxidized to form Fe^{3+}-dithiocarbamate. Reaction of ferric complex with NO forms the EPR-silent NO–Fe^{3+}-MGD complex but can be converted to an EPR detectable NO–Fe^{2+}-MGD by NO itself with 50% yield and by reductants such as ascorbate, hydroquinone, or cysteine with conversion efficiency of up to 99.9%. The use of iron carbamate complexes involves premixing of the $FeSO_4$ with excess dithiocarbamate co-ligand. Water insoluble $Fe(DETC)_2$ can be introduced as suspension with serum albumin[262] and has been reported to measure NO in porcine aorta with high sensitivity of 10 pmol/mL. Detection of NO in blood vessels as well as human umbilical endothelial cells has been successfully demonstrated using colloidal Fe^{2+}-DETC prepared by mixing DETC and Fe^{2+} in concentrated Krebs-HEPES solution.[263,264] Compared when using NN as probe for NO, Fe dithiocbamate complexes are better probes for the detection of NO due to the stability of the adducts formed. Cautions should be observed however since iron complexes also have been shown to detect HNO, nitrite and *S*-nitrosothiols. Dithiocarbamates have

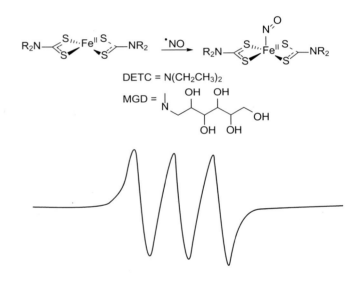

Figure 1.48 Complexation of nitric oxide with iron (II) dithiocarbamates, Fe-DETC, and Fe-MGD, giving a triplet EPR signal.

potential to chelate metals and may act as enzyme inhibitors. Through the use of NOS inhibitors, the triplet signal can be integrated to represent NOS-derived NO.[265] Nitric oxide can be trapped by hemoglobin/myoglobin (Hb) as well and can be detected using EPR but with the disadvantage of cooling the sample to ~100 K to allow observation of the signal thus making Hb impractical for real time monitoring of NO production but proved useful in determining NO production in tissues. However, deoxygenated ferrous haem forms HbNO and is detectable by EPR at normal conditions and the complex formed is very stable.[265]

1.5.2 In Vivo

Formation of reactive species *in vivo* are conventionally determined through analysis of biomarkers by using various methods such as histochemical, immunocytochemical, or EPR imaging. It should be noted that samples taken *in vivo* can be analyzed using the same techniques mentioned above employed for the formation of reactive species in *in vitro*.

1.5.2.1 Histochemical
Protein carbonyls are biomarkers of protein oxidation and their detection can be accomplished by their derivatization using dinitrophenyldrazine to form the protein-bound hydrazone and by using the anti-2,4-dintirophenyl antibody. Another approach is through the use of biotin-hydrazide in which the protein-bound acyl hydrazone is detected by the enzyme-linked avidin or streptavidin.[266]

1.5.2.2 Immunocytochemical Methods
Nitrotyrosine, lipid peroxidation end products, and DNA damage can be visualized in tissues using monoclonal or polyclonal antibodies for nitrotyrosine, HNE and 8-OHG, respectively.[266,267] Nitrated tyrosine has been considered as biochemical marker of ONOO-induced damage to proteins and lipids. By employing two dimensional polyacrylamide gel electrophoresis (2DE) and western blotting, coupled with mass spectroscopy, targets of protein nitration and HNE modification have been determined in protein systems.[220]

1.5.2.3 Low Frequency EPR Imaging
The availability of low frequency EPR instrumentation could limit the application of radical imaging to many investigators, however, provides direct visualization of probe response to ROS formation or O_2 concentrations in whole animals.[268,269] The method involves the use of low frequency, highly sensitive spectrometers, operating between 200 MHz and 1.5 GHz and paramagnetic probes. As mentioned earlier, probes such as nitroxides[252] and trityls react with ROS and show characteristic spectral behavior, that is, signal formation or its disappearance, respectively. Since ROS production has direct correlation with O_2 consumption, probes that respond to the pO_2 are very desirable such as trityl,[270] charcoal[271] and pthalocyanines,[272,273] the latter two are stable enough from being metabolized. Significant advancements have already been achieved in the development of highly sensitive detectors, data acquisition and analysis modalities. *In vivo* imaging of NO have also been achieved using commonly use iron-dithiocarbamate spin traps.[274]

1.5.2.4 In Vivo EPR Spin Tapping-Ex Vivo Measurement
In vivo spin trapping of radical metabolites have been extensively employed using the commonly used spin traps, DMPO, PBN or POBN. However, due to the susceptibility of the radical adducts to be reduced to diamagnetic hydroxylamine species, post-treatment of the samples are needed to re-oxidize the hydroxylamine back to the EPR-detectable nitroxide using mild oxidants such as potassium ferricyanide or bubbling with O_2. Carbon- or S-centered radicals have been detected from blood or biles samples after systemic injection of the xenobiotics. Spin traps have been extensively employed for the detection of transient radicals in animals.[253,275,276]

REFERENCES

1. Lelieveld, J. Atmospheric chemistry: A missing sink for radicals. *Nature* **2010**, *466*, 925–926.
2. Baumgarten, M., Muellen, K. Radical ions: Where organic chemistry meets materials sciences. *Top. Curr. Chem.* **1994**, *169*, 1–103.
3. Halliwell, B., Gutteridge, J.M.C. *Free Radicals in Biology and Medicine*, 4th ed. Oxford University Press, New York, 2007.
4. Gomberg, M. An instance of trivalent carbon: Triphenylmethyl. *J. Am. Chem. Soc.* **1900**, *22*, 757–771.
5. Adam, F.C., Weissman, S.I. Electron spin resonance and electronic structure of triphenylmethyl. *J. Am. Chem. Soc.* **1958**, *80*, 2057–2059.
6. Kazzaz, J.A., Xu, J., Palaia, T.A., Mantell, L., Fein, A.M., Horowitz, S. Cellular oxygen toxicity. Oxidant injury without apoptosis. *J. Biol. Chem.* **1996**, *271*, 15182–15186.
7. Song, Y., Wagner, B.A., Lehmler, H.J., Buettner, G.R. Semiquinone radicals from oxygenated polychlorinated biphenyls: Electron paramagnetic resonance studies. *Chem. Res. Toxicol.* **2008**, *21*(7), 1359–1367.
8. Fu, H., Zhou, Y.-H., Chen, W.-L., Deqing, Z.-G., Tong, M.-L., Ji, L.-N., Mao, Z.-W. Complexation, structure, and superoxide dismutase activity of the imidazolate-bridged dinuclear copper moiety with beta-cyclodextrin and its guanidinium-containing derivative. *J. Am. Chem. Soc.* **2006**, *128*(15), 4924–4925.
9. Valgimigli, L., Amorati, R., Fumo, M.G., DiLabio, G.A., Pedulli, G.F., Ingold, K.U., Pratt, D.A. The unusual reaction of semiquinone radicals with molecular oxygen. *J. Org. Chem.* **2008**, *73*(5), 1830–1841.
10. Sawyer, D.T., Valentine, J.S. How super is superoxide? *Acc. Chem. Res.* **1981**, *14*(12), 393–400.

11. Evans, M.G., Uri, N. Dissociation constant of hydrogen peroxide and the electron affinity of the HO2 radical. *J. Chem. Soc. Faraday Trans.* **1949**, *45*, 224–230.
12. Liu, G.-F., Filipovic, M., Ivanovic-Burmazovic, I., Beuerle, F., Witte, P., Hirsch, A. High catalytic activity of dendritic C60 monoadducts in metal-free superoxide dismutation. *Angew. Chem. Int. Ed Engl.* **2008**, *47*(21), 3991–3994.
13. Ali, S.S., Hardt, J.I., Quick, K.L., Sook Kim-Han, J., Erlanger, B.F., Huang, T.-T., Epstein, C.J., Dugan, L.L. A biologically effective fullerene (C60) derivative with superoxide dismutase mimetic properties. *Free Radic. Biol. Med.* **2004**, *37*(8), 1191–1202.
14. Osuna, S., Swart, M., Sola, M. On the mechanism of action of fullerene derivatives in superoxide dismutation. *Chem. Eur. J.* **2010**, *16*(10), 3207–3214, S3207/1–S3207/33.
15. Tietze, D., Tischler, M., Voigt, S., Imhof, D., Ohlenschlaeger, O., Goerlach, M., Buntkowsky, G. Development of a functional cis-prolyl bond biomimetic and mechanistic implications for nickel superoxide dismutase. *Chem. Eur. J.* **2010**, *16*(25), 7572–7578.
16. Batinic-Haberle, I., Spasojevic, I., Stevens, R.D., Hambright, P., Neta, P., Okado-Matsumoto, A., Fridovich, I. New class of potent catalysts of O2.bul.-dismutation. Mn(III) ortho-methoxyethylpyridyl- and di-ortho-methoxyethylimidazolylporphyrins. *Dalton Trans.* **2004**, *11*, 1696–1702.
17. Riley, D.P., Weiss, R.H. Manganese macrocyclic ligand complexes as mimics of superoxide dismutase. *J. Am. Chem. Soc.* **1994**, *116*(1), 387–388.
18. Nagano, T., Hirano, T., Hirobe, M. Superoxide dismutase mimics based on iron in vivo. *J. Biol. Chem.* **1989**, *264*(16), 9243–9249.
19. Ivanovic-Burmazovic, I., Eldik, R.V. Metal complex-assisted activation of small molecules. From NO to superoxide and peroxides. *Dalton Trans.* **2008**, *39*, 5259–5275.
20. Spasojevic, I., Batinic-Haberle, I., Reboucas, J.S., Idemori, Y.M., Fridovich, I. Electrostatic contribution in the catalysis of O2.bul.-dismutation by superoxide dismutase mimics. MnIIITE-2-PyP5+ versus MnIIIBr8T-2-PyP. *J. Biol. Chem.* **2003**, *278*(9), 6831–6837.
21. Krishna, M.C., Grahame, D.A., Samuni, A., Mitchell, J.B., Russo, A. Oxoammonium cation intermediate in the nitroxide-catalyzed dismutation of superoxide. *Proc. Natl. Acad. Sci. U.S.A.* **1992**, *89*(12), 5537–5541.
22. Krishna, M.C., Russo, A., Mitchell, J.B., Goldstein, S., Dafni, H., Samuni, A. Do nitroxide antioxidants act as scavengers of superoxide radical or as SOD mimics? *J. Biol. Chem.* **1996**, *271*(42), 26026–26031.
23. Chern, C.-I., DiCosimo, R., De Jesus, R., San Filippo, J., Jr. A study of superoxide reactivity. Reaction of potassium superoxide with alkyl halides and tosylates. *J. Am. Chem. Soc.* **1978**, *100*(23), 7317–7327.
24. Davico, G.E., Bierbaum, V.M. Reactivity and secondary kinetic isotope effects in the SN2 reaction mechanism: Dioxygen radical anion and related nucleophiles. *J. Am. Chem. Soc.* **2000**, *122*(8), 1740–1748.
25. Das, A.B., Nagy, P., Abbott, H.F., Winterbourn, C.C., Kettle, A.J. Reactions of superoxide with the myoglobin tyrosyl radical. *Free Radic. Biol. Med.* **2004**, *48*(11), 1540–1547.
26. Winterbourn, C.C., Parsons-Mair, H.N., Gebicki, S., Gebicki, J.M., Davies, M.J. Requirements for superoxide-dependent tyrosine hydroperoxide formation in peptides. *Biochem. J.* **2004**, *381*(Pt 1), 241–248.
27. Field, S.M., Villamena, F.A. Theoretical and experimental studies of tyrosyl hydroperoxide formation in the presence of H-bond donors. *Chem. Res. Toxicol.* **2008**, *21*(10), 1923–1932.
28. Sawyer, D.T., Stamp, J.J., Menton, K.A. Reactivity of superoxide ion with ethyl pyruvate, a-diketones, and benzil in dimethylformamide. *J. Org. Chem.* **1983**, *48*, 3733–3736.
29. Rene, A., Abasq, M.-L., Hauchard, D., Hapiot, P. How do phenolic compounds react toward superoxide ion? A simple electrochemical method for evaluating antioxidant capacity. *Anal. Chem.* **2010**, *82*, 8703–8710.
30. Salazar, R., Navarrete-Encina, P.A., Squella, J.A., Camargo, C., Nunez-Vergara, L.J. Reactivity of C4-indolyl substituted 1,4-dihydropyridines toward superoxide anion (O2-) in dimethylsulfoxide. *J. Phys. Org. Chem.* **2009**, *22*, 569–577.
31. Winterbourn, C.C., Metodiewa, D. Reactivity of biologically important thiol compounds with superoxide and hydrogen peroxide. *Free Radic. Biol. Med.* **1999**, *27*(3/4), 322–328.
32. Cardey, B., Foley, S., Enescu, M. Mechanism of thiol oxidation by the superoxide radical. *J. Phys. Chem. A* **2007**, *111*(50), 13046–13052.
33. Sutton, V.R., Stubna, A., Patschkowski, T., Munck, E., Beinert, H., Kiley, P.J. Superoxide destroys the [2Fe-2S]2+ cluster of FNR from *Escherichia coli*. *Biochemistry* **2004**, *43*(3), 791–798.
34. Flint, D.H., Tuminello, J.F., Emptage, M.H. The inactivation of Fe-S cluster containing hydro-lyases by superoxide. *J. Biol. Chem.* **1993**, *268*(30), 22369–22376.
35. Sawyer, D.T. The chemistry of dioxygen species (oxygen, super oxide radical anion, peroxide radical and hydrogen peroxide) and their activation by transition metals. *Int. Rev. Exp. Pathol.* **1990**, *31*, 109–131.
36. Cheng, Z., Li, Y. What is responsible for the initiating chemistry of iron-mediated lipid peroxidation: An update. *Chem. Rev.* **2007**, *107*(3), 748–766.
37. Czapski, G., Bielski, B.H.J. The formation and decay of H_2O_3 and HO_2 in electronirradiated aqueous solutions. *J. Phys. Chem.* **1963**, *67*(10), 2180–2184.
38. Haber, F., Weiss, J. The catalytic decomposition of hydrogen peroxide by iron salts. *Proc. R. Soc. London, Ser. A* **1934**, *147*, 332.
39. Perez-Benito, J.F. Iron(III)-hydrogen peroxide reaction: Kinetic evidence of a hydroxyl-mediated chain mechanism. *J. Phys. Chem. A* **2004**, *108*, 4853–4858.
40. Furtmuller, P.G., Obinger, C., Hsuanyu, Y., Dunford, H.B. Mechanism of reaction of myeloperoxidase with hydrogen

peroxide and chloride ion. *Eur. J. Biochem.* **2000**, *267*(19), 5858–5864.

41. Kettle, A.J., Winterbourn, C.C. A kinetic analysis of the catalase activity of myeloperoxidase. *Biochemistry* **2001**, *40*(34), 10204–10212.

42. Silanikove, N., Shapiro, F., Silanikove, M., Merin, U., Leitner, G. Hydrogen peroxide-dependent conversion of nitrite to nitrate as a crucial feature of bovine milk catalase. *J. Agric. Food Chem.* **2009**, *57*(17), 8018–8025.

43. Robinson, K.M., Beckman, J.S. Synthesis of peroxynitrite from nitrite and hydrogen peroxide. *Methods Enzymol.* **2005**, *396*, 207–214.

44. Oury, T.D., Tatro, L., Ghio, A.J., Piantadosi, C.A. Nitration of tyrosine by hydrogen peroxide and nitrite. *Free Radic. Res.* **1995**, *23*, 537–547.

45. Adimora, N.J., Jones, D.P., Kemp, M.L. A model of redox kinetics implicates the thiol proteome in cellular hydrogen peroxide responses. *Antioxid. Redox Signal.* **2010**, *13*(6), 731–743.

46. Scrivens, G., Gilbert, B.C., Lee, T.C.P. EPR studies of the copper-catalysed oxidation of thiols with peroxides. *J. Chem. Soc. Perkin Trans. 2* **1995**, *5*, 955–963.

47. Koppenol, W.H., Liebman, J.F. The oxidizing nature of the hydroxyl Radical. A comparison with the ferryl ion (FeO2+). *J. Phys. Chem.* **1984**, *88*, 99–101.

48. Li, H., Lawson, J.A., Reilly, M., Adiyaman, M., Hwang, S.W., Rokach, J., FitzGerald, G.A. Quantitative high performance liquid chromatography/tandem mass spectrometric analysis of the four classes of F(2)-isoprostanes in human urine. *Proc. Natl. Acad. Sci. U.S.A.* **1999**, *96*(23), 13381–13386.

49. Song, Y., Buettner, G.R., Parkin, S., Wagner, B.A., Robertson, L.W., Lehmler, H.J. Chlorination increases the persistence of semiquinone free radicals derived from polychlorinated biphenyl hydroquinones and quinones. *J. Org. Chem.* **2008**, *73*(21), 8296–8304.

50. Buxton, G.V., Greenstock, C.L., Helman, W.P., Ross, A.B. Critical review of rate constants for reactions of hydrated electrons, hydrogen atoms and hydroxyl radicals (OH/O) in aqueous solution. *J. Phys. Chem. Ref. Data* **1988**, *17*, 513–886.

51. Adhikari, S., Sprinz, H., Brede, O. Thiyl radical induced isomerization of unsaturated fatty acids: Determination of equilibrium constants. *Res. Chem. Intermed.* **2001**, *27*, 549–559.

52. Alam, M.S., Rao, B.S.M., Janata, E. •OH reactions with aliphatic alcohols: Evaluation of kinetics by direct optical absorption measurement. A pulse radiolysis study. *Radiat. Phys. Chem.* **2003**, *67*, 723–728.

53. Galano, A., Alvarez-Idaboy, J.R., Bravo-Pérez, G., Ruiz-Santoyo, M.E. Gas phase reactions of C1-C4 alcohols with the OH radical: A quantum mechanical approach. *Phys. Chem. Chem. Phys.* **2002**, *4*, 4648–4662.

54. Rachmilovich-Calis, S., Meyerstein, N., Meyerstein, D. A mechanistic study of the effects of antioxidants on the formation of malondialdehyde-like products in the reaction of hydroxyl radicals with deoxyribose. *Chem. Eur. J.* **2009**, *15*(31), 7717–7723.

55. Esterbauer, H., Schaur, R.J., Zollner, H. Chemistry and biochemistry of 4-hydroxynonenal, malonaldehyde and related aldehydes. *Free Radic. Biol. Med.* **1991**, *11*(1), 81–128.

56. Onyango, A.N., Baba, N. New hypotheses on the pathways of formation of malondialdehyde and isofurans. *Free Radic. Biol. Med.* **2010**, *49*(10), 1594–1600.

57. Bucknall, T., Edwards, H.E., Kemsley, K.G., Moore, J.S., Phillips, G.O. The formation of malonaldehyde in irradiated carbohydrates. *Carbohydr. Res.* **1978**, *62*, 49–59.

58. Gligorovski, S., Herrmann, H. Kinetics of reactions of OH with organic carbonyl compounds in aqueous solution. *Phys. Chem. Chem. Phys.* **2004**, *6*, 4118–4126.

59. Alvarez-Idaboy, J.R., Cruz-Torres, A., Galano, A., Ruiz-Santoyo, M.E. Structure-reactivity relationship in ketones + OH reactions: A quantum mechanical and TST. *J. Phys. Chem. A* **2004**, *108*, 2740–2749.

60. Alvarez-Idaboy, J.R., Mora-Diez, N., Boyd, R.J., Vivier-Bunge, A. On the importance of prereactive complexes in molecule-radical reactions: Hydrogen abstraction from aldehydes by OH. *J. Am. Chem. Soc.* **2001**, *123*, 2018–2024.

61. Ervens, B., Gligorovski, S., Herrmann, H. Temperature-dependent rate constants for hydroxyl radical reactions with organic compounds in aqueous solutions. *Phys. Chem. Chem. Phys.* **2003**, *5*, 1811–1824.

62. Morris, E.D., Niki, H. Reactivity of hydroxyl radicals with olefins. *J. Phys. Chem.* **1971**, *75*, 3640–3641.

63. Atkinson, R. Kinetics and mechanisms of the gas-phase reactions of the hydroxyl radical with organic compounds under atmospheric conditions. *Chem. Rev.* **1985**, *85*, 69–201.

64. Poole, J.S., Shi, X., Hadad, C.M., Platz, M.S. Reaction of hydroxyl radical with aromatic hydrocarbons in non-aqueous solutions: A laser flash photolysis study in acetonitrile. *J. Phys. Chem. A* **2005**, *109*, 2547–2551.

65. DeMatteo, M.P., Poole, J.S., Shi, X., Sachdeva, R., Hatcher, P.G., Hadad, C.M., Platz, M.S. On the electrophilicity of hydroxyl radical: A laser flash photolysis and computational study. *J. Am. Chem. Soc.* **2005**, *127*(19), 7094–7109.

66. Geeta, S., Rao, B.S.M., Mohan, H., Mittal, J.P. Radiation-induced oxidation of substituted benzaldehydes: A pulse radiolysis study. *J. Phys. Org. Chem.* **2004**, *17*, 194–198.

67. Enescu, M., Cardey, B. Mechanism of cysteine oxidation by a hydroxyl radical: A theoretical study. *ChemPhysChem.* **2006**, *7*(4), 912–919.

68. Cruz-Torres, A., Galano, A. On the mechanism of gas-phase reaction of C1-C3 aliphatic thiols + OH radicals. *J. Phys. Chem. A* **2007**, *111*(8), 1523–1529.

69. Karoui, H., Hogg, N., Frejaville, C., Tordo, P., Kalyanaraman, B. Characterization of sulfur-centered radical intermediates formed during the oxidation of thiols and sulfite by peroxynitrite. ESR-spin trapping and oxygen uptake studies. *J. Biol. Chem.* **1996**, *271*(11), 6000–6009.

70. DeRosa, M.C., Crutchley, R.J. Photosensitized singlet oxygen and its applications. *Coord. Chem. Rev.* **2002**, *233–234*, 351–371.
71. Wilkinson, F., Helman, W.P., Ross, A.B. Rate constants for the decay and reactions of the lowest electronically excited singlet state of molecular oxygen in solution. An expanded and revised compilation. *J. Phys. Chem. Ref. Data* **1995**, *24*, 663–1021.
72. Greer, A. Christopher Foote's discovery of the role of singlet oxygen [1O2 (1Delta g)] in photosensitized oxidation reactions. *Acc. Chem. Res.* **2006**, *39*(11), 797–804.
73. Sun, S., Bao, Z., Ma, H., Zhang, D., Zheng, X. Singlet oxygen generation from the decomposition of alpha-linolenic acid hydroperoxide by cytochrome c and lactoperoxidase. *Biochemistry* **2007**, *46*(22), 6668–6673.
74. Escobar, J.A., Vasquez-Vivar, J., Cilento, G. Free radicals and excited species in the metabolism of indole-3-acetic acid and its ethyl ester by horseradish peroxidase and by neutrophils. *Photochem. Photobiol.* **1992**, *55*(6), 895–902.
75. King, M.M., Lai, E.K., McCay, P.B. Singlet oxygen production associated with enzyme-catalyzed lipid peroxidation in liver microsomes. *J. Biol. Chem.* **1975**, *250*(16), 6496–6502.
76. Kiryu, C., Makiuchi, M., Miyazaki, J., Fujinaga, T., Kakinuma, K. Physiological production of singlet molecular oxygen in the myeloperoxidase-H2O2-chloride system. *FEBS Lett.* **1999**, *443*(2), 154–158.
77. Medeiros, M.H., Wefers, H., Sies, H. Generation of excited species catalyzed by horseradish peroxidase or hemin in the presence of reduced glutathione and H2O2. *Free Radic. Biol. Med.* **1987**, *3*(2), 107–110.
78. Zacharia, I.G., Deen, W.M. Diffusivity and solubility of nitric oxide in water and saline. *Ann. Biomed. Eng.* **2005**, *33*, 214–222.
79. Zhao, Y.L., Bartberger, M.D., Goto, K., Shimada, K., Kawashima, T., Houk, K.N. Theoretical evidence for enhanced NO dimerization in aromatic hosts: Implications for the role of the electrophile (NO)(2) in nitric oxide chemistry. *J. Am. Chem. Soc.* **2005**, *127*(22), 7964–7965.
80. Lewis, R.S., Deen, W.M. Kinetics of the reaction of nitric oxide with oxygen in aqueous solutions. *Chem. Res. Toxicol.* **1994**, *7*, 568–574.
81. Goldstein, S., Czapski, G. Kinetics of nitric oxide autoxidation in aqueous solution in the absence and presence of various reductants. The nature of the oxidizing intermediates. *J. Am. Chem. Soc.* **1995**, *117*, 12078–12084.
82. Czapski, G., Holcman, J., Bielski, B.H.J. Reactivity of nitric oxide with simple short-lived radicals in aqueous solutions. *J. Am. Chem. Soc.* **1994**, *116*, 11465–11469.
83. Batt, L., Rattray, G.N. The reaction of methoxy radicals with nitric oxide and nitrogen dioxide. *Int. J. Chem. Kinet.* **1979**, *11*, 1183–1196.
84. Goldstein, S., Lind, J., Merenyi, G. Reaction of organic peroxyl radicals with NO2 and NO in aqueous solution: Intermediacy of organic peroxynitrate and peroxynitrite species. *J. Phys. Chem. A* **2004**, *108*, 1719–1725.
85. Ford, P.C., Lorkovic, I.M. Mechanistic aspects of the reactions of nitric oxide with transition-metal complexes. *Chem. Rev.* **2002**, *102*(4), 993–1018.
86. Scheidt, W.R., Lee, Y.J., Hatano, K. Preparation and structural characterization of nitrosyl complexes of ferric porphyrinates. Molecular structure of aquonitrosyl(meso-tetraphenylporphinato)iron(III) perchlorate and nitrosyl(octaethylporphinato)iron(III) perchlorate. *J. Am. Chem. Soc.* **1984**, *106*, 3191–3198.
87. Enami, S., Hoffmann, M.R., Colussi, A.J. Absorption of inhaled NO(2). *J. Phys. Chem. B* **2009**, *113*(23), 7977–7981.
88. Huie, R.E. The reaction kinetics of NO_2. *Toxicology* **1994**, *89*, 193–216.
89. Dutton, A.S., Fukuto, J.M., Houk, K.N. Theoretical reduction potentials for nitrogen oxides from CBS-QB3 energetics and (C)PCM solvation calculations. *Inorg. Chem.* **2005**, *44*(11), 4024–4028.
90. Napolitano, A., Camera, E., Picardo, M., d'Ischia, M. Acid-promoted reactions of ethyl linoleate with nitrite ions: Formation and structural characterization of isomeric nitroalkene, nitrohydroxy, and novel 3-nitro-1,5-hexadiene and 1,5-dinitro-1, 3-pentadiene products. *J. Org. Chem.* **2000**, *65*(16), 4853–4860.
91. d'Ischia, M., Napolitano, A., Manini, P., Panzella, L. Secondary targets of nitrite-derived reactive nitrogen species: Nitrosation/nitration pathways, antioxidant defense mechanisms and toxicological implications. *Chem. Res. Toxicol.* **2011**, *24*(12), 2071–2092.
92. Solar, S., Solar, W., Getoff, N. Reactivity of hydroxyl with tyrosine in aqueous solution studied by pulse radiolysis. *J. Phys. Chem.* **1984**, *88*, 2091–2095.
93. Prütz, W.A., Mönig, H., Butler, J., Land, E.J. Reactions of nitrogen dioxide in aqueous model systems: Oxidation of tyrosine units in peptides and proteins. *Arch. Biochem. Biophys.* **1985**, *243*, 125–134.
94. Deeb, R.S., Resnick, M.J., Mittar, D., McCaffrey, T., Hajjar, D.P., Upmacis, R.K. Tyrosine nitration in prostaglandin H(2) synthase. *J. Lipid Res.* **2002**, *43*(10), 1718–1726.
95. Surmeli, N.B., Litterman, N.K., Miller, A.F., Groves, J.T. Peroxynitrite mediates active site tyrosine nitration in manganese superoxide dismutase. Evidence of a role for the carbonate radical anion. *J. Am. Chem. Soc.* **2010**, *132*(48), 17174–17185.
96. Abriata, L.A., Cassina, A., Tortora, V., Marin, M., Souza, J.M., Castro, L., Vila, A.J., Radi, R. Nitration of solvent-exposed tyrosine 74 on cytochrome c triggers heme iron-methionine 80 bond disruption. Nuclear magnetic resonance and optical spectroscopy studies. *J. Biol. Chem.* **2009**, *284*(1), 17–26.
97. Jourd'heuil, D., Jourd'heuil, F.L., Feelisch, M. Oxidation and nitrosation of thiols at low micromolar exposure to nitric oxide. Evidence for a free radical mechanism. *J. Biol. Chem.* **2003**, *278*(18), 15720–15726.

98. Hofstetter, D., Nauser, T., Koppenol, W.H. The glutathione thiyl radical does not react with nitrogen monoxide. *Biochem. Biophys. Res. Commun.* **2007**, *360*(1), 146–148.

99. Goldstein, S., Lind, J., Merenyi, G. Chemistry of peroxynitrites as compared to peroxynitrates. *Chem. Rev.* **2005**, *105*(6), 2457–2470.

100. Merenyi, G., Lind, J., Eriksen, T.E. The reactivity of superoxide (O2-) and its ability to induce chemiluminescence with luminol. *Photochem. Photobiol.* **1985**, *41*(2), 203–208.

101. Szabo, C., Ischiropoulos, H., Radi, R. Peroxynitrite: Biochemistry, pathophysiology and development of therapeutics. *Nat. Rev. Drug Discov.* **2007**, *6*(8), 662–680.

102. Pearce, L.L., Martinez-Bosch, S., Manzano, E.L., Winnica, D.E., Epperly, M.W., Peterson, J. The resistance of electron-transport chain Fe-S clusters to oxidative damage during the reaction of peroxynitrite with mitochondrial complex II and rat-heart pericardium. *Nitric Oxide* **2009**, *20*(3), 135–142.

103. Rubbo, H., Denicola, A., Radi, R. Peroxynitrite inactivates thiol-containing enzymes of Trypanosoma cruzi energetic metabolism and inhibits cell respiration. *Arch. Biochem. Biophys.* **1994**, *308*(1), 96–102.

104. Mere'nyi, G.b., Lind, J., Czapski, G., Goldstein, S. Direct determination of the Gibbs' energy of formation of peroxynitrous acid. *Inorg. Chem.* **2003**, *42*, 3796–3800.

105. Lymar, S.V., Hurst, J.K. Rapid reaction between peroxonitrite ion and carbon dioxide: Implications for biological activity. *J. Am. Chem. Soc.* **1995**, *117*, 8867–8868.

106. Bonini, M.G., Siraki, A.G., Atanassov, B.S., Mason, R.P. Immunolocalization of hypochlorite-induced, catalase-bound free radical formation in mouse hepatocytes. *Free Radic. Biol. Med.* **2007**, *42*(4), 530–540.

107. Zhang, R., Goldstein, S., Samuni, A. Kinetics of superoxide-induced exchange among nitroxide antioxidants and their oxidized and reduced forms. *Free Radic. Biol. Med.* **1999**, *26*, 1245–1252.

108. Massari, J., Tokikawa, R., Zanolli, L., Tavares, M.F.M., Assuncao, N.A., Bechara, E.J.H. Acetyl radical production by the methylglyoxal-peroxynitrite system: A possible route for L-lysine acetylation. *Chem. Res. Toxicol.* **2010**, *23*(11), 1762–1770.

109. Royer, L.O., Knudsen, F.S., De Oliveira, M.A., Tavares, M.F.M., Bechara, E.J.H. Peroxynitrite-initiated oxidation of acetoacetate and 2-methylacetoacetate esters by oxygen: Potential sources of reactive intermediates in keto acidoses. *Chem. Res. Toxicol.* **2004**, *17*(12), 1725–1732.

110. Imaram, W., Gersch, C., Kim, K.M., Johnson, R.J., Henderson, G.N., Angerhofer, A. Radicals in the reaction between peroxynitrite and uric acid identified by electron spin resonance spectroscopy and liquid chromatography mass spectrometry. *Free Radic. Biol. Med.* **2010**, *49*(2), 275–281.

111. Imaram, W., Johnson, R.J., Angerhofer, A. ESR spin trapping of the reaction between urate and peroxynitrite: The hydrogen adduct. *Appl. Magn. Reson.* **2009**, *37*(1–4), 463–472.

112. Olson, L.P., Bartberger, M.D., Houk, K.N. Peroxynitrate and peroxynitrite: A complete basis set investigation of similarities and differences between these NO_x species. *J. Am. Chem. Soc.* **2003**, *125*(13), 3999–4006.

113. Goldstein, S., Saha, A., Lymar, S.V., Czapski, G. Oxidation of peroxynitrite by inorganic radicals: A pulse radiolysis study. *J. Am. Chem. Soc.* **1998**, *120*, 5549–5554.

114. Koppenol, W.H. Thermodynamics of reactions involving nitrogen-oxygen compounds. *Methods Enzymol.* **1996**, *268*, 7–12.

115. Trujillo, M., Radi, R. Peroxynitrite reaction with the reduced and the oxidized forms of lipoic acid: New insights into the reaction of peroxynitrite with thiols. *Arch. Biochem. Biophys.* **2002**, *397*(1), 91–98.

116. Radi, R., Beckman, J.S., Bush, K.M., Freeman, B.A. Peroxynitrite oxidation of sulfhydryls. The cytotoxic potential of superoxide and nitric oxide. *J. Biol. Chem.* **1991**, *266*, 4244–4250.

117. Dubuisson, M., Vander Stricht, D., Clippe, A., Etienne, F., Nauser, T., Kissner, R., Koppenol, W.H., Rees, J.F., Knoops, B. Human peroxiredoxin 5 is a peroxynitrite reductase. *FEBS Lett.* **2004**, *571*, 161–165.

118. Alvarez, B., Ferrer-Sueta, G., Freeman, B.A., Radi, R. Kinetics of peroxynitrite reaction with amino acids and human serum albumin. *J. Biol. Chem.* **1999**, *274*(2), 842–848.

119. Jursic, B.S. Reliability of hybrid density theory-semiempirical approach for evaluation of bond dissociation energies. *J. Chem. Soc. Perkin Trans. 2* **1999**, *2*, 369–372.

120. Hoffman, M.Z., Hayon, E. Pulse radiolysis study of sulfhydryl compounds in aqueous solution. *J. Phys. Chem Ref. Data* **1973**, *77*, 990–996.

121. Folkes, L.K., Trujillo, M., Bartesaghi, S., Radi, R., Wardman, P. Kinetics of reduction of tyrosine phenoxyl radicals by glutathione. *Arch. Biochem. Biophys.* **2011**, *506*, 242–249.

122. Ferreri, C., Kratzsch, S., Landic, L., Bredeb, O. Thiyl radicals in biosystems: Effects on lipid structures and metabolisms. *Cell. Mol. Life Sci.* **2005**, *62*, 834–847.

123. Chatgilialoglu, C., Altieri, A., Fischer, H. The kinetics of thiyl radical-induced reactions of monounsaturated fatty acid esters. *J. Am. Chem. Soc.* **2002**, *124*(43), 12816–12823.

124. Schöneich, C., Dillinger, U., Bruchhausen, F.v., Asmus, K.-D. Oxidation of polyunsaturated fatty acids and lipids through thiyl and sulfonyl radicals: Reaction kinetics, and influence of oxygen and structure of thiyl radicals. *Arch. Biochem. Biophys.* **1992**, *292*, 456–467.

125. Schoneich, C., Asmus, K.-D., Bonifacic, M. Determination of absolute rate constants for the reversible hydrogen-atom transfer between thiyl radicals and alcohols or ethers. *J. Chem. Soc. Faraday Trans.* **1995**, *91*, 1923–193.

126. Nauser, T., Schoneich, C. Thiyl radicals abstract hydrogen atoms from the $^\alpha$C-H bonds in model peptides: Absolute rate constants and effect of amino acid structure. *J. Am. Chem. Soc.* **2003**, *125*, 2042–2043.

127. Schoneich, C. Cysteine residues as catalysts for covalent peptide and protein modification: A role for thiyl radicals? *Biochem. Soc. Trans.* **2011**, *39*(5), 1254–1259.

128. Hofstetter, D., Nauser, T., Koppenol, W.H. Hydrogen exchange equilibria in glutathione radicals: Rate constants. *Chem. Res. Toxicol.* **2010**, *23*(10), 1596–1600.

129. Pogocki, D., Schoneich, C. Thiyl radicals abstract hydrogen atoms from carbohydrates: Reactivity and selectivity. *Free Radic. Biol. Med.* **2001**, *31*(1), 98–107.

130. Madej, E., Folkes, L.K., Wardman, P., Czapski, G., Goldstein, S. Thiyl radicals react with nitric oxide to form S-nitrosothiols with rate constants near the diffusion-controlled limit. *Free Radic. Biol. Med.* **2008**, *44*, 2013–2018.

131. Sevilla, M.D., Becker, D., Swarts, S., Herrington, J. Sulfinyl radical formation from the reaction of cysteine and glutathione thiyl radicals with molecular oxygen. *Biochem. Biophys. Res. Commun.* **1987**, *144*, 1037–1042.

132. Mönig, J., Asmus, K.D., Forni, L.G., Willson, R.L. On the reaction of molecular oxygen with thiyl radicals: A re-examination. *Int. J. Radiat. Biol. Relat. Stud. Phys. Chem. Med.* **1987**, *52*, 589–602.

133. Tamba, M., Simone, G., Quintiliani, M. Interactions of thiyl free radicals with oxygen: A pulse radiolysis study. *Int. J. Radiat. Biol.* **1986**, *50*, 595–600.

134. Zhao, R., Lind, J., Merbnyi, G., Eriksen, T.E. Kinetics of one-electron oxidation of thiols and abstraction by thiyl radicals from α-amino C-H bonds. *J. Am. Chem. Soc.* **1994**, *116*, 12010–12015.

135. Hayon*, M.Z.H.a.E. One-electron reduction of the disulfide linkage in aqueous solution. Formation, protonation, and decay kinetics of the RSSR anion radical. *J. Am. Chem. Soc.* **1972**, *94*, 7950–7957.

136. Quintiliani, M., Badiello, R., Tamba, M., Esfandi, A., Gorin, G. Radiolysis of glutathione in oxygen-containing solutions of pH 7. *Int. J. Radiat. Biol.* **1977**, *32*, 195–202.

137. Voronkov, M.G., Deryagin, E.N. Thermal reactions of thiyl radicals. *Russ. Chem. Rev.* **1990**, *59*, 778–791.

138. Guo, W., Pleasants, J., Rabenstein, D.L. Nuclear magnetic resonance studies of thiol/disulfide chemistry. 2. Kinetics of symmetrical thiol/disulfide interchange reactions. *J. Org. Chem.* **1990**, *55*, 373–376.

139. Trivedi, M.V., Laurence, J.S., Siahaan, T.J. The role of thiols and disulfides on protein stability. *Curr. Protein Pept. Sci.* **2009**, *10*(6), 614–625.

140. Giles, G.I., Tasker, K.M., Jacob, C. Hypothesis: The role of reactive sulfur species in oxidative stress. *Free Radic. Biol. Med.* **2001**, *31*(10), 1279–1283.

141. Racker, E. Glutathione reductase from baker's yeast and beef liver. *J. Biol. Chem.* **1955**, *217*, 855–866.

142. Collet, J.F., Messens, J. Structure, function, and mechanism of thioredoxin proteins. *Antioxid. Redox Signal.* **2010**, *13*(8), 1205–1216.

143. Singh, R., Lamoureux, G.V., Lees, W.J., Whitesides, G.M. Reagents for rapid reduction of disulfide bonds. *Methods Enzymol.* **1995**, *251*, 167–173.

144. Pattison, D.I., Davies, M.J. Reactions of myeloperoxidase-derived oxidants with biological substrates: Gaining chemical insight into human inflammatory diseases. *Curr. Med. Chem.* **2006**, *13*(27), 3271–3290.

145. Khan, A.U., Kasha, M. Chemiluminescence arising from simultaneous transitions in pairs of singlet oxygen molecules. *J. Am. Chem. Soc.* **1970**, *92*, 3293–3300.

146. Long, C.A., Bielski, B.H.J. Rate of reaction of superoxide radical with chloride-containing species. *J. Phys. Chem.* **1980**, *84*, 555–557.

147. Zuo, Z., Katsumura, Y., Ueda, K., Ishigur, K. Reactions between some inorganic radicals and oxychlorides studied by pulse radiolysis and laser photolysis. *J. Chem. Soc. Faraday Trans.* **1997**, *93*, 1885–1891.

148. Miyamoto, S., Martinez, G.R., Rettori, D., Augusto, O., Medeiros, M.H., Di Mascio, P. Linoleic acid hydroperoxide reacts with hypochlorous acid, generating peroxyl radical intermediates and singlet molecular oxygen. *Proc. Natl. Acad. Sci. U.S.A.* **2006**, *103*(2), 293–298.

149. Eiserich, J.P., Cross, C.E., Jones, A.D., Halliwell, B., van der Vliet, A. Formation of nitrating and chlorinating species by reaction of nitrite with hypochlorous acid. A novel mechanism for nitric oxide-mediated protein modification. *J. Biol. Chem.* **1996**, *271*(32), 19199–19208.

150. Eiserich, J.P., Hristova, M., Cross, C.E., Jones, A.D., Freeman, B.A., Halliwell, B., van der Vliet, A. Formation of nitric oxide-derived inflammatory oxidants by myeloperoxidase in neutrophils. *Nature* **1998**, *391*(6665), 393–397.

151. Fogelman, K.D., Walker, D.M., Margerum, D.W. Non-metal redox kinetics: Hypochlorite and hypochlorous acid reactions with sulfite. *Inorg. Chem.* **1989**, *28*, 986–993.

152. Chesney, J.A., Mahoney, J.R.J., Eaton, J.W. A spectrophotometric assay for chlorine-containing compounds. *Anal. Biochem.* **1991**, *196*, 262–266.

153. Folkes, L.K., Candeias, L.P., Wardma, P. Kinetics and mechanisms of hypochlorous acid reactions. *Arch. Biochem. Biophys.* **1995**, *323*, 120–126.

154. Hawkins, C.L., Davies, M.J. Reaction of HOCl with amino acids and peptides: EPR evidence for rapid rearrangement and fragmentation reactions of nitrogen-centred radicals. *J. Chem. Soc. Perkin Trans. 2* **1998**, *9*, 1937–1945.

155. Peskin, A.V., Winterbourn, C.C. Kinetics of the reactions of hypochlorous acid and amino acid chloramines with thiols, methionine, and ascorbate. *Free Radic. Biol. Med.* **2001**, *30*(5), 572–579.

156. Carr, A.C., Winterbourn, C.C. Oxidation of neutrophil glutathione and protein thiols by myeloperoxidase-derived hypochlorous acid. *Biochem. J.* **1997**, *327*, 275–281.

157. Peskin, A.V., Turner, R., Maghzal, G.J., Winterbourn, C.C., Kettle, A.J. Oxidation of methionine to dehydromethionine

158. Fu, S., Wang, H., Davies, M., Dean, R. Reactions of hypochlorous acid with tyrosine and peptidyl-tyrosyl residues give dichlorinated and aldehydic products in addition to 3-chlorotyrosine. *J. Biol. Chem.* **2000**, *275*(15), 10851–10858.

159. Prutz, W.A., Kissner, R., Nauser, T., Koppenol, W.H. On the oxidation of cytochrome c by hypohalous acids. *Arch. Biochem. Biophys.* **2001**, *389*(1), 110–122.

160. Winterbourn, C.C., Berg, J.J.M.v.D., Roitman, E., Kuypers, F.A. Chlorohydrin formation from unsaturated fatty acids reacted with hypochlorous acid. *Arch. Biochem. Biophys.* **1992**, *296*, 547–555.

161. Hawkins, C.L., Davies, M.J. Hypochlorite-induced damage to nucleosides: Formation of chloramines and nitrogen-centered radicals. *Chem. Res. Toxicol.* **2001**, *14*(8), 1071–1081.

162. Stanley, N.R., Pattison, D.I., Hawkins, C.L. Ability of hypochlorous acid and N-chloramines to chlorinate DNA and its constituents. *Chem. Res. Toxicol.* **2010**, *23*(7), 1293–1302.

163. Whiteman, M., Jenner, A., Halliwell, B. Hypochlorous acid-induced base modifications in isolated calf thymus DNA. *Chem. Res. Toxicol.* **1997**, *10*(11), 1240–1246.

164. Prutz, W.A., Kissner, R., Koppenol, W.H., Ruegger, H. On the irreversible destruction of reduced nicotinamide nucleotides by hypohalous acids. *Arch. Biochem. Biophys.* **2000**, *380*(1), 181–191.

165. Koppenol, W.H., Stanbury, D.M., Bounds, P.L. Electrode potentials of partially reduced oxygen species, from dioxygen to water. *Free Radic. Biol. Med.* **2010**, *49*(3), 317–322.

166. Koppenol, W.H., Butler, J. Energetics of interconversion reactions of oxyradicals. *Adv. Free Radic. Biol. Med.* **1985**, *1*, 91–131.

167. Pfeiffer, S., Mayer, B., Janoschek, R. Gibbs energies of reactive species involved in peroxynitrite chemistry calculated by density functional theory. *J. Mol. Struct.* **2003**, *623*, 95–103.

168. Lagager, T., Sehested, K. Formation and decay of peroxynitrous acid: A pulse radiolysis study. *J. Phys. Chem.* **1993**, *97*, 6664–6669.

169. Buettner, G.R. The pecking order of free radicals and antioxidants: Lipid peroxidation, α-tocopherol, and ascorbate. *Arch. Biochem. Biophys.* **1993**, *300*, 535–543.

170. Simic, M.G., Jovanovi, S.V. Antioxidation mechanisms of uric acid? *J. Am. Chem. Soc.* **1989**, *111*, 5718–5182.

171. Lancaster, J.R., Jr. Nitroxidative, nitrosative, and nitrative stress: Kinetic predictions of reactive nitrogen species chemistry under biological conditions. *Chem. Res. Toxicol.* **2006**, *19*(9), 1160–1174.

172. Hotamisligil, G.S. Endoplasmic reticulum stress and the inflammatory basis of metabolic disease. *Cell* **2010**, *140*(6), 900–917.

173. Jiang, Z., Hu, Z., Zeng, L., Lu, W., Zhang, H., Li, T., Xiao, H. The role of the Golgi apparatus in oxidative stress: Is this organelle less significant than mitochondria? *Free Radic. Biol. Med.* **2011**, *50*, 907–917.

174. Quinn, M.T., Gauss, K.A. Structure and regulation of the neutrophil respiratory burst oxidase: Comparison with nonphagocyte oxidases. *J. Leukoc. Biol.* **2004**, *76*, 760–781.

175. Berry, C.E., Hare, J.M. Xanthine oxidoreductase and cardiovascular disease: Molecular mechanisms and pathophysiological implications. *J. Physiol.* **2004**, *555*(Pt 3), 589–606.

176. Cadenas, E., Davies, K.J.A. Mitochondrial free radical generation, oxidative stress, and aging. *Free Radic. Biol. Med.* **2000**, *29*, 222–230.

177. Barja, G. Mitochondrial oxygen radical generation and leak: Sites of production in states 4 and 3, organ specificity, and relation to aging and longevity. *J. Bioenerg. Biomembr.* **1999**, *31*(4), 347–366.

178. Rifkind, J.M., Ramasamy, S., Manoharan, P.T., Nagababu, E., Mohanty, J.G. Redox reactions of hemoglobin. *Antioxid. Redox Signal.* **2004**, *6*, 657–666.

179. Wei, C.-C., Crane, B.R., Stuehr, D.J. Tetrahydrobiopterin radical enzymology. *Chem. Rev.* **2003**, *103*, 2365–2384.

180. Surawatanawong, P., Tye, J.W., Hall, M.B. Density functional theory applied to a difference in pathways taken by the enzymes cytochrome P450 and superoxide reductase: Spin states of ferric hydroperoxo intermediates and hydrogen bonds from water. *Inorg. Chem.* **2010**, *49*, 188–198.

181. Kuthan, H., Tsuji, H., Graf, H., Ullrich, V., Werringloer, J., Estabrook, R.W. Generation of superoxide anion as a source of hydrogen peroxide in a reconstituted monooxygenase system. *FEBS Lett.* **1978**, *91*, 343–345.

182. Roy, P., Roy, S.K., Mitra, A., Kulkarni, A.P. Superoxide generation by lipoxygenase in the presence of NADH and NADPH. *Biochem. Biophys. Acta, Lipids Lipid Metab.* **1994**, *1214*, 171–179.

183. Gross, E., Sevier, C.S., Heldman, N., Vitu, E., Bentzur, M., Kaiser, C.A., Thorpe, C., Fass, D. Generating disulfides enzymatically: Reaction products and electron acceptors of the endoplasmic reticulum thiol oxidase Ero1p. *Proc. Natl. Acad. Sci. U.S.A.* **2006**, *103*(2), 299–304.

184. Richter, B.H.J.B.a.H.W. A study of the superoxide radical chemistry by stopped-flow radiolysis and radiation induced oxygen consumption. *J. Am. Chem. Soc.* **1977**, *99*, 3019–3023.

185. Holroyd, R.A., Bielski, B.H.J. Photochemical generation of superoxide radicals in aqueous solutions. *J. Am. Chem. Soc.* **1978**, *100*, 5796–5800.

186. Hochanadel, C.J. Photolysis of dilute H_2O_2 solution in the presence of dissolved H_2 and O_2. Evidence relating to the nature of the OH radical and the H atom produced in the radiolysis of H_2O. *Radiat. Res.* **1962**, *17*, 286–301.

187. Joshi, A., Yang, G.C. Spin trapping of radicals generated in the UV photolysis of alkyl disulfides. *J. Org. Chem.* **1981**, *46*, 3736–3738.

188. Misik, V., Miyoshi, N., Riesz, P. EPR spin-trapping study of the sonolysis of H_2O/D_2O mixtures: Probing the tem-

peratures of cavitation regions. *J. Phys. Chem.* **1995**, *99*, 3605–3611.

189. Makino, K., Mossoba, M.M., Riesz, P. Chemical effects of ultrasound on aqueous solutions. Formation of hydroxyl radicals and hydrogen atoms. *J. Phys. Chem.* **1983**, *87*, 1369–1377.

190. Misik, V., Riesz, P. Free radical intermediates in sonodynamic therapy. *Ann. N. Y. Acad. Sci.* **2000**, *899*, 335–348.

191. Draper, W.M., Crosby, D.G. Photochemical generation of superoxide radical anion in water. *J. Agric. Food Chem.* **1983**, *31*, 734–737.

192. Maurette, M.-T., Oliveros, E., Infelta, P.P., Ramsteiner, K., Braun, A.M. Singlet oxygen and superoxide: Experimental differentiation and analysis. *Helv. Chim. Acta* **1983**, *66*, 722–732.

193. Yamakoshi, Y., Sueyoshi, S., Fukuhara, K., Miyata, N. OH and O_2 generation in aqueous C60 and C70 solutions by photoirradiation: An EPR study. *J. Am. Chem. Soc.* **1998**, *120*, 12363–12364.

194. Garg, S., Rose, A.L., Waite, T.D. Production of reactive oxygen species on photolysis of dilute aqueous quinone solutions. *Photochem. Photobiol.* **2007**, *83*(4), 904–913.

195. Matsumoto, F., Okajima, T., Uesugi, S., Koura, N., Ohsaka, T. Electrogeneration of superoxide ion and its mechanism at thiol-modified Au electrodes in alkaline aqueous solution. *Electrochemistry* **2003**, *71*, 266–273.

196. Merritt, M.V., Sawyer, D.T. Electrochemical studies of the reactivity of superoxide ion with several alkyl halides in dimethyl sulfoxide. *J. Org. Chem.* **1970**, *35*, 2157–2159.

197. Maricle, D.L., Hodgson, W.G. Reduction of oxygen to superoxide in aprotic solvents. *Anal. Chem.* **1965**, *37*, 1562–1565.

198. Martiz, B., Keyrouz, R., Gmouh, S., Vaultier, M., Jouikov, V. Superoxide-stable ionic liquids: New and efficient media for electrosynthesis of functional siloxanes. *Chem. Commun.* **2004**, *6*, 674–675.

199. Bohle, D.S., Sagan, E.S., Koppenol, W.H., Kissner, R. Tetramethylammonium salts of superoxide and peroxynitrite. *Inorg. Synth.* **2004**, *34*, 36–42.

200. McElroy, A.D., Hashman, J.S. Synthesis of tetramethylammonium superoxide. *Inorg. Chem.* **1964**, *3*, 1798–1799.

201. Bolton, J.L., Trush, M.A., Penning, T.M., Dryhurst, G., Monks, T.J. Role of quinones in toxicology. *Chem. Res. Toxicol.* **2000**, *13*, 135–160.

202. Wang, P.G., Xian, M., Tang, X., Wu, X., Wen, Z., Cai, T., Janczuk, A.J. Nitric oxide donors: Chemical activities and biological applications. *Chem. Rev.* **2002**, *102*(4), 1091–1134.

203. Pryor, W.A., Cueto, R., Jin, X., Koppenol, W.H., Ngu-Schwemlein, M., Squadrito, G.L., Uppu, P.L., Uppu, R.M. A practical method for preparing peroxynitrite solutions of low ionic strength and free of hydrogen peroxide. *Free Radic. Biol. Med.* **1995**, *18*(1), 75–83.

204. Uppu, R.M. Synthesis of peroxynitrite using isoamyl nitrite and hydrogen peroxide in a homogeneous solvent system. *Anal. Biochem.* **2006**, *354*(2), 165–168.

205. Wahl, R.U.R. Decomposition mechanism of 3-N-morpholinosydnonimine (SIN-1)—a density functional study on intrinsic structures and reactivities. *J. Mol. Model.* **2004**, *10*, 121–129.

206. Wardman, P. Fluorescent and luminescent probes for measurement of oxidative and nitrosative species in cells and tissues: Progress, pitfalls, and prospects. *Free Radic. Biol. Med.* **2007**, *43*(7), 995–1022.

207. Zhao, H., Joseph, J., Fales, H.M., Sokoloski, E.A., Levine, R.L., Vasquez-Vivar, J., Kalyanaraman, B. Detection and characterization of the product of hydroethidine and intracellular superoxide by HPLC and limitations of fluorescence. *Proc. Natl. Acad. Sci. U.S.A.* **2005**, *102*(16), 5727–5732.

208. Miller, E.W., Chang, C.J. Fluorescent probes for nitric oxide and hydrogen peroxide in cell signaling. *Curr. Opin. Chem. Biol.* **2007**, *11*(6), 620–625.

209. Sikora, A., Zielonka, J., Lopez, M., Joseph, J., Kalyanaraman, B. Direct oxidation of boronates by peroxynitrite: Mechanism and implications in fluorescence imaging of peroxynitrite. *Free Radic. Biol. Med.* **2009**, *47*(10), 1401–1407.

210. Nagano, T. Bioimaging probes for reactive oxygen species and reactive nitrogen species. *J. Clin. Biochem. Nutr.* **2009**, *45*(2), 111–124.

211. Fontijn, A., Sabadell, A.J., Ronco, R.J. Homogeneous chemiluminescent measurement of nitric oxide with ozone: Implications for continuous selective monitoring of gaseous air pollutants. *Anal. Chem.* **1970**, *42*, 575–579.

212. Afanasev, I. Detection of superoxide in cells, tissues and whole organisms. *Front. Biosci.* **2009**, *E1*, 153–160.

213. Archer, S. Measurement of nitric oxide in biological models. *FASEB J.* **1993**, *7*, 349–360.

214. Kleinbongard, P., Rassaf, T., Dejam, A., Kerber, S., Kelm, M. Griess method for nitrite measurement of aqueous and protein-containing samples. *Methods Enzymol.* **2002**, *359*, 158–168.

215. Thomas, E.L., Grisham, M.B., Jefferson, M.M. Preparation and characterization of chloramines. *Methods Enzymol.* **1986**, *132*, 569–585.

216. Dypbukt, J.M., Bishop, C., Brooks, W.M., Thong, B., Eriksson, H., Kettle, A.J. A sensitive and selective assay for chloramine production by myeloperoxidase. *Free Radic. Biol. Med.* **2005**, *39*(11), 1468–1477.

217. Floyd, R.A., West, M.S., Eneff, K.L., Hogsett, W.E., Tingey, D.T. Hydroxyl free radical mediated formation of 8-hydroxyguanine in isolated DNA. *Arch. Biochem. Biophys.* **1988**, *262*(1), 266–272.

218. Freinbichler, W., Bianchi, L., Colivicchi, M.A., Ballini, C., Tipton, K.F., Linert, W., Corte, L.D. The detection of hydroxyl radicals *in vivo*. *J. Inorg. Biochem.* **2008**, *102*(5–6), 1329–1333.

219. Rebrin, I., Bregere, C., Gallaher, T.K., Sohal, R.S. Detection and characterization of peroxynitrite-induced modifications of tyrosine, tryptophan, and methionine residues

220. Sultana, R., Reed, T., Butterfield, D.A. Detection of 4-hydroxy-2-nonenal- and 3-nitrotyrosine-modified proteins using a proteomics approach. *Methods Mol. Biol.* **2009**, *519*, 351–361.

221. Butt, Y.K., Lo, S.C. Detecting nitrated proteins by proteomic technologies. *Methods Enzymol.* **2008**, *440*, 17–31.

222. Nourooz-Zadeh, J., Tajaddini-Sarmadi, J., Ling, K.L., Wolff, S.P. Low-density lipoprotein is the major carrier of lipid hydroperoxides in plasma. Relevance to determination of total plasma lipid hydroperoxide concentrations. *Biochem. J.* **1996**, *313*(Pt 3), 781–786.

223. Fukuzawa, K., Fujisaki, A., Akai, K., Tokumura, A., Terao, J., Gebicki, J.M. Measurement of phosphatidylcholine hydroperoxides in solution and in intact membranes by the ferric-xylenol orange assay. *Anal. Biochem.* **2006**, *359*(1), 18–25.

224. Gieseg, S.P., Pearson, J., Firth, C.A. Protein hydroperoxides are a major product of low density lipoprotein oxidation during copper, peroxyl radical and macrophage-mediated oxidation. *Free Radic. Res.* **2003**, *37*(9), 983–991.

225. Winterbourn, C.C., Pichorner, H., Kettle, A.J. Myeloperoxidase-dependent generation of a tyrosine peroxide by neutrophils. *Arch. Biochem. Biophys.* **1997**, *338*(1), 15–21.

226. O'Donnell, V.B., Eiserich, J.P., Chumley, P.H., Jablonsky, M.J., Krishna, N.R., Kirk, M., Barnes, S., Darley-Usmar, V.M., Freeman, B.A. Nitration of unsaturated fatty acids by nitric oxide-derived reactive nitrogen species peroxynitrite, nitrous acid, nitrogen dioxide, and nitronium ion. *Chem. Res. Toxicol.* **1999**, *12*(1), 83–92.

227. Janero, D.R. Malondialdehyde and thiobarbituric acid-reactivity as diagnostic indices of lipid peroxidation and peroxidative tissue injury. *Free Radic. Biol. Med.* **1990**, *9*(6), 515–540.

228. Lawson, J.A., Rokach, J., FitzGerald, G.A. Isoprostanes: Formation, analysis and use as indices of lipid peroxidation in vivo. *J. Biol. Chem.* **1999**, *274*(35), 24441–24444.

229. Mason, R.P. Using anti-5,5-dimethyl-1-pyrroline N-oxide (anti-DMPO) to detect protein radicals in time and space with immuno-spin trapping. *Free Radic. Biol. Med.* **2004**, *36*(10), 1214–1223.

230. Ramirez, D.C., Chen, Y.R., Mason, R.P. Immunochemical detection of hemoglobin-derived radicals formed by reaction with hydrogen peroxide: Involvement of a protein-tyrosyl radical. *Free Radic. Biol. Med.* **2003**, *34*(7), 830–839.

231. Detweiler, C.D., Lardinois, O.M., Deterding, L.J., de Montellano, P.R., Tomer, K.B., Mason, R.P. Identification of the myoglobin tyrosyl radical by immuno-spin trapping and its dimerization. *Free Radic. Biol. Med.* **2005**, *38*(7), 969–976.

232. Ramirez, D.C., Gomez Mejiba, S.E., Mason, R.P. Mechanism of hydrogen peroxide-induced Cu,Zn-superoxide dismutase-centered radical formation as explored by immuno-spin trapping: The role of copper- and carbonate radical anion-mediated oxidations. *Free Radic. Biol. Med.* **2005**, *38*(2), 201–214.

233. Ehrenshaft, M., Mason, R.P. Protein radical formation on thyroid peroxidase during turnover as detected by immuno-spin trapping. *Free Radic. Biol. Med.* **2006**, *41*(3), 422–430.

234. Ranguelova, K., Suarez, J., Magliozzo, R.S., Mason, R.P. Spin trapping investigation of peroxide- and isoniazid-induced radicals in *Mycobacterium tuberculosis* catalase-peroxidase. *Biochemistry* **2008**, *47*(43), 11377–11385.

235. Ramirez, D.C., Mejiba, S.E., Mason, R.P. Immuno-spin trapping of DNA radicals. *Nat. Methods* **2006**, *3*(2), 123–127.

236. Ramirez, D.C., Gomez-Mejiba, S.E., Mason, R.P. Immuno-spin trapping analyses of DNA radicals. *Nat. Protoc.* **2007**, *2*(3), 512–522.

237. Gomez-Mejiba, S.E., Zhai, Z., Akram, H., Deterding, L.J., Hensley, K., Smith, N., Towner, R.A., Tomer, K.B., Mason, R.P., Ramirez, D.C. Immuno-spin trapping of protein and DNA radicals: "Tagging" free radicals to locate and understand the redox process. *Free Radic. Biol. Med.* **2009**, *46*(7), 853–865.

238. Kutala, V.K., Villamena, F.A., Ilangovan, G., Maspoch, D., Roques, N., Veciana, J., Rovira, C., Kuppusamy, P. Reactivity of superoxide anion radical with a perchlorotriphenylmethyl (trityl) radical. *J. Phys. Chem. B* **2008**, *112*(1), 158–167.

239. Rizzi, C., Samouilov, A., Kutala, V.K., Parinandi, N.L., Zweier, J.L., Kuppusamy, P. Application of a trityl-based radical probe for measuring superoxide. *Free Radic. Biol. Med.* **2003**, *35*(12), 1608–1618.

240. Ballester, M., Riera-Figueras, J., Castaner, J., Badfa, C., Monso, J.M. Inert carbon free radicals. I. Perchlorodiphenylmethyl and perchlorotriphenylmethyl radical series. *J. Am. Chem. Soc.* **1971**, *93*(9), 2215–2225.

241. Dikalov, S., Skatchkov, M., Bassenge, E. Quantification of peroxynitrite, superoxide, and peroxyl radicals by a new spin trap hydroxylamine 1-hydroxy-2,2,6,6-tetramethyl-4-oxo-piperidine. *Biochem. Biophys. Res. Commun.* **1997**, *230*(1), 54–57.

242. Saito, K., Takeshita, K., Anzai, K., Ozawa, T. Pharmacokinetic study of acyl-protected hydroxylamine probe, 1-acetoxy-3-carbamoyl-2,2,5,5-tetramethylpyrrolidine, for *in vivo* measurements of reactive oxygen species. *Free Radic. Biol. Med.* **2004**, *36*(4), 517–525.

243. Dikalov, S., Grigor'ev, I.A., Voinov, M., Bassenge, E. Detection of superoxide radicals and peroxynitrite by 1-hydroxy-4-phosphonooxy-2,2,6,6-tetramethylpiperidine: Quantification of extracellular superoxide radicals formation. *Biochem. Biophys. Res. Commun.* **1998**, *248*(2), 211–215.

244. Zang, L.Y., Misra, B.R., van Kuijk, F.J., Misra, H.P. EPR studies on the kinetics of quenching singlet oxygen. *Biochem. Mol. Biol. Int.* **1995**, *37*(6), 1187–1195.

245. Villamena, F.A., Das, A., Nash, K.M. Potential implication of the chemical properties and bioactivity of nitrone spin traps for therapeutics. *Future Med. Chem.* **2012**, *4*(9), 1171–1207.
246. Britigan, B.E., Pou, S., Rosen, G.M., Lilleg, D.M., Buettner, G.R. Hydroxyl radical is not a product of the reaction of xanthine oxidase and xanthine. The confounding problem of adventitious iron bound to xanthine oxidase. *J. Biol. Chem.* **1990**, *265*(29), 17533–17538.
247. Du, G., Mouithys-Mickalad, A., Sluse, F.E. Generation of superoxide anion by mitochondria and impairment of their functions during anoxia and reoxygenation *in vitro*. *Free Radic. Biol. Med.* **1998**, *25*(9), 1066–1074.
248. Nohl, H., Jordan, W., Hegner, D. Identification of free hydroxyl radicals in respiring rat heart mitochondria by spin trapping with the nitrone DMPO. *FEBS Lett.* **1981**, *123*(2), 241–244.
249. Bannister, J.V., Bellavite, P., Serra, M.C., Thornalley, P.J., Rossi, F. An EPR study of the production of superoxide radicals by neutrophil NADPH oxidase. *FEBS Lett.* **1982**, *145*(2), 323–326.
250. Shi, H., Timmins, G., Monske, M., Burdick, A., Kalyanaraman, B., Liu, Y., Clement, J.-L., Burchiel, S., Liu, K.J. Evaluation of spin trapping agents and trapping conditions for detection of cell-generated reactive oxygen species. *Arch. Biochem. Biophys.* **2005**, *437*(1), 59–68.
251. Britigan, B.E., Cohen, M.S., Rosen, G.M. Detection of the production of oxygen-centered free radicals by human neutrophils using spin trapping techniques: A critical perspective. *J. Leukoc. Biol.* **1987**, *41*(4), 349–362.
252. Hirata, H., He, G., Deng, Y., Salikhov, I., Petryakov, S., Zweier, J.L. A loop resonator for slice-selective *in vivo* EPR imaging in rats. *J. Magn. Reson.* **2008**, *190*(1), 124–134.
253. Jiang, J.J., Liu, K.J., Jordan, S.J., Swartz, H.M., Mason, R.P. Detection of free radical metabolite formation using *in vivo* EPR spectroscopy: Evidence of rat hemoglobin thiyl radical formation following administration of phenylhydrazine. *Arch. Biochem. Biophys.* **1996**, *330*(2), 266–270.
254. Kadiiska, M.B., Burkitt, M.J., Xiang, Q.H., Mason, R.P. Iron supplementation generates hydroxyl radical *in vivo*. An ESR spin-trapping investigation. *J. Clin. Invest.* **1995**, *96*(3), 1653–1657.
255. Bolli, R., Jeroudi, M.O., Patel, B.S., DuBose, C.M., Lai, E.K., Roberts, R., McCay, P.B. Direct evidence that oxygen-derived free radicals contribute to postischemic myocardial dysfunction in the intact dog. *Proc. Natl. Acad. Sci. U.S.A.* **1989**, *86*(12), 4695–4699.
256. Goldstein, S., Russo, A., Samuni, A. Reactions of PTIO and carboxy-PTIO with .bul.NO, .bul.NO2, and O2-.bul. *J. Biol. Chem.* **2003**, *278*(51), 50949–50955.
257. Samuni, U., Samuni, Y., Goldstein, S. On the distinction between nitroxyl and nitric oxide using nitronyl nitroxides. *J. Am. Chem. Soc.* **2010**, *132*(24), 8428–8432.
258. Mordvintcev, P., Mulsch, A., Busse, R., Vanin, A.F. On-line detection of nitric oxide formation in liquid aqueous phase by EPR. *Anal. Biochem.* **1991**, *199*, 142.
259. Vanin, A.F., Huisman, A., Van Faassen, E.E. Iron dithiocarbamate as spin trap for nitric oxide detection: Pitfalls and successes. *Methods Enzymol.* **2002**, *359* (*Nitric Oxide–Part D*), 27–42.
260. Vanin, A.F. Iron diethyldithiocarbamate as spin trap for nitric oxide detection. *Methods Enzymol.* **1999**, *301*, 269–279.
261. Mikoyan, V.D., Kubrina, L.N., Serezhenkov, V.A., Stukan, R.A., Vanin, A.F. Complexes of Fe(II) with diethyldithiocarbamate or N-methyl-D-glucamine dithiocarbamate as traps of nitric oxide in animal tissues: Comparative investigations. *Biochim. Biophys. Acta* **1997**, *1336*, 225–234.
262. Tsuchiya, K., Takasugi, M., Minakuchi, K., Fukuzawa, K. Sensitive quantification of nitric oxide by EPR spectroscopy. *Free Radic. Biol. Med.* **1996**, *21*, 733–737.
263. Kleschyov, A.L., Munzel, T. Advanced spin trapping of vascular nitric oxide using colloid iron dithiocarbamate. *Methods Enzymol.* **2002**, *359*, 43–51.
264. Kleschyov, A.L., Hanke Mollnau, H., Oelze, M., Meinertz, T., Huang, Y., Harrison, D.G., Munzel, T. Spin trapping of vascular nitric oxide using colloidal Fe(II)-diethyldithiocarbamate. *Biochim. Biophys. Res. Commun.* **2000**, *275*, 672–677.
265. Hogg, N. Detection of nitric oxide by electron paramagnetic resonance spectroscopy. *Free Radic. Biol. Med.* **2010**, *49*(2), 122–129.
266. Moreira, P.I., Sayre, L.M., Zhu, X., Nunomura, A., Smith, M.A., Perry, G. Detection and localization of markers of oxidative stress by *in situ* methods: Application in the study of Alzheimer disease. *Methods Mol. Biol.* **2010**, *610*, 419–434.
267. Celes, M.R., Torres-Duenas, D., Prado, C.M., Campos, E.C., Moreira, J.E., Cunha, F.Q., Rossi, M.A. Increased sarcolemmal permeability as an early event in experimental septic cardiomyopathy: A potential role for oxidative damage to lipids and proteins. *Shock* **2010**, *33*(3), 322–331.
268. Kuppusamy, P., Zweier, J.L. Cardiac applications of EPR imaging. *NMR Biomed.* **2004**, *17*(5), 226–239.
269. Jackson, S.K., Thomas, M.P., Smith, S., Madhani, M., Rogers, S.C., James, P.E. *In vivo* EPR spectroscopy: Biomedical and potential diagnostic applications. *Faraday Discuss.* **2004**, *126*, 103–117; discussion 169–183.
270. Liu, Y., Villamena, F.A., Sun, J., Wang, T.Y., Zweier, J.L. Esterified trityl radicals as intracellular oxygen probes. *Free Radic. Biol. Med.* **2009**, *46*(7), 876–883.
271. He, G., Shankar, R.A., Chzhan, M., Samouilov, A., Kuppusamy, P., Zweier, J.L. Noninvasive measurement of anatomic structure and intraluminal oxygenation in the gastrointestinal tract of living mice with spatial and spectral EPR imaging. *Proc. Natl. Acad. Sci. U.S.A.* **1999**, *96*(8), 4586–4591.
272. Pandian, R.P., Chacko, S.M., Kuppusamy, M.L., Rivera, B.K., Kuppusamy, P. Evaluation of lithium naphthalocyanine (LiNc) microcrystals for biological EPR oximetry. *Adv. Exp. Med. Biol.* **2011**, *701*, 29–36.

273. Khan, N., Williams, B.B., Hou, H., Li, H., Swartz, H.M. Repetitive tissue pO2 measurements by electron paramagnetic resonance oximetry: Current status and future potential for experimental and clinical studies. *Antioxid. Redox Signal.* **2007**, *9*(8), 1169–1182.

274. Berliner, L.J., Fujii, H. *In vivo* spin trapping of nitric oxide. *Antioxid. Redox Signal.* **2004**, *6*(3), 649–656.

275. Sentjurc, M., Mason, R.P. Inhibition of radical adduct reduction and reoxidation of the corresponding hydroxylamines in *in vivo* spin trapping of carbon tetrachloride-derived radicals. *Free Radic. Biol. Med.* **1992**, *13*(2), 151–160.

276. Chamulitrat, W., Jordan, S.J., Mason, R.P. Fatty acid radical formation in rats administered oxidized fatty acids: *In vivo* spin trapping investigation. *Arch. Biochem. Biophys.* **1992**, *299*(2), 361–367.

277. Kirsch, M., Lehnig, M., Korth, H.-G., Sustmann, R., de Groot, H. Inhibition of peroxynitrite-induced nitration of tyrosine by glutathione in the presence of carbon dioxide through both radical repair and peroxynitrate formation. *Chem. Eur. J.* **2001**, *7*(15), 3313–3320.

2

LIPID PEROXIDATION AND NITRATION

Sean S. Davies and Lilu Guo

OVERVIEW

At first glance, peroxidation and nitrosylation of polyunsaturated fatty acids (PUFAs) may appear to form a bewildering array of nearly random products. The overall goal of this chapter is to review the reaction pathways that lead to well-established products of lipid peroxidation and nitrosylation, so that the underlying patterns in species formation becomes clear. Understanding these pathways and their major products should permit a more coherent examination of the role of lipid peroxidation and nitrosylation products in physiology and disease.

2.1 PEROXIDATION OF PUFAs

Saturated fatty acids (e.g., palmitic and stearic acid) and monounsaturated fatty acids (e.g., oleic acid [OA]) do not undergo peroxidation during physiologically relevant conditions. In contrast, PUFAs such as linoleic (C18:2), linolenic (C18:3), arachidonic (C20:4), eicosapentaenoic (C20:5), and docosahexaenoic acid (C22:6) readily undergo peroxidation. What differentiates PUFA from the other fatty acids is the energy required for a radical (typically a hydroxyl or peroxyl radical) to abstract a hydrogen from the acyl chain to form a lipid radical, the critical first step in peroxidation. The potential energy for abstraction of hydrogen (removing both the proton and the electron by breaking the carbon–hydrogen bond) from an alkane chain is quite high (1900 mV).[1] For this reason, formation of radicals from a saturated fatty acid is extremely unlikely under physiological conditions. The potential energy for abstraction of an allylic hydrogen adjacent to single double bond (e.g., from a monounsaturated fatty acid) is 960 mV,[1] which is still too great for abstraction to occur physiologically. In contrast, the potential energy for abstraction of the methylene hydrogen directly adjacent to two double bonds (i.e., the bis-allylic hydrogen of a PUFA) is only 600 mV.[1] The difference in energy of abstraction for this type of hydrogen can be easily rationalized by the ability of the five neighboring carbons (the pentadienyl moiety) to share the unpaired electron due to conjugation (Fig. 2.1).

Because hydrogen abstraction occurs at the bis-allylic hydrogens, being cognizant of the position of these bis-allylic hydrogens is essential to understanding the pattern of peroxidation products that emerge. For linoleic acid (LA), abstraction occurs at carbon 11; for α-linolenic acid, abstraction occurs at carbon 11 or 14; for γ-linolenic acid, at carbon 8 or 11; and for arachidonic acid, at carbon 7, 10, or 13. In general, each of the bis-allylic hydrogens have an equal probability of abstraction, so that the number of possible species produced by peroxidation increases with the number of bis-allylic hydrogen in the PUFA.

The rate of hydrogen abstraction ($k = 50\ M^{-1}\ s^{-1}$) controls the rate of lipid peroxidation as the next step, the addition of molecular oxygen, is extremely rapid ($3 \times 10^8\ M^{-1}\ s^{-1}$). Why is the addition of oxygen so rapid? Radical to radical bond formation does not have to overcome spin restrictions, so the activation energy for this reaction is very low and thus the reaction rate is

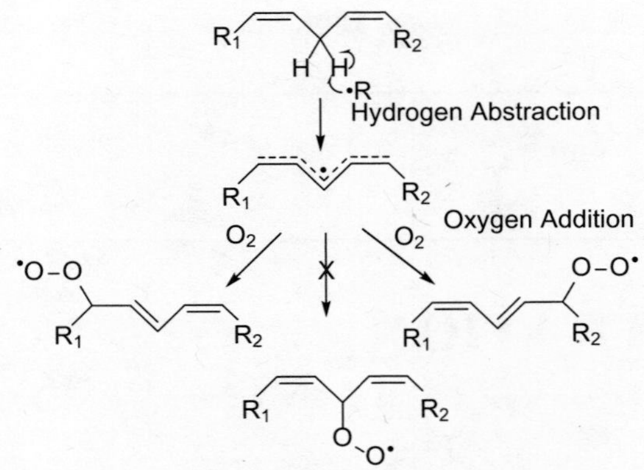

Figure 2.1 Abstraction of hydrogen by free radical and addition of oxygen forms lipid peroxyl radical.

very fast. Molecular oxygen in its ground state is a diradical and thus primed to react with nearby lipid radicals. Given the relative abundance of molecular oxygen *in vivo*, once the initial lipid radical forms, the generation of the lipid peroxyl follows almost instantaneously.

One critical point for understanding the downstream products is that oxygen generally does not add at the same carbon position where the hydrogen abstraction occurred (i.e., the carbon at the center of the pentadienyl unit). Rather, oxygen adds at carbons at the outer positions of the pentadienyl unit, two carbons away (in either direction) from the central carbon where the abstraction occurred. Because the unpaired electron is shared across all five carbons of the pentadienyl unit, stable addition of oxygen occurs at the most energetically favorable positions. Why at the outer positions and not the central position? Addition of oxygen at either of the outer carbons generates a conjugated diene, while addition at the central carbon gives only two isolated double bonds (Fig. 2.1). Perhaps for this reason, while peroxyl radical does form at the central carbon, this reaction is highly reversible and is therefore generally negligible. The formation of a conjugated diene during lipid peroxidation has been exploited to measure rates of peroxidation *in vitro*, using the UV absorbance of the conjugated dienes (~232 nm). Counting two carbons away from the bis-allylic hydrogen position allows us to predict what has been well-verified experimentally, that formation of peroxyl radicals primarily occurs at carbon 9 or 13 for LA and at carbon 5, 8, 9, 11, 12, or 15 for arachidonic acid. In the absence of other factors in the local environment, the peroxyl radical forms with equal probability at each of these positions, so that the total yield of each regioisomer is very similar as long as all of the downstream products are accounted for. As we will see in the next sections, the position of the initial peroxyl radical determines the specific products that are formed, and knowing that only specific peroxyl radicals can form helps to rationalize these various products.

One final essential point about the formation of the peroxyl radical by nonenzymatic mechanisms is the lack of stereoselectivity. Because the pentadienyl radical is planar, in the absence of constraints imposed by the active site of an enzyme, molecular oxygen can attack with equal probability from either side. Thus, the presence of racemic mixtures (i.e., equal abundance of *S* and *R* isomers) of lipid peroxides in a biological sample is a clear signature of nonenzymatic peroxidation rather than enzymatic peroxidation (e.g., by a lipoxygenase).

Once formed, the lipid peroxyl radical can undergo numerous secondary reactions depending on the specific PUFA being oxidized and on the local environment. In general, the greater the total number of pentadienyl units in the PUFA, the greater the likelihood of oxidation and the greater number of final possible lipid peroxidation products. Thus, LA oxidizes to a somewhat lesser extent and gives rise to fewer species than linolenic or arachidonic acid.

2.1.1 Hydroperoxy Fatty Acid Isomers (HpETEs and HpODEs)

Lipid peroxyl radicals generally form in a sea of other membrane-associated PUFAs (i.e., PUFA esterified in phospholipids) providing a ready supply of easily abstractable bis-allylic hydrogens nearby. Thus, the peroxyl radical can abstract a hydrogen from one of these adjacent PUFA or other hydrogen donor such as alpha-tocopherol (vitamin E) to form a hydroperoxy fatty acids (Fig. 2.2). When the peroxidized lipid is arachidonic acid or LA, the resulting product is named hydroperoxy eicosatetraenoic acid (HpETE) or hydroperoxy octadecadienoic acid (HpODE), respectively. The exact regioisomer formed can be further designated by the carbon position of the hydroperoxy group (e.g., 15-HpETE). Collectively, the family of isomers generated by nonenzymatic lipid peroxidation can be referred to as isoHpETEs or isoHpODEs.

In addition to their formation by nonenzymatic free radical mechanisms, HpETEs and HpODEs are formed in great abundance by various lipoxygenases. These iron-containing oxidoreductase enzymes generally form a single regio- and stereoisomer of HpETE or HpODE and are named based on their products (e.g., 15-lipoxygenase generates 15-S-HpETE). The HpETEs and HpODEs generated by lipoxygenases can serve as substrates for secondary peroxidation reactions, so that the genetic ablation of these enzymes can lower the level of other peroxidation products *in vivo*.[2]

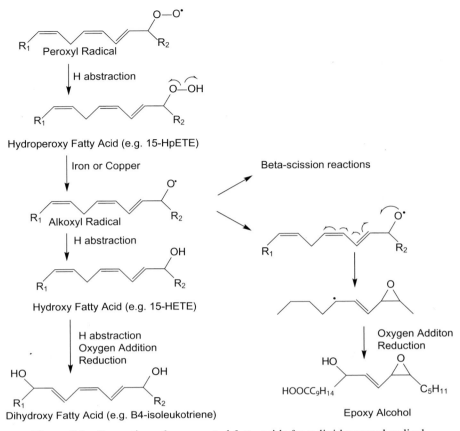

Figure 2.2 Formation of oxygenated fatty acids from lipid peroxyl radical.

2.1.2 Hydroxy Fatty Acids (HETEs and HODEs)

Metals such as iron or copper reduce hydroperoxy fatty acids to their respective hydroxy fatty acids via a two-step mechanism. First, the oxygen–oxygen bond of the peroxyl group undergoes scission to generate an alkoxyl radical. Second, the alkoxyl radical abstracts hydrogen from another molecule of PUFA to form the hydroxy fatty acid (Fig. 2.2). Hydroxy fatty acids derived from arachidonic acid and LA are named hydroxyeicosatetraenoic acids (HETEs) and hydroxyoctadecaenoic acids (HODEs), respectively, and specific isomers are designated by the position of the hydroxyl group in the same manner as with HpETEs.

HETEs and HODEs produced by lipoxygenases are biologically active.[3–7] Although nonenzymatically formed, isoHETEs and isoHODEs are typically esterified to phospholipids and cholesterol; once these products are hydrolyzed, they share the same biological activity of their enzymatically formed counterparts (e.g., 15S-HETE). The other regio- and stereoisomers may be biologically active as well.

2.1.3 Isoleukotrienes

IsoHETEs also serve as intermediates for the generation of other biologically active lipids. Because iso-HETEs retain pentadienyl moieties, they can undergo another round of hydrogen abstraction and peroxidation (Fig. 2.2). For instance, when arachidonic acid containing phospholipids are exposed to oxidants, compounds with two peroxyl moieties are readily detected, likely representing stepwise peroxidation at two independent sites. After reduction of the peroxidation products with $SnCl_2$, the resulting dihydroxy compounds display a strong absorbance at 270 nm with vibronic absorption at 260 nm and 280 nm.[8] This absorbance pattern is characteristic of a conjugated triene, and mass spectrometry confirms the presence of leukotriene B_4-like isomers (B_4-isoleukotrienes). Most interestingly, when these B_4-isoleukotrienes are released from the phospholipid by base hydrolysis, they show relatively potent activity for calcium release in neutrophils, and this effect can be blocked by LTB4 receptor antagonists.[8] Thus, isoleukotrienes show similar biological activities as their enzymatically generated counterparts. From chemical considerations, it seems reasonable to also expect that trihydroxy fatty like epi-lipoxins would also form during lipid peroxidation. However, there does not yet appear to be clear evidence for completely nonenzymatic formation of lipoxin-like compounds.

2.1.4 Epoxy Alcohols

The alkoxyl radical intermediate can also generate other products including beta-scission products (discussed later in this chapter) and epoxyalcohols (Fig. 2.2). Formation of epoxyalcohols occurs when the alkoxyl radical attacks the adjacent double bond to form an epoxide and another molecule of oxygen subsequently adds to epoxy radical. Reduction of this epoxyperoxyl radical generates the final epoxyalcohol. The nonenzymatic formation of these epoxyalcohols have been characterized using the reaction of 15-HpETEs with hematin, and four epoxyalcohols were characterized from this regioisomer.[9] Epoxyalcohols (or their metabolites) are ligands for the PPARα receptor.[10]

The formation of these other alkoxyl products from peroxy fatty acids may be limited *in vivo* by the enzymatic two-electron reduction of peroxy fatty acids to hydroxy fatty acids by nonheme peroxidases and reductases such as gluthatione peroxidase-4,[11] gluthathione *S*-transferases,[12] and peroxiredoxins.[13–15] Such enzymatic reductions likely protect the cell from the damage caused by the alkoxyl radicals.

The formation of isoHETEs, isoHODEs, isoleukotrienes, and epoxyalcohols highlights what will be a recurring theme in this chapter, which is that nonenzymatic lipid peroxidation generates isomeric families of many (if not all) of the well-established, enzymatically derived bioactive lipids. It is interesting to speculate whether the predominance of bioactive lipids in inflammatory cell signaling reflects an evolutionary process. For instance, receptors that were activated by peroxidized lipid may have allowed primitive cells to identify that they were being harmed by local conditions, to respond by altering this environment or moving away, and thus to improve their survival. Over time, enzymes may have evolved to catalyze the formation of specific isomers of these peroxidized lipids to accelerate the cellular response to danger. Regardless of the natural history of these responses, there is substantial evidence that cellular responses to the bioactive lipids generated by nonenzymatic lipid peroxidation play important roles in inflammatory responses and disease.

2.2 CYCLIC ENDOPEROXIDES AND THEIR PRODUCTS

In addition to the relatively simple reaction represented by formation of oxygenated PUFAs, lipid peroxyl radicals can generate complex cyclic compounds *via* intramolecular reactions. In theory, the double bond adjacent to the nascent peroxyl group is a ready target for 4-*exo* cyclization to form a highly strained, four-membered endoperoxide ring (i.e., a 1,2-dioxetane) (Fig. 2.3). This intermediate was proposed initially for a number of prominent lipoxidation products.[16] However, evidence supporting this mechanism has been weak. Instead, evidence supports 5-*exo* cyclization to form a 1,2-dioxolane moieties.[17] 5-*Exo* cyclization is possible when the lipid undergoing peroxidation contains an octatrienyl moiety (e.g., linolenic acid, arachidonic acid, eicosapentaenoic acid, and docosahexaenoic acid). Formation of the peroxyl radical must occur at one of the internal positions within the octratrienyl moiety (i.e., on the fourth or fifth carbon of the octatrienyl unit) in order to allow the peroxyl group to react with the nonconjugated double bond. The resulting 1,2-dioxolanylcarbinyl radical is relatively unstrained compared to a 1,2-dioxetanylcarbinyl radical, making it more stable. The monocyclic peroxides formed in this manner account for at least 25% of the oxidation products of linoleate methyl esters.[18] In addition to monocyclic peroxides, the dioxolanylcarbinyl radical can undergo transformation to more complex molecules by three distinct pathways: the isoprostane pathway, the diepoxide pathway, and the serial cyclic endoperoxide pathway (Fig. 2.3). The yield of products from each pathway depends on temperature and oxygen concentration.[18–21] Endoperoxide products account for at least 50% of oxidation products of linoleate methyl esters.[18]

2.2.1 Isoprostanes

Isoprostanes may well be the most studied of the endoperoxide products of lipid peroxidation. Formation of a cyclopentane (prostane) ring from the dioxolanylcarbinyl radical is the first dedicated step in the formation of these compounds (Fig. 2.3). The resulting bicyclic endoperoxide radical adds an additional molecule of oxygen to form prostaglandin G_2-like compounds. Reduction of the peroxyl radical yields prostaglandin H_2 (PGH_2)-like compounds, given the trivial name of H_2-isoprostanes. The four most common PUFA that give rise to prostane ring containing molecules are linolenic acid, arachidonic acid, eicosapentaenoic acid, and docosahexaenoic acid. Peroxidation of LA has not been shown to form a prostane ring, presumably because it lacks the octatrienyl moeity required to form a 1,2-dioxilane ring.

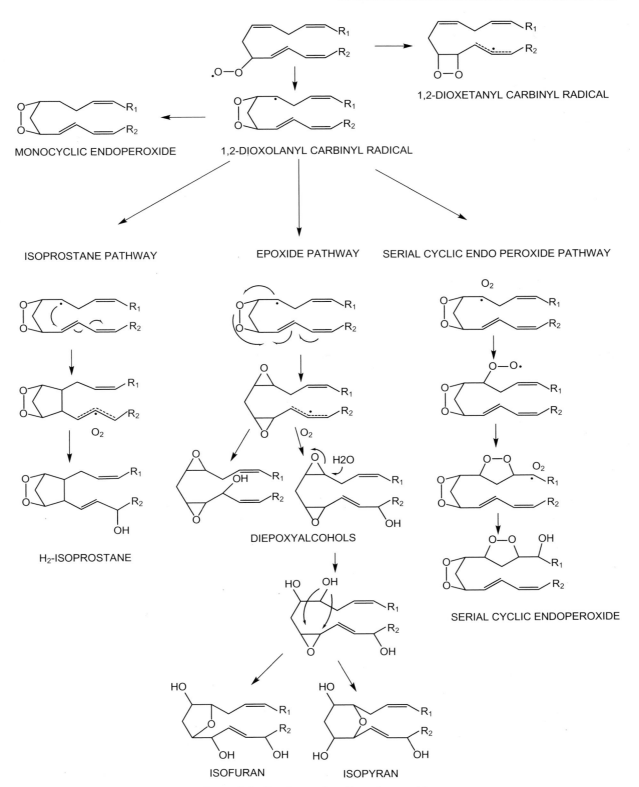

Figure 2.3 Products of cyclic endoperoxides.

2.2.1.1 Isoprostane Regio- and Stereoisomers Peroxidation of PUFA generates a large number of isoprostane regio- and stereoisomers. As discussed previously, peroxidation of arachidonic acid leads to formation of a peroxyl radical at six possible carbon positions (carbon 5, 8, 9, 11, 12, and 15) (Fig. 2.4). Peroxyls at carbon 5 or 15 do not have an appropriately situated double bond to form a dioxolane ring, so they do not give rise to isoprostanes directly. The remaining four peroxyl regiosomers each give rise to an individual bicyclic endoperoxide regioisomer. Each bicyclic endoperoxide regioisomers has four stereocenters (marked with * in Fig. 2.4), so that there are 16 potential stereoisomers for each of the four regioisomers. Thus, there are 64 potential regio- and stereoisomers of H_2-isoprostanes created by the peroxidation of arachidonic acid. When necessary to identify a particular regioisomer of H_2-isoprostane, individual isomers are designated using a numbering system based on the carbon position of the hydroxyl group and the stereochemistry of the two side chains relative to the endoperoxide ring.[22] By this nomenclature system, the regioisomer that has the hydroxyl group at carbon 15 and its acyl side chains in *trans* configuration relative to the endoperoxide (but in *cis* configuration relative to the cyclopentane ring) is designated as 15-H_{2t}-isoprostane (Fig. 2.4). Other nomenclature systems for isoprostanes have also been utilized as well[23] and in the prostaglandin nomenclature this same species is designated 8-epi-PGH_2. The use of multiple nomenclature systems is unfortunate, but perhaps inevitable as each system has differing strengths and weaknesses.

Although there are 64 isomers of H_2-isoprostanes, isomers where the acyl side chains are in the *cis* configuration relative to the cyclopentane ring are formed in greater abundance than isomers where the acyl chains are in the *trans* configuration (Fig. 2.4 inset). Thus, the configuration of most isoprostanes differs from that of prostaglandins that are formed enzymatically by prostaglandin synthases (also known as cycloxygenases,) where the two side chains are in the *trans*-configuration. Why do free radical reactions, which do not have enzymatically directed stereochemistry, predominately yield *cis* side chains instead of equal proportions of *cis* and *trans* side chains? The transition state for formation of the *cis* side chain configuration has lower free energy than that for formation of the *trans* configuration,[24] resulting in a faster rate constant for the *cis* configuration, particularly at lower temperatures.[18,24] The conformational difference between the vast majority of isoprostanes and their prostaglandin counterparts has been routinely exploited for their separation in various chromatographic systems and leads to differences in their potency for various receptors. However, it is important to keep in mind that isoprostanes with *trans* side chains are still formed in significant amounts (~8% at 37°C).[24] Thus, lipid peroxidation via the isoprostane pathway can be a significant source of prostaglandins in some situations (e.g., $PGF_{2\alpha}$ measured in the urine).[24]

2.2.1.2 F_2-Isoprostanes The endoperoxide ring of H_2-isoprostanes rearranges to form a number of products (Fig. 2.5). Complete reduction of the endoperoxide of H_2-isoprostanes generates 64 regio- and stereoisomers of PGF_2 (designated as F_2-isoprostanes). Measurement of F_2-isoprostane concentrations in plasma and urine is now considered the gold standard for monitoring oxidative stress *in vivo*,[25] and changes in F_2-isoprostane concentrations have been measured in a large number of disease-related conditions in humans and other animals. The suitability of these compounds for monitoring oxidative stress derives from their chemical stability, the relative ease of selective measurement, and their dramatic increases in response to oxidative insults *in vivo*.

In addition to their usefulness as biomarkers, F_2-isoprostanes are also biologically active. For instance, 15-F_{2t}-isoprostane is a potent vasoconstrictor and acts via the thromboxane receptor.[26] Eight other F_2-isoprostane isomers have been synthesized, including 12-F_{2t}-IsoP and 5-F_{2t}-isoprostane, and six out of these eight isomers were also vasoconstrictors.[27] Interestingly, they also induced thromboxane synthesis, so they may potentially be acting at an independent receptor in addition to binding at the thromboxane receptor.

2.2.1.3 Other Major Isoprostane Products Although F_2-isoprostanes are the most well-studied products of the isoprostane pathway, other products of this pathway may be more important in terms of direct contribution to pathophysiology of oxidative stress. Synthetase enzymes enzymatically rearrange PGH_2 to form five major prostaglandin products: PGF_2, PGE_2, PGD_2, PGI_2 (prostacyclin), and thromboxane A_2. Nonenzymatic rearrangement of PGH_2 also forms two isomers of γ-ketoaldehydes, levuglandin E_2, and D_2. Peroxidation of arachidonic acid has been demonstrated to form isomeric families of most of these prostaglandin products, including F_2-, D_2-/E_2-isoprostanes,[28] isothromboxanes,[29] and isolevuglandins (also called isoketals).[30,31] To date, no evidence for isomers of the sixth product, I_2-isoprostanes, has been published, although attempts have been made to isolate it or its metabolite.

The exact yield of each product from nonenzymatic rearrangement of H_2-isoprostane depends on the redox state under which the H_2-isoprostane forms. *In vitro* oxidation of arachidonate in typical buffers yields

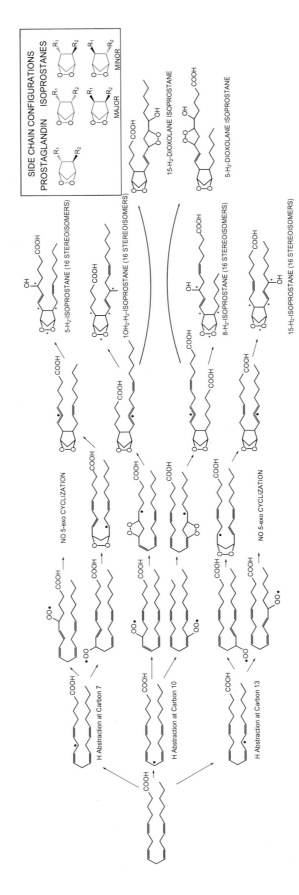

Figure 2.4 Generation of H_2-isoprostane regio- and stereo-isomers.

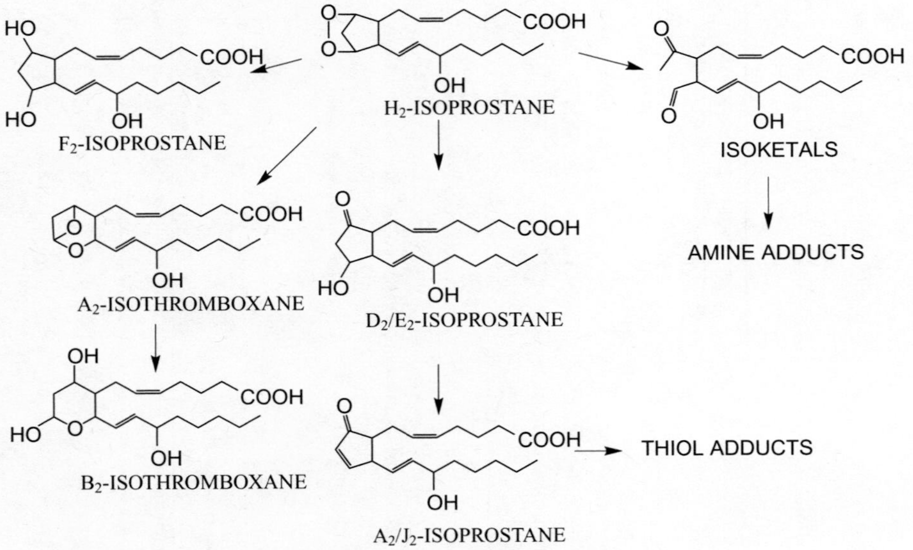

Figure 2.5 Major products of H_2-isoprostane pathway.

predominantly D_2/E_2-isoprostanes, but lipid peroxidation within the more reducing environments of cells and tissues strongly favor F_2-isoprostanes. For example, the yield of the five products in the liver of rats after carbon tetrachloride treatment was found to be 672 ng g^{-1} F_2-isoprostane, 161 ng g^{-1} D_2/E_2-isoprostane, 102 ng g^{-1} isothromboxane, and 27.4 ng g^{-1} isolevuglandins (measured as protein adduct).[29,32] Addition of 100 μM α-tocopherol, a reductant, to oxidizing synaptosomes significantly decreased the sum total of isoprostanes formed, but dramatically increased the ratio of F_2-isoprostanes to D_2/E_2-isoprostanes.[33] It may be important to note, however, that measurement of D_2/E_2-isoprostanes and isolevuglandins is less straightforward in biological systems than *in vitro*. In the case of D_2/E_2-isoprostanes, this is because they undergo dehydration to J_2/A_2-isoprostanes that react with cellular thiols to form adducts. Thus, measuring free levels of D_2/E_2-isoprostanes or even J_2/A_2-isoprostanes does not accurately reflect their initial abundance. Such consideration is even more significant when considering isolevuglandins, which react almost instantaneously with proteins.[31] Although measuring isolevuglandin protein adducts does provide an estimate of the initial isolevuglandin formed, isolevuglandins were recently discovered to form phosphatidylethanolamine adducts in greater abundance than protein adducts,[34,35] so that an accurate estimate of the total isolevuglandins formed would require measuring both types of adducts.

The biological ramification of nonenzymatic formation of prostaglandin-like products remains to be fully explored. In addition to the F_2-isoprostane isomers, 15-E_2-isoprostane has also been shown to be a potent vasoconstrictor,[28] but nothing is known about the bioactivity of isothromboxanes. In terms of pathophysiology, the most important isoprostane product may be the isolevuglandins. These highly reactive γ-ketoaldehydes are highly toxic when added exogenously to cells.[36–38] Potential mechanisms underlying their cytotoxicity include their ability to inhibit the proteasome,[37] induce mitochondrial transition pore formation,[39] activate p38 MAPK,[40] and form PE adducts that may disrupt membrane integrity.[34] They also inhibit ion channel function.[32,41,42] The importance of isolevuglandins to the deleterious effects of lipid peroxidation has gained further support by experiments where selective scavengers of these compounds provided protection against cytotoxicity and ion channel dysfunction induced by oxidizing agents.[42,43] Because at least one of these selective scavengers, salicylamine, is orally bioavailable,[44] evaluation of the efficacy of this compound in animal models of disease related to lipid peroxidation is currently underway.

2.2.1.4 Minor Isoprostane Products In addition to the major products of the isoprostane pathway, several isoprostane species with additional cyclized oxygen rings have been reported. Two G_2-isoprostanes (8-G_2-isoprostane and 12-G_2-isoprostane) have appropriately spaced double bonds to be able to generate a second dioxolane ring, and these dioxolane-isoprostane species have been demonstrated to form *in vitro*[20] (Fig. 2.4). Presumably F-, D/E-ring type compounds, as well as the isolevuglandins, can form from these H_2-dioxolane-isoprostanes.[20] Formation of the dioxolane-isoprostane species has been used to rationalize the greater abundance of 5-and 15-series isoprostanes found *in vitro* and *in vivo* compared to 8- and 12-series isoprostanes. Besides dioxolane isoprostanes, Berliner and coworkers have also characterized isoprostane compounds with additional epoxide rings that form esterified to phospholipids during the oxidation of LDL.[45,46] Interestingly, many of these epoxyisoprostane compounds induced inflammatory cytokine secretion from endothelial cells.

2.2.2 Diepoxide Pathway Products

Besides isoprostanes, the 1,2-dioxolanylcarbinyl radical is the precursor to other molecules as well, including diepoxide containing compounds and their derivatives (Fig. 2.3). Here, a concerted reaction of the carbinyl radical with the endoperoxide ring leads to scission of the oxygenoxygen bond to form two epoxide rings. Acceptance of a second molecule of oxygen gives a diepoxy, peroxyl radical product which can be reduced to a diepoxyalcohol. The two epoxide rings can undergo ring opening to potentially yield trihydroxy, epoxide products or pentahydroxy products. The biological activity of these various products has not been well characterized.

2.2.2.1 Isofurans and Related Compounds The trihydroxy, epoxide products of the diepoxide pathway were recently proposed as an alternative intermediate[19] for the formation of isofurans, a potential marker of oxidative damage at high oxygen.[21] The isofurans were first put forward as sensitive markers of oxygen tension in 2002, when Fessel et al. reported finding a family of compounds by GC/MS with mass 16 amu greater than F_2-isoprostanes during the oxidation of arachidonic acid at high oxygen concentration.[21] When these same compounds were analyzed as [$^2H_9$] TMS derivatives, rather than as the [$^1H_9$] TMS derivatives, the mass was increased by another 27 amu, consistent with the presence of three hydroxyl groups. Catalytic hydrogenation shifted the mass of the compounds 4 amu, consistent with two double bonds. The underivatized mass of these compounds was consistent with a molecular formula of $C_{20}H_{34}O_6$ and thus four double bond equivalents, so that the molecule appeared to contain either a ketone or a cyclized oxygen ring. Treatment with sodium borohydride or with methoxyamine hydrochloride did not alter the mass of the molecule, ruling out a ketone. Acid treatment also did not alter the mass of the molecule, so that the compounds could not be trihydroxy, epoxy compounds. Although the mass matched that for isothromboxane B_2, the acetal ring of isothromboxane B_2 should spontaneously open to form a ketone, so that the lack of reactivity with methoxyamine ruled out this structure. A molecule with a substituted tetrahydrofuran ring (i.e., an isofuran) is consistent with this data, as is a molecule with a substituted hydropyran ring (Fig. 2.3). Furan ring compounds have previously been shown to form by enzymatic mechanisms.[47] Based on experiments with ^{18}O-labeled water and ^{18}O oxygen, two mechanisms for the formation of isofurans were proposed by Fessel et al.[21] One mechanism involved cyclic peroxide cleavage, and the second mechanism involved epoxide hydrolysis. More recently, Jahn et al. proposed that formation of isofurans that acquired oxygen from water proceeded via the diepoxide pathway (Fig. 2.3), rather than the originally proposed epoxide hydrolysis pathway.[19] More studies are necessary to determine which of these proposed pathways account for the formation of these compounds.

While no biological activity has been attributed to isofurans to date, they have been used as biomarkers in conjunction with isoprostanes in cases where oxygen tension is expected to be markedly high. *In vitro* experiments showed that both F_2-isoprostanes and isofurans levels increase linearly with oxygen concentrations up to 21% oxygen, but above 21% only isofuran levels continue to increase.[21] Although oxygen tension in the vast majority of tissues is <5%, in cases where supplemental oxygen is used such as surgery, high performance aircraft, or deep diving operations, isofurans are likely to change to a greater extent than F_2-isoprostanes in exposed tissues like the lung. Additionally, isofuran levels are elevated to a greater extent than F_2-isoprostanes in the substantia nigra of patients who died from Parkinson's disease,[48] potentially reflecting the greater oxygen concentration present in diseased tissue due to mitochondrial dysfunction. If further validation does indeed establish a relationship between isofuran/isoprostane ratio and mitochondrial dysfunction, this may be a very useful tool for verifying mitochondrial dysfunction in many conditions where such dysfunction has been proposed.

2.2.3 Serial Cyclic Endoperoxides

Another set of products of the 1,2-dioxolanylcarbinyl radical intermediate that show sensitivity to oxygen tension are the serial cyclic endoperoxides (Fig. 2.3). These species can be generated when the 1,2-dioxolanylcarbinyl adds a second oxygen to form a peroxyl radical properly positioned to undergo a second 5-*exo* cyclization. Addition of the second oxygen must compete with the cyclization reaction, so that higher oxygen concentrations favor formation of the serial cyclic endoperoxides.[49–51] The bioactivity of these serial cyclic endoperoxides has not been explored.

2.3 FRAGMENTED PRODUCTS OF LIPID PEROXIDATION

Up to this point, we have described lipid peroxidation species that form by the addition of one or more oxygen to the PUFA. In addition to oxygenation and cyclization reactions, peroxidation results in fragmentation of acyl chains. Fragmentation products include some of the earliest and most well-studied products of lipid peroxidation, such as malondialdehyde, acrolein, and 4-hydroxy-2-nonenal, as well as highly potent ligands for receptors such as oxidatively fragmented phospholipids.

2.3.1 Short-Chain Alkanes, Aldehydes, and Acids

One of the predominant fragmentation reactions during lipid peroxidation is beta scission. After formation of a lipid hydroperoxide, reductive fragmentation by metals such as iron or copper leads to formation of an alkoxyl radical as described previously in Section 2.1.2. This alkoxyl radical can instigate the cleavage of the carbon–carbon bond on the side opposite from the conjugated double bond, leaving behind an alkanyl radical fragment as well as an aldehydic fragment (Fig. 2.6). Two electron oxidation of the aldehydic fragment yields a carboxylate, so that both aldehyde and acid fragments are detected *in vitro* and *in vivo*. For the alkanyl radical, abstraction of a hydrogen from a nearby PUFA converts the radical into a stable alkane product. In addition to these fragmentation products, fragmentation products that correspond to scission on the conjugated diene side of the alkoxyl radical can also be readily observed. These products have also been rationalized by a beta-scission reaction. However, because this mechanism requires the formation of a vinyl radical, which is not very favorable, mechanisms involving adjacent peroxy groups[52] or a dioxetene mechanisms[53] have also been proposed for the beta-scission-like fragmentation on the diene side.

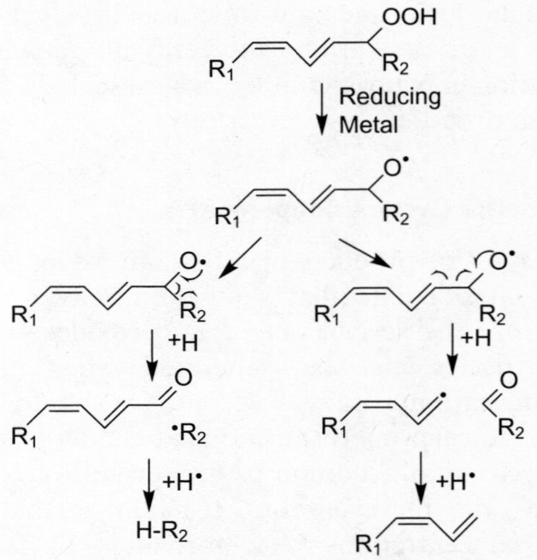

Whatever the uncertainty in the precise mechanism of fragmentation, products predicted from beta-scission-like fragmentation can be readily detected during lipid peroxidation. Measurement of pentane and ethane, the ω-terminal fragments from the alkoxyl radical formed at the carbon 15 and carbon 18 position of arachidonic and eicosapentaenoic acid, respectively, have long been used as a marker of lipid peroxidation *in vivo*.[54] For each PUFA, specific fragmentation products can be predicted based on the site of the initial peroxyl radical formation. For instance, oxidation of arachidonic acid is predicted to generate an alkoxyl radical at six different positions (carbon 5, 8, 9, 11, 12, and 15) leading to a total of 12 different aldehydes and 12 complimentary alkane fragments of variable lengths. Relatively little effort has gone into trying to measure each of these products *in vivo* or to determine their bioactivities, but some of the products have known bioactivities, often discovered in other contexts. For instance, beta scission from the carbon 5 alkoxyl radical of arachidonic acid forms butyric acid and from the carbon 9 alkoxyl radical of linoleate forms azelaic acid. Butyric acid is well known as an inhibitor of histone deacetylases.[55] Azelaic acid has antiproliferative and cytotoxic effects which has led to its use as a treatment for comedonal and inflammatory acne, as well as various cutaneous hyperpigmentary disorders.[56]

2.3.2 Oxidatively Fragmented Phospholipids

One area where the extent of formation and biological activity of beta-scission products have been more extensively characterized is oxidatively fragmented phospholipids. In particular, the beta-scission products of LA and arachidonic acid esterified to phosphatidylcholines (PC) generate potent ligands for a number of critical leukocyte receptors including the platelet-activating factor (PAF) receptor, the peroxisomal proliferator-activated receptor γ (PPARγ), and CD36.

Activation of the PAF receptor results in platelet aggregation, leukocyte activation, vasodilation, and secretion of inflammatory cytokines.[57,58] Oxidation of low-density lipoprotein (LDL) or 1-alkyl-2-arachidonyl-PC results in a number of beta-scission products, including 1-alkyl-2-butanoyl-PC (C4-PAF) and 1-alkyl-2-oxovaleroyl-PC that are potent agonists for the PAF receptor.[59–61] Oxidation of the 1-palmitoyl-2-arachidonyl-PC give rise to similar fragmentation products, including 1-palmitoyl-2-oxovaleroyl-PC (POVPC) and 1-palmitoyl-

Figure 2.6 Fragmentation of polyunsaturated fatty acids by beta-scission generates aldehydes and alkanes.

2-glutaryl-PC (PGPC), which are also PAF receptor agonists even though with less potency than their alkyl analogs.[59,60,62,63] Importantly, these products have been detected in plasma, oxidized cells, low-density lipoprotein, and atherosclerotic lesions.[64–67]

Although oxidatively fragmented phospholipid clearly activate the PAF receptor, authentic PAF cannot replicate all of the activities of these lipids, suggesting that oxidatively fragmented phospholipid act through other receptors or mechanisms as well when they stimulate cytokine secretion from endothelial cells, monocyte binding to endothelial cells, and endothelial cell death.[66,68] Oxidation of 1-alkyl-2-linoleic-PC gives 1-alkyl-2-azelaoyl-PC (azPAF) which is a relatively poor agonist of the PAF receptor but stimulates PPARγ.[69] More recently, azPAF has been also shown to act at mitochondria to induce swelling and cytochrome c release and these effects required the mitochondrial protein Bid.[70,71] Whether these effects are a general property of all oxidatively fragmented phospholipids with ω-carboxylate moieties is not known, but at least one other such species (PGPC) was able to induce swelling, while oxidatively fragmented phospholipids lacking the ω-carboxylate moiety such as C4-PAF and POVPC were much weaker agonists.[71] Other receptors which have been implicated in the activity of oxidatively fragmented phospholipids included the lectin-like oxidized low-density-lipoprotein receptor (LOX-1),[72] the VEGF receptor 2,[73] and the EP2 receptor.[74]

In addition to the PAF receptor and lipid receptors, oxidatively fragmented phospholipids are substrates for uptake by various scavenger receptors of macrophages including CD36 and SRBI.[69,75] Characterization of structural requirements for optimal binding by CD36 found that the best ligands possessed chain lengths of 6–11 carbons and a gamma-hydroxy(or oxo)-α,β-unsaturated carbonyl.[76] The potential mechanism underlying the formation of these hydroxyalkenals will be discussed in Section 2.3.4, but the essential point here is that monocytes and macrophages appear to be extremely well equipped to detect and take up oxidatively fragmented phospholipids, suggesting the biological importance of these molecules. The mechanisms of action for these oxidized phospholipids needs further investigation and such investigation will likely be very informative in increasing our understanding of how lipid peroxidation contributes to atherosclerosis and other diseases.

2.3.3 PAF Acetylhydrolase

Because of the potential role of oxidatively fragmented phospholipids in inflammation, the discovery of enzymes that degrade these products is not surprising. Perhaps the most important of these enzymes are the group VII phospholipase A_2 that make up the PAF acetylhydrolase (PAFAH) family.[77] There are a least four isoforms of PAFAH,[78,79] and two of these isoforms (PAFAH1 and PAFAH2) degrade both authentic PAF and oxidized phospholipids. The plasma form (PAFAH1) was the first enzyme discovered and was originally characterized as the enzyme that degraded PAF but not normal length alkylacyl or diacyl PC.[80] Later, it was found that this enzyme could also cleave longer sn-2 chains if they bore also an ω-oxo moieties or multiple hydroxyl groups.[81] Thus PAFAHs cleave C4-PAF, POVPC, PGPC, and F_2-isoprostanes esterified phospholipids, and probably many of the oxidatively fragmented phospholipids recognized by CD36. The effects of modulating PAFAH activity support a critical role for oxidatively fragmented phospholipids in oxidative injury. For example, overexpression of PAFAH2 protects against cell death induced by oxidants,[82] and genetic ablation of PAFAH2 increases cell death and injury during carbon tetrachloride-induced hepatic injury.[83]

2.3.4 Hydroxyalkenals

Perhaps the most studied of all lipid peroxidation products is 4-hydroxy-2-nonenal (HNE), a highly abundant hydroxyalkenal fragmentation product that results from peroxidation of both arachidonic acid and LA. Esterbauer first reported the formation of HNE in peroxidized liver microsomal lipids in 1980.[84] Extensive studies have demonstrated that HNE reacts with cellular thiols, particularly cysteines, and to a lesser extent with lysines and histidines.[85,86] Modification of proteins by HNE has significant effects on protein function and may therefore lead to cellular dysfunction. Increases in HNE-protein levels have been demonstrated in numerous conditions and more recently, the major proteins adducted by HNE have been cataloged.[87]

Interestingly, as intensively as HNE has been studied (or perhaps precisely because of how intensively it has been studied), the exact mechanism of its formation has been a matter of some controversy. Originally, Esterbauer proposed a dioxetane mechanism for its formation (Fig. 2.7).[88,89]

Subsequently, Porter and Pryor proposed two other mechanisms for HNE formation.[90] One mechanism required cyclization to form a hydroperoxy dihydropyran intermediate that then underwent beta scission to form HNE.[90] However, this mechanism seemed less favorable because the presence of a double bond greatly constrains the likelihood of the cyclization. The other, more favored mechanism, postulated the formation of an epoxyhydroperoxide that then underwent Hock rearrangement and cleavage in the presence of acid to

60 LIPID PEROXIDATION AND NITRATION

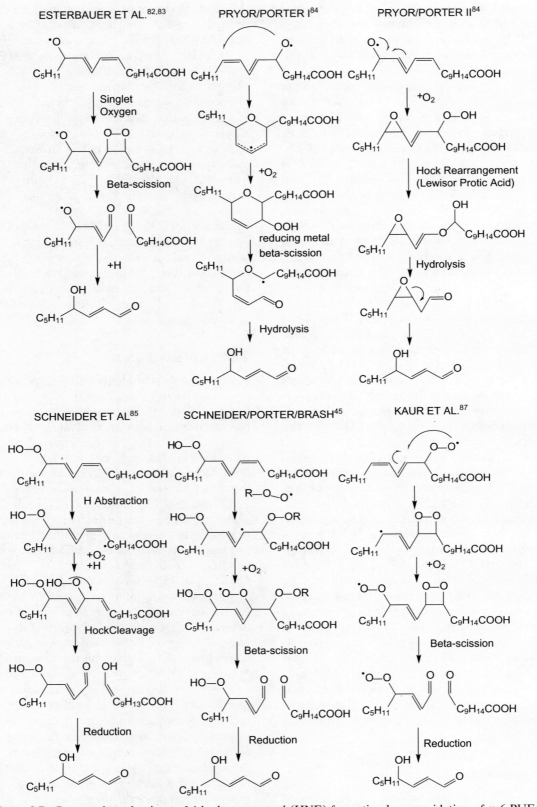

Figure 2.7 Proposed mechanisms of 4-hydroxynonenal (HNE) formation by peroxidation of ω-6 PUFA.

yield the 3,4 epoxynonal. This epoxynonal had already been definitely shown to undergo rearrangement to HNE. While these mechanism seemed reasonable, subsequent studies pointed to 4-hydroperoxy-2-nonenal (HPNE) being the immediate precursor of HNE. Therefore, Schneider et al. performed further analytical studies on various intermediates and proposed another mechanism that still utilized Hock cleavage, but that included a dihydroperoxy intermediate prior to formation of HPNE.[91] Subsequently, Schneider et al. analyzed the products of synthetic dihydroperoxides,[92] leading them to modify this mechanism to one that included formation of lipid peroxy dimer intermediates.[52] Still other mechanisms have also been put forward; for instance, Kaur et al. proposed a mechanism that proceeded through a dioxetane intermediate prior to fragmentation.[53] It is of value to note that each of these mechanisms applies basic reaction steps for which there is substantial scientific evidence. The difficulty lies in combining these potential choices of reaction pathways into a cohesive mechanism that best fits the available data on HNE intermediates. It may well be that multiple pathways lead to the formation of HNE, and that it is the multiplicity of formation routes that make HNE one of the major fragmentation products of lipid peroxidation.

The same mechanism(s) that lead to HNE formation should in principle lead to the formation of many other hydroxyalkenals. For instance, fragmentation of ω-3 fatty acids like eicosapentaenoic acid and docosahexaenoic acid, instead of ω-6 fatty acids like LA and arachidonic acid, lead to formation of hydroxyhexanal, which has many of the same properties as HNE. Fragmentation of esterified linolenic acid or arachidonic acid by the same mechanisms as for HNE, but in the inverse direction relative to the acyl chain, will generate esterified hydroxyalkenals. As discussed in the previous section, phospholipid esterified hydroxyalkenals and their more oxidized acid analogs are ligands for CD36.

A great many bioactivities have been attributed to hydroxyalkenals, with low levels inducing cell proliferation and higher levels (>20 μM) inducing cell death.[93] Cell signaling activated by HNE includes activation of NRF2-EpRE signaling pathways,[94] protein kinase C,[95] MAP kinase pathways,[96] tyrosine kinase receptors like EGF receptor,[97] and calcium signaling.[98] Cytotoxic concentrations of HNE and related hydroxyalkenals induce proteasome inhibition[99] and the mitochondrial permeability transition pore.[100]

In addition to hydroxyalkenals, the same fragmentation of LA that produces HNE will also produce 9-oxononanic acid (ONA). The two aldehydes differ in reactivity with cellular nucleophiles because the α,β-unsaturation of HNE allows it to undergo Michael addition with thiols in addition to forming Schiff base adducts with primary amines. While ONA can form a Schiff base, the ONA Schiff base adduct will not undergo further reaction to form pyrrole adducts as the HNE Schiff base adduct can. What role ONA or its analogs play in the biology of oxidative injury remains to be more fully elucidated.

2.3.5 Malondialdehyde

One of the very first fragmentation products of lipid peroxidation identified was the 1,3-dialdehyde, malondialdehyde (MDA). One well-established intermediate in the formation of MDA is bicyclic endoperoxides (e.g., H_2-isoprostanes and PGH_2), with enzymatic conversion of PGH_2 to thromboxane A_2 by thromboxane synthetase generating 1 mole of MDA and 1 mole of thromboxane A_2 for every 2 mole of PGH_2.[101,102] Free ferric iron (or the ferric thiol cluster in the case of thromboxane synthase) assists in the fragmentation of the ring, which likely proceeds through hemolytic scission of the oxygen–oxygen bond of the endoperoxide (Fig. 2.8).[103] Other mechanisms may also generate MDA from other lipid hydroperoxide precursors.[16,104] MDA reacts readily with thiobarbituric acid (TBA) to form a pinkish chromophore, and assaying TBA reactivity has long been used to detect MDA and lipid peroxidation in biological systems. While commonly used, it is important to remember that the TBA reactivity assay is not selective for MDA, because many other lipid aldehydes including HNE can react with TBA to form the same chromophore. Furthermore, as MDA can be formed from spermine,[105] the TBA reactivity assay is also not an entirely specific indicator of lipid peroxidation.

In solution with neutral or alkaline pH, MDA generally exists in its enolate form, which is not very reactive. However, at pH < 4.5, MDA exists as both β-hydroxy acrolein and its dicarbonyl form and is quite reactive with both thiols and amines. More than one MDA molecule generally contributes to the final stable adduct. Many of the same biological activities that have been characterized for HNE have also been shown to be induced by MDA.

2.3.6 Acrolein

One final aldehydic product of beta-scission-like fragmentation that deserves mention is the α,β-unsaturated aldehyde acrolein. Acrolein forms during incomplete combustion of gasoline, coal, wood, and plastics,[16] as well as the peroxidation of various PUFAs.[106] Although the exact mechanism of acrolein formation from peroxidation of PUFA has not been well characterized, its

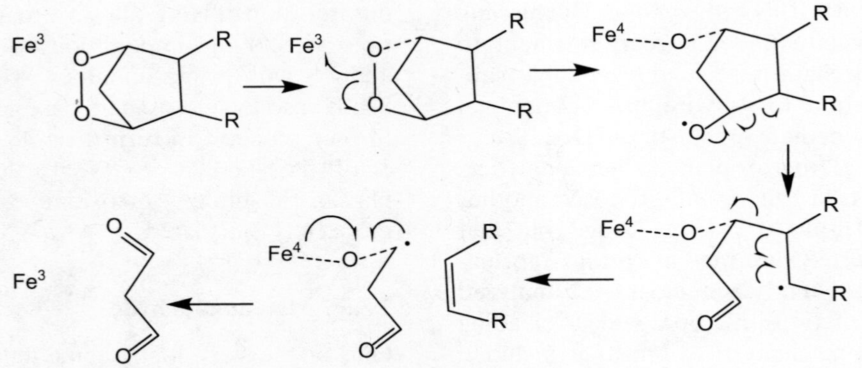

Figure 2.8 Potential mechanism of iron-mediated malondialdehyde formation from bicyclic endoperoxides.

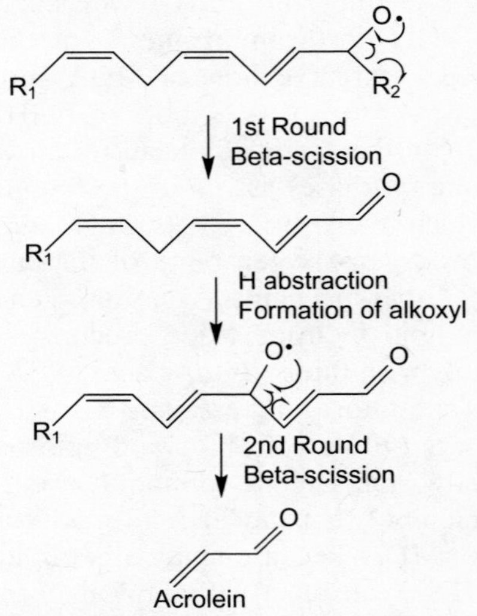

Figure 2.9 Potential mechanism of acrolein formation by peroxidation of PUFA.

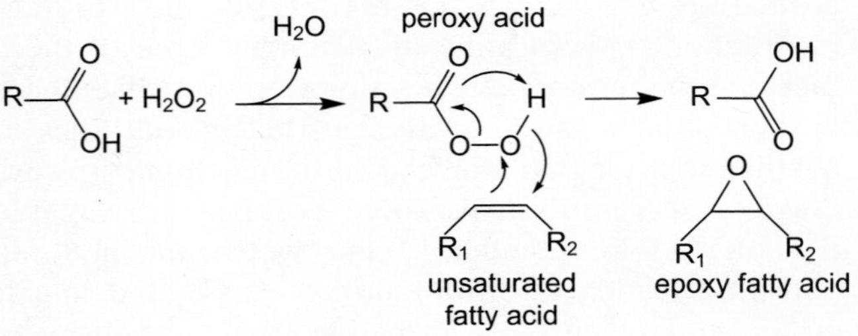

Figure 2.10 Formation of epoxy fatty acid by peroxyacids.

formation can be rationalized by two consecutive rounds of beta-scission-like fragmentations (Fig. 2.9).

Acrolein reacts with gluthatione approximately 150 times faster than 4-hydroxynonenal, making it by far the strongest electrophile of all the α,β-unsaturated aldehydes formed by lipid peroxidation.[16] Acrolein also reacts with similar rapidity to the cysteine residues of proteins. When acrolein reacts with protein thiols, the initial Michael adduct will also rapidly react with a second thiol to form a highly stable thiazolidine derivative, making acrolein a potent protein cross-linker. Reactions with amines proceed through Michael addition as well, but are slower.[107] Besides adducts to gluthathione and protein, acrolein adducts to DNA[108–110] and aminophospholipid[111] have also been detected in biological systems. Given acrolein's propensity to adduct to key macromolecules, it is not surprising to find that it inhibits the activity of a number of proteins with active-site cysteines and potently induces cell death in a large number of cell types.[16]

2.4 EPOXY FATTY ACIDS

One final family of oxygenated lipid product that bears brief discussion are the epoxy fatty acids. Unlike the peroxidation reactions previously discussed that form through free radical (one-electron) mechanisms, simple epoxidation proceeds through two-electron mechanisms and any unsaturated fatty acid can serve as a potential substrate. Both peroxy acids or hydroperoxy fatty acids can serve as the oxygen donor for these nonenzymatic reactions (Fig. 2.10). In addition to nonenzymatic formation of epoxy fatty acids, various cytochrome P450 epoxygenases such as CYP 2J3 oxygenate arachidonic acid to form epoxyeicosatrienoic acids (EETs).[112] EETs have several bioactivities relevant to vascular function,[113] including being an endothelium-derived hyperpolarizing factor.[114]

2.5 LIPID NITROSYLATION

Up to this point, we have focused on peroxidation reactions with fatty acids. In addition to modification by oxygen, lipid radicals can also undergo modification by reactive nitrogen species. Like peroxidation, nitrosylation of lipids also generates highly bioactive compounds.

2.5.1 Formation of Reactive Nitrogen Species

Nitric oxide ($^{\bullet}$NO) is generated *in vivo* by the action of various nitric oxide synthetases and is a key regulator of vascular tone. Like oxygen, $^{\bullet}$NO in its ground state is a free radical, and will readily undergo reactions with other radicals. In particular, $^{\bullet}$NO reacts with the superoxide radical ($O_2^{\bullet-}$) at near diffusion rates to form the potent nitrating agent peroxynitrite ($ONOO^-$) (Fig. 2.11). At neutral pH, $ONOO^-$ is protonated to $ONOOH$, which can undergo homolysis to form two highly reactive radicals: the nitrogen dioxide radical ($^{\bullet}NO_2$) and the hydroxyl radical ($^{\bullet}OH$). $ONOO^-$ can also react with CO_2 to form $ONOOCO_2^-$, which will also undergo homolysis to generate $^{\bullet}NO_2$ as well as $CO_3^{\bullet-}$. $^{\bullet}NO_2$ can also form directly by the two-electron oxidation of $^{\bullet}NO$.

2.5.2 Lipid Nitration Reactions

Formation of $^{\bullet}NO_2$ is a key event in lipid nitrosylation because $^{\bullet}NO_2$ can directly attack the double bonds of alkene chains. For this reason, nitrosylation differs from peroxidation, because even fatty acids like OA that lack bis-allylic hydrogens can be nitrosylated. Attack of the $^{\bullet}NO_2$ at either carbon of the double bond generates the nitrosyl lipid radical (Fig. 2.12). If $^{\bullet}NO_2$ concentration are sufficiently high, a second subsequent addition occurs to generate dinitro-fatty acids. In the presence of

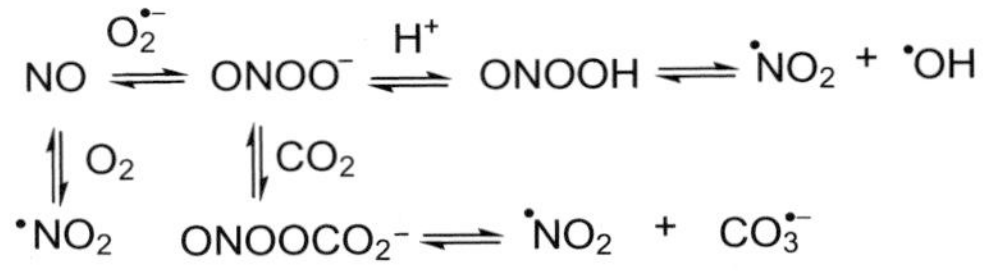

Figure 2.11 Formation of lipid-nitrating species from nitric oxide.

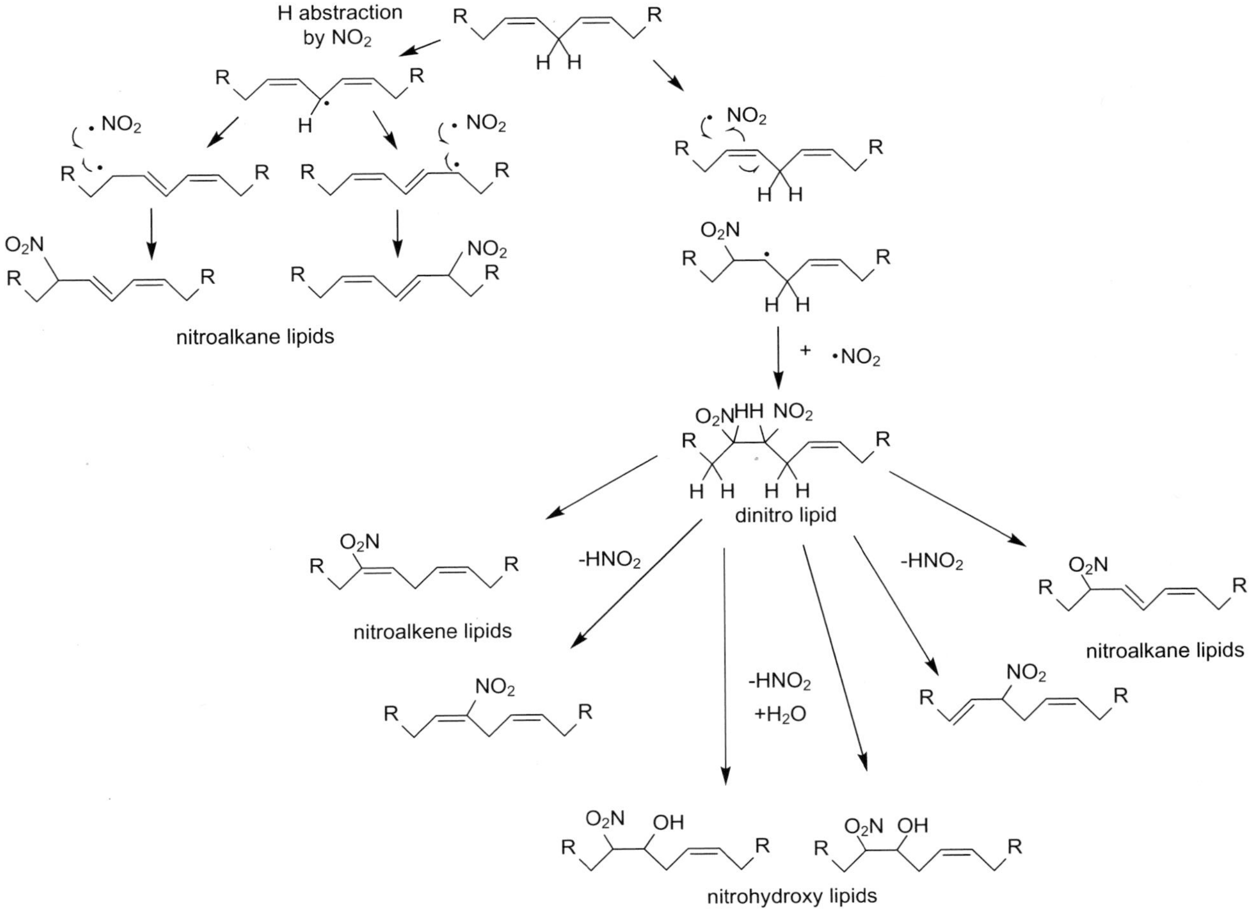

Figure 2.12 Mechanisms of nitrated lipids formation.

molecular oxygen, the dinitro-fatty acid can also undergo rearrangement to form nitro-nitrate compounds. The dinitro-fatty acid is somewhat unstable, and elimination of HNO_2 from either position generates nitrated fatty acid (NO_2–FA) species, including trans-nitroalkane, cis-nitroalkene, and trans-nitroalkenes (Fig. 2.12). Elimination of HNO_2 coupled with addition of water gives rise to nitrohydroxy fatty acid species ($NO_2(OH)$–FA).

The formation of nitroalkene NO_2–FAs is of particular interest because these products are good electrophiles that can reversibly react with cellular thiols via Michael addition.[115] The formation of nitroalkene fatty acids was verified by synthesis of authentic standards using nitronium tetrafluoroborate (NO_2BF_4), which provides a reliable synthetic route to nitroalkene NO_2–FA.[116–118]

Other mechanisms for the formation of nitrated lipids beside those described above have been proposed as well. For instance, •NO can react with lipid peroxyl radicals (LOO•) to form an unstable lipid peroxynitrite (LOONO), followed by rapid rearrangements of the caged radicals to generate the alkyl nitrate ($LONO_2$).[117] Another mechanism for PUFA with bis-allylic hydrogens is hydrogen abstraction by hydroxyl radical, followed by addition of •NO_2 to generate a NO_2–FA with a trans-, cis-conjugated double bond (Fig. 2.12).

Because NO_2–FAs form by radical reactions, multiple regio- and stereoisomers are generated, just as in the case of lipid peroxidation. With nitrosylation, an even greater number of regioisomers can form. For instance, with LA, peroxidation occurs at carbon 9 and 13, while nitrosylation can occur on carbons 9, 10, 12, and 13 and while no peroxidation of OA occurs, nitrosylation occurs at carbons 9 and 10. Jain et al. explored the yield of various nitrated OA regioisomers under several experimental conditions.[119] They found that 10-NO_2-OA formed at about a 1.2:1 ratio to 9-NO_2-OA and that 9,10-$NO_2(OH)$-OA formed at 4:1 ratio with 10,9-$NO_2(OH)$-OA in normoxic conditions. Greater levels of all products were formed under hypoxic conditions, but the same isomers remained dominant. Further studies may be useful to elucidate the mechanisms underlying the predominance of these species.

2.5.3 Detection of Lipid Nitration In Vivo

The formation of NO_2–FAs in vivo has been confirmed by mass spectrometry analysis spectrometry techniques.[120,121] The accurate detection and quantitation of various NO_2-FA in biological tissues by mass spectrometry requires appropriate internal standards and adequate separation methods. Synthesis of nitrated lipid regioisomers has been undertaken by a variety of method.[116,122–124] Currently, a stereospecific route[125] is often utilized to synthesize nitroalkene standards including NO_2-LA[126] and its $^{13}C/^{15}N$ isotopes,[127] and nitroarachidonic acid.[128] Using these standards, a compound with the mass of NO_2-LA (m/z 324) that co-eluted with the synthetic isotopic internal standards and whose fragmentation gave a product ion at m/z of 46 consistent with generation of a nitro group was shown to be present in plasma.[121] Levels of NO_2-LA were higher in hyperlipidemic subjects than normolipidemic subjects. Subsequent analysis by Baker et al. confirmed these results using both GC/MS and LC/MS/MS and demonstrated the presence in plasma of several regioisomers of both esterified and unesterified NO_2-LA.[129] Nitration of the linoleate esterified to cholesterol was subsequently detected in human plasma and lipoproteins.[130] NO_2-OA and NO_2-arachidonic acid have also been detected in vivo.[131]

The actual levels of nitrated lipid available in vivo is unclear. Initially, levels of free NO_2-FA in plasma were stated to be in the 300–600 nM range. Subsequent analysis suggested that free NO_2-FA was actually present only at very low nM concentrations because most NO_2-FA was adducted to cellular thiols, including proteins and gluthatione, and only disassociated from the thiols during the assay.[115,132,133] The biological implications of the sequestered NO_2-FA are currently being investigated. Exact quantitation of NO_2-FA is also complicated because oxidation and degradation of NO_2-FAs occurs readily.[120] Therefore, comparing the NO_2-FA levels for various conditions may be more accurate than estimating absolute levels.

2.5.4 Bioactivities of Nitrated Lipids

In general, the effects of NO_2–FAs appear to be anti-inflammatory and protective, in contrast to lipid peroxidation products which tend to be proinflammatory. Anti-inflammatory effects include suppression of cytokine expression and platelet aggregation, and upregulation of anti-inflammatory proteins such as hemeoxygenase-1.[134–138] Evidence suggests that NO_2–FAs transduce their effects through multiple mechanisms, including modifying transcription factors, interacting with specific receptors, and releasing •NO. For instance, NO_2–FAs inhibition of LPS-induced pro-inflammatory cytokine release appears to be dependent on alkylation of transcription factor NF-κB p65 as well as suppression of STAT.[136,137] NO_2-LA alkylation of the cysteine residues of Keap1 appears to be responsible for nuclear translocation of the Nrf-2 transcription factor and subsequent gene expression in vascular smooth muscle cells produced by NO_2-LA.[138] In addition to binding transcription factors, NO_2–FAs are potent agonists of PPARs,[131,139] especially PPARγ, and the binding of

NO$_2$-LA to PPARγ is sufficiently stable for X-ray analysis of the crystal structure.[140] Finally, NO$_2$–FAs release •NO when metabolized by smooth muscle cells.[126,128] Whether •NO release from NO$_2$-FA has an important physiological function is not clear, but •NO release likely mediates some of the pharmacological effects of NO$_2$–FAs. The physiology and mechanisms of action of nitrated lipids is likely to be a fruitful area of discovery for quite some time.

SUMMARY

Peroxidation and nitrosylation of lipids generates a wide range of biologically active molecules. The general mechanisms underlying the formation of these products have been worked out, although fresh insights into the details continue to be brought forward. Because peroxidation and nitrosylation of lipids proceed through nonenzymatic pathways, a great number of isomers and functionally related compounds are formed for each type of molecule. This multiplicity of compounds complicates both quantitation and characterization of biological activity. Nevertheless, the bioactivities and *in vivo* levels of a large number of these compounds have been characterized. Based on these studies, both peroxidized and nitrated lipids may make important contributions to physiology and pathophysiology.

REFERENCES

1. Koppenol, W.H. Oxyradical reactions: From bond-dissociation energies to reduction potentials. *FEBS Lett.* **1990**, *264*(2), 165–167.
2. Cyrus, T., Pratico, D., Zhao, L., Witztum, J.L., Rader, D.J., Rokach, J., FitzGerald, G.A., Funk, C.D. Absence of 12/15-lipoxygenase expression decreases lipid peroxidation and atherogenesis in apolipoprotein E-deficient mice. *Circulation* **2001**, *103*(18), 2277–2282.
3. Yu, K., Bayona, W., Kallen, C.B., Harding, H.P., Ravera, C.P., McMahon, G., Brown, M., Lazar, M.A. Differential activation of peroxisome proliferator-activated receptors by eicosanoids. *J. Biol. Chem.* **1995**, *270*, 23975–23983.
4. Naruhn, S., Meissner, W., Adhikary, T., Kaddatz, K., Klein, T., Watzer, B., Müller-Brüsselbach, S., Müller, R. 15-Hydroxyeicosatetraenoic acid is a preferential peroxisome proliferator-activated receptor β/δ agonist. *Mol. Pharmacol.* **2010**, *77*(2), 171–184.
5. Kliewer, S.A., Sundseth, S.S., Jones, S.A., Brown, P.J., Wisely, G.B., Koble, C.S., Devchand, P., Wahli, W., Willson, T.M., Lenhard, J.M., Lehmann, J.M. Fatty acids and eicosanoids regulate gene expression through direct interactions with peroxisome proliferator-activated receptors α and γ. *Proc. Natl. Acad. Sci. U.S.A.* **1997**, *94*(9), 4318–4323.
6. Shappell, S.B., Gupta, R.A., Manning, S., Whitehead, R., Boeglin, W.E., Schneider, C., Case, T., Price, J., Jack, G.S., Wheeler, T.M., Matusik, R.J., Brash, A.R., DuBois, R.N. 15S-hydroxyeicosatetraenoic acid activates peroxisome proliferator-activated receptor γ and inhibits proliferation in PC3 prostate carcinoma cells. *Cancer Res.* **2001**, *61*(2), 497–503.
7. Moreno, J.J. New aspects of the role of hydroxyeicosatetraenoic acids in cell growth and cancer development. *Biochem. Pharmacol.* **2009**, *77*(1), 1–10.
8. Harrison, K.A., Murphy, R.C. Isoleukotrienes are biologically active free radical products of lipid peroxidation. *J. Biol. Chem.* **1995**, *270*(29), 17273–17278.
9. Chang, M.S., Boeglin, W.E., Guengerich, F.P., Brash, A.R. Cytochrome P450-dependent transformations of 15R- and 15S-hydroperoxyeicosatetraenoic acids: Stereoselective formation of epoxy alcohol productsâ€. *Biochemistry* **1996**, *35*(2), 464–471.
10. Yu, Z., Schneider, C., Boeglin, W.E., Brash, A.R. Epidermal lipoxygenase products of the hepoxilin pathway selectively activate the nuclear receptor PPARalpha. *Lipids* **2007**, *42*(6), 491–497.
11. Thomas, J.P., Maiorino, M., Ursini, F., Girotti, A.W. Protective action of phospholipid hydroperoxide glutathione peroxidase against membrane-damaging lipid peroxidation. *In situ* reduction of phospholipid and cholesterol hydroperoxides. *J. Biol. Chem.* **1990**, *265*(1), 454–461.
12. Yang, Y., Cheng, J.-Z., Singhal, S.S., Saini, M., Pandya, U., Awasthi, S., Awasthi, Y.C. Role of glutathione S-transferases in protection against lipid peroxidation: Overexpression of hGSTA2-2 in K562 cell protects against hydrogen peroxide-induced apoptosis and inhibit JNK and caspase 3 activation. *J. Biol. Chem.* **2001**, *276*(22), 19220–19230.
13. Kang, S.W., Baines, I.C., Rhee, S.G. Characterization of a mammalian peroxiredoxin that contains one conserved cysteine. *J. Biol. Chem.* **1998**, *273*(11), 6303–6311.
14. Pak, J.H., Manevich, Y., Kim, H.S., Feinstein, S.I., Fisher, A.B. An antisense oligonucleotide to 1-cys peroxiredoxin causes lipid peroxidation and apoptosis in lung epithelial cells. *J. Biol. Chem.* **2002**, *277*(51), 49927–49934.
15. Koo, K.H., Lee, S., Jeong, S.Y., Kim, E.T., Kim, H.J., Kim, K., Song, K., Chae, H.Z. Regulation of thioredoxin peroxidase activity by C-terminal truncation. *Arch. Biochem. Biophys.* **2002**, *397*(2), 312–318.
16. Esterbauer, H., Schaur, R.J., Zollner, H. Chemistry and biochemistry of 4-hydroxynonenal, malonaldehyde and related aldehydes. *Free Radic. Biol. Med.* **1991**, *11*(1), 81–128.
17. Yin, H., Havrilla, C.M., Gao, L., Morrow, J.D., Porter, N.A. Mechanisms for the formation of isoprostane endoperoxides from arachidonic acid. *J. Biol. Chem.* **2003**, *278*(19), 16720–16725.
18. O'Connor, D.E., Mihelich, E.D., Coleman, M.C. Stereochemical course of the autooxidative cyclization of lipid hydroperoxides to prostaglandin-like bicyclic endoperoxides. *J. Am. Chem. Soc.* **1984**, *106*(12), 3577–3584.

19. Jahn, U., Galano, J.M., Durand, T. Beyond prostaglandins—chemistry and biology of cyclic oxygenated metabolites formed by free-radical pathways from polyunsaturated fatty acids. *Angew. Chem. Int. Ed. Engl.* **2008**, *47*(32), 5894–5955.

20. Yin, H., Morrow, J.D., Porter, N.A. Identification of a novel class of endoperoxides from arachidonate autoxidation. *J. Biol. Chem.* **2004**, *279*(5), 3766–3776.

21. Fessel, J.P., Porter, N.A., Moore, K.P., Sheller, J.R., Roberts, L.J., II. Discovery of lipid peroxidation products formed *in vivo* with a substituted tetrahydrofuran ring (isofurans) that are favored by increased oxygen tension. *Proc. Natl. Acad. Sci. U.S.A.* **2002**, *99*(26), 16713–16718.

22. Taber, D.F., Morrow, J.D., Roberts, L.J., 2nd. A nomenclature system for the isoprostanes. *Prostaglandins* **1997**, *53*(2), 63–67.

23. Rokach, J., Khanapure, S.P., Hwang, S.W., Adiyaman, M., Lawson, J.A., FitzGerald, G.A. Nomenclature of isoprostanes: A proposal. *Prostaglandins* **1997**, *54*(6), 853–873.

24. Yin, H., Gao, L., Tai, H.-H., Murphey, L.J., Porter, N.A., Morrow, J.D. Urinary prostaglandin F2{alpha} is generated from the isoprostane pathway and not the cyclooxygenase in humans. *J. Biol. Chem.* **2007**, *282*(1), 329–336.

25. Kadiiska, M.B., Gladen, B.C., Baird, D.D., Germolec, D., Graham, L.B., Parker, C.E., Nyska, A., Wachsman, J.T., Ames, B.N., Basu, S., Brot, N., Fitzgerald, G.A., Floyd, R.A., George, M., Heinecke, J.W., Hatch, G.E., Hensley, K., Lawson, J.A., Marnett, L.J., Morrow, J.D., Murray, D.M., Plastaras, J., Roberts, L.J., 2nd, Rokach, J., Shigenaga, M.K., Sohal, R.S., Sun, J., Tice, R.R., Van Thiel, D.H., Wellner, D., Walter, P.B., Tomer, K.B., Mason, R.P., Barrett, J.C. Biomarkers of oxidative stress study II: Are oxidation products of lipids, proteins, and DNA markers of CCl(4) poisoning? *Free Radic. Biol. Med.* **2005**, *38*(6), 698–710.

26. Morrow, J.D., Minton, T.A., Roberts, L.J., 2nd. The F2-isoprostane, 8-epi-prostaglandin F2 alpha, a potent agonist of the vascular thromboxane/endoperoxide receptor, is a platelet thromboxane/endoperoxide receptor antagonist. *Prostaglandins* **1992**, *44*(2), 155–163.

27. Hou, X., Roberts, L.J., 2nd, Gobeil, F., Jr., Taber, D., Kanai, K., Abran, D., Brault, S., Checchin, D., Sennlaub, F., Lachapelle, P., Varma, D., Chemtob, S. Isomer-specific contractile effects of a series of synthetic f2-isoprostanes on retinal and cerebral microvasculature. *Free Radic. Biol. Med.* **2004**, *36*(2), 163–172.

28. Morrow, J.D., Minton, T.A., Mukundan, C.R., Campbell, M.D., Zackert, W.E., Daniel, V.C., Badr, K.F., Blair, I.A., Roberts, L.J., 2nd. Free radical-induced generation of isoprostanes *in vivo*. Evidence for the formation of D-ring and E-ring isoprostanes. *J. Biol. Chem.* **1994**, *269*(6), 4317–4326.

29. Morrow, J.D., Awad, J.A., Wu, A., Zackert, W.E., Daniel, V.C., Roberts, L.J., 2nd. Nonenzymatic free radical-catalyzed generation of thromboxane-like compounds (isothromboxanes) *in vivo*. *J. Biol. Chem.* **1996**, *271*(38), 23185–23190.

30. Salomon, R.G., Subbanagounder, G., Singh, U., O'Neil, J., Hoff, H.F. Oxidation of low-density lipoproteins produces levuglandin-protein adducts. *Chem. Res. Toxicol.* **1997**, *10*(7), 750–759.

31. Brame, C.J., Salomon, R.G., Morrow, J.D., Roberts, L.J., 2nd. Identification of extremely reactive gamma-ketoaldehydes (isolevuglandins) as products of the isoprostane pathway and characterization of their lysyl protein adducts. *J. Biol. Chem.* **1999**, *274*(19), 13139–13146.

32. Brame, C.J., Boutaud, O., Davies, S.S., Yang, T., Oates, J.A., Roden, D., Roberts, L.J., 2nd. Modification of proteins by isoketal-containing oxidized phospholipids. *J. Biol. Chem.* **2004**, *279*(14), 13447–13451.

33. Montine, T.J., Montine, K.S., Reich, E.E., Terry, E.S., Porter, N.A., Morrow, J.D. Antioxidants significantly affect the formation of different classes of isoprostanes and neuroprostanes in rat cerebral synaptosomes. *Biochem. Pharmacol.* **2003**, *65*(4), 611–617.

34. Sullivan, C.B., Matafonova, E., Roberts, L.J., II, Amarnath, V., Davies, S.S. Isoketals form cytotoxic phosphatidylethanolamine adducts in cells. *J. Lipid Res.* **2010**, *51*(5), 999–1009.

35. Li, W., Laird, J.M., Lu, L., Roychowdhury, S., Nagy, L.E., Zhou, R., Crabb, J.W., Salomon, R.G. Isolevuglandins covalently modify phosphatidylethanolamines *in vivo*: Detection and quantitative analysis of hydroxylactam adducts. *Free Radic. Biol. Med.* **2009**, *47*(11), 1539–1552.

36. Davies, S.S. Modulation of protein function by isoketals and levuglandins. *Subcell Biochem.* **2008**, *49*, 49–70.

37. Davies, S.S., Amarnath, V., Montine, K.S., Bernoud-Hubac, N., Boutaud, O., Montine, T.J., Roberts, L.J., 2nd. Effects of reactive gamma-ketoaldehydes formed by the isoprostane pathway (isoketals) and cyclooxygenase pathway (levuglandins) on proteasome function. *FASEB J.* **2002**, *16*(7), 715–717.

38. Schmidley, J.W., Dadson, J., Iyer, R.S., Salomon, R.G. Brain tissue injury and blood-brain barrier opening induced by injection of LGE2 or PGE2. *Prostaglandins Leukot. Essent. Fatty Acids* **1992**, *47*(2), 105–110.

39. Stavrovskaya, I.G., Baranov, S.V., Guo, X., Davies, S.S., Roberts, L.J., 2nd, Kristal, B.S. Reactive gamma-ketoaldehydes formed via the isoprostane pathway disrupt mitochondrial respiration and calcium homeostasis. *Free Radic. Biol. Med.* **2010**, *49*(4), 567–579.

40. Bernoud-Hubac, N., Alam, D.A., Lefils, J., Davies, S.S., Amarnath, V., Guichardant, M., Roberts Ii, L.J., Lagarde, M. Low concentrations of reactive [gamma]-ketoaldehydes prime thromboxane-dependent human platelet aggregation via p38-MAPK activation. *Biochim. Biophys. Acta.* **2009**, *1791*(4), 307–313.

41. Fukuda, K., Davies, S.S., Nakajima, T., Ong, B.H., Kupershmidt, S., Fessel, J., Amarnath, V., Anderson, M.E., Boyden, P.A., Viswanathan, P.C., Roberts, L.J., 2nd, Balser, J.R. Oxidative mediated lipid peroxidation recapitulates proarrhythmic effects on cardiac sodium channels. *Circ. Res.* **2005**, *97*(12), 1262–1269.

42. Nakajima, T., Davies, S.S., Matafonova, E., Potet, F., Amarnath, V., Tallman, K.A., Serwa, R.A., Porter, N.A., Balser, J.R., Kupershmidt, S., Roberts, L.J., 3rd. Selective gamma-ketoaldehyde scavengers protect Nav1.5 from oxidant-induced inactivation. *J. Mol. Cell. Cardiol.* **2010**, *48*(2), 352–359.

43. Davies, S.S., Brantley, E.J., Voziyan, P.A., Amarnath, V., Zagol-Ikapitte, I., Boutaud, O., Hudson, B.G., Oates, J.A., Roberts L.J., II. Pyridoxamine analogues scavenge lipid-derived gamma-ketoaldehydes and protect against H(2)O(2)-mediated cytotoxicity. *Biochemistry* **2006**, *45*(51), 15756–15767.

44. Zagol-Ikapitte, I., Matafonova, E., Amarnath, V., Bodine, C.L., Boutaud, O., Tirona, R.G., Oates, J.A., Roberts, L.J., II, Davies, S.S. Determination of the pharmacokinetics and oral bioavailability of salicylamine, a potent gamma-ketoaldehyde scavenger, by LC/MS/MS. *Pharmaceutics* **2010**, *2*(1), 18–29.

45. Watson, A.D., Subbanagounder, G., Welsbie, D.S., Faull, K.F., Navab, M., Jung, M.E., Fogelman, A.M., Berliner, J.A. Structural identification of a novel pro-inflammatory epoxyisoprostane phospholipid in mildly oxidized low density lipoprotein. *J. Biol. Chem.* **1999**, *274*(35), 24787–24798.

46. Subbanagounder, G., Wong, J.W., Lee, H., Faull, K.F., Miller, E., Witztum, J.L., Berliner, J.A. Epoxyisoprostane and epoxycyclopentenone phospholipids regulate monocyte chemotactic protein-1 and interleukin-8 synthesis. Formation of these oxidized phospholipids in response to interleukin-1beta. *J. Biol. Chem.* **2002**, *277*(9), 7271–7281.

47. Pace-Asciak, C. Plyhydroxyl cyclic ethers formed from triiated arachidonic acid by acetone powders of sheep seminal vesicles. *Biochemistry* **1971**, *10*, 3664–3669.

48. Fessel, J.P., Hulette, C., Powell, S., Roberts, L.J., 2nd, Zhang, J. Isofurans, but not F2-isoprostanes, are increased in the substantia nigra of patients with Parkinson's disease and with dementia with Lewy body disease. *J. Neurochem.* **2003**, *85*(3), 645–650.

49. Khan, J.A., Porter, N.A. Serial cyclization of an arachidonic hydroperoxide. *Angew. Chem. Int. Ed. Engl.* **1982**, *21*, 217–218.

50. Yin, H., Porter, N.A. New insights regarding the autoxidation of polyunsaturated fatty acids. *Antioxid. Redox Signal.* **2005**, *7*(1–2), 170–184.

51. Yin, H., Porter, N.A. Identification of intact lipid peroxides by Ag+ coordination ion-spray mass spectrometry (CIS-MS). *Methods Enzymol.* **2007**, *433*, 193–211.

52. Schneider, C., Porter, N.A., Brash, A.R. Routes to 4-hydroxynonenal: Fundamental issues in the mechanisms of lipid peroxidation. *J. Biol. Chem.* **2008**, *283*(23), 15539–15543.

53. Kaur, K., Salomon, R.G., O'Neil, J., Hoff, H.F. Carboxyalkyl)pyrroles in human plasma and oxidized low-density lipoproteins. *Chem. Res. Toxicol.* **1997**, *10*(12), 1387–1396.

54. Knutson, M.D., Handelman, G.J., Viteri, F.E. Methods for measuring ethane and pentane in expired air from rats and humans. *Free Radic. Biol. Med.* **2000**, *28*(4), 514–519.

55. Kruh, J. Effects of sodium butyrate, a new pharmacological agent, on cells in culture. *Mol. Cell. Biochem.* **1982**, *42*(2), 65–82.

56. Fitton, A., Goa, K.L. Azelaic acid. A review of its pharmacological properties and therapeutic efficacy in acne and hyperpigmentary skin disorders. *Drugs* **1991**, *41*(5), 780–798.

57. Zimmerman, G.A., McIntyre, T.M., Prescott, S.M., Stafforini, D.M. The platelet-activating factor signaling system and its regulators in syndromes of inflammation and thrombosis. *Crit. Care Med.* **2002**, *30*(5 Suppl.), S294–S301.

58. Yost, C.C., Weyrich, A.S., Zimmerman, G.A. The platelet activating factor (PAF) signaling cascade in systemic inflammatory responses. *Biochimie* **2010**, *92*(6), 692–697.

59. Marathe, G.K., Davies, S.S., Harrison, K.A., Silva, A.R., Murphy, R.C., Castro-Faria-Neto, H., Prescott, S.M., Zimmerman, G.A., McIntyre, T.M. Inflammatory platelet-activating factor-like phospholipids in oxidized low density lipoproteins are fragmented alkyl phosphatidylcholines. *J. Biol. Chem.* **1999**, *274*(40), 28395–28404.

60. Heery, J.M., Kozak, M., Stafforini, D.M., Jones, D.A., Zimmerman, G.A., McIntyre, T.M., Prescott, S.M. Oxidatively modified LDL contains phospholipids with platelet-activating factor-like activity and stimulates the growth of smooth muscle cells. *J. Clin. Invest.* **1995**, *96*(5), 2322–2330.

61. Chen, R., Chen, X., Salomon, R.G., McIntyre, T.M. Platelet activation by low concentrations of intact oxidized LDL particles involves the PAF receptor. *Arterioscler. Thromb. Vasc. Biol.* **2009**, *29*(3), 363–371.

62. Smiley, P.L., Stremler, K.E., Prescott, S.M., Zimmerman, G.A., McIntyre, T.M. Oxidatively fragmented phosphatidylcholines activate human neutrophils through the receptor for platelet-activating factor. *J. Biol. Chem.* **1991**, *266*(17), 11104–11110.

63. Pegorier, S., Stengel, D., Durand, H., Croset, M., Ninio, E. Oxidized phospholipid: POVPC binds to platelet-activating-factor receptor on human macrophages. Implications in atherosclerosis. *Atherosclerosis* **2006**, *188*(2), 433–443.

64. Patel, K.D., Zimmerman, G.A., Prescott, S.M., McIntyre, T.M. Novel leukocyte agonists are released by endothelial cells exposed to peroxide. *J. Biol. Chem.* **1992**, *267*(21), 15168–15175.

65. Ravandi, A., Babaei, S., Leung, R., Monge, J.C., Hoppe, G., Hoff, H., Kamido, H., Kuksis, A. Phospholipids and oxophospholipids in atherosclerotic plaques at different stages of plaque development. *Lipids* **2004**, *39*(2), 97–109.

66. Watson, A.D., Leitinger, N., Navab, M., Faull, K.F., Horkko, S., Witztum, J.L., Palinski, W., Schwenke, D., Salomon, R.G., Sha, W., Subbanagounder, G., Fogelman, A.M.,

Berliner, J.A. Structural identification by mass spectrometry of oxidized phospholipids in minimally oxidized low density lipoprotein that induce monocyte/endothelial interactions and evidence for their presence in vivo. *J. Biol. Chem.* **1997**, *272*(21), 13597–13607.

67. Yang, L., Latchoumycandane, C., McMullen, M.R., Pratt, B.T., Zhang, R., Papouchado, B.G., Nagy, L.E., Feldstein, A.E., McIntyre, T.M. Chronic alcohol exposure increases circulating bioactive oxidized phospholipids. *J. Biol. Chem.* **2010**, *285*(29): 22211–22220.

68. Huber, J., Vales, A., Mitulovic, G., Blumer, M., Schmid, R., Witztum, J.L., Binder, B.R., Leitinger, N. Oxidized membrane vesicles and blebs from apoptotic cells contain biologically active oxidized phospholipids that induce monocyte-endothelial interactions. *Arterioscler. Thromb. Vasc. Biol.* **2002**, *22*(1), 101–107.

69. Davies, S.S., Pontsler, A.V., Marathe, G.K., Harrison, K.A., Murphy, R.C., Hinshaw, J.C., Prestwich, G.D., Hilaire, A.S., Prescott, S.M., Zimmerman, G.A., McIntyre, T.M. Oxidized alkyl phospholipids are specific, high affinity peroxisome proliferator-activated receptor γ ligands and agonists. *J. Biol. Chem.* **2001**, *276*(19), 16015–16023.

70. Chen, R., Feldstein, A.E., McIntyre, T.M. Suppression of mitochondrial function by oxidatively truncated phospholipids is reversible, aided by bid, and suppressed by Bcl-XL. *J. Biol. Chem.* **2009**, *284*(39), 26297–26308.

71. Chen, R., Yang, L., McIntyre, T.M. Cytotoxic phospholipid oxidation products. *J. Biol. Chem.* **2007**, *282*(34), 24842–24850.

72. Dunn, S., Vohra, R.S., Murphy, J.E., Homer-Vanniasinkam, S., Walker, J.H., Ponnambalam, S. The lectin-like oxidized low-density-lipoprotein receptor: A pro-inflammatory factor in vascular disease. *Biochem. J.* **2008**, *409*(2), 349–355.

73. Zimman, A., Mouillesseaux, K.P., Le, T., Gharavi, N.M., Ryvkin, A., Graeber, T.G., Chen, T.T., Watson, A.D., Berliner, J.A. Vascular endothelial growth factor receptor 2 plays a role in the activation of aortic endothelial cells by oxidized phospholipids. *Arterioscler. Thromb. Vasc. Biol.* **2007**, *27*(2), 332–338.

74. Berliner, J.A., Leitinger, N., Tsimikas, S. The role of oxidized phospholipids in atherosclerosis. *J. Lipid Res.* **2009**, *50*(Suppl.), S207–S212.

75. Gao, D., Ashraf, M.Z., Kar, N.S., Lin, D., Sayre, L.M., Podrez, E.A. Structural basis for the recognition of oxidized phospholipids in oxidized low density lipoproteins by class B scavenger receptors CD36 and SR-BI. *J. Biol. Chem.* **2010**, *285*(7), 4447–4454.

76. Podrez, E.A., Poliakov, E., Shen, Z., Zhang, R., Deng, Y., Sun, M., Finton, P.J., Shan, L., Gugiu, B., Fox, P.L., Hoff, H.F., Salomon, R.G., Hazen, S.L. Identification of a novel family of oxidized phospholipids that serve as ligands for the macrophage scavenger receptor CD36. *J. Biol. Chem.* **2002**, *277*(41), 38503–38516.

77. Stafforini, D.M. Biology of platelet-activating factor acetylhydrolase (PAF-AH, lipoprotein associated phospholipase A2). *Cardiovasc. Drugs Ther.* **2009**, *23*(1), 73–83.

78. Scott, B.T., Olson, N., Long, G.L., Bovill, E.G. Novel isoforms of intracellular platelet activating factor acetylhydrolase (PAFAH1b2) in human testis; encoded by alternatively spliced mRNAs. *Prostaglandins Other Lipid Mediat.* **2008**, *85*(3–4), 69–80.

79. Hattori, K., Adachi, H., Matsuzawa, A., Yamamoto, K., Tsujimoto, M., Aoki, J., Hattori, M., Arai, H., Inoue, K. cDNA cloning and expression of intracellular platelet-activating factor (PAF) acetylhydrolase II. Its homology with plasma PAF acetylhydrolase. *J. Biol. Chem.* **1996**, *271*(51), 33032–33038.

80. Stafforini, D.M., Prescott, S.M., McIntyre, T.M. Human plasma platelet-activating factor acetylhydrolase. Purification and properties. *J. Biol. Chem.* **1987**, *262*(9), 4223–4230.

81. Stremler, K.E., Stafforini, D.M., Prescott, S.M., Zimmerman, G.A., McIntyre, T.M. An oxidized derivative of phosphatidylcholine is a substrate for the platelet-activating factor acetylhydrolase from human plasma. *J. Biol. Chem.* **1989**, *264*(10), 5331–5334.

82. Matsuzawa, A., Hattori, K., Aoki, J., Arai, H., Inoue, K. Protection against oxidative stress-induced cell death by intracellular platelet-activating factor-acetylhydrolase II. *J. Biol. Chem.* **1997**, *272*(51), 32315–32320.

83. Kono, N., Inoue, T., Yoshida, Y., Sato, H., Matsusue, T., Itabe, H., Niki, E., Aoki, J., Arai, H. Protection against oxidative stress-induced hepatic injury by intracellular type II platelet-activating factor acetylhydrolase by metabolism of oxidized phospholipids in vivo. *J. Biol. Chem.* **2008**, *283*(3), 1628–1636.

84. Benedetti, A., Comporti, M., Esterbauer, H. Identification of 4-hydroxynonenal as a cytotoxic product originating from the peroxidation of liver microsomal lipids. *Biochim. Biophys. Acta* **1980**, *620*(2), 281–296.

85. Benedetti, A., Esterbauer, H., Ferrali, M., Fulceri, R., Comporti, M. Evidence for aldehydes bound to liver microsomal protein following CCl4 or BrCCl3 poisoning. *Biochim. Biophys. Acta* **1982**, *711*(2), 345–356.

86. Uchida, K., Stadtman, E.R. Covalent attachment of 4-hydroxynonenal to glyceraldehyde-3-phosphate dehydrogenase. A possible involvement of intra- and intermolecular cross-linking reaction. *J. Biol. Chem.* **1993**, *268*(9), 6388–6393.

87. Vila, A., Tallman, K.A., Jacobs, A.T., Liebler, D.C., Porter, N.A., Marnett, L.J. Identification of protein targets of 4-hydroxynonenal using click chemistry for ex vivo biotinylation of azido and alkynyl derivatives. *Chem. Res. Toxicol.* **2008**, *21*(2), 432–444.

88. Esterbauer, H., Zollner, H. Methods for determination of aldehydic lipid peroxidation products. *Free Radic. Biol. Med.* **1989**, *7*(2), 197–203.

89. Esterbauer, H., Zollner, H., Schaur, R.J. Aldehydes formed by lipid peroxidation: Mechanisms of formation, occurrence, and determination. In: *Membrane Lipid Oxidation*. ed. C. Vigo-Pelfrey. CRC Press, Boca Raton, FL, 1990:239–268.

90. Pryor, W.A., Porter, N.A. Suggested mechanisms for the production of 4-hydroxy-2-nonenal from the autoxida-

tion of polyunsaturated fatty acids. *Free Radic. Biol. Med.* **1990**, *8*(6), 541–543.

91. Schneider, C., Tallman, K.A., Porter, N.A., Brash, A.R. Two distinct pathways of formation of 4-hydroxynonenal. *J. Biol. Chem.* **2001**, *276*(24), 20831–20838.

92. Schneider, C., Boeglin, W.E., Yin, H., Ste, D.F., Hachey, D.L., Porter, N.A., Brash, A.R. Synthesis of dihydroperoxides of linoleic and linolenic acids and studies on their transformation to 4-hydroperoxynonenal. *Lipids* **2005**, *40*(11), 1155–1162.

93. Forman, H.J., Fukuto, J.M., Miller, T., Zhang, H., Rinna, A., Levy, S. The chemistry of cell signaling by reactive oxygen and nitrogen species and 4-hydroxynonenal. *Arch. Biochem. Biophys.* **2008**, *477*(2), 183–195.

94. Zhang, H., Court, N., Forman, H.J. Submicromolar concentrations of 4-hydroxynonenal induce glutamate cysteine ligase expression in HBE1 cells. *Redox Rep.* **2007**, *12*(1), 101–106.

95. Nitti, M., Domenicotti, C., d'Abramo, C., Assereto, S., Cottalasso, D., Melloni, E., Poli, G., Biasi, F., Marinari, U.M., Pronzato, M.A. Activation of PKC-beta isoforms mediates HNE-induced MCP-1 release by macrophages. *Biochem. Biophys. Res. Commun.* **2002**, *294*(3), 547–552.

96. Soh, Y., Jeong, K.S., Lee, I.J., Bae, M.A., Kim, Y.C., Song, B.J. Selective activation of the c-Jun N-terminal protein kinase pathway during 4-hydroxynonenal-induced apoptosis of PC12 cells. *Mol. Pharmacol.* **2000**, *58*(3), 535–541.

97. Liu, W., Akhand, A.A., Kato, M., Yokoyama, I., Miyata, T., Kurokawa, K., Uchida, K., Nakashima, I. 4-Hydroxynonenal triggers an epidermal growth factor receptor-linked signal pathway for growth inhibition. *J. Cell Sci.* **1999**, *112*(Pt 14), 2409–2417.

98. Carini, R., Bellomo, G., Paradisi, L., Dianzani, M.U., Albano, E. 4-Hydroxynonenal triggers Ca2+ influx in isolated rat hepatocytes. *Biochem. Biophys. Res. Commun.* **1996**, *218*(3), 772–776.

99. Ferrington, D.A., Kapphahn, R.J. Catalytic site-specific inhibition of the 20S proteasome by 4-hydroxynonenal. *FEBS Lett.* **2004**, *578*(3), 217–223.

100. Kristal, B.S., Park, B.K., Yu, B.P. 4-Hydroxyhexenal is a potent inducer of the mitochondrial permeability transition. *J. Biol. Chem.* **1996**, *271*(11), 6033–6038.

101. Pryor, W.A., Stanley, J.P. A suggested mechanism for the production of malonaldehyde during the autoxidation of polyunsaturated fatty acids. Nonenyzmatic production of prostaglandin endoperoxides during autooxidation. *J. Org. Chem.* **1975**, *40*, 3615–3617.

102. McMillan, R.M., MacIntyre, D.E., Booth, A., Gordon, J.L. Malonaldehyde formation in intact platelets is catalysed by thromboxane synthase. *Biochem. J.* **1978**, *176*(2), 595–598.

103. Hecker, M., Ullrich, V. On the mechanism of prostacyclin and thromboxane A2 biosynthesis. *J. Biol. Chem.* **1989**, *264*(1), 141–150.

104. Frankel, E.N., Neff, W.E. Formation of malonaldehyde from lipid oxidation products. *Biochim. Biophys. Acta* **1983**, *754*, 264–270.

105. Quash, G., Ripoll, H., Gazzolo, L., Doutheau, A., Saba, A., Gore, J. Malondialdehyde production from spermine by homogenates of normal and transformed cells. *Biochimie* **1987**, *69*(2), 101–108.

106. Uchida, K., Kanematsu, M., Morimitsu, Y., Osawa, T., Noguchi, N., Niki, E. Acrolein is a product of lipid peroxidation reaction. *J. Biol. Chem.* **1998**, *273*, 16058–16066.

107. Witz, G. Biological interactions of alpha,beta-unsaturated aldehydes. *Free Radic. Biol. Med.* **1989**, *7*(3), 333–349.

108. Chung, F.L., Young, R., Hecht, S.S. Formation of cyclic 1,N2-propanodeoxyguanosine adducts in DNA upon reaction with acrolein or crotonaldehyde. *Cancer Res.* **1984**, *44*(3), 990–995.

109. Shapiro, R., Sodum, R.S., Everett, D.W., Kundu, S.K. Reactions of nucleosides with glyoxal and acrolein. *IARC Sci. Publ.* **1986**, *70*, 165–173.

110. Foiles, P.G., Akerkar, S.A., Chung, F.L. Application of an immunoassay for cyclic acrolein deoxyguanosine adducts to assess their formation in DNA of *Salmonella typhimurium* under conditions of mutation induction by acrolein. *Carcinogenesis* **1989**, *10*(1), 87–90.

111. Zemski Berry, K.A., Murphy, R.C. Characterization of acrolein-glycerophosphoethanolamine lipid adducts using electrospray mass spectrometry. *Chem. Res. Toxicol.* **2007**, *20*(9), 1342–1351.

112. Spector, A.A., Fang, X., Snyder, G.D., Weintraub, N.L. Epoxyeicosatrienoic acids (EETs): Metabolism and biochemical function. *Prog. Lipid Res.* **2004**, *43*(1), 55–90.

113. Fleming, I., Busse, R. Endothelium-derived epoxyeicosatrienoic acids and vascular function. *Hypertension* **2006**, *47*(4), 629–633.

114. Campbell, W.B., Gebremedhin, D., Pratt, P.F., Harder, D.R. Identification of epoxyeicosatrienoic acids as endothelium-derived hyperpolarizing factors. *Circ. Res.* **1996**, *78*(3), 415–423.

115. Batthyany, C., Schopfer, F.J., Baker, P.R., Duran, R., Baker, L.M., Huang, Y., Cervenansky, C., Branchaud, B.P., Freeman, B.A. Reversible post-translational modification of proteins by nitrated fatty acids *in vivo*. *J. Biol. Chem.* **2006**, *281*(29), 20450–20463.

116. Napolitano, A., Camera, E., Picardo, M., d'Ischia, M. Acid-promoted reactions of ethyl linoleate with nitrite ions: Formation and structural characterization of isomeric nitroalkene, nitrohydroxy, and novel 3-nitro-1,5-hexadiene and 1,5-dinitro-1, 3-pentadiene products. *J. Org. Chem.* **2000**, *65*(16), 4853–4860.

117. O'Donnell, V.B., Eiserich, J.P., Chumley, P.H., Jablonsky, M.J., Krishna, N.R., Kirk, M., Barnes, S., Darley-Usmar, V.M., Freeman, B.A. Nitration of unsaturated fatty acids by nitric oxide-derived reactive nitrogen species peroxynitrite, nitrous acid, nitrogen dioxide, and nitronium ion. *Chem. Res. Toxicol.* **1999**, *12*(1), 83–92.

118. Olah, G.A., Malhotra, R., Narang, S.C. *Nitration: Methods and Mechanisms*. VCH Publishing Inc., New York, 1989.

119. Jain, K., Siddam, A., Marathi, A., Roy, U., Falck, J.R., Balazy, M. The mechanism of oleic acid nitration by -NO2. *Free Radic. Biol. Med.* **2008**, *45*(3), 269–283.

120. Nadtochiy, S.M., Baker, P.R., Freeman, B.A., Brookes, P.S. Mitochondrial nitroalkene formation and mild uncoupling in ischaemic preconditioning: Implications for cardioprotection. *Cardiovasc. Res.* **2009**, *82*(2), 333–340.

121. Lima, E.S., Di Mascio, P., Rubbo, H., Abdalla, D.S. Characterization of linoleic acid nitration in human blood plasma by mass spectrometry. *Biochemistry* **2002**, *41*(34), 10717–10722.

122. Napolitano, A., Camera, E., Picardo, M., d'Ishida, M. Reactions of hydro(pero)xy derivatives of polyunsaturated fatty acids/esters with nitrite ions under acidic conditions. Unusual nitrosative breakdown of methyl 13-hydro(pero)xyoctadeca-9,11-dienoate to a novel 4-nitro-2-oximinoalk-3-enal product. *J. Org. Chem.* **2002**, *67*(4), 1125–1132.

123. Napolitano, A., Crescenzi, O., Camera, E., Giudicianni, I., Picardo, M., d'Ischia, M. The acid-promoted reaction of ethyl linoleate with nitrite. New insights from 15N-labeling and peculiar reactivity of a model skipped diene. *Tetrahedron* **2004**, *58*, 5061–5067.

124. O'Donnell, V.B., Eiserich, J.P., Bloodsworth, A., Chumley, P.H., Kirk, M., Barnes, S., Darley-Usmar, V.M., Freeman, B.A. Nitration of unsaturated fatty acids by nitric oxide-derived reactive species. *Methods Enzymol.* **1999**, *301*, 454–470.

125. Hayama, T., Tomoda, S., Takeuchi, Y., Nomura, Y. Synthesis of conjugated nitroalkenes via nitrosenylation of alkenes. *Tetrahedron Lett.* **1982**, *23*(45), 4733–4734.

126. Lim, D.G., Sweeney, S., Bloodsworth, A., White, C.R., Chumley, P.H., Krishna, N.R., Schopfer, F., O'Donnell, V.B., Eiserich, J.P., Freeman, B.A. Nitrolinoleate, a nitric oxide-derived mediator of cell function: Synthesis, characterization, and vasomotor activity. *Proc. Natl. Acad. Sci. U.S.A.* **2002**, *99*(25), 15941–15946.

127. Ferreira, A.M., Ferrari, M.I., Trostchansky, A., Batthyany, C., Souza, J.M., Alvarez, M.N., Lopez, G.V., Baker, P.R., Schopfer, F.J., O'Donnell, V., Freeman, B.A., Rubbo, H. Macrophage activation induces formation of the anti-inflammatory lipid cholesteryl-nitrolinoleate. *Biochem. J.* **2009**, *417*(1), 223–234.

128. Trostchansky, A., Souza, J.M., Ferreira, A., Ferrari, M., Blanco, F., Trujillo, M., Castro, D., Cerecetto, H., Baker, P.R., O'Donnell, V.B., Rubbo, H. Synthesis, isomer characterization, and anti-inflammatory properties of nitro-arachidonate. *Biochemistry* **2007**, *46*(15), 4645–4653.

129. Baker, P.R.S., Schopfer, F.J., Sweeney, S., Freeman, B.A. Red cell membrane and plasma linoleic acid nitration products: Synthesis, clinical identification, and quantitation. *Proc. Natl. Acad. Sci. U.S.A.* **2004**, *101*(32), 11577–11582.

130. Lima, E.S., Di Mascio, P., Abdalla, D.S. Cholesteryl nitrolinoleate, a nitrated lipid present in human blood plasma and lipoproteins. *J. Lipid Res.* **2003**, *44*(9), 1660–1666.

131. Baker, P.R., Lin, Y., Schopfer, F.J., Woodcock, S.R., Groeger, A.L., Batthyany, C., Sweeney, S., Long, M.H., Iles, K.E., Baker, L.M., Branchaud, B.P., Chen, Y.E., Freeman, B.A. Fatty acid transduction of nitric oxide signaling: Multiple nitrated unsaturated fatty acid derivatives exist in human blood and urine and serve as endogenous peroxisome proliferator-activated receptor ligands. *J. Biol. Chem.* **2005**, *280*(51), 42464–42475.

132. Baker, P.R., Schopfer, F.J., O'Donnell, V.B., Freeman, B.A. Convergence of nitric oxide and lipid signaling: Anti-inflammatory nitro-fatty acids. *Free Radic. Biol. Med.* **2009**, *46*(8), 989–1003.

133. Tsikas, D., Zoerner, A., Mitschke, A., Homsi, Y., Gutzki, F.M., Jordan, J. Specific GC-MS/MS stable-isotope dilution methodology for free 9- and 10-nitro-oleic acid in human plasma challenges previous LC-MS/MS reports. *J. Chromatogr. B Analyt. Technol. Biomed. Life Sci.* **2009**, *877*(26), 2895–2908.

134. Coles, B., Bloodsworth, A., Clark, S.R., Lewis, M.J., Cross, A.R., Freeman, B.A., O'Donnell, V.B. Nitrolinoleate inhibits superoxide generation, degranulation, and integrin expression by human neutrophils: Novel anti-inflammatory properties of nitric oxide-derived reactive species in vascular cells. *Circ. Res.* **2002**, *91*(5), 375–381.

135. Coles, B., Bloodsworth, A., Eiserich, J.P., Coffey, M.J., McLoughlin, R.M., Giddings, J.C., Lewis, M.J., Haslam, R.J., Freeman, B.A., O'Donnell, V.B. Nitrolinoleate inhibits platelet activation by attenuating calcium mobilization and inducing phosphorylation of vasodilator-stimulated phosphoprotein through elevation of cAMP. *J. Biol. Chem.* **2002**, *277*(8), 5832–5840.

136. Cui, T., Schopfer, F.J., Zhang, J., Chen, K., Ichikawa, T., Baker, P.R., Batthyany, C., Chacko, B.K., Feng, X., Patel, R.P., Agarwal, A., Freeman, B.A., Chen, Y.E. Nitrated fatty acids: Endogenous anti-inflammatory signaling mediators. *J. Biol. Chem.* **2006**, *281*(47), 35686–35698.

137. Ichikawa, T., Zhang, J., Chen, K., Liu, Y., Schopfer, F.J., Baker, P.R., Freeman, B.A., Chen, Y.E., Cui, T. Nitroalkenes suppress lipopolysaccharide-induced signal transducer and activator of transcription signaling in macrophages: A critical role of mitogen-activated protein kinase phosphatase 1. *Endocrinology* **2008**, *149*(8), 4086–4094.

138. Villacorta, L., Zhang, J., Garcia-Barrio, M.T., Chen, X.L., Freeman, B.A., Chen, Y.E., Cui, T. Nitro-linoleic acid inhibits vascular smooth muscle cell proliferation via the Keap1/Nrf2 signaling pathway. *Am. J. Physiol. Heart Circ. Physiol.* **2007**, *293*(1), H770–H776.

139. Schopfer, F.J., Lin, Y., Baker, P.R., Cui, T., Garcia-Barrio, M., Zhang, J., Chen, K., Chen, Y.E., Freeman, B.A. Nitrolinoleic acid: An endogenous peroxisome proliferator-activated receptor gamma ligand. *Proc. Natl. Acad. Sci. U.S.A.* **2005**, *102*(7), 2340–2345.

140. Li, Y., Zhang, J., Schopfer, F.J., Martynowski, D., Garcia-Barrio, M.T., Kovach, A., Suino-Powell, K., Baker, P.R., Freeman, B.A., Chen, Y.E., Xu, H.E. Molecular recognition of nitrated fatty acids by PPAR gamma. *Nat. Struct. Mol. Biol.* **2008**, *15*(8), 865–867.

3

PROTEIN POSTTRANSLATIONAL MODIFICATION

JAMES L. HOUGLAND, JOSEPH DARLING, AND SUSAN FLYNN

OVERVIEW

Proteins play central and essential roles in the vast majority of biological processes. Protein function (catalysis, ligand binding, signaling, etc.) is transduced through molecular interactions involving amino acid side chains. These side chains present a wide variety of functional groups, spanning from simple hydrocarbons (e.g., alanine, valine, leucine) to thiols (cysteine), alcohols (serine, threonine), and amines (lysine). While some of these side chains are largely unaffected by oxidative stress, a subset of amino acids can be rapidly modified by reactive oxygen and nitrogen species. Furthermore, oxidative stress leads to the formation of electrophilic molecules that can subsequently react with amino acids bearing nucleophilic side chains. Through these two pathways, oxidative stress can result in a variety of posttranslational modifications (PTMs) that alter protein structure and function.

In this chapter, we focus on the protein modifications that occur due to reactive species generated during oxidative stress. The reader should note that oxidative stress also leads to changes in the distribution of PTMs such as phosphorylation and acetylation as part of signal transduction and gene regulation pathways (discussed in Section 3.4 of this chapter); as these modifications are not directly caused (in a chemical sense) by reactive species arising from oxidative stress, they will not be covered herein. The goal of this chapter is to present an overview of the modifications that can occur at each amino acid due to oxidative stress and a survey of the methods currently in use to detect oxidative stress-related PTMs.

3.1 OXIDATIVE STRESS-RELATED PTMs: OXIDATION REACTIONS

In the presence of reactive oxygen and nitrogen species, a number of amino acid side chains participate in oxidation reactions that often involve amino acid-centered radicals. The vast majority of oxidation chemistry observed to date occurs with the sulfur-containing amino acids cysteine and methionine, as the low oxidation potential of sulfur renders these amino acids particularly susceptible to redox chemistry.[1]

3.1.1 Cysteine

In the last decade, the roles of cysteine oxidation in controlling protein function and mediating cellular signaling have become a topic of great interest. The central role of cysteine in cellular redox chemistry arises from its low oxidation potential and resulting low barriers for undergoing oxidation.[2,3] In light of their susceptibility to oxidation under biologically relevant conditions, cysteine residues in many proteins are proposed to serve as members of a complex redox signaling network, or "cellular thiolstat," which regulates enzymatic and protein activity through reversible cysteine modifications.[4]

Under oxidative stress conditions, cysteine can encounter both one-electron (e.g., $O_2^{\bullet-}$, $HO^{\bullet}$, $^{\bullet}NO$) and two-electron (e.g., H_2O_2) oxidants. One-electron oxidation of the cysteine thiol leads to formation of the thiyl radical ($RS^{\bullet}$), which can then undergo subsequent radical chemistry such as hydrogen abstraction or radical coupling (Fig. 3.1a).[5] Alternatively, reaction with two-electron oxidants usually involves nucleophilic

---

Molecular Basis of Oxidative Stress: Chemistry, Mechanisms, and Disease Pathogenesis, First Edition. Edited by Frederick A. Villamena.
© 2013 John Wiley & Sons, Inc. Published 2013 by John Wiley & Sons, Inc.

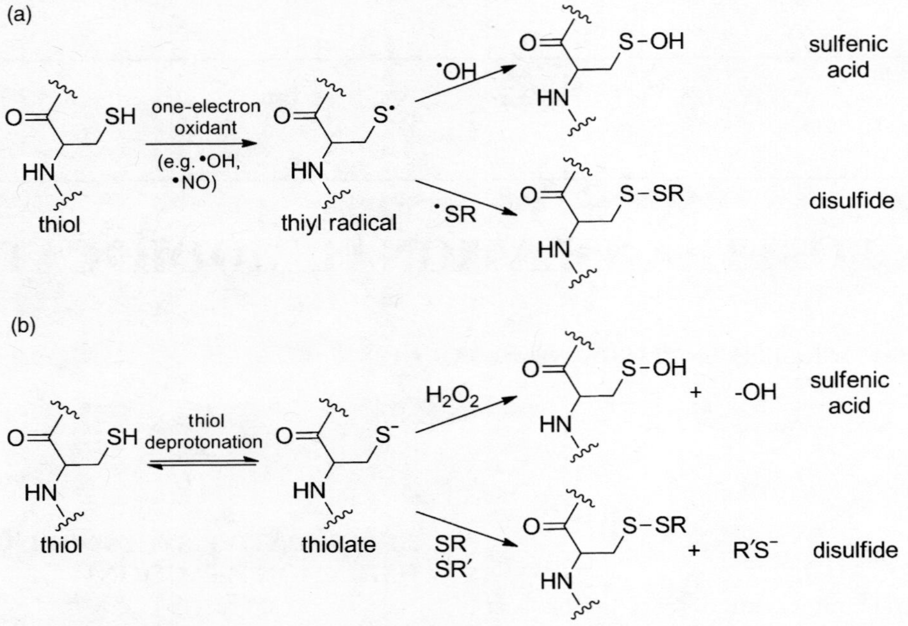

Figure 3.1 Cysteine oxidation pathways in the presence of reactive oxygen and nitrogen species. (a) One electron oxidation of cysteine thiol. The thiol hydrogen is abstracted by a radical, resulting in formation a sulfur-centered thiyl radical. The thiyl radical can then undergo radical coupling (reactions with hydroxyl and thiyl radicals shown), or abstract a hydrogen to reform a thiol. (b) Two electron oxidation of the cysteine thiol. The thiol side chain can deprotonate to form the thiolate anion, a strong nucleophile that can attack the electrophilic centers of reactive species (disulfide and hydrogen peroxide shown) resulting in cysteine oxidation.

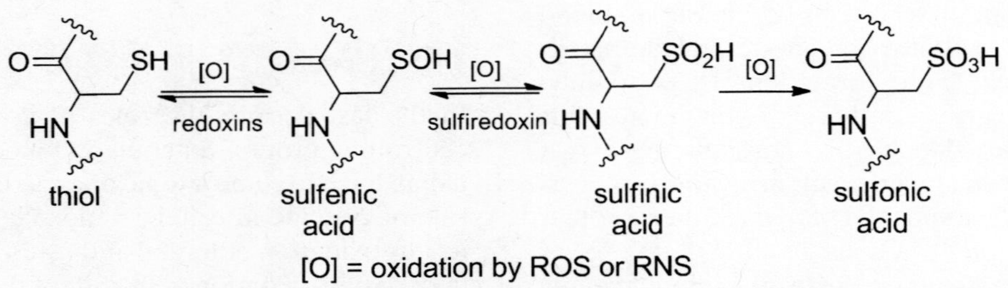

Figure 3.2 Cysteine oxidation leading to cysteine-oxygen adducts. Oxidation of cysteine can result in formation of sulfenic, sulfinic, and sulfonic acids. The first two of these oxidation products, sulfenic and sulfinic acid, can be enzymatically reduced back to cysteine under physiological conditions.

attack of the thiolate anion (RS^-) on the electrophilic center in the oxidant to yield a new covalent bond between the cysteine thiol sulfur atom and an atom with equal or higher electronegativity than sulfur (e.g., oxygen, nitrogen, sulfur) (Fig. 3.1b). Following initial oxidation by these reactive species, oxidized cysteines can then participate in a variety of reactions as outlined below.

3.1.1.1 Formation of Sulfur–Oxygen Adducts: Sulfenic, Sulfinic, and Sulfonic Acids
Exposure of cysteine to reactive oxygen species results in rapid oxidation of the side chain thiol to sulfenic acid (Fig. 3.2), which forms readily via either one-electron or two electron-oxidation reactions (Fig. 3.1).

Following one-electron oxidation, the resulting thiyl radical can couple with hydroxyl radicals to form sulfenic acids.[6] Formation of sulfenic acids through two-electron oxidations involves peroxides, peroxynitrites, and other molecules containing an electrophilic oxygen center.[7–9] Once formed, sulfenic acids exhibit both electrophilic and nucleophilic character and are considered highly reactive.[9,10] Sulfenic acids can react readily with other thiols to form disulfides with release of water (see below), with these thiols usually either cysteine side chains or small molecular thiols such as glutathione. Due to this reactivity, persistent (long-lived) sulfenic acids often require local environments that limit the access of other thiols. Sulfenic acids have been observed in crystal structures of multiple proteins, indicating that this modification is stable within the context of certain protein environments (see Reference 11 and references therein). Sulfenic acids can also react with amide nitrogens to form cyclic sulfenamides, as observed in the crystal structures of protein tyrosine phosphatase (PTP1B) and receptor protein tyrosine phosphatase (RPTP-α), described in detail later.[12,13]

Sulfenic acid formation plays a key role in the mechanism of the peroxiredoxins, a class of antioxidant enzymes that control peroxide levels within the cell.[14,15] There are three subgroups of peroxiredoxins (2-Cys, atypical 2-Cys, and 1-Cys), all of which contain an active-site cysteine that is oxidized by the peroxide substrate to a sulfenic acid in the first reaction step (Fig. 3.3).[16] This oxidized cysteine must then be reduced back to a thiol to reactivate the enzyme, which occurs by disulfide formation and subsequent reduction by thioredoxin (Trx) or other electron donors.[17–20] These enzymes exemplify the balancing act involved when cysteines participate in redox chemistry. For example, cysteine oxidation is an essential step in the peroxiredoxin catalytic cycle while overoxidation of the same active-site cysteine leads to enzyme deactivation.

In the presence of strong oxidants, the sulfur atom in a sulfenic acid (-SOH) can be further oxidized to a sulfinic (-SO$_2$H) or sulfonic (-SO$_3$H) acid (Fig. 3.2). These further oxidations have been historically considered irreversible modifications, as sulfinic and sulfonic acids cannot be reduced by common cellular reductants such as Trx or glutathione. Recent studies have identified an ATP-dependent enzyme sulfiredoxin (Srx) that reduces a sulfinic acid at the active site of peroxiredoxin to cysteine,[21–24] but as yet, no general sulfinic or sulfonic acid reductases have been identified. Irreversible oxidation of cysteines can lead to protein conformation changes, aggregation, and degradation,[9] and the presence of sulfonic acid residues would indicate a severely oxidizing environment.

3.1.1.2 Formation of Sulfur–Nitrogen Adducts: S-Nitrosothiols and Sulfonamides
Cysteine thiols react with reactive nitrogen species such as nitric oxide (NO), nitrous acid (HNO$_2$), and peroxynitrous acid (ONOOH) to form S-nitrosothiols (Fig. 3.4).

The predominant one-electron pathway for S-nitrosothiol formation involves NO, a free radical that is generated by nitric oxide synthases (NOSs) from arginine, NADPH, and oxygen.[25] NO can react directly with a thiyl radical (RS•) to form an S-nitrosothiol through radical coupling. In addition to this one-electron pathway, oxidative mechanisms for S-nitrosothiol formation include two-electron reactions of cysteine thiols with nitrous acid, NO, and peroxynitrite.[26–30] Cysteine thiolates can also undergo trans-S-nitrosylation through nucleophilic attack on either a S-nitrosylated protein or a small molecule S-nitrosothiol such as S-nitrosoglutathione, S-nitrosocysteine, and S-nitrosohomocysteine.[31–36] NO also forms metal–nitrosyl complexes and N$_2$O$_3$ that can act as S-nitrosylating agents.[37,38] The reactivity of a cysteine side chain toward S-nitrosylation can be regulated by several factors, such as nearby amino acid sequences that increase the reactivity of a target cysteine toward NO and the presence of adapter proteins near NOSs that aid in localizing target proteins near the site of NO production.[39,40]

Once formed, S-nitrosothiols can participate in a range of subsequent modifications. S-nitrosylation is reversible, with S-denitrosylation catalyzed by enzymes such as Trx, protein disulfide isomerase (PDI), and

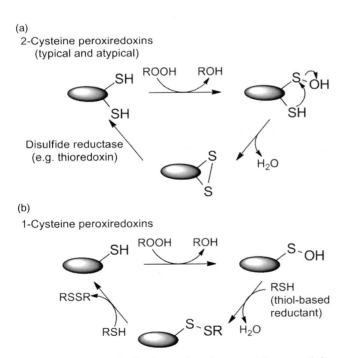

Figure 3.3 Peroxiredoxin-catalyzed peroxidase activity involves reversible sulfenic acid formation at the catalytic cysteine residue. (a) Mechanism for 2-cysteine peroxiredoxins, with the catalytic cysteine cycling between thiol, sulfenic acid, and internal protein disulfide in the catalytic cycle. (b) In 1-cysteine peroxiredoxins, the sulfenic acid formed at the catalytic cysteine residue is resolved by reduction involving small-molecule thiol-based reducing agents such as glutathione.

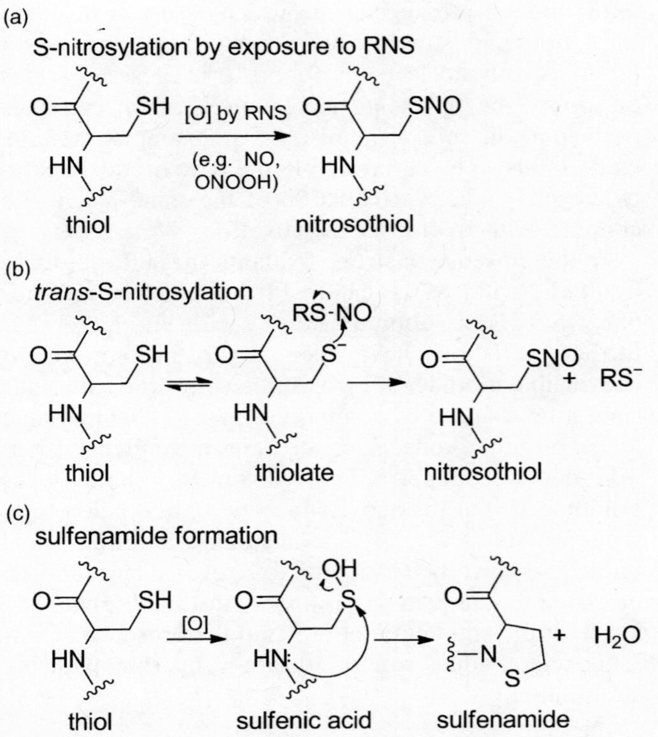

Figure 3.4 Formation of cysteine-nitrogen adducts under conditions of oxidative stress. (a) In the presence of RNS, cysteine can be directly nitrosylated through both one- and two-electron pathways to form nitrosothiol. (b) A thiolate anion can undergo *trans*-S-nitrosylation by nucleophilic attack on small-molecule nitrosothiols such as S-nitrosoglutahione. (c) Sulfenamide formation requires backbone amide nitrogen attack on a sulfenic acid side chain.

alcohol dehydrogenase class III (ADH).[41–45] These enzymes operate by distinct mechanisms: Trx and PDI catalyze *trans*-S-nitrosylation reactions to yield a cysteine thiol on the substrate protein and an S-nitrosylated enzyme; ADH employs a more complex reductive mechanism involving NADH to yield a disulfide and NH_3. S-nitrosothiols can also hydrolyze to form sulfenic acids or react with another cysteine to form disulfides,[46] mirroring the reactivity of sulfenic acids described earlier.

In addition to S-nitrosylation, cysteine thiols can form cyclic sulfenamides through reaction of a sulfenic acid with a backbone amide nitrogen when fixed in close proximity (Fig. 3.4). This modification was first observed in PTP1B[13,47] and has since been detected in a number of other proteins such as the RPTP-α and the peroxide sensor OhrR in *Bacillus subtilis*.[12,48] While in this context sulfenamide formation leads to loss of phosphatase activity in PTP1B, sulfenamide modifications are reversible and may serve a role in protecting cysteine side chains from irreversible oxidation beyond the sulfenic acid state.[49]

3.1.1.3 Formation of Sulfur–Sulfur Adducts: Disulfides and S-Glutathionylation
Formation of disulfide bonds involving cysteine side chains is perhaps the most commonly known oxidation reaction in protein biochemistry. Following from the one-electron oxidation chemistry of the cysteine thiol group described above, disulfides can form through radical coupling of two thiyl radicals (Fig. 3.1a).[28,50] More commonly under conditions of oxidative stress, however, the cysteine thiol initially forms an activated oxidized species such as sulfenic acid or S-nitrosothiol which undergoes subsequent nucleophilic attack by another thiol to form a disulfide (Fig. 3.5a).[51–53]

Intramolecular disulfides can form when an oxidation-activated cysteine side chain lies near another cysteine residue in the proper geometry to allow nucleophilic attack on the oxidized side chain. While internal protein disulfides have traditionally been considered to provide stability and rigidity to protein structure, there is growing evidence that a second group of disulfides exist that serve as redox-sensitive switches for controlling protein structure and function.[54,55] Oxidative-stress induced disulfide formation could also contribute to protein misfolding, with PDIs serving as a redox-sensitive chaperone to aid in proper protein folding.[56]

Oxidized cysteine thiols also form intermolecular disulfides with small molecule thiols, in particular the tripeptide glutathione (glutathionylation). Glutathione is the predominant reducing agent available within mammalian cells, with cellular concentrations in the low millimolar range.[57] The vast majority of glutathione within cell exists in its reduced form (GSH) rather than the oxidized disulfide-bonded dimer (GSSG).[58] The glutathione thiol acts as a nucleophile at cysteine thiols that have been oxidized to sulfenic acids, nitrosothiols, or other activated species to yield a mixed protein–glutathione disulfide (Fig. 3.5a). This disulfide can be subsequently reduced by Trx or glutaredoxin, an enzyme with specificity for gluthionyl disulfides.[59–62] Given the reservoir of reduced glutathione available within cells, glutathionylation serves as a defense against irreversible cysteine oxidation by converting reactive intermediates to stable disulfides (Fig. 3.5b).

Glutathionylation can serve as a redox-sensitive regulator of protein function, with the ability to both inhibit and activate protein activity depending on the specific context.[63] For example, glutathionylation of actin through a sulfenic acid intermediate impairs actin polymerization.[64–66] Disulfide formation modulates the activity of a large number of kinases, phosphatases, transcription factors, and other proteins involved in cell sig-

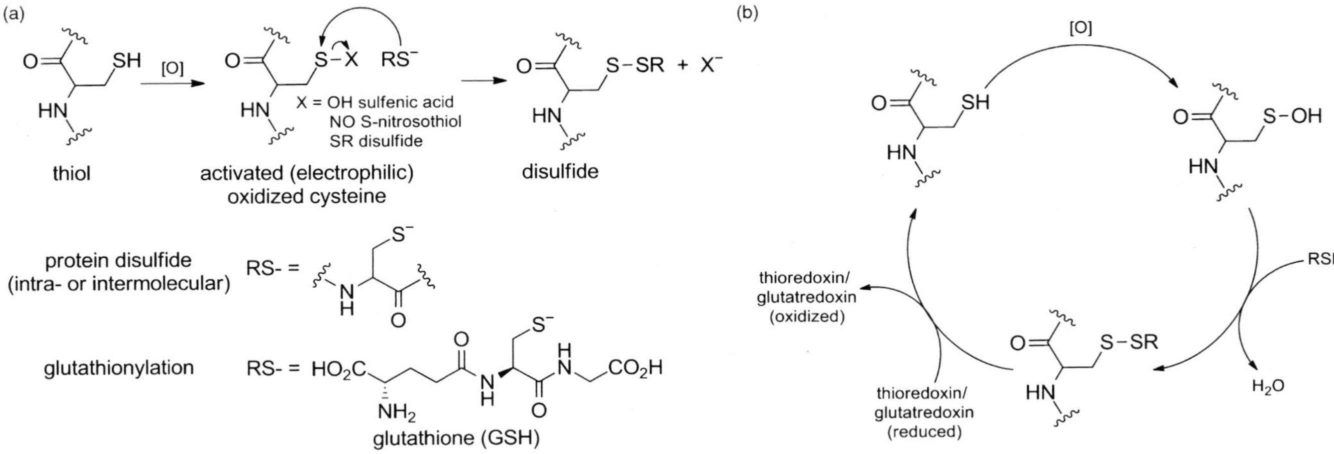

Figure 3.5 Disulfide formation plays a key role in cellular cysteine redox chemistry. (a) Formation of protein disulfides and glutationylation (protein-glutathione mixed disulfide). While disulfides can form by radical coupling between thiyl radicals, most protein disulfides form by two-electron thiol/thiolate attack on a cysteine residue that has been previously activated by oxidation. (b) The cycle of cysteine oxidation and reduction within the cell, with oxidation by ROS and RNS followed by disulfide formation and subsequent reduction to free thiols.

naling and gene expression.[49] Reversible signaling through protein glutathionylation and degluthiolation is also proposed to regulate processes such as apoptosis, mitochondrial redox state, and redox homeostasis in photosynthetic organisms.[67–69] Glutathionylation also plays a role in regulating peroxiredoxin activity and protecting the peroxiredoxin catalytic cysteines from overoxidation, thereby modulating the cell's ability to react to increases in peroxides and other ROSs.[15,68]

3.1.1.4 Redoxins: Enzymes Catalyzing Cysteine Reduction

In many cases, cysteine oxidation leads to either intramolecular disulfide formation with another cysteine within the oxidized protein or intermolecular disulfide formation with either a cysteine from another protein or the thiol group of reduced glutathione. There are two families of small proteins, the Trxs and glutaredoxins (Grxs), whose major function is to reduce disulfide bonds within proteins.[61,62,70] This reduction involves a two-step reaction in both enzymes (Fig. 3.6): (1) a catalytic cysteine residue within a CxxC motif in Trx or Grx attacks the disulfide bond, leading to release of a free cysteine side chain and formation of a new disulfide bond between the protein substrate and Trx/Grx; (2) a second cysteine within Trx/Grx attacks the newly formed disulfide to release the protein substrate and form an intramolecular disulfide within the enzyme. In addition to Trxs and Grxs with di-cysteine CxxC sequences, there are also Grx isoforms that contain a single catalytic cysteine in a CxxS sequence. These monocysteinic Grxs cannot catalyze complete reduction of the target protein and require an external thiol reductant such as reduced glutathione to release the reduced protein.

The most significant difference between Trx and Grx involves the reduction pathway leading to enzyme reactivation (Fig. 3.6b).[62,71] Following reduction of a protein disulfide by Trx, the resulting disulfide within the Trx active site is reduced by Trx reductase using NADPH as a reducing cofactor. In contrast, Grx is reduced nonenzymatically by glutathione to yield the reactivated enzyme and oxidized glutathione (GSSG). The enzymes also differ in their substrate preferences, with Trx exhibiting broad reactivity with protein disulfides and Grx showing selectivity for substrates containing glutathionyl-mixed disulfides leading to protein deglutathionylation.[72] In general, reactions catalyzed by Grx will involve glutathione at some point in the reaction cycle. These differences in reduction mechanisms and substrate selectivity suggest the Trxs and Grxs play complementary roles in maintaining protein redox homeostasis in the presence of oxidative stress.

In addition to catalyzing deglutathionylation, Grx can also catalyze protein glutathionylation by using oxidized glutathione as a substrate and a reduced protein as the reducing equivalent.[72–74] This reaction occurs primarily under oxidizing conditions, where a low GSH/GSSG ration exists. Grx has also been observed to scavenge glutathione radicals, suggesting the potential that it serves as a protein-based antioxidant.[73] By catalyzing both the forward and reverse steps in protein glutathionylation, Grx aids in the regulation of protein activity by gluthiolylation-based redox signaling as described

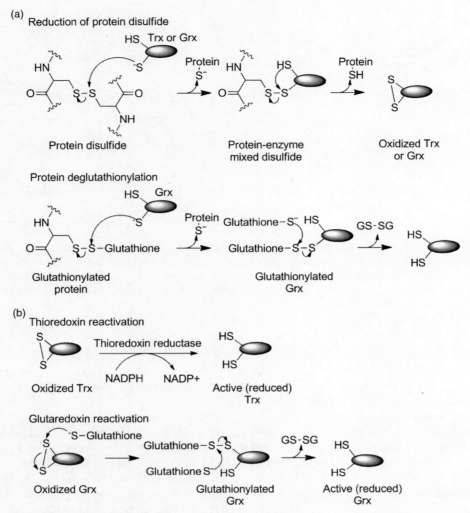

Figure 3.6 Thioredoxin and glutaredoxin catalyze disulfide reduction. (a) Protein disulfides can be reduced by either thioredoxin or glutaredoxin, with deglutathionylation primarily catalyzed by glutaredoxin. In both enzymes, substrate disulfide reduction results in formation of a new disulfide bond involving the catalytic cysteine leading to enzyme deactivation. (b) Thioredoxin is reactivated by thioredoxin reductase, which uses NADPH as an electron source to reduce the disulfide involving the thioredoxin catalytic cysteine. In contrast, glutaredoxin is reactivated by reduction involving free glutathione.

above. Changes in Grx expression or regulation of the cellular GSH/GSSG reservoir could alter protein gluthionylation affecting protein function and signaling; such changes may contribute to the progression and severity of diseases linked to oxidative stress.[75]

3.1.2 Methionine

Cysteine is generally considered to be the amino acid most susceptible to chemical modification by reactive oxygen and nitrogen species under conditions of oxidative stress. However, the presence of these reactive species can also lead to oxidative modification/damage of the other amino acids. While not observed as frequently as cysteine oxidation products, these other oxidation reactions are also biologically relevant. After cysteine, methionine is the amino acid most susceptible to modifications caused by oxidative stress due to its easily oxidized thioether side chain ($R-S-CH_3$). The methionine thioether can react with ROS/RNS (e.g., $HO^{\bullet}, NO^{\bullet}$) to produce a sulfoxide at the thioether sulfur atom, yielding methionine S-oxide. Formation of methionine S-oxide is observed during times of increased oxidative stress,[76,77] with this modification reversible through the action of methionine sulfoxide reductases (Msrs) as described later. Methionine can be further oxidized to the sulfone, but sulfone oxidation is an irreversible modification that requires nonbiologically relevant oxidation conditions (Fig. 3.7).[76,78]

In contrast to cysteine, oxidation of methionine involves the creation of a new stereocenter. The sulfur of the methionine thioether is a prochiral center, with oxidation leading to two diastereomeric products, methionine-S-sulfoxide (Met–S–SO) and methionine-R-sulfoxide (Met-R-SO) (Fig. 3.7).[79–82] Different classes of Msrs have evolved to catalyze reactions with these diastereomers. Methionine sulfoxide reductase A (MsrA) catalyzes reduction of Met–S–SO in the contexts of both intact proteins and free amino acids, with methionine sulfoxide reductase B (MsrB) catalyzing Met-R-SO reduction in intact proteins while exhibiting significantly reduced activity with free Met-R-SO.[79,81,83,84] A third enzyme, free methionine sulfoxide reductase (fMsr), also catalyzes reduction of Met-R-SO but only as a free amino acid.[85] In contrast to MsrA and MsrB, fMsr is found only in single-celled organisms such as bacteria and lower eukaryotes. Due to its efficient scavenging of free MetSO, fMsr is important for proper cell oxidative stress regulation in these organisms.[85]

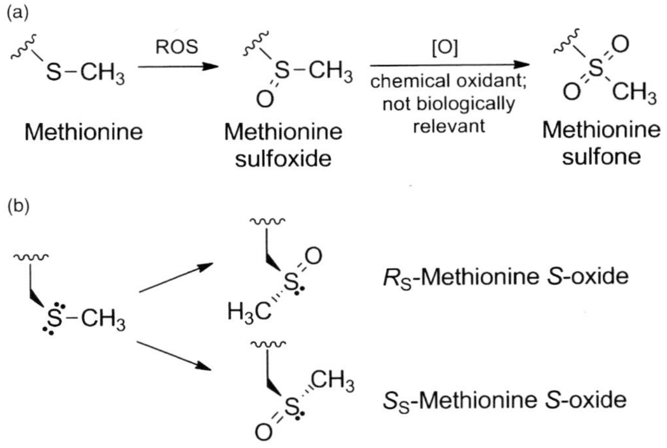

Both MsrA and MsrB are proposed to reduce S-methionine S-oxide to a thioether through a multistep catalytic mechanism involving a covalent enzyme intermediate (Fig. 3.8).[79,84] In the first step, the thiol side chain of a catalytic cysteine residue attacks the S-oxide sulfur atom, followed by release of the sulfoxide oxygen atom as water and formation of an enzyme-substrate

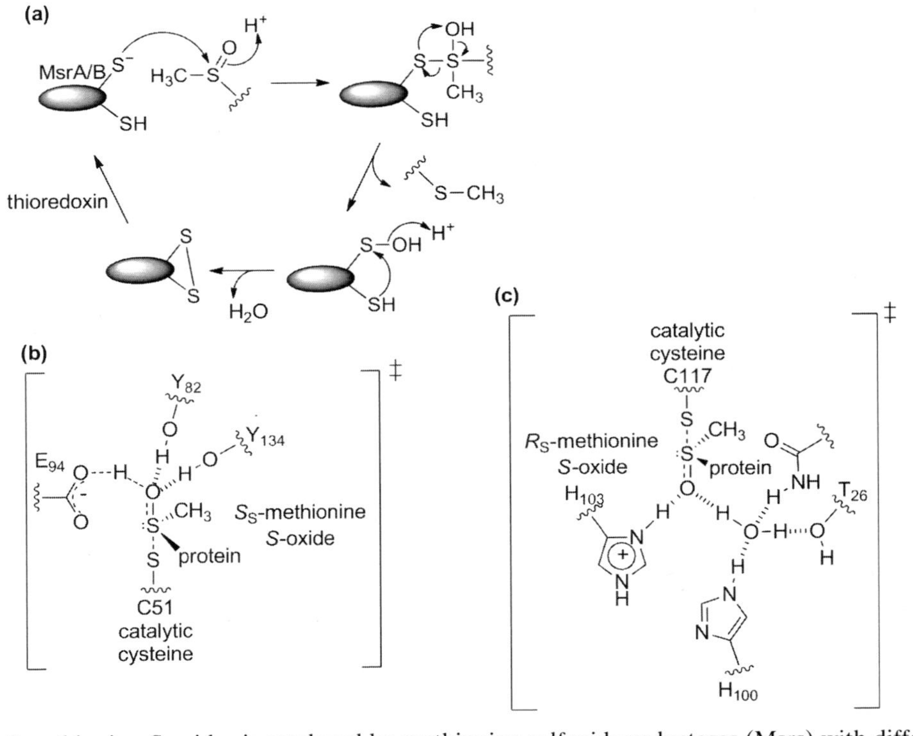

Figure 3.7 Methionine oxidation creates a new chiral center. (a) Under biological conditions, methionine only undergoes a single oxidation to methionine sulfoxide. (b) Methionine oxidation leads to a mixture of RS- and SS-methionine S-oxide stereoisomers.

Figure 3.8 Reduction of methionine S-oxides is catalyzed by methionine sulfoxide reductases (Msrs) with different stereoselectivities. (a) Proposed mechanism for methionine S-oxide reduction by MsrA and MsrB. (b and c) The stereoselectivities of MsrA for S_S- methionine S-oxide (panel B) and MsrB for R_S-methionine S-oxide (panel c) arise from distinct arrays of hydrogen bonding to the substrates within the MsrA and MsrB active sites. Figure adapted from Reference 83.

mixed disulfide. Methionine release follows water attack on the catalytic cysteine, which converts this cysteine to a sulfenic acid (S–OH). Following methionine release, a second active-site cysteine attacks the newly formed sulfenic acid to form an intramolecular disulfide bond. Finally, the newly formed disulfide bond is reduced by Trx to regenerate active enzyme.

The stereoselectivities of MsrA and MsrB for Met–S-SO and Met-R-SO, respectively, arise from differences in active-site interactions that stabilize the sulfurane intermediate that forms following attack by the catalytic cysteine (Fig. 3.8b–c).[79,83,86] In MsrA, this stabilization involves hydrogen bonding to two tyrosine residues and an aspartic acid whose geometry is complementary to the intermediate arising from Met–S-SO.[87] The active site in MsrB differs in terms of both active-site amino acids and active-site architecture, with the MsrB crystal structure revealing an array of several imidazole residues that provide hydrogen bond stabilization to the Met-R-SO reduction intermediate.

Similar to cysteine, the ease of both methionine oxidation and subsequent enzyme-catalyzed reduction allows methionine to serve as a buffer of oxidative stress within the cell. Through cycles of methionine oxidation and reduction, ROS and RNS are scavenged to protect other amino acids from irreversible oxidation by these reactive intermediates.[88] In studies that support the role of the methionine oxidation-reduction cycle in protecting against oxidative stress, knocking out MsrA in mice, yeast, and bacteria leads to increased susceptibility of these organisms to oxidative stress.[89–93] In contrast, overexpressing MsrA increases resistance to oxidative stress in mammalian cells.[94–96]

In addition to serving a general protective role against oxidative stress, methionine oxidation can also directly impact protein function. For example, oxidation of several methionine residues within the voltage dependent K+ channel ShC/B increases channel activity.[97] Methionine oxidation due to oxidative stress can also affect the actin cytoskeleton, with oxidation of several structurally important methionine residues shown to block actin polymerization and decrease actin stability.[98]

3.1.3 Oxidation of Aromatic Amino Acids

3.1.3.1 Tyrosine The side chain of tyrosine easily undergoes oxidative additions in the presence of ROS and RNS, with addition typically occurring at positions adjacent (*ortho*) to the phenol hydroxyl group.[99,100] In the presence of radicals derived from either ROS (e.g., hydrogen peroxide, H_2O_2) or RNS (e.g., peroxynitrous acid, ONO_2H) sources, a hydrogen atom can be abstracted from the tyrosine ring to form the tyrosinyl

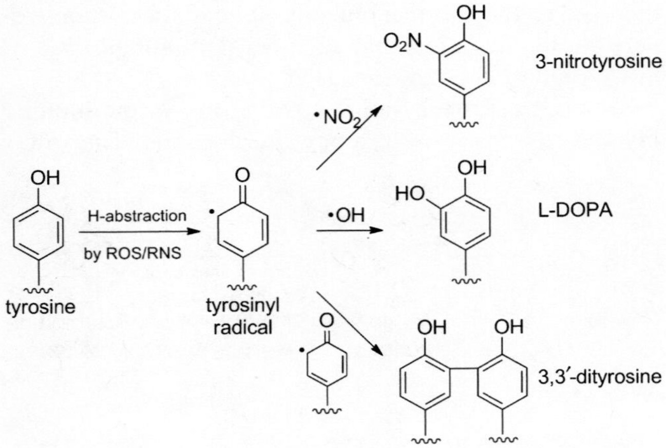

Figure 3.9 Tyrosine oxidation leads to multiple products. In the presence of ROS and RNS, tyrosine undergoes a one-electron oxidation to the tyrosinyl radical, with radical coupling reactions yielding 3-nitrotyrosine, L-3,4-dihydroxylphenylalanine (L-DOPA), and 3,3′-dityrosine.

radical (Fig. 3.9).[101,102] Coupling of tyrosinyl and hydroxyl radicals yields the neurotransmitter precursor L-3,4 dihydroxyphenylalanine (L-DOPA), with this tyrosine derivative also produced in the absence of oxidative stress by the enzyme tyrosine hydroxylase.[99,103,104] Tyrosinyl radical can react with nitrogen dioxide (NO_2) to form 3-nitrotyrosine (Fig. 3.9)[105]; alternatively, tyrosinyl radicals can also dimerize to form 3,3′-dityrosine.[105,106] As these modifications are considered biologically irreversible, tyrosine oxidation products are characterized as biomarkers for oxidative stress.[107]

Tyrosine modifications, particularly formation of 3-nitrotyrosine, can impact the regulation of protein structure and function by other PTMs. For example, protein phosphorylation frequently occurs on the phenol oxygen of tyrosine residues.[108,109] Tyrosine phosphorylation can induce changes in protein conformation and activity, such as in protein tyrosine kinase.[110] Tyrosine nitration can block subsequent phosphorylation, thereby eliminating an important signal transduction mechanism for controlling protein activity. In another enzyme, glutamine synthetase (GS), tyrosine residues are adenylylated to induce a conformational change which alters both enzyme structure and activity.[111,112] During oxidative stress resulting from exposure to peroxynitrate, several tyrosine residues in glutamine synthetase which are normally adenylylation sites are nitrated. While nitration alone led to the same changes in protein function that accompany adenylylation, subsequent adenylylation of proteins bearing 3-nitrotyrosine residues completely abolished catalytic activity.[113] As

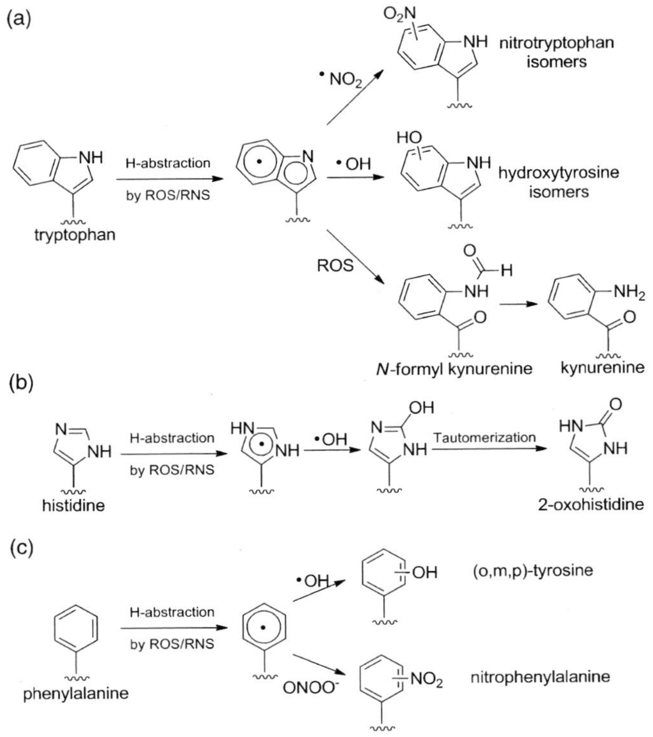

Figure 3.10 Aromatic amino acid oxidation. (a) The indole ring of tryptophan is easily oxidized to form multiple nitrotryptophans and hydroxytryptophans regioisomers, and indole ring fragmentation is also observed. (b) Histidine oxidation primarily yields 2-oxohistidine. (c) Similar to tryptophan, phenylalanine yields multiple nitro- and hydroxyl-regioisomers upon oxidation by ROS and RNS.

tophan oxidation products have been studied *in vitro*, nitrotryptophan and its derivatives have been observed *in vivo*.[122,123] There are currently a limited number of known protein tryptophan nitration sites,[118] but this list is expected to grow with continuing studies.

3.1.3.3 Histidine In the presence of ROS and RNS, histidine can react with hydroxyl radicals to yield 2-oxohistidine through radical addition followed by tautomerization (Fig. 3.10b). This modification has been observed in several proteins, such as Cu,Zn-superoxide dismutatase and the bacterial transcription factor PerR.[124,125] In PerR, metal-catalyzed oxidation of histidines by hydrogen peroxide alters metal coordination within this protein. This allows PerR to serve as a sensor for detecting low levels of intracellular peroxide as an indicator of oxidative stress.[124]

3.1.3.4 Phenylalanine While not as reactive as tyrosine, the phenyl side chain of phenylalanine can also be oxidized in the presence of ROS/RNS. Exposure to hydroxide radicals leads to formation of *para*-, *meta*-, and *ortho*-tyrosine (Fig. 3.10c), and nitrophenylalanine can form in the presence of RNS such as peroxynitrite.[99,126–128]

3.1.4 Oxidation of Aliphatic Amino Acids

Amino acids with aliphatic side chains are also susceptible to oxidative damage, but these modifications are relatively rare when compared to those of cysteine, methionine, and the aromatic amino acids. These modifications are often metal- or enzyme-catalyzed reactions, as opposed to the uncatalyzed reactions with ROS and RNS possible in the case of cysteine, methionine, and the aromatic amino acids. Reflecting the more rigorous conditions required for oxidation of aliphatic amino acids, these modifications are largely irreversible under biologically relevant conditions and if observed could suggest the presence of high levels of oxidative stress.

In oxidative environments with metal-catalyzed ROS production, lysine can be oxidized to allysine, an aldehyde derivative of lysine (Fig. 3.11).[129] This reaction can also occur naturally in the production of collagen and elastin through the action of lysyl oxidase. Under similar oxidative conditions, proline and arginine form glutamic semialdehyde via a metal-catalyzed pathway involving hydroxyl radicals.[129,130] Proline can also be converted to 2-pyrrolidone in the presence of ROS, leading to peptide backbone cleavage.[131] Similar reactions involving proline and asparagine side chains in the transcription factor HIF utilize hydroxylation of these amino acids as a sensor/readout for cellular oxygen levels.[132,133]

these examples illustrate, irreversible tyrosine nitration can alter or block signaling pathways used to control protein structure and function.

3.1.3.2 Tryptophan The indole ring of tryptophan is susceptible to oxidation by both ROS and RNS (Fig. 3.10a). Exposure to ROS results in tryptophan hydroxylation at multiple ring positions and also fragmentation of the indole ring to form *N*-formylkynurenine and kynurenine.[99,114–116] One of these oxidative products, 5-hydroxytryptophan, is also formed by the enzyme tryptophan hydroxylase in the absence of oxidative stress.[117] Tryptophan can also be nitrated in the presence of RNS to form nitrotryptophan, with the nitro group potentially attached at multiple positions on the tryptophan side chain.[118–120] Tryptophan nitration can also impact protein activity, as formation of nitrotryptophan within human Cu,Zn-superoxide dismutase leads to a drop in catalytic activity.[121] While the majority of tryp-

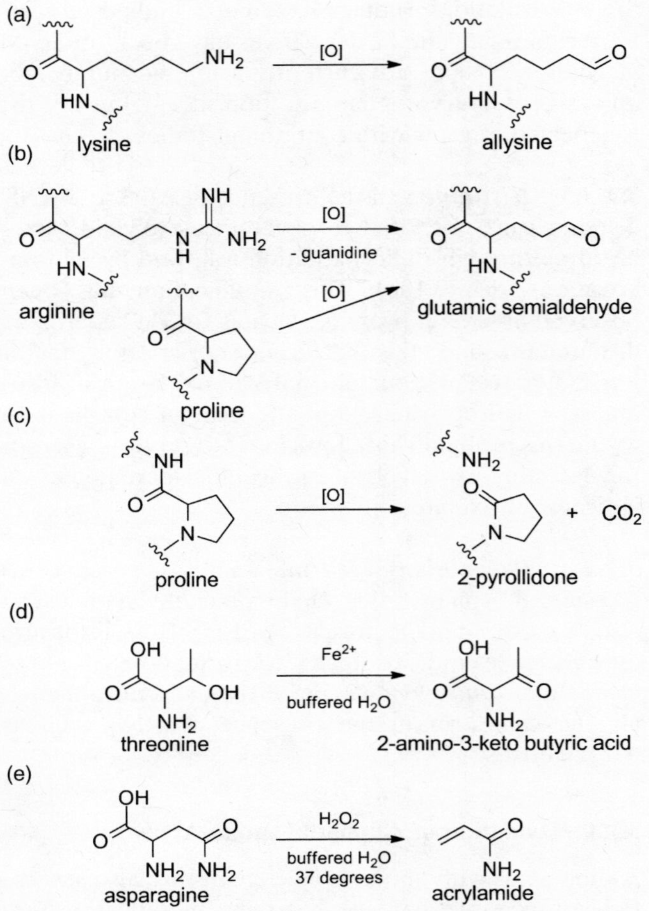

Figure 3.11 Oxidation of aliphatic amino acids yields reactive carbonyls and peptide backbone cleavage. Under strongly oxidizing conditions, the following amino acid oxidations have been observed: (a) Lysine oxidation to allysine. (b) Arginine and proline conversion to glutamic semialdehyde. (c) Oxidation of proline to 2-pyrollidone, with cleavage of the peptide backbone. (d) Threonine side chain oxidation to yield 2-amino-3-ketobutyric acid in the presence of metal-catalyzed hydroxide radical formation. (e) Asparagine conversion to acrylamide by H_2O_2 under physiological conditions.

3.2 AMINO ACID MODIFICATION BY OXIDATION-PRODUCED ELECTROPHILES

3.2.1 Electrophiles Formed by Oxidative Stress

Under oxidative stress conditions, carbonyl groups (ketones and aldehydes) can be introduced into proteins by side chain oxidation (e.g., allysine from lysine, 2-amino-3-ketobutyric acid from threonine) or oxidative backbone cleavage as described above. These carbonyls can serve as electrophiles in subsequent reactions. Reactive carbonyls can also be formed in carbohydrates and lipids in cells under high oxidative stress, as described in Chapter 2 and Chapter 4 of this book. These reactive electrophiles can then react with nucleophilic amino acid side chains, resulting in protein carbonylation.

Protein carbonylation reactions are generally considered biologically irreversible, although one recent review suggests a potential role for protein decarbonylation in cell signaling.[136] Protein carbonylation has been proposed as a biomarker for oxidative stress, as formation of protein carbonyls requires more severe oxidative conditions than cysteine or methionine oxidation.[137–140] Many human diseases, such as Alzheimer's disease, cystic fibrosis, and chronic obstructive pulmonary disease (COPD), are associated with protein carbonylation.[138] In some of these diseases, elevated levels of protein carbonylation correlate with disease severity and progression. For example, high levels of autoantibodies to carbonylation products have been found in titers of patients with COPD, suggesting that high levels of protein carbonylation in cells involved in COPD may contribute to this disease.[141]

3.2.2 Carbonylation Reactions with Amino Acids

Carbonylation of amino acids by small-molecule electrophiles occurs by two distinct chemical mechanisms, depending on the nature of the carbonyl electrophile. Lipid peroxidation can lead to formation of advanced lipidation end products (ALE) containing α,β-unsaturated carbonyls such as 4-hydroxy-2-nonenal (NHE), malondialdehyde, and acrolein (see Chapter 2 and references therein). These molecules are efficient substrates for conjugate (Michael) additions by lysine, cysteine, and histidine side chains (Fig. 3.12a).[142–145] Lysine side chains also undergo glycation and glycoxidation reactions with reducing sugars or their oxidation products (advanced glycation end products, [AGEs]), respectively, to form imine linkages that subsequently convert to α-amino ketones via Amadori rearrangements (Fig. 3.12b).[146–152] In both cases, these reactions result in the formation of protein adducts containing reactive carbonyl functional groups.

Several other amino acids are thought to be rarely oxidized under physiologically relevant oxidative stress, with evidence for the formation of oxidative products from these amino acids largely derived from *in vitro* studies (Fig. 3.11). When free threonine is oxidized under biologically relevant conditions involving metal ion-catalyzed ROS production, the secondary alcohol of the threonine side chain is converted to a ketone, yielding 2-amino-3-keto butyric acid.[134] Oxidation of asparagine under physiologically relevant conditions can result in the formation of acrylamide, a known carcinogen.[135]

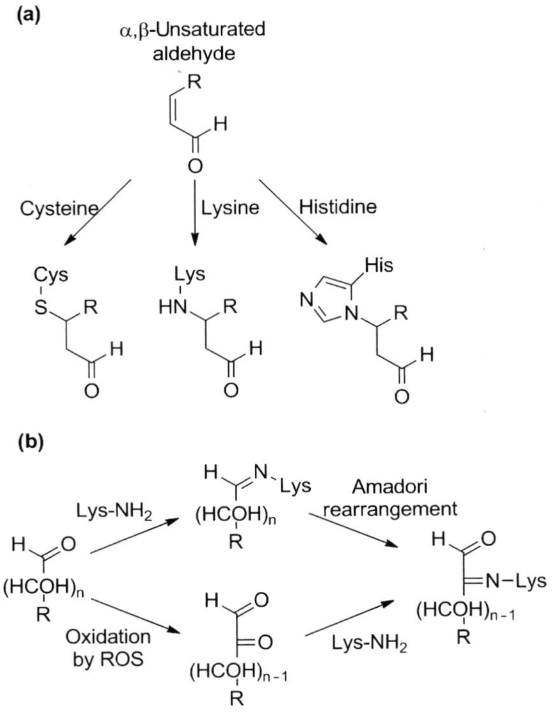

Figure 3.12 Electrophilic reactions leading to protein carbonylation. (a) Cysteine, lysine, and histidine side chain nucleophiles can perform Michael additions to α,β-unsaturated carbonyls formed by lipid oxidation (e.g., 4-hydroxy-2-nonenal). (b) Lysine carbonylation through glycation and glycoxylation reactions.

As protein carbonyls can be produced by both direct amino acid oxidation (see earlier discussion) and electrophilic addition to amino acid side chains, the presence of protein carbonyls must be interpreted as a broad and nonspecific sign of oxidative stress. Given that formation of protein carbonyls is common during oxidative stress and that this modification generates a reactive carbonyl, significant efforts have been invested in developing analytical and biochemical approaches for detecting these modifications as a means of quantifying oxidative stress within biological samples.[153–156] Moreover, given the relative stability of protein carbonyls compared to other markers of oxidative stress such as the products of lipid peroxidation, protein carbonyl content is considered a reliable marker for protein oxidation and the presence of oxidative stress.[138]

3.3 DETECTION OF OXIDATIVE-STRESS RELATED PTMs

Oxidative PTMs can lead to protein damage and negatively impact cellular health. However, these modifications also play important roles in signaling pathways that read out the redox state of the cell, particularly the array of reversible modifications that occur at cysteine residues.[9,49,157] Understanding how these modifications lead to cell signaling requires first identifying the proteins, and the specific amino acids within those proteins, that undergo oxidative modifications. Identifying specific signaling pathways involving modifications arising from oxidative stress then requires determining how the distribution and extent of these modifications changes in response to changes in the cellular environment.

Among the array of methods that have been used to identify oxidative PTMs, mass spectrometry and chemoselective modifications have proven to be the most robust and specific tools for identifying these modifications within the proteome. The following section provides a brief overview of the current state of both techniques, with more detailed descriptions available in the cited references.

3.3.1 Mass Spectrometry

Mass spectrometry is rapidly becoming the most powerful and widespread technology available for detecting protein PTMs. Changes in protein molecular weight can provide evidence for chemical modifications, with tandem mass spectrometry (MS-MS) techniques potentially leading to identification of the specific modification sites; for reviews of mass spectrometry techniques for posttranslational proteomics, please see references 158–161. The mass changes that accompany oxidative modifications, ranging from a mass loss of 2 daltons during disulfide formation to addition of >100 daltons for some protein carbonylation modifications, are easily detectable by modern mass spectrometers. In most applications, the protein of interest is digested into small peptides by one or more proteases (most commonly trypsin), and the resulting peptides are analyzed by either matrix-assisted laser desorption ionization (MALDI) or electrospray ionization (ESI) mass spectrometry.[158] This analysis can be performed on the combined proteolytic digest in the case of simple samples or, alternatively, following peptide separation by liquid chromatography.

However, investigation of PTMs caused by oxidative stress using mass spectrometry presents three major challenges. The first barrier is sensitivity, as these modifications exist only on a small fraction of cellular proteins, and only a subset of any given protein target. Therefore, the signal from a modified protein isoform must be deconvoluted from the background arising from the parent protein. Identifying the nature of the modification presents the second challenge, as many oxidative modifications lead to similar changes in molecular weight—for example, formation of a sulfenic acid, 4-hydroxyleucine, and 2-oxohistidine all lead to a mass increase of 16 daltons. While MS-MS techniques coupled with computational data searching algorithms can theoretically identify the residue(s) that are modified, the large number of possible modifications and the potential heterogeneity of modifications within the protein of interest make this approach expensive in terms of both labor and time.[156]

The last challenge in using mass spectrometry to identify oxidative stress-related PTMs arises from the chemical stability of the modifications themselves. Many oxidation modifications are not chemically reactive and are compatible with mass spectrometry, such as most aromatic and alkyl amino acid hydroxylations. Disulfide bonds are also reasonably stable under nonreducing conditions, allowing for detection of both intramolecular protein disulfide cross-linking and glutathionylation.[66,162,163] However, cysteine modifications such as sulfenic acid formation or S-nitrosylation are chemically reactive and can be lost during sample preparation or sample ionization. To aid in detection of these modifications, proteins can be treated with reagents that selectively modify reactive oxidized amino acids to generate stable adducts for subsequent identification by mass spectrometry (see later discussion).

3.3.2 Chemoselective Functionalization

Chemical labeling of oxidized amino acids is performed primarily for two reasons. First, in the case of unstable oxidation products such as sulfenic acids or nitrosothiols, chemical functionalization converts these reactive species into a more stable functional group that can survive subsequent experimental manipulations. As most of these modifications only occur on a subset of potential amino acid sites at any one time, chemical "trapping" of reactive species can also aid in accumulating modified proteins for subsequent identification. Second, chemical labeling can lead to changes in protein properties (mass, fluorescence/absorbance, chromatographic retention time, etc.) that aid in identification of proteins bearing oxidative PTMs. Finally, quantitative mapping of oxidation-related modifications can be achieved using isotopically labeled chemical probes, with protein samples modified with reagents containing either "heavy" (usually ^{13}C or ^{2}H) or "light" (^{12}C or ^{1}H) isotopes allowing the proteins from the different samples to be distinguished by small changes in the mass of the label. This approach allows for quantitative proteomics to potentially identify changes in oxidative PTMs upon protein or cellular exposure to oxidative stress.[156,158,164–166]

3.3.3 Cysteine Modifications

As reflected by the layout of this chapter, oxidative cysteine modifications constitute a large fraction of the known PTMs resulting from oxidative stress. The appreciation of cysteine's role in oxidative biology and redox signaling, coupled with the chemical reactivities provided by the various cysteine derivatives, has led to the development of a large range of chemical tools for identifying cysteines that have undergone oxidation. The following sections outline the most prevalent forms of chemoselective modifications used for detection of oxidized cysteines; the reader is referred to References 167 and 168 for a more comprehensive treatment.

3.3.3.1 Sulfenic Acids As noted above, sulfenic acids are highly reactive molecules that exhibit both nucleophilic and electrophilic reactivity. This reactivity has been harnessed in the development of probes that utilize the chemoselective reaction of sulfenic acids with dimedone (5,5-dimethyl-1,3-cyclohexadione) and dimedone derivatives (Fig. 3.13a).[169] While originally serving as a tag for identifying sulfenic acids through immunoblotting or mass spectrometry of cell lysates, these reagents have been expanded to allow fluorescent or biotin labeling for sulfenic acid detection and protein labeling within intact cells.[170,171] Dimedone-based chemistry incorporating isotopic tags allows for quantitative proteomics analysis of sulfenic acid modifications within complex protein mixtures using peptide-based proteomics tools.[164,165]

3.3.3.2 Cysteine-Nitrosothiols Chemical detection of protein S-nitrosylation is commonly accomplished using the biotin-switch method developed by Jaffrey and coworkers.[172] This three-step process results in attachment of a biotin tag at the nitrosylated cysteine. Free cysteine thiols are first blocked as methyl disulfides by treatment with methyl methanethiosulfonate while native protein disulfides remained unchanged. The S–NO bond in the nitrosothiol is then reduced by treatment with ascorbate to form a thiol, which is then is ligated to a thiol-reactive biotin conjugate (Fig. 3.13b). Attachment of biotin allows for subsequent enrichment and recovery using streptavidin beads, followed by protein analysis using mass spectrometry and other methods.[173] Variations on this approach are being explored to enhance and expand detection of S-nitrosylated proteins, with an increasing number of these proteins being identified in multiple organisms.[174,175]

3.3.3.3 Cysteine-Glutathionylation Unlike S-nitrosylation, cysteine glutathionylation results in a mixed disulfide that

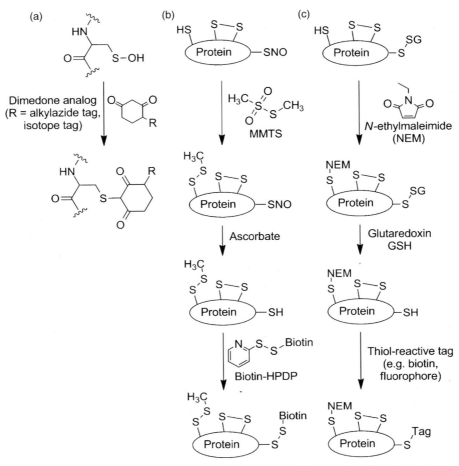

Figure 3.13 Chemoselective detection of oxidized cysteines within proteins. (a) Functionalization of sulfenic acids through reaction with dimedone analogs. (b) "Biotin switch" method for detecting nitrosothiols. (c) Glutaredoxin-catalyzed approach for specifically tagging glutathionylated cysteines. Please see the text for a detailed description of the specific reaction steps in each method.

will exhibit similar chemical reactivity to other disulfide bonds (if any) within the protein of interest. However, protein-glutathione mixed disulfides can be selectively reduced by glutaredoxins to liberate a free cysteine residue.[176] This selectivity lends itself to a multistep protocol for detecting protein gluthiolation similar to that used for nitrosothiols (Fig. 3.13c): (1) block free cysteines with a thiol-reactive alkylating reagent such as N-ethylmaleimide (NEM); (2) specific reduction of protein-glutathione disulfides using glutaredoxin; and (3) labeling the newly reduced cysteine by a thiol-selective probe. Several groups have used this approach to label glutathiolated proteins within cells with biotin for streptavidin-based detection.[176–178] Employing a thiol-reactive fluorescent dye, Zhang and coworkers have demonstrated a high-sensitivity method for detecting glutathiolated proteins using capillary gel electrophoresis with fluorescence detection.[179] This approach for chemoselectively labeling gluthiolated cysteine residues will also aid in determining glutathionylation sites using mass spectrometry.

3.3.4 Protein Carbonylation

Protein carbonylation can be detected by reacting proteins with 2,4-dinitrophenylhydrazine (2,4-DNP), which forms a colored hydrazone derivative with reactive aldehydes and ketones (Fig. 3.14). These adducts can be subsequently detected either spectrophotometrically or with an anti-DNP antibody, and DNP functionalization can be combined with mass spectrometry to identify carbonylated proteins.[140,180] A related molecule, Girard's Reagent P [1-(carboxymethyl)pyridinium chloride hydrazide], combines a carbonyl-reactive hydrazide group with a quaternary amine center that allows for isolation by cation exchange chromatography.[181] Isotopically

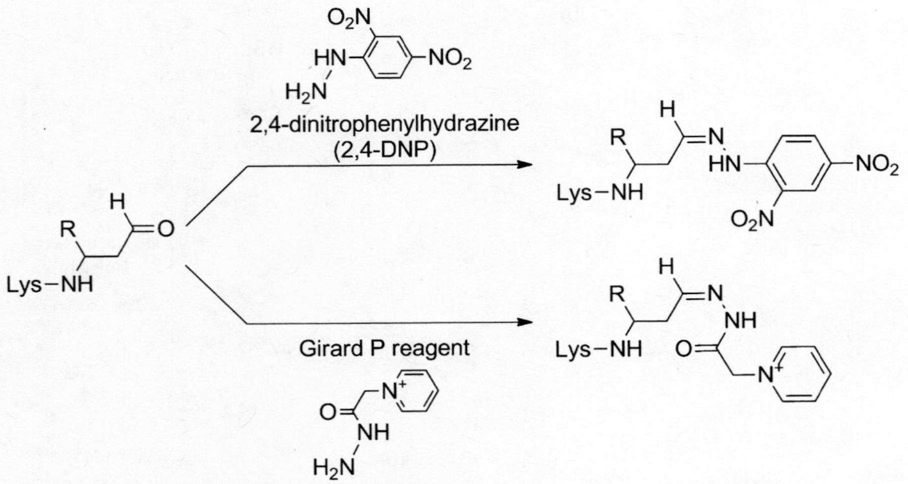

84 PROTEIN POSTTRANSLATIONAL MODIFICATION

Figure 3.14 Detection of protein carbonylation through hydrazone formation. Protein carbonyls formed through amino acid oxidation or electrophilic conjugation to amino acid side chains (shown) can be chemoselectively labeled with hydrazine reagents such as 2,4-DNP or the Girard P reagent.

labeled carbonylation-reactive probes have also recently been developed to quantify carbonylated proteins in a variety of systems.[156,166] Recent studies have also demonstrated fluorescence labeling and HPLC analysis of protein-bound aldehyde groups using reductive amination with para-aminobenzoic acid, an approach that provides higher sensitivity than previous assays.[182] Identifying the specific location and chemical nature of the modified amino acid is challenging, but methods to accomplish this task are rapidly progressing in both utility and sensitivity.[183]

3.4 ROLE OF PTMs IN CELLULAR REDOX SIGNALING

Cells contain multiple signaling mechanisms based on protein PTMs, such as phosphorylation and dephosphorylation within signal cascades or the histone modifications (e.g., acetylation, phosphorylation, methylation, ubiquitylation) that play roles in epigenetic control of gene expression. In recent years, the reversible oxidation of sulfur containing amino acids—particularly cysteine—has been acknowledged as a mechanism for redox signaling within the cell (reviewed in References 4, 9, 157, 168, and 184–186). Cellular signaling through cysteine oxidation exhibits one key distinction from other biological signaling pathways in that cysteine oxidation can occur spontaneously under oxidative stress without the need for enzymatic catalysis. In this way, redox signaling represents a direct functional linkage between the cellular environment and control of protein structure and function. Cysteine reactivity with oxidants can vary over 10^7-fold depending on the local environment, providing a mechanism for sensing the local oxidation potential based on cysteine susceptibility to oxidation.[187] This system of redox-active cysteines, termed the "cellular thiolstat" by Jacob and coworkers, is proposed to serve as mechanism for the cell to sense and respond to changes in the intracellular redox environment.[4]

Many enzymes involved in cell signaling, metabolism, and oxidative stress defense pathways contain cysteine residues whose oxidation can modulate catalytic activity. The simplest case of this control mechanism manifests in enzymes that utilize cysteine side chains as part of their catalytic machinery. Oxidation of the active-site cysteine results in enzyme deactivation, linking enzyme activity to the cellular redox state. One example of such regulation are the peroxiredoxins (discussed above), which help sequester H_2O_2 and organic peroxides. The peroxiredoxin active-site cysteine transits through a sulfenic acid intermediate during its catalytic cycle, which must be subsequently reduced to reactivate the enzyme. An excess of the peroxide substrate can lead to inactivation of some peroxiredoxins, suggesting that the sulfenic acid intermediate can serve as a redox-sensitive switch (reviewed in References 6, 58, and 188). In another example, PTPs use a catalytic cysteine as a nucleophile to attack a phosphorylated protein, resulting in formation of a phosphocysteine intermediate.[189] Oxidation of this cysteine to a sulfenic acid by H_2O_2 results in enzyme inactivation, thereby modulating protein phosphorylation and dephosphorylation involved in signal transduction.[190–194] Cysteine oxidation outside the active site can also control enzyme activity, such as in the cases of protein kinase A (PKA) and cGMP-dependent protein kinase (PKG)I-α. Interestingly, cysteine oxidation in these two enzymes results in opposite effects with PKA undergoing glutathionylation and inactivation whereas PKGI-α is activated through the formation of an intramolecular disulfide.[6,195–197] Many other examples of enzyme regulation through cysteine oxidation have been identified (reviewed in Reference 186), with more expected as methods for detecting oxidized cysteines within the proteome improve.

Cysteine oxidation is also implicated in controlling signaling mediated by transcription factors and other proteins involved in control of gene expression. Glutathionylation of nuclear factor-kappa B (NF-κB) altered DNA binding in *in vitro* assays, and the tumor suppressor ability of human p53 is inhibited by glutathionylation under oxidative stress.[198,199] Proteins can also be activated by cysteine oxidation, as in the case of the small GTPase p21ras wherein glutathionylation leads to increased signaling activity and activation of the downstream ERK and Akt signaling pathways.[200,201]

In light of the growing evidence that reversible cysteine oxidation plays a key role in regulating cellular function, three hierarchical challenges present themselves: identifying the specific proteins involved in redox signaling, determining the cysteines within those proteins whose oxidation transduces redox chemistry into signaling, and the nature of this transduction (e.g., changes in protein structure, alteration of protein–protein interactions, activation/inactivation of enzyme activity). The scale of this challenge is daunting, as evidence suggests that a significant fraction of the 214,000 cysteines encoded in the human genome may be functional.[185] However, recent advances in identifying proteins containing oxidized cysteine residues using mass spectrometry and chemoselective functionalization (see earlier discussion) offer assurance that the tools needed for deciphering cysteine-based redox signaling will be developed in the near future, if these methods are not already available. Characterizing these signaling pathways is essential to understanding how oxidative stress, and the cell's response to this stress, contributes to degradation of cell function involved in aging and disease development and progression.

SUMMARY

Proteins undergo a wide range of PTMs under conditions of oxidative stress. Once thought to be simply the obligatory by-products of oxidizing species present within the cell, in recent years the richness of chemical diversity provided by oxidative protein modifications has received new attention and appreciation. These modifications can be divided into two broad classes, those that can be reversed/reduced under physiological conditions and those that are irreversible oxidations that persist for the lifetime of the modified protein. Reversible oxidative modifications most often involve cysteine and methionine, utilizing the rich reduction and oxidation chemistry inherent to sulfur. These reversible modifications are proposed to play a central role in cell signaling, providing a mechanism for transducing the oxidation state of the cellular environment into changes in protein structure and function. In contrast, irreversible modifications resulting from an excess of oxidizing species may contribute to cell aging and disease progression. Sensing and maintaining the oxidative balance within the cell is essential to allow productive use of ROS and RNS signaling and metabolism while minimizing cellular damage caused by these highly reactive species.

With the chemistry underlying oxidative PTMs firmly established, the focus now shifts to identifying which proteins undergo oxidative modification and characterizing the impact of these modifications on protein structure and function. The continuing development of new chemical and analytical tools is essential to this endeavor, constituting the first step toward connecting specific protein modifications to changes in cellular health and functions. Identifying the sites and effects of irreversible oxidative PTMs that are observed in diseases such as neurodegenerative disorders and cardiovascular disease will aid in developing a molecular-level understanding of the pathogenic processes that underlie these conditions. The roles of oxidative PTMs in controlling cellular function are just now becoming well established, with studies in the near future holding great promise for further understanding of this form of biological signaling.

REFERENCES

1. Storz, G., Imlay, J.A. Oxidative stress. *Curr. Opin. Microbiol.* **1999**, *2*(2), 188–194.
2. Roos, G., Messens, J. Protein sulfenic acid formation: From cellular damage to redox regulation. *Free Radic. Biol. Med.* **2011**, *51*(2), 314–326.
3. Winterbourn, C.C. Superoxide as an intracellular radical sink. *Free Radic. Biol. Med.* **1993**, *14*(1), 85–90.
4. Jacob, C. Redox signalling via the cellular thiolstat. *Biochem. Soc. Trans.* **2011**, *39*(5), 1247–1253.
5. Zhao, R., Lind, J., Merenyi, G., Eriksen, T.E. Kinetics of one-electron oxidation of thiols and hydrogen abstraction by thiyl radicals from alpha-amino C-H bonds. *J. Am. Chem. Soc.* **1994**, *116*(26), 12010–12015.

6. Winterbourn, C.C., Hampton, M.B. Thiol chemistry and specificity in redox signaling. *Free Radic. Biol. Med.* **2008**, *45*(5), 549–561.
7. Giles, N.M., Giles, G.I., Jacob, C. Multiple roles of cysteine in biocatalysis. *Biochem. Biophys. Res. Commun.* **2003**, *300*(1), 1–4.
8. Harman, L.S., Carver, D.K., Schreiber, J., Mason, R.P. One- and two-electron oxidation of reduced glutathione by peroxidases. *J. Biol. Chem.* **1986**, *261*(4), 1642–1648.
9. Reddie, K.G., Carroll, K.S. Expanding the functional diversity of proteins through cysteine oxidation. *Curr. Opin. Chem. Biol.* **2008**, *12*(6), 746–754.
10. Poole, L.B., Karplus, P.A., Claiborne, A. Protein sulfenic acids in redox signaling. *Annu. Rev. Pharmacol. Toxicol.* **2004**, *44*, 325–347.
11. Salsbury, F.R., Jr., Knutson, S.T., Poole, L.B., Fetrow, J.S. Functional site profiling and electrostatic analysis of cysteines modifiable to cysteine sulfenic acid. *Protein Sci.* **2008**, *17*(2), 299–312.
12. Yang, J., Groen, A., Lemeer, S., Jans, A., Slijper, M., Roe, S.M., den Hertog, J., Barford, D. Reversible oxidation of the membrane distal domain of receptor PTPalpha is mediated by a cyclic sulfenamide. *Biochemistry* **2007**, *46*(3), 709–719.
13. Salmeen, A., Andersen, J.N., Myers, M.P., Meng, T.C., Hinks, J.A., Tonks, N.K., Barford, D. Redox regulation of protein tyrosine phosphatase 1B involves a sulphenyl-amide intermediate. *Nature* **2003**, *423*(6941), 769–773.
14. Rhee, S.G., Chae, H.Z., Kim, K. Peroxiredoxins: A historical overview and speculative preview of novel mechanisms and emerging concepts in cell signaling. *Free Radic. Biol. Med.* **2005**, *38*(12), 1543–1552.
15. Rhee, S.G., Yang, K.S., Kang, S.W., Woo, H.A., Chang, T.S. Controlled elimination of intracellular H(2)O(2): Regulation of peroxiredoxin, catalase, and glutathione peroxidase via post-translational modification. *Antioxid. Redox Signal.* **2005**, *7*(5–6), 619–626.
16. Wood, Z.A., Schroder, E., Robin Harris, J., Poole, L.B. Structure, mechanism and regulation of peroxiredoxins. *Trends Biochem. Sci.* **2003**, *28*(1), 32–40.
17. Nogoceke, E., Gommel, D.U., Kiess, M., Kalisz, H.M., Flohe, L. A unique cascade of oxidoreductases catalyses trypanothione-mediated peroxide metabolism in *Crithidia fasciculata*. *Biol. Chem.* **1997**, *378*(8), 827–836.
18. Poole, L.B., Reynolds, C.M., Wood, Z.A., Karplus, P.A., Ellis, H.R., Li Calzi, M. AhpF and other NADH:peroxiredoxin oxidoreductases, homologues of low Mr thioredoxin reductase. *Eur. J. Biochem.* **2000**, *267*(20), 6126–6133.
19. Bryk, R., Lima, C.D., Erdjument-Bromage, H., Tempst, P., Nathan, C. Metabolic enzymes of mycobacteria linked to antioxidant defense by a thioredoxin-like protein. *Science* **2002**, *295*(5557), 1073–1077.
20. Huang, H.H., Day, L., Cass, C.L., Ballou, D.P., Williams, C.H., Jr., Williams, D.L. Investigations of the catalytic mechanism of thioredoxin glutathione reductase from *Schistosoma mansoni*. *Biochemistry* **2011**, *50*(26), 5870–5882.
21. Biteau, B., Labarre, J., Toledano, M.B. ATP-dependent reduction of cysteine-sulphinic acid by *S. cerevisiae* sulphiredoxin. *Nature* **2003**, *425*(6961), 980–984.
22. Chang, T.S., Jeong, W., Woo, H.A., Lee, S.M., Park, S., Rhee, S.G. Characterization of mammalian sulfiredoxin and its reactivation of hyperoxidized peroxiredoxin through reduction of cysteine sulfinic acid in the active site to cysteine. *J. Biol. Chem.* **2004**, *279*(49), 50994–51001.
23. Jonsson, T.J., Johnson, L.C., Lowther, W.T. Structure of the sulphiredoxin-peroxiredoxin complex reveals an essential repair embrace. *Nature* **2008**, *451*(7174), 98–101.
24. Jonsson, T.J., Murray, M.S., Johnson, L.C., Lowther, W.T. Reduction of cysteine sulfinic acid in peroxiredoxin by sulfiredoxin proceeds directly through a sulfinic phosphoryl ester intermediate. *J Biol Chem* **2008**, *283*(35), 23846–23851.
25. Alderton, W.K., Cooper, C.E., Knowles, R.G. Nitric oxide synthases: Structure, function and inhibition. *Biochem. J.* **2001**, *357*(Pt 3), 593–615.
26. Hughes, M.N. Relationships between nitric oxide, nitroxyl ion, nitrosonium cation and peroxynitrite. *Biochim. Biophys. Acta* **1999**, *1411*(2–3), 263–272.
27. Denicola, A., Freeman, B.A., Trujillo, M., Radi, R. Peroxynitrite reaction with carbon dioxide/bicarbonate: Kinetics and influence on peroxynitrite-mediated oxidations. *Arch. Biochem. Biophys.* **1996**, *333*(1), 49–58.
28. Wardman, P., von Sonntag, C. Kinetic factors that control the fate of thiyl radicals in cells. *Methods Enzymol.* **1995**, *251*, 31–45.
29. Madej, E., Folkes, L.K., Wardman, P., Czapski, G., Goldstein, S. Thiyl radicals react with nitric oxide to form S-nitrosothiols with rate constants near the diffusion-controlled limit. *Free Radic. Biol. Med.* **2008**, *44*(12), 2013–2018.
30. Kharitonov, V.G., Sundquist, A.R., Sharma, V.S. Kinetics of nitrosation of thiols by nitric oxide in the presence of oxygen. *J. Biol. Chem.* **1995**, *270*(47), 28158–28164.
31. Stamler, J.S., Meissner, G. Physiology of nitric oxide in skeletal muscle. *Physiol. Rev.* **2001**, *81*(1), 209–237.
32. Pawloski, J.R., Hess, D.T., Stamler, J.S. Export by red blood cells of nitric oxide bioactivity. *Nature* **2001**, *409*(6820), 622–626.
33. Foster, M.W., McMahon, T.J., Stamler, J.S. S-nitrosylation in health and disease. *Trends Mol. Med.* **2003**, *9*(4), 160–168.
34. Mitchell, D.A., Marletta, M.A. Thioredoxin catalyzes the S-nitrosation of the caspase-3 active site cysteine. *Nat. Chem. Biol.* **2005**, *1*(3), 154–158.
35. Mitchell, D.A., Erwin, P.A., Michel, T., Marletta, M.A. S-nitrosation and regulation of inducible nitric oxide synthase. *Biochemistry* **2005**, *44*(12), 4636–4647.
36. Inoue, K., Akaike, T., Miyamoto, Y., Okamoto, T., Sawa, T., Otagiri, M., Suzuki, S., Yoshimura, T., Maeda, H. Nitro-

sothiol formation catalyzed by ceruloplasmin: Implication for cytoprotective mechanism *in vivo*. *J. Biol. Chem.* **1999**, *274*(38), 27069–27075.

37. Kim, Y.M., Chung, H.T., Simmons, R.L., Billiar, T.R. Cellular non-heme iron content is a determinant of nitric oxide-mediated apoptosis, necrosis, and caspase inhibition. *J. Biol. Chem.* **2000**, *275*(15), 10954–10961.

38. Boese, M., Mordvintcev, P.I., Vanin, A.F., Busse, R., Mulsch, A. S-nitrosation of serum albumin by dinitrosyl-iron complex. *J. Biol. Chem.* **1995**, *270*(49), 29244–29249.

39. Stamler, J.S., Toone, E.J., Lipton, S.A., Sucher, N.J. (S)NO signals: Translocation, regulation, and a consensus motif. *Neuron* **1997**, *18*(5), 691–696.

40. Paige, J.S., Jaffrey, S.R. Pharmacologic manipulation of nitric oxide signaling: Targeting NOS dimerization and protein-protein interactions. *Curr. Top. Med. Chem.* **2007**, *7*(1), 97–114.

41. Jensen, D.E., Belka, G.K., Du Bois, G.C. S-nitrosoglutathione is a substrate for rat alcohol dehydrogenase class III isoenzyme. *Biochem. J.* **1998**, *331*(Pt 2), 659–668.

42. Nikitovic, D., Holmgren, A. S-nitrosoglutathione is cleaved by the thioredoxin system with liberation of glutathione and redox regulating nitric oxide. *J. Biol. Chem.* **1996**, *271*(32), 19180–19185.

43. Sliskovic, I., Raturi, A., Mutus, B. Characterization of the S-denitrosation activity of protein disulfide isomerase. *J. Biol. Chem.* **2005**, *280*(10), 8733–8741.

44. Stoyanovsky, D.A., Tyurina, Y.Y., Tyurin, V.A., Anand, D., Mandavia, D.N., Gius, D., Ivanova, J., Pitt, B., Billiar, T.R., Kagan, V.E. Thioredoxin and lipoic acid catalyze the denitrosation of low molecular weight and protein S-nitrosothiols. *J. Am. Chem. Soc.* **2005**, *127*(45), 15815–15823.

45. Sengupta, R., Ryter, S.W., Zuckerbraun, B.S., Tzeng, E., Billiar, T.R., Stoyanovsky, D.A. Thioredoxin catalyzes the denitrosation of low-molecular mass and protein S-nitrosothiols. *Biochemistry* **2007**, *46*(28), 8472–8483.

46. Hess, D.T., Matsumoto, A., Kim, S.O., Marshall, H.E., Stamler, J.S. Protein S-nitrosylation: Purview and parameters. *Nat. Rev. Mol. Cell Biol.* **2005**, *6*(2), 150–166.

47. van Montfort, R.L., Congreve, M., Tisi, D., Carr, R., Jhoti, H. Oxidation state of the active-site cysteine in protein tyrosine phosphatase 1B. *Nature* **2003**, *423*(6941), 773–777.

48. Lee, J.W., Soonsanga, S., Helmann, J.D. A complex thiolate switch regulates the *Bacillus subtilis* organic peroxide sensor OhrR. *Proc. Natl. Acad. Sci. U.S.A.* **2007**, *104*(21), 8743–8748.

49. Paulsen, C.E., Carroll, K.S. Orchestrating redox signaling networks through regulatory cysteine switches. *ACS Chem. Biol.* **2010**, *5*(1), 47–62.

50. Bulleid, N.J., Ellgaard, L. Multiple ways to make disulfides. *Trends Biochem. Sci.* **2011**, *36*(9), 485–492.

51. Turell, L., Botti, H., Carballal, S., Ferrer-Sueta, G., Souza, J.M., Duran, R., Freeman, B.A., Radi, R., Alvarez, B. Reactivity of sulfenic acid in human serum albumin. *Biochemistry* **2008**, *47*(1), 358–367.

52. Jourd'heuil, D., Jourd'heuil, F.L., Feelisch, M. Oxidation and nitrosation of thiols at low micromolar exposure to nitric oxide. Evidence for a free radical mechanism. *J. Biol. Chem.* **2003**, *278*(18), 15720–15726.

53. Claiborne, A., Yeh, J.I., Mallett, T.C., Luba, J., Crane, E.J., 3rd, Charrier, V., Parsonage, D. Protein-sulfenic acids: Diverse roles for an unlikely player in enzyme catalysis and redox regulation. *Biochemistry* **1999**, *38*(47), 15407–15416.

54. Wouters, M.A., Fan, S.W., Haworth, N.L. Disulfides as redox switches: From molecular mechanisms to functional significance. *Antioxid. Redox Signal.* **2010**, *12*(1), 53–91.

55. Yang, Y., Song, Y., Loscalzo, J. Regulation of the protein disulfide proteome by mitochondria in mammalian cells. *Proc. Natl. Acad. Sci. U.S.A.* **2007**, *104*(26), 10813–10817.

56. Wang, C., Yu, J., Huo, L., Wang, L., Feng, W., Wang, C.C. Human protein-disulfide isomerase is a redox-regulated chaperone activated by oxidation of domain a'. *J. Biol. Chem.* **2012**, *287*(2), 1139–1149.

57. Meister, A. Glutathione metabolism and its selective modification. *J. Biol. Chem.* **1988**, *263*(33), 17205–17208.

58. Forman, H.J., Zhang, H., Rinna, A. Glutathione: Overview of its protective roles, measurement, and biosynthesis. *Mol. Aspects Med.* **2009**, *30*(1–2), 1–12.

59. Gravina, S.A., Mieyal, J.J. Thioltransferase is a specific glutathionyl mixed disulfide oxidoreductase. *Biochemistry* **1993**, *32*(13), 3368–3376.

60. Yang, Y., Jao, S., Nanduri, S., Starke, D.W., Mieyal, J.J., Qin, J. Reactivity of the human thioltransferase (glutaredoxin) C7S, C25S, C78S, C82S mutant and NMR solution structure of its glutathionyl mixed disulfide intermediate reflect catalytic specificity. *Biochemistry* **1998**, *37*(49), 17145–17156.

61. Holmgren, A. Thioredoxin and glutaredoxin systems. *J. Biol. Chem.* **1989**, *264*(24), 13963–13966.

62. Meyer, Y., Buchanan, B.B., Vignols, F., Reichheld, J.P. Thioredoxins and glutaredoxins: Unifying elements in redox biology. *Annu. Rev. Genet.* **2009**, *43*, 335–367.

63. Hill, B.G., Bhatnagar, A. Protein S-glutathiolation: Redox-sensitive regulation of protein function. *J. Mol. Cell. Cardiol.* **2011**, J. Mol. Cell. Cardiol. **2012**, *52*(3), 559–567.

64. Wang, J., Boja, E.S., Tan, W., Tekle, E., Fales, H.M., English, S., Mieyal, J.J., Chock, P.B. Reversible glutathionylation regulates actin polymerization in A431 cells. *J. Biol. Chem.* **2001**, *276*(51), 47763–47766.

65. Johansson, M., Lundberg, M. Glutathionylation of beta-actin via a cysteinyl sulfenic acid intermediary. *BMC Biochem.* **2007**, *8*, 26.

66. Hill, B.G., Ramana, K.V., Cai, J., Bhatnagar, A., Srivastava, S.K. Measurement and identification of S-glutathiolated proteins. *Methods Enzymol.* **2010**, *473*, 179–197.

67. Anathy, V., Roberson, E.C., Guala, A.S., Godburn, K.E., Budd, R.C., Janssen-Heininger, Y.M. Redox-based regulation of apoptosis: S-glutathionylation as a regulatory mechanism to control cell death. *Antioxid. Redox Signal.* **2012**, *16*(6), 496–505.

68. Murphy, M.P. Mitochondrial thiols in antioxidant protection and redox signaling: Distinct roles for glutathionylation and other thiol modifications. *Antioxid. Redox Signal.* **2012**, *16*(6), 476–495.

69. Zaffagnini, M., Bedhomme, M., Marchand, C.H., Morisse, S., Trost, P., Lemaire, S.D. Redox regulation in photosynthetic organisms: Focus on glutathionylation. *Antioxid. Redox Signal.* **2012**, *16*(6), 567–586.

70. Cheng, Z., Zhang, J., Ballou, D.P., Williams, C.H., Jr. Reactivity of thioredoxin as a protein thiol-disulfide oxidoreductase. *Chem. Rev.* **2011**, *111*(9), 5768–5783.

71. Gallogly, M.M., Starke, D.W., Mieyal, J.J. Mechanistic and kinetic details of catalysis of thiol-disulfide exchange by glutaredoxins and potential mechanisms of regulation. *Antioxid. Redox Signal.* **2009**, *11*(5), 1059–1081.

72. Gallogly, M.M., Starke, D.W., Leonberg, A.K., Ospina, S.M., Mieyal, J.J. Kinetic and mechanistic characterization and versatile catalytic properties of mammalian glutaredoxin 2: Implications for intracellular roles. *Biochemistry* **2008**, *47*(42), 11144–11157.

73. Starke, D.W., Chock, P.B., Mieyal, J.J. Glutathione-thiyl radical scavenging and transferase properties of human glutaredoxin (thioltransferase). Potential role in redox signal transduction. *J. Biol. Chem.* **2003**, *278*(17), 14607–14613.

74. Mieyal, J.J., Starke, D.W., Gravina, S.A., Hocevar, B.A. Thioltransferase in human red blood cells: Kinetics and equilibrium. *Biochemistry* **1991**, *30*(36), 8883–8891.

75. Mieyal, J.J., Gallogly, M.M., Qanungo, S., Sabens, E.A., Shelton, M.D. Molecular mechanisms and clinical implications of reversible protein S-glutathionylation. *Antioxid. Redox Signal.* **2008**, *10*(11), 1941–1988.

76. Vogt, W. Oxidation of methionyl residues in proteins: Tools, targets, and reversal. *Free Radic. Biol. Med.* **1995**, *18*(1), 93–105.

77. Bonvini, E., Bougnoux, P., Stevenson, H.C., Miller, P., Hoffman, T. Activation of the oxidative burst in human monocytes is associated with inhibition of methionine-dependent methylation of neutral lipids and phospholipids. *J. Clin. Invest.* **1984**, *73*(6), 1629–1637.

78. Arterburn, J.B., Nelson, S.L. Rhenium-catalyzed oxidation of sulfides with phenyl sulfoxide. *J. Org. Chem.* **1996**, *61*(7), 2260–2261.

79. Boschi-Muller, S., Gand, A., Branlant, G. The methionine sulfoxide reductases: Catalysis and substrate specificities. *Arch. Biochem. Biophys.* **2008**, *474*(2), 266–273.

80. Lee, B.C., Dikiy, A., Kim, H.Y., Gladyshev, V.N. Functions and evolution of selenoprotein methionine sulfoxide reductases. *Biochim. Biophys. Acta* **2009**, *1790*(11), 1471–1477.

81. Lee, B.C., Gladyshev, V.N. The biological significance of methionine sulfoxide stereochemistry. *Free Radic. Biol. Med.* **2011**, *50*(2), 221–227.

82. Stadtman, E.R. Protein oxidation and aging. *Free Radic. Res.* **2006**, *40*(12), 1250–1258.

83. Boschi-Muller, S., Olry, A., Antoine, M., Branlant, G. The enzymology and biochemistry of methionine sulfoxide reductases. *Biochim. Biophys. Acta* **2005**, *1703*(2), 231–238.

84. Lim, J.C., You, Z., Kim, G., Levine, R.L. Methionine sulfoxide reductase A is a stereospecific methionine oxidase. *Proc. Natl. Acad. Sci. U.S.A.* **2011**, *108*(26), 10472–10477.

85. Lin, Z., Johnson, L.C., Weissbach, H., Brot, N., Lively, M.O., Lowther, W.T. Free methionine-(R)-sulfoxide reductase from *Escherichia coli* reveals a new GAF domain function. *Proc. Natl. Acad. Sci. U.S.A.* **2007**, *104*(23), 9597–9602.

86. Lowther, W.T., Weissbach, H., Etienne, F., Brot, N., Matthews, B.W. The mirrored methionine sulfoxide reductases of *Neisseria gonorrhoeae* pilB. *Nat. Struct. Biol.* **2002**, *9*(5), 348–352.

87. Ranaivoson, F.M., Antoine, M., Kauffmann, B., Boschi-Muller, S., Aubry, A., Branlant, G., Favier, F. A structural analysis of the catalytic mechanism of methionine sulfoxide reductase A from *Neisseria meningitidis*. *J. Mol. Biol.* **2008**, *377*(1), 268–280.

88. Levine, R.L., Mosoni, L., Berlett, B.S., Stadtman, E.R. Methionine residues as endogenous antioxidants in proteins. *Proc. Natl. Acad. Sci. U.S.A.* **1996**, *93*(26), 15036–15040.

89. Moskovitz, J., Berlett, B.S., Poston, J.M., Stadtman, E.R. The yeast peptide-methionine sulfoxide reductase functions as an antioxidant *in vivo*. *Proc. Natl. Acad. Sci. U.S.A.* **1997**, *94*(18), 9585–9589.

90. Moskovitz, J., Bar-Noy, S., Williams, W.M., Requena, J., Berlett, B.S., Stadtman, E.R. Methionine sulfoxide reductase (MsrA) is a regulator of antioxidant defense and lifespan in mammals. *Proc. Natl. Acad. Sci. U.S.A.* **2001**, *98*(23), 12920–12925.

91. Moskovitz, J., Rahman, M.A., Strassman, J., Yancey, S.O., Kushner, S.R., Brot, N., Weissbach, H. *Escherichia coli* peptide methionine sulfoxide reductase gene: Regulation of expression and role in protecting against oxidative damage. *J. Bacteriol.* **1995**, *177*(3), 502–507.

92. Douglas, T., Daniel, D.S., Parida, B.K., Jagannath, C., Dhandayuthapani, S. Methionine sulfoxide reductase A (MsrA) deficiency affects the survival of *Mycobacterium smegmatis* within macrophages. *J. Bacteriol.* **2004**, *186*(11), 3590–3598.

93. St John, G., Brot, N., Ruan, J., Erdjument-Bromage, H., Tempst, P., Weissbach, H., Nathan, C. Peptide methionine sulfoxide reductase from *Escherichia coli* and *Mycobacterium tuberculosis* protects bacteria against oxidative damage from reactive nitrogen intermediates. *Proc. Natl. Acad. Sci. U.S.A.* **2001**, *98*(17), 9901–9906.

94. Moskovitz, J., Flescher, E., Berlett, B.S., Azare, J., Poston, J.M., Stadtman, E.R. Overexpression of peptide-methionine sulfoxide reductase in *Saccharomyces cerevisiae* and human T cells provides them with high resistance

to oxidative stress. *Proc. Natl. Acad. Sci. U.S.A.* **1998**, *95*(24), 14071–14075.

95. Yermolaieva, O., Xu, R., Schinstock, C., Brot, N., Weissbach, H., Heinemann, S.H., Hoshi, T. Methionine sulfoxide reductase A protects neuronal cells against brief hypoxia/reoxygenation. *Proc. Natl. Acad. Sci. U.S.A.* **2004**, *101*(5), 1159–1164.

96. Kantorow, M., Hawse, J.R., Cowell, T.L., Benhamed, S., Pizarro, G.O., Reddy, V.N., Hejtmancik, J.F. Methionine sulfoxide reductase A is important for lens cell viability and resistance to oxidative stress. *Proc. Natl. Acad. Sci. U.S.A.* **2004**, *101*(26), 9654–9659.

97. Ciorba, M.A., Heinemann, S.H., Weissbach, H., Brot, N., Hoshi, T. Modulation of potassium channel function by methionine oxidation and reduction. *Proc. Natl. Acad. Sci. U.S.A.* **1997**, *94*(18), 9932–9937.

98. Dalle-Donne, I., Rossi, R., Giustarini, D., Gagliano, N., Di Simplicio, P., Colombo, R., Milzani, A. Methionine oxidation as a major cause of the functional impairment of oxidized actin. *Free Radic. Biol. Med.* **2002**, *32*(9), 927–937.

99. Maskos, Z., Rush, J.D., Koppenol, W.H. The hydroxylation of phenylalanine and tyrosine: A comparison with salicylate and tryptophan. *Arch. Biochem. Biophys.* **1992**, *296*(2), 521–529.

100. Gieseg, S.P., Simpson, J.A., Charlton, T.S., Duncan, M.W., Dean, R.T. Protein-bound 3,4-dihydroxyphenylalanine is a major reductant formed during hydroxyl radical damage to proteins. *Biochemistry* **1993**, *32*(18), 4780–4786.

101. Pacher, P., Beckman, J.S., Liaudet, L. Nitric oxide and peroxynitrite in health and disease. *Physiol. Rev.* **2007**, *87*(1), 315–424.

102. Gunaydin, H., Houk, K.N. Mechanisms of peroxynitrite-mediated nitration of tyrosine. *Chem. Res. Toxicol.* **2009**, *22*(5), 894–898.

103. Nagatsu, T., Levitt, M., Udenfriend, S. Tyrosine hydroxylase: The initial step in norepinephrine biosynthesis. *J. Biol. Chem.* **1964**, *239*, 2910–2917.

104. Fitzpatrick, P.F. Steady-state kinetic mechanism of rat tyrosine hydroxylase. *Biochemistry* **1991**, *30*(15), 3658–3662.

105. Lymar, S.V., Jiang, Q., Hurst, J.K. Mechanism of carbon dioxide-catalyzed oxidation of tyrosine by peroxynitrite. *Biochemistry* **1996**, *35*(24), 7855–7861.

106. Grossi, L. Evidence of an electron-transfer mechanism in the peroxynitrite-mediated oxidation of 4-alkylphenols and tyrosine. *J. Org. Chem.* **2003**, *68*(16), 6349–6353.

107. DiMarco, T., Giulivi, C. Current analytical methods for the detection of dityrosine, a biomarker of oxidative stress, in biological samples. *Mass Spectrom. Rev.* **2007**, *26*(1), 108–120.

108. Jia, M., Mateoiu, C., Souchelnytskyi, S. Protein tyrosine nitration in the cell cycle. *Biochem. Biophys. Res. Commun.* **2011**, *413*(2), 270–276.

109. Hunter, T. Protein kinases and phosphatases: The yin and yang of protein phosphorylation and signaling. *Cell* **1995**, *80*(2), 225–236.

110. Johnson, L.N., Lewis, R.J. Structural basis for control by phosphorylation. *Chem. Rev.* **2001**, *101*(8), 2209–2242.

111. Ginsburg, A., Yeh, J., Hennig, S.B., Denton, M.D. Some effects of adenylylation on the biosynthetic properties of the glutamine synthetase from *Escherichia coli*. *Biochemistry* **1970**, *9*(3), 633–649.

112. Stadtman, E.R., Ginsburg, A., Ciardi, J.E., Yeh, J., Hennig, S.B., Shapiro, B.M. Multiple molecular forms of glutamine synthetase produced by enzyme catalyzed adenylation and deadenylylation reactions. *Adv. Enzyme Regul.* **1970**, *8*, 99–118.

113. Berlett, B.S., Friguet, B., Yim, M.B., Chock, P.B., Stadtman, E.R. Peroxynitrite-mediated nitration of tyrosine residues in *Escherichia coli* glutamine synthetase mimics adenylylation: Relevance to signal transduction. *Proc. Natl. Acad. Sci. U.S.A.* **1996**, *93*(5), 1776–1780.

114. Armstrong, R.C., Swallow, A.J. Pulse- and gamma-radiolysis of aqueous solutions of tryptophan. *Radiat. Res.* **1969**, *40*(3), 563–579.

115. Winchester, R.V., Lynn, K.R. X- and gamma-radiolysis of some tryptophan dipeptides. *Int. J. Radiat. Biol. Relat. Stud. Phys. Chem. Med.* **1970**, *17*(6), 541–548.

116. Kikugawa, K., Kato, T., Okamoto, Y. Damage of amino acids and proteins induced by nitrogen dioxide, a free radical toxin, in air. *Free Radic. Biol. Med.* **1994**, *16*(3), 373–382.

117. Fitzpatrick, P.F. Mechanism of aromatic amino acid hydroxylation. *Biochemistry* **2003**, *42*(48), 14083–14091.

118. Nuriel, T., Hansler, A., Gross, S.S. Protein nitrotryptophan: Formation, significance and identification. *J. Proteomics.* **2011**, *74*(11), 2300–2312.

119. Yamakura, F., Ikeda, K. Modification of tryptophan and tryptophan residues in proteins by reactive nitrogen species. *Nitric Oxide* **2006**, *14*(2), 152–161.

120. Pietraforte, D., Minetti, M. One-electron oxidation pathway of peroxynitrite decomposition in human blood plasma: Evidence for the formation of protein tryptophan-centred radicals. *Biochem. J.* **1997**, *321*(Pt 3), 743–750.

121. Yamakura, F., Matsumoto, T., Fujimura, T., Taka, H., Murayama, K., Imai, T., Uchida, K. Modification of a single tryptophan residue in human Cu,Zn-superoxide dismutase by peroxynitrite in the presence of bicarbonate. *Biochim. Biophys. Acta* **2001**, *1548*(1), 38–46.

122. Ishii, Y., Ogara, A., Katsumata, T., Umemura, T., Nishikawa, A., Iwasaki, Y., Ito, R., Saito, K., Hirose, M., Nakazawa, H. Quantification of nitrated tryptophan in proteins and tissues by high-performance liquid chromatography with electrospray ionization tandem mass spectrometry. *J. Pharm. Biomed. Anal.* **2007**, *44*(1), 150–159.

123. Rebrin, I., Bregere, C., Kamzalov, S., Gallaher, T.K., Sohal, R.S. Nitration of tryptophan 372 in succinyl-CoA:3-ketoacid CoA transferase during aging in rat heart mitochondria. *Biochemistry* **2007**, *46*(35), 10130–10144.

124. Lee, J.W., Helmann, J.D. The PerR transcription factor senses H_2O_2 by metal-catalysed histidine oxidation. *Nature* **2006**, *440*(7082), 363–367.

125. Uchida, K., Kawakishi, S. Identification of oxidized histidine generated at the active site of Cu,Zn-superoxide dismutase exposed to H2O2: Selective generation of 2-oxo-histidine at the histidine 118. *J. Biol. Chem.* **1994**, *269*(4), 2405–2410.

126. Wells-Knecht, M.C., Huggins, T.G., Dyer, D.G., Thorpe, S.R., Baynes, J.W. Oxidized amino acids in lens protein with age: Measurement of o-tyrosine and dityrosine in the aging human lens. *J. Biol. Chem.* **1993**, *268*(17), 12348–12352.

127. van der Vliet, A., O'Neill, C.A., Halliwell, B., Cross, C.E., Kaur, H. Aromatic hydroxylation and nitration of phenylalanine and tyrosine by peroxynitrite. Evidence for hydroxyl radical production from peroxynitrite. *FEBS Lett.* **1994**, *339*(1–2), 89–92.

128. Ferger, B., Themann, C., Rose, S., Halliwell, B., Jenner, P. 6-Hydroxydopamine increases the hydroxylation and nitration of phenylalanine *in vivo*: Implication of peroxynitrite formation. *J. Neurochem.* **2001**, *78*(3), 509–514.

129. Requena, J.R., Chao, C.C., Levine, R.L., Stadtman, E.R. Glutamic and aminoadipic semialdehydes are the main carbonyl products of metal-catalyzed oxidation of proteins. *Proc. Natl. Acad. Sci. U.S.A.* **2001**, *98*(1), 69–74.

130. Amici, A., Levine, R.L., Tsai, L., Stadtman, E.R. Conversion of amino acid residues in proteins and amino acid homopolymers to carbonyl derivatives by metal-catalyzed oxidation reactions. *J. Biol. Chem.* **1989**, *264*(6), 3341–3346.

131. Uchida, K., Kato, Y., Kawakishi, S. A novel mechanism for oxidative cleavage of prolyl peptides induced by the hydroxyl radical. *Biochem. Biophys. Res. Commun.* **1990**, *169*(1), 265–271.

132. Bruick, R.K., McKnight, S.L. A conserved family of prolyl-4-hydroxylases that modify HIF. *Science* **2001**, *294*(5545), 1337–1340.

133. Freedman, S.J., Sun, Z.Y., Kung, A.L., France, D.S., Wagner, G., Eck, M.J. Structural basis for negative regulation of hypoxia-inducible factor-1alpha by CITED2. *Nat. Struct. Biol.* **2003**, *10*(7), 504–512.

134. Taborsky, G. Oxidative modification of proteins in the presence of ferrous ion and air:. Effect of ionic constituents of the reaction medium on the nature of the oxidation products. *Biochemistry* **1973**, *12*(7), 1341–1348.

135. Tareke, E., Heinze, T.M., Gamboa da Costa, G., Ali, S. Acrylamide formed at physiological temperature as a result of asparagine oxidation. *J. Agric. Food Chem.* **2009**, *57*(20), 9730–9733.

136. Wong, C.M., Marcocci, L., Liu, L., Suzuki, Y.J. Cell signaling by protein carbonylation and decarbonylation. *Antioxid. Redox Signal.* **2010**, *12*(3), 393–404.

137. Berlett, B.S., Stadtman, E.R. Protein oxidation in aging, disease, and oxidative stress. *J. Biol. Chem.* **1997**, *272*(33), 20313–20316.

138. Dalle-Donne, I., Giustarini, D., Colombo, R., Rossi, R., Milzani, A. Protein carbonylation in human diseases. *Trends Mol. Med.* **2003**, *9*(4), 169–176.

139. Dalle-Donne, I., Rossi, R., Giustarini, D., Milzani, A., Colombo, R. Protein carbonyl groups as biomarkers of oxidative stress. *Clin Chim Acta* **2003**, *329*(1–2), 23–38.

140. Yan, L.J. Analysis of oxidative modification of proteins. *Curr. Protoc. Protein Sci.* **2009**. DOI: 10.1002/0471140864.ps1404s56.

141. Kirkham, P.A., Caramori, G., Casolari, P., Papi, A.A., Edwards, M., Shamji, B., Triantaphyllopoulos, K., Hussain, F., Pinart, M., Khan, Y., Heinemann, L., Stevens, L., Yeadon, M., Barnes, P.J., Chung, K.F., Adcock, I.M. Oxidative stress-induced antibodies to carbonyl-modified protein correlate with severity of chronic obstructive pulmonary disease. *Am. J. Respir. Crit. Care Med.* **2011**, *184*(7), 796–802.

142. Nadkarni, D.V., Sayre, L.M. Structural definition of early lysine and histidine adduction chemistry of 4-hydroxynonenal. *Chem. Res. Toxicol.* **1995**, *8*(2), 284–291.

143. Uchida, K., Stadtman, E.R. Covalent attachment of 4-hydroxynonenal to glyceraldehyde-3-phosphate dehydrogenase: A possible involvement of intra- and intermolecular cross-linking reaction. *J. Biol. Chem.* **1993**, *268*(9), 6388–6393.

144. Friguet, B., Stadtman, E.R., Szweda, L.I. Modification of glucose-6-phosphate dehydrogenase by 4-hydroxy-2-nonenal: Formation of cross-linked protein that inhibits the multicatalytic protease. *J. Biol. Chem.* **1994**, *269*(34), 21639–21643.

145. Stadtman, E.R., Levine, R.L. Free radical-mediated oxidation of free amino acids and amino acid residues in proteins. *Amino Acids* **2003**, *25*(3–4), 207–218.

146. Cerami, A., Vlassara, H., Brownlee, M. Glucose and aging. *Sci. Am.* **1987**, *256*(5), 90–96.

147. Wolff, S.P., Dean, R.T. Glucose autoxidation and protein modification. The potential role of "autoxidative glycosylation" in diabetes. *Biochem. J.* **1987**, *245*(1), 243–250.

148. Grandhee, S.K., Monnier, V.M. Mechanism of formation of the Maillard protein cross-link pentosidine: Glucose, fructose, and ascorbate as pentosidine precursors. *J. Biol. Chem.* **1991**, *266*(18), 11649–11653.

149. Kristal, B.S., Yu, B.P. An emerging hypothesis: Synergistic induction of aging by free radicals and Maillard reactions. *J. Gerontol.* **1992**, *47*(4), B107–B114.

150. Wells-Knecht, M.C., Thorpe, S.R., Baynes, J.W. Pathways of formation of glycoxidation products during glycation of collagen. *Biochemistry* **1995**, *34*(46), 15134–15141.

151. Reddy, S., Bichler, J., Wells-Knecht, K.J., Thorpe, S.R., Baynes, J.W. N epsilon-(carboxymethyl)lysine is a dominant advanced glycation end product (AGE) antigen in tissue proteins. *Biochemistry* **1995**, *34*(34), 10872–10878.

152. Wells-Knecht, K.J., Zyzak, D.V., Litchfield, J.E., Thorpe, S.R., Baynes, J.W. Mechanism of autoxidative glycosylation: Identification of glyoxal and arabinose as intermediates in the autoxidative modification of proteins by glucose. *Biochemistry* **1995**, *34*(11), 3702–3709.

153. Shacter, E. Quantification and significance of protein oxidation in biological samples. *Drug Metab. Rev.* **2000**, *32*(3–4), 307–326.

154. Conrad, C.C., Choi, J., Malakowsky, C.A., Talent, J.M., Dai, R., Marshall, P., Gracy, R.W. Identification of protein carbonyls after two-dimensional electrophoresis. *Proteomics* **2001**, *1*(7), 829–834.
155. Castegna, A., Aksenov, M., Aksenova, M., Thongboonkerd, V., Klein, J.B., Pierce, W.M., Booze, R., Markesbery, W.R., Butterfield, D.A. Proteomic identification of oxidatively modified proteins in Alzheimer's disease brain. Part I: Creatine kinase BB, glutamine synthase, and ubiquitin carboxy-terminal hydrolase L-1. *Free Radic. Biol. Med.* **2002**, *33*(4), 562–571.
156. Moller, I.M., Rogowska-Wrzesinska, A., Rao, R.S. Protein carbonylation and metal-catalyzed protein oxidation in a cellular perspective. *J. Proteomics.* **2011**, *74*(11), 2228–2242.
157. Poole, L.B., Nelson, K.J. Discovering mechanisms of signaling-mediated cysteine oxidation. *Curr. Opin. Chem. Biol.* **2008**, *12*(1), 18–24.
158. Spickett, C.M., Pitt, A.R. Protein oxidation: Role in signalling and detection by mass spectrometry. *Amino Acids* **2012**, *42*(1), 5–21.
159. Johnson, H., Eyers, C.E. Analysis of post-translational modifications by LC-MS/MS. *Methods Mol. Biol.* **2010**, *658*, 93–108.
160. Jensen, O.N. Interpreting the protein language using proteomics. *Nat. Rev. Mol. Cell Biol.* **2006**, *7*(6), 391–403.
161. Zhao, Y., Jensen, O.N. Modification-specific proteomics: Strategies for characterization of post-translational modifications using enrichment techniques. *Proteomics* **2009**, *9*(20), 4632–4641.
162. Mauri, P., Toppo, S., De Palma, A., Benazzi, L., Maiorino, M., Ursini, F. Identification by MS/MS of disulfides produced by a functional redox transition. *Methods Enzymol.* **2010**, *473*, 217–225.
163. Lindahl, M., Mata-Cabana, A., Kieselbach, T. The disulfide proteome and other reactive cysteine proteomes: Analysis and functional significance. *Antioxid. Redox Signal.* **2011**, *14*(12), 2581–2642.
164. Seo, Y.H., Carroll, K.S. Quantification of protein sulfenic acid modifications using isotope-coded dimedone and iododimedone. *Angew. Chem. Int. Ed Engl.* **2011**, *50*(6), 1342–1345.
165. Truong, T.H., Garcia, F.J., Seo, Y.H., Carroll, K.S. Isotope-coded chemical reporter and acid-cleavable affinity reagents for monitoring protein sulfenic acids. *Bioorg. Med. Chem. Lett.* **2011**, *21*(17), 5015–5020.
166. Mirzaei, H., Regnier, F. Identification and quantification of protein carbonylation using light and heavy isotope labeled Girard's P reagent. *J. Chromatogr. A* **2006**, *1134*(1–2), 122–133.
167. Leonard, S.E., Carroll, K.S. Chemical "omics" approaches for understanding protein cysteine oxidation in biology. *Curr. Opin. Chem. Biol.* **2011**, *15*(1), 88–102.
168. Jacob, C., Battaglia, E., Burkholz, T., Peng, D., Bagrel, D., Montenarh, M. Control of oxidative posttranslational cysteine modifications: From intricate chemistry to widespread biological and medical applications. *Chem. Res. Toxicol.* **2012**, *25*(3), 588–604.
169. Benitez, L.V., Allison, W.S. The inactivation of the acyl phosphatase activity catalyzed by the sulfenic acid form of glyceraldehyde 3-phosphate dehydrogenase by dimedone and olefins. *J. Biol. Chem.* **1974**, *249*(19), 6234–6243.
170. Poole, L.B., Klomsiri, C., Knaggs, S.A., Furdui, C.M., Nelson, K.J., Thomas, M.J., Fetrow, J.S., Daniel, L.W., King, S.B. Fluorescent and affinity-based tools to detect cysteine sulfenic acid formation in proteins. *Bioconjug. Chem.* **2007**, *18*(6), 2004–2017.
171. Reddie, K.G., Seo, Y.H., Muse Iii, W.B., Leonard, S.E., Carroll, K.S. A chemical approach for detecting sulfenic acid-modified proteins in living cells. *Mol. Biosyst.* **2008**, *4*(6), 521–531.
172. Jaffrey, S.R., Snyder, S.H. The biotin switch method for the detection of S-nitrosylated proteins. *Sci. STKE* **2001**, *2001*(86), pl1.
173. Jaffrey, S.R., Erdjument-Bromage, H., Ferris, C.D., Tempst, P., Snyder, S.H. Protein S-nitrosylation: A physiological signal for neuronal nitric oxide. *Nat. Cell Biol.* **2001**, *3*(2), 193–197.
174. Murray, C.I., Uhrigshardt, H., O'Meally, R.N., Cole, R.N., Van Eyk, J.E. Identification and quantification of S-nitrosylation by cysteine reactive tandem mass tag switch assay. *Mol. Cell. Proteomics* **2012**, *11*(2), M111 013441.
175. Uehara, T., Nishiya, T. Screening systems for the identification of S-nitrosylated proteins. *Nitric Oxide* **2011**, *25*(2), 108–111.
176. Aesif, S.W., Janssen-Heininger, Y.M., Reynaert, N.L. Protocols for the detection of s-glutathionylated and s-nitrosylated proteins *in situ*. *Methods Enzymol.* **2010**, *474*, 289–296.
177. Aesif, S.W., Anathy, V., Havermans, M., Guala, A.S., Ckless, K., Taatjes, D.J., Janssen-Heininger, Y.M. In situ analysis of protein S-glutathionylation in lung tissue using glutaredoxin-1-catalyzed cysteine derivatization. *Am. J. Pathol.* **2009**, *175*(1), 36–45.
178. Reynaert, N.L., Ckless, K., Guala, A.S., Wouters, E.F., van der Vliet, A., Janssen-Heininger, Y.M. *In situ* detection of S-glutathionylated proteins following glutaredoxin-1 catalyzed cysteine derivatization. *Biochim. Biophys. Acta* **2006**, *1760*(3), 380–387.
179. Zhang, C., Rodriguez, C., Circu, M.L., Aw, T.Y., Feng, J. S-glutathionyl quantification in the attomole range using glutaredoxin-3-catalyzed cysteine derivatization and capillary gel electrophoresis with laser-induced fluorescence detection. *Anal. Bioanal. Chem.* **2011**, *401*(7), 2165–2175.
180. Guo, J., Prokai, L. To tag or not to tag: A comparative evaluation of immunoaffinity-labeling and tandem mass spectrometry for the identification and localization of posttranslational protein carbonylation by 4-hydroxy-2-nonenal, an end-product of lipid peroxidation. *J. Proteomics.* **2011**, *74*(11), 2360–2369.

181. Mirzaei, H., Regnier, F. Affinity chromatographic selection of carbonylated proteins followed by identification of oxidation sites using tandem mass spectrometry. *Anal. Chem.* **2005**, *77*(8), 2386–2392.

182. Akagawa, M., Suyama, K., Uchida, K. Fluorescent detection of alpha-aminoadipic and gamma-glutamic semialdehydes in oxidized proteins. *Free Radic. Biol. Med.* **2009**, *46*(6), 701–706.

183. Madian, A.G., Regnier, F.E. Proteomic identification of carbonylated proteins and their oxidation sites. *J. Proteome Res.* **2010**, *9*(8), 3766–3780.

184. Jones, D.P. Redox sensing: Orthogonal control in cell cycle and apoptosis signalling. *J. Intern. Med.* **2010**, *268*(5), 432–448.

185. Jones, D.P., Go, Y.M. Mapping the cysteine proteome: Analysis of redox-sensing thiols. *Curr. Opin. Chem. Biol.* **2011**, *15*(1), 103–112.

186. Klomsiri, C., Karplus, P.A., Poole, L.B. Cysteine-based redox switches in enzymes. *Antioxid. Redox Signal.* **2011**, *14*(6), 1065–1077.

187. Nagy, P., Winterbourn, C.C. Redox chemistry of biological thiols. In: *Advances in Molecular Toxicology*, ed. J.C. Fishbein. Elsevier, New York, 2010. Pp. 183–222.

188. Hall, A., Karplus, P.A., Poole, L.B. Typical 2-Cys peroxiredoxins—structures, mechanisms and functions. *FEBS J.* **2009**, *276*(9), 2469–2477.

189. Denu, J.M., Dixon, J.E. Protein tyrosine phosphatases: Mechanisms of catalysis and regulation. *Curr. Opin. Chem. Biol.* **1998**, *2*(5), 633–641.

190. Denu, J.M., Tanner, K.G. Specific and reversible inactivation of protein tyrosine phosphatases by hydrogen peroxide: Evidence for a sulfenic acid intermediate and implications for redox regulation. *Biochemistry* **1998**, *37*(16), 5633–5642.

191. Seth, D., Rudolph, J. Redox regulation of MAP kinase phosphatase 3. *Biochemistry* **2006**, *45*(28), 8476–8487.

192. Sohn, J., Rudolph, J. Catalytic and chemical competence of regulation of cdc25 phosphatase by oxidation/reduction. *Biochemistry* **2003**, *42*(34), 10060–10070.

193. Tonks, N.K. Protein tyrosine phosphatases: From genes, to function, to disease. *Nat. Rev. Mol. Cell Biol.* **2006**, *7*(11), 833–846.

194. Chiarugi, P., Buricchi, F. Protein tyrosine phosphorylation and reversible oxidation: Two cross-talking post-translation modifications. *Antioxid. Redox Signal.* **2007**, *9*(1), 1–24.

195. Cross, J.V., Templeton, D.J. Regulation of signal transduction through protein cysteine oxidation. *Antioxid. Redox Signal.* **2006**, *8*(9–10), 1819–1827.

196. Burgoyne, J.R., Madhani, M., Cuello, F., Charles, R.L., Brennan, J.P., Schroder, E., Browning, D.D., Eaton, P. Cysteine redox sensor in PKGIa enables oxidant-induced activation. *Science* **2007**, *317*(5843), 1393–1397.

197. Humphries, K.M., Deal, M.S., Taylor, S.S. Enhanced dephosphorylation of cAMP-dependent protein kinase by oxidation and thiol modification. *J. Biol. Chem.* **2005**, *280*(4), 2750–2758.

198. Qanungo, S., Starke, D.W., Pai, H.V., Mieyal, J.J., Nieminen, A.L. Glutathione supplementation potentiates hypoxic apoptosis by S-glutathionylation of p65-NFkappaB. *J. Biol. Chem.* **2007**, *282*(25), 18427–18436.

199. Velu, C.S., Niture, S.K., Doneanu, C.E., Pattabiraman, N., Srivenugopal, K.S. Human p53 is inhibited by glutathionylation of cysteines present in the proximal DNA-binding domain during oxidative stress. *Biochemistry* **2007**, *46*(26), 7765–7780.

200. Clavreul, N., Adachi, T., Pimental, D.R., Ido, Y., Schoneich, C., Cohen, R.A. S-glutathiolation by peroxynitrite of p21ras at cysteine-118 mediates its direct activation and downstream signaling in endothelial cells. *FASEB J.* **2006**, *20*(3), 518–520.

201. Adachi, T., Pimentel, D.R., Heibeck, T., Hou, X., Lee, Y.J., Jiang, B., Ido, Y., Cohen, R.A. S-glutathiolation of Ras mediates redox-sensitive signaling by angiotensin II in vascular smooth muscle cells. *J. Biol. Chem.* **2004**, *279*(28), 29857–29862.

4

DNA OXIDATION

Dessalegn B. Nemera, Amy R. Jones, and Edward J. Merino

OVERVIEW

Reactive oxygen species (ROS) and the reactions that utilize it are essential for cellular growth and signaling.[1] However, excess oxidative equivalents lead to disruption of cellular function via deleterious modification of DNA, proteins, and other molecules.[2,3] Nucleic acid oxidation has existed in the literature ever since the early part of the twentieth century when it was shown that virus particles (mainly composed of oligoribonucleotides) could be inactivated via photooxidation.[4] Among the biomolecules, the oxidation of DNA plays a central role in oxidation-induced stress. Damage to DNA, which includes formation of DNA lesions, without repair, can lead directly to the inhibition of DNA replication.[5,6] Without the ability to replicate DNA, a cell will enter a variety of cell death pathways.[7] An additional and more problematic alternative is that many DNA oxidative lesions will induce mutagenesis.[8] Some of these mutations can lead to the genetic alterations associated with cancers and several other disease states.[9–13] It is important to note that cells possess several DNA repair pathways that combat the damage from oxidative stress.[14]

This chapter will first discuss DNA oxidation within the context of the cell. Then, several key oxidative adducts will be examined. The assessment of specific oxidative lesions will cover topics including their formation, repair, and their association with disease, as well as biologically relevant oxidants. This chapter will give a broad overview of these two topics so that a reader can gain an understanding of the role of DNA in oxidative stress. As the reader obtains this knowledge, each topic can then be explored in greater specific detail from the more focused reviews we will later discuss.

4.1 THE CONTEXT OF CELLULAR DNA OXIDATION

DNA is a reactive molecule that can be modified via various mechanisms, including alkylation, spontaneous degradation, reduction, and oxidation.[15,16] There are two ways to think about DNA oxidation chemistry. One view divides lesions based on the particular structure it forms. For example, one could determine the structure of a C8-guanine oxidation product, such as 8-oxo-7,8-dihydro-2′-deoxyguanosine, and in turn elucidate its impact. In this view, the effect of a specific lesion is analyzed.[8] An alternative view is to take a given oxidant or toxin and determine the spectrum of lesions formed. The variety of lesions reveals the impact of the oxidant or toxin.[17] Although both types are valid, this chapter uses the former.

DNA is a relatively poor substrate for oxidation in the context of an entire cell. As an example, the oxidation potential required to oxidize 2′-deoxyguanosine to a guanine radical is 1.3 V.[18,19] It has been shown that some of the guanine radical intermediates in the oxidative reactions can be quite stable, having a lifetime of up to several seconds.[20] The guanine radical undergoes several divergent mechanistic pathways to produce the variety of lesions discussed in Section 4.2.[21] The remaining nucleotides are much harder to oxidize with

Molecular Basis of Oxidative Stress: Chemistry, Mechanisms, and Disease Pathogenesis, First Edition. Edited by Frederick A. Villamena.
© 2013 John Wiley & Sons, Inc. Published 2013 by John Wiley & Sons, Inc.

potentials of 1.4, 1.6, and 1.7 V for 2′-deoxyadenosine, 2′-deoxycytidine, and 2′-deoxythymidine.[19] Each of these adducts has its own unique chemistry. Proteins and amino acids are easier to oxidize. For instance, the one-electron oxidation of cysteine can be accomplished at as low as 0.7 V.[22]

The vast difference in potential between oxidizing DNA and other biomolecules is further exacerbated by two key cellular realities. First, proteins comprise more than 50% of the nonwater biomaterial in a cell. In comparison, DNA comprises a relatively minor portion (approximately 10%) of the cell's biomaterial.[23] This means that once a reactive species is generated, it is less likely to react with DNA than a protein. Second, DNA is sequestered and protected. DNA in a cell is located within the nucleus, and a small portion, ~1% of its total weight, is located in the mitochondrion. Transport into the nucleus is actively controlled and some small, charged, reactive molecules have problems passing through the membrane.[24] In contrast, mitochondrial DNA is located near the electron transport chain, a major producer of endogenous ROS. An additional defensive strategy is used in this case. Mitochondrial DNA is packaged into nucleoid particles by histone-like mitochondrial transcription factor A.[25] Interestingly, these mitochondrial nucleoid particles also contain identifiable antioxidant enzymes, such as superoxide dismutase and glutathione peroxidases. These antioxidant enzymes likely serve to quench free radicals and oxidants that approach the DNA particle.[26] Nuclear DNA is highly packaged by histone proteins. A reactive species that succeeds in making it to the nuclear DNA would likely collide and react with these proteins that contain easy to oxidize amino acids. It has recently been shown that histones are posttranscriptionally modified, in order to stop gene expression, in the presence of oxidized DNA.[27] These complex strategies demonstrate the lengths a cell will undergo to protect DNA and also underscore the importance of limiting DNA oxidation.

Given the difficulty of oxidizing DNA and the many defense mechanisms involved, one may ask, "Why does DNA oxidation matter?" The answer is simple: any modification that bypasses all of these fail-safes has the potential to be sustained in future cellular divisions or to directly induce cytotoxicity.[28] The importance of DNA oxidation is not in its abundance but rather in its permanence.

4.2 OXIDATION OF OLIGONUCLEOTIDES

Oxidation of oligonucleotides has been used to determine the effect of various oxidants on genomic DNA. Since the advent of primer extension via polymerases, DNA has been incubated with oxidants and reactive species to investigate any damage that may happen.[29] The goal of these experiments is to visualize what and where nucleotide oxidative modification is occurring. Many DNA lesions do not arrest replication, and therefore, DNA is treated with an agent that induces a strand break at the site of damage to facilitate visualization. Induction of strand breaks is accomplished by alkaline or glycosylase treatment.[30] Because many oxidation products are not substrates for glycosylases, while other lesions are not broken by alkaline treatment, a recent study described the employment of an optimized mixture of the two.[31] A technique termed ligation-mediated polymerase chain reaction (LM-PCR), which allows a precise sequence mapping of DNA lesions, is employed to elucidate damage in genomic DNA that is in low copy number.[32] Mitochondrial DNA can be directly visualized using primer extension.[33] An alternative approach is to hydrolyze the large, cellular oligonucleotides into nucleosides.[34] There are several enzymes that can be used to accomplish this hydrolysis, the most common being phosphodiesterases. The nucleotides are then treated with a phosphatase to generate an uncharged nucleoside that is amenable to high-performance liquid chromatography (HPLC) separation and identification by either mass spectrometry (MS) or electrochemistry.[35,36] For MS analysis, an isotopic standard is used to give an absolute concentration of the oxidized nucleoside. The sequence dependence of any damage is lost. Thus, the total endogenous damage level is obtained. These two experiments, which give alternative but complementary results, have produced much data which have shown the specific sequences damaged on genomic and mitochondrial DNA.[37,38]

Since guanine has the lowest oxidation potential, induction of oxidative stress has yielded selective modification of guanine residues.[33,39,40] Guanine-specific oxidation of DNA has been observed with primer extension, using methylene blue photooxidation,[41] riboflavin photooxidation,[42] and Fenton chemistry,[30] to name a few systems. Of note is the fact that guanine-specific oxidation from toxic metals, like chromium and arsenic, are observed.[43,44] In fact, metal-based oxidative damage is more DNA sequence specific than metal specific.[45]

Guanine oxidation and its products are generally differentiated based on the mechanism of oxidation induced by different ROS. Recently, a means to conceptually rationalize guanine oxidation has been published.[46] The mechanism of oxidation of a few ROS will be briefly discussed. Hydroxyl radical, the most potent of the reactive oxygen forms, adds to double bonds at C8 or C5.[47] This leads to the formation of two types of radical species, with each having divergent properties and half-lives. The addition of hydroxyl radical at C5 causes the loss of a net water equivalent and forms a guanine radical/radical cation. This guanine-based radical is prone to radical–radical combination. Hydroxyl radical addition at C8 can lead to 8-oxo-7,8-dihydro-2′-deoxyguanosine (8O-dG) via a series of steps. Singlet oxygen adds via cycloaddition.[48] The cycloaddition generates an endoperoxide (i.e., the two oxygen atoms are covalently bound to guanine) that rearranges into a C8-hydroperoxide (only one oxygen atom is connected to guanine at C8). The second mechanism of guanine oxidation is electron abstraction from the highest occupied orbital. Many exogenous oxidants utilize this mechanism of electron abstraction to form guanine radical–radical cations.[49] Once an electron is abstracted, the guanine radical can react with superoxide or other radicals in ultrarapid radical–radical combination reactions or lose a second electron and undergo attack by nucleophiles.[50,51] These observations are validated by experiments showing that superoxide dismutase coincubation with transiently generated guanine radicals leads to an increased radical lifetime.[50]

Many of the guanine oxidation mechanisms begin to coalesce at a later stage. The common intermediates form either –OH or –OOH addition products at either C5 or C8.[46] When –OH is added to C8, the molecule tautomerizes into 8-oxo-7,8-dihydro-2′-deoxyguanosine. The C8 hydroperoxide is reduced to remove the terminal oxygen of the hydroperoxyl group.[48] The reduction forms 8-oxo-7,8-dihydro-2′-deoxyguanosine. It should be noted that the production of other lesions is possible by reduction. For example, after the addition of –OH to the C8 ring, an opening at the N9-C8 bond followed by reduction forms formamidopyrimidine derivatives that contain a C8-aldehyde.[6] The case for the remaining –OH and –OOH addition at C5 is much more complex. These intermediates eventually form imidazolone, spiroiminodihydantoin, and guanidinohydantoin based on specific conditions and the type of DNA being examined.[52–54] Spiroiminodihydantoin and guanidinohydantoin exist in stereoisomers and various possible tautomers.[55] Many of these derivatives further react to form a myriad of products. The key C5 and C8 intermediates in guanine oxidation are also precursors to other oxidation products. Replacement of the –OH in the key C8 and C5 guanine intermediates with the amine group of a protein's lysine leads to the formation of a crosslink.[56] This example of a cross-link can be isolated from several model systems.[57,58] Replacement of ROS with reactive nitrogen species leads to the formation of guanine-nitrogen derivatives like 8-nitroguanine.[59] At present, it is unclear which of the above adducts are occurring *in vivo* and in what relative yield. Of these lesions, 8-oxo-7,8-dihydro-2′-deoxyguanosine and formamidopyrimidine derivatives have been observed *in vivo*.[60] Additionally, spiroiminodihydantoin has been identified in bacteria lacking DNA repair that has been treated with chromium-based oxidants.[61]

Though it had been previously proposed by several authors such as Simon and Van Vunakis (1962) that guanine oxidation is critical to disease, it was not until 1986 when it was discovered that 8-oxo-7,8-dihydro-2′-deoxyguanosine is both formed and repaired in gamma-irradiated mice.[62] In the early 1990s, many groups set out and are still working on detection of 8-oxo-7,8-dihydro-2′-deoxyguanosine as a diagnostic marker of disease states.[63,64] This field of study has uncovered an important aspect of guanine oxidation: many of the oxidation products are easier to oxidize than guanine. For example, the oxidation potential of 8-oxo-7,8-dihydro-2′-deoxyguanosine is 0.5 V more favorable compared to deoxyguanosine (dG), corresponding to a ΔG of -13 kcals/mol.[65] In fact, bulk electrochemical oxidation experiments using dG concurrently lead to oxidation of 8-oxo-7,8-dihydro-2′-deoxyguanosine to form hyperoxidation products.[66] Prior critical work by Burrows showed that the products formed via several oxidation mechanisms of 8O-dG include guanidinohydantoin and spiroiminodihydantoin bases.[67] The oxidation of 8-oxo-7,8-dihydro-2′-deoxyguanosine into the various products is an active area of research. One way to conceptualize this oxidation of 8O-dG is similar to that proposed for guanine except that the C8 position is occupied, leaving only the C4–C5 double bond primed for reaction with oxidants.[46] These lesions have been found to occur with several oxidants, including the toxic metal arsenic, chromium, copper, iron, rose bengal, and riboflavin.[43,68]

Genome-wide studies on the formation of oxidative lesions show that G-content is a strong predictor of overall oxidative DNA damage.[69] It is also well known that the sequence context of a guanine modulates its oxidation propensity. The ability of sequences flanking guanine to affect its oxidation has been examined (i.e., 5′-NGN).[70] The easiest to oxidize is 5′-GGG. The order of the remaining sequences is 5′-CGG, 5′-AGG, 5′-GGA, 5′-TGG, 5′-GGT, 5′-GGC, 5′-CGA, 5′-AGA, 5′-TGA, 5′-CGT, 5′-AGT, 5′-CGC, 5′-TGT, 5′-AGC, 5′-TGC. In this series, all the sequences containing GG are easier to oxidize than those that do not. Experiments have shown that the 5′-guanine of the sequence GG is the most electron-donating site.[71] Thus, exposure of an oxidant generally leads to preferential damage at this sequence. This pattern of DNA oxidation has been verified not only from agents that abstract an electron from guanine but also Fenton reagents that produce hydroxyl radical.[42] These experiments used an innovative method to deconvolute hydroxyl radical reactions that occur with the ribose (see later discussion) and those that damage at guanine. Oxidized DNA was pretreated with the enzyme ExoII to remove ribose-based strand break products. Experiments on cellular DNA also show this trend in guanine oxidation.[30] Human cells treated with an oxidant led to damage at guanine repeats. One report, however, has shown the opposite tendency in guanine mutation, indicating that damage on genomic DNA may be more complex than previously thought.[72]

The damage at guanine repeats is caused by a phenomenon known as DNA-mediated charge transport, which occurs because the bases on DNA are stacked like an array of coins.[73,74] Holes, guanine radicals formed via electron abstraction on the DNA, are produced via oxidative events. These holes migrate extremely rapidly to other guanines in the oligonucleotide. This is compared to the slower quenching of these radicals that leads to the formation of oxidized lesions.[75] In this same manner, adjacent DNA bases can stabilize hole formation, giving rise to the aforementioned pattern of damage. Therefore, hole migration equilibrates with greater occupancy at the most stable sites, causing more oxidative lesions at these DNA sequences, making them prone to damage. The phenomenon is termed funneling.[76] It has been hypothesized that this funneling is a means through which the cell preserves DNA. It has been shown, via computational methods, that the DNA sequences 5′-GG and 5′-GGG are localized to non-protein-coding sequences in DNA.[77] In fact, these computational methods show that non-protein-coding regions are 50-fold more likely to be oxidized than DNA in coding regions within the human genome. It is theorized that when a hole is made, the hole migrates out of coding regions to preserve the integrity of these sequences. Experiments support this hypothesis by demonstrating that charge transport is able to occur over 34 nanometers.[78]

Many biologically important DNA sequences also contain oxidation prone regions. The ends of noncircular genomic DNA contain repeats, known as telomeres, that differentiate it from damaged DNA.[79] Once the repeats are removed, senescence follows. The telomeres are maintained by telomerase. Human telomeres possess the sequence 5′-TTAGGG and are selectively oxidized in the context of large DNAs.[80] Another important guanine-rich sequence is a critical mitochondrial replication element termed a conserved sequence block.[81] The sequence of the conserved block, which is 5′-GGGGGGGTGGGGG, is clearly oxidation prone. When functioning mitochondria are placed under oxidative stress, this sequence is selectively damaged and mutated.[38,82] These experiments also showed that the oxidation of DNA led to the cross-linking of proteins in the mitochondria. It was suggested that funneling could be used biologically as a means to ensure replication of undamaged mitochondrial DNA. Several reports show that DNA-based holes can be transferred to bound proteins. The proteins are then oxidized at thiols or iron–sulfur clusters.[83,84] It is proposed that transfer of holes between two bound repair proteins through the DNA is used as a means to determine if a section of genome is damaged DNA. This is because the traveling holes cannot pass damaged DNA.

The oxidation of DNA is more complex than just selective oxidation of guanine and 8-oxo-7,8-dihydro-2′-deoxyguanosine. Though guanine is the easiest base to oxidize, other bases are also oxidized in the presence of biological oxidants. For example, highly reactive oxidizing agents like the hydroxyl radical, E_0 ~1.9 V,[85] are capable of reacting with the DNA backbone to initiate direct strand cleavage or oxidize other DNA bases.[86] Historically, reactions of hydroxyl radical with the DNA backbone to initiate direct strand cleavage were used to elucidate site-specific protein contacts from DNA. The reaction pathway involves hydrogen atom abstraction to generate a ribose-based radical.[87] Common products from this type of process include 2′-deoxyribonolactone, base propenal, base propenoate, 3-formyl phosphate, erythorose abasic site, 3′-phosphoglycolate aldehyde, 2-deoxypentos-4-ulose abasic site, and nucleoside-5′-aldehyde.[88,89] Alternatively, a net hydroxyl addition can occur. At the 4′-carbon position, the additional hydroxyl leads to a hemiacetal that epimerizes to change the stereo configuration of the ribose.[90] Oxyradicals can also react with the nucleobase and form lesions that lead to strand cleavage.[91] Backbone reactivity has limited sequence specificity, but it is a common product. Another biologically important oxidative process is gamma irradiation. Gamma irradiation can lead to pyrimidine dimers, though these reactions tend to occur via cycloaddition.[92] Adenine oxidation is observed upon gamma irradiation.[93] The major adenine oxidation products observed are 8-oxo-7,8-dihydro-2′-deoxyadenosine and 2-hydroxy-2′-deoxyadenosine.[94]

4.3 EXAMINATION OF SPECIFIC OXIDATIVE LESIONS

4.3.1 8-Oxo-7,8-Dihydro-2′-Deoxyguanosine

For decades, considerable effort has been devoted to study the formation and properties of guanine oxidation products (Fig. 4.1). This section will focus specifically on 8-oxo-7,8-dihydro-2′-deoxyguanosine, or 8O-dG, one of the most well-studied DNA lesions. The structure of each specific lesion in this chapter section is listed in Figure 4.2. The discovery of the lesion, mechanism of its formation, repair potential, and the associations of this lesion with a variety of pathological events will be discussed. Lastly, this review will present a brief example

EXAMINATION OF SPECIFIC OXIDATIVE LESIONS 97

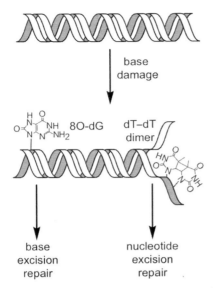

Figure 4.1 Repair of DNA base damage. Reactive oxygen species can modify the easy-to-oxidize DNA nucleobases. The most commonly oxidized nucleobase is guanine. Reactive oxygen-induced nucleobase damage can be repaired by two pathways. First, many lesions are specifically corrected by the base excision repair pathway. An example lesion is 1 8-Oxo-7,8-Dihydro-2′-deoxyguanosine. Some lesions are helix distorting, like a cyclobutane dT–dT dimer. These types of lesions are corrected by the nucleotide excision repair pathway.

of several pathogenic and disease states related to such lesions in the recent literature. Since the discovery of 8O-dG, numerous papers in the scientific literature have examined this lesion. A Pubmed search for 8-oxo-7,8-dihydro-2′-deoxyguanosine, as well as its common, however, incorrect names, 8-hydroxydeoxyguanosine and 8-oxo-2′-deoxyguanosine, yields a total of 4746 results.

The formation of 8O-dG in DNA was first reported at the Cancer Research Institute of Tokyo in 1984.[95] Kasai and Nishimura discovered 8O-dG in DNA while isolating new mutagens present in broiled foods.[96] The 8O-dG lesion was first detected by HPLC. In 1986, Floyd et al. showed a sensitive analytical detection method for 8O-dG in cellular DNA by electrochemical detection following HPLC.[97] However, the analysis of 8O-dG as a major and ubiquitous oxidation product of DNA in the urine of experimental animals was performed in 1989.[98] This was done by examining urine samples of mice for 8O-dG. The adduct was separated by solid phase extraction for concentration.[98] Next, the separated 8O-dG was purified and analyzed by gradient reverse phase HPLC combined with an electrochemical detection system.[98] Using this technique, the urine

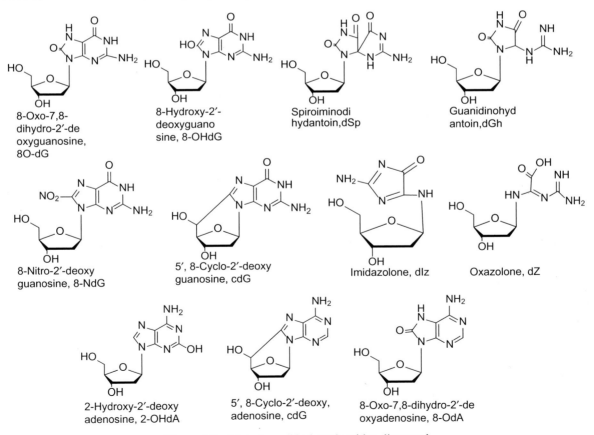

Figure 4.2 List of modified nucleosides discussed.

sample obtained from a mouse provides *in vivo* measurements of global oxidative damage to DNA. Later, several teams used different analytical techniques to quantify 8O-dG in response to ionizing radiation, pathogenic states, and other stressors. 8O-dG has been detected and analyzed with high sensitivity by gas-chromatography–mass spectrometry (GC-MS),[99,100] liquid chromatography/tandem mass spectrometry (LC/MS),[101,102] and enzyme-linked immunosorbent assay (ELISA).[103,104] Additionally, several indirect measurements to quantify 8O-dG are possible. For example, single cell gel electrophoresis of DNA, treated with glycosylases that recognize specific oxidized lesions, is possible.[105] In addition to these analytical methods, novel, sensitive, reliable, and low-cost analytical methods are being developed for the first time.[106] One new method uses capillary electrophoresis with amperometric detection to quantify the level of 8O-dG in urine.[107] These methods are currently important for the determination and analysis of 8O-dG, both in tissues and other biological fluids such as serum.[108,109] 8O-dG is a critical biomarker of aging and diseases.[110,111] The accurate determination of 8O-dG in these cases is limited by problematic 8O-dG chemistry, with each method of detection yielding divergent concentrations, and, in some methods, high deviations causing poor precision.[34] Many analytical chemists are currently seeking to reach a consensus.[112] The basal 8O-dG level is ~1–10 modified DNA base per million base pairs.[35] This level of adduct corresponds to several thousand 8O-dG molecules per cell at any given time.

The chemistry of 8O-dG is such that it is easy to oxidize, moderately stable, and makes thermodynamically favorable interactions with bases in a DNA duplex.[113] The oxidation potential of 8O-dG is 0.5 V more favorable than that of guanine base.[65,74] 8O-dG is observed to increase in concentration by Fenton-type reagents, cigarette smoke, tar, asbestos, gamma rays, and under many other conditions.[114–117] For example, reaction of hydroxyl radical with guanine leads to addition at C8 to generate a C4 radical.[47] This radical can then lose an electron and a proton to form 8O-dG, or it can rearrange to form a derivative known as formamidopyrimidine. An alternative pathway is the addition of superoxide, a radical, to an already formed guanine radical or guanine radical cation.[46] In aerobic cells, superoxide is continuously formed as a part of oxidative phosphorylation.[118] Because 8O-dG formation occurs in various oxidative conditions, it is used as a general marker of oxidative stress.[119] In the 1990s, the analysis of 8O-dG was advanced as an attractive diagnostic marker for several disease states in multiple cells. However, the researchers ran into a significant roadblock: quantification of 8O-dG was difficult and often unreliable.[120,121] The basis for this roadblock is rooted in the chemistry of 8O-dG.

Several research teams have determined the formation of 8O-dG under various conditions.[114] Although 8O-dG is a stable lesion, it has been found that its formation is highly influenced by pH. Studies by Tsou et al. focused on 8O-dG formation in calf thymus DNA treated with a chromium-based oxidant. Their findings indicated that the highest level of 8O-dG was observed at pH 7–8, and development of the lesion was unfavorable at pH <6.[122] An additional issue to consider with 8O-dG is its ease of oxidation.[67] Artifactual oxidation of 8O-dG was examined by sending authentic samples to several labs.[123] Results indicated problems with quantification that have since been reversed by the addition of antioxidants to solutions used to collect DNA. It has been established that 8O-dG is easy to oxidize with a potential of ~0.8 V. It should be noted that this potential is slightly more favorable than the oxidation of tyrosine and close to the oxidation of a –SH to a disulfide.[22] The ease of oxidation becomes important when DNA is collected. The first step is lysis, which releases several oxidants, like iron (II) and copper (II) Fenton reagents. Recall that a few thousand 8O-dG lesions are in a cell at any given time, but the amount of iron (II) and copper (II) is much greater on a molar basis. Release of these oxidants can result in an overestimation of 8O-dG levels due to guanine oxidation from these agents during lysis. Alternatively, an underestimation due to the loss of 8O-dG by further oxidation may be possible. Oxidation of 8O-dG leads to the formation of a variety of lesions, including guanidinohydantoin, spiroiminodihydantoin, oxazolone, and several others.[54,67,124]

Once formed, 8O-dG is a strong mutagen. 8O-dG can form Hoogsteen-type base pairing with adenine.[125] Base pairing is accomplished through the syn ribose conformation that allows the protonated N7 and O6 to alternatively hydrogen bond to adenine.[126] Miscoding results in a G:C→T:A transversion mutation.[127] The potential deleterious effects of this lesion are curtailed by base excision repair and mismatch repair of the transversion mutation.[128] In general, base excision repair is characterized by the excision of a base residue via hydrolysis of the *N*-glycosyl bond. This function is termed glycosylase activity. Some members of the glycosylase family also contain lyase activity which nicks the DNA backbone to give a strand break that is almost ready for DNA replication. The nicked DNA then requires the removal of 5′-terminal deoxyribose-phosphate residue, repair of the DNA strand by a polymerase, and ligation. In humans, there are several of these glycosylases, including 8-oxo-guanine glycosylase, hNEIL1-3, and several families dealing with DNA methylation or pyrimidine damage.[129] The reader should note that the names for glycosylases are different depending on the specific organism. 8O-dG is specifically recognized by 8-oxoguanine glycosylases termed hOOG in humans. The base excision repair system has overlapping substrate specificity and redundancy with hNEIL1 showing anti-8O-dG activity.[130] Recently, though controversial, it was reported that the loss of 8-oxoguanine glycosylases function correlates with aggressive forms of breast cancer.[131] Castaing et al. and Michael et al. showed that bacterial 8-Oxoguanine glycosylases remove 8O-dG from 8O-dG:X containing duplex with the following relative preference: X = C > T > G >> A.[132,133] Glycosylases must flip the nucleotide out of the DNA stack into the active site of the protein. This would suggest thermodynamic stability and a role in recognition. The conformation, the melting behavior, and the thermal stability of DNA duplexes containing 8O-dG have been determined by means of spectroscopic and calorimetric methods.[134] It was found that a single 8O-dG does not alter the global DNA duplex conformation, and the thermal properties depend rather on the base opposite the lesion. For example, an oligonucleotide that had a single 8O-dG:C base pair is 3.2°C less thermally stable than a duplex with a G:C complex. When dA is opposite 8O-dG, the thermal stability actually increases. Therefore, the presence of 8O-dG within the base stack has only a minor thermodynamic effect in DNA structure.

Kinetic changes to a DNA duplex are caused by 8O-dG. Recent NMR analysis has begun to further elucidate the dynamic nature of 8O-dG in oligonucleotides.[135] Double-stranded DNA with a single 8O-dG was compared to that of the perfect duplex. Temperature-dependent NMRs show that despite the largely similar global thermodynamic details, there are profound local differences between the two duplexes. This study revealed that 8O-dG changes the local hydrophilicity and the ability of the major groove to bind cations. These differences are proposed to be essential identification and repair of these lesions since an 8O-dG reduces the penalty for the lesion to be extrahelical.

There are two other lines of defense against 8O-dG. The human protein MTH1 prevents incorporation of 8O-dG-5′-triphosphate into DNA by hydrolyzing the nucleotide to 8O-dG-5′-monophosphate.[136,137] The hydrolysis removes the damaged nucleotide from the nucleotide pool so that it cannot be incorporated into DNA by polymerases. Haghdoost et al. added interfering RNAs that target and remove the *MutY* mRNA transcript. This study revealed that the nucleotide pool is a significant mode of incorporation for 8O-dG in DNA. The DNA glycosylase MUTY provides a second line of defense by removing the inappropriate adenine from 8-OxoG:A base pair.[129]

Because of 8O-dG's central role in DNA oxidation, it is strongly correlated with several pathogenic and diseased states. Cell genomes are under continuous assault by agents that increase the concentration of 8O-dG.[138] 8O-dG has been found to increase in concentration in a variety of diseases, including several types of cancers[114] and other chronic diseases such as diabetes,[139] heart disease,[140] and neurodegenerative diseases.[141] One consequence of elevated reactive oxygen and increase in 8O-dG level is the initiation, promotion, and malignant conversion stages of carcinogenesis.[142] Several studies have examined elevated levels of 8O-dG in various types of cancer such as leukemia,[143] lung,[144] and breast[145] cancer. Yang et al. demonstrated that the pretherapy levels of urinary 8O-dG were found to be higher in patients with lymphoma, acute leukemia, and myelodysplastic syndrome than in normal controls.[146] Recent studies also highlight the correlation between type 2 diabetic patients and 8O-dG. Diabetic patients have significantly higher concentration of 8O-dG in their urine,[147] serum,[139] blood cells,[148] pancreas,[149] muscles,[150] and kidneys.[151] Al-Aubaidy et al. also confirmed that serum 8O-dG levels were significantly higher in both obese and diabetic patients compared with controls.[146] Similar studies of patients with heart failure show an analogous correlation compared to control subjects.[152] The presence of 8O-dG lesion plays a critical role in the pathogenesis of neurodegenerative disorders, including Alzheimer's, Parkinson's, and Huntington's disease.[153,154] Recent work has shown the formation of 8O-dG in trinucleotide repeats alters structure and may be important in Huntington's disease.[155]

4.3.2 Lesions on Ribose Bases Including Apurinic or Apyrimidinic Sites

Although multiple lesions taking place on bases have been identified, damage at the 2′-deoxyribose (dR) ring is also a common occurrence (Fig. 4.3).[156] Additionally, several oxidative base lesions possess weakened glycosidic bonds that cause formation of apurinic/apyrimidininc (AP) sites after hydrolytic cleavage.[157] These AP sites are processed by endonucleases generating "unblocked" single-strand breaks (SSBs), that is, those which possess no modifications. Conversely, oxidation at dR can lead to fragmentation and the formation of "blocked" lesions that require further processing in order for repair to occur.[158] Production of both dR lesions and AP sites is quantifiable in human cells.[159] Quantification and detection of AP and dR oxidation products is accomplished by many analytical techniques, including MALDI-TOF-MS and assays similar to ELISA.[160] One method for the detection of dR oxidation in DNA and cells was developed by Chan et al. in

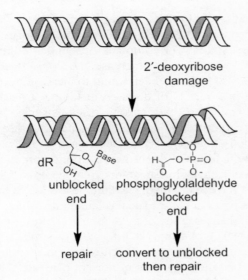

Figure 4.3 Repair of 2′-deoxyribose damage. Reactive oxygen species can abstract hydrogens from 2′-deoxyribose to form many lesions. In addition, some modified bases lose the nucleobase to form an abasic site. Modified 2′-deoxyribose adducts are grouped into two classes. First, some lesions are unblocked. These lesions can be directly repaired. Some lesions are blocked ends. An example blocked end is shown by the modified phosphoglyolaldehyde. These adducts require conversion of the DNA strand to an unblocked site by endonucleases. Once an unblocked site is generated, the DNA strand can be repaired.

which a chemically sensitive isotope-dilution GC-MS was utilized.[161] Once an AP residue has formed, the 1′-position has the same oxidation state as an aldehyde. Therefore, the AP site is capable of reacting with amines and hydrazines to form Schiff base products.[162] Recent studies by Greenberg et al. and Atamna et al. show that biotin containing probes and aldehyde reactive probes, respectively, are capable of detecting abasic sites.[163] In this review, the chemistry and detection of dR H-abstraction is explored. An overview of the blocked repair is also described.

Oxidation of dR occurs through hydrogen atom abstraction.[161,164] The abstraction forms a sugar-based radical, and each has a unique degradation product with either an aldehyde, ketone, or sugar-ring fragmentation.[88,89] The seven dR hydrogen atoms are reactive toward oxidants, each having different probabilities of being abstracted from duplex DNA.[89] The differential abstraction propensity is governed by both proximity to oxygen atoms and the accessibility of each hydrogen.[89] Computational models from Aydogan et al. show that highly reactive hydroxyl radicals abstract hydrogens based on their solvent accessibility.[165] The preference for hydrogen abstractions is as follows in descending order: 5′-hydrogen, 4′-hydrogen, 2′-hydrogen 3′-hydrogen, and 1′-hydrogen with the exception of 2′-hydrogen and 3′-hydrogen being equal.[166] The 1′-hydrogen is buried in the minor groove of DNA and, therefore, is solvent inaccessible. This position can be oxidized by oxidants that override the normal H-abstraction preference. For example, oxidation by phenanthroline copper (I) complexes is shown to produce a 2′-deoxyribonolactone product that corresponds to the difficult C1′- hydrogen abstraction.[88] This leads to an unfavorable hydrogen abstraction.[167] The C2′-position of 2-deoxyribose leads to the formation of the 2-phosphoglycolaldehyde residue by an indirect oxidation mechanism involving an erythrose intermediate. However, the 3′-hydrogen abstraction leads to the formation of a phosphoglycolaldehye residue directly from the oxidation.[168] Thus, the DNA strand is cleaved, resulting in an SSB.[89] These lesions are known to form photooxidants, including rhodium (III) complexes.[169] The high solvent accessibility and low bond dissociation energy make the 4′-hydrogen a major target for abstraction.[170] Additionally, the anticancer natural product, bleomycin, exerts its cytotoxic activity by means of 4′-hydrogen atom abstraction.[171] The chemistry of 4′-oxidation can follow several pathways, depending on the experimental conditions. For example, Levin et al. showed that treating cells with bleomycin leads to the formation of a 2′-deoxy-pentos-4-ulose dR modification and a free base.[171] Another path accessed by chemical oxidation or ionizing radiation of dR C4 leads to strand breaks terminated by base propenal, 5′-phosphate, and 3′-phosphoglycolate moieties.[171] The two 5′-hydrogens are highly solvent accessible.[89] After abstraction, the 5′-radical reacts with purine bases to form 5′,8-cyclopurines. The products are 5′-8-cyclo-2′-deoxyadenosine and 5′-8-cyclo-2′-deoxyguanosine.[172] Notably, several important enediyne antibiotics, such as calicheamicin and neocarzinostatin, form diradicals that are capable of abstracting a 5′-hydrogen.[173] The abstraction generates highly cytotoxic strand breaks. A common means to examine single dR oxidation products is to synthesize nucleosides and oligonucleosides with photooxidation or oxidation-sensitive groups that lead to selective formation of the dR radical. This is due to the mixture of dR radicals formed by common oxidants. For example, a recent study examines the formation of a 3′-dR radical via photodegradation of a 3′-phenyl selenide.[174] Hydrogen abstractions can also form complex lesions such as tandem damage sites or cross-links (reviewed in Section 4.3.3). Two dR hydrogen abstraction events occurring nearby, however, on opposite strands, lead to double-strand breaks (DSBs).[175] Such DNA DSBs are considered the most cytotoxic and deleterious type of DNA lesions produced in human cells.[176]

In addition, dR-based radicals can react with the same or opposite strand to generate an oxidative DNA–DNA cross-link.[177]

Lesions of dR are repaired effectively through a variety of DNA repair pathways. Many of the SSBs discussed previously leave unnatural fragments and are considered blocked ends. The blocked SSBs, if not repaired, can be converted into highly cytotoxic double-stranded break by DNA replication. The blocked ends must be processed into an unblocked end. The damaged DNA strand is cleaved by specific endonucleases which generate an unblocked end with a gap in the DNA helix, ready for polymerization.[178] One pathway, nucleotide excision repair, is responsible for the clearance of a large variety of DNA lesions that are bulky, helix-distorting, or blocking RNA transcription, which includes dR modifications.[179] The mechanism of nucleotide excision repair begins with the recognition of a DNA lesion by XPC complexes. Next, the excision of a lesion is accomplished by XPG and XPF associated with ERCC1. The resulting gap is then filled, and the nick is sealed. Several products, including cycloaddition adducts, large 3′-blocks, and cross-links, are considered bulky lesions. Thus, nucleotide excision repair is vital to the processing of these lesions.[180] Results from the Fisher group indicate that the function of the XPF–ERCC1 complex in the repair of oxidative damage is to trim a 3′-blocked terminus. Such activity leads to the generation of a 3′-hydroxyl end, allowing for completion of SSB repair or DSB repair.[180] Additionally, it has recently been shown that, as a last resort, translesion polymerases can add nucleotides past AP sites.[181] There are specialized translesion polymerases that replace arrested replicative polymerases at the lesion to allow bypass. The purpose of this polymerase is to seal gaps containing lesions and process closely spaced lesions on opposite DNA strands.[181]

Several methods and procedures have been developed to quantify major products of abasic and dR oxidation. Dedon et al. developed a technique to quantify the 1′-hydrogen abstraction product, 2-deoxyribonolactone, and the 5′-hydrogen abstraction product, nucleoside 5′-aldehyde, by GC-MS with isotopomeric internal standards.[161] Thus, both the least and the most common abstraction positions can be probed under biological conditions. As a positive control for 5′-abstraction products, the enediyne calicheamicin was used, producing nucleoside-5′-aldehydes at a rate of ~564 lesions per 10^6 μM nucleotides. The 2-deoxyribonolactone adduct occurred at only 10 lesions per 10^6 μM nucleotides. Methods for measuring AP sites in cells include aldehyde-reactive probes, accelerator MS, ^{32}P-postlabeling assays, and plasmid nicking assays.[164,182,183] AP sites can be measured using ^{14}C-labeled methoxyamine.[184] Other dR lesions like 2-phosphoryl-1,4-dioxobutane are benzylhydroxylamine reactive. Robert et al. developed a method based on MS detection of abasic sites that have been prelabeled with O-4-nitrobenzylhydroxylamine, followed by isolation and detection with HPLC-ESI-MS-MS.[185] Collins et al. reported a method for the quantification of a 3′-phosphoglycoaldehyde in cells with pentafluorobenzylhydroxylamine followed by GC-MS.[164] Consequently, many of the known dR lesions can be reliably quantified in cells. Burrows et al. recently quantified the number of lesions that occurred on oligonucleotides and from nucleosides that arise from different oxidation systems. The number and type of damage was attenuated by the type of oxidation system.[186]

4.3.3 Novel Types of Ribose and Guanine Oxidative Lesions and Future Outlook

4.3.3.1 Tandem Lesions ROS can induce the formation of not only single-nucleobase lesions, but also lesions in close proximity. There are two types of close proximity lesions, tandem and clustered lesions. Tandem lesions are two, adjacent, oxidized DNA residues.[187] Clustered lesions are formed when two or more damaged bases are positioned within 20 base pairs of each other.[188] Importantly, an early study by Sutherland et al. demonstrated that, upon ionizing radiation, up to 20% of DNA damage was in the form of DSBs.[189] These data pose strong evidence in support of clustered and tandem lesions (Fig. 4.4). This is due to the fact that the large size of genomic DNA makes it statistically unlikely that lesions would form within 20 base pairs of each

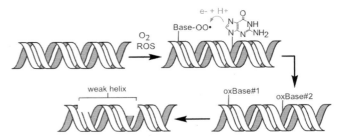

Figure 4.4 Tandem and clustered lesions. Oxidation of DNA at base #1 in the presence of oxygen forms a DNA-bound radical capable of abstracting a nearby electron at base #2. This abstraction leads to an endoperoxide at base #1 and a guanine radical cation at base #2. Both these lesions are converted into oxidized DNA lesions. Problematically, if repaired simultaneously or spontaneous formation of an abasic site occurs, a weak helix is formed. The weak helix can break apart, giving rise to highly cytotoxic double-strand breaks. The mechanism of tandem lesion formation is shown in Reference 194.

other. DSBs are formed during repair when a gap on each strand in close proximity is generated. Thus, the two strands can be easily melted, forming a DSB.[190] Several forms of insults produce these lesions.[175,191] The reason for high levels of clustered and tandem lesions is based on chemical mechanism. For example, formation of the C8-OOH intermediate (see Section 4.2) requires reduction. The terminal oxygen atom can either attack a nearby DNA base or abstract an electron to produce a radical at an adjacent DNA site.[191]

Both *in vitro* and *in vivo* results support that tandem and clustered lesions cause mutations.[192] These lesions are strong replication blocks and are more difficult to repair than when they are present alone.[193] Although repair by the base excision repair mechanisms is difficult, tandem lesions are efficiently recognized by nucleotide excision repair and translesion synthesis enzymes.[194] Since most tandem lesions are alkaline-labile, treatment with NaOH results in strand scission and enhances cleavage at the radical sites of DNA.[195] Quantitative measurement of the efficient formation of tandem lesions in DNA can be performed by several analytical techniques. These techniques include liquid chromatography coupled to ESI-MS-MS and MALDI-TOF-MS.[196]

4.3.3.2 Hyperoxidized Guanine

Guanine oxidation results predominantly in the formation of 8O-dG as well as a variety of other products.[46] A diverse array of 8O-dG oxidation products, including cyanuric acid, oxaluric acid, and 2,5-diaminoimidazalone, have been characterized due to the ease of 8O-dG oxidation (Fig. 4.5).[197] Importantly, each of these novel lesions, if they occur in cells, will impart a distinct mutational and cytotoxicity profile. Two lesion structures, spiroiminodihydantoin and guanidinohydantoin, have recently been under investigation. It should be noted that each of these two adducts possesses a novel stereocenter that leads to the formation of several isomers.[55] Spiroiminodihydantoin lesions are observed at neutral pH on nucleotides, while guanidinohydantoin lesions predominate at low pH and in oligonucleotides.[198–200] The structure of spiroiminodihydantoin was determined by HPLC, UV-vis spectroscopy, ESI-MS-MS, and ^{13}C NMR analysis.[201] Melting temperature analysis indicates that duplexes containing both spiroiminodihydantoin and guanidinohydantoin are destabilized relative to guanine.[199] Sugden et al. detected spiroiminodihydantoin lesions *in vivo* for the first time in bacteria lacking DNA repair. This finding indicates that these lesions are likely occurring in cells.[61] Both spiroiminodihydantoin and guanidinohydantoin lesions possess high mutation frequency.[202] In fact, this study found essentially 100% mutation formation. For example, a recent crystal structure showed that guanidinohydantoin adopts a syn con-

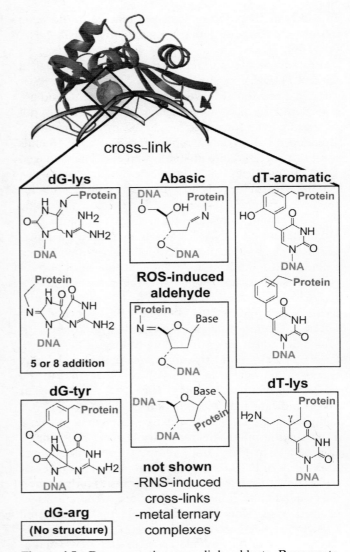

Figure 4.5 Representative cross-link adducts. Representative oxidatively induced DNA–protein adducts. Eleven possible cross-link structures are shown (single isomer).

formation that allows for dATP to be inserted opposite the oxidized base.[203] In addition, a single substitution to a phenylalanine from a tyrosine leads to misincorporation of guanidinohydantoin-ribose triphosphates by a polymerase.[204] Mutations occurring as spiroiminodihydantoin and guanidinohydantoin lesions are recognized by *Escherichia coli* base excision repair enzymes like FPG, NEI, NTH.[205,206] The eukaryotic repair enzyme NEIL1 can excise spiroiminodihydantoin and guanidinohydantoin lesions.[207] In addition, a study by Liu et al. illustrates that NEIL3 glycosylase can recognize spiroiminodihydantoin *in vitro* and *in vivo*. Current research aims to elucidate which hyperoxidized guanines occur in cells at detectable levels.

4.3.3.3 Oxidative Cross-Links Interactions between DNA and proteins are essential for proper cellular function. Cross-links are of intense interest because they are present at high levels when several repair pathways are compromised, as occurs in Fanconi anemia.[208] In addition, environmental toxins that generate ROS significantly elevate the level of cross-linking.[209] The formation of DNA–protein cross-links is possible because both nuclear and mitochondrial DNA are closely associated with proteins such as histones, single-stranded DNA-binding proteins, transcription factors, and other DNA-binding proteins. The close association means the effective concentration of protein is very high relative to water. Thus, cross-linking is an outcome when reactive oxygen is present.[210] Cross-links arrest replication and require repair due to their large size.[211] Early work supported the involvement of proteolytic degradation to yield a smaller DNA-peptide adduct.[212] More recently, nucleotide excision repair, via the UVrABC complex, has been shown to be capable of removing the smaller DNA–peptide cross-link.[213] Another mechanism, homologous recombination, has been implicated in cross-link repair since homologous recombination-deficient cells show more cross-links.[214] The bulky and unstable nature of these lesions disrupts DNA metabolism by posing significant blocks to replication and transcription.[193]

To date, 12 cross-link structures have been shown (Figure 4.5). The left panel in Figure 4.5 shows guanine-based cross-links. Guanine is the most oxidation prone base, and as such, it can participate in many types of cross-links.[215] Reactive oxygen generates 8O-dG. Subsequent oxidation causes the formation of spiroiminodihydantoin or guanidinohydantoin oxidation products after reaction with water.[216] Instead of water, a lysine can be added, at C5 or C8, to give a cross-link.[56] The left middle panel shows a cross-link observed between tyrosine and guanine.[217] Finally, the lowest left panel shows a dG–arginine cross-link has been observed with no structural information.[56] The next set of cross-links are those deriving from aldehyde additions (center panels). Upon oxidation by superoxide or hydroxyl radical, an AP site can be made. The AP site reacts with the ε-amine of lysine to form a Schiff base,[218] as shown in the top center panel. Several cross-links are formed due to cleavage of biomolecules at other positions (center middle). For example, oxidation of a dR leads to the scission of the backbone and liberation of a 5′-aldehyde[219] that can react with a protein. These cross-links are traditionally ascribed to hydroxyl radical-mediated oxidation. It should be noted that ROS can generate aldehyde-containing metabolites such as acetaldehyde, formaldehyde, and hydroxynonenals (oxidized lipids) that form cross-links (lower center middle).[220–222] The right panel shows dT-based cross-links that are hydroxyl radical specific. Dizdaroglu et al. identified a dT-aromatic cross-link in 1985.[223] Thymidine cross-links have also been observed to tryptophan, to histone, and to phenylalanine.[224–226] Mechanistically, electrons are abstracted from both the amino acid and the thymidine. The two radicals then combine to form the cross-link. Not shown are reactive nitrogen-based cross-links and metal-bound cross-links.[227–229] We were able to form a cross-link between the model protein, Ribonuclease A, and a guanine-rich DNA sequence designed to predominantly form cross-links in the presence of various oxidants, including rose bengal, riboflavin, Cu(II), and Fe(II).[57,68,230] Moreover, the group showed how the guanine oxidation product forms a cross-link with the side chain of amino acids such as lysine and tyrosine. The major forms of these lesions are detected by gel electrophoresis and MS.[231] Clearly, cross-links are a large and complex subset of DNA lesions. Their examination is ongoing.

FUTURE OUTLOOK OF DNA OXIDATIVE LESIONS

In the 100 years since DNA oxidation has been known, much progress has been made. Today, most lesions have known structures, biochemistry, and cellular effects. In this last section, a flavor for the direction of the reactive oxygen DNA damage field is illustrated. Though this is only a small sampling of research that goes on, some important observations can be made. Importantly, research is moving into the cell to determine which lesions are detectable and most important as far as reactive oxygen. The reader is reminded that the cell is a complex environment with many compounds that can alter the mechanism of lesion formation. Inside a cell, the concentrations of DNA and protein are extremely high, while the cell is not a dilute aqueous solution with freely diffusing molecules. In this complex cellular environment, tandem lesions seem to be strongly favored, and proteins may be able to participate in cross-link formation. Highly accurate, precise, and technically simple detection of DNA lesion formation in a cells may pave the way for new diagnostic markers for many disease states.[95]

REFERENCES

1. D'Autreaux, B., Toledano, M.B. ROS as signalling molecules: Mechanisms that generate specificity in ROS homeostasis. *Nat. Rev. Mol. Cell Biol.* **2007**, *8*, 813–824.

2. Apel, K., Hirt, H. Reactive oxygen species: Metabolism, oxidative stress, and signal transduction. *Annu. Rev. Plant Biol.* **2004**, *55*, 373–399.

3. Valko, M., Leibfritz, D., Moncol, J., Cronin, M.T.D., Mazur, M., Telser, J. Free radicals and antioxidants in normal physiological functions and human disease. *Int. J. Biochem. Cell Biol.* **2007**, *39*, 44–84.

4. Simon, M.I., Van Vunakis, H. The photodynamic reaction of methylene blue with deoxyribonucleic acid. *J. Mol. Biol.* **1962**, *4*, 488–499.

5. Neeley, W.L., Essigmann, J.M. Mechanisms of formation, genotoxicity, and mutation of guanine oxidation products. *Chem. Res. Toxicol.* **2006**, *19*, 491–505.

6. Burrows, C.J., Muller, J.G. Oxidative nucleobase modifications leading to strand scission. *Chem. Rev.* **1998**, *98*, 1109–1151.

7. Martindale, J.L., Holbrook, N.J. Cellular response to oxidative stress: Signaling for suicide and survival. *J. Cell. Physiol.* **2002**, *192*, 1–15.

8. Wang, D., Kreutzer, D.A., Essigmann, J.M. Mutagenicity and repair of oxidative DNA damage: Insights from studies using defined lesions. *Mutat. Res.* **1998**, *400*, 99–115.

9. Pleasance, E.D., Cheetham, R.K., Stephens, P.J., McBride, D.J., Humphray, S.J., Greenman, C.D., Varela, I., Lin, M.L., Ordonez, G.R., Bignell, G.R., Ye, K., Alipaz, J., Bauer, M.J., Beare, D., Butler, A., Carter, R.J., Chen, L.N., Cox, A.J., Edkins, S., Kokko-Gonzales, P.I., Gormley, N.A., Grocock, R.J., Haudenschild, C.D., Hims, M.M., James, T., Jia, M.M., Kingsbury, Z., Leroy, C., Marshall, J., Menzies, A., Mudie, L.J., Ning, Z.M., Royce, T., Schulz-Trieglaff, O.B., Spiridou, A., Stebbings, L.A., Szajkowski, L., Teague, J., Williamson, D., Chin, L., Ross, M.T., Campbell, P.J., Bentley, D.R., Futreal, P.A., Stratton, M.R. A comprehensive catalogue of somatic mutations from a human cancer genome. *Nature* **2010**, *463*, 191–U173.

10. Forbes, J.M., Coughlan, M.T., Cooper, M.E. Oxidative stress as a major culprit in kidney disease in diabetes. *Diabetes* **2008**, *57*, 1446–1454.

11. Martin, K.R., Barrett, J.C. Reactive oxygen species as double-edged swords in cellular processes: Low-dose cell signaling versus high-dose toxicity. *Hum. Exp. Toxicol.* **2002**, *21*, 71–75.

12. Sallmyr, A., Fan, J., Rassool, F.V. Genomic instability in myeloid malignancies: Increased reactive oxygen species (ROS), DNA double strand breaks (DSBs) and error-prone repair. *Cancer Lett.* **2008**, *270*, 1–9.

13. Pljesa-Ercegovac, M., Mimic-Oka, J., Dragicevic, D., Savic-Radojevic, A., Opacic, M., Pljesa, S., Radosavljevic, R., Simic, T. Altered antioxidant capacity in human renal cell carcinoma: Role of glutathione associated enzymes. *Urol. Oncol.* **2008**, *26*, 175–181.

14. Barzilai, A., Yamamoto, K.I. DNA damage responses to oxidative stress. *DNA Repair (Amst.)* **2004**, *3*, 1109–1115.

15. Lindahl, T. Instability and decay of the primary structure of DNA. *Nature* **1993**, *362*, 709–715.

16. Shrivastav, N., Li, D.Y., Essigmann, J.M. Chemical biology of mutagenesis and DNA repair: Cellular responses to DNA alkylation. *Carcinogenesis* **2010**, *31*, 59–70.

17. Lonkar, P., Dedon, P.C. Reactive species and DNA damage in chronic inflammation: Reconciling chemical mechanisms and biological fates. *Int. J. Cancer* **2011**, *128*, 1999–2009.

18. Steenken, S., Jovanovic, S.V. How easily oxidizable is DNA? One-electron reduction potentials of adenosine and guanosine radicals in aqueous solution. *J. Am. Chem. Soc.* **1997**, *119*, 617–618.

19. Seidel, C.A.M., Schulz, A., Sauer, M.H.M. Nucleobase-specific quenching of fluorescent dyes: 1. Nucleobase one-electron redox potentials and their correlation with static and dynamic quenching efficiencies. *J. Phys. Chem.* **1996**, *100*, 5541–5553.

20. Hildenbrand, K., Schultefrohlinde, D. ESR-spectra of radicals of single-stranded and double-stranded DNA in aqueous-solution—Implications for OH-induced strand breakage. *Free Radic. Res. Commun.* **1990**, *11*, 195–206.

21. Munk, B.H., Burrows, C.J., Schlegel, H.B. Exploration of mechanisms for the transformation of 8-hydroxy guanine radical to FAPyG by density functional theory. *Chem. Res. Toxicol.* **2007**, *20*, 432–444.

22. Stubbe, J., Nocera, D.G., Yee, C.S., Chang, M.C.Y. Radical initiation in the class I ribonucleotide reductase: Long-range proton-coupled electron transfer? *Chem. Rev.* **2003**, *103*, 2167–2201.

23. Davies, M.J. The oxidative environment and protein damage. *Biochim. Biophys. Acta* **2005**, *1703*, 93–109.

24. Salvador, A., Sousa, J., Pinto, R.E. Hydroperoxyl, superoxide and pH gradients in the mitochondrial matrix: A theoretical assessment. *Free Radic. Biol. Med.* **2001**, *31*, 1208–1215.

25. Wiesner, R.J., Zsurka, G., Kunz, W.S. Mitochondrial DNA damage and the aging process: Facts and imaginations. *Free Radic. Res.* **2006**, *40*, 1284–1294.

26. Kienhofer, J., Haussler, D.J.F., Ruckelshausen, F., Muessig, E., Weber, K., Pimentel, D., Ullrich, V., Burkle, A., Bachschmid, M.M. Association of mitochondrial antioxidant enzymes with mitochondrial DNA as integral nucleoid constituents. *FASEB J.* **2009**, *23*, 2034–2044.

27. Khobta, A., Anderhub, S., Kitsera, N., Epe, B. Gene silencing induced by oxidative DNA base damage: Association with local decrease of histone H4 acetylation in the promoter region. *Nucleic Acids Res.* **2010**, *38*, 4285–4295.

28. Beckman, K.B., Ames, B.N. The free radical theory of aging matures. *Physiol. Rev.* **1998**, *78*, 547–581.

29. Galas, D.J., Schmitz, A. DNAase footprinting: Simple method for detection of protein-DNA binding specificity. *Nucleic Acids Res.* **1978**, *5*, 3157–3170.

30. Rodriguez, H., Drouin, R., Holmquist, G.P., Oconnor, T.R., Boiteux, S., Laval, J., Doroshow, J.H., Akman, S.A. Mapping of copper hydrogen peroxide-induced DNA-damage at nucleotide resolution in human genomic

DNA by ligation-mediated polymerase chain reaction. *J. Biol. Chem.* **1995**, *270*, 17633–17640.
31. Dong, M., Vongchampa, V., Gingipalli, L., Cloutier, J.F., Kow, Y.W., O'Connor, T., Dedon, P.C. Development of enzymatic probes of oxidative and nitrosative DNA damage caused by reactive nitrogen species. *Mutat. Res.* **2006**, *594*, 120–134.
32. Tornaletti, S., Pfeifer, G.P. Slow repair of pyrimidine dimers at p53 mutation hotspots in skin cancer. *Science* **1994**, *263*, 1436–1438.
33. Merino, E.J., Barton, J.K. Oxidation by DNA charge transport damages conserved sequence block II, a regulatory element in mitochondrial DNA. *Biochemistry* **2007**, *46*, 2805–2811.
34. Ravanat, J.L. Measuring oxidized DNA lesions as biomarkers of oxidative stress: An analytical challenge. *FABAD J. Pharm. Sci.* **2005**, *30*, 100–113.
35. Boysen, G., Collins, L.B., Liao, S.K., Luke, A.M., Pachkowski, B.F., Watters, J.L., Swenberg, J.A. Analysis of 8-oxo-7,8-dihydro-2′-deoxyguanosine by ultra high pressure liquid chromatography-heat assisted electrospray ionization-tandem mass spectrometry. *J. Chromatogr. B* **2010**, *878*, 375–380.
36. Kelly, M.C., White, B., Smyth, M.R. Separation of oxidatively damaged DNA nucleobases and nucleosides on packed and monolith C18 columns by HPLC-UV-EC. *J. Chromatogr. B* **2008**, *863*, 181–186.
37. Driggers, W.J., Holmquist, G.P., LeDoux, S.P., Wilson, G.L. Mapping frequencies of endogenous oxidative damage and the kinetic response to oxidative stress in a region of rat mtDNA. *Nucleic Acids Res.* **1997**, *25*, 4362–4369.
38. Merino, E.J., Barton, J.K. DNA oxidation by charge transport in mitochondria. *Biochemistry* **2008**, *47*, 1511–1517.
39. Nunez, M.E., Noyes, K.T., Barton, J.K. Oxidative charge transport through DNA in nucleosome core particles. *Chem. Biol.* **2002**, *9*, 403–415.
40. Cloutier, J.F., Drouin, R., Weinfeld, M., O'Connor, T.R., Castonguay, A. Characterization and mapping of DNA damage induced by reactive metabolites of 4-(methylnitrosamino)-1-(3-pyridyl)-1-butanone (NNK) at nucleotide resolution in human genomic DNA. *J. Mol. Biol.* **2001**, *313*, 539–557.
41. Meniel, V., Waters, R. Spontaneous and photosensitiser-induced DNA single-strand breaks and formamidopyrimidine-DNA glycosylase sensitive sites at nucleotide resolution in the nuclear and mitochondrial DNA of *Saccharomyces cerevisiae*. *Nucleic Acids Res.* **1999**, *27*, 822–830.
42. Margolin, Y., Shafirovich, V., Geacintov, N.E., Demott, M.S., Dedon, P.C. DNA sequence context as a determinant of the quantity and chemistry of guanine oxidation produced by hydroxyl radicals and one-electron oxidants. *J. Biol. Chem.* **2008**, *283*, 35569–35578.
43. Sugden, K.D., Campo, C.K., Martin, B.D. Direct oxidation of guanine and 7,8-dihydro-8-oxoguanine in DNA by a high-valent chromium complex: A possible mechanism for chromate genotoxicity. *Chem. Res. Toxicol.* **2001**, *14*, 1315–1322.
44. Wen, W.H., Che, W.J., Lu, L., Yang, J., Gao, X.F., Wen, J.H., Heng, Z.C., Cao, S.Q., Cheng, H.R. Increased damage of exon 5 of p53 gene in workers from an arsenic plant. *Mutat. Res.* **2008**, *643*(36–40).
45. Rodriguez, H., Holmquist, G.P., Dagostino, R., Keller, J., Akman, S.A. Metal ion-dependent hydrogen peroxide-induced DNA damage is more sequence specific than metal specific. *Cancer Res.* **1997**, *57*, 2394–2403.
46. Pratviel, G., Meunier, B. Guanine oxidation: One- and two-electron reactions. *Chem. Eur. J.* **2006**, *12*, 6018–6030.
47. Candeias, L.P., Steenken, S. Reaction of HO• with guanine derivatives in aqueous solution: Formation of two different redox-active OH-adduct radicals and their unimolecular transformation reactions. Properties of G(-H). *Chem. Eur. J.* **2000**, *6*, 475–484.
48. Sheu, C., Foote, C.S. Reactivity toward singlet oxygen of a 7,8-dihydro-8-oxoguanosine (8-hydroxyguanosine) formed by photooxidation of a guanosine derivative. *J. Am. Chem. Soc.* **1995**, *117*, 6439–6442.
49. Kobayashi, K., Tagawa, S. Direct observation of guanine radical cation deprotonation in duplex DNA using pulse radiolysis. *J. Am. Chem. Soc.* **2003**, *125*, 10213–10218.
50. Misiaszek, R., Crean, C., Joffe, A., Geacintov, N.E., Shafirovich, V. Oxidative DNA damage associated with combination of guanine and superoxide radicals and repair mechanisms via radical trapping. *J. Biol. Chem.* **2004**, *279*, 32106–32115.
51. Candeias, L.P., Steenken, S. Structure and acid-base properties of one-electron-oxidized deoxyguanosine, guanosine, and 1-methylguanosine. *J. Am. Chem. Soc.* **1989**, *111*, 1094–1099.
52. Vialas, C., Pratviel, G., Claparols, C., Meunier, B. Efficient oxidation of 2′-deoxyguanosine by Mn-TMPyP/KHSO(5) to imidazolone diz without formation of 8-Oxo-dG. *J. Am. Chem. Soc.* **1998**, *120*, 11548–11553.
53. Niles, J.C., Wishnok, J.S., Tannenbaum, S.R. Spiroiminodihydantoin is the major product of the 8-oxo-7,8-dihydroguanosine reaction with peroxynitrite in the presence of thiols and guanosine photooxidation by methylene blue. *Org. Lett.* **2001**, *3*, 963–966.
54. Ye, Y., Muller, J.G., Luo, W.C., Mayne, C.L., Shallop, A.J., Jones, R.A., Burrows, C.J. Formation of C-13-, N-15-, and O-18-labeled guanidinohydantoin from guanosine oxidation with singlet oxygen: Implications for structure and mechanism. *J. Am. Chem. Soc.* **2003**, *125*, 13926–13927.
55. Gremaud, J.N., Martin, B.D., Sugden, K.D. Influence of substrate complexity on the diastereoselective formation of spiroiminodihydantoin and guanidinohydantoin from chromate oxidation. *Chem. Res. Toxicol.* **2010**, *23*, 379–385.
56. Xu, X.Y., Muller, J.G., Ye, Y., Burrows, C.J. DNA-protein cross-links between guanine and lysine depend on the mechanism of oxidation for formation of c5 vs c8 guanosine adducts. *J. Am. Chem. Soc.* **2008**, *130*, 703–709.

57. Solivio, M.J., Joy, T.J., Sallans, L., Merino, E.J. Copper generated reactive oxygen leads to formation of lysine-DNA adducts. *J. Inorg. Biochem.* **2010**, *104*, 1000–1005.
58. Kurbanyan, K., Nguyen, K.L., To, P., Rivas, E.V., Lueras, A.M.K., Kosinski, C., Steryo, M., Gonzalez, A., Mah, D.A., Stemp, E.D.A. DNA-protein cross-linking via guanine oxidation: Dependence upon protein and photosensitizer. *Biochemistry* **2003**, *42*, 10269–10281.
59. Niles, J.C., Wishnok, J.S., Tannenbaum, S.R. Peroxynitrite-induced oxidation and nitration products of guanine and 8-oxoguanine: Structures and mechanisms of product formation. *Nitric Oxide* **2006**, *14*, 109–121.
60. Douki, T., Martini, R., Ravanat, J.L., Turesky, R.J., Cadet, J. Measurement of 2,6-diamino-4-hydroxy-5-formamidopyrimidine and 8-oxo-7,8-dihydroguanine in isolated DNA exposed to gamma radiation in aqueous solution. *Carcinogenesis* **1997**, *18*, 2385–2391.
61. Hailer, M.K., Slade, P.G., Martin, B.D., Sugden, K.D. Nei deficient *Escherichia coli* are sensitive to chromate and accumulate the oxidized guanine lesion spiroiminodihydantoin. *Chem. Res. Toxicol.* **2005**, *18*, 1378–1383.
62. Kasai, H., Crain, P.F., Kuchino, Y., Nishimura, S., Ootsuyama, A., Tanooka, H. Formation of 8-hydroxyguanine moiety in cellular DNA by agents producing oxygen radicals and evidence for its repair. *Carcinogenesis* **1986**, *7*, 1849–1851.
63. Fraga, C.G., Shigenaga, M.K., Park, J.W., Degan, P., Ames, B.N. Oxidative damage to DNA during aging—8-hydroxy-2'-deoxyguanosine in rat organ DNA and urine. *Proc. Natl. Acad. Sci. U.S.A.* **1990**, *87*, 4533–4537.
64. Murtas, D., Piras, F., Minerba, L., Ugalde, J., Floris, C., Maxia, C., Demurtas, P., Perra, M.T., Sirigu, P. Nuclear 8-hydroxy-2'-deoxyguanosine as survival biomarker in patients with cutaneous melanoma. *Oncol. Rep.* **2010**, *23*, 329–335.
65. Stover, J.S., Ciobanu, M., Cliffel, D.E., Rizzo, C.J. Chemical and electrochemical oxidation of c8-arylamine adducts of 2'-deoxyguanosine. *J. Am. Chem. Soc.* **2007**, *129*, 2074–2081.
66. Goyal, R.N., Jain, N., Garg, D.K. Electrochemical and enzymic oxidation of guanosine and 8-hydroxyguanosine and the effects of oxidation products in mice. *Bioelectrochem. Bioenerg.* **1997**, *43*, 105–114.
67. Luo, W.C., Muller, J.G., Rachlin, E.M., Burrows, C.J. Characterization of spiroiminodihydantoin as a product of one-electron oxidation of 8-oxo-7,8-dihydroguanosine. *Org. Lett.* **2000**, *2*, 613–616.
68. Alp, O., Merino, E.J., Caruso, J.A. Arsenic-induced protein phosphorylation changes in hela cells. *Anal. Bioanal. Chem.* **2010**, *398*, 2099–2107.
69. Ohno, M., Miura, T., Furuichi, M., Tominaga, Y., Tsuchimoto, D., Sakumi, K., Nakabeppu, Y. A genome-wide distribution of 8-oxoguanine correlates with the preferred regions for recombination and single nucleotide polymorphism in the human genome. *Genome Res.* **2006**, *16*, 567–575.
70. Saito, I., Nakamura, T., Nakatani, K., Yoshioka, Y., Yamaguchi, K., Sugiyama, H. Mapping of the hot spots for DNA damage by one-electron oxidation: Efficacy of GG doublets and GGG triplets as a trap in long-range hole migration. *J. Am. Chem. Soc.* **1998**, *120*, 12686–12687.
71. Saito, I., Takayama, M., Sugiyama, H., Nakatani, K. Photoinduced DNA cleavage via electron-transfer—demonstration that guanine residues located 5' to guanine are the most electron-donating sites. *J. Am. Chem. Soc.* **1995**, *117*, 6406–6407.
72. Margolin, Y., Cloutier, J.F., Shafirovich, V., Geacintov, N.E., Dedon, P.C. Paradoxical hotspots for guanine oxidation by a chemical mediator of inflammation. *Nat. Chem. Biol.* **2006**, *2*, 365–366.
73. Merino, E.J., Boal, A.K., Barton, J.K. Biological contexts for DNA charge transport chemistry. *Curr. Opin. Chem. Biol.* **2008**, *12*, 229–237.
74. Hall, D.B., Holmlin, R.E., Barton, J.K. Oxidative DNA damage through long-range electron transfer. *Nature* **1996**, *382*, 731–735.
75. Wan, C.Z., Fiebig, T., Kelley, S.O., Treadway, C.R., Barton, J.K., Zewail, A.H. Femtosecond dynamics of DNA-mediated electron transfer. *Proc. Natl. Acad. Sci. U.S.A.* **1999**, *96*, 6014–6019.
76. Delaney, S., Barton, J.K. Charge transport in DNA duplex/quadruplex conjugates. *Biochemistry* **2003**, *42*, 14159–14165.
77. Freidman, K.A., Heller, A. Guanosine distribution and oxidation resistance in eight eukaryotic genomes. *J. Am. Chem. Soc.* **2004**, *126*, 2368–2371.
78. Slinker, J.D., Muren, N.B., Renfrew, S.E., Barton, J.K. DNA charge transport over 34 nm. *Nat. Chem.* **2011**, *3*, 228–233.
79. Hoeijmakers, J.H.J. Genome maintenance mechanisms for preventing cancer. *Nature* **2001**, *411*, 366–374.
80. Oikawa, S., Tada-Oikawa, S., Kawanishi, S. Site-specific DNA damage at the GGG sequence by UVA involves acceleration of telomere shortening. *Biochemistry* **2001**, *40*, 4763–4768.
81. Pham, X.H., Farge, G., Shi, Y.H., Gaspari, M., Gustafsson, C.M., Falkenberg, M. Conserved sequence box II directs transcription termination and primer formation in mitochondria. *J. Biol. Chem.* **2006**, *281*, 24647–24652.
82. Merino, E.J., Davis, M.L., Barton, J.K. Common mitochondrial DNA mutations generated through DNA-mediated charge transport. *Biochemistry* **2009**, *48*, 660–666.
83. Genereux, J.C., Boal, A.K., Barton, J.K. DNA-mediated charge transport in redox sensing and signaling. *J. Am. Chem. Soc.* **2010**, *132*, 891–905.
84. Augustyn, K.E., Merino, E.J., Barton, J.K. A role for DNA-mediated charge transport in regulating p53: Oxidation of the DNA-bound protein from a distance. *Proc. Natl. Acad. Sci. U.S.A.* **2007**, *104*, 18907–18912.
85. Sies, H. Strategies of antioxidant defense. *Eur. J. Biochem.* **1993**, *215*, 213–219.

86. Adhikary, A., Becker, D., Collins, S., Koppen, J., Sevilla, M.D. C5'- and c3'-sugar radicals produced via photoexcitation of one-electron oxidized adenine in 2'-deoxyadenosine and its derivatives. *Nucleic Acids Res.* **2006**, *34*, 1501–1511.
87. Jung, K.Y., Kodama, T., Greenberg, M.M. Repair of the major lesion resulting from c5'-oxidation of DNA. *Biochemistry* **2011**, *50*, 6273–6279.
88. Dedon, P.C. The chemical toxicology of 2-deoxyribose oxidation in DNA. *Chem. Res. Toxicol.* **2008**, *21*, 206–219.
89. Pogozelski, W.K., Tullius, T.D. Oxidative strand scission of nucleic acids: Routes initiated by hydrogen abstraction from the sugar moiety. *Chem. Rev.* **1998**, *98*, 1089–1107.
90. Kim, J., Kreller, C.R., Greenberg, M.M. Preparation and analysis of oligonucleotides containing the c4'-oxidized abasic site and related mechanistic probes. *J. Org. Chem.* **2005**, *70*, 8122–8129.
91. Adam, W., Arnold, M.A., Nau, W.M., Pischel, U., Saha-Moller, C.R. Structure-dependent reactivity of oxyfunctionalized acetophenones in the photooxidation of DNA: Base oxidation and strand breaks through photolytic radical formation (spin trapping, EPR spectroscopy, transient kinetics) versus photosensitization (electron transfer, hydrogen-atom abstraction). *Nucleic Acids Res.* **2001**, *29*, 4955–4962.
92. DeRosa, M.C., Sancar, A., Barton, J.K. Electrically monitoring DNA repair by photolyase. *Proc. Natl. Acad. Sci. U.S.A.* **2005**, *102*, 10788–10792.
93. Mori, T., Hori, Y., Dizdaroglu, M. DNA-base damage generated in-vivo in hepatic chromatin of mice upon whole-body gamma-irradiation. *Int. J. Radiat. Biol.* **1993**, *64*, 645–650.
94. Frelon, S., Douki, T., Cadet, J. Radical oxidation of the adenine moiety of nucleoside and DNA: 2-hydroxy-2'-deoxyadenosine is a minor decomposition product. *Free Radic. Res.* **2002**, *36*, 499–508.
95. Kasai, H., Hayami, H., Yamaizumi, Z., Saito, H., Nishimura, S. Detection and identification of mutagens and carcinogens as their adducts with guanosine derivatives. *Nucleic Acids Res.* **1984**, *12*, 2127–2136.
96. Kasai, H., Nishimura, S. Hydroxylation of deoxyguanosine at the c-8 position by ascorbic-acid and other reducing agents. *Nucleic Acids Res.* **1984**, *12*, 2137–2145.
97. Floyd, R.A., Watson, J.J., Wong, P.K., Altmiller, D.H., Rickard, R.C. Hydroxyl free radical adduct of deoxyguanosine: Sensitive detection and mechanisms of formation. *Free Radic. Res. Commun.* **1986**, *1*, 163–172.
98. Shigenaga, M.K., Gimeno, C.J., Ames, B.N. Urinary 8-hydroxy-2'-deoxyguanosine as a biological marker of in vivo oxidative DNA damage. *Proc. Natl. Acad. Sci. U.S.A.* **1989**, *86*, 9697–9701.
99. Lin, H.S., Jenner, A.M., Ong, C.N., Huang, S.H., Whiteman, M., Halliwell, B. A high-throughput and sensitive methodology for the quantification of urinary 8-hydroxy-2'-deoxyguanosine: Measurement with gas chromatography-mass spectrometry after single solid-phase extraction. *Biochem. J.* **2004**, *380*, 541–548.
100. Halliwell, B., Dizdaroglu, M. The measurement of oxidative damage to DNA by HPLC and GC/MS techniques. *Free Radic. Res. Commun.* **1992**, *16*, 75–87.
101. Zhang, F., Stott, W.T., Clark, A.J., Schisler, M.R., Grundy, J.J., Gollapudi, B.B., Bartels, M.J. Quantitation of 8-hydroxydeoxyguanosine in DNA by liquid chromatography/positive atmospheric pressure photoionization tandem mass spectrometry. *Rapid Commun. Mass Spectrom.* **2007**, *21*, 3949–3955.
102. Hu, C.W., Wu, M.T., Chao, M.R., Pan, C.H., Wang, C.J., Swenberg, J.A., Wu, K.Y. Comparison of analyses of urinary 8-hydroxy-2'-deoxyguanosine by isotope-dilution liquid chromatography with electrospray tandem mass spectrometry and by enzyme-linked immunosorbent assay. *Rapid Commun. Mass Spectrom.* **2004**, *18*, 505–510.
103. Toyokuni, S., Tanaka, T., Hatton, Y., Nishiyama, Y., Yoshida, A., Uchida, K., Hiai, H., Ochi, H., Osawa, T. Quantitative immunohistochemical determination of 8-hydroxy-2'-deoxyguanosine by a monoclonal antibody n45.1: Its application to ferric nitrilotriacetate-induced renal carcinogenesis model. *Lab. Invest.* **1997**, *76*, 365–374.
104. Yin, B.Y., Whyatt, R.M., Perera, F.P., Randall, M.C., Cooper, T.B., Santella, R.M. Determination of 8-hydroxydeoxyguanosine by an immunoaffinity chromatography-monoclonal antibody-based ELISA. *Free Radic. Biol. Med.* **1995**, *18*, 1023–1032.
105. Smith, C.C., O'Donovan, M.R., Martin, E.A. Hogg1 recognizes oxidative damage using the comet assay with greater specificity than FPG or ENDOIII. *Mutagenesis* **2006**, *21*, 185–190.
106. Mei, S.R., Yao, Q.H., Wu, C.Y., Xu, G.W. Determination of urinary 8-hydroxy-2'-deoxyguanosine by two approaches—capillary electrophoresis and gums: An assay for *in vivo* oxidative DNA damage in cancer patients. *J. Chromatogr. B* **2005**, *827*, 83–87.
107. Li, M.-J., Zhang, J.-B., Li, W.-L., Chu, Q.-C., Ye, J.-N. Capillary electrophoretic determination of DNA damage markers: Content of 8-hydroxy-2'-deoxyguanosine and 8-nitroguanine in urine. *J. Chromatogr. B* **2011**, *879*, 3818–3822.
108. Bogdanov, M.B., Beal, M.F., McCabe, D.R., Griffin, R.M., Matson, W.R. A carbon column-based liquid chromatography electrochemical approach to routine 8-hydroxy-2'-deoxyguanosine measurements in urine and other biologic matrices: A one-year evaluation of methods. *Free Radic. Biol. Med.* **1999**, *27*, 647–666.
109. Haghdoost, S., Czene, S., Naslund, I., Skog, S., Harms-Ringdahl, M. Extracellular 8-oxo-dg as a sensitive parameter for oxidative stress *in vivo* and *in vitro*. *Free Radic. Res.* **2005**, *39*, 153–162.
110. Barja, G., Herrero, A. Oxidative damage to mitochondrial DNA is inversely related to maximum life span in

the heart and brain of mammals. *FASEB J.* **2000**, *14*, 312–318.

111. Loft, S., Poulsen, H.E. Cancer risk and oxidative DNA damage in man. *J. Mol. Med.* **1996**, *74*, 297–312.

112. Moller, P., Cooke, M.S., Collins, A., Olinski, R., Rozalski, R., Loft, S. Harmonising measurements of 8-oxo-7,8-dihydro-2′-deoxyguanosine in cellular DNA and urine. *Free Radic. Res.* **2012**, *46*, 541–553.

113. Valko, M., Morris, H., Cronin, M.T.D. Metals, toxicity and oxidative stress. *Curr. Med. Chem.* **2005**, *12*, 1161–1208.

114. Valavanidis, A., Vlachogianni, T., Fiotakis, C. 8-hydroxy-2′-deoxyguanosine (8-ohdg): A critical biomarker of oxidative stress and carcinogenesis. *J. Environ. Sci. Health C* **2009**, *27*, 120–139.

115. Oikawa, S., Kawanishi, S. Distinct mechanisms of site-specific DNA damage induced by endogenous reductants in the presence of iron(III) and copper(II). *Biochim. Biophys. Acta* **1998**, *1399*, 19–30.

116. Pilger, A., Germadnik, D., Riedel, K., Meger-Kossien, I., Scherer, G., Rudiger, H.W. Longitudinal study of urinary 8-hydroxy-2′-deoxyguanosine excretion in healthy adults. *Free Radic. Res.* **2001**, *35*, 273–280.

117. Floyd, R.A., West, M.S., Eneff, K.L., Schneider, J.E. Methylene-blue plus light mediates 8-hydroxyguanine formation in DNA. *Arch. Biochem. Biophys.* **1989**, *273*, 106–111.

118. King, M.S., Sharpley, M.S., Hirst, J. Reduction of hydrophilic ubiquinones by the flavin in mitochondrial NADH: Ubiquinone oxidoreductase (complex I) and production of reactive oxygen species. *Biochemistry* **2009**, *48*, 2053–2062.

119. Cadet, J., Douki, T., Ravanat, J.L. Oxidatively generated base damage to cellular DNA. *Free Radic. Biol. Med.* **2010**, *49*, 9–21.

120. Cadet, J., D'Ham, C., Douki, T., Pouget, J.P., Ravanat, J.L., Sauvaigo, S. Facts and artifacts in the measurement of oxidative base damage to DNA. *Free Radic. Res.* **1998**, *29*, 541–550.

121. Dizdaroglu, M. Facts about the artifacts in the measurement of oxidative DNA base damage by gas chromatography mass spectrometry. *Free Radic. Res.* **1998**, *29*, 551–563.

122. Tsou, T.C., Chen, C.L., Liu, T.Y., Yang, J.L. Induction of 8-hydroxydeoxyguanosine in DNA by chromium(III) plus hydrogen peroxide and its prevention by scavengers. *Carcinogenesis* **1996**, *17*, 103–108.

123. Gedik, C.M., Collins, A. ESCODD establishing the background level of base oxidation in human lymphocyte DNA: Results of an interlaboratory validation study. *FASEB J.* **2005**, *19*, 82–84.

124. Slade, P.G., Priestley, N.D., Sugden, K.D. Spiroiminodihydantoin as an oxo-atom transfer product of 8-oxo-2′-deoxyguanosine oxidation by chromium(V). *Org. Lett.* **2007**, *9*, 4411–4414.

125. Bruner, S.D., Norman, D.P.G., Verdine, G.L. Structural basis for recognition and repair of the endogenous mutagen 8-oxoguanine in DNA. *Nature* **2000**, *403*, 859–866.

126. Brieba, L.G., Eichman, B.F., Kokoska, R.J., Doublie, S., Kunkel, T.A., Ellenberger, T. Structural basis for the dual coding potential of 8-oxoguanosine by a high-fidelity DNA polymerase. *EMBO J.* **2004**, *23*, 3452–3461.

127. Berdis, A.J. Mechanisms of DNA polymerases. *Chem. Rev.* **2009**, *109*, 2862–2879.

128. Zharkov, D.O. Base excision DNA repair. *Cell. Mol. Life Sci.* **2008**, *65*, 1544–1565.

129. David, S.S., O'Shea, V.L., Kundu, S. Base-excision repair of oxidative DNA damage. *Nature* **2007**, *447*, 941–950.

130. Morland, I., Rolseth, V., Luna, L., Rognes, T., Bjoras, M., Seeberg, E. Human DNA glycosylases of the bacterial Fpg/MutM superfamily: An alternative pathway for the repair of 8-oxoguanine and other oxidation products in DNA. *Nucleic Acids Res.* **2002**, *30*, 4926–4936.

131. Karihtala, P., Kauppila, S., Puistola, U., Jukkola-Vuorinen, A. Absence of the DNA repair enzyme human 8-oxoguanine glycosylase is associated with an aggressive breast cancer phenotype. *Br. J. Cancer* **2011**, *106*, 344–347.

132. Amara, P., Serre, L., Castaing, B., Thomas, A. Insights into the DNA repair process by the formamidopyrimidine-DNA glycosylase investigated by molecular dynamics. *Protein Sci.* **2004**, *13*, 2009–2021.

133. Castaing, B., Geiger, A., Seliger, H., Nehls, P., Laval, J., Zelwer, C., Boiteux, S. Cleavage and binding of a DNA fragment containing a single 8-oxoguanine by wild-type and mutant Fpg proteins. *Nucleic Acids Res.* **1993**, *21*, 2899–2905.

134. Plum, G.E., Grollman, A.P., Johnson, F., Breslauer, K.J. Influence of the oxidatively damaged adduct 8-oxodeoxyguanosine on the conformation, energetics, and thermodynamic stability of a DNA duplex. *Biochemistry* **1995**, *34*, 16148–16160.

135. Singh, S.K., Szulik, M.W., Ganguly, M., Khutsishvili, I., Stone, M.P., Marky, L.A., Gold, B. Characterization of DNA with an 8-oxoguanine modification. *Nucleic Acids Res.* **2011**, *39*, 6789–6801.

136. Haghdoost, S., Sjolander, L., Czene, S., Hanns-Ringdahl, M. The nucleotide pool is a significant target for oxidative stress. *Free Radic. Biol. Med.* **2006**, *41*, 620–626.

137. Hazra, T.K., Izumi, T., Maidt, L., Floyd, R.A., Mitra, S. The presence of two distinct 8-oxoguanine repair enzymes in human cells: Their potential complementary roles in preventing mutation. *Nucleic Acids Res.* **1998**, *26*, 5116–5122.

138. Svilar, D., Goellner, E.M., Almeida, K.H., Sobol, R.W. Base excision repair and lesion-dependent subpathways for repair of oxidative DNA damage. *Antioxid. Redox Signal.* **2011**, *14*, 2491–2507.

139. Pan, H.-Z., Zhang, H., Chang, D., Li, H., Sui, H. The change of oxidative stress products in diabetes mellitus and diabetic retinopathy. *Br. J. Ophthalmol.* **2008**, *92*, 548–551.

140. Suzuki, S., Shishido, T., Ishino, M., Katoh, S., Sasaki, T., Nishiyama, S., Miyashita, T., Miyamoto, T., Nitobe, J., Watanabe, T., Takeishi, Y., Kubota, I. 8-Hydroxy-2′-deoxyguanosine is a prognostic mediator for cardiac event. *Eur. J. Clin. Invest.* **2011**, *41*, 759–766.

141. Evans, M.D., Dizdaroglu, M., Cooke, M.S. Oxidative DNA damage and disease: Induction, repair and significance. *Mutat. Res.* **2004**, *567*, 1–61.

142. Navasumrit, P., Arayasiri, M., Hiang, O.M.T., Leechawengwongs, M., Promvijit, J., Choonvisase, S., Chantchaemsai, S., Nakngam, N., Mahidol, C., Ruchirawat, M. Potential health effects of exposure to carcinogenic compounds in incense smoke in temple workers. *Chem. Biol. Interact.* **2008**, *173*, 19–31.

143. Fujihara, J., Hasegawa, M., Kanai, R., Agusa, T., Iwata, H., Tanabe, S., Yasuda, T., Yamaguchi, S., Takeshita, H. 8-Hydroxy-2′-deoxyguanosine and arsenic compounds in urine and serum of a 4-year-old child suffering from acute promyelocytic leukemia during treatment with arsenic trioxide. *Forensic Toxicol.* **2011**, *29*, 65–68.

144. Kawahara, A., Azuma, K., Hattori, S., Nakashima, K., Basaki, Y., Akiba, J., Takamori, S., Aizawa, H., Yanagawa, T., Izumi, H., Kohno, K., Kono, S., Kage, M., Kuwano, M., Ono, M. The close correlation between 8-hydroxy-2′-deoxyguanosine and epidermal growth factor receptor activating mutation in non-small cell lung cancer. *Hum. Pathol.* **2010**, *41*, 951–959.

145. Beketic-Oreskovic, L., Ozretic, P., Rabbani, Z.N., Jackson, I.L., Sarcevic, B., Levanat, S., Maric, P., Babic, I., Vujaskovic, Z. Prognostic significance of carbonic anhydrase IX (CA-IX), endoglin (CD105) and 8-hydroxy-2′-deoxyguanosine (8-OHdG) in breast cancer patients. *Pathol. Oncol. Res.* **2011**, *17*, 593–603.

146. Yang, Y., Tian, Y., Yan, C., Jin, X., Tang, J., Shen, X. Determinants of urinary 8-hydroxy-2′-deoxyguanosine in Chinese children with acute leukemia. *Environ. Toxicol.* **2009**, *24*, 446–452.

147. Goodarzi, M.T., Navidi, A.A., Rezaei, M., Babahmadi-Rezaei, H. Oxidative damage to DNA and lipids: Correlation with protein glycation in patients with type 1 diabetes. *J. Clin. Lab. Anal.* **2010**, *24*, 72–76.

148. Wu, L.L., Chiou, C.C., Chang, P.Y., Wu, J.T. Urinary 8-OHdG: A marker of oxidative stress to DNA and a risk factor for cancer, atherosclerosis and diabetes. *Clin. Chim. Acta* **2004**, *339*, 1–9.

149. Al-Aubaidy, H.A., Jelinek, H.F. Oxidative DNA damage and obesity in type 2 diabetes mellitus. *Eur. J. Endocrinol.* **2011**, *164*, 899–904.

150. Al-Aubaidy, H.A., Jelinek, H.F. Hydroxy-2-deoxyguanosine identifies oxidative DNA damage in a rural prediabetes cohort. *Redox Rep.* **2010**, *15*, 155–160.

151. Kakimoto, M., Inoguchi, T., Sonta, T., Yu, H.Y., Imamura, M., Etoh, T., Hashimoto, T., Nawata, H. Accumulation of 8-hydroxy-2′-deoxyguanosine and mitochondrial DNA deletion in kidney of diabetic rats. *Diabetes* **2002**, *51*, 1588–1595.

152. Rossner, P., Sram, R.J. Immunochemical detection of oxidatively damaged DNA. *Free Radic. Res.* **2011**, 2012 Apr; *46*(4), 492–522. doi: 10.3109/10715762.2011.632415.

153. Isobe, C., Abe, T., Terayama, Y. Levels of reduced and oxidized coenzyme Q-10 and 8-hydroxy-2′-deoxyguanosine in the CSF of patients with Alzheimer's disease demonstrate that mitochondrial oxidative damage and/or oxidative DNA damage contributes to the neurodegenerative process. *J. Neurol.* **2010**, *257*, 399–404.

154. Carlesi, C., Caldarazzo Ienco, E., Piazza, S., Lo Gerfo, A., Alessi, R., Pasquali, L., Siciliano, G. Oxidative stress modulation in neurodegenerative diseases. *Mediterr. J. Nutr. Metab.* **2011**, *4*, 219–225.

155. Volle, C.B., Jarem, D.A., Delaney, S. Trinucleotide repeat DNA alters structure to minimize the thermodynamic impact of 8-oxo-7,8-dihydroguanine. *Biochemistry* **2012**, *51*, 52–62.

156. Strauss, P.R., O'Regan, N.E. *DNA Damage and Repair*, Vol. 3, 1st ed. Humana Press, Totowa, NJ, 2001.

157. Dutta, S., Chowdhury, G., Gates, K.S. Interstrand cross-links generated by abasic sites in duplex DNA. *J. Am. Chem. Soc.* **2007**, *129*, 1852.

158. Nilsen, L., Forstrøm, R.J., Bjørås, M., Alseth, I. AP endonuclease independent repair of abasic sites in *Schizosaccharomyces pombe*. *Nucleic Acids Res.* **2012** March; *40*(5), 2000–2009.

159. Sczepanski, J.T., Wong, R.S., McKnight, J.N., Bowman, G.D., Greenberg, M.M. Rapid DNA-protein cross-linking and strand scission by an abasic site in a nucleosome core particle. *Proc. Natl. Acad. Sci. U.S.A.* **2010**, *107*, 22475–22480.

160. Boturyn, D., Constant, J.F., Defrancq, E., Lhomme, J., Barbin, A., Wild, C.P. A simple and sensitive method for *in vitro* quantitation of abasic sites in DNA. *Chem. Res. Toxicol.* **1999**, *12*, 476–482.

161. Chan, W., Chen, B., Wang, L., Taghizadeh, K., Demott, M.S., Dedon, P.C. Quantification of the 2-deoxyribonolactone and nucleoside 5′-aldehyde products of 2-deoxyribose oxidation in DNA and cells by isotope-dilution gas chromatography mass spectrometry: Differential effects of gamma-radiation and Fe(2+)-EDTA. *J. Am. Chem. Soc.* **2010**, *132*, 6145–6153.

162. Xue, L., Greenberg, M.M. Facile quantification of lesions derived from 2′-deoxyguanosine in DNA. *J. Am. Chem. Soc.* **2007**, *129*, 7010–7011.

163. Atamna, H., Cheung, I., Ames, B.N. A method for detecting abasic sites in living cells: Age-dependent changes in base excision repair. *Proc. Natl. Acad. Sci. U.S.A.* **2000**, *97*, 686–691.

164. Dedon, P.C. Oxidation and deamination of DNA by endogenous sources. in *Chemical Carcinogenesis*, ed. T. Penning. Springer, New York, pp. 209–225.

165. Aydogan, B., Marshall, D.T., Swarts, S.G., Turner, J.E., Boone, A.J., Richards, N.G., Bolch, W.E. Site-specific OH attack to the sugar moiety of DNA: A comparison of experimental data and computational simulation. *Radiat. Res.* **2002**, *157*, 38–44.

166. Balasubramanian, B., Pogozelski, W.K., Tullius, T.D. DNA strand breaking by the hydroxyl radical is governed by the accessible surface areas of the hydrogen atoms of the DNA backbone. *Proc. Natl. Acad. Sci. U.S.A.* **1998**, *95*, 9738–9743.

167. Sy, D., Savoye, C., Begusova, M., Michalik, V., Charlier, M., SpotheimMaurizot, M. Sequence-dependent variations of DNA structure modulate radiation-induced strand breakage. *Int. J. Radiat. Biol.* **1997**, *72*, 147–155.

168. Sugiyama, H., Tsutsumi, Y., Fujimoto, K., Saito, I. Photoinduced deoxyribose-c2′ oxidation in DNA—alkali-dependent cleavage of erythrose-containing sites via a retroaldol reaction. *J. Am. Chem. Soc.* **1993**, *115*, 4443–4448.

169. Sitlani, A., Long, E.C., Pyle, A.M., Barton, J.K. DNA photocleavage by phenanthrenequinone diimine complexes of rhodium(III)—shape-selective recognition and reaction. *J. Am. Chem. Soc.* **1992**, *114*, 2303–2312.

170. Chowdhury, G., Guengerich, F.P. Tandem mass spectrometry-based detection of c4′-oxidized abasic sites at specific positions in DNA fragments. *Chem. Res. Toxicol.* **2009**, *22*, 1310–1319.

171. Levin, J.D., Demple, B. *In vitro* detection of endonuclease IV-specific DNA damage formed by bleomycin *in vivo*. *Nucleic Acids Res.* **1996**, *24*, 885–889.

172. Jaruga, P., Birincioglu, M., Rodriguez, H., Dizdaroglu, M. Mass spectrometric assays for the tandem lesion 8,5′-cyclo-2′-deoxyguanosine in mammalian DNA. *Biochemistry* **2002**, *41*, 3703–3711.

173. Shiraki, T., Uesugi, M., Sugiura, Y. C-1′ hydrogen abstraction of deoxyribose in DNA strand scission by dynemicin-A. *Biochem. Biophys. Res. Commun.* **1992**, *188*, 584–589.

174. Audat, S.A.S., Love, C.T., Al-Oudat, B.A.S., Bryant-Friedrich, A.C. Synthesis of c3′ modified nucleosides for selective generation of the c3′-deoxy-3′-thymidinyl radical: A proposed intermediate in LEE-induced DNA damage. *J. Org. Chem.* **2012**, *77*, 3829–3837.

175. Box, H.C., Dawidzik, J.B., Budzinski, E.E. Free radical-induced double lesions in DNA. *Free Radic. Biol. Med.* **2001**, *31*, 856–868.

176. Helleday, T., Lo, J., van Gent, D.C., Engelward, B.P. DNA double-strand break repair: From mechanistic understanding to cancer treatment. *DNA Repair (Amst.)* **2007**, *6*, 923–935.

177. Hong, H., Cao, H., Wang, Y. Formation and genotoxicity of a guaninecytosine intrastrand cross-link lesion *in vivo*. *Nucleic Acids Res.* **2007**, *35*, 7118–7127.

178. Boiteux, S., Guillet, M. Abasic sites in DNA: Repair and biological consequences in *Saccharomyces cerevisiae*. *DNA Repair (Amst.)* **2004**, *3*, 1–12.

179. van Hoffen, A., Balajee, A.S., van Zeeland, A.A., Mullenders, L.H.F. Nucleotide excision repair and its interplay with transcription. *Toxicology* **2003**, *193*, 79–90.

180. Fisher, L.A., Samson, L., Bessho, T. Removal of reactive oxygen species-induced 3′-blocked ends by XPF-ERCC1. *Chem. Res. Toxicol.* **2011**, *24*, 1876–1881.

181. Villani, G., Hubscher, U., Gironis, N., Parkkinen, S., Pospiech, H., Shevelev, I., di Cicco, G., Markkanen, E., Syvaoja, J.E., Le Gac, N.T. *In vitro* gap-directed translesion DNA synthesis of an abasic site involving human DNA polymerases epsilon, lambda, and beta. *J. Biol. Chem.* **2011**, *286*, 32094–32104.

182. Zhou, X.F., Liberman, R.G., Skipper, P.L., Margolin, Y., Tannenbaum, S.R., Dedon, P.C. Quantification of DNA strand breaks and abasic sites by oxime derivatization and accelerator mass spectrometry: Application to gamma-radiation and peroxynitrite. *Anal. Biochem.* **2005**, *343*, 84–92.

183. Weinfeld, M., Liuzzi, M., Paterson, M.C. Response of phage-T4 polynucleotide kinase toward dinucleotides containing apurinic sites—design of a p-32 postlabeling assay for apurinic sites in DNA. *Biochemistry* **1990**, *29*, 1737–1743.

184. Talpaertborle, M., Liuzzi, M. Reaction of apurinic/apyrimidinic sites with methoxyamine-c-14—a method for the quantitative assay of AP sites in DNA. *Biochim. Biophys. Acta* **1983**, *740*, 410–416.

185. Roberts, K.P., Sobrino, J.A., Payton, J., Mason, L.B., Turesky, R.J. Determination of apurinic/apyrimidinic lesions in DNA with high-performance liquid chromatography and tandem mass spectrometry. *Chem. Res. Toxicol.* **2006**, *19*, 300–309.

186. Fleming, A.M., Muller, J.G., Ji, I.S., Burrows, C.J. Characterization of 2′-deoxyguanosine oxidation products observed in the Fenton-like system Cu(II)/H2O2/reductant in nucleoside and oligodeoxynucleotide contexts. *Org. Biomol. Chem.* **2011**, *9*, 3338–3348.

187. Jiang, Y., Wang, Y., Wang, Y. *In vitro* replication and repair studies of tandem lesions containing neighboring thymidine glycol and 8-oxo-7,8-dihydro-2′-deoxyguanosine. *Chem. Res. Toxicol.* **2009**, *22*, 574–583.

188. Malyarchuk, S., Castore, R., Harrison, L. DNA repair of clustered lesions in mammalian cells: Involvement of non-homologous end-joining. *Nucleic Acids Res.* **2008**, *36*, 4872–4882.

189. Sutherland, B.M., Bennett, P.V., Sidorkina, O., Laval, J. Clustered damages and total lesions induced in DNA by ionizing radiation: Oxidized bases and strand breaks. *Biochemistry* **2000**, *39*, 8026–8031.

190. Harrison, L., Brame, K.L., Geltz, L.E., Landry, A.M. Closely opposed apurinic/apyrimidinic sites are converted to double strand breaks in *Escherichia coli* even in the absence of exonuclease III, endonuclease IV, nucleotide excision repair and AP lyase cleavage. *DNA Repair (Amst.)* **2006**, *5*, 324–335.

191. Bourdat, A.G., Douki, T., Frelon, S., Gasparutto, D., Cadet, J. Tandem base lesions are generated by hydroxyl radical within isolated DNA in aerated aqueous solution. *J. Am. Chem. Soc.* **2000**, *122*, 4549–4556.

192. Gentil, A., Le Page, F., Cadet, J., Sarasin, A. Mutation spectra induced by replication of two vicinal oxidative DNA lesions in mammalian cells. *Mutat. Res.* **2000**, *452*, 51–56.

193. Eot-Houllier, G., Eon-Marchais, S., Gasparutto, D., Sage, E. Processing of a complex multiply damaged DNA site by human cell extracts and purified repair proteins. *Nucleic Acids Res.* **2005**, *33*, 260–271.
194. Bergeron, F., Auvre, F., Radicella, J.P., Ravanat, J.-L. HO• radicals induce an unexpected high proportion of tandem base lesions refractory to repair by DNA glycosylases. *Proc. Natl. Acad. Sci. U.S.A.* **2010**, *107*, 5528–5533.
195. Gu, C.N., Wang, Y.S. LC-MS/MS identification and yeast polymerase eta bypass of a novel gamma-irradiation-induced intrastrand cross-link lesion G[8-5]C. *Biochemistry* **2004**, *43*, 6745–6750.
196. Carter, K.N., Greenberg, M.M. Tandem lesions are the major products resulting from a pyrimidine nucleobase radical. *J. Am. Chem. Soc.* **2003**, *125*, 13376–13378.
197. Adam, W., Arnold, M.A., Grune, M., Nau, W.M., Pischel, U., Saha-Moller, C.R. Spiroiminodihydantoin is a major product in the photooxidation of 2′-deoxyguanosine by the triplet states and oxyl radicals generated from hydroxyacetophenone photolysis and dioxetane thermolysis. *Org. Lett.* **2002**, *4*, 537–540.
198. Suzuki, T., Friesen, M.D., Ohshima, H. Identification of products formed by reaction of 3′,5′-di-*O*-acetyl-2′-deoxyguanosine with hypochlorous acid or a myeloperoxidase-H2O2 system. *Chem. Res. Toxicol.* **2003**, *16*, 382–389.
199. Kornyushyna, O., Berges, A.M., Muller, J.G., Burrows, C.J. In vitro nucleotide misinsertion opposite the oxidized guanosine lesions spiroiminodihydantoin and guanidinohydantoin and DNA synthesis past the lesions using *E. coli* DNA polymerase I. *Biochemistry* **2002**, *41*, 15304–15314.
200. Niles, J.C., Wishnok, J.S., Tannenbaum, S.R. Spiroiminodihydantoin and guanidinohydantoin are the dominant products of 8-oxoguanosine oxidation at low fluxes of peroxynitrite: Mechanistic studies with 18O. *Chem. Res. Toxicol.* **2004**, *17*, 1510–1519.
201. Ravanat, J.L., Cadet, J. Reaction of singlet oxygen with 2′-deoxyguanosine and DNA—isolation and characterization of the main oxidation products. *Chem. Res. Toxicol.* **1995**, *8*, 379–388.
202. Henderson, P.T., Delaney, J.C., Gu, F., Tannenbaum, S.R., Essigmann, J.M. Oxidation of 7,8-dihydro-8-oxoguanine affords lesions that are potent sources of replication errors in vivo. *Biochemistry* **2002**, *41*, 914–921.
203. Aller, P., Ye, Y., Wallace, S.S., Burrows, C.J., Doublie, S. Crystal structure of a replicative DNA polymerase bound to the oxidized guanine lesion guanidinohydantoin. *Biochemistry* **2010**, *49*, 2502–2509.
204. Beckman, J., Wang, M., Blaha, G., Wang, J., Konigsberg, W.H. Substitution of Ala for Tyr567 in RB69 DNA polymerase allows dAMP and dGMP to be inserted opposite guanidinohydantoin. *Biochemistry* **2010**, *49*, 8554–8563.
205. Hazra, T.K., Muller, J.G., Manuel, R.C., Burrows, C.J., Lloyd, R.S., Mitra, S. Repair of hydantoins, one electron oxidation product of 8-oxoguanine, by DNA glycosylases of *Escherichia coli*. *Nucleic Acids Res.* **2001**, *29*, 1967–1974.
206. Leipold, M.D., Muller, J.G., Burrows, C.J., David, S.S. Removal of hydantoin products of 8-oxoguanine oxidation by the *Escherichia coli* DNA repair enzyme, Fpg. *Biochemistry* **2000**, *39*, 14984–14992.
207. Hailer, M.K., Slade, P.G., Martin, B.D., Rosenquist, T.A., Sugden, K.D. Recognition of the oxidized lesions spiroiminodihydantoin and guanidinohydantoin in DNA by the mammalian base excision repair glycosylases NEIL1 and NEIL2. *DNA Repair (Amst.)* **2005**, *4*, 41–50.
208. Niedernhofer, L.J., Lalai, A.S., Hoeijmakers, J.H.J. Fanconi anemia (cross)linked to DNA repair. *Cell* **2005**, *123*, 1191–1198.
209. Macfie, A., Hagan, E., Zhitkovich, A. Mechanism of DNA-protein cross-linking by chromium. *Chem. Res. Toxicol.* **2010**, *23*, 341–347.
210. VanderVeen, L.A., Harris, T.M., Jen-Jacobson, L., Marrlett, L.J. Formation of DNA-protein cross-links between gamma-hydroxypropanodeoxyguanosine and EcoRI. *Chem. Res. Toxicol.* **2008**, *21*, 1733–1738.
211. Ide, H., Shoulkamy, M.I., Nakano, T., Miyamoto-Matsubara, M., Salem, A.M.H. Repair and biochemical effects of DNA-protein crosslinks. *Mutat. Res.* **2011**, *711*, 113–122.
212. Quievryn, G., Zhitkovich, A. Loss of DNA-protein cross-links from formaldehyde-exposed cells occurs through spontaneous hydrolysis and an active repair process linked to proteosome function. *Carcinogenesis* **2000**, *21*, 1573–1580.
213. Minko, I.G., Kurtz, A.J., Croteau, D.L., Van Houten, B., Harris, T.M., Lloyd, R.S. Initiation of repair of DNA-polypeptide cross-links by the UvrABC nuclease. *Biochemistry* **2005**, *44*, 3000–3009.
214. Nakano, T., Morishita, S., Katafuchi, A., Matsubara, M., Horikawa, Y., Terato, H., Salem, A.M.H., Izumi, S., Pack, S.P., Makino, K., Ide, H. Nucleotide excision repair and homologous recombination systems commit differentially to the repair of DNA-protein crosslinks. *Mol. Cell* **2007**, *28*, 147–158.
215. Johansen, M.E., Muller, J.G., Xu, X.Y., Burrows, C.J. Oxidatively induced DNA-protein cross-linking between single-stranded binding protein and oligodeoxynucleotides containing 8-oxo-7,8-dihydro-2′-deoxyguanosine. *Biochemistry* **2005**, *44*, 5660–5671.
216. Ghude, P., Schallenberger, M.A., Fleming, A.M., Muller, J.G., Burrows, C.J. Comparison of transition metal-mediated oxidation reactions of guanine in nucleoside and single-stranded oligodeoxynucleotide contexts. *Inorganica Chim. Acta* **2011**, *369*, 240–246.
217. Xu, X.Y., Fleming, A.M., Muller, J.G., Burrows, C.J. Formation of tricyclic 4.3.3.0 adducts between 8-oxoguanosine and tyrosine under conditions of oxidative DNA-protein cross-linking. *J. Am. Chem. Soc.* **2008**, *130*, 10080.
218. Guan, L.R., Greenberg, M.M. An oxidized abasic lesion as an intramolecular source of DNA adducts. *Aust. J. Chem.* **2011**, *64*, 438–442.

219. Zaidi, R., Bryant-Friedrich, A.C. The effect of reductant levels on the formation of damage lesions derived from a 2-deoxyribose radical in ssDNA. *Radiat. Res.* **2012**, *177*, 565–572.

220. Huang, H., Wang, H., Kozekova, A., Rizzo, C.J., Stone, M.P. Formation of a N2-dG:N2-dG carbinolamine DNA cross-link by the *trans*-4-hydroxynonenal-derived (6S,8R,11S) 1,N2-dG adduct. *J. Am. Chem. Soc.* **2011**, *133*, 16101–16110.

221. Cheng, G., Shi, Y.L., Sturla, S.J., Jalas, J.R., McIntee, E.J., Villalta, P.W., Wang, M.Y., Hecht, S.S. Reactions of formaldehyde plus acetaldehyde with deoxyguanosine and DNA: Formation of cyclic deoxyguanosine adducts and formaldehyde cross-links. *Chem. Res. Toxicol.* **2003**, *16*, 145–152.

222. Voitkun, V., Zhitkovich, A. Analysis of DNA-protein crosslinking activity of malondialdehyde *in vitro*. *Mutat. Res.* **1999**, *424*, 97–106.

223. Simic, M.G., Dizdaroglu, M. Formation of radiation-induced cross-links between thymine and tyrosine—possible model for cross-linking of DNA and proteins by ionizing-radiation. *Biochemistry* **1985**, *24*, 233–236.

224. Mitrasinovic, P.M. Cross-linking between thymine and indolyl radical: Possible mechanisms for cross-linking of DNA and tryptophan-containing peptides. *Bioconjug. Chem.* **2005**, *16*, 588–597.

225. Shi, W.-Q., Hu, J., Zhao, W., Su, X.-Y., Cai, H., Zhao, Y.-F., Li, Y.-M. Identification of radiation-induced cross-linking between thymine and tryptophan by electrospray ionization-mass spectrometry. *J. Mass Spectrom.* **2006**, *41*, 1205–1211.

226. Sun, G.X., Fecko, C.J., Nicewonger, R.B., Webb, W.W., Begley, T.P. DNA-protein cross-linking: Model systems for pyrimidine-aromatic amino acid cross-linking. *Org. Lett.* **2006**, *8*, 681–683.

227. Nakano, T., Terato, H., Asagoshi, K., Masaoka, A., Mukuta, M., Ohyama, Y., Suzuki, T., Makino, K., Ide, H. DNA-protein cross-link formation mediated by oxanine—a novel genotoxic mechanism of nitric oxide-induced DNA damage. *J. Biol. Chem.* **2003**, *278*, 25264–25272.

228. Nakano, T., Ouchi, R., Kawazoe, J., Pack, S.P., Makino, K., Ide, H. T7 RNA polymerases backed up by covalently trapped proteins catalyze highly error prone transcription. *J. Biol. Chem.* **2012**, *287*, 6562–6572.

229. Zhitkovich, A., Voitkun, V., Costa, M. Formation of the amino acid-DNA complexes by hexavalent and trivalent chromium *in vitro*: Importance of trivalent chromium and the phosphate group. *Biochemistry* **1996**, *35*, 7275–7282.

230. Solivio, M.J., Nemera, D.B., Sallans, L., Merino, E.J. Biologically relevant oxidants cause bound proteins to readily oxidatively cross-link at guanine. *Chem. Res. Toxicol.* **2012**, *25*, 326–336.

231. Zhitkovich, A., Costa, M. A simple, sensitive assay to detect DNA-protein cross-links in intact cells and in vivo. *Carcinogenesis* **1992**, *13*, 1485–1489.

5

DOWNREGULATION OF ANTIOXIDANTS AND PHASE 2 PROTEINS

Hong Zhu, Jianmin Wang, Arben Santo, and Yunbo Li

OVERVIEW

Aerobic organisms are constantly exposed to reactive oxygen species (ROS) and other related reactive species. These reactive species are able to cause oxidative damage to biomolecules, leading to cell and tissue injury. Therefore, a number of endogenous antioxidants and phase 2 proteins have been evolved to counteract the detrimental effects of ROS and related reactive species. A balance between ROS generation and their detoxification by antioxidants and phase 2 proteins is maintained under normal physiological conditions. However, disruption of this balance by increased ROS production can lead to oxidative stress and tissue injury. Similarly, downregulation of antioxidants and phase 2 proteins by chemical agents may also contribute to oxidative stress and tissue injury under certain conditions. This chapter begins with a brief description of how antioxidants and phase 2 proteins are defined, followed by a survey of the major cellular antioxidants and phase 2 proteins, their role in protecting against ROS, as well as their molecular regulation. A discussion of the chemical agents capable of inhibiting or downregulating endogenous antioxidants and phase 2 proteins is then presented, followed by conclusions and perspectives.

5.1 DEFINITIONS OF ANTIOXIDANTS AND PHASE 2 PROTEINS

5.1.1 Antioxidants

The term antioxidant has been defined in various ways in the literature. One way to define it is that antioxidant is any substance that can prevent, reduce, or repair the ROS-induced damage of a target biomolecule.[1] Antioxidants protect ROS and other related reactive species-induced damage by three general mechanisms, as listed below:

- Inhibition of ROS generation
- Scavenging of ROS already formed
- Repair of ROS-induced damage.

5.1.2 Phase 2 Proteins

Phase 1 and phase 2 reactions are related to biotransformation of xenobiotics. Phase 1 biotransformation reactions include oxidation, reduction, and hydrolysis. Phase 2 biotransformation involves primarily conjugation reactions, such as conjugation with endogenous cellular ligands glutathione (GSH) and glucuronic acid. Glutathione

---

Molecular Basis of Oxidative Stress: Chemistry, Mechanisms, and Disease Pathogenesis, First Edition. Edited by Frederick A. Villamena.
© 2013 John Wiley & Sons, Inc. Published 2013 by John Wiley & Sons, Inc.

S-transferase (GST) and uridine diphosphate (UDP)-glucuronosyltransferase catalyze conjugation with GSH and glucuronic acid, respectively. These enzymes, along with many others involved in phase 2 biotransformation reactions of xenobiotics, are classically referred to as phase 2 proteins or enzymes.

Recently, the term phase 2 proteins is expanded to include not only the above conjugation enzymes but also NAD(P)H:quinone oxidoreductase (NQO), epoxide hydrolase, dihydrodiol dehydrogenase, γ-glutamylcysteine ligase (GCL), heme oxygenase-1, leukotriene B4 dehydrogenase, aflatoxin B1 dehydrogenase, and ferritin.[2] Some of the above phase 2 proteins, such as GCL, heme oxygenase-1, and ferritin are typically classified as antioxidants. Thus, the compound term antioxidative/phase 2 proteins is frequently encountered in the literature.

5.2 ROLES IN OXIDATIVE STRESS

5.2.1 Superoxide Dismutase

There are three isoforms of superoxide dismutase (SOD) in mammals: (1) copper, zinc superoxide dismutase (Cu,ZnSOD or SOD1), (2) manganese superoxide dismutase (MnSOD or SOD2), and (3) extracellular superoxide dismutase (ECSOD or SOD3). Cu,ZnSOD is present mainly in cytosol. MnSOD exists in mitochondrial matrix. ECSOD is associated with plasma membrane or present in extracellular space. All three isoforms of SODs catalyze dismutation of superoxide ($O_2^{\cdot-}$) to form hydrogen peroxide (H_2O_2) and molecular oxygen with a similar reaction rate constant (Eq. 5.1):

$$2O_2^{\cdot-} + 2H^+ \xrightarrow{SOD} H_2O_2 + O_2 \quad (5.1)$$

The role of SODs has been demonstrated in genetically manipulated animals. Targeted disruption of SODs results in oxidative tissue degeneration in various animal models. On the other hand, transgenic overexpression of each of the three isozymes of SODs in animal models leads to protection against various pathophysiological processes associated with oxidative stress. For example, overexpression of Cu,ZnSOD, MnSOD, or ECSOD in mice protects against tissue ischemia-reperfusion injury and hyperoxia-induced lung injury.[3]

5.2.2 Catalase

Catalase (CAT) is a heme-containing enzyme primarily present in peroxisomes. It is best known for its ability to catalyze the decomposition of H_2O_2, a product of SOD enzymatic activity, to form water and molecular oxygen (Eq. 5.2). Catalase also possesses peroxidase and oxidase activities toward a number of substrates, including alcohols.[4] Equation 5.3 illustrates the peroxidase activity of catalase in converting alcohol (AH_2) to aldehyde (A):

$$2H_2O_2 \xrightarrow{CAT} H_2O + O_2 \quad (5.2)$$

$$H_2O_2 + AH_2 \xrightarrow{CAT} A + 2H_2O \quad (5.3)$$

The role of catalase in antioxidant defense is dependent on the types of tissues and the models of oxidative tissue injury. In general, animals deficient in catalase are more susceptible to oxidative stress-mediated tissue injury. Conversely, genetic overexpression of catalase in mice renders these animals' increased resistance to various pathophysiological processes involving oxidative stress, such as aging, atherosclerosis, cardiomyopathy, and tissue ischemia-reperfusion injury.[5–7]

5.2.3 GSH and GSH-Related Enzymes

5.2.3.1 GSH GSH is a tripeptide (γ-glutamylcysteinylglycine) synthesized from three amino acids via two successive enzymatic reactions in cytoplasm.[8] The first step involves a combination of cysteine and glutamate to produce γ-glutamylcysteine. This reaction is catalyzed by GCL, also known as γ-glutamylcysteine synthetase (Eq. 5.4). The next step involves the enzyme glutathione synthetase (GS), which catalyzes the addition of glycine to the dipeptide to form γ-glutamylcysteinylglycine (GSH) (Eq. 5.5). Both steps require coupled ATP hydrolysis:

$$\begin{aligned}&\text{Cysteine} + \text{Glutamate} + \text{ATP} \\ &\xrightarrow{GCL} \gamma\text{-Glutamylcysteine} + \text{ADP} + \text{Pi}\end{aligned} \quad (5.4)$$

$$\begin{aligned}&\gamma\text{-Glutamylcysteine} + \text{Glycine} + \text{ATP} \\ &\xrightarrow{GS} \gamma\text{-Glutamylcysteinylglycine} + \text{ADP} + \text{Pi}\end{aligned} \quad (5.5)$$

GCL is the rate-limiting enzyme of GSH biosynthesis. This enzyme consists of two subunits, the heavy catalytic subunit designated as GCLC and the light modifier subunit designated as GCLM. As the name indicates, GCLM modulates the activity of GCL and affects the steady-state levels of GSH in mammalian cells.

In mammalian cells and tissues, GSH is mainly involved in four types of biochemical reactions: (1) reaction with ROS, (2) reaction with electrophiles, (3) reaction with other nonenzymatic antioxidants, and (4) protein deglutathionylation.[9] These chemical properties largely contribute to the beneficial effects of GSH observed in various disease conditions.

The biological activities of GSH have been investigated by modulating its levels in cells or tissues. Increasing cellular or tissue GSH affords protection against ROS- and electrophile-elicited injury in various disease models. Elevation of tissue GSH levels also leads to protection against experimental carcinogenesis. In contrast, depletion of cellular or tissue GSH sensitizes animals to various pathophysiological processes involving oxidative and electrophilic stress.[10] These include neurodegeneration, tissue ischemia-reperfusion injury, and xenobiotic/drug-induced toxicity.

5.2.3.2 Glutathione Peroxidase

Glutathione peroxidase (GPx) refers to a family of multiple isozymes. In mammalian tissues, there are six GPx isozymes, namely, GPx1, 2, 3, 4, 5, and 6.[11] GPx1, 2, 3, and 4 are selenoproteins. All of the GPx isozymes are able to catalyze the reduction of H_2O_2 or organic hydroperoxides (LOOH) to water or corresponding alcohols (LOH) using GSH as an electron donor (Eq. 5.6 and Eq. 5.7). During the reactions, GSH is oxidized to glutathione disulfide (GSSG):

$$H_2O_2 + 2GSH \xrightarrow{GPx} 2H_2O + GSSG \quad (5.6)$$

$$LOOH + 2GSH \xrightarrow{GPx} LOH + GSSG + H_2O \quad (5.7)$$

The biological activities of individual GPx isozymes have been studied in transgenic overexpression and gene knockout animal models. For example, GPx1 knockout ($GPx1^{-/-}$) mice are more sensitive to tissue injury induced by redox cycling chemicals, including paraquat and doxorubicin. Knockout of GPx1 gene also sensitizes the mice to tissue ischemia-reperfusion injury, angiotensin II-induced vascular oxidative stress, and development of atherosclerosis.[12–14] Mice with overexpression of GPx1 are more resistant to oxidative tissue injury induced by redox cycling chemicals. GPx3 may act as one of the major scavenger of ROS in plasma. Mice overexpressing GPx3 are resistant to lipopolysaccharide-induced endotoxemia, inflammation, and oxidative stress.[15] The GPx3-overexpressing mice manifest better control of ROS levels under high body temperatures when compared with wild-type mice. In addition, GPx3 plays a role in regulating the bioavailability of vascular nitric oxide, a critical antioxidative and anti-inflammatory molecule in the cardiovascular system.[16] Taken together, extensive experimental evidence demonstrates an important role for GPx isozymes in protecting tissues from oxidative stress injury.

5.2.3.3 Glutathione Reductase

As noted above, GSH, upon reaction with ROS, is oxidized to GSSG. In mammalian cells, the ratios of intracellular GSH to GSSG are high, usually in the range of 10:1 to 100:1. Maintenance of such high ratios of intracellular GSH to GSSG is essential for normal cellular activities, including redox signaling. Glutathione reductase (GR) reduces GSSG to GSH by using NADPH as a cofactor (Eq. 5.8) and is critical for maintaining the high ratios of intracellular GSH to GSSG[17]:

$$GSSG + NADPH + H^+ \xrightarrow{GR} 2GSH + NADP^+ \quad (5.8)$$

By maintaining the high ratios of GSH to GSSG, GR plays an important role in detoxification of ROS as well as in the regulation of cellular redox homeostasis. GR-deficient mice are shown to be more susceptible to ROS-induced tissue injury. Inhibition of GR activity in animals or cultured cells by chemical inhibitors or small interfering RNA techniques results in increased sensitivity to oxidative injury. On the other hand, transgenic overexpression of GR in animal models is reported to extend survival under hyperoxic conditions.[18] Increased expression of GR in macrophages also decreases atherosclerotic lesion formation and vascular oxidative stress in low-density lipoprotein receptor-deficient mice.[19]

5.2.3.4 GST

GST is a general term for a superfamily of enzymes that catalyze the conjugation of GSH to a wide variety of xenobiotics.[20] Some GST isozymes also regulate a number of cellular processes via nonenzymatic reactions. This superfamily of enzymes consists of three major families widely distributed in mammalian tissues: (1) cytosolic GSTs, (2) mitochondrial GSTs, and (3) microsomal GSTs, which are now referred to as membrane-associated proteins in eicosanoid and glutathione (MAPEG) metabolism. Among these families, cytosolic GSTs are the most extensively studied enzymes involved in detoxification of xenobiotics and ROS, and play important roles in protecting mammalian cells and tissues from electrophilic and oxidative stress.[20] These enzymes catalyze the conjugation reactions of GSH with various electrophilic xenobiotics, including reactive aldehydes and quinone compounds to form less reactive conjugates (xenobiotic-GS) (Eq. 5.9). Some GST isozymes also exhibit GPx activity, catalyzing reduction of organic hydroperoxide (LOOH) to form alcohol (LOH) (Eq. 5.10):

$$\text{Xenobiotic} + GSH \xrightarrow{GST} \text{Xenobiotic-GS} \quad (5.9)$$

$$LOOH + 2GSH \xrightarrow{GST} LOH + GSSG + H_2O \quad (5.10)$$

The biological activities of GST isozymes can be summarized into the following three categories: (1) protection against toxicity of electrophilic xenobiotics and ROS via enzymatic reactions, (2) protection against chemical carcinogenesis through detoxification of electrophilic carcinogens, and (3) regulation of cellular processes such as cell signaling and apoptosis via nonenzymatic reactions.[21]

5.2.4 NAD(P)H:Quinone Oxidoreductase

NAD(P)H:quinone oxidoreductase (NQO) refers to a family of flavoproteins that include two members, NQO1 and NQO2 in mammals.[22] NQO1 and NQO2 stand for NAD(P)H:quinone oxidoreductase 1 and NRH:quinone oxidoreductase 2, respectively. NQO2 uses dihydronicotinamide riboside (NRH) rather than NAD(P)H as an electron donor. NQO1 has received much more extensive studies than does NQO2 in mammalian systems.

The biological functions of NQO1 include[23] (1) two electron reductions and detoxification of reactive quinone compounds and derivatives, (2) maintenance of the endogenous lipid-soluble antioxidants α-tocopherol and ubiquinone in their reduced and active forms, (3) direct scavenging of superoxide, and (4) stabilization of the tumor suppressor p53 protein.

5.2.5 Heme Oxygenase

Heme oxygenase (HO) is the rate-limiting enzyme in heme catabolism. In mammals, two major isoforms of HO exist: the inducible form HO-1 and the constitutively expressed form HO-2.[24] HO-1 is expressed at low levels in most tissues under physiological conditions, with the exception of the spleen. HO-2 is constitutively expressed in mammalian tissues under physiological conditions.

Both HO-1 and HO-2 catalyze the degradation of heme to biliverdin with concurrent release of iron (Fe^{2+}) and carbon monoxide (CO). This reaction requires molecular oxygen as well as reducing equivalents from NADPH cytochrome P450 reductase. The biliverdin generated in the HO reaction is reduced to bilirubin by biliverdin reductase.

Since heme is a pro-oxidant, its degradation by HO minimizes the oxidative stress induced by excess heme. Both biliverdin and bilirubin possess antioxidant activities as well as other cytoprotective functions, such as anti-inflammation and antiproliferation. Carbon monoxide is also known to exert anti-inflammatory, antiproliferative, and vasodilatory activities.[24] Due to the pro-oxidant activity of iron (e.g., participation in Fenton reaction, leading to the formation of hydroxyl radical), the iron released from HO-catalyzed reaction would seem to lead to detrimental effects. However, induction of HO activity and release of iron are usually associated with concurrent induction of ferritin, an iron-chelating protein. Induction of ferritin thus minimizes the pro-oxidant potential of the released iron. Extensive studies have established HO enzymes as critical antioxidant defenses in mammalian systems.

5.2.6 Ferritin

Ferritin is a 24-subunit protein whose principal role in mammals is the storage of iron in a nontoxic, but bioavailable, form. The assembled ferritin molecule, often referred to as a nanocage, can store up to 4500 atoms of iron. Iron in the ferritin nanocage is insoluble and is most likely redox inactive. Ferritin has a central role in the control of cellular iron homeostasis. Sequestration of iron ions by ferritin is also an important mechanism for controlling iron-mediated oxidative damage.[25]

5.2.7 UDP-Glucuronosyltransferase

UDP-glucuronosyltransferase (UGT) is a superfamily of phase 2 biotransformation enzymes which include UGT1 and UGT2 subfamilies.[26] These enzymes are predominately located in the endoplastic reticulum of liver and other tissues, such as kidney, gastrointestinal tract, lungs, skin, and brain. UGTs catalyze the conjugation of glucuronyl group from uridine 5′-diphosphoglucuronic acid (UDP-GA) with endogenous and exogenous substrates, generating glucuronide products that are more water soluble, less toxic, and more readily excreted. Glucuronidation is responsible for the elimination of a diverse range of compounds such as steroids, drugs, and environmental toxicants, and is thus generally considered as a major detoxification mechanism.

5.3 MOLECULAR REGULATION

5.3.1 General Consideration

The critical involvement of ROS in both physiology and pathophysiology makes it necessary to regulate endogenous antioxidants and phase 2 proteins so as to control the undesired effects of these reactive species while permitting their physiological functions. There are multiple mechanisms involved in the regulation of mammalian antioxidant and phase 2 genes. Among them, the nuclear factor-erythroid 2 p45-related factor 2 (Nrf2) appears to play the most important role in the regulation of a wide variety of antioxidant and phase 2 genes as well as other novel cytoprotective factors in mammals, including humans.[27]

5.3.2 Nrf2 Signaling

Nrf2 is a member of the vertebrate Cap"n"Collar (CNC) transcription factor subfamily of basic leucine zipper (bZip) transcription factors. Other members of the CNC subfamily of transcription factors include Nrf1, Nrf3, and p45 NF-E2. It has become established that Nrf2 plays a central role in regulating both the constitutive and inducible expression of a wide variety of mammalian antioxidant and phase 2 genes. Nrf2 activation occurs under a variety of stress conditions, including exposure to mild oxidative or electrophilic stress. Various classes of chemical inducers are known to upregulate endogenous antioxidants and phase 2 proteins via activating Nrf2 both *in vitro* and *in vivo*.[28]

Nrf2 normally resides in the cytosolic compartment through association with a cytosolic actin-binding protein, Keap1 (Kelch-like ECH-associated protein 1), which is also known as INrf2 (inhibitor of Nrf2). Keap1 plays a central role in the regulation of Nrf2 activity. Keap1 exists as dimers inside the cells and functions as a substrate linker protein for interaction of Cul3/Rbx1-based E3-ubiquitin ligase complex with Nrf2, leading to continuous ubiquitination of Nrf2 and its proteasomal degradation. Hence, the continuous degradation of Nrf2 under basal conditions keeps the Nrf2 level low and thereby the low basal levels of Nrf2-regulated antioxidant and phase 2 genes. When the cells encounter the stress, such as exposure to oxidants or chemical inducers, Nrf2 dissociates from Keap1, becomes stabilized, and translocates into the nuclei. Inside the nuclei, Nrf2 interacts with other protein factors, including small Maf (sMaf), and binds to the antioxidant response element, leading to increased transcription of antioxidant and phase 2 genes.[29] Several potential mechanisms have been proposed to explain the dissociation of Nrf2 from Keap1 under stress conditions. These include modifications of cysteine residues of Keap1 and phosphorylation of Nrf2.[29]

5.3.3 Other Regulators

In addition to Nrf2, several other transcription factors and signaling pathways are also found to regulate antioxidant and phase 2 gene expression under certain conditions. These include Nrf1, Nrf3, NF-κB, AP-1, AhR, p53, and cAMP-response element-binding protein. In general, as compared with Nrf2 these transcription factors and signaling molecules have limited roles in regulating mammalian antioxidant and phase 2 genes. They usually participate in the regulation of certain genes under particular conditions or in specific types of cells.

5.4 INDUCTION IN CHEMOPREVENTION

5.4.1 Chemical Inducers

It is established that Nrf2 plays a major role in mediating the upregulation of cellular antioxidants and phase 2 proteins by various chemical inducers. In the field of chemoprotection or chemoprevention, several classes of chemicals are commonly used for induction of cellular antioxidants and phase 2 proteins to protect against disease conditions associated with oxidative and electrophilic stress. These chemoprotective agents include the following four chemical classes: (1) phenolic compounds, (2) dithiolethiones, (3) isothiocyanates, and (4) triperpenoids.[28]

5.4.2 Chemoprotection

The chemical agents listed above in Section 5.1 have been demonstrated to provide chemoprotective effects in various animal models, including oxidative and electrophilic tissue injury, inflammatory disorders, and chemical carcinogenesis.[28,29] Studies in $Nrf2^{-/-}$ mouse models show that the above chemoprotective effects occur primarily via activation of Nrf2 and the subsequent induction of the Nrf2-regulated antioxidants and phase 2 proteins. It is worth to note that the coordinated actions of various antioxidants and phase 2 proteins are essential for effective detoxification of ROS and other related reactive species that participate in tissue injury. Hence, the coordinated induction of Nrf2-regulated antioxidant and phase 2 defenses by chemoprotective agents may represent an effective strategy for intervention of disease conditions involving an oxidative stress component. Such a strategy would be particularly useful when tissue antioxidants and phase 2 proteins are downregulated or compromised due to pathophysiological conditions.

5.5 DOWNREGULATION

As stated above, cellular antioxidants and phase 2 proteins can be upregulated by chemical inducers. They can also be downregulated or inactivated by chemical agents, including drugs and environmental toxicants. Downregulation or inhibition of cellular antioxidants and phase 2 proteins may also cause oxidative stress, contributing to tissue injury and disease pathophysiology.

5.5.1 Selective Chemical Inhibitors

A number of chemical agents have been shown to selectively inhibit antioxidant enzymes and phase 2 proteins, and as such, many of them are frequently used as chemical tools for studying the biological activities of antioxidants and phase 2 proteins. These include the SOD inhibitor *N*,*N*-diethyldithiocarbamate, catalase inhibitor 3-amino-1,2,4-triazole, GSH synthesis inhibitor buthionine sulfoximine (BSO), GPx inhibitor mercaptosuccinate, GST inhibitor sulfasalazine, NQO1 inhibitor dicumarol, and HO inhibitors metalloporphyrins. There have been a number of studies showing that some of the above inhibitors may be able to induce oxidative stress in biological systems at least partially via their inhibition of the respective antioxidants and phase 2 proteins.

5.5.1.1 N,N-Diethyldithiocarbamate
Dithiocarbamates are an important class of compounds that possess a diverse set of chemical properties and biological activities that are utilized in pesticidal, industrial, and therapeutic applications. Among dithiocarbamates, N,N-diethyldithiocarbamate has been extensively studied with respect to its inhibition of SOD. N,N-diethyldithiocarbamate chelates copper ion and as such inactivates Cu,ZnSOD both *in vitro* and *in vivo*. Multiple studies demonstrated that inhibition of Cu,ZnSOD by N,N-diethyldithiocarbamate led to oxidative stress and tissue injury.[30,31] It should be noted that N,N-diethyldithiocarbamate may also affect other enzymes and cause oxidative stress via different mechanisms. Indeed, treatment with N,N-diethyldithiocarbamate was shown to induce cellular copper ion accumulation, leading to copper-dependent oxygen radical generation.[32,33]

5.5.1.2 3-Amino-1,2,4-Triazole
Although azide, cyanide, peroxynitrite, and hypochlorious acid all inhibit catalase, they are nonselective. In contrast, 3-amino-1,2,4-triazole has been characterized as a highly selective inhibitor of catalase in both cultured cells and animal models. Treatment of both cell cultures and experimental animals with this compound has been shown to selectively inhibit catalase activity leading to oxidative stress, as revealed by increased levels of H_2O_2 and oxidative damage of biomolecules.[34,35] Inhibition of cellular catalase by 3-amino-1,2,4-triazole was also found to sensitize the cells to exogenous ROS-induced cytotoxicity.[36,37]

5.5.1.3 BSO
BSO is a highly selective inhibitor of GCL, the key enzyme in GSH biosynthesis (Section 5.2.3.1). Treatment of both cell cultures and animals with BSO has been conclusively shown to deplete cellular or tissue GSH. Depletion of cellular GSH by BSO was found to increase ROS levels and induce oxidative damage, including apoptosis in various types of cells, supporting the notion that GSH is an important endogenous antioxidant defense under basal conditions. Administration of BSO to animals could also cause oxidative stress and tissue degeneration under basal conditions, and sensitize the animals to a wide variety of pathophysiological conditions, such as tissue ischemia-reperfusion injury and chemically induced oxidative tissue degeneration.[38,39]

5.5.1.4 Sulfasalazine
Sulfasalazine is a commonly used mesalamine analog for the treatment of inflammatory bowel disease. Sulfasalazine was reported to inhibit GST activity in cancer cells and sensitize these cells to anticancer drug-induced cytotoxicity.[40] Administration of sulfasalazine to animals was shown to inhibit GST activity and cause oxidative damage[41,42]; however, the causal role of GST inhibition in the elicited oxidative tissue injury was not clear. In this regard, treatment of animals with sulfasalazine also resulted in decreased levels/activities of other antioxidants, including SOD and GSH.[42] Inhibition of these antioxidants might also contribute to the oxidative stress induced by sulfasalazine in various animal models.

5.5.1.5 Dicumarol
Dicumarol is a selective inhibitor of NQO1 though it may also affect other enzymes such as mitochondrial electron transport enzyme complexes at high concentrations. As discussed earlier, NQO1 is an important antioxidant enzyme with multiple activities against oxidative stress. For example, NQO1 is able to maintain α-tocopherol and ubiquinone in their reduced forms which are effective antioxidants. Inhibition of NQO1 by dicumarol in cultured cells was shown to diminish the antioxidant activity of ubiquinone and sensitize the cells to ROS-induced cytotoxicity.[43,44] In cancer cells, inhibition of NQO1 by dicumarol also led to increased formation of superoxide and induction of cytotoxicity.[45,46] These observations suggested that inhibition of NQO1 could provoke oxidative stress in cells.

5.5.2 Drugs and Environmental Toxic Agents

In addition to the selective inhibitors of antioxidants and phase 2 proteins, a number of other chemical agents, including drugs and environmental toxicants, have also been shown to induce oxidative stress possibly via inhibition of endogenous antioxidant and phase 2 defenses. For example, administration of therapeutic doses of doxorubicin to mice resulted in decreased levels/activities of myocardial antioxidants, including GSH, GR, and GPx.[47] Such deceased antioxidant defenses were thought to contribute to oxidative cardiomyopathy associated with doxorubicin treatment. However, it remains unclear how doxorubicin treatment causes decreased levels/activities of the aforementioned antioxidants in myocardium. Another example of drug-mediated downregulation of antioxidants and phase 2 proteins is butulinic acid. This compound is purified from *Pulsatilla chinensis*, and has been found to have selective inhibitory effects on hepatitis B virus. It was recently reported that butulinic acid inhibited hepatitis B virus replication in hepatocytes by downregulation of MnSOD expression with subsequent ROS generation and mitochondrial damage in the infected cells.[48] It was further shown that MnSOD expression was suppressed by butulinic acid-induced cAMP-response element-binding protein dephosphorylation at Ser133, which subsequently prevented MnSOD transcription.[48]

Exposure to environmental toxicants, especially heavy metals, has been repeatedly reported to lead to decreased levels/activities of various antioxidants and phase 2 proteins.[49-51] Inhibition of the above antioxidant and phase 2 defenses was often accompanied by oxidative stress and tissue injury. It has been suggested that inhibition of the antioxidants and phase 2 proteins by heavy metals and other environmental toxicants may disrupt the normal balance between antioxidant defenses and generation of ROS from various cellular metabolic pathways, leading to accumulation of ROS. Although there are extensive studies showing the inhibitory effects of heavy metals and other environmental toxicants on antioxidants and phase 2 proteins, the exact underlying mechanisms remain poorly understood. This is in contrast to our understanding of the molecular mechanisms involved in the upregulation of these cellular defenses by chemoprotective agents (Section 5). Nevertheless, several potential mechanisms have been proposed to account for the decreased levels/activities of antioxidants and phase 2 proteins following exposure to heavy metals and other toxic chemicals. As illustrated in Figure 5.1, these potential mechanisms include (1) direct inhibition or depletion of the antioxidants and phase 2 proteins due to chemical–chemical interactions, (2) posttranscriptional inhibition of the synthesis of antioxidants and phase 2 proteins, and (3) disruption of the signaling pathways and transcriptional mechanisms involved in the regulation of antioxidant and phase 2 genes.[52,53]

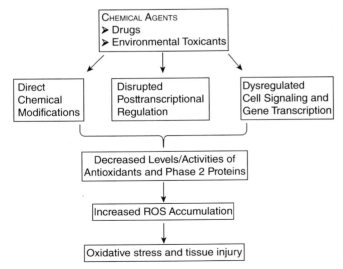

Figure 5.1 Potential mechanisms by which drugs and environmental toxicants downregulate antioxidants and phase 2 proteins, resulting in oxidative stress and tissue injury.

CONCLUSIONS AND PERSPECTIVES

Under physiological conditions, production of ROS and the functionality of endogenous antioxidants and phase 2 proteins are balanced, and thus, no obvious oxidative damage occurs. However, this balance could be disturbed, leading to overt ROS accumulation by either stimulation of ROS production or inhibition/downregulation of antioxidants and phase 2 proteins. While increased ROS generation is a well-recognized mechanism of oxidative stress, inhibition or downregulation of antioxidants and phase 2 proteins as a mechanism leading to oxidative stress has not been extensively investigated. Several classes of chemicals have been found to either inhibit or downregulate endogenous antioxidants and phase 2 proteins, resulting in oxidative stress and tissue/cell injury. However, the molecular pathways leading to inactivation or downregulation of the above endogenous defenses remain largely unknown. Future studies should focus on identification of additional chemical agents capable of inhibiting or downregulating antioxidants and phase 2 enzymes, the underlying molecular mechanisms, and the causal involvement in oxidative stress and tissue injury. In addition, future studies should also focus on identification of nonchemical agents that may cause downregulation of antioxidants and phase 2 proteins and subsequent oxidative stress. In this context, it was reported that the antioxidants MnSOD and HO-1 could be downregulated by viral infections, which might explain the increased susceptibility of the infected cells to oxidative stress injury.[54,55] Together, these studies would provide important insight into the molecular mechanisms of oxidative stress in human disease conditions, and contribute to the development of mechanistically based strategies for disease intervention.

REFERENCES

1. Halliwell, B. Biochemistry of oxidative stress. *Biochem. Soc. Trans.* **2007**, *35*(Pt 5), 1147–1150.
2. Talalay, P. Chemoprotection against cancer by induction of phase 2 enzymes. *Biofactors* **2000**, *12*(1–4), 5–11.
3. Miao, L., St Clair, D.K. Regulation of superoxide dismutase genes: Implications in disease. *Free Radic. Biol. Med.* **2009**, *47*(4), 344–356.
4. Vetrano, A.M., Heck, D.E., Mariano, T.M., Mishin, V., Laskin, D.L., Laskin, J.D. Characterization of the oxidase activity in mammalian catalase. *J. Biol. Chem.* **2005**, *280*(42), 35372–35381.
5. Kozower, B.D., Christofidou-Solomidou, M., Sweitzer, T.D., Muro, S., Buerk, D.G., Solomides, C.C., Albelda, S.M., Patterson, G.A., Muzykantov, V.R. Immunotargeting of catalase to the pulmonary endothelium alleviates oxidative

5. stress and reduces acute lung transplantation injury. *Nat. Biotechnol.* **2003**, *21*(4), 392–398.

6. Schriner, S.E., Linford, N.J., Martin, G.M., Treuting, P., Ogburn, C.E., Emond, M., Coskun, P.E., Ladiges, W., Wolf, N., Van Remmen, H., Wallace, D.C., Rabinovitch, P.S. Extension of murine life span by overexpression of catalase targeted to mitochondria. *Science* **2005**, *308*(5730), 1909–1911.

7. Dai, D.F., Santana, L.F., Vermulst, M., Tomazela, D.M., Emond, M.J., MacCoss, M.J., Gollahon, K., Martin, G.M., Loeb, L.A., Ladiges, W.C., Rabinovitch, P.S. Overexpression of catalase targeted to mitochondria attenuates murine cardiac aging. *Circulation* **2009**, *119*(21), 2789–2797.

8. Lu, S.C. Regulation of glutathione synthesis. *Mol. Aspects Med.* **2009**, *30*(1–2), 42–59.

9. Mieyal, J.J., Gallogly, M.M., Qanungo, S., Sabens, E.A., Shelton, M.D. Molecular mechanisms and clinical implications of reversible protein S-glutathionylation. *Antioxid. Redox Signal.* **2008**, *10*(11), 1941–1988.

10. Meister, A. Glutathione deficiency produced by inhibition of its synthesis, and its reversal; applications in research and therapy. *Pharmacol. Ther.* **1991**, *51*(2), 155–194.

11. Margis, R., Dunand, C., Teixeira, F.K., Margis-Pinheiro, M. Glutathione peroxidase family—an evolutionary overview. *FEBS J.* **2008**, *275*(15), 3959–3970.

12. Shiomi, T., Tsutsui, H., Matsusaka, H., Murakami, K., Hayashidani, S., Ikeuchi, M., Wen, J., Kubota, T., Utsumi, H., Takeshita, A. Overexpression of glutathione peroxidase prevents left ventricular remodeling and failure after myocardial infarction in mice. *Circulation* **2004**, *109*(4), 544–549.

13. Chrissobolis, S., Didion, S.P., Kinzenbaw, D.A., Schrader, L.I., Dayal, S., Lentz, S.R., Faraci, F.M. Glutathione peroxidase-1 plays a major role in protecting against angiotensin II-induced vascular dysfunction. *Hypertension* **2008**, *51*(4), 872–877.

14. Torzewski, M., Ochsenhirt, V., Kleschyov, A.L., Oelze, M., Daiber, A., Li, H., Rossmann, H., Tsimikas, S., Reifenberg, K., Cheng, F., Lehr, H.A., Blankenberg, S., Forstermann, U., Munzel, T., Lackner, K.J. Deficiency of glutathione peroxidase-1 accelerates the progression of atherosclerosis in apolipoprotein E-deficient mice. *Arterioscler. Thromb. Vasc. Biol.* **2007**, *27*(4), 850–857.

15. Mirochnitchenko, O., Prokopenko, O., Palnitkar, U., Kister, I., Powell, W.S., Inouye, M. Endotoxemia in transgenic mice overexpressing human glutathione peroxidases. *Circ. Res.* **2000**, *87*(4), 289–295.

16. Loscalzo, J. Nitric oxide insufficiency, platelet activation, and arterial thrombosis. *Circ. Res.* **2001**, *88*(8), 756–762.

17. Meister, A. Glutathione metabolism and its selective modification. *J. Biol. Chem.* **1988**, *263*(33), 17205–17208.

18. Mockett, R.J., Sohal, R.S., Orr, W.C. Overexpression of glutathione reductase extends survival in transgenic *Drosophila melanogaster* under hyperoxia but not normoxia. *FASEB J.* **1999**, *13*(13), 1733–1742.

19. Qiao, M., Kisgati, M., Cholewa, J.M., Zhu, W., Smart, E.J., Sulistio, M.S., Asmis, R. Increased expression of glutathione reductase in macrophages decreases atherosclerotic lesion formation in low-density lipoprotein receptor-deficient mice. *Arterioscler. Thromb. Vasc. Biol.* **2007**, *27*(6), 1375–1382.

20. Hayes, J.D., Flanagan, J.U., Jowsey, I.R. Glutathione transferases. *Annu. Rev. Pharmacol. Toxicol.* **2005**, *45*, 51–88.

21. Oakley, A.J. Glutathione transferases: New functions. *Curr. Opin. Struct. Biol.* **2005**, *15*(6), 716–723.

22. Vasiliou, V., Ross, D., Nebert, D.W. Update of the NAD(P)H:quinone oxidoreductase (NQO) gene family. *Hum. Genomics* **2006**, *2*(5), 329–335.

23. Ross, D. Quinone reductases multitasking in the metabolic world. *Drug Metab. Rev.* **2004**, *36*(3–4), 639–654.

24. Abraham, N.G., Kappas, A. Pharmacological and clinical aspects of heme oxygenase. *Pharmacol. Rev.* **2008**, *60*(1), 79–127.

25. Arosio, P., Ingrassia, R., Cavadini, P. Ferritins: A family of molecules for iron storage, antioxidation and more. *Biochim. Biophys. Acta* **2009**, *1790*(7), 589–599.

26. Nzila, A.M., Mberu, E.K., Nduati, E., Ross, A., Watkins, W.M., Sibley, C.H. Genetic diversity of *Plasmodium falciparum* parasites from Kenya is not affected by antifolate drug selection. *Int. J. Parasitol.* **2002**, *32*(12), 1469–1476.

27. Itoh, K., Chiba, T., Takahashi, S., Ishii, T., Igarashi, K., Katoh, Y., Oyake, T., Hayashi, N., Satoh, K., Hatayama, I., Yamamoto, M., Nabeshima, Y. An Nrf2/small Maf heterodimer mediates the induction of phase II detoxifying enzyme genes through antioxidant response elements. *Biochem. Biophys. Res. Commun.* **1997**, *236*(2), 313–322.

28. Kensler, T.W., Wakabayashi, N. Nrf2: Friend or foe for chemoprevention? *Carcinogenesis* **2010** *31*(1), 90–99.

29. Hayes, J.D., McMahon, M. NRF2 and KEAP1 mutations: Permanent activation of an adaptive response in cancer. *Trends Biochem. Sci.* **2009**, *34*(4), 176–188.

30. Lushchak, V.I., Bagnyukova, T.V., Lushchak, O.V., Storey, J.M., Storey, K.B. Diethyldithiocarbamate injection induces transient oxidative stress in goldfish tissues. *Chem. Biol. Interact.* **2007**, *170*(1), 1–8.

31. Viquez, O.M., Lai, B., Ahn, J.H., Does, M.D., Valentine, H.L., Valentine, W.M. N,N-diethyldithiocarbamate promotes oxidative stress prior to myelin structural changes and increases myelin copper content. *Toxicol. Appl. Pharmacol.* **2009**, *239*(1), 71–79.

32. Viquez, O.M., Valentine, H.L., Amarnath, K., Milatovic, D., Valentine, W.M. Copper accumulation and lipid oxidation precede inflammation and myelin lesions in N,N-diethyldithiocarbamate peripheral myelinopathy. *Toxicol. Appl. Pharmacol.* **2008**, *229*(1), 77–85.

33. Valentine, H.L., Viquez, O.M., Valentine, W.M. Peripheral nerve and brain differ in their capacity to resolve N,N-diethyldithiocarbamate-mediated elevations in copper and oxidative injury. *Toxicology* **2010**, *274*, 10–17.

34. Koepke, J.I., Wood, C.S., Terlecky, L.J., Walton, P.A., Terlecky, S.R. Progeric effects of catalase inactivation in

human cells. *Toxicol. Appl. Pharmacol.* **2008**, *232*(1), 99–108.
35. Bagnyukova, T.V., Vasylkiv, O.Y., Storey, K.B., Lushchak, V.I. Catalase inhibition by amino triazole induces oxidative stress in goldfish brain. *Brain Res.* **2005**, *1052*(2), 180–186.
36. Kim, S.Y., Lee, S.M., Park, J.W. Antioxidant enzyme inhibitors enhance singlet oxygen-induced cell death in HL-60 cells. *Free Radic. Res.* **2006**, *40*(11), 1190–1197.
37. Yang, E.S., Park, J.W. Antioxidant enzyme inhibitors enhance peroxynitrite-induced cell death in U937 cells. *Mol. Cell. Biochem.* **2007**, *301*(1–2), 61–68.
38. Martensson, J., Jain, A., Stole, E., Frayer, W., Auld, P.A., Meister, A. Inhibition of glutathione synthesis in the newborn rat: A model for endogenously produced oxidative stress. *Proc. Natl. Acad. Sci. U.S.A.* **1991**, *88*(20), 9360–9364.
39. Biswas, S.K., Rahman, I. Environmental toxicity, redox signaling and lung inflammation: The role of glutathione. *Mol. Aspects Med.* **2009**, *30*(1–2), 60–76.
40. Awasthi, S., Sharma, R., Singhal, S.S., Herzog, N.K., Chaubey, M., Awasthi, Y.C. Modulation of cisplatin cytotoxicity by sulphasalazine. *Br. J. Cancer* **1994**, *70*(2), 190–194.
41. Alonso, V., Linares, V., Belles, M., Albina, M.L., Sirvent, J.J., Domingo, J.L., Sanchez, D.J. Sulfasalazine induced oxidative stress: A possible mechanism of male infertility. *Reprod. Toxicol.* **2009**, *27*(1), 35–40.
42. Linares, V., Alonso, V., Albina, M.L., Belles, M., Sirvent, J.J., Domingo, J.L., Sanchez, D.J. Lipid peroxidation and antioxidant status in kidney and liver of rats treated with sulfasalazine. *Toxicology* **2009**, *256*(3), 152–156.
43. Kishi, T., Takahashi, T., Mizobuchi, S., Mori, K., Okamoto, T. Effect of dicumarol, a Nad(P)h: Quinone acceptor oxidoreductase 1 (DT-diaphorase) inhibitor on ubiquinone redox cycling in cultured rat hepatocytes. *Free Radic. Res.* **2002**, *36*(4), 413–419.
44. Chan, T.S., O'Brien, P.J. Hepatocyte metabolism of coenzyme Q1 (ubiquinone-5) to its sulfate conjugate decreases its antioxidant activity. *Biofactors* **2003**, *18*(1–4), 207–218.
45. Cullen, J.J., Hinkhouse, M.M., Grady, M., Gaut, A.W., Liu, J., Zhang, Y.P., Weydert, C.J., Domann, F.E., Oberley, L.W. Dicumarol inhibition of NADPH:quinone oxidoreductase induces growth inhibition of pancreatic cancer via a superoxide-mediated mechanism. *Cancer Res.* **2003**, *63*(17), 5513–5520.
46. Lewis, A., Ough, M., Li, L., Hinkhouse, M.M., Ritchie, J.M., Spitz, D.R., Cullen, J.J. Treatment of pancreatic cancer cells with dicumarol induces cytotoxicity and oxidative stress. *Clin. Cancer Res.* **2004**, *10*(13), 4550–4558.
47. Gustafson, D.L., Swanson, J.D., Pritsos, C.A. Modulation of glutathione and glutathione dependent antioxidant enzymes in mouse heart following doxorubicin therapy. *Free Radic. Res. Commun.* **1993**, *19*(2), 111–120.
48. Yao, D., Li, H., Gou, Y., Zhang, H., Vlessidis, A.G., Zhou, H., Evmiridis, N.P., Liu, Z. Betulinic acid-mediated inhibitory effect on hepatitis B virus by suppression of manganese superoxide dismutase expression. *FEBS J.* **2009**, *276*(9), 2599–2614.
49. Jurczuk, M., Brzoska, M.M., Moniuszko-Jakoniuk, J., Galazyn-Sidorczuk, M., Kulikowska-Karpinska, E. Antioxidant enzymes activity and lipid peroxidation in liver and kidney of rats exposed to cadmium and ethanol. *Food Chem. Toxicol.* **2004**, *42*(3), 429–438.
50. Ognjanovic, B.I., Markovic, S.D., Pavlovic, S.Z., Zikic, R.V., Stajn, A.S., Saicic, Z.S. Effect of chronic cadmium exposure on antioxidant defense system in some tissues of rats: Protective effect of selenium. *Physiol. Res.* **2008**, *57*(3), 403–411.
51. Shaikh, Z.A., Vu, T.T., Zaman, K. Oxidative stress as a mechanism of chronic cadmium-induced hepatotoxicity and renal toxicity and protection by antioxidants. *Toxicol. Appl. Pharmacol.* **1999**, *154*(3), 256–263.
52. Kamiya, T., Hara, H., Yamada, H., Imai, H., Inagaki, N., Adachi, T. Cobalt chloride decreases EC-SOD expression through intracellular ROS generation and p38-MAPK pathways in COS7 cells. *Free Radic. Res.* **2008**, *42*(11–12), 949–956.
53. Anwar-Mohamed, A., El-Kadi, A.O. Down-regulation of the detoxifying enzyme NAD(P)H:quinone oxidoreductase 1 by vanadium in Hepa 1c1c7 cells. *Toxicol. Appl. Pharmacol.* **2009**, *236*(3), 261–269.
54. Berger, M.M., Jia, X.Y., Legay, V., Aymard, M., Tilles, J.G., Lina, B. Nutrition- and virus-induced stress represses the expression of manganese superoxide dismutase *in vitro*. *Exp. Biol. Med.* **2004**, *229*(8), 843–849.
55. Abdalla, M.Y., Britigan, B.E., Wen, F., Icardi, M., McCormick, M.L., LaBrecque, D.R., Voigt, M., Brown, K.E., Schmidt, W.N. Down-regulation of heme oxygenase-1 by hepatitis C virus infection *in vivo* and by the *in vitro* expression of hepatitis C core protein. *J. Infect. Dis.* **2004**, *190*(6), 1109–1118.

6

MITOCHONDRIAL DYSFUNCTION

Yeong-Renn Chen

OVERVIEW

Mitochondria are the powerhouses of the living cell, producing most of the cell's energy by oxidative phosphorylation. The process of energy transduction requires the orchestrated action of four major respiratory enzyme complexes and ATP synthase (ATPase). High-resolution structure is now available for the hydrophilic domain and part of hydrophobic domain from complex I, complete structures of complexes II–IV, and F_1-ATPase. In addition, mitochondria play a central role in the regulation of programmed cell death. Mitochondria trigger apoptosis by rupture of electron transport and energy metabolism, by releasing (cytochrome c) and activating (caspase) proteins that mediate apoptosis, and by altering cellular redox potential via reactive oxygen species (ROS) production. The abovementioned mechanism can help to explain a variety of disease pathogenesis caused by mitochondrial defects or mitochondrial dysfunction. In this chapter we focus on the mechanism of oxygen free radical(s) production by mitochondrial electron transport chain (ETC), redox alterations, and how these mechanisms control disease process.

6.1 MITOCHONDRIA AND SUBMITOCHONDRIAL PARTICLES

The classical mitochondrial cross-section is obtained from thin sections viewed under the electron microscope.[1] Their shape is not fixed but can change continuously in the cell, and the appearance and density of cristae can be quite different in mitochondria isolated from different tissues. Thus, heart mitochondria tend to have a greater surface area of cristae than liver and others due to required periods of high respiratory activity.

The outer mitochondrial membrane possesses proteins, termed porins, which act as nonspecific pores for solutes of molecular weight less than 10 kDa, and is therefore freely permeable to ions and most metabolites (Fig. 6.1). The mitochondrial porin is also called voltage-dependent anion channel (VDAC). It should be emphasized that there is no potential gradient across the permeable outer membrane, and the voltage dependency is only seen in synthetic reconstitution experiments.

The inner membrane hosts the ETC and ATP synthase, which is energy transducing. When mitochondrial preparations are negatively stained with phosphotungstate, the component of ATPase where adenine nucleotide and phosphate bind can been seen as knobs on the matrix face (N-side or the negative compartment; Fig. 6.1).

The enzymes of the citric acid cycle are in the matrix, except for succinate dehydrogenase, which is bound to the N face of inner membrane. The protein concentration in the matrix can approach 500 mg mL^{-1}; thus, the consistency of the inner mitochondrial compartment is more gel-like rather than a dilute medium. NAD$^+$ and NADP$^+$ of the matrix pools are separate from those in the cytosol, while matrix ADP and ATP communicate with the cytoplasm through the adenine nucleotide translocator (ANT) of inner membrane (Fig. 6.2 and Fig. 6.7).

Molecular Basis of Oxidative Stress: Chemistry, Mechanisms, and Disease Pathogenesis, First Edition. Edited by Frederick A. Villamena.
© 2013 John Wiley & Sons, Inc. Published 2013 by John Wiley & Sons, Inc.

124 MITOCHONDRIAL DYSFUNCTION

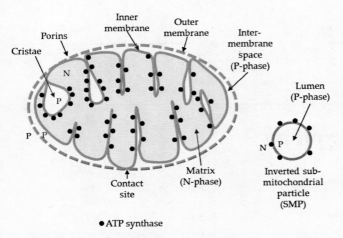

Figure 6.1 Schematic representation of a typical mitochondria and submitochondrial particle (SMP). P and N refer to the positive and negative compartments. SMP is inverted and inside out.

Mitochondria are usually prepared by gentle homogenization of the tissue in isotonic sucrose (for osmotic support and to minimize aggregation) followed by differential centrifugation to separate mitochondria from nuclei, cell debris, and microsome (fragmented endoplasmic reticulum [ER]). This method is effective with fragile tissues such as liver and brain. Preparation of mitochondria from tougher tissue such as heart must either first be incubated with the protease nagarse or briefly exposed to a blender to break the muscle fibers. Yeast mitochondria are isolated following digestion of the cell wall with snail-gut enzyme.

Ultrasonic disintegration or mechanical pressing (such as French Press) of mitochondria produces inverted submitochondrial particles (SMPs) (inside-out vesicle, Fig. 6.1). Because these have the substrate binding sites for both the respiratory chain and the F_1-ATPase on the outside, they have been much exploited

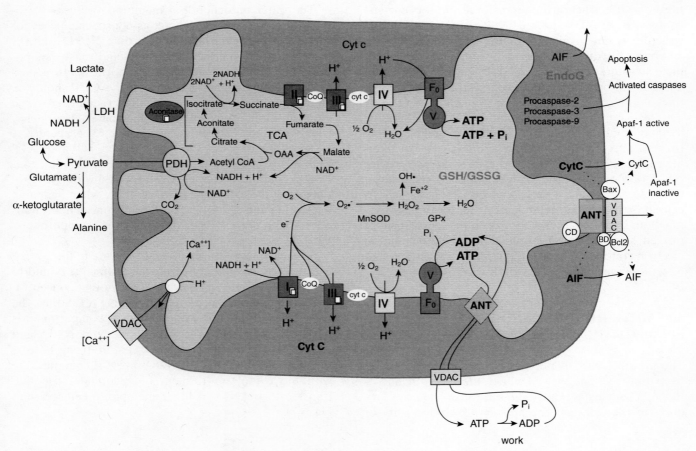

Figure 6.2 Schematic representation illustrating the relationship of oxidative phosphorylation, production of reactive oxygen species, and initiation of programmed cell death or apoptosis. (Adapted with permission from *Science* **1999**, *283*, 1482–1488. Copyright 1999 AAAS.) See color insert.

for investigation into the mechanism of energy transduction and oxygen free radical generation.

6.2 ENERGY TRANSDUCTION

Mitochondrial oxidative phosphorylation plays a major role in three important physiological or pathophysiological aspects: (1) energy transduction, (2) production of reactive oxygen species, and (3) regulation of cellular apoptosis.[2,3] Mitochondrial energy transduction for ATP synthesis is driven by reoxidation of NADH and $FADH_2$, which is carried out by the ETC located within the inner membrane. ETC mediates a stepwise electron flow from NADH or succinate to molecular oxygen through a series of electron carriers, including complex I, II, ubiquinone, complex III, cytochrome c, and complex IV. Electron flow mediated by ETC can drive proton translocation from the matrix side to the cytoplasmic side. ATP synthesis is then catalyzed by F_1,F_0-ATPase driven by the flow of protons back across the membrane. This process is called oxidative phosphorylation (Fig. 6.2).

When the Clark oxygen electrode was first being applied to mitochondrial studies, Chance and Williams proposed a convention following the typical order of addition of agents during an experiment, which allows to define so-called respiratory "states" and determine the P/O ratios. In this experiment, mitochondria were added to an oxygen electrode chamber, followed by a mixture of glutamate and malate or succinate as substrate. Respiration is slow (state 2 respiration) due to low amount of ADP and that proton circuit is not completed by H^+ re-entry through ATPase. A limited amount of ADP is added, allowing ATPase to synthesize ATP coupled to proton reentry across the membrane, which is defined as "state 3 respiration." Oxygen uptake is accelerated during state 3 respiration, and the total oxygen uptake is effectively used for ATP synthesis. When ADP is exhausted, respiration slows down and finally, anoxia is attained, which is state 4 respiration. State 4 respiration is not ADP-dependent, and technically can be attained by the ATPase inhibitor, oligomycin.

6.3 MITOCHONDRIAL STRESS

Under most physiological conditions, electron transport is tightly coupled to oxidative phosphorylation (Fig. 6.2).[2,3] In other words, electrons do not usually flow through the ETC to O_2 unless ADP is simultaneously phosphorylated to ATP. Oxidative phosphorylation is the major endogenous source of reactive oxygen species such as $O_2^{\bullet-}$ (superoxide anion radical), H_2O_2 (hydrogen peroxide), and $HO^{\bullet}$ (hydroxyl radical), which are toxic by-products of respiration (Fig. 6.2). The one-electron reduction of O_2 yields $O_2^{\bullet-}$. Therefore, $O_2^{\bullet-}$ generation by mitochondria is mainly derived from the electron leakage from ETC. For example, under the physiological conditions of state 4 respiration, the oxygen tension in mitochondria is low; oxygen consumption by ETC does not meet the needs of oxidative phosphorylation. A decrease in the rate of mitochondrial phosphorylation can increase the electron leakage from the ETC and subsequent production of $O_2^{\bullet-}$. Superoxide anion radical is converted or detoxified to hydrogen peroxide (H_2O_2) by the mitochondrial Mn superoxide dismutase (MnSOD or SOD2 isoform), and H_2O_2 is further converted to H_2O by glutathione peroxidase (GPx) in the presence of a reductant (Eq. 6.1 and Eq. 6.2). In mitochondria, the reductant is glutathione (GSH), and the produced oxidized glutathione (GSSG) is recycled to GSH by glutathione reductase in an NADPH-dependent catalysis. Normally, mitochondria produce 0.6–1.0 nmol of H_2O_2 min^{-1} mg^{-1} protein, accounting for 2% oxygen uptake, under the conditions of state 4 respiration.[4] The produced H_2O_2 may modestly induce mitochondrial oxidative stress, or diffuse to cytosol, acting as a signaling molecule to trigger important physiological response, such as vessel dilation:

$$O_2^{\bullet-} + O_2^{\bullet-} \rightarrow H_2O_2 + O_2 \quad (6.1)$$

$$H_2O_2 + 2GSH \rightarrow 2H_2O + GSSG \quad (6.2)$$

$$H_2O_2 + Fe^{2+} \rightarrow Fe^{3+} + HO^- + HO^{\bullet} \quad (6.3)$$

$$H_2O_2 + O_2^{\bullet-} \rightarrow O_2 + HO^- + HO^{\bullet} \quad (6.4)$$

The proteins of mitochondrial ETC are rich in metal cofactors such as heme (complexes II, III, and IV) and iron–sulfur (Fe–S) clusters (complexes I, II, and III). Oxidative stress in mitochondria can be greatly enhanced in the presence of reduced transition metal because H_2O_2 can be converted to the highly reactive hydroxyl radical via the Fe^{2+}-dependent Fenton reaction (Eq. 6.3). Alternatively, hydroxyl radical also can be produced via the mechanism of Fe^{3+}-catalyzed Haber–Weiss mechanism (Eq. 6.4). Under patho-physiological conditions of certain diseases, decreasing the function of ETC and oxidative phosphorylation can increase mitochondrial ROS production. Acute ROS exposure can inactivate the iron–sulfur centers of complexes I, II, and III, and tricarboxylic acid (TCA) cycle aconitase, resulting in a shutdown of mitochondrial energy production, and chronic ROS exposure can result in oxidative damage to mitochondrial and cellular proteins, lipids, and nucleic acids. Therefore, oxidative stress induced by mitochondrial ROS overproduction is closely linked to disease pathogenesis.

6.4 SUPEROXIDE RADICAL ANION GENERATION AS MEDIATED BY ETC AND DISEASE PATHOGENESIS

Under these conditions, a decrease in the rate of mitochondrial oxidative phosphorylation can increase the oxygen free radicals, in the form of $O_2^{\bullet-}$, from the early stages of ETC.[2-8] Two segments of the ETC have been widely hypothesized to be responsible for $O_2^{\bullet-}$ generation. One, on the NADH dehydrogenase (NDH) (or flavin subcomplex) of complex I, operates via electron leakage from the reduced flavin mononucleotide (FMN). The other, on complex III, mediates $O_2^{\bullet-}$ production through Q-cycle mechanism, in which electron leakage presumably results from auto-oxidation of ubisemiquinone and reduced cytochrome b. Furthermore, there is an increasing body of evidence that links overproduction of $O_2^{\bullet-}$ with a defect in complex II. Under ischemic conditions, phosphorylation of complex IV may also increase electron leakage and $O_2^{\bullet-}$ production. Therefore, mediation of $O_2^{\bullet-}$ production by complex II or complex IV is highly relevant in disease conditions.

6.4.1 Mediation of $O_2^{\bullet-}$ Generation by Complex I

Mitochondrial complex I (EC 1.6.5.3. NADH: ubiquinone oxidoreductase [NQR]) is the first energy-conserving segment of the ETC.[9-15] There are two reasons why complex I is the least well-understood component of the mammalian mitochondrial ETC. First, it is very large, about the same size as the large subunit of a ribosome. Purified bovine heart complex I contains up to 45 different polypeptides with a total molecular mass approaching 980 kDa. With the use of chaotropic anions such as perchlorate, complex I can be resolved into three fractions (or three subcomplexes): a flavoprotein fraction (Fp), an iron-sulfur protein fraction (Ip), and a hydrophobic fraction (Hp). The Fp contains the catalytic activity of NDH. Second, the redox centers, apart from the FMN, are iron–sulfur centers, which cannot be studied by optical spectroscopy but instead require the more difficult technique of low-temperature electron paramagnetic resonance (EPR).[13] The complex I catalyzes the transfer of two electrons from NADH to ubiquinone in a reaction that is coupled with the translocation of four protons across the membrane, and that is inhibited by rotenone and piericidin A. The redox centers of complex I that are involved in mediation of two-electron transfer include a noncovalent FMN, eight iron–sulfur clusters, and ubiquinone. Current evidence suggests that the proton translocation stoichiometry is $4H^+/2e^-$.

In addition to its functions of electron-transfer and energy transduction, the catalysis of complex I provides the major source of oxygen free radicals in mitochondria. Studies using isolated mitochondria indicate $O_2^{\bullet-}$ and H_2O_2 production by complex I is mainly controlled by $NADH/NAD^+$ redox coupling and succinate-induced reverse electron transfer. Two segments of complex I are widely recognized to be responsible for enzyme-mediated $O_2^{\bullet-}$ generation: one is involved in the cofactor of FMN and FMN-binding moiety at the 51 kDa polypeptide (FMN-binding subunit), and the other is located on the ubiquinone-binding site that mediates ubiquinone reduction.

6.4.1.1 The Role of FMN Moiety
The mechanism of $O_2^{\bullet-}$ generation is likely derived from direct reaction of reduced FMN ($FMNH_2$) or reduced FMN semiquinone radical ($FMNH^{\bullet-}$) with molecular oxygen.[14,15] Based on investigations using the models of intact complex I and its Fp subcomplex (NDH) with EPR spin-trapping technique, enzyme-mediated $O_2^{\bullet-}$ generation by intact complex I or NDH can be inhibited by a general inhibitor of flavoprotein, diphenyleneiodonium chloride (DPI). Addition of free FMN can increase enzyme-mediated $O_2^{\bullet-}$ generation by intact complex I or NDH. However, FMN addition could not reverse the inhibition by either DPI treatment or heat denaturation of either intact complex I or NDH. These results support the involvement of FMN and FMN-binding protein moiety in the mediation of $O_2^{\bullet-}$ generation by the Fp of complex I. Further evidence has been provided by redox titration of mitochondrial $O_2^{\bullet-}$ generation using mitochondrial inner membrane (SMP). The redox potential of $O_2^{\bullet-}$ production at the site of complex I was determined to be ~ -295 mV. The midpoint potential of the superoxide-producing site at complex I clearly differs from the values of iron–sulfur clusters but resembles the value of FMN (~ -310 mV), also supporting the FMN moiety as a potential site of superoxide generation.

6.4.1.2 The Role of Ubiquinone-Binding Domain
The second site associated with complex I-mediated $O_2^{\bullet-}$ generation has been proposed to be the site of ubiquinone reduction, presumably due to the formation of an unstable ubisemiquinone radical (SQ) that is the source of $O_2^{\bullet-}$.[16,17] The ubisemiquinone radical is formed via incomplete reduction of ubiquinone under the conditions of enzyme turnover. Evidence of SQ involvement has been investigated using EPR detection of enzymatic system in the presence or absence of spin trap. SQ can be categorized as a stable ubisemiquinone radical that is EPR detectable, and as an unstable ubisemiquinone radical that is not detected by EPR due to a short life at room temperature. Unstable ubisemiquinone radical is thus a source of $O_2^{\bullet-}$ under physiological conditions. A nitrone spin trap such as DEPMPO or DMPO normally does not trap unstable SQ due to a

less reactive nature of SQ. It is thus anticipated that $O_2^{\bullet-}$ generated from unstable SQ can be trapped by DMPO or DEPMPO and ready for EPR detection. Studies using isolated complex I and EPR spin trapping indicates that enzyme-mediated $O_2^{\bullet-}$ generation driven by NADH was enhanced twofold in the presence of ubiquinone-1 (Q_1) compared with that in the absence of Q_1. More than 80% of the enhanced $O_2^{\bullet-}$ generation can be inhibited by rotenone, thus supporting (but not necessarily proving) a $O_2^{\bullet-}$ generation mechanism involving the reduction of ubiquinone to an unstable semiquinone radical. The formation of an unstable semiquinone radical under enzyme turnover conditions is likely mediated through the ubiquinone-binding site of the Hp subcomplex in complex I.

The formation of a stable ubisemiquinone-1 (SQ_1) has been detected by direct EPR under enzyme turnover conditions (or steady state) in the presence of intact complex I, NADH, and Q_1. The signal of formed SQ_1 can be enhanced by rotenone or piericidin A. Presumably, binding of rotenone or piericidin A to the ubiquinone-binding site of complex I decreases the formation of unstable SQ1 and subsequently increases stable SQ_1 at steady state, thus reducing $O_2^{\bullet-}$ generation mediated by ubiquinone-binding domain.

Ohnishi et al.[16] have reported the presence of two distinct semiquinone species with different spin relaxation times in complex I *in situ* based on a study using rat heart SMP. They are SQ_{Nf} (semiquinone with fast relaxing time) and SQ_{Ns} (semiquinone with slow relaxing time). The SQ_1 detected during steady state at room temperature can be logically assigned to be SQ_{Ns}, because the signal is not quenched by rotenone or piericidin A. The unstable semiquinone as a source of $O_2^{\bullet-}$ is likely to be SQ_{Nf}, which is highly sensitive to rotenone or piericidin A.

6.4.1.3 The Role of Iron–Sulfur Clusters
Mammalian complex I hosts as many as eight iron–sulfur clusters, which are essential redox centers controlling the electron transfer during enzyme turnover.[12,13,17,18] Iron–sulfur clusters identified with the use of EPR include the binuclear clusters N1a and N1b and the tetranuclear clusters N2, N3, N4, and N5. Sequence comparisons suggest that complex I contains two more tetranuclear clusters, N6a and N6b, in subunit TYKY. Based on the X-ray structure of the hydrophilic domain of bacterial respiratory complex I from *Thermus thermophilus*, the organization of iron–sulfur clusters in complex I has been elucidated. The main route for electron transfer within the enzyme is likely to be NADH→FMN→N3→N1b→N4→N5→N6a→N6b→N2→ubiquinone. N1a (2Fe-2S center) of the 24 kDa subunit is not reduced by NADH and has the lowest midpoint potential (~ –370 mV).

N1a may act as an antioxidant to protect the enzyme from oxidative damage. The iron–sulfur clusters in complex I should play a secondary role in the formation of $O_2^{\bullet-}$ by complex I. Studies using intact mitochondria from rat brain indicates rotenone-insensitive $O_2^{\bullet-}$ generation by complex I is far less sensitive to CMB (*p*-chloromercuribenzoate, iron–sulfur cluster blocker) treatment than DPI treatment.[14] However, several research groups have proposed that in intact mitochondria and SMP, most of the $O_2^{\bullet-}$ may be generated around the N2/ubiquinone site, suggesting that electron leakage at the N2 center is coupled to incomplete reduction of ubiquinone.

The 75-kDa subunit hosts three iron–sulfur clusters (N1b, N4, N5) of mammalian complex I. Residues 77–97 of the 75-kDa mature protein, $_{100}$WNILTNSEKTKKAREGVMEFL$_{120}$ (p75), may exhibit a β-sheet-α-turn-helix (blue region in the homolog model of 75-kDa subunit in Fig. 6.3A). A similar structure (^{91}MVVDTLSDVVREAQAGMVEFT^{111} in the Nqo3 of 2FUG) is observed in *T. thermophilus*. The nearest distances between the major α-helix of p75 and the iron–sulfur clusters are about 11.2 Å (from M_{117} (Cα) to N5), 11.6 Å (to N1b), and 14.3 Å (to N4) based on the homolog model (Fig. 6.3A). The peptide of p75 can be designed as a B cell epitope. The polyclonal antibody generated against p75 is named Ab75. Binding of Ab75 antibody to complex I resulted in the inhibition of $O_2^{\bullet-}$ generation by 35% as assayed by EPR spin-trapping with DEPMPO.

6.4.1.4 The Role of Cysteinyl Redox Domains
In mitochondria, the generation of $O_2^{\bullet-}$ and the oxidants derived from it can act as a redox signal in triggering cellular events such apoptosis, proliferation, and senescence.[17,19–24] The mitochondrial redox pool is enriched in GSH with a high physiological concentration (in millimolar range). Overproduction of $O_2^{\bullet-}$ and $O_2^{\bullet-}$-derived oxidants increases the ratio of GSSG to GSH. Moreover, the proteins of the mitochondrial ETC are rich in protein thiols. Complex I is the major component of the ETC to host protein thiols, which comprise structural thiols involved in the ligands of iron–sulfur clusters and the reactive/regulatory thiols that are thought to function in antioxidant defense and redox signaling. Physiologically, the complex I-derived regulatory thiols have been implicated in the regulation of respiration, nitric oxide utilization, and redox status of mitochondria. It has been well documented that the 51-kDa and 75-kDa subunits of the complex I hydrophilic domain are two of the major polypeptides that host regulatory thiols.

The C_{206} (Cys$_{206}$) moiety of the 51-kDa subunit plays a unique role as a reactive thiol in oxidative damage to complex I. The C_{206} of the 51-kDa subunit is also involved

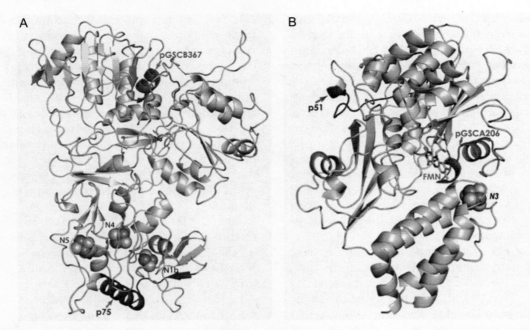

Figure 6.3 (A) Homology model of the 75-kDa subunit using the crystal structure of *T. thermophilus* (2FUG) as a template. Arrows show the domains of pGSCB367 and p75, denoted by red and blue ribbons. (B) Homology model of the 51-kDa subunit. Arrows show the domains of pGSCA206 and p51, denoted by red and blue ribbons. (Adapted with permission from *J. Pept. Sci.* **2011**, *96*, 207–221. Copyright 2011 John Wiley & Sons.) See color insert.

in site-specific S-glutathionylation (via binding of glutathione). The peptide identified to form protein thiyl radical and GS-binding is $_{200}$GAGAYIC^{206}GEETALIESIEGK$_{219}$, which is highly conserved in the bacterial, fungal, and mammalian enzymes (90% sequence identity to the bacterial enzyme). An X-ray crystal structure of the hydrophilic domain of respiratory complex I from *T. thermophilus* indicates that this conserved cysteine (Cys$_{182}$ in *T. thermophilus*) is only 6 Å from the FMN, which is consistent with the role of C$_{206}$ as a redox-sensitive thiol and FMN's serving as a source of O$_2^{\bullet-}$.[12]

The polypeptide of 75 kDa is the other subunit of complex I to be involved in redox modification via S-glutathionylation. Based on the liquid chromatography–tandem mass spectrometry (LC-MS-MS) analysis, S-glutathionylation of C$_{367}$ can be induced by oxidized GSSG through protein thiol disulfide exchange. C$_{554}$ and C$_{727}$ were S-glutathionylated when the complex I of bovine heart mitochondria was oxidatively stressed by diamide. The GS-binding peptides identified include $_{361}$VDSDTLC^{367}TEEVFPTAGAGTDLR$_{382}$, $_{544}$MLFLLGADGGC^{554}ITR$_{557}$, and $_{713}$AVTEGAHAVEEPSIC^{727}. Although the identified GS-binding domains are not conserved in the bacterial and fungal enzymes (only 27.3% sequence identity in the bacterial enzyme), they are highly conserved in the mammalian enzymes.

Based on an EPR spin-trapping study, GSSG-induced glutathionylation of complex I at the 51-kDa and the 75-kDa subunits affects the O$_2^{\bullet-}$ generation activity of complex I by marginally decreasing electron leakage and increasing electron transfer efficiency. High dosage of GSSG or diamide-induced glutathionylation tends to decrease the catalytic function of complex I and increase enzyme-mediated O$_2^{\bullet-}$ generation. Binding of antibodies against the peptide of ^{200}GAGAYIC$_{206}$GEETALIESIEGK^{219} (pGSCA206 in Fig. 6.3B) decreases complex I-mediated O$_2^{\bullet-}$ generation by 37%. Since FMN serves as a source of O$_2^{\bullet-}$, binding of antibodies may prevent molecular oxygen from accessing FMN, resulting in a subsequent decrease in O$_2^{\bullet-}$ production. In addition, binding of the antibodies against the peptide of ^{361}VDSDTLC$_{367}$TEEVFPTAGAGTDLR^{382} (pGSCB367 in Fig. 6.3A) inhibits O$_2^{\bullet-}$ production by complex I by 57%. The distance of pGSCB367 to iron–sulfur clusters is relatively long; thus, binding of antibodies to the 75-kDa subunit of complex I likely triggers long range conformational changes in the 75-kDa polypeptide to reduce O$_2$ interactions. Binding of either antibody does not affect the electron transfer activity of complex I. Therefore, the redox domains involved in the glutathionylation are responsible for modulating electron leakage for O$_2^{\bullet-}$ production by complex I.

Other types of complex I-derived redox modifications, such as protein tyrosine nitration, have been detected in the postischemic myocardium and aging process. Several redox domains targeted by tyrosine nitration have been identified, but their roles in the regulation of $O_2^{\bullet -}$ production remain unclear and have yet to be defined.

6.4.1.5 Complex I, Free Radicals, and Parkinsonism

Two important complex I inhibitors, 1-methyl-4-phenylpyridinium (MPP^+) and rotenone, have been used to induce permanent symptoms of Parkinson's disease by destroying neurons in the substantia nigra.[25,26] They have been used in the study of animal disease model. MPTP (1-methyl-4-phenyl-1,2,3,6-tetrahydropyridine) is the precursor of MPP^+ and a potential neurotoxin. Injection of MPTP causes a rapid onset of Parkinsonism. MPTP is not toxic. However, MPTP is a lipophilic compound that can cross the blood–brain barrier. Once inside the brain, MPTP is metabolized to toxic anion MPP^+ by monoamine oxidase-B of glial cells. MPP^+ primarily kills dopamine-producing neurons of the substantia nigra. MPP^+ inhibits complex I of the ETC and leads to overproduction of oxygen free radicals and other free radicals that contribute to neuronal death in dopaminergic cells. Like MPP^+, rotenone also interferes with complex I and induces the buildup of free radicals, causing cell death.

6.4.2 Mediation of $O_2^{\bullet -}$ Generation by Complex II

Mitochondrial complex II (EC 1.3.5.1. succinate:ubiquinone oxidoreductase) is a key membrane complex in the TCA cycle that catalyzes the oxidation of succinate to fumarate in the mitochondrial matrix.[27–32] Succinate oxidation is coupled to reduction of ubiquinone at the mitochondrial inner membrane as one part of the respiratory ETC. Complex II mediates electron transfer from succinate to ubiquinone through the prosthetic groups of flavin adenine nucleotide (FAD); [2Fe-2S] (S1), [4Fe-4S] (S2), and [3Fe-4S] (S3); and heme b. The enzyme is composed of two parts: a soluble succinate dehydrogenase (SDH) and a membrane-anchoring protein fraction. SDH contains two protein subunits, a 70-kDa protein with a covalently bound FAD, and a 30-kDa iron–sulfur protein hosting S1, S2, and S3 iron–sulfur clusters. The membrane-anchoring protein fraction contains two hydrophobic polypeptides (14 kDa and 9 kDa) with heme b binding.

The catalysis of complex II is believed to contribute to superoxide generation in mitochondria. Two regions of the enzyme complex are hypothesized to be responsible for generating the $O_2^{\bullet -}$. One is located on the FAD cofactor and is modulated by $FADH^{\bullet}$ semiquinone, while the other is likely located on the ubiquinone-binding site which acts in the mediation of ubiquinone reduction. Defects in the complex II leading to $O_2^{\bullet -}$ overproduction have been linked with the pathogenesis of certain diseases.

6.4.2.1 The Role of FAD Moiety

The generation of $O_2^{\bullet -}$ by the FAD moiety of complex II may arise from $FADH_2$ auto-oxidation or $FADH^{\bullet -}$ auto-oxidation.[27,31,32] Evidence has been documented in the mammalian and bacterial enzyme via use of the inhibitor, 2-thenoyltrifluoroacetone (TTFA). However, until now there has been no direct evidence supported by EPR. TTFA is a classical inhibitor for the ubiquinone reduction of complex II by occupying its ubiquinone-binding sites. Therefore, binding of TTFA to the enzyme induces electron accumulation at the early stage of complex II. Based on the EPR spin-trapping study using isolated complex II and a supercomplex (SCR) hosting complex II and complex III, the inhibitory effect of TTFA on the $O_2^{\bullet -}$ generation by complex II or SCR indicates that $FADH_2$ auto-oxidation mediated by FAD-binding moiety partially contributes to the $O_2^{\bullet -}$ production. The production of $O_2^{\bullet -}$ by complex II or SCR is minimized under the conditions of enzyme turnover in the presence of Q_2 or cyt c.

6.4.2.2 The Role of Ubiquinone-Binding Site

Mammalian complex II contains at least two ubiquinone-binding sites (namely, Qp and Qd).[28–30] Qp is on the matrix side, and Qd is near the intermembrane space site. Because the ubiquinone-binding site in complex II is close to the TTFA-binding site, EPR signal derived from ubisemiquinone of complex II is sensitive to TTFA. The location of Qp site and its quinone-binding pocket can be revealed by comparison of mammalian complex II (pdb 1ZOY) structure with the superimposed *Escherichia coli* SQR structure (pdb 1NEK). The quinone-binding pocket at the Qp site is formed by second helix of cybL, second helix of cybS, and the [3Fe-4S] ligating region. The residues in the quinone-binding site, His B216, Pro B169, Trp B172, Trp B173, Ile B218, Ser C42, Ile C43, Arg C46, Tyr D91, and Asp D90, are strictly conserved among mammalian complex II and *E. coli* SQR (His B207, Pro B160, Trp B163, Trp B164, Ile B209, Ser C27, Ile C28, Arg C31, Tyr D83, and Asp D82 in Fig. 6.4). These conserved residues form the quinone-binding environment (Fig. 6.4). Under physiological conditions, ubiquinone reduction mediated by complex II is well controlled via quinone-binding environment. Alteration(s) of quinone-binding environment results in incomplete reduction of ubiquinone, leading to increasing ROS production.

6.4.2.3 Mutations of Complex II Are Related with Mitochondrial Diseases

Mutations of complex II cause electron leakage from the prosthetic groups in the electron transfer chain, producing ROS and leading to tumor formation and neurodegeneration.[29,30,33–36] Electrons could leak from the FAD group and the Qp site.

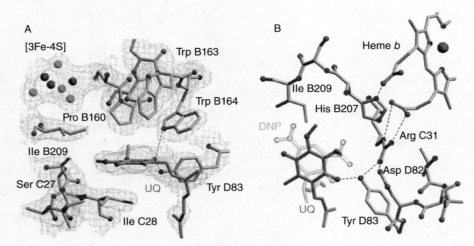

Figure 6.4 (A) Hydrophobic residues in the ubiquinone-binding site of *E. coli* SQR. (B) Polar interactions in the ubiquinone-binding site of *E. coli* SQR. (Adapted with permission from *Science* **2003**, *299*, 700–704. Copyright 2003 AAAS.) See color insert.

However, mutations at the residue around the iron–sulfur clusters or heme *b* environment may also cause electron leakage for ROS production. For example, (1) mutation of Cys to Tyr at position B73 expands [2Fe-2S] (S1 center) ligating pocket and destroys its ligation. In this instance, electrons cannot be transferred from FAD to [4Fe-4S] (S2 center) but can leak into the matrix, thus producing ROS and resulting in pheochromocytoma symptom.[36] (2) Mutation of His to Leu at D79 destroys the ligation of heme *b*, thus abolishing ubiquinone reduction at Qp site.[33] Consequently, electron is accumulated at [3Fe-4S] (S3 center), resulting in leakage into the membrane and ROS production, which manifests itself as head-and-neck paraganglioma. (3) Mutation of Pro B169 (Pro B160 in the *E. coli* SQR, Fig. 6.4) to Arg alters the quinone-binding environment via change of [3Fe-4S] ligation conformation and destruction of Qp site, increasing ROS production and causing hereditary paraganglioma in humans.[34] (4) Mutation of Tyr D91 (Tyr D83 in the *E. coli* SQR, Fig. 6.4) to Cys alters the quinone-binding environment via destruction of binding of ubiquinone at the Qp site, resulting in ROS production and leading to head-and-neck paraganglioma in humans.[35] (5) In *Caenorhabditis elegans* (*mev-1*), mutation of the SDHC (equivalent to CybL of human complex II) or the residue equivalent to Ile C28 confers hypersensitivity to hyperoxia and to oxidative stress.[29] The mutation affects the Qp site, leads to ROS overproduction, the accumulation of aging markers, and a shortened life span.

6.4.2.4 Mitochondrial Complex II in Myocardial Infarction
In the animal disease model of myocardial ischemia and reperfusion injury, oxidative impairment of the electron transfer activity of complex II is marked in the region of myocardial infarction.[32,37,38] The injury of complex II is closely related to the mitochondrial dysfunction or loss of FAD-linked oxygen consumption in the postischemic heart. Further evaluation of the redox biochemistry of complex II indicated alternations of oxidative posttranslational modification are marked in the postischemic myocardium, including deglutathionylation of 70-kDa FAD-binding subunit (loss of GSH binding) and an increase in the level of protein tyrosine nitration of the 70 kDa subunit. Therefore, oxidative inactivation and the consequence of redox modifications may be served as the disease marker of myocardial infarction.

6.4.3 Mediation of $O_2^{\cdot-}$ Generation by Complex III

6.4.3.1 The Q-Cycle Mediated by Complex III
Mitochondrial complex III catalyzes the electron transfer from ubiquinol (QH_2) to ferricytochrome *c*, which is coupled to proton translocation for ATP synthesis.[31,39–43] The redox centers of complex III consist of QH_2, cytochrome b_L (low potential *b* or b_{566}), b_H (high potential *b* or b_{562}), and c_1, and the Rieske iron–sulfur cluster (RISP). The electron transfer from QH_2 to cytochrome *c* follows the Q-cycle mechanism.

In the Q-cycle mechanism, there are two semiquinones formed in different parts of the cycle (Fig. 6.5A). An unstable semiquinone ($Q_o^{\cdot-}$) is formed near the cytoplasmic site. The other semiquinone ($Q_i^{\cdot-}$), formed near the matrix side (negative site or inner site), is a stable semiquinone and EPR detectable. During the

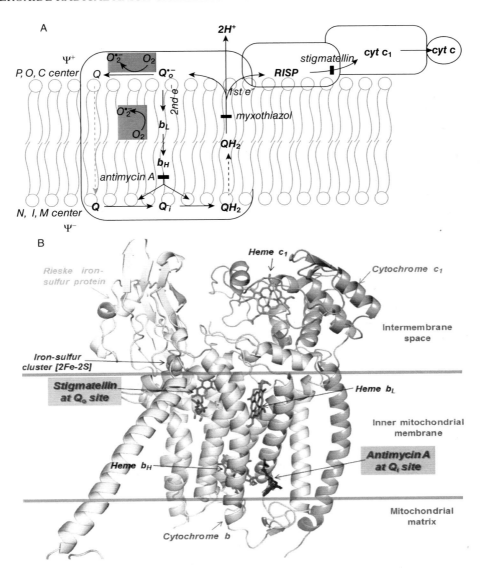

Figure 6.5 (A) Superoxide generation mediated by the Q-cycle mechanism in the complex III. The scheme is adapted from Reference 31 with modification. Gray areas symbolize the reactions involved in $O_2^{\bullet-}$ production. P, O, and C represent positive, outside, and cytoplasmic side, respectively. N, I, and M stands for negative, inside, and matrix side, respectively. (B) X-ray structure (pdb 1PPJ) of complex III shows Q_o site (occupied by Q_o site inhibitor, stigmatellin in green color) is located immediately next to the intermembrane space. Structure also shows the Q_i site (occupied by Q_i site inhibitor, antimycin A in blue color) is located next to the matrix site. The subunits of cytochrome b, cytochrome c_1, and Rieske iron–sulfur protein (RISP) are denoted by cyan, pale green, and yellow ribbons, respectively. See color insert.

enzyme turnover of complex III, one electron from QH_2 is sequentially transferred to RISP, then cyt c_1, and then ferricytochrome c. QH_2 contains two electrons, but cyt c only accepts one electron. This leaves an unstable semiquinone formed at the cytoplamic site ($Q_o^{\bullet-}$). One electron from unstable semiquinone is transferred to low-potential heme b (b_L), and then transferred to high-potential heme b (b_H). One electron from b_H is then transferred to ubiquinone (Q) to form a stable semiquinone at the matrix side. The stable semiquinone can accept one electron from a second single turnover to complete the cycle, forming QH_2.

6.4.3.2 Role of Q Cycle in $O_2^{\bullet-}$ Generation
The Q-cycle mechanism provides an important theoretical basis to explain why complex III is an endogenous

source of ROS in mitochondria.[31,40,43] The semiquinone formed near the cytoplasmic space (or outer site, or positive site in Figure 6.5A) is an unstable semiquinone radical, namely $Q_o^{\bullet-}$. $Q_o^{\bullet-}$ is a source of $O_2^{\bullet-}$ due to its poor stability. Antimycin, myxothiazol, and stigmatellin are inhibitors of complex III. Antimycin acts on Q_i site, preventing the formation of the relatively stable $Q_i^{\bullet-}$. In the presence of this inhibitor, a reduction of cytochrome b occurs by accepting an electron from the Q_o as the other electron from QH_2 passes to the RISP and onward to cytochromes c_1 and c. Therefore, antimycin inhibition increases the formation of $Q_o^{\bullet-}$ and subsequent $O_2^{\bullet-}$ production. Myxothiazol blocks the formation of $Q_o^{\bullet-}$, while stigmatellin inhibits electron transfer to the cytochrome c_1, thus decreasing the formation of $Q_o^{\bullet-}$ and subsequent $O_2^{\bullet-}$ production (Figure 6.5A).

6.4.3.3 The Role of Cytochrome b_L in $O_2^{\bullet-}$ Generation

As indicated in Figure 6.5A of Q-cycle mechanism, low potential heme b (cytochrome b_L or b_{566}) accepts one electron from unstable semiquinone.[31,44] The midpoint potential of b_L is −50 mV that tends to give away the electron. *In vitro* evidence has supported the notion of ferrocytochrome b_L to be a source of $O_2^{\bullet-}$. Evidence has shown the involvement of $O_2^{\bullet-}$ generation during intramolecular electron transfer from ferrocytochrome b to ferricytochrome c_1.

6.4.3.4 Bidirectionality of Superoxide Release as Mediated by Complex III

Since the anionic form of $O_2^{\bullet-}$ is too strongly negatively charged to readily pass the inner membrane, $O_2^{\bullet-}$ production mediated by complex III can exhibit a distinct membrane sidedness or "topology."[41] The Q_o site, one of the sources of $O_2^{\bullet-}$, is located near the intermembrane space (Fig. 6.5B). The close proximity of the Q_o site to the intermembrane space can result in invariable release of a fraction of complex III-derived $O_2^{\bullet-}$ to the cytoplasmic side of the inner membrane. Experimental evidence has supported that ~50% of total electron leak in mitochondria lacking CuZnSOD accounts for extramitochondrial superoxide release. The remaining ~50% of electron leak is due to superoxide released to the matrix. Hydrogen peroxide derived from dismutation of $O_2^{\bullet-}$ in the mitochondrial matrix can induce aconitase inhibition. The inhibition of aconitase by H_2O_2 in the matrix can be relieved by the Q_o site inhibitor stigmatellin. These results indicate that superoxide is released to both sides of the inner mitochondrial membrane by the semiquinone at the Q_o site of complex III. In contrast, complex I-mediated superoxide is exclusively released into the matrix and that no detectable levels escape from intact mitochondria, indicating the hydrophilic domain as the major site of electron leak.

6.4.4 Complex IV

Mitochondrial complex IV is cytochrome c oxidase (CcO) that catalyzes the sequential transfer of four electrons from the reduced cytochrome c (or ferrocytochrome c) to O_2 forming $2H_2O$.[45–48] The reduction of O_2 to H_2O provides the driving force for protein translocation and ATP synthesis. High-resolution structure of the bovine heart complex IV, in both fully oxidized and fully reduced states, has been obtained. The catalytic function of complex IV is mainly controlled by subunit I and subunit II in which both subunits are encoded by mitochondria DNA (Fig. 6.6). The redox centers of complex IV includes two Cu_A (in a cluster with sulfur atoms in subunit II), two a-type heme (heme a and heme a_3 are located in subunit I), and Cu_B (in subunit I). The subunit II, in addition to two transmembrane helices, has a globular domain, folded as a β-barrel, which projects into the intermembrane space. This is the location of two CuA that undergoes a one-electron oxidation–reduction reaction. Binuclear Cu_A center receives electrons, one at a time, from ferrocytochrome c.

Heme a and heme a_3 of complex IV, located approximately 15 Å below the P-surface of the membrane bilayer, are sandwiched between some of the 12 transmembrane α-helices of subunit I. Heme a, slightly closer to the Cu_A center, is the electron acceptor from Cu_A. Heme a_3 is within a few angstroms of heme a. One axial coordination position to the iron of heme a_3 is not occupied by amino acid chain. This is the position where

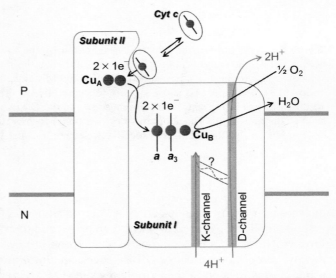

Figure 6.6 Schematic representation of subunits I and II of complex IV. Copper atoms are represented by green spheres, and the iron atoms of the heme a and heme a₃ are represented by brown spheres.

oxygen binds before its reduction to water. Heme a_3 is also the site of binding of several inhibitors, including cyanide, azide, NO, and CO. Cu_B, the third copper, is immediately adjacent to heme a_3; it has three histidine ligands, suggesting a fourth coordination position may be occupied by a reaction product during a specific stage of oxygen reduction reaction. One of three ligands, histidine-240 (H_{240} in bovine complex IV), is cross-linked to a nearby specific tyrosine residue (Y_{244} in bovine complex IV) through a covalent bond.[47] This specific tyrosine is proved in participation to the formation of protein-tyrosyl radical intermediate during enzyme turnover or under oxidant stress induced by H_2O_2. Cu_B and nearby heme a_3 form the so-called binuclear center which tightly controls oxygen binding, the formation of intermediates (oxy, compound P, and compound F), and reduction of oxygen to H_2O. Electron leakage for $O_2^{\bullet-}$ generation is not likely occurring during the enzyme turnover of complex IV or under normal physiological conditions. However, $O_2^{\bullet-}$ production can be mediated by complex IV under certain pathophysiological conditions. The activity of complex IV is modulated in response to O_2 tension. Submitochondrial particle exposed to hypoxic conditions *in vitro* show reduced complex IV activity. Hypoxia exposure of murine monocyte macrophage results in cAMP-mediated inhibition of complex IV activity. The inhibition of complex IV is accompanied by phosphorylation mediated by protein kinase A (PKA) and markedly increased $O_2^{\bullet-}$ generation by complex IV. Conditions of myocardial ischemia also activate mitochondrial PKA, enhancing phosphorylation of complex IV, vastly increasing reactive oxygen species production by complex IV, and augmenting ischemic and subsequent reperfusion injuries. Therefore, specific inhibitors of PKA have been proposed to render cardiac protection against ischemia/reperfusion injury via scavenging the pro-oxidant activity of complex IV.

SUMMARY

Mitochondrial superoxide anion radical is generated at complex III and complex I. The former site likely releases $O_2^{\bullet-}$ into the intermembrane space, while $O_2^{\bullet-}$ from complex I may be released into the matrix (Fig. 6.7). In addition, $O_2^{\bullet-}$ may be membrane permeable in its protonated form HO_2. Under the pathological conditions of certain diseases, overproduction of $O_2^{\bullet-}$ can be mediated by complex II (pheochromocytoma symptom, head-and-neck paraganglioma, ischemia/reperfusion injury) and complex IV (ischemia and ischemia/reperfusion injury). $O_2^{\bullet-}$ mediated by the ETC is mainly controlled by the moieties of flavin-binding (complexes I and II), ubiquinone-binding (complexes I, II, and III), cytochrome b_L (complex III), and possible iron–surfur clusters (complex I).

Mitochondria have developed antioxidant defense to detoxify $O_2^{\bullet-}$ and $O_2^{\bullet-}$-derived oxidant. Superoxide dismutase 2 (SOD2) and glutathione couple play the critical role in scavenging the reactive oxygen species produced in mitochondria (Fig. 6.7). Using glutathione

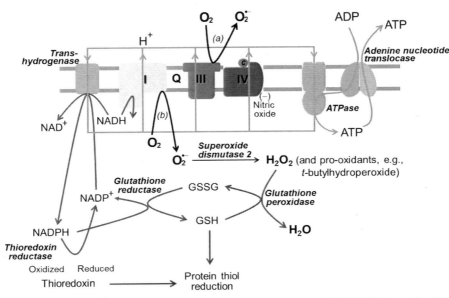

Figure 6.7 Generation of superoxide in mitochondria and interactions among GSH/GSSG couple, NADPH/NADP$^+$ couple, NADH/NAD$^+$ couple, and reactive oxygen species.

peroxidase, GSH detoxifies H_2O_2 generated by SOD2, and other pro-oxidants such as *tert*-butylhydroperoxide or lipid hydroperoxide.

NADPH/NADP$^+$ and NADH/NAD$^+$ are two other important redox couples. The ratio of NADPH/NADP$^+$ is much higher than NADH/NAD$^+$ in the mitochondrial matrix. This disequilibrium with highly reduced NADP pool is maintained by transhydrogenase and NADP-linked cytosolic isocitrate dehydrogenase (Fig. 6.7). NADPH reduces GSSG to GSH via glutathione reductase and reduces the oxidized thioredoxin 2 via thioredoxin reductase. GSH and reduced thioredoxin 2 maintain protein thiols in reduced state.[49,50]

REFERENCES

1. Frey, T.G., Mannella, C.A. The internal structure of mitochondria. *Trends Biochem. Sci.* **2000**, *25*(7), 319–324.
2. Wallace, D.C. A mitochondrial paradigm of metabolic and degenerative diseases, aging, and cancer: A dawn for evolutionary medicine. *Annu. Rev. Genet.* **2005**, *39*, 359–407.
3. Wallace, D.C. Mitochondrial diseases in man and mouse. *Science* **1999**, *283*(5407), 1482–1488.
4. Turrens, J.F., Boveris, A. Generation of superoxide anion by the NADH dehydrogenase of bovine heart mitochondria. *Biochem. J.* **1980**, *191*(2), 421–427.
5. Turrens, J.F., Alexandre, A., Lehninger, A.L. Ubisemiquinone is the electron donor for superoxide formation by complex III of heart mitochondria. *Arch. Biochem. Biophys.* **1985**, *237*(2), 408–414.
6. Nohl, H., Jordan, W. The mitochondrial site of superoxide formation. *Biochem. Biophys. Res. Commun.* **1986**, *138*(2), 533–539.
7. Cadenas, E., Boveris, A., Ragan, C.I., Stoppani, A.O. Production of superoxide radicals and hydrogen peroxide by NADH-ubiquinone reductase and ubiquinol-cytochrome c reductase from beef-heart mitochondria. *Arch. Biochem. Biophys.* **1977**, *180*(2), 248–257.
8. Boveris, A., Oshino, N., Chance, B. The cellular production of hydrogen peroxide. *Biochem. J.* **1972**, *128*(3), 617–630.
9. Hirst, J., Carroll, J., Fearnley, I.M., Shannon, R.J., Walker, J.E. The nuclear encoded subunits of complex I from bovine heart mitochondria. *Biochim. Biophys. Acta* **2003**, *1604*(3), 135–150.
10. Yagi, T., Matsuno-Yagi, A. The proton-translocating NADH-quinone oxidoreductase in the respiratory chain: The secret unlocked. *Biochemistry* **2003**, *42*(8), 2266–2274.
11. Walker, J.E. The NADH:ubiquinone oxidoreductase (complex I) of respiratory chains. *Q. Rev. Biophys.* **1992**, *25*(3), 253–324.
12. Sazanov, L.A., Hinchliffe, P. Structure of the hydrophilic domain of respiratory complex I from *Thermus thermophilus*. *Science* **2006**, *311*(5766), 1430–1436.
13. Ohnishi, T. Iron-sulfur cluster/semiquinones in complex I. *Biochim. Biophys. Acta*. **1998** *1364*(2), 186–206.
14. Kudin, A.P., Bimpong-Buta, N.Y., Vielhaber, S., Elger, C.E., Kunz, W.S. Characterization of superoxide-producing sites in isolated brain mitochondria. *J. Biol. Chem.* **2004**, *279*(6), 4127–4135.
15. Chen, Y.R., Chen, C.L., Zhang, L., Green-Church, K.B., Zweier, J.L. Superoxide generation from mitochondrial NADH dehydrogenase induces self-inactivation with specific protein radical formation. *J. Biol. Chem.* **2005**, *280*(45), 37339–37348.
16. Magnitsky, S., Toulokhonova, L., Yano, T., Sled, V.D., Hagerhall, C., Grivennikova, V.G., Burbaev, D.S., Vinogradov, A.D., Ohnishi, T. EPR characterization of ubisemiquinones and iron-sulfur cluster N2, central components of the energy coupling in the NADH-ubiquinone oxidoreductase (complex I) in situ. *J. Bioenerg. Biomembr.* **2002**, *34*(3), 193–208.
17. Chen, J., Chen, C.L., Rawale, S., Chen, C.A., Zweier, J.L., Kaumaya, P.T., Chen, Y.R. Peptide-based antibodies against glutathione-binding domains suppress superoxide production mediated by mitochondrial complex I. *J. Biol. Chem.* **2010**, *285*(5), 3168–3180.
18. Kang, P.T., Yun, J., Kaumaya, P.P., Chen, Y.R. Design and use of peptide-based antibodies decreasing superoxide production by mitochondrial complex I and complex II. *Biopolymers* **2011**, *96*(2), 207–221.
19. Hurd, T.R., Requejo, R., Filipovska, A., Brown, S., Prime, T.A., Robinson, A.J., Fearnley, I.M., Murphy, M.P. Complex I within oxidatively stressed bovine heart mitochondria is glutathionylated on Cys-531 and Cys-704 of the 75-kDa subunit: Potential role of CYS residues in decreasing oxidative damage. *J. Biol. Chem.* **2008**, *283*(36), 24801–24815.
20. Hurd, T.R., Costa, N.J., Dahm, C.C., Beer, S.M., Brown, S.E., Filipovska, A., Murphy, M.P. Glutathionylation of mitochondrial proteins. *Antioxid. Redox Signal.* **2005**, *7*(7–8), 999–1010.
21. Filipovska, A., Kelso, G.F., Brown, S.E., Beer, S.M., Smith, R.A.J., Murphy, M.P. Synthesis and characterization of a triphenylphosphonium-conjugated peroxidase mimetic. *J. Biol. Chem.* **2005**, *280*, 24113–24126.
22. Costa, N.J., Dahm, C.C., Hurrell, F., Taylor, E.R., Murphy, M.P. Interactions of mitochondrial thiols with nitric oxide. *Antioxid. Redox Signal.* **2003**, *5*(3), 291–305.
23. Taylor, E.R., Hurrell, F., Shannon, R.J., Lin, T.K., Hirst, J., Murphy, M.P. Reversible glutathionylation of complex I increases mitochondrial superoxide formation. *J. Biol. Chem.* **2003**, *278*(22), 19603–19610.
24. Chen, C.L., Zhang, L., Yeh, A., Chen, C.A., Green-Church, K.B., Zweier, J.L., Chen, Y.R. Site-specific S-glutathiolation of mitochondrial NADH ubiquinone reductase. *Biochemistry* **2007**, *46*(19), 5754–5765.
25. Cicchetti, F., Drouin-Ouellet, J., Gross, R.E. Environmental toxins and Parkinson's disease: What have we learned from pesticide-induced animal models? *Trends Pharmacol. Sci.* **2009**, *30*(9), 475–483.

26. Langston, J.W., Ballard, P., Tetrud, J.W., Irwin, I. Chronic Parkinsonism in humans due to a product of meperidine-analog synthesis. *Science* **1983**, *219*(4587), 979–980.
27. Messner, K.R., Imlay, J.A. Mechanism of superoxide and hydrogen peroxide formation by fumarate reductase, succinate dehydrogenase, and aspartate oxidase. *J. Biol. Chem.* **2002**, *277*(45), 42563–42571.
28. Yankovskaya, V., Horsefield, R., Tornroth, S., Luna-Chavez, C., Miyoshi, H., Leger, C., Byrne, B., Cecchini, G., Iwata, S. Architecture of succinate dehydrogenase and reactive oxygen species generation. *Science* **2003**, *299*(5607), 700–704.
29. Guo, J. and Lemire, B.D. The ubiquinone-binding site of the Saccharomyces cerevisiae succinate-ubiquinone oxidireductase is a source of superoxide. *J. Biol. Chem.* **2003**, *278* (48), 47629–47635.
30. Sun, F., Huo, X., Zhai, Y., Wang, A., Xu, J., Su, D., Bartlam, M., Rao, Z. Crystal structure of mitochondrial respiratory membrane protein complex II. *Cell* **2005**, *121*(7), 1043–1057.
31. Chen, Y.R., Chen, C.L., Yeh, A., Liu, X., Zweier, J.L. Direct and indirect roles of cytochrome b in the mediation of superoxide generation and NO catabolism by mitochondrial succinate-cytochrome c reductase. *J. Biol. Chem.* **2006**, *281*(19), 13159–13168.
32. Chen, Y.R., Chen, C.L., Pfeiffer, D.R., Zweier, J.L. Mitochondrial complex II in the post-ischemic heart: Oxidative injury and the role of protein S-glutathionylation. *J. Biol. Chem.* **2007**, *282*(45), 32640–32654.
33. Baysal, B.E., Ferrell, R.E., Willett-Brozick, J.E., Lawrence, E.C., Myssiorek, D., Bosch, A., van der Mey, A., Taschner, P.E., Rubinstein, W.S., Myers, E.N., Richard, C.W., 3rd, Cornelisse, C.J., Devilee, P., Devlin, B. Mutations in SDHD, a mitochondrial complex II gene, in hereditary paraganglioma. *Science* **2000**, *287*(5454), 848–851.
34. Astuti, D., Douglas, F., Lennard, T.W., Aligianis, I.A., Woodward, E.R., Evans, D.G., Eng, C., Latif, F., Maher, E.R. Germline SDHD mutation in familial phaeochromocytoma. *Lancet* **2001**, *357*(9263), 1181–1182.
35. Milunsky, J.M., Maher, T.A., Michels, V.V., Milunsky, A. Novel mutations and the emergence of a common mutation in the SDHD gene causing familial paraganglioma. *Am. J. Med. Genet.* **2001**, *100*(4), 311–314.
36. Neumann, H.P., Bausch, B., McWhinney, S.R., Bender, B.U., Gimm, O. Franke, G., Schipper, J., Klisch, J., Altehoefer, C., Zerres, K., et al. Germ-line mutations in nonsyndromic pheochromocytoma. *N. Engl. J. Med.* **2002**, *346*, 1459–1466.
37. Chen, C.L., Chen, J., Rawale, S., Varadharaj, S., Kaumaya, P.P., Zweier, J.L., Chen, Y.R. Protein tyrosine nitration of the flavin subunit is associated with oxidative modification of mitochondrial complex II in the post-ischemic myocardium. *J. Biol. Chem.* **2008**, *283*(41), 27991–28003.
38. Zhang, L., Chen, C.L., Kang, P.T., Garg, V., Hu, K., Green-Church, K.B., Chen, Y.R. Peroxynitrite-mediated oxidative modifications of complex II: Relevance in myocardial infarction. *Biochemistry* **2010**, *49*(11), 2529–2539.
39. Zhang, Z., Huang, L., Shulmeister, V.M., Chi, Y.I., Kim, K.K., Hung, L.W., Crofts, A.R., Berry, E.A., Kim, S.H. Electron transfer by domain movement in cytochrome bc1. *Nature* **1998**, *392*(6677), 677–684.
40. Muller, F.L., Roberts, A.G., Bowman, M.K., Kramer, D.M. Architecture of the Q_o site of the cytochrome bc1 complex probed by superoxide production. *Biochemistry* **2003**, *42*(21), 6493–6499.
41. Muller, F.L., Liu, Y., Van Remmen, H. Complex III releases superoxide to both sides of the inner mitochondrial membrane. *J. Biol. Chem.* **2004**, *279*(47), 49064–49073.
42. Berry, E.A., Guergova-Kuras, M., Huang, L.S., Crofts, A.R. Structure and function of cytochrome bc complexes. *Annu. Rev. Biochem.* **2000**, *69*, 1005–1075.
43. Zhang, L., Yu, L., Yu, C.A. Generation of superoxide anion by succinate-cytochrome c reductase from bovine heart mitochondria. *J. Biol. Chem.* **1998**, *273*(51), 33972–33976.
44. Gong, X., Yu, L., Xia, D., Yu, C.A. Evidence for electron equilibrium between the two hemes bL in the dimeric cytochrome bc1 complex. *J. Biol. Chem.* **2005**, *280*(10), 9251–9257.
45. Tsukihara, T., Aoyama, H., Yamashita, E., Tomizaki, T., Yamaguchi, H., Shinzawa-Itoh, K., Nakashima, R., Yaono, R., Yoshikawa, S. The whole structure of the 13-subunit oxidized cytochrome c oxidase at 2.8 A. *Science* **1996**, *272*(5265), 1136–1144.
46. Proshlyakov, D.A., Pressler, M.A., Babcock, G.T. Dioxygen activation and bond cleavage by mixed-valence cytochrome c oxidase. *Proc. Natl. Acad. Sci. U.S.A.* **1998**, *95*(14), 8020–8025.
47. Yoshikawa S., Shinzawa-Itoh K., Nakashima R., Yaono R., Yamashita E., Inoue N., Yao M., Fei M.J., Libeu C.P., Mizushima T., Yamaguchi H., Tomizaki T., Tsukihara T. Redox-coupled crystal structural changes in bovine heart cytochrome c oxidase. *Science* **1998**, *280*(5370), 1723–1729.
48. Prabu, S.K., Anandatheerthavarada, H.K., Raza, H., Srinivasan, S., Spear, J.F., Avadhani, N.G. Protein kinase A-mediated phosphorylation modulates cytochrome c oxidase function and augments hypoxia and myocardial ischemia-related injury. *J. Biol. Chem.* **2006**, *281*(4), 2061–2070.
49. Tanaka, T., Nakamura, H., Nishiyama, A., Hosoi, F., Masutani, H., Wada, H., Yodoi, J. Redox regulation by thioredoxin superfamily; protection against oxidative stress and aging. *Free Radic. Res.* **2000**, *33*(6), 851–855.
50. Schafer, F.Q., Buettner, G.R. Redox environment of the cell as viewed through the redox state of the glutathione disulfide/glutathione couple. *Free Radic. Biol. Med.* **2001**, *30*(11), 1191–1212.

7

NADPH OXIDASES: STRUCTURE AND FUNCTION

MARK T. QUINN

OVERVIEW

NADPH oxidases are transmembrane, multiprotein enzyme complexes that catalyze the formation of reactive oxygen species (ROS). Although originally identified in phagocytic cells, NADPH oxidases have subsequently been shown to be present in a wide variety of tissues and contribute to a multitude of cellular processes. In this chapter, key structural and functional features of the phagocyte and nonphagocyte NADPH oxidases are described, including a consideration of their roles in physiological and pathophysiological events.

7.1 INTRODUCTION

The NADPH oxidases are a family of multi-subunit enzymes that catalyze the formation of ROS, such as superoxide anion ($O_2^{\bullet-}$) and hydrogen peroxide (H_2O_2). The classical NADPH oxidase was first characterized in phagocytic leukocytes, such as macrophages and neutrophils, and it was originally thought that the enzyme was restricted to leukocytes and for use in host defense.[1] However, more recent studies have demonstrated that similar NADPH oxidases are present in a wide variety of nonphagocytic cells and tissues (reviewed in References 2 and 3). While these enzymes are distinct from the phagocyte NADPH oxidase and serve in functional roles other than host defense, their structural features are often similar to those of their phagocyte counterpart. Understanding the physiological functions of these enzymes and their potential roles in pathophysiological events is currently an area of great interest. This chapter provides an overview of the structural and functional features of the phagocyte and nonphagocyte NADPH oxidases.

7.2 PHAGOCYTE NADPH OXIDASE STRUCTURE

Over time since their discovery, the names given to the various phagocyte NADPH oxidase components has been revised; however, the currently accepted nomenclature for the phagocyte oxidase-specific components includes the suffix *phox*, which refers to *ph*agocyte *ox*idase.[4] The one exception is gp91^{phox}, which was more recently renamed NOX2.[5] Thus, the core NADPH oxidase enzyme is composed of a flavocytochrome b (heterodimer of p22^{phox} and NOX2), two oxidase-specific cytosolic proteins (p47^{phox} and p67^{phox}), and a low-molecular-weight GTPase (Rac1/2). An additional oxidase-specific cytosolic protein (p40^{phox}) and a second low-molecular-weight GTPase (Rap1A) have also been shown to play roles in regulating oxidase activity; however, their specific functions are still being defined. Overall, the *phox* proteins have been highly conserved among the various species studied to date, confirming the extreme importance of this system in mammalian host defense.[6,7] Details of each of the individual components are summarized as follows.

Molecular Basis of Oxidative Stress: Chemistry, Mechanisms, and Disease Pathogenesis, First Edition. Edited by Frederick A. Villamena.
© 2013 John Wiley & Sons, Inc. Published 2013 by John Wiley & Sons, Inc.

7.2.1 Flavocytochrome b

The first NADPH oxidase component to be identified was a nonmitochondrial cytochrome b,[8] and it was originally proposed that this cytochrome b and a flavoprotein formed the phagocyte NADPH oxidase.[9] Subsequent studies demonstrated that neutrophil cytochrome b was actually a heterodimer of a 22-kDa nonglycosylated protein (α-chain, later named p22^{phox}) and a 91-kDa glycoprotein (β-chain, later named gp91^{phox} and then NOX2).[10–12] Based on the experimentally determined heme content of purified cytochrome b, Parkos et al.[13] proposed that more than one heme was present per cytochrome b molecule, and additional biochemical studies by a number of groups demonstrated that neutrophil cytochrome b was a bihistidinyl, multi-heme cytochrome with closely spaced hemes.[14–16] This model was refined even further by Cross et al.[17] who reported that neutrophil cytochrome b contained two nonidentical hemes with midpoint redox potentials of −225 and −265 mV, respectively. Hydropathy analyses of p22^{phox} and gp91^{phox} indicated the presence of two to three and four to six transmembrane regions in these proteins, respectively,[13,18] and it is now generally thought that p22^{phox} contains two transmembrane domains, while gp91^{phox}/NOX2 contains six transmembrane domains (reviewed in Reference 19) (Fig. 7.1). In both proteins, the C-terminus and N-terminus appear to be located inside the cytoplasm.

The presence of two nonidentical hemes fits with the proposed function of cytochrome b, which is to transport electrons across the membrane, either extracellularly or into the phagosome, and the location of these hemes has been a major topic of investigation in the field. Currently, the general consensus is that both hemes are coordinated by gp91^{phox}/NOX2. This model was originally based on sequence similarities between gp91^{phox}/NOX2 and yeast iron reductase Fre1, and it was proposed[20] that both hemes are stacked within gp91^{phox}/NOX2, between transmembrane helices III and V, and coordinated by histidines 101, 115, 209, and 222 (Fig. 7.1). Additional support for this model came from mutagenesis studies showing that replacement of these histidines with leucine or arginine resulted in lost or significantly decreased heme spectrum and from data showing that missense mutations in any of these residues resulted in the X-linked form of a rare genetic disorder known as chronic granulomatous disease (CGD), which is characterized by defective NADPH oxidase activity.[21] Furthermore, p22^{phox} contains only one invariant histidine (His94), and mutagenesis of this residue showed that a histidine at this position was not required for cytochrome b function, further suggesting that heme was not shared between subunits in this

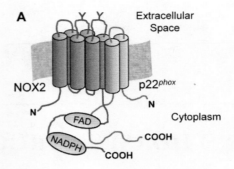

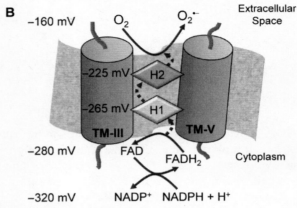

Figure 7.1 Model of phagocyte flavocytochrome b and proposed electron flow. Panel A. Flavocytochrome b is a heterodimer of gp91$^{\mathbf{phox}}$/NOX2 and p22^{phox}. The proposed sites of NOX2 glycosylation (Y) and general location of binding domains for other redox components (FAD, NADPH) are indicated. Panel B. Flavocytochrome b transfers electrons across the plasma membrane or phagosomal membrane. Redox components involved in this process. Flavocytochrome b hemes (H1 are H2) are coordinated by NOX2 transmembrane helices (TM-III and TM-V), resulting in a pathway for electron transfer across the membrane. The redox midpoint potentials at pH 7.0 (Em) shown for each step in the pathway indicate that the transfer electrons from NADPH to O_2 is energetically favorable. See color insert.

system.[22] Finally, the more recent discovery of NOX2 homologs demonstrated conservation of the general structure proposed for NOX2, including conservation of transmembrane domains and location of the hemes within these NOX homologs (see further details later).

As discussed above, early studies suggested that the NADPH oxidase included a separate flavoprotein, and a number of candidates were proposed (reviewed in Reference 23). However, discrepancies in these studies with what was known about cytochrome b led to the proposal that the flavin adenine dinucleotide (FAD)-binding site might actually be present within cytochrome b, and three groups concurrently provided data demonstrating that cytochrome b was indeed a flavocytochrome b (aka flavocytochrome b_{558}).[24–26] Likewise, these groups also suggested that in addition to binding FAD, flavocytochrome b might also contain an NADPH-binding domain.[24–26] Indeed, subsequent studies by Koshkin and Pick[27] showed that relipidated flavocytochrome b *alone* could generate $O_2^{\bullet-}$, indicating that flavocytochrome b was able to functionally bind both FAD and NADPH. Thus, NOX2 contains a membrane domain that consists of six transmembrane domains that coordinate two nonidentical hemes and a cytosolic dehydrogenase domain that contains binding sites for FAD and NADPH (Fig. 7.1).

It is widely agreed that flavocytochrome b is the only catalytic component of the oxidase and functions to transfer electrons from NADPH via FAD and two hemes to ultimately reduce O_2 (Fig. 7.1). On the other hand, flavocytochrome b cannot function independently in the cell and requires assembly of several cytosolic proteins for enzymatic activity, including $p47^{phox}$, $p67^{phox}$, and Rac GTPase (see below). This compartmentalization of NADPH oxidase components in different cellular locations is essential for maintaining the enzyme in an inactive state in resting cells. When cells are activated, the cytosolic components required for NADPH activity translocate very rapidly to the plasma membrane or phagosomal membranes where they assemble with flavocytochrome b_{558} to form the active enzyme. This process that is tightly regulated by a number of protein–protein interactions and phosphorylation events are described as follows.

7.2.2 $p47^{phox}$

The existence of additional proteins required for NADPH oxidase activity in addition to flavocytochrome b was first noticed in neutrophils from individuals with CGD (reviewed in Reference 28). For example, Segal et al.[29] reported that neutrophils from patients with autosomal recessive CGD failed to phosphorylate a 44-kDa protein and suggested this protein could be one of these oxidase-related proteins. Through genetic analyses, complementation experiments, and cell-free assays, two groups concurrently identified this protein (now known as $p47^{phox}$), which was previously named neutrophil cytosolic factor 1 (NCF1).[30,31] $p47^{phox}$ is a highly basic protein and contains a number of potential phosphorylation sites, tandem Src homology 3 (SH3) domains, a C-terminal proline-rich domain, and an N-terminal *phox* homology (PX) domain that plays a role in phosphoinositide binding[32–34] (Fig. 7.2).

In resting neutrophils, $p47^{phox}$ is located in the cytosol, either in a free form or in a large complex consisting of $p47^{phox}$ and other cytosolic oxidase proteins.[35–37] Follow-

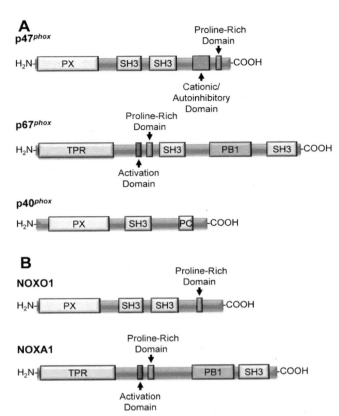

Figure 7.2 Structural models of the NADPH oxidase cytosolic cofactors. Major functional domains are indicated on these scale models of the phagocyte NADPH oxidase cytosolic proteins (Panel A) and the nonphagocyte homologs identified (Panel B). Src homology 3 (SH3) domains are present in all cytosolic cofactors and play a role in protein–protein binding interactions by binding to proline-rich domains within and between proteins. The N-terminal *phox* homology (PX) domains present in $p47^{phox}$, $p40^{phox}$, and NOXO1 participate in phosphoinositide binding interactions involved in targeting to cellular locaction. The $p47^{phox}$ auto-inhibitory region (AIR) mediates an intramolecular interaction to sequester the tandem SH3 domains, which become exposed during activation. In addition, phosphorylation of $p47^{phox}$ releases an intramolecular autoinhibitory interaction between the PX domain and second SH3 domain. In $p67^{phox}$ and NOXA1, the N-terminal tetratricopeptide repeat (TPR) motifs mediate Rac GTPase binding, whereas the activation domains in these proteins participate in the regulation of electron flow from NADPH to FAD. The $p67^{phox}$ Phox and Bem (PB1) domain plays a role in binding to the $p40^{phox}$ N-terminal *Phox* and Cdc (PC) domain, whereas the role of the NOXA1 PB1 domain is undefined. See text for additional details. See color insert.

ing neutrophil activation, $p47^{phox}$ translocates, either alone or complexed with other oxidase cofactors, to the membrane and associates with the nascent NADPH oxidase complex.[35,36,38] $p47^{phox}$ seems to be the first cytosolic component to interact with flavocytochrome b during oxidase assembly,[39,40] and its association with flavocytochrome b is a prerequisite for translocation of the other cytosolic cofactors (p47^{phox} and/or p40^{phox}), which are described later.[39–41] Indeed, p47^{phox} appears to function as an organizing cofactor during oxidase activation, resulting in correct orientation and assembly of the cofactors.[42] It has also been suggested that binding of p47^{phox} functions to facilitate electron transfer from FAD to the proximal heme of flavocytochrome b.[43]

Because p47^{phox} association with flavocytochrome b seems to be one of the key oxidase-activating events, a significant amount of research has focused on understanding why this interaction is absent in resting cells. This research has shown that the tandem p47^{phox} SH3 domains bind intramolecularly to a proline-rich C-terminal region of the resting, nonphosphorylated protein to keep the protein in an "autoinhibited" state, resulting in the designation of this C-terminal region of p47^{phox} as the autoinhibitory region (AIR)[44–46] (Fig. 7.2). During oxidase activation, phosphorylation of p47^{phox} results in the release of AIR interaction, resulting in exposure of the tandem SH3 domains and allowing its binding to a proline-rich domain in the cytosolic tail of p22^{phox}.[44,47,48] In addition to the release of AIR interaction, phosphorylation of p47^{phox} also releases a second intramolecular autoinhibitory interaction that prevents the PX domain from binding to membrane phosphoinositides, such as phosphatidylinositol 3,4-bisphosphate and phosphatidic acid.[49–51] It is thought that this PX domain plays a role in targeting p47^{phox} to membranes,[50,52,53] and analysis of the crystallized p47^{phox} PX domain showed that it actually contained two binding pockets, one for phosphatidylinositol 3,4-bisphosphate and the other for anionic phospholipids, such as phosphatidic acid or phosphatidylserine.[51] Recently, Li et al.[54] showed that the p47^{phox} PX domain was important for NADPH oxidase activation on the plasma membrane but did not seem to be involved in oxidase activation in phagosomal membranes. Thus, they proposed that the role of the p47^{phox} PX domain is differentially dependent on the membrane compartment involved and the stimulus.[54]

As discussed above, phosphorylation of p47^{phox} is necessary for NADPH oxidase activation, and a number of phosphorylation sites are present in this protein (reviewed in Reference 42). A wide variety of protein kinases have been shown to phosphorylate p47^{phox}, including protein kinase C (PKC),[55,56] mitogen-activated protein kinases (p38 MAPKs and extracellular signal-regulated kinase [ERK1/2]),[56–58] p21-activated kinase (PAK),[59] Akt,[60] IRAK-4,[61] casein kinase 2 (CK2),[62] src kinase,[63] and a phosphatidic acid-activated kinase.[64] While it is clear that PKC plays a dominant role in p47^{phox} activation,[65–67] participation of the aforementioned protein kinases is not as well understood, and it has been proposed that the combination of kinases involved and the specific p47^{phox} phophorylation events occurring in the cell are dependent on the nature of the agonist encountered (reviewed in Reference 42).

7.2.3 p67^{phox}

As with p47^{phox}, p67^{phox} was also first identified as a cytosolic factor missing in neutrophils from patients with autosomal recessive CGD and was named neutrophil cytosolic factor 2 (NCF2).[30,31,68] Subsequent cloning and characterization of p67^{phox} showed that it contained two SH3 domains, a proline-rich region, four N-terminal tetratricopeptide repeat (TPR) motifs, and a novel modular domain called the *P*hox and *B*em (PB1) domain because of its presence in p67^{phox} and Bem1p, a yeast scaffold protein involved in cell polarity[68–70] (Fig. 7.2). p67^{phox} plays an essential role in NADPH oxidase activation, and results from complementation studies using cytosol from CGD neutrophils indicated that p67^{phox} is the rate-limiting NADPH oxidase cofactor in neutrophil cytosol.[71]

It is generally thought that translocation of p67^{phox} is dependent on cotranslocation of p47^{phox} to the plasma membrane and prior interaction of p47^{phox} with flavocytochrome b.[39–41] Indeed, atomic force microscopy was used to directly demonstrate that p47^{phox} preceded p67^{phox} binding and actually enhanced affinity of the p67^{phox}-flavocytochrome b association during NADPH oxidase assembly.[72] While it is clear that the presence of p47^{phox} enhances binding to flavocytochrome b, p67^{phox} is able to bind directly to flavocytochrome b, and this binding is enhanced by the presence of Rac GTPase.[73] Indeed, NADPH oxidase activity can be reconstituted *in vitro* in the absence of p47^{phox} when flavocytochrome b is combined with high concentrations of p67^{phox} and Rac GTPase[74,75] or with a chimeric protein containing truncated p67^{phox} fused to Rac GTPase,[76,77] indicating that p67^{phox} plays a more direct role in NADPH oxidase activation.

The interaction between p67^{phox} and Rac GTPase has been shown to be essential for NADPH oxidase activation and appears to be mediated primarily by binding of Rac to the N-terminal TPR motifs of p67^{phox}[70,78–80] (Fig. 7.2). The p67^{phox} N-terminus also contains an activation domain that is required for NADPH oxidase activation,[81,82] and this domain appears to be important for regulating electron flow from NADPH to FAD in flavocytochrome b[83] (see Fig. 7.1). The role of p67^{phox} phosphorylation in NADPH oxidase activation is not well understood. It is clear that p67^{phox} is phosphorylated in activated neutrophils,[84] and the actual phosphorylation site has been mapped to a consensus MAPK substrate sequence in the C-terminal region of the protein.[85] p67^{phox} can also be phosphorylated by ERK2, with the primary ERK2 target localized to the N-terminus of p67^{phox}.[86] The role of these phosphorylation events has not yet been defined.

7.2.4 p40^{phox}

Compared to the other oxidase proteins, less is known about the function of p40^{phox}. Initially, p40^{phox} was found as a protein that co-immunoprecipitated with p47^{phox} and p67^{phox} from neutrophil cytosol.[87] This protein contains a single SH3 domain,[87] an N-terminal PX domain,[88] and a C-terminal *Phox* and Cdc (PC) motif[69] (Fig. 7.2). p40^{phox} is present in the cytosolic complex of *phox* proteins[87] and seems to preferentially bind to p67^{phox}.[89] Indeed, p40^{phox} has been proposed to play a role in stabilization of the cytosolic protein complex and cotranslocates to the membrane during neutrophil activation.[87] As with the other cytosolic *phox* proteins, p40^{phox} is phosphorylated during NADPH oxidase activation.[90]

Although the role of p40^{phox} in NADPH oxidase function is still being defined, recent research suggests it plays a more important role than previously appreciated. Early studies suggested that p40^{phox} was either a positive[91] or a negative[92] regulator of NADPH oxidase activity. However, subsequent research indicates that p40^{phox} is most likely a positive regulator of the NADPH oxidase. For example, studies have shown that p40^{phox} promotes oxidase activation by increasing the affinity of p47^{phox} for flavocytochrome b[93] and enhances translocation of both p47^{phox} and p67^{phox} in stimulated cells.[94] Based on *in vitro* studies, Tamura et al.[95] proposed that p40^{phox} could interact with p22^{phox} and substitute for p47^{phox} as an organizer of NADPH oxidase activation; however, the physiological relevance of this observation is not clear, as the presence of p40^{phox} does not appear to overcome the absence of p47^{phox} in individuals with CGD.

The p40^{phox} PX domain preferentially binds phosphatidylinositol 3-phosphate, suggesting differential membrane targeting of p40^{phox} compared to p47^{phox}, which prefers phosphatidylinositol 3,4-bisphosphate.[49,50,96] Thus, the p40^{phox} PX domain appears to participate in intracellular targeting to membranes that become enriched in phosphatidylinositol 3-phosphate.[97,98] For example, it has been proposed that accumulation of phosphatidylinositol 3-phosphate in the phagosomal membrane facilitates NADPH oxidase assembly by recruiting p40^{phox} and associated components (i.e., p67^{phox} and p67^{phox}) to the assembling oxidase in this area of the cell.[99] Indeed, mice deficient in p40^{phox} or expressing p40^{phox} with a PX domain mutation have reduced oxidase activity and impaired killing of phagocytosed *Staphylococcus aureus*.[100,101] Recently, Matute et al.[102] identified a new genetic subgroup of CGD with autosomal recessive mutations in the p40^{phox} PX domain and showed that these mutations resulted in defective NADPH oxidase activity. These studies also confirmed that the p40^{phox} PX domain is essential for phagocytosis-induced oxidant production in human neutrophils.

The p40^{phox} PX domain has also been shown to bind to the actin cytoskeleton, and Chen et al.[103] proposed that interaction of the PX domain with the actin cytoskeleton may stabilize NADPH oxidase proteins in resting cells, while its binding to phosphatidylinositol 3-phosphate potentiates $O_2^{\bullet-}$ production upon agonist stimulation. This finding was recently confirmed by Shao et al.,[104] who also proposed that the p40^{phox} PX domain is a dual F-actin/lipid-binding module and that interactions with actin dictate at least in part the intracellular trafficking of p40^{phox}.

Investigation of the interaction between p40^{phox} and p67^{phox} resulted in the identification of a novel protein–protein binding motif called the PC motif because it is present in p40^{phox} and Cdc24p (a guanine nucleotide exchange factor in yeast).[69] The PC motif target is the p67^{phox} PB1 domain (Fig. 7.2). PB1 domains exist in a variety of proteins and appear to provide a scaffold for PC motif binding, facilitating protein–protein interactions in a variety of biological processes, and appear to mediate the strong interaction between the p40^{phox} and p67^{phox} (reviewed in Reference 105).

7.2.5 Rac1/2

Early studies of the NADPH oxidase suggested the participation of a cytosolic guanosine triphosphate (GTP)-binding factor, and this factor was concurrently identified by two groups as the small GTP-binding protein Rac (Rac1 or Rac2, depending on the cell type and species) (reviewed in Reference 106). Subsequent analysis of Rac-deficient mice showed that neutrophils from these mice had diminished NADPH oxidase activity[107] and that Rac2 was essential when cells were activated with physiologically relevant agents, such as fMLF or IgG-opsonized particles.[108] Although both Rac1 and Rac2 are capable of reconstituting oxidase activity in *in vitro* assays, Rac1 cannot substitute completely for Rac2 *in vivo*.[109,110] For example, NADPH oxidase activity is normal in Rac1-deficient cells but is markedly diminished in Rac2-deficient cells, indicating that Rac2 is essential for oxidase activation under physiological situations.[109,110] Further confirmation of the importance of Rac2 in human neutrophil NADPH oxidase function was provided by clinical studies showing that a patient with decreased neutrophil NADPH oxidase activity and other neutrophil functional defects had an inhibitory (dominant-negative) mutation in Rac2.[111,112]

When neutrophils are activated, Rac2 translocates to the membrane with similar kinetics as the translocation of p47^{phox} and p67^{phox},[113,114] although Rac translocation is independent of the other cytosolic oxidase components.[115,116] Thus, Rac2 does not directly mediate translocation of p47^{phox} and p67^{phox} but may enhance NADPH oxidase assembly through an indirect mechanism involving activation of PAK-mediated phosphorylation of p47^{phox}.[59] Rac has been shown to interact with p67^{phox},[78] as well as with flavocytochrome b.[115] Recently, Kao et al.[117] identified a Rac2 binding site in the C-terminus of NOX2 and showed that this region was required for Rac2 binding and was independent of cytosolic oxidase factors. Furthermore, mutational analysis indicated that the specific residues required for Rac-dependent NADPH oxidase activity were conserved in other Rac-regulated NOX enzymes. Thus, this research demonstrated a direct regulatory interaction of Rac2 with NOX2 to promote NADPH oxidase activation.

7.2.6 Rap1A

Early studies on the identity of the NADPH oxidase-associated GTPase suggested that Rap1A might be involved because of its association with flavocytochrome b.[118,119] However, subsequent studies described earlier indicated that the GTPase required for NADPH oxidase activity was Rac2. Nevertheless, a number of observations suggest that Rap1A may still play a role in NADPH oxidase regulation. For example, cells transfected with a dominant inhibitory (17N) mutant of Rap1A had significantly reduced NADPH oxidase activity compared to control cells, supporting a regulatory role for Rap1A.[120,121] More recently, Li et al.[122] generated Rap1A-deficient mice and reported that neutrophils from these mice had reduced formyl peptide-stimulated $O_2^{\bullet-}$ production as well as a weaker initial response to phorbol ester. They concluded that because only the transient formyl peptide response and the early stages of the phorbol ester response were attenuated, Rap1A may serve a function in enhancing complex assembly. Nevertheless, further studies are necessary to define the exact role of Rap1A in this complex system.

7.3 PHAGOCYTE ROS PRODUCTION

Activation of the phagocyte NADPH oxidase results in the generation of $O_2^{\bullet-}$; however, subsequent biochemical events can convert this radical into a much more potent microbicidal oxidant species, as described below.

7.3.1 Superoxide Anion ($O_2^{\bullet-}$)

The phagocyte NADPH oxidase catalyzes oxidation of NADPH and the univalent reduction of O_2, resulting in the formation of $O_2^{\bullet-}$,[1] as shown in Equation 7.1 (Fig. 7.3):

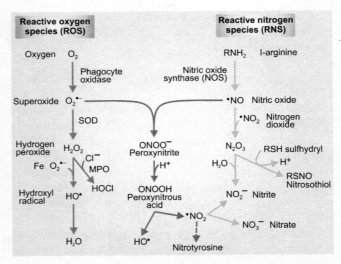

Figure 7.3 Production of reactive oxygen species by phagocytes and potential reaction with reactive nitrogen species. As shown on the left, the initial product of the phagocyte NADPH oxidase is superoxide anion ($O_2^{\bullet-}$), which can be converted to other reactive oxygen species. Superoxide dismutase (SOD) dismutates $O_2^{\bullet-}$ into hydrogen peroxide (H_2O_2), which can be utilized by myeloperoxidase to form hypochlorous acid (HOCl) or can be reduced by $O_2^{\bullet-}$ in the iron-catalyzed Haber–Weiss reaction to produce hydroxyl radical (HO$^{\bullet}$). HO$^{\bullet}$ can also be formed by the reduction of HOCl with $O_2^{\bullet-}$ (not shown). As shown on the right, nitric oxide ($^{\bullet}$NO) is produced by the activity of nitric oxide synthase (NOS) and can lead to additional reactive nitrogen species (RNS). $^{\bullet}$NO can react with $NO_2^{\bullet}$ to form dinitrogen trioxide (N_2O_3), which can react with cysteine sulfydryls, resulting in S-nitrosylation (RSNO), or with water to form nitrite (NO_2^-). NO_2^- can be oxidized by H_2O_2 in a reaction catalyzed by MPO to produce nitrogen dioxide $NO_2^{\bullet}$ radicals (not shown). As shown in the center, the simultaneous presence of $^{\bullet}$NO and $O_2^{\bullet-}$ leads to a rapid reaction, resulting in the formation of peroxynitrite ($ONOO^-$). At physiological pH, $ONOO^-$ exists in equilibrium with peroxynitrous acid (ONOOH), which can spontaneously decompose to form HO$^{\bullet}$ and nitrogen dioxide ($NO_2^{\bullet}$). $NO_2^{\bullet}$ can nitrate tyrosine residues on proteins, resulting in nitrotyrosine. In addition, NO_2^- and nitrate (NO_3^-) can be formed by the decomposition of $NO_2^{\bullet}$.

$$NADPH + 2O_2 \rightarrow NADP^+ + 2O_2^{\bullet-} + H^+ \quad (7.1)$$

As such, activation of the phagocyte NADPH oxidase results in a rapid and substantial increase in neutrophil O_2 consumption, which is known as the respiratory burst.[123,124] While significant amounts of $O_2^{\bullet-}$ are generated by the phagocyte respiratory burst, it is relatively unstable at physiological pH, has limited reactivity toward biological molecules, and exhibits minimal antibacterial activity.[125,126] On the other hand, high concentrations of $O_2^{\bullet-}$ (~5–10 mM s^{-1}) can accumulate in phagosomes, where the pH is low, and in this environment, $O_2^{•-}$ has been suggested to be directly toxic to some pathogens by virtue of its ability to oxidize iron–sulfur clusters required by important bacterial enzymes.[127] More generally, however, $O_2^{•-}$ represents the precursor to more potent microbicidal oxidants, which appear to play important roles in host defense.[128]

7.3.2 Hydrogen Peroxide (H_2O_2)

At physiological pH, $O_2^{•-}$ is rapidly converted to H_2O_2 by spontaneous or enzymatic dismutation,[129] as shown in Equation 7.2 (Fig. 7.3):

$$2\,O_2^{•-} + 2H^+ \rightarrow H_2O_2 + O_2 \quad (7.2)$$

Enzymatic dismutation of $O_2^{•-}$ is catalyzed by superoxide dismutase (SOD).[126] This abundant antioxidant enzyme is extremely efficient and plays an important role in protecting cells and tissues from ROS.[130] Additional antioxidant enzymes are also present to detoxify H_2O_2. For example, catalase catalyzes the dismutation of H_2O_2, resulting in the formation of H_2O (Eq. 7.3):

$$2H_2O_2 \;-\!catalase \rightarrow 2H_2O + O_2 \quad (7.3)$$

Likewise, H_2O_2 can be detoxified through a multistep reduction reaction catalyzed by glutathione (GSH) peroxidase, resulting in oxidized glutathione (GSSG) (Eq. 7.4):

$$H_2O_2 + 2GSH \;-\!GSH\ peroxidase \rightarrow 2H_2O + GSSG \quad (7.4)$$

7.3.3 Hypochlorous Acid (HOCl)

Phagocyte activation not only results in the production of ROS, but it also leads to the release of cytoplasmic granule contents into the phagosome or into the extracellular space. Fusion of neutrophil azurophil granules results in the release of high concentrations of myeloperoxidase (MPO), which comprises up to 5% of the proteins present in neutrophils.[131] MPO is also found in monocytes, but its expression is normally lost when monocytes mature into tissue macrophages.[131] Note that there are some exceptions to this pattern, and MPO expression has been reported in macrophages from atherosclerotic lesions[132] and in microglia from individuals with multiple sclerosis[133] and Alzheimer's disease.[134]

MPO utilizes H_2O_2 and Cl^- ions to form hypochlorous acid (HOCl), as described in Equation 7.5 (Fig. 7.3):

$$Cl^- + H_2O_2 + H^+ \rightarrow HOCl + H_2O \quad (7.5)$$

Although MPO can also utilize other halide ions besides Cl^-, it is generally accepted that Cl^- is the most physiologically relevant ion *in vivo*.[135] Furthermore, it is clear that the H_2O_2 utilized by MPO to produce HOCl is derived from NADPH oxidase-generated $O_2^{•-}$, as MPO is unable to contribute to the host defense in the absence of NADPH oxidase activity.[136]

HOCl is a potent oxidant and is cytotoxic to a wide spectrum of pathogens, including bacteria, viruses, and fungi.[131,137] Indeed, the MPO–H_2O_2–halide system seems to be the most efficient oxygen-dependent microbicidal mechanism in neutrophils, and much of the H_2O_2 generated by activated neutrophils participates in HOCl formation.[138] HOCl can enter microorganisms and is involved in a variety of oxidation and chlorination reactions, resulting in pathogen killing.[137] For example, HOCl can oxidize α-amino acids to form reactive aldehydes and tyrosyl radicals that can attack important biomolecules.[139,140] HOCl can also oxidize heme groups and iron–sulfur centers.[141] The primary targets for HOCl-mediated chlorination are primary amines, resulting in the formation of chloramines. Other HOCl targets include pyridine nucleotides, unsaturated lipids, and cholesterol.[137,142] Chloramines have been reported to play an important role in regulating the inflammatory response and, therefore, may represent a critical product of this pathway.[143]

Note, however, that while HOCl is an efficient microbicidal agent, it is not absolutely required for host defense. MPO deficiencies are relatively common, and individuals with MPO deficiency generally do not seem to have an increased incidence of infection, even though they cannot generate HOCl.[144] Indeed, MPO-deficient neutrophils appear to utilize compensatory MPO-independent, but oxygen-dependent antimicrobial systems.[131]

7.3.4 Hydroxyl Radical (HO•)

HO• can be formed by the metal-catalyzed reduction of H_2O_2 by $O_2^{•-}$, which is commonly known as the Haber–Weiss reaction or $O_2^{•-}$-assisted Fenton reaction,[128] as shown in Equation 7.6, Equation 7.7, and Equation 7.8 (Fig. 7.3):

$$O_2^{•-} + Fe^{3+} \rightarrow O_2 + Fe^{2+} \quad (7.6)$$

$$\underline{H_2O_2 + Fe^{2+} \rightarrow HO^{•} + OH^- + Fe^{3+}} \quad (7.7)$$

$$O_2^{•-} + H_2O_2 \rightarrow HO^{•} + OH^- + O_2 \quad (7.8)$$

This reaction requires the presence of redox-active transition metals, such as iron or copper. Here, NADPH oxidase-derived $O_2^{•-}$ is required both as a source of

H_2O_2 and as an Fe^{3+} reducing-agent. $HO^\bullet$ is an extremely powerful and highly reactive oxidant that can attack most organic compounds at a diffusion-controlled rate.[145] While the generation of $HO^\bullet$ inside microbes likely contributes to pathogen killing, it has been proposed that $HO^\bullet$ generated in the phagosome probably reacts with the many phagosomal constituents before it ever has a chance to reach targets in the pathogen.[138] Thus, there remains some question as to the importance of $HO^\bullet$ in phagocyte host defense. Indeed, analysis of activated neutrophils showed that <1% of the $O_2^{\bullet-}$ formed was converted to $HO^\bullet$.[146] In addition, the cell is maintained in a highly reduced state and contains a number of efficient reducing agents that can reduce Fe^{3+} and Cu^{2+}. Based on this feature, it has been suggested that $O_2^{\bullet-}$ is unlikely to serve as a metal-reducing agent.[147] Furthermore, intracellular and extracellular iron is highly regulated by iron-binding proteins. For example, delivery of lactoferrin to the phagosome through fusion of specific granules would likely bind free iron and limit its participation in this reaction,[148,149] and iron bound to this protein does not appear to serve as a catalyst for this reaction.[150] Thus, the nature of the iron or copper complexes required for metal-catalyzed $HO^\bullet$ formation *in vivo* is currently unknown, although one possibility is that they may be provided by the cell itself.[151]

Alternatively, if $HO^\bullet$ is essential for phagocyte microbicidal activity, metal ion-independent mechanisms could be involved in its generation. One such mechanism involves the reduction of HOCl by $O_2^{\bullet-}$, resulting in the formation of $HO^\bullet$, as shown in Equation 7.9:

$$O_2^{\bullet-} + HOCl \rightarrow HO^\bullet + O_2 + Cl^- \quad (7.9)$$

This transition metal ion-independent mechanism of $HO^\bullet$ formation represents an important pathway in neutrophils, and $HO^\bullet$ generated by this reaction has been shown to damage a number of biomolecules, including nucleic acid and proteins.[152,153] In addition, Saran et al.[154] proposed this $HO^\bullet$ might be stabilized by reacting with Cl^- to form an equilibrium with $HOCl^{\bullet-}$ and, if an appropriate target is not reached, could lead to the formation of chlorine radicals ($Cl^\bullet$), which are also highly reactive, as shown in Equation 7.10:

$$HO^\bullet + Cl^- \leftrightarrow HOCl^{\bullet-} + H^+ \leftrightarrow Cl^\bullet + H_2O \quad (7.10)$$

Overall, the actual role of $HO^\bullet$ in phagocyte microbicidal activity *in vivo* is still unclear and remains to be determined.

7.3.5 Singlet Oxygen ($^1O_2^*$)

The generation of high concentrations of ROS during neutrophil activation provides an ideal environment for the formation of additional secondary metabolites. An additional metabolite formed by HOCl-mediated oxidation of H_2O_2 is $^1O_2^*$,[155] as shown in Equation 7.11:

$$HOCl + H_2O_2 \rightarrow {}^1O_2^* + HCl + H_2O \quad (7.11)$$

$^1O_2^*$ is a highly reactive and relatively long-lived ROS and has been implicated as a bactericidal oxidant in the phagosome,[155] although this conclusion has been debated.[141] Intracellular production of $^1O_2^*$ by neutrophils has been demonstrated, and it has been reported that phagocytosing neutrophils generate significant amounts of $^1O_2^*$.[156] In support of this idea, bacteria expressing lycopene (a $^1O_2^*$ quencher) were found to be relatively more resistant to neutrophil killing than wild-type bacteria.[157] Although the role of $^1O_2^*$ in bacterial killing is currently unknown, it has been proposed to cause DNA damage, lipid peroxidation, and protein oxidation,[158] all of which are toxic to microbes.

7.3.6 Nitric Oxide ($^\bullet NO$) and Peroxynitrite ($OONO^-$)

$^\bullet NO$ plays important regulatory roles in a number of physiological processes involving the cardiovascular and neuronal systems; however, $^\bullet NO$ is also important in innate immunity and has been widely implicated in the inflammatory response (reviewed in Reference 159). Most evidence suggests that the microbicidal effects of $^\bullet NO$ are due to the formation of $ONOO^-$, which results from the rapid reaction between $^\bullet NO$ and $O_2^{\bullet-}$. Indeed, $^\bullet NO$ can even outcompete SOD for $O_2^{\bullet-}$, making this one of the fastest known reactions.[160] $ONOO^-$ exists in equilibrium with its conjugate acid OONOH at physiological pH and can decompose to nitrate (NO_3^-),[159] as described in Equation 7.12 (Fig. 7.3):

$$^\bullet NO + O_2^{\bullet-} \rightarrow ONOO^- \leftrightarrow OONOH \rightarrow NO_3^- \quad (7.12)$$

In the presence of an oxidizable substrate, $ONOO^-$ decomposition also yields nitrite (NO_2^-), which can be converted into additional toxic metabolites that contribute to the microbicidal effects of $ONOO^-$[161] (Fig. 7.3). For example, NO_2^- can be oxidized by H_2O_2 in a reaction catalyzed by MPO to produce nitrogen dioxide $NO_2^\bullet$ radicals (see Eq. 7.13) that can nitrate tyrosine and other aromatic compounds[161,162] and promote lipid peroxidation[163]:

$$NO_2^- + H_2O_2 + H^+ \xrightarrow{MPO} NO_2^\bullet + HO^- + H_2O \quad (7.13)$$

Significant evidence indicates that $ONOO^-$ plays an important role in host defense[164] and in the pathogenesis of inflammatory disease.¹⁵⁹ ONOO⁻ is a potent oxidant that can attack a wide variety of biological tissues, and ONOO⁻-mediated tissue damage has been implicated in a number of inflammatory diseases.¹⁵⁹ ONOO⁻-mediated tissue damage has been reported to result from its ability to initiate lipid peroxidation,¹⁶⁵ oxidize protein sulfhydryl groups,¹⁶⁶ and nitrate protein tyrosine residues.¹⁶⁷

7.4 PHAGOCYTE NADPH OXIDASE FUNCTION

TABLE 7.1 Defects in Phagocyte NADPH Oxidase Components Associated with Chronic Granulomatous Disease

Protein	Gene	Chromosome	Frequency
gp91^{phox} (NOX2)	CYBB	Xp21.1	69%
p22^{phox}	CYBA	16p24	5%
p40^{phox}	NCF4	22q13.1	Unknown
p47^{phox}	NCF1	7q11.23	20%
p67^{phox}	NCF2	1q25	6%

See References 188, 472, and 473.

The phagocyte NADPH oxidase is expressed in neutrophils, eosinophils, basophils, mast cells, monocytes, and macrophages, and plays an essential role in host defense against pathogenic microorganisms (reviewed in References 168 and 169). Note, however, that the NADPH oxidase is regulated differently in various types of phagocytes. For example, while monocyte/macrophages and neutrophils express the same NADPH oxidase components, there are significant differences in NADPH oxidase responses between these cell types.¹⁷⁰⁻¹⁷² Activated monocyte/macrophages exhibit a gradual increase in ROS production which peaks about 1 hour after stimulation,¹⁷³ whereas the response in neutrophils is very rapid, peaking in 2–10 minutes.¹⁷⁴ In addition, monocytes are capable of mounting an additional response after recovery from the initial response, which is typically not the case for neutrophils.¹⁷⁵ Lastly, stimuli that activate the monocyte/macrophage NADPH oxidase do not necessarily activate the neutrophil NADPH oxidase (reviewed in References 171 and 176). These differences likely reflect the distinct roles of monocyte/macrophages and neutrophils in chronic versus acute inflammatory responses, respectively.¹⁷⁴,¹⁷⁷,¹⁷⁸

The phagocyte NADPH oxidase is activated when cells encounter a variety of soluble and particulate stimuli, including cytokines, chemokines, microbes and microbial products, viruses, and other foreign antigens.¹⁷⁹ Depending on the stimulus, NADPH oxidase assembly and activation can occur at the plasma membrane or in the phagosomal membrane. The generation of ROS during phagocytosis is determined in part by the type of receptors involved in pathogen recognition, and phagocytic cells utilize a variety of receptors to recognize molecules expressed by bacteria and fungi.¹⁸⁰ Among these, pattern recognition receptors, such as Toll-like receptors (TLRs), play an important role by recognizing microbes or microbial products.¹⁸¹ Engagement of TLRs has been reported to prime phagocytes for enhanced responses to subsequent stimuli, such as phagocytosis itself, and priming results in increased production of microbicidal ROS.¹⁸² Priming can induce partial phosphorylation of p47^{phox} and translocation of flavocytochrome b to the membrane without actual NADPH oxidase activation.¹⁸³ Neutrophil priming is thought to play a key role in host defense and host survival (reviewed in Reference 184). Indeed, phagocytes obtained from mice with deficient TLR function have an impaired bactericidal function, which was proposed to result from a defect in TLR-dependent priming of the partial phosphorylation of p47^{phox}.¹⁸⁵

The importance of the phagocyte NADPH oxidase in host defense is clearly evident in individuals with CGD, which occurs with an incidence of 1/200,000–1/250,000 (reviewed in Reference 28). The majority of CGD mutations that have been identified result in defective gp91^{phox}/NOX2, p22^{phox}, p47^{phox}, or p67^{phox} and result in absent or substantially diminished NADPH oxidase activity, depending on the mutation (Table 7.1).²⁸ Recently, an additional CGD group with defects in p40^{phox} was added¹⁰² (Table 7.1). As a result of NADPH oxidase insufficiency, individuals with CGD experience severe, recurrent bacterial and fungal infections and can develop granulomas that form from accumulated phagocytes containing ingested bacteria.²⁸ In addition to their impaired microbicidal function, CGD phagocytes exhibit abnormal apoptosis/turnover, which is also due to defective NADPH oxidase function.¹⁸⁶

Infections in individuals with CGD are caused mostly by catalase-positive organisms, such as *Aspergillus* spp. or *Staphylococcus* spp., and are often life-threatening.¹⁸⁷ However, recent advances in diagnosis and medical care have significantly improved prognosis, and the survival of individuals with CGD has dramatically improved.¹⁸⁸ For example, the development of improved therapeutic interventions for CGD, such as treatment with interferon-γ or itraconazole, has greatly improved prognosis of this disease.¹⁸⁹ Although neutrophils from CGD patients are unable to generate ROS or have substantially diminished ROS-generating capacity, they are still able to kill a number of pathogens, presumably through the action of other antimicrobial components,¹⁹⁰ and neutrophils from individuals with CGD have been shown to upregulate expression of other host defense proteins.[186] In addition, it has been proposed that the microbicidal capacity of CGD neutrophils depends, to some degree, on H_2O_2 produced by the pathogen itself.[191] In support of this idea, bacteria that do not produce H_2O_2 seem to be resistant to killing by CGD phagocytes.[192] On the other hand, some bacteria express catalase and can detoxify H_2O_2, suggesting that catalase-producing microbes would have enhanced resistance to killing by CGD phagocytes.[191] This appears to be generally true, as catalase-positive organisms are responsible for many of the infections observed in CGD individuals, whereas infections due to catalase-negative organisms are less common.[193] Note, however, that this paradigm is not absolute, as subsequent studies have shown that certain catalase-deficient organisms can be virulent in mouse models of CGD.[194] In addition, Kottilil et al.[195] reported the presence of catalase-negative *Hemophilus* spp. infection in individuals with CGD.

While phagocytes generate ROS during host defense against pathogens, these oxidants can also damage host tissues and, when inappropriately regulated, contribute to inflammatory disease. Indeed, ROS have been shown to contribute to the tissue damage associated with arthritic diseases, such as rheumatoid arthritis (RA),[196,197] and acute inflammatory arthritis.[198] Joint synovial fluid of RA patients contains large numbers of neutrophils,[199] and higher levels of ROS are present in the synovial fluid of arthritic compared to nonarthritic patients.[200] In addition, phagocytes from patients with arthritis produce higher levels of ROS than cells from healthy individuals.[201] Indeed, phagocytes from individuals with arthritis appear to be primed because of their exposure to tumor necrosis factor-α *in vivo*.[202] Furthermore, treatments that inhibit ROS production can suppress the development of inflammation and symptoms associated with arthritis.[203–206]

In addition to its role in host defense and the inflammatory response, the phagocyte NADPH oxidase also plays a role in turning off this response and in resolution of the inflammatory response by inducing neutrophil apoptosis[207] (Fig. 7.4). Apoptosis is essential for neutrophil homeostasis,[208] and its importance in the resolution phase of infections is evident in diseases where tissue destruction and inflammation are associated with prolonged neutrophil activation or cell lysis (reviewed in Reference 209). NADPH oxidase-derived ROS have been implicated in spontaneous neutrophil apoptosis,[210,211] and ROS are required for bacteria-induced neutrophil apoptosis and/or phagocytosis-induced cell death.[212] Indeed, inhibition of NADPH oxidase activation has been shown to prevent phagocytosis-induced neutrophil apoptosis.[213] In part, NADPH oxidase-derived ROS appear to promote cleavage of caspases 3 and 8 following phagocytosis, leading to cell apoptosis[213] (Fig. 7.4). Studies on neutrophils from individuals with CGD further demonstrate the importance of NADPH oxidase-derived ROS in neutrophil apoptosis, as CGD neutrophils exhibit impaired apoptosis and accumulate to form granulomas.[186,214,215] ROS directly or indirectly modulate expression of pro-apoptotic factors that are induced during phagocytosis,[186] and ROS have been shown to be essential for appropriate clearance of neutrophils from inflammatory sites as inflammation is resolved[215] (Fig. 7.4). Indeed, recent studies suggest that ROS can suppress autoimmunity, inflammation, and arthritis (reviewed in Reference 216). For example, inflammation and bone damage are worse in $p47^{phox}$- and $gp91^{phox}$-deficient mice when compared to wild-type mice with experimentally induced arthritis.[217,218] Likewise, individuals with CGD have an increased frequency of autoimmune diseases and hyperinflammation.[219] Based on the role of ROS in the resolution of inflammation described earlier,[220] excessive inflammation in CGD could be due to inability to efficiently terminate inflammatory responses.[219] In support of this idea, it has been proposed that NADPH oxidase-derived ROS can reduce arthritis by regulating arthritogenic T cells and limiting expansion of T-cell dependent autoimmune responses directed to self-antigens.[221,222] Furthermore, macrophage NADPH oxidase activity has recently been shown to be important for the induction of regulatory T cells, which are important in controlling inflammation.[223] Thus, it may be that physiological levels of ROS are required to prevent arthritis and other autoimmune diseases, while excessive ROS production can contribute directly to disease pathogenesis (Fig. 7.4).

7.5 NONPHAGOCYTE NADPH OXIDASE STRUCTURE

For some time after discovery of the phagocyte NADPH oxidase, it was thought that this system was specific only to phagocytic leukocytes. However, more recent research has revealed the presence of analogous NADPH oxidase enzymes in nonphagocyte tissues (reviewed in References 3, 19, and 224–227) (Table 7.2). These enzymes are functionally distinct from the phagocyte NADPH oxidase, serve distinct physiological roles, and respond to a variety of factors, such as growth factors, cytokines, and hormones, and mechanical inputs, such as shear stress and cyclic stretch (reviewed in References 3, 19, 224–227). Characterization of these tissues not only demonstrated the expression of various classical phagocyte oxidase proteins, but also resulted in the identification of novel proteins homologous to those found in phagocytes (see details later). Thus, the assembly of

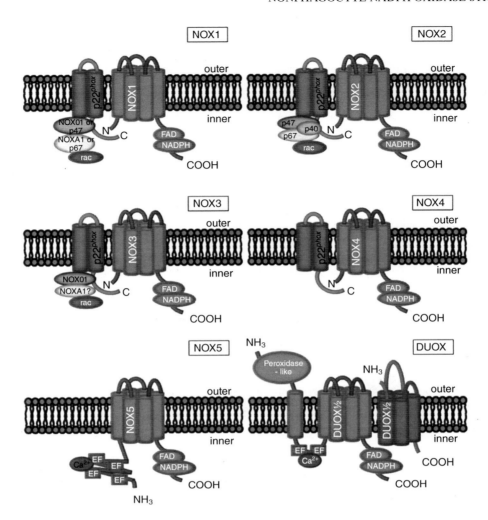

Figure 7.4 NOX family members and their proposed regulatory subunits. NOX proteins are believed to contain six transmembrane domains based on hydrophobicity analysis (seven for DUOX1/2). NADPH oxidase activity occurs when NADPH binds to the cytosolic tail of NOX, where it transfers electrons to FAD. From FAD, the electrons are transferred to the heme centers (refer to Fig. 7.1) and ultimately to oxygen near the outer membrane surface, resulting in $O_2^{\bullet-}$ formation. The transmembrane subunit $p22^{phox}$ is associated with both active and inactive forms of NOX1–NOX4. During activation, NOX1/$p22^{phox}$ is believed to primarily assemble with the cytosolic subunits NOXO1, NOXA1, and Rac GTPase; however, $p47^{phox}$ and $p67^{phox}$ can substitute in some situations for NOXO1 and NOXA1, respectively. NOX2/$p22^{phox}$ activation involves assembly with $p47^{phox}$, $p67^{phox}$, $p40^{phox}$, and Rac2 GTPase. NOX3/$p22^{phox}$ activation is less well defined, but is believed to primarily involve NOXO1 and Rac GTPase in the inner ear. NOX4/$p22^{phox}$ is constitutively active in the absence of cytosolic cofactors. NOX5 and DUOX1/2 activation involves Ca^{2+} binding to EF-hand domains in the cytosol. In addition, DUOX1/2 require the association of DUOXA1/2, respectively, for localization to the plasma membrane. (Adapted from *Free Radic. Biol. Med.* **2009**, *47*(9), 1239–1253. Copyright 2009 Elsevier.) See color insert.

various combinations of classical phagocyte oxidase proteins and unique nonphagocyte oxidase proteins results in an array of tissue-specific NADPH oxidase systems.

7.5.1 NOX1

Characterization of the nonphagocyte NADPH oxidases focused initially on a search for the presence of homologs of $gp91^{phox}$ because of the key role it plays in electron transport, as described earlier. The first $gp91^{phox}$ homolog identified was designated as mitogenic oxidase 1 (Mox-1)[228] or NADPH oxidase homolog 1 (NOH-1)[229] (Fig. 7.5). At this point, the nomenclature for $gp91^{phox}$ and its homologs was revised to avoid anticipated confusion as additional homologs were found. Thus, Mox-1/NOH-1 was renamed NADPH oxidase 1 (NOX1), while

TABLE 7.2 Expression of Phagocyte and Nonphagocyte NADPH Oxidase Proteins

Oxidase Protein	Tissue Expression
NOX1	Vascular smooth muscle, gastric pit (species-dependent), colon epithelium, prostate, uterus, placenta, osteoclasts, retina, neurons
NOX2 (gp91^{phox})	Phagocytes, lymphocytes, vascular smooth muscle, fibroblasts, endothelium, skeletal muscle, neurons, lung, carotid body, kidney, liver, heart, colon
NOX3	Fetal tissue, inner ear
NOX4	Fetal tissue, kidney, pancreas, placenta, ovary, testis, carotid body, vascular smooth muscle, melanocytes, osteoclasts, eye, lung, kidney, thyroid, monocyte/macrophages, prostate, neurons, vascular endothelium, cardiomyocytes, lung fibroblasts
NOX5	Fetal tissue, lymphocytes, spleen, testis, ovary, placenta, pancreas, stomach, mammary glands, colon, vascular endothelium, vascular smooth muscle
DUOX1	Thyroid, salivary glands, colon, rectum, airway epithelium
DUOX2	Thyroid, salivary glands, colon, rectum, airway epithelium, pancreas, prostate, intestinal epithelium, rectal mucosa
p22^{phox}	Phagocytes, lymphocytes, testis, placenta, ovary, kidney, liver, lung, spleen, pancreas, skeletal muscle, neurons, eye, vascular smooth muscle, fibroblasts, endothelium, lung, carotid body, kidney, melanocytes, osteoclasts, inner ear, thyroid
p40^{phox}	Phagocytes
p47^{phox}	Phagocytes, lymphocytes, testis, placenta, ovary, kidney, liver, lung, spleen, pancreas, skeletal muscle, neurons, eye, vascular smooth muscle, fibroblasts, endothelium, lung, carotid body, kidney
p67^{phox}	Phagocytes, lymphocytes, testis, placenta, ovary, kidney, liver, lung, spleen, pancreas, skeletal muscle, neurons, eye, vascular smooth muscle, fibroblasts, endothelium, lung carotid body, kidney,
NOXO1 (p41^{nox})	Colon, liver, small intestine, gastric mucosal cells, cochlea, liver, pancreas, thymus, testis, inner ear
NOXA1 (p51^{nox})	Colon, uterus, salivary gland, small intestine, stomach, lung, thyroid, liver, kidney, pancreas, spleen, prostate, testis, ovary, bronchial epithelial cells, vascular smooth muscle cells

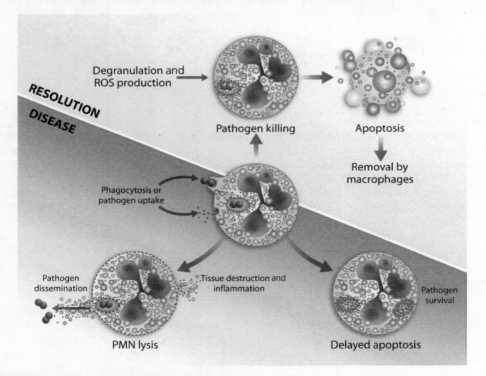

Figure 7.5 A paradigm illustrating the role of ROS in the resolution of infection. Production of ROS by the phagocyte NADPH oxidase is essential for host defense against pathogens and is a key feature of the inflammatory response. However, the presence of ROS is also required for inducing neutrophil apoptosis during resolution of the inflammatory response. In the absence of phagocyte NADPH oxidase activity, neutrophil apoptosis is impaired, leading to excessive inflammation and inflammatory tissue damage. (Adapted from *Clin. Sci.* **2006**, *111*(1), 1–20. Copyright 2006 Biochemical Society.) See color insert.

gp91^{phox} was renamed NOX2.[230] The sequence of NOX1 is quite similar to that of NOX2 (60% identity), and NOX1 has a similar size and domain structure as NOX2[228,229] (Fig. 7.5). Likewise, NOX1 requires association with p22^{phox} to be active.[231,232] Thus, p22^{phox} plays a functional role in both phagocyte and nonphogocyte NADPH oxidases. Indeed, when p22^{phox} was first identified, it was found to be expressed in a variety of nonphagocytic cells, although its role in these cells was a mystery at the time.[13] Since initial studies in NOX1-transfected cells showed that NOX1 alone produced very low levels of ROS,[228] it was proposed that NOX1 might require additional cytosolic cofactors for proper function. In support of this idea, Geiszt et al.[233] reported that NOX1 activity could be substantially enhanced by coexpression of p47^{phox} and p67^{phox} and verified that these *phox* proteins could functionally associate with NOX1. On the other hand, it was clear that alternative cofactors must exist in NOX1-expressing tissues, and three groups concurrently identified homologs of both p47^{phox} and p67^{phox} in mouse[234] and human[235,236] colon. These p47^{phox} and p67^{phox} homologs were subsequently designated as NOX organizer 1 (NOXO1) and NOX activator 1 (NOXA1) due to their proposed roles in oxidase organization and activation, respectively, which is analogous to the roles of p47^{phox} and p67^{phox}[234–236] (Fig. 7.2; see further details about NOXO1 and NOXA1 later). In addition, subsequent studies have demonstrated that Rac1 plays a role in regulation of the NOX1 system and binds to the TPR domain of NOXA1, which is analogous to the interaction of Rac2 with p67^{phox}.[237,238] Thus, it appears that the most efficient NOX1-based NADPH oxidase system is composed of NOX1/p22^{phox}, NOXO1, NOXA1, and Rac.

NOX1 is not expressed in phagocytes but is highly expressed in colon epithelial cells.[228,229] It is also expressed at lower levels in other tissues and cell types, such as vascular smooth muscle cells, endothelial cells, prostate, uterus, placenta, osteoclasts, and retinal pericytes (reviewed in Reference 19) (Table 7.2).

7.5.2 NOX3

Based on sequence similarity to NOX1 and NOX2, additional homologs of comparable size and domain structure were soon identified.[239] Structurally, NOX3 appears to contain a similar number of transmembrane domains, analogous FAD- and NADPH-binding domains, and similar placement of heme-coordinating histidines[239] (Fig. 7.5). NOX3 forms a functional heterodimer with p22^{phox},[240] and mutations in the proline-rich region of p22^{phox} result in dominant inhibition of ROS production by NOX3 and other NOX-based enzymes.[241] NOX3 is expressed almost exclusively in the inner ear vestibular system,[242] although low levels of expression have been noted in some fetal tissues[243] (Table 7.2). Nevertheless, it is generally thought that the primary role of NOX3 is in the formation of otoconia in the inner ear, and mutations in NOX3 were found to lead to the vestibular defects that were observed in mutant mice.[242,243]

Although NOX3/p22^{phox} can generate ROS constitutively in the absence of cytosolic cofactors,[240] this activity is substantially enhanced by the presence of regulatory subunits. NOX3/p22^{phox} activity is significantly enhanced by NOXO1 and, to a lesser extent, by p47^{phox}.[240] Since inactivation of NOXO1 in mice results in the same phenotype as that found in mice with NOX3 mutations, it is generally thought that NOXO1 is physiologically the most important of these cofactors.[244] The role of NOXA1 in the NOX3-based oxidase system is not clear, as some studies suggest NOXA1 is important,[238,243] while others do not.[240,245] Rac GTPase also appears to play a role in the NOX3-based system, although the contribution of Rac to this system appears to be variable and may depend on which combination of the different cofactors is present.[246] For example, studies in transfected cell lines indicated that Rac played a positive role in NOX3 activation in the presence of p47^{phox} and either p67^{phox} or NOXA1, whereas Rac failed to enhance NOX3 activity when p47^{phox} was replaced with NOXO1.[246] Since NOXO1 seems to be the most important regulator of NOX3 activity, these studies suggest that Rac may not be absolutely required.

7.5.3 NOX4

NOX4 was originally identified in the kidney and was first designated as Renox.[247] This NOX2 homolog was subsequently identified in kidney tissue by two other groups and renamed NOX4[239,248] (Fig. 7.5). Although NOX4 does not exhibit the high degree of similarity that is present among NOX1–NOX3, it does require p22^{phox} to be functionally active.[232,249] On the other hand, the role of p22^{phox} in NOX4 function seems to be distinct from that of NOX1–NOX3, and p22^{phox} mutations that abolished NOX2 and NOX3 function *in vivo* had no effect on NOX4 activity when expressed in lung cancer cells.[250] In addition, p22^{phox} mutations that inhibited NOX1–NOX3 maturation did not impair the association of p22^{phox} with NOX4.[250] In contrast to NOX3, NOX4 is expressed in a wide variety of tissues, including the renal cortex, vascular smooth muscle and endothelium, cardiomyocytes, ovary, eye, skeletal muscle, testis, osteoclasts, and prostate (reviewed in Reference 19) (Table 7.2). In addition, recent studies indicate NOX4 is expressed in thyroid tissue[251] and in human monocytes and mature macrophages.[252]

NOX4/p22^{phox} does not appear to require any additional cytosolic oxidase cofactors for activity and apparently generates ROS constitutively.[247–249] However, not all studies are consistent with NOX4 being constitutively active, and it has been reported that this NADPH oxidase can be activated by different signals, such as lipopolysaccharide,[253] insulin,[254] and angiotensin II[255] under certain conditions. The role of Rac GTPase in NOX4/p22^{phox} is also a matter of debate. Rac does not appear to be required in cell lines transfected with NOX4[249] but has been reported to regulate NOX4 activation in cells endogenously expressing NOX4, such as mesangial cells.[255] On the other hand, analysis of the Rac-binding site in NOX proteins revealed that specific residues required for Rac-dependent NADPH oxidase activity are present in NOX1–NOX3 but are not conserved in NOX4,[117] indicating that if Rac is involved in NOX4 function, it utilizes an unconventional site of interaction. Recently, Lyle et al.[256] reported that Poldip2 was a novel p22^{phox}-binding protein and that Poldip2 associates with p22^{phox} to activate NOX4 in vascular smooth muscle cells. Whether Poldip2 plays a global role in NOX4 regulation remains to be determined.

7.5.4 NOX5

As described above, NOX1–NOX4 are characterized by similar transmembrane catalytic regions. While this catalytic core is also present in NOX5, NOX5 contains an additional N-terminal extension that contains four EF hand-like, Ca^{2+}-binding motifs and is activated in a Ca^{2+}-dependent manner[257] (Fig. 7.5). NOX5 was found to be expressed in spleen, lymph nodes, and testis[257] (Table 7.2). More recent work indicates that NOX5 is functionally expressed in vascular endothelial cells.[227,258] NOX5 does not require association with p22^{phox} for activity, as knockdown of p22^{phox} had no effect on NOX5,[241] but is localized to the plasma membrane through binding of an N-terminal polybasic region with phosphatidylinositol (4,5)-bisphosphate.[259] NOX5 does not require additional cytosolic cofactors[260] and also functions independently of Rac GTPase.[261] Thus, it has been proposed that the N-terminal Ca^{2+}-binding domain may function in a similar capacity as the cytosolic regulatory subunits to activate electron transfer.[260] In the case of NOX5, it is proposed that Ca^{2+} binding to the EF hand domains induces a conformational change in the NOX5 N-terminus that facilitates intramolecular protein–protein binding interactions with the catalytic core, resulting in enzymatic activity[260] (Fig. 7.5).

7.5.5 DUOX1 and DUOX2

Shortly after the cloning of NOX1, two novel NOX2 homologs were identified in thyroid tissue and were designated as thyroid oxidase 1 and 2 (ThOX1 and ThOX2).[262,263] These proteins are much longer than the other NOX proteins because of the addition of large N-terminal extensions containing two proximal EF hand motifs that function in Ca^{2+} binding, an additional transmembrane helix, and a distal peroxidase homology domain (Fig. 7.5). Based on the presence of both NADPH oxidase and peroxidase homology domains, these proteins were renamed dual oxidase 1 and 2 (DUOX1 and DUOX2).[230] Both DUOX proteins are expressed in thyroid tissue, airway epithelial cells, and prostate tissue. In addition, DUOX2 is also present in salivary glands, rectal mucosa, and intestinal epithelium (reviewed in Reference 3) (Table 7.2). DUOXes require maturation factors known as DUOXA1 and DUOXA2 for proper expression and membrane localization[264]; however, cytosolic cofactors and p22^{phox} do not seem to be required for function of these NADPH oxidase homologs. Thus, it has been suggested that DUOXA1 and DUOXA2 may function in stabilization of the DUOX membrane complexes, which is analogous to the role of p22^{phox} in NOX complexes.[265]

7.5.6 NOXO1

The search for a p47^{phox} homolog resulted in the identification of p41 or p41^{nox}, based on its predicted molecular weight, which was subsequently renamed NOXO1.[234–236] NOXO1 is expressed primarily in colon and testis, but is also found at low levels in pancreas, liver, thymus, uterus, inner ear, and small intestine[234–236] (Table 7.2). NOXO1 shares 23% amino acid sequence identity with p47^{phox}; however, there is a very high degree of homology in the placement of their functional domains. Both proteins contain similarly placed PX domains, tandem SH3 domains, and proline-rich domains (Fig. 7.2). In contrast, NOXO1 lacks the AIR that regulates exposure of the p47^{phox} tandem SH3 domains during activation.[234–236] This region of p47^{phox} and NOXO1 plays a role in binding to p22^{phox}.[44,47,48] Thus, it appears that the NOXO1 may be able to associate with p22^{phox} constitutively, resulting in constitutive ROS production by systems utilizing this cofactor, which include NOX1- and NOX3-based NADPH oxidase systems. However, this issue is still unclear, as only low levels of constitutive activity have been observed.[234–236] Recently, Dutta and Rittinger[266] reported that the region C-terminal to the tandem SH3 domains of NOXO1 interferes with binding to p22^{phox} and may thereby prevent the formation of a fully active oxidase complex in unstimulated cells. They also proposed that the inhibitory effect of the C-terminal tail is less pronounced with NOXO1 than with p47^{phox}, suggesting a possible explanation of why the system using NOXO1 is capable of producing low levels of $O_2^{\bullet-}$ in a constitutive manner.[266]

Gianni et al.[267] recently reported that the c-Src substrate proteins Tks4 (tyrosine kinase substrate with four SH3 domains) and Tks5 are functional members of a NOXO1 superfamily. They showed that Tks proteins could interact with NOXA1 to support NOX1/p22^{phox} and NOX3/p22^{phox} activity in reconstituted cellular systems and played a role in the formation of functional invadopodia in human colon cancer cells.[267,268] Whether additional NOX organizers exist remains to be determined.

7.5.7 NOXA1

The search for a p67^{phox} homolog resulted in the identification of p51 or p51^{nox}, based on its predicted molecular weight, which was subsequently renamed NOXA1.[234–236] NOXA1 is highly expressed in colon and seems to be expressed in a wider range of tissues than NOXO1[234–236] (Table 7.2). NOXA1 exhibits 28% amino acid sequence identity with p67^{phox} but is significantly shorter due to the absence of the first SH3 domain and a small region near the second SH3 domain[234–236] (Fig. 7.2). NOXA1 contains TPR motifs for interaction with Rac GTPase, an activation domain, a PB1 domain, and a C-terminal SH3 domain that are analogous to p67^{phox}[234–236] (Fig. 7.2). Analogous to the interaction between p67^{phox} and p47^{phox}, the C-terminal SH3 domain of NOXA1 binds to the proline-rich region in the C-terminus of NOXO1[236]; however, this interaction is weaker than the p67^{phox}–p47^{phox} interaction, indicating that the molecular features of the NOXA1–NOXO1 interaction may be significantly different.[266] In addition, the NOXA1 PB1 domain has characteristics distinct from that of p67^{phox} and does not support binding with p40^{phox}.[236] Recently, Kroviarski et al.[269] reported that NOXA1 was phosphorylated by MAPK, PKC, and PKA and that phosphorylation of NOXA1 decreased its binding to NOX1 and Rac1. Based on these studies, they proposed a role for NOXA1 phosphorylation in controlling excessive activation of NOX1-dependent NADPH oxidases.[269]

7.6 NONPHAGOCYTE ROS PRODUCTION

As with the NOX2-based NADPH oxidase, the primary product of NOX1, NOX3, and NOX5 appears to be $O_2^{\bullet-}$, which results from the oxidation of NADPH and the univalent reduction of O_2 (Fig. 7.1). Note, however, that nonphagocyte NADPH oxidase systems generally produce much lower amounts of ROS, which is consistent with their functional roles in signaling and cellular homeostasis.[226] In addition, a number of nonphagocytic cell types have been shown to express multiple NOX enzymes that localize to different subcellular compartments (Fig. 7.6 and Table 7.2).

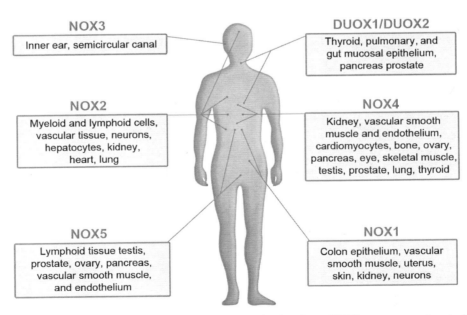

Figure 7.6 Schematic representation of the various regions of the body where NOX systems are located. Tissues reported to express NOX homologs throughout the human body are indicated. See text for further details. (Adapted from *Clin. Sci.* **2006**, *111*(1), 1–20. Copyright 2006 Biochemical Society.)

As discussed above, NOX4/p22^{phox} appears to be constitutively active, without requirement for additional cytosolic components. However, in contrast to other NOX systems, the primary product of NOX4/p22^{phox} is H_2O_2.[249] Generation of H_2O_2 requires NADPH and FAD, and recent studies indicate that the cytosolic dehydrogenase domain of NOX4 is constitutively "turned on" compared with the dehydrogenase domains of NOX1–NOX3, which explains the constitutive activity observed for this enzyme.[270] NOX4 is localized primarily in the endoplasmic reticulum[271,272] and plays a role in endoplasmic reticulum stress responses.[273]

DUOX1 and DUOX2 generate H_2O_2 in a Ca^{2+}-dependent manner, and it has been proposed that the N-terminal peroxidase homology domain plays a role in H_2O_2 formation (reviewed in Reference 3) (Fig. 7.5). Mature DUOX enzymes coexpressed with their respective maturation factors do not appear to generate $O_2^{\bullet-}$, despite the homology with NOX2.[264,265] However, it has also been suggested that DOUX enzymes initially generate a $O_2^{\bullet-}$ intermediate, which is rapidly dismutated within the peroxidase homology domain of the DUOX molecule so that free $O_2^{\bullet-}$ is never released.[274]

7.7 FUNCTIONS OF NONPHAGOCYTE NADPH OXIDASES

NADPH oxidases play important functions in many organ systems and cell types, thereby contributing to normal physiological processes and, when in excess or dysregulated, to the pathogenesis of various human diseases (Fig. 7.6). The roles of nonphagocyte NADPH oxidases in a variety of organs and tissues are described as follows.

7.7.1 Cardiovascular System

A number of NADPH oxidases have been reported to be important in the cardiovascular system, and they have been shown to play important roles in physiological processes, such as blood pressure regulation, as well as pathophysiological events, including hypertension, restenosis, inflammation, and atherosclerosis (recently reviewed in References 226 and 227) (Fig. 7.6). The functions of NADPH oxidase-generated ROS in the cardiovascular system seem to be quite complex, involving multiple NADPH oxidases and a range of signaling and regulatory mechanisms. NADPH oxidase-dependent ROS production has been observed throughout the vascular system, and NADPH oxidase expression has been observed in adventitial fibroblasts, endothelial cells, and smooth muscle cells.[226,227] In the vascular system, ROS have been shown to participate in the regulation of $^{\bullet}$NO bioavailability, cell growth and differentiation, cell migration, cell contraction, and fibrosis.[226,227]

One of the key roles of ROS in the cardiovascular system is in the regulation of vascular tone and blood pressure.[275] As discussed above, $O_2^{\bullet-}$ can rapidly react with $^{\bullet}$NO and thus neutralize the biological properties of this important mediator of vascular homeostasis.[276] $^{\bullet}$NO (aka endothelial-derived relaxation factor) is generated by endothelial nitric oxide synthase (eNOS), and generation of $^{\bullet}$NO by endothelial cells is required for relaxation of vascular smooth muscle cells and vasodilation.[277] It is now clear from a number of studies that vascular tone is regulated by the balance between $^{\bullet}$NO and $O_2^{\bullet-}$ and that oxidative stress resulting in excessive levels of ROS can upset this balance, leading to vascular diseases (reviewed in References 226, 275, and 278). In addition to NADPH oxidase activity, xanthine oxidase also represents an important source of vascular ROS that contributes to vascular dysfunction.[279] Additionally, it is thought that oxidative stress due to increased vascular ROS can lead to tetrahydrobiopterin (BH_4) oxidation and thus uncoupling of eNOS, which requires BH_4 as a cofactor.[280] Consequently, uncoupled eNOS produces $O_2^{\bullet-}$ rather than $^{\bullet}$NO, and it has been reported that eNOS-derived $O_2^{\bullet-}$ contributes to the development of vascular diseases, such as atherosclerosis and hypertension (reviewed in References 275 and 280).

In addition to the regulation of $^{\bullet}$NO bioavailability, vascular ROS also have direct effects on vascular tissues, independent of $^{\bullet}$NO. For example, ROS have been shown to induce vascular smooth muscle cell proliferation, which contributes to vascular disease pathogenesis.[281,282] ROS also play important roles in regulation of redox-sensitive intracellular targets, such as protein kinases, protein tyrosine phosphatases, and transcription factors, during vascular cell responses to hormones and growth factors (reviewed in Reference 283).

It is now known that a number of NADPH oxidase enzymes participate in vascular ROS production, including those utilizing NOX1, NOX2, NOX4, and NOX5.[226] Indeed, SOD mimetics, antioxidants, and inhibitors of NADPH oxidase assembly have been shown to reduce hypertension (reviewed in Reference 284). The first NADPH oxidase to be identified in vascular tissue was a NOX2-dependent system present in aortic adventitial fibroblasts.[285] This NADPH oxidase was found to be constitutively active, but its activity could also be enhanced by angiotensin II.[285] Subsequent work has shown that NOX2 is expressed in all vascular wall cells except vascular smooth muscle cells from large arteries (reviewed in Reference 226). NOX2-dependent NADPH oxidase activity is elevated in animal models of hypertension, such as renovascular hypertension and angiotensin II-induced hypertension, supporting a role for vascular-derived $O_2^{\cdot-}$ in regulating $^\cdot$NO bioavailability.[286] Indeed, the development of hypertension was diminished in $p47^{phox}$-deficient mice[287] and in some models of hypertension using NOX2-deficient mice.[288] Note, however, that ROS in hypertension induced by chronic angiotensin II administration seem to be NOX2-independent, suggesting a role for other NOX homologs.[289] Since knockdown of $p22^{phox}$ expression in vivo also alleviated angiotensin II-induced hypertension,[290] it can be concluded that activity of this NOX homolog is $p22^{phox}$-dependent. In addition to its role in hypertension, NOX2-dependent NADPH oxidase activity has also been implicated in angiotensin II-induced vascular hypertrophy,[291] cardiac hypertrophy,[292] aortic stenosis,[293] angioplasty-induced neointimal hyperplasia,[294] ischemia-induced angiogenesis,[295] and aldosterone-induced inflammation of the heart.[296] In atherosclerosis, NOX2 is upregulated, and its expression correlates with lesion severity.[297] Likewise, the development of atherosclerotic lesions was substantially reduced in NOX2-deficient mice.[298]

Besides NOX2, additional NOX homologs have been identified in vascular tissue, including NOX1, which is expressed in endothelial cells, smooth muscle cells, and adventitial fibroblasts[299,300]; NOX4, which is constitutively expressed in large vessel smooth muscle cells and endothelium[301,302]; and NOX5, which is expressed in vascular smooth muscle cells and endothelial cells.[303,304]

NOX1-dependent ROS production has been reported to play a role in hypertension, and NOX1 overexpression in smooth muscle cells enhanced angiotensin II-dependent hypertension.[305] Likewise, angiotensin II-dependent hypertension was reduced in NOX1-deficient animals.[306] NOX1 has also been implicated in proliferative vascular disease and appears to play a role in abnormal vascular growth induced by oxidized low-density lipoprotein[307] and advanced glycation end products.[308] NOX1 expression is increased in hypercholesterolemia and has been reported to contribute to vascular lesion formation, which is due in part to its role in foam cell formation.[309,310] In support of the role of NOX1 in atherosclerosis, Niu et al.[311] recently demonstrated that NOXA1 expression was significantly increased in aortas and atherosclerotic lesions of ApoE-deficient mice compared with wild-type mice and that NOXA1 was present in intimal and medial smooth muscle cells of human early carotid atherosclerotic lesions.

NOX4 is abundantly expressed in vascular cells and appears to play important roles in cell differentiation, cell growth, and cell migration (reviewed in Reference 226). NOX4 is expressed at higher levels than other NOX proteins in vascular endothelial cells and may be the primary NADPH oxidase system in these cells.[271,312] NOX1 and NOX4 appear to be localized in different cellular compartments in vascular smooth muscle cells, with NOX1 primarily localized in surface domains and NOX4 concentrated in focal adhesions.[301] In endothelial cells, NOX4 is localized in the endoplasmic reticulum and possibly the nucleus.[272,313] Thus, NADPH oxidase compartmentalization may facilitate contributions to distinct signaling events within or between vascular cells. In addition, recent studies suggest that the relative levels or activities of different NOX homologs in a given cell may change, depending on physiological events. For example, ROS production is found in the plaque shoulder of atherosclerotic arteries, and analysis of the plaque shoulder demonstrated the presence of NOX2 and NOX4, which are expressed in plaque-associated macrophages and vascular smooth muscle cells.[297] On the other hand, NOX4 expression does not appear to be upregulated in parallel with the increase in NOX2 expression during atherosclerosis, suggesting increased NOX2 expression is a better correlate of disease pathogenesis.[297] NOX2 and NOX4 expression is enhanced in vessels from patients with coronary artery disease, suggesting a role for these oxidases in this disease.[314] Interestingly, NOX1, NOX2, and NOX4 are upregulated sequentially during restenosis after balloon injury of the carotid artery; however, the kinetics were different for each NOX protein, with NOX1 upregulation at 3 days, followed by increased NOX2 at 7–15 days, and finally NOX4 upregulation at >15 days postinjury,[315] suggesting differential roles in vascular repair. Recently, ROS production by NOX4 was found to be required for oxidized low-density lipoprotein-induced macrophage death, further implicating this NOX homolog in atherogenesis.[252]

Not much is known regarding the functions of NOX5 in the vasculature. NOX5 is expressed in vascular smooth muscle cells and endothelial cells and appears to play a role in vascular cell proliferation.[303,304] Overexpression of NOX5 in intact vessels increased eNOS activity and $O_2^{\cdot-}$ production, with a net loss in $^\cdot$NO bioavailability.[316] NOX5 expression has been reported to be upregulated in atherosclerosis and may play a role in lesion development by contributing to vascular oxidative stress.[258]

The importance of phagocyte NADPH oxidase proteins in regulation of the vascular system suggests that individuals with CGD should exhibit cardiovascular abnormalities, including problems in blood pressure regulation. Surprisingly, there appears to be little effect of these defects in ROS production on blood pressure regulation, regardless of the defect. Indeed, a recent study of CGD patients with NOX2 deficiency showed these patients had enhanced arterial dilatation but normal blood pressure values[317] Thus, it appears that the absence of vascular abnormalities in CGD patients may be due, in part, to the compensatory effects of other NOX homologs. For example, NOX1 has been shown to participate in regulation of blood pressure, and hypertension induced by angiotensin II is decreased or absent in NOX1-deficient mice.[306,318] On the other hand, chronic angiotensin II-dependent hypertension can occur independently of both NOX1 and NOX2,[289,319] suggesting involvement of a different NOX homolog. One candidate for a potential compensatory oxidase is NOX4, which is highly expressed in the vasculature and can function in the absence of the regulatory cytosolic factors that are missing in autosomal forms of CGD[232,249] (Table 7.1). Note however, that NOX4 requires association with p22^{phox},[232,249] which is defective in certain forms of CGD[28] (Table 7.1). Another possibility is NOX5, which has recently been shown to be regulated by angiotensin II[261] and does not require additional cofactors for activity.[260] Further research is needed to determine if the various NADPH oxidase enzymes can functionally compensate for each other and the mechanisms involved.

NADPH oxidase-dependent ROS production is also important in cardiac tissues, and NOX2 and NOX4 appear to be the primary systems involved (reviewed in Reference 19). NOX4 has been reported to play an important role in cardiomyocyte differentiation,[320] whereas NOX2 is the main NOX form in adult heart tissue.[19] NOX2 is upregulated in cardiac tissue after myocardial infarction[321] or ischemia.[322] On the other hand, myocardial damage was not found to be decreased in mice with a defective NOX2-based NADPH oxidase.[323] Recently, Kuroda et al.[324] reported that NOX4 is expressed in mitochondria of cardiac myocytes and is a major source of mitochondrial oxidative stress during pressure overload and contributes to heart failure. Conversely, Zhang et al.[325] reported that NOX4 mediates protection against chronic load-induced stress in mouse hearts by enhancing angiogenesis. Thus, further work will be needed to reconcile these observations.

7.7.2 Renal System

ROS play important roles in renal tissues, and NADPH oxidase expression has been observed in a number of cells in the kidney (reviewed in References 19 and 326) (Fig. 7.6). ROS have been implicated in the regulation of Na$^+$ transport, renovascular tone, tubuloglomerular feedback, and renal oxygenation (reviewed in References 327 and 328). For example, ROS increased salt absorption in the thick ascending limb of Henle's loop by modulation of Na$^+$/H$^+$ exchangers.[329] In addition, enhanced tubuloglomerular feedback observed in spontaneously hypertensive rats (SHRs) has been shown to be due to the removal of •NO by increased O$_2$•$^-$ production in the juxtaglomerular apparatus (JGA).[330] Furthermore, ROS have been reported to exert a tonic regulatory action on renal medullary blood flow in the renal outer medulla.[331] During hypoxia, renal medullary function is enhanced by ROS through regulation of hypoxia-inducible factor-1α (HIF-1α) expression.[328]

All NADPH oxidase cytosolic subunits and three of the NOX homologs (NOX1, NOX2, and NOX4) are expressed in the kidney and have been localized to the renal arterioles, glomeruli, macula densa, podocytes, and distal nephron.[326] NOX4 (aka Renox) was originally discovered in kidney and appears to be the most abundant NOX expressed in kidney, with high levels of expression being observed in the renal cortex.[247,248] Comparatively lower levels of NOX4 are expressed in glomerular mesangial cells, where it has been proposed to play a role in angiotensin II-induced activation of Akt/protein kinase B.[255] NOX2-dependent ROS production in podocytes has been proposed to play a role in the regulation of renal glomerular blood flow, and these cells have been shown to express NOX2, p22^{phox}, p47^{phox}, p67^{phox}.[332] Indeed, analysis of renal function in NOX2-deficient mice confirmed a role for this oxidase in the maintenance of renal vascular tone.[333]

While ROS play essential roles in normal renal function, oxidant stress has also been implicated in pathological conditions related to abnormal kidney function, including hypertension and diabetes. For example, NOX protein expression is upregulated in the SHR kidney, suggesting a role in renovascular hypertension.[334] Furthermore, it was shown that oxidant stress in spontaneous hypertension was due not only to upregulation of NOX1 and p22^{phox}, but also to loss of extracellular SOD.[335] Likewise, knockdown of p22^{phox} expression resulted in decreased oxidative stress and inhibited angiotensin II-induced hypertension.[290] Dopamine is an important regulator of blood pressure, sodium balance, and renal and adrenal function, and abnormal signaling of D$_1$-like receptors (D$_1$ and D$_5$) is involved in animal models of hypertension and in humans with essential hypertension (reviewed in Reference 336). Interestingly, activation of D$_1$-like receptors has been shown to inhibit expression of NOX2, p47^{phox}, and NOX4, leading to decreased ROS production in kidney and brain.[337] In addition, recent studies suggest D$_1$-like receptors may decrease renal NADPH oxidase activity by disrupting the assembly of NADPH oxidase components in cell membrane microdomains.[338] Thus, the ability of D$_1$-like receptor stimulation to decrease ROS production may explain, in part, the antihypertensive action of dopamine.

ROS also play a role in diabetes, and NOX4-dependent ROS production is evident in the kidney during early stages of diabetes and has been reported to contribute to renal hypertrophy.³³⁹ In addition, high glucose has been reported to enhance PKC activity in mesangial cells, leading to increased ROS associated with diabetic glomerulopathy.³⁴⁰ NADPH oxidase activity is enhanced in diabetic nephropathy, and has been reported to be involved in glomerular cell apoptosis, endothelial dysfunction, phagocyte adherence, and impaired coagulation in the kidney.³⁴¹ In support of this idea, treatment of rats with apocynin, an NADPH oxidase inhibitor, prevented proteinuria and glomerulopathy associated with experimental diabetes.³⁴² Likewise, treatment of animals with the antioxidant α-lipoic acid reduced NADPH oxidase protein expression, as well as kidney damage associated with diabetes.³⁴³ Similarly, combined treatment of rats with the antioxidant probucol and an angiotensin II type I receptor antagonist arrested proteinuria and disease progression in mesangioproliferative glomerulonephritis.³⁴⁴

7.7.3 Pulmonary System

ROS play distinct physiological roles in the pulmonary system and participate in airway and vasculature remodeling, O_2 sensing, and mucosal defense (reviewed in References 345 and 346) (Fig. 7.6). Airway epithelial cells express both DUOX1 and DUOX2[347,348] (Table 7.2). In these cells, DUOX proteins are expressed on the apical surface, where they generate H_2O_2 that is utilized by lactoperoxidase to form microbicidal hypothiocyanite, which contributes to mucosal host defense against airway pathogens (reviewed in Reference 3). Airway DUOX expression is differentially regulated by cytokines, including IL-4 and IFN-γ.[348]

In airway smooth muscle cells and pulmonary artery smooth muscle cells, ROS play an important role in signaling events associated with cell proliferation. For example, proliferation of airway smooth muscle cells involves a $p22^{phox}$-dependent NADPH oxidase and ROS-dependent activation of NF-κB.[349] Likewise, proliferation in human pulmonary artery smooth muscle cells induced by transforming growth factor β (TGF-β) is dependent on NOX4, which may be the dominant NOX isoform in these cells. NADPH oxidase expression has also been described in human lung endothelial cells, and its activation was upregulated by hyperoxia.[350] Subsequent studies demonstrated a role for both NOX4 and NOX2 in hyperoxia-induced ROS generation and migration of human lung endothelial cells.[351] Furthermore, knockdown of either NOX4 or NOX2 resulted in upregulated expression of the remaining NOX protein, suggesting activation of compensatory mechanisms when either one of these NOX homologs was missing.[351] NOX4-dependent NADPH oxidase activity has been identified in lung fibroblasts, and ROS production was found to be upregulated in these cells by rhinovirus infection.[352] In addition to NOX4 and $p22^{phox}$, expression of $p47^{phox}$ and $p67^{phox}$ but not NOX2 was observed in these cells.[352] Since NOX4 does not require these cytosolic cofactors, their role in lung fibroblast function is not clear. On the other hand, orally bioavailable NOX4 inhibitors have been suggested as potential new treatments for idiopathic pulmonary fibrosis.[353]

NADPH oxidases play an important role in O_2 sensing throughout the body through their involvement in pulmonary chemoreceptors (i.e., neuroepithelial bodies [NEB]), carotid body, erythropoietin (EPO)-producing cells in the kidney, and other organ systems (reviewed in Reference 354). Based on the expression of NOX2, $p22^{phox}$, $p47^{phox}$, and $p67^{phox}$ in the carotid body and airway chemoreceptor cells, it has been proposed that a NOX2-dependnt oxidase is involved in O_2 sensing in these tissues.[355,356] Indeed, mice deficient in NOX2 have defective O_2 sensing in pulmonary NEB[357] and defective respiratory control.[358] On the other hand, it has also been reported that inhibition of NADPH oxidase activity does not block hypoxia sensing in carotid body chemoreceptor cells.[359] In addition, normal hypoxia-induced gene expression was found in NOX2-deficient cells,[360] and normal O_2 sensing was observed in pulmonary artery smooth muscle cells and carotid body isolated from NOX2-deficient mice.[361,362] In more recent studies, a role for additional NOX homologs has been suggested in O_2 sensing. For example, NOX4 has been proposed to be an oxygen sensor, leading to regulation of TASK-1 activity in HEK293 cells.[363] Recent analysis of NADPH oxidase protein expression and K^+-sensitive O_2 sensing channels (Kv) showed that in rat lung, NOX2, NOX4, $p22^{phox}$, Kv3.3, and TASK1 localized to the apical plasma membrane of NEB cells and that knockdown of NOX2, but not NOX4, in NEB inhibited responses to hypoxia.[364] In addition, co-immunoprecipitation studies showed NOX2 molecular complexes with Kv but not with TASK, while NOX4 associated with TASK1 but not with Kv channel proteins, suggesting that unique combination of NOX and O_2-sensing channels play diverse physiological NEB functions.[364]

In addition to their physiological functions, NADPH oxidases have been proposed to play roles in the pathology of asthma and pulmonary vascular diseases (reviewed in References 365 and 366). ROS production is increased in pulmonary hypertension, and hypoxia-induced activation of ROS has been reported to contribute to pulmonary artery vasoconstriction.[367] Indeed, treatment of isolated rat pulmonary arteries with NADPH oxidase inhibitors blocked hypoxia-induced vasoconstriction.[368] Likewise, NADPH oxidase-dependent ROS production has been implicated in the vascular changes associated with chronic hypoxia-induced pulmonary artery hypertension, and both NOX2[369] and NOX4[370] systems have been implicated in this process. Pulmonary fibrosis is another disease associated with excessive ROS production in the lung, and it has been proposed that NOX4 is involved in myofibroblast activation and fibrogenic responses to lung injury.[371] In addition, the absence of ROS production in p47^{phox}-deficient mice protected them against bleomycin-induced pulmonary fibrosis.[372] Thus, additional NOX homologs besides NOX4 are likely involved. ROS contribute to the inflammation associated with cystic fibrosis, and recent studies suggest that ROS production by DUOX enzymes may contribute to the pathogenesis of this disease.[373]

NADPH oxidase-derived ROS have been implicated in lung injury associated with asthma and acute lung injury (reviewed in References 374 and 375). In addition to the involvement of ROS generated by pulmonary airway cells,[374] ROS produced by inflammatory phagocytes are also involved in actual pulmonary injury (reviewed in Reference 376). Indeed, excessive phagocyte-derived ROS has been implicated in lung injury during asthma, allergic rhinitis, and adult respiratory distress syndrome (ARDS).[376,377] For example, lung microvascular injury due to phagocyte-generated ROS is absent in lung microvessels from NOX2- and p47^{phox}-deficient mice.[378] In addition, chronic ethanol ingestion increases ROS production and NOX2 expression in the lung, possibly contributing to ARDS.[379] Treatment with antioxidants or overexpression of SOD were reported to reduce airway hyperresponsiveness,[380] supporting a role for ROS in asthma. •NO deficiency has been shown to contribute to allergen-induced airway hyperreactivity, and it was proposed that an oxidant–antioxidant imbalance contributed to this response due to removal of •NO by $O_2^{•-}$.[365] However, subsequent studies demonstrated that the •NO deficiency associated with airway hyperreactivity was due to downregulation of constitutive NOS activity and not due to quenching by $O_2^{•-}$.[381] Thus, the specific roles played by NADPH oxidases in the pulmonary system can also depend on the nature of underlying conditions.

7.7.4 Central Nervous System

NADPH oxidase-generated ROS play important roles in central nervous system function, with NOX2 being the predominant NOX, although NOX1 and NOX4 have also been identified in neurons (reviewed in Refererence 382). Microglia are macrophage-like cells in the central nervous system and, like other phagocytes, generate ROS using a NOX2-based NADPH oxidase.[383] Microglial ROS are thought to participate in central nervous system host defense[384] and in the regulation of microglial proliferation.[385] Astrocytes also express a NOX2-based oxidase, which plays a role in astrocyte responses to inflammatory agents[386] and astrocyte intracellular signaling.[387] NADPH oxidases are expressed in neurons, including systems involving NOX1,[388] NOX2,[389,390] and NOX4.[391] NOX1 is upregulated by nerve growth factor, and increased NOX1-dependent ROS production has been proposed to negatively regulate neuronal differentiation by suppressing excessive neurite outgrowth.[388] NOX2, p22^{phox}, p40^{phox}, p47^{phox}, and p67^{phox} are present in sympathetic neurons, and ROS generated by this system contributes to neuronal apoptosis during nerve growth factor deprivation.[389] ROS are also important signaling molecules in the central nervous system, and NOX2-dependent NADPH oxidase activity has been reported to be involved in angiotensin II signaling in the nucleus tractus solitarius[392] and hypothalamic cardiovascular regulatory nuclei.[393] Similarly, NOX2-dependent NADPH oxidase activity is required for N-methyl D-aspartate (NMDA) receptor signaling in the hippocampus.[394] However, prolonged NOX2-dependent glutamate release has been suggested to lead to neuroadaptive downregulation of NMDA receptor subunits.[395] Finally, ROS have been proposed to play a role in synaptic plasticity and memory (reviewed in Reference 396).

As in the vascular system, NADPH oxidases have been shown to participate in the regulation of cerebral vascular tone (reviewed in Reference 397). NADPH oxidase-generated ROS have been reported to reduce •NO bioavailability in cerebral vessels, and chronic exposure to nicotine was shown to increase p47^{phox} expression and ROS production in the parietal cortex, leading to impaired NOS-dependent vasodilation of pial arterioles.[398] In addition, NOX4 has been proposed to play a role in regulating cerebral vasodilation, and cerebral artery NOX4 expression and activity are increased during chronic hypertension.[399]

Elevated oxidative stress is characteristic of normal brain aging, as well as disease states, such as Alzheimer's disease, Parkinson's disease, ischemic injury, and stroke (reviewed in References 400 and 401). For example, microglial ROS production has been proposed to contribute to inflammatory neurodegeneration associated with Alzheimer's disease and Parkinson's disease (reviewed in References 402 and 403). Indeed, activation of microglial ROS has been shown to contribute to death of neighboring neurons,[404] and $ONOO^-$ generated by reaction of microglial ROS with •NO causes neuronal apoptosis.[405] Alzheimer's disease is associated with oxidant stress,[406] and NADPH oxidase-derived ROS contribute to the protein and lipid oxidation associated with this disease.[407] In patients with Parkinson's disease, ROS production is enhanced,[408] and NOX2-dependent ROS production contributes to inflammation-induced toxicity to dopaminergic neurons.[409]

ROS have been shown to play an important role in brain injury associated with ischemia/reperfusion and stroke. For example, phagocyte NOX2-dependent ROS have been shown to contribute to inflammatory and ischemic central nervous system injury, and inhibition of leukocyte recruitment to the central nervous system reduced experimental cerebral injury.[410] Likewise, depletion of neutrophils was found to be neuroprotective in hypoxic-ischemic brain injury.[411] Indeed, antioxidants have been shown to prevent ROS-induced lipid peroxidation during ischemic injury,[412] and brain-penetrating antioxidants are effective neuroprotective agents in central nervous system ischemic injury.[413] In addition to NOX2, NOX1 has also been shown to contribute to ischemic injury in experimental stroke, and genetic deletion of NOX1 resulted in decreased lesion size and improved neurological outcome.[414] However, ROS-mediated direct cellular injury did not appear to be involved in the protective effect achieved by NOX1 deficiency, suggesting involvement of other NOX homologs.[414] One likely candidate is NOX4, and Kleinschnitz et al.[415] recently reported that cerebral vascular NOX4 expression was induced during ischemic stroke and caused oxidative stress and death of nerve cells after a stroke. They also demonstrated that brain damage was reduced and neurological function was improved in NOX4-deficient animals or by treatment with a NOX inhibitor after stroke injury.[415] Thus, their work suggests that NOX4 could be a novel therapeutic target for reducing brain injury due to stroke.

Phagocyte ROS also play an important role in inflammatory and ischemic central nervous system injury. For example, inhibition of leukocyte recruitment to the central nervous system has been shown to reduce experimental cerebral injury.[410] Likewise, depletion of neutrophils was found to be neuroprotective in hypoxic-ischemic brain injury.[411] Antioxidants have been shown to prevent ROS-induced lipid peroxidation during ischemic injury,[412] and brain-penetrating antioxidants are effective neuroprotective agents in central nervous system ischemic injury.[413]

7.7.5 Gastrointestinal System

Resident and recruited phagocytes are essential for gastrointestinal defense against pathogens and represent a source of ROS in the gut.[416] However, it is now clear that additional NOX-based enzymes contribute to healthy gut immunity (reviewed in Reference 417) (Fig. 7.6). Expression of NOX1 is abundant in colon and gastric pit epithelial cells,[228,229,418] and NOX1 is induced during epithelial cell differentiation, with the highest levels of NOX1 expressed on crypt cells.[233] NOXO1 and NOXA1 are also expressed in colon epithelial cells, where they are required as cofactors for regulating NOX1 activity.[234–236] In humans, NOX1 and cofactors are expressed in the colon, but not in gastric tissues.[417] Thus, there appears to be species-specific difference in localization of this NADPH oxidase in the gastrointestinal system.

NOX1 appears to be important in normal colon function, and ROS generated by a NOX1-dependent oxidase in colon mucosa has been reported to promote serotonin biosynthesis, which is important in regulating secretion and motility.[419] In addition, NOX1-based NADPH oxidase activity has been shown recently to modulate WNT and NOTCH1 signaling to control the fate of proliferative progenitor cells in the colon.[420] The NOX1-based NADPH oxidase also has been shown to contribute to gut immunity in various models. For example, treatment of guinea pig gastric mucosal cells with *Helicobacter pylori* lipopolysaccharide independently activated Rac1, as well as transcription of NOX1 and NOXO1.[421] Upregulation of the NOX1-based NADPH oxidase in intestinal epithelial cells appears to be mediated by TLR5 signaling.[422] Indeed, bacterial flagellin has been shown to activate TLR5 and induce epithelial proinflammatory gene expression.[423] Although TLR4 has been suggested to play a role in regulation of inflammation caused by *H. pylori* infection,[424] TLR4-mediated activation of NOX1 in colon epithelial cells was not observed.[422] On the other hand, stimulation of TLR4 by lipopolysaccharide induced a time- and dose-dependent contractile dysfunction in colon smooth muscle cells, which was associated with oxidative imbalance due to increased ROS production.[425] NOX1 is induced by a number of inflammatory cytokines, which is consistent with a role in mucosal host defense. For example, interferon-γ activated NOX1 transcription and upregulated ROS production by colon epithelial cells through a signal transducers and activators of transcription 1 (STAT1)-dependent pathway.[426] In addition, IL-1β, IL-18, and TNF-α were also found to induce NOX1 in T84 colon cancer cells.[417]

In addition to NOX1, recent studies suggest DUOX2 contributes to gastrointestinal tract host defense. In *Drosophila*, *duox2* is expressed throughout the digestive tract,[427] and mortality due to infection is greatly increased by silencing Duox2 in adult flies.[428] In humans, DUOX2 is expressed in the distal gastrointestinal tract, especially in the cecum, sigmoidal colon, and rectal glands.[347,427] In these tissues, DUOX2 expression has been localized to enterocytes within the apical membrane of the brush border[427] and to the lower half of the rectal glands.[347] In the gastrointestinal system, DUOX2 has been proposed to serve as a source of H_2O_2 that is utilized by lactoperoxidase to generate microbicidal thiocyanite for mucosal defense (reviewed in References

429 and 430). In support of this idea, Flores et al.[431] recently showed that zebrafish DUOX is highly expressed in intestinal epithelial cells and that knockdown of DUOX impaired larval capacity to control enteric *Salmonella* infection.

Oxidant stress has been proposed to play a role in inflammatory tissue damage associated with gastrointestinal diseases, such as Crohn's disease and ulcerative colitis (reviewed in References 417, 432, and 433). Abnormalities in vascular structure and function, in addition to inflammatory lesions in the gut, may be initiated and exacerbated by ROS production mediated by gastrointestinal NADPH oxidases, including those utilizing NOX1, NOX2, and possibly NOX5. For example, injury of gut mucosal tissues by nonsteroidal anti-inflammatory drugs appears to be mediated, in part, by phagocyte-generated ROS.[434] In addition, NOX2 is upregulated in mucosal macrophages from intestinal tissue of patients with Crohn's disease.[435] Conversely, significantly increased gastritis was observed in NOX2-deficient mice treated with dextran-sodium-sulfate (DSS) to induce colitis,[436] and gut inflammation due to DSS treatment was found to be similar in $p47^{phox}$-deficient and wild-type mice.[437] In addition, phagocytes from patients with Crohn's disease generate lower levels of ROS and exhibit impaired inflammatory responses compared to phagocytes from healthy individuals.[438,439] These diminished responses have been attributed to decreased phagocyte NOX2 expression.[440] Thus, it is clear that nonphagocyte NADPH oxidases must be involved and may represent important sources of ROS in inflammatory bowel diseases. Messages for NOX2 and NOX5 are expressed in human stomach biopsies, although only NOX2 message is upregulated as a function of age and inflammation.[441] Likewise, messages for NOX1 and NOX5 are present in colonic lymphocytes from healthy individuals, while lymphocytes in lesions from patients with Crohn's disease or ulcerative colitis express only NOX1.[441] Thus, there is not much evidence for involvement of NOX5 in these tissues, and NOX5 protein may not be expressed.

7.7.6 Hepatic System

ROS are generated by a number of cell types in the liver (reviewed in Reference 442). Hepatocytes generate ROS in response to various stimuli, and NADPH oxidase components are expressed in hepatocytes.[19] NOX2 seems to be the main NOX protein expressed in hepatocytes,[443,444] and ROS generation seems to be involved in hepatocyte apoptosis.[445,446] For example, hydrophobic bile salts were found to activate the hepatocyte NADPH oxidase, leading to apoptosis in these cells.[446] Likewise, NADPH oxidase-dependent generation of ROS is involved in apoptotic cell death of HepG2 human hepatoma cells exposed to capsaicin.[445] TGF-β induces oxidative stress that mediates fetal hepatocyte apoptosis, and it has been proposed that mitochondrial and extramitochondrial NADPH oxidase-dependent ROS contribute to this process.[447] On the other hand, it appears that certain regenerating adult hepatocytes are resistant to ROS-mediated apoptosis.[448] Thus, the impact of ROS on hepatocytes also appears to be dependent on hepatocyte phenotype.

The liver plays an important role in innate immunity, and specialized tissue macrophages called Kupffer cells play an important role in this process (reviewed in Reference 449). Kupffer cells are resident macrophage of the liver and, like all phagocytes, utilize a NOX2-based NADPH oxidase to generate ROS in defense against pathogens.[450] Kupffer cells generate ROS in response to a variety of inflammatory stimuli, including lipopolysaccharide, carcinogens, and zymosan.[451,452] ROS production is absent in Kupffer cells from $p47^{phox}$-deficient mice, supporting a role for the NOX2-dependent oxidase system.[452] Kupffer cell ROS production has also been reported to contribute to defense against pathogens during liver infection.[453]

Hepatic stellate cells are normally quiescent cells located in the perisinusoidal space of the liver but become activated in response to chronic liver injury and develop a myofibroblast-like phenotype associated with increased proliferation and collagen synthesis.[454] Activated hepatic stellate cells generate ROS in response to various stimuli, including angiotensin II,[455] platelet-derived growth factor,[456] and leptin.[457] NOX2 is expressed in hepatic stellate cells,[455,456] and it has been reported that NOX2-dependent ROS production participates in angiotensin II signaling during liver fibrosis.[455] Indeed, hepatic fibrosis was inhibited in $p47^{phox}$-deficient mice, further supporting a role for ROS in hepatic stellate cell proliferation and fibrosis during liver injury.[455,456]

Increased oxidant stress is a characteristic of chronic liver diseases, and this is due to enhanced NADPH oxidase activity and decreased activity of antioxidant defenses (reviewed in References 458 and 459). Indeed, oxidant stress is associated with alcohol-induced liver disease, hepatitis C virus infection, and fibrosis associated with chronic liver injury.[458] In addition, excessive ROS generation by Kupffer cells has been reported to be a factor in hepatocarcinogenesis.[460] Thus, it is evident that hepatic NADPH oxidase-generated ROS can be involved in both physiological and pathological events in the liver and that disruption of the redox balance toward oxidative stress can lead to inflammatory events and tissue injury. Indeed, antioxidants have been the focus of anticancer therapies in the liver. For example, treatment with vitamin E was shown to inhibit NADPH oxidase-dependent transformation in a murine liver cancer model,[461] and SOD mimetics have been considered for treatment of colon and liver malignancies.[462]

7.7.7 Thyroid Gland

Thyrocytes generate H_2O_2, which is essential for thyroid hormone formation, and early studies suggested that an NADPH oxidase was responsible for H_2O_2 production in these cells.[463] However, features of the thyroid NADPH oxidase were distinct from the phagocyte NADPH oxidase system, and the ensuing search for this enzyme resulted in the identification of two thyroid oxidase proteins, which are now known as DUOX1 and DUOX2[262,263] (Fig. 7.6). Thyroid peroxidase utilizes H_2O_2 to oxidize iodide, which is then incorporated into thyroglobulin. H_2O_2 is also required in the subsequent coupling reaction leading to formation of thyroxine (T_4) and triiodothyronine (T_3) (reviewed in Reference 464). In addition, recent studies suggest thyroperoxidase exhibits a catalase-like effect and protects DUOX from inhibition by H_2O_2.[465] Genetic mutations in DUOX2 have been shown to cause congenital hypothyroidism, demonstrating the importance of DUOX in thyroid function.[466–468] In addition, mutations in DUOXA2 have also been associated with congenital hypothyroidism.[469] Interestingly, DUOX1/DUOXA1 does not appear to be able to substitute for defects in DUOX2/DUOXA2, despite their structural similarities,[466] suggesting alternative compensatory mechanisms to explain the transient hypothyroidism observed in patients with defective DUOX2 protein.[470]

Recently, NOX4/p22^{phox} expression was demonstrated in normal thyroid tissue and was found to be positively regulated by thyroid-stimulating hormone.[251] The intracellular NOX4/p22^{phox} localization suggested a role in cytoplasmic redox signaling,[251] but this localization also indicated that NOX4/p22^{phox} could not substitute for DUOX. In addition, increased thyroid NOX4/p22^{phox} activity was found to be associated with papillary thyroid carcinoma, the most common thyroid cancer, suggesting ROS might be related to a higher proliferation rate and tumor progression in this tissue.[251]

SUMMARY

Virtually all aerobic organisms generate ROS through cellular metabolism and utilize these molecules to maintain cellular homeostasis. However, excess ROS favor oxidant stress, leading to pathological processes, and ROS-mediated damage has been linked to the aging process itself.[471] Although ROS are produced by a variety of intracellular mechanisms, one of the predominant sources of extracellular ROS is a family of multisubunit enzymes known as the NADPH oxidases. These enzymes transfer electrons across cell membranes to catalyze the univalent reduction of molecular oxygen (O_2), forming $O_2^{\bullet-}$ or, in some cases, H_2O_2 (reviewed in Reference 3 and 19). The first NADPH oxidase to be characterized was found in phagocytic leukocytes, where it was shown to play an essential role in innate immunity and inflammation (reviewed in Reference 169). This system is composed of membrane and cytosolic proteins that are sequestered from each other in resting phagocytes, and highly regulated events involving phosphorylation, translocation, and multiple conformational changes are required for enzyme assembly and ROS production. Such regulation is essential because of the potential toxicity of ROS to host tissue, and ROS-mediated tissue damage is evident in inflammatory conditions where inappropriate or excessive levels of ROS are produced. On the other hand, regulation of phagocyte ROS production plays an important role in resolution of inflammatory responses, and absence of ROS is also detrimental. Thus, there is clearly an important balance of ROS levels required to support immune defenses, while still facilitating appropriate resolution of inflammatory responses.

While it was thought for some time that the NADPH oxidase was a specific feature of phagocytic leukocytes, subsequent research by a number of researchers demonstrated that analogous NADPH oxidase enzymes were present in nonphagocyte tissues and contributed to a wide variety of different physiological functions (reviewed in References 3, 19, and 224–227). These enzymes are functionally distinct from the phagocyte NADPH oxidase and are assembled from various combinations of classical phagocyte oxidase proteins and unique nonphagocyte oxidase proteins, resulting in an array of tissue-specific NADPH oxidase systems. These nonphagocyte NADPH oxidases play important functions in many organ systems and cell types, where they participate in normal physiological processes and, when in excess or dysregulated, can contribute to oxidative stress and the pathogenesis of various human diseases, such as hypertension, atherosclerosis, diabetes, neurodegeneration, and cancer. Overall, expansion of our understanding of the NADPH oxidase family has contributed to an increased appreciation of their function in physiological and pathophysiological processes.

ACKNOWLEDGMENTS

This work was supported in part by National Institutes of Health grants P01 GM103500 and the Montana State University Agricultural Experimental Station.

REFERENCES

1. Babior, B.M., Kipnes, R.S., Curnutte, J.T. Biological defense mechanisms: Production by leukocytes of superoxide, a potential bactericidal agent. *J. Clin. Invest.* **1973**, *52*, 741–744.

2. Brown, D.I., Griendling, K.K. Nox proteins in signal transduction. *Free Radic. Biol. Med.* **2009**, *47*(9), 1239–1253.

3. Leto, T.L., Morand, S., Hurt, D., Ueyama, T. Targeting and regulation of reactive oxygen species generation by Nox family NADPH oxidases. *Antioxid. Redox Signal.* **2009**, *11*(10), 2607–2619.

4. Babior, B.M. The respiratory burst oxidase and the molecular basis of chronic granulomatous disease. *Am. J. Hematol.* **1991**, *37*, 263–266.

5. Lambeth, J.D., Cheng, G., Arnold, R.S., Edens, W.A. Novel homologs of gp91phox. *Trends Biochem. Sci.* **2000**, *25*(10), 459–461.

6. Kawahara, T., Quinn, M.T., Lambeth, J.D. Molecular evolution of the reactive oxygen-generating NADPH oxidase (Nox/Duox) family of enzymes. *BMC Evol. Biol.* **2007**, *7*, 109.

7. Kawahara, T., Lambeth, J.D. Molecular evolution of phox-related regulatory subunits for NADPH oxidase enzymes. *BMC Evol. Biol.* **2007**, *7*, 178.

8. Segal, A.W., Jones, O.T.G. Novel cytochrome b system in phagocytic vacuoles of human granulocytes. *Nature* **1978**, *276*, 515–517.

9. Babior, B.M. The respiratory burst of phagocytes. *J. Clin. Invest.* **1984**, *73*, 599–601.

10. Parkos, C.A., Allen, R.A., Cochrane, C.G., Jesaitis, A.J. Purified cytochrome b from human granulocyte plasma membrane is composed of two polypeptides with relative molecular weights of 91,000 and 22,000. *J. Clin. Invest.* **1987**, *80*, 732–742.

11. Parkos, C.A., Allen, R.A., Cochrane, C.G., Jesaitis, A.J. The quaternary structure of the plasma membrane b-type cytochrome of human granulocytes. *Biochim. Biophys. Acta* **1988**, *932*, 71–83.

12. Huang, J., Hitt, N.D., Kleinberg, M.E. Stoichiometry of p22-phox and gp91-phox in phagocyte cytochrome b_{558}. *Biochemistry* **1995**, *34*(51), 16753–16757.

13. Parkos, C.A., Dinauer, M.C., Walker, L.E., Allen, R.A., Jesaitis, A.J., Orkin, S.H. Primary structure and unique expression of the 22-kilodalton light chain of human neutrophil cytochrome b. *Proc. Natl. Acad. Sci. U.S.A.* **1988**, *85*, 3319–3323.

14. Iizuka, T., Kanegasaki, S., Makino, R., Tanaka, T., Ishimura, Y. Studies on neutrophil b-type cytochrome *in situ* by low temperature absorbance spectroscopy. *J. Biol. Chem.* **1985**, *260*, 12049–12053.

15. Hurst, J.K., Loehr, T.M., Curnutte, J.T., Rosen, H. Resonance raman and electron paramagnetic resonance structural investigations of neutrophil cytochrome b558. *J. Biol. Chem.* **1991**, *266*, 1627–1634.

16. Ueno, I., Fujii, S., Ohya-Nishiguchi, H., Iizuka, T., Kanegasaki, S. Characterization of neutrophil b-type cytochrome *in situ* by electron paramagnetic resonance spectroscopy. *FEBS Lett.* **1991**, *281*, 130–132.

17. Cross, A.R., Rae, J., Curnutte, J.T. Cytochrome b_{245} of the neutrophil superoxide-generating system contains two nonidentical hemes: Potentiometric studies of a mutant form of gp91^{phox}. *J. Biol. Chem.* **1995**, *270*(29), 17075–17077.

18. Royer-Pokora, B., Kunkel, L.M., Monaco, A.P., Goff, S.C., Newburger, P.E., Baehner, R.L., Cole, F.S., Curnutte, J.T., Orkin, S.H. Cloning the gene for an inherited human disorder—chronic granulomatous disease—on the basis of its chromosomal location. *Nature* **1986**, *322*, 32–38.

19. Bedard, K., Krause, K.H. The NOX family of ROS-generating NADPH oxidases: Physiology and pathophysiology. *Physiol. Rev.* **2007**, *87*(1), 245–313.

20. Finegold, A.A., Shatwell, K.P., Segal, A.W., Klausner, R.D., Dancis, A. Intramembrane bis-heme motif for transmembrane electron transport conserved in a yeast iron reductase and the human NADPH oxidase. *J. Biol. Chem.* **1996**, *271*(49), 31021–31024.

21. Heyworth, P.G., Curnutte, J.T., Rae, J., Noack, D., Roos, D., Van Koppen, E., Cross, A.R. Hematologically important mutations: X-linked chronic granulomatous disease (second update). *Blood Cells Mol. Dis.* **2001**, *27*(1), 16–26.

22. Biberstine-Kinkade, K.J., Yu, L., Stull, N., LeRoy, B., Cross, A.R., Dinauer, M.C. Mutagenesis of p22^{phox} histidine 94: A histidine in this position is not required for flavocytochrome b_{558} function. *J. Biol. Chem.* **2002**, *277*(33), 30368–30374.

23. Vignais, P.V. The superoxide-generating NADPH oxidase: Structural aspects and activation mechanism. *Cell. Mol. Life Sci.* **2002**, *59*(9), 1428–1459.

24. Rotrosen, D., Yeung, C.L., Leto, T.L., Malech, H.L., Kwong, C.H. Cytochrome b_{558}: The flavin-binding component of the phagocyte NADPH oxidase. *Science* **1992**, *256*, 1459–1462.

25. Sumimoto, H., Sakamoto, N., Nozaki, M., Sakaki, Y., Takeshige, K., Minakami, S. Cytochrome b_{558}, a component of the phagocyte NADPH oxidase, is a flavoprotein. *Biochem. Biophys. Res. Commun.* **1992**, *186*, 1368–1375.

26. Segal, A.W., West, I., Wientjes, F., Nugent, J.H.A., Chavan, A.J., Haley, B., Garcia, R.C., Rosen, H., Scarce, G. Cytochrome b_{245} is a flavocytochrome containing FAD and the NADPH-binding site of the microbicidal oxidase of phagocytes. *Biochem. J.* **1992**, *284*, 781–788.

27. Koshkin, V., Pick, E. Generation of superoxide by purified and relipidated cytochrome b_{559} in the absence of cytosolic activators. *FEBS Lett.* **1993**, *327*, 57–62.

28. Holland, S.M. Chronic granulomatous disease. *Clin. Rev. Allergy Immunol.* **2010**, *38*(1), 3–10.

29. Segal, A.W., Heyworth, P.G., Cockcroft, S., Barrowman, M.M. Stimulated neutrophils from patients with autosomal recessive chronic granulomatous disease fail to phos-

phorylate a Mr-44,000 protein. *Nature* **1985**, *316*, 547–549.
30. Volpp, B.D., Nauseef, W.M., Clark, R.A. Two cytosolic neutrophil oxidase components absent in autosomal chronic granulomatous disease. *Science* **1988**, *242*, 1295–1297.
31. Nunoi, H., Rotrosen, D., Gallin, J.I., Malech, H.L. Two forms of autosomal chronic granulomatous disease lack distinct neutrophil cytosolic factors. *Science* **1988**, *242*, 1298–1301.
32. Volpp, B.D., Nauseef, W.M., Donelson, J.E., Moser, D.R., Clark, R.A. Cloning of the cDNA and functional expression of the 47-kilodalton cytosolic component of human neutrophil respiratory burst oxidase. *Proc. Natl. Acad. Sci. U.S.A.* **1989**, *86*, 7195–7199.
33. Lomax, K.J., Leto, T.L., Nunoi, H., Gallin, J.I., Malech, H.L. Recombinant 47-kilodalton cytosolic factor restores NADPH oxidase in chronic granulomatous disease. *Science* **1989**, *245*, 409–412.
34. Hiroaki, H., Ago, T., Ito, T., Sumimoto, H., Kohda, D. Solution structure of the PX domain, a target of the SH3 domain. *Nat. Struct. Biol.* **2001**, *8*(6), 526–530.
35. Park, J.W., Ma, M.C., Ruedi, J.M., Smith, R.M., Babior, B.M. The cytosolic components of the respiratory burst oxidase exist as a M(r) similar to 240,000 complex that acquires a membrane-binding site during activation of the oxidase in a cell-free system. *J. Biol. Chem.* **1992**, *267*, 17327–17332.
36. Iyer, S.S., Pearson, D.W., Nauseef, W.M., Clark, R.A. Evidence for a readily dissociable complex of p47phox and p67phox in cytosol of unstimulated human neutrophils. *J. Biol. Chem.* **1994**, *269*, 22405–22411.
37. Lapouge, K., Smith, S.J.M., Groemping, Y., Rittinger, K. Architecture of the p40-p47-p67^{phox} complex in the resting state of the NADPH oxidase—A central role for p67^{phox}. *J. Biol. Chem.* **2002**, *277*(12), 10121–10128.
38. El Benna, J., Ruedi, J.M., Babior, B.M. Cytosolic guanine nucleotide-binding protein Rac2 operates *in vivo* as a component of the neutrophil respiratory burst oxidase. Transfer of Rac2 and the cytosolic oxidase components p47^{phox} and p67^{phox} to the submembranous actin cytoskeleton during oxidase activation. *J. Biol. Chem.* **1994**, *269*, 6729–6734.
39. Heyworth, P.G., Curnutte, J.T., Nauseef, W.M., Volpp, B.D., Pearson, D.W., Rosen, H., Clark, R.A. Neutrophil nicotinamide adenine dinucleotide phosphate oxidase assembly. Translocation of p47-phox and p67-phox requires interaction between p47-phox and cytochrome b$_{558}$. *J. Clin. Invest.* **1991**, *87*, 352–356.
40. Kleinberg, M.E., Malech, H.L., Rotrosen, D. The phagocyte 47-kilodalton cytosolic oxidase protein is an early reactant in activation of the respiratory burst oxidase. *J. Biol. Chem.* **1990**, *265*, 15577–15583.
41. Uhlinger, D.J., Taylor, K.L., Lambeth, J.D. p67-*phox* enhances the binding of p47-*phox* to the human neutrophil respiratory burst oxidase complex. *J. Biol. Chem.* **1994**, *269*, 22095–22098.
42. El-Benna, J., Dang, P.M., Gougerot-Pocidalo, M.A., Marie, J.C., Braut-Boucher, F. p47^{phox}, the phagocyte NADPH oxidase/NOX2 organizer: Structure, phosphorylation and implication in diseases. *Exp. Mol. Med.* **2009**, *41*(4), 217–225.
43. Cross, A.R., Curnutte, J.T. The cytosolic activating factors p47^{phox} and p67^{phox} have distinct roles in the regulation of electron flow in NADPH oxidase. *J. Biol. Chem.* **1995**, *270*, 6543–6548.
44. de Mendez, I., Homayounpour, N., Leto, T.L. Specificity of p47^{phox} SH3 domain interactions in NADPH oxidase assembly and activation. *Mol. Cell. Biol.* **1997**, *17*(4), 2177–2185.
45. Groemping, Y., Rittinger, K. Activation and assembly of the NADPH oxidase: A structural perspective. *Biochem. J.* **2005**, *386*, 401–416.
46. Yuzawa, S., Suzuki, N.N., Fujioka, Y., Ogura, K., Sumimoto, H., Inagaki, F. Crystallization and preliminary crystallographic analysis of the autoinhibited form of the tandem SH3 domain of p47^{phox}. *Acta Crystallogr. D Biol. Crystallogr.* **2003**, *59*(Pt 8), 1479–1480.
47. Ago, T., Nunoi, H., Ito, T., Sumimoto, H. Mechanism for phosphorylation-induced activation of the phagocyte NADPH oxidase protein p47^{phox}—Triple replacement of serines 303, 304, and 328 with aspartates disrupts the SH3 domain-mediated intramolecular interaction in p47^{phox}, thereby activating the oxidase. *J. Biol. Chem.* **1999**, *274*(47), 33644–33653.
48. Ogura, K., Nobuhisa, I., Yuzawa, S., Takeya, R., Torikai, S., Saikawa, K., Sumimoto, H., Inagaki, F. NMR solution structure of the tandem Src homology 3 domains of p47^{phox} complexed with a p22^{phox}-derived proline-rich peptide. *J. Biol. Chem.* **2006**, *281*(6), 3660–3668.
49. Kanai, F., Liu, H., Field, S.J., Akbary, H., Matsuo, T., Brown, G.E., Cantley, L.C., Yaffe, M.B. The PX domains of p47phox and p40phox bind to lipid products of PI(3) K. *Nat. Cell Biol.* **2001**, *3*(7), 675–678.
50. Ago, T., Takeya, R., Hiroaki, H., Kuribayashi, F., Ito, T., Kohda, D., Sumimoto, H. The PX domain as a novel phosphoinositide-binding module. *Biochem. Biophys. Res. Commun.* **2001**, *287*(3), 733–738.
51. Karathanassis, D., Stahelin, R.V., Bravo, J., Perisic, O., Pacold, C.M., Cho, W.W., Williams, R.L. Binding of the PX domain of p47^{phox} to phosphatidylinositol 3,4-bisphosphate and phosphatidic acid is masked by an intramolecular interaction. *EMBO J.* **2002**, *21*(19), 5057–5068.
52. Zhan, Y., Virbasius, J.V., Song, X., Pomerleau, D.P., Zhou, G.W. The p40^{phox} and p47^{phox} PX domains of NADPH oxidase target cell membranes via direct and indirect recruitment by phosphoinositides. *J. Biol. Chem.* **2002**, *277*, 4512–4518.
53. Stahelin, R.V., Burian, A., Bruzik, K.S., Murray, D., Cho, W.W. Membrane binding mechanisms of the PX domains of NADPH oxidase p40^{phox} and p47^{phox}. *J. Biol. Chem.* **2003**, *278*(16), 14469–14479.

54. Li, X.J., Marchal, C.C., Stull, N.D., Stahelin, R.V., Dinauer, M.C. p47^{phox} Phox domain regulates plasma membrane but not phagosome neutrophil NADPH oxidase activation. *J. Biol. Chem.* **2010**, *285*(45), 35169–35179.

55. Wolfson, M., McPhail, L.C., Nasrallah, V.N., Snyderman, R. Phorbol myristate acetate mediates redistribution of protein kinase C in human neutrophils: Potential role in the activation of the respiratory burst enzyme. *J. Immunol.* **1985**, *135*, 2057–2062.

56. El Benna, J., Faust, L.P., Babior, B.M. The phosphorylation of the respiratory burst oxidase component p47^{phox} during neutrophil activation: Phosphorylation of sites recognized by protein kinase C and by proline-directed kinases. *J. Biol. Chem.* **1994**, *269*, 23431–23436.

57. El Benna, J., Faust, L.P., Johnson, J.L., Babior, B.M. Phosphorylation of the respiratory burst oxidase subunit p47^{phox} as determined by two-dimensional phosphopeptide mapping—Phosphorylation by protein kinase C, protein kinase A, and a mitogen-activated protein kinase. *J. Biol. Chem.* **1996**, *271*(11), 6374–6378.

58. Dewas, C., Fay, M., Gougerot-Pocidalo, M.A., El Benna, J. The mitogen-activated protein kinase extracellular signal-regulated kinase 1/2 pathway is involved in formyl-methionyl-leucyl-phenylalanine-induced p47^{phox} phosphorylation in human neutrophils. *J. Immunol.* **2000**, *165*(9), 5238–5244.

59. Knaus, U.G., Morris, S., Dong, H.-J., Chernoff, J., Bokoch, G.M. Regulation of human leukocyte p21-activated kinases through G protein-coupled receptors. *Science* **1995**, *269*(5221), 221–223.

60. Didichenko, S.A., Tilton, B., Hemmings, B.A., Ballmer-Hofer, K., Thelen, M. Constitutive activation of protein kinase B and phosphorylation of p47^{phox} by a membrane-targeted phosphoinositide 3-kinase. *Curr. Biol.* **1996**, *6*(10), 1271–1278.

61. Pacquelet, S., Johnson, J.L., Ellis, B.A., Brzezinska, A.A., Lane, W.S., Munafo, D.B., Catz, S.D. Cross-talk between IRAK-4 and the NADPH oxidase. *Biochem. J.* **2007**, *403*, 451–461.

62. Park, H.S., Lee, S.M., Lee, J.H., Kim, Y.S., Bae, Y.S., Park, J.W. Phosphorylation of the leucocyte NADPH oxidase subunit p47^{phox} by casein kinase 2: Conformation-dependent phosphorylation and modulation of oxidase activity. *Biochem. J.* **2001**, *358*(3), 783–790.

63. Chowdhury, A.K., Watkins, T., Parinandi, N.L., Saatian, B., Kleinberg, M.E., Usatyuk, P.V., Natarajan, V. Src-mediated tyrosine phosphorylation of p47(phox) in hyperoxia-induced activation of NADPH oxidase and generation of reactive oxygen species in lung endothelial cells. *J. Biol. Chem.* **2005**, *280*(21), 20700–20711.

64. Waite, K.A., Wallin, R., Qualliotine-Mann, D., McPhail, L.C. Phosphatidic acid-mediated phosphorylation of the NADPH oxidase component p47-*phox*: Evidence that phosphatidic acid may activate a novel protein kinase. *J. Biol. Chem.* **1997**, *272*(24), 15569–15578.

65. Reeves, E.P., Dekker, L.V., Forbes, L.V., Wientjes, F.B., Grogan, A., Pappin, D.J.C., Segal, A.W. Direct interaction between p47^{phox} and protein kinase C: Evidence for targeting of protein kinase C by p47^{phox} in neutrophils. *Biochem. J.* **1999**, *344*(3), 859–866.

66. Dang, P.M.C., Fontayne, A., Hakim, J., El Benna, J., Périanin, A. Protein kinase C ζ phosphorylates a subset of selective sites of the NADPH oxidase component p47^{phox} and participates in formyl peptide-mediated neutrophil respiratory burst. *J. Immunol.* **2001**, *166*(2), 1206–1213.

67. Fontayne, A., Dang, P.M.C., Gougerot-Pocidalo, M.A., El Benna, J. Phosphorylation of p47^{phox} sites by PKC α, βII, δ, and ξ: Effect on binding to p22^{phox} and on NADPH oxidase activation. *Biochemistry* **2002**, *41*(24), 7743–7750.

68. Leto, T.L., Lomax, K.J., Volpp, B.D., Nunoi, H., Sechler, J.M.G., Nauseef, W.M., Clark, R.A., Gallin, J.I., Malech, H.L. Cloning of a 67-kD neutrophil oxidase factor with similarity to a noncatalytic region of p60-src. *Science* **1990**, *248*, 727–730.

69. Nakamura, R., Sumimoto, H., Mizuki, K., Hata, K., Ago, T., Kitajima, S., Takeshige, K., Sakaki, Y., Ito, T. The PC motif: A novel and evolutionarily conserved sequence involved in interaction between p40^{phox} and p67^{phox}, SH3 domain-containing cytosolic factors of the phagocyte NADPH oxidase. *Eur. J. Biochem.* **1998**, *251*(3), 583–589.

70. Koga, H., Terasawa, H., Nunoi, H., Takeshige, K., Inagaki, F., Sumimoto, H. Tetratricopeptide repeat (TPR) motifs of p67^{phox} participate in interaction with the small GTPase Rac and activation of the phagocyte NADPH oxidase. *J. Biol. Chem.* **1999**, *274*(35), 25051–25060.

71. Vergnaud, S., Paclet, M.H., El Benna, J., Pocidalo, M.A., Morel, F. Complementation of NADPH oxidase in p67-phox-deficient CGD patients—p67-phox/p40-phox interaction. *Eur. J. Biochem.* **2000**, *267*(4), 1059–1067.

72. Paclet, M.H., Coleman, A.W., Vergnaud, S., Morel, F. p67-phox-mediated NADPH oxidase assembly: Imaging of cytochrome b_{558} liposomes by atomic force microscopy. *Biochemistry* **2000**, *39*(31), 9302–9310.

73. Dang, P.M.C., Cross, A.R., Babior, B.M. Assembly of the neutrophil respiratory burst oxidase: A direct interaction between p67^{phox} and cytochrome b_{558}. *Proc. Natl. Acad. Sci. U.S.A.* **2001**, *98*(6), 3001–3005.

74. Freeman, J.L., Lambeth, J.D. NADPH oxidase activity is independent of p47^{phox} *in vitro*. *J. Biol. Chem.* **1996**, *271*(37), 22578–22582.

75. Koshkin, V., Lotan, O., Pick, E. The cytosolic component p47^{phox} is not a *sine qua non* participant in the activation of NADPH oxidase but is required for optimal superoxide production. *J. Biol. Chem.* **1996**, *271*(48), 30326–30329.

76. Alloul, N., Gorzalczany, Y., Itan, M., Sigal, N., Pick, E. Activation of the superoxide-generating NADPH oxidase by chimeric proteins consisting of segments of the cytosolic component p67^{phox} and the small GTPase Rac1. *Biochemistry* **2001**, *40*(48), 14557–14566.

77. Miyano, K., Ogasawara, S., Han, C.H., Fukuda, H., Tamura, M. A fusion protein between Rac and p67^{phox}

(1-210) reconstitutes NADPH oxidase with higher activity and stability than the individual components. *Biochemistry* **2001**, *40*(46), 14089–14097.

78. Diekmann, D., Abo, A., Johnston, C., Segal, A.W., Hall, A. Interaction of Rac with p67^{phox} and regulation of phagocytic NADPH oxidase activity. *Science* **1994**, *265*, 531–533.

79. Lapouge, K., Smith, S.J.M., Walker, P.A., Gamblin, S.J., Smerdon, S.J., Rittinger, K. Structure of the TPR domain of p67^{phox} in complex with Rac•GTP. *Mol. Cell* **2000**, *6*, 899–907.

80. Grizot, S., Fieschi, F., Dagher, M.C., Pebay-Peyroula, E. The active N-terminal region of p67^{phox}—Structure at 1.8 Å resolution and biochemical characterizations of the A128V mutant implicated in chronic granulomatous disease. *J. Biol. Chem.* **2001**, *276*(24), 21627–21631.

81. Han, C.-H., Freeman, J.L.R., Lee, T.H., Motalebi, S.A., Lambeth, J.D. Regulation of the neutrophil respiratory burst oxidase—Identification of an activation domain in p67^{phox}. *J. Biol. Chem.* **1998**, *273*(27), 16663–16668.

82. Maehara, Y., Miyano, K., Yuzawa, S., Akimoto, R., Takeya, R., Sumimoto, H. A conserved region between the TPR and activation domains of p67phox participates in activation of the phagocyte NADPH oxidase. *J. Biol. Chem.* **2010**, *285*(41), 31435–31445.

83. Nisimoto, Y., Motalebi, S., Han, C.H., Lambeth, J.D. The p67^{phox} activation domain regulates electron flow from NADPH to flavin in flavocytochrome b_{558}. *J. Biol. Chem.* **1999**, *274*(33), 22999–23005.

84. El Benna, J., Dang, P.M., Gaudry, M., Fay, M., Morel, F., Hakim, J., Gougerot-Pocidalo, M.A. Phosphorylation of the respiratory burst oxidase subunit p67^{phox} during human neutrophil activation—Regulation by protein kinase C-dependent and independent pathways. *J. Biol. Chem.* **1997**, *272*(27), 17204–17208.

85. Forbes, L.V., Truong, O., Wientjes, F.B., Moss, S.J., Segal, A.W. The major phosphorylation site of the NADPH oxidase component p67^{phox} is Thr^{233}. *Biochem. J.* **1999**, *338*(1), 99–105.

86. Dang, P.M.C., Morel, F., Gougerot-Pocidalo, M.A., El Benna, J. Phosphorylation of the NADPH oxidase component p67^{PHOX} by ERK2 and P38MAPK: Selectivity of phosphorylated sites and existence of an intramolecular regulatory domain in the tetratricopeptide-rich region. *Biochemistry* **2003**, *42*(15), 4520–4526.

87. Wientjes, F.B., Hsuan, J.J., Totty, N.F., Segal, A.W. p40^{phox}, a third cytosolic component of the activation complex of the NADPH oxidase to contain *src* homology 3 domains. *Biochem. J.* **1993**, *296*, 557–561.

88. Ponting, C.P. Novel domains in NADPH oxidase subunits, sorting nexins, and PtdIns 3-kinases: Binding partners of SH3 domains? *Protein Sci.* **1996**, *5*(11), 2353–2357.

89. Wientjes, F.B., Panayotou, G., Reeves, E., Segal, A.W. Interactions between cytosolic components of the NADPH oxidase: p40^{phox} interacts with both p67^{phox} and p47^{phox}. *Biochem. J.* **1996**, *317*, 919–924.

90. Bouin, A.P., Grandvaux, N., Vignais, P.V., Fuchs, A. p40^{phox} is phosphorylated on threonine 154 and serine 315 during activation of the phagocyte NADPH oxidase—Implication of a protein kinase C type kinase in the phosphorylation process. *J. Biol. Chem.* **1998**, *273*(46), 30097–30103.

91. Tsunawaki, S., Kagara, S., Yoshikawa, K., Yoshida, L.S., Kuratsuji, T., Namiki, H. Involvement of p40^{phox} in activation of phagocyte NADPH oxidase through association of its carboxyl-terminal, but not its amino-terminal, with p67^{phox}. *J. Exp. Med.* **1996**, *184*(3), 893–902.

92. Sathyamoorthy, M., de Mendez, I., Adams, A.G., Leto, T.L. p40^{phox} down-regulates NADPH oxidase activity through interactions with its SH3 domain. *J. Biol. Chem.* **1997**, *272*(14), 9141–9146.

93. Cross, A.R. p40^{phox} participates in the activation of NADPH oxidase by increasing the affinity of p47^{phox} for flavocytochrome b_{558}. *Biochem. J.* **2000**, *349*(1), 113–117.

94. Kuribayashi, F., Nunoi, H., Wakamatsu, K., Tsunawaki, S., Sato, K., Ito, T., Sumimoto, H. The adaptor protein p40^{phox} as a positive regulator of the superoxide-producing phagocyte oxidase. *EMBO J.* **2002**, *21*(23), 6312–6320.

95. Tamura, M., Shiozaki, L., Ono, S., Miyano, K., Kunihiro, S., Sasaki, T. p40(phox) as an alternative organizer to p47(phox) in Nox2 activation: A new mechanism involving an interaction with p22(phox). *FEBS Lett.* **2007**, *581*(23), 4533–4538.

96. Bravo, J., Karathanassis, D., Pacold, C.M., Pacold, M.E., Ellson, C.D., Anderson, K.E., Butler, P.J., Lavenir, I., Perisic, O., Hawkins, P.T., Stephens, L., Williams, R.L. The crystal structure of the PX domain from p40^{phox} bound to phosphatidylinositol 3-phosphate. *Mol. Cell* **2001**, *8*(4), 829–839.

97. Ellson, C.D., Anderson, K.E., Morgan, G., Chilvers, E.R., Lipp, P., Stephens, L.R., Hawkins, P.T. Phosphatidylinositol 3-phosphate is generated in phagosomal membranes. *Curr. Biol.* **2001**, *11*(20), 1631–1635.

98. Ellson, C.D., Andrews, S., Stephens, L.R., Hawkins, P.T. The PX domain: A new phosphoinositide-binding module. *J. Cell. Sci.* **2002**, *115*, 1099–1105.

99. Ellson, C.D., Gobert-Gosse, S., Anderson, K.E., Davidson, K., Erdjument-Bromage, H., Tempst, P., Thuring, J.W., Cooper, M.A., Lim, Z.Y., Holmes, A.B., Gaffney, P.R., Coadwell, J., Chilvers, E.R., Hawkins, P.T., Stephens, L.R. PtdIns(3)P regulates the neutrophil oxidase complex by binding to the PX domain of p40^{phox}. *Nat. Cell Biol.* **2001**, *3*(7), 679–682.

100. Ellson, C.D., Davidson, K., Ferguson, G.J., O'Connor, R., Stephens, L.R., Hawkins, P.T. Neutrophils from p40$^{phox-/-}$ mice exhibit severe defects in NADPH oxidase regulation and oxidant-dependent bacterial killing. *J. Exp. Med.* **2006**, *203*(8), 1927–1937.

101. Ellson, C., Davidson, K., Anderson, K., Stephens, L.R., Hawkins, P.T. PtdIns3P binding to the PX domain of p40^{phox} is a physiological signal in NADPH oxidase activation. *EMBO J.* **2006**, *25*(19), 4468–4478.

102. Matute, J.D., Arias, A.A., Wright, N.A., Wrobel, I., Waterhouse, C.C., Li, X.J., Marchal, C.C., Stull, N.D., Lewis,

D.B., Steele, M., Kellner, J.D., Yu, W., Meroueh, S.O., Nauseef, W.M., Dinauer, M.C. A new genetic subgroup of chronic granulomatous disease with autosomal recessive mutations in p40^{phox} and selective defects in neutrophil NADPH oxidase activity. *Blood* **2009**, *114*(15), 3309–3315.

103. Chen, J., He, R., Minshall, R.D., Dinauer, M.C., Ye, R.D. Characterization of a mutation in the Phox homology domain of the NADPH oxidase component p40^{phox} identifies a mechanism for negative regulation of superoxide production. *J. Biol. Chem.* **2007**, *282*(41), 30273–30284.

104. Shao, D., Segal, A.W., Dekker, L.V. Subcellular localisation of the p40^{phox} component of NADPH oxidase involves direct interactions between the Phox homology domain and F-actin. *Int. J. Biochem. Cell Biol.* **2010**, *42*(10), 1736–1743.

105. Matute, J.D., Arias, A.A., Dinauer, M.C., Patino, P.J. p40^{phox}: The last NADPH oxidase subunit. *Blood Cells Mol. Dis.* **2005**, *35*(2), 291–302.

106. Bokoch, G.M., Zhao, T.M. Regulation of the phagocyte NADPH oxidase by Rac GTPase. *Antioxid. Redox Signal.* **2006**, *8*(9–10), 1533–1548.

107. Roberts, A.W., Kim, C., Zhen, L., Lowe, J.B., Kapur, R., Petryniak, B., Spaetti, A., Pollock, J.D., Borneo, J.B., Bradford, G.B., Atkinson, S.J., Dinauer, M.C., Williams, D.A. Deficiency of the hematopoietic cell-specific Rho family GTPase Rac2 is characterized by abnormalities in neutrophil function and host defense. *Immunity* **1999**, *10*(2), 183–196.

108. Kim, C., Dinauer, M.C. Rac2 is an essential regulator of neutrophil nicotinamide adenine dinucleotide phosphate oxidase activation in response to specific signaling pathways. *J. Immunol.* **2001**, *166*(2), 1223–1232.

109. Glogauer, M., Marchal, C.C., Zhu, F., Worku, A., Clausen, B.E., Foerster, I., Marks, P., Downey, G.P., Dinauer, M., Kwiatkowski, D.J. Rac1 deletion in mouse neutrophils has selective effects on neutrophil functions. *J. Immunol.* **2003**, *170*(11), 5652–5657.

110. Gu, Y., Filippi, M.D., Cancelas, J.A., Siefring, J.E., Williams, E.P., Jasti, A.C., Harris, C.E., Lee, A.W., Prabhakar, R., Atkinson, S.J., Kwiatkowski, D.J., Williams, D.A. Hematopoietic cell regulation by Rac1 and Rac2 guanosine triphosphatases. *Science* **2003**, *302*(5644), 445–449.

111. Williams, D.A., Tao, W., Yang, F.C., Kim, C., Gu, Y., Mansfield, P., Levine, J.E., Petryniak, B., Derrow, C.W., Harris, C., Jia, B.Q., Zheng, Y., Ambruso, D.R., Lowe, J.B., Atkinson, S.J., Dinauer, M.C., Boxer, L. Dominant negative mutation of the hematopoietic-specific Rho GTPase, Rac2, is associated with a human phagocyte immunodeficiency. *Blood* **2000**, *96*(5), 1646–1654.

112. Ambruso, D.R., Knall, C., Abell, A.N., Panepinto, J., Kurkchubasche, A., Thurman, G., Gonzalez-Aller, C., Hiester, A., deBoer, M., Harbeck, R.J., Oyer, R., Johnson, G.L., Roos, D. Human neutrophil immunodeficiency syndrome is associated with an inhibitory Rac2 mutation. *Proc. Natl. Acad. Sci. U.S.A.* **2000**, *97*(9), 4654–4659.

113. Quinn, M.T., Evans, T., Loetterle, L.R., Jesaitis, A.J., Bokoch, G.M. Translocation of Rac correlates with NADPH oxidase activation: Evidence for equimolar translocation of oxidase components. *J. Biol. Chem.* **1993**, *268*, 20983–20987.

114. Abo, A., Webb, M.R., Grogan, A., Segal, A.W. Activation of NADPH oxidase involves the dissociation of p21^{rac} from its inhibitory GDP/GTP exchange protein (rhoGDI) followed by its translocation to the plasma membrane. *Biochem. J.* **1994**, *298*, 585–591.

115. Heyworth, P.G., Bohl, B.P., Bokoch, G.M., Curnutte, J.T. Rac translocates independently of the neutrophil NADPH oxidase components p47^{phox} and p67^{phox}: Evidence for its interaction with flavocytochrome b_{558}. *J. Biol. Chem.* **1994**, *269*, 30749–30752.

116. Dorseuil, O., Quinn, M.T., Bokoch, G.M. Dissociation of Rac translocation from p47-phox/p67-phox movements in human neutrophils by tyrosine kinase inhibitors. *J. Leukoc. Biol.* **1995**, *58*, 108–113.

117. Kao, Y.Y., Gianni, D., Bohl, B., Taylor, R.M., Bokoch, G.M. Identification of a conserved Rac-binding site on NADPH oxidases supports a direct GTPase regulatory mechanism. *J. Biol. Chem.* **2008**, *283*(19), 12736–12746.

118. Quinn, M.T., Parkos, C.A., Walker, L., Orkin, S.H., Dinauer, M.C., Jesaitis, A.J. Association of a ras-related protein with cytochrome b of human neutrophils. *Nature* **1989**, *342*, 198–200.

119. Bokoch, G.M., Quilliam, L.A., Bohl, B.P., Jesaitis, A.J., Quinn, M.T. Inhibition of Rap1A binding to cytochrome b_{558} of NADPH oxidase by phosphorylation of Rap1A. *Science* **1991**, *254*, 1794–1796.

120. Maly, F.-E., Quilliam, L.A., Dorseuil, O., Der, C.J., Bokoch, G.M. Activated or dominant inhibitory mutants of Rap1A decrease the oxidative burst of Epstein-Barr virus-transformed human B lymphocytes. *J. Biol. Chem.* **1994**, *269*, 18743–18746.

121. Gabig, T.G., Crean, C.D., Mantel, P.L., Rosli, R. Function of wild-type or mutant Rac2 and Rap1a GTPases in differentiated HL60 cell NADPH oxidase activation. *Blood* **1995**, *85*, 804–811.

122. Li, Y., Yan, J.L., De, P., Chang, H.C., Yamauchi, A., Christopherson, K.W., Paranavitana, N.C., Peng, X.D., Kim, C., Munugulavadla, V., Kapur, R., Chen, H.Y., Shou, W.N., Stone, J.C., Kaplan, M.H., Dinauer, M.C., Durden, D.L., Quilliam, L.A. Rap1a null mice have altered myeloid cell functions suggesting distinct roles for the closely related Rap1a and 1b proteins. *J. Immunol.* **2007**, *179*(12), 8322–8331.

123. Sbarra, A.J., Karnovsky, M.L. The biochemical basis of phagocytosis: I. Metabolic changes during the ingestion of particles by polymorphonuclear leukocytes. *J. Biol. Chem.* **1959**, *234*(6), 1355–1362.

124. Rossi, F., Zatti, M. Changes in the metabolic pattern of polymorphonuclear leukocytes during phagocytosis. *Br. J. Exp. Pathol.* **1964**, *45*, 548–559.

125. Rosen, G.M., Pou, S., Ramos, C.L., Cohen, M.S., Britigan, B.E. Free radicals and phagocytic cells. *FASEB J.* **1995**, *9*, 200–209.

126. Fridovich, I. Oxygen toxicity: A radical explanation. *J. Exp. Biol.* **1998**, *201*(8), 1203–1209.
127. Liochev, S.I., Fridovich, I. The role of O_2^- in the production of HO: *In vitro* and *in vivo*. *Free Radic. Biol. Med.* **1994**, *16*(1), 29–33.
128. Halliwell, B., Gutteridge, J.M.C. *Free Radicals in Biology and Medicine*, 2nd ed. Oxford University Press, Inc., New York, 1989.
129. Fridovich, I. Superoxide anion radical (O_2^-), superoxide dismutases, and related matters. *J. Biol. Chem.* **1997**, *272*(30), 18515–18517.
130. McCord, J.M. Superoxide dismutase in aging and disease: An overview. *Meth. Enzymol.* **2002**, *349*, 331–341.
131. Klebanoff, S.J. Myeloperoxidase: Friend and foe. *J. Leukoc. Biol.* **2005**, *77*(5), 598–625.
132. Daugherty, A., Dunn, J.L., Rateri, D.L., Heinecke, J.W. Myeloperoxidase, a catalyst for lipoprotein oxidation, is expressed in human atherosclerotic lesions. *J. Clin. Invest.* **1994**, *94*(1), 437–444.
133. Nagra, R.M., Becher, B., Tourtellotte, W.W., Antel, J.P., Gold, D., Paladino, T., Smith, R.A., Nelson, J.R., Reynolds, W.F. Immunohistochemical and genetic evidence of myeloperoxidase involvement in multiple sclerosis. *J. Neuroimmunol.* **1997**, *78*(1–2), 97–107.
134. Reynolds, W.F., Rhees, J., Maciejewski, D., Paladino, T., Sieburg, H., Maki, R.A., Masliah, E. Myeloperoxidase polymorphism is associated with gender specific risk for Alzheimer's disease. *Exp. Neurol.* **1999**, *155*(1), 31–41.
135. Klebanoff, S.J. Myeloperoxidase. *Proc. Assoc. Am. Physicians* **1999**, *111*, 383–389.
136. Aratani, Y., Kura, F., Watanabe, H., Akagawa, H., Takano, Y., Suzuki, K., Dinauer, M.C., Maeda, N., Koyama, H. Relative contributions of myeloperoxidase and NADPH-oxidase to the early host defense against pulmonary infections with *Candida albicans* and *Aspergillus fumigatus*. *Med. Mycol.* **2002**, *40*(6), 557–563.
137. Winterbourn, C.C. Biological reactivity and biomarkers of the neutrophil oxidant, hypochlorous acid. *Toxicology* **2002**, *181*, 223–227.
138. Halliwell, B. Phagocyte-derived reactive species: Salvation or suicide? *Trends Biochem. Sci.* **2006**, *31*(9), 509–515.
139. Hazen, S.L., Hsu, F.F., D'Avignon, A., Heinecke, J.W. Human neutrophils employ myeloperoxidase to convert α-amino acids to a battery of reactive aldehydes: A pathway for aldehyde generation at sites of inflammation. *Biochemistry* **1998**, *37*(19), 6864–6873.
140. Heinecke, J.W. Tyrosyl radical production by myeloperoxidase: A phagocyte pathway for lipid peroxidation and dityrosine cross-linking of proteins. *Toxicology* **2002**, *177*(1), 11–22.
141. Albrich, J.M., McCarthy, C.A., Hurst, J.K. Biological reactivity of hypochlorous acid: Implications for microbicidal mechanisms of leukocyte myeloperoxidase. *Proc. Natl. Acad. Sci. U.S.A.* **1981**, *78*(1), 210–214.
142. Heinecke, J.W. Mechanisms of oxidative damage by myeloperoxidase in atherosclerosis and other inflammatory disorders. *J. Lab. Clin. Med.* **1999**, *133*(4), 321–325.
143. Marcinkiewicz, J. Neutrophil chloramines: Missing links between innate and acquired immunity. *Immunol. Today* **1997**, *18*(12), 577–580.
144. Winterbourn, C.C., Vissers, M.C., Kettle, A.J. Myeloperoxidase. *Curr. Opin. Hematol.* **2000**, *7*(1), 53–58.
145. Halliwell, B. Superoxide-dependent formation of hydroxyl radicals in the presence of iron chelates: Is it a mechanism for hydroxyl radical production in biochemical systems? *FEBS Lett.* **1978**, *92*(2), 321–326.
146. Ramos, C.L., Pou, S., Britigan, B.E., Cohen, M.S., Rosen, G.M. Spin trapping evidence for myeloperoxidase-dependent hydroxyl radical formation by human neutrophils and monocytes. *J. Biol. Chem.* **1992**, *267*(12), 8307–8312.
147. Winterbourn, C.C. Superoxide as an intracellular radical sink. *Free Radic. Biol. Med.* **1993**, *14*(1), 85–90.
148. Segal, A.W. How neutrophils kill microbes. *Annu. Rev. Immunol.* **2005**, *23*, 197–223.
149. Britigan, B.E., Coffman, T.J., Buettner, G.R. Spin trapping evidence for the lack of significant hydroxyl radical production during the respiration burst of human phagocytes using a spin adduct resistant to superoxide-mediated destruction. *J. Biol. Chem.* **1990**, *265*(5), 2650–2656.
150. Aruoma, O.I., Halliwell, B. Superoxide-dependent and ascorbate-dependent formation of hydroxyl radicals from hydrogen peroxide in the presence of iron. Are lactoferrin and transferrin promoters of hydroxyl-radical generation? *Biochem. J.* **1987**, *241*(1), 273–278.
151. Gannon, D.E., Varani, J., Phan, S.H., Ward, J.H., Kaplan, J., Till, G.O., Simon, R.H., Ryan, U.S., Ward, P.A. Source of iron in neutrophil-mediated killing of endothelial cells. *Lab. Invest.* **1987**, *57*(1), 37–44.
152. Shen, Z., Wu, W., Hazen, S.L. Activated leukocytes oxidatively damage DNA, RNA, and the nucleotide pool through halide-dependent formation of hydroxyl radical. *Biochemistry* **2000**, *39*(18), 5474–5482.
153. Hawkins, C.L., Rees, M.D., Davies, M.J. Superoxide radicals can act synergistically with hypochlorite to induce damage to proteins. *FEBS Lett.* **2002**, *510*(1–2), 41–44.
154. Saran, M., Beck-Speier, I., Fellerhoff, B., Bauer, G. Phagocytic killing of microorganisms by radical processes: Consequences of the reaction of hydroxyl radicals with chloride yielding chlorine atoms. *Free Radic. Biol. Med.* **1999**, *26*(3–4), 482–490.
155. Krinsky, N.I. Singlet excited oxygen as a mediator of the antibacterial action of leukocytes. *Science* **1974**, *186*(4161), 363–365.
156. Steinbeck, M.J., Khan, A.U., Karnovsky, M.J. Intracellular singlet oxygen generation by phagocytosing neutrophils in response to particles coated with a chemical trap. *J. Biol. Chem.* **1992**, *267*(19), 13425–13433.
157. Tatsuzawa, H., Maruyama, T., Hori, K., Sano, Y., Nakano, M. Singlet oxygen ($^1\Delta_g O_2$) as the principal oxidant in

157. myeloperoxidase-mediated bacterial killing in neutrophil phagosome. *Biochem. Biophys. Res. Commun.* **1999**, *262*(3), 647–650.

158. Sies, H., De Groot, H. Role of reactive oxygen species in cell toxicity. *Toxicol. Lett.* **1992**, *64–65*, 547–551.

159. Pacher, P., Beckman, J.S., Liaudet, L. Nitric oxide and peroxynitrite in health and disease. *Physiol. Rev.* **2007**, *87*(1), 315–424.

160. Pryor, W.A., Squadrito, G.L. The chemistry of peroxynitrite: A product from the reaction of nitric oxide with superoxide. *Am. J. Physiol. Lung Cell Mol. Physiol.* **1995**, *12*, L699–L722.

161. Klebanoff, S.J. Reactive nitrogen intermediates and antimicrobial activity: Role of nitrite. *Free Radic. Biol. Med.* **1993**, *14*(4), 351–360.

162. Eiserich, J.P., Hristova, M., Cross, C.E., Jones, A.D., Freeman, B.A., Halliwell, B., Van der Vliet, A. Formation of nitric oxide derived inflammatory oxidants by myeloperoxidase in neutrophils. *Nature* **1998**, *391*(6665), 393–397.

163. Byun, J., Mueller, D.M., Fabjan, J.S., Heinecke, J.W. Nitrogen dioxide radical generated by the myeloperoxidase-hydrogen peroxide-nitrite system promotes lipid peroxidation of low density lipoprotein. *FEBS Lett.* **1999**, *455*(3), 243–246.

164. Kaplan, S.S., Lancaster, J.R., Basford, R.E., Simmons, R.L. Effect of nitric oxide on staphylococcal killing and interactive effect with superoxide. *Infect. Immun.* **1996**, *64*, 69–76.

165. Rubbo, H., Radi, R., Trujillo, M., Telleri, R., Kalyanaraman, B., Barnes, S., Kirk, M., Freeman, B.A. Nitric oxide regulation of superoxide and peroxynitrite-dependent lipid peroxidation. Formation of novel nitrogen-containing oxidized lipid derivatives. *J. Biol. Chem.* **1994**, *269*, 26066–26075.

166. Radi, R., Beckman, J.S., Bush, K.M., Freeman, B.A. Peroxynitrite oxidation of sulfhydryls. The cytotoxic potential of superoxide and nitric oxide. *J. Biol. Chem.* **1991**, *266*(7), 4244–4250.

167. Crow, J.P., Beckman, J.S. Reactions between nitric oxide, superoxide, and peroxynitrite: Footprints of peroxynitrite *in vivo*. *Adv. Pharmacol.* **1995**, *34*, 17–43.

168. Fang, F.C. Antimicrobial reactive oxygen and nitrogen species: Concepts and controversies. *Nat. Rev. Microbiol.* **2004**, *2*(10), 820–832.

169. Nauseef, W.M. Nox enzymes in immune cells. *Semin. Immunopathol.* **2008**, *30*(3), 195–208.

170. Johansson, A., Jesaitis, A.J., Lundqvist, H., Magnusson, K.-E., Sjölin, C., Karlsson, A., Dahlgren, C. Different subcellular localization of cytochrome *b* and the dormant NADPH-oxidase in neutrophils and macrophages: Effect on the production of reactive oxygen species during phagocytosis. *Cell. Immunol.* **1995**, *161*, 61–71.

171. Forman, H.J., Torres, M. Reactive oxygen species and cell signaling: Respiratory burst in macrophage signaling. *Am. J. Respir. Crit. Care Med.* **2002**, *166*(12), S4–S8.

172. Zhao, X., Carnevale, K.A., Cathcart, M.K. Human monocytes use Rac1, not Rac2, in the NADPH oxidase complex. *J. Biol. Chem.* **2003**, *278*(42), 40788–40792.

173. Cathcart, M.K., McNally, A.K., Morel, D.W., Chisolm, G.M., III. Superoxide anion participation in human monocyte-mediated oxidation of low-density lipoprotein and conversion of low-density lipoprotein to a cytotoxin. *J. Immunol.* **1989**, *142*(6), 1963–1969.

174. Dewald, B., Payne, T.G., Baggiolini, M. Activation of NADPH oxidase of human neutrophils: Potentiation of chemotactic peptide by a diacylglycerol. *Biochem. Biophys. Res. Commun.* **1984**, *125*(1), 367–373.

175. Carlyon, J.A., Latif, D.A., Pypaert, M., Lacy, P., Fikrig, E. *Anaplasma phagocytophilum* utilizes multiple host evasion mechanisms to thwart NADPH oxidase-mediated killing during neutrophil infection. *Infect. Immun.* **2004**, *72*(8), 4772–4783.

176. Cathcart, M.K. Regulation of superoxide anion production by NADPH oxidase in monocytes/macrophages—Contributions to atherosclerosis. *Arterioscler. Thromb. Vasc. Biol.* **2004**, *24*(1), 23–28.

177. Doherty, D.E., Downey, G.P., Worthen, G.S., Haslett, C., Henson, P.M. Monocyte retention and migration in pulmonary inflammation. Requirement for neutrophils. *Lab. Invest.* **1988**, *59*(2), 200–213.

178. Gabay, C. Interleukin-6 and chronic inflammation. *Arthritis Res. Ther.* **2006**, *8*(Suppl. 2), S3.

179. Cross, A.R., Segal, A.W. The NADPH oxidase of professional phagocytes—prototype of the NOX electron transport chain systems. *Biochim. Biophys. Acta* **2004**, *1657*(1), 1–22.

180. Akira, S., Takeda, K. Toll-like receptor signalling. *Nat. Rev. Immunol.* **2004**, *4*(7), 499–511.

181. Beutler, B., Hoebe, K., Du, X., Ulevitch, R.J. How we detect microbes and respond to them: The Toll-like receptors and their transducers. *J. Leukoc. Biol.* **2003**, *74*(4), 479–485.

182. Hayashi, F., Means, T.K., Luster, A.D. Toll-like receptors stimulate human neutrophil function. *Blood* **2003**, *102*(7), 2660–2669.

183. El-Benna, J., Dang, P.M., Gougerot-Pocidalo, M.A. Priming of the neutrophil NADPH oxidase activation: Role of p47phox phosphorylation and NOX2 mobilization to the plasma membrane. *Semin. Immunopathol.* **2008**, *30*(3), 279–289.

184. Swain, S.D., Rohn, T.T., Quinn, M.T. Neutrophil priming in host defense: Role of oxidants as priming agents. *Antioxid. Redox Signal.* **2002**, *4*(1), 69–83.

185. Laroux, F.S., Romero, X., Wetzler, L., Engel, P., Terhorst, C. Cutting edge: MyD88 controls phagocyte NADPH oxidase function and killing of gram-negative bacteria. *J. Immunol.* **2005**, *175*(9), 5596–5600.

186. Kobayashi, S.D., Voyich, J.M., Braughton, K.R., Whitney, A.R., Nauseef, W.M., Malech, H.L., DeLeo, F.R. Gene expression profiling provides insight into the pathophysi-

ology of chronic granulomatous disease. *J. Immunol.* **2004**, *172*(1), 636–643.
187. Bylund, J., Goldblatt, D., Speert, D.P. Chronic granulomatous disease: From genetic defect to clinical presentation. *Adv. Exp. Med. Biol.* **2005**, *568*, 67–87.
188. Winkelstein, J.A., Marino, M.C., Johnston, R.B., Jr., Boyle, J., Curnutte, J., Gallin, J.I., Malech, H.L., Holland, S.M., Ochs, H., Quie, P., Buckley, R.H., Foster, C.B., Chanock, S.J., Dickler, H. Chronic granulomatous disease. Report on a national registry of 368 patients. *Medicine (Baltimore)* **2000**, *79*(3), 155–169.
189. Rosenzweig, S.D., Holland, S.M. Phagocyte immunodeficiencies and their infections. *J. Allergy Clin. Immunol.* **2004**, *113*(4), 620–626.
190. Ganz, T. Antimicrobial polypeptides. *J. Leukoc. Biol.* **2004**, *75*, 34–38.
191. Segal, B.H., Leto, T.L., Gallin, J.I., Malech, H.L., Holland, S.M. Genetic, biochemical, and clinical features of chronic granulomatous disease. *Medicine (Baltimore)* **2000**, *79*(3), 170–200.
192. Saito, M., Ohga, S., Endoh, M., Nakayama, H., Mizunoe, Y., Hara, T., Yoshida, S. H_2O_2-nonproducing *Streptococcus pyogenes* strains: Survival in stationary phase and virulence in chronic granulomatous disease. *Microbiology* **2001**, *147*, 2469–2477.
193. Holland, S.M., Gallin, J.I. Evaluation of the patient with recurrent bacterial infections. *Annu. Rev. Med.* **1998**, *49*, 185–199.
194. Messina, C.G.M., Reeves, E.P., Roes, E., Segal, A.W. Catalase negative *Staphylococcus aureus* retain virulence in mouse model of chronic granulomatous disease. *FEBS Lett.* **2002**, *518*(1–3), 107–110.
195. Kottilil, S., Malech, H.L., Gill, V.J., Holland, S.M. Infections with *Haemophilus* species in chronic granulomatous disease: Insights into the interaction of bacterial catalase and H_2O_2 production. *Clin. Immunol.* **2003**, *106*(3), 226–230.
196. Robinson, J., Watson, F., Bucknall, R.C., Edwards, S.W. Activation of neutrophil reactive-oxidant production by synovial fluid from patients with inflammatory joint disease—Soluble and insoluble immunoglobulin aggregates activate different pathways in primed and unprimed cells. *Biochem. J.* **1992**, *286*, 345–351.
197. Edwards, S.W., Hallett, M.B. Seeing the wood for the trees: The forgotten role of neutrophils in rheumatoid arthritis. *Immunol. Today* **1997**, *18*, 320–324.
198. Leirisalo-Repo, M., Lauhio, A., Repo, H. Chemotaxis and chemiluminescence responses of synovial fluid polymorphonuclear leucocytes during acute reactive arthritis. *Ann. Rheum. Dis.* **1990**, *49*, 615–619.
199. Weissmann, G., Serhan, C., Korchak, H.M., Smolen, J.E. Neutrophils: Release of mediators of inflammation with special reference to rheumatoid arthritis. *Ann. N. Y. Acad. Sci.* **1982**, *389*, 11–24.
200. Lunec, J., Halloran, S.P., White, A.G., Dormandy, T.L. Free-radical oxidation (peroxidation) products in serum and synovial fluid in rheumatoid. *J. Rheumatol.* **1981**, *8*, 233–245.
201. Biemond, P., Swaak, A.J., Penders, J.M., Beindorff, C.M., Koster, J.F. Superoxide production by polymorphonuclear leucocytes in rheumatoid arthritis and osteoarthritis: *In vivo* inhibition by the antirheumatic drug piroxicam due to interference with the activation of the NADPH-oxidase. *Ann. Rheum. Dis.* **1986**, *45*(3), 249–255.
202. Miesel, R., Hartung, R., Kroeger, H. Priming of NADPH oxidase by tumor necrosis factor alpha in patients with inflammatory and autoimmune rheumatic diseases. *Inflammation* **1996**, *20*(4), 427–438.
203. 't Hart, B.A., Simons, J.M., Knaan-Shanzer, S., Bakker, N.P.M., LaBadie, R.P. Antiarthritic activity of the newly developed neutrophil oxidative burst antagonist apocynin. *Free Radic. Biol. Med.* **1990**, *9*, 127–131.
204. Hougee, S., Hartog, A., Sanders, A., Graus, Y.M., Hoijer, M.A., Garssen, J., van den Berg, W.B., van Beuningen, H.M., Smit, H.F. Oral administration of the NADPH-oxidase inhibitor apocynin partially restores diminished cartilage proteoglycan synthesis and reduces inflammation in mice. *Eur. J. Pharmacol.* **2006**, *531*(1–3), 264–269.
205. Cronstein, B.N., Eberle, M.A., Gruber, H.E., Levin, R.I. Methotrexate inhibits neutrophil function by stimulating adenosine release from connective tissue cells. *Proc. Natl. Acad. Sci. U.S.A.* **1991**, *88*, 2441–2445.
206. Miesel, R., Kurpisz, M., Kröger, H. Suppression of inflammatory arthritis by simultaneous inhibition of nitric oxide synthase and NADPH oxidase. *Free Radic. Biol. Med.* **1996**, *20*(1), 75–81.
207. Quinn, M.T., Ammons, M.C., DeLeo, F.R. The expanding role of NADPH oxidases in health and disease: No longer just agents of death and destruction. *Clin. Sci.* **2006**, *111*(1), 1–20.
208. Savill, J. Apoptosis in resolution of inflammation. *J. Leukoc. Biol.* **1997**, *61*(4), 375–380.
209. Kennedy, A.D., DeLeo, F.R. Neutrophil apoptosis and the resolution of infection. *Immunol. Res.* **2009**, *43*(1–3), 25–61.
210. Kasahara, Y., Iwai, K., Yachie, A., Ohta, K., Konno, A., Seki, H., Miyawaki, T., Taniguchi, N. Involvement of reactive oxygen intermediates in spontaneous and CD95(Fas/APO-1)-mediated apoptosis of neutrophils. *Blood* **1997**, *89*(5), 1748–1753.
211. Gardai, S., Whitlock, B.B., Helgason, C., Ambruso, D., Fadok, V., Bratton, D., Henson, P.M. Activation of SHIP by NADPH oxidase-stimulated Lyn leads to enhanced apoptosis in neutrophils. *J. Biol. Chem.* **2002**, *277*(7), 5236–5246.
212. Watson, R.W., Redmond, H.P., Wang, J.H., Condron, C., Bouchier-Hayes, D. Neutrophils undergo apoptosis following ingestion of *Escherichia coli*. *J. Immunol.* **1996**, *156*(10), 3986–3992.
213. Zhang, B., Hirahashi, J., Cullere, X., Mayadas, T.N. Elucidation of molecular events leading to neutrophil apoptosis following phagocytosis: Cross-talk between caspase 8,

213. reactive oxygen species, and MAPK/ERK activation. *J. Biol. Chem.* **2003**, *278*(31), 28443–28454.

214. Coxon, A., Rieu, P., Barkalow, F.J., Askari, S., Sharpe, A.H., von Adrian, U.H., Arnaout, M.A., Mayadas, T.N. A novel role for the β2 integrin CD11b/CD18 in neutrophil apoptosis: A homeostatic mechanism in inflammation. *Immunity* **1996**, *5*(6), 653–666.

215. Hampton, M.B., Vissers, M.C.M., Keenan, J.I., Winterbourn, C.C. Oxidant-mediated phosphatidylserine exposure and macrophage uptake of activated neutrophils: Possible impairment in chronic granulomatous disease. *J. Leukoc. Biol.* **2002**, *71*(5), 775–781.

216. Sareila, O., Kelkka, T., Pizzolla, A., Hultqvist, M., Holmdahl, R. NOX2 complex derived ROS as immune regulators. *Antioxid. Redox Signal.* **2010**, *15*(8), 2197–2208.

217. Van de Loo, F.A.J., Bennink, M.B., Arntz, O.J., Smeets, R.L., Lubberts, E., Joosten, L.A.B., van Lent, P.L.E.M., Coenen-de Roo, C.J.J., Cuzzocrea, S., Segal, B.H., Holland, S.M., van den Berg, W.B. Deficiency of NADPH oxidase components p47^{phox} and gp91^{phox} caused granulomatous synovitis and increased connective tissue destruction in experimental arthritis models. *Am. J. Pathol.* **2003**, *163*(4), 1525–1537.

218. Batsalova, T., Dzhambazov, B., Klaczkowska, D., Holmdahl, R. Mice producing less reactive oxygen species are relatively resistant to collagen glycopeptide vaccination against arthritis. *J. Immunol.* **2010**, *185*(5), 2701–2709.

219. Schappi, M.G., Jaquet, V., Belli, D.C., Krause, K.H. Hyperinflammation in chronic granulomatous disease and anti-inflammatory role of the phagocyte NADPH oxidase. *Semin. Immunopathol.* **2008**, *30*(3), 255–271.

220. Serhan, C.N., Savill, J. Resolution of inflammation: The beginning programs the end. *Nat. Immunol.* **2005**, *6*(12), 1191–1197.

221. Hultqvist, M., Olofsson, P., Holmberg, J., Backstrom, B.T., Tordsson, J., Holmdahl, R. Enhanced autoimmunity, arthritis, and encephalomyelitis in mice with a reduced oxidative burst due to a mutation in the *Ncf1* gene. *Proc. Natl. Acad. Sci. U.S.A.* **2004**, *101*, 12646–12651.

222. Olofsson, P., Holmberg, J., Tordsson, J., Lu, S., Akerstrom, B., Holmdahl, R. Positional identification of Ncf1 as a gene that regulates arthritis severity in rats. *Nat. Genet.* **2003**, *33*(1), 25–32.

223. Kraaij, M.D., Savage, N.D., van der Kooij, S.W., Koekkoek, K., Wang, J., van den Berg, J.M., Ottenhoff, T.H., Kuijpers, T.W., Holmdahl, R., van Kooten, C., Gelderman, K.A. Induction of regulatory T cells by macrophages is dependent on production of reactive oxygen species. *Proc. Natl. Acad. Sci. U.S.A.* **2010**, *107*(41), 17686–17691.

224. Lambeth, J.D., Kawahara, T., Diebold, B. Regulation of Nox and Duox enzymatic activity and expression. *Free Radic. Biol. Med.* **2007**, *43*(3), 319–331.

225. Sumimoto, H. Structure, regulation and evolution of Nox-family NADPH oxidases that produce reactive oxygen species. *FEBS J.* **2008**, *275*(13), 3249–3277.

226. Lassegue, B., Griendling, K.K. NADPH oxidases: Functions and pathologies in the vasculature. *Arterioscler. Thromb. Vasc. Biol.* **2010**, *30*(4), 653–661.

227. Sedeek, M., Hebert, R.L., Kennedy, C.R., Burns, K.D., Touyz, R.M. Molecular mechanisms of hypertension: Role of Nox family NADPH oxidases. *Curr. Opin. Nephrol. Hypertens.* **2009**, *18*(2), 122–127.

228. Suh, Y.A., Arnold, R.S., Lassegue, B., Shi, J., Xu, X., Sorescu, D., Chung, A.B., Griendling, K.K., Lambeth, J.D. Cell transformation by the superoxide-generating oxidase Mox1. *Nature* **1999**, *401*, 79–82.

229. Bánfi, B., Maturana, A., Jaconi, S., Arnaudeau, S., Laforge, T., Sinha, B., Ligeti, E., Demaurex, N., Krause, K.H. A mammalian H$^+$ channel generated through alternative splicing of the NADPH oxidase homolog *NOH*-1. *Science* **2000**, *287*(5450), 138–142.

230. Lambeth, J.D. NOX enzymes and the biology of reactive oxygen. *Nat. Rev. Immunol.* **2004**, *4*(3), 181–189.

231. Hanna, I.R., Hilenski, L.L., Dikalova, A., Taniyama, Y., Dikalov, S., Lyle, A., Qum, M.T., Lassègue, B., Griendling, K.K. Functional association of NOX1 with P22PHOX in vascular smooth muscle cells. *Free Radic. Biol. Med.* **2004**, *37*(10), 1542–1549.

232. Ambasta, R.K., Kumar, P., Griendling, K.K., Schmidt, H.H., Busse, R., Brandes, R.P. Direct interaction of the novel Nox proteins with p22^{phox} is required for the formation of a functionally active NADPH oxidase. *J. Biol. Chem.* **2004**, *279*(44), 45935–45941.

233. Geiszt, M., Lekstrom, K., Brenner, S., Hewitt, S.M., Dana, R., Malech, H.L., Leto, T.L. NAD(P)H oxidase 1, a product of differentiated colon epithelial cells, can partially replace glycoprotein 91^{phox} in the regulated production of superoxide by phagocytes. *J. Immunol.* **2003**, *171*(1), 299–306.

234. Bánfi, B., Clark, R.A., Steger, K., Krause, K.H. Two novel proteins activate superoxide generation by the NADPH oxidase NOX1. *J. Biol. Chem.* **2003**, *278*(6), 3510–3513.

235. Geiszt, M., Lekstrom, K., Witta, J., Leto, T.L. Proteins homologous to p47^{phox} and p67^{phox} support superoxide production by NAD(P)H oxidase 1 in colon epithelial cells. *J. Biol. Chem.* **2003**, *278*(22), 20006–20012.

236. Takeya, R., Ueno, N., Kami, K., Taura, M., Kohjima, M., Izaki, T., Nunoi, H., Sumimoto, H. Novel human homologues of p47^{phox} and p67^{phox} participate in activation of superoxide-producing NADPH oxidases. *J. Biol. Chem.* **2003**, *278*(27), 25234–25246.

237. Cheng, G., Diebold, B.A., Hughes, Y., Lambeth, J.D. Nox1-dependent reactive oxygen generation is regulated by Rac1. *J. Biol. Chem.* **2006**, *281*(26), 17718–17726.

238. Ueyama, T., Geiszt, M., Leto, T.L. Involvement of Rac1 in activation of multicomponent Nox1- and Nox3-based NADPH oxidases. *Mol. Cell. Biol.* **2006**, *26*(6), 2160–2174.

239. Cheng, G., Cao, Z., Xu, X., Van Meir, E.G., Lambeth, J.D. Homologs of gp91*phox*: Cloning and tissue expression of Nox3, Nox4, and Nox5. *Gene* **2001**, *269*(1–2), 131–140.

240. Ueno, N., Takeya, R., Miyano, K., Kikuchi, H., Sumimoto, H. The NADPH oxidase Nox3 constitutively produces superoxide in a p22^{phox}-dependent manner. *J. Biol. Chem.* **2005**, *280*(24), 23328–23339.

241. Kawahara, T., Ritsick, D., Cheng, G.J., Lambeth, J.D. Point mutations in the proline-rich region of p22^{phox} are dominant inhibitors of Nox1- and Nox2-dependent reactive oxygen generation. *J. Biol. Chem.* **2005**, *280*(36), 31859–31869.

242. Paffenholz, R., Bergstrom, R.A., Pasutto, F., Wabnitz, P., Munroe, R.J., Jagla, W., Heinzmann, U., Marquardt, A., Bareiss, A., Laufs, J., Russ, A., Stumm, G., Schimenti, J.C., Bergstrom, D.E. Vestibular defects in head-tilt mice result from mutations in *Nox3*, encoding an NADPH oxidase. *Genes Dev.* **2004**, *18*(5), 486–491.

243. Banfi, B., Malgrange, B., Knisz, J., Steger, K., Dubois-Dauphin, M., Krause, K.H. NOX3, a superoxide-generating NADPH oxidase of the inner ear. *J. Biol. Chem.* **2004**, *279*(44), 46065–46072.

244. Kiss, P.J., Knisz, J., Zhang, Y.Z., Baltrusaitis, J., Sigmund, C.D., Thalmann, R., Smith, R.J.H., Verpy, E., Banfi, B. Inactivation of *NADPH oxidase organizer 1* results in severe imbalance. *Curr. Biol.* **2006**, *16*(2), 208–213.

245. Cheng, G.J., Ritsick, D., Lambeth, J.D. Nox3 Regulation by NOXO1, p47^{phox} and p67^{phox}. *J. Biol. Chem.* **2004**, *279*, 34250–34255.

246. Miyano, K., Sumimoto, H. Role of the small GTPase Rac in p22(phox)-dependent NADPH oxidases. *Biochimie* **2007**, *89*(9), 1133–1144.

247. Geiszt, M., Kopp, J.B., Várnai, P., Leto, T.L. Identification of Renox, an NAD(P)H oxidase in kidney. *Proc. Natl. Acad. Sci. U.S.A.* **2000**, *97*(14), 8010–8014.

248. Shiose, A., Kuroda, J., Tsuruya, K., Hirai, M., Hirakata, H., Naito, S., Hattori, M., Sakaki, Y., Sumimoto, H. A novel superoxide-producing NAD(P)H oxidase in kidney. *J. Biol. Chem.* **2001**, *276*(2), 1417–1423.

249. Martyn, K.D., Frederick, L.M., von Löhneysen, K., Dinauer, M.C., Knaus, U.G. Functional analysis of Nox4 reveals unique characteristics compared to other NADPH oxidases. *Cell. Signal.* **2006**, *18*(1), 69–82.

250. von Löhneysen, K., Noack, D., Jesaitis, A.J., Dinauer, M.C., Knaus, U.G. Mutational analysis reveals distinct features of the Nox4-p22^{phox} complex. *J. Biol. Chem.* **2008**, *283*(50), 35273–35282.

251. Weyemi, U., Caillou, B., Talbot, M., Ameziane-El-Hassani, R., Lacroix, L., Lagent-Chevallier, O., Al Ghuzlan, A., Roos, D., Bidart, J.M., Virion, A., Schlumberger, M., Dupuy, C. Intracellular expression of reactive oxygen species-generating NADPH oxidase NOX4 in normal and cancer thyroid tissues. *Endocr. Relat. Cancer* **2010**, *17*(1), 27–37.

252. Lee, C.F., Qiao, M., Schroder, K., Zhao, Q., Asmis, R. Nox4 is a novel inducible source of reactive oxygen species in monocytes and macrophages and mediates oxidized low density lipoprotein-induced macrophage death. *Circ. Res.* **2010**, *106*(9), 1489–1497.

253. Park, H.S., Chun, J.N., Jung, H.Y., Choi, C., Bae, Y.S. Role of NADPH oxidase 4 in lipopolysaccharide-induced proinflammatory responses by human aortic endothelial cells. *Cardiovasc. Res.* **2006**, *72*(3), 447–455.

254. Mahadev, K., Motoshima, H., Wu, X., Ruddy, J.M., Arnold, R.S., Cheng, G., Lambeth, J.D., Goldstein, B.J. The NAD(P)H oxidase homolog Nox4 modulates insulin-stimulated generation of H_2O_2 and plays an integral role in insulin signal transduction. *Mol. Cell. Biol.* **2004**, *24*(5), 1844–1854.

255. Gorin, Y., Ricono, J.M., Kim, N.H., Bhandari, B., Choudhury, G.G., Abboud, H.E. Nox4 mediates angiotensin II-induced activation of Akt/protein kinase B in mesangial cells. *Am. J. Physiol. Renal Physiol.* **2003**, *285*(2), F219–F229.

256. Lyle, A.N., Deshpande, N.N., Taniyama, Y., Seidel-Rogol, B., Pounkova, L., Du, P., Papaharalambus, C., Lassegue, B., Griendling, K.K. Poldip2, a novel regulator of Nox4 and cytoskeletal integrity in vascular smooth muscle cells. *Circ. Res.* **2009**, *105*(3), 249–259.

257. Bánfi, B., Molnár, G., Maturana, A., Steger, K., Hegedûs, B., Demaurex, N., Krause, K.H. A Ca^{2+}-activated NADPH oxidase in testis, spleen, and lymph nodes. *J. Biol. Chem.* **2001**, *276*(40), 37594–37601.

258. Guzik, T.J., Chen, W., Gongora, M.C., Guzik, B., Lob, H.E., Mangalat, D., Hoch, N., Dikalov, S., Rudzinski, P., Kapelak, B., Sadowski, J., Harrison, D.G. Calcium-dependent NOX5 nicotinamide adenine dinucleotide phosphate oxidase contributes to vascular oxidative stress in human coronary artery disease. *J. Am. Coll. Cardiol.* **2008**, *52*(22), 1803–1809.

259. Kawahara, T., Lambeth, J.D. Phosphatidylinositol (4,5)-bisphosphate modulates Nox5 localization via an N-terminal polybasic region. *Mol. Biol. Cell* **2008**, *19*(10), 4020–4031.

260. Banfi, B., Tirone, F., Durussel, I., Knisz, J., Moskwa, P., Molnar, G.Z., Krause, K.H., Cox, J.A. Mechanism of Ca^{2+} activation of the NADPH oxidase 5 (NOX5). *J. Biol. Chem.* **2004**, *279*(18), 18583–18591.

261. Montezano, A.C., Burger, D., Paravicini, T.M., Chignalia, A.Z., Yusuf, H., Almasri, M., He, Y., Callera, G.E., He, G., Krause, K.H., Lambeth, D., Quinn, M.T., Touyz, R.M. Nicotinamide adenine dinucleotide phosphate reduced oxidase 5 (Nox5) regulation by angiotensin II and endothelin-1 is mediated via calcium/calmodulin-dependent, rac-1-independent pathways in human endothelial cells. *Circ. Res.* **2010**, *106*(8), 1363–1373.

262. Dupuy, C., Ohayon, R., Valent, A., Noël-Hudson, M.S., Dème, D., Virion, A. Purification of a novel flavoprotein involved in the thyroid NADPH oxidase—Cloning of the porcine and human cDNAs. *J. Biol. Chem.* **1999**, *274*(52), 37265–37269.

263. De Deken, X., Wang, D.T., Many, M.C., Costagliola, S., Libert, F., Vassart, G., Dumont, J.E., Miot, F. Cloning of two human thyroid cDNAs encoding new members of the NADPH oxidase family. *J. Biol. Chem.* **2000**, *275*(30), 23227–23233.

264. Grasberger, H., Refetoff, S. Identification of the maturation factor for dual oxidase—Evolution of an eukaryotic operon equivalent. *J. Biol. Chem.* **2006**, *281*(27), 18269–18272.

265. Morand, S., Ueyama, T., Tsujibe, S., Saito, N., Korzeniowska, A., Leto, T.L. Duox maturation factors form cell surface complexes with Duox affecting the specificity of reactive oxygen species generation. *FASEB J.* **2009**, *23*, 1205–1218.

266. Dutta, S., Rittinger, K. Regulation of NOXO1 activity through reversible interactions with p22 and NOXA1. *PLoS ONE* **2010**, *5*(5), e10478.

267. Gianni, D., Dermardirossian, C., Bokoch, G.M. Direct interaction between Tks proteins and the N-terminal proline-rich region (PRR) of NoxA1 mediates Nox1-dependent ROS generation. *Eur. J. Cell Biol.* **2010**, *90*(2–3), 164–171.

268. Gianni, D., Taulet, N., Dermardirossian, C., Bokoch, G.M. c-Src-mediated phosphorylation of NoxA1 and Tks4 induces the reactive oxygen species (ROS)-dependent formation of functional invadopodia in human colon cancer cells. *Mol. Biol. Cell* **2010**, *21*(23), 4287–4298.

269. Kroviarski, Y., Debbabi, M., Bachoual, R., Perianin, A., Gougerot-Pocidalo, M.A., El-Benna, J., Dang, P.M. Phosphorylation of NADPH oxidase activator 1 (NOXA1) on serine 282 by MAP kinases and on serine 172 by protein kinase C and protein kinase A prevents NOX1 hyperactivation. *FASEB J.* **2010**, *24*(6), 2077–2092.

270. Nisimoto, Y., Jackson, H.M., Ogawa, H., Kawahara, T., Lambeth, J.D. Constitutive NADPH-dependent electron transferase activity of the Nox4 dehydrogenase domain. *Biochemistry* **2010**, *49*(11), 2433–2442.

271. Van Buul, J.D., Fernandez-Borja, M., Anthony, E.C., Hordijk, P.L. Expression and localization of NOX2 and NOX4 in primary human endothelial cells. *Antioxid. Redox Signal.* **2005**, *7*(3–4), 308–317.

272. Chen, K., Kirber, M.T., Xiao, H., Yang, Y., Keaney, J.F. Regulation of ROS signal transduction by NADPH oxidase 4 localization. *J. Cell Biol.* **2008**, *181*(7), 1129–1139.

273. Wu, R.F., Ma, Z., Liu, Z., Terada, L.S. Nox4-derived H_2O_2 mediates endoplasmic reticulum signaling through local Ras activation. *Mol. Cell. Biol.* **2010**, *30*(14), 3553–3568.

274. Ameziane-El-Hassani, R., Morand, S., Boucher, J.L., Frapart, Y.M., Apostolou, D., Agnandji, D., Gnidehou, S., Ohayon, R., Noel-Hudson, M.S., Francon, J., Lalaoui, K., Virion, A., Dupuy, C. Dual oxidase-2 has an intrinsic Ca^{2+}-dependent H_2O_2-generating activity. *J. Biol. Chem.* **2005**, *280*(34), 30046–30054.

275. Briones, A.M., Touyz, R.M. Oxidative stress and hypertension: Current concepts. *Curr. Hypertens. Rep.* **2010**, *12*(2), 135–142.

276. Beckman, J.S., Koppenol, W.H. Nitric oxide, superoxide, and peroxynitrite: The good, the bad, and the ugly. *Am. J. Physiol. Cell Physiol.* **1996**, *271*(5), C1424–C1437.

277. Moncada, S., Higgs, E.A. The discovery of nitric oxide and its role in vascular biology. *Br. J. Pharmacol.* **2006**, *147*(Suppl. 1), S193–S201.

278. Muller, G., Morawietz, H. Nitric oxide, NAD(P)H oxidase, and atherosclerosis. *Antioxid. Redox Signal.* **2009**, *11*, 1711–1731.

279. Viel, E.C., Benkirane, K., Javeshghani, D., Touyz, R.M., Schiffrin, E.L. Xanthine oxidase and mitochondria contribute to vascular superoxide anion generation in DOCA-salt hypertensive rats. *Am. J. Physiol. Heart Circ. Physiol.* **2008**, *295*(1), H281–H288.

280. Vasquez-Vivar, J. Tetrahydrobiopterin, superoxide, and vascular dysfunction. *Free Radic. Biol. Med.* **2009**, *47*(8), 1108–1119.

281. Jeong, H.Y., Jeong, H.Y., Kim, C.D. $p22^{phox}$-derived superoxide mediates enhanced proliferative capacity of diabetic vascular smooth muscle cells. *Diabetes Res. Clin. Pract.* **2004**, *64*(1), 1–10.

282. Weber, D.S., Rocic, P., Mellis, A.M., Laude, K., Lyle, A.N., Harrison, D.G., Griendling, K.K. Angiotensin II-induced hypertrophy is potentiated in mice overexpressing p22(phox) in vascular smooth muscle. *Am. J. Physiol. Heart Circ. Physiol.* **2005**, *288*(1), H37–H42.

283. Brandes, R.P., Kreuzer, J. Vascular NADPH oxidases: Molecular mechanisms of activation. *Cardiovasc. Res.* **2005**, *65*(1), 16–27.

284. Cifuentes, M.E., Pagano, P.J. Targeting reactive oxygen species in hypertension. *Curr. Opin. Nephrol. Hypertens.* **2006**, *15*(2), 179–186.

285. Pagano, P.J., Clark, J.K., Cifuentes-Pagano, M.E., Clark, S.M., Callis, G.M., Quinn, M.T. Localization of a constitutively active, phagocyte-like NADPH oxidase in rabbit aortic adventitia: Enhancement by angiotensin II. *Proc. Natl. Acad. Sci. U.S.A.* **1997**, *94*(26), 14483–14488.

286. Landmesser, U., Harrison, D.G., Drexler, H. Oxidant stress: A major cause of reduced endothelial nitric oxide availability in cardiovascular disease. *Eur. J. Clin. Pharmacol.* **2006**, *62*(Suppl. 13), 13–19.

287. Grote, K., Ortmann, M., Salguero, G., Doerries, C., Landmesser, U., Luchtefeld, M., Brandes, R.P., Gwinner, W., Tschernig, T., Brabant, E.G., Klos, A., Schaefer, A., Drexler, H., Schieffer, B. Critical role for $p47^{phox}$ in renin-angiotensin system activation and blood pressure regulation. *Cardiovasc. Res.* **2006**, *71*(3), 596–605.

288. Jung, O., Schreiber, J.G., Geiger, H., Pedrazzini, T., Busse, R., Brandes, R.P. gp91phox-containing NADPH oxidase mediates endothelial dysfunction in renovascular hypertension. *Circulation* **2004**, *109*(14), 1795–1801.

289. Touyz, R.M., Mercure, C., He, Y., Javeshghani, D., Yao, G.Y., Callera, G.E., Yogi, A., Lochard, N., Reudelhuber, T.L. Angiotensin II-dependent chronic hypertension and cardiac hypertrophy are unaffected by gp91phox-containing NADPH oxidase. *Hypertension* **2005**, *45*(4), 530–537.

290. Modlinger, P., Chabrashvili, T., Gill, P.S., Mendonca, M., Harrison, D.G., Griendling, K.K., Li, M., Raggio, J., Wellstein, A., Chen, Y.F., Welch, W.J., Wilcox, C.S. RNA silencing *in vivo* reveals role of $p22^{phox}$ in rat angiotensin slow pressor response. *Hypertension* **2006**, *47*(2), 238–244.

291. Wang, H.D., Xu, S., Johns, D.G., Du, Y., Quinn, M.T., Cayatte, A.J., Cohen, R.A. Role of NADPH oxidase in the vascular hypertrophic and oxidative stress response to angiotensis II in mice. *Circ. Res.* **2001**, *88*, 947–953.

292. Bendall, J.K., Cave, A.C., Heymes, C., Gall, N., Shah, A.M. Pivotal role of a gp91^{phox}-containing NADPH oxidase in angiotensin II-induced cardiac hypertrophy in mice. *Circulation* **2002**, *105*, 293–296.

293. Sindhu, R.K., Roberts, C.K., Ehdaie, A., Zhan, C.D., Vaziri, N.D. Effects of aortic coarctation on aortic antioxidant enzymes and NADPH oxidase protein expression. *Life Sci.* **2005**, *76*(8), 945–953.

294. Dourron, H.M., Jacobson, G.M., Park, J.L., Liu, J.H., Reddy, D.J., Scheel, M.L., Pagano, P.J. Perivascular gene transfer of NADPH oxidase inhibitor suppresses angioplasty-induced neointimal proliferation of rat carotid artery. *Am. J. Physiol. Heart Circ. Physiol.* **2005**, *288*(2), H946–H953.

295. Tojo, T., Ushio-Fukai, M., Yamaoka-Tojo, M., Ikeda, S., Patrushev, N., Alexander, R.W. Role of gp91^{phox} (Nox2)-containing NAD(P)H oxidase in angiogenesis in response to hindlimb ischemia. *Circulation* **2005**, *111*(18), 2347–2355.

296. Sun, Y., Zhang, J.K., Lu, L., Chen, S.S., Quinn, M.T., Weber, K.T. Aldosterone-induced inflammation in the rat heart—Role of oxidative stress. *Am. J. Pathol.* **2002**, *161*(5), 1773–1781.

297. Sorescu, D., Weiss, D., Lassegue, B., Clempus, R.E., Szocs, K., Sorescu, G.P., Valppu, L., Quinn, M.T., Lambeth, J.D., Vega, J.D., Taylor, W.R., Griendling, K.K. Superoxide production and expression of nox family proteins in human atherosclerosis. *Circulation* **2002**, *105*(12), 1429–1435.

298. Judkins, C.P., Diep, H., Broughton, B.R., Mast, A.E., Hooker, E.U., Miller, A.A., Selemidis, S., Dusting, G.J., Sobey, C.G., Drummond, G.R. Direct evidence of a role for Nox2 in superoxide production, reduced nitric oxide bioavailability, and early atherosclerotic plaque formation in ApoE-/- mice. *Am. J. Physiol. Heart Circ. Physiol.* **2010**, *298*(1), H24–H32.

299. Lassègue, B., Sorescu, D., Szöcs, K., Yin, Q.Q., Akers, M., Zhang, Y., Grant, S.L., Lambeth, J.D., Griendling, K.K. Novel gp91^{phox} homologues in vascular smooth muscle cells—Nox1 mediates angiotensin II-induced superoxide formation and redox-sensitive signaling pathways. *Circ. Res.* **2001**, *88*(9), 888–894.

300. Csanyi, G., Taylor, W.R., Pagano, P.J. NOX and inflammation in the vascular adventitia. *Free Radic. Biol. Med.* **2009**, *47*(9), 1254–1266.

301. Hilenski, L.L., Clempus, R.E., Quinn, M.T., Lambeth, J.D., Griendling, K.K. Distinct subcellular localizations of Nox1 and Nox4 in vascular smooth muscle cells. *Arterioscler. Thromb. Vasc. Biol.* **2004**, *24*, 677–683.

302. Ellmark, S.H.M., Dusting, G.J., Fui, M.N.T., Guzzo-Pernell, N., Drummond, G.R. The contribution of Nox4 to NADPH oxidase activity in mouse vascular smooth muscle. *Cardiovasc. Res.* **2005**, *65*(2), 495–504.

303. Jay, D.B., Papaharalambus, C.A., Seidel-Rogol, B., Dikalova, A.E., Lassegue, B., Griendling, K.K. Nox5 mediates PDGF-induced proliferation in human aortic smooth muscle cells. *Free Radic. Biol. Med.* **2008**, *45*(3), 329–335.

304. BelAiba, R.S., Djordjevic, T., Petry, A., Diemer, K., Bonello, S., Banfi, B., Hess, J., Pogrebniak, A., Bickel, C., Gorlach, A. NOX5 variants are functionally active in endothelial cells. *Free Radic. Biol. Med.* **2007**, *42*(4), 446–459.

305. Dikalova, A., Clempus, R., Lassegue, B., Cheng, G.J., Mccoy, J., Dikalov, S., Martin, A.S., Lyle, A., Weber, D.S., Weiss, D., Taylor, R., Schmidt, H.H.H.W., Owens, G.K., Lambeth, J.D., Griendling, K.K. Nox1 overexpression potentiates angiotensin II-induced hypertension and vascular smooth muscle hypertrophy in transgenic mice. *Circulation* **2005**, *112*(17), 2668–2676.

306. Gavazzi, G., Banfi, B., Deffert, C., Fiette, L., Schappi, M., Herrmann, F., Krause, K.H. Decreased blood pressure in NOX1-deficient mice. *FEBS Lett.* **2006**, *580*(2), 497–504.

307. Yin, C.C., Huang, K.T. H_2O_2 but not O_2^- elevated by oxidized LDL enhances human aortic smooth muscle cell proliferation. *J. Biomed. Sci.* **2007**, *14*(2), 245–254.

308. San Martin, A., Foncea, R., Laurindo, F.R., Ebensperger, R., Griendling, K.K., Leighton, F. Nox1-based NADPH oxidase-derived superoxide is required for VSMC activation by advanced glycation end-products. *Free Radic. Biol. Med.* **2007**, *42*(11), 1671–1679.

309. Lee, J.G., Lim, E.J., Park, D.W., Lee, S.H., Kim, J.R., Baek, S.H. A combination of Lox-1 and Nox1 regulates TLR9-mediated foam cell formation. *Cell. Signal.* **2008**, *20*(12), 2266–2275.

310. Park, D.W., Baek, K., Kim, J.R., Lee, J.J., Ryu, S.H., Chin, B.R., Baek, S.H. Resveratrol inhibits foam cell formation via NADPH oxidase 1-mediated reactive oxygen species and monocyte chemotactic protein-1. *Exp. Mol. Med.* **2009**, *41*(3), 171–179.

311. Niu, X.L., Madamanchi, N.R., Vendrov, A.E., Tchivilev, I., Rojas, M., Madamanchi, C., Brandes, R.P., Krause, K.H., Humphries, J., Smith, A., Burnand, K.G., Runge, M.S. Nox activator 1: A potential target for modulation of vascular reactive oxygen species in atherosclerotic arteries. *Circulation* **2010**, *121*(4), 549–559.

312. Ago, T., Kitazono, T., Ooboshi, H., Iyama, T., Han, Y.H., Takada, J., Wakisaka, M., Ibayashi, S., Utsumi, H., Iida, M. Nox4 as the major catalytic component of an endothelial NAD(P)H oxidase. *Circulation* **2004**, *109*(2), 227–233.

313. Goettsch, C., Goettsch, W., Muller, G., Seebach, J., Schnittler, H.J., Morawietz, H. Nox4 overexpression activates reactive oxygen species and p38 MAPK in human endothelial cells. *Biochem. Biophys. Res. Commun.* **2009**, *380*(2), 355–360.

314. Guzik, T.J., Sadowski, J., Guzik, B., Jopek, A., Kapelak, B., Przybylowski, P., Wierzbicki, K., Korbut, R., Harrison, D.G., Channon, K.M. Coronary artery superoxide production and nox isoform expression in human coronary

artery disease. *Arterioscler. Thromb. Vasc. Biol.* **2006**, *26*(2), 333–339.

315. Szocs, K., Lassegue, B., Sorescu, D., Hilenski, L.L., Valppu, L., Couse, T.L., Wilcox, J.N., Quinn, M.T., Lambeth, J.D., Griendling, K.K. Upregulation of Nox-based NAD(P)H oxidases in restenosis after carotid injury. *Arterioscler. Thromb. Vasc. Biol.* **2002**, *22*(1), 21–27.

316. Zhang, Q., Malik, P., Pandey, D., Gupta, S., Jagnandan, D., Belin de Chantemele, E., Banfi, B., Marrero, M.B., Rudic, R.D., Stepp, D.W., Fulton, D.J. Paradoxical activation of endothelial nitric oxide synthase by NADPH oxidase. *Arterioscler. Thromb. Vasc. Biol.* **2008**, *28*(9), 1627–1633.

317. Violi, F., Sanguigni, V., Carnevale, R., Plebani, A., Rossi, P., Finocchi, A., Pignata, C., De Mattia, D., Martire, B., Pietrogrande, M.C., Martino, S., Gambineri, E., Soresina, A.R., Pignatelli, P., Martino, F., Basili, S., Loffredo, L. Hereditary deficiency of gp91^{phox} is associated with enhanced arterial dilatation: Results of a multicenter study. *Circulation* **2009**, *120*(16), 1616–1622.

318. Matsuno, K., Yamada, H., Iwata, K., Jin, D., Katsuyama, M., Matsuki, M., Takai, S., Yamanishi, K., Miyazaki, M., Matsubara, H., Yabe-Nishimura, C. Nox1 is involved in angiotensin II-mediated hypertension—A study in Nox1-deficient mice. *Circulation* **2005**, *112*(17), 2677–2685.

319. Yogi, A., Mercure, C., Touyz, J., Callera, G.E., Montezano, A.C.I., Aranha, A.B., Tostes, R.C., Reudelhuber, T., Touyz, R.M. Renal redox-sensitive signaling, but not blood pressure, is attenuated by Nox1 knockout in angiotensin II-dependent chronic hypertension. *Hypertension* **2008**, *51*(2), 500–506.

320. Li, J., Stouffs, M., Serrander, L., Banfi, B., Bettiol, E., Charnay, Y., Steger, K., Krause, K.H., Jaconi, M.E. The NADPH oxidase NOX4 drives cardiac differentiation: Role in regulating cardiac transcription factors and MAP kinase activation. *Mol. Biol. Cell* **2006**, *17*(9), 3978–3988.

321. Krijnen, P.A., Meischl, C., Hack, C.E., Meijer, C.J., Visser, C.A., Roos, D., Niessen, H.W. Increased Nox2 expression in human cardiomyocytes after acute myocardial infarction. *J. Clin. Pathol.* **2003**, *56*(3), 194–199.

322. Meischl, C., Krijnen, P.A.J., Sipkens, J.A., Cillessen, S.A.G.M., Munoz, I.G., Okroj, M., Ramska, M., Muller, A., Visser, C.A., Musters, R.J.P., Simonides, W.S., Hack, C.E., Roos, D., Niessen, H.W.M. Ischemia induces nuclear NOX2 expression in cardiomyocytes and subsequently activates apoptosis. *Apoptosis* **2006**, *11*(6), 913–921.

323. Hoffmeyer, M.R., Jones, S.P., Ross, C.R., Sharp, B., Grisham, M.B., Laroux, F.S., Stalker, T.J., Scalia, R., Lefer, D.J. Myocardial ischemia/reperfusion injury in NADPH oxidase-deficient mice. *Circ. Res.* **2000**, *87*(9), 812–817.

324. Kuroda, J., Ago, T., Matsushima, S., Zhai, P., Schneider, M.D., Sadoshima, J. NADPH oxidase 4 (Nox4) is a major source of oxidative stress in the failing heart. *Proc. Natl. Acad. Sci. U.S.A.* **2010**, *107*(35), 15565–15570.

325. Zhang, M., Brewer, A.C., Schroder, K., Santos, C.X., Grieve, D.J., Wang, M., Anilkumar, N., Yu, B., Dong, X., Walker, S.J., Brandes, R.P., Shah, A.M. NADPH oxidase-4 mediates protection against chronic load-induced stress in mouse hearts by enhancing angiogenesis. *Proc. Natl. Acad. Sci. U.S.A.* **2010**, *107*(42), 18121–18126.

326. Wilcox, C.S. Oxidative stress and nitric oxide deficiency in the kidney: A critical link to hypertension? *Am. J. Physiol. Regul. Integr. Comp. Physiol.* **2005**, *289*(4), R913–R935.

327. Wilcox, C.S. Redox regulation of the afferent arteriole and tubuloglomerular feedback. *Acta Physiol. Scand.* **2003**, *179*(3), 217–223.

328. Zou, A.P., Cowley, A.W. Reactive oxygen species and molecular regulation of renal oxygenation. *Acta Physiol. Scand.* **2003**, *179*(3), 233–241.

329. Juncos, R., Hong, N.J., Garvin, J.L. Differential effects of superoxide on luminal and basolateral Na+/H+ exchange in the thick ascending limb. *Am. J. Physiol. Regul. Integr. Comp. Physiol.* **2006**, *290*(1), R79–R83.

330. Welch, W.J., Tojo, A., Wilcox, C.S. Roles of NO and oxygen radicals in tubuloglomerular feedback in SHR. *Am. J. Physiol. Renal Physiol.* **2000**, *278*, F769–F776.

331. Zou, A.P., Li, N., Cowley, A.W., Jr. Production and actions of superoxide in the renal medulla. *Hypertension* **2001**, *37*(2 Pt 2), 547–553.

332. Greiber, S., Munzel, T., Kastner, S., Muller, B., Schollmeyer, P., Pavenstadt, H. NAD(P)H oxidase activity in cultured human podocytes: Effects of adenosine triphosphate. *Kidney Int.* **1998**, *53*, 654–663.

333. Haque, M.Z., Majid, D.S.A. Assessment of renal functional phenotype in mice lacking gp91^{PHOX} subunit of NAD(P)H oxidase. *Hypertension* **2004**, *43*(2 Pt 2), 335–340.

334. Chabrashvili, T., Tojo, A., Onozato, M.L., Kitiyakara, C., Quinn, M.T., Fujita, T., Welch, W.J., Wilcox, C.S. Expression and cellular localization of classic NADPH oxidase subunits in the spontaneously hypertensive rat kidney. *Hypertension* **2002**, *39*(2 Pt 1), 269–274.

335. Adler, S., Huang, H. Oxidant stress in kidneys of spontaneously hypertensive rats involves both oxidase overexpression and loss of extracellular superoxide dismutase. *Am. J. Physiol. Renal Physiol.* **2004**, *287*(5), F907–F913.

336. Wang, X., Villar, V.A., Armando, I., Eisner, G.M., Felder, R.A., Jose, P.A. Dopamine, kidney, and hypertension: Studies in dopamine receptor knockout mice. *Pediatr. Nephrol.* **2008**, *23*(12), 2131–2146.

337. Yang, Z.W., Asico, L.D., Yu, P.Y., Wang, Z., Jones, J.E., Escano, C.S., Wang, X.Y., Quinn, M.T., Sibley, D.R., Romero, G.G., Felder, R.A., Jose, P.A. D_5 dopamine receptor regulation of reactive oxygen species production, NADPH oxidase, and blood pressure. *Am. J. Physiol. Regul. Integr. Comp. Physiol.* **2006**, *290*(1), R96–R104.

338. Li, H., Han, W., Villar, V.A., Keever, L.B., Lu, Q., Hopfer, U., Quinn, M.T., Felder, R.A., Jose, P.A., Yu, P. D1-like receptors regulate NADPH oxidase activity and subunit expression in lipid raft microdomains of renal proximal tubule cells. *Hypertension* **2009**, *53*(6), 1054–1061.

339. Gorin, Y., Block, K., Hernandez, J., Bhandari, B., Wagner, B., Barnes, J.L., Abboud, H.E. Nox4 NAD(P)H oxidase

mediates hypertrophy and fibronectin expression in the diabetic kidney. *J. Biol. Chem.* **2005**, *280*(47), 39616–39626.

340. Xia, L., Wang, H., Goldberg, H.J., Munk, S., Fantus, I.G., Whiteside, C.I. Mesangial cell NADPH oxidase upregulation in high glucose is protein kinase C dependent and required for collagen IV expression. *Am. J. Physiol. Renal Physiol.* **2006**, *290*(2), F345–F356.

341. Li, J.M., Shah, A.M. ROS generation by nonphagocytic NADPH oxidase: Potential relevance in diabetic nephropathy. *J. Am. Soc. Nephrol.* **2003**, *14*, S221–S226.

342. Asaba, K., Tojo, A., Onozato, M.L., Goto, A., Quinn, M.T., Fujita, T., Wilcox, C.S. Effects of NADPH oxidase inhibitor in diabetic nephropathy. *Kidney Int.* **2005**, *67*(5), 1890–1898.

343. Bhatti, F., Mankhey, R.W., Asico, L., Quinn, M.T., Welch, W.J., Maric, C. Mechanisms of antioxidant and prooxidant effects of α-lipoic acid in the diabetic and nondiabetic kidney. *Kidney Int.* **2005**, *67*(4), 1371–1380.

344. Kondo, S., Shimizu, M., Urushihara, M., Tsuchiya, K., Yoshizumi, M., Tamaki, T., Nishiyama, A., Kawachi, H., Shimizu, F., Quinn, M.T., Lambeth, D.J., Kagami, S. Addition of the antioxidant probucol to angiotensin II type I receptor antagonist arrests progressive mesangioproliferative glomerulonephritis in the rat. *J. Am. Soc. Nephrol.* **2006**, *17*(3), 783–794.

345. Fischer, H. Mechanisms and function of DUOX in epithelia of the lung. *Antioxid. Redox Signal.* **2009**, *11*(10), 2453–2465.

346. Perez-Vizcaino, F., Cogolludo, A., Moreno, L. Reactive oxygen species signaling in pulmonary vascular smooth muscle. *Respir. Physiol. Neurobiol.* **2010**, *174*(3), 212–220.

347. Geiszt, M., Witta, J., Baffi, J., Lekstrom, K., Leto, T.L. Dual oxidases represent novel hydrogen peroxide sources supporting mucosal surface host defense. *FASEB J.* **2003**, *17*(9), NIL362–NIL375.

348. Harper, R.W., Xu, C., Eiserich, J.P., Chen, Y., Kao, C.Y., Thai, P., Setiadi, H., Wu, R. Differential regulation of dual NADPH oxidases/peroxidases, Duox1 and Duox2, by Th1 and Th2 cytokines in respiratory tract epithelium. *FEBS Lett.* **2005**, *579*(21), 4911–4917.

349. Brar, S.S., Kennedy, T.P., Sturrock, A.B., Huecksteadt, T.P., Quinn, M.T., Murphy, T.M., Chitano, P., Hoidal, J.R. NADPH oxidase promotes NF-κB activation and proliferation in human airway smooth muscle. *Am. J. Physiol. Lung Cell Mol. Physiol.* **2002**, *282*(4), L782–L795.

350. Parinandi, N.L., Kleinberg, M.A., Usatyuk, P.V., Cummings, R.J., Pennathur, A., Cardounel, A.J., Zweier, J.L., Garcia, J.G., Natarajan, V. Hyperoxia-induced NAD(P)H oxidase activation and regulation by MAP kinases in human lung endothelial cells. *Am. J. Physiol. Lung Cell Mol. Physiol.* **2003**, *284*(1), L26–L38.

351. Pendyala, S., Gorshkova, I.A., Usatyuk, P.V., He, D., Pennathur, A., Lambeth, J.D., Thannickal, V.J., Natarajan, V. Role of Nox4 and Nox2 in hyperoxia-induced reactive oxygen species generation and migration of human lung endothelial cells. *Antioxid. Redox Signal.* **2009**, *11*(4), 747–764.

352. Dhaunsi, G.S., Paintlia, M.K., Kaur, J., Turner, R.B. NADPH oxidase in human lung fibroblasts. *J. Biomed. Sci.* **2004**, *11*(5), 617–622.

353. Laleu, B., Gaggini, F., Orchard, M., Fioraso-Cartier, L., Cagnon, L., Houngninou-Molango, S., Gradia, A., Duboux, G., Merlot, C., Heitz, F., Szyndralewiez, C., Page, P. First in class, potent, and orally bioavailable NADPH oxidase isoform 4 (Nox4) inhibitors for the treatment of idiopathic pulmonary fibrosis. *J. Med. Chem.* **2010**, *53*(21), 7715–7730.

354. Wolin, M.S., Ahmad, M., Gupte, S.A. Oxidant and redox signaling in vascular oxygen sensing mechanisms: Basic concepts, current controversies, and potential importance of cytosolic NADPH. *Am. J. Physiol. Lung Cell Mol. Physiol.* **2005**, *289*(2), L159–L173.

355. Kummer, W., Acker, H. Immunohistochemical demonstration of four subunits of neutrophil NAD(P)H oxidase in type I cells of carotid body. *J. Appl. Physiol.* **1995**, *78*, 1904–1909.

356. Youngson, C., Nurse, C., Yeger, H., Curnutte, J.T., Vollmer, C., Wong, V., Cutz, E. Immunocytochemical localization of O_2^--sensing protein (NADPH oxidase) in chemoreceptor cells. *Microsc. Res. Tech.* **1997**, *37*(1), 101–106.

357. Fu, X.W., Wang, D.S., Nurse, C.A., Dinauer, M.C., Cutz, E. NADPH oxidase is an O_2 sensor in airway chemoreceptors: Evidence from K^+ current modulation in wild-type and oxidase-deficient mice. *Proc. Natl. Acad. Sci. U.S.A.* **2000**, *97*(8), 4374–4379.

358. Kazemian, P., Stephenson, R., Yeger, H., Cutz, E. Respiratory control in neonatal mice with NADPH oxidase deficiency. *Respir. Physiol.* **2001**, *126*(2), 89–101.

359. Obeso, A., Gómez-Niño, A., Gonzalez, C. NADPH oxidase inhibition does not interfere with low P_{O_2} transduction in rat and rabbit CB chemoreceptor cells. *Am. J. Physiol. Cell Physiol.* **1999**, *276*(3), C593–C601.

360. Wenger, R.H., Marti, H.H., Schuerer-Maly, C.C., Kvietikova, I., Bauer, C., Gassmann, M., Maly, F.E. Hypoxic induction of gene expression in chronic granulomatous disease-derived B-cell lines: Oxygen sensing is independent of the cytochrome b_{558}-containing nicotinamide adenine dinucleotide phosphate oxidase. *Blood* **1996**, *87*(2), 756–761.

361. Archer, S.L., Reeve, H.L., Michelakis, E., Puttagunta, L., Waite, R., Nelson, D.P., Dinauer, M.C., Weir, E.K. O_2 sensing is preserved in mice lacking the gp91 phox subunit of NADPH oxidase. *Proc. Natl. Acad. Sci. U.S.A.* **1999**, *96*(14), 7944–7949.

362. He, L., Chen, J., Dinger, B., Sanders, K., Sundar, K., Hoidal, J., Fidone, S. Characteristics of carotid body chemosensitivity in NADPH oxidase-deficient mice. *Am. J. Physiol. Cell Physiol.* **2002**, *282*, C27–C33.

363. Lee, Y.M., Kim, B.J., Chun, Y.S., So, I., Choi, H., Kim, M.S., Park, J.W. NOX4 as an oxygen sensor to regulate TASK-1 activity. *Cell. Signal.* **2006**, *18*(4), 499–507.

364. Cutz, E., Pan, J., Yeger, H. The role of NOX2 and "novel oxidases" in airway chemoreceptor O_2 sensing. *Adv. Exp. Med. Biol.* **2009**, *648*, 427–438.

365. Rahman, I., Biswas, S.K., Kode, A. Oxidant and antioxidant balance in the airways and airway diseases. *Eur. J. Pharmacol.* **2006**, *533*(1–3), 222–239.

366. Sanders, K.A., Hoidal, J.R. The NOX on pulmonary hypertension. *Circ. Res.* **2007**, *101*(3), 224–226.

367. Jones, R.D., Morice, A.H. Hydrogen peroxide—An intracellular signal in the pulmonary circulation: Involvement in hypoxic pulmonary vasoconstriction. *Pharmacol. Ther.* **2000**, *88*(2), 153–161.

368. Jones, R.D., Thompson, J.S., Morice, A.H. The NADPH oxidase inhibitors iodonium diphenyl and cadmium sulphate inhibit hypoxic pulmonary vasoconstriction in isolated rat pulmonary arteries. *Physiol. Res.* **2000**, *49*(5), 587–596.

369. Liu, J.Q., Zelko, I.N., Erbynn, E.M., Sham, J.S.K., Folz, R.J. Hypoxic pulmonary hypertension: Role of superoxide and NADPH oxidase (gp91^{phox}). *Am. J. Physiol. Lung Cell Mol. Physiol.* **2006**, *290*(1), L2–L10.

370. Mittal, M., Roth, M., Konig, P., Hofmann, S., Dony, E., Scherer, A., Schermuly, R.T., Ghofrani, H.A., Kwapiszewska, G., Kummer, W., Schmidt, H.H.H.W., Seeger, W., Hanze, J., Grimminger, F., Fink, L., Weissmann, N. Hypoxia driven upregulation of the catalytic NADPH oxidase subunit NOX4 in the pulmonary arterial hypertension. *Pathol. Res. Pract.* **2007**, *203*(5), 324.

371. Hecker, L., Vittal, R., Jones, T., Jagirdar, R., Luckhardt, T.R., Horowitz, J.C., Pennathur, S., Martinez, F.J., Thannickal, V.J. NADPH oxidase-4 mediates myofibroblast activation and fibrogenic responses to lung injury. *Nat. Med.* **2009**, *15*(9), 1077–1081.

372. Manoury, B., Nenan, S., Leclerc, O., Guenon, I., Boichot, E., Planquois, J.M., Bertrand, C.P., Lagente, V. The absence of reactive oxygen species production protects mice against bleomycin-induced pulmonary fibrosis. *Respir. Res.* **2005**, *6*, 11.

373. Pongnimitprasert, N., El-Benna, J., Foglietti, M.J., Gougerot-Pocidalo, M.A., Bernard, M., Braut-Boucher, F. Potential role of the "NADPH oxidases" (NOX/DUOX) family in cystic fibrosis. *Ann. Biol. Clin. (Paris)* **2008**, *66*(6), 621–629.

374. Hoidal, J.R., Brar, S.S., Sturrock, A.B., Sanders, K.A., Dinger, B., Fidone, S., Kennedy, T.P. The role of endogenous NADPH oxidases in airway and pulmonary vascular smooth muscle function. *Antioxid. Redox Signal.* **2003**, *5*(6), 751–758.

375. Auten, R.L., Davis, J.M. Oxygen toxicity and reactive oxygen species: The devil is in the details. *Pediatr. Res.* **2009**, *66*(2), 121–127.

376. Abraham, E. Neutrophils and acute lung injury. *Crit. Care Med.* **2003**, *31*(4 Suppl.), S195–S199.

377. Moraes, T.J., Zurawska, J.H., Downey, G.P. Neutrophil granule contents in the pathogenesis of lung injury. *Curr. Opin. Hematol.* **2006**, *13*(1), 21–27.

378. Gao, X.P., Standiford, T.J., Rahman, A., Newstead, M., Holland, S.M., Dinauer, M.C., Liu, Q.H., Malik, A.B. Role of NADPH oxidase in the mechanism of lung neutrophil sequestration and microvessel injury induced by gram-negative sepsis: Studies in p47$^{phox-/-}$ and gp91$^{phox-/-}$ mice. *J. Immunol.* **2002**, *168*(8), 3974–3982.

379. Polikandriotis, J.A., Rupnow, H.L., Elms, S.C., Clempus, R.E., Campbell, D.J., Sutliff, R.L., Brown, L.A., Guidot, D.M., Hart, C.M. Chronic ethanol ingestion increases superoxide production and NADPH oxidase expression in the lung. *Am. J. Respir. Cell Mol. Biol.* **2006**, *34*(3), 314–319.

380. Henricks, P.A., Nijkamp, F.P. Reactive oxygen species as mediators in asthma. *Pulm. Pharmacol. Ther.* **2001**, *14*(6), 409–420.

381. De Boer, J., May, F., Pouw, H., Zaagsma, J., Meurs, H. Effects of endogenous superoxide anion and nitric oxide on cholinergic constriction of normal and hyperreactive guinea pig airways. *Am. J. Respir. Crit. Care Med.* **1998**, *158*(6), 1784–1789.

382. Sorce, S., Krause, K.H. NOX enzymes in the central nervous system: From signaling to disease. *Antioxid. Redox Signal.* **2009**, *11*(10), 2481–2504.

383. Sankarapandi, S., Zweier, J.L., Mukherjee, G., Quinn, M.T., Huso, D.L. Measurement and characterization of superoxide generation in microglial cells: Evidence for an NADPH oxidase-dependent pathway. *Arch. Biochem. Biophys.* **1998**, *353*(2), 312–321.

384. Lavigne, M.C., Malech, H.L., Holland, S.M., Leto, T.L. Genetic requirement of p47*phox* for superoxide production by murine microglia. *FASEB J.* **2001**, *15*(2), 285–287.

385. Mander, P.K., Jekabsone, A., Brown, G.C. Microglia proliferation is regulated by hydrogen peroxide from NADPH oxidase. *J. Immunol.* **2006**, *176*(2), 1046–1052.

386. Abramov, A.Y., Jacobson, J., Wientjes, F., Hothersall, J., Canevari, L., Duchen, M.R. Expression and modulation of an NADPH oxidase in mammalian astrocytes. *J. Neurosci.* **2005**, *25*(40), 9176–9184.

387. Pawate, S., Shen, Q., Fan, F., Bhat, N.R. Redox regulation of glial inflammatory response to lipopolysaccharide and interferongamma. *J. Neurosci. Res.* **2004**, *77*(4), 540–551.

388. Ibi, M., Katsuyama, M., Fan, C., Iwata, K., Nishinaka, T., Yokoyama, T., Yabe-Nishimura, C. NOX1/NADPH oxidase negatively regulates nerve growth factor-induced neurite outgrowth. *Free Radic. Biol. Med.* **2006**, *40*(10), 1785–1795.

389. Tammariello, S.P., Quinn, M.T., Estus, S. NADPH oxidase contributes directly to oxidative stress and apoptosis in nerve growth factor-deprived sympathetic neurons. *J. Neurosci.* **2000**, *20*(1), RC53–NIL16.

390. Tejada-Simon, M.V., Serrano, F., Villasana, L.E., Kanterewicz, B.I., Wu, G.Y., Quinn, M.T., Klann, E. Synaptic localization of a functional NADPH oxidase in the mouse hippocampus. *Mol. Cell. Neurosci.* **2005**, *29*(1), 97–106.

391. Vallet, P., Charnay, Y., Steger, K., Ogier-Denis, E., Kovari, E., Herrmann, F., Michel, J.P., Szanto, I. Neuronal expres-

sion of the NADPH oxidase NOX4, and its regulation in mouse experimental brain ischemia. *Neuroscience* **2005**, *132*(2), 233–238.
392. Wang, G., Anrather, J., Huang, J., Speth, R.C., Pickel, V.M., Iadecola, C. NADPH oxidase contributes to angiotensin II signaling in the nucleus tractus solitarius. *J. Neurosci.* **2004**, *24*(24), 5516–5524.
393. Erdös, B., Broxson, C.S., King, M.A., Scarpace, P.J., Tümer, N. Acute pressor effect of central angiotensin II is mediated by NAD(P)H-oxidase-dependent superoxide production in the hypothalamic cardiovascular regulatory nuclei. *J. Hypertens.* **2006**, *24*(1), 109–116.
394. Kishida, K.T., Pao, M., Holland, S.M., Klann, E. NADPH oxidase is required for NMDA receptor-dependent activation of ERK in hippocampal area CA1. *J. Neurochem.* **2005**, *94*(2), 299–306.
395. Sorce, S., Schiavone, S., Tucci, P., Colaianna, M., Jaquet, V., Cuomo, V., Dubois-Dauphin, M., Trabace, L., Krause, K.H. The NADPH oxidase NOX2 controls glutamate release: A novel mechanism involved in psychosis-like ketamine responses. *J. Neurosci.* **2010**, *30*(34), 11317–11325.
396. Kishida, K.T., Klann, E. Sources and targets of reactive oxygen species in synaptic plasticity and memory. *Antioxid. Redox Signal.* **2007**, *9*(2), 233–244.
397. Faraci, F.M. Reactive oxygen species: Influence on cerebral vascular tone. *J. Appl. Physiol.* **2006**, *100*(2), 739–743.
398. Fang, Q., Sun, H., Arrick, D.M., Mayhan, W.G. Inhibition of NADPH oxidase improves impaired reactivity of pial arterioles during chronic exposure to nicotine. *J. Appl. Physiol.* **2006**, *100*(2), 631–636.
399. Paravicini, T.M., Chrissobolis, S., Drummond, G.R., Sobey, C.G. Increased NADPH-oxidase activity and Nox4 expression during chronic hypertension is associated with enhanced cerebral vasodilatation to NADPH in vivo. *Stroke* **2004**, *35*(2), 584–589.
400. Warner, D.S., Sheng, H., Batinic-Haberle, I. Oxidants, antioxidants and the ischemic brain. *J. Exp. Biol.* **2004**, *207*(Pt 18), 3221–3231.
401. Lucas, S.-M., Rothwell, N.J., Gibson, R.M. The role of inflammation in CNS injury and disease. *Br. J. Pharmacol.* **2006**, *147*(Suppl. 1), S232–S240.
402. Lull, M.E., Block, M.L. Microglial activation and chronic neurodegeneration. *Neurotherapeutics* **2010**, *7*(4), 354–365.
403. Sawada, M., Imamura, K., Nagatsu, T. Role of cytokines in inflammatory process in Parkinson's disease. *J. Neural Transm. Suppl.* **2006**, 70, 373–381.
404. Qin, B., Cartier, L., Dubois-Dauphin, M., Li, B., Serrander, L., Krause, K.H. A key role for the microglial NADPH oxidase in APP-dependent killing of neurons. *Neurobiol. Aging* **2006**, *27*(11), 1577–1587.
405. Brown, G.C., Neher, J.J. Inflammatory neurodegeneration and mechanisms of microglial killing of neurons. *Mol. Neurobiol.* **2010**, *41*(2–3), 242–247.
406. Butterfield, D.A., Howard, B., Yatin, S., Koppal, T., Drake, J., Hensley, K., Aksenov, M., Aksenova, M., Subramaniam, R., Varadarajan, S., Harris-White, M.E., Pedigo, N.W., Jr., Carney, J.M. Elevated oxidative stress in models of normal brain aging and Alzheimer's disease. *Life Sci.* **1999**, *65*(18–19), 1883–1892.
407. Zekry, D., Epperson, T.K., Krause, K.H. A role for NOX NADPH oxidases in Alzheimer's disease and other types of dementia? *IUBMB Life* **2003**, *55*(6), 307–313.
408. Kalra, J., Rajput, A.H., Mantha, S.V., Chaudhary, A.K., Prasad, K. Oxygen free radical producing activity of polymorphonuclear leukocytes in patients with Parkinson's disease. *Mol. Cell. Biochem.* **1992**, *112*(2), 181–186.
409. Wu, D.C., Teismann, P., Tieu, K., Vila, M., Jackson-Lewis, V., Ischiropoulos, H., Przedborski, S. NADPH oxidase mediates oxidative stress in the 1-methyl-4-phenyl-1,2,3,6-tetrahydropyridine model of Parkinson's disease. *Proc. Natl. Acad. Sci. U.S.A.* **2003**, *100*(10), 6145–6150.
410. Bowes, M.P., Rothlein, R., Fagan, S.C., Zivin, J.A. Monoclonal antibodies preventing leukocyte activation reduce experimental neurologic injury and enhance efficacy of thrombolytic therapy. *Neurology* **1995**, *45*, 815–819.
411. Hudome, S., Palmer, C., Roberts, R.L., Mauger, D., Housman, C., Towfighi, J. The role of neutrophils in the production of hypoxic-ischemic brain injury in the neonatal rat. *Pediatr. Res.* **1997**, *41*(5), 607–616.
412. Green, A.R., Ashwood, T. Free radical trapping as a therapeutic approach to neuroprotection in stroke: Experimental and clinical studies with NXY-059 and free radical scavengers. *Curr. Drug Targets CNS Neurol. Disord.* **2005**, *4*, 109–118.
413. Hall, E.D. Brain attack. Acute therapeutic interventions. Free radical scavengers and antioxidants. *Neurosurg. Clin. N. Am.* **1997**, *8*(2), 195–206.
414. Kahles, T., Kohnen, A., Heumueller, S., Rappert, A., Bechmann, I., Liebner, S., Wittko, I.M., Neumann-Haefelin, T., Steinmetz, H., Schroeder, K., Brandes, R.P. NADPH oxidase Nox1 contributes to ischemic injury in experimental stroke in mice. *Neurobiol. Dis.* **2010**, *40*(1), 185–192.
415. Kleinschnitz, C., Grund, H., Wingler, K., Armitage, M.E., Jones, E., Mittal, M., Barit, D., Schwarz, T., Geis, C., Kraft, P., Barthel, K., Schuhmann, M.K., Herrmann, A.M., Meuth, S.G., Stoll, G., Meurer, S., Schrewe, A., Becker, L., Gailus-Durner, V., Fuchs, H., Klopstock, T., de Angelis, M.H., Jandeleit-Dahm, K., Shah, A.M., Weissmann, N., Schmidt, H.H. Post-stroke inhibition of induced NADPH oxidase type 4 prevents oxidative stress and neurodegeneration. *PLoS Biol.* **2010**, *8*(9), e1000479.
416. Sherman, M.A., Kalman, D. Initiation and resolution of mucosal inflammation. *Immunol. Res.* **2004**, *29*(1–3), 241–252.
417. Rokutan, K., Kawahara, T., Kuwano, Y., Tominaga, K., Nishida, K., Teshima-Kondo, S. Nox enzymes and oxidative stress in the immunopathology of the gastrointestinal tract. *Semin. Immunopathol.* **2008**, *30*(3), 315–327.

418. Teshima, S., Kutsumi, H., Kawahara, T., Kishi, K., Rokutan, K. Regulation of growth and apoptosis of cultured guinea pig gastric mucosal cells by mitogenic oxidase 1. *Am. J. Physiol. Gastrointest. Liver Physiol.* **2000**, *279*(6), G1169–G1176.
419. Kojim, S., Ikeda, M., Shibukawa, A., Kamikawa, Y. Modification of 5-hydroxytryptophan-evoked 5-hydroxytryptamine formation of guinea pig colonic mucosa by reactive oxygen species. *Jpn. J. Pharmacol.* **2002**, *88*(1), 114–118.
420. Coant, N., Ben, M.S., Pedruzzi, E., Guichard, C., Treton, X., Ducroc, R., Freund, J.N., Cazals-Hatem, D., Bouhnik, Y., Woerther, P.L., Skurnik, D., Grodet, A., Fay, M., Biard, D., Lesuffleur, T., Deffert, C., Moreau, R., Groyer, A., Krause, K.H., Daniel, F., Ogier-Denis, E. NADPH oxidase 1 modulates WNT and NOTCH1 signaling to control the fate of proliferative progenitor cells in the colon. *Mol. Cell. Biol.* **2010**, *30*(11), 2636–2650.
421. Kawahara, T., Kohjima, M., Kuwano, Y., Mino, H., Teshima-Kondo, S., Takeya, R., Tsunawaki, S., Wada, A., Sumimoto, H., Rokutan, K. *Helicobacter pylori* lipopolysaccharide activates Rac1 and transcription of NADPH oxidase Nox1 and its organizer NOXO1 in guinea pig gastric mucosal cells. *Am. J. Physiol. Cell Physiol.* **2005**, *288*(2), C450–C457.
422. Kawahara, T., Kuwano, Y., Teshima-Kondo, S., Takeya, R., Sumimoto, H., Kishi, K., Tsunawaki, S., Hirayama, T., Rokutan, K. Role of nicotinamide adenine dinucleotide phosphate oxidase 1 in oxidative burst response to Toll-like receptor 5 signaling in large intestinal epithelial cells. *J. Immunol.* **2004**, *172*(5), 3051–3058.
423. Gewirtz, A.T., Navas, T.A., Lyons, S., Godowski, P.J., Madara, J.L. Cutting edge: Bacterial flagellin activates basolaterally expressed TLR5 to induce epithelial proinflammatory gene expression. *J. Immunol.* **2001**, *167*(4), 1882–1885.
424. Amieva, M.R., El-Omar, E.M. Host-bacterial interactions in *Helicobacter pylori* infection. *Gastroenterology* **2008**, *134*(1), 306–323.
425. Scirocco, A., Matarrese, P., Petitta, C., Cicenia, A., Ascione, B., Mannironi, C., Ammoscato, F., Cardi, M., Fanello, G., Guarino, M.P., Malorni, W., Severi, C. Exposure of Toll-like receptors 4 to bacterial lipopolysaccharide (LPS) impairs human colonic smooth muscle cell function. *J. Cell. Physiol.* **2010**, *223*(2), 442–450.
426. Kuwano, Y., Kawahara, T., Yamamoto, H., Teshima-Kondo, S., Tominaga, K., Masuda, K., Kishi, K., Morita, K., Rokutan, K. Interferon-γ activates transcription of NADPH oxidase 1 gene and upregulates production of superoxide anion by human large intestinal epithelial cells. *Am. J. Physiol. Cell Physiol.* **2006**, *290*(2), C433–C443.
427. El Hassani, R.A., Benfares, N., Caillou, B., Talbot, M., Sabourin, J.C., Belotte, V., Morand, S., Gnidehou, D., Agnandji, D., Ohayon, R., Kaniewski, J., Noel-Hudson, M.S., Bidart, J.M., Schlumberger, M., Virion, A., Dupuy, C. Dual oxidase2 is expressed all along the digestive tract. *Am. J. Physiol. Gastrointest. Liver Physiol.* **2005**, *288*(5), G933–G942.
428. Ha, E.M., Oh, C.T., Bae, Y.S., Lee, W.J. A direct role for dual oxidase in *Drosophila* gut immunity. *Science* **2005**, *310*(5749), 847–850.
429. Allaoui, A., Botteaux, A., Dumont, J.E., Hoste, C., De Deken, X. Dual oxidases and hydrogen peroxide in a complex dialogue between host mucosae and bacteria. *Trends Mol. Med.* **2009**, *15*(12), 571–579.
430. Bae, Y.S., Choi, M.K., Lee, W.J. Dual oxidase in mucosal immunity and host-microbe homeostasis. *Trends Immunol.* **2010**, *31*(7), 278–287.
431. Flores, M.V., Crawford, K.C., Pullin, L.M., Hall, C.J., Crosier, K.E., Crosier, P.S. Dual oxidase in the intestinal epithelium of zebrafish larvae has anti-bacterial properties. *Biochem. Biophys. Res. Commun.* **2010**, *400*(1), 164–168.
432. Laroux, F.S., Pavlick, K.P., Wolf, R.E., Grisham, M.B. Dysregulation of intestinal mucosal immunity: Implications in inflammatory bowel disease. *News Physiol. Sci.* **2001**, *16*, 272–277.
433. Asquith, M., Powrie, F. An innately dangerous balancing act: Intestinal homeostasis, inflammation, and colitis-associated cancer. *J. Exp. Med.* **2010**, *207*(8), 1573–1577.
434. Tanaka, J., Yuda, Y., Yamakawa, T. Mechanism of superoxide generation system in indomethacin-induced gastric mucosal injury in rats. *Biol. Pharm. Bull.* **2001**, *24*(2), 155–158.
435. Hausmann, M., Spöttl, T., Andus, T., Rothe, G., Falk, W., Schölmerich, J., Herfarth, H., Rogler, G. Subtractive screening reveals up-regulation of NADPH oxidase expression in Crohn's disease intestinal macrophages. *Clin. Exp. Immunol.* **2001**, *125*(1), 48–55.
436. Blanchard, T.G., Yu, F., Hsieh, C.L., Redline, R.W. Severe inflammation and reduced bacteria load in murine *Helicobacter* infection caused by lack of phagocyte oxidase activity. *J. Infect. Dis.* **2003**, *187*(10), 1609–1615.
437. Krieglstein, C.F., Cerwinka, W.H., Laroux, F.S., Salter, J.W., Russell, J.M., Schuermann, G., Grisham, M.B., Ross, C.R., Granger, D.N. Regulation of murine intestinal inflammation by reactive metabolites of oxygen and nitrogen: Divergent roles of superoxide and nitric oxide. *J. Exp. Med.* **2001**, *194*(9), 1207–1218.
438. Marks, D.J., Harbord, M.W., MacAllister, R., Rahman, F.Z., Young, J., Al-Lazikani, B., Lees, W., Novelli, M., Bloom, S., Segal, A.W. Defective acute inflammation in Crohn's disease: A clinical investigation. *Lancet* **2006**, *367*(9511), 668–678.
439. Curran, F.T., Allan, R.N., Keighley, M.R. Superoxide production by Crohn's disease neutrophils. *Gut* **1991**, *32*(4), 399–402.
440. Solis-Herruzo, J.A., Fernandez, B., Vilalta-Castell, E., Muñoz-Yagüe, M.T., Hernandez-Muñoz, I., De la Torre-Merino, M.P., Balsinde, J. Diminished cytochrome *b* content and toxic oxygen metabolite production in circulating neutrophils from patients with Crohn's disease. *Dig. Dis. Sci.* **1993**, *38*, 1631–1637.
441. Salles, N., Szanto, I., Herrmann, F., Armenian, B., Stumm, M., Stauffer, E., Michel, J.P., Krause, K.H. Expression of

mRNA for ROS-generating NADPH oxidases in the aging stomach. *Exp. Gerontol.* **2005**, *40*(4), 353–357.

442. Guichard, C., Moreau, R., Pessayre, D., Epperson, T.K., Krause, K.H. NOX family NADPH oxidases in liver and in pancreatic islets: A role in the metabolic syndrome and diabetes? *Biochem. Soc. Trans.* **2008**, *36*(Pt 5), 920–929.

443. Kikuchi, H., Hikage, M., Miyashita, H., Fukumoto, M. NADPH oxidase subunit, gp91^{phox} homologue, preferentially expressed in human colon epithelial cells. *Gene* **2000**, *254*(1–2), 237–243.

444. Reinehr, R., Becker, S., Eberle, A., Grether-Beck, S., Haussinger, D. Involvement of NADPH oxidase isoforms and Src family kinases in CD95-dependent hepatocyte apoptosis. *J. Biol. Chem.* **2005**, *280*(29), 27179–27194.

445. Lee, Y.S., Kang, Y.S., Lee, J.S., Nicolova, S., Kim, J.A. Involvement of NADPH oxidase-mediated generation of reactive oxygen species in the apototic cell death by capsaicin in HepG2 human hepatoma cells. *Free Radic. Res.* **2004**, *38*(4), 405–412.

446. Reinehr, R., Becker, S., Keitel, V., Eberle, A., Grether-Beck, S., Haussinger, D. Bile salt-induced apoptosis involves NADPH oxidase isoform activation. *Gastroenterology* **2005**, *129*(6), 2009–2031.

447. Herrera, B., Murillo, M.M., Alvarez-Barrientos, A., Beltran, J., Fernandez, M., Fabregat, I. Source of early reactive oxygen species in the apoptosis induced by transforming growth factor-β in fetal rat hepatocytes. *Free Radic. Biol. Med.* **2004**, *36*(1), 16–26.

448. Herrera, B., Alvarez, A.M., Beltran, J., Valdes, F., Fabregat, I., Fernandez, M. Resistance to TGF-β-induced apoptosis in regenerating hepatocytes. *J. Cell. Physiol.* **2004**, *201*(3), 385–392.

449. Parker, G.A., Picut, C.A. Liver immunobiology. *Toxicol. Pathol.* **2005**, *33*(1), 52–62.

450. Kono, H., Rusyn, I., Yin, M., Gabele, E., Yamashina, S., Dikalova, A., Kadiiska, M.B., Connor, H.D., Mason, R.P., Segal, B.H., Bradford, B.U., Holland, S.M., Thurman, R.G. NADPH oxidase-derived free radicals are key oxidants in alcohol-induced liver disease. *J. Clin. Invest.* **2000**, *106*(7), 867–872.

451. Bhatnagar, R., Schirmer, R., Ernst, M., Decker, K. Superoxide release by zymosan-stimulated rat Kupffer cells *in vitro*. *Eur. J. Biochem.* **1981**, *119*(1), 171–175.

452. Rusyn, I., Yamashina, S., Segal, B.H., Schoonhoven, R., Holland, S.M., Cattley, R.C., Swenberg, J.A., Thurman, R.G. Oxidants from nicotinamide adenine dinucleotide phosphate oxidase are involved in triggering cell proliferation in the liver due to peroxisome proliferators. *Cancer Res.* **2000**, *60*(17), 4798–4803.

453. LaCourse, R., Ryan, L., North, R.J. Expression of NADPH oxidase-dependent resistance to listeriosis in mice occurs during the first 6 to 12 hours of liver infection. *Infect. Immun.* **2002**, *70*(12), 7179–7181.

454. Senoo, H. Structure and function of hepatic stellate cells. *Med. Electron Microsc.* **2004**, *37*(1), 3–15.

455. Bataller, R., Schwabe, R.F., Choi, Y.H., Yang, L., Paik, Y.H., Lindquist, J., Qian, T., Schoonhoven, R., Hagedorn, C.H., Lemasters, J.J., Brenner, D.A. NADPH oxidase signal transduces angiotensin II in hepatic stellate cells and is critical in hepatic fibrosis. *J. Clin. Invest.* **2003**, *112*(9), 1383–1394.

456. Adachi, T., Togashi, H., Suzuki, A., Kasai, S., Ito, J., Sugahara, K., Kawata, S. NAD(P)H oxidase plays a crucial role in PDGF-induced proliferation of hepatic stellate cells. *Hepatology* **2005**, *41*(6), 1272–1281.

457. De Minicis, S., Seki, E., Oesterreicher, C., Schnabl, B., Schwabe, R.F., Brenner, D.A. Reduced nicotinamide adenine dinucleotide phosphate oxidase mediates fibrotic and inflammatory effects of leptin on hepatic stellate cells. *Hepatology* **2008**, *48*(6), 2016–2026.

458. De Minicis, S., Brenner, D.A. Oxidative stress in alcoholic liver disease: Role of NADPH oxidase complex. *J. Gastroenterol. Hepatol.* **2008**, *23*(Suppl. 1), S98–S103.

459. Muriel, P. Role of free radicals in liver diseases. *Hepatol. Int.* **2009**, *3*(4), 526–536.

460. Teufelhofer, O., Parzefall, W., Kainzbauer, E., Ferk, F., Freiler, C., Knasmüller, S., Elbling, L., Thurman, R., Schulte-Hermann, R. Superoxide generation from Kupffer cells contributes to hepatocarcinogenesis: Studies on NADPH oxidase knockout mice. *Carcinogenesis* **2005**, *26*(2), 319–329.

461. Calvisi, D.F., Ladu, S., Hironaka, K., Factor, V.M., Thorgeirsson, S.S. Vitamin E down-modulates iNOS and NADPH oxidase in c-Myc/TGF-alpha transgenic mouse model of liver cancer. *J. Hepatol.* **2004**, *41*(5), 815–822.

462. Laurent, A., Nicco, C., Chereau, C., Goulvestre, C., Alexandre, J., Alves, A., Levy, E., Goldwasser, F., Panis, Y., Soubrane, O., Weill, B., Batteux, F. Controlling tumor growth by modulating endogenous production of reactive oxygen species. *Cancer Res.* **2005**, *65*(3), 948–956.

463. Virion, A., Michot, J.L., Deme, D., Kaniewski, J., Pommier, J. NADPH-dependent H_2O_2 generation and peroxidase activity in thyroid particular fraction. *Mol. Cell. Endocrinol.* **1984**, *36*(1–2), 95–105.

464. Ohye, H., Sugawara, M. Dual oxidase, hydrogen peroxide and thyroid diseases. *Exp. Biol. Med. (Maywood)* **2010**, *235*(4), 424–433.

465. Fortunato, R.S., Lima de Souza, E.C., Ameziane-el Hassani, R., Boufraqech, M., Weyemi, U., Talbot, M., Lagente-Chevallier, O., Pires de Carvalho, D., Bidart, J.M., Schlumberger, M., Dupuy, C. Functional consequences of dual oxidase-thyroperoxidase interaction at the plasma membrane. *J. Clin. Endocrinol. Metab* **2010**, *95*(12), 5403–5411.

466. Moreno, J.C., Bikker, H., Kempers, M.J., van Trotsenburg, A.S., Baas, F., de Vijlder, J.J., Vulsma, T., Ris-Stalpers, C. Inactivating mutations in the gene for thyroid oxidase 2 (THOX2) and congenital hypothyroidism. *N. Engl. J. Med.* **2002**, *347*(2), 95–102.

467. Park, S.M., Chatterjee, V.K. Genetics of congenital hypothyroidism. *J. Med. Genet.* **2005**, *42*(5), 379–389.

468. Grasberger, H., De Deken, X., Miot, F., Pohlenz, J., Refetoff, S. Missense mutations of dual oxidase 2 (DUOX2) implicated in congenital hypothyroidism have impaired trafficking in cells reconstituted with DUOX2 maturation factor. *Mol. Endocrinol.* **2007**, *21*(6), 1408–1421.

469. Zamproni, I., Grasberger, H., Cortinovis, F., Vigone, M.C., Chiumello, G., Mora, S., Onigata, K., Fugazzola, L., Refetoff, S., Persani, L., Weber, G. Biallelic inactivation of the dual oxidase maturation factor 2 (DUOXA2) gene as a novel cause of congenital hypothyroidism. *J. Clin. Endocrinol. Metab.* **2008**, *93*(2), 605–610.

470. Maruo, Y., Takahashi, H., Soeda, I., Nishikura, N., Matsui, K., Ota, Y., Mimura, Y., Mori, A., Sato, H., Takeuchi, Y. Transient congenital hypothyroidism caused by biallelic mutations of the dual oxidase 2 gene in Japanese patients detected by a neonatal screening program. *J. Clin. Endocrinol. Metab.* **2008**, *93*(11), 4261–4267.

471. Finkel, T., Holbrook, N.J. Oxidants, oxidative stress and the biology of ageing. *Nature* **2000**, *408*(6809), 239–247.

472. Roos, D., Kuhns, D.B., Maddalena, A., Roesler, J., Lopez, J.A., Ariga, T., Avcin, T., de Boer, M., Bustamante, J., Condino-Neto, A., Di Matteo, G., He, J., Hill, H.R., Holland, S.M., Kannengiesser, C., Koker, M.Y., Kondratenko, I., van Leeuwen, K., Malech, H.L., Marodi, L., Nunoi, H., Stasia, M.J., Ventura, A.M., Witwer, C.T., Wolach, B., Gallin, J.I. Hematologically important mutations: X-linked chronic granulomatous disease (third update). *Blood Cells Mol. Dis.* **2010**, *45*(3), 246–265.

473. Roos, D., Kuhns, D.B., Maddalena, A., Bustamante, J., Kannengiesser, C., de Boer, M., van Leeuwen, K., Koker, M.Y., Wolach, B., Roesler, J., Malech, H.L., Holland, S.M., Gallin, J.I., Stasia, M.J. Hematologically important mutations: The autosomal recessive forms of chronic granulomatous disease (second update). *Blood Cells Mol. Dis.* **2010**, *44*(4), 291–299.

8

CELL SIGNALING AND TRANSCRIPTION

IMRAN REHMANI, FANGE LIU, AND AIMIN LIU

OVERVIEW

Cell signaling and transcription is a tightly regulated process, integrating the activities of multiple interlinked pathways to respond in a precise manner to changes in the cellular environment. Many pathological conditions can be traced to defects in one or more of these regulatory mechanisms, which illustrates the requirement for exactness in which, when, and how genes are expressed. Considering this, it may seem unusual to consider the role of reactive oxygen species (ROS) in cell signaling. ROS were initially characterized as the undesirable by-products of aerobic metabolism and earned a reputation for being destructive molecules that necessitated the existence of antioxidant enzymes to keep them from accumulating. How could these molecules, which literally have "reactive" in their name, have a role in a process that requires tight regulation and specificity?

As decades of surprising research has revealed, ROS are an indispensable aspect of cell signaling and are mediators of cross talk between different pathways. While they are reactive and capable of damaging biological molecules, their chemical properties give them a preference for specific molecules and structural features with which they react. The conservation and recurrence of these redox-sensitive structural features throughout nature has made ROS relevant to practically every level of eukaryotic cell signaling, from the mediation of growth factor receptor activity on the cell's plasma membrane to the regulation of transcription factors in the nucleus. ROS have come a long way from their initial characterization as a threat to life and will likely continue to be involved at the forefront of medical and biochemical research for many years to come.

This chapter highlights some of the best-characterized mechanisms of ROS and oxygen signaling, especially those metal-dependent, in both prokaryotes and eukaryotes. We will start by briefly introducing the reactive properties of ROS and the most common structural features seen in redox-sensitive proteins. The types of simple redox and oxygen sensors found in bacteria will be reviewed with emphasis on a recently characterized transcription factor with a novel metal-catalysis mechanism, followed by a more in-depth discussion of the signaling systems found in metazoans (multicellular eukaryotes lacking a cell wall). Finally, the roles of redox and oxygen signaling in our understanding of disease will be highlighted.

8.1 COMMON MECHANISMS OF REDOX SIGNALING

Superoxide ($O_2^{\bullet-}$), the product of a one-electron reduction of oxygen, constitutes a majority of the ROS produced both enzymatically or as a by-product of aerobic metabolism. However, while $O_2^{\bullet-}$ has been demonstrated to be capable of participating in redox signaling, its physiological relevance has been vastly overshadowed by its dismutation product, hydrogen peroxide (H_2O_2). The rate that $O_2^{\bullet-}$ reacts with most molecules does not compare to the diffusion-limited rate of its

Molecular Basis of Oxidative Stress: Chemistry, Mechanisms, and Disease Pathogenesis, First Edition. Edited by Frederick A. Villamena.
© 2013 John Wiley & Sons, Inc. Published 2013 by John Wiley & Sons, Inc.

dismutation by superoxide dismutase (SOD).[1] Thus, any direct modification of a biological molecule by $O_2^{\bullet-}$ has to occur in proximity to the site of its production. The mechanisms of $O_2^{\bullet-}$ participating in signaling are not as well characterized as those for H_2O_2, but the ability of $O_2^{\bullet-}$ to act as a nucleophile allows it to participate in reactions that would be much less likely with H_2O_2.[2]

Unlike the negatively charged $O_2^{\bullet-}$ molecule, the lack of a charge on an H_2O_2 molecule allows it to diffuse through the plasma membranes of cells and organelles with relative ease. H_2O_2 also has a much longer half-life than $O_2^{\bullet-}$ and is less reactive due to its requirement that its dioxygen bond be broken before it can accept electrons.[3] Both ROS can be rapidly generated by eukaryotic cells and quickly removed by antioxidant enzymes when it is no longer needed. In addition, the oxidative products of ROS signaling are often reversible,[4] allowing their target proteins to be switched on and off depending on intracellular conditions. Even though ROS can have deleterious effects on cells and is not directly recognized by any protein, these reactive properties make them effective cellular messengers.

While H_2O_2 has the potential to be a strong oxidant, its activation energy for most of its reactions is too high to be a threat to most molecules.[5] However, transition metals and cysteines in protein environments that lower the pK_a of their thiol side chain both have characteristics that make them prime targets for H_2O_2 oxidation.

Transition metals are reactive with both oxygen and ROS. Proteins containing one or more metal centers, or metalloproteins, have long been associated with the biological roles of oxygen and ROS. Catalase and SOD, two of the most prominent antioxidant enzymes that scavenge hydrogen peroxide and $O_2^{\bullet-}$, respectively, both use bound transition metals to interact with their ROS targets. There are also metalloproteins that interact with or serve as transcription factors. These proteins, through the reactivity and environment of their metal center, are sensitive to changes in intracellular concentrations of oxygen or ROS and alter transcription accordingly.

The properties of metalloproteins that make them effective catalysts in reactions involving oxygen or ROS also make them appealing targets for ROS inflicting oxidative damage. Proteins containing iron–sulfur (Fe–S) clusters, which serve as the redox-sensitive component of multiple prokaryotic transcription factors, are vulnerable to oxidation by $O_2^{\bullet-}$ due to the electrostatic attraction from the prosthetic group's positive charge.[3] While this is a desirable feature in a protein meant to respond to changing redox conditions, it is also responsible for loss of function of the citric acid cycle enzymes containing solvent-exposed Fe–S clusters during oxidative stress.[6]

Along with bound transition metals, cysteine residues are among the most common ways for a protein to interact with H_2O_2. Hydrogen peroxide has a tendency to react with the thiol group of cysteine residues, modifying the protein's structure or function through oxidation of the cysteine to a sulfenic acid (S–OH) that can form a disulfide bond with a neighboring cysteine residue on itself or another subunit or protein. The formation of one or more disulfide bonds can change the conformation of a protein or, if the cysteine is part of an enzyme's active site, lead to enzyme inactivation. The reduced/oxidized ratio of the reactive thiols in these proteins, and therefore their activity, is determined by the redox environment of the cell.[5]

Normally, the reaction between H_2O_2 and an exposed thiol group is too slow for it to be significant unless the thiol is deprotonated, forming a thiolate anion. The pK_a of the thiol side chain of cysteine is around 8.3. At physiological pH it is unlikely to lose its proton. However, the pK_a of an amino acid side chain can be dramatically influenced by its protein environment. Thiol groups with shifted pK_a in protein favor the presence of a thiolate anion and have reaction rates with H_2O_2 that are magnitudes higher than what they would be with a normal cysteine.[5]

Sulfenic acid formation and disulfide bonds are easily reduced back to their original form by glutaredoxins or thioredoxins.[7] However, in cases where the levels of $[H_2O_2]$ are particularly high and a sulfenic acid has a chance to be oxidized further before it forms a bond with a nearby cysteine, it forms a sulfinic acid (S–O_2H) or a sulfonic acid (S–O_3H) that are essentially irreversible (Fig. 8.1).

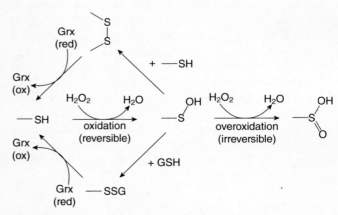

Figure 8.1 Redox modification of protein thiols and disulfide bonds by reactions with H_2O_2, GSH, and glutaredoxin (Grx) enzymes.

Sulfenic acids formed from peroxide oxidation can also bond covalently to other molecules containing thiol groups, forming a mixed disulfide. The modification of an oxidized thiol with glutathione, a process called glutathiolation, is a common reversible posttranslational modification observed in cells experiencing oxidative stress.[8] Like disulfide bond formation, glutathiolation can affect the structure or function of a protein and acts as another redox-regulated switch.

8.2 REDOX AND OXYGEN-SENSITIVE TRANSCRIPTION FACTORS IN PROKARYOTES

The formation of ROS is believed to be responsible for the toxicity of oxygen. Organisms living in aerobic environments have to utilize oxygen, and ROS are inevitably produced in the process of metabolism. The remarkable catalytic efficiency (up to 4×10^7 s^{-1}) and near-ubiquitous nature of the antioxidant enzymes catalase and SOD among aerobic organisms show how crucial the removal of ROS is for survival in aerobic environments.[9] These antioxidant enzymes are constitutively expressed in both prokaryotes and eukaryotes to maintain sufficiently low concentrations of endogenously produced ROS.

In addition to the ROS produced as a by-product of aerobic metabolism, prokaryotes are subject to oxidative stress from external sources to a much higher degree than metazoans.[10] ROS are commonly used by other organisms as a defense mechanism against bacterial infection. These defenses can come in the form of a direct release of ROS or indirectly through compounds that cause the bacteria to generate lethal amounts of ROS. The fluoroquinolone class of antibiotics has been seen to have this effect.[11] Of course, as with most antimicrobial strategies nature and technology have developed, bacteria have evolved mechanisms to help withstand these oxidative attacks and adapt to transient increases in ROS. The constitutive defenses against ROS found in prokaryotes are supplemented with inducible responses to an overexposure to ROS due to exogenous sources. These inducible responses are triggered by redox-sensitive transcription factors containing either metal ions or reactive cysteines. Since the properties of each ROS give them a preference for reacting with certain structural features or protein environments, ROS-sensitive transcription factors that contain these structural features are likely to be oxidized before other proteins. This helps ensure that they are among the first molecules in the cell to react with the ROS, making them effective early sensors of oxidative stress. In addition, it provides the specificity required for a functional signaling pathway and ensures that the correct genes are activated in response to a specific ROS rather than the general condition of oxidative stress.

8.2.1 Fe–S Cluster Proteins

Fe–S are a ubiquitous and ancient prosthetic group used for a variety of purposes, including electron transfer, catalysis, and regulation of gene transcription. Tetrahedral iron ions are typically bound by cysteine residues or inorganic sulfur and can be arranged into a variety of structures with varying amounts of iron and sulfur, contributing to their functional versatility.[12] Their redox properties are heavily influenced by their protein environment, allowing for a wide range of redox potentials even between proteins that have the same Fe–S cluster shape.[12] This unique property allows Fe–S proteins to have a redox potential fine-tuned for their function, which could potentially be shifted by conformation changes.

Facultative anaerobes, organisms capable of surviving in both aerobic and anaerobic environments, use the fumarate and nitrate reduction (FNR) transcription factor to determine whether or not oxygen is present in their environment.[13] Many of the gene products needed for anaerobic survival, such as those that utilize alternative terminal electron acceptors in lieu of oxygen, would be useless in aerobic conditions and need to be switched off to conserve resources. FNR is a homodimer with each subunit containing a [4Fe-4S] cluster that it requires for dimerization. When in its dimerized form, it activates genes needed for anaerobic survival.[14] However, when exposed to oxygen, the [4Fe-4S] cluster is converted to a [2Fe-2S] cluster that produces a conformation change, making the transcription factor unable to dimerize or bind DNA.[15] Inactive apo-FNR can be regenerated back to its active form by taking up a new [4Fe-4S] cluster.[16] However, the rate of degradation of the [4Fe-4S] clusters in the presence of oxygen outpaces their uptake, and the apo form of FNR dominates in aerobic conditions.[17]

SoxR is another example of an Fe–S-containing transcription factor that senses elevated concentrations of $O_2^{\bullet-}$ or nitric oxide (NO) and initiates the organism's antioxidant defenses in response to the oxidation of its [2Fe-2S] cluster.[18] SoxR does not directly activate the transcription of antioxidant genes. Instead, it regulates the expression of the *soxS* gene producing the protein of the same name, which goes onward to induce its target genes in response to the oxidative attack.[19]

Both the reduced and oxidized form of SoxR bind DNA with similar affinity.[20] However, when in its reduced form ([2Fe-2S]$^{1+}$), it is unable to promote transcription of SoxS.[21] Upon oxidation of its Fe–S cluster to [2Fe-2S]$^{2+}$, the increased negative charge of the cluster causes electrostatic repulsion that alters the conformation of the protein into its active form.[22]

Although the link between SoxR activation and elevated $O_2^{\bullet-}$ levels have been demonstrated using multiple $O_2^{\bullet-}$ generating compounds,[23] there is also evidence suggesting that SoxR might not be a direct sensor of $O_2^{\bullet-}$ or even required for activation at all.[24] SoxR activity in *Pseudomonas aeruginosa* has been successfully induced in laboratory using a redox-active compound in anaerobic conditions where $O_2^{\bullet-}$ formation would be impossible.[25] Also, it has been observed that the [2Fe-2S] cluster in FNR is destroyed by $O_2^{\bullet-}$,[26] suggesting that direct contact between the SoxR [2Fe-2S] cluster and $O_2^{\bullet-}$ could render the protein inactive. An alternative explanation is that guanine radicals, an early marker of oxidative stress, are the activators of SoxR activity.[24]

8.2.2 Prokaryotic Hydrogen Peroxide Sensors: Proteins Utilizing Reactive Thiols

In spite of its lower reactivity than other ROS, the toxicity and mutagenic effects of H_2O_2 to cells are well known. Catalase, one of the primary scavengers of H_2O_2, is almost ubiquitous among aerobic organisms. Catalase-deficient mutants of bacteria have a decreased survival rate unless they are placed in an anaerobic environment.[27] The peroxiredoxin family of proteins, one of the most highly expressed proteins in cells, also assists in removing H_2O_2.

Many of the observed toxic effects of H_2O_2 are not a result of what the molecule is, but rather what it is capable of becoming. H_2O_2 readily reacts with ferrous iron and undergoes what is referred to as a Fenton reaction that produces a hydroxide ion (OH^-) and the much more reactive and damaging hydroxyl radical ($HO\bullet$). Once produced, $HO\bullet$ reacts indiscriminately with surrounding molecules at a diffusion-limited rate, causing oxidative damage at the site of its production. The observed mutagenic effects of H_2O_2 are caused by Fenton reactions with metal ions associated with DNA, causing oxidative lesions.[28] Given the difficulty of intercepting an $HO\bullet$ radical before it reacts, detecting and scavenging excess H_2O_2 and limiting the levels of free iron that could potentially participate in a Fenton reaction is a much more effective strategy for avoiding damage from $HO\bullet$ than removing it directly. Sensor proteins containing reactive cysteine residues that are susceptible to H_2O_2 help accomplish this task.

The prokaryotic transcription factor OxyR takes advantage of this mechanism by using reactive cysteines that are extremely sensitive to changing levels of H_2O_2.[18] OxyR, similar to SoxR, exists in either an oxidized or reduced form where only the oxidized form is capable of activating transcription. Upon oxidation by H_2O_2, the Cys199 residue of OxyR becomes a sulfenic acid that forms a disulfide bond with Cys208 and converts the protein to its active form.[29] Once active, it binds the promoter regions of over 20 genes, including *ahpCF* (alkyl hydroperoxide reductase), *katG* (catalase), *dps* (an iron-binding protein), *fur* (ferric uptake regulator protein), *grxA* (glutaredoxin), and *trxC* (thioredoxin).[30] This occurs through direct interaction with RNA polymerase α-subunits.[31] The oxidation and disulfide bond formation by Cys199 is reversible by glutaredoxin, inactivating the protein and creating a negative feedback mechanism where OxyR can shut itself off with its own gene products upon reestablishing normal level of H_2O_2 concentrations.[29]

It is important to note the types of genes induced by OxyR activation in response to H_2O_2 detection. The logic behind the upregulation of AhpCF and catalase, both antioxidant enzymes, during oxidative stress is self-explanatory. Glutaredoxin and thioredoxin reduce the disulfide bonds that form from H_2O_2 oxidation. However, Fur and Dps have no reducing or antioxidant action. Instead, they regulate or sequester iron. Because the formation of $HO\bullet$ radicals from the interaction of H_2O_2 and iron mediates much of the toxicity of both substances, linking the regulation of free intracellular iron concentrations to H_2O_2 concentrations is an effective strategy that ameliorates the damage caused by oxidative stress from both fronts.

8.2.3 PerR: A Unique Metalloprotein Sensor of Hydrogen Peroxide

PerR, a metalloprotein found in the gram-positive *Bacillus subtilis*, has attracted attention due to its unique mechanism for sensing H_2O_2 that uses a bound metal ion.[32] Named after the regulon it regulates, PerR is a metal-dependent homodimeric protein with two metal-binding sites per subunit. Rather than being an activator of transcription like OxyR, it instead acts as a repressor that is inactivated by oxidation. In its reduced and active conformation, PerR is capable of binding to specific DNA sequences and repressing transcription of genes responsible for defenses against H_2O_2 stress.[33]

The PerR regulon contains the *katA*, *ahpCF*, and *mrgA* genes, which are responsible for the production of catalase, alkylhydroperoxide reductase, and Dps, respectively. The same genes are found within the OxyR regulon in other bacteria. OxyR and PerR have a significant amount of crossover between the genes they regulate, indicating that the response to H_2O_2 stress in bacteria has been conserved even as the proteins regulating them have not. PerR loses its ability to bind DNA upon oxidation by H_2O_2 and releases its repression of its regulon. The resulting upregulation of H_2O_2 scavenging enzymes along with proteins that regulate intracellular concentrations of iron mimics the damage control strategy regulated by other H_2O_2 sensors.

Although the functions and gene targets of OxyR and PerR are very similar, they belong to different protein families and have different reaction mechanisms. While OxyR interacts with H_2O_2 through reactive cysteines that form disulfide bonds upon oxidation and alter the protein's structure and ability to bind DNA, PerR participates in a Fenton-type reaction with H_2O_2 through the iron(II) ion bound to its active site.[32] The result of this reaction is the oxidation of histidine residues within its binding site that drastically alters the protein's conformation.[34]

PerR is a member of the Fur family, a family of transcription factors that serve as regulators of intracellular iron concentrations. Notably, Fur is regulated by both OxyR and PerR. PerR shares structural homology with other Fur proteins such as its homodimer structure, N-terminal DNA-binding domains and C-terminal dimerization domains, structural Zn(II) binding site containing two conserved CXXC motifs, and similar DNA binding sequences.[35]

The active site of PerR is capable of binding either an Mn(II) or Fe(II) ion with five protein ligands (His37, His91, His93, Asp85, and Asp104) that are all required for the protein's function. Mn(II) and Fe(II), having similar sizes and coordination geometries, compete for the binding site. Other divalent metals have been seen to bind, but only when their concentrations exceed that of iron and manganese, making them less relevant in physiological conditions.[33] Another site, located in the C-terminal domain responsible for the dimerization of the subunits, binds a zinc ion using four cysteine residues in a tetrahedral formation.[36] This zinc ion and its four cysteine ligands, while essential for the stability of the dimer, has not been demonstrated to play a direct role in the detection of ROS or its transcriptional activity. Thus, it will not be discussed further.

As mentioned previously, PerR uses its bound Fe^{2+} ion to produce an HO• radical via Fenton chemistry. This radical oxidizes one of two histidine residues in its binding site, His37 or His91, forming a 2-oxo-histidine product that has been shown using both mass spectrometry and crystallography. Unlike proteins that use cysteine residues to respond to H_2O_2 which are often enzymatically reduced and recycled back to their original form, the oxidation of PerR and its dissociation from DNA is irreversible, and the protein is degraded after a single use. Interestingly, PerR regulates its own production. This results in an increased production of PerR during peroxide stress that restores normal levels of gene repression without requiring a separate enzyme to recycle PerR to its active form.

Well before PerR was identified and characterized as a metalloprotein transcription factor, it was established that both metals were required for the repression of the PerR regulon.[35] Mn(II), previously seen to have a protective effect against oxidative stress in bacteria, had the unexpected effect of potentiating the damage caused in B. subtilis by high H_2O_2 concentrations. This phenomenon was later explained by its ability to substitute for iron in the regulatory site of PerR and its inability to participate in Fenton reactions, rendering the protein insensitive to increasing levels of H_2O_2. The competition between Fe(II) and Mn(II) for the active site of PerR and the differences in H_2O_2 sensitivity between the two ions gives PerR a secondary role as an iron sensor, similar to its Fur family members. Higher concentration of Fe(II) would result in it occupying a higher percentage of active sites, making them more sensitive to H_2O_2 levels. Since the toxicity of H_2O_2 is linked to its interaction with Fe(II), it would follow that the sensitivity of an effective H_2O_2 sensor would be linked to the intracellular concentration of Fe(II).

Like other Fur proteins, PerR's function as a transcriptional repressor is metal-dependent. PerR requires both its DNA-binding domains to bind DNA, requiring that the two domains assume a caliper-like conformation in its active form when the conserved His37 residue is bound to the metal ion found in the regulatory site (Fig. 8.2).[34] Unlike the other four ligands of the regulatory metal (His91, His93, Asp85, and Asp104), His37 resides in the N-terminal region and pulls the DNA-binding domains into their active conformation when it is bound to the metal.

When oxidized by the HO• radical produced by the reaction of H_2O_2 with the Fe(II) ion, His37 or His91 is irreversibly modified to 2-oxo-histidine.[32] 2-Oxo-histidine is unable to bind to the metal, causing the DNA-binding domains to take a more planar conformation that is unable to bind DNA.[36] Both the oxidized protein and the apo-protein have the inactive planar conformation, demonstrating that the metal ion bound to His37 is required for the active conformation.[34,36] The other possibility, oxidation of His91, drastically reduces the regulatory site's affinity for metal which results in the inactive conformation as His37 no longer has metal with which to bind.[34] Thus, while the two possible oxidation products of the HO• radical produced by the Fenton reaction have different consequences, the end result of inactivation is the same. While similar mechanisms using metal-catalyzed oxidation have been described in the inactivation of iron-containing SOD and NADP-isocitrate dehydrogenase,[37,38] this is the first known protein to utilize what is normally considered a

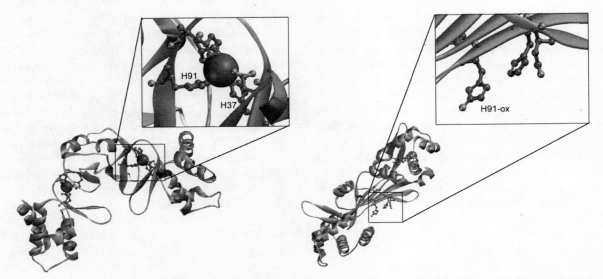

Figure 8.2 The structure of the regulatory metal site in PerR (left) and 2-oxo-histidine in oxidized PerR protein (right) (PDB codes: 3F8N and 2RGV, respectively). See color insert.

harmful and unpredictable molecule as a method of cell signaling.

The iron(II) or manganese(II) bound to the regulatory site is held by its three histidine residues and two aspartate residues in a distorted square pyramidal formation, with His37 and His91 forming the base with the two aspartate residues, and His97 binding at the apical position toward the interior of the protein.[39] The other apical position facing the outside of the protein is available for H_2O_2 to interact with the metal, forming the HO• radical that oxidizes His37 or His91 causing a change in conformation. His97, which has not been seen to form a 2-oxo-histidine product upon oxidation by H_2O_2, is thought to be protected from oxidation by its apical position on the metal opposite of the available site where H_2O_2 interacts and the HO• radical is produced.[39]

Because each ROS requires a different set of defense mechanisms and the levels of one type are not necessarily linked to the levels of others during oxidative stress, each ROS sensor should be specific for what it is supposed to detect. The peroxide response of *B. subtilis* is regulated by multiple transcription factors with different characteristics, thus allowing the differentiation between H_2O_2 and organic peroxides.[40] PerR accomplishes its specificity for H_2O_2 by having a hydrophilic environment surrounding its regulatory site, which consists of four carbonyls, a hydroxyl group from Thr88, and an amino group from Lys101.[39] OhrR, the transcriptional regulator in *B. subtilis* responsible for responding to organic peroxides, has a more hydrophobic environment surrounding its active site containing thiol groups that would otherwise be targeted by H_2O_2.

Just as PerR has a unique mechanism for sensing and producing a cellular response to H_2O_2, its 3-His-2-Asp coordination is also interesting and rarely seen in nonheme iron(II) proteins. Most nonheme iron(II) proteins have a 2-His-1-Asp/Glu coordination. Many members of the Cupin superfamily have been seen to have a 3-His-1-Asp/Glu motif with two water molecules that complete their octahedral environment. Iron-containing SOD, which as mentioned before is also capable of being inactivated via Fenton reactions, also has five ligands similar to PerR with the exception that one of the aspartate ligands is replaced by a solvent ion.[41] This suggests that the 3-His-2-Asp motif might favor the metal-catalyzed oxidation mechanism seen in PerR. The rate constant of the Fenton-type reaction is increased by magnitudes when bound to anionic ligands,[42] which potentially explains the utility of having two aspartates in the active site rather than the one found in most other nonheme iron(II) proteins.

8.2.4 Summary

Prokaryotes employ a range of strategies for adapting to changes in oxygen or ROS levels. The strategies used by each transcription factor to change its conformation in response to oxidative stress reflect the different ways ROS commonly cause unwanted damage to proteins.

Solvent-exposed Fe–S clusters, as seen in aconitase, are often easily destroyed by $O_2^{\bullet-}$.[6] However, it is this vulnerability to oxidation that makes FNR and SoxR effective sensors. The $OH^{\bullet}$ radical produced by a Fenton reaction is highly reactive and is the basis of much of the damage H_2O_2 causes but is harnessed by PerR and restricted to oxidizing only specific residues. The prokaryotic sensors using Fe–S clusters do not have any known homologs in humans, although Fe–S proteins in humans that have been proposed to act as redox sensors do exist.[43] However, as will be discussed later, thiol modification is a recurring theme in ROS signaling in higher organisms.

From the above discussions, a general idea of how oxygen and ROS can be used as signals for transcription is presented. In what follows, the much more intricate mechanisms of oxygen and redox sensing will be discussed. While prokaryotes and eukaryotes share similarities in the antioxidant enzymes they use to remove ROS, how eukaryotes use ROS in signaling is very different from what we have discussed so far.

8.3 REDOX SIGNALING IN METAZOANS

Like prokaryotes, eukaryotic cells require sensors for detecting elevated levels of ROS or other redox changes. However, the cellular responses observed in eukaryotes in response to a shift in a cell's redox potential are somewhat different from what has been observed in bacteria. As mentioned previously, the majority of the elevated ROS levels bacteria experience are due to other organisms trying to kill them. Therefore, the first priority of bacteria encountering elevated ROS levels is to eliminate them through the ways that have been described. Metazoans, having more control over their environment than bacteria, are more likely to experience elevated ROS levels as a result of increased endogenous production. Elevated ROS due to environmental factors such as radiation or exposure to pollutants can be a factor at times, but a majority of the ROS found in all cell types are produced by the cell itself.[44]

The differences in the relationships bacteria and metazoans have with ROS reflect the differences in their redox-sensing mechanisms. Redox sensors in bacteria, usually being activated during an attack by another organism, generally play a defensive role by directly modifying the expression of specific genes. However, far from being only a toxin as they are in prokaryotic cells, endogenously generated ROS have important roles in eukaryotic cellular signaling. Rather than indicating an attack by another organism, shifts in redox conditions within a eukaryotic cell are often the result of increased ROS production induced by changes in one or more signaling pathways. Along with indicating a change in the signaling of other pathways, elevated ROS levels also serve as a mechanism for further inducing change by altering the activity of eukaryotic redox sensors. As a result, redox-sensitive proteins in eukaryotes are linked to a diverse array of cellular functions and more often are upstream modulators of transcription rather than the direct actors seen in prokaryotes.

8.3.1 Primary Sources of ROS in Eukaryotic Redox Signaling

The electron transport chain in mitochondria is responsible for most of the ROS found in cells.[44] The premature oxidation of oxygen to $O_2^{\bullet-}$ is primarily seen in complexes I[45] and III,[46] although complexes II and IV have also been observed to produce a small amount of $O_2^{\bullet-}$. There is not a complete consensus on the amount of electrons passing through the respiratory chain that leak out and end up producing $O_2^{\bullet-}$, but 1–2% is the most common estimate. Complexes I through III release $O_2^{\bullet-}$ in the mitochondrial matrix, but only complex III releases it into the intermembrane space as well.[46] $O_2^{\bullet-}$, carrying a negative charge, is unable to travel across membranes into the cytosol until it is dismutated to H_2O_2 or transported through a voltage-dependent anion channel (VDAC). Mitochondria contain SOD in both its matrix and intermembrane space, permitting the export of its generated ROS in the form of H_2O_2. Intermembrane $O_2^{\bullet-}$ produced by complex III can also be exported into the cytosol by VDACs.[47,48]

For a long time, mitochondrial ROS were viewed as only a by-product of oxidative phosphorylation. That view has been challenged with evidence of mitochondrial ROS being inducible and participating in cell signaling.[49] In particular, mitochondrial ROS and the mitochondria itself are closely linked to Ca^{2+} homeostasis, signaling induced by hypoxia, and cell proliferation and death. Each of these topics will be discussed separately later in this chapter, where the participation of mitochondrial ROS will be explained in further detail.

Unlike mitochondrial ROS, which are produced by enzymes with other purposes, the multiprotein enzyme complex NADPH oxidase has the sole function of generating $O_2^{\bullet-}$. NADPH oxidase was initially found in phagocytes where their generation of $O_2^{\bullet-}$ was induced during phagocytosis. Since then, multiple nitrogen oxide (NOX) isoforms have been found in many other cell types in eukaryotes.[50]

The catalytic subunit of phagocytic NADPH oxidase is a glycoprotein known as gp91^{phox} or NOX2, which contains six transmembrane domains observed in all other NOX isoforms. Within the transmembrane segment of gp91^{phox}/NOX2 are two asymmetrical heme groups bound to His residues that are conserved across other NOX homologs.[50] The hemes, along with a bound flavin adenine nucleotide (FAD) on the cytosolic side of the membrane that serves as the initial electron acceptor, allow the transfer of an electron from a cytosolic NADPH through the plasma membrane to an oxygen molecule bound to the extracellular side of the protein. gp91^{phox}/NOX2 associates with another transmembrane subunit, p22^{phox}, which it requires for its stability.[50] Three other subunits, p40^{phox}, p47^{phox}, and p67^{phox}, are required for its induction and exist as a protein complex in the cytosol until they are recruited.[51] The NADPH oxidase complex is assembled upon activation by the small GTPase Rac1, which interacts with the activation domain of p67^{phox}.[52]

It was later found that gp91^{phox}/NOX2 was present in nonphagocytic cells. In addition, four other nonphagocytic homologs of gp91^{phox}/NOX2 have been identified: NOX1, NOX3, NOX4, and NOX5.[50] The "dual oxidases," DUOX1 and DUOX2, are also related to the NOX family but instead generate H_2O_2 instead of $O_2^{\bullet-}$.[53] The homologs are expressed differently from one cell type to another, vary in cellular localization, and have different subunits required for their activity. NOX1, found primarily in colon epithelial cells and vascular smooth muscle cells (VSMCs), was the first homolog discovered and shares the most similarities with NOX2. Like NOX2, NOX1 is inducible and requires Rac1 for its activity when interacting with p47^{phox} and p67^{phox}. However, its active complex can include NOXO1 and NOXA1 subunits instead of their p47^{phox} and p67^{phox} homologs.[54] From there, the NOX homologs begin to deviate from the characteristics of their original family member. NOX3 is also flexible in the cytosolic subunits it can be activated with, being able to interchange p47^{phox}/p67^{phox} with NOXO1/NOXA1, but has a moderate amount of activity even when it is lacking p47^{phox}/NOXA1 or Rac.[55] NOX4, while still requiring p22^{phox} for its stability, produces $O_2^{\bullet-}$ constitutively and does not require a complex formation with the cytosolic subunits for its function.[56] NOX5 does not require interaction with any subunits, including p22^{phox}, and is dependent on calcium for its induction.[57]

NADPH oxidase has been linked to multiple signaling pathways, allowing the generation of a ROS signal in response to stimuli that increase NADPH oxidase expression or activity. Signaling molecules, like growth factors and hormones, as well as physical stimuli like shear stress,[58] are capable of stimulating NADPH oxidase activity and producing a transient burst of ROS that can modify the activity of other signaling proteins and transcription factors. Some of the signaling cascades associated with growth factors are not just facilitated by this increase in ROS concentration but absolutely require it. Many effects of growth factors are able to be blocked with NADPH inhibitors.[59]

8.3.2 The Floodgate Hypothesis

The relationship between a cell's redox environment and its signaling creates additional consequences for oxidative stress than their potential to damage biological molecules. If the redox equilibrium of the cell is not maintained through antioxidant defenses, it can disrupt the regulation of signaling pathways and contribute to disease. However, there is no chemical difference between the "bad" ROS produced during oxidative stress and the "good" ROS that are produced in response to biological or physical stimuli. The antioxidant defenses that maintain redox equilibrium by removing ROS would also block the generation of ROS associated with cell signaling unless there was a way to distinguish between oxidative stress and signaling. A rule of thumb is that the "good" ROS are tightly regulated whereas the "bad" ROS are produced in an uncontrolled manner. The "floodgate" hypothesis attempts to provide an explanation for how this is managed.[60]

2-Cys peroxiredoxins (2-Cys Prxs), another group of redox-sensitive proteins utilizing reactive thiols, act alongside the glutaredoxin system to maintain a cell's redox condition at equilibrium. 2-Cys Prxs reduce peroxides, resulting in the oxidation of a reactive thiol to a sulfenic acid intermediate that quickly forms a disulfide bond with a nearby cysteine. Similar to glutathione, the oxidized disulfide form of 2-Cys Prxs are recycled by a reductase enzyme. 2-Cys Prxs are ubiquitous in both prokaryotes and eukaryotes and are the primary method of peroxide removal in some organisms, but some eukaryotic 2-Cys Prxs have structural features that stabilize their sulfenic acid intermediate. As a result, the sulfenic acid on these 2-Cys Prxs are capable of being oxidized by a second peroxide molecule to a sulfinic acid.[60] This overoxidized 2-Cys Prx is inactive and unable to be recycled back to the reduced form of the protein by a reductase. Therefore, a burst of ROS associated with signaling that could cause the overoxidation of 2-Cys Prxs would temporarily disable part of the cell's antioxidant defenses and allow a transient change in the cell's redox conditions that would be sufficient for the proliferation of the signal. The remaining antioxidant enzymes would remove the excess ROS while the overoxidized 2-Cys Prxs are recycled by the ATP-dependent enzyme sulfiredoxin,[61] restoring the cell's redox environment and antioxidant defenses to their original state. Overexpressing 2-Cys Prx to levels that are not significantly overoxidized by the signaling ROS bursts has been observed to inhibit peroxide-mediated signaling.[62] Only organisms utilizing peroxide signaling have been found to have 2-Cys Prxs capable of being overoxidized along with expression of sulfiredoxin to allow recovery after overoxidation. 2-Cys Prx oxidation is generally only seen in eukaryotes with the notable exception of certain cyanobacteria due to their evolutionary relationship with chloroplasts, which are the site of 2-Cys Prx overoxidation in plants.[63]

8.3.3 Redox Regulation of Kinase and Phosphatase Activity

Protein phosphorylation is often the switch that determines whether a signaling protein is active or inactive. ROS are able to modify the activity of the enzymes that control this process, kinases and phosphatases, allowing them to influence the signaling activity of many important proteins.

One of the largest families of kinases is receptor tyrosine kinase (RTK). RTKs act as extracellular receptors for growth factors and hormones. The binding of their ligand induces autophosphorylation of their intracellular C-terminal region, allowing them to convert the extracellular signal to an intracellular response by phosphorylating other proteins.[64] RTKs are often the initiators of signaling cascades that have been linked to cell adhesion, proliferation, differentiation, and apoptosis.[64] An important group of redox-sensitive proteins, protein tyrosine phosphatases (PTPs), negatively regulate RTK activity by undoing the phosphorylation catalyzed by the RTK. The active site of all PTPs is highly conserved and contains a reactive low pK_a cysteine residue that is a thiolate anion at physiological pH.[65] This nucleophilic thiolate is necessary for the phosphatase activity of the PTP and is susceptible to inactivation by oxidation.[4] Stimulation of an RTK receptor by its ligand frequently induces a local increase in ROS through Rac-mediated activation of NADPH oxidase, which inhibits its PTP and allows transduction of the extracellular signal.

PI3K/PKB: The PI3K/Protein kinase B (PKB/Akt) signaling pathway regulates multiple pathways associated with cell proliferation and apoptosis. PKB activation upregulates the expression of survival and proliferation genes through activation of NF-κB while inhibiting activation of caspases.[66] Phosphoinositide 3-kinase (PI3K) is activated by RTK receptors for various growth factors and hormones, such as angiotensin II[67] and EGFR[68] as well as direct treatment of H_2O_2.[67] Once active, it catalyzes the phosphorylation of phosphatidylinositol 4,5-bisphosphate (PIP_2) to form phosphatidylinositol 3,4,5-triphosphate (PIP_3). PIP_3 then interacts with phosphoinositide dependent kinase 1 (PDK1) to activate PKB, which goes on to phosphorylate multiple protein targets that has the overall effect of activating cell proliferation and inhibiting apoptosis. PI3K signaling is negatively regulated by the PTP PTEN (phosphatase with tensin homology), which counteracts the activity of PI3K by removing a phosphate group from PIP_3 and reverting it to PIP_2.[69] When PTEN is inactivated by H_2O_2 oxidation, PIP_3 accumulates in the cell and PKB activity increases.[69] In addition to assisting in the activation of PKB, PIP_3 is also capable of inducing NADPH oxidase activity.[70] Thus, accumulated PIP_3 is able to amplify and prolong its proliferative signal by increasing the amount of ROS in its local environment and further inhibiting PTEN.

PKB activity can also be regulated by ROS through PI3K-independent mechanisms. During H_2O_2 stress, Hsp27 associates with PKB and induces its activity.[71] However, large concentrations of H_2O_2 degrade PKB and eventually result in apoptosis.[72]

PKC: Protein kinase C (PKC) is another mediator of receptor signaling and transcription that is associated with immunity,[73] smooth muscle contraction,[74] and cell proliferation and apoptosis.[75] Most of its isoforms are activated by diacylglycerol (DAG) and Ca^{2+}. DAG, in addition to IP_3, is formed by the cleavage of PIP_2 by phospholipase Cγ1 (PLC-γ1). IP_3 opens Ca^{2+} channels that increase intracellular levels of calcium, further activating most PKC isoforms. PLC-γ1 activity, and by extension PKC activity, has been shown to be upregulated by H_2O_2.[76]

PKC contains many cysteine residues that can be directly modified by ROS. Generally, these oxidative modifications have a favorable effect on PKC. Treatment of hepatocytes with ROS-generating compounds that promoted disulfide bonding was shown to increase the V_{max} of PKC.[77] But, similar to PKB, H_2O_2 concentrations beyond a certain level have an inhibitory effect upon PKC.[78]

MAPK: The mitogen-activated protein kinase (MAPK) consists of the ERK, JNK, and p38 kinases. ERK mediates growth factor signaling while JNK and p38 are associated with the proliferation of survival signals induced by cellular stress. MAPK signaling cascades are commonly activated by RTKs in a process mediated by the small G protein Ras.[79] Each group of MAPKs is activated by direct H_2O_2 treatment.[80] The increase in intracellular ROS associated with RTK-mediated signaling appears to be the primary method MAPKs are induced by cytokines. For example, the overexpression of catalase has been shown to block the angiotensin II-mediated induction of p38.[81]

8.3.4 Communication between ROS and Calcium Signaling

Calcium ions are possibly the most widely used second messengers in intracellular signaling. Similar to ROS, cytoplasmic levels of Ca^{2+} are maintained at a low level through multiple homeostatic mechanisms and are only elevated in response to specific stimuli.[82] The primary intracellular store of Ca^{2+} is the sarcoplasmic reticulum (SR), which pumps Ca^{2+} ions into its lumen using ATP-dependent transporters and releases them into the cytoplasm through receptor-mediated channels. Extracellular Ca^{2+} can also be transported into the cell through voltage- and receptor-operated calcium channels. Many of these transporters have reactive thiols similar to the redox-sensitive proteins that have been discussed so far, potentially providing a way for these two widespread messaging systems to communicate with each other.

Influx and efflux of Ca^{2+} from the SR are major contributors to the cytosolic concentration of Ca^{2+}, which is 10–50 µM as opposed to the mM range concentrations found in the SR.[83] ER/SR Ca^{2+}-ATPases (SERCA) pump Ca^{2+} from the cytoplasm into the SR and are inhibited by both $O_2^{\cdot-}$ and H_2O_2.[84] Plasma membrane Ca^{2+}-ATPases (PMCAs) are also inhibited by ROS, but only at significantly higher concentrations.[84] In contrast, two receptors that mediate the release of Ca^{2+} from the SR in response to ryanodine and the signaling molecule 1,4,5-inositol-triphosphate (IP_3) are activated by elevated H_2O_2 concentrations. Ryanodine receptors (RyRs) have between 40 and 50 reactive cysteines on both the luminal and transmembrane sides of its four subunits that are able to be oxidized, forming either disulfide bonds[85] or mixed disulfides with glutathione.[86] RyRs have less activity in reducing conditions where all thiols are unmodified. When roughly half the reactive thiols in RyRs are oxidized, its activity is stimulated due to conformation changes brought by inter-subunit disulfide cross-linking and blocked interaction with its inhibitors.[87] However, the receptor irreversibly loses function when more than ~70% of its thiols are oxidized during prolonged or severe oxidative stress.[87] IP_3 receptors (IP_3Rs) and RyRs share a high amount of sequence homology, particularly in regions containing reactive cysteines, and are similarly activated by H_2O_2.[88] H_2O_2 further stimulates Ca^{2+} release via IP_3 receptors by activating phospholipase Cγ1 (PLC-γ1), which cleaves PIP_2 to form IP_3 and DAG.[89]

Ca^{2+} signaling can also influence ROS homeostasis. Depending on the context, Ca^{2+} signaling is capable of both positive and negative regulation of ROS production.[90] Ca^{2+} stimulates TCA cycle activity and oxidative phosphorylation, increasing the output of mitochondrial ROS proportional to the increase in oxygen consumption.[90] NOX5, which contains N-terminal EF-hands not present in NOX1 through NOX4, is also dependent upon Ca^{2+} for its activity.[57]

8.3.5 Redox Modulation of Transcription Factors

Transcription factors and other nuclear proteins, often the last step of a signaling pathway before a stimulus is converted into a change in gene transcription, are indirectly affected by redox conditions through activation or inhibition of the signaling mechanisms that have been mentioned so far. In addition to inducing changes in transcription via participation in signaling cascades, ROS are also able to directly oxidize and modify the activity of transcription factors. Both direct and indirect mechanisms of redox mediated changes in transcription will be illustrated by the knowledge learned from two transcription factors, p53 and NF-κB.

p53: p53 is a key regulator of the cellular stress response.[91] Out of the genes activated by H_2O_2, one-third are regulated by p53 and are primarily related to cell cycle arrest and apoptosis.[92] Mutations in p53 resulting in a loss of function, resulting in the unchecked proliferation of damaged DNA, are common in many cancerous cells. Basal expression of p53 has been found to be required for expression of glutathione peroxidase and other antioxidant enzymes.[93] However, activation of p53 by DNA damage results in the suppression of SOD[94] and the production of multiple pro-oxidant enzymes,[95] inducing oxidative stress and apoptosis.

The DNA-binding domain of p53 contains redox active cysteines that affect its binding affinity.[96] When oxidized, p53 takes an inactive conformation that is unable to bind to DNA.[96] Ref-1, a bifunctional enzyme that repairs DNA and also utilizes thioredoxin to reduce oxidized transcription factors, stabilizes p53 activity during oxidative stress.[97] Oxidative stress can also potentially activate p53. p38α, an MAPK that is activated by ROS, induces p53 activity which further contributes to oxidative stress and creates a feedback loop leading to cell death.[98]

NF-κB: The discovery of the NF-κB family of transcription factors has spawned tens of thousands of published articles since its initial characterization 25 years ago. NF-κB is a regulator of over 100 genes and the point of convergence for many signaling pathways.[99] NF-κB's activation is normally associated with cell proliferation and survival, although links to apoptosis have also been established.[100] The NF-κB family is composed of p50, p52, RelA/p65, RelB, and c-Rel,[99] which form both homo- and heterodimers that bind to specific DNA sequences (κB sites). The Rel subunits contain transactivation domains, allowing them to activate transcription. p50 and p52 lack these domains and therefore block transcription unless it exists in a dimer with a Rel subunit. NF-κB is sequestered in the cytosol by IκB complexes.[99] In response to various stimuli, IκB is phosphorylated by the IKK family and degraded, enabling the translocation of the NF-κB dimer to the nucleus where it can bind to its gene targets.

ROS have been observed to modify both NF-κB and the proteins that contribute to its activation. The binding of p50 to DNA is inhibited by the H_2O_2 oxidation and subsequent glutathiolation of its Cys 62 residue.[101] Glutathiolation of Cys189 of IκB has also been shown to inhibit its phosphorylation and degradation.[102] Taken alone, these results would lead to the conclusion that ROS negatively regulate NF-κB activity. However, generation of H_2O_2 and activation of NF-κB are both hallmarks of inflammation, and multiple studies have demonstrated that H_2O_2 positively regulates NF-κB activity when administered with TNF-α.[103] In addition, the TNF-α-induced phosphorylation of Ser 276 on RelA/p65, required for activation of some of NF-κB's target genes, is a ROS-dependent process that is inhibited by antioxidants.[104] PKB activation promotes NF-κB activity through induction of mTOR,[105] but PKB is also both activated and inhibited by ROS depending on the concentration.

The synergy between H_2O_2 and TNF-α in activating NF-κB demonstrates the reciprocal relationship between redox environment and other cellular conditions. As we have seen, cellular conditions and behavior are heavily influenced by the cell's redox environment. But, in many instances, the effects of a ROS signal can have different or even opposite effects during different conditions or lengths of exposure. The varying affinities of κB sites for NF-κB also provides an example of how the ROS-mediated activation of a transcription factor could affect specific genes more than others in spite of being regulated by the same protein. Genes with lower affinity κB sites were upregulated more by H_2O_2 treatment than genes with higher affinity κB sites that were already saturated without the H_2O_2 stimulation.[103]

Hydrogen peroxide and iron are known to activate NF-κB, but the activation mechanisms are resolved. A few coactivators of NF-κB has been identified, including Brad 1, Jab1, Pirin, and Tip60.[106] Among them, Pirin is a Fe-containing nuclear protein present in all human tissues. This metalloprotein is found to enhance the DNA binding ability of the NF-κB proteins. The three-dimension structure of Pirin determined by X-ray crystal structure shows that Pirin belongs to the Cupin protein superfamily and contains an iron ion bound by a 3-His-1-Glu ligand motif,[107] which is similar to the metal-containing proteins involved in gene transcription regulations discussed in this chapter. Although the mechanism by which Pirin regulates NF-κB signaling pathway is presently unknown and whether or not its regulation is associated with H_2O_2 is yet unclear, the finding of an iron-containing nuclear activator of NF-κB is a significant advance and offers a clue for understanding how iron ions may participate in the NF-κB transcription regulation.

8.3.6 Summary

Multicellular eukaryotes, not having to experience the constant threat of elevated ROS levels from their environment that prokaryotes do, have been able to evolve signaling mechanisms that co-opt their endogenously produced ROS as a way to facilitate cross talk between different pathways. The thiol modification mechanism of H_2O_2 signaling is seen in both prokaryotes and eukaryotes, but the transcriptional responses of prokaryotes are focused on removing the H_2O_2 while the response in mammals has not been shown to include antioxidant enzymes.[92] Multiple antioxidant genes (MnSOD, glutathione reductase 1, and thioredoxin), along with proteins that regulate intracellular metal concentrations, a response that resembles the prokaryotic response to oxidative stress, are regulated by NF-κB activation.[108] However, genes for pro-oxidants xanthine oxidase and NOX also contain κB sites.[108] Unlike the consistent antioxidant responses to any change in ROS concentration that have been characterized in prokaryotes, any upregulation of antioxidant genes in metazoans is likely to be based on many other factors and conditions in addition to the concentration of ROS within the cell.

However, the lack of a definite inducible antioxidant response in mammals should not be mistaken as a lack of an overall strategy against oxidative stress. Moderately elevated concentrations of ROS generally promote the activation of signaling pathways leading to proliferation and survival and the inhibition of those that initiate apoptosis. While these elevations in ROS can be the product of normal signaling, they can also be due to oxidative stress. In this situation, the anti-apoptotic signals from the ROS can double as a survival mechanism that helps the cell endure the stress.[76] PKB is an example of a pathway that is activated by both growth factors via PTEN/PI3K regulation and by stress responses mediated by Hsp26. PKC also has a role in resisting cellular stress. Mammalian cells with knocked out PLC-γ1, which contributes to the activation of PKC by ROS, are only able to tolerate a fraction of the H_2O_2 concentration that wild-type cells can comfortably survive in.[76] While low or transient exposure to ROS promotes survival, severe or chronic ROS exposure has the opposite effect. Overexposure to ROS inhibits many of these pro-survival responses that promote signaling associated with apoptosis.

8.4 OXYGEN SENSING IN METAZOANS

Since the introduction of oxygen into our atmosphere, life has developed ways to not only cope with this potentially toxic element but to utilize it. Oxidative phosphorylation confers a large advantage to aerobic organisms, allowing them to extract 18 times more ATP from a single molecule of glucose than they would anaerobically. Molecular oxygen is also a necessary substrate for a variety of metabolic pathways. With the importance of oxygen to normal physiological and cellular function, the means for a cell to detect an inadequate supply of oxygen and translate the pathological condition into a signal that results in the adjustment of the oxygen requirements of the cell is necessary.

As we discussed in prokaryotes, oxygen sensing is a relatively straightforward process involving a single switch-like transcription factor that experiences alterations in its ability to bind DNA or dimerize into its active form upon exposure to oxygen. These oxygen-sensing proteins are efficient for their intended purpose but can only operate in a strictly "on" or "off" manner, requiring strictly anaerobic conditions to be turned "on." While this is suitable for single-celled organisms, larger and more complicated organisms are more likely to experience levels of oxygen that are merely lower than normal (a condition known as hypoxia) rather than completely anaerobic conditions. Hypoxia can be acute or chronic, as well as mild or more severe, and each of these conditions would require the organism to adapt in different ways. Larger organisms have multiple types of tissue with different physiological oxygen concentrations and requirements. This increased complexity in physiology and lifestyles obviously requires a more complex and nuanced mechanism that is capable of differentiating between different degrees of hypoxia and responding accordingly to each. The central regulators of this oxygen-sensing pathway in metazoans are hypoxia-inducible factors (HIFs), transcription factors that upregulate the expression of genes associated with glycolysis, angiogenesis, and red blood cell production in response to hypoxia.[109] These processes assist in transitioning to a low-oxygen environment where anaerobic metabolism is relied upon more and additional means of delivering blood and oxygen are needed. Classic examples of this can been seen in the adaptations that occur in people living at higher altitudes with less oxygen and in the Pasteur effect where glycolysis and glucose-consumption are upregulated in conditions with limited oxygen.[110]

8.4.1 HIF

HIF is a heterodimeric protein. Three types of α subunits (HIF-1α, HIF-2α, and HIF-3α) exist and require interaction with an HIF-β subunit for activity, with each subunit containing a basic-helix-loop-helix (bHLH) domain seen in other transcription factors.[111] HIF-β, also referred to as aryl hydrocarbon receptor nuclear translocator (ARNT), is a ubiquitous and constitutively expressed nuclear protein. It is not an exclusive binding partner of HIF-α nor does HIF-β participate exclusively in hypoxic signaling. HIF-α is a cytosolic protein subject to much tighter oxygen-dependent regulation at both the transcriptional and posttranscriptional levels that render it nearly undetectable during normal cellular conditions, unlike the constitutive HIF-β which is not affected by oxygen levels.[112] The HIF-α/β complex recognizes conserved sequences called hypoxic response elements (HREs) located near its target genes. Recognition of HREs and the enhanced expression seen during hypoxia requires the dimerization of the HIF-α and HIF-β subunits, which is contingent on HIF-α being permitted by the cell to accumulate to levels above what is found during normoxia.[113]

Even though HIF-α has a central role in the regulation of hypoxia-sensitive genes, it cannot detect or react with oxygen itself. When looking at biological reactions, molecular oxygen is not a reactive molecule in most situations. As a stable diradical at triplet state ($S = 1$) with two unpaired electrons, it can only accept the transfer of a single electron at a time.[3] While oxygen cannot react with molecules with paired electrons, it is more than willing to accept unpaired electrons from transition metals. In fact, in any metabolic or signaling pathway where oxygen is utilized, a metalloprotein is usually not far away. In the case of oxygen sensing, the metalloproteins are HIF-α prolyl hydroxylase domain (PHD) enzymes and the asparaginyl hydroxylase factor-inhibiting HIF (FIH), both capable of using oxygen as a substrate to posttranslationally modify and inhibit HIF-α. Both enzymes belong to a family of Fe(II)/α-ketoglutarate (α-KG) dependent proteins.[114] Their requirement of an iron-bound O_2 substrate provides the oxygen-sensing component of HIF signaling. When deprived of oxygen in hypoxic conditions they are unable to block HIF-α function, allowing it to accumulate and dimerize with HIF-β, and its target genes are expressed until oxygen levels and PHD/FIH activity returns to normal.

8.4.2 PHD Enzymes

The PHD acts as the first line of HIF-α regulation, the hydroxylation of HIF-α leading to its degradation and

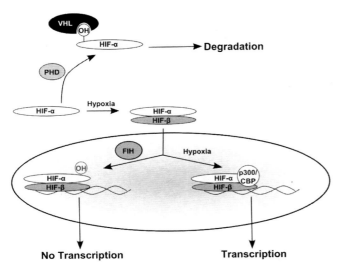

Figure 8.3 Cartoon of HIF-α regulation by hydroxylase activity of PHD and FIH on transcription.

subsequently the low levels of HIF-α found during normoxia (Fig. 8.3). Three different PHDs (PHD1, PHD2, PHD3) capable of regulating HIF-α, each encoded by their own gene, have been found in mammals.[115] All three isoforms are capable of hydroxylating HIF-α and suppressing hypoxia-regulated genes *in vitro*. However, PHD2 has been shown to play the most prominent role in the suppression of HIF-α *in vivo*.[116] All three PHDs specifically target and hydroxylate prolines within a conserved LxxLAP motif. When interacting with HIF-α at sufficient oxygen levels, two conserved proline residues (Pro402 and/or Pro564) within the oxygen-dependent degradation domain (ODDD) of HIF-α are converted to hydroxyprolines (Hyp). Upon formation of the Hyp residues, HIF-α becomes capable of being recognized by the von-Hippel Lindau (VHL) segment of the E3 ubiquitin ligase complex, resulting in its ubiquination and degradation via the proteasome.[117] The Hyp residues increase HIF-α's affinity for VHL by over a thousandfold. Hyp, along with having differences in conformation preferences, is more capable of hydrogen bonding than proline with its additional oxygen. The hydroxyl group of Hyp564 forms hydrogen bonds with Ser111 and His115 in VHL, stabilizing the interaction between the two proteins.[118]

The three different PHD proteins have similar structures and overlapping functions, but their locations within the cell and their patterns of expression suggest that they may also have specific roles in hypoxia signaling.[119] PHD1 is exclusively located in the nucleus while PHD2 is almost entirely located in the cytosol and PHD3 is distributed evenly between both.[119] Across multiple cell lines, PHD1 mRNA was not found to be affected by oxygen concentration while expression of PHD2 and PHD3 were generally upregulated during hypoxia.[115] In particular, PHD3 levels are dramatically higher in hypoxic conditions and are believed to contribute more to the regulation of HIF-α levels during hypoxia than PHD2.[115] The inducibility of PHD2 and PHD3 by the transcription factor they regulate suggests a feedback loop mechanism where an increase in the levels of PHD2 and PHD3 during hypoxia assists the cell in rapidly degrading the accumulated HIF-α and shutting off hypoxia-regulated gene expression upon the restoration of normal oxygen levels. Rate of degradation of HIF-α upon reoxygenation after different lengths of hypoxia was proportional to how long the cells were in hypoxic conditions, due to the higher levels of PHD2 and PHD3 expressed in the longer period of hypoxia.[120]

Prolyl-4-hydroxylases (P4H) have been well-characterized due to their role in both oxygen sensing and the hydroxylation and stabilization of collagen.[121] As with all P4H hydroxylation reactions, the P4H responsible for the hydroxylation of collagen requires O_2 but is not disabled by hypoxia as PHD is. If it were otherwise, chronic hypoxia would lead to the instability of collagen, a symptom of scurvy. The differences in oxygen requirements between PHDs and other P4Hs are explainable by their differences in O_2 affinity. PHD, by having a K_m value for O_2 that is near the concentration present in the atmosphere, is more sensitive to shifts in O_2 levels than other P4Hs that have much lower K_m values.[122] Other studies have challenged this theory, claiming that PHD's K_m value for O_2 is within the range of other α-KG dependent enzymes (although still sufficiently above the estimate of intracellular concentrations of O_2 for it to be an effective sensor of oxygen), and other factors that govern α-KG reaction rates such as the size of the bound substrate, iron/α-KG concentrations, or redox conditions contribute to PHD's efficacy as an oxygen sensor.[123]

8.4.3 FIH

On top of the PHD's regulation of HIF-α stability, the asparaginyl hydroxylase FIH provides an additional layer of regulation that inhibits HIF-α from inducing its target genes. However, instead of targeting the ODDD of HIF-α and tagging it for degradation, FIH modifies an asparagine residue (Asn803) on its C-terminal transactivation domain (C-TAD) that makes the transcription factor unable to recruit its coactivators, p300/CBP, required for it to activate transcription of many of its target genes.[124] PHD and FIH work together in inhibiting HIF-α, with either one of their activities sufficient to affect its activity (Fig. 8.3). Mutations of the target proline residue in the ODDD of HIF-α displayed a higher stability during normoxia as one would expect from a mutant that was unable to be targeted by PHD, but only displayed a fraction of the transcriptional activity seen during hypoxia. Only by mutating both the ODDD proline and Asn803 within the C-TAD was full activity restored during normoxia.[124]

FIH shares an inhibitory function with PHD, but the functions of the two proteins should not be thought of as being redundant. FIH targets the C-TAD domain of HIF-α, but HIF-1α and HIF-2α contain a second transactivation domain closer to the N-terminal (N-TAD) that is not recognized by FIH, while HIF-3α lacks a C-TAD altogether. Many, but not all, of the hypoxia-regulated genes require a functional C-TAD domain that is capable of recruiting p300/CBP, making them resistant to being silenced by FIH hydroxylation.[125] FIH also has a higher affinity for O_2 than PHD, thus requiring less oxygen for its function.[126] This difference in affinities allows for an oxygen concentration where PHD is inactivated, allowing HIF-α to accumulate, but FIH continues to hydroxylate the C-TAD and inhibit the expression of certain genes.[126] The different O_2 concentrations that PHD and FIH can operate at and the different genes they can potentially silence contribute to the versatility and sensitivity of this oxygen-sensing mechanism.

8.4.4 Factors Influencing Fe(II)/α-KG Dependent Enzymes

PHD and FIH, other than having different residues they hydroxylate, share very similar mechanisms and structural characteristics that can be seen in other Fe(II)/α-KG dependent oxygenases.[114] α-KG dependent enzymes bind an Fe(II) ion using two histidines and one aspartate or glutamate residue in an octahedral coordination with the remaining sites occupied by water, a well-characterized motif seen in other dioxygenases.[127] α-KG and O_2 bind to the Fe(II) ion, displacing the water molecules, while the substrate to be oxidized binds on a separate site. The co-substrate α-KG is converted to succinate and CO_2 and released, leaving behind a highly reactive Fe(IV)=O intermediate that hydroxylates the bound substrate.[114] The O_2 substrate is split apart and incorporated into the succinate product and the hydroxylated substrate (Fig. 8.4).

As mentioned previously, the O_2 requirement in this reaction is absolute and is the underlying mechanism behind the oxygen sensitivity of the HIF pathway. However, the other substrate, α-KG, and the oxidation state of the iron ion are equally necessary for the

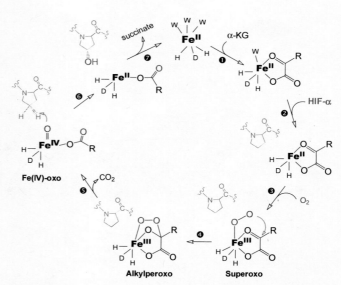

Figure 8.4 α-Ketoglutarate and Fe-dependent hydroxylation of HIF-α by prolyl hydroxylase. The mechanistic cycle shown in this model includes the following steps: (1) binding of co-substrate α-KG and replacing of the two water ligands at the Fe(II) center, (2) binding of the primary substrate HIF-α resulted in a conformational change and lost of the third water ligand, (3) binding of O_2 to the five-coordinate Fe(II) center and production of a superoxo anion intermediate, (4) formation of an alkylperoxy intermediate, (5) O–O bond cleavage leads to an Fe(IV)-oxo intermediate while oxidative decarboxylation of the bound co-substrate produces CO_2 and a bound succinate, (6) oxygen transfer from the high-valent Fe(IV)-oxo intermediate to a proline residue of HIF-α, and (7) succinate release resumes the initial 2-His-1-Asp ferrous center. See color insert.

enzyme's activity and open up other possibilities for altering PHD/FIH activity and HIF-α stability.

α-KG, an intermediate of the citric acid cycle, has been found to compete with two other citric acid cycle intermediates, succinate and fumarate, for binding to α-KG enzymes.[128] Studies examining the effects of oxygen and glucose deprivation on HIF-α stability have observed a link between metabolism and HIF-α degradation.[129,130] Cells deprived of glucose during hypoxia, producing a low α-KG/fumarate ratio, was found to have a lower rate of HIF-α degradation upon reoxygenation than cells only deprived of oxygen, providing a role of metabolism and glucose availability in the modulation of HIF activity.[129]

The Fe(II) ion is also required for the activity of α-KG dependent enzymes, HIF and FIH not being an exception. Iron chelators have been observed to mimic the effects of hypoxia, leading to the accumulation of

HIF-α due to the loss of HIF and FIH function.[131] The oxidation state of iron is also an important factor in the functionality of α-KG dependent enzymes. The ferrous state of the iron ion is required for activity and is typically regenerated at the end of each reaction cycle. However, occasionally, the oxidative decarboxylation of α-KG by the Fe ion is not coupled to an oxidation of a substrate, leaving behind an inactive Fe(III) ion that requires reduction by the antioxidant ascorbate (also known as vitamin C).[132] Ascorbate is considered as a requirement for the function of many, but not all, α-KG dependent enzymes. P4Hs are among the α-KG dependent enzymes requiring ascorbate.[133] The disease scurvy, caused by collagen instability due to a lack of P4H activity, is linked to a lack of dietary ascorbate. PHD has been found to have higher activity in the presence of higher concentrations of ascorbate, accompanied by a reduction in HIF-α levels.[134]

The oxidation state of Fe(II) can also be influenced by the accumulation of H_2O_2, a ROS that is capable of oxidizing Fe(II) to form Fe(III). H_2O_2 has been implicated as an activator of angiogenesis,[135] a process which is enhanced by hypoxia and HIF activation. Increased H_2O_2 levels has a stabilizing effect on HIF-α accompanied by a higher proportion of PHD2 in the Fe(III) oxidation state.[136] The effect of H_2O_2 on HIF-α stability was reversed by ascorbate treatment.[136] ROS, having close ties to many pathophysiological conditions linked to hypoxia such as cancer and inflammation, have been found to play a significant role in the HIF pathway.

8.4.5 ROS and Oxygen Sensing

The function of ROS in the regulation of HIF-α has provided fertile ground for investigation and debate over the past decade. It is generally agreed upon that ROS have an effect on HIF-α stability, but the sources of the ROS and the signaling pathways and mechanisms involved have yet to be completely elucidated. The importance of mitochondrial ROS in HIF-α stability has been argued on both sides. Some studies have claimed that the ROS produced from the electron transport chain contribute to HIF-α accumulation.[137] However, both explanations support that a functional electron transport chain contributes to the stability of HIF-α.

Another source of ROS that has been investigated in HIF signaling is NADPH oxidase. There is accumulating evidence that hypoxia, HIF-α accumulation, and the expression of NOX4 are linked. The gene for NOX4 has been found to contain an HRE and is upregulated in response to higher HIF-α expression, increasing levels of ROS resulting in an increase of pulmonary arterial smooth muscle cell proliferation during hypoxia.[138]

One of the more significant cell signaling proteins that HIF is connected to both directly and via ROS signaling is NF-κB. A link between hypoxia and NF-κB has been known since the 1990s.[139] More recently, IκB kinase B, an upstream activator of NF-κB, has been identified as another substrate of PHD that has reduced activity upon hydroxylation,[140] providing a hypoxic activation mechanism for NF-κB that is similar to HIF-α. Interestingly, it was also found that the expression of HIF-α is directly regulated by NF-κB.[141,142] These findings are significant in two ways. First, it suggests that the accumulation of HIF-α during hypoxia occurs not only through its stabilization but also through its upregulated expression, a revelation further supported by recent studies demonstrating that inhibition of NF-κB signaling reduces the induction of HIF-α's target genes.[143] Second, the direct link between NF-κB and HIF-α provides further insight into how ROS generated in response to different physiological conditions can modulate HIF-α expression. ROS can regulate the activation of NF-κB in numerous ways and varies greatly by cell type.[144] The proinflammatory cytokine TNF-α, an inducer of NF-κB and H_2O_2 accumulation via increased NADPH oxidase activity, has been found to increase HIF-α stability and DNA-binding affinity in nonhypoxic conditions using a ROS-dependent mechanism that can be inhibited by antioxidant treatment.[145] In addition, NF-κB activity has also been linked to the upregulation of NOX1 and NOX4 in smooth muscle cell (SMC),[146] illustrating the potential for cross talk between the HIF and NF-κB signaling pathways.

8.4.6 Summary

Far from the simple oxygen-sensitive proteins found in bacteria that have a simple switch-like function, the effects of hypoxia on cellular signaling are still under investigation, with new possible pathways and mechanisms activated in response to low oxygen levels being revealed all the time. Substrates for PHD and FIH other than HIF-α continue to be discovered, providing new possible roles for the hydroxylases and the VHL complex. One interesting substrate of FIH, the ankyrin repeat domains of proteins in the NF-κB pathway, is still being investigated.[147] The conventional wisdom that HIF activation requires hypoxia has been effectively challenged with evidence that nonhypoxic induction of HIF-α is also possible in response to stimulation by cytokines and growth factors during inflammation,[145] made possible by HIF's interdependence with NF-κB and the effects of ROS generation on both pathways (Fig. 8.5). The study of oxygen-sensing metalloproteins, the pathways these metalloproteins interact with, and

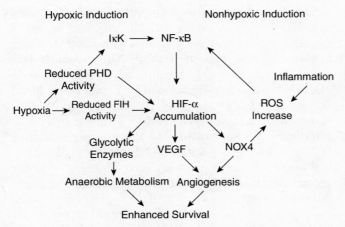

Figure 8.5 Hypoxic and nonhypoxic induction of hypoxia-inducible factor (HIF).

their dependence on cellular redox conditions has exploded over the past decade and shows no sign of slowing down.

8.5 MEDICAL SIGNIFICANCE OF REDOX AND OXYGEN-SENSING PATHWAYS

With the large contributions that oxygen levels and ROS make to cell signaling, it is no surprise that dysregulation of both are associated with pathological conditions. ROS, while a necessary part of our cell signaling network, remain a double-edged sword. Defects in signaling pathways caused by disease can result in an increase in ROS production and the development of oxidative stress and vice versa, causing the disease or contributing to its effects. The role of ROS in signaling has become increasingly more relevant for the study of many diseases, including Alzheimer's disease,[148] Parkinson's disease,[149] and diabetes.[150] In particular, the understanding of cancer and vascular disease have benefited greatly from the study of hypoxia and oxidative stress.

8.5.1 Cancer

Perhaps more than any other category of disease, the study of redox and oxygen signaling has intertwined with the research and treatment of cancer. In 1924, German biochemist Otto Warburg first hypothesized that the growth of cells within a tumor was being driven by an increased rate of anaerobic metabolism. Almost a century of research has supported his hypothesis, and we now know that the hypoxic environment of tumors is responsible for what is now known as the "Warburg effect."[151] An increased rate of glycolysis is not the only way an increase in HIF levels promotes tumorgenesis. Vascular endothelial growth factor (VEGF), an angiogenic growth factor, is upregulated by HIF and results in the vascularization seen in tumors.[152]

Acting in conjunction with hypoxia, high levels of ROS contribute to the faulty cellular signaling that enables the proliferation and survival of cancerous cells. Elevated ROS concentrations are a common feature of tumors.[153] VEGF stimulates the activation of NOX1 and NOX2, causing an increase in ROS.[154] Signaling pathways and transcription factors activated by ROS signaling, such as the PI3K/PKB pathway and NF-κB, are associated with cell proliferation and the suppression of apoptosis that when activated can contribute to oncogenesis.[66,155] ROS and NF-κB also further promote the stability and expression of HIF-1, assisting the hypoxic survival mechanisms utilized by cancerous cells. Hypoxia and ROS both contribute to the proliferation of the signals of the other, creating a dangerous feedback loop.

With elevated HIF and ROS levels being widely associated with cancerous cells, some anticancer strategies have focused inhibiting HIF and ROS activity. Both the inhibition of HIF-1 and treatment of antioxidants have been demonstrated to reduce the rates of growth and angiogenesis in tumors.[156,157] Interestingly, going in the other direction and increasing ROS concentrations of tumor cells to a level where their signaling effects shift from proliferation to apoptosis may also have therapeutic potential. Cancer cells, while generally more resistant to oxidative stress than normal cells due to selection and the activation of pro-survival signaling pathways, still have to maintain ROS concentrations within a specific range to avoid cell death. Because cancer cells are producing large amounts of signaling-derived $O_2^{\bullet-}$ and their SOD activity is often reduced, they are more vulnerable to being killed by SOD inhibition than a noncancerous cell.[158] 2-Methoxyestradiol (2-ME), an inhibitor of SOD, has been shown to induce apoptosis in multiple cancer cell lines as well as showing efficacy in animal studies with minimal side effects upon noncancerous cells.[159] This therapy, while promising, still has issues with low bioavailability.[159]

8.5.2 Vascular Pathophysiology

H_2O_2 is an important mediator of vascular function at both the physiological and cellular level. H_2O_2 is involved in the hyperpolarization and relaxation of smooth muscle in response to acetylcholine.[160] In addition, angiotensin II is a key regulator of cell growth, apoptosis, and migration in vascular cells and relies upon NOX1 and NOX4 for much of its function.[161] NOX1 and NOX4 are the primary sources of $O_2^{\bullet-}$ in endothelial and VSMCs, and elevated expression of both enzymes are frequently associated with atherosclerosis and hypertension.[162] While normal levels of ROS have a protective, dilatory effect upon the vasculature, excessive $O_2^{\bullet-}$ production beyond the ability of SOD to scavenge it induces vasoconstriction.[163] Increased ROS levels also upregulate expression of adhesion molecules vascular cell adhesion molecule 1(VCAM-1) and intercellular adhesion molecule 1(ICAM-1), promoting the accumulation of inflammatory molecules that eventually lead to lesion formation.[163] In addition, oxidative stress contributes to vascular remodeling, which results in thicker vessel walls and a reduction in lumen size, and contributes to hypertension.[164]

Studies examining the efficacy of antioxidant therapy on patients with vascular diseases have been extensive and with mixed rates of success. While some studies have demonstrated improvement in vascular health in patients treated with antioxidants,[165] others show no significant improvement in health.[166] The success of a treatment likely depends on the length of the treatment, the patient, the antioxidant used, and at what stage of the disease the treatment was given.[166] In the case of atherosclerosis, NOX and ROS levels are higher during the beginning of lesion formation than the end, making early treatment more effective.[163]

Inhibitors of NADPH oxidase may also have therapeutic potential.[167] Upregulated NADPH oxidase expression and activity are commonly found in patients with poor vascular health. Direct targeting of NADPH oxidase, the source of the harmful ROS, could be more effective than the nonspecific scavenging of ROS by antioxidants.

CONCLUDING REMARKS

Metal-dependent oxidation/reduction (redox) reactions intimately associated with the regulation of gene transcription have become elucidated and increasingly appreciated. Inducible regulation of gene expression, by redox reactions involving metal ions, may be one of the most widespread mechanisms for translation of signals. The importance of a cell's redox environment and metal (e.g., Ca, Fe, and Zn) concentrations to its signaling and transcription profile are outlined in this chapter. When stepping back from the labyrinth of signaling pathways that ROS can affect and trying to answer the question "What is the overall effect of elevated ROS concentrations in our cells?", we are left with the answer of "It depends." Our relationship with these molecules is much more nuanced than what the prokaryotes have. As a result, our redox-sensitive proteins have evolved beyond the simple detection and elimination systems seen in bacteria and instead are able to harness their reactivity and integrate them into signals that promote growth rather than death.

A potential pitfall of studying ROS signaling is making sweeping generalizations on what ROS do or how they are produced based on observations from a single cell type. The five NOX isoforms, each regulated and induced by different stimuli, have variable expression between different types of cells. Thus, a signal or transcription factor that induces an NADPH oxidase-generated increase of ROS in one type of cell may not do the same in another. The relationship between ROS and Ca^{2+} homeostasis allows for even more variation in the outcomes of redox signaling in different cells. The signaling effects of ROS have been seen to depend heavily on cell type, ROS concentrations, the length of exposure, and the overall condition of the cell and activity of its other signaling pathways. As we have seen in many pathways, adjustments to any of those parameters can mean the difference between a signal that encourages cell proliferation and one that results in apoptosis or necrosis. This can sometimes complicate the investigation of ROS signaling and lead to conflicts between the results of different investigators who used different cell lines or experimental methods. However, it is possible that our cells would not be as fine-tuned to their individual environments and purpose without these variations in the cause and effect of ROS signals between cell types.

REFERENCES

1. Forman, H.J., Maiorino, M., Ursini, F. Signaling functions of reactive oxygen species. *Biochemistry* **2010**, *49*(5), 835–842.
2. Afanas'ev, I.B. Signaling functions of free radicals superoxide & nitric oxide under physiological & pathological conditions. *Mol. Biotechnol.* **2007**, *37*(1), 2–4.
3. Imlay, J.A. Pathways of oxidative damage. *Annu. Rev. Microbiol.* **2003**, *57*, 395–418.
4. Denu, J.M., Tanner, K.G. Specific and reversible inactivation of protein tyrosine phosphatases by hydrogen peroxide: Evidence for a sulfenic acid intermediate and implications for redox regulation. *Biochemistry* **1998**, *37*(16), 5633–5642.
5. Winterbourn, C.C., Hampton, M.B. Thiol chemistry and specificity in redox signaling. *Free Radic. Biol. Med.* **2008**, *45*(5), 549–561.
6. Gardner, P.R., Fridovich, I. Superoxide sensitivity of the *Escherichia coli* aconitase. *J. Biol. Chem.* **1991**, *266*(29), 19328–19333.

7. Berndt, C., Lillig, C.H., Holmgren, A. Thiol-based mechanisms of the thioredoxin and glutaredoxin systems: Implications for diseases in the cardiovascular system. *Am. J. Physiol. Heart Circ. Physiol.* **2007**, *292*(3), H1227–H1236.

8. Hill, B.G., Bhatnagar, A. Role of glutathiolation in preservation, restoration and regulation of protein function. *IUBMB Life* **2007**, *59*(1), 21–26.

9. McCord, J.M., Keele, B.B., Fridovich, I. An enzyme-based theory of obligate anaerobiosis: The physiological function of superoxide dismutase. *Proc. Natl. Acad. Sci. U.S.A.* **1971**, *68*(5), 1024–1027.

10. Imlay, J.A. Cellular defenses against superoxide and hydrogen peroxide. *Annu. Rev. Biochem.* **2008**, *77*, 755–776.

11. Dwyer, D.J., Kohanski, M.A., Collins, J.J. Role of reactive oxygen species in antibiotic action and resistance. *Curr. Opin. Microbiol.* **2009**, *12*(5), 482–489.

12. Kiley, P.J., Beinert, H. The role of Fe-S proteins in sensing and regulation in bacteria. *Curr. Opin. Microbiol.* **2003**, *6*(2), 181–185.

13. Crack, J.C., Jervis, A.J., Gaskell, A.A., White, G.F., Green, J., Thomson, A.J., Le Brun, N.E. Signal perception by FNR: The role of the iron-sulfur cluster. *Biochem. Soc. Trans.* **2008**, *36*(Pt 6), 1144–1148.

14. Melville, S.B., Gunsalus, R.P. Isolation of an oxygen-sensitive FNR protein of *Escherichia coli*: Interaction at activator and repressor sites of FNR-controlled genes. *Proc. Natl. Acad. Sci. U.S.A.* **1996**, *93*(3), 1226–1231.

15. Khoroshilova, N., Popescu, C., Munck, E., Beinert, H., Kiley, P.J. Iron-sulfur cluster disassembly in the FNR protein of *Escherichia coli* by O2: [4Fe-4S] to [2Fe-2S] conversion with loss of biological activity. *Proc. Natl. Acad. Sci. U.S.A.* **1997**, *94*(12), 6087–6092.

16. Dibden, D.P., Green, J. *In vivo* cycling of the *Escherichia coli* transcription factor FNR between active and inactive states. *Microbiology* **2005**, *151*(Pt 12), 4063–4070.

17. Reinhart, F., Achebach, S., Koch, T., Unden, G. Reduced apo-fumarate nitrate reductase regulator (ApoFNR) as the major form of FNR in aerobically growing *Escherichia coli*. *J. Bacteriol.* **2008**, *190*(3), 879–886.

18. Zheng, M., Storz, G. Redox sensing by prokaryotic transcription factors. *Biochem. Pharmacol.* **2000**, *59*(1), 1–6.

19. Wu, J., Weiss, B. Two-stage induction of the soxRS (superoxide response) regulon of *Escherichia coli*. *J. Bacteriol.* **1992**, *174*(12), 3915–3920.

20. Hidalgo, E., Demple, B. An iron-sulfur center essential for transcriptional activation by the redox-sensing SoxR protein. *EMBO J.* **1994**, *13*(1), 138–146.

21. Gaudu, P., Weiss, B. SoxR, a [2Fe-2S] transcription factor, is active only in its oxidized form. *Proc. Natl. Acad. Sci. U.S.A.* **1996**, *93*(19), 10094–10098.

22. Watanabe, S., Kita, A., Kobayashi, K., Miki, K. Crystal structure of the [2Fe-2S] oxidative-stress sensor SoxR bound to DNA. *Proc. Natl. Acad. Sci. U.S.A.* **2008**, *105*(11), 4121–4126.

23. Nunoshiba, T., Hidalgo, E., Amabile Cuevas, C.F., Demple, B. Two-stage control of an oxidative stress regulon: The *Escherichia coli* SoxR protein triggers redox-inducible expression of the soxS regulatory gene. *J. Bacteriol.* **1992**, *174*(19), 6054–6060.

24. Lee, P.E., Demple, B., Barton, J.K. DNA-mediated redox signaling for transcriptional activation of SoxR. *Proc. Natl. Acad. Sci. U.S.A.* **2009**, *106*(32), 13164–13168.

25. Dietrich, L.E.P., Price-Whelan, A., Petersen, A., Whiteley, M., Newman, D.K. The phenazine pyocyanin is a terminal signalling factor in the quorum sensing network of *Pseudomonas aeruginosa*. *Mol. Microbiol.* **2006**, *61*(5), 1308–1321.

26. Sutton, V.R., Stubna, A., Patschkowski, T., Münck, E., Beinert, H., Kiley, P.J. Superoxide destroys the [2Fe-2S]2+ cluster of FNR from *Escherichia coli*. *Biochemistry* **2003**, *43*(3), 791–798.

27. Loewen, P.C. Isolation of catalase-deficient *Escherichia coli* mutants and genetic mapping of katE, a locus that affects catalase activity. *J. Bacteriol.* **1984**, *157*(2), 622–626.

28. Imlay, J.A., Chin, S.M., Linn, S. Toxic DNA damage by hydrogen peroxide through the Fenton reaction *in vivo* and *in vitro*. *Science* **1988**, *240*(4852), 640–642.

29. Zheng, M., Aslund, F., Storz, G. Activation of the OxyR transcription factor by reversible disulfide bond formation. *Science* **1998**, *279*(5357), 1718–1721.

30. Zheng, M., Wang, X., Templeton, L.J., Smulski, D.R., LaRossa, R.A., Storz, G. DNA microarray-mediated transcriptional profiling of the *Escherichia coli* response to hydrogen peroxide. *J. Bacteriol.* **2001**, *183*(15), 4562–4570.

31. Tao, K., Fujita, N., Ishihama, A. Involvement of the RNA polymerase alpha subunit C-terminal region in co-operative interaction and transcriptional activation with OxyR protein. *Mol. Microbiol.* **1993**, *7*(6), 859–864.

32. Lee, J.-W., Helmann, J.D. The PerR transcription factor senses H2O2 by metal-catalysed histidine oxidation. *Nature* **2006**, *440*(7082), 363–367.

33. Herbig, A.F., Helmann, J.D. Roles of metal ions and hydrogen peroxide in modulating the interaction of the *Bacillus subtilis* PerR peroxide regulon repressor with operator DNA. *Mol. Microbiol.* **2001**, *41*(4), 849–859.

34. Traore, D.A.K., El Ghazouani, A., Jacquamet, L., Borel, F., Ferrer, J.-L., Lascoux, D., Ravanat, J.-L., Jaquinod, M., Blondin, G., Caux-Thang, C., Duarte, V., Latour, J.-M. Structural and functional characterization of 2-oxo-histidine in oxidized PerR protein. *Nat. Chem. Biol.* **2009**, *5*(1), 53–59.

35. Bsat, N., Herbig, A., Casillas-Martinez, L., Setlow, P., Helmann, J.D. *Bacillus subtilis* contains multiple Fur homologs: Identification of the iron uptake (Fur) and peroxide regulon (PerR) repressors. *Mol. Microbiol.* **1998**, *29*(1), 189–198.

36. Traore, D.A.K., El Ghazouani, A., Ilango, S., Dupuy, J., Jacquamet, L., Ferrer, J.-L., Caux-Thang, C., Duarte, V., Latour, J.-M. Crystal structure of the apo-PerR-Zn protein from *Bacillus subtilis*. *Mol. Microbiol.* **2006**, *61*(5), 1211–1219.

37. Beyer, W.F., Jr., Fridovich, I. Effect of hydrogen peroxide on the iron-containing superoxide dismutase of *Escherichia coli*. *Biochemistry* **1987**, *26*(5), 1251–1257.

38. Murakami, K., Tsubouchi, R., Fukayama, M., Ogawa, T., Yoshino, M. Oxidative inactivation of reduced NADP-generating enzymes in *E. coli*: Iron-dependent inactivation with affinity cleavage of NADP-isocitrate dehydrogenase. *Arch. Microbiol.* **2006**, *186*(5), 385–392.

39. Jacquamet, L., Traore, D.A.K., Ferrer, J.L., Proux, O., Testemale, D., Hazemann, J.L., Nazarenko, E., El Ghazouani, A., Caux-Thang, C., Duarte, V., Latour, J.M. Structural characterization of the active form of PerR: Insights into the metal-induced activation of PerR and fur proteins for DNA binding. *Mol. Microbiol.* **2009**, *73*(1), 20–31.

40. Helmann, J.D., Wu, M.F., Gaballa, A., Kobel, P.A., Morshedi, M.M., Fawcett, P., Paddon, C. The global transcriptional response of *Bacillus subtilis* to peroxide stress is coordinated by three transcription factors. *J. Bacteriol.* **2003**, *185*(1), 243–253.

41. Cooper, J.B., McIntyre, K., Badasso, M.O., Wood, S.P., Zhang, Y., Garbe, T.R., Young, D. X-ray structure analysis of the iron-dependent superoxide dismutase from *Mycobacterium tuberculosis* at 2.0 Angstroms resolution reveals novel dimer-dimer interactions. *J. Mol. Biol.* **1995**, *246*(4), 531–544.

42. Park, S., You, X., Imlay, J.A. Substantial DNA damage from submicromolar intracellular hydrogen peroxide detected in Hpx- mutants of *Escherichia coli*. *Proc. Natl. Acad. Sci. U.S.A.* **2005**, *102*(26), 9317–9322.

43. Lillig, C.H., Berndt, C., Vergnolle, O., Lonn, M.E., Hudemann, C., Bill, E., Holmgren, A. Characterization of human glutaredoxin 2 as iron-sulfur protein: A possible role as redox sensor. *Proc. Natl. Acad. Sci. U.S.A.* **2005**, *102*(23), 8168–8173.

44. Cadenas, E., Davies, K.J.A. Mitochondrial free radical generation, oxidative stress, and aging. *Free Radic. Biol. Med.* **2000**, *29*(3–4), 222–230.

45. Grivennikova, V.G., Vinogradov, A.D. Generation of superoxide by the mitochondrial complex I. *Biochim. Biophys. Acta* **2006**, *1757*(5–6), 553–561.

46. Muller, F.L., Liu, Y., Van Remmen, H. Complex III releases superoxide to both sides of the inner mitochondrial membrane. *J. Biol. Chem.* **2004**, *279*(47), 49064–49073.

47. Han, D., Antunes, F., Canali, R., Rettori, D., Cadenas, E. Voltage-dependent anion channels control the release of the superoxide anion from mitochondria to cytosol. *J. Biol. Chem.* **2003**, *278*(8), 5557–5563.

48. Okado-Matsumoto, A., Fridovich, I. Subcellular distribution of superoxide dismutases (SOD) in rat liver. *J. Biol. Chem.* **2001**, *276*(42), 38388–38393.

49. Hamanaka, R.B., Chandel, N.S. Mitochondrial reactive oxygen species regulate cellular signaling and dictate biological outcomes. *Trends Biochem. Sci.* **2010**, *35*(9), 505–513.

50. Sumimoto, H. Structure, regulation and evolution of Nox-family NADPH oxidases that produce reactive oxygen species. *FEBS J.* **2008**, *275*(13), 3249–3277.

51. Bedard, K., Krause, K.-H. The NOX family of ROS-generating NADPH oxidases: Physiology and pathophysiology. *Physiol. Rev.* **2007**, *87*(1), 245–313.

52. Bokoch, G.M., Diebold, B.A. Current molecular models for NADPH oxidase regulation by Rac GTPase. *Blood* **2002**, *100*(8), 2692–2695.

53. Donko, A., Peterfi, Z., Sum, A., Leto, T., Geiszt, M. Dual oxidases. *Philos. Trans. R. Soc. Lond., B, Biol. Sci.* **2005**, *360*(1464), 2301–2308.

54. Takeya, R., Ueno, N., Kami, K., Taura, M., Kohjima, M., Izaki, T., Nunoi, H., Sumimoto, H. Novel human homologues of p47phox and p67phox participate in activation of superoxide-producing NADPH oxidases. *J. Biol. Chem.* **2003**, *278*(27), 25234–25246.

55. Cheng, G., Ritsick, D., Lambeth, J.D. Nox3 regulation by NOXO1, p47phox, and p67phox. *J. Biol. Chem.* **2004**, *279*(33), 34250–34255.

56. Martyn, K.D., Frederick, L.M., von Loehneysen, K., Dinauer, M.C., Knaus, U.G. Functional analysis of Nox4 reveals unique characteristics compared to other NADPH oxidases. *Cell. Signal.* **2006**, *18*(1), 69–82.

57. Jagnandan, D., Church, J.E., Banfi, B., Stuehr, D.J., Marrero, M.B., Fulton, D.J.R. Novel mechanism of activation of NADPH oxidase 5. *J. Biol. Chem.* **2007**, *282*(9), 6494–6507.

58. Hwang, J., Saha, A., Boo, Y.C., Sorescu, G.P., McNally, J.S., Holland, S.M., Dikalov, S., Giddens, D.P., Griendling, K.K., Harrison, D.G., Jo, H. Oscillatory shear stress stimulates endothelial production of O_2^- from p47phox-dependent NAD(P)H oxidases, leading to monocyte adhesion. *J. Biol. Chem.* **2003**, *278*(47), 47291–47298.

59. Ammendola, R., Ruocchio, M.R., Chirico, G., Russo, L., De Felice, C., Esposito, F., Russo, T., Cimino, F. Inhibition of NADH/NADPH oxidase affects signal transduction by growth factor receptors in normal fibroblasts. *Arch. Biochem. Biophys.* **2002**, *397*(2), 253–257.

60. Wood, Z.A., Poole, L.B., Karplus, P.A. Peroxiredoxin evolution and the regulation of hydrogen peroxide signaling. *Science* **2003**, *300*(5619), 650–653.

61. Biteau, B., Labarre, J., Toledano, M.B. ATP-dependent reduction of cysteine-sulphinic acid by *S. cerevisiae* sulphiredoxin. *Nature* **2003**, *425*(6961), 980–984.

62. Kang, S.W., Chae, H.Z., Seo, M.S., Kim, K., Baines, I.C., Rhee, S.G. Mammalian peroxiredoxin isoforms can reduce hydrogen peroxide generated in response to growth factors and tumor necrosis factor-α. *J. Biol. Chem.* **1998**, *273*(11), 6297–6302.

63. Pascual, M.B., Mata-Cabana, A., Florencio, F.J., Lindahl, M., Cejudo, F.J. Overoxidation of 2-Cys peroxiredoxin in prokaryotes. *J. Biol. Chem.* **2010**, *285*(45), 34485–34492.

64. Chiarugi, P., Cirri, P. Redox regulation of protein tyrosine phosphatases during receptor tyrosine kinase signal transduction. *Trends Biochem. Sci.* **2003**, *28*(9), 509–514.
65. Denu, J.M., Dixon, J.E. Protein tyrosine phosphatases: Mechanisms of catalysis and regulation. *Curr. Opin. Chem. Biol.* **1998**, *2*(5), 633–641.
66. Cully, M., You, H., Levine, A.J., Mak, T.W. Beyond PTEN mutations: The PI3K pathway as an integrator of multiple inputs during tumorigenesis. *Nat. Rev. Cancer* **2006**, *6*(3), 184–192.
67. Ushio-Fukai, M., Alexander, R.W., Akers, M., Yin, Q., Fujio, Y., Walsh, K., Griendling, K.K. Reactive oxygen species mediate the activation of Akt/protein kinase B by angiotensin II in vascular smooth muscle cells. *J. Biol. Chem.* **1999**, *274*(32), 22699–22704.
68. Wang, X., McCullough, K.D., Franke, T.F., Holbrook, N.J. Epidermal growth factor receptor-dependent Akt activation by oxidative stress enhances cell survival. *J. Biol. Chem.* **2000**, *275*(19), 14624–14631.
69. Leslie, N.R., Bennett, D., Lindsay, Y.E., Stewart, H., Gray, A., Downes, C.P. Redox regulation of PI 3-kinase signalling via inactivation of PTEN. *EMBO J.* **2003**, *22*(20), 5501–5510.
70. Welch, H.C.E., Coadwell, W.J., Ellson, C.D., Ferguson, G.J., Andrews, S.R., Erdjument-Bromage, H., Tempst, P., Hawkins, P.T., Stephens, L.R. P-Rex1, a PtdIns(3,4,5)P3- and G^{23}-regulated guanine-nucleotide exchange factor for Rac. *Cell* **2002**, *108*(6), 809–821.
71. Konishi, H., Matsuzaki, H., Tanaka, M., Takemura, Y., Kuroda, S., Ono, Y., Kikkawa, U. Activation of protein kinase B (Akt/RAC-protein kinase) by cellular stress and its association with heat shock protein Hsp27. *FEBS Lett.* **1997**, *410*(2–3), 493–498.
72. Martín, D., Salinas, M., Fujita, N., Tsuruo, T., Cuadrado, A. Ceramide and reactive oxygen species generated by H2O2 induce caspase-3-independent degradation of Akt/protein kinase B. *J. Biol. Chem.* **2002**, *277*(45), 42943–42952.
73. Spitaler, M., Cantrell, D.A. Protein kinase C and beyond. *Nat. Immunol.* **2004**, *5*(8), 785–790.
74. Rasmussen, H., Takuwa, Y., Park, S. Protein kinase C in the regulation of smooth muscle contraction. *FASEB J.* **1987**, *1*(3), 177–185.
75. Musashi, M., Ota, S., Shiroshita, N. The role of protein kinase C isoforms in cell proliferation and apoptosis. *Int. J. Hematol.* **2000**, *72*(1), 12–19.
76. Wang, X.T., McCullough, K.D., Wang, X.J., Carpenter, G., Holbrook, N.J. Oxidative stress-induced phospholipase C-gamma 1 activation enhances cell survival. *J. Biol. Chem.* **2001**, *276*(30), 28364–28371.
77. Kass, G.E., Duddy, S.K., Orrenius, S. Activation of hepatocyte protein kinase C by redox-cycling quinones. *Biochem. J.* **1989**, *260*(2), 499–507.
78. Gopalakrishna, R., Anderson, W.B. Ca2+- and phospholipid-independent activation of protein kinase C by selective oxidative modification of the regulatory domain. *Proc. Natl. Acad. Sci. U.S.A.* **1989**, *86*(17), 6758–6762.
79. Chakraborti, S., Chakraborti, T. Oxidant-mediated activation of mitogen-activated protein kinases and nuclear transcription factors in the cardiovascular system: A brief overview. *Cell. Signal.* **1998**, *10*(10), 675–683.
80. Guyton, K.Z., Liu, Y., Gorospe, M., Xu, Q., Holbrook, N.J. Activation of mitogen-activated protein kinase by HO. *J. Biol. Chem.* **1996**, *271*(8), 4138–4142.
81. Ushio-Fukai, M., Alexander, R.W., Akers, M., Griendling, K.K. p38 mitogen-activated protein kinase is a critical component of the redox-sensitive signaling pathways activated by angiotensin II. *J. Biol. Chem.* **1998**, *273*(24), 15022–15029.
82. Marín, J., Encabo, A., Briones, A., García-Cohen, E.-C., Alonso, M.J. Mechanisms involved in the cellular calcium homeostasis in vascular smooth muscle: Calcium pumps. *Life Sci.* **1998**, *64*(5), 279–303.
83. Bygrave, F.L., Benedetti, A. What is the concentration of calcium ions in the endoplasmic reticulum? *Cell Calcium* **1996**, *19*(6), 547–551.
84. Grover, A.K., Samson, S.E., Fomin, V.P. Peroxide inactivates calcium pumps in pig coronary artery. *Am. J. Physiol. Heart Circ. Physiol.* **1992**, *263*(2), H537–H543.
85. Han, H.M., Wei, R.S., Lai, F.A., Yin, C.C. Molecular nature of sulfhydryl modification by hydrogen peroxide on type 1 ryanodine receptor. *Acta Pharmacol. Sin.* **2006**, *27*(7), 888–894.
86. Hidalgo, C., Sanchez, G., Barrientos, G., Aracena-Parks, P. A transverse tubule NADPH oxidase activity stimulates calcium release from isolated triads via ryanodine receptor type 1 S-glutathionylation. *J. Biol. Chem.* **2006**, *281*(36), 26473–26482.
87. Zissimopoulos, S., Lai, F.A. Redox regulation of the ryanodine receptor/calcium release channel. *Biochem. Soc. Trans.* **2006**, *34*(Pt 5), 919–921.
88. Zheng, Y., Shen, X. H2O2 directly activates inositol 1,4,5-trisphosphate receptors in endothelial cells. *Redox Rep.* **2005**, *10*(1), 29–36.
89. González-Pacheco, F.R., Caramelo, C., Castilla, M.Á., Deudero, J.J.P., Arias, J., Yagüe, S., Jiménez, S., Bragado, R., Álvarez-Arroyo, M.V. Mechanism of vascular smooth muscle cells activation by hydrogen peroxide: Role of phospholipase C gamma. *Nephrol. Dial. Transplant.* **2002**, *17*(3), 392–398.
90. Yan, Y., Wei, C.L., Zhang, W.R., Cheng, H.P., Liu, J. Crosstalk between calcium and reactive oxygen species signaling. *Acta Pharmacol. Sin.* **2006**, *27*(7), 821–826.
91. Liu, G., Chen, X. Regulation of the p53 transcriptional activity. *J. Cell. Biochem.* **2006**, *97*(3), 448–458.
92. Desaint, S., Luriau, S., Aude, J.-C., Rousselet, G., Toledano, M.B. Mammalian antioxidant defenses are not inducible by H2O2. *J. Biol. Chem.* **2004**, *279*(30), 31157–31163.
93. Sablina, A.A., Budanov, A.V., Ilyinskaya, G.V., Agapova, L.S., Kravchenko, J.E., Chumakov, P.M. The antioxidant

function of the p53 tumor suppressor. *Nat. Med.* **2005**, *11*(12), 1306–1313.

94. Drane, P., Bravard, A., Bouvard, V., May, E. Reciprocal down-regulation of p53 and SOD2 gene expression-implication in p53 mediated apoptosis. *Oncogene* **2001**, *20*(4), 430–439.

95. Rivera, A., Maxwell, S.A. The p53-induced gene-6 (proline oxidase) mediates apoptosis through a calcineurin-dependent pathway. *J. Biol. Chem.* **2005**, *280*(32), 29346–29354.

96. Rainwater, R., Parks, D., Anderson, M., Tegtmeyer, P., Mann, K. Role of cysteine residues in regulation of p53 function. *Mol. Cell. Biol.* **1995**, *15*(7), 3892–3903.

97. Seemann, S., Hainaut, P. Roles of thioredoxin reductase 1 and APE/Ref-1 in the control of basal p53 stability and activity. *Oncogene* **2005**, *24*(24), 3853–3863.

98. Bragado, P., Armesilla, A., Silva, A., Porras, A. Apoptosis by cisplatin requires p53 mediated p38alpha MAPK activation through ROS generation. *Apoptosis* **2007**, *12*(9), 1733–1742.

99. Oeckinghaus, A., Ghosh, S. The NF-κB family of transcription factors and its regulation. *Cold Spring Harb. Perspect. Biol.* **2009**, *1*(4), a000034.

100. Barkett, M., Gilmore, T.D. Control of apoptosis by Rel/NF-kappaB transcription factors. *Oncogene* **1999**, *18*(49), 6910–6924.

101. Pineda-Molina, E., Klatt, P., Vázquez, J., Marina, A., García de Lacoba, M., Pérez-Sala, D., Lamas, S. Glutathionylation of the p50 subunit of NF-κB: A mechanism for redox-induced inhibition of DNA binding. *Biochemistry* **2001**, *40*(47), 14134–14142.

102. Kil, I.S., Kim, S.Y., Park, J.W. Glutathionylation regulates IkappaB. *Biochem. Biophys. Res. Commun.* **2008**, *373*(1), 169–173.

103. Oliveira-Marques, V., Marinho, H.S., Cyrne, L., Antunes, F. Modulation of NF-κB-dependent gene expression by H_2O_2: A major role for a simple chemical process in a complex biological response. *Antioxid. Redox Signal.* **2009**, *11*(9), 2043–2053.

104. Jamaluddin, M., Wang, S., Boldogh, I., Tian, B., Brasier, A.R. TNF-alpha-induced NF-kappaB/RelA Ser(276) phosphorylation and enhanceosome formation is mediated by an ROS-dependent PKAc pathway. *Cell. Signal.* **2007**, *19*(7), 1419–1433.

105. Dan, H.C., Cooper, M.J., Cogswell, P.C., Duncan, J.A., Ting, J.P., Baldwin, A.S. Akt-dependent regulation of NF-{kappa}B is controlled by mTOR and Raptor in association with IKK. *Genes Dev.* **2008**, *22*(11), 1490–1500.

106. Dechend, R., Hirano, F., Lehmann, K., Heissmeyer, V., Ansieau, S., Wulczyn, F.G., Scheidereit, C., Leutz, A. The Bcl-3 oncoprotein acts as a bridging factor between NF-κB/Rel and nuclear co-regulators. *Oncogene* **1999**, *18*(22), 3316–3323.

107. Pang, H., Bartlam, M., Zeng, Q., Miyatake, H., Hisano, T., Miki, K., Wong, L.L., Gao, G.F., Rao, Z. Crystal structure of human pirin: An iron-binding nuclear protein and transcription cofactor. *J. Biol. Chem.* **2004**, *279*(2), 1491–1498.

108. Morgan, M.J., Liu, Z.-G. Crosstalk of reactive oxygen species and NF-[kappa]B signaling. *Cell Res.* **2011**, *21*(1), 103–115.

109. Fandrey, J. Hypoxia-inducible gene expression. *Respir. Physiol.* **1995**, *101*(1), 1–10.

110. Seagroves, T.N., Ryan, H.E., Lu, H., Wouters, B.G., Knapp, M., Thibault, P., Laderoute, K., Johnson, R.S. Transcription factor HIF-1 is a necessary mediator of the pasteur effect in mammalian cells. *Mol. Cell. Biol.* **2001**, *21*(10), 3436–3444.

111. Wang, G.L., Jiang, B.H., Rue, E.A., Semenza, G.L. Hypoxia-inducible factor 1 is a basic-helix-loop-helix-PAS heterodimer regulated by cellular O_2 tension. *Proc. Natl. Acad. Sci. U.S.A.* **1995**, *92*(12), 5510–5514.

112. Hirota, K., Semenza, G.L. Regulation of hypoxia-inducible factor 1 by prolyl and asparaginyl hydroxylases. *Biochem. Biophys. Res. Commun.* **2005**, *338*(1), 610–616.

113. Huang, L.E., Gu, J., Schau, M., Bunn, H.F. Regulation of hypoxia-inducible factor 1alpha is mediated by an O_2-dependent degradation domain via the ubiquitin-proteasome pathway. *Proc. Natl. Acad. Sci. U.S.A.* **1998**, *95*(14), 7987–7992.

114. Schofield, C.J., Zhang, Z. Structural and mechanistic studies on 2-oxoglutarate-dependent oxygenases and related enzymes. *Curr. Opin. Struct. Biol.* **1999**, *9*(6), 722–731.

115. Appelhoff, R.J., Tian, Y.M., Raval, R.R., Turley, H., Harris, A.L., Pugh, C.W., Ratcliffe, P.J., Gleadle, J.M. Differential function of the prolyl hydroxylases PHD1, PHD2, and PHD3 in the regulation of hypoxia-inducible factor. *J. Biol. Chem.* **2004**, *279*(37), 38458–38465.

116. Berra, E., Benizri, E., Ginouves, A., Volmat, V., Roux, D., Pouyssegur, J. HIF prolyl-hydroxylase 2 is the key oxygen sensor setting low steady-state levels of HIF-1alpha in normoxia. *EMBO J.* **2003**, *22*(16), 4082–4090.

117. Jaakkola, P., Mole, D.R., Tian, Y.M., Wilson, M.I., Gielbert, J., Gaskell, S.J., Kriegsheim, A., Hebestreit, H.F., Mukherji, M., Schofield, C.J., Maxwell, P.H., Pugh, C.W., Ratcliffe, P.J. Targeting of HIF-alpha to the von Hippel-Lindau ubiquitylation complex by O_2-regulated prolyl hydroxylation. *Science* **2001**, *292*(5516), 468–472.

118. Hon, W.C., Wilson, M.I., Harlos, K., Claridge, T.D., Schofield, C.J., Pugh, C.W., Maxwell, P.H., Ratcliffe, P.J., Stuart, D.I., Jones, E.Y. Structural basis for the recognition of hydroxyproline in HIF-1 alpha by pVHL. *Nature* **2002**, *417*(6892), 975–978.

119. Metzen, E., Berchner-Pfannschmidt, U., Stengel, P., Marxsen, J.H., Stolze, I., Klinger, M., Huang, W.Q., Wotzlaw, C., Hellwig-Burgel, T., Jelkmann, W., Acker, H., Fandrey, J. Intracellular localisation of human HIF-1{alpha} hydroxylases: Implications for oxygen sensing. *J. Cell Sci.* **2003**, *116*(7), 1319–1326.

120. Berra, E., Richard, D.E., Gothie, E., Pouyssegur, J. HIF-1-dependent transcriptional activity is required for

oxygen-mediated HIF-1alpha degradation. *FEBS Lett.* **2001**, *491*(1–2), 85–90.

121. Gorres, K.L., Raines, R.T. Prolyl 4-hydroxylase. *Crit. Rev. Biochem. Mol. Biol.* **2010**, *45*(2), 106–124.

122. Hirsila, M., Koivunen, P., Gunzler, V., Kivirikko, K.I., Myllyharju, J. Characterization of the human prolyl 4-hydroxylases that modify the hypoxia-inducible factor. *J. Biol. Chem.* **2003**, *278*(33), 30772–30780.

123. Ehrismann, D., Flashman, E., Genn, D.N., Mathioudakis, N., Hewitson, K.S., Ratcliffe, P.J., Schofield, C.J. Studies on the activity of the hypoxia-inducible-factor hydroxylases using an oxygen consumption assay. *Biochem. J.* **2007**, *401*(1), 227–234.

124. Lando, D., Peet, D.J., Whelan, D.A., Gorman, J.J., Whitelaw, M.L. Asparagine hydroxylation of the HIF transactivation domain a hypoxic switch. *Science* **2002**, *295*(5556), 858–861.

125. Dayan, F., Roux, D., Brahimi-Horn, M.C., Pouyssegur, J., Mazure, N.M. The oxygen sensor factor-inhibiting hypoxia-inducible factor-1 controls expression of distinct genes through the bifunctional transcriptional character of hypoxia-inducible factor-1α. *Cancer Res.* **2006**, *66*(7), 3688–3698.

126. Koivunen, P., Hirsila, M., Gunzler, V., Kivirikko, K.I., Myllyharju, J. Catalytic properties of the asparaginyl hydroxylase (FIH) in the oxygen sensing pathway are distinct from those of its prolyl 4-hydroxylases. *J. Biol. Chem.* **2004**, *279*(11), 9899–9904.

127. Hegg, E.L., Que, L., Jr. The 2-His-1-carboxylate facial triad—An emerging structural motif in mononuclear non-heme iron(II) enzymes. *Eur. J. Biochem.* **1997**, *250*(3), 625–629.

128. Selak, M.A., Armour, S.M., MacKenzie, E.D., Boulahbel, H., Watson, D.G., Mansfield, K.D., Pan, Y., Simon, M.C., Thompson, C.B., Gottlieb, E. Succinate links TCA cycle dysfunction to oncogenesis by inhibiting HIF-alpha prolyl hydroxylase. *Cancer Cell* **2005**, *7*(1), 77–85.

129. Serra-Perez, A., Planas, A.M., Nunez-O'Mara, A., Berra, E., Garcia-Villoria, J., Ribes, A., Santalucia, T. Extended ischemia prevents HIF1alpha degradation at reoxygenation by impairing prolyl-hydroxylation: Role of Krebs cycle metabolites. *J. Biol. Chem.* **2010**, *285*(24), 18217–18224.

130. Vordermark, D., Kraft, P., Katzer, A., Bolling, T., Willner, J., Flentje, M. Glucose requirement for hypoxic accumulation of hypoxia-inducible factor-1alpha (HIF-1alpha). *Cancer Lett.* **2005**, *230*(1), 122–133.

131. Epstein, A.C., Gleadle, J.M., McNeill, L.A., Hewitson, K.S., O'Rourke, J., Mole, D.R., Mukherji, M., Metzen, E., Wilson, M.I., Dhanda, A., Tian, Y.M., Masson, N., Hamilton, D.L., Jaakkola, P., Barstead, R., Hodgkin, J., Maxwell, P.H., Pugh, C.W., Schofield, C.J., Ratcliffe, P.J. C. elegans EGL-9 and mammalian homologs define a family of dioxygenases that regulate HIF by prolyl hydroxylation. *Cell* **2001**, *107*(1), 43–54.

132. De Jong, L., Albracht, S.P.J., Kemp, A. Prolyl 4-hydroxylase activity in relation to the oxidation state of enzyme-bound iron: The role of ascorbate in peptidyl proline hydroxylation. *Biochim. Biophys. Acta* **1982**, *704*(2), 326–332.

133. Kivirikko, K.I., Myllyharju, J. Prolyl 4-hydroxylases and their protein disulfide isomerase subunit. *Matrix Biol.* **1998**, *16*(7), 357–368.

134. Knowles, H.J., Raval, R.R., Harris, A.L., Ratcliffe, P.J. Effect of ascorbate on the activity of hypoxia-inducible factor in cancer cells. *Cancer Res.* **2003**, *63*(8), 1764–1768.

135. Ushio-Fukai, M. Redox signaling in angiogenesis: Role of NADPH oxidase. *Cardiovasc. Res.* **2006**, *71*(2), 226–235.

136. Gerald, D., Berra, E., Frapart, Y.M., Chan, D.A., Giaccia, A.J., Mansuy, D., Pouysségur, J., Yaniv, M., Mechta-Grigoriou, F. JunD reduces tumor angiogenesis by protecting cells from oxidative stress. *Cell* **2004**, *118*(6), 781–794.

137. Brunelle, J.K., Bell, E.L., Quesada, N.M., Vercauteren, K., Tiranti, V., Zeviani, M., Scarpulla, R.C., Chandel, N.S. Oxygen sensing requires mitochondrial ROS but not oxidative phosphorylation. *Cell Metab.* **2005**, *1*(6), 409–414.

138. Diebold, I., Petry, A., Hess, J., Gorlach, A. The NADPH oxidase subunit NOX4 is a new target gene of the hypoxia-inducible factor-1. *Mol. Biol. Cell* **2010**, *21*(12), 2087–2096.

139. Koong, A.C., Chen, E.Y., Giaccia, A.J. Hypoxia causes the activation of nuclear factor kappa B through the phosphorylation of I kappa B alpha on tyrosine residues. *Cancer Res.* **1994**, *54*(6), 1425–1430.

140. Cummins, E.P., Berra, E., Comerford, K.M., Ginouves, A., Fitzgerald, K.T., Seeballuck, F., Godson, C., Nielsen, J.E., Moynagh, P., Pouyssegur, J., Taylor, C.T. Prolyl hydroxylase-1 negatively regulates IκB kinase-β, giving insight into hypoxia-induced NFκB activity. *Proc. Natl. Acad. Sci. U.S.A.* **2006**, *103*(48), 18154–18159.

141. Bonello, S., Zahringer, C., BelAiba, R.S., Djordjevic, T., Hess, J., Michiels, C., Kietzmann, T., Gorlach, A. Reactive oxygen species activate the HIF-1alpha promoter via a functional NFkappaB site. *Arterioscler. Thromb. Vasc. Biol.* **2007**, *27*(4), 755–761.

142. van Uden, P., Kenneth, N.S., Rocha, S. Regulation of hypoxia-inducible factor-1alpha by NF-kappaB. *Biochem. J.* **2008**, *412*(3), 477–484.

143. Fitzpatrick, S.F., Tambuwala, M.M., Bruning, U., Schaible, B., Scholz, C.C., Byrne, A., O'Connor, A., Gallagher, W.M., Lenihan, C.R., Garvey, J.F., Howell, K., Fallon, P.G., Cummins, E.P., Taylor, C.T. An intact canonical NF-{kappa}B pathway is required for inflammatory gene expression in response to hypoxia. *J. Immunol.* **2011**, *186*(2), 1091–1096.

144. Gloire, G., Legrand-Poels, S., Piette, J. NF-[kappa]B activation by reactive oxygen species: Fifteen years later. *Biochem. Pharmacol.* **2006**, *72*(11), 1493–1505.

145. Haddad, J.J., Land, S.C. A non-hypoxic, ROS-sensitive pathway mediates TNF-[alpha]-dependent regulation of HIF-1[alpha]. *FEBS Lett.* **2001**, *505*(2), 269–274.

146. Manea, A., Tanase, L.I., Raicu, M., Simionescu, M. Transcriptional regulation of NADPH oxidase isoforms, Nox1 and Nox4, by nuclear factor-[kappa]B in human aortic smooth muscle cells. *Biochem. Biophys. Res. Commun.* **2010**, *396*(4), 901–907.

147. Cockman, M.E., Webb, J.D., Ratcliffe, P.J. FIH-dependent asparaginyl hydroxylation of ankyrin repeat domain-containing proteins. *Ann. N. Y. Acad. Sci.* **2009**, *1177*, 9–18.

148. Christen, Y. Oxidative stress and Alzheimer disease. *Am. J. Clin. Nutr.* **2000**, *71*(2), 621s–629s.

149. Jenner, P. Oxidative stress in Parkinson's disease. *Ann. Neurol.* **2003**, *53*(Suppl. 3), S26–S36; discussion S36–S38.

150. Maritim, A.C., Sanders, R.A., Watkins, J.B., 3rd. Diabetes, oxidative stress, and antioxidants: A review. *J. Biochem. Mol. Toxicol.* **2003**, *17*(1), 24–38.

151. Semenza, G.L. HIF-1 mediates the Warburg effect in clear cell renal carcinoma. *J. Bioenerg. Biomembr.* **2007**, *39*(3), 231–234.

152. Semenza, G.L. Hypoxia-inducible factor 1 and cancer pathogenesis. *IUBMB Life* **2008**, *60*(9), 591–597.

153. Szatrowski, T.P., Nathan, C.F. Production of large amounts of hydrogen peroxide by human tumor cells. *Cancer Res.* **1991**, *51*(3), 794–798.

154. Ushio-Fukai, M., Nakamura, Y. Reactive oxygen species and angiogenesis: NADPH oxidase as target for cancer therapy. *Cancer Lett.* **2008**, *266*(1), 37–52.

155. Dolcet, X., Llobet, D., Pallares, J., Matias-Guiu, X. NF-kB in development and progression of human cancer. *Virchows Arch.* **2005**, *446*(5), 475–482.

156. Cai, T., Fassina, G., Morini, M., Aluigi, M.G., Masiello, L., Fontanini, G., D'Agostini, F., De Flora, S., Noonan, D.M., Albini, A. N-acetylcysteine inhibits endothelial cell invasion and angiogenesis. *Lab. Invest.* **1999**, *79*(9), 1151–1159.

157. Onnis, B., Rapisarda, A., Melillo, G. Development of HIF-1 inhibitors for cancer therapy. *J. Cell. Mol. Med.* **2009**, *13*(9A), 2780–2786.

158. Huang, P., Feng, L., Oldham, E.A., Keating, M.J., Plunkett, W. Superoxide dismutase as a target for the selective killing of cancer cells. *Nature* **2000**, *407*(6802), 390–395.

159. Kirches, E., Warich-Kirches, M. 2-methoxyestradiol as a potential cytostatic drug in gliomas? *Anticancer Agents Med. Chem.* **2009**, *9*(1), 55–65.

160. Matoba, T., Shimokawa, H., Nakashima, M., Hirakawa, Y., Mukai, Y., Hirano, K., Kanaide, H., Takeshita, A. Hydrogen peroxide is an endothelium-derived hyperpolarizing factor in mice. *J. Clin. Invest.* **2000**, *106*(12), 1521–1530.

161. Touyz, R.M. Reactive oxygen species and angiotensin II signaling in vascular cells—Implications in cardiovascular disease. *Braz. J. Med. Biol. Res.* **2004**, *37*(8), 1263–1273.

162. Griendling, K.K., Sorescu, D., Ushio-Fukai, M. NAD(P)H oxidase: Role in cardiovascular biology and disease. *Circ. Res.* **2000**, *86*(5), 494–501.

163. Taniyama, Y., Griendling, K.K. Reactive oxygen species in the vasculature: Molecular and cellular mechanisms. *Hypertension* **2003**, *42*(6), 1075–1081.

164. Fortuño, A., José, G.S., Moreno, M.U., Díez, J., Zalba, G. Oxidative stress and vascular remodelling. *Exp. Physiol.* **2005**, *90*(4), 457–462.

165. Shargorodsky, M., Debby, O., Matas, Z., Zimlichman, R. Effect of long-term treatment with antioxidants (vitamin C, vitamin E, coenzyme Q10 and selenium) on arterial compliance, humoral factors and inflammatory markers in patients with multiple cardiovascular risk factors. *Nutr. Metab.* **2010**, *7*(1), 55.

166. Griendling, K.K., FitzGerald, G.A. Oxidative stress and cardiovascular injury: Part II: Animal and human studies. *Circulation* **2003**, *108*(17), 2034–2040.

167. Williams, H.C., Griendling, K.K. NADPH oxidase inhibitors: New antihypertensive agents? *J. Cardiovasc. Pharmacol.* **2007**, *50*(1), 9–16.

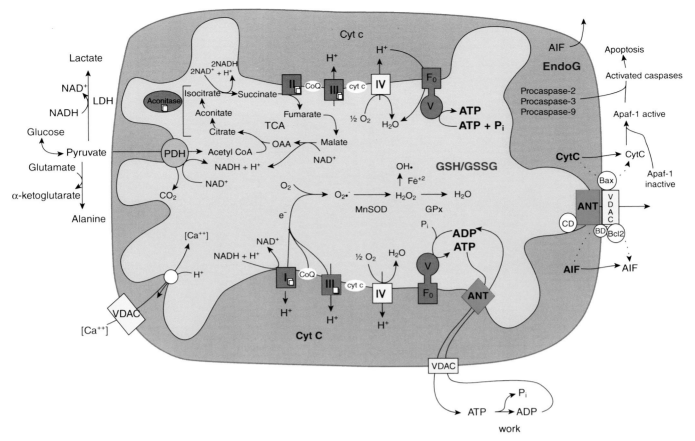

Figure 6.2 Schematic representation illustrating the relationship of oxidative phosphorylation, production of reactive oxygen species, and initiation of programmed cell death or apoptosis. (Adapted with permission from *Science* **1999**, *283*, 1482–1488. Copyright 1999 AAAS.)

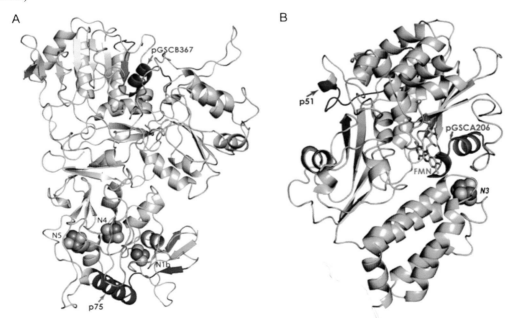

Figure 6.3 (A) Homology model of the 75-kDa subunit using the crystal structure of *T. thermophilus* (2FUG) as a template. Arrows show the domains of pGSCB367 and p75, denoted by red and blue ribbons. (B) Homology model of the 51-kDa subunit. Arrows show the domains of pGSCA206 and p51, denoted by red and blue ribbons. (Adapted with permission from *J. Pept. Sci.* **2011**, *96*, 207–221. Copyright 2011 John Wiley & Sons.)

Molecular Basis of Oxidative Stress: Chemistry, Mechanisms, and Disease Pathogenesis, First Edition. Edited by Frederick A. Villamena.
© 2013 John Wiley & Sons, Inc. Published 2013 by John Wiley & Sons, Inc.

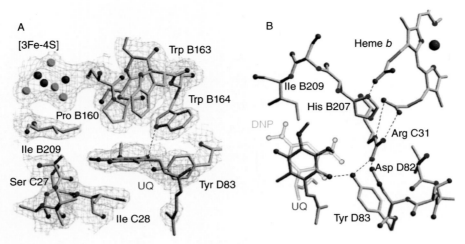

Figure 6.4 (A) Hydrophobic residues in the ubiquinone-binding site of *E. coli* SQR. (B) Polar interactions in the ubiquinone-binding site of *E. coli* SQR. (Adapted with permission from *Science* **2003**, *299*, 700–704. Copyright 2003 AAAS.)

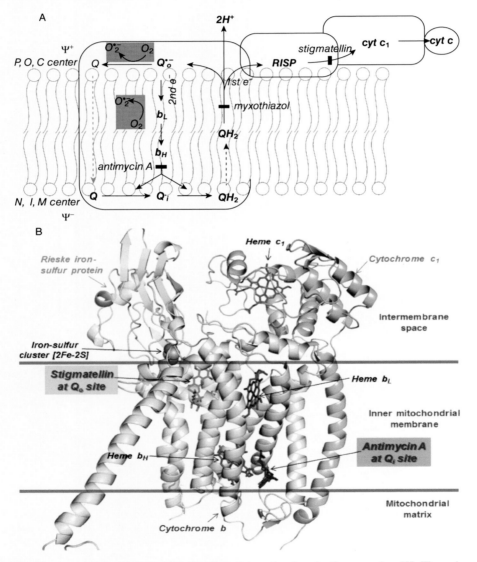

Figure 6.5 (A) Superoxide generation mediated by the Q-cycle mechanism in the complex III. The scheme is adapted from Reference 31 with modification. Gray areas symbolize the reactions involved in $O_2^{\bullet-}$ production. P, O, and C represent positive, outside, and cytoplasmic side, respectively. N, I, and M stands for negative, inside, and matrix side, respectively. (B) X-ray structure (pdb 1PPJ) of complex III shows Q_o site (occupied by Q_o site inhibitor, stigmatellin in green color) is located immediately next to the intermembrane space. Structure also shows the Q_i site (occupied by Q_i site inhibitor, antimycin A in blue color) is located next to the matrix site. The subunits of cytochrome *b*, cytochrome c_1, and Rieske iron–sulfur protein (RISP) are denoted by cyan, pale green, and yellow ribbons, respectively.

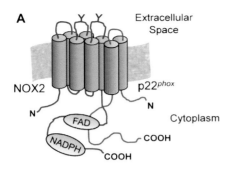

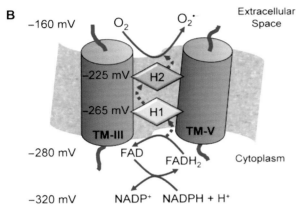

Figure 7.1 Model of phagocyte flavocytochrome b and proposed electron flow. Panel A. Flavocytochrome b is a heterodimer of gp91^{phox}/NOX2 and p22^{phox}. The proposed sites of NOX2 glycosylation (Y) and general location of binding domains for other redox components (FAD, NADPH) are indicated. Panel B. Flavocytochrome b transfers electrons across the plasma membrane or phagosomal membrane. Redox components involved in this process. Flavocytochrome b hemes (H1 are H2) are coordinated by NOX2 transmembrane helices (TM-III and TM-V), resulting in a pathway for electron transfer across the membrane. The redox midpoint potentials at pH 7.0 (Em) shown for each step in the pathway indicate that the transfer electrons from NADPH to O$_2$ is energetically favorable.

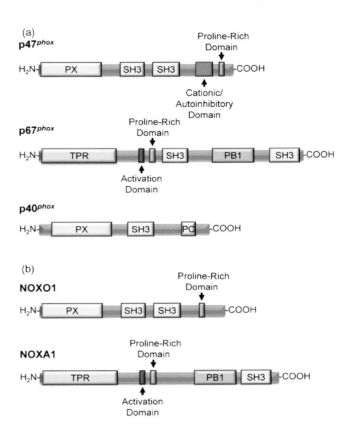

Figure 7.2 Structural models of the NADPH oxidase cytosolic cofactors. Major functional domains are indicated on these scale models of the phagocyte NADPH oxidase cytosolic proteins (Panel A) and the nonphagocyte homologs identified (Panel B). Src homology 3 (SH3) domains are present in all cytosolic cofactors and play a role in protein–protein binding interactions by binding to proline-rich domains within and between proteins. The N-terminal *phox* homology (PX) domains present in p47^{phox}, p40^{phox}, and NOXO1 participate in phosphoinositide binding interactions involved in targeting to cellular locaction. The p47^{phox} auto-inhibitory region (AIR) mediates an intramolecular interaction to sequester the tandem SH3 domains, which become exposed during activation. In addition, phosphorylation of p47^{phox} releases an intramolecular autoinhibitory interaction between the PX domain and second SH3 domain. In p67^{phox} and NOXA1, the N-terminal tetratricopeptide repeat (TPR) motifs mediate Rac GTPase binding, whereas the activation domains in these proteins participate in the regulation of electron flow from NADPH to FAD. The p67^{phox} Phox and Bem (PB1) domain plays a role in binding to the p40^{phox} N-terminal *Phox* and Cdc (PC) domain, whereas the role of the NOXA1 PB1 domain is undefined. See text for additional details.

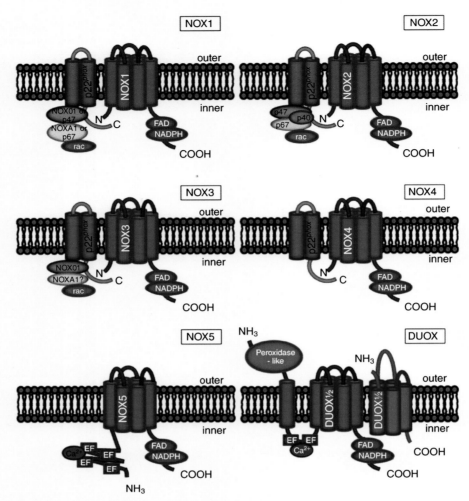

Figure 7.4 NOX family members and their proposed regulatory subunits. NOX proteins are believed to contain six transmembrane domains based on hydrophobicity analysis (seven for DUOX1/2). NADPH oxidase activity occurs when NADPH binds to the cytosolic tail of NOX, where it transfers electrons to FAD. From FAD, the electrons are transferred to the heme centers (refer to Fig. 7.1) and ultimately to oxygen near the outer membrane surface, resulting in $O_2^{\cdot-}$ formation. The transmembrane subunit $p22^{phox}$ is associated with both active and inactive forms of NOX1–NOX4. During activation, NOX1/$p22^{phox}$ is believed to primarily assemble with the cytosolic subunits NOXO1, NOXA1, and Rac GTPase; however, $p47^{phox}$ and $p67^{phox}$ can substitute in some situations for NOXO1 and NOXA1, respectively. NOX2/$p22^{phox}$ activation involves assembly with $p47^{phox}$, $p67^{phox}$, $p40^{phox}$, and Rac2 GTPase. NOX3/$p22^{phox}$ activation is less well defined, but is believed to primarily involve NOXO1 and Rac GTPase in the inner ear. NOX4/$p22^{phox}$ is constitutively active in the absence of cytosolic cofactors. NOX5 and DUOX1/2 activation involves Ca^{2+} binding to EF-hand domains in the cytosol. In addition, DUOX1/2 require the association of DUOXA1/2, respectively, for localization to the plasma membrane. (Adapted from *Free Radic. Biol. Med.* **2009**, *47*(9), 1239–1253. Copyright 2009 Elsevier).

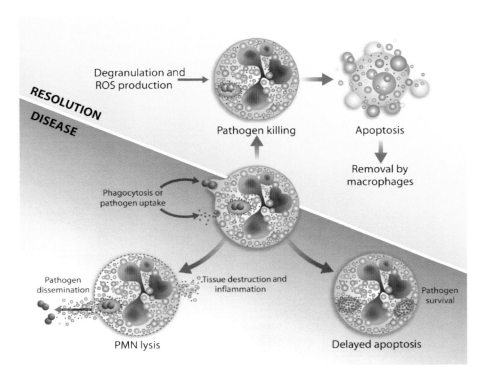

Figure 7.5 A paradigm illustrating the role of ROS in the resolution of infection. Production of ROS by the phagocyte NADPH oxidase is essential for host defense against pathogens and is a key feature of the inflammatory response. However, the presence of ROS is also required for inducing neutrophil apoptosis during resolution of the inflammatory response. In the absence of phagocyte NADPH oxidase activity, neutrophil apoptosis is impaired, leading to excessive inflammation and inflammatory tissue damage. (Adapted from *Clin. Sci.* **2006**, *111*(1), 1–20. Copyright 2006 Biochemical Society.)

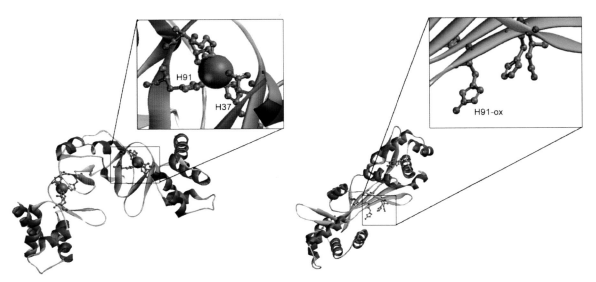

Figure 8.2 The structure of the regulatory metal site in PerR (left) and 2-oxo-histidine in oxidized PerR protein (right) (PDB codes: 3F8N and 2RGV, respectively).

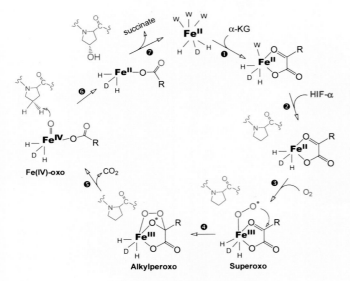

Figure 8.4 α-Ketoglutarate and Fe-dependent hydroxylation of HIF-α by prolyl hydroxylase. The mechanistic cycle shown in this model includes the following steps: (1) binding of co-substrate α-KG and replacing of the two water ligands at the Fe(II) center, (2) binding of the primary substrate HIF-α resulted in a conformational change and lost of the third water ligand, (3) binding of O_2 to the five-coordinate Fe(II) center and production of a superoxo anion intermediate, (4) formation of an alkylperoxy intermediate, (5) O–O bond cleavage leads to an Fe(IV)-oxo intermediate while oxidative decarboxylation of the bound co-substrate produces CO_2 and a bound succinate, (6) oxygen transfer from the high-valent Fe(IV)-oxo intermediate to a proline residue of HIF-α, and (7) succinate release resumes the initial 2-His-1-Asp ferrous center.

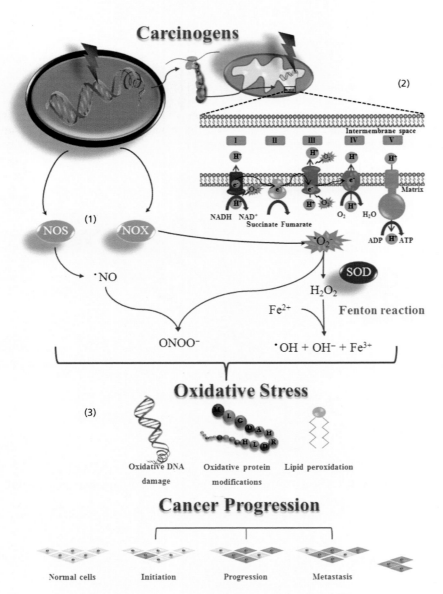

Figure 9.1 Cancer cells exhibit increased generation of ROS and RNS compared with normal cells, which provides favorable conditions for cancer initiation, progression, or metastasis. This increased generation of RS is given particularly by alterations in the protein levels of RS generating enzymes such as NOX and NOS (1). In the mitochondria, $O_2^{\bullet -}$ is formed by monoelectronic reduction of O_2 by complexes I and III of the respiratory chain. Mutations in nuclear or mitochondrial genes encoding components of the ETC can lead to an increase in ROS generation by inhibition of electron transfer (2). Although oxidative DNA damage is a central element in the transformation process toward malignancy, it is also evident that RS can promote/predispose to cancer development by oxidative damage/modification to lipids and proteins/enzymes (3).

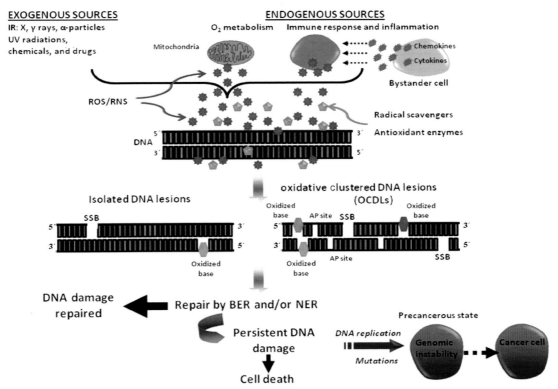

Figure 9.2 Accumulation and persistence of oxidatively induced DNA damage can lead to cancer. Cells in every organism are exposed to various oxidizing and damaging agents ranging from exogenous sources like environmental, medical/diagnostic ionizing and nonionizing radiations (X- or γ-rays, α-particles, UV radiation) or chemicals like benzo[a]pyrene, to intracellular (endogenous) sources of oxidative stress primarily produced by O_2 metabolism, immune responses, and inflammation. The final outcome of the production of ROS, RNS, is the generation of oxidation by-products, including DNA lesions and adducts. Cellular defense against DNA damage consists of endogenous radical scavengers like GSH, antioxidant enzymes like SOD, catalase, PRDX, and GPXs, as well as DNA repair pathways. The major types of DNA damage include SSBs and oxidized bases like 8-oxodG and thymine glycol in a clustered (OCDLs) or isolated formation. These lesions are thought to be processed primarily by BER, although in some cases the involvement of NER cannot be excluded. Cells can bypass DNA damage and enter DNA replication. Lesions that escape repair and enter replication are expected to be primarily repaired by MMR. The participation of additional repair mechanisms like GG-NER or TC-NER as subpathways of NER can also be involved depending on the efficiency of the major repair pathways to correct the damage, or the location of the DNA lesions that are in actively transcribed DNA strands (TC-NER). Alternatively, the presence of unrepaired DNA lesions can activate cell death pathways. Chronic exposure to DNA lesions can lead to mutations and genomic instability (precancerous state) and eventually to malignant transformations (cancerous state).

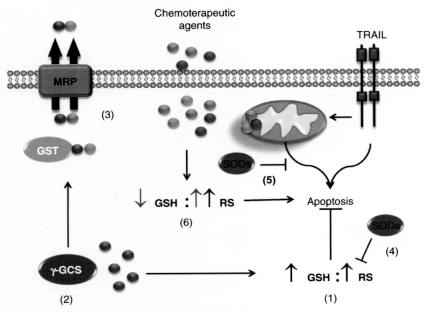

Figure 9.3 Apoptotic deregulation and multidrug resistance in cancer cells. In cancer cells, deregulation of apoptotic pathways is associated with cellular redox imbalance. High levels of both ROS/RNS and high intracellular GSH impair the activation of apoptotic enzymes by oxidative posttranslational modifications and/or scavenge pro-apoptotic oxidative damage (1). An increase in γ-GCS levels in cancer cells augments GSH concentration, whose metabolism also contributes to multidrug resistance against chemotherapeutic agents (2). Increased expression of GSTs and MRPs is a common feature of transformed cells that are associated with high resistance to chemotherapy (3). SODs exert antitumorigenic effects by reduction of RS levels in cancer cells (4). Interestingly, SODs also protect against the activation of death receptor (TRAIL) and mitochondrial pathways of apoptosis induced by chemotherapeutic agents, which depend on the generation of RS (5). Thus, chemotherapeutic approaches involve both the depletion of intracellular GSH and/or increased oxidative damage (6).

9

OXIDATIVE STRESS AND REDOX SIGNALING IN CARCINOGENESIS

Rodrigo Franco, Aracely Garcia-Garcia, Thomas B. Kryston, Alexandros G. Georgakilas, Mihalis I. Panayiotidis, and Aglaia Pappa

OVERVIEW

Most cells in an adult multicellular organism are not dividing and must receive mitogenic signals in order to proliferate. For a multicellular organism to survive, it is necessary to regulate the number of cells proliferating in the interest of the whole organism. A tumor arouses when tissue growth exceeds that of normal tissue in an uncoordinated fashion. Carcinogenesis is a multistage process of conversion of a normal cell to a malignant state as a result of a series of mutations, cell divisions, and changes in the gene expression pattern. Carcinogenesis leads to a self-replicating transformed cell which conveys the generation of a larger number of cells with an increasing accumulation of abnormalities. Numerous factors contribute to the initiation, development, and progression of cancer, including those of genetic, environmental, and dietary influences. Oxidative stress and reactive species (RS) participate at all stages of the malignancy/carcinogenic process aroused from both inherited and environmental factors.[1]

9.1 REDOX ENVIRONMENT AND CANCER

Cancer cells often exhibit increased generation of RS compared with normal cells, which is thought to provide favorable conditions for cancer cell growth, genetic instability, and survival. This increased generation of RS is given particularly by alterations in the protein levels of RS-generating enzymes and/or mutations in the mitochondrial electron transport chain (ETC) components (Fig. 9.1).

9.1.1 Pro-Oxidant Environment and Endogenous Sources of RS in Cancer

9.1.1.1 Reactive Oxygen Species (ROS)-Generating NADPH Oxidases and Cancer Cancer cells have increased levels of ROS, which are associated with increased nicotinamide adenine dinucleotide phosphate (NADPH)-oxidases (NOX) activity/expression[2] (Fig. 9.1). Cancer cells, constitutively expressing the activated form of the oncoprotein Ras present significantly elevated levels of $O_2^{\bullet-}$, which correlates with resistance to apoptosis. NOX1 has been postulated as the central mediator of oncogenic Ras-MEK-ERK signaling by regulation of growth and transcription factors.[3,4] Other oncogenic signaling proteins such as Src regulate NOX1-dependent ROS generation.[5]

Several other NOX enzymes are also implicated in cellular transformation. The Epstein–Barr virus (EBV) nuclear antigen (EBNA), the only viral protein expressed in all EBV-carrying malignancies, induces

---

Molecular Basis of Oxidative Stress: Chemistry, Mechanisms, and Disease Pathogenesis, First Edition. Edited by Frederick A. Villamena.
© 2013 John Wiley & Sons, Inc. Published 2013 by John Wiley & Sons, Inc.

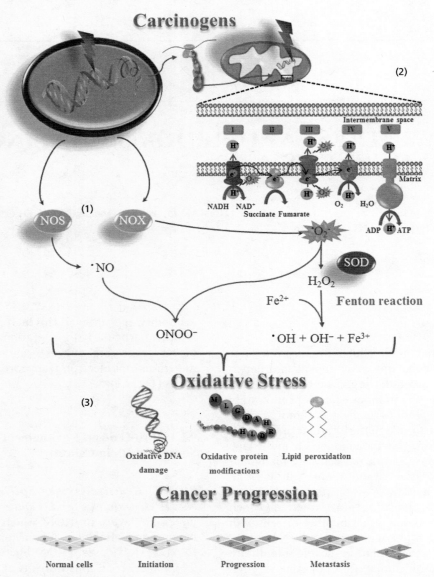

Figure 9.1 Cancer cells exhibit increased generation of ROS and RNS compared with normal cells, which provides favorable conditions for cancer initiation, progression, or metastasis. This increased generation of RS is given particularly by alterations in the protein levels of RS generating enzymes such as NOX and NOS (1). In the mitochondria, $O_2^{\bullet-}$ is formed by monoelectronic reduction of O_2 by complexes I and III of the respiratory chain. Mutations in nuclear or mitochondrial genes encoding components of the ETC can lead to an increase in ROS generation by inhibition of electron transfer (2). Although oxidative DNA damage is a central element in the transformation process toward malignancy, it is also evident that RS can promote/predispose to cancer development by oxidative damage/modification to lipids and proteins/enzymes (3). See color insert.

chromosomal aberrations and DNA damage by transcriptional induction of NOX2 and increased ROS production.[6] NOX4-generated ROS are associated with melanoma cell growth[7] and pancreatic cancer cell resistance to apoptosis by promotion of Akt survival signaling.[8] Furthermore, overexpression of NOX4 induces cellular senescence, tumorigenic transformation, and cell migration.[9,10] NOX5 is overexpressed in esophageal adenocarcinoma cells and mediates acid-induced ROS production contributing to cell proliferation and decreased apoptosis.[11–14] In contrast to the increased overexpression of NOXs in cancer cells, the dual NADPH oxidase/peroxidase, DUOX, is silenced by promoter hypermethylation in lung cancer.[15]

9.1.1.2 Mitochondria Mutations, Oxidative Stress, and Cancer

In cancer cells, mutations in nuclear or mitochondrial genes encoding components of the ETC induce an increase in ROS generation by inhibition of electron transfer, leading to the accumulation of electrons that can be captured by O_2 (Fig. 9.1). ETC gene defects are directly linked to some hereditary cancers. Mutations in the mitochondrial fumarate hydratase (FH) and in mitochondrial succinate dehydrogenase complex II subunits SDHD (cytb membrane subunit) and SDHC (coenzyme Q-binding membrane subunit), and SDHB (iron–sulfur), which impede electron flux through complex II increasing ROS production, are associated with distinct carcinomas.[16] Mitochondrial DNA (mtDNA) mutations that inhibit oxidative phosphorylation and impede electron flow down the ETC also increase ROS production and have been described in cancer cells. Mutations in the mtDNA of the cytochrome oxidase subunit I gene (COI) are also associated with prostate cancer.[16]

9.1.1.3 Nitric Oxide Synthases (NOS) and Cancer

Nitric oxide (NO) is synthesized by NOS, which catalyze the oxidation of L-arginine resulting in the formation of NO and L-citrulline. Three different NOS have been described, two of which are constitutively expressed primarily in neurons (nNOS) and endothelial cells (eNOS). A third class of NOS called inducible NOS (iNOS) can be upregulated in response to different cytokines. Depending on the tumor type and stage, the expression levels and activities of NOS are increased compared with normal tissues. NO is involved at different stages of carcinogenesis, including tumor cell invasion, proliferation, and angiogenesis. The role of NO in cancer is quite complex, as both protective and cytotoxic effects have been demonstrated.[17] iNOS mediates neoplastic transformation in models of oncogene- and chemical-induced carcinogenesis, and its expression together with elevated nitrotyrosine levels in human metastatic melanoma tumors correlate with poor patient survival.[18] Furthermore, iNOS is linked to angiogenesis by regulation of vascular endothelial growth factor (VEGF) receptor levels.[19] In clear contrast, iNOS has also been shown to inhibit tumor growth and metastasis.[20,21]

iNOS has received most of the attention in cancer research because it can be induced by a variety of cytokines and can produce micromolar levels of NO, which damage DNA and modify protein structure/function. However, eNOS and nNOS also participate in the carcinogenic processes. Increased eNOS expression has been observed in the vasculature of human tumor tissues when compared with normal tissue samples.[22] Studies on human colon cancer samples suggest that high levels of eNOS expression correlate with tumor cell vascular invasion.[23] nNOS expression is increased in glial tumors, which correlates with the proliferative potential, suggesting a role of nNOS with malignancy.[24]

9.1.1.4 Other Sources of RS in Cancer

Increased myeloperoxidase (MPO) and lipid peroxide-modified proteins have been recently described in the plasma of patients with gynecological malignancies.[25] MPO polymorphisms are associated with the etiology of lung, larynx, pancreatic and breast cancer, and leukemia[26–28] Cyclooxygenase 1 (COX1) overexpression mediates tumorigenic transformation of immortalized ECV endotheial cells,[29] while COX2 overexpression contributes to DNA oxidation and DNA-adduct formation from lipidperoxide bioproducts.[30,31] COX2 inhibitors reduce tumor growth and metastasis of breast cancer.[32] However, whether the protective effect of COX inhibition on carcinogenesis is associated to its pro-oxidant effects is unclear.[33] Increased xantine oxidase (XO)-mediated ROS induces the expression of hypoxia-inducible factor-1α (HIF-1α) in cancer cells, which is involved in angiogenesis and, thus, cancer tumor growth.[34]

9.1.2 Alterations in Antioxidant Systems and Cancer

Cells have evolved intrinsic antioxidant mechanisms to detoxify RS generated under both physiological and pathological conditions. Under normal/basal conditions, a balance exists between both the activities of enzymatic antioxidant defenses and the intracellular levels of non-enzymatic antioxidants, which is essential for the survival of organisms and their health. In addition to their ability to scavenge RS, antioxidants appear to modulate cell signaling pathways by redox-dependent mechanisms. Thus, antioxidants can modulate cancer at the cellular level by regulation of cell proliferation and cell death; at the tissue level by inhibiting tumor invasion, angiogenesis, and inflammation; and also by mediating detoxification processes.[35,36]

9.1.2.1 Glutathione (GSH) and Glutathione-Dependent Enzymes in Cancer

Elevated GSH levels are observed in various types of cancerous cells and solid tumors[37,38] which are a consequence of high γ-glutamylcysteine synthetase (γ-GCS) expression.[39] Genetic polymorphisms of human γ-GCS are also associated with increased GSH levels and drug sensitivity in tumor cell lines.[40] Cysteine availability is rate limiting for the synthesis of GSH. Several types of malignant cells, such as leukemias and lymphomas, are incapable of synthesizing cysteine; thus, its uptake from the microenvironment is essential for their growth and viability.

Because cysteine derived from the breakdown of proteins and peptides is unstable, its main extracellular source is the dipeptide cystine. Cystine uptake occurs by a Na^+-independent heteroexchange mechanism-denominated x_c^-. The x_c^- transporter is present in numerous transformed cell lines and types of cancer tissue from patients, and its inhibition reduces GSH content and cancer cell growth,[41–43] while its increased expression is associated with cancer chemoresistance.[44] Glutamine via glutaminase-mediated glutamate synthesis is also one of the precursors for synthesis of GSH via the glutaminase-mediated glutamate synthesis. Recently, the myc oncogene was shown to regulate mitochondrial glutaminase expression, and either glutamine withdrawal or glutamine synthesis knockdown were shown to decrease cancer cell proliferation, which was associated with alterations in GSH content.[45,46]

Changes in the intracellular thiol–disulfide (GSH/GSSG) balance are considered major determinants in the redox status/signaling of the cell. In tumor cells, the GSH/GSSG ratio remains constant despite the high levels of RS and oxidative stress, which might be associated with increased GSSG efflux.[47] GSH reductase (GR), which reduces GSSG back to GSH, requires NADPH as the electron donor reductant, and glucose-6-phosphate dehydrogenase (G6PD) is indispensable for the regeneration of NADPH from $NADP^+$.[48] G6PD is overexpressed in different cancer tissues[49] and its knockdown decreases cancer cell proliferation and renders cells more sensitive to oxidative stress and chemotherapy.[50] In contrast, both elevated or decreased GR expression/enzymatic levels have been detected in transformed cells and tumor tissues.[51,52]

Glutathione peroxidases (GPXs) are selenoproteins that reduce peroxides. A clinical trial demonstrated that daily supplementation with selenium significantly reduces mortality associated with lung, prostate, and colon cancers, which suggests a role for GPXs in selenium's protective effects.[53] GPX1 preferred substrate is H_2O_2, and its expression is decreased or repressed in cancer cell lines and tumors.[54,55] GPX1 overexpression reduces tumor cell growth,[56] while its polymorphisms are associated with cancer.[57] Other GPX isoforms are also linked to the carcinogenic process. Early inactivation of GPX2, the intestinal and extracellular GPX enzyme, contributes to squamos cell carcinoma formation induced by UV radiation.[58] Interestingly, while GPX2 activity inhibits tumor cell migration and invasion, it can also promote the growth of transformed cells.[59] GPX3 is inactivated in prostate cancers, and its overexpression induces the suppression of colony formation and anchorage-independent growth.[60] GPX4 (also known as phospholipid hydroperoxide glutathione peroxidase, PHGPX) was shown to reduce pancreatic cancer growth.[61]

9.1.2.2 Catalase

Catalase overexpression, which mediates the decomposition of H_2O_2 to water and O_2 has been shown to inhibit tumor cell growth[62] and reverse malignant cell tumorigenicity.[63] In contrast, catalase can also mediate cancer cell resistance to oxidant-induced cell death and DNA damaging agents.[64–66] Similarly, glucocorticoids (commonly used in the treatment of lymphoid neoplasms) were shown to induce lymphoid cell apoptosis by downregulation of catalase.[67] Although catalase is known for its antioxidant properties, a recent study shows that formation of ROS induced by high energy ultraviolet B light was potentiated by catalase, suggesting a pro-oxidant action of this enzyme.[68]

9.1.2.3 Superoxide Dismutases (SODs)

SODs, metalloenzymes that catalyze the conversion of $O_2^{\bullet-}$ to H_2O_2, have been largely implicated in carcinogenesis. Both pro- and antitumoral effects have been demonstrated for mitochondrial MnSOD (SOD2).[69] Reduced MnSOD levels are detected in a variety of cancer cells, and this is correlated with mutations in the DNA sequence of its promoter region.[70] Promoter methylation and histone modifications are also implicated in the decreased expression of MnSOD in multiple myeloma and breast carcinoma cell lines.[71,72] Increased MnSOD levels have been associated with glioblastoma patient survival and prognosis,[73] while polymorphisms are linked with increased cancer risk and prognosis.[28,74] The tumor suppressor activity of MnSOD has been extensively reported. For example, MnSOD overexpression suppresses the malignant phenotype and growth of a variety of cancer cells.[75,76] MnSOD overexpression also suppresses radiation-induced neoplastic transformation.[77] In agreement with the catalytic conversion of $O_2^{\bullet-}$ to H_2O_2 by MnSOD, the tumor suppressive effects of MnSOD expression are reported to be mediated by H_2O_2 and thus, inhibited by GPX and catalase.[78,79] Distinct signaling cascades mediate the pro- and antitumoral effects of MnSOD. Tumor cell growth inhibition by MnSOD is associated with cellular senescence mediated by mitochondrial membrane depolarization and upregulation of p53.[80] Tumor suppression induced by overexpressing MnSOD is also related to the modulation of activated-protein 1 (AP-1) and nuclear factor-kappa B (NF-κB) signaling.[81,82] MnSOD activity also suppresses HER2/neu oncogene expression[83] and reduces HIF-1α accumulation and VEGF expression, suggesting a role for MnSOD in angiogenesis.[84] Finally, the C/EBP homologous protein (CHOP/GADD153), which is involved in DNA damage repair, was reported to be upregulated by overexpression of MnSOD.[85]

In contrast to the antitumoral effects of MnSOD described earlier, an increase in MnSOD levels has been found in brain tumors and mesothelioma cells, which correlates with increased oxidative stress.[86,87] Elevated MnSOD levels in distinct cancer cells correlate with chemoresistance[88–90] and enhanced invasiveness and migration of tumor cells.[91] Similarly, leukemia cell death induced by estrogen derivatives is associated with the inhibition of SOD activity and increased ROS generation.[92]

Cu,ZnSOD (SOD1) also participates in the carcinogenic processes. Mice deficient in Cu,ZnSOD present elevated incidence of liver cancer and high mutation frequency.[93] While Cu,ZnSOD overexpression suppresses cancer cell growth,[94] it confers radioresistance to human glioma cells.[95] Inhibition of Cu,ZnSOD exerts antiangiogenic and antitumor effects.[96]

9.1.2.4 Peroxiredoxins (PRDXs)
PRDXs reduce intracellular peroxides (H_2O_2 and organic peroxides) by using the peroxide reactivity of the cysteine sulfur atom and the thioredoxin (TRX) system as the electron donor. Aberrant expression of PRDXs is found in various kinds of cancer and some members of the PRDX family are thought to be biomarkers of cancer progression and prognosis.[97] Increased PRDX1 expression is found in human tumors and cancer cells,[98,99] which is associated with nuclear factor (erythroid-derived 2)-like 2 (Nrf2) activation by hypoxia in tumor microenvironment.[100] Both PRDX1 and TRX1 overexpression are associated with human breast carcinoma,[101] while PRDX5 overexpression in mammary tissue is linked to poor prognosis in breast cancer.[102] Several PRDX isoforms have also been shown to be overexpressed in cancer.[103,104] Furthermore, autoantibodies against PRDX6 were identified as serum markers in esophageal squamous cell carcinoma.[105] In contrast, PRDX2 silencing by aberrant methylation of promoter CpG islands is observed in human malignant melanomas.[106]

PRDX1 knockdown predisposes mice to malignant cancer development, which is associated with increased sensitivity to oxidative DNA damage.[107] PRDX1 interacts and regulates c-Myc transcriptional activity and protects against oxidative DNA damage.[108] c-Myc-mediated neoplastic transformation was also reported to involve transcriptional regulation of PRDX3,[109] while PRDX3 silencing inhibits cell proliferation in breast cancer cells.[110] More recently, the inhibition of Ras- or human epidermal growth factor receptor 2 (ErbB-2)-induced tumorigenesis by PRDX1 was reported to be mediated by its protective effect against oxidant-induced inactivation of the phosphatase and tensin homolog (PTEN).[111] PRDX1 and PRDX5 protect against ionizing radiation and oxidative DNA damage in cancer cells.[112,113]

9.1.2.5 Heme Oxygenase (HO)
The enzymatic activity of HO results in decreased oxidative stress due to removal of heme, a potent pro-oxidant. HO-1 overexpression is reported in a variety of transformed cell types.[114,115] HO-1 polymorphisms are associated with increased risk of cancer.[116,117] Survival of chronic myeloid leukemia cells mediated by the breakpoint cluster region (BCR)/V-abl Abelson murine leukemia viral oncogene homolog 1 (ABL) complex is associated with regulation of HO-1 expression,[118] while in acute myeloid leukemia cells, increased HO-1 expression is regulated by the transcriptional repressor Bach1.[119] HO-1 overexpression increases the viability, proliferation, metastasis, and angiogenic potential of melanoma cells, while it also decreases survival of tumor-bearing mice.[120] Overexpression of HO-1 potentiates pancreatic cancer tumor growth, angiogenesis, and metastasis.[121] On the other hand, HO-1 silencing and/or inhibition reduces prostate cancer cell proliferation, survival, and invasion *in vitro*, as well as prostate tumor growth, and lymph node and lung metastases *in vivo*.[122]

9.1.2.6 TRX/TRX Reductase System
TRXs reduce or bind to proteins modifying their activity by thiol–disulfide exchange reactions. The TRX redox system also comprises TRX reductases (TRs), which transfer reducing equivalents from NADPH to TRXs. While TRX1/TR1 system is localized in the cytoplasm, TRX2/TR2 is mitochondrial. TRX1 is overexpressed in a variety of cancer cell types, including pancreatic ductal carcinoma.[123] Aggressive invasive mammary carcinomas and advanced malignant melanomas overexpress TRX1 and TR1,[124] which are also increased in thyroid cancer cells.[125] TR1 and the TRX1-interacting protein (TXNIP) are also associated with breast cancer prognosis.[126] TRX1 overexpression increases cell proliferation HIF-1α, VEGF production, and angiogenesis,[127] while a dominant-negative TRX1 mutant reverses the transformed phenotype of human breast cancer cells.[128] Accordingly, TRX1 knockdown has been shown to reverse malignant characteristics in cancer cells.[129,130] Protein disulfide isomerase (PDI), a TRX family member found in the endoplasmic reticulum (ER), is strongly expressed in glioma cells, and its inhibition prevents tumor cell migration and invasion.[131]

9.2 OXIDATIVE MODIFICATIONS TO BIOMOLECULES AND CARCINOGENESIS

In physiological settings, ROS/reactive nitrogen species (RNS) exhibit substrate specificity. Until recently, the study of the biomolecular alterations caused by RS and its specific role in carcinogenesis has been undertaken.

DNA damage caused by RS is a central element in the transformation process toward malignancy. However, it is evident that RS can promote/predispose to cancer development by oxidatively induced damage and modifications to lipids, and proteins/enzymes as well (Fig. 9.1).

9.2.1 Oxidative Posttranslational Protein Modifications in Cancer

Oxidative protein modifications regulate the activity of a wide variety of proteins, such as kinases, phosphatases, proteases (caspases), molecular adaptors and chaperones, and transcription factors known to be associated with cellular transformation. Cys, Met, Trp, and Tyr residues are prone to specific oxidative modification. In general, oxidative protein modifications can be classified as reversible and irreversible. Highly RS such as hypochlorous acid (HOCL), peroxynitrite (ONOO$^-$), and hydroxyl radical (HO$^•$) are thought to lead to the irreversible formation of oxidized residues such as 3-nitrotyrosine and protein carbonyls (PCOs). Physiological oxidants such as NO, $O_2^{•-}$, and H_2O_2, are implicated in reversible protein modifications at the Cys level (nitrosylation, hydroxylation, glutathionylation, disulfide bond formation). A wide variety of enzymes regulate these posttranslational modifications, including sulfiredoxins (SRX), TRXs, glutaredoxins (GRXs), and methionine sulfoxide reductases (MSRs).[132–134]

9.2.1.1 Protein Carbonylation Protein amino acid oxidation of Lys, Arg, Pro, or Thr, or secondary reaction of Cys, His, or Lys residues with reactive carbonyl compounds lead to the formation of PCO derivatives (aldehydes and ketones), which are considered indicative of severe oxidative stress. Increased PCO content is observed in tumor tissues and cancer cells,[135,136] while several carcinogens induce PCO accumulation.[137,138] Recently, chrysotile asbestos was shown to induce carbonylation of G6PD.[139] However, the exact relevance of protein carbonylation to the carcinogenic process is still unclear.

9.2.1.2 Protein Nitration Tyr nitration is a covalent protein modification derived from the reaction of proteins with nitrating agents, particularly ONOO$^-$.[140] Cigarette smoke increases stress and RNS, resulting in nitration of plasma proteins.[141] Increased ONOO$^-$-mediated Tyr nitration and Tyr phosphorylation of c-Src kinase have been found in human pancreatic ductal adenocarcinoma.[142]

9.2.1.3 Protein Nitrosylation or Nitrosation Protein (S)-nitrosylation, which refers to the reaction of NO with cysteine residues in proteins, is emerging as one of the most important mechanisms mediating the biological effects of NO. The only mechanism described for protein S-nitrosylation *in vivo* is the reversible transnitrosylation of protein thiols by glutathione-SNO (GSNO) or S-nitrosoglutathione.[134] Estrogen-induced breast cancer cell invasion has been shown to be mediated by Src nitrosylation.[143] GSNO is the substrate for glutathione-dependent formaldehyde dehydrogenase class III alcohol dehydrogenase (ADH III), also named GSNO reductase (GSNOR). Deregulated nitrosylation induced by GSNOR deficiency promotes hepatocellular carcinoma through proteosomal degradation of DNA-repair enzymes and increased oxidative DNA damage.[144] NO releasing nonsteroidal anti-inflammatory drugs (NSAID) or NO-NSAIDs reduced NF-κB protein levels through its nitrosylation, leading to increased caspase-dependent apoptosis in human colon cancer cells.[145]

9.2.1.4 Protein Glutathionylation Protein (S-)glutathionylation refers to the formation of a protein mixed disulfide between the Cys of GSH and a Cys moiety of a protein. A significant increase of glutathionylated proteins in red blood cells and plasma concentrations is found in cigarette smokers,[146] while GSH, GSSG, and glutathionylated protein levels were shown to be increased during oral carcinogenesis.[147] Protein–SSG linkages are removed by changes in the intracellular GSH/GSSG balance and/or the activity of GRX enzymes. Increased GRX1 levels are found in pancreatic ductal carcinoma tissues.[123] Alternative transcript variants of mitochondrial GRX2 encoding nonmitochondiral GRX2B and GRX2C have been reported in various cancer cell lines, but their significance to cellular transformation has not been studied.[148]

9.2.1.5 Methionine Sulfoxide Methionine is one of the most sensitive amino acid residues to oxidation by ROS and RNS. Addition of an extra O_2 oxidizes methionine to methionine sulfoxide (MetSO), and a strong oxidant might further oxidize MetSO to methionine sulfone (MetSO$_2$).[149] Two different stereoisomers of methionine sulfoxide, L-Met-S-SO and L-Met-R-SO, can be found at physiological conditions. Methionine oxidation is reverted by MetSO reductases. MsrA is specific to the S-stereoisomer of MetSO, whereas MsrB catalytically reduces the R-stereoisomer of MetSO.[150] Reduced levels of MsrA are associated with increased proliferation and extracellular matrix degradation in human breast cancer cells, and this was associated with increased PI3K signaling pathways and upregulation of VEGF.[151]

9.2.2 Lipid Peroxidation (LPO) and Cancer

LPO generates a number of lipid hydroperoxide products, such as malondialdehyde, 4-hydroperoxy-2-nonenal, 4-oxo-2-nonenal, and 4-hydroxy-2-nonenal (4HNE). These aldehyde products react with individual nucleotides and nucleophilic amino acids, thus inducing several signaling effects. Reactive aldehydes produced by LPO, such as 4-HNE, malondialdehyde, acrolein, and crotonaldehyde, react with DNA bases or generate bifunctional intermediates forming exocyclic DNA adducts. Modification of DNA bases by these electrophiles yields promutagenic exocyclic adducts, which contribute to the mutagenic and carcinogenic effects associated with oxidative stress-induced LPO.[152] Increased levels of lipid peroxide products such as malondialdehyde and 4-HNE are present in different types of cancer.[153–155]

9.2.3 Oxidative DNA Damage and Carcinogenesis

Cells of the human body are continuously challenged by oxidation attacks originated extracellularly (exogenous sources) or intracellularly (endogenous sources). Exogenous sources of oxidation relate to specific exposures of the organism to ionizing radiations like X-, γ- or cosmic rays and α-particles from radon decay or oxidizing chemicals like benzo[a]pyrene and UV solar light, while endogenous sources usually associate with metabolic, cell signaling, or inflammation processes.[156–158] Oxidation attack to DNA is induced primarily but not exclusively by RS. However, not all ROS or RNS react with the deoxyribose moieties or nucleobases at the same rate. For example, the HO• will react with all four DNA nucleobases (adenine, thymine, gunanine, and cytosine), while singlet oxygen ($^1O_2$) will react only with guanines.[156,159] In another case, HOCl, an inflammation release product, will only halogenate amino-substituted nucleobases like guanine.[160]

Oncogenic changes induce a chronic inflammatory microenvironment within tumors and surrounding tissues including presence of inflammatory cells and inflammatory mediators, such as chemokines, cytokines, and prostaglandins,[161] as well as elevated levels of endogenous oxidative stress and ROS production.[2,162] These ROS, produced either directly by tumors or indirectly via inflammatory responses, cause DNA damage in healthy neighboring cells[163] (Fig. 9.2). ROS interact with the DNA and disrupt its normal synthesis and repair. Tumor growth elevates oxidative stress and DNA damage accumulation in the organism.[158,164,165] The chronic inflammatory microenvironment within and surrounding tumor area[161] has many oncogenic effects, including elevated levels of intracellular stress and ROS/RNS.[158,166] These RS can result in bystander DNA damage similar to cellular inflammatory responses.[167] Although most ROS have a short half-life and cause damage locally, H_2O_2 has a relatively long half-life and can travel long distances, causing DNA damage at distant sites.[168]

9.2.3.1 Types of Oxidatively Induced DNA Damage
DNA lesions that usually arise from the persistent endogenous oxidative stress are apurinic/apyrimidinic (abasic; AP) DNA sites, oxidized purines and pyrimidines, single-stranded DNA breaks (SSBs), and double-stranded DNA breaks (DSBs) (Fig. 9.2). DNA damage can be induced directly by the interaction of DNA with various RS or indirectly as the result of DNA repair. An oxidized base will be excised after its detection by a DNA glycosylas, and an AP site will be created. In many cases, the chemical structure of the initial DNA lesions and the ones created *de novo* as "repair intermediates" can be different. Two of the most abundant endogenous DNA base modifications are the 8-oxo-7,8-dihydro-2′-deoxyguanosine (8-oxdG) and 2,6-diamino-4-hydroxy-5-formamidopyrimidine (FapydG). Both of them originate from the addition of the HO• to the C8 position of the guanine ring producing a C8-OH adduct radical, which can be either oxidized to 8-oxodG or reduced to give the ring-opened product FapydG.[157] FapydG is currently considered as the most prevalent guanine-derived lesion formed under low O_2 conditions that is hypoxia.[169,170] The hypoxic conditions not only alter the type and level of DNA lesions but in most cases decrease their yields. For this reason, hypoxia or even anoxia (lack of O_2) are considered very important in radiation biology and radiotherapy since they are present in all solid tumors resulting in resistance of the tumor to radiation treatment.[171] Although the creation of an altered base or base loss is not expected to result into a significant destabilization of the DNA molecule, a localized perturbation of the stacking forces, hydrogen bonds and interaction with water molecules and/or positive ions like Na^+ surrounding the DNA double helix is expected.[172,173] It is generally accepted that this localized destabilization and conformational changes of the DNA at the site of the DNA lesion are part of the recognition mechanisms used by the DNA glycosylases to detect the altered guanine 8-oxodG or FapydG.[173,174] Interaction of HO• with pyrimidines (thymine and cytosine) at positions 5 or 6 of the ring can produce several base lesions and two of the most abundant and well-known products, 5,6-dihydroxy-5,6-dihydrothymine (thymine glycol; Tg) and 5,6-dihydroxy-5,6-dihydrocytosine (cytosine glycol). 8-oxodG and Tg are often chosen as reliable markers of oxidative stress in human cancer patients. 8-oxodG has

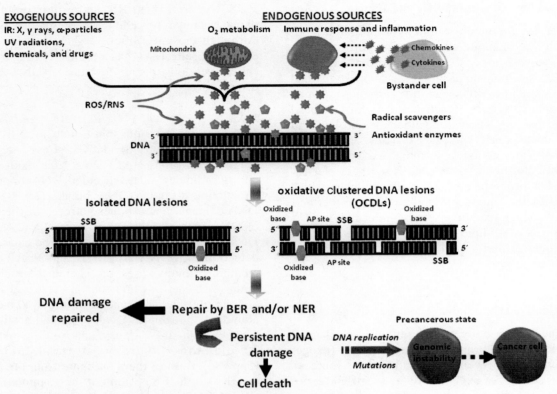

Figure 9.2 Accumulation and persistence of oxidatively induced DNA damage can lead to cancer. Cells in every organism are exposed to various oxidizing and damaging agents ranging from exogenous sources like environmental, medical/diagnostic ionizing and nonionizing radiations (X- or γ-rays, α-particles, UV radiation) or chemicals like benzo[a]pyrene, to intracellular (endogenous) sources of oxidative stress primarily produced by O₂ metabolism, immune responses, and inflammation. The final outcome of the production of ROS, RNS, is the generation of oxidation by-products, including DNA lesions and adducts. Cellular defense against DNA damage consists of endogenous radical scavengers like GSH, antioxidant enzymes like SOD, catalase, PRDX, and GPXs, as well as DNA repair pathways. The major types of DNA damage include SSBs and oxidized bases like 8-oxodG and thymine glycol in a clustered (OCDLs) or isolated formation. These lesions are thought to be processed primarily by BER, although in some cases the involvement of NER cannot be excluded. Cells can bypass DNA damage and enter DNA replication. Lesions that escape repair and enter replication are expected to be primarily repaired by MMR. The participation of additional repair mechanisms like GG-NER or TC-NER as subpathways of NER can also be involved depending on the efficiency of the major repair pathways to correct the damage, or the location of the DNA lesions that are in actively transcribed DNA strands (TC-NER). Alternatively, the presence of unrepaired DNA lesions can activate cell death pathways. Chronic exposure to DNA lesions can lead to mutations and genomic instability (precancerous state) and eventually to malignant transformations (cancerous state). See color insert.

been reported in genomic, mitochondrial, and telomeric DNA and RNA.[175] Two other pyrimidine lesions often detected in cancer patients as the result of the interaction of the HO• with the methyl group of the thymine are the 5-(hydroxymethyl)uracil and 5-formyluracil.[157] Although all these purine and pyrimidine oxidation products are not considered lethal for the cell, they are usually noncoding and are highly mutagenic.[176] The same stands for AP sites. AP sites are considered very common DNA lesions resulting from the hydrolysis of the N-glycosidic bond of nucleotides in the DNA, which releases the DNA base but leaves the phosphodiester backbone intact. They occur spontaneously or enzymatically as repair intermediates after the removal (excision) of the damage base by a DNA glycosylase in the base excision repair pathway (BER).[177] AP sites are not lethal unless they are present in high levels and block DNA polymerases and therefore have a high mutagenic potential.[178,179]

Interaction of DNA with HO• can result in strand breakage SSBs through hydrogen abstraction from the 2-deoxyribose and ribose leading to the formation of carbon-based radicals which, under the presence of O₂, can be converted to peroxyl radicals (ROO•). The ROO•

can also abstract hydrogen atoms from sugar moieties, and reaction of the sugar radicals leads to DNA chain breaks. ROO• are also implicated in LPO-mediated DNA damage and carcinogenesis, especially under the presence of O_2.[180] In many human fibroblastic cell lines, 8-oxodG is the main product of DNA oxidation by ROO•, which also induce tandem mutations.[180] H_2O_2 production in the cells leads to a variety of DNA damage ranging from oxidized bases to strand breaks via formation of the highly reactive HO•.[181] The ability of these radicals to travel up to 15 Å prior to reacting with DNA can lead to simultaneous attack of HO• to DNA, causing two neighboring SSBs; that is, a DSB.[182] Exposure of mammalian cells to low-moderate H_2O_2 concentrations although producing SSBs is not efficient to kill cells, therefore suggesting the existence of "locally multiply damage sites" (LMDSs).[182] Only under much higher H_2O_2 concentrations is cell killing observed due to the induction of clustered DNA lesions (DSBs and other non-DSB lesions) of high complexity.

If closely spaced damaged DNA sites (SSBs, oxidized bases, and/or AP sites) occur in two or more bases within a few helical turns, DNA damage is said to be clustered. Strong experimental evidence was recently provided for the existence of non-DSB clusters.[182,183] Since then, a few laboratories have showed the presence of non-DSB DNA clustered lesions in human cells or tissues and their accumulation under persistent oxidative stress or DNA repair deficiencies.[184] The direct association between these DNA lesions and the occurrence of endogenous or exogenous oxidative stress leads to the most properly known idea of oxidatively induced bistranded clustered DNA lesions (OCDLs).[185]

9.2.3.2 Base and Nucleotide Excision Repair in Oxidative DNA Damage Processing

The pool of oxidative base lesions, AP sites, and SSBs induced by endogenous sources or ionizing radiation and chemicals (alkylating agents) are predominantly repaired by BER and, to a lesser extent, by the nucleotide excision repair (NER). BER is a multistep procedure involving the sequential action of several proteins. The BER pathways are initiated by a DNA glycosylase (in human cells: mainly hOGG1, hNTH1, NEIL1, and NEIL2) that recognizes and hydrolytically cleaves and removes the altered base, giving rise to an AP site. The AP site is then processed by AP endonuclease (APE1), which incises the DNA strand 5′ to the baseless sugar. Then, DNA polymerase β catalyzes the β-elimination of the 5′-sugar phosphate residue and at the same time fills one-nucleotide gap (single-nucleotide patch BER pathway). Finally, the nick is sealed by the DNA ligase III/XRCC1 complex.[186] This repair process results in the removal and replacement of a single damaged nucleotide with a normal one.[187] DNA glycosylases (OGG1, NTH1, NEIL1, and NEIL2) are bifunctional as they have an associated β-lyase (OGG1 and NTH1) or β,δ-lyase activity (NEIL1-2). This intrinsic AP lyase activity incises 3′ AP sites to the baseless sugar, leaving a 3′-(2,3-didehydro-2,3-dideoxyribose) terminus that is then removed by AP endonuclease. The gap is then filled by DNA polymerase, and the nick is sealed by DNA ligase. In the case of a single nucleotide filling and final ligation step performed by LIG3/XRCC1 complex, we refer to short-patch BER (SP-BER). Alternatively, when BER synthesis extends beyond one nucleotide and is performed with the incorporation of 2–10 nucleotides, we refer to long-patch BER (LP-BER). In this case, the enzymes and proteins that participate are also associated with DNA replication: DNA polymerase δ (POLδ) or polymerase ε (POLε), replication factor C (RFC) and the proliferating cell nuclear antigen (PCNA), which acts as a cofactor for both POLδ and POLε, flap endonuclease (FEN1), and ligase I (LIG1).[157] However, major human glycosylases possess strong glycosylase activity relative to their lyase activity, especially in the presence of an AP endonuclease.[188,189]

Glycosylases NEIL1 and NEIL2, the nuclease Artemis and the variant form of histone H2A called H2AX have also been implicated in the repair of oxidative DNA damage in addition to DSB repair.[190,191] Bifunctional glycosylases NEIL1 and NEIL2 participate in APE1-independent processing of oxidative lesions induced by ionizing radiation and ROS, where an SSB with a 3′-phosphate blocking end is produced by β,δ-elimination, which is removed by the polynucleotide kinase/phosphatase (PNK).[190,192] When an SSB is directly induced by ROS, the poly [ADP-ribose] polymerase 1 enzyme (PARP1) is implicated in its repair. Automodification of PARP by poly ADP-ribosylation results in its inhibition.[193] One of the important functions of PARP in promoting BER is its ability to perform ADP-ribosylation of histone proteins allowing decondensation of the chromosome.[194–197] PARP1 is composed of three domains of which one is referred to as the DNA binding domain containing two zinc finger domains. It binds sites with SSB through its N-terminal zinc fingers and recruits XRCC1, DNA ligase III, DNA polymerase β and PNK to the nick.[198]

Because OCDLs are highly mutagenic and resistant to repair, they have the highest biological significance[199–201] and thus, a lot of emphasis has been given in the last 10 years in the repair pathways involved in their processing.[158,184,202] Many fundamental questions about the repair of clustered DNA lesions still remain unanswered, such as (1) to what extent does the presence of one lesion affect the recognition and simultaneous processing of the opposite neighboring lesion(s); (2) is there a hierarchy in repair of closely spaced DNA lesions;[203] (3) are there any strategies to avoid DSB formation that trigger cell death or genomic instability; and (4) which repair proteins process these lesions and what mechanisms are used to eliminate their presence?

Some major inferences can be made based on *in vitro* studies that consider the effect of neighboring lesions on the excision/repair of a base damage or AP site by the corresponding enzyme or cell extract.[204] First, excision of 8-oxo-7,8-dihydroguanine (8-oxoG) or thymine glycol (Tg) by human glycosylase hOGG1 or whole cell extracts is retarded by the presence of an AP site or SSB on the opposite strand up to five bases in the 3′ or 5′ direction, whereas a second base lesion has little or no influence on the rate of excision of the lesion.[205,206] In addition, DNA damage clusters containing two tandem 8-oxoG lesions opposing an AP site show delayed processing of the AP site in nuclear extracts and an elevated mutation frequency after transformation into wild-type or mutY *Escherichia coli*.[207] Second, hAPE1, expected to be the major AP endonuclease dealing with abasic clusters in human cells, is significantly retarded in the processing of an AP site in the presence of an opposing AP site or SSB one to three bases in the 3′ direction, whereas no inhibition was observed in the presence of a base damage.[208,209]

To what extent DSBs resulting from repair intermediates (mainly opposing SSBs) are produced in human cells is still a major open question. The general suggestion is that clustered DNA lesions in human cells may have a significantly increased lifetime compared to isolated ones, if they are repaired at all.[210] Simultaneous processing of both lesions constituting an oxybase or abasic cluster can lead to a DSB,[211,212] a phenomenon often called "abortive excision repair."[213] Considering that repair of DSB frequently results in deletions and loss of genetic information, it is possible that the BER mechanism includes features that reduce the probability of DSB formation during repair of closely spaced DNA lesions. A number of studies indicate that there should be a certain hierarchy in the processing of clustered DNA lesions in order to minimize the possibility of DSB creation.[210,214] The key role in the biochemical basis for this mechanism of "DSB avoidance" seems to be played by the creation of an AP site or SSB (nt gap). Once an AP site or SSB is created in the processing of clustered DNA lesions, then this event is expected to significantly inhibit the processing of the opposing oxybase lesion or AP site by DNA glycosylase or AP endonuclease.[184,203]

Recent evidence supports the role of non-BER repair proteins in the processing of OCDLs and, in general, oxidative lesions.[215,216] Different laboratories have implicated nonhomologous end joining (NHEJ) components in the repair of non-DSB oxidative clustered lesions.[217–219] DNA-PK activation has been shown by SSBs induced in plasmid DNA[220] or by the radiomimetic drug bleomycin,[221] which induces both DSBs as well as non-DSB-clustered DNA damage.[222] In addition, DNA-PK, BRCA1 and MSH2 deficiencies are associated with compromised processing of single and complex oxidative DNA lesions.[216,223–226]

Non-DSB clusters have a decreased ability of inducing cell death unless they are converted to DSBs.[184] These bistranded base clusters are compatible with normal cell survival and DNA replication unless they are induced at very high levels. In the case of environmentally induced radiation injury in the DNA, a relatively low level of these lesions is expected to be induced; therefore, cells are expected to survive and proceed to cell division.[210] Thus, every round of DNA replication is expected not only to decrease the number of these bistranded lesions by conversion to unistranded but also increase the overall mutation level in surviving cells.[224,227] Consequently, the biological significance of clusters, especially in the case of environmentally induced damage, is high since the accumulation of OCDLs combined with cellular survival can make cells with a high level of chromosomal aberrations and instability prone to transformation. There is an association between deficient non-DSB cluster repair and accumulation of chromatid breaks in BRCA1-deficient breast cancer cells, suggesting a role of BRCA1 in the mitigation of radiotoxicity, chromosomal instability, and human cancer etiology.[225,226]

Although BER is expected to be the predominant repair system involved in the processing of oxidatively induced DNA damage, experimental evidence supports the actual involvement of NER as a backup system in the repair of minor base damages induced by alkylating and oxidizing agents in a variety of biological systems ranging from bacterial up to human cells and especially mitochondria.[157,228] In addition, two NER-subpathways, the global genome repair (GG-NER; active on DNA lesions across the genome) and the transcription-coupled repair (TC-NER; present in actively transcribed genes), are hypothesized to be involved in the elimination of base lesions that escape BER or NER.[229,230] As shown in Figure 9.2, when DNA lesions escape repair pathways, including mismatch repair (MMR) pathway, the cell can enter DNA replication through the existence of specific DNA polymerases that can bypass the damage,[231,232] which may result in the accumulation of mutations and genomic instability (precancerous state). The chronic exposure to persistent DNA damage and mutations, especially in key genes controlling cell proliferation, cell cycle, and apoptosis, increases the possibility of a malignant transformation (cancerous state).

9.3 MEASUREMENT OF OXIDATIVE DNA DAMAGE IN HUMAN CANCER

Despite cell's defensive mechanisms against oxidative injury, different studies suggest an accumulation of oxidative DNA lesions like 8-oxodG, AP sites, and others in the range of hundreds to thousands/cell/day primarily due to inefficiency of the DNA repair systems.[233–235] DNA repair polymorphisms in genes coding for pivotal DNA repair enzymes and proteins like OGG1, NTH1, XRCC1, and others, as well as decline of repair enzymes activity with age[235–237] lead to a deficient BER system, leaving a window to damage persistence after chronic exposure to oxidative stress, mutations, and susceptibility to cancer.[238–240] The additional perturbation of antioxidant and DNA repair systems due to chronic exposure to viral infections adds up a factor of cellular vulnerability to oxidative injury.[241] Increasing evidence using OGG1 null animals shows an accumulation of 8-oxodG in their blood, urine, and tissues but not a significantly shorter life span. In addition, the animals do not exhibit any severe pathophysiological symptom including tumor formation, and surprisingly, they show resistance to inflammation.[175] Recent evidence from OGG1$^{-/-}$ mice suggests an actual association of the glycosylase OGG1 deficiency in the control of inflammatory responses. Specifically, OGG1 deficiency has a protective role against inflammation and genotoxicity induced by *Helicobacter pylori*,[242] and also presents lower levels of the chemokine MIP-1α and Th1 cytokines IL-12 and TNF-α, but higher levels of protective Th2 cytokines IL-4 and IL-10, in the pancreas of OGG-1$^{-/-}$ mice compared with the levels measured in wild-type mice.[243]

Due to the importance of accurate measurement of endogenous DNA damage, different laboratories have employed a variety of techniques for the detection of single oxidative DNA lesions like 8-oxodG, Tg, and AP sites, as well as clustered ones. Substantial experimental evidence suggests the presence of "background" levels of oxidatively generated DNA lesions (e.g., abasic sites or oxidized bases) in human or animal cells and tissues at values ranging from 100–10,000 lesions/Gb.[235,244–246] There are very limited data on the possible accumulation of OCDL in human cells or tissues.[247,248] Endogenous OCDLs can be detected in nonirradiated mice tissues at a steady state of ~0.5–0.9 clusters/Mb and at elevated levels in the human breast cancer line MCF-7 compared to the nonmalignant MCF-10A.[185] Recent studies[249] showed the accumulation of OCDLs in human tumor tissues compared to adjacent normal tissue in a variety of human cancer patients accompanied by the presence of high levels of γ-H2AX foci, an indicator of DSBs.[250] Overall, there is still a discrepancy between the data provided by either HPLC-electrochemical detection or HPLC–tandem mass spectrometry (HPLC-MS/MS), and those inferred from the assessment of formamidopyrimidine (Fpg)-sensitive sites by either the comet assay or the alkaline elution technique.[251,252]

There is a lot of controversy and debate about the steady-state levels of 8-oxo-7,8-dihydroguanine (8-oxoGua) and in general about the oxidized bases steady-state level.[158,184] The measurement of 8-oxoGua or its related 2′-deoxyribonucleoside (8-oxodGuo) is often used as an indicator of oxidation reactions to cellular DNA. However, several of the methods used for detecting 8-oxoGua in cellular DNA give rise to erroneous conclusions due to artifactual oxidation reactions for chromatographic methods and a lack of specificity for the immunoassays.[251,252] Several of the abovementioned characteristic studies (also in Table 9.1) that would support the implication of persistent oxidative stress and damage in human cancer and tumors may suffer in some cases from a lack of accuracy and a relative overestimation of DNA lesions. Nevertheless, there is an accepted idea of using several oxidative DNA lesions (like 8-oxodG, Tg, AP sites, and potentially OCDLs) as novel biomarkers of oxidative stress and susceptibility to cancer. The many well-documented cases of higher levels of DNA damage in tumor and cancer cells compared to nonmalignant controls definitely reveals a great potential in the usage of oxidative DNA damage biomarkers toward prognostic and curative applications in cancer and inflammation as shown in Table 9.1.[253] Although no knowledge of a definite mechanism exists on the occurrence of elevated DNA damage in the presence of cancerous cells or a tumor in the organism, all the above discussed mechanisms of deficient DNA repair and/or antioxidant systems as well as the induction of inflammatory responses may be involved.

9.4 EPIGENETIC INVOLVEMENT IN OXIDATIVE STRESS-INDUCED CARCINOGENESIS

Cancer is a multistage process that involves changes in the transcriptional levels of various genes associated with a number of critical cellular processes absolutely important in tumor development, including those of proliferation, senescence, angiogenesis, metastasis, and so on. In addition, there are both genotoxic and nongenotoxic mechanisms contributing to malignant transformation. To this end, by "genotoxic mechanisms" we often refer to changes in genomic DNA sequences that ultimately lead to mutations, whereas by "nongenotoxic" we include mechanisms capable of modulating

TABLE 9.1 Cases with Documented Elevated Levels of Oxidative DNA Damage Repair in Human Cancer

Type of Cancer	Study Model	Findings
Breast	Human mammary tissues: normal, benign hyperplasia (BH), ductal carcinoma in situ (DCIS), and invasive breast cancer (IBC)	A number of redox-related proteins, DNA repair proteins, and damage markers overexpressed in human breast cancer tissue.[363]
Breast	Human breast cancer patients	Significantly higher ($P < 0.0001$) levels of 8-OH-dG in DNA from tumor compared to nonmalignant adjacent tissue.[364]
Breast	Human breast cancer patients and cell lines	Significantly elevated levels of 8-OH-dG ($P < 0.001$) in malignant breast tissue (invasive ductal carcinoma). Significantly greater levels ($P = 0.007$) in estrogen receptor positive (ORP) versus ORP negative malignant tissue and cancer cell lines.[365]
Breast	Human breast cancer cell lines	Defective DNA repair of 8-hydroxyguanine in mitochondria of MCF-7 and MDA-MB-468 human breast cancer cell lines.[366] Reduced repair of 8-hydroxyguanine in the human breast cancer cell line, HCC1937.[367] Accumulation of oxidatively induced DNA damage in human breast cancer cell lines following treatment with H_2O_2.[368]
Breast	Human breast cancer cell lines	Higher levels of OCDLs in human breast cancer cell line MCF-7 compared to nonmalignant MCF-10A.[185]
Breast	Breast cancer patients	Mean levels of 5-hydroxymethyl-2′-deoxyuridine were significantly higher in blood of women who had high risk or invasive breast lesions versus women with benign lesions.[369]
Cervical cancer	Cervical tissues in human patients	Levels of 8-OH-dG significantly increased ($P < 0.001$) in DNA from low-grade and high-grade levels of dysplasia, compared to normal, although this did not correlate with human papillomavirus status.[370]
Colorectal cancer	Sporadic colorectal tumors patients	Malondialdehyde and 8-OH-dG levels were twofold higher in colorectal tumors compared to normal mucosa ($P < 0.005$). Seven of ten DNA tumor samples (70%) showing higher values of 8-OH-dG also had genetic alterations at different chromosomal loci.[371]
Colorectal cancer	Colorectal tumor patients	Significantly higher levels of 8-OH-dG in nuclear DNA of primary adenocarcinoma, compared to surrounding nontumorous tissue ($P < 0.005$).[372] 8-OH-dG-specific lyase activity and expression were significantly upregulated in carcinoma; a proportional association between 8-OH-dG levels and either 8-OH-dG lyase activity ($P < 0.05$) or expression ($P < 0.05$) present.[373]
Colorectal cancer	Colon cancer patients	Significantly elevated levels of 8-OH-dG lymphocyte DNA in colorectal cancer patients, compared to controls accompanied by reduced levels of antioxidant vitamins[374]
Colorectal	Colorectal cancer patients	Immunostaining for pATM, γH2AX and pChk2 revealed that all were significantly expressed during tumor progression in advanced carcinoma (vs. normal tissue for pATM ($P < 0.05$); vs. normal and adenoma for γH2AX ($P < 0.05$); and vs. normal tissue for pChk2 ($P < 0.05$)). Western blot analysis of γH2AX and pChk2 revealed that their level increased gradually during tumor progression and was maximal in advanced carcinoma (vs. normal tissue; $P < 0.05$).[375]
Gastric	Gastric cancer patients	Significantly higher levels of 8-OH-dG in DNA from tumor-adjacent and tumor adenocarcinoma tissues than in normal tissue ($P < 0.001$) of gastric cancer patients.[376]

TABLE 9.1 (*Continued*)

Type of Cancer	Study Model	Findings
Gastric	Human patients with chronic gastritis and gastric cancer.	Levels of 8-OH-dG significantly elevated in DNA from chronic atrophic gastritis ($P = 0.0009$), intestinal metaplasia ($P = 0.035$), and *Helicobacter pylori*-infected ($P = 0.001$) tissues, compared to unaffected controls.[377]
Gynecologic cancer	Female cancer patients	Significantly higher ($P \leq 0.05$) levels of urinary 8-OH-dG in patients with gynecological cancer compared to control subjects.[378]
Hepatocellular carcinoma (HCC)	HCC patients	Significantly ($P < 0.005$) elevated levels of 8-OH-dG in DNA from peritumoral tissue compared to tumor tissue in HCC. In contrast, patients with hepatic metastases (non-HCC) or end-stage alcoholic liver disease showed no differences between the corresponding two regions.[379]
Acute lymphoblastic leukemia (ALL)	Human lympocytes from ALL patients and controls	Lymphocyte DNA levels of FapyGua, 8-OH-Gua, FapyAde, 8-OH-Ade, 5-OH-Cyt, 5-OH-5-MeHyd, and 5-OH-Hyd significantly ($P < 0.05$) elevated in ALL compared to control subjects.[380]
Adult T-cell leukemia lymphoma; lymphoma, acute leukemia, and myelodysplastic syndrome	Human leukemia and lymphoma patients	Significant difference in levels of urinary 8-OH-dG between adult T-cell leukemia/lymphoma and controls ($P < 0.05$); no significant difference in levels of urinary 8-OH-dG between lymphoma, acute leukemia, and myelodysplastic syndrome.[381]
Lung	Lung cancer patients	Lymphocyte DNA levels of 8-OH-dG significantly elevated ($P < 0.05$) compared to controls.[382]
Lung	Lung cancer patients	An increase in urinary 8-OH-dG/creatinine was found in non-small-cell carcinoma (non-SCC) patients during the course of radiotherapy. SCC patients showed higher levels of urinary 8-OH-dG/creatinine than the controls ($P < 0.05$).[383]
Lung	Patients with lung squamous cell carcinoma (SCC)	In a pilot study of five subjects, levels of 8-OH-Ade elevated in tumor tissue of all SCC patients versus controls; levels of 8-OH-Gua elevated in 4/5b patients; levels of FapyGua elevated in 3 patients; 5-OHMeUra, 5-OH-Ura, 5-OH-Cyt, 2-OH-Ade levels elevated in 3/5 patients; 5-OH-Hyd, 5,6-diOH-Ura, FapyAde (DNA)-levels elevated in only 1/5 or 2/5 patients. Antioxidant enzyme (GPx, SOD, and CAT) levels were lower in cancerous tissues.[384]
Liver, ovary, kidney, breast, and colon	Tumor and adjacent normal tissues from human cancer patients.	Higher OCDLs in tumor versus normal tissues, importance of endogenous non-DSB clusters in human cancer and their potential use as cancer biomarkers.[249]
Nasopharyngeal	Human nasopharyngeal carcinoma (NPC) cells	All cases of NPC were positive for 8-NitroG, 8-OHdG, and 94.7% were positive for INOS. NPC samples exhibited significantly more intense staining for 8-NitroG, 8-OHdG, and iNOS than those of chronic nasopharyngitis. Pathological stimulation of nasopharyngeal tissue, caused by bacterial, viral, or parasitic inflammation, may lead to nitrative and oxidative DNA lesions, caused by NO.[328]
Prostate	Male prostate cancer patients	Significant increased risk was observed for individuals who carried 1 or 2 copies of the variant allele of the XRCC-1 Arg399Gln polymorphism, compared with those who only harbored the wild-type allele. Variability in the capacity of repairing oxidative DNA damage influences susceptibility to prostate cancer.[385]
Renal cell carcinoma (RCC)	RCC patients	A 54% higher content of 8-OHdG was found in RCC than in the corresponding nontumorous kidney, suggesting that the DNA of RCC is more exposed to ROS than is the DNA of nontumorous kidneys.[386]

gene expression without directly affecting the DNA sequence itself.[254] The term "epigenetics" refers to altered gene expression levels without directly affecting primary DNA nucleotide sequences. It involves alterations in DNA methylation patterns and specific histone modifications (methylations, acetylations, deacetylations, etc.), both of which can contribute to transcriptional inactivation.[255] In this respect, epigenetically induced modulation of gene expression can be viewed as a nongenotoxic mechanism for contributing toward malignant transformation.

In general, DNA methylation and chromatin are interconnected in ways that genes will either be transcribed or not. The process is initiated by DNA methyltransferases (DNMTs), bringing together the DNA methylation machinery to the chromatin itself through recruitment of histone deacetylases (HDACs) and other chromatin-binding proteins to promoter sites. Thus, the chromatin's acetylation status regulates transcriptional activity in a way where acetylation maintains a transcriptional active chromatin whereas deacetylation maintains a transcriptional inactive state.[256,257] In mammalian cells, DNA methyltransferases 1 (DNMT1), 3a (DNMT3a), and 3b (DNMT3b) are critical enzymes involved in DNA methylation by means of either maintenance (DNMT 1) and/or *de novo* methylation (DNMT 3a and DNMT 3b). Thus, they could potentially contribute not only to increased promoter methylation status (through *de novo* methylation) but also ensure inheritance of gene silencing (through maintenance of methylation), both of which could account for the acquisition of a malignant transformation phenotype.[258,259] In addition, many genes in tumor cells contain alterations in their DNA methylation patterns, including global hypomethylation (occurring early in the progression phase of neoplasia) and regional hypermethylation of normally unmethylated CpG islands (regions of 0.5–4.0 Kb length in genomic DNA). Both events can result in the transcriptional silencing of tumor suppressor gene expression and the transcriptional activation of oncogenes, respectively. Finally, genome-wide hypomethylation has been shown to increase mutation rates, thus leading to genome instability.[260] It is therefore evident that alterations in DNA methylation patterns underlie aberrant gene expression that is associated with malignant transformation.

ROS are implicated in the multistage process of malignant transformation by means of both genotoxic and nongenotoxic mechanisms. In terms of genotoxic mechanisms, ROS modulate cell proliferation by regulating various cell cycle proteins, and cell death processes (senescence), by acting either as an antisenescence stimulus[261] or through suppression of apoptosis.[262] On the other hand, in terms of nongenotoxic (epigenetic) mechanisms, ROS cause a wide range of DNA damage by-products which interfere with the ability of DNA to function as a substrate for the DNMTs, resulting in global hypomethylation.[263] The presence of one such major by-product (8-oxodG) in CpG islands strongly inhibits methylation of adjacent cytosine residues[264,265] and interferes with the ability of restriction nucleases to cleave DNA.[266] The presence of O^6-methylguanine, another potentially mutagenic DNA lesion, inhibits the binding of DNA methyltransferases and therefore can lead to hypomethylation by means of inhibiting methylation of adjacent cytosine molecules.[267,268] Alternatively, O^6-methylguanine can be spontaneously mispaired with thymine and contribute to DNA hypomethylation.[269] MnSOD and metallothionein are found silenced via promoter DNA hypermethylation events in pancreatic[270] and breast carcinomas[271] as well as in human myeloma cells.[272] Finally, major phase II xenobiotic metabolizing enzymes such as NADPH: quinone oxidoreductase 1 (NQO1) and glutathione-S-transferase P1 (GSTP1) are also inactivated via promoter hypermethylation in hepatocellular carcinoma,[273,274] breast, renal,[275] and prostate cancers,[276] respectively. These studies clearly demonstrate the association between repressed expression of key antioxidant enzymes and tumor development by means of promoter hypermethylation-induced gene silencing.

9.5 DEREGULATION OF CELL DEATH PATHWAYS BY OXIDATIVE STRESS IN CANCER PROGRESSION

Cell death is classified by biochemical and morphological criteria. According to the recommended classification of cell death, three distinct types of pathways can be defined, which are apoptosis, necrosis, and autophagy, although there are numerous examples in which cell death displays mixed features. Impairment and/or alteration of cell death pathways are a central element in cancer progression (Fig. 9.3).[277,278]

9.5.1 Apoptosis

Apoptosis or programmed cell death is a ubiquitous homeostatic process involved in numerous biological phenomena. Under physiological conditions, apoptosis is critical not only in the turnover of cells in tissues but also during normal development and senescence, while its impairment occurs during cancer progression.[279] Apoptosis is a highly organized program induced by a myriad of stimuli. It is characterized by the progressive activation of precise pathways leading to specific biochemical and morphological alterations in individual cells without involvement of an inflammatory response. Early stages (initiation phase) of apoptosis are charac-

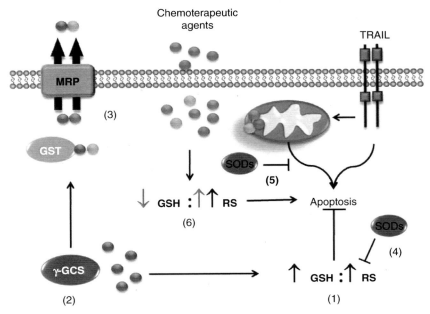

Figure 9.3 Apoptotic deregulation and multidrug resistance in cancer cells. In cancer cells, deregulation of apoptotic pathways is associated with cellular redox imbalance. High levels of both ROS/RNS and high intracellular GSH impair the activation of apoptotic enzymes by oxidative posttranslational modifications and/or scavenge pro-apoptotic oxidative damage (1). An increase in γ-GCS levels in cancer cells augments GSH concentration, whose metabolism also contributes to multidrug resistance against chemotherapeutic agents (2). Increased expression of GSTs and MRPs is a common feature of transformed cells that are associated with high resistance to chemotherapy (3). SODs exert antitumorigenic effects by reduction of RS levels in cancer cells (4). Interestingly, SODs also protect against the activation of death receptor (TRAIL) and mitochondrial pathways of apoptosis induced by chemotherapeutic agents, which depend on the generation of RS (5). Thus, chemotherapeutic approaches involve both the depletion of intracellular GSH and/or increased oxidative damage (6). See color insert.

terized by initiator caspase activation, cell shrinkage, loss of plasma membrane lipid asymmetry, and chromatin condensation. Later during the execution phase, apoptosis is characterized by activation of executioner caspases and endonucleases, apoptotic body formation, and cell fragmentation.[280]

The role of oxidative stress in apoptosis has been largely studied.[281,282] Redox imbalance in cancer cells impairs the activation of the apoptotic machinery. Increased $O_2^{\bullet-}$ and NO formation in cancer cells is associated with increased resistance to apoptosis.[283,284] Accordingly, inhibition and/or downregulation of $O_2^{\bullet-}$-generating enzymes such as NOXs increase cellular sensitivity to apoptosis.[261,285] An interesting observation is the fact that both increased ROS/RNS and increased GSH are common hallmarks of cancer cells. Increased intracellular GSH might be acting as a protective mechanism of transformed cells against the deleterious effects of increased oxidative stress. This was evidenced by reports showing that overexpression of the Ha-ras oncogene increases the levels of the $O_2^{\bullet-}$ and H_2O_2, as well as GSH, which was associated with increased cell growth and proliferation. Interestingly, GSH depletion induced apoptosis in Ha-ras-transformed cells by increasing oxidative damage.[286] Similarly, GSH depletion in human neuroblastoma cells triggers oxidative stress and apoptosis.[287]

The signaling cascades that regulate the progression of apoptosis have been extensively studied and characterized, and both extrinsic and intrinsic pathways have been described for the activation of apoptosis. Induction of apoptosis via the extrinsic pathway is triggered by the activation of death receptors such as those activated by the TNF-related apoptosis-inducing ligand or TRAIL (DR4, DR5). Activation of DR4 and DR5 leads to the formation of the death-inducing signaling complex (DISC) by the recruitment of the Fas-associated death domain (FADD) and initiator caspase 8 (and in some cases caspase 10) and the cellular FLICE-inhibitory protein (cFLIP$_L$). Initiator caspase 8 amplifies the apoptotic cascade by activation of executioner caspases (3, 6, and 7). In cells that have lower levels of DISC formation and thus reduced caspase 8 activation (Type II cells), the progression of apoptosis relies on an amplification loop induced by caspase 8-dependent cleavage of the Bcl-2-family protein Bid (Bcl-2 interacting domain), translocation to the mitochondria, and subsequent release of cytochrome c (cyt c).[288]

TRAIL has great therapeutic potential, as it is a rather specific apoptosis-inducing cytokine in cancer cells.[289] However, its therapeutic use has been hampered by the observation that cancer cells present resistance to TRAIL-induced apoptosis. TRAIL-induced activation of caspase 8 and release of cyt *c* and Smac (second mitochondria-derived activator of caspases)/DIABLO is mediated by oxidative stress.[290,291] TRAIL-induced apoptosis is also potentiated by uncoupling of the mitochondrial oxidative phosphorylation, generation of ROS,[292] and GSH depletion.[293] Paradoxically, ROS generation by NOX4 has been shown to impair TRAIL-induced executioner caspase 3 activation by oxidation of the catalytic cysteine residue.[294] TRAIL-induced ROS also suppress apoptosis by upregulation of the anti-apoptotic Bcl-2 member MCL-1,[295] while RNS sensitize prostate carcinoma cells to TRAIL-induced apoptosis by inhibition of the DR5 transcription repressor Ying Yang 1.[296]

The intrinsic pathway of apoptosis, also referred to as the mitochondrial pathway, is activated by a wide variety of chemotherapeutic agents. The mechanisms by which these stimuli trigger apoptosis differ between them, but they all convey the release of pro-apoptotic proteins from the mitochondria including cyt *c*. However, the exact mechanisms mediating cyt *c* release are still controversial. Distinct mitochondrial released proteins such as AIF (apoptosis inducing factor), EndoG (Endonuclease G), Omi/HtrA2 (HtrA serine peptidase 2), and Smac/DIABLO participate in the mitochondrial pathway of apoptosis. The intrinsic pathway is also regulated by the Bcl-2 family of proteins. The BH3 (Bcl-2 homology domain 3)-only proteins Bad (Bcl-2-associated death promoter), Bid, Bim, Bik (Bcl-2-interacting killer), NOXA, and PUMA (p53 upregulated modulator of apoptosis) regulate the anti-apoptotic Bcl-2 proteins Bcl-2 and Bcl-xL (B-cell lymphoma-extra large) to promote apoptosis. Bcl-2 and Bcl-xL inhibit Bax (Bcl-2–associated X protein) and Bak (Bcl-2 homologous antagonist/killer). BH3-only proteins derepress Bax and Bak by direct binding and inhibition of Bcl-2 and other anti-apoptotic family members. Bax and Bak activation mediates the release of cyt *c* by its translocation to the mitochondria. Released cyt *c* leads to the recruitment of APAF1 (apoptotic protease activating factor 1) into the apoptosome and activation of caspase-9. Once activated, initiator caspases converge in the cleavage/activation of executioner caspases 3, 6, and 7, which further cleave different cellular substrates leading to the organized demise of the cell.[297]

GSH is a major regulator of apoptotic signaling pathways.[298,299] Elevated GSH levels in cancerous cells and solid tumors are directly associated with resistance to chemotherapeutic agents by inhibition of apoptotic pathways.[37,38,300] GSH depletion can predispose transformed cells to apoptosis or directly trigger cell death by induction of the permeability transition pore and/or the activation of executioner caspases.[301,302] A reduction in the GSH content is necessary for the formation of the apoptosome.[303] GSH depletion also activates the mitochondrial apoptotic initiator Bax through its oxidation-dependent dimerization[304] and enhances apoptosis induced by Bax overexpression in lung cancer cells.[305] GSH depletion triggers cyt *c* release, and it has been proposed that released cyt *c* needs to be oxidized for its pro-apoptotic action, which would require cytosolic GSH levels to be depleted.[306,307] Recently, it was demonstrated that in cancer cells, cyt *c* is reduced and held inactive by increased GSH content generated as a result of glucose metabolism via the pentose phosphate pathway.[308]

Overexpression of members of the Bcl-2 family of pro-survival proteins is commonly associated with cancer pathogenesis.[309] The anti-apoptotic role of Bcl-2 has been linked to the regulation of GSH content in different cellular compartments.[310] Overexpression of Bcl-2 increases GSH levels and inhibits mitochondrial-induced cell death elicited by GSH-depleting reagents.[311,312] Recent studies suggest that Bcl-2 can regulate the mitochondrial GSH pool by a direct interaction with GSH via the BH3 groove.[313] Accordingly, depletion of intracellular GSH overcomes Bcl-2-mediated resistance to apoptosis.[300,314] However, these effects appear to be cell type-specific and context-dependent.[315] The anti-apoptotic effect of Bcl-xl is also attributed to the regulation of GSH homeostasis by preventing GSH loss.[316] Bcl-2 is itself regulated by oxidative posttranslational modifications. Increased levels of NOS in cancer cells mediate Bcl-2 nitrosylation, inhibiting its ubiquitination and proteasomal degradation, which renders cells resistant to chemotherapeutic agents.[317] In contrast, NO sensitizes prostate cancer cells to TRAIL-induced apoptosis by inhibition of NF-κB activity and Bcl-xL expression.[318]

Other pathways of apoptosis, such as ER stress and DNA damage, have been described, which can be dependent or independent of a cross talk with the mitochondrial pathway. The ER is highly sensitive to perturbations in cellular energy levels, redox state, and/or Ca^{2+} concentration. Such perturbations result in the accumulation and aggregation of unfolded proteins, which are toxic to cells triggering the unfolded protein response. Unfolded proteins cause the chaperone glucose-regulated protein (GRP78) to release the inositol-requiring kinase 1 (IRE1α), the PRKR-like ER kinase (PERK), and the activating transcription factor 6 (ATF6). IRE1α binds to TRAF2 (TNF-receptor associated factor 2) and ASK1 (apoptosis signaling-regulating kinase 1), inducing the activation of stress activated kinases (SAPKs). Induction of CHOP via SAPKs, ATF6,

PERK-induced ATF4, and/or IRE1α-induced X-box-binding protein 1 (XBP1) regulates the expression of Bim and Bcl-2 proteins.[319] In an environment with limited nutrients, cancer cells utilize cellular regulatory pathways to facilitate adaptation and promote tumor growth and survival. The ER senses nutrient deprivation triggering signaling pathways that promote adaptive strategies. Thus, ER stress-signaling pathways represent potential antineoplastic targets and regulate redox homeostasis. CHOP overexpression downregulates GSH levels and induces increased ROS formation sensitizing cells to ER-stress.[320] PERK-dependent activation of Nrf-2 regulates intracellular GSH levels, which might protect against oxidative stress induced during unfolded protein response.[321] Ablation of PERK in mammary carcinoma cells results in impaired regeneration of intracellular antioxidants and accumulation of ROS, suggesting that PERK signaling is involved in tumor initiation and expansion.[322] The ER stress-linked transcriptional factor XBP1 also protects against oxidative stress by upregulating catalase expression.[323]

It is well known that specific DNA lesions, including O^6-methylguanine, base N-alkylations, bulky DNA adducts, DNA cross-links, and DSBs trigger apoptosis. BER deficiency in the presence of N-alkylations and MMR mediated by O^6-methylguanine lesions result in secondary DNA lesions which interfere with replication leading to secondary DSB lesions. DSBs are detected by ATM (ataxia telangiectasia mutated gene) and ATR (ataxia telangiectasia and Rad3 related) proteins, which signal downstream to p53. ATM can also be directly activated by oxidative stress, which involves the formation of a disulfide-cross-linked dimer of ATM.[324] p53 is a transcription factor with tumor suppressive properties. p53 induces transcriptional activation of pro-apoptotic factors that can activate death receptor and mitochondrial pathways.[325] However, nontranscriptional regulation of apoptosis by p53 and p53-independent pathways have also been described. Oxidative and nitrosative stress modulates p53 activity,[326] and a decrease in the ability to bind consensus DNA sequence has been reported in glutathionylated p53,[327] which is abundant in cancer cell lines.[328] NO is capable of modulating p53 by inhibition of Hdm2 binding through protein nitrosylation, which acts as a negative regulator of p53.[329] In contrast, TRX1 enhances DNA binding activity of p53.[330] Moreover, p53-dependent apoptosis is reported to involve alterations in the expression levels of MnSOD and oxidative stress.[331,332]

9.5.2 Autophagy

Autophagy is a major catabolic pathway by which eukaryotic cells degrade and recycle macromolecules and organelles. It plays an essential role in differentiation, development, and cellular response to stress. Autophagy can be activated during amino acid deprivation and has been associated with cancer progression. Autophagy is initiated by the surrounding of cytoplasmic constituents by a phagophore, which forms a closed double-membrane structure, called autophagosome. The autophagosome subsequently fuses with a lysosome to become an autolysosome, whose content is degraded by acidic lysosomal hydrolases. Autophagic cell death is morphologically defined by massive autophagic vacuolization of the cytoplasm in the absence of chromatin condensation.[333] Several studies suggest that autophagy is one of the survival processes of cancer cells in response to starvation, hypoxia, and chemotherapeutic agents. In contrast, in human breast, ovarian, and prostate tumors, the autophagy gene Beclin-1 is monoallelically deleted.[334] Furthermore, Beclin-1 also acts as a tumor suppressor.[335] The mechanisms by which autophagy inhibits tumorigenesis are unknown but might involve prevention of oxidative damage and mutagenesis through the removal of ROS generated by mitochondria.[336] Many anticancer agents and DNA-damaging conditions induce autophagy, but it is unknown whether autophagy acts as a survival signaling to counteract the stress, or if it participates in the tumor cell killing.[337,338] However, there is evidence that oxidative stress induces autophagic cell death in transformed and cancer cells.[339]

9.5.3 Redox Regulation of Drug Resistance in Cancer Cells

GSH metabolism participates in chemotherapy-induced tumor resistance to apoptotic cell death. Increased expression of glutathione-S-transferases (GST), transporters for GSH/GSH-conjugates, as well as high GSH concentration, are common features of transformed cells that are associated with high resistance to chemotherapeutic agent-induced apoptosis (Fig. 9.3).[340] GSTs catalyze nucleophilic attack by GSH on nonpolar compounds that contain an electrophilic atom of carbon, nitrogen, or sulfur. Overexpression of GSTs and increased levels of GSH are often associated with an increased resistance to apoptosis-induced by cancer chemotherapeutic drugs.[341] The most highly expressed GST isoenzyme in various human cancerous and precancerous tissues is GSTP1.[342] Several studies clearly identify a relationship between GSH/GSTs levels and the extent of drug-induced apoptosis like that induced by doxorubicin and cisplatin.[38,343] Several cancer chemotherapeutic agents such as adriamycin, 1,3-bis(2-chloroethyl)-1-nitrosourea (BCNU), carmustine, ethacrynic acid, and melphalan, are also detoxified by GSTs. GST isoenzyme expression increases during cancer development as evidenced by GSTP1 upregulation during chemical liver carcinogenesis. GST polymorphisms are linked to specific cancer types.[341,344] For example, GSTP1 polymorphisms are associated with the chemotherapy response in patients with metastatic colorectal cancer or multiple myeloma.[345]

Because GSTs are overexpressed in many tumors, prodrug therapy has been designed using inactive agents that are converted to cytotoxins by GST activity. TLK286 is a GSH analogue in which an inactive alkylating agent is linked to a modified GSH backbone. GSTT catalyzed elimination of TLK286 yields aziridinium ring moieties that can alkylate DNA having toxic activity against a variety of tumors and tumor cell lines where GSTT is overexpressed.[346,347] A variety of GST inhibitors modulate drug resistance by sensitizing tumor cells to anticancer drugs. Ethacrynic acid and its derivatives, which inhibit the activity of GST by binding directly to the substrate-binding site and depleting GSH, reduce resistance against drug-induced apoptosis in different tumoral cells.[348,349] TLK199 is a peptidomimetic inhibitor of GSTT which has a GSH backbone in which a benzyl group is added to the sulfur atom and glycine is replaced with phenylglycine. It inhibits GSTT and enhances the toxicity of different chemotherapeutic agents while also reversing the resistance of MRP1 overexpressing cell lines.[346]

GSTs protect against a variety of chemical carcinogens. Cytosolic GSTs exhibit genetic polymorphisms that increase susceptibility to carcinogenesis and inflammatory disease. Epoxides derived from environmental carcinogens, including aflatoxin B1, 1-nitropyrene, 4-nitroquinoline, polycyclic aromatic hydrocarbons (PAHs), and styrene, are detoxified by GST. Carcinogenic heterocyclic amines, produced by cooking protein-rich food, are also detoxified by GST isoenzymes.[342,345] Interestingly, drug resistance is exhibited in cells expressing certain isoforms of GSTs even when that specific selecting drug is not an enzyme substrate. This is explained by recent findings reporting the ability of GSTs to play critical roles in kinase signaling and protein glutathionylation.[341]

GSH/GSH-conjugate transporters also act as important contributors to the carcinogenic process. When antineoplastic or chemotherapeutic drugs enter cancer cells, they are conjugated with GSH and are excreted through GSH pumps of the MRP family of transporters. It is known that MRP transporters are highly expressed in malignant cells.[350] Thus, inhibitors of MRP-mediated transport of GSH-conjugates have been largely used as a common adjuvant in anticancer therapy.[351–353] The multidrug resistance family of transporters ABCC/MRPs act as transporters of GSH, GSSG, and GSH adducts, and require the hydrolysis of ATP for its transport activity.[351] MRP1 is capable of transporting the GSH conjugate of the chemotherapeutic agent doxorubicin. In addition, MRP1 and MRP2 transporters have the ability to transport vincristine and doxorubicin in their unmodified state. Such transport is dependent on or stimulated by GSH.[354,355] The mechanism by which GSH stimulates MRP transport activity is not fully understood. GSH is a poor substrate for both MRP1 and MRP2. In contrast, GSSG is more efficiently transported by MRP1, suggesting a role for MRP proteins in maintaining GSH/GSSG balance.[350] The transport of heavy metals by MRP1 and MRP2 is also dependent on GSH. For example, MRP1 transports the GSH-conjugate arsenic triglutathione, which can be formed enzymatically by GSTP1.[356,357] Another potential approach for antineoplastic therapy is the modulation of GSH efflux through its transporters or pumps. Recently, it has been shown that transformed cells are sensitized to cell death when intracellular GSH is depleted through stimulation of GSH efflux pumps (MRP, CFTR, and SLCO proteins),[358–360] whose expression varies in different cancer cell lines.[361,362]

CONCLUSIONS AND PERSPECTIVE

As highlighted in Fig. 9.4, RS, oxidative stress, and redox-dependent signaling cascades participate at all stages of the malignancy/carcinogenic process aroused from both inherited and environmental factors. The carcinogenic process can be triggered when oxidative DNA damage transforms or immortalizes cells by promoting uncontrolled cell proliferation and/or impaired activation of cell death pathways. Nonoxidative DNA damage might also mediate cellular transformation by promoting a redox imbalance through (1) increased generation of RS associated with alterations in protein levels of RS-generating enzymes (NOX and NOS) and/or mutations in the mitochondrial ETC; and (2) increased levels of intracellular antioxidant defenses, particularly GSH, which also participates in multidrug resistance against chemotherapeutic approaches. Increased RS formation in cancer cells can regulate cell proliferation/survival pathways such as those mediated by MAPKs and PI3K/AKT, and also impair the activation of cell death pathways (apoptosis or autophagy). High intracellular GSH might then exert a protective mechanism against the deleterious effects of increased RS formation and/or chemotherapeutic agents.

DNA damage caused by RS is a central element in the transformation process toward malignancy. Chronic exposure to DNA lesions can lead to mutations and genomic instability (precancerous state) and eventually to malignant transformations (cancerous state), while

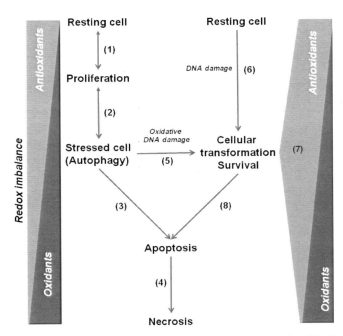

Figure 9.4 Redox imbalance in carcinogenesis: a simplified hypothesis. Resting cells are characterized by a reduced environment with a high concentration of intracellular antioxidants (GSH) that counteract the formation of RS during normal cell metabolism. Cell proliferation is associated with a regulated increase in RS, whose effects are reversed by the reducing environment of the cell (1). In response to "mild" oxidative stress (defined as an imbalance between production of reactive metabolites, oxidants, and their elimination by antioxidant systems), stressed cells activate counteractive measures that might involve induction of transcriptional regulated expression of antioxidant systems or the removal of RS-producing mitochondria by autophagy (2). If oxidative stress persists and/or exceeds antioxidant defenses, programmed cell death or apoptosis is triggered (3), which can further develop into necrotic cell death upon "excessive" oxidative damage (4). The carcinogenic process is triggered when oxidative DNA damage transforms or immortalizes cells by promoting uncontrolled cell proliferation and/or impaired activation of cell death pathways (5). Nonoxidative DNA damage might also mediate cellular transformation by upregulation of enzymes involved in RS formation (NOXs or NOS) and antioxidant (GSH) synthesis (6). Accordingly, transformed cells are characterized not only by an increase in RS formation but also by an increase in antioxidant defenses such as GSH, which might act as a protective mechanism against the deleterious effects of RS and/or chemotherapeutic agents (7). Thus, redox-based chemotherapeutic approaches involve either the depletion of antioxidant defenses in cancer cells or enhancement of oxidative damage in order to overcome anti-apoptotic defenses and activate cell death pathways (8).

impaired DNA repair mechanisms have been reported in transformed cells. However, it is evident that RS can promote/predispose to cancer development by oxidative damage/modification to lipids and proteins/enzymes. Impairment and/or alteration of cell death pathways are also a central element in cancer progression, and chemotherapeutic approaches involve either the depletion of antioxidant defenses in cancer cells or enhancement of oxidative damage in order to overcome anti-apoptotic defenses and activate cell death pathways.

ACKNOWLEDGMENTS

This work was supported by the National Institutes of Health Grant P20RR17675 Centers of Biomedical Research Excellence (COBRE), the Layman Award, and the Life Sciences Grant Program of the University of Nebraska-Lincoln (R. Franco); the Marie Curie International Reintegration Grants within the 6th and the 7th European Community Framework Programmes (MIRG-CT-2006-036585) (A. Pappa) and (PIRG05-GA-2009-249315) (M.I. Panayiotidis), respectively; and a 2009/2010 ECU Research/Creative Activity Award (A.G. Georgakilas).

REFERENCES

1. Klaunig, J.E., Kamendulis, L.M., Hocevar, B.A. Oxidative stress and oxidative damage in carcinogenesis. *Toxicol. Pathol.* **2010**, *38*(1), 96–109.
2. Szatrowski, T.P., Nathan, C.F. Production of large amounts of hydrogen peroxide by human tumor cells. *Cancer Res.* **1991**, *51*(3), 794–798.
3. Irani, K., Xia, Y., Zweier, J.L., Sollott, S.J., Der, C.J., Fearon, E.R., Sundaresan, M., Finkel, T., Goldschmidt-Clermont, P.J. Mitogenic signaling mediated by oxidants in Ras-transformed fibroblasts. *Science* **1997**, *275*(5306), 1649–1652.
4. Mitsushita, J., Lambeth, J.D., Kamata, T. The superoxide-generating oxidase Nox1 is functionally required for Ras oncogene transformation. *Cancer Res.* **2004**, *64*(10), 3580–3585.
5. Gianni, D., Bohl, B., Courtneidge, S.A., Bokoch, G.M. The involvement of the tyrosine kinase c-Src in the regulation of reactive oxygen species generation mediated by NADPH oxidase-1. *Mol. Biol. Cell* **2008**, *19*(7), 2984–2994.
6. Gruhne, B., Sompallae, R., Marescotti, D., Kamranvar, S.A., Gastaldello, S., Masucci, M.G. The Epstein-Barr virus nuclear antigen-1 promotes genomic instability via induction of reactive oxygen species. *Proc. Natl. Acad. Sci. U.S.A.* **2009**, *106*(7), 2313–2318.
7. Brar, S.S., Kennedy, T.P., Sturrock, A.B., Huecksteadt, T.P., Quinn, M.T., Whorton, A.R., Hoidal, J.R. An NAD(P)H

oxidase regulates growth and transcription in melanoma cells. *Am. J. Physiol. Cell Physiol.* **2002**, *282*(6), C1212–C1224.

8. Mochizuki, T., Furuta, S., Mitsushita, J., Shang, W.H., Ito, M., Yokoo, Y., Yamaura, M., Ishizone, S., Nakayama, J., Konagai, A., Hirose, K., Kiyosawa, K., Kamata, T. Inhibition of NADPH oxidase 4 activates apoptosis via the AKT/apoptosis signal-regulating kinase 1 pathway in pancreatic cancer PANC-1 cells. *Oncogene* **2006**, *25*(26), 3699–3707.

9. von Lohneysen, K., Noack, D., Wood, M.R., Friedman, J.S., Knaus, U.G. Structural insights into Nox4 and Nox2: Motifs involved in function and cellular localization. *Mol. Cell. Biol.* **2010**, *30*(4), 961–975.

10. Graham, K.A., Kulawiec, M., Owens, K.M., Li, X., Desouki, M.M., Chandra, D., Singh, K.K. NADPH oxidase 4 is an oncoprotein localized to mitochondria. *Cancer Biol. Ther.* **2010**, *10*(3), 223–231.

11. Si, J., Behar, J., Wands, J., Beer, D.G., Lambeth, D., Chin, Y.E., Cao, W. STAT5 mediates PAF-induced NADPH oxidase NOX5-S expression in Barrett's esophageal adenocarcinoma cells. *Am. J. Physiol. Gastrointest. Liver Physiol.* **2008**, *294*(1), G174–G183.

12. Si, J., Fu, X., Behar, J., Wands, J., Beer, D.G., Souza, R.F., Spechler, S.J., Lambeth, D., Cao, W. NADPH oxidase NOX5-S mediates acid-induced cyclooxygenase-2 expression via activation of NF-kappaB in Barrett's esophageal adenocarcinoma cells. *J. Biol. Chem.* **2007**, *282*(22), 16244–16255.

13. Fu, X., Beer, D.G., Behar, J., Wands, J., Lambeth, D., Cao, W. cAMP-response element-binding protein mediates acid-induced NADPH oxidase NOX5-S expression in Barrett esophageal adenocarcinoma cells. *J. Biol. Chem.* **2006**, *281*(29), 20368–20382.

14. Hong, J., Behar, J., Wands, J., Resnick, M., Wang, L.J., Delellis, R.A., Lambeth, D., Cao, W. Bile acid reflux contributes to development of esophageal adenocarcinoma via activation of phosphatidylinositol-specific phospholipase Cgamma2 and NADPH oxidase NOX5-S. *Cancer Res.* **2010**, *70*(3), 1247–1255.

15. Luxen, S., Belinsky, S.A., Knaus, U.G. Silencing of DUOX NADPH oxidases by promoter hypermethylation in lung cancer. *Cancer Res.* **2008**, *68*(4), 1037–1045.

16. Brandon, M., Baldi, P., Wallace, D.C. Mitochondrial mutations in cancer. *Oncogene* **2006**, *25*(34), 4647–4662.

17. Olson, S.Y., Garban, H.J. Regulation of apoptosis-related genes by nitric oxide in cancer. *Nitric Oxide* **2008**, *19*(2), 170–176.

18. Ekmekcioglu, S., Ellerhorst, J., Smid, C.M., Prieto, V.G., Munsell, M., Buzaid, A.C., Grimm, E.A. Inducible nitric oxide synthase and nitrotyrosine in human metastatic melanoma tumors correlate with poor survival. *Clin. Cancer Res.* **2000**, *6*(12), 4768–4775.

19. Konopka, T.E., Barker, J.E., Bamford, T.L., Guida, E., Anderson, R.L., Stewart, A.G. Nitric oxide synthase II gene disruption: Implications for tumor growth and vascular endothelial growth factor production. *Cancer Res.* **2001**, *61*(7), 3182–3187.

20. Le, X., Wei, D., Huang, S., Lancaster, J.R., Jr., Xie, K. Nitric oxide synthase II suppresses the growth and metastasis of human cancer regardless of its up-regulation of protumor factors. *Proc. Natl. Acad. Sci. U.S.A.* **2005**, *102*(24), 8758–8763.

21. Wei, D., Richardson, E.L., Zhu, K., Wang, L., Le, X., He, Y., Huang, S., Xie, K. Direct demonstration of negative regulation of tumor growth and metastasis by host-inducible nitric oxide synthase. *Cancer Res.* **2003**, *63*(14), 3855–3859.

22. Fukumura, D., Kashiwagi, S., Jain, R.K. The role of nitric oxide in tumour progression. *Nat. Rev. Cancer* **2006**, *6*(7), 521–534.

23. Wang, L., Shi, G.G., Yao, J.C., Gong, W., Wei, D., Wu, T.T., Ajani, J.A., Huang, S., Xie, K. Expression of endothelial nitric oxide synthase correlates with the angiogenic phenotype of and predicts poor prognosis in human gastric cancer. *Gastric Cancer* **2005**, *8*(1), 18–28.

24. Tanriover, N., Ulu, M.O., Isler, C., Durak, H., Oz, B., Uzan, M., Akar, Z. Neuronal nitric oxide synthase expression in glial tumors: Correlation with malignancy and tumor proliferation. *Neurol. Res.* **2008**, *30*(9), 940–944.

25. Song, M., Santanam, N. Increased myeloperoxidase and lipid peroxide-modified protein in gynecological malignancies. *Antioxid. Redox Signal.* **2001**, *3*(6), 1139–1146.

26. Cascorbi, I., Henning, S., Brockmoller, J., Gephart, J., Meisel, C., Muller, J.M., Loddenkemper, R., Roots, I. Substantially reduced risk of cancer of the aerodigestive tract in subjects with variant–463A of the myeloperoxidase gene. *Cancer Res.* **2000**, *60*(3), 644–649.

27. Ahn, J., Gammon, M.D., Santella, R.M., Gaudet, M.M., Britton, J.A., Teitelbaum, S.L., Terry, M.B., Neugut, A.I., Josephy, P.D., Ambrosone, C.B. Myeloperoxidase genotype, fruit and vegetable consumption, and breast cancer risk. *Cancer Res.* **2004**, *64*(20), 7634–7639.

28. Ambrosone, C.B., Ahn, J., Singh, K.K., Rezaishiraz, H., Furberg, H., Sweeney, C., Coles, B., Trovato, A. Polymorphisms in genes related to oxidative stress (MPO, MnSOD, CAT) and survival after treatment for breast cancer. *Cancer Res.* **2005**, *65*(3), 1105–1111.

29. Narko, K., Ristimaki, A., MacPhee, M., Smith, E., Haudenschild, C.C., Hla, T. Tumorigenic transformation of immortalized ECV endothelial cells by cyclooxygenase-1 overexpression. *J. Biol. Chem.* **1997**, *272*(34), 21455–21460.

30. Nikolic, D., van Breemen, R.B. DNA oxidation induced by cyclooxygenase-2. *Chem. Res. Toxicol.* **2001**, *14*(4), 351–354.

31. Lee, S.H., Williams, M.V., Dubois, R.N., Blair, I.A. Cyclooxygenase-2-mediated DNA damage. *J. Biol. Chem.* **2005**, *280*(31), 28337–28346.

32. Connolly, E.M., Harmey, J.H., O'Grady, T., Foley, D., Roche-Nagle, G., Kay, E., Bouchier-Hayes, D.J. Cyclooxygenase inhibition reduces tumour growth and metas-

tasis in an orthotopic model of breast cancer. *Br. J. Cancer* **2002**, *87*(2), 231–237.

33. Menter, D.G., Schilsky, R.L., DuBois, R.N. Cyclooxygenase-2 and cancer treatment: Understanding the risk should be worth the reward. *Clin. Cancer Res.* **2010**, *16*(5), 1384–1390.

34. Griguer, C.E., Oliva, C.R., Kelley, E.E., Giles, G.I., Lancaster, J.R., Jr., Gillespie, G.Y. Xanthine oxidase-dependent regulation of hypoxia-inducible factor in cancer cells. *Cancer Res.* **2006**, *66*(4), 2257–2263.

35. Valko, M., Leibfritz, D., Moncol, J., Cronin, M.T., Mazur, M., Telser, J. Free radicals and antioxidants in normal physiological functions and human disease. *Int. J. Biochem. Cell Biol.* **2007**, *39*(1), 44–84.

36. Mates, J.M., Sanchez-Jimenez, F.M. Role of reactive oxygen species in apoptosis: Implications for cancer therapy. *Int. J. Biochem. Cell Biol.* **2000**, *32*(2), 157–170.

37. Estrela, J.M., Ortega, A., Obrador, E. Glutathione in cancer biology and therapy. *Crit. Rev. Clin. Lab. Sci.* **2006**, *43*(2), 143–181.

38. Roshchupkina, G.I., Bobko, A.A., Bratasz, A., Reznikov, V.A., Kuppusamy, P., Khramtsov, V.V. *In vivo* EPR measurement of glutathione in tumor-bearing mice using improved disulfide biradical probe. *Free Radic. Biol. Med.* **2008**, *45*, 312–320.

39. Yao, K.S., Godwin, A.K., Johnson, S.W., Ozols, R.F., O'Dwyer, P.J., Hamilton, T.C. Evidence for altered regulation of gamma-glutamylcysteine synthetase gene expression among cisplatin-sensitive and cisplatin-resistant human ovarian cancer cell lines. *Cancer Res.* **1995**, *55*(19), 4367–4374.

40. Walsh, A.C., Feulner, J.A., Reilly, A. Evidence for functionally significant polymorphism of human glutamate cysteine ligase catalytic subunit: Association with glutathione levels and drug resistance in the National Cancer Institute tumor cell line panel. *Toxicol. Sci.* **2001**, *61*(2), 218–223.

41. Chung, W.J., Lyons, S.A., Nelson, G.M., Hamza, H., Gladson, C.L., Gillespie, G.Y., Sontheimer, H. Inhibition of cystine uptake disrupts the growth of primary brain tumors. *J. Neurosci.* **2005**, *25*(31), 7101–7110.

42. Lo, M., Ling, V., Wang, Y.Z., Gout, P.W. The xc-cystine/glutamate antiporter: A mediator of pancreatic cancer growth with a role in drug resistance. *Br. J. Cancer* **2008**, *99*(3), 464–472.

43. Gout, P.W., Buckley, A.R., Simms, C.R., Bruchovsky, N. Sulfasalazine, a potent suppressor of lymphoma growth by inhibition of the x(c)- cystine transporter: A new action for an old drug. *Leukemia* **2001**, *15*(10), 1633–1640.

44. Huang, Y., Dai, Z., Barbacioru, C., Sadee, W. Cystine-glutamate transporter SLC7A11 in cancer chemosensitivity and chemoresistance. *Cancer Res.* **2005**, *65*(16), 7446–7454.

45. Gao, P., Tchernyshyov, I., Chang, T.C., Lee, Y.S., Kita, K., Ochi, T., Zeller, K.I., De Marzo, A.M., Van Eyk, J.E., Mendell, J.T., Dang, C.V. c-Myc suppression of miR-23a/b enhances mitochondrial glutaminase expression and glutamine metabolism. *Nature* **2009**, *458*(7239), 762–765.

46. Collins, C.L., Wasa, M., Souba, W.W., Abcouwer, S.F. Determinants of glutamine dependence and utilization by normal and tumor-derived breast cell lines. *J. Cell. Physiol.* **1998**, *176*(1), 166–178.

47. Navarro, J., Obrador, E., Carretero, J., Petschen, I., Avino, J., Perez, P., Estrela, J.M. Changes in glutathione status and the antioxidant system in blood and in cancer cells associate with tumour growth *in vivo*. *Free Radic. Biol. Med.* **1999**, *26*(3–4), 410–418.

48. Ho, H.Y., Cheng, M.L., Chiu, D.T. Glucose-6-phosphate dehydrogenase: From oxidative stress to cellular functions and degenerative diseases. *Redox Rep.* **2007**, *12*(3), 109–118.

49. Dessi, S., Batetta, B., Pani, P., Barra, S., Miranda, F., Puxeddu, P. Glucose-6-phosphate dehydrogenase (G6PD) activity in tumoral tissues of G6PD-deficient subjects affected by larynx carcinoma. *Cancer Lett.* **1990**, *53*(2–3), 159–162.

50. Tome, M.E., Johnson, D.B., Samulitis, B.K., Dorr, R.T., Briehl, M.M. Glucose 6-phosphate dehydrogenase overexpression models glucose deprivation and sensitizes lymphoma cells to apoptosis. *Antioxid. Redox Signal.* **2006**, *8*(7–8), 1315–1327.

51. el-Sharabasy, M.M., el-Dosoky, I., Horria, H., Khalaf, A.H. Elevation of glutathione, glutathione-reductase and nucleic acids in both normal tissues and tumour of breast cancer patients. *Cancer Lett.* **1993**, *72*(1–2), 11–15.

52. Di Ilio, C., Sacchetta, P., Angelucci, S., Zezza, A., Tenaglia, R., Aceto, A. Glutathione peroxidase and glutathione reductase activities in cancerous and non-cancerous human kidney tissues. *Cancer Lett.* **1995**, *91*(1), 19–23.

53. Combs, G.F., Jr. Status of selenium in prostate cancer prevention. *Br. J. Cancer* **2004**, *91*(2), 195–199.

54. Esworthy, R.S., Baker, M.A., Chu, F.F. Expression of selenium-dependent glutathione peroxidase in human breast tumor cell lines. *Cancer Res.* **1995**, *55*(4), 957–962.

55. Hasegawa, Y., Takano, T., Miyauchi, A., Matsuzuka, F., Yoshida, H., Kuma, K., Amino, N. Decreased expression of glutathione peroxidase mRNA in thyroid anaplastic carcinoma. *Cancer Lett.* **2002**, *182*(1), 69–74.

56. Liu, J., Hinkhouse, M.M., Sun, W., Weydert, C.J., Ritchie, J.M., Oberley, L.W., Cullen, J.J. Redox regulation of pancreatic cancer cell growth: Role of glutathione peroxidase in the suppression of the malignant phenotype. *Hum. Gene Ther.* **2004**, *15*(3), 239–250.

57. Zhuo, P., Goldberg, M., Herman, L., Lee, B.S., Wang, H., Brown, R.L., Foster, C.B., Peters, U., Diamond, A.M. Molecular consequences of genetic variations in the glutathione peroxidase 1 selenoenzyme. *Cancer Res.* **2009**, *69*(20), 8183–8190.

58. Walshe, J., Serewko-Auret, M.M., Teakle, N., Cameron, S., Minto, K., Smith, L., Burcham, P.C., Russell, T., Strutton, G., Griffin, A., Chu, F.F., Esworthy, S., Reeve, V., Saunders,

N.A. Inactivation of glutathione peroxidase activity contributes to UV-induced squamous cell carcinoma formation. *Cancer Res.* **2007**, *67*(10), 4751–4758.

59. Banning, A., Kipp, A., Schmitmeier, S., Lowinger, M., Florian, S., Krehl, S., Thalmann, S., Thierbach, R., Steinberg, P., Brigelius-Flohe, R. Glutathione peroxidase 2 inhibits cyclooxygenase-2-mediated migration and invasion of HT-29 adenocarcinoma cells but supports their growth as tumors in nude mice. *Cancer Res.* **2008**, *68*(23), 9746–9753.

60. Yu, Y.P., Yu, G., Tseng, G., Cieply, K., Nelson, J., Defrances, M., Zarnegar, R., Michalopoulos, G., Luo, J.H. Glutathione peroxidase 3, deleted or methylated in prostate cancer, suppresses prostate cancer growth and metastasis. *Cancer Res.* **2007**, *67*(17), 8043–8050.

61. Liu, J., Du, J., Zhang, Y., Sun, W., Smith, B.J., Oberley, L.W., Cullen, J.J. Suppression of the malignant phenotype in pancreatic cancer by overexpression of phospholipid hydroperoxide glutathione peroxidase. *Hum. Gene Ther.* **2006**, *17*(1), 105–116.

62. Hanke, N.T., Finch, J.S., Bowden, G.T. Loss of catalase increases malignant mouse keratinocyte cell growth through activation of the stress activated JNK pathway. *Mol. Carcinog.* **2008**, *47*(5), 349–360.

63. Finch, J.S., Tome, M.E., Kwei, K.A., Bowden, G.T. Catalase reverses tumorigenicity in a malignant cell line by an epidermal growth factor receptor pathway. *Free Radic. Biol. Med.* **2006**, *40*(5), 863–875.

64. Kasugai, I., Yamada, M. High production of catalase in hydrogen peroxide-resistant human leukemia HL-60 cell lines. *Leuk. Res.* **1992**, *16*(2), 173–179.

65. Smith, P.S., Zhao, W., Spitz, D.R., Robbins, M.E. Inhibiting catalase activity sensitizes 36B10 rat glioma cells to oxidative stress. *Free Radic. Biol. Med.* **2007**, *42*(6), 787–797.

66. Bai, J., Cederbaum, A.I. Catalase protects HepG2 cells from apoptosis induced by DNA-damaging agents by accelerating the degradation of p53. *J. Biol. Chem.* **2003**, *278*(7), 4660–4667.

67. Tome, M.E., Baker, A.F., Powis, G., Payne, C.M., Briehl, M.M. Catalase-overexpressing thymocytes are resistant to glucocorticoid-induced apoptosis and exhibit increased net tumor growth. *Cancer Res.* **2001**, *61*(6), 2766–2773.

68. Vetrano, A.M., Heck, D.E., Mariano, T.M., Mishin, V., Laskin, D.L., Laskin, J.D. Characterization of the oxidase activity in mammalian catalase. *J. Biol. Chem.* **2005**, *280*(42), 35372–35381.

69. Oberley, L.W. Anticancer therapy by overexpression of superoxide dismutase. *Antioxid. Redox Signal.* **2001**, *3*(3), 461–472.

70. Xu, Y., Fang, F., Dhar, S.K., Bosch, A., St Clair, W.H., Kasarskis, E.J., St Clair, D.K. Mutations in the SOD2 promoter reveal a molecular basis for an activating protein 2-dependent dysregulation of manganese superoxide dismutase expression in cancer cells. *Mol. Cancer Res.* **2008**, *6*(12), 1881–1893.

71. Hurt, E.M., Thomas, S.B., Peng, B., Farrar, W.L. Integrated molecular profiling of SOD2 expression in multiple myeloma. *Blood* **2007**, *109*(9), 3953–3962.

72. Hitchler, M.J., Oberley, L.W., Domann, F.E. Epigenetic silencing of SOD2 by histone modifications in human breast cancer cells. *Free Radic. Biol. Med.* **2008**, *45*(11), 1573–1580.

73. Ria, F., Landriscina, M., Remiddi, F., Rosselli, R., Iacoangeli, M., Scerrati, M., Pani, G., Borrello, S., Galeotti, T. The level of manganese superoxide dismutase content is an independent prognostic factor for glioblastoma. Biological mechanisms and clinical implications. *Br. J. Cancer* **2001**, *84*(4), 529–534.

74. Wang, L.I., Miller, D.P., Sai, Y., Liu, G., Su, L., Wain, J.C., Lynch, T.J., Christiani, D.C. Manganese superoxide dismutase alanine-to-valine polymorphism at codon 16 and lung cancer risk. *J. Natl. Cancer Inst.* **2001**, *93*(23), 1818–1821.

75. Church, S.L., Grant, J.W., Ridnour, L.A., Oberley, L.W., Swanson, P.E., Meltzer, P.S., Trent, J.M. Increased manganese superoxide dismutase expression suppresses the malignant phenotype of human melanoma cells. *Proc. Natl. Acad. Sci. U.S.A.* **1993**, *90*(7), 3113–3117.

76. Zhong, W., Oberley, L.W., Oberley, T.D., St Clair, D.K. Suppression of the malignant phenotype of human glioma cells by overexpression of manganese superoxide dismutase. *Oncogene* **1997**, *14*(4), 481–490.

77. St Clair, D.K., Wan, X.S., Oberley, T.D., Muse, K.E., St Clair, W.H. Suppression of radiation-induced neoplastic transformation by overexpression of mitochondrial superoxide dismutase. *Mol. Carcinog.* **1992**, *6*(4), 238–242.

78. Li, S., Yan, T., Yang, J.Q., Oberley, T.D., Oberley, L.W. The role of cellular glutathione peroxidase redox regulation in the suppression of tumor cell growth by manganese superoxide dismutase. *Cancer Res.* **2000**, *60*(14), 3927–3939.

79. Rodriguez, A.M., Carrico, P.M., Mazurkiewicz, J.E., Melendez, J.A. Mitochondrial or cytosolic catalase reverses the MnSOD-dependent inhibition of proliferation by enhancing respiratory chain activity, net ATP production, and decreasing the steady state levels of H_2O_2. *Free Radic. Biol. Med.* **2000**, *29*(9), 801–813.

80. Behrend, L., Mohr, A., Dick, T., Zwacka, R.M. Manganese superoxide dismutase induces p53-dependent senescence in colorectal cancer cells. *Mol. Cell. Biol.* **2005**, *25*(17), 7758–7769.

81. Li, J.J., Oberley, L.W., Fan, M., Colburn, N.H. Inhibition of AP-1 and NF-kappaB by manganese-containing superoxide dismutase in human breast cancer cells. *FASEB J.* **1998**, *12*(15), 1713–1723.

82. Zhao, Y., Xue, Y., Oberley, T.D., Kiningham, K.K., Lin, S.M., Yen, H.C., Majima, H., Hines, J., St Clair, D. Overexpression of manganese superoxide dismutase suppresses tumor formation by modulation of activator protein-1 signaling in a multistage skin carcinogenesis model. *Cancer Res.* **2001**, *61*(16), 6082–6088.

83. Chuang, T.C., Liu, J.Y., Lin, C.T., Tang, Y.T., Yeh, M.H., Chang, S.C., Li, J.W., Kao, M.C. Human manganese superoxide dismutase suppresses HER2/neu-mediated breast cancer malignancy. *FEBS Lett.* **2007**, *581*(23), 4443–4449.

84. Wang, M., Kirk, J.S., Venkataraman, S., Domann, F.E., Zhang, H.J., Schafer, F.Q., Flanagan, S.W., Weydert, C.J., Spitz, D.R., Buettner, G.R., Oberley, L.W. Manganese superoxide dismutase suppresses hypoxic induction of hypoxia-inducible factor-1alpha and vascular endothelial growth factor. *Oncogene* **2005**, *24*(55), 8154–8166.

85. Li, Z., Khaletskiy, A., Wang, J., Wong, J.Y., Oberley, L.W., Li, J.J. Genes regulated in human breast cancer cells overexpressing manganese-containing superoxide dismutase. *Free Radic. Biol. Med.* **2001**, *30*(3), 260–267.

86. Kinnula, V.L., Torkkeli, T., Kristo, P., Sormunen, R., Soini, Y., Paakko, P., Ollikainen, T., Kahlos, K., Hirvonen, A., Knuutila, S. Ultrastructural and chromosomal studies on manganese superoxide dismutase in malignant mesothelioma. *Am. J. Respir. Cell Mol. Biol.* **2004**, *31*(2), 147–153.

87. Cobbs, C.S., Levi, D.S., Aldape, K., Israel, M.A. Manganese superoxide dismutase expression in human central nervous system tumors. *Cancer Res.* **1996**, *56*(14), 3192–3195.

88. Yeung, B.H., Wong, K.Y., Lin, M.C., Wong, C.K., Mashima, T., Tsuruo, T., Wong, A.S. Chemosensitisation by manganese superoxide dismutase inhibition is caspase-9 dependent and involves extracellular signal-regulated kinase 1/2. *Br. J. Cancer* **2008**, *99*(2), 283–293.

89. Hur, G.C., Cho, S.J., Kim, C.H., Kim, M.K., Bae, S.I., Nam, S.Y., Park, J.W., Kim, W.H., Lee, B.L. Manganese superoxide dismutase expression correlates with chemosensitivity in human gastric cancer cell lines. *Clin. Cancer Res.* **2003**, *9*(15), 5768–5775.

90. Hirose, K., Longo, D.L., Oppenheim, J.J., Matsushima, K. Overexpression of mitochondrial manganese superoxide dismutase promotes the survival of tumor cells exposed to interleukin-1, tumor necrosis factor, selected anticancer drugs, and ionizing radiation. *FASEB J.* **1993**, *7*(2), 361–368.

91. Connor, K.M., Hempel, N., Nelson, K.K., Dabiri, G., Gamarra, A., Belarmino, J., Van De Water, L., Mian, B.M., Melendez, J.A. Manganese superoxide dismutase enhances the invasive and migratory activity of tumor cells. *Cancer Res.* **2007**, *67*(21), 10260–10267.

92. Huang, P., Feng, L., Oldham, E.A., Keating, M.J., Plunkett, W. Superoxide dismutase as a target for the selective killing of cancer cells. *Nature* **2000**, *407*(6802), 390–395.

93. Busuttil, R.A., Garcia, A.M., Cabrera, C., Rodriguez, A., Suh, Y., Kim, W.H., Huang, T.T., Vijg, J. Organ-specific increase in mutation accumulation and apoptosis rate in CuZn-superoxide dismutase-deficient mice. *Cancer Res.* **2005**, *65*(24), 11271–11275.

94. Zhang, Y., Zhao, W., Zhang, H.J., Domann, F.E., Oberley, L.W. Overexpression of copper zinc superoxide dismutase suppresses human glioma cell growth. *Cancer Res.* **2002**, *62*(4), 1205–1212.

95. Gao, Z., Sarsour, E.H., Kalen, A.L., Li, L., Kumar, M.G., Goswami, P.C. Late ROS accumulation and radiosensitivity in SOD1-overexpressing human glioma cells. *Free Radic. Biol. Med.* **2008**, *45*(11), 1501–1509.

96. Donate, F., Juarez, J.C., Burnett, M.E., Manuia, M.M., Guan, X., Shaw, D.E., Smith, E.L., Timucin, C., Braunstein, M.J., Batuman, O.A., Mazar, A.P. Identification of biomarkers for the antiangiogenic and antitumour activity of the superoxide dismutase 1 (SOD1) inhibitor tetrathiomolybdate (ATN-224). *Br. J. Cancer* **2008**, *98*(4), 776–783.

97. Zhang, B., Wang, Y., Su, Y. Peroxiredoxins, a novel target in cancer radiotherapy. *Cancer Lett.* **2009**, *286*(2), 154–160.

98. Yanagawa, T., Ishikawa, T., Ishii, T., Tabuchi, K., Iwasa, S., Bannai, S., Omura, K., Suzuki, H., Yoshida, H. Peroxiredoxin I expression in human thyroid tumors. *Cancer Lett.* **1999**, *145*(1–2), 127–132.

99. Kim, J.H., Bogner, P.N., Baek, S.H., Ramnath, N., Liang, P., Kim, H.R., Andrews, C., Park, Y.M. Up-regulation of peroxiredoxin 1 in lung cancer and its implication as a prognostic and therapeutic target. *Clin. Cancer Res.* **2008**, *14*(8), 2326–2333.

100. Kim, Y.J., Ahn, J.Y., Liang, P., Ip, C., Zhang, Y., Park, Y.M. Human prx1 gene is a target of Nrf2 and is up-regulated by hypoxia/reoxygenation: Implication to tumor biology. *Cancer Res.* **2007**, *67*(2), 546–554.

101. Cha, M.K., Suh, K.H., Kim, I.H. Overexpression of peroxiredoxin I and thioredoxin1 in human breast carcinoma. *J. Exp. Clin. Cancer Res.* **2009**, *28*, 93.

102. Seo, M.J., Liu, X., Chang, M., Park, J.H. GATA-binding protein 1 is a novel transcription regulator of peroxiredoxin. *Int. J. Oncol.* **2012**, *40*(3), 655–664.

103. Kinnula, V.L., Lehtonen, S., Sormunen, R., Kaarteenaho-Wiik, R., Kang, S.W., Rhee, S.G., Soini, Y. Overexpression of peroxiredoxins I, II, III, V, and VI in malignant mesothelioma. *J. Pathol.* **2002**, *196*(3), 316–323.

104. Karihtala, P., Mantyniemi, A., Kang, S.W., Kinnula, V.L., Soini, Y. Peroxiredoxins in breast carcinoma. *Clin. Cancer Res.* **2003**, *9*(9), 3418–3424.

105. Fujita, Y., Nakanishi, T., Hiramatsu, M., Mabuchi, H., Miyamoto, Y., Miyamoto, A., Shimizu, A., Tanigawa, N. Proteomics-based approach identifying autoantibody against peroxiredoxin VI as a novel serum marker in esophageal squamous cell carcinoma. *Clin. Cancer Res.* **2006**, *12*(21), 6415–6420.

106. Furuta, J., Nobeyama, Y., Umebayashi, Y., Otsuka, F., Kikuchi, K., Ushijima, T. Silencing of peroxiredoxin 2 and aberrant methylation of 33 CpG islands in putative promoter regions in human malignant melanomas. *Cancer Res.* **2006**, *66*(12), 6080–6086.

107. Neumann, C.A., Krause, D.S., Carman, C.V., Das, S., Dubey, D.P., Abraham, J.L., Bronson, R.T., Fujiwara, Y., Orkin, S.H., Van Etten, R.A. Essential role for the

107. ...peroxiredoxin Prdx1 in erythrocyte antioxidant defence and tumour suppression. *Nature* **2003**, *424*(6948), 561–565.

108. Egler, R.A., Fernandes, E., Rothermund, K., Sereika, S., de Souza-Pinto, N., Jaruga, P., Dizdaroglu, M., Prochownik, E.V. Regulation of reactive oxygen species, DNA damage, and c-Myc function by peroxiredoxin 1. *Oncogene* **2005**, *24*(54), 8038–8050.

109. Wonsey, D.R., Zeller, K.I., Dang, C.V. The c-Myc target gene PRDX3 is required for mitochondrial homeostasis and neoplastic transformation. *Proc. Natl. Acad. Sci. U.S.A.* **2002**, *99*(10), 6649–6654.

110. Chua, P.J., Lee, E.H., Yu, Y., Yip, G.W., Tan, P.H., Bay, B.H. Silencing the peroxiredoxin III gene inhibits cell proliferation in breast cancer. *Int. J. Oncol.* **2010**, *36*(2), 359–364.

111. Cao, J., Schulte, J., Knight, A., Leslie, N.R., Zagozdzon, A., Bronson, R., Manevich, Y., Beeson, C., Neumann, C.A. Prdx1 inhibits tumorigenesis via regulating PTEN/AKT activity. *EMBO J.* **2009**, *28*(10), 1505–1517.

112. Kropotov, A., Serikov, V., Suh, J., Smirnova, A., Bashkirov, V., Zhivotovsky, B., Tomilin, N. Constitutive expression of the human peroxiredoxin V gene contributes to protection of the genome from oxidative DNA lesions and to suppression of transcription of noncoding DNA. *FEBS J.* **2006**, *273*(12), 2607–2617.

113. Gao, M.C., Jia, X.D., Wu, Q.F., Cheng, Y., Chen, F.R., Zhang, J. Silencing Prx1 and/or Prx5 sensitizes human esophageal cancer cells to ionizing radiation and increases apoptosis via intracellular ROS accumulation. *Acta Pharmacol. Sin.* **2011**, *32*(4), 528–536.

114. Sass, G., Leukel, P., Schmitz, V., Raskopf, E., Ocker, M., Neureiter, D., Meissnitzer, M., Tasika, E., Tannapfel, A., Tiegs, G. Inhibition of heme oxygenase 1 expression by small interfering RNA decreases orthotopic tumor growth in livers of mice. *Int. J. Cancer* **2008**, *123*(6), 1269–1277.

115. Liu, Z.M., Chen, G.G., Ng, E.K., Leung, W.K., Sung, J.J., Chung, S.C. Upregulation of heme oxygenase-1 and p21 confers resistance to apoptosis in human gastric cancer cells. *Oncogene* **2004**, *23*(2), 503–513.

116. Chang, K.W., Lee, T.C., Yeh, W.I., Chung, M.Y., Liu, C.J., Chi, L.Y., Lin, S.C. Polymorphism in heme oxygenase-1 (HO-1) promoter is related to the risk of oral squamous cell carcinoma occurring on male areca chewers. *Br. J. Cancer* **2004**, *91*(8), 1551–1555.

117. Sawa, T., Mounawar, M., Tatemichi, M., Gilibert, I., Katoh, T., Ohshima, H. Increased risk of gastric cancer in Japanese subjects is associated with microsatellite polymorphisms in the heme oxygenase-1 and the inducible nitric oxide synthase gene promoters. *Cancer Lett.* **2008**, *269*(1), 78–84.

118. Mayerhofer, M., Florian, S., Krauth, M.T., Aichberger, K.J., Bilban, M., Marculescu, R., Printz, D., Fritsch, G., Wagner, O., Selzer, E., Sperr, W.R., Valent, P., Sillaber, C. Identification of heme oxygenase-1 as a novel BCR/ABL-dependent survival factor in chronic myeloid leukemia. *Cancer Res.* **2004**, *64*(9), 3148–3154.

119. Miyazaki, T., Kirino, Y., Takeno, M., Samukawa, S., Hama, M., Tanaka, M., Yamaji, S., Ueda, A., Tomita, N., Fujita, H., Ishigatsubo, Y. Expression of heme oxygenase-1 in human leukemic cells and its regulation by transcriptional repressor Bach1. *Cancer Sci.* **2010**, *101*(6), 1409–1416.

120. Was, H., Cichon, T., Smolarczyk, R., Rudnicka, D., Stopa, M., Chevalier, C., Leger, J.J., Lackowska, B., Grochot, A., Bojkowska, K., Ratajska, A., Kieda, C., Szala, S., Dulak, J., Jozkowicz, A. Overexpression of heme oxygenase-1 in murine melanoma: Increased proliferation and viability of tumor cells, decreased survival of mice. *Am. J. Pathol.* **2006**, *169*(6), 2181–2198.

121. Sunamura, M., Duda, D.G., Ghattas, M.H., Lozonschi, L., Motoi, F., Yamauchi, J., Matsuno, S., Shibahara, S., Abraham, N.G. Heme oxygenase-1 accelerates tumor angiogenesis of human pancreatic cancer. *Angiogenesis* **2003**, *6*(1), 15–24.

122. Alaoui-Jamali, M.A., Bismar, T.A., Gupta, A., Szarek, W.A., Su, J., Song, W., Xu, Y., Xu, B., Liu, G., Vlahakis, J.Z., Roman, G., Jiao, J., Schipper, H.M. A novel experimental heme oxygenase-1-targeted therapy for hormone-refractory prostate cancer. *Cancer Res.* **2009**, *69*(20), 8017–8024.

123. Nakamura, H., Bai, J., Nishinaka, Y., Ueda, S., Sasada, T., Ohshio, G., Imamura, M., Takabayashi, A., Yamaoka, Y., Yodoi, J. Expression of thioredoxin and glutaredoxin, redox-regulating proteins, in pancreatic cancer. *Cancer Detect. Prev.* **2000**, *24*(1), 53–60.

124. Lincoln, D.T., Ali Emadi, E.M., Tonissen, K.F., Clarke, F.M. The thioredoxin-thioredoxin reductase system: Over-expression in human cancer. *Anticancer Res.* **2003**, *23*(3B), 2425–2433.

125. Lincoln, D.T., Al-Yatama, F., Mohammed, F.M., Al-Banaw, A.G., Al-Bader, M., Burge, M., Sinowatz, F., Singal, P.K. Thioredoxin and thioredoxin reductase expression in thyroid cancer depends on tumour aggressiveness. *Anticancer Res.* **2010**, *30*(3), 767–775.

126. Cadenas, C., Franckenstein, D., Schmidt, M., Gehrmann, M., Hermes, M., Geppert, B., Schormann, W., Maccoux, L.J., Schug, M., Schumann, A., Wilhelm, C., Freis, E., Ickstadt, K., Rahnenfuhrer, J., Baumbach, J.I., Sickmann, A., Hengstler, J.G. Role of thioredoxin reductase 1 and thioredoxin interacting protein in prognosis of breast cancer. *Breast Cancer Res.* **2010**, *12*(3), R44.

127. Welsh, S.J., Bellamy, W.T., Briehl, M.M., Powis, G. The redox protein thioredoxin-1 (Trx-1) increases hypoxia-inducible factor 1alpha protein expression: Trx-1 overexpression results in increased vascular endothelial growth factor production and enhanced tumor angiogenesis. *Cancer Res.* **2002**, *62*(17), 5089–5095.

128. Gallegos, A., Gasdaska, J.R., Taylor, C.W., Paine-Murrieta, G.D., Goodman, D., Gasdaska, P.Y., Berggren, M., Briehl, M.M., Powis, G. Transfection with human thioredoxin increases cell proliferation and a dominant-negative mutant thioredoxin reverses the transformed phenotype

of human breast cancer cells. *Cancer Res.* **1996**, *56*(24), 5765–5770.

129. Yoo, M.H., Xu, X.M., Carlson, B.A., Patterson, A.D., Gladyshev, V.N., Hatfield, D.L. Targeting thioredoxin reductase 1 reduction in cancer cells inhibits self-sufficient growth and DNA replication. *PLoS ONE* **2007**, *2*(10), e1112.

130. Yoo, M.H., Xu, X.M., Carlson, B.A., Gladyshev, V.N., Hatfield, D.L. Thioredoxin reductase 1 deficiency reverses tumor phenotype and tumorigenicity of lung carcinoma cells. *J. Biol. Chem.* **2006**, *281*(19), 13005–13008.

131. Goplen, D., Wang, J., Enger, P.O., Tysnes, B.B., Terzis, A.J., Laerum, O.D., Bjerkvig, R. Protein disulfide isomerase expression is related to the invasive properties of malignant glioma. *Cancer Res.* **2006**, *66*(20), 9895–9902.

132. Janssen-Heininger, Y.M., Mossman, B.T., Heintz, N.H., Forman, H.J., Kalyanaraman, B., Finkel, T., Stamler, J.S., Rhee, S.G., van der Vliet, A. Redox-based regulation of signal transduction: Principles, pitfalls, and promises. *Free Radic. Biol. Med.* **2008**, *45*(1), 1–17.

133. Stadtman, E.R., Levine, R.L. Free radical-mediated oxidation of free amino acids and amino acid residues in proteins. *Amino Acids* **2003**, *25*(3–4), 207–218.

134. Foster, M.W., Hess, D.T., Stamler, J.S. Protein S-nitrosylation in health and disease: A current perspective. *Trends Mol. Med.* **2009**, *15*(9), 391–404.

135. Morabito, F., Cristani, M., Saija, A., Stelitano, C., Callea, V., Tomaino, A., Minciullo, P.L., Gangemi, S. Lipid peroxidation and protein oxidation in patients affected by Hodgkin's lymphoma. *Mediators Inflamm.* **2004**, *13*(5–6), 381–383.

136. Tanriverdi, T., Hanimoglu, H., Kacira, T., Sanus, G.Z., Kemerdere, R., Atukeren, P., Gumustas, K., Canbaz, B., Kaynar, M.Y. Glutathione peroxidase, glutathione reductase and protein oxidation in patients with glioblastoma multiforme and transitional meningioma. *J. Cancer Res. Clin. Oncol.* **2007**, *133*(9), 627–633.

137. Wang, T.C., Jan, K.Y., Wang, A.S., Gurr, J.R. Trivalent arsenicals induce lipid peroxidation, protein carbonylation, and oxidative DNA damage in human urothelial cells. *Mutat. Res.* **2007**, *615*(1–2), 75–86.

138. Colombo, G., Aldini, G., Orioli, M., Giustarini, D., Gornati, R., Rossi, R., Colombo, R., Carini, M., Milzani, A., Dalle-Donne, I. Water-soluble alpha,beta-unsaturated aldehydes of cigarette smoke induce carbonylation of human serum albumin. *Antioxid. Redox Signal.* **2010**, *12*(3), 349–364.

139. Ogasawara, Y., Ishii, K. Exposure to chrysotile asbestos causes carbonylation of glucose 6-phosphate dehydrogenase through a reaction with lipid peroxidation products in human lung epithelial cells. *Toxicol. Lett.* **2010**, *195*(1), 1–8.

140. Ischiropoulos, H. Protein tyrosine nitration—An update. *Arch. Biochem. Biophys.* **2009**, *484*(2), 117–121.

141. Pignatelli, B., Li, C.Q., Boffetta, P., Chen, Q., Ahrens, W., Nyberg, F., Mukeria, A., Bruske-Hohlfeld, I., Fortes, C., Constantinescu, V., Ischiropoulos, H., Ohshima, H. Nitrated and oxidized plasma proteins in smokers and lung cancer patients. *Cancer Res.* **2001**, *61*(2), 778–784.

142. MacMillan-Crow, L.A., Greendorfer, J.S., Vickers, S.M., Thompson, J.A. Tyrosine nitration of c-SRC tyrosine kinase in human pancreatic ductal adenocarcinoma. *Arch. Biochem. Biophys.* **2000**, *377*(2), 350–356.

143. Rahman, M.A., Senga, T., Ito, S., Hyodo, T., Hasegawa, H., Hamaguchi, M. S-nitrosylation at cysteine 498 of c-Src tyrosine kinase regulates nitric oxide-mediated cell invasion. *J. Biol. Chem.* **2010**, *285*(6), 3806–3814.

144. Wei, W., Li, B., Hanes, M.A., Kakar, S., Chen, X., Liu, L. S-nitrosylation from GSNOR deficiency impairs DNA repair and promotes hepatocarcinogenesis. *Sci. Transl. Med.* **2010**, *2*(19), 19ra13.

145. Chattopadhyay, M., Goswami, S., Rodes, D.B., Kodela, R., Velazquez, C.A., Boring, D., Crowell, J.A., Kashfi, K. NO-releasing NSAIDs suppress NF-kappaB signaling *in vitro* and *in vivo* through S-nitrosylation. *Cancer Lett.* **2010**, *298*(2), 204–211.

146. Muscat, J.E., Kleinman, W., Colosimo, S., Muir, A., Lazarus, P., Park, J., Richie, J.P., Jr. Enhanced protein glutathiolation and oxidative stress in cigarette smokers. *Free Radic. Biol. Med.* **2004**, *36*(4), 464–470.

147. Huang, Z., Komninou, D., Kleinman, W., Pinto, J.T., Gilhooly, E.M., Calcagnotto, A., Richie, J.P., Jr. Enhanced levels of glutathione and protein glutathiolation in rat tongue epithelium during 4-NQO-induced carcinogenesis. *Int. J. Cancer* **2007**, *120*(7), 1396–1401.

148. Lonn, M.E., Hudemann, C., Berndt, C., Cherkasov, V., Capani, F., Holmgren, A., Lillig, C.H. Expression pattern of human glutaredoxin 2 isoforms: Identification and characterization of two testis/cancer cell-specific isoforms. *Antioxid. Redox Signal.* **2008**, *10*(3), 547–557.

149. Levine, R.L., Mosoni, L., Berlett, B.S., Stadtman, E.R. Methionine residues as endogenous antioxidants in proteins. *Proc. Natl. Acad. Sci. U.S.A.* **1996**, *93*(26), 15036–15040.

150. Boschi-Muller, S., Azza, S., Sanglier-Cianferani, S., Talfournier, F., Van Dorsselear, A., Branlant, G. A sulfenic acid enzyme intermediate is involved in the catalytic mechanism of peptide methionine sulfoxide reductase from *Escherichia coli*. *J. Biol. Chem.* **2000**, *275*(46), 35908–35913.

151. De Luca, A., Sanna, F., Sallese, M., Ruggiero, C., Grossi, M., Sacchetta, P., Rossi, C., De Laurenzi, V., Di Ilio, C., Favaloro, B. Methionine sulfoxide reductase A downregulation in human breast cancer cells results in a more aggressive phenotype. *Proc. Natl. Acad. Sci. U.S.A.* **2010**, *107*(43), 18628–18633.

152. Nair, U., Bartsch, H., Nair, J. Lipid peroxidation-induced DNA damage in cancer-prone inflammatory diseases: A review of published adduct types and levels in humans. *Free Radic. Biol. Med.* **2007**, *43*(8), 1109–1120.

153. Otamiri, T., Sjodahl, R. Increased lipid peroxidation in malignant tissues of patients with colorectal cancer. *Cancer* **1989**, *64*(2), 422–425.

154. Batcioglu, K., Mehmet, N., Ozturk, I.C., Yilmaz, M., Aydogdu, N., Erguvan, R., Uyumlu, B., Genc, M., Karagozler, A.A. Lipid peroxidation and antioxidant status in stomach cancer. *Cancer Invest.* **2006**, *24*(1), 18–21.

155. Boyd, N.F., McGuire, V. The possible role of lipid peroxidation in breast cancer risk. *Free Radic. Biol. Med.* **1991**, *10*(3–4), 185–190.

156. Cadet, J., Douki, T., Ravanat, J.-L. Oxidatively generated base damage to cellular DNA. *Free Radic. Biol. Med.* **2010**. doi: 10.1016/j.freeradbiomed.2010.03.025.

157. Altieri, F., Grillo, C., Maceroni, M., Chichiarelli, S. DNA damage and repair: From molecular mechanims to health implications. *Antioxid. Redox Signal.* **2008**, *10*, 891–930.

158. Sedelnikova, O.A., Redon, C.E., Dickey, J.S., Nakamura, A.J., Georgakilas, A.G., Bonner, W.M. Role of oxidatively induced DNA lesions in human pathogenesis. *Mutat. Res.* **2010**, *704*, 152–159.

159. Cadet, J., Berger, M., Douki, T., Ravanat, J.L. Oxidative damage to DNA: Formation, measurement and biological significance. *Rev. Physiol. Biochem. Pharmacol.* **2006**, *131*, 1–87.

160. Cadet, J., Douki, T., Gasparutto, D., Ravanat, J.L., Wagner, J.R. Chemical reactions of the radical cations of nucleobases in isolated and cellular DNA: Formation of single-base lesions. In: *Radical and Radical Ion Reactivity in Nucleic Acid Chemistry*, Vol. 3., ed. M.M. Greenberg. John Wiley and Sons, Inc., Hoboken, NJ, 2010, pp. 69–97.

161. Mantovani, A., Allavena, P., Sica, A., Balkwill, F. Cancer-related inflammation. *Nature* **2008**, *454*(7203), 436–444.

162. De Bont, R., van Larebeke, N. Endogenous DNA damage in humans: A review of quantitative data. *Mutagenesis* **2004**, *19*, 169–185.

163. Hussain, S.P., Hofseth, L.J., Harris, C.C. Radical causes of cancer. *Nat. Rev. Cancer* **2003**, *3*, 276–285.

164. Redon, C., Naf, D., Dickey, J.S., Kareva, I., Flood, B., Nowsheen, S., Georgakilas, A.G., Bonner, W.M., Sedelnikova, O.A. In vivo tumor growth induces DNA damage in distal organs. *AACR Annual Meeting*, San Diego, CA, April 2008, 2008; Proceedings of the American Association for Cancer Research, San Diego, CA, 2008, p. 1888.

165. Georgakilas, A.G., Nowsheen, S., Wukovich, R.L., Kalogerinis, P.T., Redon, C., Dickey, J.S., Naf, D., Bonner, W.M., Sedelnikova, O.A. Accumulation of complex DNA lesions in tumors and bystander tissues. *Proceedings of the 55th Annual Meeting of the Radiation Research Society*, Savannah, GA, 2009; RRS, Ed. Radiation Research Society, Savannah, GA, 2009, pp. 161–162.

166. Bonner, W.M., Redon, C.E., Dickey, J.S., Nakamura, A.J., Sedelnikova, O.A., Solier, S., Pommier, Y. [gamma]H2AX and cancer. *Nat. Rev. Cancer* **2008**, *8*(12), 957–967.

167. Dickey, J.S., Baird, B.J., Redon, C.E., Sokolov, M.V., Sedelnikova, O.A., Bonner, W.M. Intercellular communication of cellular stress monitored by {gamma}-H2AX induction. *Carcinogenesis* **2009**, *30*(10), 1686–1695.

168. Sokolov, M.V., Dickey, J.S., Bonner, W.M., Sedelnikova, O.A. γ-H2AX in bystander cells: Not just a radiation-triggered event, a cellular response to stress mediated by intercellular communication. *Cell Cycle* **2007**, *6*, 2210–2212.

169. Douki, T., Martini, R., Ravanat, J.L., Turesky, R.J., Cadet, J. Measurement of 2,6-diamino-4-hydroxy-5-formamidopyrimidine and 8-oxo-7,8-dihydroguanine in isolated DNA exposed to gamma radiation in aqueous solution. *Carcinogenesis* **1997**, *18*(12), 2385–2391.

170. Pouget, J.P., Douki, T., Richard, M.J., Cadet, J. DNA damage induced in cells by γ and UVA radiation as measured by HPLC/GC–MS and HPLC–EC and comet assay. *Chem. Res. Toxicol.* **2000**, *13*(7), 541–549.

171. Vaupel, P., Kelleher, D.K., Hoeckel, M. Oxygen status of malignant tumors: Pathogenesis of hypoxia and significance for tumor therapy. *Semin. Oncol.* **2001**, *28*, 29–35.

172. Konsta, A.A., Visvardis, E.E., Haveles, K.S., Georgakilas, A.G., Sideris, E.G. Detecting radiation-induced DNA damage: From changes in dielectric properties to programmed cell death. *J. Non Cryst. Solids* **2003**, *305*, 295–302.

173. Jiranusornkul, S., Laughton, C.A. Destabilization of DNA duplexes by oxidative damage at guanine: Implications for lesion recognition and repair. *J. R. Soc. Interface* **2008**, *5*(Suppl. 3), 191–198.

174. Spassky, A., Angelov, D. Influence of the local helical conformation on the guanine modifications generated from one-electron DNA oxidationâ€. *Biochemistry Mosc.* **1997**, *36*(22), 6571–6576.

175. Radak, Z., Boldogh, I. 8-Oxo-7,8-dihydroguanine: Links to gene expression, aging, and defense against oxidative stress. *Free Radic. Biol. Med.* **2010**, *49*(4), 587–596.

176. Wallace, S.S. Biological consequences of free radical-damaged DNA bases. *Free Radic. Biol. Med.* **2002**, *33*, 1–14.

177. Sancar, A., Sancar, G.B. DNA repair enzymes. *Annu. Rev. Biochem.* **1988**, *57*(1), 29–67.

178. Yu, S.L., Lee, S.K., Johnson, R.E., Prakash, L., Prakash, S. The stalling of transcription at abasic sites is highly mutagenic. *Mol. Cell. Biol.* **2003**, *23*, 382–388.

179. Wilson, D.M.III, Barsky, D. The major human abasic endonuclease: Formation, consequences and repair of abasic lesions in DNA. *Mutat. Res.* **2001**, *485*, 283–307.

180. Lim, P., Wuenschell, G.E., Holland, V., Lee, D.-H., Pfeifer, G.P., Rodriguez, H., Termini, J. Peroxyl radical mediated oxidative DNA base damage: Implications for lipid peroxidation induced mutagenesis. *Biochemistry Mosc.* **2004**, *43*(49), 15339–15348.

181. Pogozelski, W.K., Tullius, T.D. Oxidative strand scission of nucleic acids: Routes initiated by hydrogen abstraction from the sugar moiety. *Chem. Rev.* **1998**, *98*(3), 1089–1108.

182. Ward, J.F., Blakely, W.F., Joner, E.I. Mammialian cells are not killed by DNA single-strand breaks caused by hydroxyl radicals from hydrogen peroxide. *Radiat. Res.* **1985**, *103*, 383–392.

183. Sutherland, B., Bennett, P.V., Sidorkina, O., Laval, J. DNA damage clusters induced by ionizing radiation in isolated DNA and in human cells. *Proc. Natl. Acad. Sci. U.S.A.* **2000**, *97*, 103–108.
184. Georgakilas, A.G. Processing of DNA damage clusters in human cells: Current status of knowledge. *Mol. Biosyst.* **2008**, *4*, 30–35.
185. Gollapalle, E., Wong, R., Adetolu, R., Tsao, D., Francisco, D., Sigounas, G., Georgakilas, A.G. Detection of oxidative clustered DNA lesions in X-irradiated mouse skin tissues and human MCF-7 breast cancer cells. *Radiat. Res.* **2007**, *167*, 207–216.
186. Slupphaug, G., Kavli, B., Krokan, H.E. The interacting pathways for prevention and repair of oxidative DNA damage. *Mutat. Res.* **2003**, *531*, 231–251.
187. Lindahl, T., Wood, R.D. Quality control of DNA repair. *Science* **1999**, *286*, 1897–1905.
188. Demple, B., DeMott, M.S. Dynamics and diversions in base excision DNA repair of oxidized abasic lesions. *Oncogene* **2002**, *21*, 8926–8934.
189. Ocampo, M.T.A., Chaung, W., Marenstein, D.R., Chan, M.K., Altamirano, A., Basu, A.K., Boorstein, R.J., Cunningham, R.P., Teebor, G.W. Targeted deletion of mNth1 reveals a novel DNA repair enzyme activity. *Mol. Cell. Biol.* **2002**, *22*, 6111–6121.
190. Wiederhold, L., Leppard, J.B., Kedar, P., Karimi-Busheri, F., Rasouli-Nia, A., Weinfeld, M., Tomkinson, A.E., Izumi, T., Prasad, R., Wilson, S.H., Mitra, S., Hazra, T.K. AP endonuclease-independent DNA base excision repair in human cells. *Mol. Cell* **2004**, *15*, 209–220.
191. Riballo, E., Kuhne, M., Rief, N., Doherty, A., Smith, G.C.M., Recio, M.J., Reis, C., Dahm, K., Fricke, A., Krempler, A., Parker, A.R., Jackson, S.P., Gennery, A., Jeggo, P.A., Lobrich, M. A pathway of double-strand break rejoining dependent upon ATM, Artemis and proteins locating to H2AX foci. *Mol. Cell* **2004**, *16*, 715–724.
192. Das, A., Wiederhold, L., Leppard, J.B., Kedar, P., Prasad, R., Wang, H., Boldogh, I., Karimi-Busheri, F., Weinfeld, M., Tomkinson, A.E., Wilson, S.H., Mitra, S., Hazra, T.K. NEIL2-initiated, APE-independent repair of oxidized bases in DNA: Evidence for a repair complex in human cells. *DNA Repair (Amst.)* **2006**, *5*(12), 1439–1448.
193. Nguewa, P.A., Fuertes, M.A., Valladares, B., Alonso, C., Pérez, J.M. Poly(ADP-Ribose) polymerases: Homology, structural domains and functions. Novel therapeutical applications. *Prog. Biophys. Mol. Biol.* **2005**, *88*(1), 143–172.
194. Satoh, M.S., Lindahl, T. Role of poly(ADP-ribose) formation in DNA repair. *Nature* **1992**, *356*(6367), 356–358.
195. de Murcia, G., Menissier de Murcia, J. Poly(ADP-ribose) polymerase: A molecular nick-sensor. *Trends Biochem. Sci.* **1994**, *19*(4), 172–176.
196. D'Amours, D., Desnoyers, S., D'Silva, I., Poirier, G.G. Poly(ADP-ribosyl)ation reactions in the regulation of nuclear functions. *Biochem. J.* **1999**, *342*(2), 249–268.
197. Perrin, D., Gras, S.P., van Hille, B.T., Hill, B.T. Expression in yeast and purification of functional recombinant human poly(ADP-RIBOSE)polymerase(PARP). Comparative pharmacological profile with that of the rat enzyme. *J. Enzyme Inhib. Med. Chem.* **2000**, *15*(5), 461–469.
198. Sukhanova, M.V., Khodyreva, S.N., Lebedeva, N.A., Prasad, R., Wilson, S.H., Lavrik, O.I. Human base excision repair enzymes apurinic/apyrimidinic endonuclease1 (APE1), DNA polymerase {beta} and poly(ADP-ribose) polymerase 1: Interplay between strand-displacement DNA synthesis and proofreading exonuclease activity. *Nucleic Acids Res.* **2005**, *33*(4), 1222–1229.
199. Pearson, C.G., Shikazono, N., Thacker, J., O'Neill, P. Enhanced mutagenic potential of 8-oxo-7,8-dihydroguanine when present within a clustered DNA damage site. *Nucleic Acids Res.* **2004**, *32*, 263–270.
200. Malyarchuk, S.G., Youngblodd, R., Landry, A.M., Quillin, E., Harrison, L. The mutation frequency of 8-oxo-7,8-dihydroguanine (8-oxodG) situated in a multiply damaged site: Comparison of a single and two closely spaced 8-oxodG in *Escherichia coli*. *DNA Repair (Amst.)* **2003**, *152*, 1–11.
201. Malyarchuk, S., Brame, K.L., Youngblood, R., Shi, R., Harrison, L. Two clustered 8-oxo-7,8-dihydroguanine (8-oxodG) lesions increase the point mutation frequency of 8-oxodG, but do not result in double strand breaks or deletions in *Escherichia coli*. *Nucleic Acids Res.* **2004**, *32*, 5721–5731.
202. Hada, M., Georgakilas, A.G. Formation of clustered DNA damage after High-LET irradiation: A review. *J. Radiat. Res.* **2008**, *49*, 203–210.
203. Dianov, G.L., O'Neill, P., Goodhead, D.T. Securing genome stability by orchestrating DNA repair: Removal of radiation-induced clustered lesions in DNA. *Bioessays* **2001**, *23*, 745–749.
204. Weinfeld, M., Rasouli-Nia, A., Chaudhry, M.A., Britten, R.A. Response of base excision repair enzymes to complex DNA lesions. *Radiat. Res.* **2001**, *156*, 584–589.
205. David-Cordonnier, M.H., Boiteux, S., O'Neill, P. Efficiency of excision of 8-oxo-guanine within DNA clustered damage by XRS5 nuclear extracts and purified human OGG1 protein. *Biochemistry Mosc.* **2001**, *40*, 11811–11818.
206. Eot-Houllier, G., Gonera, M., Gasparutto, D., Giustranti, C., Sage, E. Interplay between DNA N-glycosylases/AP lyases at multiply damaged sites and biological consequences. *Nucleic Acids Res.* **2007**, *35*, 3355–3366.
207. Eccles, L.J., Lomax, M.E., O'Neill, P. Hierarchy of lesion processing governs the repair, double-strand break formation and mutability of three-lesion clustered DNA damage. *Nucleic Acids Res.* **2010**, *38*(4), 1123–1134.
208. David-Cordonnier, M.H., Cunniffe, S.M.T., Hickson, I.D., O'Neill, P. Efficiency of incision of an AP site within clustered DNA damage by the major human AP endonuclease. *Biochemistry Mosc.* **2002**, *41*, 634–642.
209. Budworth, H., Dianov, G.L. Mode of inhibition of short-patch base excision repair by thymine glycol within clustered DNA lesions. *J. Biol. Chem.* **2003**, *278*, 9378–9381.

210. Georgakilas, A.G., Bennett, P.V., Wilson, D.M., III, Sutherland, B.M. Processing of bistranded abasic DNA clusters in gamma-irradiated human hematopoietic cells. *Nucleic Acids Res.* **2004**, *32*, 5609–5620.

211. Bellon, S., Shikazono, N., Cunniffe, S., Lomax, M., O'Neill, P. Processing of thymine glycol in a clustered DNA damage site: Mutagenic or cytotoxic. *Nucleic Acids Res.* **2009**, *37*, 4430–4440.

212. Kozmin, S.G., Sedletska, Y., Reynaud-Angelin, A., Gasparutto, D., Sage, E. The formation of double-strand breaks at multiply damaged sites is driven by the kinetics of excision/incision at base damage in eukaryotic cells. *Nucleic Acids Res.* **2009**, *37*, 1767–1777.

213. Blaisdell, J.O., Wallace, S. Abortive base-excision repair of radiation-induced clustered DNA lesions in *Escherichia coli*. *Proc. Natl. Acad. Sci. U.S.A.* **2001**, *98*, 7426–7430.

214. Malyarchuk, S., Harrison, L. DNA repair of clustered uracils in HeLa cells. *J. Mol. Biol.* **2005**, *345*, 731–743.

215. Georgakilas, A.G. A possible role of repair proteins BRCA1 and DNA-PK in the processing of oxidative DNA damage. *J. Biochem. Technol.* **2008**, *1*(1), 9–11.

216. Peddi, P., Loftin, C.W., Dickey, J.S., Hair, J.M., Burns, K.J., Aziz, K., Francisco, D.C., Panayiotidis, M.I., Sedelnikova, O.A., Bonner, W.M., Winters, T.A., Georgakilas, A.G. DNA-PKcs deficiency leads to persistence of oxidatively-induced clustered DNA lesions in human tumor cells. *Free Radic. Biol. Med.* **2010**, *48*, 1435–1443.

217. Gulston, M., de Lara, C., Jenner, T., Davis, E., O'Neill, P. Processing of clustered DNA damage generates additional double-strand breaks in mammalian cells post-irradiation. *Nucleic Acids Res.* **2004**, *32*, 1602–1609.

218. Malyarchuk, S., Castore, R., Harrison, L. DNA repair of clustered lesions in mammalian cells: Involvement of non-homologous end-joining. *Nucleic Acids Res.* **2008**, *36*, 4872–4882.

219. Hashimoto, M., Donald, C.D., Yannone, S.M., Chen, D.J., Roy, R., Kow, Y.W. A possible role of Ku in mediating sequential repair of closely opposed lesions. *J. Biol. Chem.* **2001**, *276*, 12827–12831.

220. Plumb, M.A., Smith, G.C.M., Cunniffe, S.M.T., O'Neill, P. DNA-PK activation by ionizing radiation-induced DNA single-strand breaks. *Int. J. Radiat. Biol.* **1999**, *75*, 553–561.

221. Mårtensson, S., Nygren, J., Osheroff, N., Hammarsten, O. Activation of the DNA-dependent protein kinase by drug-induced and radiation-induced DNA strand breaks. *Radiat. Res.* **2003**, *160*, 291–301.

222. Regulus, P., Duroux, B., Bayle, P.A., Favier, A., Cadet, J., Ravanat, J.L. Oxidation of the sugar moiety of DNA by ionizing radiation or bleomycin could induce the formation of a cluster DNA lesion. *Proc. Natl. Acad. Sci. U.S.A.* **2007**, *104*, 14032–14037.

223. Peddi, P., Francisco, D.C., Cecil, A., Hair, J.M., Panayiotidis, M.I., Georgakilas, A.G. Deficient processing of clustered DNA damage in human breast cancer cells MCF-7 with silenced DNA-PKcs expression. *Cancer Lett.* **2008**, *269*, 174–183.

224. Holt, S.M., Scemama, J.L., Panayiotidis, M.I., Georgakilas, A.G. Compromised repair of clustered DNA damage in the human acute lymphoblastic leukemia MSH2-deficient NALM-6 cells. *Mutat. Res.* **2009**, *674*, 123–130.

225. Georgakilas, A.G., Aziz, K., Ziech, D., Georgakila, S., Panayiotidis, M.I. BRCA1 involvement in toxicological responses and human cancer etiology. *Toxicol. Lett.* **2009**, *188*, 77–83.

226. Hair, J.M., Terzoudi, G.I., Hatzi, V.I., Lehockey, K.A., Srivastava, D., Wang, W., Pantelias, G.E., Georgakilas, A.G. BRCA1 role in the mitigation of radiotoxicity and chromosomal instability through repair of clustered DNA lesions. *Chem. Biol. Interact.* **2010**. doi: 10.1016/j.cbi.2010.03.046.

227. Tsao, D., Tabrizi, I., Dingfelder, M., Stewart, R.D., Georgakilas, A.G. Induction and processing of oxidative clustered DNA lesions and double strand breaks induced by high-LET 56Fe space radiation in human monocytes. *Proceedings of the 53th Annual Meeting of the Radiation Research Society*, 2006. Philadelphia, PA, Nov 5–8, 2006, p. 53.

228. Gros, L., Saparbaev, M.K., Laval, J. Enzymology of the repair of free radicals-induced DNA damage. *Oncogene* **2002**, *21*, 8905–8925.

229. Pastoriza Gallego, M., Sarasin, A. Transcription-coupled repair of 8-oxoguanine in human cells and its deficiency in some DNA repair diseases. *Biochimie* **2003**, *85*(11), 1073–1082.

230. Frosina, G. The current evidence for defective repair of oxidatively damaged DNA in Cockayne syndrome. *Free Radic. Biol. Med.* **2007**, *43*(2), 165–177.

231. Avkin, S., Adar, S., Blander, G., Livneh, Z. Quantitative measurement of translesion replication in human cells: Evidence of bypass of abasic sites by a replicative DNA polymerase. *Proc. Natl. Acad. Sci. U.S.A.* **2002**, *99*, 3764–3769.

232. Kamiya, H., Yamaguchi, A., Suzuki, T., Harashima, H. Roles of specialized DNA polymerases in mutagenesis by 8-hydroxyguanine in human cells. *Mutat. Res.* **2010**, *686*(1–2), 90–95.

233. Cappelli, E., Hazra, T., Hill, J.W., Slupphaug, G., Bogliolo, M., Frosina, G. Rates of base excision repair are not solely dependent on levels of initiating enzymes. *Carcinogenesis* **2001**, *22*, 387–393.

234. Wagner, J.R., Hu, C.C., Ames, B.N. Endogenous oxidative damage of deoxycytidine in DNA. *Proc. Natl. Acad. Sci. U.S.A.* **1992**, *89*, 3380–3384.

235. Atamna, H., Cheung, I., Ames, B.N. A method of detecting abasic sites in living cells: Age-dependent changes in base excision repair. *Proc. Natl. Acad. Sci. U.S.A.* **2000**, *97*, 686–691.

236. Maynard, S., Schurman, S.H., Harboe, C., de Souza-Pinto, N.C., Bohr, V.A. Base excision repair of oxidative DNA damage and association with cancer and aging. *Carcinogenesis* **2009**, *30*(1), 2–10.

237. Møller, P., Løhr, M., Folkmann, J.K., Mikkelsen, L., Loft, S. Aging and oxidatively damaged nuclear DNA in animal organs. *Free Radic. Biol. Med.* **2010**, *48*(10), 1275–1285.

238. Bhatti, P., Struewing, J.P., Alexander, B.H., Hauptmann, M., Bowen, L., Mateus-Pereira, L.H., Pineda, M.A., Simon, S.L., Weinstock, R.M., Rosenstein, M., Stovall, M., Preston, D.L., Linet, M.S., Doody, M.M., Sigurdson, A.J. Polymorphisms in DNA repair genes, ionizing radiation exposure and risk of breast cancer in U.S. radiologic technologists. *Int. J. Cancer* **2008**, *122*(1), 177–182.

239. Joseph, T., Kusumakumary, P., Chacko, P., Abraham, A., Pillai, M.R. DNA repair gene XRCC1 polymorphisms in childhood acute lymphoblastic leukemia. *Cancer Lett.* **2005**, *217*, 17–24.

240. Tudek, B. Base excision repair modulation as a risk factor for human cancers. *Mol. Aspects Med.* **2007**, *28*(3–4), 258–275.

241. Georgakilas, A.G., Mosley, W., Georgakila, S., Zeich, D., Panayiotidis, M.I. Viral-induced human carcinogenesis: An oxidative stress perspective. *Mol. Biosyst.* **2010**, *6*, 1162–1172.

242. Touati, E., Michel, V., Thiberge, J.M., Avé, P., Huerre, M., Bourgade, F., Klungland, A., Labigne, A. Deficiency in OGG1 protects against inflammation and mutagenic effects associated with *H. pylori* infection in mouse. *Helicobacter* **2006**, *11*(5), 494–505.

243. Mabley, J.G., Pacher, P., Deb, A., Wallace, R., Elder, R.H., Szabo, C. Potential role for 8-oxoguanine DNA glycosylase in regulating inflammation. *FASEB J.* **2005**, *19*(2), 290–292.

244. Nakamura, J., Swenberg, A.J. Endogenous apurinic/apyrimidinic sites in genomic DNA of mammalian tissues. *Cancer Res.* **1999**, *59*, 2522–2526.

245. Klungland, A., Rosewell, I., Hollenbach, S., Larsen, E., Daly, G., Epe, B., Seeberg, E., Lindahl, T., Barnes, D.E. Accumulation of premutagenic DNA lesions in mice defective in removal of oxidative base damage. *Proc. Natl. Acad. Sci. U.S.A.* **1999**, *96*, 13300–13305.

246. Cadet, J., Douki, T., Frelon, S., Sauvaigo, S., Pouget, J.P., Ravanat, J.L. Assessment of oxidative base damage to isolated and cellular DNA by HPLC-MS/MS measurement. *Free Radic. Biol. Med.* **2002**, *33*, 441–449.

247. Bennett, P.V., Cuomo, N.L., Paul, S., Tafrov, S.T., Sutherland, B.M. Endogenous DNA damage clusters in human skin, 3-D model and cultured skin cells. *Free Radic. Biol. Med.* **2005**, *39*, 832–839.

248. Bennett, P., Ishchenko, A.A., Laval, J., Paap, B., Sutherland, B.M. Endogenous DNA damage clusters in human hematopoietic stem and progenitor cells. *Free Radic. Biol. Med.* **2008**, *45*(9), 1352–1359.

249. Nowsheen, S., Wukovich, R.L., Aziz, K., Kalogerinis, P.T., Richardson, C.C., Panayiotidis, M.I., Bonner, W.M., Sedelnikova, O.A., Georgakilas, A.G. Accumulation of oxidatively induced clustered DNA lesions in human tumor tissues. *Mutat. Res.* **2009**, *674*(1–2), 131–136.

250. Sedelnikova, O.A., Bonner, W.M. GammaH2AX in cancer cells: A potential biomarker for cancer diagnostics, prediction and recurrence. *Cell Cycle* **2006**, *5*, 231–240.

251. Gedik, C.M., Collins, A., ESCODD (European Standards Committee on Oxidative DNA Damage). Establishing the background level of base oxidation in human lymphocyte DNA: Results of an interlaboratory validation study. *FASEB J.* **2005**, *19*, 82–84.

252. Collins, A.R., Cadet, J., Möller, L., Poulsen, H.E., Viña, J. Are we sure we know how to measure 8-oxo-7,8-dihydroguanine in DNA from human cells? *Arch. Biochem. Biophys.* **2004**, *423*, 57–65.

253. Evans, D.M., Dizdaroglu, M., Cooke, M.S. Oxidative DNA damage and disease: Induction, repair and significance. *Mutat. Res.* **2004**, *567*, 1–61.

254. Klaunig, J.E., Kamendulis, L.M. The role of oxidative stress in carcinogenesis. *Annu. Rev. Pharmacol. Toxicol.* **2004**, *44*, 239–267.

255. Fuks, F. DNA methylation and histone modifications: Teaming up to silence genes. *Curr. Opin. Genet. Dev.* **2005**, *15*(5), 490–495.

256. Baylin, S.B. DNA methylation and gene silencing in cancer. *Nat. Clin. Pract. Oncol.* **2005**, *2*(Suppl. 1), S4–S11.

257. Esteller, M. DNA methylation and cancer therapy: New developments and expectations. *Curr. Opin. Oncol.* **2005**, *17*(1), 55–60.

258. Bird, A. The essentials of DNA methylation. *Cell* **1992**, *70*(1), 5–8.

259. Hitchler, M.J., Domann, F.E. An epigenetic perspective on the free radical theory of development. *Free Radic. Biol. Med.* **2007**, *43*(7), 1023–1036.

260. Klaunig, J.E., Kamendulis, L.M., Xu, Y. Epigenetic mechanisms of chemical carcinogenesis. *Hum. Exp. Toxicol.* **2000**, *19*(10), 543–555.

261. Vaquero, E.C., Edderkaoui, M., Pandol, S.J., Gukovsky, I., Gukovskaya, A.S. Reactive oxygen species produced by NAD(P)H oxidase inhibit apoptosis in pancreatic cancer cells. *J. Biol. Chem.* **2004**, *279*(33), 34643–34654.

262. Urbano, A., Lakshmanan, U., Choo, P.H., Kwan, J.C., Ng, P.Y., Guo, K., Dhakshinamoorthy, S., Porter, A. AIF suppresses chemical stress-induced apoptosis and maintains the transformed state of tumor cells. *EMBO J.* **2005**, *24*(15), 2815–2826.

263. Wachsman, J.T. DNA methylation and the association between genetic and epigenetic changes: Relation to carcinogenesis. *Mutat. Res.* **1997**, *375*(1), 1–8.

264. Turk, P.W., Laayoun, A., Smith, S.S., Weitzman, S.A. DNA adduct 8-hydroxyl-2′-deoxyguanosine (8-hydroxyguanine) affects function of human DNA methyltransferase. *Carcinogenesis* **1995**, *16*(5), 1253–1255.

265. Weitzman, S.A., Turk, P.W., Milkowski, D.H., Kozlowski, K. Free radical adducts induce alterations in DNA cytosine methylation. *Proc. Natl. Acad. Sci. U.S.A.* **1994**, *91*(4), 1261–1264.

266. Turk, P.W., Weitzman, S.A. Free radical DNA adduct 8-OH-deoxyguanosine affects activity of Hpa II and Msp

I restriction endonucleases. *Free Radic. Res.* **1995**, *23*(3), 255–258.

267. Hepburn, P.A., Margison, G.P., Tisdale, M.J. Enzymatic methylation of cytosine in DNA is prevented by adjacent O6-methylguanine residues. *J. Biol. Chem.* **1991**, *266*(13), 7985–7987.

268. Tan, N.W., Li, B.F. Interaction of oligonucleotides containing 6-O-methylguanine with human DNA (cytosine-5-)-methyltransferase [published erratumm appears in Biochemistry 1992 Aug 4;31(30):7008]. *Biochemistry Mosc.* **1990**, *29*(39), 9234–9240.

269. Xiao, W., Samson, L. In vivo evidence for endogenous DNA alkylation damage as a source of spontaneous mutation in eukaryotic cells. *Proc. Natl. Acad. Sci. U.S.A.* **1993**, *90*(6), 2117–2121.

270. Hurt, E.M., Thomas, S.B., Peng, B., Farrar, W.L. Molecular consequences of SOD2 expression in epigenetically silenced pancreatic carcinoma cell lines. *Br. J. Cancer* **2007**, *97*(8), 1116–1123.

271. Hitchler, M.J., Wikainapakul, K., Yu, L., Powers, K., Attatippaholkun, W., Domann, F.E. Epigenetic regulation of manganese superoxide dismutase expression in human breast cancer cells. *Epigenetics* **2006**, *1*(4), 163–171.

272. Hodge, D.R., Peng, B., Pompeia, C., Thomas, S., Cho, E., Clausen, P.A., Marquez, V.E., Farrar, W.L. Epigenetic silencing of manganese superoxide dismutase (SOD-2) in KAS 6/1 human multiple myeloma cells increases cell proliferation. *Cancer Biol. Ther.* **2005**, *4*(5), 585–592.

273. Tada, M., Yokosuka, O., Fukai, K., Chiba, T., Imazeki, F., Tokuhisa, T., Saisho, H. Hypermethylation of NAD(P)H: Quinone oxidoreductase 1 (NQO1) gene in human hepatocellular carcinoma. *J. Hepatol.* **2005**, *42*(4), 511–519.

274. Zhong, S., Tang, M.W., Yeo, W., Liu, C., Lo, Y.M., Johnson, P.J. Silencing of GSTP1 gene by CpG island DNA hypermethylation in HBV-associated hepatocellular carcinomas. *Clin. Cancer Res.* **2002**, *8*(4), 1087–1092.

275. Esteller, M., Corn, P.G., Urena, J.M., Gabrielson, E., Baylin, S.B., Herman, J.G. Inactivation of glutathione S-transferase P1 gene by promoter hypermethylation in human neoplasia. *Cancer Res.* **1998**, *58*(20), 4515–4518.

276. Millar, D.S., Ow, K.K., Paul, C.L., Russell, P.J., Molloy, P.L., Clark, S.J. Detailed methylation analysis of the glutathione S-transferase pi (GSTP1) gene in prostate cancer. *Oncogene* **1999**, *18*(6), 1313–1324.

277. Kroemer, G., Galluzzi, L., Vandenabeele, P., Abrams, J., Alnemri, E.S., Baehrecke, E.H., Blagosklonny, M.V., El-Deiry, W.S., Golstein, P., Green, D.R., Hengartner, M., Knight, R.A., Kumar, S., Lipton, S.A., Malorni, W., Nunez, G., Peter, M.E., Tschopp, J., Yuan, J., Piacentini, M., Zhivotovsky, B., Melino, G. Classification of cell death: Recommendations of the Nomenclature Committee on Cell Death 2009. *Cell Death Differ.* **2009**, *16*(1), 3–11.

278. Golstein, P., Kroemer, G. Cell death by necrosis: Towards a molecular definition. *Trends Biochem. Sci.* **2007**, *32*(1), 37–43.

279. Hotchkiss, R.S., Strasser, A., McDunn, J.E., Swanson, P.E. Cell death. *N. Engl. J. Med.* **2009**, *361*(16), 1570–1583.

280. Galluzzi, L., Maiuri, M.C., Vitale, I., Zischka, H., Castedo, M., Zitvogel, L., Kroemer, G. Cell death modalities: Classification and pathophysiological implications. *Cell Death Differ.* **2007**, *14*(7), 1237–1243.

281. West, J.D., Marnett, L.J. Endogenous reactive intermediates as modulators of cell signaling and cell death. *Chem. Res. Toxicol.* **2006**, *19*(2), 173–194.

282. Ryter, S.W., Kim, H.P., Hoetzel, A., Park, J.W., Nakahira, K., Wang, X., Choi, A.M. Mechanisms of cell death in oxidative stress. *Antioxid. Redox Signal.* **2007**, *9*(1), 49–89.

283. Pervaiz, S., Ramalingam, J.K., Hirpara, J.L., Clement, M.V. Superoxide anion inhibits drug-induced tumor cell death. *FEBS Lett.* **1999**, *459*(3), 343–348.

284. Salvucci, O., Carsana, M., Bersani, I., Tragni, G., Anichini, A. Antiapoptotic role of endogenous nitric oxide in human melanoma cells. *Cancer Res.* **2001**, *61*(1), 318–326.

285. Clement, M.V., Hirpara, J.L., Pervaiz, S. Decrease in intracellular superoxide sensitizes Bcl-2-overexpressing tumor cells to receptor and drug-induced apoptosis independent of the mitochondria. *Cell Death Differ.* **2003**, *10*(11), 1273–1285.

286. Chuang, J.I., Chang, T.Y., Liu, H.S. Glutathione depletion-induced apoptosis of Ha-ras-transformed NIH3T3 cells can be prevented by melatonin. *Oncogene* **2003**, *22*(9), 1349–1357.

287. Marengo, B., De Ciucis, C., Verzola, D., Pistoia, V., Raffaghello, L., Patriarca, S., Balbis, E., Traverso, N., Cottalasso, D., Pronzato, M.A., Marinari, U.M., Domenicotti, C. Mechanisms of BSO (L-buthionine-S,R-sulfoximine)-induced cytotoxic effects in neuroblastoma. *Free Radic. Biol. Med.* **2008**, *44*(3), 474–482.

288. Guicciardi, M.E., Gores, G.J. Life and death by death receptors. *FASEB J.* **2009**, *23*(6), 1625–1637.

289. Merino, D., Lalaoui, N., Morizot, A., Solary, E., Micheau, O. TRAIL in cancer therapy: Present and future challenges. *Expert Opin. Ther. Targets* **2007**, *11*(10), 1299–1314.

290. Perez-Cruz, I., Carcamo, J.M., Golde, D.W. Caspase-8 dependent TRAIL-induced apoptosis in cancer cell lines is inhibited by vitamin C and catalase. *Apoptosis* **2007**, *12*(1), 225–234.

291. Mohr, A., Buneker, C., Gough, R.P., Zwacka, R.M. MnSOD protects colorectal cancer cells from TRAIL-induced apoptosis by inhibition of Smac/DIABLO release. *Oncogene* **2008**, *27*(6), 763–774.

292. Izeradjene, K., Douglas, L., Tillman, D.M., Delaney, A.B., Houghton, J.A. Reactive oxygen species regulate caspase activation in tumor necrosis factor-related apoptosis-inducing ligand-resistant human colon carcinoma cell lines. *Cancer Res.* **2005**, *65*(16), 7436–7445.

293. Meurette, O., Lefeuvre-Orfila, L., Rebillard, A., Lagadic-Gossmann, D., Dimanche-Boitrel, M.T. Role of intracellular glutathione in cell sensitivity to the apoptosis induced by tumor necrosis factor {alpha}-related apoptosis-inducing ligand/anticancer drug combinations. *Clin. Cancer Res.* **2005**, *11*(8), 3075–3083.

294. Choi, K., Ryu, S.W., Song, S., Choi, H., Kang, S.W., Choi, C. Caspase-dependent generation of reactive oxygen species in human astrocytoma cells contributes to resistance to TRAIL-mediated apoptosis. *Cell Death Differ.* **2010**, *17*(5), 833–845.

295. Son, J.K., Varadarajan, S., Bratton, S.B. TRAIL-activated stress kinases suppress apoptosis through transcriptional upregulation of MCL-1. *Cell Death Differ.* **2010**, *17*(8), 1288–1301.

296. Huerta-Yepez, S., Vega, M., Escoto-Chavez, S.E., Murdock, B., Sakai, T., Baritaki, S., Bonavida, B. Nitric oxide sensitizes tumor cells to TRAIL-induced apoptosis via inhibition of the DR5 transcription repressor Yin Yang 1. *Nitric Oxide* **2009**, *20*(1), 39–52.

297. Wang, C., Youle, R.J. The role of mitochondria in apoptosis. *Annu. Rev. Genet.* **2009**, *43*, 95–118.

298. Franco, R., Cidlowski, J.A. Apoptosis and glutathione: Beyond an antioxidant. *Cell Death Differ.* **2009**, *16*(10), 1303–1314.

299. Franco, R., Schoneveld, O.J., Pappa, A., Panayiotidis, M.I. The central role of glutathione in the pathophysiology of human diseases. *Arch. Physiol. Biochem.* **2007**, *113*(4–5), 234–258.

300. Yoshida, A., Takemura, H., Inoue, H., Miyashita, T., Ueda, T. Inhibition of glutathione synthesis overcomes Bcl-2-mediated topoisomerase inhibitor resistance and induces nonapoptotic cell death via mitochondrial-independent pathway. *Cancer Res.* **2006**, *66*(11), 5772–5780.

301. Armstrong, J.S., Jones, D.P. Glutathione depletion enforces the mitochondrial permeability transition and causes cell death in Bcl-2 overexpressing HL60 cells. *FASEB J.* **2002**, *16*(10), 1263–1265.

302. Varghese, J., Khandre, N.S., Sarin, A. Caspase-3 activation is an early event and initiates apoptotic damage in a human leukemia cell line. *Apoptosis* **2003**, *8*(4), 363–370.

303. Sato, T., Machida, T., Takahashi, S., Iyama, S., Sato, Y., Kuribayashi, K., Takada, K., Oku, T., Kawano, Y., Okamoto, T., Takimoto, R., Matsunaga, T., Takayama, T., Takahashi, M., Kato, J., Niitsu, Y. Fas-mediated apoptosome formation is dependent on reactive oxygen species derived from mitochondrial permeability transition in Jurkat cells. *J. Immunol.* **2004**, *173*(1), 285–296.

304. D'Alessio, M., De Nicola, M., Coppola, S., Gualandi, G., Pugliese, L., Cerella, C., Cristofanon, S., Civitareale, P., Ciriolo, M.R., Bergamaschi, A., Magrini, A., Ghibelli, L. Oxidative Bax dimerization promotes its translocation to mitochondria independently of apoptosis. *FASEB J.* **2005**, *19*(11), 1504–1506.

305. Honda, T., Coppola, S., Ghibelli, L., Cho, S.H., Kagawa, S., Spurgers, K.B., Brisbay, S.M., Roth, J.A., Meyn, R.E., Fang, B., McDonnell, T.J. GSH depletion enhances adenoviral bax-induced apoptosis in lung cancer cells. *Cancer Gene Ther.* **2004**, *11*(4), 249–255.

306. Ghibelli, L., Coppola, S., Fanelli, C., Rotilio, G., Civitareale, P., Scovassi, A.I., Ciriolo, M.R. Glutathione depletion causes cytochrome c release even in the absence of cell commitment to apoptosis. *FASEB J.* **1999**, *13*(14), 2031–2036.

307. Hancock, J.T., Desikan, R., Neill, S.J. Does the redox status of cytochrome C act as a fail-safe mechanism in the regulation of programmed cell death? *Free Radic. Biol. Med.* **2001**, *31*(5), 697–703.

308. Vaughn, A.E., Deshmukh, M. Glucose metabolism inhibits apoptosis in neurons and cancer cells by redox inactivation of cytochrome c. *Nat. Cell Biol.* **2008**, *10*(12), 1477–1483.

309. Lessene, G., Czabotar, P.E., Colman, P.M. BCL-2 family antagonists for cancer therapy. *Nat. Rev. Drug Discov.* **2008**, *7*(12), 989–1000.

310. Voehringer, D.W., Meyn, R.E. Redox aspects of Bcl-2 function. *Antioxid. Redox Signal.* **2000**, *2*(3), 537–550.

311. Ellerby, L.M., Ellerby, H.M., Park, S.M., Holleran, A.L., Murphy, A.N., Fiskum, G., Kane, D.J., Testa, M.P., Kayalar, C., Bredesen, D.E. Shift of the cellular oxidation-reduction potential in neural cells expressing Bcl-2. *J. Neurochem.* **1996**, *67*(3), 1259–1267.

312. Kane, D.J., Sarafian, T.A., Anton, R., Hahn, H., Gralla, E.B., Valentine, J.S., Ord, T., Bredesen, D.E. Bcl-2 inhibition of neural death: Decreased generation of reactive oxygen species. *Science* **1993**, *262*(5137), 1274–1277.

313. Zimmermann, A.K., Loucks, F.A., Schroeder, E.K., Bouchard, R.J., Tyler, K.L., Linseman, D.A. Glutathione binding to the Bcl-2 homology-3 domain groove: A molecular basis for Bcl-2 antioxidant function at mitochondria. *J. Biol. Chem.* **2007**, *282*(40), 29296–29304.

314. Rudin, C.M., Yang, Z., Schumaker, L.M., VanderWeele, D.J., Newkirk, K., Egorin, M.J., Zuhowski, E.G., Cullen, K.J. Inhibition of glutathione synthesis reverses Bcl-2-mediated cisplatin resistance. *Cancer Res.* **2003**, *63*(2), 312–318.

315. Schor, N.F., Rudin, C.M., Hartman, A.R., Thompson, C.B., Tyurina, Y.Y., Kagan, V.E. Cell line dependence of Bcl-2-induced alteration of glutathione handling. *Oncogene* **2000**, *19*(3), 472–476.

316. Bojes, H.K., Datta, K., Xu, J., Chin, A., Simonian, P., Nunez, G., Kehrer, J.P. Bcl-xL overexpression attenuates glutathione depletion in FL5.12 cells following interleukin-3 withdrawal. *Biochem. J.* **1997**, *325*(Pt 2), 315–319.

317. Chanvorachote, P., Nimmannit, U., Stehlik, C., Wang, L., Jiang, B.H., Ongpipatanakul, B., Rojanasakul, Y. Nitric oxide regulates cell sensitivity to cisplatin-induced apoptosis through S-nitrosylation and inhibition of Bcl-2 ubiquitination. *Cancer Res.* **2006**, *66*(12), 6353–6360.

318. Huerta-Yepez, S., Vega, M., Jazirehi, A., Garban, H., Hongo, F., Cheng, G., Bonavida, B. Nitric oxide sensitizes prostate carcinoma cell lines to TRAIL-mediated apoptosis via inactivation of NF-kappa B and inhibition of Bcl-xl expression. *Oncogene* **2004**, *23*(29), 4993–5003.

319. Kim, I., Xu, W., Reed, J.C. Cell death and endoplasmic reticulum stress: Disease relevance and therapeutic opportunities. *Nat. Rev. Drug Discov.* **2008**, *7*(12), 1013–1030.

320. McCullough, K.D., Martindale, J.L., Klotz, L.O., Aw, T.Y., Holbrook, N.J. Gadd153 sensitizes cells to endoplasmic reticulum stress by down-regulating Bcl2 and perturbing the cellular redox state. *Mol. Cell. Biol.* **2001**, *21*(4), 1249–1259.

321. Cullinan, S.B., Diehl, J.A. PERK-dependent activation of Nrf2 contributes to redox homeostasis and cell survival following endoplasmic reticulum stress. *J. Biol. Chem.* **2004**, *279*(19), 20108–20117.

322. Bobrovnikova-Marjon, E., Grigoriadou, C., Pytel, D., Zhang, F., Ye, J., Koumenis, C., Cavener, D., Diehl, J.A. PERK promotes cancer cell proliferation and tumor growth by limiting oxidative DNA damage. *Oncogene* **2010**, *29*(27), 3881–3895.

323. Liu, Y., Adachi, M., Zhao, S., Hareyama, M., Koong, A.C., Luo, D., Rando, T.A., Imai, K., Shinomura, Y. Preventing oxidative stress: A new role for XBP1. *Cell Death Differ.* **2009**, *16*(6), 847–857.

324. Guo, Z., Kozlov, S., Lavin, M.F., Person, M.D., Paull, T.T. ATM activation by oxidative stress. *Science* **2010**, *330*(6003), 517–521.

325. Roos, W.P., Kaina, B. DNA damage-induced cell death by apoptosis. *Trends Mol. Med.* **2006**, *12*(9), 440–450.

326. Calmels, S., Hainaut, P., Ohshima, H. Nitric oxide induces conformational and functional modifications of wild-type p53 tumor suppressor protein. *Cancer Res.* **1997**, *57*(16), 3365–3369.

327. Velu, C.S., Niture, S.K., Doneanu, C.E., Pattabiraman, N., Srivenugopal, K.S. Human p53 is inhibited by glutathionylation of cysteines present in the proximal DNA-binding domain during oxidative stress. *Biochemistry Mosc.* **2007**, *46*(26), 7765–7780.

328. Yusuf, M.A., Chuang, T., Bhat, G.J., Srivenugopal, K.S. Cys-141 glutathionylation of human p53: Studies using specific polyclonal antibodies in cancer samples and cell lines. *Free Radic. Biol. Med.* **2010**, *49*(5), 908–917.

329. Schonhoff, C.M., Daou, M.C., Jones, S.N., Schiffer, C.A., Ross, A.H. Nitric oxide-mediated inhibition of Hdm2-p53 binding. *Biochemistry Mosc.* **2002**, *41*(46), 13570–13574.

330. Ueno, M., Masutani, H., Arai, R.J., Yamauchi, A., Hirota, K., Sakai, T., Inamoto, T., Yamaoka, Y., Yodoi, J., Nikaido, T. Thioredoxin-dependent redox regulation of p53-mediated p21 activation. *J. Biol. Chem.* **1999**, *274*(50), 35809–35815.

331. Hussain, S.P., Amstad, P., He, P., Robles, A., Lupold, S., Kaneko, I., Ichimiya, M., Sengupta, S., Mechanic, L., Okamura, S., Hofseth, L.J., Moake, M., Nagashima, M., Forrester, K.S., Harris, C.C. p53-induced up-regulation of MnSOD and GPx but not catalase increases oxidative stress and apoptosis. *Cancer Res.* **2004**, *64*(7), 2350–2356.

332. Li, P.F., Dietz, R., von Harsdorf, R. p53 regulates mitochondrial membrane potential through reactive oxygen species and induces cytochrome c-independent apoptosis blocked by Bcl-2. *EMBO J.* **1999**, *18*(21), 6027–6036.

333. He, C., Klionsky, D.J. Regulation mechanisms and signaling pathways of autophagy. *Annu. Rev. Genet.* **2009**, *43*, 67–93.

334. Qu, X., Yu, J., Bhagat, G., Furuya, N., Hibshoosh, H., Troxel, A., Rosen, J., Eskelinen, E.L., Mizushima, N., Ohsumi, Y., Cattoretti, G., Levine, B. Promotion of tumorigenesis by heterozygous disruption of the beclin 1 autophagy gene. *J. Clin. Invest.* **2003**, *112*(12), 1809–1820.

335. Liang, X.H., Jackson, S., Seaman, M., Brown, K., Kempkes, B., Hibshoosh, H., Levine, B. Induction of autophagy and inhibition of tumorigenesis by beclin 1. *Nature* **1999**, *402*(6762), 672–676.

336. Eng, C.H., Abraham, R.T. The autophagy conundrum in cancer: Influence of tumorigenic metabolic reprogramming. *Oncogene* **2011**, *30*(47), 4687–4696.

337. Azad, M.B., Chen, Y., Gibson, S.B. Regulation of autophagy by reactive oxygen species (ROS): Implications for cancer progression and treatment. *Antioxid. Redox Signal.* **2009**, *11*(4), 777–790.

338. Rodriguez-Rocha, H., Garcia-Garcia, A., Panayiotidis, M.I., Franco, R. DNA damage and autophagy. *Mutat. Res.* **2011**, *711*(1–2), 158–166.

339. Chen, Y., McMillan-Ward, E., Kong, J., Israels, S.J., Gibson, S.B. Oxidative stress induces autophagic cell death independent of apoptosis in transformed and cancer cells. *Cell Death Differ.* **2008**, *15*(1), 171–182.

340. Yang, P., Ebbert, J.O., Sun, Z., Weinshilboum, R.M. Role of the glutathione metabolic pathway in lung cancer treatment and prognosis: A review. *J. Clin. Oncol.* **2006**, *24*(11), 1761–1769.

341. McIlwain, C.C., Townsend, D.M., Tew, K.D. Glutathione S-transferase polymorphisms: Cancer incidence and therapy. *Oncogene* **2006**, *25*(11), 1639–1648.

342. Sau, A., Pellizzari Tregno, F., Valentino, F., Federici, G., Caccuri, A.M. Glutathione transferases and development of new principles to overcome drug resistance. *Arch. Biochem. Biophys.* **2010**, *500*(2), 116–122.

343. Balendiran, G.K., Dabur, R., Fraser, D. The role of glutathione in cancer. *Cell Biochem. Funct.* **2004**, *22*(6), 343–352.

344. Parl, F.F. Glutathione S-transferase genotypes and cancer risk. *Cancer Lett.* **2005**, *221*(2), 123–129.

345. Hayes, J.D., Flanagan, J.U., Jowsey, I.R. Glutathione transferases. *Annu. Rev. Pharmacol. Toxicol.* **2005**, *45*, 51–88.

346. Hamilton, D., Batist, G. Glutathione analogues in cancer treatment. *Curr. Oncol. Rep.* **2004**, *6*(2), 116–122.

347. Zhao, G., Wang, X. Advance in antitumor agents targeting glutathione-S-transferase. *Curr. Med. Chem.* **2006**, *13*(12), 1461–1471.

348. Oakley, A.J., Lo Bello, M., Mazzetti, A.P., Federici, G., Parker, M.W. The glutathione conjugate of ethacrynic acid can bind to human pi class glutathione transferase P1-1 in two different modes. *FEBS Lett.* **1997**, *419*(1), 32–36.

349. Singh, S.V., Xu, B.H., Maurya, A.K., Mian, A.M. Modulation of mitomycin C resistance by glutathione transferase

inhibitor ethacrynic acid. *Biochim. Biophys. Acta* **1992**, *1137*(3), 257–263.
350. Deeley, R.G., Westlake, C., Cole, S.P. Transmembrane transport of endo- and xenobiotics by mammalian ATP-binding cassette multidrug resistance proteins. *Physiol. Rev.* **2006**, *86*(3), 849–899.
351. Cole, S.P., Deeley, R.G. Transport of glutathione and glutathione conjugates by MRP1. *Trends Pharmacol. Sci.* **2006**, *27*(8), 438–446.
352. Szakacs, G., Paterson, J.K., Ludwig, J.A., Booth-Genthe, C., Gottesman, M.M. Targeting multidrug resistance in cancer. *Nat. Rev. Drug Discov.* **2006**, *5*(3), 219–234.
353. Lorico, A., Rappa, G., Finch, R.A., Yang, D., Flavell, R.A., Sartorelli, A.C. Disruption of the murine MRP (multidrug resistance protein) gene leads to increased sensitivity to etoposide (VP-16) and increased levels of glutathione. *Cancer Res.* **1997**, *57*(23), 5238–5242.
354. Loe, D.W., Almquist, K.C., Deeley, R.G., Cole, S.P. Multidrug resistance protein (MRP)-mediated transport of leukotriene C4 and chemotherapeutic agents in membrane vesicles. Demonstration of glutathione-dependent vincristine transport. *J. Biol. Chem.* **1996**, *271*(16), 9675–9682.
355. Loe, D.W., Deeley, R.G., Cole, S.P. Characterization of vincristine transport by the M(r) 190,000 multidrug resistance protein (MRP): Evidence for cotransport with reduced glutathione. *Cancer Res.* **1998**, *58*(22), 5130–5136.
356. Leslie, E.M., Haimeur, A., Waalkes, M.P. Arsenic transport by the human multidrug resistance protein 1 (MRP1/ABCC1). Evidence that a tri-glutathione conjugate is required. *J. Biol. Chem.* **2004**, *279*(31), 32700–32708.
357. Liu, J., Chen, H., Miller, D.S., Saavedra, J.E., Keefer, L.K., Johnson, D.R., Klaassen, C.D., Waalkes, M.P. Overexpression of glutathione S-transferase II and multidrug resistance transport proteins is associated with acquired tolerance to inorganic arsenic. *Mol. Pharmacol.* **2001**, *60*(2), 302–309.
358. Franco, R., Cidlowski, J.A. SLCO/OATP-like transport of glutathione in FasL-induced apoptosis: Glutathione efflux is coupled to an organic anion exchange and is necessary for the progression of the execution phase of apoptosis. *J. Biol. Chem.* **2006**, *281*(40), 29542–29557.
359. Benlloch, M., Ortega, A., Ferrer, P., Segarra, R., Obrador, E., Asensi, M., Carretero, J., Estrela, J.M. Acceleration of glutathione efflux and inhibition of gamma-glutamyltranspeptidase sensitize metastatic B16 melanoma cells to endothelium-induced cytotoxicity. *J. Biol. Chem.* **2005**, *280*(8), 6950–6959.
360. Trompier, D., Chang, X.B., Barattin, R., du Moulinet D'Hardemare, A., Di Pietro, A., Baubichon-Cortay, H. Verapamil and its derivative trigger apoptosis through glutathione extrusion by multidrug resistance protein MRP1. *Cancer Res.* **2004**, *64*(14), 4950–4956.
361. Cui, Y., Konig, J., Nies, A.T., Pfannschmidt, M., Hergt, M., Franke, W.W., Alt, W., Moll, R., Keppler, D. Detection of the human organic anion transporters SLC21A6 (OATP2) and SLC21A8 (OATP8) in liver and hepatocellular carcinoma. *Lab. Invest.* **2003**, *83*(4), 527–538.
362. Perez-Tomas, R. Multidrug resistance: Retrospect and prospects in anti-cancer drug treatment. *Curr. Med. Chem.* **2006**, *13*(16), 1859–1876.
363. Curtis, C.D., Thorngren, D.L., Nardulli, A.M. Immunohistochemical analysis of oxidative stress and DNA repair proteins in normal mammary and breast cancer tissues. *BMC Cancer* **2010**, *10*, 9.
364. Matsui, A., Ikeda, T., Enomoto, K., Hosoda, K., Nakashima, H., Omae, K., Watanabe, M., Hibi, T., Kitajima, M. Increased formation of oxidative DNA damage 8-hydroxy-2'-deoxyguanosine in human breast cancer tissue and its relationship to GSTP1 and COMT genotypes. *Cancer Lett.* **2000**, *151*, 87–95.
365. Musarrat, J., Arezina-Wilson, J., Wani, A.A. Prognostic and aetiological relevance of 8-hydroxyguanosine in human breast carcinogenesis. *Eur. J. Cancer* **1996**, *32A*(7), 1209–1214.
366. Mambo, E., Nyaga, S.G., Bohr, V.A., Evans, M.K. Defective DNA repair of 8-hydroxyguanine in mitochondria of MCF-7 and MDA-MB-468 human breast cancer cell lines. *Cancer Res.* **2002**, *62*, 1349–1355.
367. Nyaga, S.G., Lohani, A., Jaruga, P., Trzeciak, A.R., Dizdaroglu, M., Evans, M.K. Reduced repair of 8-hydroxyguanine in the human breast cancer cell line, HCC1937. *BMC Cancer* **2006**, *6*, 297–312.
368. Nyaga, S.G., Jaruga, P., Lohani, A., Dizdaroglu, M., Evans, M.K. Accumulation of oxidatively induced DNA damage in human breast cancer cell lines following treatment with hydrogen peroxide. *Cell Cycle* **2007**, *6*, 1472–1478.
369. Djuric, Z., Heilbrun, L.K., Lababidi, S., Berzinkas, E., Simon, M.S., Kosir, M.A. Levels of 5-hydroxymethyl-2'-deoxyuridine in DNA from blood of women scheduled for breast biopsy. *Cancer Epidemiol. Biomarkers Prev.* **2001**, *10*, 147–149.
370. Romano, G., Sgambato, A., Mancini, R., Capelli, G., Giovagnoli, M.R., Flamini, G., Boninsegna, A., Vecchione, A., Cittadini, A. 8-hydroxy-2'-deoxyguanosine in cervical cells: Correlation with grade of dysplasia and human papillomavirus infection. *Carcinogenesis* **2000**, *21*(6), 1143–1147.
371. Oliva, M.R., Ripoll, F., Muniz, P., Iradi, A., Trullenque, R., Valls, V., Drehmer, E., Saez, G.T. Genetic alterations and oxidative metabolism in sporadic colorectal tumors from a Spanish community. *Mol. Carcinog.* **1997**, *18*(4), 232–243.
372. Kondo, S., Toyokuni, S., Iwasa, Y., Tanaka, T., Onodera, H., Hiai, H., Imamura, M. Persistent oxidative stress in human colorectal carcinoma, but not in adenoma. *Free Radic. Biol. Med.* **1999**, *27*(3–4), 401–410.
373. Kondo, S., Toyokuni, S., Tanaka, T., Hiai, H., Onodera, H., Kasai, H., Imamura, M. Overexpression of the hOGG1 Gene and high 8-Hydroxy-2'-deoxyguanosine (8-OHdG) lyase activity in human colorectal carcinoma: Regulation mechanism of the 8-OHdG level in DNA. *Clin. Cancer Res.* **2000**, *6*(4), 1394–1400.

374. Gackowski, D., Banaszkiewicz, Z., Rozalski, R., Jawien, A., Olinski, R. Persistent oxidative stress in colorectal carcinoma patients. *Int. J. Cancer* **2002**, *101*(4), 395–397.

375. Oka, K., Tanaka, T., Enoki, T., Yoshimura, K., Ohshima, M., Kubo, M., Murakami, T., Gondou, T., Minami, Y., Takemoto, Y., Harada, E., Tsushimi, T., Li, T.S., Traganos, F., Darzynkiewicz, Z., Hamano, K. DNA damage signaling is activated during cancer progression in human colorectal carcinoma. *Cancer Biol. Ther.* **2010**, *9*, 246–252.

376. Lee, B.M., Jang, J.J., Kim, H.S. Benzo[a]pyrene diol-epoxide-I-DNA and oxidative DNA adducts associated with gastric adenocarcinoma. *Cancer Lett.* **1998**, *125*(1–2), 61–68.

377. Farinati, F., Cardin, R., Degan, P., Rugge, M., Mario, F.D., Bonvicini, P., Naccarato, R. Oxidative DNA damage accumulation in gastric carcinogenesis. *Gut* **1998**, *42*(3), 351–356.

378. Yamamoto, T., Hosokawa, K., Tamura, T., Kanno, H., Urabe, M., Honjo, H. Urinary 8-hydroxy-2'-deoxyguanosine (8-OHdG) levels in women with or without gynecologic cancer. *J. Obstet. Gynaecol. Res.* **1996**, *22*, 359–363.

379. Schwarz, K.B., Kew, M., Klein, A., Abrams, R.A., Sitzmann, J., Jones, L., Sharma, S., Britton, R.S., Di Bisceglie, A.M., Groopman, J. Increased hepatic oxidative DNA damage in patients with hepatocellular carcinoma. *Dig. Dis. Sci.* **2001**, *46*(10), 2173–2178.

380. Senturker, S., Karahalil, B., Inal, M., Yilmaz, H., Muslumanoglu, H., Gedikoglu, G., Dizdaroglu, M. Oxidative DNA base damage and antioxidant enzyme levels in childhood acute lymphoblastic leukemia. *FEBS Lett.* **1997**, *416*(3), 286–290.

381. Honda, M., Yamada, Y., Tomonaga, M., Ichinose, H., Kamihira, S. Correlation of urinary 8-hydroxy-2'-deoxyguanosine (8-OHdG), a biomarker of oxidative DNA damage, and clinical features of hematological disorders: A pilot study. *Leuk. Res.* **2000**, *24*(6), 461–468.

382. Vulimiri, S.V., Wu, X., Baer-Dubowska, W., de Andrade, M., Detry, M., Spitz, M.R., DiGiovanni, J. Analysis of aromatic DNA adducts and 7,8-dihydro-8-oxo-2'-deoxyguanosine in lymphocyte DNA from a case-control study of lung cancer involving minority populations. *Mol. Carcinog.* **2000**, *27*(1), 34–46.

383. Erhola, M., Yoyokumi, S., Okada, K., Tanaka, T., Hiai, H., Ochi, H., Uchida, K., Osawa, T., Nieminen, M.M., Alho, H., Kellokumpu-Lehtinen, P. Biomarker evidence of DNA oxidation in lung cancer patients: Association of urinary 8-hydroxy-2'-deoxyguanosine excretion with radiotherapy, chemotherapy, and response to treatment. *FEBS Lett.* **1997**, *409*, 287–291.

384. Jaruga, P., Zastawny, T.H., Skokowski, J., Dizdaroglu, M., Olinski, R. Oxidative DNA base damage and antioxidant enzyme activities in human lung cancer. *FEBS Lett.* **1994**, *341*(1), 59–64.

385. Zhang, J., Dhakal, I.B., Greene, G., Lang, N.P., Kadlubar, F.F. Polymorphisms in hOGG1 and XRCC1 and risk of prostate cancer: Effects modified by plasma antioxidants. *Urology* **2010**, *75*(4), 779–785.

386. Okamoto, K., Toyokuni, S., Uchida, K., Ogawa, O., Takenewa, J., Kakehi, Y., Kinoshita, H., Hattori-Nakakuki, Y., Hiai, H., Yoshida, O. Formation of 8-hydroxy-2'-deoxyguanosine and 4-hydroxy-2-nonenal-modified proteins in human renal-cell carcinoma. *Int. J. Cancer* **1994**, *58*(6), 825–829.

10

NEURODEGENERATION FROM DRUGS AND AGING-DERIVED FREE RADICALS

ANNMARIE RAMKISSOON, AARON M. SHAPIRO, MARGARET M. LONIEWSKA, AND PETER G. WELLS

OVERVIEW

This chapter focuses upon the potential neurotoxic and neurodegenerative effects of reactive oxygen species (ROS) and free radical intermediates formed either from drugs or from endogenous brain chemicals such as neurotransmitters, and the various protective mechanisms that neutralize ROS or repair ROS-mediated damage, in animal models, with some discussion of potential human relevance. In particular, we comment upon long-term free radical-mediated neurodegenerative effects of amphetamine analogs like methamphetamine (METH) and ecstasy following adult exposure, and related neurodevelopmental deficits following *in utero* exposure of the developing embryo and fetus. Also discussed is a similar but endogenous ROS-dependent mechanism of neurodegeneration with aging, which results from an imbalance in the activity of pathways for ROS formation versus ROS detoxification and repair of cellular macromolecules. The ultimate risk of neurodegeneration is mitigated by protective central nervous system (CNS) substrates and enzymes like antioxidants and antioxidative enzymes, and by enzymes and proteins involved in the repair of oxidatively damaged DNA, many of which are regulated by ROS-sensing mechanisms, including nuclear factor erythroid 2-related factor 2 (Nrf2).

The term "neurotoxicity" in this chapter refers to toxicity as a consequence of exposure to a toxic agent that results in pathological effects, including oxidatively damaged cellular macromolecules, swollen organelles, shrunken cell membranes, and autophagic vacuoles that may lead to cell death or nerve terminal degeneration. This may result in an increase in the expression of glial fibrillary acidic protein (GFAP), which is a marker of astrogliosis. In METH studies in rodent models, loss of presynaptic nerve terminals has been identified by silver staining to obtain evidence of METH-induced neuronal damage as identified by morphological signs of axonal degeneration.[1,2] This has been correlated to a reduction in neurotransmitters or enzymes for their synthesis, or transporters present in the nerve terminal, which are associated with functional deficits (discussed later). These biochemical changes and functional deficits may reflect receptor-mediated mechanisms and may be reversible within days as the drug is eliminated. However, damage to critical components may be only partially reversible and can persist for months after drug exposure[3,4] suggesting a role for ROS in the toxicity.

This chapter will discuss the mechanisms of drug-initiated and age-related ROS-mediated neurodegeneration, the various protective enzymes and substrates, and the role of Nrf2 in responding to ROS-mediated insult to alleviate toxicity.

10.1 ROS FORMATION

10.1.1 Introduction to ROS

ROS, such as superoxide anions, hydrogen peroxide (H_2O_2), and hydroxyl radicals are produced due to the incomplete reduction of oxygen during aerobic

Molecular Basis of Oxidative Stress: Chemistry, Mechanisms, and Disease Pathogenesis, First Edition. Edited by Frederick A. Villamena.
© 2013 John Wiley & Sons, Inc. Published 2013 by John Wiley & Sons, Inc.

metabolism.[5] Superoxide anion, the product of a one-electron reduction of oxygen, is the precursor of most ROS and a mediator in oxidative chain reactions. Dismutation of superoxide anions, either spontaneously or through a reaction catalyzed by superoxide dismutases (SODs), produces H_2O_2. This may be fully reduced to water or partially reduced to hydroxyl radicals, one of the strongest oxidants in nature. The formation of hydroxyl radicals is also catalyzed by reduced transition metals. Important radicals are summarized in Table 10.1. These ROS are highly reactive as they contain an unpaired electron and can oxidize molecular targets such as proteins, lipids, and DNA in a process known as oxidative stress.[5] Such damage, if not repaired, can accumulate over time and can lead to loss of cellular function and even cell death (Fig. 10.1).[6,7] The brain is especially susceptible to oxidative stress due to the high rate of oxygen consumption, low antioxidant levels, and ROS-generating enzymatic reactions.[5] ROS have been implicated in many neurodegenerative diseases such as Alzheimer's disease (AD),[8] Parkinson's disease (PD),[9] and multiple sclerosis (MS).[10]

10.1.2 CNS Sources of ROS

There are numerous sources of ROS in the brain that can lead to oxidative damage to macromolecules and neurotoxicity. These are summarized in Figure 10.2.

10.1.2.1 Mitochondria
Mitochondria play a central role in the survival and death of neurons. Mitochondria exert multiple influences on neuronal function, including the generation of adenosine-5′-triphosphate (ATP) and sequestering of calcium. The mitochondrial respiratory chain is also the major site for the generation of superoxide anions[11] and H_2O_2.[12,13] The electron transport chain, which is embedded in the inner membrane of the mitochondria, consists of five multiprotein complexes. Although molecular oxygen is reduced to water in complex IV by a sequential four-electron transfer, a

TABLE 10.1 Important Reactive Species

Reactive Species		Notes
Superoxide anion	$O_2^{\bullet-}$	Free radical, source of H_2O_2, $t_{1/2} = 1 \times 10^{-6}$ s
Hydrogen peroxide	H_2O_2	Oxidizing agent, source of $OH^{\bullet}$, $t_{1/2}$ = min
Hydroxyl radical	$HO^{\bullet}$	Very reactive, $t_{1/2} = 1 \times 10^{-9}$ s
Nitric oxide radical	NO	Free radical, reacts with $O_2^{\bullet-}$
Peroxynitrite	$ONOO^-$	Can decompose to $HO^{\bullet}$
Ferrous iron	Fe^{2+}	Reacts with H_2O_2, oxidation leads to $O_2^{\bullet-}$
Ferric iron	Fe^{3+}	Oxidized form of ferrous iron

Sources: References 5 and 458.

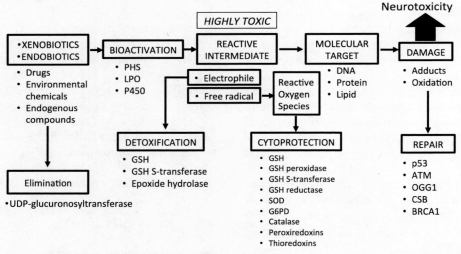

Figure 10.1 Enzymatic pathways involved in reactive intermediate-mediated neurotoxicity. Susceptibility to the neurotoxic effects involves the balance between drug bioactivation, elimination, detoxification, and pathways of cytoprotection and repair. Under normal conditions or during low oxidative stress, cells are able to detoxify endogenous and xenobiotic reactive intermediates and ROS with appropriate enzymes: GSH reductase, GSH peroxidase, G6PD, SOD, catalase, peroxiredoxins. However, when bioactivation exceeds detoxification, high levels of reactive intermediates can lead to cellular damage. Such damage can be repaired by p53, ATM, OGG1, CSB, Trx. If not repaired, molecular damage and/or alteration of signal transduction can lead to neurotoxicity. ATM, ataxia telangiectasia mutated; CSB, Cockayne syndrome B; G6PD, glucose-6-phosphate dehydrogenase; GSH, glutathione; LPO, lipoxygenase; OGG1, oxoguanine glycosylase 1; P450, cytochromes 450; PHS, prostaglandin H synthase; ROS, reactive oxygen species; SOD, superoxide dismutase. Modified from Reference 6.

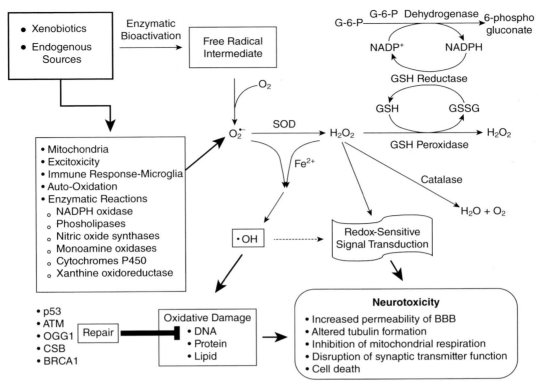

Figure 10.2 Sources of reactive oxygen species (ROS) in the brain. When pro-oxidants exceed the antioxidative and repair mechanisms, oxidative damage can occur leading to neurotoxicity. ATM, ataxia telangiectasia mutated; CSB, Cockayne syndrome B; Fe, iron; G-6-P, glucose-6-phosphate; GSH, glutathione; GSSG, GSH disulfide; H_2O_2, hydrogen peroxide; HO•, hydroxyl radical; LPO, lipoxygenase; NADP+, nicotinamide adenine dinucleotide phosphate; $O_2^{•-}$, superoxide anion, OGG1, oxoguanine glycosylase 1; P450, cytochromes P450; PHS, prostaglandin H synthase; SOD, superoxide dismutase. Modified from Reference 213.

portion can be reduced by a one-electron addition that occurs in complex I[11] and also in complex III,[12,14] resulting in the formation of ROS. The mitochondrial matrix, however, contains the antioxidant enzyme manganese superoxide dismutase (MnSOD, SOD2) that may combat the high rate of superoxide production in the mitochondrial inner membrane.[5] The production of H_2O_2 by mitochondria appears to account for 1–2% of the total oxygen consumed *in vitro*.[15] As a result, the steady state concentrations of superoxide anions and H_2O_2 in the mitochondrial matrix have been estimated to be around 1×10^{-10} M and 5×10^{-9} M, respectively.[16] Alterations in mitochondrial function can play an important role in increasing ROS steady-state levels and contribute to neurotoxicity.

10.1.2.2 Nicotinamide Adenine Dinucleotide Phosphate Hydrogen (NADPH) Oxidase (NOX)
NOX is found in the brain. There are seven different isoforms, including NOXs 1 to 5 and dual oxidases (DUOX) DUOX1 and DUOX2. Little is known about the role of NOX5, DUOX1, and DUOX2 in the brain.[17] All NOX family members are transmembrane proteins that transport electrons across biological membranes to reduce oxygen to superoxide.[18] When the complex is active, it generates superoxide by transferring an electron from NADPH in the cytosol to oxygen on the luminal or extracellular space. Superoxide is the primary product of the electron transfer, but H_2O_2 is also generated.[19] The NOX isoforms have been detected in various brain regions.[20–22] Most studied has been NOX2 which can be found in regions including the cortex, hippocampus,[21] and striatum.[22] Within these regions, NOX enzymes have been investigated with respect to inflammatory processes regulated by NOX2 in microglia.[23,24] However, NOXs are also present in astrocytes[25] and in mouse and human neurons.[22,26] There is also increasing evidence for a role of NOX2 in neurodegeneration, including in AD and amyotrophic lateral sclerosis (ALS).[27,28]

10.1.2.3 Phospholipase A2 (PLA2)
PLA2 enzymes are esterases that cleave the acyl ester bond at the sn-2

position of membrane phospholipids to produce free fatty acids, for example arachidonic acid (AA).[29] These enzymes are broadly classified into groups, including secretory phospholipase A2 (sPLA2), cytosolic phospholipase A2 (cPLA2), plasmalogen-selective phospholipase A2 (PlsEtn-PLA2), and calcium-independent phospholipase A2 (iPLA2).[29–31] They are present in various regions of the brain and are expressed in astrocytes[32,33] and in neurons.[29,31,34] Enzymatic activity is regulated by calcium, ROS, and neurotransmitters, and activation releases AA that is metabolized by cyclooxygenases and lipoxygenases, in the process generating ROS (discussed in detail later). cPLA2 activities may contribute to neurotoxicity in AD and MS,[32,35] and cPLA2-deficient mice are resistant to 1-methyl-4-phenyl-1,2,3,6-tetrahydropyridine (MPTP)-induced neurotoxicity, an animal model for PD.[36]

10.1.2.4 Nitric Oxide Synthases (NOSs)

There are four members of the NOS family. A constitutive isoform of NOS called neuronal NOS (nNOS) is found in neurons.[37,38] Another isoform is inducible NOS (iNOS) in which inflammatory mediators such as lipopolysaccharide (LPS) and cytokines cause its expression in microglia and astrocytes[39,40] and possibly in neurons.[41] Other isoforms include endothelial NOS (eNOS) and mitochondrial NOS (mtNOS).[42–44] L-arginine is used by NOS to produce NO and citrulline in a process requiring NADPH and O_2.[45] Although all NOS isoforms can potentially produce superoxide anions, iNOS is the most likely to produce superoxide anions *in vivo* due to L-arginine depletion during inflammation.[46] Nitric oxide (NO) is mainly used for guanylate cyclase activation with the subsequent production of cyclic guanosine-3′,5′-monophosphate (cGMP).[47] NO, however, can react with superoxide anions forming peroxynitrite anion ($ONOO^-$), which is a very reactive anion that can oxidize proteins.[48] The NO free radical also triggers apoptosis when it binds to cytochrome *c* oxidase and induces the formation of superoxide anions in the mitochondria, generating $ONOO^-$.[49] These enzymes are involved in various neurodegenerative diseases including AD.[50] In MS, NOS may be induced, and there is evidence of oxidative stress where $ONOO^-$ is believed to contribute to the cellular damage.[51,52]

10.1.2.5 Monoamine Oxidase (MAO)

During catecholamine metabolism, ROS can also be generated through enzymatic reactions. For example, MAOs, which are associated with mitochondrial membranes, catalyze the oxidation of amines to their corresponding aldehydes and ammonia with the formation of H_2O_2 as a by-product.[53] There are two isoforms of this enzyme, MAO-A and MAO-B. MAO-A is in neurons in the catecholinergic cell areas.[54] MAO-B is contained in serotonergic neurons in the median raphe and in astrocytes.[54,55] MAO-B is selectively inhibited by L-deprenyl,[56] while MAO-A is selectively inhibited by clorgyline.[57] Both forms oxidize dopamine (DA), tyramine, and octopamine.[58] MAO-B is also responsible for the oxidation of MPTP to MPP+ which damages dopaminergic neurons.[59] Alterations in MAO-B activity have been implicated in PD,[60] and patients with PD have elevated MAO-B activity in the substantia nigra.[61]

10.1.2.6 Cytochromes P450 (CYPs)

CYPs are a superfamily of heme-containing monooxygenases that metabolize a large number of compounds. CYPs are involved in the biosynthetic pathways of steroid and bile acid production, and most CYPs metabolize xenobiotics. CYPs carry out the oxidation of carbon and nitrogen groups usually resulting in the addition of an –OH.[62] In catalyzing the metabolism of a drug, CYPs use NADPH to reduce O_2, leading to the production of H_2O_2 and superoxide anion radicals. CYP2E1 metabolism of a number of substrates is known to lead to increased ROS.[63] CYP2E1 mRNA has been detected in several mammalian brain regions.[64,65] It has also been proposed that selective localization of CYP2E1 in DA-containing neurons may contribute to nigrostriatal toxicity in chemically induced PD.[66] While it is uncertain whether other CYPs may contribute to ROS generation, the expression of CYPs in the brain is 1–2% of that in the liver.[67,68] CYP2D6 is an isozyme involved in the metabolism of many drugs active in the CNS, such as antidepressants and antipsychotics. This enzyme is coded by a polymorphic gene, with 7% of the Caucasian population showing no enzymatic activity ("poor metabolizers"). Approximately 20–30% of Caucasians carry one active and one inactive allele, and show intermediate enzyme activity (here referred to as "intermediate metabolizers"). Individuals carrying two active alleles are "extensive metabolizers."[69] Whether this enzyme directly forms ROS in the brain is uncertain.

10.1.2.7 Xanthine Oxidoreductase

Guanine degrades into xanthine, and during ATP catabolism under hypoxic conditions, hypoxanthine can be formed.[70] Xanthine oxidoreductase is a widely distributed enzyme that catalyzes the oxidation of hypoxanthine to xanthine and of xanthine to uric acid.[70] The enzyme occurs in two forms, xanthine dehydrogenase and xanthine oxidase. During these reactions, H_2O_2 and superoxide anions are produced. Only xanthine dehydrogenase is capable of reducing nicotinamide adenine dinucleotide (NAD+). Both forms can reduce molecular oxygen, although xanthine oxidase is more effective.[71] In the reoxidation of fully reduced xanthine oxidase, the first two steps each involve transfer of two electrons to oxygen, generating H_2O_2. Xanthine oxidase then transfers its remaining electrons in separate steps, with each electron independently reducing O_2 to produce superoxide anions.[72] Xanthine dehydrogenase and xanthine oxidase can be interconverted by means of sulfhydryl reagents. When xanthine dehydrogenase is treated with proteases, like trypsin, it is irreversibly transformed into xanthine oxidase.[72] Reversible conversion occurs due to conditions that oxidize thiol groups of Cys535 and Cys992, exposure to sulfhydryl agents and exposure to anaerobic conditions.[72] Xanthine oxidase is localized to the vascular endothelium of brain.[73] It is thought to play a neurotoxic role during ischemia-reperfusion injury.[74,75] During ischemia, ATP is broken down into hypoxanthine and upon reoxygenation, xanthine oxidase converts the excessive hypoxanthine to xanthine, thereby generating ROS.[72]

10.1.2.8 Excitotoxicity
Excitotoxicity is a phenomenon whereby prolonged activation of excitatory amino acid receptors leads to cell death.[76,77] Glutamate has been identified as the principal transmitter mediating fast excitatory synaptic responses in the vertebrate brain. Glutamate distribution within the brain is extensive. Glutamate is present at concentrations of 5–15 µmol/g weight of wet tissue in humans, and regional distribution of glutamate in the brain is similar among species from rat to human.[78,79] Glutamate receptors are divided into ionotropic, which are ligand-gated ion channels, and metabotropic receptors that are linked to G-proteins.[80] These are further subdivided, for example, the ionotropic receptors are characterized by their selective affinity for the specific agonists N-methyl-D-aspartate (NMDA), α-amino-3-hydroxy-5-methylisoxazole-4-propionic acid (AMPA), and kainic acid.[81] Excitotoxicity can result in excessive intracellular calcium, generation of free radicals, and activation of the mitochondrial permeability transition and secondary excitotoxicity. Excitotoxicity more directly is toxicity that is related to calcium influx subsequent to glutamate receptor activation, and it is greatly attenuated in the absence of calcium.[82] Excitotoxicity leads to ROS production, and the oxidative agents in turn promote excitotoxic mechanisms, usually by disrupting calcium homeostasis and activating calcium-dependent proteases.[82,83] Termination of the excitatory action of glutamate is mediated by a high-affinity uptake system on both pre- and post-synaptic neuronal cell membranes and the membranes of adjacent glial cells.[84] Glutamate transporter 1, which is located on astrocytes,[85] is responsible for most of the total glutamate transport.[84] Excitotoxicity not only affects neurons, but also astrocytes and oligodendrocytes as well.,[86,87] Through these various mechanisms, excitotoxicity has been proposed to explain the pathology characteristic of neurodegenerative diseases such as AD, Huntington's disease (HD), and ALS.[88–90]

10.1.2.9 Immune Response Microglia
Microglia are considered the local immune cells of the brain and are present in various regions.[91] Microglial density varies by brain region in the adult human and in adult mice; these cells are in the gray matter, with the highest concentrations being found in the hippocampus, olfactory telencephalon, basal ganglia, and substantia nigra.[91,92] Microglia are normally in a resting state and become activated in response to injury.[93,94] The resting microglial cells are activated by detecting LPS, beta-amyloid (Aβ), thrombin, interferon-gamma (IFN-γ), and other proinflammatory cytokines.[95] A crucial function of microglia is their ability to generate significant immune responses. For example, microglia initiate responses including the production of cytokines, chemokines, ROS, and NO.[96,97] Activated microglia release interleukin (IL)-1, tumor necrosis factor-alpha (TNF-α), and chemokines for lymphocyte recruitment.[95,98,99] Phagocytic and cytotoxic functions of microglia are also triggered during CNS injury. During these processes, significant amounts of ROS can be produced. When activated, microglia can contribute to ROS production via NOX enzymes as discussed previously.[23,24] The brain microglia can generate significant quantities of superoxide anion and NO.[97] However, species differences in their generation have been observed, with mouse microglia generating large amounts of NO when stimulated. In contrast, human and hamster microglia do not produce measurable amounts of NO under the same stimulation conditions, but both human and hamster microglia generate significantly more superoxide anion than rat microglia.[100] Neurotoxins such as MPTP as well as LPS can also overactivate microglia, leading to increased ROS generation and neuronal death.[101,102] Microglia have been shown to play a role in AD, PD, and HD.[103–105]

10.1.3 Prostaglandin H Synthases (PHSs)

10.1.3.1 Role of Prostaglandin Synthesis and Their Receptors
Eicosanoids comprise a class of bioactive lipid mediators derived from the metabolism of polyunsaturated fatty acids by PHSs and lipoxygenases leading to prostanoids and leukotrienes, respectively.[106,107] Prostanoids can be further divided into prostaglandins (PGs) and thromboxanes (TXs). The most typical actions are the relaxation and contraction of various types of smooth muscles. They also modulate neuronal activity by either inhibiting or stimulating neurotransmitter release or inducing central actions such as fever and sleep induction.[108] PGs also regulate secretion and

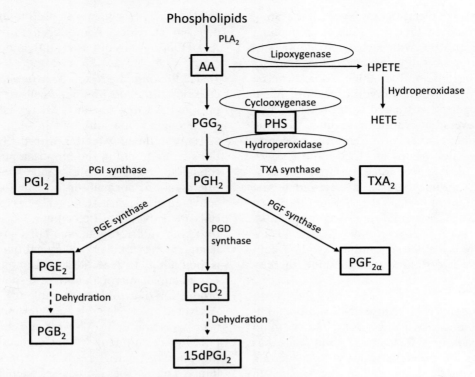

Figure 10.3 Biosynthesis of prostaglandins. Arachidonic acid (AA) is released from phospholipids by phospholipase A2 and is used in both prostaglandin synthesis and the lipoxygenase–eicosanoid pathway. Cyclooxygenase and hydroperoxidase are components of prostaglandin H synthase (PHS). PG, prostaglandin; PGG2, prostaglandin G2; PGH2, prostaglandin H2; HPETE, hydroperoxyeicosatetraenoic acid; HETE, hydroxyeicosatetraenoic acid.

motility in the gastrointestinal tract as well as transport of ions and water in the kidney.[108] AA serves as the metabolic precursor for eicosanoid synthesis. AA is not available in large quantities in the free acid form, but is stored in the backbone of membrane phospholipids. To be used for biosynthesis, PLA2 (reviewed previously) liberates AA from phospholipids in the membrane.[109] This is the rate-limiting step in eicosanoid synthesis.[110]

5-Lipoxygenase (5-LOX) performs the initial enzymatic step in leukotriene synthesis,[111] creating 5-hydroperoxyeicosatetraenoic acid (5-HPETE) by incorporating one molecular oxygen at the C5 position of AA. Depending on cellular conditions, 5-HPETE has a number of potential metabolic fates. It can be secreted in its peroxide form, reduced to 5-hydroxyeicosatetraenoic acid (5-HETE), or undergo a catalytic rearrangement in the 5-LOX active site to form leukotriene A4.

PHSs, also known as cyclooxygenases (COXs), consist of two isozymes, PHS-1 (COX-1) and PHS-2 (COX-2). PHSs contain two distinct active sites, a COX site and a hydroperoxidase site, both of which use the same tyrosyl radical and heme-iron for catalysis. The COX site incorporates molecular O_2 at the 11- and 15-carbon on AA to form the hydroperoxy endoperoxide PGG_2. The hydroperoxidase site reduces the peroxide to the corresponding alcohol, the hydroxy endoperoxide PGH_2, which is the substrate for various PG synthases.[112,113] PGH_2 can form a number of different bioactive products through the action of PG synthases (Fig. 10.3). This includes a number of important signaling molecules, including PGI_2 (also known as prostacyclin), PGD_2, PGE_2, $PGF_{2\alpha}$, and thromboxane A_2 (TXA_2). PGI_2 is formed by prostacyclin synthase, a member of the CYP superfamily.[114] PGI_2 binds the G protein-coupled receptor (GPCR) PGI_2 receptor (IP),[115] as well as the transcription factors peroxisome proliferator-activated receptor (PPAR)α, PPARδ, and PPARγ.[116] PGD_2 is major product of rat brain homogenate formed by PGD synthases.[117] PGD_2 has two known receptors, PGD_2 receptor 1 (DP1) and chemoattractant receptor-homologous, molecules expressed on T helper-2 cells (CRTH2) (DP2).[118] The effects of PGE_2 have been implicated in many biological processes.[119] PGE_2 synthesis occurs through three unique enzymes, the cytosolic PGE synthase (cPGES), microsomal PGE synthase-1 (mPGES-1), and mPGES-2.[119] $PGF_{2\alpha}$, made by PGF synthase, has been implicated in a number of physiological processes and disease states.[120] Only one PGF2α-specific receptor has been cloned,[120] a GPCR termed PGF receptor (FP), which upon binding ligand results in an elevation of intracellular calcium. TXA$_2$ biosynthesis is catalyzed by a member of the CYP superfamily, thromboxane A synthase. It also binds a specific GPCR, termed the thromboxane A receptor (TP), which leads to increased intracellular calcium.[121] Cyclopentenone PGs are a family of molecules that are formed by dehydration of PGE$_2$ and PGD$_2$.[122] Dehydration of PGE$_2$ leads to PGB$_2$. PGD$_2$ dehydration leads to 15-deoxy-12,14-prostaglandin J2 (15d-PGJ$_2$). 15d-PGJ$_2$ has been identified as a high-affinity ligand for the transcription factor PPARγ as well as a less potent activator of PPARα and PPARδ.[123]

There are nine types and subtypes of receptor for prostanoids, designated PGD receptor (DP1) and the CRTH2 or (DP2), EP1, EP2, EP3, and EP4 subtypes of PGE receptor, PGF receptor (FP), PGI receptor (IP), and TXA$_2$ receptor (TP).[108] All of these prostanoid receptors are GPCRs, and their main signal transduction pathways leads to a rise in intracellular cyclic adenosine monophosphate (cAMP) and/or increases in calcium. Their functions and expression are presented in Table 10.2.

In addition to their synthesis, extracellular levels of PGs also depend on transport processes, which are regulated by PG transporter (an influx transporter) and the multidrug resistance-associated protein 4 (an efflux transporter).[124] Also, inactivation in the cytoplasm can occur through hydroxyprostaglandin dehydrogenase (also known 15-PGDH). 15-PGDH is highly expressed in normal tissues but is lacking in human colon.[125] Lack of 15-PGDH expression in tumors results in increased endogenous PGE2 levels.[125]

10.1.3.2 Genetics of PHS

Genes Despite their close structural and functional similarities, the PHS isozymes are encoded by different genes that are differentially regulated, leading to distinct expression patterns and biological functions. Sheep PHS-1 was determined by cDNA cloning in 1989[126–128] followed by cloning and sequence analysis of human PHS-1.[129] It is located on chromosome 9.[129] The human and mouse genes for PHS-1 are approximately 22 kilobases (kb) in length with 11 exons and 10 introns and are transcribed as a 2.8 kb mRNA. PHS-1 is a glycoprotein that in processed form has 576 amino acids with an apparent molecular mass of 70 kilodaltons (kDa).

The PHS-1 gene promoter lacks TATA and CAAT boxes but is GC-rich, and contains multiple transcription start sites.[130] These promoter features are usually characteristic of housekeeping genes that are constitutively expressed under basal conditions. Within the 5′ flanking region of the human PHS-1 promoter there are three functional specificity protein 1 (Sp1)-binding sites at −610, −111, and −89 relative to the ATG start site. Reporter gene assays have demonstrated that the Sp1 sites at −610 and −111 are functionally important in maintaining basal constitutive expression of PHS-1.[131]

PHS-2 was discovered in 1991 as a primary response gene.[132,133] The gene for PHS-2 is approximately 8.3 kb long with 10 exons, and is transcribed as 4.6, 4.0, and 2.8 kb mRNA variants. The human PHS-2 gene is located on chromosome 1. The cDNA for COX-2 encodes a polypeptide that with the signal peptide region sequence contains 604 amino acids and shares 61% homology with the human COX-1 polypeptide.[134] The gene structures of PHS-1 and PHS-2 demonstrate conservation of exon–intron junctions.[135] Unlike PHS-1, sequence analysis of the 5′ flanking region of the human PHS-2 gene has identified several potential transcriptional regulatory elements, including a peroxisome proliferator response element (PPRE), two cyclic AMP response elements (CRE), a sterol response element (SRE), two nuclear factor-kappa B (NF-κB) sites, an Sp1 site, a CAAT enhancer-binding protein (C/EBP, or nuclear factor for interleukin-6 expression [NF-IL6]) motif, two activator protein 2 (AP-2) sites, an E-box, and a TATA box. The promoter regions of PHS-2 genes have sequences of typical immediate early genes such as c-fos and c-jun.[107,136]

A comparison of the genes for PHS-1 and PHS-2 showed that the first and last exons differ in size. There is a 42-base deletion in exon 1 of the human PHS-2 gene, which encodes a smaller signal peptide than the PHS-1 gene.[135] PHS-2 exon 1 encodes the signal peptide region and is only 14% identical in amino acid sequence to the corresponding exon of PHS-1.[137] Furthermore, exon 10 of the PHS-2 gene has a larger 3′ untranslated sequence and a 54-base insert in the protein-coding region, which encodes 18 PHS-2-specific amino acids. Exon 3 of the human PHS-2 gene contains an additional three-nucleotide insert, which codes for a proline residue absent in exon 4 of human PHS-1.[135] For PHS-2, the human and mouse genes have similar structures in genomic organization. Sequence comparison showed the first 200 base pairs (bp) of the human PHS-2 promoter share 67% and 65% identity with that of mouse and rat, respectively.[138] There are also some interspecies differences in the sequences of the human and mouse PHS-2 genes. For example, the mouse PHS-2 promoter has one NF-κB motif and two C/EBP sites instead of the two NF-κB sites and one C/EBP motif found in the human PHS-2 promoter. However, for PHS-1, human

TABLE 10.2 Prostanoid Receptor Subtypes, Expression, and Functions

Receptor	Ligand	Expression	Function
EP	PGE_2		
EP1		• Cerebral cortex, hippocampus, Purkinje cells • Kidney, lung, stomach • Gastrointestinal tract	• Acute signs of inflammation • Bone formation • Cerebral blood flow
EP2		• Least abundant, inducible • Neurons in forebrain, hypothalamus • Induced in uterus, implantation • Low in gastrointestinal tract	• Role in synaptic plasticity • Role in cerebral blood flow
EP3		• Widely distributed • Kidney, uterus • mRNA in neurons of cortex, hippocampus, midbrain • Monominergic neurons of the substantia nigra • Smooth muscle of gastrointestinal tract	• Fever generation
EP4		• Forebrain neurons • Kidney • Uterus	• Anti-apoptotic effects • Inflammatory and anti-inflammatory effects
DP1	PGD_2	• Low levels in human tissue • Low in lung, stomach, uterus • Leptomeninges of brain	• Immune response • Sleep Induction
DP2 (CRTH2)	PGD_2	• Th2 cell specific • Lymphocytes	• Allergy • Chemotaxis
FP	$PGF_{2\alpha}$	• High in corpus luteum, ovaries	• Luteolysis in pregnancy
IP	PGI_2	• Neurons in dorsal root ganglion • Megakaryocytes • Smooth muscle of arteries • Kidney	• Mediation of pain • Inflammation • Vasodilator
TP	TXA_2	• Vasculature of lung, kidney, heart • Thymus, Spleen	• Regulation of immunity • Hemodynamics • Vasoconstrictor • Bronchoconstrictor

Sources: References 108, 616–627.

and mouse genes share approximately 60% sequence identity in the 230-bp 5′ flanking region.[130]

A new member of the PHS family, cyclooxygenase-3 (COX-3), also known as COX-1b, has been identified and characterized in canine tissues.[139] Canine COX-3 mRNA is identical to the PHS-1 mRNA, except that the intron-1 is retained. In canines, COX-3 is 90 nucleotides in length and represents an in-frame insertion into the portion of the PHS-1 open reading frame encoding the N-terminal hydrophobic signal peptide.[139] Since the normal start codon resides in exon 1, and the 90-bp intron-1 sequence maintains the open reading frame, canine COX-3 mRNA creates an enzymatically active PHS-1-related peptide containing a 30-amino acid insertion near the N-terminus.[139] Recently, COX-3 mRNA has been detected in tissues from rat,[140] mouse,[141] and humans.[139,142] It does not appear that a full-length, catalytically active form of COX-3 exists in humans. Retention of intron-1, however, which is 98 bp in rat and mouse and 94 bp in human should lead to a shift in the reading frame and to the synthesis of a protein very different from PHS-1 and possibly without enzymatic activity.[142]

Transcriptional Regulation In the human PHS-1 promoter there are three functional Sp1-binding sites at −610, −111, and −89 relative to the ATG start site. Reporter gene assays have demonstrated that the Sp1 sites at −610 and −111 are functionally important in maintaining basal expression of PHS-1.[131] Deletion of either site leads to a reduction of 50% in basal transcription, and with deletion of both sites leading to a reduction of about 75%. There is an AP-1 site located in intron 8 of the PHS-1 gene that is highly conserved across species and that interacts with the −111 Sp1 site of the promoter to regulate induced expression of PHS-1 in MEG-01 cells.[143] PHS-1 gene expression is controlled and can be upregulated by tumor-promoting phorbol esters or growth factors as seen in some cell lines (Table 10.3). As discussed previously, the PHS-2 5′

TABLE 10.3 Transcriptional Activation of PHS-1

Activator	Cell Type	Reference
Tumor promoter (i.e., PMA)	Epithelial cells	628
	Megakaryoblasts	629
Cytokines (i.e., IL-1β IL-2 TNFα)	Fibroblasts	630
Growth factors (i.e., TGFβ VEGF)	Fibroblasts	631
	Vascular endothelial cells	632

IL-1β, interleukin-1β; IL-2; interleukin-2; PMA, Phorbol-12-myristate-13-acetate; TGF, tumor growth factor; TNFα, tumor necrosis factor alpha; VEGF, vascular endothelial growth factor.

TABLE 10.4 Transcriptional Activation of PHS-2

Activator	Cell Type	Reference
Tumor promoter (i.e., PMA)	Fibroblasts	633
	Endothelial cells	634
	Epithelial cells	635
	Macrophages	634
	Osteoblasts	636
Cytokines (i.e., IL-1β TNFα)	Endothelial cells	637,638
	Fibroblasts	639,640
Growth factors (i.e., EGF PDGF)	Fibroblasts	132
LPS	Macrophages	641,642
	Endothelial cells	643
Shear stress	Osteoblasts	644
	Endothelial cells	645
Hormones (i.e., LH GnRH)	Granulosa cells	646
Oncogenes (i.e., v-src)	Fibroblasts	647

EGF, epidermal growth factor; GnRH, gonadotrophin-releasing hormone; IL-1β, interleukin-1β; LH, luteinizing hormone; LPS, lipopolysaccharide; PDGF, platelet-derived growth factor; PMA, Phorbol-12-myristate-13-acetate; TNFα, tumor necrosis factor alpha; v-src, rous sarcoma virus.

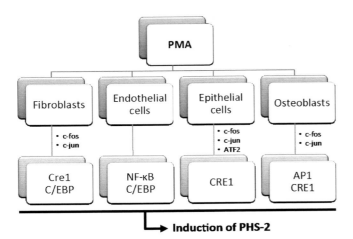

Figure 10.4 Cell-dependent PHS-2-activation. The figure illustrates the numerous functional regulatory elements in the PHS-2 gene promoters that can be involved in the transcriptional regulation of expression when exposed to PMA. Only certain of these pathways are operative in individual cell types. AP, activator protein; ATF, activating transcription factor; C/EBP, a CAAT enhancer-binding protein; CRE, cyclic AMP response elements; NF-κB, nuclear factor kappa B; PMA, phorbol-12-myristate-13-acetate.

untranslated region has many *cis*-acting regulatory elements, which suggests complex regulation of the gene by a number of signaling pathways, including the mitogen-activated protein kinase (MAPK). Depending on the cell type and the stimulus, distinct combinations of cis-regulatory elements can be utilized to activate PHS-2 transcription (Fig. 10.4). PHS-2 inducers of transcriptional activation range from growth factors and hormones to shear stress (Table 10.4).

PPARs interfere with the transcriptional activation of the PHS-2 gene. Repression by PPARα results from interference with NF-κB signaling pathways. PPARγ activated by ligands can block both AP-1 and NF-κB-mediated gene expression of PHS-2.[144] The mechanism of glucocorticoid-mediated repression of PHS-2 gene expression also involves suppression of AP-1 and NF-κB-dependent transcription, but also has posttranscriptional mechanisms of repression, possibly involving the regulation PHS-2 mRNA stability.[145]

Posttranscriptional Regulation While the open reading frame is conserved in both PHS-1 and -2, the promoter or 5′ untranslated region (discussed above) and the 3′ untranslated region are divergent. The 3′ untranslated region of the PHS-2 mRNA is approximately 1.5 kb longer than that of the PHS-1 transcript and contains 23 copies of the Adenine (A)- and Uridine (U)-rich AUUUA motif that has been associated with RNA instability and may participate in posttranscriptional regulation of COX-2 expression.[134,137] AUUUA motifs may also contribute to the different length of mRNA transcripts. PHS-1 contains only 1 AUUUA motif, contributing to stable mRNA.[146]

The N-terminal active-site region of the exon 10-encoded polypeptide is similar (57%) between the two isoenzymes; however, hCox-2 contains a unique 18-amino acid insertion in the C-terminal region which contains a potential N-linked glycosylation site.[137] COX-1 can be glycosylated at three sites, whereas COX-2 has four functional N-glycosylation sites. The last glycosylation site of COX-2 (Asn-594) is variably glycosylated.[137] A role for this addition is not well established, but it may be a marker for PGHS-2 for rapid proteolysis or provide a signal for subcellular trafficking. Some inducers of PHS transcription can act to stabilize the AU rich regions through activation of the c-Jun N-terminal kinases (JNK) and mitogen-activated protein kinase kinase kinase (MEKK1) pathways.[147,148] IL-1 and TNF-α, for example, can both activate signal pathways and stabilize mRNA of PHS-2 through interaction with AU-rich elements, thereby increasing the half-life of mRNA from 1 to 4 hours.[145,149] Conversely, glucocorticoids such as dexamethasone can destabilize mRNA, effectively acting as a PHS-2 inhibitor.[145]

10.1.3.3 Primary Protein Structures of PHSs The cDNA for PHS-2 encodes a polypeptide that before cleavage of the signal sequence contains 604 amino acids, and is 61% identical to the sequence of the human PHS-1 polypeptide.[134] Crystallographic structures of PHS-2 show striking similarity with PHS-1.[150,151] The PHS enzymes are glycosylated, integral membrane proteins with globular catalytic domains. After posttranslational processing in the endoplasmic reticulum, the mature PHS-1 and PHS-2 proteins have apparent molecular masses of 67–72 kDa and exist as homodimers, which bind 1 mole of heme per mole monomer.

PHS-1 and PHS-2 have many different domains, starting from the amino terminus with the signal peptide, the epidermal growth factor (EGF)-like region, the membrane-binding domain (MBD), and the catalytic domain with distinct peroxidase and cyclooxygenase sites.[152] Each region has important functions and may vary between the isozymes as summarized in Table 10.5. Additional structural features include dimerization domains through which PHS-1 and PHS-2 dimers are held together via hydrophobic interactions, hydrogen bonding, and salt bridges between the dimerization domains of each monomer.[150–152] Heterodimerization of PHS-1 and PHS-2 subunits does not occur.[151]

Asparagine (N)-linked polysaccharides are dispersed at several points along the polypeptide. Potential sites for N-linked glycosylation are conserved at residues 68,

Table 10.5 Comparison of Domain Regions of PHS-1 and 2[152]

Domain	Function	Isozyme Differences
Amino terminal signal peptide	• Directs polypetides to lumen of ER and nuclear envelope • 57–65% conserved between human and mouse[152]	• Length: 22–26 aa (PHS-1) 17 aa (PHS-2) • Larger hydrophobic core in PHS-1 and translocates faster to ER compared to PHS-2[133]
Epidermal growth factor (EGF) domain	• 50 aa at the N terminus • Has 3 disulfides interlocking • 1 disulfide linking Cys37 to Cys159 to attach EGF domain to the catalytic domain	• Highly conserved in both PHSs
Membrane-binding domain (MBD)	• Anchors enzyme to lipid bilayer • Forms the mouth of a hydrophobic channel that leads to the cox site	• Amino acid sequence sharing only 33% identity in this region between isozymes
Catalytic domain	• 80% of the protein contains the cyclooxygenase and peroxidase sites	• Ile-523 in PHS-1 is a valine in PHS -2. • Ile-434 in PHS -1 is a valine in PHS-2.
Cyclooxygenase	• Converts AA to PGG2 via oxygenation reactions	• Increases the volume of the PHS-2 cyclooxygenase site by 25% over that in PHS-1[151] • His 513 in PHS-1 is an Arg in PHS-2 and is required for the time-dependent inhibition of PHS-2[150] • PHS-2 competes more effectively arachidonic acid[165] • Arg120 critical residue for arachidonic acid binding in PHS-1, unessential in binding substrate in PHS-2[648]
Peroxidase	• Reduces PGG2 to PGH2 • Contains heme prosthetic group	• Low conservation of side chain structure near the peroxidase site in residues 445–457 • Structural stability of the peroxidase active site greater in PHS-1 than in PHS-2[649] • PHS-1 catalyzes a two-electron reduction of hydroperoxidase substrates almost exclusively, whereas PHS-2 catalyzes 60% two-electron and 40% one-electron reductions[160]

104, 144, and 410 in PHS-1; PHS-2 lacks the site at 104 but has two additional consensus sites in the C-terminal insert at residues 579 and 591 (murine PHS-2 numbering).[153] Glycosylation of asparagine 410 in PHS-1 is essential for cyclooxygenase and peroxidase activities, probably by promoting proper protein folding.[153] Blocking glycosylation destroys the activity of both isoforms. At the carboxy terminus of the catalytic domain of PHS-1 and PHS-2 are sequences that act as a signal for retention of proteins in the lumen of the endoplasmic reticulum and nuclear envelope. PHS-2 appears to be relatively more concentrated within the nuclear envelope.[154–156] The major isozyme differences in the primary structure are that PHS-2 has a shorter signal peptide and an 18-amino acid C-terminal insertion. Deletion of the entire insertion site has little effect on the cyclooxygenase activity of human PHS-2.[157] The catalytic domain also contains a major structural landmark called the Arg277 loop, which when cleaved destroys peroxidase activity in PHS-1 but not in PHS-2. The primary sequence in the Arg277 loop region is much less conserved between the isoforms than in the overall sequence, with only 25% identity between the human isoforms.[152]

10.1.3.4 PHS Enzymology

PHS-1 and PHS-2 contain both cyclooxygenase and hydroperoxidase activity and are involved in AA metabolism as discussed previously. The peroxidase reaction occurs at a heme-containing active site located near the protein surface, while the cyclooxygenase reaction occurs in a hydrophobic channel in the core of the enzyme. An activated cyclooxygenase component with a crucial tyrosyl radical at Tyr385 is required to initiate hydrogen abstraction from AA.[158] The Tyr385 radical is actually formed through a process dependent on the hydroperoxidase activity of PHS (Fig. 10.5).[158] The first step involves a two-electron reduction of the hydroperoxide substrate to the alcohol product, which is supported by the two-electron oxidation of the resting ferric heme (FeIII) to a oxyferryl (FeIV) protoporphyrin cation radical (Compound 1) (Fig. 10.5).[159] The process of a 1-electron reduction of the protoporphyrin cation radical leads to an oxyferryl group (FeIV) plus a neutral protoporphyrin IX (Compound 2) and a Tyr385 tyrosyl radical in the cyclooxygenase active site.[107,160–163] The first step of the cyclooxygenase reaction is then the hydrogen abstraction from C13 of AA to form a radical involving C11–C15 (Fig. 10.5). O_2 is attacked by the C11 radical, and cyclization

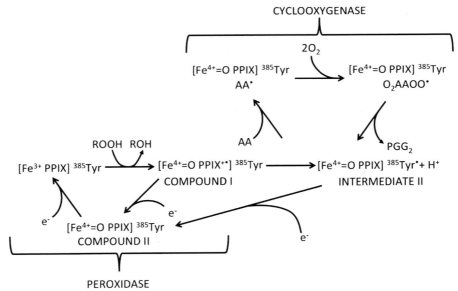

Figure 10.5 Cyclooxygenase and peroxidase catalysis by PHSs. A two-electron oxidation of the heme group by a hydroperoxide leads to compound I with iron as Fe^{4+} and protoporphyrin as a cation radical. This radical oxidizes ^{385}Tyr of cyclooxygenase generating a protein tyrosyl radical. This leads to hydrogen abstraction from C13 of AA to form a radical involving C11–C15. Through cyclization reactions and $2O_2$ additions, PGG2 is formed and as in step 1 of the peroxidase reaction, hydroperoxidase site reduces PGG2 to the corresponding alcohol, PGH2. Fe^{3+} PPIX, ferric iron protoporphyrin IX (heme); ROOH, alkyl hydroperoxide; ROH, alcohol; AA, arachidonic acid; $Fe^{4+}=O$ PPIX, oxyferryl heme; Compound I, an oxyferryl group ($Fe(IV)=O$) plus a protoporphyrin IX radical cation; intermediate II, an oxyferryl group plus a neutral protoporphyrin IX plus a Tyr385 tyrosyl radical; compound II, an oxyferryl group plus a neutral protoporphyrin IX; intermediate III, a spectral intermediate of a heme group with a protein radical located on an amino acid side chain other than Tyr385. (Adapted from *Prostaglandins Other Lipid Mediat.* **2002**, *68–69*, 115–128. Copyright 2002 Elsevier.)

forms a 9,11-dioxo bridge and leaves a carbon-centered radical at C8. Cyclization then occurs between C8 and C12 to generate an allyl radical at C15, in which a second O_2 can give rise to a hydroperoxy radical at C15. The PGG2 radical is reduced to form PGG2.[107,160–164] The hydroperoxidase site reduces PGG2 to the corresponding alcohol, PGH2, and the process can regenerate tyrosyl radicals. However, these radicals can also cross-link and lead to inactivation of PHSs. It is during the process of reduction that one- or two-electron oxidation of endogenous and exogenous compounds can occur (discussed later).

The hydroperoxidase activity can function independently of ongoing cyclooxygenase catalysis. In contrast, the cyclooxygenase reaction is peroxide-dependent and requires that the heme group at the hydroperoxidase site undergo a two-electron oxidation to form the tyrosyl radical.[160] *In vitro* studies have shown that the catalytic sites of the cyclooxygenase and hydroperoxidase are active for less than 2 minutes, which may be due to the various radicals formed and their involvement in enzyme inactivation through internal protein cross-linking.[161,165]

In terms of kinetic properties, PHS-1 and PHS-2 have similar K_m values for arachidonate (5 μM) and O_2 (5 μM).[107,166,167] Cyclooxygenase turnover rates (3500 mol/min of arachidonate per mole of dimer) are similar as well.[166] There are, however, some isozyme-dependent substrate specificities. AA with its 20-carbon chain and four *cis* double bonds (i.e., 20:4) is a fatty acid substrate for oxygenation, and 20:3 is 30–50% as effective as 20:4.[167] For COX-1, 18:2 and α-18:3 are poor substrates, but they are better substrates for COX-2.[167] Both substrates are converted to monohydroperoxide products. Furthermore, the concentration of peroxide needed to activate and sustain cyclooxygenase activity was approximately 2 nM for COX-2 and 20 nM for COX-1; therefore, COX-2 may be catalytically active at much lower concentrations of hydroperoxide.[168] The catalytic activities of PHS-1 and PHS-2 isozymes also respond differently to acetylsalicylic acid (ASA, aspirin) treatment, where the cyclooxygenase activity of PHS-1 was completely inhibited, whereas ASA-treated COX-2 converted AA to 15-HPETE reduced to 15-HETE.[169] ASA treatment leads to acetylation of the serine residue in the cyclooxygenase active site that block PHS-1 oxygenation, but the larger active site in PHS-2 allows AA to bind after ASA treatment. The presence of the acetyl group alters the conformation of the AA so that the product of oxygenation is 15-HPETE rather than PGG2.[169,170]

10.1.3.5 Inhibition of PHSs The main PHS inhibitors are the nonsteroidal anti-inflammatory drugs (NSAIDs) (Fig. 10.6). In 1971, John Vane used a cell-free homogenate of guinea pig lung to demonstrate that aspirin, indomethacin, and salicylate, all popular NSAIDs, were inhibitors of PHS, which constituted the mechanism of action of these drugs.[171] These classical PHS inhibitors are not selective and inhibit both PHS-1

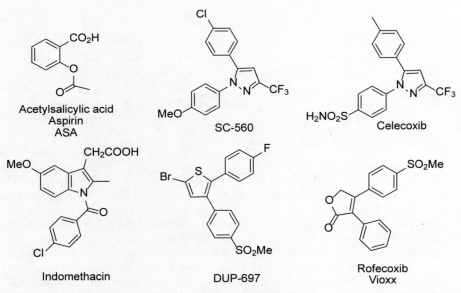

Figure 10.6 Examples of PHS inhibitors. Aspirin is a covalent modifier of PHS-1 and PHS-2. Indomethacin is a time-dependent PHS inhibitor. Selective PHS-2 inhibitors are celecoxib, rofecoxib (Vioxx), and DUP-697 while SC-560 is a selective PHS-1 inhibitor.

and PHS-2. They also target the cyclooxygenase component of PHS. NSAIDs are widely prescribed as analgesics and anti-inflammatory agents.

There are different classes of PHS inhibitors.[172] The differences among them are based on their selectivity for the different isozymes, PHS-1 and PHS-2.[173] These classes are (1) ASA: shown to trigger a covalent acetylation in the enzyme, irreversibly blocking its activity; (2) competitively acting NSAIDs such as indomethacin, naproxen, and ibuprofen; and (3) PHS-specific inhibitors (coxibs).[172–174]

ASA is a covalent modifier of PHS-1 and PHS-2 as it acetylates serine 530 of PHS-1.[175,176] As mentioned previously, because the catalytic pocket of the channel in cyclooxygenase is larger in PHS-2 than in PHS-1, access of ASA to the Ser530 of PHS-2 is reduced due to a lack of stabilization in the binding pocket, and acetylation efficiency in PHS-2 is limited. This accounts for the lowered sensitivity of PHS-2 compared with PHS-1 to inhibition by ASA.[175,176] Therapeutically, this 10- to 100-fold greater selective inhibition of PHS-1 over PHS-2 by low-dose aspirin is employed in the prophylactic treatment of thromboembolic disease and myocardial infarction as ASA can inhibit PHS-1 in platelets and the vascular endothelium at low doses.[177] Another PHS-1-specific inhibitor is SC-560.

Other NSAIDs inhibit PHS-1 and PHS-2 by competing with AA for binding in the COX active site. However, NSAIDs significantly differ from each other in whether they bind the COX active site in a time-dependent or independent manner. Some NSAIDs like ibuprofen have very rapid reversible binding and therefore are not time dependent.[178] Conversely, NSAIDs such as indomethacin and diclofenac are time dependent, in that they require typically seconds to minutes to bind the COX active site. Once bound, however, these drugs typically bind with high affinity and may require hours to be washed out of the active site.[178] Carboxyl-containing NSAIDs form a salt bridge between the carboxylate of the NSAID and the Arg120 moiety, which provides a positive charge that binds the negative charges of carboxylic acid substrates.[175,179] These inhibitors block entry of AA to the COX active site.

Selective inhibitors of PHS-2 were introduced in 1999. The first NSAIDs to be introduced as selective PHS-2 inhibitors were celecoxib (Celebrex) and rofecoxib (Vioxx). In place of the carboxyl group of the NSAIDs, the structure of celecoxib contains a sulfonamide group and that of rofecoxib contains a methylsulfone, as does DUP-697.[150,180] Each of these compounds is a weak time-independent inhibitor of PHS-1 but a potent time-dependent inhibitor of PHS-2. Like time-dependent carboxyl-containing NSAIDs, time dependence for celecoxib and rofecoxib requires these compounds to enter and be stabilized in the catalytic pocket of the COX component of PHS.[150,180] However, because these drugs lack a carboxyl group, stabilization of binding for both of these drugs does not require Arg120. Instead, the combination of hydrophobic and hydrogen bonding interactions stabilizes binding. Specifically, the selective, time-dependent inhibition of PHS-2 is due to the insertion of the methylsulfonyl or sulfonamide group of the inhibitor past Val-523 in PHS-2 and into the side pocket. This is precluded in PHS-1 by the extra steric bulk of Ile-523.[180,181] The sulfur-containing phenyl rings of these drugs bind into the side pocket of the cyclooxygenase catalytic channel of PHS-2 but interact weakly with the active site of PHS-1.[150]

Previously it was believed that the inhibition of PHS-2 mediated the therapeutic actions of NSAIDs, while the inhibition of PHS-1 caused unwanted side effects, particularly in the gastrointestinal tract. PHS-1 is the major PHS isoform expressed in platelets and gastric mucosa. NSAID toxicity in the gastrointestinal mucosa, leading to ulceration and bleeding, is the result of inhibition of PHS-1 activity in platelets and a reduction in prostanoids important for protecting the stomach from erosion and ulceration.[182] Coxibs selective for PHS-2 were developed to reduce the incidence of serious upper gastrointestinal toxicity associated with the administration of nonselective NSAIDs, and hence inhibition of PHS-1-derived prostanoids. However, the reduced incidence of serious gastrointestinal adverse effects compared to nonselective NSAIDs has been countered by an increased incidence (i.e., 1% in placebo group vs. 3.4% in celecoxib group) of myocardial infarction and stroke detected in placebo-controlled trials involving celecoxib and rofecoxib.[183,184] Therefore, chronic administration of the selective PHS-2 inhibitors for prophylactic purposes may also carry certain risks.

10.1.3.6 Cellular Localization and CNS Expression of PHSs PHS-1 is a constitutive isoform that is widely distributed in various cell types and is thought to mediate physiological responses. PHS-2 is rapidly induced in several cell types in response to various stimuli, such as neuronal activity, cytokines, and proinflammatory molecules.[185–187] Both PHS-1 and PHS-2 are expressed under physiological conditions in some organs, such as brain, kidney, heart, liver, spleen, and small intestine.[187,188] PHS-1 can also be induced during T-cell development.[189] Several lines of evidence suggest that PHS-1 also has a role in inflammation and, like PHS-2, can be upregulated in certain conditions (see later discussion).

In the human brain, PHS-1 mRNA has been found in regions including the hippocampus, midfrontal cortex, amygdala, substantia nigra, thalamus, occipital cortex, motor cortex, caudate, and the cerebellum.[188] PHS-1 protein is constitutively expressed in both glia and neurons. For example, PHS-1 immunoreactivity was present in microglial cells in gray and white matter in the hippocampus and cortex.[190,191] In rat and ovine brain, PHS-1 immunoreactivity is enriched in midbrain, pons, and medulla.[192]

Mouse peripheral dorsal root ganglion neurons also appear to constitutively express PHS-1 exclusively, and lack detectable PHS-2 expression, either under basal conditions or during peripheral inflammatory states.[193] Recent studies also have indicated a proinflammatory role of PHS-1 in the pathophysiology of acute and chronic LPS-induced neurotoxicity and brain injury,[194–197] and have found increased PHS-1 immunopositive microglia in association with amyloid plaques in AD.[190,191,198]

Several studies have shown the presence of PHS-2 mRNA and protein in different brain regions such as cerebral cortex, substantia nigra, caudate, thalamus, hippocampus, and amygdala.[187,188,199] PHS-2 immunoreactivity is localized to the perinuclear regions and seems to be primarily neuronal.[187,188,190,200,201] It may not be detected in glia under physiologic conditions, except in radial glia of the spinal cord.[202] However, astrocytes and microglia can express PHS-2 after exposure to proinflammatory mediators *in vitro* or following CNS injury *in vivo*.[201,203,204] In hippocampal and cortical glutamatergic neurons, PHS-2 has a central role in synaptic activity and long-term synaptic plasticity.[192,199,205] Within neurons, PHS-2 immunoreactivity has been localized to postsynaptic sites and dendritic and axonal domains of neurons.[199] The dendritic spine is a neuronal structure that can modulate large fluctuations in calcium and is believed to function in altering the efficiency of transmission at excitatory synapses. The level of neuronal PHS-2 expression within the CNS appears to be coupled to excitatory neuronal activity as PHS-2 protein expression in the brain and spinal cord is upregulated by seizure activity and peripheral inflammation.[187,206,207]

10.1.3.7 PHS in ROS Generation, Aging, and Neurotoxicity

As discussed previously, co-oxidation of endogenous and exogenous substrates can occur during the catalytic process of PG biosynthesis by PHSs. When the hydroperoxidase site reduces PGG2 to the corresponding alcohol, PGH2, the process can generate tyrosyl radicals and 1- or 2-electron oxidation of endogenous and exogenous compounds (Fig. 10.7A).[208] Peroxyl radical-mediated bioactivation can also result when there is a direct transfer of the hydroperoxide oxygen to the co-substrate. This occurs during the bioactivation of compounds such as aflatoxin B1 and benzo[*a*]pyrene-7,8-dihydrodiol, the latter of which is bioactivated to the reactive 9,10-epoxide intermediate that is the proximate carcinogen (Fig. 10.7B).[208–210] Compounds can also be oxidized by PHS to C, N, or S free radicals that trap O_2 forming peroxyl radical; for example, retinoic acid is oxidized to carbon-centered radicals that react with O_2 to form peroxyl free radicals (Fig. 10.7C). Peroxyl radicals are stable oxy radicals and are able to diffuse some distance from the site of their generation to form ROS and lead to toxicity.[208,209] A phenylbutazone carbon-centered radical can be formed by PHS hydroperoxidase,[211] and heterocyclic amines can be bioactivated by PHS to oxide intermediates that covalently bind to DNA, which can initiate cancer.[212]

As shown in Figure 10.8, one-electron oxidations of endogenous or exogenous substrates can lead to formation of free radicals that generate ROS and oxidize macromolecules such as protein, lipid, RNA, or DNA.[213] A variety of reducing compounds can serve as peroxi-

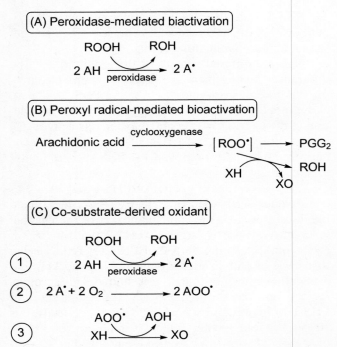

Figure 10.7 Mechanisms of PHS bioactivation of substrates. (A) In peroxidase-mediated bioactivation, as ROOH (hydroperoxide) is reduced to ROH, cosubstrate (AH) can be oxidized to a free radical (A·). (B) In peroxyl radical-mediated bioactivation, the peroxyl radical generated from cyclooxygenase (ROO·) can transfer oxygen to the substrate (X). (C) A cosubstrate-derived oxidant can be formed as the substrate free radical can trap then transfer oxygen to other compounds. Adapted from reference 745.

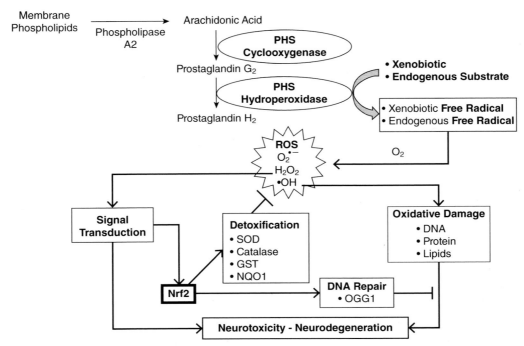

Figure 10.8 Postulated bioactivation of endogenous substrates to a free radical intermediate by PHS which generates ROS to activate Nrf2. Cyclooxygenase and hydroperoxidase are the components of PHS. Arachidonic acid released from membrane phospholipids by phospholipase A2 serves as the co-substrate in the cyclooxygenase-dependent pathway, generating a hydroperoxide (PGG2), which can then be reduced by hydroperoxidase to an alcohol. In the peroxidase pathway, xenobiotics or endogenous compounds may serve as the reducing co-substrate, themselves being oxidized to a reactive free radical intermediate that can initiate the formation of reactive oxygen species (ROS) including superoxide anions ($O_2^{\bullet-}$), hydrogen peroxide (H_2O_2) or hydroxyl radicals (HO•). If ROS are not detoxified by enzymes such as superoxide dismutase (SOD) or catalase, they can oxidatively damage cellular macromolecules and/or alter signal transduction, thereby causing irreversible damage and neurotoxicity. Nrf2, nuclear factor erythroid 2-related factor 2; PGG2, prostaglandin G2, PHS, prostaglandin H synthase. Modified from References 222 and 615.

dase co-substrates for PHS-1 and PHS-2, promoting conversion of peroxide activators to the corresponding alcohols.[214] Addition of aromatic amines, phenols or hydroquinones, epinephrine, melatonin, and serotonin (SE) facilitate the conversion of PGG2 to PGH2.[214,215] These can lead to free radical formation only when peroxidase reductants react via a one-electron transfer. Phenols, catechols, and amines are good substrates for PHS, and many neurotransmitters, their precursors, and metabolites contain these functional groups. The neurotransmitter DA can be converted to a reactive quinone that can covalently bind to DNA and protein sulfhydryl groups.[216,217] DA quinones can also undergo one-electron reductions catalyzed by NADPH cytochrome P450 reductase (Fig. 10.9).[218] This reaction can create a redox cycling process with oxygen, leading to the formation of ROS.[218] NAD(P)H quinone oxidoreductase 1 (NQO1) can catalyze two-electron reductions hence preventing semiquinone radicals. Aside from binding covalently to protein and DNA, these compounds are able to generate ROS that react with DNA to form over 20 types of macromolecular lesions,[219] including the oxidation of 2′-deoxyguanosine in DNA by hydroxyl radicals to form 8-oxo-2′-deoxyguanosine (8-oxodG).[213] To determine the potential contribution of PHS-dependent ROS formation, the neurotransmitter DA or its precursor and metabolites were incubated *in vitro* with purified ovine PHS-1 and calf thymus DNA. DA, its L-dihydroxyphenylalanine (L-DOPA), precursor, and its dihydroxyphenylacetic acid (DOPAC) metabolite (Figure 10.10) were excellent PHS-1 substrates, resulting in PHS-1-dependent ROS formation that initiated oxidative DNA damage, selectively quantified as 8-oxodG. Most substrates generated isotropic electron spin resonance (ESR) spectra with a resolved hyperfine structure attributable to ortho-semiquinone free radical intermediates upon autoxidation at pH 6, with up to an 18-fold increase via horseradish peroxidase (HRP)-catalyzed oxidation. Remarkably, HRP-mediated oxidation of DOPAC and dihydroxymandelic acid (DHMA)

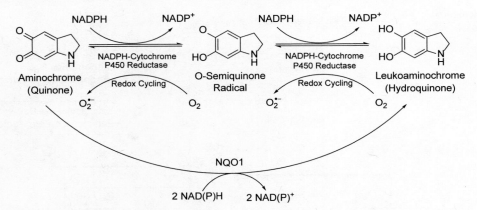

Figure 10.9 Oxidation of aminochrome by NADPH-cytochrome P450 reductase and NQO1. NADPH-cytochrome P450 reductase catalyzes one-electron reduction of aminochrome to o-semiquinone which is very reactive. The continuous NADPH oxidation and oxygen consumption leads to autoxidation. Aminochrome o-semiquinone autoxidize by reducing oxygen to superoxide radicals giving rise a redox cycling. NQO1 catalyzes two-electron reduction of aminochrome to o-hydroquinone.

produced asymmetric ESR spectra characteristic of an immobilized radical, possibly due to free radical intermediates and melanin or melanin-like polymers. These results show that the precursors and metabolites of endogenous neurotransmitters, while inactive in receptor-binding assays, may actually play an important role in free radical formation. Additionally, ROS generated by PHS-catalyzed bioactivation produce oxidative DNA damage in the CNS, which may initiate neurodegeneration associated with aging.[220] The biological consequences of PHS-catalyzed neurotransmitter bioactivation were evaluated in cells expressing human PHS-1 (hPHS-1) or hPHS-2. hPHS-1 and hPHS-2 cells incubated with DA, L-DOPA, DOPAC, or homovanillic acid (HVA) exhibited increased cytotoxicity compared to untransfected cells, and cytotoxicity was increased further by exogenous AA, which increased hPHS activity. Preincubation with catalase, which detoxifies ROS, or acetylsalicylic acid, an inhibitor of hPHS-1 and hPHS-2, reduced the cytotoxicity caused by DA, L-DOPA, DOPAC, and HVA in hPHS-1 and hPHS-2 cells both with and without AA. Protein oxidation was increased in hPHS-1 and hPHS-2 cells exposed to DA or L-DOPA, and further increased by AA addition. DNA oxidation was enhanced earlier and at lower substrate concentrations than protein oxidation in both hPHS-1 and hPHS-2 cells by DA, L-DOPA, DOPAC, and HVA, and further enhanced by AA addition. hPHS-2 cells appeared more susceptible than hPHS-1 cells, while untransfected CHO-K1 cells were less susceptible. Thus, isozyme-specific, hPHS-dependent oxidative damage and cytotoxicity caused by neurotransmitters, their precursors, and metabolites may contribute to neurodegeneration associated with aging.[221]

The amphetamines methamphetamine (METH, Speed), 3,4-methylenedioxymethamphetamine (MDMA; Ecstasy), and its major metabolite 3,4-methylenedioxyamphetamine (MDA) can be bioactivated by mouse brain PHS-1 to free radical intermediates that generate ROS and oxidatively damage brain DNA leading to neurodegeneration.[3,4] Also, PHS-catalyzed bioactivation, ROS formation, and embryonic DNA oxidation have been implicated in the teratogenicity of numerous xenobiotics including phenytoin, thalidomide, benzo[a]pyrene, and METH.[222–227]

There are several methodological issues of importance in determining the *in vivo* relevance of PHS-catalyzed bioactivation. The first is that bioactivation by the COX-2/COX-1 ratio for a particular xenobiotic or endobiotic will vary according to whether it is measured in intact cells, cellular homogenates, purified enzymes, or recombinant proteins expressed in bacterial, insect, or animal cells.[152,212,228,229] Secondly, the adverse consequences of PHS-catalyzed bioactivation also varies when measured in different types of cells derived from various species depending on their potential for antioxidative processes and repair.[192,196,223,229,230a] To investigate the role of PHS-mediated bioactivation of amphetamines, cells expressing hPHS-1 or hPHS-2 were used as described above. Both METH and MDA (250 uM–1000 uM) caused concentration-independent cytotoxicity in hPHS-1 cells, suggesting maximal bioactivation at the lowest concentration. In hPHS-2 cells, with half the activity of hPHS-1 cells, METH (250 uM–1000 uM) cytotoxicity was less than that for hPHS-1 cells, but was increased by exogenous AA, which increased hPHS activity. While 10 uM MDA and METH were not cytotoxic, at 100 uM both analogs caused AA-dependent and drug concentration-dependent increases in cytotoxicity and DNA oxidation in both hPHS-1/2 cells. The hPHS-2 isozyme appeared to provide more efficacious bioactivation of these amphetamine analogs at lower concentrations. Acetylsalicylic acid, an irreversible inhibitor of both hPHS-1 and hPHS-2, blocked cytotoxicity and DNA oxidation in both cell lines, and untransfected CHO-K1 cells lacking PHS activity were similarly resistant. Accordingly, isozyme-dependent hPHS-catalyzed bioactivation of METH and MDA can cause oxidative macromolecular damage and cytotoxicity, which may contribute to their neurotoxicity.[230b]

10.1.3.8 PHS in Neurodegenerative Diseases A number of studies have investigated the expression of PHSs in neurodegenerative diseases, especially those involving extensive neuroinflammatory effects that are thought to be mediated by PHS-2 (Table 10.6). ROS generation and downstream effectors (i.e., PG and/or TX synthases and their respective receptors) that are responsible for the deleterious and/or protective effects of PHS activation in neurodegenerative diseases remain an area of active research. However, research on the expression of PHS in neurodegenerative diseases have been conflicting, for example in some cases showing PHS induction or decreases in lesions of AD.[190,198,231] Of particular interest are the use of NSAIDs and specific PHS-2 inhibitors and their potential neuroprotective effects. The use of NSAIDs to slow pathology of neurodegenerative diseases or aid in prevention of cognitive decline has been inconsistent. COX-2 inhibitor rofecoxib failed to slow cognitive decline in patients with mild-to-moderate AD.[232] In contrast, others showed that cyclooxygenase-2 inhibition improves β-amyloid-associated suppression of memory and synaptic plasticity.[233] The observed inconsistencies may be due to any or all of the following: (1) Treatment of neurodegenerative diseases may require long-term prophylactic dosing, and PHS-2 inhibitors are associated with increased cardiovascular toxicities as discussed previously. (2) The protective effects of NSAIDs may be via non-PHS-inhibitory mechanisms, such as activation of PPARs or through second messenger systems, suggesting that selective inhibition of COX-2 may not be the optimal therapeutic strategy.[234] (3) Furthermore, treatment may need to change as the disease progresses as the stage of the disease may modulate protein expression. (4) The observed variability in PHS expression in the studies may be related to disease stage as well.

PHS has been shown to be involved with many neurodegenerative diseases as investigated in both patients and using animal models (Table 10.6). AD is the most common cause of dementia with neurodegeneration in the elderly. It is clinically characterized by a progressive memory loss and other cognitive impairments. The characteristics of AD include deposits of amyloid fibrils in senile plaques, presence of abnormal tau protein filaments in neurofibrillary tangles, and extensive neuronal degeneration and loss in regions such as the frontal cortex and hippocampus. AD brains also exhibit several additional pathological abnormalities, including reactive gliosis, microglial activation, and chronic inflammatory processes.[188,235] Recently β-amyloid-induced neuronal apoptosis has been associated with COX-2 upregulation directly through the activation of NF-κB.[234] Generally, chronic therapy with the PHS-2 inhibitors rofecoxib or naproxen do not aid in decreasing cognitive decline; however, nonselective PHS inhibitors may prove helpful.[232,236]

MS is a demyelinating disease of the brain characterized by perivascular infiltration of lymphocytes and macrophages into the brain. Glutamate-mediated excitotoxic death of oligodendrocytes has also been reported to contribute to the pathogenesis of demyelinating diseases.[237,238] Since inflammation is associated with demyelination, oligodendrocyte death, axonal damage and, ultimately, neuronal loss, numerous studies have investigated a potential role for PHS (Table 10.6).

PD is a neurodegenerative disorder that results in the loss of dopaminergic transmission in the substantia nigra and striatum, which leads to rigidity, resting tremors, and slowness of movement. Idiopathic PD accounts for the majority (>90%) of the cases. The remaining cases are mostly familial PD forms that are correlated with mutations of genes such as synuclein and parkin.[239] Idiopathic PD, unlike the familial form occurring earlier, usually begins in the fifth decade of life and progresses over long periods of time (10–20 years). Biochemical analyses have implicated mitochondrial dysfunction as a mechanism in idiopathic cases of PD.[240] Since PD progression has an inflammatory pathology, a potential role for PHS has been investigated (Table 10.6).

ALS is characterized by the progressive loss of motor neurons, typically resulting in death within 5 years of onset. It has been suggested that inflammatory-related processes may promote motor neuron death. Although the sporadic form of ALS is the most frequent, 5–10% of cases are familial, being associated with several genes. Missense mutations in the gene encoding for the Cu,Zn superoxide dismutase (SOD1) account for a familial form of ALS linked to chromosome 21q and present in 20% of the inherited cases.[241] Mutant SOD1 produces motor neuron injury by a toxic gain of function and several hypotheses exist, including aberrant free radical handling, abnormal protein aggregation, and increased susceptibility to excitotoxicity, although the exact mechanism of action is unclear.[242] Transgenic mice expressing the human mutant SOD1 with a phenotype that mimics

TABLE 10.6 Summary of PHS-Mediated Effects in Neurodegenerative Diseases

Disease/Model	PHS-Mediated Evidence	References
Multiple Sclerosis (MS)		
• MS patients	• PHS-2-positive cells were present in all chronic active lesions • Associated with cells expressing the macrophage/ microglial marker CD64, associated with activated macrophages • Possible oligodendroglial excitotoxic death	650, 651
• Mouse experimental autoimmune encephalomyelitis (EAE)	• PHS-2 expression is confined within infiltrating macrophages • PHS-2 induction in astrocytes during relapse phase • PHS inhibitor indomethacin suppressed active EAE	652 653
• Rat model of delayed-type hypersensitivity leading to demyelination	• PHS-2 expression was restricted to major infiltrating neutrophils and phagocytes • Macrophages and/or endothelial cells near lesion • Neuronal PHS-2 not affected • No obvious PHS-2 staining in astrocytes and microglia	654
Alzheimer's disease (AD) • Patients with AD	• PHS-2 mRNA levels reported as either decreased or increased • Increased PHS-2 in neurons • PHS-2-positive neurons decreased with the severity of dementia • Early AD, an increase in PHS-2 • PHS-1 expressed by microglial cells in association with amyloid deposits • Increased PHS-1 • PHS-2 inhibitors (rofecoxib or naproxen) failed to slow cognitive decline • NSAIDs decrease the severity of cognitive symptoms • Indomethacin appeared to protect the degree of cognitive decline	190,191,655,656 657,658 191,659 232,236 660,661
• Mouse models (i.e., overexpressing amyloid precursor protein [APP])	• PHS-2 inhibitors may protect against AD by blocking the PHS-2-mediated PGE2 response at synapses • Overexpressing human PHS-2 show an increase in amyloid plaques • Amyloid plaques are surrounded by a few PHS-2 immunoreactive astrocytes • A subset of NSAIDs, such as ibuprofen, have been shown to reduce serum levels of amyloid, a primary component of senile plaques in AD • Celebrex (PHS-2 inhibitor) did not decrease amyloid load in APP transgenic mice • Indomethacin, but not nimesulide, showed a significant reduction in the amyloid burden	233 662 663 664 665 666
Parkinson's disease (PD) • PD patients	• Increased expression of PHS-2 in activated microglial cells in the substantia nigra • Unchanged neuronal and astroglial PHS-2 expression • Moderate PHS-1 immunoreactivity in neurons and glia	667
• PD patients Postmortem analysis • PD mouse models (MPTP) rodent model	• PHS-2 is specifically induced in substantia nigra dopaminergic neurons • PGE2 levels were increased in both human and mouse tissues • PHS-2 KO mice exhibited resistance to dopaminergic neuron degeneration due to MPTP • Increased levels of PHS-2 generates toxic dopamine-quinone species leading to dopaminergic neuronal degeneration in substantia nigra neurons • Selective PHS-2 inhibitor rofecoxib and paracoxib neuroprotective effect on tyrosine hydroxylase expression, motor and cognitive functions in MPTP-rat model	668 668,669 670 669 671
Amyotrophic lateral sclerosis (ALS) • ALS patients postmortem analysis	• PHS-2 mRNA and protein were increased in spinal cords, localized to both neurons and glial cells	672
• ALS patients	• Increased levels of PGE2 in the CSF • Celecoxib for 12 months to ALS subjects did not slow the decline in muscle strength or affect survival	673,674 675
• ALS mouse model (i.e., transgenic expressing human mutated SOD1)	• PHS-2 mRNA and protein were increased in spinal cords, localized to both neurons and glial cells • Celecoxib delayed the onset of disease, reduced spinal neurodegeneration and glial activation • PHS-1 KO does not improve preservation of motor neurons and survival of transgenic mutant mice • Selective PHS-2 inhibitors (e.g., celecoxib and rofecoxib) improve motor performance, extend survival, and reduce CSF levels of PGE2 • Nimesulide decreased spinal cord PGE2 levels and delayed the onset of ALS type motor impairment in mice • Nimesulide did not affect the onset of end-stage disease	676 677 678 679 680

clinical and pathological characteristics of the human disease have been developed and have been used to study PHS-mediated effects (Table 10.6).

10.1.4 Amphetamines

10.1.4.1 History and Uses
The amphetamine analogs MDMA (Ecstasy), MDA (active major metabolite of MDMA), and METH (Speed) are synthetic substances that are common drugs of abuse, as they promote the release of neurotransmitters and induce euphoria and hallucinations. They are illicit drugs classified under Controlled Drug Acts in both the United States and Canada. Amphetamine derivatives intended for recreational use have been referred to as "designer drugs" because they are designed to circumvent existing legal restrictions.[243]

These amphetamine analogs have a chiral center at the alpha carbon, and thus exist as a pair of optical isomers; for example, d-METH (corresponding to the configuration of S-(+)) and l-METH (corresponding to the configuration of R-(−)). Their pharmacological profiles are stereoselectively distinct. The d-enantiomer is the dominant CNS stimulant and is five times more biologically active than the l-enantiomer, which has greater sympathomimetic activity.[244] The l-enantiomer is also formed as a metabolite of selegiline, an anti-Parkinsonian drug.[245] These drugs belong to a class of sympathomimetic drugs called phenylethylamines and are structurally similar to many endogenous neurotransmitters such as DA, SE, and epinephrine as well as their metabolites (Fig. 10.10).

Amphetamines were drugs originally developed as synthetics used as substitutes for ephedrine. Japanese scientists synthesized METH in 1919. In 1932, the Smith, Kline and French pharmaceutical company introduced these drugs in over-the-counter inhalers for asthma and congestion. In the 1930s, the American Medical Association approved the use of amphetamines under names like benzedrine (d/l-amphetamine) for treatment of a range of disorders such as narcolepsy, depression, PD, attention deficit disorder, and even as an appetite

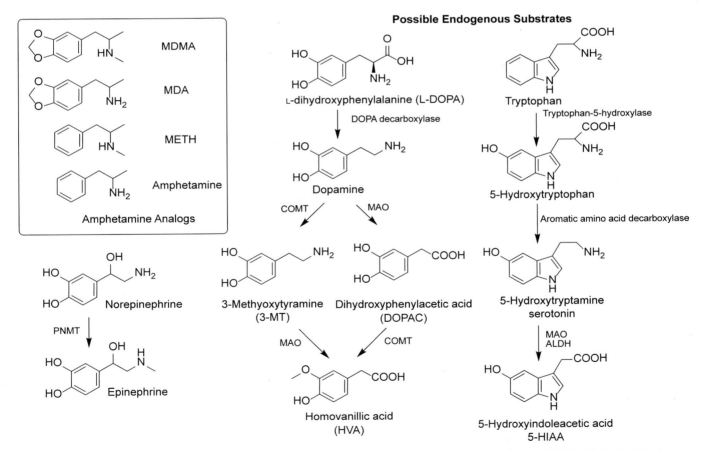

Figure 10.10 Amphetamine, its analogs and neurotransmitters, their precursors and metabolites. ALDH, aldehyde dehydrogenase; COMT, catechol-O-methyl-transferase; MAO, monoamine oxidase; PNMT, phenylethanolamine-N-methyltransferase.

suppressant. The therapeutic dose for these drugs in tablet form is typically 5–10 mg.[243,246,247] The first reported misuse of amphetamine was in 1937 when it was used by students in Minnesota to avoid sleep during examination periods. Thereafter, both amphetamine and METH were widely used both clinically and illicitly during the Second World War by the Americans, Germans, and Japanese, and became a serious problem in postwar Japan. Increasing popularity of METH as a drug of abuse within the United States led to its illicit production in the 1960s, and by the 1970s, laws were passed to make METH illegal to possess without a prescription.[243]

First synthesized by Merck in 1912, MDMA was patented in 1914 but never marketed.[248] While possibly one of the drugs used as a stimulant during the World Wars, it gained recognition in 1965 when Alexander Shulgin manufactured MDMA in his laboratory, but it was not until the 1970s–1980s that MDMA was first used recreationally and in psychotherapy and was said to increase patient self-esteem and facilitate therapeutic communication.[248,249] In 1985, the U.S. Drug Enforcement Administration classified MDMA as a Schedule 1 drug due to its high abuse potential and evidence that MDA, a related compound and major MDMA metabolite, induced serotonergic nerve terminal degeneration in rat brain.[250] However, since the mid 1980s, MDMA has been a popular recreational drug at "raves," causing a state of euphoria, allowing the user to socialize and dance all night.[251] The absence of commercially produced METH and MDMA/MDA led to the clandestine production of these drugs. Furthermore, the ease of obtaining precursors for METH synthesis, such as ephedrine and pseudoephedrine found in cough medication, resulted in the production of the higher quality d-METH and by the early 1980s, METH became more easily synthesized and readily available for abuse. Despite efforts to limit production of the precursors, METH and MDMA use and the associated behavioral problems and addiction have always been and remain a concern, especially among young people.[243,247,248,251]

Amphetamine marketed as Adderall and Dexedrine is prescribed for the treatment of narcolepsy and attention deficit-hyperactivity disorder (ADHD), while METH is indicated for the treatment of ADHD and the short-term treatment of obesity. METH prescriptions are very rare.[252] The recommended starting dose for METH treatment of ADHD is 5 mg/day in individuals who are at least 6 years old. The maximum recommended dose for ADHD is 25 mg/day. For obesity, the recommended METH dose is 5 mg before a meal.[252] During illicit use, METH can be taken through a variety of routes, including ingestion, injection, nasal insufflation, and inhalation (smoking).[253–255]

10.1.4.2 Pharmacokinetics Following ingestion, METH is absorbed across the gastrointestinal tract. Controlled studies with therapeutic formulations (5–10 mg) have indicated t_{max} values ranging from 3 to 6 hours postingestion.[253–256] In chronic METH abusers, the average plasma concentration ranges from 150 µg/L (ng/mL) to 1700 µg/L (ng/mL) (1 µM–10 µM), while the concentration of its amphetamine metabolite ranges from 30 µg/L to 300 µg/L.[257] Following intranasal administration of METH, peak plasma concentrations do not occur until approximately 3–4 hours postexposure.[258]

Inhalation of METH via smoking provides a bioavailability that ranges from 67% to 90%, with the differences depending on smoking technique and the temperature of the flame.[258] Following oral ingestion, 67% may be absorbed.[258] METH is very lipophilic and distributes extensively across the blood–brain barrier (BBB).[259] METH is also distributed into breast milk, appearing in the milk within minutes of intravenous use.[260] Because of its low molecular weight and high lipid solubility, there is also significant transfer of METH from maternal to fetal blood.[261] METH and amphetamine show no differences in their effect on DA release in the striatum, elimination rates, or other pharmacokinetic properties.[262]

The related amphetamine analogs MDMA and MDA are readily absorbed from the intestinal tract and reach their peak concentration in the plasma about 2 hours after oral administration.[263] Doses of 50 mg, 75 mg, and 125 mg to healthy human volunteers produced peak blood concentrations of 106 ng/mL, 131 ng/mL, and 236 ng/mL of MDMA, respectively. These concentrations are relatively low, in part because the drug passes readily into the tissues, and much of it is bound to tissue constituents.[263] Most of the cases of serious toxicity or fatality with MDMA have involved blood levels ranging from 0.5 mg/L to 10 mg/L (500–10,000 ng/mL); that is, up to 40 times higher than the usual recreational range.[247]

The mean value for the half-life of MDMA in humans ranges between 9 and 12 hours,[247,253,254,256,264] and is not measurably altered by the route of drug administration.[253,254,258] The half-life in rodents is only 70 minutes to 3 hours, which is significantly lower than that in humans.[246,265] To approximate human plasma concentrations, mice typically are administered four doses of METH (5–20 mg/kg i.p.), with a 2- to 6-hour interval between each dose.[246,265] This dosing regimen causes significant neurotoxicity in mice and achieves a plasma concentration in mice similar to that resulting from a human METH binge pattern of self-administration.[1,3,246,265–267] These concentrations are similar to plasma concentrations of METH in humans after chronic use in the range 0.176–1.743 mg/L (176–1743 ng/mL)[257,268]; however, concentrations in rats average 10 times higher in the brain versus plasma.[268]

Human abuse patterns vary from single day usage to regular users where Cho et al. reported a dose range of 20–250 mg or more per "hit" in METH abusers with total daily doses of up to several grams, which is substantially greater than the doses used normally in controlled clinical experiments.[246,269] Further estimates of plasma concentrations come from impaired drivers testing positive for METH, in whom plasma METH concentrations were typically 300–550 ng/mL, with plasma concentrations up to 1665 ng/mL in nonfatal cases.[262,268,270]

10.1.4.3 Distribution
Postmortem analysis in humans has shown METH distributes to many different tissues, including brain, liver, and kidney.[271] METH is homogeneously distributed within the brain of chronic human users in the globus pallidus, caudate, hippocampus, and temporal cortex.[272] Human brain levels of METH range from 44–100 nmol/g brain tissue, but can be as high as 200 nmol/g.[257,272] Studies with [11C] d-METH tracing in human brain have shown a relatively rapid distribution across brain regions with high and persistent uptake in both subcortical and cortical areas, slow clearance from gray matter, and no observable clearance from white matter regions. METH uptake correlated with dopamine transporter (DAT) availability in the striatum but not the cerebellum over the time course of the study.[273] The highest peak uptake of METH was in the putamen after i.v. administration, with 7–8% of the injected dose accumulating in the brain within 10 minutes. Estimates from this study suggest that a typical human dose of 30 mg in a METH user would result in a brain accumulation of about 2.5 mg of METH (14 µM).[273] This is similar to findings in the nonhuman primate brain[274] and to studies in rat brain after i.v. METH administration.[259,274,275] Levels of METH are relatively uniformly distributed throughout the brain of experimental animals administered a single dose of the drug[274–277] with concentrations in the frontal cortex, striatum, and cerebellum of 65 ± 3, 55 ± 5, and 46 ± 2 nmol/g, respectively.[275] Male Sprague-Dawley rats that received a pharmacologically active METH i.v. bolus dose (1.0 mg/kg) showed distribution of METH into brain and other tissues. This study also revealed that the highest concentrations were observed in the kidney, liver, brain, and heart with a delayed peak concentration in the spleen. The METH metabolite amphetamine also distributes extensively into these tissues and could significantly contribute to the pharmacological effects after administration of METH.[259]

METH uptake has also been reported to be higher in the striatum versus other brain areas of chronically treated animals.[278] Theoretically, the extracellular level of drug in the brain is important for determining its psychopharmacological action via receptor/transporters located on neuronal cells. Plasma concentrations of MDMA and METH after recreational doses are usually in the range of 1–10 µM in humans[257,263]; however, concentrations in brain may be substantially higher. Studies in rats with a dose of 5–10 mg/kg of MDMA or METH achieve extracellular striatal concentrations of 10–100 µM,[262,279] which are usually on average 10 times higher than plasma concentrations. Intracellular concentrations may also be elevated as active transporters may concentrate these drugs inside the neuronal terminal and in the brains of tolerant abusers during high-dose binges. Other studies have shown that the DAT plays an important role in METH neurotoxicity, as DAT knockout (KO) mice are less susceptible than wild-type controls to METH-initiated neurotoxicity.[280] Postmortem tissue is devoid of these active transport systems, so METH levels determined in these samples may be significantly different from those in living subjects.

As mentioned previously, METH is also distributed into breast milk,[260] and there is significant transfer from maternal to fetal blood.[261] The 40 mg/kg dose used in the studies from our laboratory gives a concentration in fetal brain similar to that in METH-exposed infants.[261,281,282] Won et al. measured maternal and fetal brain levels of METH and amphetamine after s.c. injection of mouse dams with 40 mg/kg d-METH hydrochloride on gestational day (GD) 14. In maternal striatum, METH levels peaked at approximately 510 ng/mg protein 1 hour after injection, while fetal striatum had 99 ng/mg protein in the striatum, approximately 102 ng/mg protein in the brainstem and 57 ng/mg protein in the cortex which peaked at 1 hour after injection, indicating that the drug can accumulate in fetal brain.[282]

10.1.4.4 Metabolism by Cytochromes P450 (CYPs) and Elimination
METH and amphetamine can be excreted unchanged, but the amount and disposition of the metabolites are influenced by urinary pH.[283,284] With pKa values of approximately 9.9 for the parent drugs, at normal physiological pH, these compounds are primarily in their ionized form. In the urine under acidic conditions, the drugs are ionized and primarily secreted unchanged, with insignificant reabsorption by the kidneys. Conversely, alkaline urine converts more of these drugs to their neutral form, which are readily reabsorbed by the kidneys, thereby increasing the half-lives of the drugs.[283,284]

About one-half of METH in humans is excreted unchanged, and the remainder undergoes CYP2D6-catalyzed N-demethylation to amphetamine, which can then be hydroxylated (Fig. 10.11).[253,284] METH also is oxidized by CYP2D6 to 4-hydroxy-METH, which is the predominant metabolite constituting almost 50% of all metabolites excreted in the urine.[253,284] These metabolites also accumulate in the striatum after administration of the parent drug.[262] Other minor metabolites include norephedrine and 4-hydroxynorephedrine.

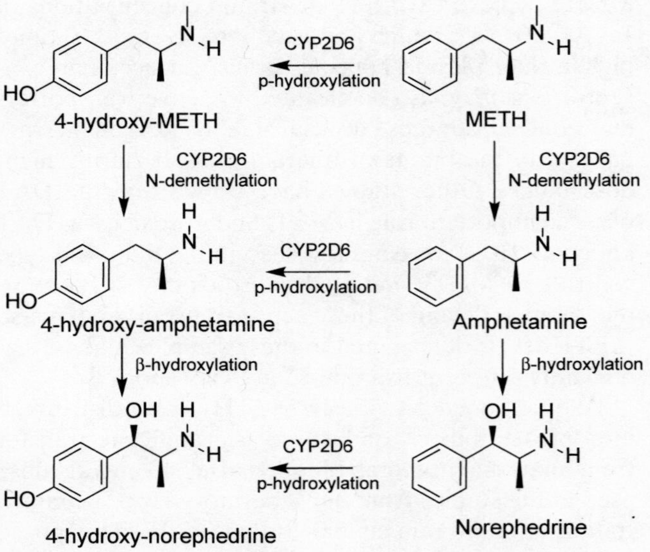

MDMA is N-demethylated to MDA by CYP1A2. MDA can be further metabolized by CYP2D6/CYP3A4 to the catechol intermediate HHA (3,4-dihydroxyamphetamine), and finally O-methylated by catechol-O-methyltransferase (COMT) to 4-hydroxy-3-methoxyamphetamine (HMA) (Fig. 10.12).[285,286] MDMA can also be O-demethylenated by CYP2D6 to 3,4-dihydroxymethamphetamine (HHMA), followed by O-methylation by COMT to 4-hydroxy-3-methoxymethamphetamine (HMMA). These metabolites are believed to be downstream products formed after the opening of the methylendioxyphenyl ring, a process that is mainly catalyzed by CYP2D6, with low-affinity contributions from CYP1A2, CYP2B6, and CYP3A4.[285,286]

Systemic metabolism of MDMA may play a role in its neurotoxicity. This was concluded from the observation that direct injection of Ecstasy into the brain failed to reproduce the neurotoxic effects seen after systemic administration.[279] Metabolites such as HHMA and HHA are easily oxidized to their corresponding quinones by CYPs, and possibly by PHSs; these quinones can form adducts with glutathione (GSH) or conjugation with sulfate or glucuronide and other thiol-containing compounds,[287,288] including proteins critical for neural function and survival. The GSH-derived conjugated metabolites are low in urine.[289]

Catechol metabolites of METH, MDMA, and MDA can be oxidized by CYPs to reactive quinones that redox cycle to form semiquinone radicals that generate ROS.[287] While it is uncertain whether CYPs contribute

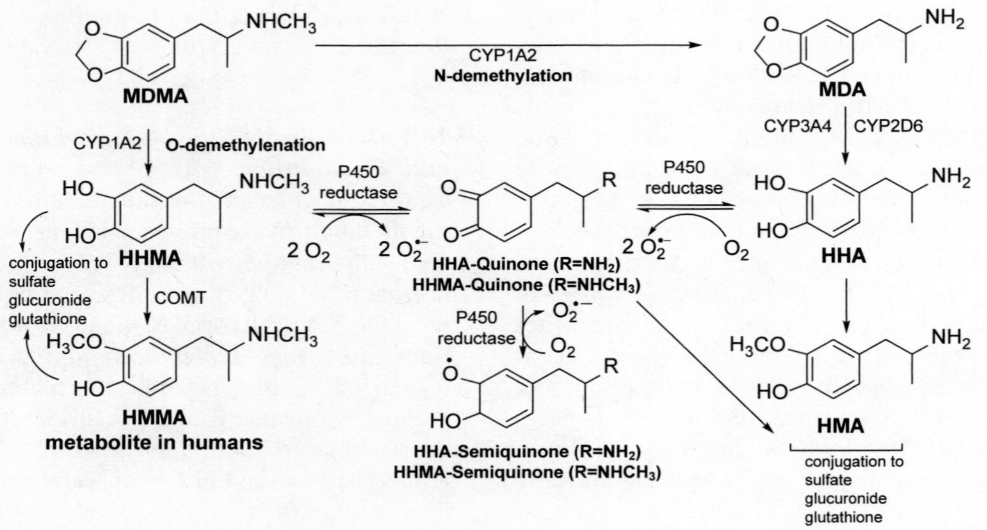

Figure 10.11 Metabolism of methamphetamine by cytochromes P450. Methamphetamine (METH) can be demethylated to amphetamine by CYP2D6. 4-Hydroxylation of METH is catalyzed by CYP2D6 to form 4-hydroxy-derivatives while β-hydroxylase generates norephedrine-derivatives.

Figure 10.12 MDMA metabolism by CYPs and P450 reductase. HHA, 3,4-dihydroxyamphetamine; HHMA, 3,4-dihydroxymethamphetamine; HMA, 4-hydroxy-3-methoxyamphetamine; HMMA, 4-hydroxy-3-methoxymethamphetamine; MDA, 3,4-methylenedioxyamphetamine; MDMA, 3,4-methylenedioxymethamphetamine.

to ROS generation, the expression of CYPs in the brain is only around 1–2% of that in the liver.[68] It is also reported that MDMA is a potent competitive inhibitor of CYP2D6 in human liver microsomes.[290,291] Furthermore, CYP2D6 mRNA transcripts are widely expressed in different brain regions, but constitute only 3% of expression in the liver. At the protein level, CYP2D6 is present in the human frontal lobe, hippocampus, and cerebellum.[64] This enzyme is also coded by a polymorphic gene with some subjects expressing no activity[69]; however, it is not known whether CYP2D6 in human brain can bioactivate amphetamines to neurotoxic intermediates. Evidence from rats suggest that this may not be the case, since rats deficient in CYP2D1, the rat homolog of CYP2D6 in humans, remain susceptible to MDMA neurotoxicity.[292] The activity of NADPH cytochrome P450 reductase, which converts quinones to semiquinone radicals, has high levels in the putamen–pallidum region where amphetamine analogs and their metabolites accumulate. However, this activity was low in human brain microsomes, representing only 8% of the value reported in rat brain microsomes.[293] Therefore, other mechanisms of ROS generation may be present in the brain that can contribute to the metabolism and neurotoxicity of amphetamine analogs.

10.1.4.5 Receptor-Mediated Pharmacological Actions of METH

METH can cause reversible, receptor-mediated effects in both the peripheral and central nervous systems. In the periphery, clinical manifestations are usually via alpha- and beta-adrenergic receptor-mediated sympathomimetic effects.[294,295] METH leads to the increased release of the key neurotransmitters by several different processes in the brain. METH enters the presynaptic terminals by both passive diffusion across the lipid membrane and through the plasma membrane catecholamine-uptake transporters such as DA, norepinephrine (NE), and serotonin transporters (SERT) (Fig. 10.13).[294,295] Within the cytosol, METH enters the presynaptic vesicle via the membrane-bound vesicular monoamine transporter-2 (VMAT-2) and facilitates the redistribution of the monoamines into the cytosol by disrupting the pH gradient that drives the accumulation of the monoamines within the vesicles.[294,295] This contributes to elevated neurotransmitter concentrations within the cytosol, leading to increased movement into the synapse via the plasma membrane transporters which change from an influx to efflux state; for example, the DAT reverses the direction of DA transport causing the transporter to move DA from the cytoplasm into the synapse.[280,296,297] The detailed mechanisms by which this occurs are unclear. The net effect of these mechanisms is to acutely increase the levels of neurotransmitters in the synaptic cleft, thereby increas-

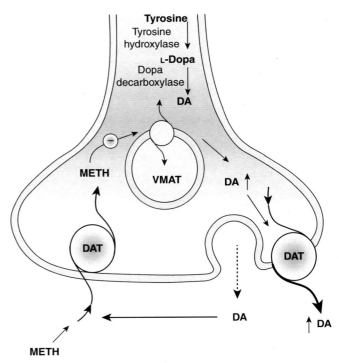

Figure 10.13 METH actions at the dopaminergic nerve terminal. METH are substrates of DAT transporters and are taken up into the cell. Once in the cell, METH interferes with the vesicular monoamine transporter (VMAT), depleting synaptic vesicles of their neurotransmitter content. As a consequence, levels of dopamine (or other transmitter amines) in the cytoplasm increase and quickly become sufficient to cause release into the synapse by reversal of the plasma membrane DAT.

ing their potential for receptor binding. Hence, METH acts as an indirect agonist at DA, NE, and SE receptors.

KI values of METH, amphetamine, and MDMA for the DAT, NE transporter (NET) and SERT are presented for human, mouse and rat in Table 10.7. KI values of uptake inhibition reflect the apparent affinities of the drugs to each transporter. Low KI values for amphetamine and METH suggest decreased neurotransmitter exchange at its transporter. The human and mouse transporters are similar in their efficacies for each of the tested drugs (KI values within a fourfold range), whereas chemical modification substantially increases the potency of MDMA for inhibiting SERT, while reducing its potency for inhibiting DAT and NET compared to METH.[297–299] In vitro studies have shown that the amphetamines are better than SE at releasing DA and NE.[297–299]

These neurotransmitters when released bind to a number of receptors that are activated to mediate

TABLE 10.7 Amphetamine Analog Binding Affinities to Uptake Transporters

Drug	Transporter	Inhibition constant (KI) (uM)	Notes
Amphetamine	rDAT	0.034	rDAT in rat synaptosomes[681]
		2.3	rDAT cultured cells expressing[682]
		0.094	rDAT rat striatal synaptosomes[298]
	mDAT	0.56	mDAT expressed in intestinal cells[299]
	hDAT	0.64	hDAT expressed in intestinal cells[299]
	rNET	0.039	rNET in rat synaptosomes[681]
	mNET	0.12	mNET expressed in intestinal cells[299]
	hNET	0.07	hNET expressed in intestinal cells[299]
	rSERT	3.8	rSERT in rat synaptosomes[681]
		8	rSERT rat striatal synaptosomes[298]
	mSERT	23	mSERT expressed in intestinal cells[299]
	hSERT	38	hSERT expressed in intestinal cells[299]
METH	rDAT	0.291	rDAT rat striatal synaptosomes[298]
	mDAT	0.47	mDAT expressed in intestinal cells[299]
	hDAT	0.46	hDAT expressed in intestinal cells[299]
		0.082	hDAT expressed in human embryonic kidney cells[683]
	mNET	0.19	mNET expressed in intestinal cells[299]
	hNET	0.11	hNET expressed in intestinal cells[299]
		0.0013	hNET expressed in human embryonic kidney cells[683]
	rSERT	9 uM	rSERT rat striatal synaptosomes[298]
METH	mSERT	9.28	mSERT expressed in intestinal cells[299]
	hSERT	31	hSERT expressed in intestinal cells[299]
		20.7	hSERT expressed in human embryonic kidney cells[683]
MDMA	rDAT	1.572	rDAT[297]
		1.53	rDAT rat striatal synaptosomes[298]
	mDAT	4.87	mDAT expressed in intestinal cells[299]
	hDAT	8.29	hDAT expressed in intestinal cells[299]
	rNET	0.462	rNET[297]
	mNET	1.75	mNET expressed in intestinal cells[299]
	hNET	1.19	hNET expressed in intestinal cells[299]
	rSERT	0.24	rSERT[297]
		2.6	rSERT rat striatal synaptosomes[298]
	mSERT	0.64	mSERT expressed in intestinal cells[299]
	hSERT	2.41	hSERT expressed in intestinal cells[299]

h, human; m, mouse; r, rat; DAT, dopamine transporter; NET, norepinephrine transporter; SERT, serotonin transporter.

complex physiological acute, chronic, and adverse responses to METH (Table 10.8). For example, hyperthermia involves the alpha1 adrenoreceptor,[300] DA-1,[301] DA-2,[302] and SE-2A receptors.[303] In ADHD, the mechanism of the therapeutic effects of amphetamine and METH is not known. It is believed that amphetamines increase levels of catecholamine in the synaptic space by blocking reuptake of NE and DA, hence increasing the effect of these neurotransmitters.[304] Interestingly, DA receptors and adrenoceptors reportedly have no affinity for amphetamines themselves.[305]

10.1.4.6 Effects of METH Abuse Effects following METH exposure can vary widely as the release of neurotransmitters like DA, NE, and SE can act to control the messaging systems of the brain for reward and pleasure, sleep, , and mood.[294,295] At doses of between 5 and 50 mg, the first effect is usually a characteristic "rush," which is believed to be the result of an initial release of high concentrations of DA in the CNS.[295] The effects are rapid when the drug is smoked or injected, appearing within 2–3 minutes after inhalation or 20 minutes after oral ingestion.[252] The immediate effects of METH are similar to the fight-or-flight response and involve increased heart and respiratory rates, blood pressure, and body temperature due to peripheral effects of adrenergic stimulation (Table 10.8).[252,294,295] The rush is often associated with euphoria and increased energy, which can last for several hours given the long half-life of METH. Low doses tend to produce a sense of height-

TABLE 10.8 Effects of METH

Acute Effects	Adverse Effects	Chronic Effects
Increased heart and respiratory rates	Hyperthermia	Insomnia
Increased blood pressure	Tremors	Paranoia
Increased body temperature	Anxiety	Hallucinations
Increased energy	Cardiac arrhythmias	Psychotic symptoms
Alertness	Acute myocardial infarction	(e.g., delusions, obsessive compulsive thoughts)
Attentiveness	Stroke	Depression, anxiety, suicide and violent behaviors
Euphoria	Death can occur	Slowed motor function
Enhanced self-esteem		Central nervous system Impairments leading to cognitive and functional decline in learning and memory
Decreased appetite		

From References 294, 295, 306, 308, 309, 312, and 684.

ened alertness, attentiveness, and energy, whereas high doses produce a sense of well-being, euphoria, and enhanced self-esteem.[252,294,295]

The adverse effects are both short-term (cardiac problems, hyperthermia, depression, confusion) and chronic (Table 10.8). With METH use, the effect is sustained for hours, placing an extended burden on the nervous, circulatory, and respiratory systems. METH causes several adverse cardiac effects.[306] In a case control study of users, only 64% of METH users showed normal heart function compared to 88% of age-matched controls. In addition, 28% of METH users showed severe cardiac dysfunction compared to 7% of age-matched controls. When used chronically, METH may cause long-term CNS consequences that result in impaired memory, impaired motor coordination, and psychiatric problems long after termination of use (Table 10.8).[307] The acute reversible effects are likely receptor-mediated and dependent upon the plasma concentration of the amphetamine analog at that time, whereas effects persisting long after the drug is gone are likely due to macromolecular damage. A study of over 1000 METH users in treatment found a high incidence of psychiatric problems, such as depression, anxiety, suicide, and violent or assaultive behaviors. Residual psychiatric symptoms included a prolonged inability to experience pleasure, as well as increased anxiety and psychotic episodes.[308,309]

Gross motor abnormalities in METH abusers, including Parkinsonism, however, are not prevalent despite the vulnerability of the striatum to METH-associated neurotoxicity.[310,311] Meta-analysis revealed moderate effects for basic motor functioning (i.e., fine-motor speed and coordination) and cognitive processing speed[307]; however, with abstinence, these effects may recover over years and such long-term studies are lacking.

Human studies are confounded especially by a lack of information on cognitive and functional tests prior to METH use, polydrug use, and variation in the amount and extent of drug exposures. Furthermore, the effects of METH may be modulated by genetic characteristics of the individual that cannot be extensively assessed. One study, for example, has shown that patients with METH psychosis lasting for 1 month or more after discontinuing METH may be associated with a polymorphism in the hDAT1 gene (SLC6A3) encoding the DAT.[312]

Several factors add complexity to understanding the stimulatory effects of amphetamines upon monoamines and the associated chronic toxicities. These factors include the multiple receptor subtypes that exist for NE, DA, and SE, each with distinct binding affinities, second-messenger effects, and CNS distribution, together with the added dimension that neuronal pathways can interact with each other; for example, monoamine neurons stimulate and/or inhibit excitatory glutamate neurons and inhibitory gamma-aminobutyric acid (GABA) neurons to alter toxicity.[313–316]

10.1.4.7 Evidence from Animal and Human Studies for Neurotoxicity METH-initiated neurotoxicity is evident in several neurotransmitter systems, but it is most notable in nigrostriatal dopaminergic pathways. Although an acute, moderate dose of METH is unlikely to reduce DA stores permanently,[317] high-doses in experimental animals (10–20 mg/kg) cause a significant reduction in levels of DA and SE and their metabolites, as well as tyrosine hydroxylase and tryptophan hydroxylase activity, enzymes involved in the synthesis of DA and other catecholamines, as well as their receptors and transporters.[1,4,318–323] In the case of long-term deficits in these nerve terminal markers (weeks to months/years), destruction of nerve terminals is evident, although catecholamine neuron cell bodies themselves do not seem to be destroyed.[1,321,323] In rodents, evidence of METH neurotoxic effects has been determined through morphological signs of axonal and nerve terminal degeneration as indicated by silver staining.[1,2] In rodents and nonhuman primates, administration of either a large single dose or repeated high doses of METH or MDMA produces long-lasting deficits in nerve terminal markers, but amphetamines also produce astrogliosis, and these markers colocalize with silver-staining, flurojade B staining, as well as caspase activation, indicating toxicity in both neuronal and nonneuronal cells.[1,320,324,325] Neurotoxic amphetamines also damage glutamatergic cortical neurons as well as gabanergic neurons in the striatum and olfactory bulb.[316,320,325–328] Furthermore, bioactivation and oxidative damage can occur in brain cells other than neurons, for example, in microglia, where the expression and activation of PHSs is associated with the neurotoxic properties of METH.[195,316,328,329]

Regionally, METH most severely affects DA terminals in the striatum; however, toxicity to the cerebral cortex, hippocampus, and olfactory bulb has been detected through decreased tyrosine hydroxylase and increased GFAP, apoptosis, and oxidative macromolecular damage.[1,320,325,326] The magnitude of the toxicity in the different brain regions is variable, and can be attributed to varied densities of DAT in these regions as well as differences in detoxification and repair pathways.[213,227,330] For example, DA levels were more severely reduced in the caudate of the striatum than in the putamen (motor area) in postmortem tissue of METH abusers, which may relate to the low prevalence of Parkinsonian motor symptoms observed in chronic METH abusers, which affects the caudate.[311] A persistent reduction in most DA markers, as found with positron emission tomography (PET) studies of human METH users, revealed decreased D2 receptors that may represent downregulation from exposure to increased synaptic DA concentrations; however, reduction in levels of DA, DAT, and tyrosine hydroxylase also have been found in postmortem striatum of chronic METH abusers.[257,331,332] The effects of METH on other markers are highly variable. VMAT2 remained unchanged in one study of human chronic abusers[257] but was elevated in recently abstinent users.[333] Other studies demonstrated modest decreases in VMAT-2 binding in PET studies.[334] In rodent models, METH decreased striatal VMAT-2 ligand binding assessed 14 days after treatment.[335]

Several studies have shown brain abnormalities in METH abusers and in animals exposed to METH that are not limited to brain regions containing DA cells and their terminals, implicating non-DA mechanisms of METH toxicity.[2,336–340] Modest decreases in SERT have been observed in human chronic METH users; however, it is uncertain if this is due to nerve terminal damage or decreased protein expression.[341,342] SE nerve terminals in various brain regions including hippocampus, prefrontal cortex, amygdala, and striatum are sensitive to the toxic effects of METH.[343] Given the effects of METH on serotonergic as well as dopaminergic systems, perhaps other GABAergic, glutaminergic systems may be similarly affected, indicating a more global pattern of degeneration and loss of neuronal connectivity in METH toxicity. Studies have shown deregulation of GABA systems with a decrease in the density of presynaptic immunolabeling for GABA 1 week postdrug, and an increase after 4 weeks.[313] Furthermore, METH administration causes dopaminergic neuronal death within the olfactory bulb of mice,[320] a region rich in dopaminergic neurons and regulatory GABA interneurons.[344,345]

METH can generate ROS such as superoxide anions, H_2O_2, and hydroxyl radicals, and can oxidize cellular macromolecules, such as proteins, lipids, and DNA.[3,4,346] By causing cumulative oxidative macromolecular damage and/or by chronically altering signal transduction in the brain, ROS may contribute to the initiation and/or progression of a number of neurodegenerative diseases and neurodegeneration associated with aging. Also, postmortem brains of chronic METH users had elevated levels of the lipid peroxidation products 4-hydroxynonenal and malondialdehyde in the caudate nucleus and, to a lesser extent, in the frontal cortex.[347] METH administration causes a marked increase in 2,3- and 2,5-dihydrobenzoic acid (the product of the reaction of salicylate with hydroxyl radicals) in striatal dialysate of rats.[348] Antioxidants (e.g., ascorbic acid or vitamin E) can attenuate METH-initiated toxicity,[349] as can the overexpression of SOD1.[350] GSH levels are also decreased by repeated administration of METH.[311] METH can cause major disruptions or even induce antioxidant enzymes in the brain.[266,326] Following a single-day administration of 10 mg/kg × 4 doses of METH to rats, SOD activity was decreased 16 hours later in the cortex, but not in the striatum, while activities of catalase and GSH peroxidase (GPx) were decreased in the striatal region.[266] However, these studies did not evaluate the time course of activities, and may have missed possible induction effects due to transcriptional responses of Nrf2 discussed later.

Experiments with PHS-1 KO mice and PHS inhibitors revealed the important role of this isozyme in the molecular mechanism of ROS generation resulting from MDMA, METH, and MDA administration, where KOs, or mice treated with the PHS-1/2 inhibitor ASA, showed decreased DNA oxidation, nerve terminal degeneration, and locomotor functional deficits when compared to PHS-normal wild-type mice or saline controls.³,⁴ METH-initiated superoxide anions might combine with NO to yield ONOO⁻, which can rapidly degrade to form •OH that oxidize proteins, lipids, and DNA. This idea is supported by the protective effect of 7-nitroindazole, an inhibitor of nNOS, in reducing METH-initiated depletion of DA and its metabolites and the loss of DAT-binding sites.³⁵¹,³⁵² Similarly, nNOS KO mice are protected against METH neurotoxicity, exhibiting less depletion of dopaminergic markers.³⁵³

A number of pathways have been implicated in METH-initiated neurotoxicity, including production of reactive oxygen and nitrogen species, hyperthermia, or triggering of an apoptotic cascade dependent upon mitochondria, but the exact processes are still unclear.³⁵⁴ Taken together, these studies suggest that a variety of different acute or chronic neuropathological processes may be associated with METH neurotoxicity, including neuronal injury or death, astrocytosis, cellular membrane alterations, and dysregulation of energy metabolism. From these studies, long-term neurotoxic effects are seen in animal models where direct degeneration has been assessed, but toxicity markers in human brain have been inconsistent with respect to assessing long-term, irreversible neurotoxicity. The complexity in assessing neurotoxicity is complicated by the fact that the markers may change regionally and over time; for example, DAT density may return to normal slowly during prolonged drug abstinence.³³² The mechanisms of recovery are unclear, but may be partly explained by the sprouting of remaining axons or a compensatory increase in monoamine levels as seen in rats treated with METH.³⁵⁵ However, even if DAT density normalizes following abstinence, cognitive deficits may still persist.³³² Also, VMAT2 may redistribute from vesicles to the plasma membrane, changing its regional localization.³⁵⁶ Importantly, direct evidence in humans for the loss of nerve terminals and/or their corresponding cell bodies or the cells surrounding these neurons, such as astrocytes and microglia, has not been provided.³²¹,³⁵⁷ Thus, multiple markers are needed in human studies, and the human studies must be complemented by more comprehensive molecular and biochemical studies in animal models to fully elucidate the molecular mechanisms leading to neurotoxicity.

10.2 PROTECTION AGAINST ROS

10.2.1 Blood Brain Barrier (BBB)

The CNS is isolated from the bloodstream by the BBB. This unique barrier is supported by the endothelial cells of the brain capillaries, which form complex tight junctions. The capillaries are also surrounded by a sheath of astrocytes end-feet, which provides a high density of cells to restrict passage from the blood to the brain.³⁵⁸⁻³⁶⁰ While the barrier makes it difficult for polar or large molecules to cross into the brain in high quantity, the BBB is quite dynamic in regulating the exchange of substances between blood and brain.³⁶¹ Transport of nutrients, ions, and hormones is required to maintain optimal conditions for neuronal and glial functions.³⁵⁸⁻³⁶¹ In addition, enzymes such as MAO, gamma-glutamyl transpeptidase, and several CYP isozymes may provide metabolic protection in the brain.²⁹³,³⁶²⁻³⁶⁴ This is especially important as a wide range of lipid-soluble molecules can diffuse though the BBB and enter the brain passively. In addition, a number of ATP-binding cassette (ABC) energy-dependent efflux transporters (ATP-binding cassette transporters) actively pump many of these agents out of the brain.³⁶⁴,³⁶⁵ There is also evidence that BBB dysfunction, either through structural changes or the alteration of transporters, may contribute to neurotoxicity associated with MS, PD, and AD.³⁶⁶⁻³⁷⁰

10.2.2 Antioxidative Enzymes and Antioxidants

Endogenous antioxidants are needed to protect the brain, as most exogenous antioxidants do not efficiently cross the BBB due to their hydrophilic nature.³⁷¹ To limit ROS-mediated damage, there are a number of endogenous antioxidant and protective xenobiotic-metabolizing enzymes present in the brain. These can include different enzymes to regenerate antioxidant potential as well as phase II detoxification enzymes. In the brain, expression of phase II detoxification enzymes occurs in astrocytes and much less in neurons.³⁷² Chemical inducers can increase detoxification proteins in both neurons and astrocytes. Furthermore, neurons in close proximity to astrocytes may obtain protective factors from the astrocytes.³⁷³

10.2.2.1 Glucose-6-Phosphate Dehydrogenase (G6PD)
This section will discuss the highly polymorphic antioxidant enzyme G6PD.

Genetics and Regulation of G6PD G6PD (D-glucose 6-phosphate: NADP oxidoreductase, EC 1.1.1.49) gene has been mapped to the telomeric region of the long arm of the X chromosome (band Xq28).³⁷⁴,³⁷⁵ G6PD is a typical X-linked gene and has helped in the development of the X-chromosome inactivation hypothesis (Lyon Law) where one of the two X chromosomes in the female cell is randomly inactivated.³⁷⁶,³⁷⁷ The gene consists of 13 exons and 12 introns and covers about 18.5 kb.³⁷⁸ Exon 1 contains no coding sequence and the ATG start site is found in exon 2. The first intron is highly conserved,³⁷⁹ while the rest of the intron sequences vary among species. The second intron is unusually long, measuring about 11 kb.[380] The promoter is embedded in a CpG island spanning from nt −1200 to intron 1, and X-chromosome inactivation is associated with the methylation of this entire island region.[381] The promoter region contains several stimulatory protein 1 (Sp1) and activator protein 2 (AP2)-binding sites[382] and an atypical TATA box (ATTAAAT) at −30 to −25 bp.[379] Deletion experiments have shown that the core promoter resides between nt −147 and +45, inhibitory sequences are located between nt −358 and −147, and upstream stimulatory sequences are present between nt −613 and −358.[382] Even though the gene is typically regarded as a "housekeeping" gene, expression has been shown to be tissue dependent and species dependent. The genetic elements that determine the rate of transcription of the gene in different cell types and in response to cellular changes are not yet fully known.[383] The mRNA product of the G6PD contains a relatively short 5′ untranslated region of 69 bp, which corresponds to all of exon 1 and part of exon 2, and a longer 3′ untranslated region of 655 bp. The gene product is about 2.4 kb in size (including the poly (A) tail).[383] Kletzien, et al. partially reviews regulation of G6PD expression by hormones, diet, and oxidative stress.[384]

Mutations in the G6PD gene are well studied and reviewed, and the most current database describes 186 mutations.[385] Historically, the number of distinct variants has fluctuated, reaching as high as 442,[386] but current sequencing advances have shown that many of the reported variants were caused by the same mutation. Most of the mutations (85%) are single nucleotide substitutions throughout the entire coding region except in exon 1, and two mutations have even been found in the intronic regions. Some of the highest prevalence rates reside in tropical Africa, the Middle East, tropical and subtropical Asia, areas of the Mediterranean, and Papua New Guinea, where the incidence of G6PD polymorphisms can approach 60% of the population.[387]

G6PD is constitutively but not uniformly expressed in all cells, with basal activity varying up to about 10-fold among different organs and tissues.[384,388] In various animal models and humans, the levels of mRNA and protein expression in the brain are constitutive in neurons and glial cells.[389–391]

G6PD Protein Structure The mature G6PD protein contains 514 amino acids and has an approximate molecular weight of 59 kDa. The gene codes for 515 amino acids, but it has been discovered that the N-terminal amino acid of the mature protein present in human erythrocytes is an acetylated alanine and the initiating methionine is cleaved and the following alanine is acetylated during posttranslational processing.[392] The active enzyme exists in an equilibrium of dimers and tetramers determined by ion concentrations and pH.[393] Each mammalian G6PD monomer contains a catalytic site and a second structural $NADP^+$-binding site,[394] the latter of which is not found in the bacterial enzyme. The role of the second structural $NADP^+$-binding site has been postulated as providing long-term stability, and many of the mutations that affect activity are found in this region.[395] Multiple sequence alignments reveal a conserved nine-residue peptide (198-RIDHYLGKE-206 in the human)[396] where the aspartate, histidine, and lysine have been shown to be important in glucose-6-phosphate (G6P) binding.[397] The consensus nucleotide-binding fingerprint, 38-GASGDLA-44, has been associated with coenzyme $NADP^+$ binding.[398]

G6PD Enzymology G6PD is the first and rate-limiting enzyme in the hexose monophosphate pathway (HMP) or the pentose phosphate pathway (PPP). In this pathway G6P is ultimately converted to ribose-5-phosphate (R5P). R5P is required for glycolysis and for DNA and RNA synthesis. During the conversion of G6P to 6-phosphogluconolactone, G6PD generates NADPH, which is essential for many reductive biosynthetic pathways, including cholesterol and fatty acid synthesis (Fig. 10.14). During cellular oxidative stress, NADPH is critical for the regeneration of reduced glutathione, catalyzed by NADPH-dependent GSH reductase, which is essential for the detoxification of reactive free radicals and lipid hydroperoxides by GPx. Another important role of NADPH is the maintenance of the catalytic activity of catalase, which is required for the detoxification of H_2O_2.[399,400] NADPH binds to catalase and prevents the formation of inactive catalase (compound II), as well as mediating the rapid reduction of catalase compound II back to its active form. Impaired catalase activity was found to contribute largely to a H_2O_2-mediated enhancement of oxidant sensitivity in G6PD-deficient erythrocytes. However, some catalase activity did remain in the G6PD-deficient cells since catalase activity did not drop below 50% of its initial level.[401] Furthermore, the amount of NADPH required for the prevention of catalase inactivation is very low (below 0.1 µM) *in vivo*, and the reduction of catalase compound II to the active form (compound I) is known to occur in the absence of NADPH, albeit at much slower rates.[399] Other enzymes and systems requiring NADPH are summarized in Table 10.9.

G6PD Deficiency and Classification (Table 10.10) G6PD deficiency refers to the condition of reduced activity of the enzyme. It has frequently been referred to as the most common enzymopathy with a prevalence worldwide of 4.9%.[402] As mentioned previously, the

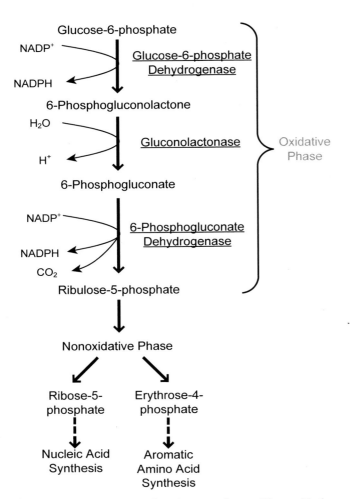

Figure 10.14 Pentose phosphate pathway. The oxidative phase of the pentose phosphate pathway (PPP) converts NADP⁺ to NADPH. The nonoxidative phase forms nucleic acids and amino acids.

TABLE 10.9 NADPH-Dependent Enzymes and Systems

Enzyme/System
Antioxidative systems (GSH reductase, catalase, thioredoxins, NQO1)
Fatty acid synthesis (β-ketoacyl-ACP reductase, enoyl-ACP reductase)
Cholesterol synthesis (3-hydroxy-3-methyl-glutaryl-CoA reductase, squalene synthase)
NADPH oxidases[18]
Nitric Oxide synthase[685]
Cytochrome p450 oxidoreductase[686]

G6PD gene is highly mutated and the World Health Organization has created a 5-class system to classify G6PD variants according to the level of enzyme activity in the red blood cell, and the clinical manifestations. Class I refers to severely deficient variants with less than

PROTECTION AGAINST ROS

TABLE 10.10 G6PD Deficiency Classification

Class	Percent Activity	Symptoms
Severe deficiency, I	<10%	Chronic nonspherocytic hemolytic anemia
Severe deficiency, II	<10%	Intermittent hemolysis
Moderate deficiency, III	10–60%	Drug induced hemolysis
Nondeficient, IV	60–150%	–
Increased activity, V	>150%	–

10% of activity that are associated with chronic nonspherocytic hemolytic anemia (CNSHA). Class II variants are severely deficient resulting in less than 10% of residual enzyme activity but are not associated with CNSHA. Class III variants are moderately deficient (10–60% of activity). Class IV are the variants resulting in normal enzyme activity (60–150%) and the Class V produce G6PD activity higher than normal (>150%).[403]

Protective Role of G6PD in Blood Cells Most individuals with a G6PD deficiency are normally asymptomatic, but can exhibit a clinical syndrome in response to an enhanced oxidative insult or exogenous factors. Although some deficiencies belong to the Class I of G6PD variants associated with CNSHA, there is no link to date between a specific genetic G6PD variant and a single clinical syndrome. It is generally believed that G6PD deficiencies constitute a problem only for mature red blood cells,[404] which lack a nucleus and the ability to increase G6PD expression in response to oxidative stress.

G6PD deficiency generally manifests as acute red blood cell hemolysis and anemia caused by oxidative stress initiated by drugs, infection, or exposure to fava beans (favism). Since in the red blood cell G6PD is the only source of NADPH, it is an essential part of the defense against oxidative stress.[405] G6PD deficiency was discovered when certain patients receiving the antimalarial drug primaquine developed acute hemolysis.[406] Since that discovery, several drugs have been linked to this manifestation,[383] making G6PD a highly important pharmacogene.[407] The other and probably most common causes of oxidative stress leading to hemolysis in G6PD-deficient patients are various types of infections, including hepatitis viruses A and B, cyclomegalovirus, pneumonia, and typhoid fever.[404] Favism is a unique type of hemolysis caused by the ingestion of the fava bean. It is believed that divicine, isouramil, and convicine are the compounds found in the beans that increase the activity of the HMP and promote hemolysis in G6PD-deficient individuals.[408] In the case of another manifestation, neonatal jaundice, it is still unclear why G6PD-deficient neonates are more susceptible to this syndrome, but these neonates may have an impaired ability to conjugate and clear bilirubin in the liver.[409] Neonatal jaundice, if not treated timely, leads to chronic bilirubin encephalopathy (kernicterus) and can leave the child mentally impaired.[409]

The widespread prevalence of malaria is believed to be the reason why G6PD deficiency is common throughout the world (especially in areas affected by outbreaks of the parasite). It is commonly believed that G6PD-deficient red blood cells are inhospitable to the survival and reproduction of the parasite *Plasmodium falciparum* responsible for malaria.[410] It has been suggested that G6PD deficiency also provides protection against *Plasmodium vivax*,[411,412] which could provide further selective pressure for the high frequency of this enzymopathy. This benefit makes G6PD mutations balanced polymorphisms, since they provide a selective advantage even though they have the potential for facilitating a disease state.[413]

G6PD Protection in Development Using a mutant G6PD-deficient mouse model, G6PD was discovered to be a developmentally critical antioxidative enzyme that protects the embryo from the pathological effects of both endogenous and xenobiotic-enhanced oxidative stress and DNA damage.[414] It was shown that G6PD-inactivated embryonic stem cells were viable but highly sensitive to oxidative stress.[405] Attempts were made to create G6PD null mice, but the resulting embryos died *in utero* by E10.5.[415]

Protective Role of G6PD in the Brain G6PD has been shown to be coordinately modulated with the expression of other antioxidative enzymes, including GPx and GSH reductase, in the developing and adult rat brain.[416,417] Moreover, NADPH generated via G6PD is necessary for the maintentance of reduced glutathione by GSH reductase. Hence, in addition to G6PD deficiencies, other antioxidative defense mechanisms may be compromised, which would increase susceptibility of the brain to ROS-mediated oxidative stress and subsequent neuronal damage. Only a few studies have examined the protective role of G6PD in the brain. One of these studied the carcinogenic implications of G6PD deficiency in brain, employing young mice that contained the X-ray-induced low efficiency allele of G6PD.[418] The brains of these G6PD-deficient male mice exhibited a decrease in the ratio of reduced glutathione to oxidized glutathione, an accumulation of promutagenic etheno DNA adducts secondary to lipid peroxidation, and an increased somatic mutation rate, suggesting an enhanced risk for brain cancer.

Selective overexpression of G6PD activity in the dopaminergic nigrostriatal system of mice was protective against the toxic effects of 1-methyl-4-phenyl-1,2,3,6-tetrahydropyridine (MPTP).[419] These findings suggest that G6PD plays a protective role in the brain. Further gene expression studies in these G6PD overexpressing mice displayed changes mainly in the expression of proteins related to antioxidant defense, detoxification, and synaptic function.[420]

Histochemical and immunohistochemical analyses revealed that the highest expression of cerebellar G6PD activity and protein was in the Purkinje cells,[389,421] and later was found to be colocalized with NADPH-dependent enzymes, including NADPH-cytochrome P450 reductase and GSH reductase.[422] G6PD activity may be particularly important in the hippocampus, where increased neuronal G6PD and sulfhydryl levels were found in patients with AD, presumably in compensation for increased oxidative stress in that region.[423] The levels of CuZnSOD mRNA and protein are particularly high in hippocampal pyramidal neurons and granular cells,[424] which in the absence of adequate G6PD-mediated detoxification may result in H_2O_2 overproduction and peroxidative damage within these cells. In addition to the cytoprotective role of G6PD against ROS-mediated oxidative damage, G6PD may be required for normal cell growth by providing NADPH for redox regulation.[405,417,425]

10.2.2.2 SOD SODs are a family of metal-containing enzymes that include Cu,ZnSOD (SOD1), MnSOD (SOD2), and extracellular Cu,ZnSOD (SOD3).[426,427] They catalyze the dismutation of superoxide anion to molecular oxygen and H_2O_2.[426] SOD1 dismutates superoxide throughout the cytoplasm and is also found in peroxisomes and the nucleus, is mainly in astrocytes and neurons. By immunohistochemistry, motor neurons of the spinal cord appear to have greater SOD1 protein expression.[426] SOD2 is inducible and located in mitochondria and is predominantly localized in neurons throughout the brain and spinal cord.[428] SOD2 KO mice suffer from postnatal neurodegeneration highlighted by cell death in the cortex and brainstem regions.[429] SOD1 KO mice do not exhibit these effects, but are more sensitive to neurotoxicity caused by MPTP.[430] Conversely, transgenics overexpressing SOD1 are protected from 3-nitropropionic acid (3-NP) and METH.[266,431] In the case of SOD3, overexpression has been found to be detrimental to neurons impairing long-term potentiation[432]; however, others have found mice overexpressing SOD3 may be protected from neurobehavioral deficits during aging.[433]

10.2.2.3 H_2O_2 Detoxifying Enzymes Detoxification of H_2O_2 is performed by antioxidant enzymes such as GPx, catalase, peroxiredoxins (Prx), and thioredoxins (Trx).

Catalase Catalase is an intracellular antioxidant enzyme that catalyzes the conversion of H_2O_2 into water and molecular oxygen. It is located in the peroxisomes and, to a lesser extent, in the cytosol, and is ubiquitously expressed in a wide variety of brain cells; however, catalase expression is relatively low in brain compared to other tissues. For example, activity of catalase is two orders of magnitude lower in brain than in kidney or liver.[434]

Prx The family of Prx enzymes consists of six antioxidant proteins that are involved in the degradation of H_2O_2, organic hydroperoxides, and peroxynitrite, and are dependent on a reactive cysteine at the active site.[435] Prx proteins reduce peroxides by employing an electron donor and redox-sensitive cysteines that undergo oxidation/reduction cycles.[436] Prx proteins are abundantly expressed in the cytosol; however, several isoforms can also be found in mitochondria, peroxisomes, nuclei, and membranes.[437] Different brain regions and cell types show distinct basal expression profiles of Prx. For example, Prx proteins 1 and 6 are expressed in glial cells, whereas Prx proteins 2–5 are localized in neurons.[436]

Trx Another family of proteins acting in conjunction with Prx are Trx, with Trx1 in the cytosol and Trx2 in the mitochondria. They bind to proteins and form intermediates that reduce protein disulfide bridges, in the process becoming oxidized themselves. Oxidized Trx proteins can be reduced through the action of Trx reductase using NADPH as a cofactor.[438] Trx1 has been detected in human brain which showed positive Trx1-like staining in white matter astrocytes.[438] The mitochondrial isoform, Trx2, is abundant and widely distributed in rat brain.[439] Brain regions showing highest expression at the RNA and protein levels include the olfactory bulb, frontal cortex, thalamus, cerebellum, and the brainstem.[439] A range of hormones, chemicals, and stress conditions have been shown to induce Trx1 expression in the brain.[438]

10.2.2.4 Heat Shock Proteins
Inducible heme oxygenase-1 (HO-1) and constitutive HO-2 belong to the family of heat shock proteins and protect brain cells from oxidative stress by degrading toxic heme into the antioxidant biliverdin, free iron, and carbon monoxide.[440] Subsequently, biliverdin is converted by biliverdin reductase into bilirubin, which also has antioxidant properties when it is oxidized back to biliverdin.[441] HOs have both antioxidative and anti-inflammatory properties, and their expression is induced by a variety of different stimuli, including their substrate heme and oxidative stress. HOs are localized in membranes, including the endoplasmic reticulum, nucleus, and plasma membrane.[442] Polymorphisms in the lengths of guanine-thymine (GT) repeats (11–40) within the HO-1 promoter appear to be an important determinant of HO-1 expression and function in humans. Long GT sequences code for a protein with lowered basal levels and reduced induction when stimulated. Robust HO-1 activity is associated with the short-GT polymorphisms, and appears to be protective against atherosclerosis-linked conditions.[443] HO-1 is expressed in unstressed brain; however, the level is minimal and HO-1 is confined to glial cells and sparse populations of neurons in the cerebellum (Purkinje cells), thalamus, hypothalamus, brain stem, hippocampal dentate gyrus, and cerebral cortex.[444,445] In contrast, HO-2 mRNA and protein are widely distributed and strongly expressed in neurons, with the highest concentrations in hippocampal pyramidal cells and dentate gyrus, olfactory epithelium and olfactory bulb, and cerebellar granule and Purkinje cell layers.[445] Following exposure to a variety of stimuli, HO-1 induction occurs in neuronal and nonneuronal brain cells, although it has been argued that astrocytes have a greater capacity than neurons for robust HO-1 expression.

10.2.2.5 NAD(P)H: Quinone Oxidoreductase
NAD(P)H: quinone oxidoreductase (NQO1) is a cytosolic flavoprotein that catalyzes the two-electron reduction of quinones to hydroquinones by using either NADPH or NADH as the donor. This prevents quinone electrophiles from participating in reactions that could lead to either sulfhydryl depletion, or to one-electron reductions that can generate semiquinones and various reactive oxygen intermediates as a result of redox cycling.[446,447] In addition, the hydroquinone products of the NQO1 reaction can be further metabolized to glucuronide and sulfate conjugates, which can then be excreted. NQO1 is mainly expressed in astrocytes and brain endothelial cells and prevents the generation of ROS.[447–449] In the human brain, NQO1 is mainly expressed in astrocytes, vascular endothelium, and in a subpopulation of dopaminergic neurons.[449]

10.2.2.6 GSH
GSH is a tripeptide called gamma-glutamyl-cysteinyl-glycine. GSH is synthesized from L-glutamate and cysteine via the enzyme gamma-glutamylcysteine synthetase (GGCS) (also known as glutamate cysteine ligase, GCLg) forming gamma-glutamylcysteine. This reaction is the rate-limiting step in GSH synthesis. Glycine is added to the C-terminal of gamma-glutamylcysteine via the enzyme GSH synthetase to then form GSH.[450] The GSH system is one of the

most important antioxidant systems in the brain. Neuronal cells contain high concentrations of GSH and display high activity of GPx and GSH reductase.[450] However, astrocytes also have important roles to play in supplying GSH substrates to neurons. This mechanism of substrate supply minimizes the neurotoxic effects of large amounts of extracellular cysteine, which can activate glutamate receptors.[451] Astrocytes synthesize and export GSH, which can then undergo transpeptidation to cysteinylglycine and gamma-glutamyl amino acid by the ecto-enzyme gamma-glutamyl transpeptidase. The cysteinylglycine generated can then be utilized by neurons to manufacture GSH, after first undergoing dipeptide cleavage to its constituent amino acids.

GSH reduces disulfide bonds formed within cytoplasmic proteins to cysteines by serving as an electron donor and in the process being oxidized to GSH disulfide (GSSG). GSH can also conjugate with electrophilic compounds, catalyzed by the glutathione S-transferases (GSTs), and these conjugates can be removed from the cell.[450] In the reaction catalyzed by GPx, the tripeptide GSH serves as an electron donor to reduce H_2O_2 to water. Besides H_2O_2, GPx also reduces organic hydroperoxides to their corresponding alcohols.[452] During this reduction of peroxides, GSH is oxidized by GPx to GS•, two molecules of which combine to form GSSG. Within cells, GSH is regenerated from GSSG in a reaction catalyzed by GSH reductase. This enzyme transfers electrons from NADPH to GSSG, thereby regenerating GSH.[453] GPx KO mice appear normal at birth, but when exposed to neurotoxins exhibit enhanced levels of neuropathology.[454] In astrocytes, when H_2O_2 clearance by catalase is inhibited, the GSH system can almost completely compensate for its detoxification.[455]

10.2.2.7 Dietary Antioxidants in the Brain Other antioxidants in the brain include vitamins. Vitamin E, also known as α-tocopherol, is a lipid-soluble antioxidant concentrated in the cell membranes. It contributes an electron to the peroxyl radical that is formed during the chain reaction of lipid peroxidation, and therefore is classified as a chainbreaking antioxidant.[5] The vitamin E radical produced when the parent compound donates an electron is unreactive, and it eventually degrades or is recycled to vitamin E by ascorbate. α-Tocopherol is important for normal brain physiology as patients with a prolonged deficiency of this vitamin suffer from neurological deficits.[456] The function of vitamin C (ascorbic acid) in the brain is a double-edged sword with respect to free radical damage.[5] In general, both the gray and white matter of the brain contain ascorbic acid, which can act as a scavenger and functions as an antioxidant in recycling the vitamin E radical back to vitamin E. Vitamin C is an effective antioxidant as it can react with most other biologically relevant radicals and oxidants to generate the ascorbyl radical, which has a low reactivity due to resonance stabilization of the unpaired electron and readily dismutates back to ascorbate.[457] In contrast, vitamin C can also be a strong prooxidant through interactions with iron.[457,458] Ascorbate can maintain metal ions in their reduced state and hence can lead to reaction of the reduced metal ions with H_2O_2, generating hydroxyl radicals.[457]

10.2.3 DNA Repair

DNA damage can exist in a number of different forms, including single-strand breaks, double-strand breaks, single base modifications, DNA adducts, and nucleotide cross-linkages. The enzymes responsible for the repair of the aforementioned lesions are numerous and take on important roles in identifying damage, excising damage sites, remodeling DNA to access damage sites, recruiting genes involved in repair, and altering cell cycle progression to either halt replication (to accommodate DNA repair) or activate apoptotic pathways. These repair pathways have mainly been studied in the context of cancer; however, there is strong support, both clinically and experimentally, for the protective role of DNA repair in the developing and adult brain following exposure to various neurotoxicants. Herein, we will discuss some key examples of DNA repair genes with specific reference to their role in preventing neurotoxicity.

10.2.3.1 Ataxia Telangiectasia Mutated (ATM) ATM is a serine/threonine kinase that detects double-strand breaks and recruits other repair proteins to the damaged site by phosphorylating key proteins, including p53, leading to the downstream activation of downstream repair genes such as Brca1.[459] Characteristic features of individuals with ataxia telangiectasia (mutated ATM) include premature aging and progressive neurodegeneration. The latter has been proposed to be the result of uncontrolled proliferation and increased genetic instability in adult neural progenitor cells.[460] The role of ROS in ataxia telangiectasia is not fully understood. However, ROS have been shown to activate ATM-mediated phosphorylation in the absence of DNA damage,[461] suggesting that ATM-deficient individuals are more susceptible to drug-initiated or age-related neurodegeneration through oxidative stress.

10.2.3.2 Oxoguanine Glycosylase 1 (Ogg1) Ogg1 is a base excision repair (BER) gene responsible for the removal of the oxidatively damaged DNA lesion, 8-oxo-2′-deoxyguanosine (8-oxodGuo). Ogg1 is responsible for the recognition of the 8-oxodGuo lesion, its excision from DNA, and coordination with other DNA repair proteins.[462] Enzyme activity levels vary with age where fetal brain excision activity is about twofold that of adults.[227] Fetal exposure to ROS-initiating chemicals like METH in Ogg1-deficient mice causes an increase in DNA oxidation in fetal brain, and leads to greater neurological deficits in adulthood, implicating this form of DNA repair in protecting against oxidative stress.[227]

10.2.3.3 Cockayne Syndrome B (CSB)
CSB is a gene that plays an important role in the transcription-coupled repair form of nucleotide excision repair (NER). NER primarily removes bulky and helix-distorting lesions that cross-link to sugars or other bases on a DNA molecule that enzymes involved in BER are unable to remove. Mutations in the CSB gene lead to Cockayne syndrome, which is characterized by increased sensitivity to ultraviolet light and premature aging. The mechanism of repair is unknown but is thought to involve binding to DNA lesions, possibly to prevent replication while recruiting other repair proteins.[463] CSB is capable of repairing oxidative lesions,[464] and deficiencies in CSB activity can lead to increased METH-initiated DNA oxidation in fetal brain and sensitivity to postnatal neurodevelopmental deficits in progeny exposed in utero to METH.[465]

10.2.3.4 Breast Cancer 1 (Brca1)
Brca1 is a DNA repair gene that, when mutated, leads to high incidences of ovarian and breast cancers.[466] Attempts to generate KO lines have been unsuccessful due to embryolethality, with severe neural tube defects characterized by disorganized neuroepithelium with signs of increased proliferation and apoptosis,[467] suggesting an important role for Brca1 in neurodevelopment. Brca1 is best known for its role in recruiting DNA repair genes involved in homologous recombination repair at sites of double-strand breaks.[468] However, a growing body of evidence indicates a role of Brca1 in other repair pathways, including nonhomologous end joining repair of double-strand breaks, NER-mediated bulky lesions,[469] and BER.[470]

10.3 NRF2 REGULATION OF PROTECTIVE RESPONSES

10.3.1 Overview
Cap"n"collar (CNC) proteins form a family of basic leucine zipper (bZip) transcription factors conserved across species from worms to humans. Most CNC factors are transcriptional activators as they contain a conserved 43-amino acid CNC domain usually located N-terminally to a DNA-binding domain. CNC transcription factors comprise the Caenorhabditis elegans SKN-1 (Skinhead family member 1),[471] the Drosophila CNC,[472] and four related vertebrate counterparts. The latter include the p45-nuclear factor erythroid-derived 2 (NF-E2) and the NF-E2-related factors Nrf1, Nrf2, and Nrf3.[473–479] These factors were originally named from studies assessing activation of β-globin gene expression through characterization of the regulatory locus control region. This regulatory region contained a tandem AP-1-NF-E2 motif, which had strong enhancer activity in erythroid cells. Investigation of the transcription factors that bind to this AP-1-NF-E2 site led to the discovery of the above CNC and bZip transcription factors. While p45-NF-E2 was localized in erythroid cells, Nrf1 and Nrf2 were expressed in many tissues (discussed below).[473,478,480] The related vertebrate transcription factors Breakpoint cluster region/Abelson murine leukemia viral oncogene homolog 1 (Bach1) and Bach2 are characterized by the additional presence of a Broad complex, Tramtrack, Bric-a-brac (BTB) protein interaction domain along with CNC homology 1; however, Bach1 and Bach2 function as transcriptional repressors.[481,482]

10.3.2 Mechanism of Action of Nrf2
The Nrf transcription factors have both physiological and stress response functions depending on the cellular environment, and are important signaling pathways that can detect oxidative stress and initiate protective mechanisms.[477,483] Nrf2 is part of a vital network in protective cellular responses in many tissues. In the absence of oxidative stress, Nrf2 binds to Keap1 (Kelch-like erythroid cell-derived protein with CNC homolog (ECH)-associated protein 1) which keeps Nrf2 in the cytosol and hence inactive (Fig. 10.15).[484] The Nrf2–Keap1 molecule is also bound to an ubiquitin ligase complex which promotes proteasomal degradation of Nrf2. In response to oxidative stress, Nrf2 is released from Keap1. Another protein in the cytosol called DJ-1, also known as PARK7 (Parkinson disease (autosomal recessive, early onset) 7), is thought to stabilize Nrf2 and prevent its association with its inhibitor Keap1, allowing Nrf2 to translocate to the nucleus. In the nucleus, Nrf2 can dimerize with other bZIP proteins such as Maf (named for the avian musculoaponeurotic fibrosarcoma oncogene).[485] Maf proteins possess the characteristic bZip domain but lack a transactivation domain.[486] Nrf2 factors do not homodimerize, but Nrf2–Maf protein heterodimers and Nrf2–Jun protein complexes can function to modulate transcription when they bind to the antioxidant response element (ARE) (also known as the electrophile response element [EpRE]), an enhancer sequence on DNA that regulates the transcription of cytoprotective components of the cell.[477,483–485,487–489] In addition to Nrf2, Nrf1

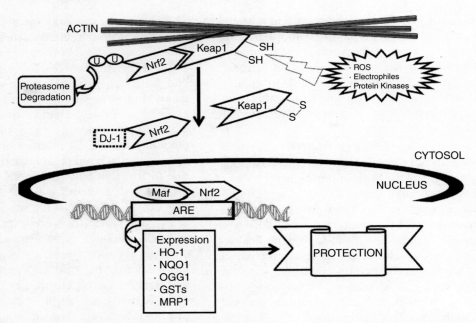

Figure 10.15 Mechanism of Nrf2 in cytoprotection. Nrf2 is frequently activated during various types of oxidative stress. Under normal conditions, Nrf2 is located in the cytoplasm as an inactive form associated with its repressor protein Keap1. The Nrf2–Keap1 molecule is also bound to an ubiquitin ligase complex which promotes proteasomal degradation of Nrf2. Oxidation of redox-sensitive cysteines in Keap1, during oxidative stress, leads to dissociation of Nrf2 from Keap1. DJ-1 is thought to stabilize Nrf2 and prevent its association with its inhibitor Keap1, thereby preventing the ubiquitination of Nrf2. This allows Nrf2 to translocate to the nucleus where it heterodimerizes with members of the small Maf protein family and binds to the antioxidant response element (ARE) that regulates the transcription of cytoprotective enzymes such as heme oxygenase (HO-1), NAD(P)H: quinone oxidoreductase (NQO1), oxoguanine glycosylase 1 (OGG1), glutathione-S-transferase (GST), and multidrug resistance-associated protein 1 (MRP1) among others listed in Table 10.14.

and Nrf3 also regulate transcription of ARE-containing genes; however, evidence from binding to the ARE suggests Nrf2 is more efficient than Nrf1 or Nrf3.[477,489] Ultimately, the binding of Nrfs depends on the presence of various nuclear proteins/cofactors that aid or repress binding to the ARE (discussed later).

10.3.3 Genetics of Nrf2

The Nrfs share many structural and functional similarities but are encoded by different genes. Chromosomal localization of the human Nrf2 gene (official symbol NFE2L2) maps to 2q31, Nrf1 (official symbol NFE2L1) maps to 17q21-3, Nrf3 (official symbol NFE2L3) maps to 7p15.2, and p45-NF-E2 (official symbol NFE2) maps to 12q13[476,479] (NCBI genome database). In mice, Nrf2 maps to chromosome 2, Nrf1 maps to chromosome 11, Nrf3 maps to chromosome 6 and p45-NF-E2 chromosome 15. These genes map close to the homeobox (hox) A-D gene cluster, suggesting a single ancestral gene for the CNC family members that may have diverged to give rise to the four closely related CNC factors.[473,476]

The Nrf2 gene consists of five exons with four introns.[490] The sequence of the first exon of the human Nrf2 gene shows 70% homology with that of the mouse Nrf2 gene, with the first intron being approximately 25 kb.[491] Alternatively, in humans, the Nrf1 gene spans 15 kb and has nine exons with two polyadenylation sites.[492] Alternative first exons, differential splicing, and alternate polyadenylation sites give rise to at least four different transcripts of Nrf1.[492,493] The promoter regions of mouse and human Nrf2 have been analyzed within the first 0.5–2 kb pair region of exon 1.[149,490,491,494,495] In the mouse promoter of Nrf2, within the first 500 bp upstream of exon 1, the promoter is GC rich with Sp1 and AP-2-binding sites but contains no TATA box nor a CCAAT box.[490] Further analysis showed the Nrf2 promoter contains two ARE sequences.[494] The genes encoding Keap1, Nrf2, and MafG (a small Maf protein) also contain the ARE, and therefore their expression is stimulated by Nrf2.[494,496,497] Studies have also determined the presence of xenobiotic response elements (XREs).[495] The XRE is the ultimate target of a protein complex that includes the ligand-activated aryl hydrocarbon receptor (AHR) that translocates to the nucleus after binding to a chaperone nuclear transporter.[498] Interestingly, the human Nrf2 promoter contains five copies of XRE-like elements in the 2-kb region of the promoter compared to only three copies in rodents.[495] Multiple single nucleotide polymorphisms exist in the promoter of human NRF2 (discussed later), and one of these (−617 C/A) significantly reduces gene expression.[499]

10.3.4 Protein Structure of Nrf2

Human and mouse Nrf2 proteins contain 605 and 597 amino acids, respectively. At the amino acid level they have 80% homology. There are six highly conserved domains called Nrf2-ECH homology (Neh)1 to Neh6 in Nrf2.[500] Their function and presence in the various CNC-related proteins are summarized in Table 10.11. The first conserved domain, Neh1, contains the CNC homology region and bZip domain, which are highly conserved in the CNC family. Neh1 is located in the C-terminal half of the molecule. The Neh2 domain is located in the proximal N-terminus. The Neh2 domain of Nrf2 binds with the Kelch domain of Keap1 in an inhibitory interaction,[477] but Nrf1 with a Neh2-like domain is not regulated by Keap1.[501,502] Furthermore, the Neh2 domain is not conserved in Nrf3.[503]

Next to Neh2 are the Neh4 and Neh5 regions, which are transcriptional activation domains. The Neh4 and Neh5 domains act cooperatively to recruit cAMP responsive element-binding protein (CREB)-binding protein (CBP) to ARE-regulated genes.[504] Within the central part of Nrf2 lies the Neh6 domain, which functions in the destabilization of Nrf2 only under conditions of oxidative stress.[505] The C-terminus of Nrf2 contains the Neh3 domain, which is highly conserved between species and is important for the transcriptional activity of the protein.[506] Deletion of the final 16 amino acids of Nrf2 gives rise to a molecule that is transcriptionally silent but translocates normally to the nucleus and binds DNA. The function of the Neh3 domain is dependent upon the VFLVPK motif, which is conserved across species and among other members of the CNC-bZIP family, including Nrf1, Nrf3, and NF-45.[506]

Nrf2 is mostly localized to the cytoplasm and upon activation translocates to the nucleus. Present in the protein domain are both nuclear localization sequence (NLS) and nuclear export signal (NES) motifs.[507] Nrf2 is also partly associated with the mitochondria under basal conditions. Conversely, Nrf1 and Nrf3 are integral membrane proteins targeted to the endoplasmic reticulum through a conserved N-terminal homology box 1 (NHB1) domain.[501,508] Nrf2 is a more potent transcriptional activator than either Nrf1 or Nrf3,[476,489] which may relate to the membrane association of Nrf1 and Nrf3. Also, as seen in Table 10.11, p45 and Nrf1 contain the transcription activation region Neh5 but not Neh4. Nrf3 does not contain regions homologous to either Neh4 or Neh5.[504] Nrf2 has both motifs, which may increase its transcriptional activity.[504]

Posttranslational modifications, including phosphorylation, acetylation, and regulated cleavage, can influence Nrf2 activity. The posttranslational phosphorylation

TABLE 10.11 Nrf Domains and Function

Protein Domain	Function	Nrf2	Nrf1	Nrf3	p45-NF
Neh1	• bZip region fused to a CNC region • Dimerizes with small MAF proteins • Binds DNA as a heterodimer • Nuclear localization sequence • Nuclear export signal	+	+	+	+
Neh2	• Negative regulatory domain of Nrf2 function • Interaction with Keap1 • Contains double glycine repeat (DLG) and a peptide sequence (ETGE) motif important for Keap1 binding • Contains lysines between DLG and ETGE that may be ubiquitinated • Contains a peptide sequence (DIDLID) element (influences degradation)	+	+		
Neh3	• Interacts with chromodomain helicase DNA-binding protein 6 (CHD6) • May be involved with transcription	+	+	+	+
Neh4	• Transactivation domain • Interacts with CREB-binding protein (CBP)	+			
Neh5	• Transactivation domain • Interacts with CBP • Nuclear export signal	+	+		+
Neh6	• Destabilization of Nrf2 under conditions of oxidative stress	+			

From References 480, 500, 504–506, 513, and 515.

of Nrf2 by a tyrosine kinase at Ser-40 can disrupt the association of Nrf2 with Keap1,[509] while phosphorylation at Tyr-568 may be required for the nuclear export of the transcription factor.[507] Acetylation of Nrf2 by its transcriptional coactivator CBP promotes Nrf2 DNA binding to ARE promoters.[510] Also, caspases can remove the N-terminal transactivation domain and convert Nrf2 to an apoptosis-promoting ARE repressor.[511] Modulation of Nrf2 activity is discussed below.

10.3.5 Regulators of Nrf2

As seen from its various protein domains, Nrf2 has many different components that can regulate its transcriptional activity. Nrf2 interactions with Keap1 and its degradation by the proteasome are important in the cytosolic compartment where Nrf2 levels are kept low, with a half-life of 10–30 minutes under physiological conditions.[512–514] In the nucleus, Nrf2 can interact with a number of cofactors forming a transcriptional complex that binds to the ARE and modulates transcription (Table 10.12).

10.3.5.1 Negative Regulation by Kelch-Like ECH-Associated Protein 1 (Keap1) Nrf2 is localized mainly in the cytoplasm bound to a repressor, KIAA0132 (also called iNrf2), the human homolog to Keap1.[500] Mammalian forms of Keap1 are metalloproteins having 624 amino acids. Mouse Keap1 shares approximately 94% amino acid sequence homology to its iNrf2 human homolog and is highly conserved among species.[500] Keap1 is homologous to the Kelch protein that binds actin in *Drosophila*.[500] Starting at the N-terminus, the Keap1 protein consists of five domains including the N-terminal region (NTR), the BTB dimerization domain, the intervening or linker region (IVR), the double glycine repeats (DGR), or Kelch repeats region in the C-terminal region.

The BTB region is involved in forming homodimers with other Keap1 proteins, as well as targeting Nrf2 for ubiquitination.[513,515,516] The IVR has a high cysteine content and serves as a redox sensor, reacting with electrophilic reactive intermediates to form adducts, or reacting with ROS to become oxidized.[517,518] Either reaction leads to conformational changes that reduce the interaction between Keap1 and Nrf2. Human Keap1 contains 27 cysteine residues, 25 of which are highly conserved across species, including mice.[484] The DRG repeats, comprising six double glycine repeats, are necessary for the interaction with the Neh2 domain of Nrf2 (i.e., motifs DLG and ETGE of the Neh2 of Nrf2), and also interacts with actin in the cytosol.[484,516,519,520] These domains of Keap1 interact through various functions (Table 10.13) to keep Nrf2 sequestered in the cytosol and inactive in the absence of chemical or pathophysiological stress. In fact, overexpression of Keap1 reduces Nrf2-mediated activation of ARE-regulated genes.[484,518] In addition, deletion of the DRG, IVR, or the C-terminal region abolished the ability of Keap1 to repress the activity of Nrf2.[519] Other studies have also shown that Keap1 has a nuclear export signal that is required for termination of Nrf2–ARE signaling by escorting the nuclear export of Nrf2.[521] However, it is still unknown how Keap1 enters the nucleus, as it has no known nuclear localization sequence.

10.3.5.2 Negative Regulation by Proteasomal Degradation Under normal conditions, Keap1, through its BTB domain, anchors the Nrf2 transcription factor within the cytoplasm, targeting it for ubiquitination and proteasomal degradation to maintain low levels of Nrf2.[513] The integrity of the proteasomal system therefore also plays an important role in regulation of Nrf2 transcriptional activity. Keap1 protein functions as a link between Nrf2 and the Cullin3-based E3-ubiquitin ligase ubiquitination complex.[515,522,523] The Neh2 domain of Nrf2 represents the region through which the repressor protein Keap1 recognizes and targets lysines in the Neh2 domain of Nrf2 to the Cullin3-based E3-ubiquitin ligase for ubiquitination and subsequent degradation by the 26S proteasome. These lysines are located between the two Kelch-binding sites on Neh2, DLG and ETGE, and a model has been proposed whereby binding of a Keap1 homodimer to these two sites allows for ubiquitination to occur. As shown in Table 10.11, another motif in Neh2 called the DIDLID element, which is a peptide sequence, may aid in the recruitment of ubiquitin ligase, and is required for the rapid turnover of Nrf2 under normal homeostatic conditions.[505]

Studies have also shown that, even in the absence of a Keap1 interaction, Nrf2 can be targeted for proteasomal degradation.[505] Under conditions of oxidative stress, the Neh2 domain directs a less rapid, Keap1-independent degradation of the bZip factor. Degradation of Nrf2 in stressed cells is predominantly mediated by the redox-insensitive Neh6 domain of Nrf2.[505] Another protein in the cytosol called DJ-1, which has been linked to PD, is also important for the stabilization of Nrf2. Proper function of DJ-1 is necessary for the proper induction of NQO1. Loss of DJ-1 leads to deficits in NQO1 expression. DJ-1 is thought to stabilize Nrf2 and prevent its association with its inhibitor Keap1, thereby preventing the ubiquitination of Nrf2.[524] Without intact DJ-1, the Nrf2 protein is unstable, and transcriptional responses may be reduced. However, the Nrf2 pathway can still be activated by tertiary butylhydroquinone (tBHQ), a potent Nrf2 activator, in DJ-1-deficient cells.[525]

TABLE 10.12 Nrf2 Interactions in the Nucleus

Cofactor	Function	Reference
Small Maf proteins: MafF, MafG and MafK (also referred to as p18)	• Nrf2–Maf dimers have been proposed to function as both positive regulators • Negative regulators of ARE-dependent gene transcription (i.e., MafK) • May form homodimers with each other to repress ARE binding of Nrf2	496,527,687–689
CBP	• Complexed with Nrf2, transcriptional coactivator • Acetylation of Nrf2 promotes its DNA binding to ARE	510
Fyn	• When phosphorylated can enter the nucleus to phosphorylate Nrf2 on Tyr-568 to promote nuclear export of Nrf2	507,510,690
Fos proteins: c-Fos, FosB, Fra-1, and Fra-2	• Downregulation of human and mouse ARE-mediated transcription	488,691
Jun proteins: c-Jun, Jun-B, and Jun-D	• Possible combinations were ineffective in repression or upregulation of hARE-mediated gene expression. • Jun–Nrf2 complexes have also been implicated as positive effectors of ARE-dependent genes • Role of c-Jun phosphorylation in activation • Heterodimerization and binding of Nrf–Jun proteins require unknown cytosolic factors • Phosphorylation of JunD at Ser-100, an activated form of JunD, is an ARE regulatory protein	488,489,692
Bach1	• Transcription repressor that competes with Nrf2 • Binds to Mafs to repress gene activation • Increased nuclear export during oxidative stress • Phosphorylation of tyrosine 486 leads to rapid nuclear export of Bach1	482,693–695
Bach2	• Oxidative stress induces nuclear accumulation • Suppresses ARE activity and promotes apoptosis • Interferes with Nrf2–Maf recognition	696–698
NFKB	• Antagonizes NRF2, but in this instance it entails competition for CBP and recruitment of histone deacetylase to MafK	699
BRG1	• Chromatin remodeling factor • Interacts with Nrf2 • Enhances ARE activity	700
SMRT	• Transcriptional repressor • Silences ARE activity possibly through binding Nrf2 protein	701
ATF	ATF1 • Transcriptional repressor of the ferritin H ARE • Blocks Nrf2 binding	702
	ATF3 • Can repress Nrf2-mediated signaling • Direct ATF3–Nrf2 protein–protein interactions that result in displacement of CBP from the ARE	703
	ATF4 • Nrf2-interacting protein possibly with other cofactor proteins increase ARE activation	704
PPAR gamma	In macrophages: PPAR–Nrf2 interactions may suppress ARE binding	705
ER	• Estrogen-bound ERα, but not ERβ, is able to repress Nrf2-mediated transcription • Physical interaction between ERα and Nrf2 • SFhERRbeta repressed Nrf2 activity possibly through physical interaction in a complex with Nrf2, not by competing for the ARE DNA-binding sites	706
RAR	• Reduces the ability of Nrf2 to mediate induction of ARE-driven genes • Evidence of Nrf2 forming a complex with RARalpha therefore preventing binding to ARE	707

ARE, antioxidant response element; ATF, activating transcription factor; Bach, Breakpoint cluster region/Abelson murine leukemia viral oncogene homolog; BRG1, Brahma-related gene 1; CBP, (CREB (cAMP Responsive Element Binding protein) Binding Protein); ER, estrogen receptor; Maf, musculoaponeurotic fibrosarcoma; NFKB, nuclear factor kappa B; PPAR gamma, peroxisome proliferator-activated receptor gamma; RAR, retinoic acid receptor; SFhERRbeta, estrogen-related receptor beta (ERR)-beta-short-form; SMRT, silencing mediator for retinoid and thyroid hormone receptors.

TABLE 10.13 Regulatory Mechanisms Provided by Keap1

Selective Evidence for Keap1 Functions	References
Directs Nrf2 for ubiquitination	
Ubiquitinated Nrf2 has been detected	515,522,540
Critical cysteine residues in Keap1, C273 and C288, C151 are required for Keap1-dependent ubiquitination of Nrf2	540
Keap1 (BTB domain) functions as a substrate adaptor for a Cullin (Cul)-dependent E3 ubiquitin ligase complex (i.e., immunoprecipitation of Keap1 with Cul3)	515,522,540
Inhibiton of cul 3 increases basal expression of Nrf2	513,515,540
Mutation of the lysine residues located between DLG and ETGE motifs in the Neh2 domain of Nrf2 inhibits Keap1-directed ubiquitination	523
Redox sensor	
Cys-151 important during oxidative stress conditions, oxidation leads to Nrf2 activation	523,540
Mutation of other cysteines within the IVR, N-terminal and C-terminal domains has limited effect on Keap1 function *in vitro*	518,540
Detection of protein disulfides. The most reactive residues of Keap1 C-257, C-273, C-288, and C-297 in the IVR	517
Interactions with cellular components	
Deletion of DGR of Keap1 results in no binding to Nrf2	519
Mutations of serine 104 (BTB domain) led to the disruption of the human Keap1 homodimers and lead to the release of Nrf2	516
Interactions with PGAM5 (a member of the phosphoglycerate mutase family) forms ternary complex containing Keap1 and Nrf2 that is localized to mitochondria. Mediates Nrf2 response to changes in mitochondrial functions	708
Scaffolding of Keap1 to the actin cytoskeleton to maintain Nrf2 in cytosol. Disruption of the actin cytoskeleton promotes nuclear entry of Nrf2	519

10.3.5.3 *Regulation of Transcriptional Complex in Nucleus*

In response to oxidative stress, Nrf2 accumulates in the nucleus, where it binds nuclear proteins that can also modulate its activity. Nrf2 activates transcription primarily as a dimer with members of the small Maf proteins, which act as coactivators.[526] Different cell types may have varying amounts of these coactivators and repressors, which provide another layer of regulation to the system and may result in cell-specific responses to oxidative stress. Gene transcription is regulated by the balance between activation and repression mechanisms in response to stimuli. The Nrf2–Maf heterodimer is the primary transcriptional complex regulating ARE-dependent gene expression; however, with excessive Maf expression, homodimers can form, which will reduce Nrf2–Maf complexes and decrease ARE activation.[486,496,527] Transcription factors, such as Jun, c-Fos, FRA-1, FRA-2, and Nrf1, also can interact with the ARE, thereby competing with Nrf2.[488,489,496,527,528] Other transcription factors, including activating transcription factor (ATF), estrogen receptor (ER), peroxisome proliferator-activated receptor (PPAR), and retinoic acid receptor (RAR) may form inhibitory complexes with Nrf2 (Table 10.12). Transcription factors can inhibit Nrf2 actions by either competing for binding to AREs or by inhibiting Nrf2 through physical complexes.

10.3.6 ARE

The ARE is a DNA sequence found within the promoter regions of numerous cytoprotective genes. The ARE core sequence, also known as the EpRE, was first identified in the promoters of Ya subunits of rat and mouse GST as 5′-GTGACnnnGC-3′, where "n" was used to denote any nucleotide.[529] The AP-1 recognition site TRE (12-O-tetradecanoylphorbol-13-acetate [TPA]-responsive element) (5′-TGACTCA-3′) and the ATF/CREB-binding sequence (5′-TGACGTCA-3′) partially overlap with the ARE sequence.[530,531] Wasserman and Fahl further characterized the ARE sequence to a "core" sequence of 5′-RTGACnnnGCR-3′ using murine GST-Ya ARE, and identified many other genes that contained the sequence in their promoters.[532] However, subsequent mutagenesis studies identified deviations from the consensus ARE and identification of ARE sequences may illustrate differences depending on the gene's promoter region and the species source.[533,534]

Venugopal and Jaiswal identified a cis-element resembling the ARE sequence in 5′ flanking regulatory region of the human NQO1 gene that was physically able to bind Nrf1 and Nrf2, which resulted in an increase in transactivation activity and NQO1 gene expression.[488] This was evaluated through reporter transgene and electrophoretic mobility shift assay experiments.[488,531] Despite being integral membrane proteins, both Nrf1 and Nrf3 can bind the ARE as well.[476,489] Furthermore, Nrf1 has a 65-kDa isoform that is nuclear.[501] However, many studies have shown that Nrf2 is more efficient at transactivation, possibly due to its extra activation domain.[476,489,535] Evidence *in vivo* was subsequently provided by Itoh and coworkers (1997) who observed an impaired constitutive and butylated hydroxyanisole (BHA)-induced expression of the phase II GSTs Ya and Yb in Nrf2-disrupted mice.[526] The Nrf2 promoter also contains two ARE-like sequences located at −492 and −754 from the start codon.[494] One motif is described as the "perfect" ARE (AREL1; TGACTccGC) consensus sequence, while the second is described as "imperfect" with one more base before the GC box (AREL2; TGACTgtgGC).[494] Cooperatively, both AREL1 and AREL2 are necessary to fully activate Nrf2 expression by Nrf2 chemical activators (discussed later).[494] Functional XRE and ARE sequences sometimes exist in proximity in promoters of several important detoxifying genes, including GST, NQO1, and Nrf2 itself.[536,537] The XRE motif is the ultimate target of a protein complex that includes a ligand-activated AHR translocating to the nucleus after binding to a chaperone nuclear transporter.[498] This provides added regulation to detoxifying genes along with the ARE sequences. It is necessary to find functional AREs in the promoter regions of genes in order to seek out genes directly regulated by the Nrf2–ARE pathway and determine how other sequences, such as XRE, interact with them. Common genes with the ARE sequences that are regulated by Nrf2 activation are listed in Table 10.14. These include many phase 2 detoxification enzymes and especially those regenerating

TABLE 10.14 Examples of Genes Containing AREs in Their Promoter Region

Gene	Species	References
NRF2 (transcription factor)	Mouse, human	494
KEAP1 (negative regulator of Nrf2)	Mouse, human	497
HO-1 (catabolizes heme to yield biliverdin)	Mouse, rat	485,512,709
	Human	
NQO1 (catalyzes two-electron reduction and detoxification of quinones)	Human	488,526,534
	Mouse	
OGG1 (DNA repair)	Human	710
MafG (transcription factor)	Mouse, human	496
SOD1 (catalyzes the dismutation of superoxide radicals)	Human	711
AhR (XRE regulation)	Mouse	712
GST Ya (catalyzing GSH conjugation to electrophiles)	Mouse, rat	529,532,537
GCS (enzyme involved in GSH synthesis)	Mouse	713,714
	Human	
GCLg (catalyzes the conjugation of cysteine with l-glutamate, involved in GSH synthesis)	Human	533,715
UGT1A1 (catalyzing phase 2 conjugation)	Mouse	715
	Human	
GPx (catalyze the reduction of H2O2 using GSH)	Human	716
MRP1 (transporter)	Mouse	717–719
	Human	
MRP2 (transporter)	Mouse	718–720
	Human	
MRP3 (transporter)	Mouse	719
MRP4 (transporter)	Mouse	719
Ferritin H	Mouse	692,721
	Human	
ETS 1 (transcription factor)	Human	722
Prx1 (reduces H2O2)	Human	723
PSMB5 (proteasome factory)	Mouse	724
Trx (catalyzes reduction of disulfides to sulfhydryls)	Human	725
		726
MEH (catalyzes epoxides to polar diols)	Mouse	727

AhR, arylhydrocarbon receptor; ETS, E-twenty six; GCLg, glutamate cysteine ligase; GCS, gamma-glutamylcysteine synthetase regulatory subunit; GPx, glutathione peroxidase; GST, glutathione *S*-transferase; MEH, microsomal epoxide hydrolase; MRP, multidrug resistance protein; NQO1, NADPH quinone oxidoreductase 1; OGG1, oxoguanine glycosylase 1; Prx1, peroxiredoxin 1; TRx1, thioredoxin reductase; Trx, thioredoxin; UGT, UDP-glucuronosyltransferases.

TABLE 10.15 Selective Examples of Nrf2 Activators

Nrf2 Activator	Cell Type	References
Sulforaphane	Mouse small intestine	579,580,728,729
	Hepatic cells	
	Intracerebral hemorrhage brain tissue	
	Blood–brain barrier tissue	
Diethylmaleate	Fibroblasts	500
Dimethyl fumarate (DMF)	Mouse astrocytes	372,500
tert-Butylhydroquinone	Human neuroblastoma cells	555,557,573,730,731
	Human neural stem cells	
	Primary cortical astrocytes	
	Mouse striatal tissue	
	Mouse cortical tissue	
Oltipraz	Hepatic and gastric tissues	552,721
	Hepatic cells and fibroblasts	
Electrophilic neurite outgrowth-promoting prostaglandin (NEPP)	Neuronal cells	542
Acrolein	Murine keratinocyte PE cells	732
B-NF (beta-naphthoflavone)	Hepatic cells and fibroblasts	721
D3T (1,2-dithiole-3-thione)	Hepatic cells and fibroblasts	576,721
	Mouse striatal tissue	
Curcumin	Renal epithelial cells	733
Cadmium	Hepatic cells	734,735
Manganese	Hepatic cells	734
Methylmercury	SH-SY5Y cells and with primary mouse hepatocytes, direct binding to Keap1	736

reduced glutathione. However, a diverse set of genes involved in antioxidation, detoxification, and repair are also transcriptionally regulated by Nrf2.

10.3.7 Activators of Nrf2

Oxidative stress is induced by a wide range of factors, including drugs, heavy metals, and ionizing radiation, as well as by endogenous processes. Oxidative stress results from the generation of ROS and reactive electrophiles, which can be toxic to cells if basal cytoprotective mechanisms are overwhelmed. These compounds may activate Nrf2 (Table 10.15) and thereby enhance the protective mechanisms discussed above that protect against toxicity initiated by both electrophilic reactive intermediates and ROS. The Nrf2 transcription system is activated by a wide variety of natural and synthetic chemical compounds. As discussed above, under normal cellular conditions, the cytosolic Keap1–Nrf2 complex is constantly degrading Nrf2. When a cell is exposed to oxidative stress, Nrf2 dissociates from the Keap1 complex, stabilizes, and translocates into the nucleus, leading to activation of ARE-mediated gene expression. Questions remain as to how Nrf2 is transcriptionally activated by such diverse chemical compounds and agents. The diversity may be related to the possibility of interactions with both Nrf2 and Keap1.

By far the most studied regulatory mechanism of Nrf2 activation is its interaction with Keap1. Keap1 has a number of reactive cysteines which may play an important role in sensing oxidative stress and then responding by either the release of Nrf2 or preventing Nrf2 degradation by the Keap1-mediated proteasomal degradation pathway.[538] Three key cysteine residues (C151, C273, and C288) were identified by both in vitro alkylation and in vivo site-directed mutagenesis assays to be important in redox sensing.[517,539] Mutation of C151 completely abolished induction of Nrf2 by many Nrf2 activators, such as sulforaphane and tBHQ.[540] Those chemical activators that are strong electrophiles can interact with Keap1 cysteines through direct alkylation or arylation, while free radicals and ROS generation can oxidize the cysteine residues resulting in the formation of disulfide bridges.[538,540] Talalay and coworkers found many inducers of environmental stress response genes that may act through Nrf2 activation, which they classified into 10 chemically distinct classes: (1) oxidizable diphenols, phenylenediamines, and quinones; (2) Michael acceptors (olefins or acetylenes conjugated to electron-withdrawing groups); (3) isothiocyanates;

(4) thiocarbamates; (5) trivalent arsenicals; (6) dithiolethiones; (7) hydroperoxides; (8) vicinal dimercaptans; (9) heavy metals; and (10) polyenes.[541] Because of the 25 conserved cysteine residues, it is possible that the sites of adduct formation or oxidative modification may vary among the different classes of inducers.[541] Electrophiles disrupt the interaction between Keap1 and the Neh2 domain of Nrf2 either by inducing phosphorylation of Nrf2 or by directly modifying cysteine residues in Keap1.[509,518] Specifically in the brain, electrophilic neurite outgrowth-promoting prostaglandin (NEPP) compounds are taken up preferentially into neurons and bind to Keap1. NEPPs prevent Keap1-mediated inactivation of Nrf2, thereby enhancing Nrf2 translocation into the nucleus of cultured neuronal cells.[542]

Another regulatory mechanism of activation of Nrf2 is phosphorylation. The MAPKs are involved in the regulation of the ARE in a Nrf2-dependent manner.[543] The extracellular signal-regulated kinases 2 and 5 (ERK2, ERK5) and JNK1 upregulate the ARE,[544-546] while the p38 MAPK appears to suppress it.[547] However, it is uncertain which cellular components involved in ARE regulation may be targets of any of these kinases. Huang et al. have reported that protein kinase C (PKC) can directly phosphorylate Nrf2, and Ser-40 appears to be a site of potential phosphorylation.[509] Activation may involve one or more of these mechanisms converging to work together depending on the chemical, cell, or tissue types and the gene of interest. Compounds like oltipraz and dimethyl fumarate (DMF),[548,549] as well as natural products like curcumin[550] and sulforaphane,[551] are effective in mouse models of experimental carcinogenesis, and their chemopreventive and neuroprotective actions are reduced or abolished in Nrf2 KOs, indicating that their activities are mediated by the induction of Nrf2.[548,552]

10.3.8 Nrf2 in Neurotoxicity and CNS Diseases

10.3.8.1 Nrf2 Expression Nrf2 is expressed in a variety of tissues, including liver, kidney, small intestines, stomach, lung, and heart.[483,490,526] Studies evaluating mouse fetal RNA and embryonic stem cell RNA revealed expression of Nrf2 as early as the blastocyst stage and continuing throughout gestation.[490] In the adult mouse brain, it is expressed in different regions, including the striatum, substantia nigra, cortex, and hippocampus.[553-557] The expression of Nrf2 has been studied in human brain tissue. In the cortex, substantia nigra and hippocampus Nrf2 is expressed in neurons and astrocytes, where it is localized in the cytoplasm and the nucleus.[554] In the motor cortex and spinal cord, Nrf2 expression was predominantly in the cytoplasm, and no nuclear labeling could be observed.[558]

Ultimately, the transcription of cytoprotective genes is determined by the activation of Nrf2, the presence of the coactivators, and the promoter elements of each gene, hence the potential for regional selectivity upon Nrf2 activation.

Of importance is the role of the astrocytes in the production of GSH, which is mainly regulated by Nrf2-mediated activation of the genes for GSH synthesis. Nrf2 and ARE activation in glial cells not only prevents oxidative damage in astrocytes, but also protects neighboring neurons, via its production and secretion of GSH and other potentially protective cofactors, from glutamate- and H_2O_2-induced neuronal cell death.[553,555,559,560] In addition, studies with the Nrf2 stabilizer DJ-1 have shown that DJ-1 knockdown in astrocytes diminishes astrocyte-mediated neuroprotection.[561] Furthermore, localization of Nrf2 in neurons, together with evidence that NEPPs prevent excitotoxicity by activating the Keap1–Nrf2–HO-1 pathway, suggests that preferential accumulation of protective compounds can occur in neurons.[542] These processes in neurons constitute a complementary mechanism for targeting areas of neurotoxicity as an alternative to mechanisms involving electrophilic compounds such as tBHQ, which activates the antioxidant-responsive element predominantly in astrocytes.

10.3.8.2 Nrf2 in Neurodegenerative Diseases Differential regulation of Nrf2 has been detected in PD and AD.[554] Studies assessing the localization of Nrf2 in the substantia nigra of PD brains demonstrated that, in addition to the cytoplasm, a strong nuclear immunoreactivity is observed in neurons.[554] In the case of AD, there is a significant reduction in nuclear Nrf2 in hippocampal neurons. Furthermore, there is less Nrf2 in nuclear fractions from the cortex in AD cases.[554] This outcome in AD is contrary to the expected Nrf2 localization in the nucleus of neurons, which should occur in oxidative stress. Potentially, inactivation of Nrf2 by proteases during the neuropathological process may contribute to its reduced nuclear localization.[511] Likewise, evidence for a role of Nrf2 and its regulation and stabilization are currently being investigated for other neurodegenerative diseases such as ALS, MS, and HD (Table 10.16). Although the Nrf2–ARE signaling route may be activated in these diseases, factors may be operating to counteract Nrf2-activated gene transcription depending on the stage of the disease.[562] For example, activation of Nrf2 that is not sufficient to inhibit the neuropathology of PD may be related to altered stabilization by DJ-1.[524] Mutations in DJ-1 are associated with autosomal recessive forms and some sporadic cases of PD.[563] Evidence is emerging that DJ-1 has antioxidant properties, through its interaction with Nrf2, and

TABLE 10.16 Evidence for Nrf2 Activation or Deregulation in Neurodegenerative Diseases

Disease	Evidence	References
AD	• Nrf2 localizes in the cytoplasm of hippocampal neurons	737
	• Not a major component of β-amyloid plaques or neurofibrillary tangles	
	• Nuclear Nrf2 expression level significantly decreased in AD cases	738
	• Cu/ZnSOD activity was significantly decreased in AD frontal and AD temporal cortex	739
	• Catalase activity was significantly decreased in AD temporal cortex	
	• Increased expression, however, decreased activity of many Nrf2-regulated antioxidant genes	740
	• Binding of APP inhibits HO activity	
	• HO-1 immunoreactivity is significantly augmented in neurons and astrocytes of the hippocampus and cerebral cortex and colocalizes to senile plaques, neurofibrillary tangles	
	• Specific localization of NQO1 staining and increased activity in astrocytes and neurites surrounding senile plaques in the frontal cortex of AD brains	741
	• DJ-1 immunoreactivity was seen in a subset of neurofibrillary tangles, neuropil threads, and neurites in extracellular plaques in AD	742
PD	• NQO1 expressed in astroglial and endothelial cells and less frequently in dopaminergic neurons in PD brains	449
	• HO-1 in Lewy bodies of affected dopaminergic neurons	740
	• Highly overexpressed in astrocytes within the substantia nigra	
	• Localization of Nrf2 in the substantia nigra of PD brain in cytoplasm and nuclear immunoreactivity is observed in neurons	554
	• Defects in DJ-1 linked to PD and may destabilize Nrf2	524,563
	• Mutations in DJ-1 are associated with autosomal recessive forms and some sporadic cases of Parkinson's disease	
MS	• HO-1 is upregulated in glial cells within multiple sclerosis plaques in spinal cord	445
	• Nrf2 is upregulated in active MS lesions, in both the nucleus and the cytoplasm of infiltrating macrophages	567
	• Nrf2 upregulated to a lesser extent in reactive astrocytes	
	• Cytoplasmic and nuclear staining of Nrf2 in macrophages and astrocytes	
	• Oligodendrocytes at the edge of MS lesions expressed relatively low levels of Nrf2	
	• DJ-1 protein expression is predominantly increased in astrocytes in both active and chronic inactive MS lesions	
	• SOD1 and SOD2, catalase, and HO-1, are upregulated in active demyelinating MS lesions especially in hypertrophic astrocytes and myelin-laden macrophages	568
	• DJ-1 levels are increased in cerebral spinal fluid of patients with relapsing–remitting MS	743
HD	• HO-1 immunopositivity were enhanced in HD brain sections	744
	• Activities of erythrocyte Cu/Zn-SOD and GPx reduced in 16 HD patients	
ALS	• Reduced mRNA expression of Nrf2 throughout the cortical layers in ALS tissues	558
	• Reduced cellular Nrf2 protein expression in both the ventral horn and the primary motor cortex in ALS tissue.	
	• Most of the Nrf2 signal was detected in pyramidal cells	
	• Keap1 protein expression in motor cortex and spinal cord was not different from controls	

AD, Alzheimer's disease; ALS, amyotrophic lateral sclerosis; APP, amyloid precursor protein; Cu/Zn-SOD, Cu/Zn-superoxide dismutase; GPx, glutathione peroxidase; HD, Huntington's disease; HO-1, heme oxygenase-1; MS, multiple sclerosis; NQO1, NAD(P)H: quinone oxidoreductase; PD, Parkinson's disease; TBARS, thiobarbituric acid-reactive substances.

protects neurons against oxidative stress-induced cell injury.[564,565] DJ-1 is homogenously expressed in all CNS regions in both human and mouse brain.[566–568] It is found in neurons of different neurotransmitter subtypes and in glial cells, such as astrocytes, microglia, and oligodendrocytes. It is also distributed throughout the cytoplasm of the soma and the proximal parts of the processes, with limited detection in the nuclei.[566] Analyses of Nrf2, Keap1, DJ-1, and Nrf2-regulated proteins in neurodegenerative diseases are providing insights into the potential involvement of ARE-mediated responses and their localization. This is summarized in Table 10.16. Protective effects of Nrf2 activation have been demonstrated in various animal models for PD,[569] AD,[570] ALS,[571] and in mouse models of HD.[553,555]

10.3.8.3 Nrf2 in Chemically Initiated Neurotoxicities

Whether activation is predominantly neuronal or in

astrocytes may depend on the origin of the pathology and/or the site of increased ROS generation, kinase activation, or electrophilic activation of Nrf2. For example, neurotoxins such as METH and MPTP that localize and generate the majority of ROS in striatal dopaminergic nerve terminals and cell bodies of the substantia nigra would be expected to generate more neurotoxicity, and hence a greater Nrf2-mediated response. Cytoprotective effects of Nrf2-induced protective mechanisms have been seen in many CNS cell types and animal models for neurodegeneration. Nrf2-driven antioxidative enzymes are able to protect primary astrocytes from H_2O_2-induced apoptosis as well as glutamate toxicity, with protection afforded by pretreatment with the Nrf2 activator tBHQ.[372,572–575]

Overexpression of Nrf2 and tBHQ-mediated and sulforaphane-mediated activation of the Nrf2–ARE pathway in astrocytes increased resistance of neurons to nonexcitotoxic glutamate toxicity.[560] In most cases, the presence of induced Nrf2 was neuroprotective, whereas Nrf2 deficiency conversely increased sensitivity to neurotoxins that cause inhibition of mitochondrial complex II (malonate or 3-NP), kainic acid, MPTP, and 6-hydroxydopamine (6-OHDA), as well as cerebral ischemia.[553,559,576–580] Protection or enhanced susceptibility were assessed in multiple brain regions such as the striatum, hippocampus, cortex, and substantia nigra, while degenerative end points assessed included decreased nerve terminal markers, neurotransmitters, caspase activation (apoptosis), neurotoxic glial activation, oxidative macromolecular damage, and even functional effects on motor coordination and learning and memory deficits.[553,559,576–581] Further evidence of Nrf2-mediated protection comes from studies which show that treatment with chemical activators of the Nrf2–ARE pathway prior to the toxic insults can reduce cellular damage in Nrf2 wild-type mice but not in Nrf2 KO mice.[555,557,576,579,580]

It has been suggested that METH-initiated neurotoxicity does not involve Nrf2, as DA depletion in the striatum at 2 weeks posttreatment was not different between Nrf2 KO mice and their wild-type controls.[582] However, studies in rat striatum indicate that METH can activate Nrf2 and regulate HO-1 expression.[326] Accordingly, it is uncertain whether ROS-mediated Nrf2 activation in specific brain regions caused by METH can modulate its neurotoxicity. Of interest is that some of the studies show regional differences in Nrf2 activation and the type of protective genes induced, as well as differences in the temporal pattern of induction, which highlights the potential for brain regional susceptibilities to neurotoxicity. For example, in a study of the involvement of the Keap1–Nrf2 system in ischemic brain injury, there was a difference in temporal changes of Nrf2-regulated TRX and HO-1 between the peri-infarct regions and regions destined to infarct.[583]

Activation of redox-sensitive transcription factors like Nrf2 can enhance the transcription of cytoprotective genes during oxidative stress. We investigated whether Nrf2 in mouse brain is activated by METH, thereby altering its neurotoxicity.[584] Multiple-day dosing with METH enhanced DNA oxidation and decreased tyrosine hydroxylase and DAT staining in the striatum, all of which were more severe in Nrf2 KO mice, as were deficits in motor coordination and olfactory discrimination. Similarly, METH increased striatal GFAP, indicating neurotoxic glial activation. Nrf2-mediated protection against METH was also observed in the glial cells and in the GABAergic system of the olfactory bulbs, whereas dopaminergic parameters were unaffected. In contrast, with a single-day dosing of METH, there were no differences between Nrf2 wild-type and Nrf2 KO mice in either basal or METH-enhanced DNA oxidation and neurotoxicity markers. Nrf2-mediated pathways accordingly protect against the neurodegenerative effects and functional deficits initiated by METH, and perhaps other ROS-enhancing neurotoxins, when repeated exposures allow time for transcriptional activation and protein induction.[584]

In rats, administration of the Nrf2 inducer sulforaphane following cortical impact injury improved behavioral performance when the treatment was initiated 1 hour, but not 6 hours, postinjury.[581] Chemical inducers of Nrf2 like tBHQ have been shown in rats to increase the levels of GSH, but not GST or NQO1, in the cortex and striatum.[555] However, under the same conditions, all of these antioxidant systems were induced in the liver, indicating a more robust Nrf2 effect in this organ than in the brain, possibly due to the contribution of hepatic metabolism in xenobiotic detoxification.[555]

Some neurotoxins themselves have been shown to activate Nrf2, for example, 6-OHDA was found to activate the Nrf2–ARE system and protect against ROS-mediated damage.[559,577] The loss of Nrf2 increased vulnerability to 6-OHDA both *in vitro* and *in vivo*, whereas upregulation of Nrf2 using tBHQ protected against 6-OHDA-induced cell death *in vitro*.[577] Similarly, selegiline (deprenyl), a drug used in the treatment of PD, stimulated Nrf2 activity as part of its cytoprotective mechanism of action.[585] Currently, clinical trials with DMF are ongoing.[548] DMF is able to activate Nrf2 and hence antioxidant enzyme production,[372] and oral administration of DMF significantly reduced the formation of new MS lesions in humans.[548]

10.3.8.4 Nrf2 in Fetal Neurodevelopmental Deficits

Preliminary evidence suggests that Nrf2 is expressed in the fetus and plays an important role in protecting the fetal brain from xenobiotic-enhanced oxidative stress.[584] METH-exposed –/– Nrf2-deficient KO fetuses were unable to increase transcription of the ROS-protective genes HO-1, NQO1, or OGG1, unlike wild-type controls, and exhibited enhanced DNA oxidation, fetal resorptions, a red phenotype with edema, and reduced fetal weight, with greater toxicity in female –/– Nrf2 fetuses. Postnatal neurodevelopmental deficits in activity and olfactory function were similarly enhanced, and the olfactory bulb GABAergic marker GAD-65 decreased, in –/– Nrf2-deficient offspring exposed *in utero* to METH, suggesting that olfactory deficits provide a potentially sensitive biomarker for long-term neurotoxicity.[586]

10.3.9 Nrf KO Mouse Models

Several types of Nrf KO mice have been generated, including double KOs of Nrf1 and Nrf2, which have revealed that although all Nrf isoforms can regulate the ARE, they are not functionally redundant.[479,490,526,587,588] Nrf3 KO mice are viable and have no obvious phenotypes.[587] Nrf1 has been shown to be important for development, as Nrf1 –/– mice die *in utero* due to anemia as a result of abnormal fetal liver erythropoiesis.[479] The viability of mice lacking Nrf2 has been taken to indicate that Nrf2 plays no role in normal mouse development.[490,526] However, in the context of oxidative stress, Nrf2-deficient fetuses may be more susceptible to developmental toxicities. Nrf2 mRNA and protein are expressed during the period of organogenesis[490]; hence, Nrf2 may become activated and modulate toxicity during *in utero* exposure to oxidative stress. Selective upregulation of Nrf2 during fetal development in instances of increased oxidative stress may be protective, as shown in our laboratory, and in studies from another laboratory where activation of Nrf2 in embryos protected against ethanol-induced embryotoxicity.[589] On the other hand, constitutive activation of Nrf2 in Keap1 –/– mice is detrimental, presumably by causing hyperkeratotic lesions that block the esophagus and increase postnatal death, possibly due to starvation.[590]

Mice lacking Nrf2 are highly sensitive to toxic stressors, some of which are neurotoxic, as discussed in the previous section. Primary astrocytes and neurons derived from Nrf2 KO mice are more sensitive than wild-type controls to oxidative damage, calcium disturbance and mitochondrial toxins,[591] and Nrf2 –/– stable cells are more sensitive to NO•-induced apoptosis.[592] Recent data suggest that Nrf2 plays an essential role in oligodendrocyte function and myelination, as Nrf2 KO mice exhibit severe myelin degeneration and oxidative damage to the myelin sheath, and aged Nrf2 KO animals develop vacuolar leukoencephalopathy and astrogliosis.[593]

In aged untreated animals, myelin degeneration in Nrf2-null mice was seen in the striatum, hippocampus, cerebellum, and corpus callosum, but importantly, the cortex and brainstem were spared, suggesting a regionally specific impact of chronic Nrf2 deficiency.[593] In addition, aged female Nrf2 –/– mice develop a multiorgan autoimmune disease similar to human systemic lupus erythematosus.[594–596] The reason for possible gender specificity remains unclear and may depend on Nrf2-regulated genes and their interactions with female-specific genes.

On the basis of experiments with Nrf2 KO mice and cells derived from these mice, there is interest in the Nrf2 system not only in neurotoxicity, but also in cancer chemoprevention, which shares some features with the rapid cell division and differentiation that occurs during embryonic and fetal development. Nrf2 has been shown to be protective against benzo[*a*]pyrene carcinogenicity.[552] Nrf2 is constitutively activated in diverse human cancers, suggesting that Nrf2-mediated regulation of the antioxidant response pathway may play a role in cancer susceptibility.[597] Cancer cells take control of the Nrf2 pathway, apparently to protect themselves from cellular stresses, which may make chemotherapies ineffective. Nrf2 may also play a role in experimental models of pulmonary diseases such as asthma,[598] inflammatory disorders such as inflammatory bowel disease,[599] acetaminophen hepatotoxicity,[600,601] and atherosclerosis.[602]

10.3.10 Evidence for Polymorphisms in the Keap1–Nrf2–ARE Pathway

Single nucleotide polymorphisms have been identified within the promoter region of the human Nrf2 gene,[491] and genetic variation here along with polymorphisms in the ARE region of Nrf2 target genes[603] may cause variability in Nrf2 activity, and hence the response to oxidative stress and neurotoxicity. Three single nucleotide polymorphisms (–650C/A, –684G/A, –686A/G) and one triplet repeat polymorphism (CCG) have been identified in the promoter region of the Nrf2 gene.[491] As discussed above, cancer cells with mutations to increase Nrf2 activity may promote their survival.[597] Conversely, inherited DNA polymorphisms can reduce the abundance of Nrf2. Promoter polymorphisms at positions –617 C/A or –651 G/A reduce the basal expression level of Nrf2, thereby resulting in attenuation of ARE-mediated gene transcription. The allelic frequencies of these inhibitory polymorphisms are as high as 20% in Europeans, 40% in Asians, and 55% in Native Americans.[499] Moreover, Marzec et al. reported that the –617 A single nucleotide polymorphism had a significantly higher risk for developing acute lung injury after major trauma.[499]

Vitiligo is an acquired pigmentary disorder and Nrf2 gene polymorphisms with the −650 C/A allele may be a risk factor associated with the development of vitiligo.[604] These polymorphisms are also significantly associated with the development of gastric mucosal inflammation and peptic ulcer diseases.[605–607] The recent analysis of Keap1 (iNrf2) in human lung cancer patients and cell lines showed that deletion, insertion, and missense mutations in functionally important domains of Keap1 result in reduction of Keap1 affinity for Nrf2 and elevated expression of cytoprotective genes, giving a selective advantage to cancerous cells.[608,609] The G430C substitution is a somatic mutation in Keap1 discovered in a human lung cancer patient.[608]

A number of somatic genetic events are important in cancerous cells, including Keap1 or Nrf2 mutations, and have been shown to provide advantages for lung and breast cancer growth as well as chemoresistance.[608–612] Studies on the epigenetic hypermethylation of the Keap1 promoter evaluated cytosine methylation in the Keap1 promoter and demonstrated that the P1 region, including 12 CpG sites, was highly methylated in lung cancer cells and tissues, but not in normal cells.[613,614] This would reduce levels of Keap1 and thereby activate Nrf2 in cancerous cells.[613,614]

Aside from the detrimental consequences of Nrf2–Keap1–ARE mutations arising in cancer cells, a deficient response in this pathway would be expected to enhance the risk of diseases and xenobiotic toxicities.

SUMMARY AND CONCLUSIONS

In this chapter, we have reviewed the endogenous sources of ROS and the various protective mechanisms in place for ROS detoxification, which under normal conditions are sufficient to maintain a balanced redox status. However, exposure to xenobiotics that cause increased ROS can challenge this delicate balance, which may leave particularly those individuals with enhanced pathways for ROS production, such as PHSs, or those with deficiencies in protective enzymes such as G6PD and others like OGG1 known to be regulated by the Nrf family, at increased risk for neurodegenerative diseases. Even endogenous sources of oxidative stress may contribute to neurodegenerative outcomes under conditions of excessive ROS production or compromised ROS detoxification or DNA repair.

The identification of new gene polymorphisms for PHSs, G6PD, and Nrf2, and their functional impacts, along with environmental modulators of these proteins, including epigenetic regulators, will provide a better understanding of not only regional susceptibility, but also individual risks for neurodegenerative outcomes, fetal damage, and neurodevelopmental deficits caused by exogenous and endogenous initiators of oxidative stress. In such cases, variations in drug abuse patterns, dietary exposures, and genetic predisposition may alter the induction and duration of enzymes like PHS involved in ROS production, and ROS-detoxifying enzymes and enzymes for the repair of DNA damage, including those regulated by Nrf2 activation, thereby varying the susceptibility to ROS-initiating compounds. Hence, genetic and environmental differences in PHS, G6PD, Nrf2, and the other ROS-related enzymes discussed herein likely constitute important determinants of neurodegenerative and neurodevelopmental risk from oxidative stress, which may be reduced by novel therapeutic strategies that limit PHS activity and/or enhance protective antioxidative/repair responses including G6PD and related ROS-protective proteins regulated by Nrf2.

ACKNOWLEDGMENTS

The authors' research was supported by grants from the Canadian Institutes of Health Research (CIHR). A.R. was supported in part by a doctoral scholarship from the CIHR and the Rx&D Health Research Foundation. A.M.S. was supported in part by a CIHR Frederick Banting and Charles Best Canada Graduate Scholarship.

REFERENCES

1. O'Callaghan, J.P., Miller, D.B. Neurotoxicity profiles of substituted amphetamines in the C57BL/6J mouse. *J. Pharmacol. Exp. Ther.* **1994**, *270*(2), 741–751.
2. Ricaurte, G.A., Guillery, R.W., Seiden, L.S., Schuster, C.R., Moore, R.Y. Dopamine nerve terminal degeneration produced by high doses of methylamphetamine in the rat brain. *Brain Res.* **1982**, *235*(1), 93–103.
3. Jeng, W., Ramkissoon, A., Parman, T., Wells, P.G., Prostaglandin, H. Synthase-catalyzed bioactivation of amphetamines to free radical intermediates that cause CNS regional DNA oxidation and nerve terminal degeneration. *FASEB J.* **2006**, *20*(6), 638–650.
4. Jeng, W., Wells, P.G. Reduced 3,4-methylenedioxymethamphetamine (MDMA, Ecstasy)-initiated oxidative DNA damage and neurodegeneration in prostaglandin H synthase-1 knockout mice. *ACS Chem Neurosci* **2010**, *1*(5), 366–380.
5. Halliwell, B., Gutteridge, J.M.C. *Free Radicals in Biology and Medicine*, 4th ed. Oxford University Press Inc., New York, 2007.
6. Wells, P.G., Winn, L.M. Biochemical toxicology of chemical teratogenesis. *Crit. Rev. Biochem. Mol. Biol.* **1996**, *31*(1), 1–40.

7. Wells, P.G., McCallum, G.P., Chen, C.S., Henderson, J.T., Lee, C.J., Perstin, J., Preston, T.J., Wiley, M.J., Wong, A.W. Oxidative stress in developmental origins of disease: Teratogenesis, neurodevelopmental deficits, and cancer. *Toxicol. Sci.* **2009**, *108*(1), 4–18.

8. Gabbita, S.P., Lovell, M.A., Markesbery, W.R. Increased nuclear DNA oxidation in the brain in Alzheimer's disease. *J. Neurochem.* **1998**, *71*(5), 2034–2040.

9. Shimura-Miura, H., Hattori, N., Kang, D., Miyako, K., Nakabeppu, Y., Mizuno, Y. Increased 8-oxo-dGTPase in the mitochondria of substantia nigral neurons in Parkinson's disease. *Ann. Neurol.* **1999**, *46*(6), 920–924.

10. Vladimirova, O., O'Connor, J., Cahill, A., Alder, H., Butunoi, C., Kalman, B. Oxidative damage to DNA in plaques of MS brains. *Mult. Scler.* **1998**, *4*(5), 413–418.

11. Turrens, J.F., Boveris, A. Generation of superoxide anion by the NADH dehydrogenase of bovine heart mitochondria. *Biochem. J.* **1980**, *191*(2), 421–427.

12. Boveris, A., Cadenas, E., Stoppani, A.O. Role of ubiquinone in the mitochondrial generation of hydrogen peroxide. *Biochem. J.* **1976**, *156*(2), 435–444.

13. Boveris, A., Oshino, N., Chance, B. The cellular production of hydrogen peroxide. *Biochem. J.* **1972**, *128*(3), 617–630.

14. Sugioka, K., Nakano, M., Totsune-Nakano, H., Minakami, H., Tero-Kubota, S., Ikegami, Y. Mechanism of O2- generation in reduction and oxidation cycle of ubiquinones in a model of mitochondrial electron transport systems. *Biochim. Biophys. Acta* **1988**, *936*(3), 377–385.

15. Chance, B., Sies, H., Boveris, A. Hydroperoxide metabolism in mammalian organs. *Physiol. Rev.* **1979**, *59*(3), 527–605.

16. Cadenas, E., Davies, K.J. Mitochondrial free radical generation, oxidative stress, and aging. *Free Radic. Biol. Med.* **2000**, *29*(3–4), 222–230.

17. Sorce, S., Krause, K.H. NOX enzymes in the central nervous system: From signaling to disease. *Antioxid. Redox Signal.* **2009**, *11*(10), 2481–2504.

18. Bedard, K., Krause, K.H. The NOX family of ROS-generating NADPH oxidases: Physiology and pathophysiology. *Physiol. Rev.* **2007**, *87*(1), 245–313.

19. Babior, B.M., Curnutte, J.T., Kipnes, B.S. Pyridine nucleotide-dependent superoxide production by a cell-free system from human granulocytes. *J. Clin. Invest.* **1975**, *56*(4), 1035–1042.

20. Infanger, D.W., Sharma, R.V., Davisson, R.L. NADPH oxidases of the brain: Distribution, regulation, and function. *Antioxid. Redox Signal.* **2006**, *8*(9–10), 1583–1596.

21. Kim, M.J., Shin, K.S., Chung, Y.B., Jung, K.W., Cha, C.I., Shin, D.H. Immunohistochemical study of p47Phox and gp91Phox distributions in rat brain. *Brain Res.* **2005**, *1040*(1–2), 178–186.

22. Serrano, F., Kolluri, N.S., Wientjes, F.B., Card, J.P., Klann, E. NADPH oxidase immunoreactivity in the mouse brain. *Brain Res.* **2003**, *988*(1–2), 193–198.

23. Block, M.L., Zecca, L., Hong, J.S. Microglia-mediated neurotoxicity: Uncovering the molecular mechanisms. *Nat. Rev. Neurosci.* **2007**, *8*(1), 57–69.

24. Lavigne, M.C., Malech, H.L., Holland, S.M., Leto, T.L. Genetic requirement of p47phox for superoxide production by murine microglia. *FASEB J.* **2001**, *15*(2), 285–287.

25. Abramov, A.Y., Jacobson, J., Wientjes, F., Hothersall, J., Canevari, L., Duchen, M.R. Expression and modulation of an NADPH oxidase in mammalian astrocytes. *J. Neurosci.* **2005**, *25*(40), 9176–9184.

26. Haorah, J., Ramirez, S.H., Floreani, N., Gorantla, S., Morsey, B., Persidsky, Y. Mechanism of alcohol-induced oxidative stress and neuronal injury. *Free Radic. Biol. Med.* **2008**, *45*(11), 1542–1550.

27. Shimohama, S., Tanino, H., Kawakami, N., Okamura, N., Kodama, H., Yamaguchi, T., Hayakawa, T., Nunomura, A., Chiba, S., Perry, G., Smith, M.A., Fujimoto, S. Activation of NADPH oxidase in Alzheimer's disease brains. *Biochem. Biophys. Res. Commun.* **2000**, *273*(1), 5–9.

28. Wu, D.C., Re, D.B., Nagai, M., Ischiropoulos, H., Przedborski, S. The inflammatory NADPH oxidase enzyme modulates motor neuron degeneration in amyotrophic lateral sclerosis mice. *Proc. Natl. Acad. Sci. U.S.A.* **2006**, *103*(32), 12132–12137.

29. Farooqui, A.A., Ong, W.Y., Horrocks, L.A. Inhibitors of brain phospholipase A2 activity: Their neuropharmacological effects and therapeutic importance for the treatment of neurologic disorders. *Pharmacol. Rev.* **2006**, *58*(3), 591–620.

30. Hirashima, Y., Farooqui, A.A., Mills, J.S., Horrocks, L.A. Identification and purification of calcium-independent phospholipase A2 from bovine brain cytosol. *J. Neurochem.* **1992**, *59*(2), 708–714.

31. Matsuzawa, A., Murakami, M., Atsumi, G., Imai, K., Prados, P., Inoue, K., Kudo, I. Release of secretory phospholipase A2 from rat neuronal cells and its possible function in the regulation of catecholamine secretion. *Biochem. J.* **1996**, *318*(Pt 2), 701–709.

32. Gattaz, W.F., Maras, A., Cairns, N.J., Levy, R., Forstl, H. Decreased phospholipase A2 activity in Alzheimer brains. *Biol. Psychiatry* **1995**, *37*(1), 13–17.

33. Stephenson, D., Rash, K., Smalstig, B., Roberts, E., Johnstone, E., Sharp, J., Panetta, J., Little, S., Kramer, R., Clemens, J. Cytosolic phospholipase A2 is induced in reactive glia following different forms of neurodegeneration. *Glia* **1999**, *27*(2), 110–128.

34. Sandhya, T.L., Ong, W.Y., Horrocks, L.A., Farooqui, A.A. A light and electron microscopic study of cytoplasmic phospholipase A2 and cyclooxygenase-2 in the hippocampus after kainate lesions. *Brain Res.* **1998**, *788*(1–2), 223–231.

35. Kalyvas, A., David, S. Cytosolic phospholipase A2 plays a key role in the pathogenesis of multiple sclerosis-like disease. *Neuron* **2004**, *41*(3), 323–335.

36. Klivenyi, P., Beal, M.F., Ferrante, R.J., Andreassen, O.A., Wermer, M., Chin, M.R., Bonventre, J.V. Mice deficient

in group IV cytosolic phospholipase A2 are resistant to MPTP neurotoxicity. *J. Neurochem.* **1998**, *71*(6), 2634–2637.

37. Cork, R.J., Perrone, M.L., Bridges, D., Wandell, J., Scheiner, C.A., Mize, R.R. A web-accessible digital atlas of the distribution of nitric oxide synthase in the mouse brain. *Prog. Brain Res.* **1998**, *118*, 37–50.

38. Bredt, D.S., Hwang, P.M., Snyder, S.H. Localization of nitric oxide synthase indicating a neural role for nitric oxide. *Nature* **1990**, *347*(6295), 768–770.

39. Murphy, S. Production of nitric oxide by glial cells: Regulation and potential roles in the CNS. *Glia* **2000**, *29*(1), 1–13.

40. Lee, S.C., Dickson, D.W., Liu, W., Brosnan, C.F. Induction of nitric oxide synthase activity in human astrocytes by interleukin-1 beta and interferon-gamma. *J. Neuroimmunol.* **1993**, *46*(1–2), 19–24.

41. Heneka, M.T., Feinstein, D.L. Expression and function of inducible nitric oxide synthase in neurons. *J. Neuroimmunol.* **2001**, *114*(1–2), 8–18.

42. Brown, G.C. Mechanisms of inflammatory neurodegeneration: iNOS and NADPH oxidase. *Biochem. Soc. Trans.* **2007**, *35*(Pt 5), 1119–1121.

43. Elfering, S.L., Sarkela, T.M., Giulivi, C. Biochemistry of mitochondrial nitric-oxide synthase. *J. Biol. Chem.* **2002**, *277*(41), 38079–38086.

44. Guix, F.X., Uribesalgo, I., Coma, M., Munoz, F.J. The physiology and pathophysiology of nitric oxide in the brain. *Prog. Neurobiol.* **2005**, *76*(2), 126–152.

45. Iyengar, R., Stuehr, D.J., Marletta, M.A. Macrophage synthesis of nitrite, nitrate, and N-nitrosamines: Precursors and role of the respiratory burst. *Proc. Natl. Acad. Sci. U.S.A.* **1987**, *84*(18), 6369–6373.

46. Xia, Y., Zweier, J.L. Superoxide and peroxynitrite generation from inducible nitric oxide synthase in macrophages. *Proc. Natl. Acad. Sci. U.S.A.* **1997**, *94*(13), 6954–6958.

47. Ignarro, L.J. Signal transduction mechanisms involving nitric oxide. *Biochem. Pharmacol.* **1991**, *41*(4), 485–490.

48. Beckman, J.S., Beckman, T.W., Chen, J., Marshall, P.A., Freeman, B.A. Apparent hydroxyl radical production by peroxynitrite: Implications for endothelial injury from nitric oxide and superoxide. *Proc. Natl. Acad. Sci. U.S.A.* **1990**, *87*(4), 1620–1624.

49. Brown, G.C. Nitric oxide and mitochondrial respiration. *Biochim. Biophys. Acta* **1999**, *1411*(2–3), 351–369.

50. Smith, M.A., Richey Harris, P.L., Sayre, L.M., Beckman, J.S., Perry, G. Widespread peroxynitrite-mediated damage in Alzheimer's disease. *J. Neurosci.* **1997**, *17*(8), 2653–2657.

51. Bagasra, O., Michaels, F.H., Zheng, Y.M., Bobroski, L.E., Spitsin, S.V., Fu, Z.F., Tawadros, R., Koprowski, H. Activation of the inducible form of nitric oxide synthase in the brains of patients with multiple sclerosis. *Proc. Natl. Acad. Sci. U.S.A.* **1995**, *92*(26), 12041–12045.

52. Liu, J.S., Zhao, M.L., Brosnan, C.F., Lee, S.C. Expression of inducible nitric oxide synthase and nitrotyrosine in multiple sclerosis lesions. *Am. J. Pathol.* **2001**, *158*(6), 2057–2066.

53. Shih, J.C. Molecular basis of human MAO A and B. *Neuropsychopharmacology* **1991**, *4*(1), 1–7.

54. Westlund, K.N., Denney, R.M., Kochersperger, L.M., Rose, R.M., Abell, C.W. Distinct monoamine oxidase A and B populations in primate brain. *Science* **1985**, *230*(4722), 181–183.

55. Levitt, P., Pintar, J.E., Breakefield, X.O. Immunocytochemical demonstration of monoamine oxidase B in brain astrocytes and serotonergic neurons. *Proc. Natl. Acad. Sci. U.S.A.* **1982**, *79*(20), 6385–6389.

56. Magyar, K., Knoll, J. Selective inhibition of the "B form" of monoamine oxidase. *Pol. J. Pharmacol. Pharm.* **1977**, *29*(3), 233–246.

57. Johnston, J.P. Some observations upon a new inhibitor of monoamine oxidase in brain tissue. *Biochem. Pharmacol.* **1968**, *17*(7), 1285–1297.

58. Youdim, M.B., Riederer, P. Dopamine metabolism and neurotransmission in primate brain in relationship to monoamine oxidase A and B inhibition. *J. Neural Transm. Gen. Sect.* **1993**, *91*(2–3), 181–195.

59. Heikkila, R.E., Manzino, L., Cabbat, F.S., Duvoisin, R.C. Protection against the dopaminergic neurotoxicity of 1-methyl-4-phenyl-1,2,5,6-tetrahydropyridine by monoamine oxidase inhibitors. *Nature* **1984**, *311*(5985), 467–469.

60. Hotamisligil, G.S., Girmen, A.S., Fink, J.S., Tivol, E., Shalish, C., Trofatter, J., Baenziger, J., Diamond, S., Markham, C., Sullivan, J., et al. Hereditary variations in monoamine oxidase as a risk factor for Parkinson's disease. *Mov. Disord.* **1994**, *9*(3), 305–310.

61. Riederer, P., Jellinger, K. Neurochemical insights into monoamine oxidase inhibitors, with special reference to deprenyl (selegiline). *Acta Neurol. Scand. Suppl.* **1983**, *95*, s43–s55.

62. Casarett, L.J., Doull, J., Klaassen, C.D. *Casarett and Doull's Toxicology: The Basic Science of Poisons*, 7th ed. McGraw-Hill, New York, 2008.

63. Mari, M., Cederbaum, A.I. CYP2E1 overexpression in HepG2 cells induces glutathione synthesis by transcriptional activation of gamma-glutamylcysteine synthetase. *J. Biol. Chem.* **2000**, *275*(20), 15563–15571.

64. Dutheil, F., Dauchy, S., Diry, M., Sazdovitch, V., Cloarec, O., Mellottee, L., Bieche, I., Ingelman-Sundberg, M., Flinois, J.P., de Waziers, I., Beaune, P., Decleves, X., Duyckaerts, C., Loriot, M.A. Xenobiotic-metabolizing enzymes and transporters in the normal human brain: Regional and cellular mapping as a basis for putative roles in cerebral function. *Drug Metab. Dispos.* **2009**, *37*(7), 1528–1538.

65. Farin, F.M., Omiecinski, C.J. Regiospecific expression of cytochrome P-450s and microsomal epoxide hydrolase in human brain tissue. *J. Toxicol. Environ. Health* **1993**, *40*(2–3), 317–335.

66. Jenner, P. Oxidative mechanisms in nigral cell death in Parkinson's disease. *Mov. Disord.* **1998**, *13*(Suppl. 1), 24–34.

67. Sasame, H.A., Ames, M.M., Nelson, S.D. Cytochrome P-450 and NADPH cytochrome c reductase in rat brain: Formation of catechols and reactive catechol metabolites. *Biochem. Biophys. Res. Commun.* **1977**, *78*(3), 919–926.

68. Warner, M., Hellmold, H., Yoshida, S., Liao, D., Hedlund, E., Gustafsson, J.A. Cytochrome P450 in the breast and brain: Role in tissue-specific activation of xenobiotics. *Mutat. Res.* **1997**, *376*(1–2), 79–85.

69. Sachse, C., Brockmoller, J., Bauer, S., Roots, I. Cytochrome P450 2D6 variants in a Caucasian population: Allele frequencies and phenotypic consequences. *Am. J. Hum. Genet.* **1997**, *60*(2), 284–295.

70. Harrison, R. Structure and function of xanthine oxidoreductase: Where are we now? *Free Radic. Biol. Med.* **2002**, *33*(6), 774–797.

71. Saito, T., Nishino, T. Differences in redox and kinetic properties between NAD-dependent and O2-dependent types of rat liver xanthine dehydrogenase. *J. Biol. Chem.* **1989**, *264*(17), 10015–10022.

72. Berry, C.E., Hare, J.M. Xanthine oxidoreductase and cardiovascular disease: Molecular mechanisms and pathophysiological implications. *J. Physiol.* **2004**, *555*(Pt 3), 589–606.

73. Betz, A.L. Identification of hypoxanthine transport and xanthine oxidase activity in brain capillaries. *J. Neurochem.* **1985**, *44*(2), 574–579.

74. Beetsch, J.W., Park, T.S., Dugan, L.L., Shah, A.R., Gidday, J.M. Xanthine oxidase-derived superoxide causes reoxygenation injury of ischemic cerebral endothelial cells. *Brain Res.* **1998**, *786*(1–2), 89–95.

75. Battelli, M.G., Buonamici, L., Virgili, M., Abbondanza, A., Contestabile, A. Simulated ischaemia-reperfusion conditions increase xanthine dehydrogenase and oxidase activities in rat brain slices. *Neurochem. Int.* **1998**, *32*(1), 17–21.

76. Lucas, D.R., Newhouse, J.P. The toxic effect of sodium L-glutamate on the inner layers of the retina. *AMA Arch. Ophthalmol.* **1957**, *58*(2), 193–201.

77. Olney, J.W., Sharpe, L.G. Brain lesions in an infant rhesus monkey treated with monsodium glutamate. *Science* **1969**, *166*(903), 386–388.

78. Erecinska, M., Silver, I.A. Metabolism and role of glutamate in mammalian brain. *Prog. Neurobiol.* **1990**, *35*(4), 245–296.

79. Perry, T.L., Yong, V.W., Bergeron, C., Hansen, S., Jones, K. Amino acids, glutathione, and glutathione transferase activity in the brains of patients with Alzheimer's disease. *Ann. Neurol.* **1987**, *21*(4), 331–336.

80. Dong, X.X., Wang, Y., Qin, Z.H. Molecular mechanisms of excitotoxicity and their relevance to pathogenesis of neurodegenerative diseases. *Acta Pharmacol. Sin.* **2009**, *30*(4), 379–387.

81. Monaghan, D.T., Bridges, R.J., Cotman, C.W. The excitatory amino acid receptors: Their classes, pharmacology, and distinct properties in the function of the central nervous system. *Annu. Rev. Pharmacol. Toxicol.* **1989**, *29*, 365–402.

82. Carriedo, S.G., Sensi, S.L., Yin, H.Z., Weiss, J.H. AMPA exposures induce mitochondrial Ca(2+) overload and ROS generation in spinal motor neurons in vitro. *J. Neurosci.* **2000**, *20*(1), 240–250.

83. Avshalumov, M.V., Rice, M.E. NMDA receptor activation mediates hydrogen peroxide-induced pathophysiology in rat hippocampal slices. *J. Neurophysiol.* **2002**, *87*(6), 2896–2903.

84. Danbolt, N.C. Glutamate uptake. *Prog. Neurobiol.* **2001**, *65*(1), 1–105.

85. Conti, F., Weinberg, R.J. Shaping excitation at glutamatergic synapses. *Trends Neurosci.* **1999**, *22*(10), 451–458.

86. Chen, C.J., Liao, S.L., Kuo, J.S. Gliotoxic action of glutamate on cultured astrocytes. *J. Neurochem.* **2000**, *75*(4), 1557–1565.

87. Oka, A., Belliveau, M.J., Rosenberg, P.A., Volpe, J.J. Vulnerability of oligodendroglia to glutamate: Pharmacology, mechanisms, and prevention. *J. Neurosci.* **1993**, *13*(4), 1441–1453.

88. Foran, E., Trotti, D. Glutamate transporters and the excitotoxic path to motor neuron degeneration in amyotrophic lateral sclerosis. *Antioxid. Redox Signal.* **2009**, *11*(7), 1587–1602.

89. Song, C., Zhang, Y., Parsons, C.G., Liu, Y.F. Expression of polyglutamine-expanded huntingtin induces tyrosine phosphorylation of N-methyl-D-aspartate receptors. *J. Biol. Chem.* **2003**, *278*(35), 33364–33369.

90. Parameshwaran, K., Dhanasekaran, M., Suppiramaniam, V. Amyloid beta peptides and glutamatergic synaptic dysregulation. *Exp. Neurol.* **2008**, *210*(1), 7–13.

91. Mittelbronn, M., Dietz, K., Schluesener, H.J., Meyermann, R. Local distribution of microglia in the normal adult human central nervous system differs by up to one order of magnitude. *Acta Neuropathol. (Berl)* **2001**, *101*(3), 249–255.

92. Lawson, L.J., Perry, V.H., Dri, P., Gordon, S. Heterogeneity in the distribution and morphology of microglia in the normal adult mouse brain. *Neuroscience* **1990**, *39*(1), 151–170.

93. Nimmerjahn, A., Kirchhoff, F., Helmchen, F. Resting microglial cells are highly dynamic surveillants of brain parenchyma in vivo. *Science* **2005**, *308*(5726), 1314–1318.

94. Fetler, L., Amigorena, S. Neuroscience. Brain under surveillance: The microglia patrol. *Science* **2005**, *309*(5733), 392–393.

95. Dheen, S.T., Kaur, C., Ling, E.A. Microglial activation and its implications in the brain diseases. *Curr. Med. Chem.* **2007**, *14*(11), 1189–1197.

96. Nakamichi, K., Saiki, M., Sawada, M., Yamamuro, Y., Morimoto, K., Kurane, I. Double-stranded RNA stimulates chemokine expression in microglia through vacuolar pH-dependent activation of intracellular signaling pathways. *J. Neurochem.* **2005**, *95*(1), 273–283.

97. Colton, C.A., Snell, J., Chernyshev, O., Gilbert, D.L. Induction of superoxide anion and nitric oxide production in cultured microglia. *Ann. N. Y. Acad. Sci.* **1994**, *738*, 54–63.

98. Hartlage-Rubsamen, M., Lemke, R., Schliebs, R. Interleukin-1beta, inducible nitric oxide synthase, and nuclear factor-kappaB are induced in morphologically distinct microglia after rat hippocampal lipopolysaccharide/interferon-gamma injection. *J. Neurosci. Res.* **1999**, *57*(3), 388–398.

99. Floden, A.M., Li, S., Combs, C.K. Beta-amyloid-stimulated microglia induce neuron death via synergistic stimulation of tumor necrosis factor alpha and NMDA receptors. *J. Neurosci.* **2005**, *25*(10), 2566–2575.

100. Colton, C., Wilt, S., Gilbert, D., Chernyshev, O., Snell, J., Dubois-Dalcq, M. Species differences in the generation of reactive oxygen species by microglia. *Mol. Chem. Neuropathol.* **1996**, *28*(1–3), 15–20.

101. Gao, H.M., Jiang, J., Wilson, B., Zhang, W., Hong, J.S., Liu, B. Microglial activation-mediated delayed and progressive degeneration of rat nigral dopaminergic neurons: Relevance to Parkinson's disease. *J. Neurochem.* **2002**, *81*(6), 1285–1297.

102. Gao, H.M., Liu, B., Zhang, W., Hong, J.S. Critical role of microglial NADPH oxidase-derived free radicals in the in vitro MPTP model of Parkinson's disease. *FASEB J.* **2003**, *17*(13), 1954–1956.

103. McGeer, P.L., Itagaki, S., Boyes, B.E., McGeer, E.G. Reactive microglia are positive for HLA-DR in the substantia nigra of Parkinson's and Alzheimer's disease brains. *Neurology* **1988**, *38*(8), 1285–1291.

104. McGeer, P.L., Itagaki, S., Tago, H., McGeer, E.G. Reactive microglia in patients with senile dementia of the Alzheimer type are positive for the histocompatibility glycoprotein HLA-DR. *Neurosci. Lett.* **1987**, *79*(1–2), 195–200.

105. Sapp, E., Kegel, K.B., Aronin, N., Hashikawa, T., Uchiyama, Y., Tohyama, K., Bhide, P.G., Vonsattel, J.P., DiFiglia, M. Early and progressive accumulation of reactive microglia in the Huntington disease brain. *J. Neuropathol. Exp. Neurol.* **2001**, *60*(2), 161–172.

106. Funk, C.D. Prostaglandins and leukotrienes: Advances in eicosanoid biology. *Science* **2001**, *294*(5548), 1871–1875.

107. Smith, W.L., DeWitt, D.L., Garavito, R.M. Cyclooxygenases: Structural, cellular, and molecular biology. *Annu. Rev. Biochem.* **2000**, *69*, 145–182.

108. Narumiya, S., Sugimoto, Y., Ushikubi, F. Prostanoid receptors: Structures, properties, and functions. *Physiol. Rev.* **1999**, *79*(4), 1193–1226.

109. Schaloske, R.H., Dennis, E.A. The phospholipase A2 superfamily and its group numbering system. *Biochim. Biophys. Acta* **2006**, *1761*(11), 1246–1259.

110. Irvine, R.F. How is the level of free arachidonic acid controlled in mammalian cells? *Biochem. J.* **1982**, *204*(1), 3–16.

111. Murphy, R.C., Gijon, M.A. Biosynthesis and metabolism of leukotrienes. *Biochem. J.* **2007**, *405*(3), 379–395.

112. Hamberg, M., Svensson, J., Wakabayashi, T., Samuelsson, B. Isolation and structure of two prostaglandin endoperoxides that cause platelet aggregation. *Proc. Natl. Acad. Sci. U.S.A.* **1974**, *71*(2), 345–349.

113. Ohki, S., Ogino, N., Yamamoto, S., Hayaishi, O. Prostaglandin hydroperoxidase, an integral part of prostaglandin endoperoxide synthetase from bovine vesicular gland microsomes. *J. Biol. Chem.* **1979**, *254*(3), 829–836.

114. Wu, K.K., Liou, J.Y. Cellular and molecular biology of prostacyclin synthase. *Biochem. Biophys. Res. Commun.* **2005**, *338*(1), 45–52.

115. Boie, Y., Rushmore, T.H., Darmon-Goodwin, A., Grygorczyk, R., Slipetz, D.M., Metters, K.M., Abramovitz, M. Cloning and expression of a cDNA for the human prostanoid IP receptor. *J. Biol. Chem.* **1994**, *269*(16), 12173–12178.

116. Kojo, H., Fukagawa, M., Tajima, K., Suzuki, A., Fujimura, T., Aramori, I., Hayashi, K., Nishimura, S. Evaluation of human peroxisome proliferator-activated receptor (PPAR) subtype selectivity of a variety of anti-inflammatory drugs based on a novel assay for PPAR delta(beta). *J. Pharmacol. Sci.* **2003**, *93*(3), 347–355.

117. Abdel-Halim, M.S., Hamberg, M., Sjoquist, B., Anggard, E. Identification of prostaglandin D2 as a major prostaglandin in homogenates of rat brain. *Prostaglandins* **1977**, *14*(4), 633–643.

118. Pettipher, R. The roles of the prostaglandin D(2) receptors DP(1) and CRTH2 in promoting allergic responses. *Br. J. Pharmacol.* **2008**, *153*(Suppl. 1), S191–S199.

119. Park, J.Y., Pillinger, M.H., Abramson, S.B. Prostaglandin E2 synthesis and secretion: The role of PGE2 synthases. *Clin. Immunol.* **2006**, *119*(3), 229–240.

120. Basu, S. Novel cyclooxygenase-catalyzed bioactive prostaglandin F2alpha from physiology to new principles in inflammation. *Med. Res. Rev.* **2007**, *27*(4), 435–468.

121. Kinsella, B.T., O'Mahony, D.J., Fitzgerald, G.A. The human thromboxane A2 receptor alpha isoform (TP alpha) functionally couples to the G proteins Gq and G11 in vivo and is activated by the isoprostane 8-epi prostaglandin F2 alpha. *J. Pharmacol. Exp. Ther.* **1997**, *281*(2), 957–964.

122. Straus, D.S., Glass, C.K. Cyclopentenone prostaglandins: New insights on biological activities and cellular targets. *Med. Res. Rev.* **2001**, *21*(3), 185–210.

123. Powell, W.S. 15-Deoxy-delta12,14-PGJ2: Endogenous PPARgamma ligand or minor eicosanoid degradation product? *J. Clin. Invest.* **2003**, *112*(6), 828–830.

124. Schuster, V.L. Prostaglandin transport. *Prostaglandins Other Lipid Mediat.* **2002**, *68-69*, 633–647.

125. Backlund, M.G., Mann, J.R., Holla, V.R., Buchanan, F.G., Tai, H.H., Musiek, E.S., Milne, G.L., Katkuri, S., DuBois, R.N. 15-Hydroxyprostaglandin dehydrogenase is downregulated in colorectal cancer. *J. Biol. Chem.* **2005**, *280*(5), 3217–3223.

126. DeWitt, D.L., Smith, W.L. Primary structure of prostaglandin G/H synthase from sheep vesicular gland determined

from the complementary DNA sequence. *Proc. Natl. Acad. Sci. U.S.A.* **1988**, *85*(5), 1412–1416.

127. Merlie, J.P., Fagan, D., Mudd, J., Needleman, P. Isolation and characterization of the complementary DNA for sheep seminal vesicle prostaglandin endoperoxide synthase (cyclooxygenase). *J. Biol. Chem.* **1988**, *263*(8), 3550–3553.

128. Yokoyama, C., Takai, T., Tanabe, T. Primary structure of sheep prostaglandin endoperoxide synthase deduced from cDNA sequence. *FEBS Lett.* **1988**, *231*(2), 347–351.

129. Yokoyama, C., Tanabe, T. Cloning of human gene encoding prostaglandin endoperoxide synthase and primary structure of the enzyme. *Biochem. Biophys. Res. Commun.* **1989**, *165*(2), 888–894.

130. Wang, L.H., Hajibeigi, A., Xu, X.M., Loose-Mitchell, D., Wu, K.K. Characterization of the promoter of human prostaglandin H synthase-1 gene. *Biochem. Biophys. Res. Commun.* **1993**, *190*(2), 406–411.

131. Xu, X.M., Tang, J.L., Chen, X., Wang, L.H., Wu, K.K. Involvement of two Sp1 elements in basal endothelial prostaglandin H synthase-1 promoter activity. *J. Biol. Chem.* **1997**, *272*(11), 6943–6950.

132. Kujubu, D.A., Fletcher, B.S., Varnum, B.C., Lim, R.W., Herschman, H.R. TIS10, a phorbol ester tumor promoter-inducible mRNA from Swiss 3T3 cells, encodes a novel prostaglandin synthase/cyclooxygenase homologue. *J. Biol. Chem.* **1991**, *266*(20), 12866–12872.

133. Xie, W.L., Chipman, J.G., Robertson, D.L., Erikson, R.L., Simmons, D.L. Expression of a mitogen-responsive gene encoding prostaglandin synthase is regulated by mRNA splicing. *Proc. Natl. Acad. Sci. U.S.A.* **1991**, *88*(7), 2692–2696.

134. Sawaoka, H., Dixon, D.A., Oates, J.A., Boutaud, O. Tristetraprolin binds to the 3′-untranslated region of cyclooxygenase-2 mRNA. A polyadenylation variant in a cancer cell line lacks the binding site. *J. Biol. Chem.* **2003**, *278*(16), 13928–13935.

135. Kosaka, T., Miyata, A., Ihara, H., Hara, S., Sugimoto, T., Takeda, O., Takahashi, E., Tanabe, T. Characterization of the human gene (PTGS2) encoding prostaglandin-endoperoxide synthase 2. *Eur. J. Biochem.* **1994**, *221*(3), 889–897.

136. Tanabe, T., Tohnai, N. Cyclooxygenase isozymes and their gene structures and expression. *Prostaglandins Other Lipid Mediat.* **2002**, *68–69*, 95–114.

137. Appleby, S.B., Ristimaki, A., Neilson, K., Narko, K., Hla, T. Structure of the human cyclo-oxygenase-2 gene. *Biochem. J.* **1994**, *302*(Pt 3), 723–727.

138. Tazawa, R., Xu, X.M., Wu, K.K., Wang, L.H. Characterization of the genomic structure, chromosomal location and promoter of human prostaglandin H synthase-2 gene. *Biochem. Biophys. Res. Commun.* **1994**, *203*(1), 190–199.

139. Chandrasekharan, N.V., Dai, H., Roos, K.L., Evanson, N.K., Tomsik, J., Elton, T.S., Simmons, D.L. COX-3, a cyclooxygenase-1 variant inhibited by acetaminophen and other analgesic/antipyretic drugs: Cloning, structure, and expression. *Proc. Natl. Acad. Sci. U.S.A.* **2002**, *99*(21), 13926–13931.

140. Kis, B., Snipes, J.A., Busija, D.W. Acetaminophen and the cyclooxygenase-3 puzzle: Sorting out facts, fictions, and uncertainties. *J. Pharmacol. Exp. Ther.* **2005**, *315*(1), 1–7.

141. Shaftel, S.S., Olschowka, J.A., Hurley, S.D., Moore, A.H., O'Banion, M.K. COX-3: A splice variant of cyclooxygenase-1 in mouse neural tissue and cells. *Brain Res. Mol. Brain Res.* **2003**, *119*(2), 213–215.

142. Dinchuk, J.E., Liu, R.Q., Trzaskos, J.M. COX-3: In the wrong frame in mind. *Immunol. Lett.* **2003**, *86*(1), 121.

143. DeLong, C.J., Smith, W.L. An intronic enhancer regulates cyclooxygenase-1 gene expression. *Biochem. Biophys. Res. Commun.* **2005**, *338*(1), 53–61.

144. Inoue, H., Tanabe, T., Umesono, K. Feedback control of cyclooxygenase-2 expression through PPARgamma. *J. Biol. Chem.* **2000**, *275*(36), 28028–28032.

145. Ristimaki, A., Narko, K., Hla, T. Down-regulation of cytokine-induced cyclo-oxygenase-2 transcript isoforms by dexamethasone: Evidence for post-transcriptional regulation. *Biochem. J.* **1996**, *318*(Pt 1), 325–331.

146. Hla, T. Molecular characterization of the 5.2 KB isoform of the human cyclooxygenase-1 transcript. *Prostaglandins* **1996**, *51*(1), 81–85.

147. Chen, C.Y., Del Gatto-Konczak, F., Wu, Z., Karin, M. Stabilization of interleukin-2 mRNA by the c-Jun NH2-terminal kinase pathway. *Science* **1998**, *280*(5371), 1945–1949.

148. Ming, X.F., Kaiser, M., Moroni, C. c-jun N-terminal kinase is involved in AUUUA-mediated interleukin-3 mRNA turnover in mast cells. *EMBO J.* **1998**, *17*(20), 6039–6048.

149. Mahboubi, K., Young, W., Ferreri, N.R. Tumour necrosis factor-alpha-dependent regulation of prostaglandin endoperoxide synthase-2. *Cytokine* **1998**, *10*(3), 175–184.

150. Kurumbail, R.G., Stevens, A.M., Gierse, J.K., McDonald, J.J., Stegeman, R.A., Pak, J.Y., Gildehaus, D., Miyashiro, J.M., Penning, T.D., Seibert, K., Isakson, P.C., Stallings, W.C. Structural basis for selective inhibition of cyclooxygenase-2 by anti-inflammatory agents. *Nature* **1996**, *384*(6610), 644–648.

151. Luong, C., Miller, A., Barnett, J., Chow, J., Ramesha, C., Browner, M.F. Flexibility of the NSAID binding site in the structure of human cyclooxygenase-2. *Nat. Struct. Mol. Biol.* **1996**, *3*(11), 927–933.

152. Kulmacz, R.J., van der Donk, W.A., Tsai, A.L. Comparison of the properties of prostaglandin H synthase-1 and -2. *Prog. Lipid Res.* **2003**, *42*(5), 377–404.

153. Otto, J.C., DeWitt, D.L., Smith, W.L. N-glycosylation of prostaglandin endoperoxide synthases-1 and -2 and their orientations in the endoplasmic reticulum. *J. Biol. Chem.* **1993**, *268*(24), 18234–18242.

154. Morita, I., Schindler, M., Regier, M.K., Otto, J.C., Hori, T., DeWitt, D.L., Smith, W.L. Different intracellular loca-

tions for prostaglandin endoperoxide H synthase-1 and -2. *J. Biol. Chem.* **1995**, *270*(18), 10902–10908.

155. Song, I., Smith, W.L. C-terminal Ser/Pro-Thr-Glu-Leu tetrapeptides of prostaglandin endoperoxide H synthases-1 and -2 target the enzymes to the endoplasmic reticulum. *Arch. Biochem. Biophys.* **1996**, *334*(1), 67–72.

156. Spencer, A.G., Woods, J.W., Arakawa, T., Singer, I.I., Smith, W.L. Subcellular localization of prostaglandin endoperoxide H synthases-1 and -2 by immunoelectron microscopy. *J. Biol. Chem.* **1998**, *273*(16), 9886–9893.

157. Guo, Q., Kulmacz, R.J. Distinct influences of carboxyl terminal segment structure on function in the two isoforms of prostaglandin H synthase. *Arch. Biochem. Biophys.* **2000**, *384*(2), 269–279.

158. Karthein, R., Dietz, R., Nastainczyk, W., Ruf, H.H. Higher oxidation states of prostaglandin H synthase. EPR study of a transient tyrosyl radical in the enzyme during the peroxidase reaction. *Eur. J. Biochem.* **1988**, *171*(1–2), 313–320.

159. Smith, W.L., Song, I. The enzymology of prostaglandin endoperoxide H synthases-1 and -2. *Prostaglandins Other Lipid Mediat.* **2002**, *68–69*, 115–128.

160. Landino, L.M., Crews, B.C., Gierse, J.K., Hauser, S.D., Marnett, L.J. Mutational analysis of the role of the distal histidine and glutamine residues of prostaglandin-endoperoxide synthase-2 in peroxidase catalysis, hydroperoxide reduction, and cyclooxygenase activation. *J. Biol. Chem.* **1997**, *272*(34), 21565–21574.

161. Rouzer, C.A., Marnett, L.J. Mechanism of free radical oxygenation of polyunsaturated fatty acids by cyclooxygenases. *Chem. Rev.* **2003**, *103*(6), 2239–2304.

162. Tsai, A., Kulmacz, R.J., Palmer, G. Spectroscopic evidence for reaction of prostaglandin H synthase-1 tyrosyl radical with arachidonic acid. *J. Biol. Chem.* **1995**, *270*(18), 10503–10508.

163. Tsai, A.L., Kulmacz, R.J. Prostaglandin H synthase: Resolved and unresolved mechanistic issues. *Arch. Biochem. Biophys.* **2010**, *493*(1), 103–124.

164. Hamberg, M., Samuelsson, B. Oxygenation of unsaturated fatty acids by the vesicular gland of sheep. *J. Biol. Chem.* **1967**, *242*(22), 5344–5354.

165. Chen, W., Pawelek, T.R., Kulmacz, R.J. Hydroperoxide dependence and cooperative cyclooxygenase kinetics in prostaglandin H synthase-1 and -2. *J. Biol. Chem.* **1999**, *274*(29), 20301–20306.

166. Kulmacz, R.J., Pendleton, R.B., Lands, W.E. Interaction between peroxidase and cyclooxygenase activities in prostaglandin-endoperoxide synthase. Interpretation of reaction kinetics. *J. Biol. Chem.* **1994**, *269*(8), 5527–5536.

167. Laneuville, O., Breuer, D.K., Xu, N., Huang, Z.H., Gage, D.A., Watson, J.T., Lagarde, M., DeWitt, D.L., Smith, W.L. Fatty acid substrate specificities of human prostaglandin-endoperoxide H synthase-1 and -2. Formation of 12-hydroxy-(9Z, 13E/Z, 15Z)- octadecatrienoic acids from alpha-linolenic acid. *J. Biol. Chem.* **1995**, *270*(33), 19330–19336.

168. Kulmacz, R.J., Wang, L.H. Comparison of hydroperoxide initiator requirements for the cyclooxygenase activities of prostaglandin H synthase-1 and -2. *J. Biol. Chem.* **1995**, *270*(41), 24019–24023.

169. Meade, E.A., Smith, W.L., DeWitt, D.L. Differential inhibition of prostaglandin endoperoxide synthase (cyclooxygenase) isozymes by aspirin and other non-steroidal anti-inflammatory drugs. *J. Biol. Chem.* **1993**, *268*(9), 6610–6614.

170. Lecomte, M., Laneuville, O., Ji, C., DeWitt, D.L., Smith, W.L. Acetylation of human prostaglandin endoperoxide synthase-2 (cyclooxygenase-2) by aspirin. *J. Biol. Chem.* **1994**, *269*(18), 13207–13215.

171. Vane, J.R. Inhibition of prostaglandin synthesis as a mechanism of action for aspirin-like drugs. *Nat. New Biol.* **1971**, *231*(25), 232–235.

172. Blobaum, A.L., Marnett, L.J. Structural and functional basis of cyclooxygenase inhibition. *J. Med. Chem.* **2007**, *50*(7), 1425–1441.

173. Capone, M.L., Tacconelli, S., Di Francesco, L., Sacchetti, A., Sciulli, M.G., Patrignani, P. Pharmacodynamic of cyclooxygenase inhibitors in humans. *Prostaglandins Other Lipid Mediat.* **2007**, *82*(1–4), 85–94.

174. Simmons, D.L., Botting, R.M., Hla, T. Cyclooxygenase isozymes: The biology of prostaglandin synthesis and inhibition. *Pharmacol. Rev.* **2004**, *56*(3), 387–437.

175. Loll, P.J., Picot, D., Garavito, R.M. The structural basis of aspirin activity inferred from the crystal structure of inactivated prostaglandin H2 synthase. *Nat. Struct. Mol. Biol.* **1995**, *2*(8), 637–643.

176. Picot, D., Loll, P.J., Garavito, R.M. The X-ray crystal structure of the membrane protein prostaglandin H2 synthase-1. *Nature* **1994**, *367*(6460), 243–249.

177. Pedersen, A.K., FitzGerald, G.A. Dose-related kinetics of aspirin. Presystemic acetylation of platelet cyclooxygenase. *N. Engl. J. Med.* **1984**, *311*(19), 1206–1211.

178. Selinsky, B.S., Gupta, K., Sharkey, C.T., Loll, P.J. Structural analysis of NSAID binding by prostaglandin H2 synthase: Time-dependent and time-independent inhibitors elicit identical enzyme conformations. *Biochemistry (Mosc)* **2001**, *40*(17), 5172–5180.

179. Mancini, J.A., Riendeau, D., Falgueyret, J.P., Vickers, P.J., O'Neill, G.P. Arginine 120 of prostaglandin G/H synthase-1 is required for the inhibition by nonsteroidal anti-inflammatory drugs containing a carboxylic acid moiety. *J. Biol. Chem.* **1995**, *270*(49), 29372–29377.

180. Gierse, J.K., McDonald, J.J., Hauser, S.D., Rangwala, S.H., Koboldt, C.M., Seibert, K. A single amino acid difference between cyclooxygenase-1 (COX-1) and -2 (COX-2) reverses the selectivity of COX-2 specific inhibitors. *J. Biol. Chem.* **1996**, *271*(26), 15810–15814.

181. Copeland, R.A., Williams, J.M., Giannaras, J., Nurnberg, S., Covington, M., Pinto, D., Pick, S., Trzaskos, J.M. Mechanism of selective inhibition of the inducible isoform of prostaglandin G/H synthase. *Proc. Natl. Acad. Sci. U.S.A.* **1994**, *91*(23), 11202–11206.

182. FitzGerald, G.A., Patrono, C. The coxibs, selective inhibitors of cyclooxygenase-2. *N. Engl. J. Med.* **2001**, *345*(6), 433–442.

183. Bresalier, R.S., Sandler, R.S., Quan, H., Bolognese, J.A., Oxenius, B., Horgan, K., Lines, C., Riddell, R., Morton, D., Lanas, A., Konstam, M.A., Baron, J.A. Cardiovascular events associated with rofecoxib in a colorectal adenoma chemoprevention trial. *N. Engl. J. Med.* **2005**, *352*(11), 1092–1102.

184. Solomon, S.D., McMurray, J.J., Pfeffer, M.A., Wittes, J., Fowler, R., Finn, P., Anderson, W.F., Zauber, A., Hawk, E., Bertagnolli, M. Cardiovascular risk associated with celecoxib in a clinical trial for colorectal adenoma prevention. *N. Engl. J. Med.* **2005**, *352*(11), 1071–1080.

185. Cao, C., Matsumura, K., Yamagata, K., Watanabe, Y. Endothelial cells of the rat brain vasculature express cyclooxygenase-2 mRNA in response to systemic interleukin-1 beta: A possible site of prostaglandin synthesis responsible for fever. *Brain Res.* **1996**, *733*(2), 263–272.

186. Matsumura, K., Cao, C., Ozaki, M., Morii, H., Nakadate, K., Watanabe, Y. Brain endothelial cells express cyclooxygenase-2 during lipopolysaccharide-induced fever: Light and electron microscopic immunocytochemical studies. *J. Neurosci.* **1998**, *18*(16), 6279–6289.

187. Yamagata, K., Andreasson, K.I., Kaufmann, W.E., Barnes, C.A., Worley, P.F. Expression of a mitogen-inducible cyclooxygenase in brain neurons: Regulation by synaptic activity and glucocorticoids. *Neuron* **1993**, *11*(2), 371–386.

188. Yasojima, K., Schwab, C., McGeer, E.G., McGeer, P.L. Up-regulated production and activation of the complement system in Alzheimer's disease brain. *Am. J. Pathol.* **1999**, *154*(3), 927–936.

189. Rocca, B., Spain, L.M., Pure, E., Langenbach, R., Patrono, C., FitzGerald, G.A. Distinct roles of prostaglandin H synthases 1 and 2 in T-cell development. *J. Clin. Invest.* **1999**, *103*(10), 1469–1477.

190. Hoozemans, J.J., Rozemuller, A.J., Janssen, I., De Groot, C.J., Veerhuis, R., Eikelenboom, P. Cyclooxygenase expression in microglia and neurons in Alzheimer's disease and control brain. *Acta Neuropathol. (Berl)* **2001**, *101*(1), 2–8.

191. Yermakova, A.V., Rollins, J., Callahan, L.M., Rogers, J., O'Banion, M.K. Cyclooxygenase-1 in human Alzheimer and control brain: Quantitative analysis of expression by microglia and CA3 hippocampal neurons. *J. Neuropathol. Exp. Neurol.* **1999**, *58*(11), 1135–1146.

192. Breder, C.D., Dewitt, D., Kraig, R.P. Characterization of inducible cyclooxygenase in rat brain. *J. Comp. Neurol.* **1995**, *355*(2), 296–315.

193. Dou, W., Jiao, Y., Goorha, S., Raghow, R., Ballou, L.R. Nociception and the differential expression of cyclooxygenase-1 (COX-1), the COX-1 variant retaining intron-1 (COX-1v), and COX-2 in mouse dorsal root ganglia (DRG). *Prostaglandins Other Lipid Mediat.* **2004**, *74*(1–4), 29–43.

194. Candelario-Jalil, E., de Oliveira, A.C., Graf, S., Bhatia, H.S., Hull, M., Munoz, E., Fiebich, B.L. Resveratrol potently reduces prostaglandin E2 production and free radical formation in lipopolysaccharide-activated primary rat microglia. *J. Neuroinflammation* **2007**, *4*, 25.

195. Choi, S.H., Langenbach, R., Bosetti, F. Genetic deletion or pharmacological inhibition of cyclooxygenase-1 attenuate lipopolysaccharide-induced inflammatory response and brain injury. *FASEB J.* **2008**, *22*(5), 1491–1501.

196. Pepicelli, O., Fedele, E., Berardi, M., Raiteri, M., Levi, G., Greco, A., Ajmone-Cat, M.A., Minghetti, L. Cyclooxygenase-1 and -2 differently contribute to prostaglandin E2 synthesis and lipid peroxidation after in vivo activation of N-methyl-D-aspartate receptors in rat hippocampus. *J. Neurochem.* **2005**, *93*(6), 1561–1567.

197. Schwab, J.M., Beschorner, R., Meyermann, R., Gozalan, F., Schluesener, H.J. Persistent accumulation of cyclooxygenase-1-expressing microglial cells and macrophages and transient upregulation by endothelium in human brain injury. *J. Neurosurg.* **2002**, *96*(5), 892–899.

198. Yermakova, A.V., O'Banion, M.K. Downregulation of neuronal cyclooxygenase-2 expression in end stage Alzheimer's disease. *Neurobiol. Aging* **2001**, *22*(6), 823–836.

199. Kaufmann, W.E., Worley, P.F., Pegg, J., Bremer, M., Isakson, P. COX-2, a synaptically induced enzyme, is expressed by excitatory neurons at postsynaptic sites in rat cerebral cortex. *Proc. Natl. Acad. Sci. U.S.A.* **1996**, *93*(6), 2317–2321.

200. Ho, L., Purohit, D., Haroutunian, V., Luterman, J.D., Willis, F., Naslund, J., Buxbaum, J.D., Mohs, R.C., Aisen, P.S., Pasinetti, G.M. Neuronal cyclooxygenase 2 expression in the hippocampal formation as a function of the clinical progression of Alzheimer disease. *Arch. Neurol.* **2001**, *58*(3), 487–492.

201. Tomimoto, H., Akiguchi, I., Wakita, H., Lin, J.X., Budka, H. Cyclooxygenase-2 is induced in microglia during chronic cerebral ischemia in humans. *Acta Neuropathol. (Berl)* **2000**, *99*(1), 26–30.

202. Ghilardi, J.R., Svensson, C.I., Rogers, S.D., Yaksh, T.L., Mantyh, P.W. Constitutive spinal cyclooxygenase-2 participates in the initiation of tissue injury-induced hyperalgesia. *J. Neurosci.* **2004**, *24*(11), 2727–2732.

203. Maslinska, D., Wozniak, R., Kaliszek, A., Modelska, I. Expression of cyclooxygenase-2 in astrocytes of human brain after global ischemia. *Folia Neuropathol.* **1999**, *37*(2), 75–79.

204. O'Banion, M.K., Miller, J.C., Chang, J.W., Kaplan, M.D., Coleman, P.D. Interleukin-1 beta induces prostaglandin G/H synthase-2 (cyclooxygenase-2) in primary murine astrocyte cultures. *J. Neurochem.* **1996**, *66*(6), 2532–2540.

205. Yang, H., Chen, C. Cyclooxygenase-2 in synaptic signaling. *Curr. Pharm. Des.* **2008**, *14*(14), 1443–1451.

206. Beiche, F., Scheuerer, S., Brune, K., Geisslinger, G., Goppelt-Struebe, M. Up-regulation of cyclooxygenase-2 mRNA in the rat spinal cord following peripheral inflammation. *FEBS Lett.* **1996**, *390*(2), 165–169.

207. Samad, T.A., Moore, K.A., Sapirstein, A., Billet, S., Allchorne, A., Poole, S., Bonventre, J.V., Woolf, C.J. Interleukin-1beta-mediated induction of Cox-2 in the CNS contributes to inflammatory pain hypersensitivity. *Nature* **2001**, *410*(6827), 471–475.
208. Eling, T.E., Curtis, J.F. Xenobiotic metabolism by prostaglandin H synthase. *Pharmacol. Ther.* **1992**, *53*(2), 261–273.
209. Marnett, L.J. Prostaglandin synthase-mediated metabolism of carcinogens and a potential role for peroxyl radicals as reactive intermediates. *Environ. Health Perspect.* **1990**, *88*, 5–12.
210. Reed, G.A., Marnett, L.J. Metabolism and activation of 7,8-dihydrobenzo[a]pyrene during prostaglandin biosynthesis. Intermediacy of a bay-region epoxide. *J. Biol. Chem.* **1982**, *257*(19), 11368–11376.
211. Hughes, M.F., Mason, R.P., Eling, T.E. Prostaglandin hydroperoxidase-dependent oxidation of phenylbutazone: Relationship to inhibition of prostaglandin cyclooxygenase. *Mol. Pharmacol.* **1988**, *34*(2), 186–193.
212. Liu, Y., Levy, G.N. Activation of heterocyclic amines by combinations of prostaglandin H synthase-1 and -2 with N-acetyltransferase 1 and 2. *Cancer Lett.* **1998**, *133*(1), 115–123.
213. Wells, P.G., McCallum, G.P., Lam, K.C., Henderson, J.T., Ondovcik, S.L. Oxidative DNA damage and repair in teratogenesis and neurodevelopmental deficits. *Birth Defects Res. C Embryo Today* **2010**, *90*(2), 103–109.
214. Markey, C.M., Alward, A., Weller, P.E., Marnett, L.J. Quantitative studies of hydroperoxide reduction by prostaglandin H synthase. Reducing substrate specificity and the relationship of peroxidase to cyclooxygenase activities. *J. Biol. Chem.* **1987**, *262*(13), 6266–6279.
215. Kulmacz, R.J. Regulation of cyclooxygenase catalysis by hydroperoxides. *Biochem. Biophys. Res. Commun.* **2005**, *338*(1), 25–33.
216. Hastings, T.G. Enzymatic oxidation of dopamine: The role of prostaglandin H synthase. *J. Neurochem.* **1995**, *64*(2), 919–924.
217. Mattammal, M.B., Strong, R., Lakshmi, V.M., Chung, H.D., Stephenson, A.H., Prostaglandin, H. Synthetase-mediated metabolism of dopamine: Implication for Parkinson's disease. *J. Neurochem.* **1995**, *64*(4), 1645–1654.
218. Segura-Aguilar, J., Metodiewa, D., Welch, C.J. Metabolic activation of dopamine o-quinones to o-semiquinones by NADPH cytochrome P450 reductase may play an important role in oxidative stress and apoptotic effects. *Biochim. Biophys. Acta* **1998**, *1381*(1), 1–6.
219. Cooke, M.S., Evans, M.D., Dizdaroglu, M., Lunec, J. Oxidative DNA damage: Mechanisms, mutation, and disease. *FASEB J.* **2003**, *17*(10), 1195–1214.
220. Goncalves, L.L., Ramkissoon, A., Wells, P.G. Prostaglandin H synthase-1-catalyzed bioactivation of neurotransmitters, their precursors, and metabolites: Oxidative DNA damage and electron spin resonance spectroscopy studies. *Chem. Res. Toxicol.* **2009**, *22*(5), 842–852.
221. Ramkissoon, A., Wells, P.G. Human prostaglandin H synthase (hPHS)-1- and hPHS-2-dependent bioactivation, oxidative macromolecular damage, and cytotoxicity of dopamine, its precursor, and its metabolites. *Free Radic Biol Med.* **2011**, *50*(2), 295–304.
222. Parman, T., Chen, G., Wells, P.G. Free radical intermediates of phenytoin and related teratogens. Prostaglandin H synthase-catalyzed bioactivation, electron paramagnetic resonance spectrometry, and photochemical product analysis. *J. Biol. Chem.* **1998**, *273*(39), 25079–25088.
223. Parman, T., Wells, P.G. Embryonic prostaglandin H synthase-2 (PHS-2) expression and benzo[a]pyrene teratogenicity in PHS-2 knockout mice. *FASEB J.* **2002**, *16*(9), 1001–1009.
224. Parman, T., Wiley, M.J., Wells, P.G. Free radical-mediated oxidative DNA damage in the mechanism of thalidomide teratogenicity. *Nat. Med.* **1999**, *5*(5), 582–585.
225. Winn, L.M., Wells, P.G. Evidence for embryonic prostaglandin H synthase-catalyzed bioactivation and reactive oxygen species-mediated oxidation of cellular macromolecules in phenytoin and benzo[a]pyrene teratogenesis. *Free Radic. Biol. Med.* **1997**, *22*(4), 607–621.
226. Jeng, W., Wong, A.W., Ting, A.K.R., Wells, P.G. Methamphetamine-enhanced embryonic oxidative DNA damage and neurodevelopmental deficits. *Free Radic. Biol. Med.* **2005**, *39*(3), 317–326.
227. Wong, A.W., McCallum, G.P., Jeng, W., Wells, P.G. Oxoguanine glycosylase 1 protects against methamphetamine-enhanced fetal brain oxidative DNA damage and neurodevelopmental deficits. *J. Neurosci.* **2008**, *28*(36), 9047–9054.
228. Liu, Y., Levy, G.N., Weber, W.W. Activation of 2-aminofluorene by prostaglandin endoperoxide H synthase-2. *Biochem. Biophys. Res. Commun.* **1995**, *215*(1), 346–354.
229. Wiese, F.W., Thompson, P.A., Kadlubar, F.F. Carcinogen substrate specificity of human COX-1 and COX-2. *Carcinogenesis* **2001**, *22*(1), 5–10.
230a. Jeng, W., Ramkissoon, A., Wells, P.G. Reduced DNA oxidation in aged prostaglandin H synthase-1 knockout mice. *Free Radic. Biol. Med.* **2011**, *50*(4), 550–556.
230b. Ramkissoon, A., Wells, P.G, Human prostaglandin H synthase (hPHS)-1 and hPHS-2 in amphetamine analog bioactivation, DNA oxidation and cytotoxicity. *Toxicol. Sci.* **2011**, *120*(1), 154–162.
231. Pasinetti, G.M., Aisen, P.S. Cyclooxygenase-2 expression is increased in frontal cortex of Alzheimer's disease brain. *Neuroscience* **1998**, *87*(2), 319–324.
232. Aisen, P.S., Schafer, K.A., Grundman, M., Pfeiffer, E., Sano, M., Davis, K.L., Farlow, M.R., Jin, S., Thomas, R.G., Thal, L.J. Effects of rofecoxib or naproxen vs placebo on Alzheimer disease progression: A randomized controlled trial. *JAMA* **2003**, *289*(21), 2819–2826.
233. Kotilinek, L.A., Westerman, M.A., Wang, Q., Panizzon, K., Lim, G.P., Simonyi, A., Lesne, S., Falinska, A., Younkin, L.H., Younkin, S.G., Rowan, M., Cleary, J., Wallis, R.A.,

233. Sun, G.Y., Cole, G., Frautschy, S., Anwyl, R., Ashe, K.H. Cyclooxygenase-2 inhibition improves amyloid-beta-mediated suppression of memory and synaptic plasticity. *Brain* **2008**, *131*(Pt 3), 651–664.

234. Jang, J.H., Surh, Y.J. Beta-amyloid-induced apoptosis is associated with cyclooxygenase-2 up-regulation via the mitogen-activated protein kinase-NF-kappaB signaling pathway. *Free Radic. Biol. Med.* **2005**, *38*(12), 1604–1613.

235. Ritchie, K., Lovestone, S. The dementias. *Lancet* **2002**, *360*(9347), 1759–1766.

236. Thal, L.J., Ferris, S.H., Kirby, L., Block, G.A., Lines, C.R., Yuen, E., Assaid, C., Nessly, M.L., Norman, B.A., Baranak, C.C., Reines, S.A. A randomized, double-blind, study of rofecoxib in patients with mild cognitive impairment. *Neuropsychopharmacology* **2005**, *30*(6), 1204–1215.

237. Martino, G., Adorini, L., Rieckmann, P., Hillert, J., Kallmann, B., Comi, G., Filippi, M. Inflammation in multiple sclerosis: The good, the bad, and the complex. *Lancet Neurol.* **2002**, *1*(8), 499–509.

238. Matute, C., Alberdi, E., Domercq, M., Perez-Cerda, F., Perez-Samartin, A., Sanchez-Gomez, M.V. The link between excitotoxic oligodendroglial death and demyelinating diseases. *Trends Neurosci.* **2001**, *24*(4), 224–230.

239. Polymeropoulos, M.H., Lavedan, C., Leroy, E., Ide, S.E., Dehejia, A., Dutra, A., Pike, B., Root, H., Rubenstein, J., Boyer, R., Stenroos, E.S., Chandrasekharappa, S., Athanassiadou, A., Papapetropoulos, T., Johnson, W.G., Lazzarini, A.M., Duvoisin, R.C., Di Iorio, G., Golbe, L.I., Nussbaum, R.L. Mutation in the alpha-synuclein gene identified in families with Parkinson's disease. *Science* **1997**, *276*(5321), 2045–2047.

240. Schapira, A.H., Cooper, J.M., Dexter, D., Clark, J.B., Jenner, P., Marsden, C.D. Mitochondrial complex I deficiency in Parkinson's disease. *J. Neurochem.* **1990**, *54*(3), 823–827.

241. Bendotti, C., Carri, M.T. Lessons from models of SOD1-linked familial ALS. *Trends Mol. Med.* **2004**, *10*(8), 393–400.

242. Bruijn, L.I., Houseweart, M.K., Kato, S., Anderson, K.L., Anderson, S.D., Ohama, E., Reaume, A.G., Scott, R.W., Cleveland, D.W. Aggregation and motor neuron toxicity of an ALS-linked SOD1 mutant independent from wild-type SOD1. *Science* **1998**, *281*(5384), 1851–1854.

243. Anglin, M.D., Burke, C., Perrochet, B., Stamper, E., Dawud-Noursi, S. History of the methamphetamine problem. *J. Psychoactive Drugs* **2000**, *32*(2), 137–141.

244. Kuczenski, R., Segal, D.S., Cho, A.K., Melega, W. Hippocampus norepinephrine, caudate dopamine and serotonin, and behavioral responses to the stereoisomers of amphetamine and methamphetamine. *J. Neurosci.* **1995**, *15*(2), 1308–1317.

245. Cho, A.K. Ice: A new dosage form of an old drug. *Science* **1990**, *249*(4969), 631–634.

246. Cho, A.K., Melega, W.P., Kuczenski, R., Segal, D.S. Relevance of pharmacokinetic parameters in animal models of methamphetamine abuse. *Synapse* **2001**, *39*(2), 161–166.

247. Kalant, H. The pharmacology and toxicology of "ecstasy" (MDMA) and related drugs. *Can. Med. Assoc. J.* **2001**, *165*(7), 917–928.

248. Gibb, J.W., Johnson, M., Stone, D., Hanson, G.R. MDMA: Historical perspectives. *Ann. N. Y. Acad. Sci.* **1990**, *600*, 601–611; discussion 611–612.

249. Grinspoon, L., Bakalar, J.B. Can drugs be used to enhance the psychotherapeutic process? *Am. J. Psychother.* **1986**, *40*(3), 393–404.

250. Ricaurte, G., Bryan, G., Strauss, L., Seiden, L., Schuster, C. Hallucinogenic amphetamine selectively destroys brain serotonin nerve terminals. *Science* **1985**, *229*(4717), 986–988.

251. Green, A.R., Mechan, A.O., Elliott, J.M., O'Shea, E., Colado, M.I. The pharmacology and clinical pharmacology of 3,4-methylenedioxymethamphetamine (MDMA, "ecstasy"). *Pharmacol. Rev.* **2003**, *55*(3), 463–508.

252. Golub, M., Costa, L., Crofton, K., Frank, D., Fried, P., Gladen, B., Henderson, R., Liebelt, E., Lusskin, S., Marty, S., Rowland, A., Scialli, J., Vore, M. NTP-CERHR Expert Panel Report on the reproductive and developmental toxicity of amphetamine and methamphetamine. *Birth Defects Res. B Dev. Reprod. Toxicol.* **2005**, *74*(6), 471–584.

253. Cook, C.E., Jeffcoat, A.R., Hill, J.M., Pugh, D.E., Patetta, P.K., Sadler, B.M., White, W.R., Perez-Reyes, M. Pharmacokinetics of methamphetamine self-administered to human subjects by smoking S-(+)-methamphetamine hydrochloride. *Drug Metab. Dispos.* **1993**, *21*(4), 717–723.

254. Cook, C.E., Jeffcoat, A.R., Sadler, B.M., Hill, J.M., Voyksner, R.D., Pugh, D.E., White, W.R., Perez-Reyes, M. Pharmacokinetics of oral methamphetamine and effects of repeated daily dosing in humans. *Drug Metab. Dispos.* **1992**, *20*(6), 856–862.

255. Huestis, M.A., Cone, E.J. Methamphetamine disposition in oral fluid, plasma, and urine. *Ann. N. Y. Acad. Sci.* **2007**, *1098*, 104–121.

256. Schepers, R.J., Oyler, J.M., Joseph, R.E., Jr., Cone, E.J., Moolchan, E.T., Huestis, M.A. Methamphetamine and amphetamine pharmacokinetics in oral fluid and plasma after controlled oral methamphetamine administration to human volunteers. *Clin. Chem.* **2003**, *49*(1), 121–132.

257. Wilson, J.M., Kalasinsky, K.S., Levey, A.I., Bergeron, C., Reiber, G., Anthony, R.M., Schmunk, G.A., Shannak, K., Haycock, J.W., Kish, S.J. Striatal dopamine nerve terminal markers in human, chronic methamphetamine users. *Nat. Med.* **1996**, *2*(6), 699–703.

258. Harris, D.S., Boxenbaum, H., Everhart, E.T., Sequeira, G., Mendelson, J.E., Jones, R.T. The bioavailability of intranasal and smoked methamphetamine. *Clin. Pharmacol. Ther.* **2003**, *74*(5), 475–486.

259. Riviere, G.J., Gentry, W.B., Owens, S.M. Disposition of methamphetamine and its metabolite amphetamine in brain and other tissues in rats after intravenous administration. *J. Pharmacol. Exp. Ther.* **2000**, *292*(3), 1042–1047.

260. Bartu, A., Dusci, L.J., Ilett, K.F. Transfer of methylamphetamine and amphetamine into breast milk following recreational use of methylamphetamine. *Br. J. Clin. Pharmacol.* **2009**, *67*(4), 455–459.
261. Stewart, J.L., Meeker, J.E. Fetal and infant deaths associated with maternal methamphetamine abuse. *J. Anal. Toxicol.* **1997**, *21*(6), 515–517.
262. Melega, W.P., Williams, A.E., Schmitz, D.A., DiStefano, E.W., Cho, A.K. Pharmacokinetic and pharmacodynamic analysis of the actions of D-amphetamine and D-methamphetamine on the dopamine terminal. *J. Pharmacol. Exp. Ther.* **1995**, *274*(1), 90–96.
263. de la Torre, R., Farre, M., Ortuno, J., Mas, M., Brenneisen, R., Roset, P.N., Segura, J., Cami, J. Non-linear pharmacokinetics of MDMA ("ecstasy") in humans. *Br. J. Clin. Pharmacol.* **2000**, *49*(2), 104–109.
264. Shappell, S.A., Kearns, G.L., Valentine, J.L., Neri, D.F., DeJohn, C.A. Chronopharmacokinetics and chronopharmacodynamics of dextromethamphetamine in man. *J. Clin. Pharmacol.* **1996**, *36*(11), 1051–1063.
265. de la Torre, R., Farre, M. Neurotoxicity of MDMA (ecstasy): The limitations of scaling from animals to humans. *Trends Pharmacol. Sci.* **2004**, *25*(10), 505–508.
266. Jayanthi, S., Ladenheim, B., Cadet, J.L. Methamphetamine-induced changes in antioxidant enzymes and lipid peroxidation in copper/zinc-superoxide dismutase transgenic mice. *Ann. N. Y. Acad. Sci.* **1998**, *844*, 92–102.
267. Kita, T., Shimada, K., Mastunari, Y., Wagner, G.C., Kubo, K., Nakashima, T. Methamphetamine-induced striatal dopamine neurotoxicity and cyclooxygenase-2 protein expression in BALB/c mice. *Neuropharmacology* **2000**, *39*(3), 399–406.
268. Melega, W.P., Cho, A.K., Harvey, D., Lacan, G. Methamphetamine blood concentrations in human abusers: Application to pharmacokinetic modeling. *Synapse* **2007**, *61*(4), 216–220.
269. McKetin, R., Kelly, E., McLaren, J. The relationship between crystalline methamphetamine use and methamphetamine dependence. *Drug Alcohol Depend.* **2006**, *85*(3), 198–204.
270. Logan, B.K. Methamphetamine and driving impairment. *J. Forensic Sci.* **1996**, *41*(3), 457–464.
271. Kojima, T., Une, I., Yashiki, M., Noda, J., Sakai, K., Yamamoto, K. A fatal methamphetamine poisoning associated with hyperpyrexia. *Forensic Sci. Int.* **1984**, *24*(1), 87–93.
272. Kalasinsky, K.S., Bosy, T.Z., Schmunk, G.A., Reiber, G., Anthony, R.M., Furukawa, Y., Guttman, M., Kish, S.J. Regional distribution of methamphetamine in autopsied brain of chronic human methamphetamine users. *Forensic Sci. Int.* **2001**, *116*(2–3), 163–169.
273. Fowler, J.S., Volkow, N.D., Logan, J., Alexoff, D., Telang, F., Wang, G.J., Wong, C., Ma, Y., Kriplani, A., Pradhan, K., Schlyer, D., Jayne, M., Hubbard, B., Carter, P., Warner, D., King, P., Shea, C., Xu, Y., Muench, L., Apelskog, K. Fast uptake and long-lasting binding of methamphetamine in the human brain: Comparison with cocaine. *Neuroimage* **2008**, *43*(4), 756–763.
274. Fowler, J.S., Kroll, C., Ferrieri, R., Alexoff, D., Logan, J., Dewey, S.L., Schiffer, W., Schlyer, D., Carter, P., King, P., Shea, C., Xu, Y., Muench, L., Benveniste, H., Vaska, P., Volkow, N.D. PET studies of d-methamphetamine pharmacokinetics in primates: Comparison with l-methamphetamine and (–)-cocaine. *J. Nucl. Med.* **2007**, *48*(10), 1724–1732.
275. O'Neil, M.L., Kuczenski, R., Segal, D.S., Cho, A.K., Lacan, G., Melega, W.P. Escalating dose pretreatment induces pharmacodynamic and not pharmacokinetic tolerance to a subsequent high-dose methamphetamine binge. *Synapse* **2006**, *60*(6), 465–473.
276. Jonsson, G., Nwanze, E. Selective (+)-amphetamine neurotoxicity on striatal dopamine nerve terminals in the mouse. *Br. J. Pharmacol.* **1982**, *77*(2), 335–345.
277. Shiue, C.Y., Shiue, G.G., Cornish, K.G., O'Rourke, M.F. Comparative PET studies of the distribution of (-)-3,4-methylenedioxy-N-[11C]methamphetamine and (-)-[11C]methamphetamine in a monkey brain. *Nucl. Med. Biol.* **1995**, *22*(3), 321–324.
278. Miyazaki, I., Asanuma, M., Diaz-Corrales, F.J., Fukuda, M., Kitaichi, K., Miyoshi, K., Ogawa, N. Methamphetamine-induced dopaminergic neurotoxicity is regulated by quinone-formation-related molecules. *FASEB J.* **2006**, *20*(3), 571–573.
279. Esteban, B., O'Shea, E., Camarero, J., Sanchez, V., Green, A.R., Colado, M.I. 3,4-Methylenedioxymethamphetamine induces monoamine release, but not toxicity, when administered centrally at a concentration occurring following a peripherally injected neurotoxic dose. *Psychopharmacology (Berl)* **2001**, *154*(3), 251–260.
280. Fumagalli, F., Gainetdinov, R.R., Valenzano, K.J., Caron, M.G. Role of dopamine transporter in methamphetamine-induced neurotoxicity: Evidence from mice lacking the transporter. *J. Neurosci.* **1998**, *18*(13), 4861–4869.
281. Bost, R.O., Kemp, P., Hnilica, V. Tissue distribution of methamphetamine and amphetamine in premature infants. *J. Anal. Toxicol.* **1989**, *13*(5), 300–302.
282. Won, L., Bubula, N., McCoy, H., Heller, A. Methamphetamine concentrations in fetal and maternal brain following prenatal exposure. *Neurotoxicol. Teratol.* **2001**, *23*(4), 349–354.
283. Beckett, A.H., Rowland, M. Urinary excretion of methylamphetamine in man. *Nature* **1965**, *206*(990), 1260–1261.
284. Caldwell, J., Dring, L.G., Williams, R.T. Metabolism of (14 C)methamphetamine in man, the guinea pig and the rat. *Biochem. J.* **1972**, *129*(1), 11–22.
285. Kreth, K., Kovar, K., Schwab, M., Zanger, U.M. Identification of the human cytochromes P450 involved in the oxidative metabolism of "Ecstasy"-related designer drugs. *Biochem. Pharmacol.* **2000**, *59*(12), 1563–1571.
286. Meyer, M.R., Peters, F.T., Maurer, H.H. The role of human hepatic cytochrome P450 isozymes in the metabolism of racemic 3,4-methylenedioxy-methamphetamine and its enantiomers. *Drug Metab. Dispos.* **2008**, *36*(11), 2345–2354.

287. Hiramatsu, M., Kumagai, Y., Unger, S.E., Cho, A.K. Metabolism of methylenedioxymethamphetamine: Formation of dihydroxymethamphetamine and a quinone identified as its glutathione adduct. *J. Pharmacol. Exp. Ther.* **1990**, *254*(2), 521–527.

288. Segura, M., Ortuno, J., Farre, M., McLure, J.A., Pujadas, M., Pizarro, N., Llebaria, A., Joglar, J., Roset, P.N., Segura, J., de La Torre, R. 3,4-Dihydroxymethamphetamine (HHMA). A major in vivo 3,4-methylenedioxymethamphetamine (MDMA) metabolite in humans. *Chem. Res. Toxicol.* **2001**, *14*(9), 1203–1208.

289. Musshoff, F. Illegal or legitimate use? Precursor compounds to amphetamine and methamphetamine. *Drug Metab. Rev.* **2000**, *32*(1), 15–44.

290. Heydari, A., Yeo, K.R., Lennard, M.S., Ellis, S.W., Tucker, G.T., Rostami-Hodjegan, A. Mechanism-based inactivation of CYP2D6 by methylenedioxymethamphetamine. *Drug Metab. Dispos.* **2004**, *32*(11), 1213–1217.

291. Wu, D., Otton, S.V., Inaba, T., Kalow, W., Sellers, E.M. Interactions of amphetamine analogs with human liver CYP2D6. *Biochem. Pharmacol.* **1997**, *53*(11), 1605–1612.

292. Colado, M.I., Camarero, J., Mechan, A.O., Sanchez, V., Esteban, B., Elliott, J.M., Green, A.R. A study of the mechanisms involved in the neurotoxic action of 3,4-methylenedioxymethamphetamine (MDMA, "ecstasy") on dopamine neurones in mouse brain. *Br. J. Pharmacol.* **2001**, *134*(8), 1711–1723.

293. Ghersi-Egea, J.F., Leninger-Muller, B., Suleman, G., Siest, G., Minn, A. Localization of drug-metabolizing enzyme activities to blood-brain interfaces and circumventricular organs. *J. Neurochem.* **1994**, *62*(3), 1089–1096.

294. Sulzer, D., Sonders, M.S., Poulsen, N.W., Galli, A. Mechanisms of neurotransmitter release by amphetamines: A review. *Prog. Neurobiol.* **2005**, *75*(6), 406–433.

295. Cruickshank, C.C., Dyer, K.R. A review of the clinical pharmacology of methamphetamine. *Addiction* **2009**, *104*(7), 1085–1099.

296. Khoshbouei, H., Wang, H., Lechleiter, J.D., Javitch, J.A., Galli, A. Amphetamine-induced dopamine efflux. A voltage-sensitive and intracellular Na+-dependent mechanism. *J. Biol. Chem.* **2003**, *278*(14), 12070–12077.

297. Rothman, R.B., Baumann, M.H. Monoamine transporters and psychostimulant drugs. *Eur. J. Pharmacol.* **2003**, *479*(1–3), 23–40.

298. Fleckenstein, A.E., Haughey, H.M., Metzger, R.R., Kokoshka, J.M., Riddle, E.L., Hanson, J.E., Gibb, J.W., Hanson, G.R. Differential effects of psychostimulants and related agents on dopaminergic and serotonergic transporter function. *Eur. J. Pharmacol.* **1999**, *382*(1), 45–49.

299. Han, D.D., Gu, H.H. Comparison of the monoamine transporters from human and mouse in their sensitivities to psychostimulant drugs. *BMC Pharmacol.* **2006**, 66.

300. Sprague, J.E., Brutcher, R.E., Mills, E.M., Caden, D., Rusyniak, D.E. Attenuation of 3,4-methylenedioxymethamphetamine (MDMA, Ecstasy)-induced rhabdomyolysis with alpha1- plus beta3-adrenoreceptor antagonists. *Br. J. Pharmacol.* **2004**, *142*(4), 667–670.

301. Mechan, A.O., Esteban, B., O'Shea, E., Elliott, J.M., Colado, M.I., Green, A.R. The pharmacology of the acute hyperthermic response that follows administration of 3,4-methylenedioxymethamphetamine (MDMA, "ecstasy") to rats. *Br. J. Pharmacol.* **2002**, *135*(1),170–180.

302. Bowyer, J.F., Davies, D.L., Schmued, L., Broening, H.W., Newport, G.D., Slikker, W., Jr., Holson, R.R. Further studies of the role of hyperthermia in methamphetamine neurotoxicity. *J. Pharmacol. Exp. Ther.* **1994**, *268*(3), 1571–1580.

303. Herin, D.V., Liu, S., Ullrich, T., Rice, K.C., Cunningham, K.A. Role of the serotonin 5-HT2A receptor in the hyperlocomotive and hyperthermic effects of (+)-3,4-methylenedioxymethamphetamine. *Psychopharmacology (Berl)* **2005**, *178*(4), 505–513.

304. Kita, T., Wagner, G.C., Nakashima, T. Current research on methamphetamine-induced neurotoxicity: Animal models of monoamine disruption. *J. Pharmacol. Sci.* **2003**, *92*(3), 178–195.

305. Kraemer, T., Maurer, H.H. Toxicokinetics of amphetamines: Metabolism and toxicokinetic data of designer drugs, amphetamine, methamphetamine, and their N-alkyl derivatives. *Ther. Drug Monit.* **2002**, *24*(2), 277–289.

306. Wijetunga, M., Seto, T., Lindsay, J., Schatz, I. Crystal methamphetamine-associated cardiomyopathy: Tip of the iceberg? *J. Toxicol. Clin. Toxicol.* **2003**, *41*(7), 981–986.

307. Scott, J.C., Woods, S.P., Matt, G.E., Meyer, R.A., Heaton, R.K., Atkinson, J.H., Grant, I. Neurocognitive effects of methamphetamine: A critical review and meta-analysis. *Neuropsychol. Rev.* **2007**, *17*(3), 275–297.

308. Cohen, J.B., Dickow, A., Horner, K., Zweben, J.E., Balabis, J., Vandersloot, D., Reiber, C. Abuse and violence history of men and women in treatment for methamphetamine dependence. *Am. J. Addict.* **2003**, *12*(5), 377–385.

309. Zweben, J.E., Cohen, J.B., Christian, D., Galloway, G.P., Salinardi, M., Parent, D., Iguchi, M. Psychiatric symptoms in methamphetamine users. *Am. J. Addict.* **2004**, *13*(2), 181–190.

310. Caligiuri, M.P., Buitenhuys, C. Do preclinical findings of methamphetamine-induced motor abnormalities translate to an observable clinical phenotype? *Neuropsychopharmacology* **2005**, *30*(12), 2125–2134.

311. Moszczynska, A., Fitzmaurice, P., Ang, L., Kalasinsky, K.S., Schmunk, G.A., Peretti, F.J., Aiken, S.S., Wickham, D.J., Kish, S.J. Why is Parkinsonism not a feature of human methamphetamine users? *Brain* **2004**, *127*(Pt 2), 363–370.

312. Ujike, H., Harano, M., Inada, T., Yamada, M., Komiyama, T., Sekine, Y., Sora, I., Iyo, M., Katsu, T., Nomura, A., Nakata, K., Ozaki, N. Nine- or fewer repeat alleles in VNTR polymorphism of the dopamine transporter gene is a strong risk factor for prolonged methamphetamine psychosis. *Pharmacogenomics J.* **2003**, *3*(4), 242–247.

313. Burrows, K.B., Meshul, C.K. High-dose methamphetamine treatment alters presynaptic GABA and glutamate immunoreactivity. *Neuroscience* **1999**, *90*(3), 833–850.
314. Lehmann, K., Lehmann, D. Transmitter balances in the olfactory cortex: Adaptations to early methamphetamine trauma and rearing environment. *Brain Res.* **2007**, *1141*, 37–47.
315. Bowyer, J.F., Scallet, A.C., Holson, R.R., Lipe, G.W., Slikker, W., Jr., Ali, S.F. Interactions of MK-801 with glutamate-, glutamine- and methamphetamine-evoked release of [3H]dopamine from striatal slices. *J. Pharmacol. Exp. Ther.* **1991**, *257*(1), 262–270.
316. Pu, C., Broening, H.W., Vorhees, C.V. Effect of methamphetamine on glutamate-positive neurons in the adult and developing rat somatosensory cortex. *Synapse* **1996**, *23*(4), 328–334.
317. Chan, P., Di Monte, D.A., Luo, J.J., DeLanney, L.E., Irwin, I., Langston, J.W. Rapid ATP loss caused by methamphetamine in the mouse striatum: Relationship between energy impairment and dopaminergic neurotoxicity. *J. Neurochem.* **1994**, *62*(6), 2484–2487.
318. Axt, K.J., Molliver, M.E. Immunocytochemical evidence for methamphetamine-induced serotonergic axon loss in the rat brain. *Synapse* **1991**, *9*(4), 302–313.
319. Bakhit, C., Morgan, M.E., Peat, M.A., Gibb, J.W. Long-term effects of methamphetamine on the synthesis and metabolism of 5-hydroxytryptamine in various regions of the rat brain. *Neuropharmacology* **1981**, *20*(12A), 1135–1140.
320. Deng, X., Ladenheim, B., Jayanthi, S., Cadet, J.L. Methamphetamine administration causes death of dopaminergic neurons in the mouse olfactory bulb. *Biol. Psychiatry* **2007**, *61*(11), 1235–1243.
321. Harvey, D.C., Lacan, G., Tanious, S.P., Melega, W.P. Recovery from methamphetamine induced long-term nigrostriatal dopaminergic deficits without substantia nigra cell loss. *Brain Res.* **2000**, *871*(2), 259–270.
322. Hotchkiss, A.J., Gibb, J.W. Long-term effects of multiple doses of methamphetamine on tryptophan hydroxylase and tyrosine hydroxylase activity in rat brain. *J. Pharmacol. Exp. Ther.* **1980**, *214*(2), 257–262.
323. Wagner, G.C., Ricaurte, G.A., Seiden, L.S., Schuster, C.R., Miller, R.J., Westley, J. Long-lasting depletions of striatal dopamine and loss of dopamine uptake sites following repeated administration of methamphetamine. *Brain Res.* **1980**, *181*(1), 151–160.
324. Yu, J., Wang, J., Cadet, J.L., Angulo, J.A. Histological evidence supporting a role for the striatal neurokinin-1 receptor in methamphetamine-induced neurotoxicity in the mouse brain. *Brain Res.* **2004**, *1007*(1–2), 124–131.
325. Jayanthi, S., Deng, X., Noailles, P.A., Ladenheim, B., Cadet, J.L. Methamphetamine induces neuronal apoptosis via cross-talks between endoplasmic reticulum and mitochondria-dependent death cascades. *FASEB J.* **2004**, *18*(2), 238–251.
326. Jayanthi, S., McCoy, M.T., Beauvais, G., Ladenheim, B., Gilmore, K., Wood, W., 3rd, Becker, K., Cadet, J.L. Methamphetamine induces dopamine D1 receptor-dependent endoplasmic reticulum stress-related molecular events in the rat striatum. *PLoS ONE* **2009**, *4*(6), e6092.
327. Deng, X., Wang, Y., Chou, J., Cadet, J.L. Methamphetamine causes widespread apoptosis in the mouse brain: Evidence from using an improved TUNEL histochemical method. *Brain Res. Mol. Brain Res.* **2001**, *93*(1), 64–69.
328. Eisch, A.J., Schmued, L.C., Marshall, J.F. Characterizing cortical neuron injury with Fluoro-Jade labeling after a neurotoxic regimen of methamphetamine. *Synapse* **1998**, *30*(3), 329–333.
329. Thomas, D.M., Kuhn, D.M. MK-801 and dextromethorphan block microglial activation and protect against methamphetamine-induced neurotoxicity. *Brain Res.* **2005**, *1050*(1–2), 190–198.
330. Chu, P.W., Seferian, K.S., Birdsall, E., Truong, J.G., Riordan, J.A., Metcalf, C.S., Hanson, G.R., Fleckenstein, A.E. Differential regional effects of methamphetamine on dopamine transport. *Eur. J. Pharmacol.* **2008**, *590*(1–3), 105–110.
331. Volkow, N.D., Chang, L., Wang, G.J., Fowler, J.S., Ding, Y.S., Sedler, M., Logan, J., Franceschi, D., Gatley, J., Hitzemann, R., Gifford, A., Wong, C., Pappas, N. Low level of brain dopamine D2 receptors in methamphetamine abusers: Association with metabolism in the orbitofrontal cortex. *Am. J. Psychiatry* **2001**, *158*(12), 2015–2021.
332. Volkow, N.D., Chang, L., Wang, G.J., Fowler, J.S., Franceschi, D., Sedler, M., Gatley, S.J., Miller, E., Hitzemann, R., Ding, Y.S., Logan, J. Loss of dopamine transporters in methamphetamine abusers recovers with protracted abstinence. *J. Neurosci.* **2001**, *21*(23), 9414–9418.
333. Boileau, I., Rusjan, P., Houle, S., Wilkins, D., Tong, J., Selby, P., Guttman, M., Saint-Cyr, J.A., Wilson, A.A., Kish, S.J. Increased vesicular monoamine transporter binding during early abstinence in human methamphetamine users: Is VMAT2 a stable dopamine neuron biomarker? *J. Neurosci.* **2008**, *28*(39), 9850–9856.
334. Johanson, C.E., Frey, K.A., Lundahl, L.H., Keenan, P., Lockhart, N., Roll, J., Galloway, G.P., Koeppe, R.A., Kilbourn, M.R., Robbins, T., Schuster, C.R. Cognitive function and nigrostriatal markers in abstinent methamphetamine abusers. *Psychopharmacology (Berl)* **2006**, *185*(3), 327–338.
335. Guilarte, T.R., Nihei, M.K., McGlothan, J.L., Howard, A.S. Methamphetamine-induced deficits of brain monoaminergic neuronal markers: Distal axotomy or neuronal plasticity. *Neuroscience* **2003**, *122*(2), 499–513.
336. Chung, A., Lyoo, I.K., Kim, S.J., Hwang, J., Bae, S.C., Sung, Y.H., Sim, M.E., Song, I.C., Kim, J., Chang, K.H., Renshaw, P.F. Decreased frontal white-matter integrity in abstinent methamphetamine abusers. *Int. J. Neuropsychopharmacol.* **2007**, *10*(6), 765–775.
337. Ernst, T., Chang, L., Leonido-Yee, M., Speck, O. Evidence for long-term neurotoxicity associated with methamphetamine abuse: A 1H MRS study. *Neurology* **2000**, *54*(6), 1344–1349.

338. Kuczenski, R., Everall, I.P., Crews, L., Adame, A., Grant, I., Masliah, E. Escalating dose-multiple binge methamphetamine exposure results in degeneration of the neocortex and limbic system in the rat. *Exp. Neurol.* **2007**, *207*(1), 42–51.

339. Thompson, P.M., Hayashi, K.M., Simon, S.L., Geaga, J.A., Hong, M.S., Sui, Y., Lee, J.Y., Toga, A.W., Ling, W., London, E.D. Structural abnormalities in the brains of human subjects who use methamphetamine. *J. Neurosci.* **2004**, *24*(26), 6028–6036.

340. Volkow, N.D., Chang, L., Wang, G.J., Fowler, J.S., Franceschi, D., Sedler, M.J., Gatley, S.J., Hitzemann, R., Ding, Y.S., Wong, C., Logan, J. Higher cortical and lower subcortical metabolism in detoxified methamphetamine abusers. *Am. J. Psychiatry* **2001**, *158*(3), 383–389.

341. Kish, S.J., Fitzmaurice, P.S., Boileau, I., Schmunk, G.A., Ang, L.C., Furukawa, Y., Chang, L.J., Wickham, D.J., Sherwin, A., Tong, J. Brain serotonin transporter in human methamphetamine users. *Psychopharmacology (Berl)* **2009**, *202*(4), 649–661.

342. Sekine, Y., Ouchi, Y., Takei, N., Yoshikawa, E., Nakamura, K., Futatsubashi, M., Okada, H., Minabe, Y., Suzuki, K., Iwata, Y., Tsuchiya, K.J., Tsukada, H., Iyo, M., Mori, N. Brain serotonin transporter density and aggression in abstinent methamphetamine abusers. *Arch. Gen. Psychiatry* **2006**, *63*(1), 90–100.

343. Seiden, L.S., Commins, D.L., Vosmer, G., Axt, K., Marek, G. Neurotoxicity in dopamine and 5-hydroxytryptamine terminal fields: A regional analysis in nigrostriatal and mesolimbic projections. *Ann. N. Y. Acad. Sci.* **1988**, *537*, 161–172.

344. Gheusi, G., Cremer, H., McLean, H., Chazal, G., Vincent, J.D., Lledo, P.M. Importance of newly generated neurons in the adult olfactory bulb for odor discrimination. *Proc. Natl. Acad. Sci. U.S.A.* **2000**, *97*(4), 1823–1828.

345. Parrish-Aungst, S., Shipley, M.T., Erdelyi, F., Szabo, G., Puche, A.C. Quantitative analysis of neuronal diversity in the mouse olfactory bulb. *J. Comp. Neurol.* **2007**, *501*(6), 825–836.

346. Jayanthi, S., Ladenheim, B., Andrews, A.M., Cadet, J.L. Overexpression of human copper/zinc superoxide dismutase in transgenic mice attenuates oxidative stress caused by methylenedioxymethamphetamine (Ecstasy). *Neuroscience* **1999**, *91*(4), 1379–1387.

347. Fitzmaurice, P.S., Tong, J., Yazdanpanah, M., Liu, P.P., Kalasinsky, K.S., Kish, S.J. Levels of 4-hydroxynonenal and malondialdehyde are increased in brain of human chronic users of methamphetamine. *J. Pharmacol. Exp. Ther.* **2006**, *319*(2), 703–709.

348. Giovanni, A., Liang, L.P., Hastings, T.G., Zigmond, M.J. Estimating hydroxyl radical content in rat brain using systemic and intraventricular salicylate: Impact of methamphetamine. *J. Neurochem.* **1995**, *64*(4), 1819–1825.

349. De Vito, M.J., Wagner, G.C. Methamphetamine-induced neuronal damage: A possible role for free radicals. *Neuropharmacology* **1989**, *28*(10), 1145–1150.

350. Hirata, H., Ladenheim, B., Carlson, E., Epstein, C., Cadet, J.L. Autoradiographic evidence for methamphetamine-induced striatal dopaminergic loss in mouse brain: Attenuation in CuZn-superoxide dismutase transgenic mice. *Brain Res.* **1996**, *714*(1–2), 95–103.

351. Di Monte, D.A., Royland, J.E., Jakowec, M.W., Langston, J.W. Role of nitric oxide in methamphetamine neurotoxicity: Protection by 7-nitroindazole, an inhibitor of neuronal nitric oxide synthase. *J. Neurochem.* **1996**, *67*(6), 2443–2450.

352. Imam, S.Z., el-Yazal, J., Newport, G.D., Itzhak, Y., Cadet, J.L., Slikker, W., Jr., Ali, S.F. Methamphetamine-induced dopaminergic neurotoxicity: Role of peroxynitrite and neuroprotective role of antioxidants and peroxynitrite decomposition catalysts. *Ann. N. Y. Acad. Sci.* **2001**, *939*, 366–380.

353. Itzhak, Y., Gandia, C., Huang, P.L., Ali, S.F. Resistance of neuronal nitric oxide synthase-deficient mice to methamphetamine-induced dopaminergic neurotoxicity. *J. Pharmacol. Exp. Ther.* **1998**, *284*(3), 1040–1047.

354. Cadet, J.L., Krasnova, I.N. Molecular bases of methamphetamine-induced neurodegeneration. *Int. Rev. Neurobiol.* **2009**, *88*, 101–119.

355. Cass, W.A., Manning, M.W. Recovery of presynaptic dopaminergic functioning in rats treated with neurotoxic doses of methamphetamine. *J. Neurosci.* **1999**, *19*(17), 7653–7660.

356. Fleckenstein, A.E., Volz, T.J., Hanson, G.R. Psychostimulant-induced alterations in vesicular monoamine transporter-2 function: Neurotoxic and therapeutic implications. *Neuropharmacology* **2009**, *56*(Suppl. 1), 133–138.

357. Davidson, C., Gow, A.J., Lee, T.H., Ellinwood, E.H. Methamphetamine neurotoxicity: Necrotic and apoptotic mechanisms and relevance to human abuse and treatment. *Brain Res. Brain Res. Rev.* **2001**, *36*(1), 1–22.

358. Reese, T.S., Karnovsky, M.J. Fine structural localization of a blood-brain barrier to exogenous peroxidase. *J. Cell Biol.* **1967**, *34*(1), 207–217.

359. LeFevre, P.G., Peters, A.A. Evidence of mediated transfer of monosaccharides from blood to brain in rodents. *J. Neurochem.* **1966**, *13*(1), 35–46.

360. Abbott, N.J., Patabendige, A.A., Dolman, D.E., Yusof, S.R., Begley, D.J. Structure and function of the blood-brain barrier. *Neurobiol. Dis.* **2010**, *37*(1), 13–25.

361. Pardridge, W.M., Triguero, D., Yang, J., Cancilla, P.A. Comparison of in vitro and in vivo models of drug transcytosis through the blood-brain barrier. *J. Pharmacol. Exp. Ther.* **1990**, *253*(2), 884–891.

362. el-Bacha, R.S., Minn, A. Drug metabolizing enzymes in cerebrovascular endothelial cells afford a metabolic protection to the brain. *Cell. Mol. Biol.* **1999**, *45*(1), 15–23.

363. DeBault, L.E., Cancilla, P.A. gamma-Glutamyl transpeptidase in isolated brain endothelial cells: Induction by glial cells in vitro. *Science* **1980**, *207*(4431), 653–655.

364. Dauchy, S., Dutheil, F., Weaver, R.J., Chassoux, F., Daumas-Duport, C., Couraud, P.O., Scherrmann, J.M.,

De Waziers, I., Decleves, X. ABC transporters, cytochromes P450 and their main transcription factors: Expression at the human blood-brain barrier. *J. Neurochem.* **2008**, *107*(6), 1518–1528.

365. Begley, D.J. ABC transporters and the blood-brain barrier. *Curr. Pharm. Des.* **2004**, *10*(12), 1295–1312.

366. Correale, J., Villa, A. The blood-brain-barrier in multiple sclerosis: Functional roles and therapeutic targeting. *Autoimmunity* **2007**, *40*(2), 148–160.

367. Desai, B.S., Monahan, A.J., Carvey, P.M., Hendey, B. Blood-brain barrier pathology in Alzheimer's and Parkinson's disease: Implications for drug therapy. *Cell Transplant.* **2007**, *16*(3), 285–299.

368. Cirrito, J.R., Deane, R., Fagan, A.M., Spinner, M.L., Parsadanian, M., Finn, M.B., Jiang, H., Prior, J.L., Sagare, A., Bales, K.R., Paul, S.M., Zlokovic, B.V., Piwnica-Worms, D., Holtzman, D.M. P-glycoprotein deficiency at the blood-brain barrier increases amyloid-beta deposition in an Alzheimer disease mouse model. *J. Clin. Invest.* **2005**, *115*(11), 3285–3290.

369. Bartels, A.L., Willemsen, A.T., Kortekaas, R., de Jong, B.M., de Vries, R., de Klerk, O., van Oostrom, J.C., Portman, A., Leenders, K.L. Decreased blood-brain barrier P-glycoprotein function in the progression of Parkinson's disease, PSP and MSA. *J. Neural Transm.* **2008**, *115*(7), 1001–1009.

370. Kortekaas, R., Leenders, K.L., van Oostrom, J.C., Vaalburg, W., Bart, J., Willemsen, A.T., Hendrikse, N.H. Blood-brain barrier dysfunction in Parkinsonian midbrain in vivo. *Ann. Neurol.* **2005**, *57*(2), 176–179.

371. Moosmann, B., Behl, C. Antioxidants as treatment for neurodegenerative disorders. *Expert Opin. Investig. Drugs* **2002**, *11*(10), 1407–1435.

372. Murphy, T.H., Yu, J., Ng, R., Johnson, D.A., Shen, H., Honey, C.R., Johnson, J.A. Preferential expression of antioxidant response element mediated gene expression in astrocytes. *J. Neurochem.* **2001**, *76*(6), 1670–1678.

373. Johnson, D.A., Andrews, G.K., Xu, W., Johnson, J.A. Activation of the antioxidant response element in primary cortical neuronal cultures derived from transgenic reporter mice. *J. Neurochem.* **2002**, *81*(6), 1233–1241.

374. Pai, G.S., Sprenkle, J.A., Do, T.T., Mareni, C.E., Migeon, B.R. Localization of loci for hypoxanthine phosphoribosyltransferase and glucose-6-phosphate dehydrogenase and biochemical evidence of nonrandom X chromosome expression from studies of a human X-autosome translocation. *Proc. Natl. Acad. Sci. U.S.A.* **1980**, *77*(5), 2810–2813.

375. Mason, P.J., Bautista, J.M., Vulliamy, T.J., Turner, N., Luzzatto, L. Human red cell glucose-6-phosphate dehydrogenase is encoded only on the X chromosome. *Cell* **1990**, *62*(1), 9–10.

376. Beutler, E., Yeh, M., Fairbanks, V.F. The normal human female as a mosaic of X-chromosome activity: Studies using the gene for C-6-PD-deficiency as a marker. *Proc. Natl. Acad. Sci. U.S.A.* **1962**, *48*, 9–16.

377. Lyon, M.F. Gene action in the X-chromosome of the mouse (*Mus musculus* L.). *Nature* **1961**, *190*, 372–373.

378. Martini, G., Toniolo, D., Vulliamy, T., Luzzatto, L., Dono, R., Viglietto, G., Paonessa, G., D'Urso, M., Persico, M.G. Structural analysis of the X-linked gene encoding human glucose 6-phosphate dehydrogenase. *EMBO J.* **1986**, *5*(8), 1849–1855.

379. Toniolo, D., Filippi, M., Dono, R., Lettieri, T., Martini, G. The CpG island in the 5′ region of the G6PD gene of man and mouse. *Gene* **1991**, *102*(2), 197–203.

380. Chen, E.Y., Cheng, A., Lee, A., Kuang, W.J., Hillier, L., Green, P., Schlessinger, D., Ciccodicola, A., D'Urso, M. Sequence of human glucose-6-phosphate dehydrogenase cloned in plasmids and a yeast artificial chromosome. *Genomics* **1991**, *10*(3), 792–800.

381. Toniolo, D., Martini, G., Migeon, B.R., Dono, R. Expression of the G6PD locus on the human X chromosome is associated with demethylation of three CpG islands within 100 kb of DNA. *EMBO J.* **1988**, *7*(2), 401–406.

382. Philippe, M., Larondelle, Y., Lemaigre, F., Mariame, B., Delhez, H., Mason, P., Luzzatto, L., Rousseau, G.G. Promoter function of the human glucose-6-phosphate dehydrogenase gene depends on two GC boxes that are cell specifically controlled. *Eur. J. Biochem.* **1994**, *226*(2), 377–384.

383. Luzzatto, L., Mehta, A., Vulliamy, T. Glucose-6-phosphate dehydrogenase deficiency. In: *The Metabolic and Molecular Bases of Inherited Disease*, 8th ed. eds. C.R. Scriver, W.S. Sly, B. Childs, A.L. Beaudet, D. Valle, K.W. Kinzler, B. Vogelstein. McGraw-Hill, New York, 2001. http://www.ommbid.com (accessed March 14, 2013).

384. Kletzien, R.F., Harris, P.K., Foellmi, L.A. Glucose-6-phosphate dehydrogenase: A "housekeeping" enzyme subject to tissue-specific regulation by hormones, nutrients, and oxidant stress. *FASEB J.* **1994**, *8*(2), 174–181.

385. Minucci, A., Moradkhani, K., Hwang, M.J., Zuppi, C., Giardina, B., Capoluongo, E. Glucose-6-phosphate dehydrogenase (G6PD) mutations database: Review of the "old" and update of the new mutations. *Blood Cells Mol. Dis.* **2012**, *48*(3), 154–165.

386. Beutler, E. G6PD deficiency. *Blood* **1994**, *84*(11), 3613–3636.

387. Sodeinde, O. Glucose-6-phosphate dehydrogenase deficiency. *Baillieres Clin. Haematol.* **1992**, *5*(2), 367–382.

388. Corcoran, C.M., Fraser, P., Martini, G., Luzzatto, L., Mason, P.J. High-level regulated expression of the human G6PD gene in transgenic mice. *Gene* **1996**, *173*(2), 241–246.

389. Biagiotti, E., Guidi, L., Capellacci, S., Ambrogini, P., Papa, S., Del Grande, P., Ninfali, P. Glucose-6-phosphate dehydrogenase supports the functioning of the synapses in rat cerebellar cortex. *Brain Res.* **2001**, *911*(2), 152–157.

390. Cammer, W., Zimmerman, T.R., Jr. Glycerolphosphate dehydrogenase, glucose-6-phosphate dehydrogenase, lactate dehydrogenase and carbonic anhydrase activities in oligodendrocytes and myelin: Comparisons between

species and CNS regions. *Brain Res.* **1982**, *282*(1), 21–26.

391. Philbert, M.A., Beiswanger, C.M., Roscoe, T.L., Waters, D.K., Lowndes, H.E. Enhanced resolution of histochemical distribution of glucose-6-phosphate dehydrogenase activity in rat neural tissue by use of a semipermeable membrane. *J. Histochem. Cytochem.* **1991**, *39*(7), 937–943.

392. Camardella, L., Damonte, G., Carratore, V., Benatti, U., Tonetti, M., Moneti, G. Glucose 6-phosphate dehydrogenase from human erythrocytes: Identification of N-acetylalanine at the N-terminus of the mature protein. *Biochem. Biophys. Res. Commun.* **1995**, *207*(1), 331–338.

393. Cohen, P., Rosemeyer, M.A. Subunit interactions of glucose-6-phosphate dehydrogenase from human erythrocytes. *Eur. J. Biochem.* **1969**, *8*(1), 8–15.

394. Au, S.W., Gover, S., Lam, V.M., Adams, M.J. Human glucose-6-phosphate dehydrogenase: The crystal structure reveals a structural NADP(+) molecule and provides insights into enzyme deficiency. *Structure* **2000**, *8*(3), 293–303.

395. Wang, X.T., Chan, T.F., Lam, V.M., Engel, P.C. What is the role of the second "structural" NADP+-binding site in human glucose 6-phosphate dehydrogenase? *Protein Sci.* **2008**, *17*(8), 1403–1411.

396. Kotaka, M., Gover, S., Vandeputte-Rutten, L., Au, S.W., Lam, V.M., Adams, M.J. Structural studies of glucose-6-phosphate and NADP+ binding to human glucose-6-phosphate dehydrogenase. *Acta Crystallogr. D Biol. Crystallogr.* **2005**, *61*(Pt 5), 495–504.

397. Cosgrove, M.S., Gover, S., Naylor, C.E., Vandeputte-Rutten, L., Adams, M.J., Levy, H.R. An examination of the role of asp-177 in the His-Asp catalytic dyad of *Leuconostoc mesenteroides* glucose 6-phosphate dehydrogenase: X-ray structure and pH dependence of kinetic parameters of the D177N mutant enzyme. *Biochemistry (Mosc)* **2000**, *39*(49), 15002–15011.

398. Levy, H.R., Vought, V.E., Yin, X., Adams, M.J. Identification of an arginine residue in the dual coenzyme-specific glucose-6-phosphate dehydrogenase from *Leuconostoc mesenteroides* that plays a key role in binding NADP+ but not NAD+. *Arch. Biochem. Biophys.* **1996**, *326*(1), 145–151.

399. Kirkman, H.N., Galiano, S., Gaetani, G.F. The function of catalase-bound NADPH. *J. Biol. Chem.* **1987**, *262*(2), 660–666.

400. Kirkman, H.N., Rolfo, M., Ferraris, A.M., Gaetani, G.F. Mechanisms of protection of catalase by NADPH. Kinetics and stoichiometry. *J. Biol. Chem.* **1999**, *274*(20), 13908–13914.

401. Scott, M.D., Wagner, T.C., Chiu, D.T. Decreased catalase activity is the underlying mechanism of oxidant susceptibility in glucose-6-phosphate dehydrogenase-deficient erythrocytes. *Biochim. Biophys. Acta* **1993**, *1181*(2), 163–168.

402. Nkhoma, E.T., Poole, C., Vannappagari, V., Hall, S.A., Beutler, E. The global prevalence of glucose-6-phosphate dehydrogenase deficiency: A systematic review and meta-analysis. *Blood Cells Mol. Dis.* **2009**, *42*(3), 267–278.

403. WHO. Glucose-6-phosphate dehydrogenase deficiency. WHO Working Group. *Bull. World Health Organ.* **1989**, *67*(6), 601–611.

404. Cappellini, M.D., Fiorelli, G. Glucose-6-phosphate dehydrogenase deficiency. *Lancet* **2008**, *371*(9606), 64–74.

405. Pandolfi, P.P., Sonati, F., Rivi, R., Mason, P., Grosveld, F., Luzzatto, L. Targeted disruption of the housekeeping gene encoding glucose 6-phosphate dehydrogenase (G6PD): G6PD is dispensable for pentose synthesis but essential for defense against oxidative stress. *EMBO J.* **1995**, *14*(21), 5209–5215.

406. Beutler, E. The hemolytic effect of primaquine and related compounds: A review. *Blood* **1959**, *14*(2), 103–139.

407. McDonagh, E.M., Thorn, C.F., Bautista, J.M., Youngster, I., Altman, R.B., Klein, T.E. PharmGKB summary: Very important pharmacogene information for G6PD. *Pharmacogenet. Genomics* **2012**, *22*(3), 219–228.

408. Arese, P., De Flora, A. Pathophysiology of hemolysis in glucose-6-phosphate dehydrogenase deficiency. *Semin. Hematol.* **1990**, *27*(1), 1–40.

409. Mason, P.J., Bautista, J.M., Gilsanz, F. G6PD deficiency: The genotype-phenotype association. *Blood Rev.* **2007**, *21*(5), 267–283.

410. Hedrick, P.W. Population genetics of malaria resistance in humans. *Heredity* **2011**, *107*(4), 283–304.

411. Carter, R., Mendis, K.N. Evolutionary and historical aspects of the burden of malaria. *Clin. Microbiol. Rev.* **2002**, *15*(4), 564–594.

412. Leslie, T., Briceno, M., Mayan, I., Mohammed, N., Klinkenberg, E., Sibley, C.H., Whitty, C.J., Rowland, M. The impact of phenotypic and genotypic G6PD deficiency on risk of *Plasmodium vivax* infection: A case-control study amongst Afghan refugees in Pakistan. *PLoS Med.* **2010**, *7*(5), e1000283.

413. Beutler, E. G6PD: Population genetics and clinical manifestations. *Blood Rev.* **1996**, *10*(1), 45–52.

414. Nicol, C.J., Zielenski, J., Tsui, L.C., Wells, P.G. An embryo-protective role for glucose-6-phosphate dehydrogenase in developmental oxidative stress and chemical teratogenesis. *FASEB J.* **2000**, *14*(1), 111–127.

415. Longo, L., Vanegas, O.C., Patel, M., Rosti, V., Li, H., Waka, J., Merghoub, T., Pandolfi, P.P., Notaro, R., Manova, K., Luzzatto, L. Maternally transmitted severe glucose 6-phosphate dehydrogenase deficiency is an embryonic lethal. *EMBO J.* **2002**, *21*(16), 4229–4239.

416. Ninfali, P., Aluigi, G., Pompella, A. Postnatal expression of glucose-6-phosphate dehydrogenase in different brain areas. *Neurochem. Res.* **1998**, *23*(9), 1197–1204.

417. Ninfali, P., Cuppini, C., Marinoni, S. Glucose-6-phosphate dehydrogenase and glutathione reductase support antioxidant enzymes in nerves and muscles of rats during nerve regeneration. *Restor. Neurol. Neurosci.* **1996**, *10*, 69–75.

418. Felix, K., Rockwood, L.D., Pretsch, W., Nair, J., Bartsch, H., Bornkamm, G.W., Janz, S. Moderate G6PD deficiency

increases mutation rates in the brain of mice. *Free Radic. Biol. Med.* **2002**, *32*(7), 663–673.

419. Mejias, R., Villadiego, J., Pintado, C.O., Vime, P.J., Gao, L., Toledo-Aral, J.J., Echevarria, M., Lopez-Barneo, J. Neuroprotection by transgenic expression of glucose-6-phosphate dehydrogenase in dopaminergic nigrostriatal neurons of mice. *J. Neurosci.* **2006**, *26*(17), 4500–4508.

420. Romero-Ruiz, A., Mejias, R., Diaz-Martin, J., Lopez-Barneo, J., Gao, L. Mesencephalic and striatal protein profiles in mice over-expressing glucose-6-phosphate dehydrogenase in dopaminergic neurons. *J. Proteomics* **2010**, *73*(9), 1747–1757.

421. Biagiotti, E., Guidi, L., Del Grande, P., Ninfali, P. Glucose-6-phosphate dehydrogenase expression associated with NADPH-dependent reactions in cerebellar neurons. *Cerebellum* **2003**, *2*(3), 178–183.

422. Ferri, P., Biagiotti, E., Ambrogini, P., Santi, S., Del Grande, P., Ninfali, P. NADPH-consuming enzymes correlate with glucose-6-phosphate dehydrogenase in Purkinje cells: An immunohistochemical and enzyme histochemical study of the rat cerebellar cortex. *Neurosci. Res.* **2005**, *51*(2), 185–197.

423. Russell, R.L., Siedlak, S.L., Raina, A.K., Bautista, J.M., Smith, M.A., Perry, G. Increased neuronal glucose-6-phosphate dehydrogenase and sulfhydryl levels indicate reductive compensation to oxidative stress in Alzheimer disease. *Arch. Biochem. Biophys.* **1999**, *370*(2), 236–239.

424. Ceballos-Picot, I., Nicole, A., Sinet, P.M. Cellular clones and transgenic mice overexpressing copper-zinc superoxide dismutase: Models for the study of free radical metabolism and aging. *EXS* **1992**, *62*, 89–98.

425. Tian, W.N., Braunstein, L.D., Pang, J., Stuhlmeier, K.M., Xi, Q.C., Tian, X., Stanton, R.C. Importance of glucose-6-phosphate dehydrogenase activity for cell growth. *J. Biol. Chem.* **1998**, *273*(17), 10609–10617.

426. Johnson, F., Giulivi, C. Superoxide dismutases and their impact upon human health. *Mol. Aspects Med.* **2005**, *26*(4–5), 340–352.

427. Marklund, S.L. Human copper-containing superoxide dismutase of high molecular weight. *Proc. Natl. Acad. Sci. U.S.A.* **1982**, *79*(24), 7634–7638.

428. Maier, C.M., Chan, P.H. Role of superoxide dismutases in oxidative damage and neurodegenerative disorders. *Neuroscientist* **2002**, *8*(4), 323–334.

429. Melov, S., Schneider, J.A., Day, B.J., Hinerfeld, D., Coskun, P., Mirra, S.S., Crapo, J.D., Wallace, D.C. A novel neurological phenotype in mice lacking mitochondrial manganese superoxide dismutase. *Nat. Genet.* **1998**, *18*(2), 159–163.

430. Zhang, J., Graham, D.G., Montine, T.J., Ho, Y.S. Enhanced N-methyl-4-phenyl-1,2,3,6-tetrahydropyridine toxicity in mice deficient in CuZn-superoxide dismutase or glutathione peroxidase. *J. Neuropathol. Exp. Neurol.* **2000**, *59*(1), 53–61.

431. Beal, M.F., Ferrante, R.J., Henshaw, R., Matthews, R.T., Chan, P.H., Kowall, N.W., Epstein, C.J., Schulz, J.B. 3-Nitropropionic acid neurotoxicity is attenuated in copper/zinc superoxide dismutase transgenic mice. *J. Neurochem.* **1995**, *65*(2), 919–922.

432. Thiels, E., Urban, N.N., Gonzalez-Burgos, G.R., Kanterewicz, B.I., Barrionuevo, G., Chu, C.T., Oury, T.D., Klann, E. Impairment of long-term potentiation and associative memory in mice that overexpress extracellular superoxide dismutase. *J. Neurosci.* **2000**, *20*(20), 7631–7639.

433. Hu, D., Serrano, F., Oury, T.D., Klann, E. Aging-dependent alterations in synaptic plasticity and memory in mice that overexpress extracellular superoxide dismutase. *J. Neurosci.* **2006**, *26*(15), 3933–3941.

434. Ho, Y.S., Magnenat, J.L., Bronson, R.T., Cao, J., Gargano, M., Sugawara, M., Funk, C.D. Mice deficient in cellular glutathione peroxidase develop normally and show no increased sensitivity to hyperoxia. *J. Biol. Chem.* **1997**, *272*(26), 16644–16651.

435. Rhee, S.G., Chae, H.Z., Kim, K. Peroxiredoxins: A historical overview and speculative preview of novel mechanisms and emerging concepts in cell signaling. *Free Radic. Biol. Med.* **2005**, *38*(12), 1543–1552.

436. Sarafian, T.A., Verity, M.A., Vinters, H.V., Shih, C.C., Shi, L., Ji, X.D., Dong, L., Shau, H. Differential expression of peroxiredoxin subtypes in human brain cell types. *J. Neurosci. Res.* **1999**, *56*(2), 206–212.

437. Hofmann, B., Hecht, H.J., Flohe, L. Peroxiredoxins. *Biol. Chem.* **2002**, *383*(3–4), 347–364.

438. Patenaude, A., Murthy, M.R., Mirault, M.E. Emerging roles of thioredoxin cycle enzymes in the central nervous system. *Cell. Mol. Life Sci.* **2005**, *62*(10), 1063–1080.

439. Rybnikova, E., Damdimopoulos, A.E., Gustafsson, J.A., Spyrou, G., Pelto-Huikko, M. Expression of novel antioxidant thioredoxin-2 in the rat brain. *Eur. J. Neurosci.* **2000**, *12*(5), 1669–1678.

440. Wagener, F.A., Volk, H.D., Willis, D., Abraham, N.G., Soares, M.P., Adema, G.J., Figdor, C.G. Different faces of the heme-heme oxygenase system in inflammation. *Pharmacol. Rev.* **2003**, *55*(3), 551–571.

441. Dore, S., Takahashi, M., Ferris, C.D., Zakhary, R., Hester, L.D., Guastella, D., Snyder, S.H. Bilirubin, formed by activation of heme oxygenase-2, protects neurons against oxidative stress injury. *Proc. Natl. Acad. Sci. U.S.A.* **1999**, *96*(5), 2445–2450.

442. Ryter, S.W., Alam, J., Choi, A.M. Heme oxygenase-1/carbon monoxide: From basic science to therapeutic applications. *Physiol. Rev.* **2006**, *86*(2), 583–650.

443. Exner, M., Minar, E., Wagner, O., Schillinger, M. The role of heme oxygenase-1 promoter polymorphisms in human disease. *Free Radic. Biol. Med.* **2004**, *37*(8), 1097–1104.

444. Baranano, D.E., Snyder, S.H. Neural roles for heme oxygenase: Contrasts to nitric oxide synthase. *Proc. Natl. Acad. Sci. U.S.A.* **2001**, *98*(20), 10996–11002.

445. Schipper, H.M., Song, W., Zukor, H., Hascalovici, J.R., Zeligman, D. Heme oxygenase-1 and neurodegeneration: Expanding frontiers of engagement. *J. Neurochem.* **2009**, *110*(2), 469–485.

446. Dinkova-Kostova, A.T., Talalay, P. NAD(P)H:quinone acceptor oxidoreductase 1 (NQO1), a multifunctional antioxidant enzyme and exceptionally versatile cytoprotector. *Arch. Biochem. Biophys.* **2010**, *501*(1), 116–123.

447. Siegel, D., Ross, D. Immunodetection of NAD(P)H:quinone oxidoreductase 1 (NQO1) in human tissues. *Free Radic. Biol. Med.* **2000**, *29*(3–4), 246–253.

448. van Horssen, J., Schreibelt, G., Bo, L., Montagne, L., Drukarch, B., van Muiswinkel, F.L., de Vries, H.E. NAD(P)H:quinone oxidoreductase 1 expression in multiple sclerosis lesions. *Free Radic. Biol. Med.* **2006**, *41*(2), 311–317.

449. van Muiswinkel, F.L., de Vos, R.A., Bol, J.G., Andringa, G., Jansen Steur, E.N., Ross, D., Siegel, D., Drukarch, B. Expression of NAD(P)H:quinone oxidoreductase in the normal and Parkinsonian substantia nigra. *Neurobiol. Aging* **2004**, *25*(9), 1253–1262.

450. Dringen, R. Metabolism and functions of glutathione in brain. *Prog. Neurobiol.* **2000**, *62*(6), 649–671.

451. Zeevalk, G.D., Razmpour, R., Bernard, L.P. Glutathione and Parkinson's disease: Is this the elephant in the room? *Biomed. Pharmacother.* **2008**, *62*(4), 236–249.

452. Ursini, F., Maiorino, M., Brigelius-Flohe, R., Aumann, K.D., Roveri, A., Schomburg, D., Flohe, L. Diversity of glutathione peroxidases. *Methods Enzymol.* **1995**, *252*, 38–53.

453. Dringen, R., Pawlowski, P.G., Hirrlinger, J. Peroxide detoxification by brain cells. *J. Neurosci. Res.* **2005**, *79*(1–2), 157–165.

454. Klivenyi, P., Andreassen, O.A., Ferrante, R.J., Dedeoglu, A., Mueller, G., Lancelot, E., Bogdanov, M., Andersen, J.K., Jiang, D., Beal, M.F. Mice deficient in cellular glutathione peroxidase show increased vulnerability to malonate, 3-nitropropionic acid, and 1-methyl-4-phenyl-1,2,5,6-tetrahydropyridine. *J. Neurosci.* **2000**, *20*(1), 1–7.

455. Liddell, J.R., Dringen, R., Crack, P.J., Robinson, S.R. Glutathione peroxidase 1 and a high cellular glutathione concentration are essential for effective organic hydroperoxide detoxification in astrocytes. *Glia* **2006**, *54*(8), 873–879.

456. Muller, D.P., Goss-Sampson, M.A. Neurochemical, neurophysiological, and neuropathological studies in vitamin E deficiency. *Crit. Rev. Neurobiol.* **1990**, *5*(3), 239–263.

457. Carr, A., Frei, B. Does vitamin C act as a pro-oxidant under physiological conditions? *FASEB J.* **1999**, *13*(9), 1007–1024.

458. Reiter, R.J. Oxidative processes and antioxidative defense mechanisms in the aging brain. *FASEB J.* **1995**, *9*(7), 526–533.

459. Andegeko, Y., Moyal, L., Mittelman, L., Tsarfaty, I., Shiloh, Y., Rotman, G. Nuclear retention of ATM at sites of DNA double strand breaks. *J. Biol. Chem.* **2001**, *276*(41), 38224–38230.

460. Allen, D.M., van Praag, H., Ray, J., Weaver, Z., Winrow, C.J., Carter, T.A., Braquet, R., Harrington, E., Ried, T., Brown, K.D., Gage, F.H., Barlow, C. Ataxia telangiectasia mutated is essential during adult neurogenesis. *Genes Dev.* **2001**, *15*(5), 554–566.

461. Guo, Z., Kozlov, S., Lavin, M.F., Person, M.D., Paull, T.T. ATM activation by oxidative stress. *Science* **2010**, *330*(6003), 517–521.

462. Huffman, J.L., Sundheim, O., Tainer, J.A. DNA base damage recognition and removal: New twists and grooves. *Mutat. Res.* **2005**, *577*(1–2), 55–76.

463. Beerens, N., Hoeijmakers, J.H.J., Kanaar, R., Vermeulen, W., Wyman, C. The CSB protein actively wraps DNA. *J. Biol. Chem.* **2005**, *280*(6), 4722–4729.

464. Kirkali, G., de Souza-Pinto, N.C., Jaruga, P., Bohr, V.A., Dizdaroglu, M. Accumulation of (5'S)-8,5'-cyclo-2'-deoxyadenosine in organs of Cockayne syndrome complementation group B gene knockout mice. *DNA Repair* **2009**, *8*(2), 274–278.

465. McCallum, G.P., Wong, A.W., Wells, P.G. Cockayne syndrome B (CSB) protects against methamphetamine-enhanced oxidative DNA damage in murine fetal brain and postnatal neurodevelopmental deficits. *Antioxid. Redox Signal.* **2011**, *14*(5), 747–756.

466. Couch, F.J., DeShano, M.L., Blackwood, M.A., Calzone, K., Stopfer, J., Campeau, L., Ganguly, A., Rebbeck, T., Weber, B.L., Jablon, L., Cobleigh, M.A., Hoskins, K., Garber, J.E. BRCA1 mutations in women attending clinics that evaluate the risk of breast cancer. *N. Engl. J. Med.* **1997**, *336*(20), 1409–1415.

467. Gowen, L.C., Avrutskaya, A.V., Latour, A.M., Koller, B.H., Leadon, S.A. BRCA1 required for transcription-coupled repair of oxidative DNA damage. *Science* **1998**, *281*(5379), 1009–1012.

468. Moynahan, M.E., Chiu, J.W., Koller, B.H., Jasin, M. Brca1 controls homology-directed DNA repair. *Mol. Cell* **1999**, *4*(4), 511–518.

469. Deng, C.X., Wang, R.H. Roles of BRCA1 in DNA damage repair: A link between development and cancer. *Hum. Mol. Genet.* **2003**, *12*(Suppl. 1), R113–R123.

470. Saha, T., Rih, J.K., Roy, R., Ballal, R., Rosen, E.M. Transcriptional regulation of the base excision repair pathway by BRCA1. *J. Biol. Chem.* **2010**, *285*(25), 19092–19105.

471. Bowerman, B., Eaton, B.A., Priess, J.R. skn-1, a maternally expressed gene required to specify the fate of ventral blastomeres in the early *C. elegans* embryo. *Cell* **1992**, *68*(6), 1061–1075.

472. Mohler, J., Vani, K., Leung, S., Epstein, A. Segmentally restricted, cephalic expression of a leucine zipper gene during *Drosophila* embryogenesis. *Mech. Dev.* **1991**, *34*(1), 3–9.

473. Chan, J.Y., Cheung, M.C., Moi, P., Chan, K., Kan, Y.W. Chromosomal localization of the human NF-E2 family of bZIP transcription factors by fluorescence in situ hybridization. *Hum. Genet.* **1995**, *95*(3), 265–269.

474. Chan, J.Y., Han, X.L., Kan, Y.W. Cloning of Nrf1, an NF-E2-related transcription factor, by genetic selection in yeast. *Proc. Natl. Acad. Sci. U.S.A.* **1993**, *90*(23), 11371–11375.

475. Chan, J.Y., Han, X.L., Kan, Y.W. Isolation of cDNA encoding the human NF-E2 protein. *Proc. Natl. Acad. Sci. U.S.A.* **1993**, *90*(23), 11366–11370.

476. Kobayashi, A., Ito, E., Toki, T., Kogame, K., Takahashi, S., Igarashi, K., Hayashi, N., Yamamoto, M. Molecular cloning and functional characterization of a new Cap"n" collar family transcription factor Nrf3. *J. Biol. Chem.* **1999**, *274*(10), 6443–6452.

477. Kobayashi, M., Itoh, K., Suzuki, T., Osanai, H., Nishikawa, K., Katoh, Y., Takagi, Y., Yamamoto, M. Identification of the interactive interface and phylogenic conservation of the Nrf2-Keap1 system. *Genes Cells* **2002**, *7*(8), 807–820.

478. Moi, P., Chan, K., Asunis, I., Cao, A., Kan, Y.W. Isolation of NF-E2-related factor 2 (Nrf2), a NF-E2-like basic leucine zipper transcriptional activator that binds to the tandem NF-E2/AP1 repeat of the beta-globin locus control region. *Proc. Natl. Acad. Sci. U.S.A.* **1994**, *91*(21), 9926–9930.

479. Chan, J.Y., Kwong, M., Lu, R., Chang, J., Wang, B., Yen, T.S., Kan, Y.W. Targeted disruption of the ubiquitous CNC-bZIP transcription factor, Nrf-1, results in anemia and embryonic lethality in mice. *EMBO J.* **1998**, *17*(6), 1779–1787.

480. Itoh, K., Igarashi, K., Hayashi, N., Nishizawa, M., Yamamoto, M. Cloning and characterization of a novel erythroid cell-derived CNC family transcription factor heterodimerizing with the small Maf family proteins. *Mol. Cell. Biol.* **1995**, *15*(8), 4184–4193.

481. Oyake, T., Itoh, K., Motohashi, H., Hayashi, N., Hoshino, H., Nishizawa, M., Yamamoto, M., Igarashi, K. Bach proteins belong to a novel family of BTB-basic leucine zipper transcription factors that interact with MafK and regulate transcription through the NF-E2 site. *Mol. Cell. Biol.* **1996**, *16*(11), 6083–6095.

482. Dhakshinamoorthy, S., Jain, A.K., Bloom, D.A., Jaiswal, A.K. Bach1 competes with Nrf2 leading to negative regulation of the antioxidant response element (ARE)-mediated NAD(P)H:quinone oxidoreductase 1 gene expression and induction in response to antioxidants. *J. Biol. Chem.* **2005**, *280*(17), 16891–16900.

483. Lee, J.M., Li, J., Johnson, D.A., Stein, T.D., Kraft, A.D., Calkins, M.J., Jakel, R.J., Johnson, J.A. Nrf2, a multi-organ protector? *FASEB J.* **2005**, *19*(9), 1061–1066.

484. Itoh, K., Wakabayashi, N., Katoh, Y., Ishii, T., Igarashi, K., Engel, J.D., Yamamoto, M. Keap1 represses nuclear activation of antioxidant responsive elements by Nrf2 through binding to the amino-terminal Neh2 domain. *Genes Dev.* **1999**, *13*(1), 76–86.

485. Alam, J., Stewart, D., Touchard, C., Boinapally, S., Choi, A.M., Cook, J.L. Nrf2, a Cap"n"Collar transcription factor, regulates induction of the heme oxygenase-1 gene. *J. Biol. Chem.* **1999**, *274*(37), 26071–26078.

486. Motohashi, H., Katsuoka, F., Shavit, J.A., Engel, J.D., Yamamoto, M. Positive or negative MARE-dependent transcriptional regulation is determined by the abundance of small Maf proteins. *Cell* **2000**, *103*(6), 865–875.

487. Motohashi, H., Yamamoto, M. Nrf2-Keap1 defines a physiologically important stress response mechanism. *Trends Mol. Med.* **2004**, *10*(11), 549–557.

488. Venugopal, R., Jaiswal, A.K. Nrf1 and Nrf2 positively and c-Fos and Fra1 negatively regulate the human antioxidant response element-mediated expression of NAD(P)H:quinone oxidoreductase1 gene. *Proc. Natl. Acad. Sci. U.S.A.* **1996**, *93*(25), 14960–14965.

489. Venugopal, R., Jaiswal, A.K. Nrf2 and Nrf1 in association with Jun proteins regulate antioxidant response element-mediated expression and coordinated induction of genes encoding detoxifying enzymes. *Oncogene* **1998**, *17*(24), 3145–3156.

490. Chan, K., Lu, R., Chang, J.C., Kan, Y.W. NRF2, a member of the NFE2 family of transcription factors, is not essential for murine erythropoiesis, growth, and development. *Proc. Natl. Acad. Sci. U.S.A.* **1996**, *93*(24), 13943–13948.

491. Yamamoto, T., Yoh, K., Kobayashi, A., Ishii, Y., Kure, S., Koyama, A., Sakamoto, T., Sekizawa, K., Motohashi, H., Yamamoto, M. Identification of polymorphisms in the promoter region of the human NRF2 gene. *Biochem. Biophys. Res. Commun.* **2004**, *321*(1), 72–79.

492. Luna, L., Johnsen, O., Skartlien, A.H., Pedeutour, F., Turc-Carel, C., Prydz, H., Kolsto, A.B. Molecular cloning of a putative novel human bZIP transcription factor on chromosome 17q22. *Genomics* **1994**, *22*(3), 553–562.

493. Luna, L., Skammelsrud, N., Johnsen, O., Abel, K.J., Weber, B.L., Prydz, H., Kolsto, A.B. Structural organization and mapping of the human TCF11 gene. *Genomics* **1995**, *27*(2), 237–244.

494. Kwak, M.K., Itoh, K., Yamamoto, M., Kensler, T.W. Enhanced expression of the transcription factor Nrf2 by cancer chemopreventive agents: Role of antioxidant response element-like sequences in the Nrf2 promoter. *Mol. Cell. Biol.* **2002**, *22*(9), 2883–2892.

495. Miao, W., Hu, L., Scrivens, P.J., Batist, G. Transcriptional regulation of NF-E2 p45-related factor (NRF2) expression by the aryl hydrocarbon receptor-xenobiotic response element signaling pathway: Direct cross-talk between phase I and II drug-metabolizing enzymes. *J. Biol. Chem.* **2005**, *280*(21), 20340–20348.

496. Katsuoka, F., Motohashi, H., Engel, J.D., Yamamoto, M. Nrf2 transcriptionally activates the mafG gene through an antioxidant response element. *J. Biol. Chem.* **2005**, *280*(6), 4483–4490.

497. Lee, O.H., Jain, A.K., Papusha, V., Jaiswal, A.K. An autoregulatory loop between stress sensors INrf2 and Nrf2 controls their cellular abundance. *J. Biol. Chem.* **2007**, *282*(50), 36412–36420.

498. Rushmore, T.H., Kong, A.N. Pharmacogenomics, regulation and signaling pathways of phase I and II drug metabolizing enzymes. *Curr. Drug Metab.* **2002**, *3*(5), 481–490.

499. Marzec, J.M., Christie, J.D., Reddy, S.P., Jedlicka, A.E., Vuong, H., Lanken, P.N., Aplenc, R., Yamamoto, T., Yamamoto, M., Cho, H.Y., Kleeberger, S.R. Functional polymorphisms in the transcription factor NRF2 in

humans increase the risk of acute lung injury. *FASEB J.* **2007**, *21*(9), 2237–2246.

500. Itoh, K., Ishii, T., Wakabayashi, N., Yamamoto, M. Regulatory mechanisms of cellular response to oxidative stress. *Free Radic. Res.* **1999**, *31*(4), 319–324.

501. Wang, W., Chan, J.Y. Nrf1 is targeted to the endoplasmic reticulum membrane by an N-terminal transmembrane domain. Inhibition of nuclear translocation and transacting function. *J. Biol. Chem.* **2006**, *281*(28), 19676–19687.

502. Zhang, Y., Crouch, D.H., Yamamoto, M., Hayes, J.D. Negative regulation of the Nrf1 transcription factor by its N-terminal domain is independent of Keap1: Nrf1, but not Nrf2, is targeted to the endoplasmic reticulum. *Biochem. J.* **2006**, *399*(3), 373–385.

503. Chenais, B., Derjuga, A., Massrieh, W., Red-Horse, K., Bellingard, V., Fisher, S.J., Blank, V. Functional and placental expression analysis of the human NRF3 transcription factor. *Mol. Endocrinol.* **2005**, *19*(1), 125–137.

504. Katoh, Y., Itoh, K., Yoshida, E., Miyagishi, M., Fukamizu, A., Yamamoto, M. Two domains of Nrf2 cooperatively bind CBP, a CREB binding protein, and synergistically activate transcription. *Genes Cells* **2001**, *6*(10), 857–868.

505. McMahon, M., Thomas, N., Itoh, K., Yamamoto, M., Hayes, J.D. Redox-regulated turnover of Nrf2 is determined by at least two separate protein domains, the redox-sensitive Neh2 degron and the redox-insensitive Neh6 degron. *J. Biol. Chem.* **2004**, *279*(30), 31556–31567.

506. Nioi, P., Nguyen, T., Sherratt, P.J., Pickett, C.B. The carboxy-terminal Neh3 domain of Nrf2 is required for transcriptional activation. *Mol. Cell. Biol.* **2005**, *25*(24), 10895–10906.

507. Jain, A.K., Jaiswal, A.K. Phosphorylation of tyrosine 568 controls nuclear export of Nrf2. *J. Biol. Chem.* **2006**, *281*(17), 12132–12142.

508. Zhang, Y., Kobayashi, A., Yamamoto, M., Hayes, J.D. The Nrf3 transcription factor is a membrane-bound glycoprotein targeted to the endoplasmic reticulum through its N-terminal homology box 1 sequence. *J. Biol. Chem.* **2009**, *284*(5), 3195–3210.

509. Huang, H.C., Nguyen, T., Pickett, C.B. Phosphorylation of Nrf2 at Ser-40 by protein kinase C regulates antioxidant response element-mediated transcription. *J. Biol. Chem.* **2002**, *277*(45), 42769–42774.

510. Sun, Z., Chin, Y.E., Zhang, D.D. Acetylation of Nrf2 by p300/CBP augments promoter-specific DNA binding of Nrf2 during the antioxidant response. *Mol. Cell. Biol.* **2009**, *29*(10), 2658–2672.

511. Ohtsubo, T., Kamada, S., Mikami, T., Murakami, H., Tsujimoto, Y. Identification of NRF2, a member of the NF-E2 family of transcription factors, as a substrate for caspase-3(-like) proteases. *Cell Death Differ.* **1999**, *6*(9), 865–872.

512. Alam, J., Killeen, E., Gong, P., Naquin, R., Hu, B., Stewart, D., Ingelfinger, J.R., Nath, K.A. Heme activates the heme oxygenase-1 gene in renal epithelial cells by stabilizing Nrf2. *Am. J. Physiol. Renal Physiol.* **2003**, *284*(4), F743–F752.

513. Furukawa, M., Xiong, Y. BTB protein Keap1 targets antioxidant transcription factor Nrf2 for ubiquitination by the Cullin 3-Roc1 ligase. *Mol. Cell. Biol.* **2005**, *25*(1), 162–171.

514. He, X., Chen, M.G., Lin, G.X., Ma, Q. Arsenic induces NAD(P)H-quinone oxidoreductase I by disrupting the Nrf2 x Keap1 x Cul3 complex and recruiting Nrf2 x Maf to the antioxidant response element enhancer. *J. Biol. Chem.* **2006**, *281*(33), 23620–23631.

515. Cullinan, S.B., Gordan, J.D., Jin, J., Harper, J.W., Diehl, J.A. The Keap1-BTB protein is an adaptor that bridges Nrf2 to a Cul3-based E3 ligase: Oxidative stress sensing by a Cul3-Keap1 ligase. *Mol. Cell. Biol.* **2004**, *24*(19), 8477–8486.

516. Zipper, L.M., Mulcahy, R.T. The Keap1 BTB/POZ dimerization function is required to sequester Nrf2 in cytoplasm. *J. Biol. Chem.* **2002**, *277*(39), 36544–36552.

517. Dinkova-Kostova, A.T., Holtzclaw, W.D., Cole, R.N., Itoh, K., Wakabayashi, N., Katoh, Y., Yamamoto, M., Talalay, P. Direct evidence that sulfhydryl groups of Keap1 are the sensors regulating induction of phase 2 enzymes that protect against carcinogens and oxidants. *Proc. Natl. Acad. Sci. U.S.A.* **2002**, *99*(18), 11908–11913.

518. Wakabayashi, N., Dinkova-Kostova, A.T., Holtzclaw, W.D., Kang, M.I., Kobayashi, A., Yamamoto, M., Kensler, T.W., Talalay, P. Protection against electrophile and oxidant stress by induction of the phase 2 response: Fate of cysteines of the Keap1 sensor modified by inducers. *Proc. Natl. Acad. Sci. U.S.A.* **2004**, *101*(7), 2040–2045.

519. Kang, M.I., Kobayashi, A., Wakabayashi, N., Kim, S.G., Yamamoto, M. Scaffolding of Keap1 to the actin cytoskeleton controls the function of Nrf2 as key regulator of cytoprotective phase 2 genes. *Proc. Natl. Acad. Sci. U.S.A.* **2004**, *101*(7), 2046–2051.

520. Tong, K.I., Katoh, Y., Kusunoki, H., Itoh, K., Tanaka, T., Yamamoto, M. Keap1 recruits Neh2 through binding to ETGE and DLG motifs: Characterization of the two-site molecular recognition model. *Mol. Cell. Biol.* **2006**, *26*(8), 2887–2900.

521. Sun, Z., Zhang, S., Chan, J.Y., Zhang, D.D. Keap1 controls postinduction repression of the Nrf2-mediated antioxidant response by escorting nuclear export of Nrf2. *Mol. Cell. Biol.* **2007**, *27*(18), 6334–6349.

522. Kobayashi, A., Kang, M.I., Okawa, H., Ohtsuji, M., Zenke, Y., Chiba, T., Igarashi, K., Yamamoto, M. Oxidative stress sensor Keap1 functions as an adaptor for Cul3-based E3 ligase to regulate proteasomal degradation of Nrf2. *Mol. Cell. Biol.* **2004**, *24*(16), 7130–7139.

523. Zhang, D.D., Lo, S.C., Cross, J.V., Templeton, D.J., Hannink, M. Keap1 is a redox-regulated substrate adaptor protein for a Cul3-dependent ubiquitin ligase complex. *Mol. Cell. Biol.* **2004**, *24*(24), 10941–10953.

524. Clements, C.M., McNally, R.S., Conti, B.J., Mak, T.W., Ting, J.P. DJ-1, a cancer- and Parkinson's disease-associated protein, stabilizes the antioxidant transcriptional master regulator Nrf2. *Proc. Natl. Acad. Sci. U.S.A.* **2006**, *103*(41), 15091–15096.

525. Gan, L., Johnson, D.A., Johnson, J.A. Keap1-Nrf2 activation in the presence and absence of DJ-1. *Eur. J. Neurosci.* **2010**, *31*(6), 967–977.

526. Itoh, K., Chiba, T., Takahashi, S., Ishii, T., Igarashi, K., Katoh, Y., Oyake, T., Hayashi, N., Satoh, K., Hatayama, I., Yamamoto, M., Nabeshima, Y. An Nrf2/small Maf heterodimer mediates the induction of phase II detoxifying enzyme genes through antioxidant response elements. *Biochem. Biophys. Res. Commun.* **1997**, *236*(2), 313–322.

527. Katsuoka, F., Motohashi, H., Ishii, T., Aburatani, H., Engel, J.D., Yamamoto, M. Genetic evidence that small maf proteins are essential for the activation of antioxidant response element-dependent genes. *Mol. Cell. Biol.* **2005**, *25*(18), 8044–8051.

528. Jeyapaul, J., Jaiswal, A.K. Nrf2 and c-Jun regulation of antioxidant response element (ARE)-mediated expression and induction of gamma-glutamylcysteine synthetase heavy subunit gene. *Biochem. Pharmacol.* **2000**, *59*(11), 1433–1439.

529. Rushmore, T.H., Morton, M.R., Pickett, C.B. The antioxidant responsive element: Activation by oxidative stress and identification of the DNA consensus sequence required for functional activity. *J. Biol. Chem.* **1991**, *266*(18), 11632–11639.

530. Dalton, T.P., Shertzer, H.G., Puga, A. Regulation of gene expression by reactive oxygen. *Annu. Rev. Pharmacol. Toxicol.* **1999**, *39*, 67–101.

531. Nguyen, T., Sherratt, P.J., Pickett, C.B. Regulatory mechanisms controlling gene expression mediated by the antioxidant response element. *Annu. Rev. Pharmacol. Toxicol.* **2003**, *43*, 233–260.

532. Wasserman, W.W., Fahl, W.E. Functional antioxidant responsive elements. *Proc. Natl. Acad. Sci. U.S.A.* **1997**, *94*(10), 5361–5366.

533. Erickson, A.M., Nevarea, Z., Gipp, J.J., Mulcahy, R.T. Identification of a variant antioxidant response element in the promoter of the human glutamate-cysteine ligase modifier subunit gene. Revision of the ARE consensus sequence. *J. Biol. Chem.* **2002**, *277*(34), 30730–30737.

534. Nioi, P., McMahon, M., Itoh, K., Yamamoto, M., Hayes, J.D. Identification of a novel Nrf2-regulated antioxidant response element (ARE) in the mouse NAD(P)H:quinone oxidoreductase 1 gene: Reassessment of the ARE consensus sequence. *Biochem. J.* **2003**, *374*(Pt 2), 337–348.

535. Biswas, M., Chan, J.Y. Role of Nrf1 in antioxidant response element-mediated gene expression and beyond. *Toxicol. Appl. Pharmacol.* **2010**, *244*(1), 16–20.

536. Favreau, L.V., Pickett, C.B. Transcriptional regulation of the rat NAD(P)H:quinone reductase gene: Identification of regulatory elements controlling basal level expression and inducible expression by planar aromatic compounds and phenolic antioxidants. *J. Biol. Chem.* **1991**, *266*(7), 4556–4561.

537. Rushmore, T.H., King, R.G., Paulson, K.E., Pickett, C.B. Regulation of glutathione S-transferase Ya subunit gene expression: Identification of a unique xenobiotic-responsive element controlling inducible expression by planar aromatic compounds. *Proc. Natl. Acad. Sci. U.S.A.* **1990**, *87*(10), 3826–3830.

538. Kobayashi, A., Kang, M.I., Watai, Y., Tong, K.I., Shibata, T., Uchida, K., Yamamoto, M. Oxidative and electrophilic stresses activate Nrf2 through inhibition of ubiquitination activity of Keap1. *Mol. Cell. Biol.* **2006**, *26*(1), 221–229.

539. Hong, F., Sekhar, K.R., Freeman, M.L., Liebler, D.C. Specific patterns of electrophile adduction trigger Keap1 ubiquitination and Nrf2 activation. *J. Biol. Chem.* **2005**, *280*(36), 31768–31775.

540. Zhang, D.D., Hannink, M. Distinct cysteine residues in Keap1 are required for Keap1-dependent ubiquitination of Nrf2 and for stabilization of Nrf2 by chemopreventive agents and oxidative stress. *Mol. Cell. Biol.* **2003**, *23*(22), 8137–8151.

541. Dinkova-Kostova, A.T., Holtzclaw, W.D., Kensler, T.W. The role of Keap1 in cellular protective responses. *Chem. Res. Toxicol.* **2005**, *18*(12), 1779–1791.

542. Satoh, T., Okamoto, S.I., Cui, J., Watanabe, Y., Furuta, K., Suzuki, M., Tohyama, K., Lipton, S.A. Activation of the Keap1/Nrf2 pathway for neuroprotection by electrophilic [correction of electrophillic] phase II inducers. *Proc. Natl. Acad. Sci. U.S.A.* **2006**, *103*(3), 768–773.

543. Yu, R., Chen, C., Mo, Y.Y., Hebbar, V., Owuor, E.D., Tan, T.H., Kong, A.N. Activation of mitogen-activated protein kinase pathways induces antioxidant response element-mediated gene expression via a Nrf2-dependent mechanism. *J. Biol. Chem.* **2000**, *275*(51), 39907–39913.

544. Keum, Y.S., Owuor, E.D., Kim, B.R., Hu, R., Kong, A.N. Involvement of Nrf2 and JNK1 in the activation of antioxidant responsive element (ARE) by chemopreventive agent phenethyl isothiocyanate (PEITC). *Pharm. Res.* **2003**, *20*(9), 1351–1356.

545. Shen, G., Hebbar, V., Nair, S., Xu, C., Li, W., Lin, W., Keum, Y.S., Han, J., Gallo, M.A., Kong, A.N. Regulation of Nrf2 transactivation domain activity: The differential effects of mitogen-activated protein kinase cascades and synergistic stimulatory effect of Raf and CREB-binding protein. *J. Biol. Chem.* **2004**, *279*(22), 23052–23060.

546. Yu, R., Lei, W., Mandlekar, S., Weber, M.J., Der, C.J., Wu, J., Kong, A.N. Role of a mitogen-activated protein kinase pathway in the induction of phase II detoxifying enzymes by chemicals. *J. Biol. Chem.* **1999**, *274*(39), 27545–27552.

547. Yu, R., Mandlekar, S., Lei, W., Fahl, W.E., Tan, T.H., Kong, A.N. p38 mitogen-activated protein kinase negatively regulates the induction of phase II drug-metabolizing enzymes that detoxify carcinogens. *J. Biol. Chem.* **2000**, *275*(4), 2322–2327.

548. Kappos, L., Gold, R., Miller, D.H., Macmanus, D.G., Havrdova, E., Limmroth, V., Polman, C.H., Schmierer, K., Yousry, T.A., Yang, M., Eraksoy, M., Meluzinova, E., Rektor, I., Dawson, K.T., Sandrock, A.W., O'Neill, G.N. Efficacy and safety of oral fumarate in patients with relapsing-remitting multiple sclerosis: A multicentre, randomised, double-blind, placebo-controlled phase IIb study. *Lancet* **2008**, *372*(9648), 1463–1472.

549. Zhang, Y., Munday, R. Dithiolethiones for cancer chemoprevention: Where do we stand? *Mol. Cancer Ther.* **2008**, *7*(11), 3470–3479.

550. Shen, G., Xu, C., Hu, R., Jain, M.R., Gopalkrishnan, A., Nair, S., Huang, M.T., Chan, J.Y., Kong, A.N. Modulation of nuclear factor E2-related factor 2-mediated gene expression in mice liver and small intestine by cancer chemopreventive agent curcumin. *Mol. Cancer Ther.* **2006**, *5*(1), 39–51.

551. Juge, N., Mithen, R.F., Traka, M. Molecular basis for chemoprevention by sulforaphane: A comprehensive review. *Cell. Mol. Life Sci.* **2007**, *64*(9), 1105–1127.

552. Ramos-Gomez, M., Kwak, M.K., Dolan, P.M., Itoh, K., Yamamoto, M., Talalay, P., Kensler, T.W. Sensitivity to carcinogenesis is increased and chemoprotective efficacy of enzyme inducers is lost in Nrf2 transcription factor-deficient mice. *Proc. Natl. Acad. Sci. U.S.A.* **2001**, *98*(6), 3410–3415.

553. Calkins, M.J., Jakel, R.J., Johnson, D.A., Chan, K., Kan, Y.W., Johnson, J.A. Protection from mitochondrial complex II inhibition in vitro and in vivo by Nrf2-mediated transcription. *Proc. Natl. Acad. Sci. U.S.A.* **2005**, *102*(1), 244–249.

554. Ramsey, C.P., Glass, C.A., Montgomery, M.B., Lindl, K.A., Ritson, G.P., Chia, L.A., Hamilton, R.L., Chu, C.T., Jordan-Sciutto, K.L. Expression of Nrf2 in neurodegenerative diseases. *J. Neuropathol. Exp. Neurol.* **2007**, *66*(1), 75–85.

555. Shih, A.Y., Imbeault, S., Barakauskas, V., Erb, H., Jiang, L., Li, P., Murphy, T.H. Induction of the Nrf2-driven antioxidant response confers neuroprotection during mitochondrial stress in vivo. *J. Biol. Chem.* **2005**, *280*(24), 22925–22936.

556. Shih, A.Y., Johnson, D.A., Wong, G., Kraft, A.D., Jiang, L., Erb, H., Johnson, J.A., Murphy, T.H. Coordinate regulation of glutathione biosynthesis and release by Nrf2-expressing glia potently protects neurons from oxidative stress. *J. Neurosci.* **2003**, *23*(8), 3394–3406.

557. Shih, A.Y., Li, P., Murphy, T.H. A small-molecule-inducible Nrf2-mediated antioxidant response provides effective prophylaxis against cerebral ischemia in vivo. *J. Neurosci.* **2005**, *25*(44), 10321–10335.

558. Sarlette, A., Krampfl, K., Grothe, C., Neuhoff, N., Dengler, R., Petri, S. Nuclear erythroid 2-related factor 2-antioxidative response element signaling pathway in motor cortex and spinal cord in amyotrophic lateral sclerosis. *J. Neuropathol. Exp. Neurol.* **2008**, *67*(11), 1055–1062.

559. Jakel, R.J., Townsend, J.A., Kraft, A.D., Johnson, J.A. Nrf2-mediated protection against 6-hydroxydopamine. *Brain Res.* **2007**, *1144*, 192–201.

560. Kraft, A.D., Johnson, D.A., Johnson, J.A. Nuclear factor E2-related factor 2-dependent antioxidant response element activation by tert-butylhydroquinone and sulforaphane occurring preferentially in astrocytes conditions neurons against oxidative insult. *J. Neurosci.* **2004**, *24*(5), 1101–1112.

561. Mullett, S.J., Hinkle, D.A. DJ-1 knock-down in astrocytes impairs astrocyte-mediated neuroprotection against rotenone. *Neurobiol. Dis.* **2009**, *33*(1), 28–36.

562. de Vries, H.E., Witte, M., Hondius, D., Rozemuller, A.J., Drukarch, B., Hoozemans, J., van Horssen, J. Nrf2-induced antioxidant protection: A promising target to counteract ROS-mediated damage in neurodegenerative disease? *Free Radic. Biol. Med.* **2008**, *45*(10), 1375–1383.

563. Wider, C., Wszolek, Z.K. Clinical genetics of Parkinson's disease and related disorders. *Parkinsonism Relat. Disord.* **2007**, *13*(Suppl. 3), S229–S232.

564. Canet-Aviles, R.M., Wilson, M.A., Miller, D.W., Ahmad, R., McLendon, C., Bandyopadhyay, S., Baptista, M.J., Ringe, D., Petsko, G.A., Cookson, M.R. The Parkinson's disease protein DJ-1 is neuroprotective due to cysteine-sulfinic acid-driven mitochondrial localization. *Proc. Natl. Acad. Sci. U.S.A.* **2004**, *101*(24), 9103–9108.

565. Lev, N., Ickowicz, D., Barhum, Y., Lev, S., Melamed, E., Offen, D. DJ-1 protects against dopamine toxicity. *J. Neural Transm.* **2009**, *116*(2), 151–160.

566. Bader, V., Ran Zhu, X., Lubbert, H., Stichel, C.C. Expression of DJ-1 in the adult mouse CNS. *Brain Res.* **2005**, *1041*(1), 102–111.

567. van Horssen, J., Drexhage, J.A., Flor, T., Gerritsen, W., van der Valk, P., de Vries, H.E. Nrf2 and DJ1 are consistently upregulated in inflammatory multiple sclerosis lesions. *Free Radic. Biol. Med.* **2010**, *49*(8), 1283–1289.

568. van Horssen, J., Schreibelt, G., Drexhage, J., Hazes, T., Dijkstra, C.D., van der Valk, P., de Vries, H.E. Severe oxidative damage in multiple sclerosis lesions coincides with enhanced antioxidant enzyme expression. *Free Radic. Biol. Med.* **2008**, *45*(12), 1729–1737.

569. Chen, P.C., Vargas, M.R., Pani, A.K., Smeyne, R.J., Johnson, D.A., Kan, Y.W., Johnson, J.A. Nrf2-mediated neuroprotection in the MPTP mouse model of Parkinson's disease: Critical role for the astrocyte. *Proc. Natl. Acad. Sci. U.S.A.* **2009**, *106*(8), 2933–2938.

570. Kanninen, K., Heikkinen, R., Malm, T., Rolova, T., Kuhmonen, S., Leinonen, H., Yla-Herttuala, S., Tanila, H., Levonen, A.L., Koistinaho, M., Koistinaho, J. Intrahippocampal injection of a lentiviral vector expressing Nrf2 improves spatial learning in a mouse model of Alzheimer's disease. *Proc. Natl. Acad. Sci. U.S.A.* **2009**, *106*(38), 16505–16510.

571. Vargas, M.R., Johnson, D.A., Sirkis, D.W., Messing, A., Johnson, J.A. Nrf2 activation in astrocytes protects against neurodegeneration in mouse models of familial amyotrophic lateral sclerosis. *J. Neurosci.* **2008**, *28*(50), 13574–13581.

572. Desagher, S., Glowinski, J., Premont, J. Astrocytes protect neurons from hydrogen peroxide toxicity. *J. Neurosci.* **1996**, *16*(8), 2553–2562.

573. Lee, J.M., Calkins, M.J., Chan, K., Kan, Y.W., Johnson, J.A. Identification of the NF-E2-related factor-2-dependent genes conferring protection against oxidative stress in primary cortical astrocytes using oligonucleotide micro-

array analysis. *J. Biol. Chem.* **2003**, *278*(14), 12029–12038.

574. Li, J., Lee, J.M., Johnson, J.A. Microarray analysis reveals an antioxidant responsive element-driven gene set involved in conferring protection from an oxidative stress-induced apoptosis in IMR-32 cells. *J. Biol. Chem.* **2002**, *277*(1), 388–394.

575. Lucius, R., Sievers, J. Postnatal retinal ganglion cells in vitro: Protection against reactive oxygen species (ROS)-induced axonal degeneration by cocultured astrocytes. *Brain Res.* **1996**, *743*(1–2), 56–62.

576. Burton, N.C., Kensler, T.W., Guilarte, T.R. In vivo modulation of the Parkinsonian phenotype by Nrf2. *Neurotoxicology* **2006**, *27*(6), 1094–1100.

577. Jakel, R.J., Kern, J.T., Johnson, D.A., Johnson, J.A. Induction of the protective antioxidant response element pathway by 6-hydroxydopamine in vivo and in vitro. *Toxicol. Sci.* **2005**, *87*(1), 176–186.

578. Kraft, A.D., Lee, J.M., Johnson, D.A., Kan, Y.W., Johnson, J.A. Neuronal sensitivity to kainic acid is dependent on the Nrf2-mediated actions of the antioxidant response element. *J. Neurochem.* **2006**, *98*(6), 1852–1865.

579. Zhao, J., Moore, A.N., Redell, J.B., Dash, P.K. Enhancing expression of Nrf2-driven genes protects the blood brain barrier after brain injury. *J. Neurosci.* **2007**, *27*(38), 10240–10248.

580. Zhao, X., Sun, G., Zhang, J., Strong, R., Dash, P.K., Kan, Y.W., Grotta, J.C., Aronowski, J. Transcription factor Nrf2 protects the brain from damage produced by intracerebral hemorrhage. *Stroke* **2007**, *38*(12), 3280–3286.

581. Dash, P.K., Zhao, J., Orsi, S.A., Zhang, M., Moore, A.N. Sulforaphane improves cognitive function administered following traumatic brain injury. *Neurosci. Lett.* **2009**, *460*(2), 103–107.

582. Pacchioni, A.M., Vallone, J., Melendez, R.I., Shih, A., Murphy, T.H., Kalivas, P.W. Nrf2 gene deletion fails to alter psychostimulant-induced behavior or neurotoxicity. *Brain Res.* **2007**, *1127*(1), 26–35.

583. Tanaka, N., Ikeda, Y., Ohta, Y., Deguchi, K., Tian, F., Shang, J., Matsuura, T., Abe, K. Expression of Keap1-Nrf2 system and antioxidative proteins in mouse brain after transient middle cerebral artery occlusion. *Brain Res.* **2011**, *1370*, 246–253.

584. Wells, P., Jeng, W., Loniewska, M., McCallum, G., Perstin, J., Ramkissoon, A., Shapiro, A. In *Proceedings of the 6th Meeting of the Canadian Oxidative Stress Consortium*, Winnipeg, Manitoba, 2009, p. 82.

585. Nakaso, K., Nakamura, C., Sato, H., Imamura, K., Takeshima, T., Nakashima, K. Novel cytoprotective mechanism of anti-Parkinsonian drug deprenyl: PI3K and Nrf2-derived induction of antioxidative proteins. *Biochem. Biophys. Res. Commun.* **2006**, *339*(3), 915–922.

586. Ramkissoon, A., Wells, P.G. Methamphetamine-initiated cognitive neurodevelopmental deficits in nuclear factor-E2-related factor 2 (Nrf2)-deficient knockout mice. *Birth Defects Res. A. Clin Mol. Teratol.* **2009**, *85*, 449 (No. P67).

587. Derjuga, A., Gourley, T.S., Holm, T.M., Heng, H.H., Shivdasani, R.A., Ahmed, R., Andrews, N.C., Blank, V. Complexity of CNC transcription factors as revealed by gene targeting of the Nrf3 locus. *Mol. Cell. Biol.* **2004**, *24*(8), 3286–3294.

588. Leung, L., Kwong, M., Hou, S., Lee, C., Chan, J.Y. Deficiency of the Nrf1 and Nrf2 transcription factors results in early embryonic lethality and severe oxidative stress. *J. Biol. Chem.* **2003**, *278*(48), 48021–48029.

589. Dong, J., Sulik, K.K., Chen, S.Y. Nrf2-mediated transcriptional induction of antioxidant response in mouse embryos exposed to ethanol in vivo: Implications for the prevention of fetal alcohol spectrum disorders. *Antioxid. Redox Signal.* **2008**, *10*(12), 2023–2033.

590. Wakabayashi, N., Itoh, K., Wakabayashi, J., Motohashi, H., Noda, S., Takahashi, S., Imakado, S., Kotsuji, T., Otsuka, F., Roop, D.R., Harada, T., Engel, J.D., Yamamoto, M. Keap1-null mutation leads to postnatal lethality due to constitutive Nrf2 activation. *Nat. Genet.* **2003**, *35*(3), 238–245.

591. Lee, J.M., Shih, A.Y., Murphy, T.H., Johnson, J.A. NF-E2-related factor-2 mediates neuroprotection against mitochondrial complex I inhibitors and increased concentrations of intracellular calcium in primary cortical neurons. *J. Biol. Chem.* **2003**, *278*(39), 37948–37956.

592. Dhakshinamoorthy, S., Porter, A.G. Nitric oxide-induced transcriptional up-regulation of protective genes by Nrf2 via the antioxidant response element counteracts apoptosis of neuroblastoma cells. *J. Biol. Chem.* **2004**, *279*(19), 20096–20107.

593. Hubbs, A.F., Benkovic, S.A., Miller, D.B., O'Callaghan, J.P., Battelli, L., Schwegler-Berry, D., Ma, Q. Vacuolar leukoencephalopathy with widespread astrogliosis in mice lacking transcription factor Nrf2. *Am. J. Pathol.* **2007**, *170*(6), 2068–2076.

594. Li, J., Stein, T.D., Johnson, J.A. Genetic dissection of systemic autoimmune disease in Nrf2-deficient mice. *Physiol. Genomics* **2004**, *18*(3), 261–272.

595. Ma, Q., Battelli, L., Hubbs, A.F. Multiorgan autoimmune inflammation, enhanced lymphoproliferation, and impaired homeostasis of reactive oxygen species in mice lacking the antioxidant-activated transcription factor Nrf2. *Am. J. Pathol.* **2006**, *168*(6), 1960–1974.

596. Yoh, K., Itoh, K., Enomoto, A., Hirayama, A., Yamaguchi, N., Kobayashi, M., Morito, N., Koyama, A., Yamamoto, M., Takahashi, S. Nrf2-deficient female mice develop lupus-like autoimmune nephritis. *Kidney Int.* **2001**, *60*(4), 1343–1353.

597. Hayes, J.D., McMahon, M. NRF2 and KEAP1 mutations: Permanent activation of an adaptive response in cancer. *Trends Biochem. Sci.* **2009**, *34*(4), 176–188.

598. Rangasamy, T., Guo, J., Mitzner, W.A., Roman, J., Singh, A., Fryer, A.D., Yamamoto, M., Kensler, T.W., Tuder, R.M., Georas, S.N., Biswal, S. Disruption of Nrf2 enhances susceptibility to severe airway inflammation and asthma in mice. *J. Exp. Med.* **2005**, *202*(1), 47–59.

599. Khor, T.O., Huang, M.T., Kwon, K.H., Chan, J.Y., Reddy, B.S., Kong, A.N. Nrf2-deficient mice have an increased susceptibility to dextran sulfate sodium-induced colitis. *Cancer Res.* **2006**, *66*(24), 11580–11584.

600. Chan, K., Han, X.D., Kan, Y.W. An important function of Nrf2 in combating oxidative stress: Detoxification of acetaminophen. *Proc. Natl. Acad. Sci. U.S.A.* **2001**, *98*(8), 4611–4616.

601. Enomoto, A., Itoh, K., Nagayoshi, E., Haruta, J., Kimura, T., O'Connor, T., Harada, T., Yamamoto, M. High sensitivity of Nrf2 knockout mice to acetaminophen hepatotoxicity associated with decreased expression of ARE-regulated drug metabolizing enzymes and antioxidant genes. *Toxicol. Sci.* **2001**, *59*(1), 169–177.

602. Collins, A.R., Lyon, C.J., Xia, X., Liu, J.Z., Tangirala, R.K., Yin, F., Boyadjian, R., Bikineyeva, A., Pratico, D., Harrison, D.G., Hsueh, W.A. Age-accelerated atherosclerosis correlates with failure to upregulate antioxidant genes. *Circ. Res.* **2009**, *104*(6), e42–e54.

603. Wang, X., Tomso, D.J., Chorley, B.N., Cho, H.Y., Cheung, V.G., Kleeberger, S.R., Bell, D.A. Identification of polymorphic antioxidant response elements in the human genome. *Hum. Mol. Genet.* **2007**, *16*(10), 1188–1200.

604. Guan, C.P., Zhou, M.N., Xu, A.E., Kang, K.F., Liu, J.F., Wei, X.D., Li, Y.W., Zhao, D.K., Hong, W.S. The susceptibility to vitiligo is associated with NF-E2-related factor2 (Nrf2) gene polymorphisms: A study on Chinese Han population. *Exp. Dermatol.* **2008**, *17*(12), 1059–1062.

605. Arisawa, T., Tahara, T., Shibata, T., Nagasaka, M., Nakamura, M., Kamiya, Y., Fujita, H., Hasegawa, S., Takagi, T., Wang, F.Y., Hirata, I., Nakano, H. The relationship between *Helicobacter pylori* infection and promoter polymorphism of the Nrf2 gene in chronic gastritis. *Int. J. Mol. Med.* **2007**, *19*(1), 143–148.

606. Arisawa, T., Tahara, T., Shibata, T., Nagasaka, M., Nakamura, M., Kamiya, Y., Fujita, H., Yoshioka, D., Okubo, M., Hirata, I., Nakano, H. Nrf2 gene promoter polymorphism and gastric carcinogenesis. *Hepatogastroenterology.* **2008**, *55*(82–83), 750–754.

607. Arisawa, T., Tahara, T., Shibata, T., Nagasaka, M., Nakamura, M., Kamiya, Y., Fujita, H., Yoshioka, D., Okubo, M., Sakata, M., Wang, F.Y., Hirata, I., Nakano, H. Nrf2 gene promoter polymorphism is associated with ulcerative colitis in a Japanese population. *Hepatogastroenterology* **2008**, *55*(82–83), 394–397.

608. Padmanabhan, B., Tong, K.I., Ohta, T., Nakamura, Y., Scharlock, M., Ohtsuji, M., Kang, M.I., Kobayashi, A., Yokoyama, S., Yamamoto, M. Structural basis for defects of Keap1 activity provoked by its point mutations in lung cancer. *Mol. Cell* **2006**, *21*(5), 689–700.

609. Singh, A., Misra, V., Thimmulappa, R.K., Lee, H., Ames, S., Hoque, M.O., Herman, J.G., Baylin, S.B., Sidransky, D., Gabrielson, E., Brock, M.V., Biswal, S. Dysfunctional KEAP1-NRF2 interaction in non-small-cell lung cancer. *PLoS Med.* **2006**, *3*(10), e420.

610. Nioi, P., Nguyen, T. A mutation of Keap1 found in breast cancer impairs its ability to repress Nrf2 activity. *Biochem. Biophys. Res. Commun.* **2007**, *362*(4), 816–821.

611. Ohta, T., Iijima, K., Miyamoto, M., Nakahara, I., Tanaka, H., Ohtsuji, M., Suzuki, T., Kobayashi, A., Yokota, J., Sakiyama, T., Shibata, T., Yamamoto, M., Hirohashi, S. Loss of Keap1 function activates Nrf2 and provides advantages for lung cancer cell growth. *Cancer Res.* **2008**, *68*(5), 1303–1309.

612. Shibata, T., Kokubu, A., Gotoh, M., Ojima, H., Ohta, T., Yamamoto, M., Hirohashi, S. Genetic alteration of Keap1 confers constitutive Nrf2 activation and resistance to chemotherapy in gallbladder cancer. *Gastroenterology* **2008**, *135*(4), 1358–1368, 68 e1–4.

613. Wang, R., An, J., Ji, F., Jiao, H., Sun, H., Zhou, D. Hypermethylation of the Keap1 gene in human lung cancer cell lines and lung cancer tissues. *Biochem. Biophys. Res. Commun.* **2008**, *373*(1), 151–154.

614. Wang, X.J., Sun, Z., Villeneuve, N.F., Zhang, S., Zhao, F., Li, Y., Chen, W., Yi, X., Zheng, W., Wondrak, G.T., Wong, P.K., Zhang, D.D. Nrf2 enhances resistance of cancer cells to chemotherapeutic drugs, the dark side of Nrf2. *Carcinogenesis* **2008**, *29*(6), 1235–1243.

615. Winn, L.M., Wells, P.G. Phenytoin-initiated DNA oxidation in murine embryo culture, and embryo protection by the antioxidative enzymes superoxide dismutase and catalase: Evidence for reactive oxygen species-mediated DNA oxidation in the molecular mechanism of phenytoin teratogenicity. *Mol. Pharmacol.* **1995**, *48*(1), 112–120.

616. Andreasson, K. Emerging roles of PGE2 receptors in models of neurological disease. *Prostaglandins Other Lipid Mediat.* **2010**, *91*(3–4), 104–112.

617. Boie, Y., Sawyer, N., Slipetz, D.M., Metters, K.M., Abramovitz, M. Molecular cloning and characterization of the human prostanoid DP receptor. *J. Biol. Chem.* **1995**, *270*(32), 18910–18916.

618. Coleman, R.A., Smith, W.L., Narumiya, S. International Union of Pharmacology classification of prostanoid receptors: Properties, distribution, and structure of the receptors and their subtypes. *Pharmacol. Rev.* **1994**, *46*(2), 205–229.

619. Hasumoto, K., Sugimoto, Y., Yamasaki, A., Morimoto, K., Kakizuka, A., Negishi, M., Ichikawa, A. Association of expression of mRNA encoding the PGF2 alpha receptor with luteal cell apoptosis in ovaries of pseudopregnant mice. *J. Reprod. Fertil.* **1997**, *109*(1), 45–51.

620. Hata, A.N., Breyer, R.M. Pharmacology and signaling of prostaglandin receptors: Multiple roles in inflammation and immune modulation. *Pharmacol. Ther.* **2004**, *103*(2), 147–166.

621. Hirata, M., Kakizuka, A., Aizawa, M., Ushikubi, F., Narumiya, S. Molecular characterization of a mouse prostaglandin D receptor and functional expression of the cloned gene. *Proc. Natl. Acad. Sci. U.S.A.* **1994**, *91*(23), 11192–11196.

622. Katsuyama, M., Sugimoto, Y., Morimoto, K., Hasumoto, K., Fukumoto, M., Negishi, M., Ichikawa, A. "Distinct cellular localization" of the messenger ribonucleic acid for prostaglandin E receptor subtypes in the mouse

uterus during pseudopregnancy. *Endocrinology* **1997**, *138*(1), 344–350.

623. Namba, T., Sugimoto, Y., Hirata, M., Hayashi, Y., Honda, A., Watabe, A., Negishi, M., Ichikawa, A., Narumiya, S. Mouse thromboxane A2 receptor: cDNA cloning, expression and northern blot analysis. *Biochem. Biophys. Res. Commun.* **1992**, *184*(3), 1197–1203.

624. Oida, H., Hirata, M., Sugimoto, Y., Ushikubi, F., Ohishi, H., Mizuno, N., Ichikawa, A., Narumiya, S. Expression of messenger RNA for the prostaglandin D receptor in the leptomeninges of the mouse brain. *FEBS Lett.* **1997**, *417*(1), 53–56.

625. Oida, H., Namba, T., Sugimoto, Y., Ushikubi, F., Ohishi, H., Ichikawa, A., Narumiya, S. In situ hybridization studies of prostacyclin receptor mRNA expression in various mouse organs. *Br. J. Pharmacol.* **1995**, *116*(7), 2828–2837.

626. Sugimoto, Y., Shigemoto, R., Namba, T., Negishi, M., Mizuno, N., Narumiya, S., Ichikawa, A. Distribution of the messenger RNA for the prostaglandin E receptor subtype EP3 in the mouse nervous system. *Neuroscience* **1994**, *62*(3), 919–928.

627. Watabe, A., Sugimoto, Y., Honda, A., Irie, A., Namba, T., Negishi, M., Ito, S., Narumiya, S., Ichikawa, A. Cloning and expression of cDNA for a mouse EP1 subtype of prostaglandin E receptor. *J. Biol. Chem.* **1993**, *268*(27), 20175–20178.

628. Kitzler, J., Hill, E., Hardman, R., Reddy, N., Philpot, R., Eling, T.E. Analysis and quantitation of splicing variants of the TPA-inducible PGHS-1 mRNA in rat tracheal epithelial cells. *Arch. Biochem. Biophys.* **1995**, *316*(2), 856–863.

629. Ueda, N., Yamashita, R., Yamamoto, S., Ishimura, K. Induction of cyclooxygenase-1 in a human megakaryoblastic cell line (CMK) differentiated by phorbol ester. *Biochim. Biophys. Acta* **1997**, *1344*(1), 103–110.

630. Kirtikara, K., Morham, S.G., Raghow, R., Laulederkind, S.J., Kanekura, T., Goorha, S., Ballou, L.R. Compensatory prostaglandin E2 biosynthesis in cyclooxygenase 1 or 2 null cells. *J. Exp. Med.* **1998**, *187*(4), 517–523.

631. Diaz, A., Chepenik, K.P., Korn, J.H., Reginato, A.M., Jimenez, S.A. Differential regulation of cyclooxygenases 1 and 2 by interleukin-1 beta, tumor necrosis factor-alpha, and transforming growth factor-beta 1 in human lung fibroblasts. *Exp. Cell Res.* **1998**, *241*(1), 222–229.

632. Bryant, C.E., Appleton, I., Mitchell, J.A. Vascular endothelial growth factor upregulates constitutive cyclooxygenase 1 in primary bovine and human endothelial cells. *Life Sci.* **1998**, *62*(24), 2195–2201.

633. Zhu, Y., Saunders, M.A., Yeh, H., Deng, W.G., Wu, K.K. Dynamic regulation of cyclooxygenase-2 promoter activity by isoforms of CCAAT/enhancer-binding proteins. *J. Biol. Chem.* **2002**, *277*(9), 6923–6928.

634. Inoue, H., Yokoyama, C., Hara, S., Tone, Y., Tanabe, T. Transcriptional regulation of human prostaglandin-endoperoxide synthase-2 gene by lipopolysaccharide and phorbol ester in vascular endothelial cells. Involvement of both nuclear factor for interleukin-6 expression site and cAMP response element. *J. Biol. Chem.* **1995**, *270*(42), 24965–24971.

635. Subbaramaiah, K., Lin, D.T., Hart, J.C., Dannenberg, A.J. Peroxisome proliferator-activated receptor gamma ligands suppress the transcriptional activation of cyclooxygenase-2. Evidence for involvement of activator protein-1 and CREB-binding protein/p300. *J. Biol. Chem.* **2001**, *276*(15), 12440–12448.

636. Okada, Y., Voznesensky, O., Herschman, H., Harrison, J., Pilbeam, C. Identification of multiple cis-acting elements mediating the induction of prostaglandin G/H synthase-2 by phorbol ester in murine osteoblastic cells. *J. Cell. Biochem.* **2000**, *78*(2), 197–209.

637. Jones, D.A., Carlton, D.P., McIntyre, T.M., Zimmerman, G.A., Prescott, S.M. Molecular cloning of human prostaglandin endoperoxide synthase type II and demonstration of expression in response to cytokines. *J. Biol. Chem.* **1993**, *268*(12), 9049–9054.

638. Wu, G., Mannam, A.P., Wu, J., Kirbis, S., Shie, J.L., Chen, C., Laham, R.J., Sellke, F.W., Li, J. Hypoxia induces myocyte-dependent COX-2 regulation in endothelial cells: Role of VEGF. *Am. J. Physiol. Heart Circ. Physiol.* **2003**, *285*(6), H2420–H2429.

639. Deng, W.G., Zhu, Y., Wu, K.K. Role of p300 and PCAF in regulating cyclooxygenase-2 promoter activation by inflammatory mediators. *Blood* **2004**, *103*(6), 2135–2142.

640. Warnock, L.J., Hunninghake, G.W. Multiple second messenger pathways regulate IL-1 beta-induced expression of PGHS-2 mRNA in normal human skin fibroblasts. *J. Cell. Physiol.* **1995**, *163*(1), 172–178.

641. Inoue, H., Nanayama, T., Hara, S., Yokoyama, C., Tanabe, T. The cyclic AMP response element plays an essential role in the expression of the human prostaglandin-endoperoxide synthase 2 gene in differentiated U937 monocytic cells. *FEBS Lett.* **1994**, *350*(1), 51–54.

642. Reddy, S.T., Herschman, H.R. Ligand-induced prostaglandin synthesis requires expression of the TIS10/PGS-2 prostaglandin synthase gene in murine fibroblasts and macrophages. *J. Biol. Chem.* **1994**, *269*(22), 15473–15480.

643. Haeffner, A., Thieblemont, N., Deas, O., Marelli, O., Charpentier, B., Senik, A., Wright, S.D., Haeffner-Cavaillon, N., Hirsch, F. Inhibitory effect of growth hormone on TNF-alpha secretion and nuclear factor-kappaB translocation in lipopolysaccharide-stimulated human monocytes. *J. Immunol.* **1997**, *158*(3), 1310–1314.

644. Ogasawara, A., Arakawa, T., Kaneda, T., Takuma, T., Sato, T., Kaneko, H., Kumegawa, M., Hakeda, Y. Fluid shear stress-induced cyclooxygenase-2 expression is mediated by C/EBP beta, cAMP-response element-binding protein, and AP-1 in osteoblastic MC3T3-E1 cells. *J. Biol. Chem.* **2001**, *276*(10), 7048–7054.

645. Okahara, K., Sun, B., Kambayashi, J. Upregulation of prostacyclin synthesis-related gene expression by shear stress in vascular endothelial cells. *Arterioscler. Thromb. Vasc. Biol.* **1998**, *18*(12), 1922–1926.

646. Morris, J.K., Richards, J.S. An E-box region within the prostaglandin endoperoxide synthase-2 (PGS-2) promoter is required for transcription in rat ovarian granulosa cells. *J. Biol. Chem.* **1996**, *271*(28), 16633–16643.
647. Xie, W., Fletcher, B.S., Andersen, R.D., Herschman, H.R. v-src Induction of the TIS10/PGS2 prostaglandin synthase gene is mediated by an ATF/CRE transcription response element. *Mol. Cell. Biol.* **1994**, *14*(10), 6531–6539.
648. Rieke, C.J., Mulichak, A.M., Garavito, R.M., Smith, W.L. The role of arginine 120 of human prostaglandin endoperoxide H synthase-2 in the interaction with fatty acid substrates and inhibitors. *J. Biol. Chem.* **1999**, *274*(24), 17109–17114.
649. Xiao, G., Chen, W., Kulmacz, R.J. Comparison of structural stabilities of prostaglandin H synthase-1 and -2. *J. Biol. Chem.* **1998**, *273*(12), 6801–6811.
650. Rose, J.W., Hill, K.E., Watt, H.E., Carlson, N.G. Inflammatory cell expression of cyclooxygenase-2 in the multiple sclerosis lesion. *J. Neuroimmunol.* **2004**, *149*(1–2), 40–49.
651. Bezzi, P., Carmignoto, G., Pasti, L., Vesce, S., Rossi, D., Rizzini, B.L., Pozzan, T., Volterra, A. Prostaglandins stimulate calcium-dependent glutamate release in astrocytes. *Nature* **1998**, *391*(6664), 281–285.
652. Deininger, M.H., Schluesener, H.J. Cyclooxygenases-1 and -2 are differentially localized to microglia and endothelium in rat EAE and glioma. *J. Neuroimmunol.* **1999**, *95*(1–2), 202–208.
653. Reder, A.T., Thapar, M., Sapugay, A.M., Jensen, M.A. Prostaglandins and inhibitors of arachidonate metabolism suppress experimental allergic encephalomyelitis. *J. Neuroimmunol.* **1994**, *54*(1–2), 117–127.
654. Minghetti, L., Hughes, P., Perry, V.H. Restricted cyclooxygenase-2 expression in the central nervous system following acute and delayed-type hypersensitivity responses to Bacillus Calmette-Guerin. *Neuroscience* **1999**, *92*(4), 1405–1415.
655. Chang, J.W., Coleman, P.D., O'Banion, M.K. Prostaglandin G/H synthase-2 (cyclooxygenase-2) mRNA expression is decreased in Alzheimer's disease. *Neurobiol. Aging* **1996**, *17*(5), 801–808.
656. Pasinetti, G.M. Cyclooxygenase and inflammation in Alzheimer's disease: Experimental approaches and clinical interventions. *J. Neurosci. Res.* **1998**, *54*(1), 1–6.
657. Ho, L., Pieroni, C., Winger, D., Purohit, D.P., Aisen, P.S., Pasinetti, G.M. Regional distribution of cyclooxygenase-2 in the hippocampal formation in Alzheimer's disease. *J. Neurosci. Res.* **1999**, *57*(3), 295–303.
658. Hoozemans, J.J., Veerhuis, R., Janssen, I., van Elk, E.J., Rozemuller, A.J., Eikelenboom, P. The role of cyclooxygenase 1 and 2 activity in prostaglandin E(2) secretion by cultured human adult microglia: Implications for Alzheimer's disease. *Brain Res.* **2002**, *951*(2), 218–226.
659. Kitamura, Y., Shimohama, S., Koike, H., Kakimura, J., Matsuoka, Y., Nomura, Y., Gebicke-Haerter, P.J., Taniguchi, T. Increased expression of cyclooxygenases and peroxisome proliferator-activated receptor-gamma in Alzheimer's disease brains. *Biochem. Biophys. Res. Commun.* **1999**, *254*(3), 582–586.
660. Etminan, M., Gill, S., Samii, A. Effect of non-steroidal anti-inflammatory drugs on risk of Alzheimer's disease: Systematic review and meta-analysis of observational studies. *Br. Med. J.* **2003**, *327*(7407), 128.
661. Rogers, J., Kirby, L.C., Hempelman, S.R., Berry, D.L., McGeer, P.L., Kaszniak, A.W., Zalinski, J., Cofield, M., Mansukhani, L., Willson, P., et al. Clinical trial of indomethacin in Alzheimer's disease. *Neurology* **1993**, *43*(8), 1609–1611.
662. Xiang, Z., Ho, L., Valdellon, J., Borchelt, D., Kelley, K., Spielman, L., Aisen, P.S., Pasinetti, G.M. Cyclooxygenase (COX)-2 and cell cycle activity in a transgenic mouse model of Alzheimer's disease neuropathology. *Neurobiol. Aging* **2002**, *23*(3), 327–334.
663. Matsuoka, Y., Picciano, M., Malester, B., LaFrancois, J., Zehr, C., Daeschner, J.M., Olschowka, J.A., Fonseca, M.I., O'Banion, M.K., Tenner, A.J., Lemere, C.A., Duff, K. Inflammatory responses to amyloidosis in a transgenic mouse model of Alzheimer's disease. *Am. J. Pathol.* **2001**, *158*(4), 1345–1354.
664. Morihara, T., Teter, B., Yang, F., Lim, G.P., Boudinot, S., Boudinot, F.D., Frautschy, S.A., Cole, G.M. Ibuprofen suppresses interleukin-1beta induction of pro-amyloidogenic alpha1-antichymotrypsin to ameliorate beta-amyloid (Abeta) pathology in Alzheimer's models. *Neuropsychopharmacology* **2005**, *30*(6), 1111–1120.
665. Jantzen, P.T., Connor, K.E., DiCarlo, G., Wenk, G.L., Wallace, J.L., Rojiani, A.M., Coppola, D., Morgan, D., Gordon, M.N. Microglial activation and beta-amyloid deposit reduction caused by a nitric oxide-releasing nonsteroidal anti-inflammatory drug in amyloid precursor protein plus presenilin-1 transgenic mice. *J. Neurosci.* **2002**, *22*(6), 2246–2254.
666. Sung, S., Yang, H., Uryu, K., Lee, E.B., Zhao, L., Shineman, D., Trojanowski, J.Q., Lee, V.M., Pratico, D. Modulation of nuclear factor-kappa B activity by indomethacin influences A beta levels but not A beta precursor protein metabolism in a model of Alzheimer's disease. *Am. J. Pathol.* **2004**, *165*(6), 2197–2206.
667. Knott, C., Stern, G., Wilkin, G.P. Inflammatory regulators in Parkinson's disease: iNOS, lipocortin-1, and cyclooxygenases-1 and -2. *Mol. Cell. Neurosci.* **2000**, *16*(6), 724–739.
668. Teismann, P., Tieu, K., Choi, D.K., Wu, D.C., Naini, A., Hunot, S., Vila, M., Jackson-Lewis, V., Przedborski, S. Cyclooxygenase-2 is instrumental in Parkinson's disease neurodegeneration. *Proc. Natl. Acad. Sci. U.S.A.* **2003**, *100*(9), 5473–5478.
669. Teismann, P., Ferger, B. Inhibition of the cyclooxygenase isoenzymes COX-1 and COX-2 provide neuroprotection in the MPTP-mouse model of Parkinson's disease. *Synapse* **2001**, *39*(2), 167–174.
670. Feng, Z.H., Wang, T.G., Li, D.D., Fung, P., Wilson, B.C., Liu, B., Ali, S.F., Langenbach, R., Hong, J.S. Cyclooxygenase-

2-deficient mice are resistant to 1-methyl-4-phenyl1, 2, 3, 6-tetrahydropyridine-induced damage of dopaminergic neurons in the substantia nigra. *Neurosci. Lett.* **2002**, *329*(3), 354–358.

671. Reksidler, A.B., Lima, M.M., Zanata, S.M., Machado, H.B., da Cunha, C., Andreatini, R., Tufik, S., Vital, M.A. The COX-2 inhibitor parecoxib produces neuroprotective effects in MPTP-lesioned rats. *Eur. J. Pharmacol.* **2007**, *560*(2–3), 163–175.

672. Yasojima, K., Tourtellotte, W.W., McGeer, E.G., McGeer, P.L. Marked increase in cyclooxygenase-2 in ALS spinal cord: Implications for therapy. *Neurology* **2001**, *57*(6), 952–956.

673. Maihofner, C., Probst-Cousin, S., Bergmann, M., Neuhuber, W., Neundorfer, B., Heuss, D. Expression and localization of cyclooxygenase-1 and -2 in human sporadic amyotrophic lateral sclerosis. *Eur. J. Neurosci.* **2003**, *18*(6), 1527–1534.

674. Almer, G., Teismann, P., Stevic, Z., Halaschek-Wiener, J., Deecke, L., Kostic, V., Przedborski, S. Increased levels of the pro-inflammatory prostaglandin PGE2 in CSF from ALS patients. *Neurology* **2002**, *58*(8), 1277–1279.

675. Cudkowicz, M.E., Shefner, J.M., Schoenfeld, D.A., Zhang, H., Andreasson, K.I., Rothstein, J.D., Drachman, D.B. Trial of celecoxib in amyotrophic lateral sclerosis. *Ann. Neurol.* **2006**, *60*(1), 22–31.

676. Almer, G., Guegan, C., Teismann, P., Naini, A., Rosoklija, G., Hays, A.P., Chen, C., Przedborski, S. Increased expression of the pro-inflammatory enzyme cyclooxygenase-2 in amyotrophic lateral sclerosis. *Ann. Neurol.* **2001**, *49*(2), 176–185.

677. Drachman, D.B., Frank, K., Dykes-Hoberg, M., Teismann, P., Almer, G., Przedborski, S., Rothstein, J.D. Cyclooxygenase 2 inhibition protects motor neurons and prolongs survival in a transgenic mouse model of ALS. *Ann. Neurol.* **2002**, *52*(6), 771–778.

678. Almer, G., Kikuchi, H., Teismann, P., Przedborski, S. Is prostaglandin E(2) a pathogenic factor in amyotrophic lateral sclerosis? *Ann. Neurol.* **2006**, *59*(6), 980–983.

679. Klivenyi, P., Kiaei, M., Gardian, G., Calingasan, N.Y., Beal, M.F. Additive neuroprotective effects of creatine and cyclooxygenase 2 inhibitors in a transgenic mouse model of amyotrophic lateral sclerosis. *J. Neurochem.* **2004**, *88*(3), 576–582.

680. Pompl, P.N., Ho, L., Bianchi, M., McManus, T., Qin, W., Pasinetti, G.M. A therapeutic role for cyclooxygenase-2 inhibitors in a transgenic mouse model of amyotrophic lateral sclerosis. *FASEB J.* **2003**, *17*(6), 725–727.

681. Rothman, R.B., Baumann, M.H., Dersch, C.M., Romero, D.V., Rice, K.C., Carroll, F.I., Partilla, J.S. Amphetamine-type central nervous system stimulants release norepinephrine more potently than they release dopamine and serotonin. *Synapse* **2001**, *39*(1), 32–41.

682. Giros, B., Caron, M.G. Molecular characterization of the dopamine transporter. *Trends Pharmacol. Sci.* **1993**, *14*(2), 43–49.

683. Eshleman, A.J., Carmolli, M., Cumbay, M., Martens, C.R., Neve, K.A., Janowsky, A. Characteristics of drug interactions with recombinant biogenic amine transporters expressed in the same cell type. *J. Pharmacol. Exp. Ther.* **1999**, *289*(2), 877–885.

684. Sekine, Y., Iyo, M., Ouchi, Y., Matsunaga, T., Tsukada, H., Okada, H., Yoshikawa, E., Futatsubashi, M., Takei, N., Mori, N. Methamphetamine-related psychiatric symptoms and reduced brain dopamine transporters studied with PET. *Am. J. Psychiatry* **2001**, *158*(8), 1206–1214.

685. Stuehr, D.J. Mammalian nitric oxide synthases. *Biochim. Biophys. Acta* **1999**, *1411*(2–3), 217–230.

686. Iyanagi, T. Molecular mechanism of phase I and phase II drug-metabolizing enzymes: Implications for detoxification. *Int. Rev. Cytol.* **2007**, *260*, 35–112.

687. Dhakshinamoorthy, S., Jaiswal, A.K. Small maf (MafG and MafK) proteins negatively regulate antioxidant response element-mediated expression and antioxidant induction of the NAD(P)H:Quinone oxidoreductase1 gene. *J. Biol. Chem.* **2000**, *275*(51), 40134–40141.

688. Marini, M.G., Chan, K., Casula, L., Kan, Y.W., Cao, A., Moi, P. hMAF, a small human transcription factor that heterodimerizes specifically with Nrf1 and Nrf2. *J. Biol. Chem.* **1997**, *272*(26), 16490–16497.

689. Wild, A.C., Moinova, H.R., Mulcahy, R.T. Regulation of gamma-glutamylcysteine synthetase subunit gene expression by the transcription factor Nrf2. *J. Biol. Chem.* **1999**, *274*(47), 33627–33636.

690. Jain, A.K., Jaiswal, A.K. GSK-3beta acts upstream of Fyn kinase in regulation of nuclear export and degradation of NF-E2 related factor 2. *J. Biol. Chem.* **2007**, *282*(22), 16502–16510.

691. Yoshioka, K., Deng, T., Cavigelli, M., Karin, M. Antitumor promotion by phenolic antioxidants: Inhibition of AP-1 activity through induction of Fra expression. *Proc. Natl. Acad. Sci. U.S.A.* **1995**, *92*(11), 4972–4976.

692. Tsuji, Y. JunD activates transcription of the human ferritin H gene through an antioxidant response element during oxidative stress. *Oncogene* **2005**, *24*(51), 7567–7578.

693. Igarashi, K., Hoshino, H., Muto, A., Suwabe, N., Nishikawa, S., Nakauchi, H., Yamamoto, M. Multivalent DNA binding complex generated by small Maf and Bach1 as a possible biochemical basis for beta-globin locus control region complex. *J. Biol. Chem.* **1998**, *273*(19), 11783–11790.

694. Kaspar, J.W., Jaiswal, A.K. Antioxidant-induced phosphorylation of tyrosine 486 leads to rapid nuclear export of Bach1 that allows Nrf2 to bind to the antioxidant response element and activate defensive gene expression. *J. Biol. Chem.* **2010**, *285*(1), 153–162.

695. Suzuki, H., Tashiro, S., Sun, J., Doi, H., Satomi, S., Igarashi, K. Cadmium induces nuclear export of Bach1, a transcriptional repressor of heme oxygenase-1 gene. *J. Biol. Chem.* **2003**, *278*(49), 49246–49253.

696. Hoshino, H., Kobayashi, A., Yoshida, M., Kudo, N., Oyake, T., Motohashi, H., Hayashi, N., Yamamoto, M., Igarashi, K. Oxidative stress abolishes leptomycin B-sensitive nuclear export of transcription repressor Bach2 that

counteracts activation of Maf recognition element. *J. Biol. Chem.* **2000**, *275*(20), 15370–15376.

697. Muto, A., Hoshino, H., Madisen, L., Yanai, N., Obinata, M., Karasuyama, H., Hayashi, N., Nakauchi, H., Yamamoto, M., Groudine, M., Igarashi, K. Identification of Bach2 as a B-cell-specific partner for small maf proteins that negatively regulate the immunoglobulin heavy chain gene 3′ enhancer. *EMBO J.* **1998**, *17*(19), 5734–5743.

698. Muto, A., Tashiro, S., Tsuchiya, H., Kume, A., Kanno, M., Ito, E., Yamamoto, M., Igarashi, K. Activation of Maf/AP-1 repressor Bach2 by oxidative stress promotes apoptosis and its interaction with promyelocytic leukemia nuclear bodies. *J. Biol. Chem.* **2002**, *277*(23), 20724–20733.

699. Liu, G.H., Qu, J., Shen, X. NF-kappaB/p65 antagonizes Nrf2-ARE pathway by depriving CBP from Nrf2 and facilitating recruitment of HDAC3 to MafK. *Biochim. Biophys. Acta* **2008**, *1783*(5), 713–727.

700. Zhang, J., Ohta, T., Maruyama, A., Hosoya, T., Nishikawa, K., Maher, J.M., Shibahara, S., Itoh, K., Yamamoto, M. BRG1 interacts with Nrf2 to selectively mediate HO-1 induction in response to oxidative stress. *Mol. Cell. Biol.* **2006**, *26*(21), 7942–7952.

701. Ki, S.H., Cho, I.J., Choi, D.W., Kim, S.G. Glucocorticoid receptor (GR)-associated SMRT binding to C/EBPbeta TAD and Nrf2 Neh4/5: Role of SMRT recruited to GR in GSTA2 gene repression. *Mol. Cell. Biol.* **2005**, *25*(10), 4150–4165.

702. Iwasaki, K., Hailemariam, K., Tsuji, Y. PIAS3 interacts with ATF1 and regulates the human ferritin H gene through an antioxidant-responsive element. *J. Biol. Chem.* **2007**, *282*(31), 22335–22343.

703. Brown, S.L., Sekhar, K.R., Rachakonda, G., Sasi, S., Freeman, M.L. Activating transcription factor 3 is a novel repressor of the nuclear factor erythroid-derived 2-related factor 2 (Nrf2)-regulated stress pathway. *Cancer Res.* **2008**, *68*(2), 364–368.

704. He, C.H., Gong, P., Hu, B., Stewart, D., Choi, M.E., Choi, A.M., Alam, J. Identification of activating transcription factor 4 (ATF4) as an Nrf2-interacting protein: Implication for heme oxygenase-1 gene regulation. *J. Biol. Chem.* **2001**, *276*(24), 20858–20865.

705. Ikeda, Y., Sugawara, A., Taniyama, Y., Uruno, A., Igarashi, K., Arima, S., Ito, S., Takeuchi, K. Suppression of rat thromboxane synthase gene transcription by peroxisome proliferator-activated receptor gamma in macrophages via an interaction with NRF2. *J. Biol. Chem.* **2000**, *275*(42), 33142–33150.

706. Zhou, W., Lo, S.C., Liu, J.H., Hannink, M., Lubahn, D.B. ERRbeta: A potent inhibitor of Nrf2 transcriptional activity. *Mol. Cell. Endocrinol.* **2007**, *278*(1–2), 52–62.

707. Wang, X.J., Hayes, J.D., Henderson, C.J., Wolf, C.R. Identification of retinoic acid as an inhibitor of transcription factor Nrf2 through activation of retinoic acid receptor alpha. *Proc. Natl. Acad. Sci. U.S.A.* **2007**, *104*(49), 19589–19594.

708. Lo, S.C., Hannink, M. PGAM5 tethers a ternary complex containing Keap1 and Nrf2 to mitochondria. *Exp. Cell Res.* **2008**, *314*(8), 1789–1803.

709. Liby, K., Hock, T., Yore, M.M., Suh, N., Place, A.E., Risingsong, R., Williams, C.R., Royce, D.B., Honda, T., Honda, Y., Gribble, G.W., Hill-Kapturczak, N., Agarwal, A., Sporn, M.B. The synthetic triterpenoids, CDDO and CDDO-imidazolide, are potent inducers of heme oxygenase-1 and Nrf2/ARE signaling. *Cancer Res.* **2005**, *65*(11), 4789–4798.

710. Dhenaut, A., Hollenbach, S., Eckert, I., Epe, B., Boiteux, S., Radicella, J.P. Characterization of hOGG1 promoter structure, expression during cell cycle and overexpression in mammalian cells. *Adv. Exp. Med. Biol.* **2001**, *500*, 613–616.

711. Park, E.Y., Rho, H.M. The transcriptional activation of the human copper/zinc superoxide dismutase gene by 2,3,7,8-tetrachlorodibenzo-p-dioxin through two different regulator sites, the antioxidant responsive element and xenobiotic responsive element. *Mol. Cell. Biochem.* **2002**, *240*(1–2), 47–55.

712. Shin, S., Wakabayashi, N., Misra, V., Biswal, S., Lee, G.H., Agoston, E.S., Yamamoto, M., Kensler, T.W. NRF2 modulates aryl hydrocarbon receptor signaling: Influence on adipogenesis. *Mol. Cell. Biol.* **2007**, *27*(20), 7188–7197.

713. Kwak, M.K., Wakabayashi, N., Itoh, K., Motohashi, H., Yamamoto, M., Kensler, T.W. Modulation of gene expression by cancer chemopreventive dithiolethiones through the Keap1-Nrf2 pathway: Identification of novel gene clusters for cell survival. *J. Biol. Chem.* **2003**, *278*(10), 8135–8145.

714. Moinova, H.R., Mulcahy, R.T. An electrophile responsive element (EpRE) regulates beta-naphthoflavone induction of the human gamma-glutamylcysteine synthetase regulatory subunit gene. Constitutive expression is mediated by an adjacent AP-1 site. *J. Biol. Chem.* **1998**, *273*(24), 14683–14689.

715. Yueh, M.F., Tukey, R.H. Nrf2-Keap1 signaling pathway regulates human UGT1A1 expression in vitro and in transgenic UGT1 mice. *J. Biol. Chem.* **2007**, *282*(12), 8749–8758.

716. Banning, A., Deubel, S., Kluth, D., Zhou, Z., Brigelius-Flohe, R. The GI-GPx gene is a target for Nrf2. *Mol. Cell. Biol.* **2005**, *25*(12), 4914–4923.

717. Hayashi, A., Suzuki, H., Itoh, K., Yamamoto, M., Sugiyama, Y. Transcription factor Nrf2 is required for the constitutive and inducible expression of multidrug resistance-associated protein 1 in mouse embryo fibroblasts. *Biochem. Biophys. Res. Commun.* **2003**, *310*(3), 824–829.

718. Kauffmann, H.M., Pfannschmidt, S., Zoller, H., Benz, A., Vorderstemann, B., Webster, J.I., Schrenk, D. Influence of redox-active compounds and PXR-activators on human MRP1 and MRP2 gene expression. *Toxicology* **2002**, *171*(2–3), 137–146.

719. Maher, J.M., Dieter, M.Z., Aleksunes, L.M., Slitt, A.L., Guo, G., Tanaka, Y., Scheffer, G.L., Chan, J.Y., Manautou, J.E., Chen, Y., Dalton, T.P., Yamamoto, M., Klaassen, C.D. Oxidative and electrophilic stress induces multidrug resistance-associated protein transporters via the nuclear factor-E2-related factor-2 transcriptional pathway. *Hepatology* **2007**, *46*(5), 1597–1610.

720. Vollrath, V., Wielandt, A.M., Iruretagoyena, M., Chianale, J. Role of Nrf2 in the regulation of the Mrp2 (ABCC2) gene. *Biochem. J.* **2006**, *395*(3), 599–609.

721. Pietsch, E.C., Chan, J.Y., Torti, F.M., Torti, S.V. Nrf2 mediates the induction of ferritin H in response to xenobiotics and cancer chemopreventive dithiolethiones. *J. Biol. Chem.* **2003**, *278*(4), 2361–2369.

722. Wilson, L.A., Gemin, A., Espiritu, R., Singh, G. ets-1 is transcriptionally up-regulated by H2O2 via an antioxidant response element. *FASEB J.* **2005**, *19*(14), 2085–2087.

723. Kim, Y.J., Ahn, J.Y., Liang, P., Ip, C., Zhang, Y., Park, Y.M. Human prx1 gene is a target of Nrf2 and is up-regulated by hypoxia/reoxygenation: Implication to tumor biology. *Cancer Res.* **2007**, *67*(2), 546–554.

724. Kwak, M.K., Wakabayashi, N., Greenlaw, J.L., Yamamoto, M., Kensler, T.W. Antioxidants enhance mammalian proteasome expression through the Keap1-Nrf2 signaling pathway. *Mol. Cell. Biol.* **2003**, *23*(23), 8786–8794.

725. Kim, Y.C., Masutani, H., Yamaguchi, Y., Itoh, K., Yamamoto, M., Yodoi, J. Hemin-induced activation of the thioredoxin gene by Nrf2. A differential regulation of the antioxidant responsive element by a switch of its binding factors. *J. Biol. Chem.* **2001**, *276*(21), 18399–18406.

726. Sakurai, A., Nishimoto, M., Himeno, S., Imura, N., Tsujimoto, M., Kunimoto, M., Hara, S. Transcriptional regulation of thioredoxin reductase 1 expression by cadmium in vascular endothelial cells: Role of NF-E2-related factor-2. *J. Cell. Physiol.* **2005**, *203*(3), 529–537.

727. Kwak, M.K., Itoh, K., Yamamoto, M., Sutter, T.R., Kensler, T.W. Role of transcription factor Nrf2 in the induction of hepatic phase 2 and antioxidative enzymes in vivo by the cancer chemoprotective agent, 3H-1, 2-dimethiole-3-thione. *Mol. Med.* **2001**, *7*(2), 135–145.

728. Shinkai, Y., Sumi, D., Fukami, I., Ishii, T., Kumagai, Y. Sulforaphane, an activator of Nrf2, suppresses cellular accumulation of arsenic and its cytotoxicity in primary mouse hepatocytes. *FEBS Lett.* **2006**, *580*(7), 1771–1774.

729. Thimmulappa, R.K., Mai, K.H., Srisuma, S., Kensler, T.W., Yamamoto, M., Biswal, S. Identification of Nrf2-regulated genes induced by the chemopreventive agent sulforaphane by oligonucleotide microarray. *Cancer Res.* **2002**, *62*(18), 5196–5203.

730. Lee, J.M., Moehlenkamp, J.D., Hanson, J.M., Johnson, J.A. Nrf2-dependent activation of the antioxidant responsive element by *tert*-butylhydroquinone is independent of oxidative stress in IMR-32 human neuroblastoma cells. *Biochem. Biophys. Res. Commun.* **2001**, *280*(1), 286–292.

731. Li, J., Johnson, D., Calkins, M., Wright, L., Svendsen, C., Johnson, J. Stabilization of Nrf2 by tBHQ confers protection against oxidative stress-induced cell death in human neural stem cells. *Toxicol. Sci.* **2005**, *83*(2), 313–328.

732. Kwak, M.K., Kensler, T.W., Casero, R.A., Jr. Induction of phase 2 enzymes by serum oxidized polyamines through activation of Nrf2: Effect of the polyamine metabolite acrolein. *Biochem. Biophys. Res. Commun.* **2003**, *305*(3), 662–670.

733. Balogun, E., Hoque, M., Gong, P., Killeen, E., Green, C.J., Foresti, R., Alam, J., Motterlini, R. Curcumin activates the haem oxygenase-1 gene via regulation of Nrf2 and the antioxidant-responsive element. *Biochem. J.* **2003**, *371*(Pt 3), 887–895.

734. Casalino, E., Calzaretti, G., Landriscina, M., Sblano, C., Fabiano, A., Landriscina, C. The Nrf2 transcription factor contributes to the induction of alpha-class GST isoenzymes in liver of acute cadmium or manganese intoxicated rats: Comparison with the toxic effect on NAD(P)H:quinone reductase. *Toxicology* **2007**, *237*(1–3), 24–34.

735. Stewart, D., Killeen, E., Naquin, R., Alam, S., Alam, J. Degradation of transcription factor Nrf2 via the ubiquitin-proteasome pathway and stabilization by cadmium. *J. Biol. Chem.* **2003**, *278*(4), 2396–2402.

736. Toyama, T., Sumi, D., Shinkai, Y., Yasutake, A., Taguchi, K., Tong, K.I., Yamamoto, M., Kumagai, Y. Cytoprotective role of Nrf2/Keap1 system in methylmercury toxicity. *Biochem. Biophys. Res. Commun.* **2007**, *363*(3), 645–650.

737. Marcus, D.L., Thomas, C., Rodriguez, C., Simberkoff, K., Tsai, J.S., Strafaci, J.A., Freedman, M.L. Increased peroxidation and reduced antioxidant enzyme activity in Alzheimer's disease. *Exp. Neurol.* **1998**, *150*(1), 40–44.

738. Omar, R.A., Chyan, Y.J., Andorn, A.C., Poeggeler, B., Robakis, N.K., Pappolla, M.A. Increased expression but reduced activity of antioxidant enzymes in Alzheimer's disease. *J. Alzheimers Dis.* **1999**, *1*(3), 139–145.

739. Takahashi, M., Dore, S., Ferris, C.D., Tomita, T., Sawa, A., Wolosker, H., Borchelt, D.R., Iwatsubo, T., Kim, S.H., Thinakaran, G., Sisodia, S.S., Snyder, S.H. Amyloid precursor proteins inhibit heme oxygenase activity and augment neurotoxicity in Alzheimer's disease. *Neuron* **2000**, *28*(2), 461–473.

740. Schipper, H.M. Heme oxygenase expression in human central nervous system disorders. *Free Radic. Biol. Med.* **2004**, *37*(12), 1995–2011.

741. SantaCruz, K.S., Yazlovitskaya, E., Collins, J., Johnson, J., DeCarli, C. Regional NAD(P)H:quinone oxidoreductase activity in Alzheimer's disease. *Neurobiol. Aging* **2004**, *25*(1), 63–69.

742. Kumaran, R., Kingsbury, A., Coulter, I., Lashley, T., Williams, D., de Silva, R., Mann, D., Revesz, T., Lees, A., Bandopadhyay, R. DJ-1 (PARK7) is associated with 3R and 4R tau neuronal and glial inclusions in neurodegenerative disorders. *Neurobiol. Dis.* **2007**, *28*(1), 122–132.

743. Hirotani, M., Maita, C., Niino, M., Iguchi-Ariga, S., Hamada, S., Ariga, H., Sasaki, H. Correlation between DJ-1 levels in the cerebrospinal fluid and the progression of disabilities in multiple sclerosis patients. *Mult. Scler.* **2008**, *14*(8), 1056–1060.

744. Browne, S.E., Ferrante, R.J., Beal, M.F. Oxidative stress in Huntington's disease. *Brain Pathol.* **1999**, *9*(1), 147–163.

745. Smith, B.J., Curtis, J.F., Eling, T.E. Bioactivation of xenobiotics by prostaglandin H synthase. *Chem. Biol. Interact.* **1991**, *79*(3), 245–264.

11

CARDIAC ISCHEMIA AND REPERFUSION

Murugesan Velayutham and Jay L. Zweier

OVERVIEW

Coronary artery disease and secondary acute myocardial infarction are among the most prevalent health problems in the world, and are leading causes of morbidity and mortality.[1] Myocardial infarction is the leading cause of heart failure.[1] Worldwide, 22 million individuals are afflicted and are living with heart failure.[2] In the United States alone, nearly 5 million people live with heart failure, and about 550,000 new cases are diagnosed annually.[2]

Current therapy for acute myocardial infarction is to reopen the occluded coronary artery and reperfuse the effected myocardium in an effort to salvage the heart muscle at risk.[3,4] However, this process is associated with oxygen radical generation and secondary reperfusion injury.[1,5,6] Research over the last two decades has focused on understanding reperfusion injury and the role of free radicals and their mechanisms of formation in this process.[5,7,8] Through understanding the process of free radical generation in the ischemic and postischemic heart, salvage of heart muscle at risk can be achieved with prevention of secondary heart failure.

11.1 OXYGEN IN THE HEART

11.1.1 Beneficial and Deleterious Effects of Oxygen in the Heart

Molecular oxygen (O_2) is essential for cellular respiration in all aerobic organisms. In humans, the consumption of oxygen by the heart is 11.6% of the total oxygen consumed by the body. The heart has the highest oxygen consumption among all the organs in the human body.[9] The oxygen consumption increases eightfold or more under maximal workload conditions.[9] Oxygen is used as an electron acceptor in mitochondria to generate chemical energy. Its reduction to water by the mitochondrial electron transport chain (ETC) helps supply the metabolic demands of human life.[10] Although oxygen is essential for cell survival, it can be toxic to cells because of its ability to generate free radicals or reactive oxygen species (ROS). Interestingly, both increases and decreases in cellular oxygen levels result in the generation of ROS.[11] ROS are known to cause cell death via oxidative stress. Oxidative stress is recognized to have a central role in the development and progression of myocardial dysfunction, but the molecular targets of ROS remain unclear.

11.1.2 Ischemia and Reperfusion

Blood flow blockage in a coronary artery leads to myocardial infarction with ischemic death of cardiomyocytes that increases as a function of the duration of ischemia.[12] Limitation of myocardial infarct size has major ramifications on clinical outcomes.[13] Therefore, timely reperfusion of acute ischemic myocardium is essential for myocardial salvage.[5,13] In patients with acute myocardial infarction, reperfusion is typically achieved through interventional procedures such as balloon angioplasty or the use of thrombolytic agents.[4] However, there is evidence that reperfusion itself contributes to oxidative damage to endothelial cells, cardiomyocytes, and myocardium.[5,7,8] This process has been

Molecular Basis of Oxidative Stress: Chemistry, Mechanisms, and Disease Pathogenesis, First Edition. Edited by Frederick A. Villamena.
© 2013 John Wiley & Sons, Inc. Published 2013 by John Wiley & Sons, Inc.

termed reperfusion injury. Reperfusion of the ischemic myocardium results in irreversible tissue injury and cell necrosis, leading to decreased cardiac performance.

Ischemia and reperfusion also occurs with surgical procedures on the heart with cardiac arrest or with heart transplantation.[14-17] Protection of the native or donor heart by a preservation solution that will ensure rapid and complete recovery of normal myocardial function is essential for successful clinical cardiovascular surgical procedures and transplantation.[18] In clinical heart transplantation, the donor hearts are stored in ice-cold cardioplegic solution, supplemented with various antioxidants, for 4–5 hours.[14,19,20] The hearts are also under cardioplegic arrest in various cardiovascular surgical procedures. The restoration of blood flow, following cardioplegic ischemia, can result in ischemia and reperfusion injury.[21]

11.1.3 Oxidative Stress and Injury

It has been reported that oxidative stress or damage is increased in heart disease patients and during surgical reperfusion of the whole heart.[22-24] Oxidative stress is defined as the imbalance of pro- and antioxidants in a biological system in favor of the former. Oxidative stress is associated with increased formation of ROS. The level of ROS is increased under various *in vivo* pathophysiological conditions such as ischemia and greatly accelerated with the sudden reintroduction of oxygen at the onset of reperfusion.[5] The levels of ROS in the myocardium are also increased in the failing human heart.[25] The increased production of ROS leads to oxidative damage of various biomolecules, cells, and tissues. The recurring or prolonged oxidative damage results in organ failure and death. Thus, ROS are an important contributing factor in the development of ischemia and reperfusion injuries.

11.2 SOURCES OF ROS DURING ISCHEMIA AND REPERFUSION

Under normal cellular physiological conditions, ROS are formed continuously in small amounts and inactivated by endogenous antioxidant enzymes and small molecule antioxidants. Thus, ROS are important mediators linking metabolic activity to various cell signaling pathways, and a number of these pathways affect cells and tissues. Under pathophysiological conditions such as cardiac ischemia and reperfusion, the generation of ROS is greatly increased. The massive production of ROS during ischemia and reperfusion in turn leads to tissue injury causing organ failure. During myocardial ischemia, sources of ROS formation have been identified. The massive burst of ROS during reperfusion has been shown to originate from several cellular sources. As described below, a number of molecular mechanisms have been shown to contribute to and mediate reperfusion injury. Several other mechanisms remain under investigation.

11.2.1 Cellular Organelles

11.2.1.1 Mitochondria In mammalian cells, mitochondria are the major source of ROS under physiological and pathophysiological conditions.[26] Under normal physiological conditions ~1–2% of the oxygen consumed by the heart is converted into ROS.[27] In the heart, ~30% of the total volume is occupied by mitochondria.[28] Electron leakage from the ETC in mitochondria occurs with partial reduction of oxygen with generation of ROS (see Fig. 11.1).[29] Under physiological conditions, in mitochondria, complex I and complex III produce ROS (see Fig. 11.1).[26,30] In addition, sulfhydryl oxidases generate disulfide bonds with the reduction of oxygen to hydrogen peroxide (H_2O_2) in the mitochondrial intermembrane space (IMS).[31] Cardiac ischemia leads to a decline in mitochondrial respiratory function that can be exacerbated upon reperfusion.[5] During ischemia (I) and reperfusion (R) in the heart, the mitochondrial production of ROS increases.[5,32]

11.2.1.2 Endoplasmic Reticulum (ER) The ER is a continuous membrane network in the cytosol. Several biochemical reactions and processes of cell biology are compartmentalized in the ER. In eukaryotes, up to 30% of newly synthesized proteins traffic through the ER.

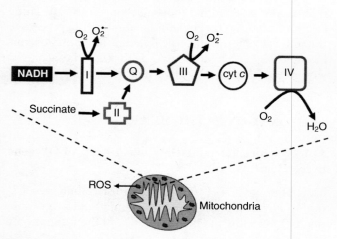

Figure 11.1 Complete and partial reduction of molecular oxygen in the mitochondrial electron transport chain. Partial (one-electron) reduction of oxygen produces superoxide radical ($O_2^{\bullet-}$).

Proteins enter this compartment as extended polypeptides and fold into their functional state in a process that is assisted by luminal chaperones. The ER is the major site for protein disulfide introduction and isomerization in eukaryotes. Oxidative protein folding in the ER lumen is one of the main sources of production of H_2O_2.[33] Recent observations suggest that the luminal H_2O_2 concentration is higher than that in the cytosol.[34] However, ER is an underappreciated source of ROS.

The ER lumen microenvironment is characteristically different from the cytosol. The major redox buffer of the ER lumen is composed of GSH and GSSG and the ratio of GSH/GSSG is nearly one hundred times lower in ER lumen as compared to cytosol.[35] The possible interference between the GSH/GSSG system and both oxidation and reduction of protein thiols underscores the importance of a tight regulation of the ER redox milieu, since changes in either direction drastically interfere with ER function. Failure of the adaptive capacity of the ER results in activation of the unfolded protein response (UPR), which increases the stress signaling pathways. Increased ER stress plays an important role in atherosclerosis.[36] In addition, ER stress-induced apoptosis is implicated in the pathophysiology of cardiovascular diseases.[37] Therefore, ER is an emerging player in redox pathophysiology. However, the role of ER stress-induced generation of ROS in oxidative damage during cardiac ischemia and reperfusion is not known.

11.2.1.3 Peroxisomes In mammals, the peroxisomes are essential cellular organelles in mammals.[38] Peroxisomes are multifunctional organelles with an important role in cellular metabolism and physiology.[38] The unique characteristic of peroxisomes is their central role in the metabolism of compounds with low solubility in water or lipids.[38] They contain more than 100 enzymes and other proteins. Mammalian peroxisomes carry out a wide range of reactions such as the oxidation of purines, L-α-hydroxy acids, fatty acids, polyamines, and amino acids.[38] Importantly, peroxisomes harbor several ROS generating and powerful antioxidant defense systems.[38] The ROS-generating systems present in peroxisomes are acyl-CoA oxidases, 2-hydroxyacid oxidase, polyamine oxidase, amino acid oxidase, xanthine oxidase (XO), and inducible nitric oxide synthase (NOS2). Peroxisomes are also equipped with antioxidant defense systems composed mainly of enzymes involved in the decomposition of H_2O_2 and superoxide radical ($O_2^{•-}$). It is predicted that peroxisomes serve as an intracellular sink for H_2O_2. The antioxidant enzymes present in peroxisomes are catalase, Cu,Zn superoxide dismutase (SOD1), peroxiredoxin V (PrxV), epoxide hydrolase, glutathione S-transferase (GST), and glutathione peroxidase (GPx). Catalase is the most abundant antioxidant enzyme in mammalian peroxisomes. Catalase decomposes H_2O_2 to water and oxygen. SOD1 dismutates $O_2^{•-}$ to H_2O_2 and oxygen.[39] Therefore, peroxisomes play an important role in the generation and decomposition of ROS. This implies considerable involvement of peroxisomes in processes leading to oxidative stress. However, the role of peroxisomes in the ROS-mediated oxidative damage in the heart remains unclear.

11.2.2 Cellular Enzymes

11.2.2.1 Xanthine Oxidoreductase (XOR) Pharmacologic and biochemical studies implicate XO as a major source of ROS in the cardiovascular system.[5,40,41] XOR exists in two forms, xanthine dehydrogenase (XDH) and XO, which are interconvertible.[42] In humans, both forms of XOR are involved in the catabolism of purine, oxidizing hypoxanthine to xanthine, and finally xanthine to its terminal catabolite, uric acid, Figure 11.2. Mammalian XDH can be readily converted to XO by oxidation of sulfhydryl residues (reversible) or by proteolysis during extraction or purification procedures (irreversible). Classically, XDH is considered to be NAD^+-dependent catalyzing the oxidative reaction of xanthine to urate with reduction of NAD^+ to NADH, whereas XO reduces oxygen with the production of $O_2^{•-}$ and H_2O_2, Figure 11.2. While the activity of XO was reported to be low in human heart compared to other species, most recently, XO has been shown to be upregulated in human dilated cardiomyopathy patients.[43] Inhibition of

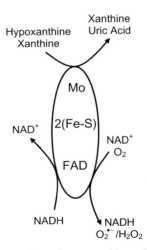

Figure 11.2 Generation of superoxide radical anion ($O_2^{•-}$) and hydrogen peroxide (H_2O_2) by xanthine oxidoreductase (XOR). XOR catalyzes the oxidation of hypoxanthine to xanthine and subsequently to uric acid. NADH is oxidized to NAD^+.

XO has been shown to improve postoperative recovery[44,45] and reduce lipid peroxidation[44] in open-heart surgery patients. Reduction of reperfusion injury in human myocardium has also been reported.[46] Recently, it has been reported that XO activity contributes to exercise-induced myocardial ischemia in patients with chronic stable angina.[47] Despite the presence of numerous supporting studies, there have also been negative reports, such that the role of XO in human cardiovascular disease remains controversial and merits further investigation.[48] XO can also simultaneously reduce both nitrite and oxygen to nitric oxide (NO) and $O_2^{\bullet-}$, respectively, in the presence of electron donors such as xanthine or NADH.[49,50] The products, NO and $O_2^{\bullet-}$, react with each other to form peroxynitrite ($ONOO^-$).

11.2.2.2 Aldehyde Oxidase (AO)
AO and XO are related enzymes in terms of their general structure, biochemical characteristics, and amino acid sequences, but their substrate specificity and inhibitor susceptibility are different.[51] Unlike XOR, AO catalyzes a diverse range of endogenous aldehydes derived from lipid peroxidation, glycation, α-amino acid oxidation, biogenic amine metabolism, and ethanol metabolism. AO not only catalyzes the oxidation of aldehydes but also the oxidation of NADH, aromatic azaheterocyclic compounds, and certain drugs of pharmacological and toxicological importance.[52,53] AO is a permanent oxidase and transfers its electrons exclusively to oxygen with concomitant formation of $O_2^{\bullet-}$ and H_2O_2 (Fig. 11.3).[53] AO is predominantly present in the liver, but it is also present in other tissues, such as lung, kidney, and heart.[54] Compared to XOR, the levels and activities of AO in freshly prepared rat liver and heart are found to be higher.[53] Hence, AO is predicted to be an important enzyme playing a major role in oxidative damage during cardiac ischemia and reperfusion.

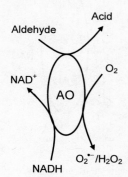

Figure 11.3 Generation of superoxide radical anion ($O_2^{\bullet-}$) and hydrogen peroxide (H_2O_2) by aldehyde oxidase (AO). AO catalyzes the oxidation of aldehydes and NADH to acids and NAD^+, respectively.

11.2.2.3 NADPH Oxidase (Nox)
In myocytes damaged by ischemia, the innate immune system plays a crucial role in both initiation and progression of the subsequent repair response, which involves immune cells such as neutrophils and macrophages.[55,56] Whether these repair mechanisms are favorable or harmful to cardiac function is partly dependent on the amount of tissue damage that is caused by inflammatory cells early after ischemia. Nox is best known for its role in neutrophil and macrophage phagocytosis.[57,58] In macrophages, mitochondrial-derived ROS (mROS) is also involved in phagocytocis.[59] There are a total of seven Nox isoforms that are present in phagocytic and nonphagocytic cells.[58] The Nox family generate ROS ($O_2^{\bullet-}$ and H_2O_2) through electron transfer from NADPH to oxygen.[58] The two Nox isoforms for which there are good data on expression and functional effects in cardiac myocytes are Nox2[60] and Nox4.[61] Nox2 is normally quiescent and is acutely activated by stimuli such as angiotensin II and endothelin-1.[58] Nox4 has a constitutive, low-level activity and seems to be regulated largely by changes in abundance.[58] Therefore, Nox4 may be regarded as an inducible isoform. The intracellular locations of the two isoforms in cardiomyocytes are distinct. Nox2 is found predominantly on the plasma membrane.[62] Nox4 is found intracellularly in ER[63] and mitochondria.[61] Nox2 may be activated by several different cardiac stresses and plays important pathophysiological roles in cardiac disease.[62] Myocardial Nox4 expression increases during hypoxia, myocardial ischemia, or *in vivo* pressure overload.[63,64] Nox2 activity in myocardium is increased in the failing human heart.[60]

11.2.2.4 NADH Oxidase(s)
NADH is an endogenous substrate of complex I in the ETC in the mitochondria.[65] The level of NADH increases during ischemia and reperfusion in isolated rabbit heart.[66] It has been demonstrated that the mitochondrial NADH-oxidase activity increases during ischemia in rat and dog hearts.[32,67] The production of $O_2^{\bullet-}$ increases with the duration of ischemia and increased oxidation of NADH.[32] It has been proposed that the activity of enzyme NADH oxidases could play an important role in the damage caused by ROS in cardiac cells during ischemic conditions.[32,67] However, the questions remain regarding the molecular enzyme responsible for the oxidation of NADH and generation of $O_2^{\bullet-}$. Various studies have shown that both XOR and AO catalyze the oxidation of NADH and generate $O_2^{\bullet-}$ and H_2O_2, (Fig. 11.2 and Fig. 11.3).[52,68,69] Recently, it has also been shown that the oxidation of NADH by Fe^{3+}cytochrome *c* (cyt *c*) in the presence of H_2O_2 increases the generation of $O_2^{\bullet-}$ (Fig. 11.4).[70]

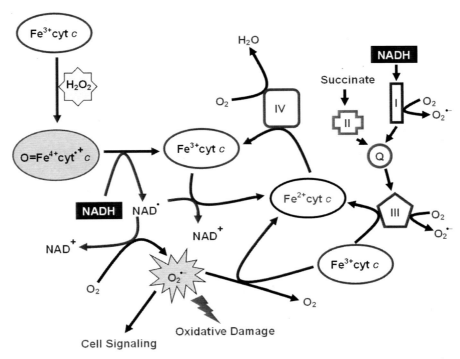

Figure 11.4 Proposed model of cytochrome c mediated removal of H_2O_2, oxidation of NADH, generation of superoxide radical ($O_2^{\bullet-}$), and alternative electron transfer pathway in mitochondria.

11.2.2.5 Cyt c A high concentration of cyt c is present in human heart.[71] In heart, ~30% of the total volume is occupied by mitochondria.[28] Cyt c is a small, globular heme protein which exists in high concentration (0.5–5 mM) in the inner membrane of mitochondria.[72] At least 15% of cyt c is tightly bound to the inner membrane, and the remainder is loosely attached to the inner membrane and can be readily mobilized.[73] Physiologically, cyt c mediates electron shuttling between cyt c reductase (complex III) and cyt c oxidase (complex IV) during mitochondrial respiration.[73] The loosely associated cyt c participates in electron transport, mediates $O_2^{\bullet-}$ removal, and prevents oxidative stress,[73,74] while the tightly bound cyt c accounts for the peroxidase activity.[75] Cyt c is also a key player in cell death through apoptosis, sometimes known as programmed cell death.[76,77] Release of cyt c from the inner mitochondrial membrane into the cytosol is a proapoptotic factor.[78,79] In the early event of apoptosis, the redox function of cyt c in the respiratory chain switches to a peroxidase function.[80]

Cardiolipin (CL) is an acidic phospholipid found primarily in the inner mitochondrial membrane, and confers fluidity and stability to the membrane.[81,82] Cyt c is anchored to the mitochondrial membrane through both electrostatic and hydrophobic interactions with cardiolipin.[81,82] Upon interaction with cardiolipin, cyt c has been shown to alter its tertiary structure and gain peroxidase activity.[80] Under conditions of oxidative and nitrative stress, the peroxidase activity of cyt c also increases.[83,84] Importantly, the oxidative stress is increased during ischemia and reperfusion in the heart.[5] Recently, it has also been shown that the peroxidase activity of cyt c increases the $O_2^{\bullet-}$ production in the presence of endogenous electron donors such as NADH (Fig. 11.4).[70] Hence, it has been suggested that cyt c could have a novel role in the deleterious effects of ischemia and postischemic reperfusion.

11.2.2.6 NOSs To date, three major NOS isoforms have been identified. Two of these (NOS1 and NOS3) are constitutive and calcium and calmodulin dependent. Cardiomyocytes constitutively express both neuronal (nNOS/NOS1) and endothelial (eNOS/NOS3) isoforms. The third isoform, iNOS/NOS2, is inducible and calcium independent and is primarily involved in inflammation. NOS converts L-arginine to L-citrulline and NO, and requires the substrates NADPH and oxygen as well as the cofactor tetrahydrobiopterin (BH_4) (Fig. 11.5). Coronary endothelium, endocardial endothelium, cardiac nerves, and cardiomyocytes of the normal heart all express constitutive NOS and have basal production of

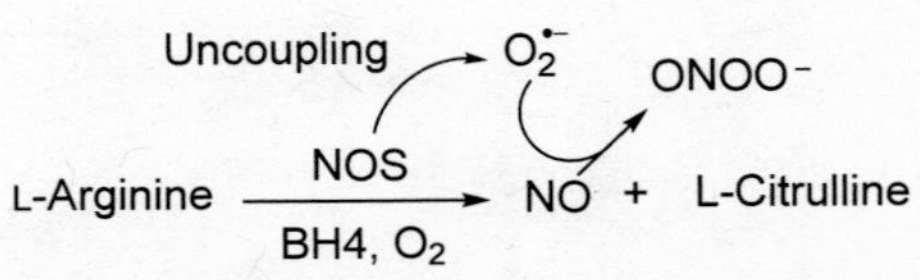

Figure 11.5 Generation of superoxide radical anion ($O_2^{\bullet-}$) and nitric oxide (NO) by nitric oxide synthase (NOS). NOS catalyzes the conversion of L-arginine to L-citruline and NO in the presence of tetrahydrobiopterin (BH_4). In the absence of BH_4 NOS uncouples and generates $O_2^{\bullet-}$. Under pathophysiological conditions, the products NO and $O_2^{\bullet-}$ react with each other to form peroxynitrite ($ONOO^-$).

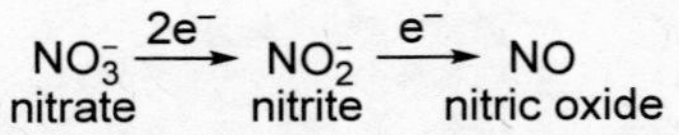

Figure 11.6 Reduction of nitrate to nitrite and subsequently to NO.

NO.[85] This process of vascular and myocardial NOS generation of NO is greatly altered by ischemia.[5] During ischemia, the rise in myocardial calcium leads to initial activation; however, subsequently, oxygen levels fall, limiting its availability as a substrate. In addition, marked intracellular acidosis occurs, with prolonged ischemia rendering the enzyme inactive. Moreover, postischemic oxidants alter NOS function and endothelial reactivity.[5]

NOSs require a tightly bound BH_4 for activity. It has been observed that NOS can become "uncoupled" and generate $O_2^{\bullet-}$ (Fig. 11.5).[5] NOS can become a potent source of $O_2^{\bullet-}$, with depletion of either the substrate L-arginine or the cofactor BH_4 triggering this fundamental alteration in NOS function.[5,86] There are other mechanisms that could trigger uncontrolled production, including release of flavin adenine nucleotide (FAD) from the enzyme, disruption of the active dimer, structural changes resulting in uncoupling of the reductase and oxygenase sites.[5] ROS production by uncoupled NOS can act as an amplifying mechanism for deleterious effects of ROS. Partially or fully uncoupled eNOS can also generate peroxynitrite (Fig. 11.5).[87,88]

11.2.2.7 Nitrate/Nitrite Reductase(s) Studies have reported that nitrate and nitrite are important sources of NO in the cardiovascular system.[89,90] Two- and one-electron reduction of nitrate and nitrite form nitrite and NO, respectively (Fig. 11.6). In mammalian tissues, nitrate can be enzymatically reduced to nitrite and then to NO under hypoxic or normoxic conditions.[91–93] Further, various enzymes such as hemoglobin, deoxymyoglobin, XOR, AO, and mitochondrial enzymes are involved in the reduction of nitrite to NO.[89–96]

11.3 MODULATION OF SUBSTRATES, METABOLITES, AND COFACTORS DURING I-R

11.3.1 ROS

The formation of ROS increases under ischemia and postischemic reperfusion.[5,97] Several studies have shown that in cardiac myocytes a distinct molecular mechanism is involved in increased ROS generation during excessive oxidative stress conditions such as ischemia and reperfusion.[98–100] In this mechanism, mitochondria can significantly amplify a low level of ROS in or around mitochondria as a cellular signal amplifier which is converted into a pathological ROS signal. It has been shown that the mitochondrial transition pore opening and ETC uncoupling are the main sources of increased ROS generation during oxidative stress. The increased generation of ROS couples mitochondria through "ROS-induced ROS-release" (RIRR) signaling. The increased generation of ROS by the mitochondrial RIRR signaling results in mitochondrial and cellular injury/death and fatal arrhythmias. Recently, it has been shown that, in the presence of H_2O_2 (ROS), the oxidation of NADH by Fe^{3+}cyt c results in the generation of $O_2^{\bullet-}$ (ROS).[70]

11.3.2 Hypoxanthine and Xanthine

During ischemia, the substrates for XOR, hypoxanthine and xanthine, accumulate.[41] In isolated ischemic hearts and hypoxic endothelial cells, ATP is metabolized to ADP, AMP, adenosine, inosine, and finally the XO substrates hypoxanthine and xanthine, which react with XO upon reoxygenation or reflow resulting in $O_2^{\bullet-}$ and H_2O_2 generation (see Fig. 11.2).[101,102] This substrate availability is of critical importance in controlling the magnitude and time course of endothelial radical generation. Thus, in the postischemic heart, increased XO formation occurs due to conversion of XDH to XO, and there is marked XO substrate formation due to ischemic ATP degradation (Fig. 11.7).[41]

11.3.3 NADH

Under physiological conditions, NADH is a substrate for complex I (NADH: ubiquinone oxidoreductase) in the ETC in mitochondria.[65] In normal tissues, the cellular levels of NADH have been measured in the range of 0.33–0.83 mM.[103] In the heart, the level of NADH increases during ischemia (reaching 1.85 mM) as a consequence of anaerobic glycolysis.[103–106] The levels of NADH in the ischemic and postischemic heart are higher than in the normal heart.[66] During postischemic reperfusion of the isolated heart, the levels of $O_2^{\bullet-}$ generation and NADH are increased and decreased,

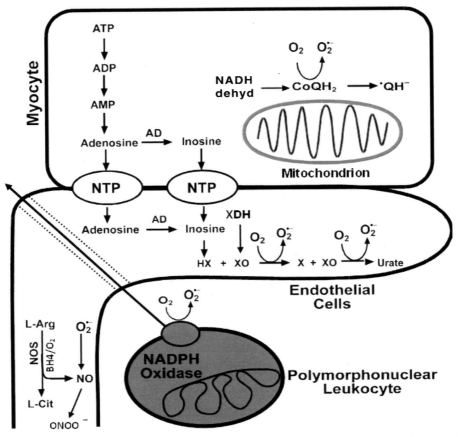

Figure 11.7 Molecular mechanisms involved in the generation of reactive oxygen species (ROS) in cardiac cells. Cellular communication involved in the generation of ROS during ischemia and reperfusion.

respectively.[5,66,106] In addition, the level of NAD^+ is increased during postischemic reperfusion.[66] Therefore, NADH is oxidized to NAD^+ during reperfusion. Recently, it was reported that the increase in NADH oxidation induces a concurrent increase in $O_2^{\bullet-}$ production during ischemia.[32]

11.3.4 BH$_4$

BH$_4$ serves as an important cofactor for aromatic amino acid hydroxylases as well as for all three NOS isoforms. BH$_4$ is highly redox-sensitive and is readily oxidized. In the ischemic and reperfused heart, there is a marked increase in oxidants.[5] BH$_4$ can be readily depleted by oxidants.[107] In isolated rat hearts, the level of BH$_4$ decreased, ≥95%, during global ischemia.[108] In addition, BH$_4$ is irreversibly oxidized in postischemic hearts.[108] Thus, reduced BH$_4$ levels, or its oxidation to dihydrobiopterin (BH$_2$) and other metabolites uncouple NOS, resulting in increased formation of $O_2^{\bullet-}$ (ROS).[108,109]

11.3.5 NO

In tissues, nitrite can be nonenzymatically reduced to NO.[110] Under acidic conditions, nitrite is reduced to NO.[110] Myocardial ischemia results in intracellular acidosis and severe hypoxia leading to a highly reduced state that subsequently leads to nitrite reduction.[110] An electron paramagnetic resonance (EPR) spectroscopic study has shown that NO is generated in isolated rat hearts subjected to ischemia in the presence of nitrite.[110] This nonenzymatic reduction of nitrite to NO results in myocardial injury with a loss of contractile function.[110]

EPR spin trapping using Fe-MGD (Fe^{2+}-MGD$_2$) has shown that the generation of NO increases in ischemic rat hearts.[5] The generation of NO increases as a function of the duration of ischemia.[5] It has been shown that the generation of NO during ischemia is largely NOS dependent by using inhibitors, such as N-nitro-L-arginine methyl ester (L-NAME) or N-monomethyl-L-arginine (L-NMMA).[5]

Nitrate/nitrite reductase reduces nitrate/nitrite to NO under anaerobic conditions.[93] The molybdenum enzymes XOR and AO can catalyze the reduction of nitrite to NO under hypoxia/anaerobic conditions.[91,92] XOR and AO can be seen as complementary to NOS. XOR and AO could promote NO-induced vasodilation in ischemia when NOS activity is low. Therefore, XOR and AO are potentially important sources of NO in ischemic biological tissues. The mitochondrial protein cyt c also reduces nitrite to NO under hypoxia.[94]

11.3.6 Peroxynitrite (ONOO⁻)

In isolated rat hearts, the production of NO and $O_2^{\bullet-}$ are increased during the early period of postischemic reperfusion.[111] Both XOR and AO increase the generation of $O_2^{\bullet-}$ during ischemia and reperfusion. The increased production of NO reacts rapidly with $O_2^{\bullet-}$ to produce highly reactive $ONOO^-$.[111] The concomitant production of NO and $O_2^{\bullet-}$ is generally associated with nitroxidative stress and protein nitration via $ONOO^-$ or heme peroxidase-dependent nitration.[112] 3-Nitrotyrosine provides a useful marker for NO-associated stress in various diseases and increased levels are found in failing hearts.[112]

$ONOO^-$ reacts with Fe^{3+}cyt c and increases the peroxidase activity.[84] The increased peroxidase activity of Fe^{3+}cyt c increases the oxidation of NADH and formation of $O_2^{\bullet-}$.[70] Moreover, $ONOO^-$ uncouples eNOS (NOS3) and increases the generation of $O_2^{\bullet-}$.[113,114] In addition, $ONOO^-$ oxidizes free and eNOS-bound BH_4.[113,114] During ischemia and reperfusion, the decreased level of BH_4 and uncoupled eNOS increase the generation of ROS.[108]

11.3.7 Free Amino Acids

Taurine, glutamine, glutamate, aspartate, and alanine are the most abundant intracellular free amino acids in the human heart.[16] The level of taurine, glutamine, and glutamate are decreased during cardiac ischemia and reperfusion.[16] The metabolic by-product of glutamate is alanine and hence, the level of alanine is increased during ischemia and reperfusion. Glutamate has an antioxidant activity in cardiomyocytes.[20] Due to its cardioprotective effects, glutamate has been used in cardioplegic buffer during cardioplegic arrest.[20]

In tissues and endothelial cells, amino acids L-arginine and asymmetric dimethylarginine (ADMA) are intrinsically present.[115,116] L-arginine, BH_4, and ADMA are substrate, cofactor, and inhibitor of eNOS, respectively. In the heart, the level of BH_4 decreases due to irreversible oxidation during ischemia and reperfusion.[108,109] With limited BH_4, both L-arginine and ADMA increase eNOS-derived $O_2^{\bullet-}$.[115] However, the molecular mechanisms and effect of amino acids on oxidative damage in cardiac ischemia and reperfusion are not presently known and require further investigation.

11.4 ROS-MEDIATED CELLULAR COMMUNICATION DURING I-R

As summarized above, it is clear that in the ischemic and reperfused heart, a number of pathways of ROS formation are activated, giving rise to high levels of oxidative stress. An isolated focus on ROS formation as detrimental is far too simplistic. Importantly, the functional communication, integrated cellular signaling, and cross talk among the various cardiac cell types (i.e., cardiomyocytes, and endothelial cells) and inflammatory cells are also critical for the generation of ROS during ischemia and reperfusion. This integrated cellular communication leads to myocardial infarction and heart failure.

XO is highly expressed within the capillary endothelium of the heart (Fig. 11.7).[5] In human aortic or venous endothelial cells, XO-mediated radical generation has been shown to be a central mechanism of oxygen radical generation upon anoxia and postischemic reoxygenation.[117,118] It has been suggested that in ischemic tissues, XDH is converted via proteolytic cleavage to XO.[40] XO inhibition with allopurinol or oxypurinol has beneficial effects on the failing heart by preventing high-energy phosphate depletion and decreasing $O_2^{\bullet-}$ production.[101] Systemic allopurinol, an XO inhibitor, reduces markers of oxidative stress and improves endothelial function in patients diagnosed with chronic heart failure.[119]

In isolated rat heart, the substrates of XOR, hypoxanthine and xanthine, are not present prior to ischemia.[41] However, in ischemic heart and hypoxic endothelial cells, ATP is metabolized to ADP, AMP, adenosine, inosine, and finally the XOR substrates hypoxanthine and xanthine (Fig. 11.7).[40,41] In the ischemic heart, it has been shown that myocytes are the main sources of XO substrates.[40,41,120] Another substrate for XO, molecular oxygen, is supplied during reperfusion of the heart.[5] It has been shown that blocking XO substrate formation with inhibition of adenosine deaminase (AD) or blocking nucleotide transport (NT) from the myocytes to the endothelium decreases ROS generation in the heart.[102] Moreover, the mitochondrial ETC is also an important source of ROS formation in myocytes.[121] Hence, the communication between myocytes and endothelium is important for the generation of ROS during ischemia and reperfusion.

Importantly, necrotic cardiomyocytes are eliminated slowly by infiltrating inflammatory cells.[122] The oxidative stress signals triggered by the endangered myocardium influence the behavior of nonimmune cells such as endothelial cells and myofibroblasts.[55] The ROS generation by myocytes and endothelial cells during ischemia activates chemotaxis of polymorphonuclear leukocytes (PMNs) in the postischemic hearts.[56] In a canine model, it has been suggested that inflammatory cells may participate in ischemia and reperfusion induced myocyte cell death.[123] The increased generation of ROS triggers PMN adhesion molecule expression and PMN adhesion.[55,56,69,124] Thus, endothelial or myocyte-derived ROS trigger PMN-mediated injury with adhesion to the endothelium followed by transmigration to the cardiac myocyte (see Fig. 11.7). In injured myocytes, leukocytes can increase the generation of ROS once the oxygen supply is restored during reperfusion. However, in view of several negative clinical studies, the importance of leukocytes in clinical reperfusion injury has been questioned.[125]

11.5 ROS AND CELL DEATH DURING ISCHEMIA AND REPERFUSION

The increased production of ROS leads to oxidative damage of various biomolecules, cells, and tissues. The increased generation of ROS during ischemia and reperfusion may injure cells by causing peroxidation of membrane lipids, denaturation of proteins including enzymes and ion channels, and strand breaks in DNA.[5] Moreover, intracellular organelles such as mitochondria are damaged during ischemia and reperfusion.[121] Mitochondria are recognized as a key player in cell death through apoptosis, necrosis, and autophagy.[121] The recurring or prolonged oxidative damage results in organ failure and death.

11.5.1 Apoptosis

Several studies have established that apoptosis, sometimes called "programmed cell death," occurs in cardiac pathologies, such as ischemia and reperfusion injury, myocardial infarction, and heart failure.[126] Myocardial apoptosis contributes significantly to ischemia and reperfusion induced myocardial injury. Mitochondria are now recognized to play a critical role in mediating apoptotic cell death. In apoptosis, a selective permeabilization of the outer mitochondrial membrane (OMM) occurs that releases cyt *c* and other apoptotic factors.[121,127] The release of cyt *c* is an early step in apoptosis.[78,79] In animal models, it has been demonstrated that apoptosis occurs during ischemia and reperfusion.[123,126,128–130] It has also been reported that apoptosis is a significant contributor to myocardial cell death as a result of reperfusion injury.[128] In addition, it has been suggested that macrophages may be involved in induction of apoptotic cell death in the later phase of reperfusion.[123] In myocardial infarction in humans, apoptosis has been observed in three different regions: (1) in the core of the ischemic myocardial area, (2) in the border zone of the infarction, and (3) in the viable myocardium, remote from the ischemic area.[131] Unfortunately, the relative contribution of apoptosis to the total amount of cardiac cell death is still not clear. In addition, whether or not preventing apoptosis would lead to functional improvement is also not known.

11.5.2 Necrosis

Mitochondria also play a key role in necrosis, involving uncontrolled cell death in response to a severe insult that leads to plasma membrane rupture and an inflammatory response. In necrosis, the opening of mitochondrial permeability transition pore (MPTP) in the inner mitochondrial membrane (IMM) plays a major role.[121] MPTP causes necrotic cell death of the heart that occurs during reperfusion after a long period of ischemia. Oxidative stress is one of the potent activators of MPTP opening. The peroxidation of lipids occurs upon reperfusion.[132] The peroxidation of lipids triggers loss of membrane integrity, necrosis, and cell death.[133] In a canine model, the majority of myocytes died during the early phase of reperfusion due to the necrotic pathway.[123] In another study, it has been shown that cell death in acute global ischemia followed by reperfusion occurs predominantly by the necrosis pathway.[134] Neutrophils and macrophages have also been associated with ischemia and reperfusion-induced necrotic myocyte cell death.[123] Reducing oxidative stress during reperfusion represents a potential target for cardioprotection.

11.5.3 Autophagy

Autophagy has evolved as a conserved process for bulk degradation and recycling of cytoplasmic components, such as long-lived proteins and organelles.[135,136] In nutrient-deprived cells, autophagy is a long term cell-survival mechanism.[136] In autophagy, amino acids and fatty acids are recycled for ATP generation.[135] However, excessive activation of autophagy can be toxic and may even induce cell death.[137] In the heart, autophagy is important for the turnover of organelles at low basal levels under normal conditions. In the heart, autophagy is upregulated in response to stresses such as ischemia/reperfusion.[135,136] Autophagy can be protective during acute myocardial ischemia and detrimental during myocardial reperfusion after a short period of ischemia.[138] Studies have suggested that autophagy may be involved in non-apoptotic cell death by ischemia/reperfusion.[136] However, it is unclear as to whether autophagy is a sign of failed cardiomyocyte repair or is a suicide pathway for failing cardiomyocytes. Suppression of oxidative stress during the reperfusion phase inhibits autophagy and prevents myocardial injury.[135] Hence, oxidative stress regulates cell survival/death through autophagy. It has also been demonstrated that a regulatory link exists between mitochondrial function and autophagy.[139] The functional role of autophagy during IR in the heart is complex. It is possible that the extent of autophagy and its pathophysiological function depend on the severity and duration of ischemia and consequent tissue damage during reperfusion in the heart. The level of autophagy may determine whether autophagy is protective or detrimental in response to ischemia and reperfusion in the heart. At present, the contribution of autophagic cell death to overall myocardial injury is unknown.

Ischemia and reperfusion leading to cardiomyocyte cell death has been reported as a combination of apoptosis, necrosis, and autophagy. Nevertheless, further research is needed for a better understanding of the role of apoptosis, necrosis, and autophagy in myocardial injury during ischemia and reperfusion and heart failure.

11.6 POTENTIAL THERAPEUTIC STRATEGIES

Cardioprotective strategies against ischemia and reperfusion are aimed at inhibiting the formation of deleterious metabolites or preserving intracellular metabolites by supplementation.

11.6.1 Inhibitors of XDH/XO (Allopurinol/Febuxostat)

Allopurinol (Zyloprim), a potent inhibitor of XO, has been shown to improve postoperative recovery[44,45] and reduce lipid peroxidation[44] in open-heart surgery patients. Reduction of reperfusion injury in human myocardium has also been reported.[46] Allopurinol is a useful, inexpensive, well-tolerated, and safe anti-ischemic drug for patients with angina.[47] Allopurinol exerts relatively weak competitive inhibition on activity of the oxidized form of XO.[140] Allopurinol is, however, rapidly oxidized by XO to the active inhibitor, oxypurinol. Oxidation of allopurinol by XO results in the generation of $O_2^{\bullet-}$ (ROS).[141] Oxypurinol binds tightly to the reduced form of XO, coordinating to the reduced molybdenum center and acting as a suicide inhibitor of XO.[140] However, oxypurinol binds only weakly to the oxidized form of XO. Moreover, oxypurinol is displaced from the enzyme during spontaneous reoxidation of molybdenum with accompanying reactivation of XO.[140]

Recently, the U.S. Food and Drug Administration (FDA) approved Uloric (febuxostat), an oral inhibitor of XO, for the chronic management of gout.[142] Uloric is structurally unrelated to allopurinol. Importantly, febuxostat is a potent and selective inhibitor of both the oxidized and reduced forms of XO. Febuxostat effectively blocks substrate access to the active site.[143] Febuxostat binding is not affected by the redox state of XO and does not induce ROS generation. Febuxostat is also a more efficient (6000 times) XO inhibitor than allopurinol.[144] Therefore, febuxostat is a promising drug for the protection of the heart from oxidative damage due to ischemia and reperfusion injury.

11.6.2 BH₄ Supplementation

In order to increase the tissue level of BH_4, supplementation with BH_4 and its various analogues have been employed.[145,146] In isolated postichemic reperfused rat hearts, BH_4 supplementation is effective in NO production and partial restoration of endothelium-dependent coronary flow.[108,146] Therefore, supplementation of BH_4 may be an important therapeutic approach to reverse endothelial dysfunction in postischemic tissues as well as in other oxidant-associated cardiovascular diseases.[109,147] However, the uptake of BH_4 by myocytes of the failing or ischemic heart needs to be explored.

Many patients with a severe deficiency of BH_4 have received BH_4 therapy and have had their quality of life greatly improved. BH_4 supplementation reduces ischemia and reperfusion induced endothelial dysfunction and injury in patients with type 2 diabetes and coronary artery disease.[148]

11.6.3 Nitrate/Nitrite Supplementation

Nitrate is abundant in many vegetables such as spinach, lettuce, and beetroot.[93] Nitrate is reduced to nitrite and further to NO by various molecular mechanisms. Dietary nitrates improve mitochondrial efficiency in humans.[149] In rats, *in vivo* administration of nitrate decreases blood pressure and enhanced postischemic blood flow.[93] A substantial amount of ingested nitrate is actively taken up from the circulation by the salivary glands and concentrated in saliva. Salivary nitrate is then efficiently reduced to nitrite by commensal bacteria in the oral cavity. The concentration of nitrite is 1000-fold higher in saliva compared to plasma.[89] Nitrite protects against ischemia and reperfusion injury in the heart.[89,90] Acidic conditions during tissue ischemia and reperfusion or chronic ischemia are optimal for the reduction of nitrite to NO.[110] Various enzymes are also involved in the reduction of nitrite to NO.[89–92,94–96,110] Hence, there is a therapeutic potential for nitrates and nitrites in conditions involving ischemia and reperfusion.[150]

11.6.4 Ischemic Preconditioning (IPC)

Brief periods of myocardial ischemia and reperfusion are protective against cell injury and death caused by subsequent prolonged ischemia, a phenomenon known as IPC.[13] IPC provides potent cardioprotection in mammalian hearts.[13,151] From earlier studies evaluating the effect of ischemic duration on the magnitude of ROS generation, it is clear that even short periods of ischemia produce a measurable but small oxidant burst upon reperfusion.[152] There is evidence that IPC may be induced by low-level oxidant formation.[13,153] These signaling ROS can be derived from mitochondria[154,155] or from other enzymes such as Nox.[155] Importantly, IPC decreases the production of ROS and infarct size in postischemic hearts.[156] IPC also preserves mitochondrial function after ischemia and reperfusion injury.[156] In addition, IPC markedly decreases NO generation during subsequent prolonged ischemia.[5,157] It has been suggested that the IPC-induced decrease of postischemic NO production would prevent myocardial hyperoxygenation upon postischemic reperfusion, in turn diminishing oxygen radical and $ONOO^-$ generation with preservation of mitochondrial function.[5]

Preconditioning triggers not only induce immediate protection against ischemia and reperfusion injury, but also provide delayed protection 24 hours or more after the initial trigger-termed delayed or late preconditioning.[158]

IPC is cardioprotective across all animal species investigated and is considered to be the most protective intervention against myocardial ischemia and reperfusion injury to date. The major limitation to the clinical use of IPC is that the intervention must be performed prior to the onset of the clinical symptoms of acute myocardial infarction. Therefore, the majority of clinical investigations that have been performed have been restricted to various cardiovascular surgical procedures, including coronary artery bypass surgery and transplantation[13] in which the ischemic period is predictable.

11.6.5 Pharmacological Preconditioning

Various pharmacological agents such as phosphodiesterase inhibitors, adenosine, bradykinin, opioid agonists, and endothelin could impart cardioprotection when administered prior to the onset of sustained myocardial ischemia.[13] The use of pharmacological agents to induce myocardial preconditioning is very clinically relevant, since the delivery of a single dose of a well characterized and safe pharmacological agent is far more feasible and often safer than using ischemia to obtain myocardial preconditioning.[13] Hence, pharmacological preconditioning is an effective method for various cardiovascular surgical procedures, including both vascular and cardiac operations and transplantation.

SUMMARY AND CONCLUSION

Over the last three decades, studies have shown that ROS formation is increased in the ischemic and reperfused heart and are important mediators of myocardial injury. Various intracellular organelles and enzymatic and nonenzymatic processes are the main sources of ROS generation during ischemia and reperfusion. ROS formation or oxidative stress is greatly increased through a series of interacting enzymes and cellular pathways. At lower levels, the ROS have an important role in regulation of normal processes such as cell signaling. The concentrations of ROS under normal conditions are maintained at nontoxic levels by a variety of antioxidant defenses and repair enzymes. The balance between antioxidant defenses and ROS production may be disrupted by inhibition of ETC in mitochondria, increased production of substrates, alterations of normal functions of organelles and enzymes, and deficient antioxidant defenses. This imbalance occurs in ischemia/reperfusion injury in which the resulting oxidative insult causes tissue damage and, eventually, cell death. The role of various molecular mechanisms in the generation of ROS during ischemia and reperfusion have been explored, however; this process and the specific molecular targets of ROS are still not fully understood. Various pharmacological agents, supplements, and treatments such as IPC can protect the heart from ischemia and reperfusion injuries. Understanding the molecular mechanisms involved in the generation of ROS during ischemia and reperfusion and specific molecular targets of ROS will facilitate development of improved therapeutic treatments leading to the prevention of myocardial injury.

REFERENCES

1. Braunersreuther, V., Jaquet, V. Reactive oxygen species in myocardial reperfusion injury: From physiopathology to therapeutic approaches. *Curr. Pharm. Biotechnol.* **2012**, *13*(1), 97–114.
2. Ott, H.C., Matthiesen, T.S., Goh, S.K., Black, L.D., Kren, S.M., Netoff, T.I., Taylor, D.A. Perfusion-decellularized matrix: Using nature's platform to engineer a bioartificial heart. *Nat. Med.* **2008**, *14*(2), 213–221.
3. Hansen, P.R. Myocardial reperfusion injury: Experimental evidence and clinical relevance. *Eur. Heart J.* **1995**, *16*(6), 734–740.
4. Ribichini, F., Wijns, W. Acute myocardial infarction: Reperfusion treatment. *Heart* **2002**, *88*(3), 298–305.

5. Zweier, J.L., Talukder, M.A. The role of oxidants and free radicals in reperfusion injury. *Cardiovasc. Res.* **2006**, *70*(2), 181–190.

6. Suleiman, M.S., Zacharowski, K., Angelini, G.D. Inflammatory response and cardioprotection during open-heart surgery: The importance of anaesthetics. *Br. J. Pharmacol.* **2008**, *153*(1), 21–33.

7. Vinten-Johansen, J., Johnston, W.E., Mills, S.A., Faust, K.B., Geisinger, K.R., DeMasi, R.J., Cordell, A.R. Reperfusion injury after temporary coronary occlusion. *J. Thorac. Cardiovasc. Surg.* **1988**, *95*(6), 960–968.

8. Yellon, D.M., Hausenloy, D.J. Myocardial reperfusion injury. *N. Engl. J. Med.* **2007**, *357*(11), 1121–1135.

9. Santos, C.X., Anilkumar, N., Zhang, M., Brewer, A.C., Shah, A.M. Redox signaling in cardiac myocytes. *Free Radic. Biol. Med.* **2011**, *50*(7), 777–793.

10. Shaw, J.M., Winge, D.R. Shaping the mitochondrion: Mitochondrial biogenesis, dynamics and dysfunction. Conference on Mitochondrial Assembly and Dynamics in Health and Disease. *EMBO Rep.* **2009**, *10*(12), 1301–1305.

11. Brahimi-Horn, M.C., Pouysségur, J. Oxygen, a source of life and stress. *FEBS Lett.* **2007**, *581*(19), 3582–3591.

12. Reimer, K.A., Lowe, J.E., Rasmussen, M.M., Jennings, R.B. The wavefront phenomenon of ischemic cell death: 1. Myocardial infarct size vs duration of coronary occlusion in dogs. *Circulation* **1977**, *56*(5), 786–794.

13. Granfeldt, A., Lefer, D.J., Vinten-Johansen, J. Protective ischaemia in patients: Preconditioning and postconditioning. *Cardiovasc. Res.* **2009**, *83*(2), 234–246.

14. Jabbour, A., Gao, L., Kwan, J., Watson, A., Sun, L., Qiu, M.R., Liu, X., Zhou, M.D., Graham, R.M., Hicks, M., Macdonald, P.S. A recombinant human neuregulin-1 peptide improves preservation of the rodent heart after prolonged hypothermic storage. *Transplantation* **2011**, *91*(9), 961–967.

15. Caputo, M., Bryan, A.J., Calafiore, A.M., Suleiman, M.S., Angelini, G.D. Intermittent antegrade hyperkalaemic warm blood cardioplegia supplemented with magnesium prevents myocardial substrate derangement in patients undergoing coronary artery bypass surgery. *Eur. J. Cardiothorac Surg.* **1998**, *14*(6), 596–601.

16. Venturini, A., Ascione, R., Lin, H., Polesel, E., Angelini, G.D., Suleiman, M.S. The importance of myocardial amino acids during ischemia and reperfusion in dilated left ventricle of patients with degenerative mitral valve disease. *Mol. Cell. Biochem.* **2009**, *330*(1–2), 63–70.

17. Taylor, D.O., Edwards, L.B., Aurora, P., Christie, J.D., Dobbels, F., Kirk, R., Rahmel, A.O., Kucheryavaya, A.Y., Hertz, M.I. Registry of the International Society for Heart and Lung Transplantation: Twenty-fifth official adult heart transplant report–2008. *J. Heart Lung Transplant.* **2008**, *27*(9), 943–956.

18. Ledingham, S.J., Katayama, O., Lachno, D.R., Yacoub, M. Prolonged cardiac preservation. Evaluation of the University of Wisconsin preservation solution by comparison with the St. Thomas' Hospital cardioplegic solutions in the rat. *Circulation* **1990**, *82*(5 Suppl.), IV351–IV358.

19. Ku, K., Kin, S., Hashimoto, M., Saitoh, Y., Nosaka, S., Iwasaki, S., Alam, M.S., Nakayama, K. The role of a hydroxyl radical scavenger (nicaraven) in recovery of cardiac function following preservation and reperfusion. *Transplantation* **1996**, *62*(8), 1090–1095.

20. Rosenkranz, E.R., Okamoto, F., Buckberg, G.D., Robertson, J.M., Vinten-Johansen, J., Bugyi, H.I. Safety of prolonged aortic clamping with blood cardioplegia: III. Aspartate enrichment of glutamate-blood cardioplegia in energy-depleted hearts after ischemic and reperfusion injury. *J. Thorac. Cardiovasc. Surg.* **1986**, *91*(3), 428–435.

21. Marenzi, G., Giorgio, M., Trinei, M., Moltrasio, M., Ravagnani, P., Cardinale, D., Ciceri, F., Cavallero, A., Veglia, F., Fiorentini, C., Cipolla, C.M., Bartorelli, A.L., Pelicci, P. Circulating cytochrome c as potential biomarker of impaired reperfusion in ST-segment elevation acute myocardial infarction. *Am. J. Cardiol.* **2010**, *106*(10), 1443–1449.

22. Tomaselli, G.F., Barth, A.S. Sudden cardio arrest: Oxidative stress irritates the heart. *Nat. Med.* **2010**, *16*(6), 648–649.

23. Rajesh, K.G., Surekha, R.H., Mrudula, S.K., Prasad, Y., Sanjib, K.S., Prathiba, N. Oxidative and nitrosative stress in association with DNA damage in coronary heart disease. *Singapore Med. J.* **2011**, *52*(4), 283–288.

24. Ferrari, R., Agnoletti, L., Comini, L., Gaia, G., Bachetti, T., Cargnoni, A., Ceconi, C., Curello, S., Visioli, O. Oxidative stress during myocardial ischaemia and heart failure. *Eur. Heart J.* **1998**, *19*(Suppl. B), B2–B11.

25. Sam, F., Kerstetter, D.L., Pimental, D.R., Mulukutla, S., Tabaee, A., Bristow, M.R., Colucci, W.S., Sawyer, D.B. Increased reactive oxygen species production and functional alterations in antioxidant enzymes in human failing myocardium. *J. Card. Fail.* **2005**, *11*(6), 473–480.

26. St-Pierre, J., Buckingham, J.A., Roebuck, S.J., Brand, M.D. Topology of superoxide production from different sites in the mitochondrial electron transport chain. *J. Biol. Chem.* **2002**, *277*(47), 44784–44790.

27. O'Rourke, B., Cortassa, S., Aon, M.A. Mitochondrial ion channels: Gatekeepers of life and death. *Physiology (Bethesda)* **2005**, *20*, 303–315.

28. Baines, C.P., Kaiser, R.A., Purcell, N.H., Blair, N.S., Osinska, H., Hambleton, M.A., Brunskill, E.W., Sayen, M.R., Gottlieb, R.A., Dorn, G.W., Robbins, J., Molkentin, J.D. Loss of cyclophilin D reveals a critical role for mitochondrial permeability transition in cell death. *Nature* **2005**, *434*(7033), 658–662.

29. Giorgio, M., Trinei, M., Migliaccio, E., Pelicci, P.G. Hydrogen peroxide: A metabolic by-product or a common mediator of ageing signals? *Nat. Rev. Mol. Cell Biol.* **2007**, *8*(9), 722–728.

30. Kowaltowski, A.J., de Souza-Pinto, N.C., Castilho, R.F., Vercesi, A.E. Mitochondria and reactive oxygen species. *Free Radic. Biol. Med.* **2009**, *47*(4), 333–343.

31. Riemer, J., Fischer, M., Herrmann, J.M. Oxidation-driven protein import into mitochondria: Insights and blind spots. *Biochim. Biophys. Acta* **2011**, *1808*(3), 981–989.
32. Matsuzaki, S., Szweda, L.I., Humphries, K.M. Mitochondrial superoxide production and respiratory activity: Biphasic response to ischemic duration. *Arch. Biochem. Biophys.* **2009**, *484*(1), 87–93.
33. Tu, B.P., Weissman, J.S. Oxidative protein folding in eukaryotes: Mechanisms and consequences. *J. Cell Biol.* **2004**, *164*(3), 341–346.
34. Enyedi, B., Varnai, P., Geiszt, M. Redox state of the endoplasmic reticulum is controlled by Ero1L-alpha and intraluminal calcium. *Antioxid. Redox Signal.* **2010**, *13*(6), 721–729.
35. Hwang, C., Sinskey, A.J., Lodish, H.F. Oxidized redox state of glutathione in the endoplasmic reticulum. *Science* **1992**, *257*(5076), 1496–1502.
36. Hotamisligil, G.S. Endoplasmic reticulum stress and atherosclerosis. *Nat. Med.* **2010**, *16*(4), 396–399.
37. Szegezdi, E., Logue, S.E., Gorman, A.M., Samali, A. Mediators of endoplasmic reticulum stress-induced apoptosis. *EMBO Rep.* **2006**, *7*(9), 880–885.
38. Antonenkov, V.D., Grunau, S., Ohlmeier, S., Hiltunen, J.K. Peroxisomes are oxidative organelles. *Antioxid. Redox Signal.* **2010**, *13*(4), 525–537.
39. McCord, J.M., Fridovich, I. Superoxide dismutase: An enzymic function for erythrocuprein (hemocuprein). *J. Biol. Chem.* **1969**, *244*(22), 6049–6055.
40. McCord, J.M. Oxygen-derived free radicals in postischemic tissue injury. *N. Engl. J. Med.* **1985**, *312*(3), 159–163.
41. Xia, Y., Zweier, J.L. Substrate control of free radical generation from xanthine oxidase in the postischemic heart. *J. Biol. Chem.* **1995**, *270*(32), 18797–18803.
42. Nishino, T., Okamoto, K., Eger, B.T., Pai, E.F., Nishino, T. Mammalian xanthine oxidoreductase: Mechanism of transition from xanthine dehydrogenase to xanthine oxidase. *FEBS J.* **2008**, *275*(13), 3278–3289.
43. Cappola, T.P., Kass, D.A., Nelson, G.S., Berger, R.D., Rosas, G.O., Kobeissi, Z.A., Marban, E., Hare, J.M. Allopurinol improves myocardial efficiency in patients with idiopathic dilated cardiomyopathy. *Circulation* **2001**, *104*(20), 2407–2411.
44. Coghlan, J.G., Flitter, W.D., Clutton, S.M., Panda, R., Daly, R., Wright, G., Ilsley, C.D., Slater, T.F. Allopurinol pretreatment improves postoperative recovery and reduces lipid peroxidation in patients undergoing coronary artery bypass grafting. *J. Thorac. Cardiovasc. Surg.* **1994**, *107*(1), 248–256.
45. Adachi, H., Motomatsu, K., Yara, I. Effect of allopurinol (zyloric) on patients undergoing open heart surgery. *Jpn. Circ. J.* **1979**, *43*(5), 395–401.
46. Gimpel, J.A., Lahpor, J.R., van der Molen, A.J., Damen, J., Hitchcock, J.F. Reduction of reperfusion injury of human myocardium by allopurinol: A clinical study. *Free Radic. Biol. Med.* **1995**, *19*(2), 251–255.
47. Noman, A., Ang, D.S., Ogston, S., Lang, C.C., Struthers, A.D. Effect of high-dose allopurinol on exercise in patients with chronic stable angina: A randomised, placebo controlled crossover trial. *Lancet* **2010**, *375*(9732), 2161–2167.
48. Berry, C.E., Hare, J.M. Xanthine oxidoreductase and cardiovascular disease: Molecular mechanisms and pathophysiological implications. *J. Physiol.* **2004**, *555*(Pt 3), 589–606.
49. Godber, B.L., Doel, J.J., Durgan, J., Eisenthal, R., Harrison, R. A new route to peroxynitrite: A role for xanthine oxidoreductase. *FEBS Lett.* **2000**, *475*(2), 93–96.
50. Millar, T.M. Peroxynitrite formation from the simultaneous reduction of nitrite and oxygen by xanthine oxidase. *FEBS Lett.* **2004**, *562*(1–3), 129–133.
51. Beedham, C. Molybdenum hydroxylases: Biological distribution and substrate-inhibitor specificity. *Prog. Med. Chem.* **1987**, *24*, 85–127.
52. Kundu, T.K., Velayutham, M., Zweier, J.L. Generation of reactive oxygen species by aldehyde oxidase through the oxidation of NADH. *Free Radic. Biol. Med.* **2010**, *49*(Suppl. 1), S96.
53. Kundu, T.K., Hille, R., Velayutham, M., Zweier, J.L. Characterization of superoxide production from aldehyde oxidase: An important source of oxidants in biological tissues. *Arch. Biochem. Biophys.* **2007**, *460*(1), 113–121.
54. Moriwaki, Y., Yamamoto, T., Higashino, K. Distribution and pathophysiologic role of molybdenum-containing enzymes. *Histol. Histopathol.* **1997**, *12*(2), 513–524.
55. Kawaguchi, M., Takahashi, M., Hata, T., Kashima, Y., Usui, F., Morimoto, H., Izawa, A., Takahashi, Y., Masumoto, J., Koyama, J., Hongo, M., Noda, T., Nakayama, J., Sagara, J., Taniguchi, S., Ikeda, U. Inflammasome activation of cardiac fibroblasts is essential for myocardial ischemia/reperfusion injury. *Circulation* **2011**, *123*(6), 594–604.
56. Arslan, F., de Kleijn, D.P., Pasterkamp, G. Innate immune signaling in cardiac ischemia. *Nat. Rev. Cardiol.* **2011**, *8*(5), 292–300.
57. Bylund, J., Brown, K.L., Movitz, C., Dahlgren, C., Karlsson, A. Intracellular generation of superoxide by the phagocyte NADPH oxidase: How, where, and what for? *Free Radic. Biol. Med.* **2010**, *49*(12), 1834–1845.
58. Lambeth, J.D. NOX enzymes and the biology of reactive oxygen. *Nat. Rev. Immunol.* **2004**, *4*(3), 181–189.
59. West, A.P., Brodsky, I.E., Rahner, C., Woo, D.K., Erdjument-Bromage, H., Tempst, P., Walsh, M.C., Choi, Y., Shadel, G.S., Ghosh, S. TLR signalling augments macrophage bactericidal activity through mitochondrial ROS. *Nature* **2011**, *472*(7344), 476–480.
60. Heymes, C., Bendall, J.K., Ratajczak, P., Cave, A.C., Samuel, J.L., Hasenfuss, G., Shah, A.M. Increased myocardial NADPH oxidase activity in human heart failure. *J. Am. Coll. Cardiol.* **2003**, *41*(12), 2164–2171.
61. Ago, T., Kuroda, J., Pain, J., Fu, C., Li, H., Sadoshima, J. Upregulation of Nox4 by hypertrophic stimuli promotes

apoptosis and mitochondrial dysfunction in cardiac myocytes. *Circ. Res.* **2010**, *106*(7), 1253–1264.

62. Cave, A.C., Brewer, A.C., Narayanapanicker, A., Ray, R., Grieve, D.J., Walker, S., Shah, A.M. NADPH oxidases in cardiovascular health and disease. *Antioxid. Redox Signal.* **2006**, *8*(5–6), 691–728.

63. Zhang, M., Brewer, A.C., Schroder, K., Santos, C.X., Grieve, D.J., Wang, M., Anilkumar, N., Yu, B., Dong, X., Walker, S.J., Brandes, R.P., Shah, A.M. NADPH oxidase-4 mediates protection against chronic load-induced stress in mouse hearts by enhancing angiogenesis. *Proc. Natl. Acad. Sci. U.S.A* **2010**, *107*(42), 18121–18126.

64. Byrne, J.A., Grieve, D.J., Bendall, J.K., Li, J.M., Gove, C., Lambeth, J.D., Cave, A.C., Shah, A.M. Contrasting roles of NADPH oxidase isoforms in pressure-overload versus angiotensin II-induced cardiac hypertrophy. *Circ. Res.* **2003**, *93*(9), 802–805.

65. Efremov, R.G., Baradaran, R., Sazanov, L.A. The architecture of respiratory complex I. *Nature* **2010**, *465*(7297), 441–445.

66. Ceconi, C., Bernocchi, P., Boraso, A., Cargnoni, A., Pepi, P., Curello, S., Ferrari, R. New insights on myocardial pyridine nucleotides and thiol redox state in ischemia and reperfusion damage. *Cardiovasc. Res.* **2000**, *47*(3), 586–594.

67. Vandeplassche, G., Hermans, C., Thone, F., Borgers, M. Stunned myocardium has increased mitochondrial NADH oxidase and ATPase activities. *Cardioscience* **1991**, *2*(1), 47–53.

68. Lee, M., Velayutham, M., Shoji, H., Yoshino, F., Hille, R., Zweier, J.L. Measurement and characterization of superoxide generation from xanthine dehydrogenase: A redox regulated pathway of radical generation in ischemic tissues. *Free Radic. Biol. Med.* **2008**, *45*(Suppl. 1), S150.

69. Harrison, R. Structure and function of xanthine oxidoreductase: Where are we now? *Free Radic. Biol. Med.* **2002**, *33*(6), 774–797.

70. Velayutham, M., Hemann, C., Zweier, J.L. Removal of H_2O_2 and generation of superoxide radical: Role of cytochrome c and NADH. *Free Radic. Biol. Med.* **2011**, *51*(1), 160–170.

71. Dallman, P.R. Cytochrome c in normal and hypertrophied heart. *Nature* **1966**, *212*(5062), 608–609.

72. Forman, H.J., Azzi, A. On the virtual existence of superoxide anions in mitochondria: Thoughts regarding its role in pathophysiology. *FASEB J.* **1997**, *11*(5), 374–375.

73. Semak, I., Naumova, M., Korik, E., Terekhovich, V., Wortsman, J., Slominski, A. A novel metabolic pathway of melatonin: Oxidation by cytochrome C. *Biochemistry* **2005**, *44*(26), 9300–9307.

74. Pereverzev, M.O., Vygodina, T.V., Konstantinov, A.A., Skulachev, V.P. Cytochrome c, an ideal antioxidant. *Biochem. Soc. Trans.* **2003**, *31*(Pt 6), 1312–1315.

75. Kagan, V.E., Borisenko, G.G., Tyurina, Y.Y., Tyurin, V.A., Jiang, J., Potapovich, A.I., Kini, V., Amoscato, A.A., Fujii, Y. Oxidative lipidomics of apoptosis: Redox catalytic interactions of cytochrome c with cardiolipin and phosphatidylserine. *Free Radic. Biol. Med.* **2004**, *37*(12), 1963–1985.

76. Liu, X., Kim, C.N., Yang, J., Jemmerson, R., Wang, X. Induction of apoptotic program in cell-free extracts: Requirement for dATP and cytochrome c. *Cell* **1996**, *86*(1), 147–157.

77. Reed, J.C. Cytochrome c: Can't live with it—can't live without it. *Cell* **1997**, *91*(5), 559–562.

78. Yang, J., Liu, X., Bhalla, K., Kim, C.N., Ibrado, A.M., Cai, J., Peng, T.I., Jones, D.P., Wang, X. Prevention of apoptosis by Bcl-2: Release of cytochrome c from mitochondria blocked. *Science* **1997**, *275*(5303), 1129–1132.

79. Kluck, R.M., Bossy-Wetzel, E., Green, D.R., Newmeyer, D.D. The release of cytochrome c from mitochondria: A primary site for Bcl-2 regulation of apoptosis. *Science* **1997**, *275*(5303), 1132–1136.

80. Kagan, V.E., Bayir, H.A., Belikova, N.A., Kapralov, O., Tyurina, Y.Y., Tyurin, V.A., Jiang, J., Stoyanovsky, D.A., Wipf, P., Kochanek, P.M., Greenberger, J.S., Pitt, B., Shvedova, A.A., Borisenko, G. Cytochrome c/cardiolipin relations in mitochondria: A kiss of death. *Free Radic. Biol. Med.* **2009**, *46*(11), 1439–1453.

81. Gonzalvez, F., Gottlieb, E. Cardiolipin: Setting the beat of apoptosis. *Apoptosis* **2007**, *12*(5), 877–885.

82. Orrenius, S., Zhivotovsky, B. Cardiolipin oxidation sets cytochrome c free. *Nat. Chem. Biol.* **2005**, *1*(4), 188–189.

83. Chen, Y.R., Deterding, L.J., Sturgeon, B.E., Tomer, K.B., Mason, R.P. Protein oxidation of cytochrome C by reactive halogen species enhances its peroxidase activity. *J. Biol. Chem.* **2002**, *277*(33), 29781–29791.

84. Abriata, L.A., Cassina, A., Tortora, V., Marin, M., Souza, J.M., Castro, L., Vila, A.J., Radi, R. Nitration of solvent-exposed tyrosine 74 on cytochrome c triggers heme iron-methionine 80 bond disruption. Nuclear magnetic resonance and optical spectroscopy studies. *J. Biol. Chem.* **2009**, *284*(1), 17–26.

85. Hare, J.M., Colucci, W.S. Role of nitric oxide in the regulation of myocardial function. *Prog. Cardiovasc. Dis.* **1995**, *38*(2), 155–166.

86. Landmesser, U., Dikalov, S., Price, S.R., McCann, L., Fukai, T., Holland, S.M., Mitch, W.E., Harrison, D.G. Oxidation of tetrahydrobiopterin leads to uncoupling of endothelial cell nitric oxide synthase in hypertension. *J. Clin. Invest.* **2003**, *111*(8), 1201–1209.

87. Vasquez-Vivar, J., Kalyanaraman, B., Martasek, P., Hogg, N., Masters, B.S., Karoui, H., Tordo, P., Pritchard, K.A., Jr. Superoxide generation by endothelial nitric oxide synthase: The influence of cofactors. *Proc. Natl. Acad. Sci. U.S.A* **1998**, *95*(16), 9220–9225.

88. Cai, H., Harrison, D.G. Endothelial dysfunction in cardiovascular diseases: The role of oxidant stress. *Circ. Res.* **2000**, *87*(10), 840–844.

89. Lundberg, J.O., Gladwin, M.T., Ahluwalia, A., Benjamin, N., Bryan, N.S., Butler, A., Cabrales, P., Fago, A., Feelisch, M., Ford, P.C., Freeman, B.A., Frenneaux, M., Friedman,

J., Kelm, M., Kevil, C.G., Kim-Shapiro, D.B., Kozlov, A.V., Lancaster, J.R., Jr., Lefer, D.J., McColl, K., McCurry, K., Patel, R.P., Petersson, J., Rassaf, T., Reutov, V.P., Richter-Addo, G.B., Schechter, A., Shiva, S., Tsuchiya, K., van Faassen, E.E., Webb, A.J., Zuckerbraun, B.S., Zweier, J.L., Weitzberg, E. Nitrate and nitrite in biology, nutrition and therapeutics. *Nat. Chem. Biol.* **2009**, *5*(12), 865–869.

90. Zweier, J.L., Li, H., Samouilov, A., Liu, X. Mechanisms of nitrite reduction to nitric oxide in the heart and vessel wall. *Nitric Oxide* **2010**, *22*(2), 83–90.

91. Li, H., Kundu, T.K., Zweier, J.L. Characterization of the magnitude and mechanism of aldehyde oxidase-mediated nitric oxide production from nitrite. *J. Biol. Chem.* **2009**, *284*(49), 33850–33858.

92. Li, H., Cui, H., Kundu, T.K., Alzawahra, W., Zweier, J.L. Nitric oxide production from nitrite occurs primarily in tissues not in the blood: Critical role of xanthine oxidase and aldehyde oxidase. *J. Biol. Chem.* **2008**, *283*(26), 17855–17863.

93. Jansson, E.A., Huang, L., Malkey, R., Govoni, M., Nihlen, C., Olsson, A., Stensdotter, M., Petersson, J., Holm, L., Weitzberg, E., Lundberg, J.O. A mammalian functional nitrate reductase that regulates nitrite and nitric oxide homeostasis. *Nat. Chem. Biol.* **2008**, *4*(7), 411–417.

94. Basu, S., Azarova, N.A., Font, M.D., King, S.B., Hogg, N., Gladwin, M.T., Shiva, S., Kim-Shapiro, D.B. Nitrite reductase activity of cytochrome c. *J. Biol. Chem.* **2008**, *283*(47), 32590–32597.

95. Kozlov, A.V., Staniek, K., Nohl, H. Nitrite reductase activity is a novel function of mammalian mitochondria. *FEBS Lett.* **1999**, *454*(1–2), 127–130.

96. van Faassen, E.E., Bahrami, S., Feelisch, M., Hogg, N., Kelm, M., Kim-Shapiro, D.B., Kozlov, A.V., Li, H., Lundberg, J.O., Mason, R., Nohl, H., Rassaf, T., Samouilov, A., Slama-Schwok, A., Shiva, S., Vanin, A.F., Weitzberg, E., Zweier, J., Gladwin, M.T. Nitrite as regulator of hypoxic signaling in mammalian physiology. *Med. Res. Rev.* **2009**, *29*(5), 683–741.

97. Valko, M., Leibfritz, D., Moncol, J., Cronin, M.T., Mazur, M., Telser, J. Free radicals and antioxidants in normal physiological functions and human disease. *Int. J. Biochem. Cell Biol.* **2007**, *39*(1), 44–84.

98. Zorov, D.B., Juhaszova, M., Sollott, S.J. Mitochondrial ROS-induced ROS release: An update and review. *Biochim. Biophys. Acta* **2006**, *1757*(5–6), 509–517.

99. Aon, M.A., Cortassa, S., O'Rourke, B. Mitochondrial oscillations in physiology and pathophysiology. *Adv. Exp. Med. Biol.* **2008**, *641*, 98–117.

100. Kurz, F.T., Aon, M.A., O'Rourke, B., Armoundas, A.A. Spatio-temporal oscillations of individual mitochondria in cardiac myocytes reveal modulation of synchronized mitochondrial clusters. *Proc. Natl. Acad. Sci. U.S.A* **2010**, *107*(32), 14315–14320.

101. Thompson-Gorman, S.L., Zweier, J.L. Evaluation of the role of xanthine oxidase in myocardial reperfusion injury. *J. Biol. Chem.* **1990**, *265*(12), 6656–6663.

102. Xia, Y., Khatchikian, G., Zweier, J.L. Adenosine deaminase inhibition prevents free radical-mediated injury in the postischemic heart. *J. Biol. Chem.* **1996**, *271*(17), 10096–10102.

103. Williamson, J.R., Corkey, B.E. Assay of citric acid cycle intermediates and related compounds—update with tissue metabolite levels and intracellular distribution. *Methods Enzymol.* **1979**, *55*, 200–222.

104. Correa, F., Garcia, N., Robles, C., Martinez-Abundis, E., Zazueta, C. Relationship between oxidative stress and mitochondrial function in the post-conditioned heart. *J. Bioenerg. Biomembr.* **2008**, *40*(6), 599–606.

105. Hill, B.G., Awe, S.O., Vladykovskaya, E., Ahmed, Y., Liu, S.Q., Bhatnagar, A., Srivastava, S. Myocardial ischaemia inhibits mitochondrial metabolism of 4-hydroxy-trans-2-nonenal. *Biochem. J.* **2009**, *417*(2), 513–524.

106. Ceconi, C., Cargnoni, A., Francolini, G., Parinello, G., Ferrari, R. Heart rate reduction with ivabradine improves energy metabolism and mechanical function of isolated ischaemic rabbit heart. *Cardiovasc. Res.* **2009**, *84*(1), 72–82.

107. Vasquez-Vivar, J. Tetrahydrobiopterin, superoxide, and vascular dysfunction. *Free Radic. Biol. Med.* **2009**, *47*(8), 1108–1119.

108. Dumitrescu, C., Biondi, R., Xia, Y., Cardounel, A.J., Druhan, L.J., Ambrosio, G., Zweier, J.L. Myocardial ischemia results in tetrahydrobiopterin (BH4) oxidation with impaired endothelial function ameliorated by BH4. *Proc. Natl. Acad. Sci. U.S.A* **2007**, *104*(38), 15081–15086.

109. Nishijima, Y., Sridhar, A., Bonilla, I., Velayutham, M., Khan, M., Terentyeva, R., Li, C., Kuppusamy, P., Elton, T.S., Terentyev, D., Gyorke, S., Zweier, J.L., Cardounel, A.J., Carnes, C.A. Tetrahydrobiopterin depletion and NOS2 uncoupling contribute to heart-failure-induced alterations in atrial electrophysiology. *Cardiovasc. Res.* **2011**, *91*(1), 71–79.

110. Zweier, J.L., Wang, P., Samouilov, A., Kuppusamy, P. Enzyme-independent formation of nitric oxide in biological tissues. *Nat. Med.* **1995**, *1*(8), 804–809.

111. Wang, P., Zweier, J.L. Measurement of nitric oxide and peroxynitrite generation in the postischemic heart. Evidence for peroxynitrite-mediated reperfusion injury. *J. Biol. Chem.* **1996**, *271*(46), 29223–29230.

112. Peluffo, G., Radi, R. Biochemistry of protein tyrosine nitration in cardiovascular pathology. *Cardiovasc. Res.* **2007**, *75*(2), 291–302.

113. Schulz, E., Jansen, T., Wenzel, P., Daiber, A., Munzel, T. Nitric oxide, tetrahydrobiopterin, oxidative stress, and endothelial dysfunction in hypertension. *Antioxid. Redox Signal.* **2008**, *10*(6), 1115–1126.

114. Chen, W., Druhan, L.J., Chen, C.A., Hemann, C., Chen, Y.R., Berka, V., Tsai, A.L., Zweier, J.L. Peroxynitrite induces destruction of the tetrahydrobiopterin and heme in endothelial nitric oxide synthase: Transition from reversible to irreversible enzyme inhibition. *Biochemistry* **2010**, *49*(14), 3129–3137.

115. Druhan, L.J., Forbes, S.P., Pope, A.J., Chen, C.A., Zweier, J.L., Cardounel, A.J. Regulation of eNOS-derived superoxide by endogenous methylarginines. *Biochemistry* **2008**, *47*(27), 7256–7263.

116. Bode-Boger, S.M., Scalera, F., Ignarro, L.J. The L-arginine paradox: Importance of the L-arginine/asymmetrical dimethylarginine ratio. *Pharmacol. Ther.* **2007**, *114*(3), 295–306.

117. Zweier, J.L., Kuppusamy, P., Lutty, G.A. Measurement of endothelial cell free radical generation: Evidence for a central mechanism of free radical injury in postischemic tissues. *Proc. Natl. Acad. Sci. U.S.A* **1988**, *85*(11), 4046–4050.

118. Zweier, J.L., Broderick, R., Kuppusamy, P., Thompson-Gorman, S., Lutty, G.A. Determination of the mechanism of free radical generation in human aortic endothelial cells exposed to anoxia and reoxygenation. *J. Biol. Chem.* **1994**, *269*(39), 24156–24162.

119. Farquharson, C.A., Butler, R., Hill, A., Belch, J.J., Struthers, A.D. Allopurinol improves endothelial dysfunction in chronic heart failure. *Circulation* **2002**, *106*(2), 221–226.

120. DeWall, R.A., Vasko, K.A., Stanley, E.L., Kezdi, P. Responses of the ischemic myocardium to allopurinol. *Am. Heart J.* **1971**, *82*(3), 362–370.

121. Halestrap, A.P. A pore way to die: The role of mitochondria in reperfusion injury and cardioprotection. *Biochem. Soc. Trans.* **2010**, *38*(4), 841–860.

122. Yaoita, H., Ogawa, K., Maehara, K., Maruyama, Y. Attenuation of ischemia/reperfusion injury in rats by a caspase inhibitor. *Circulation* **1998**, *97*(3), 276–281.

123. Zhao, Z.Q., Velez, D.A., Wang, N.P., Hewan-Lowe, K.O., Nakamura, M., Guyton, R.A., Vinten-Johansen, J. Progressively developed myocardial apoptotic cell death during late phase of reperfusion. *Apoptosis* **2001**, *6*(4), 279–290.

124. Williams, F.M. Neutrophils and myocardial reperfusion injury. *Pharmacol. Ther.* **1996**, *72*(1), 1–12.

125. Vinten-Johansen, J. Involvement of neutrophils in the pathogenesis of lethal myocardial reperfusion injury. *Cardiovasc. Res.* **2004**, *61*(3), 481–497.

126. Dispersyn, G.D., Borgers, M. Apoptosis in the heart: About programmed cell death and survival. *News Physiol. Sci.* **2001**, *16*, 41–47.

127. Petrosillo, G., Moro, N., Ruggiero, F.M., Paradies, G. Melatonin inhibits cardiolipin peroxidation in mitochondria and prevents the mitochondrial permeability transition and cytochrome c release. *Free Radic. Biol. Med.* **2009**, *47*(7), 969–974.

128. Fliss, H., Gattinger, D. Apoptosis in ischemic and reperfused rat myocardium. *Circ. Res.* **1996**, *79*(5), 949–956.

129. Cheng, Y., Zhu, P., Yang, J., Liu, X., Dong, S., Wang, X., Chun, B., Zhuang, J., Zhang, C. Ischaemic preconditioning-regulated miR-21 protects heart against ischaemia/reperfusion injury via anti-apoptosis through its target PDCD4. *Cardiovasc. Res.* **2010**, *87*(3), 431–439.

130. Kajstura, J., Cheng, W., Reiss, K., Clark, W.A., Sonnenblick, E.H., Krajewski, S., Reed, J.C., Olivetti, G., Anversa, P. Apoptotic and necrotic myocyte cell deaths are independent contributing variables of infarct size in rats. *Lab. Invest.* **1996**, *74*(1), 86–107.

131. Borgers, M., Voipio-Pulkki, L., Izumo, S. Apoptosis. *Cardiovasc. Res.* **2000**, *45*(3), 525–527.

132. Manning, A.S., Hearse, D.J. Reperfusion-induced arrhythmias: Mechanisms and prevention. *J. Mol. Cell. Cardiol.* **1984**, *16*(6), 497–518.

133. Park, J.L., Lucchesi, B.R. Mechanisms of myocardial reperfusion injury. *Ann. Thorac. Surg.* **1999**, *68*(5), 1905–1912.

134. Freude, B., Masters, T.N., Robicsek, F., Fokin, A., Kostin, S., Zimmermann, R., Ullmann, C., Lorenz-Meyer, S., Schaper, J. Apoptosis is initiated by myocardial ischemia and executed during reperfusion. *J. Mol. Cell. Cardiol.* **2000**, *32*(2), 197–208.

135. Hariharan, N., Zhai, P., Sadoshima, J. Oxidative stress stimulates autophagic flux during ischemia/reperfusion. *Antioxid. Redox Signal.* **2011**, *14*(11), 2179–2190.

136. Nishida, K., Kyoi, S., Yamaguchi, O., Sadoshima, J., Otsu, K. The role of autophagy in the heart. *Cell Death Differ.* **2009**, *16*(1), 31–38.

137. Levine, B., Kroemer, G. Autophagy in the pathogenesis of disease. *Cell* **2008**, *132*(1), 27–42.

138. Matsui, Y., Takagi, H., Qu, X., Abdellatif, M., Sakoda, H., Asano, T., Levine, B., Sadoshima, J. Distinct roles of autophagy in the heart during ischemia and reperfusion: Roles of AMP-activated protein kinase and Beclin 1 in mediating autophagy. *Circ. Res.* **2007**, *100*(6), 914–922.

139. Graef, M., Nunnari, J. Mitochondria regulate autophagy by conserved signaling pathways. *EMBO J.* **2011**, *30*, 2101–2114.

140. Takano, Y., Hase-Aoki, K., Horiuchi, H., Zhao, L., Kasahara, Y., Kondo, S., Becker, M.A. Selectivity of febuxostat, a novel non-purine inhibitor of xanthine oxidase/xanthine dehydrogenase. *Life Sci.* **2005**, *76*(16), 1835–1847.

141. Galbusera, C., Orth, P., Fedida, D., Spector, T. Superoxide radical production by allopurinol and xanthine oxidase. *Biochem. Pharmacol.* **2006**, *71*(12), 1747–1752.

142. Adams, J.U. New relief for gout. *Nat. Biotechnol.* **2009**, *27*(4), 309–311.

143. Okamoto, K., Nishino, T. Crystal structures of mammalian xanthine oxidoreductase bound with various inhibitors: Allopurinol, febuxostat, and FYX-051. *J. Nippon Med. Sch.* **2008**, *75*(1), 2–3.

144. Okamoto, K., Eger, B.T., Nishino, T., Kondo, S., Pai, E.F., Nishino, T. An extremely potent inhibitor of xanthine oxidoreductase. Crystal structure of the enzyme-inhibitor complex and mechanism of inhibition. *J. Biol. Chem.* **2003**, *278*(3), 1848–1855.

145. Sawabe, K., Wakasugi, K.O., Hasegawa, H. Tetrahydrobiopterin uptake in supplemental administration: Elevation of tissue tetrahydrobiopterin in mice following uptake of the exogenously oxidized product 7,8-

dihydrobiopterin and subsequent reduction by an antifolate-sensitive process. *J. Pharmacol. Sci.* **2004**, *96*(2), 124–133.
146. Moens, A.L., Kietadisorn, R., Lin, J.Y., Kass, D. Targeting endothelial and myocardial dysfunction with tetrahydrobiopterin. *J. Mol. Cell. Cardiol.* **2011**, *51*(4), 559–563.
147. Moens, A.L., Kass, D.A. Therapeutic potential of tetrahydrobiopterin for treating vascular and cardiac disease. *J. Cardiovasc. Pharmacol.* **2007**, *50*(3), 238–246.
148. Settergren, M., Bohm, F., Malmstrom, R.E., Channon, K.M., Pernow, J. L-arginine and tetrahydrobiopterin protects against ischemia/reperfusion-induced endothelial dysfunction in patients with type 2 diabetes mellitus and coronary artery disease. *Atherosclerosis* **2009**, *204*(1), 73–78.
149. Larsen, F.J., Schiffer, T.A., Borniquel, S., Sahlin, K., Ekblom, B., Lundberg, J.O., Weitzberg, E. Dietary inorganic nitrate improves mitochondrial efficiency in humans. *Cell Metab.* **2011**, *13*(2), 149–159.
150. Lundberg, J.O., Weitzberg, E., Gladwin, M.T. The nitrate-nitrite-nitric oxide pathway in physiology and therapeutics. *Nat. Rev. Drug Discov.* **2008**, *7*(2), 156–167.
151. Yellon, D.M., Alkhulaifi, A.M., Pugsley, W.B. Preconditioning the human myocardium. *Lancet* **1993**, *342*(8866), 276–277.
152. Zweier, J.L., Kuppusamy, P., Williams, R., Rayburn, B.K., Smith, D., Weisfeldt, M.L., Flaherty, J.T. Measurement and characterization of postischemic free radical generation in the isolated perfused heart. *J. Biol. Chem.* **1989**, *264*(32), 18890–18895.
153. Yellon, D.M., Downey, J.M. Preconditioning the myocardium: From cellular physiology to clinical cardiology. *Physiol. Rev.* **2003**, *83*(4), 1113–1151.
154. Juhaszova, M., Zorov, D.B., Kim, S.H., Pepe, S., Fu, Q., Fishbein, K.W., Ziman, B.D., Wang, S., Ytrehus, K., Antos, C.L., Olson, E.N., Sollott, S.J. Glycogen synthase kinase-3beta mediates convergence of protection signaling to inhibit the mitochondrial permeability transition pore. *J. Clin. Invest.* **2004**, *113*(11), 1535–1549.
155. Kimura, S., Zhang, G.X., Nishiyama, A., Shokoji, T., Yao, L., Fan, Y.Y., Rahman, M., Suzuki, T., Maeta, H., Abe, Y. Role of NAD(P)H oxidase- and mitochondria-derived reactive oxygen species in cardioprotection of ischemic reperfusion injury by angiotensin II. *Hypertension* **2005**, *45*(5), 860–866.
156. Quarrie, R., Cramer, B.M., Lee, D.S., Steinbaugh, G.E., Erdahl, W., Pfeiffer, D.R., Zweier, J.L., Crestanello, J.A. Ischemic preconditioning decreases mitochondrial proton leak and reactive oxygen species production in the postischemic heart. *J. Surg. Res.* **2011**, *165*(1), 5–14.
157. Csonka, C., Szilvassy, Z., Fulop, F., Pali, T., Blasig, I.E., Tosaki, A., Schulz, R., Ferdinandy, P. Classic preconditioning decreases the harmful accumulation of nitric oxide during ischemia and reperfusion in rat hearts. *Circulation* **1999**, *100*(22), 2260–2266.
158. Wang, Y., Kodani, E., Wang, J., Zhang, S.X., Takano, H., Tang, X.L., Bolli, R. Cardioprotection during the final stage of the late phase of ischemic preconditioning is mediated by neuronal NO synthase in concert with cyclooxygenase-2. *Circ. Res.* **2004**, *95*(1), 84–91.

12

ATHEROSCLEROSIS: OXIDATION HYPOTHESIS

CHANDRAKALA ALUGANTI NARASIMHULU, DMITRY LITVINOV, XUETING JIANG, ZHAOHUI YANG, AND SAMPATH PARTHASARATHY

OVERVIEW

Peroxidation of polyunsaturated fatty acid (PUFA) containing lipids has been known for a long time. Numerous studies have documented that peroxidized lipids as well as products derived from their decomposition, particularly aldehydes, have deleterious biological properties. This concept has been exemplified in the study of atherosclerosis. A plethora of *in vitro* and animal studies, as well as human epidemiological and correlatory studies have supported the notion that oxidative processes may contribute to the disease process. Yet the negative outcome of human clinical trials with α-tocopherol and other antioxidants has convinced even staunch supporters of the hypothesis to take a step backward and reconsider reasons of their failure and suggest alternative approaches.

In this chapter, based on our recent studies, we point out that lipid peroxidation-derived aldehydes are readily oxidized to carboxylic acids, and many enzyme systems that are suggested to be involved in the oxidation of lipids themselves are capable of accelerating this conversion. Presence of antioxidants prevented such conversions suggesting that toxic aldehydes that are not only pro-atherogenic but also pro-inflammatory and could accumulate under such conditions. Considering the literature that human fatty streak foam cell lesions abound even early in life and that the antioxidant trials were conducted in adult clinical population late in life, it is likely that antioxidants could have interfered with the oxidative clearance of toxic aldehydes and exacerbated plaque vulnerability.

12.1 LIPID PEROXIDATION

Lipid peroxidation has been a topic of interest for decades. Countless studies have documented that the presence of PUFA in esterified lipids make them vulnerable for oxidative damage. Edible oil industry has long been aware of the sensitivity of plant-derived oils to oxidation and the plethora of side products that could be formed if PUFA containing oils are left to become rancid. Elegant chemistry has elucidated the position, stereochemistry, and the mechanisms of oxidation as well as the products generated from almost all PUFAs that are commonly present in cooking oils, cell membranes, and lipoproteins. Innumerable oxidation systems, from exposure to air and ozone to complex enzyme systems have been described in the literature.[1-11] Basically, they can be divided into two groups: those that generate oxidants such as oxygen free radicals and those which directly act on the fatty acids. NADPH oxidase and xanthine oxidase (XAO) are examples of the former, while lipoxygenases represent the latter. Oxidation of lipids by the former is nonselective and could cause oxidation at random double bond positions depending on how the lipids are presented. True lipid oxygenases are highly specific enzymes that catalyze the oxygenation in a site and stereospecific manner. There are also peroxidases, for example, myeloperoxidase (MPO), which utilize hydrogen peroxide and generate products that are capable of oxidizing a variety of biological molecules, including lipids. In general, the type of fatty acid, its esterification status and to which molecule it is esterified to, oxygen tension, presence/absence

Molecular Basis of Oxidative Stress: Chemistry, Mechanisms, and Disease Pathogenesis, First Edition. Edited by Frederick A. Villamena.
© 2013 John Wiley & Sons, Inc. Published 2013 by John Wiley & Sons, Inc.

of redox metals, the nature and specificity of the enzymes, and the composition of the surrounding milieu could influence the oxidation of PUFA. While the vulnerability of PUFA is extensively discussed about, monounsaturated fatty acids as well as the carboxylic group itself can undergo oxidation. In addition, oxidation also affects other non-fatty acid-containing lipids such as cholesterol.

The realization that lipid peroxides are relatively unstable and gave rise to secondary decomposition products led to the identification of a variety of end products. These include carbonyl compounds such as aldehydes and ketones, hydrocarbons, epoxides, alcohols, carboxylic acids, and several types of polymerized products. In addition, the chemical reactivity of some of these products led to their interaction with many different types of biological molecules, for example, amine- and thiol-containing macromolecules with the formation of novel adducts. Of these, the aldehydes are considered biologically important.

For reasons that are not obvious, it was always felt that oxidative stress is harmful and that products of oxidative stress are etiological in many disease processes. The etiology of almost every major human pathological condition has been linked to oxidative stress. In addition, many beneficial cultural, social, and nutritional trends around the world have been attributed to antioxidative aspects. For example, aspects of Mediterranean diet, yoga, vegetarianism, curry powder, red wine and chocolate consumption, physical activity, zinc, selenium or lack of it in the soil, and so on (Fig. 12.1) have been attributed to their abilities to affect oxidative stress and thereby influence the disease process. Likewise, there are constant attempts to link proven therapies and lifestyle modalities to their potential antioxidant effects. Exercise, statins,[12,13] many antihypertensive drugs, drugs that contain a phenolic hydroxyl group or thiol function,[14-29] and so on have been suggested to have antioxidant effects. On the contrary, exposure to radiation, environmental pollutants and toxins, exposure to carcinogenic chemicals, exposure to viruses, bacteria, or other pathogens are suggested to be oxidative, and their physiological effects have been suggested to be linked to such stress. One would expect from the long list of antioxidant compounds that we consume that the human race would be free of chronic diseases and would have a very long life expectancy.

12.2 OXIDATION HYPOTHESIS OF ATHEROSCLEROSIS

Atherosclerosis is the major manifestation of cardiovascular diseases (CVDs) and is also one of the major risk factors for heart failure. For a long time, the disease was considered as a natural consequence of the aging process. However, recent evidence indicates that young adults and even neonates may have substantial atherosclerosis,[30-32] suggesting that the aging process itself might promote clinically relevant effects of atherosclerosis such as calcification, inflammation, thrombosis, and poor heart function.[33-35] Diabetes,[36-46] hypertension,[47-51] smoking,[52-58] environmental pollutants,[59] lack of physical activity/obesity,[60] food rich in cholesterol leading to hypercholesterolemia, consumption of fried food, certain chemical pollutants, and so on (Fig. 12.2) are suggested to increase oxidative stress and perhaps exacerbate the progression of the early atherosclerotic lesions to advanced lesions.

The role of lipoproteins in atherosclerosis development has been a topic of interest for several decades. It is now established beyond doubt that high levels of low-density lipoprotein (LDL) and low levels of high-density lipoprotein (HDL) contribute significantly to the development and progression of cardiovascular diseases.[33, 61-65] While the former might promote progression, the latter might be intricately involved in not only preventing the progression but also in promoting regression.

12.2.1 The Oxidized LDL (Ox-LDL)

The biochemical processes that contribute to the formation of early atherosclerotic lesions, the fatty streak lesions, are still under debate. The LDL oxidation hypothesis was put forward in the eighties to explain the formation of fatty streak lesions.[34,35,66-68] Countless reviews and over 5000 articles have appeared on the

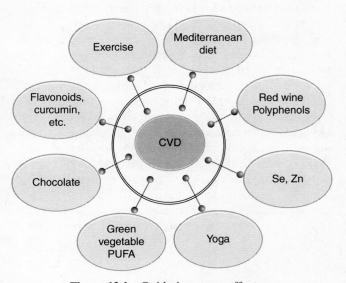

Figure 12.1 Oxidative stress affectors.

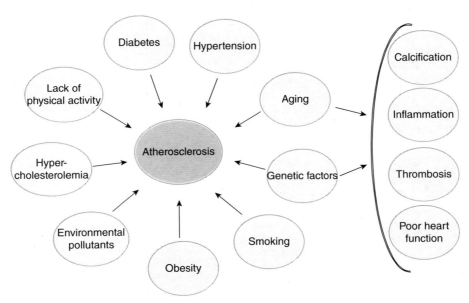

Figure 12.2 Atherosclerosis-causing agents.

TABLE 12.1 Biologically Relevant, Lipid-Peroxidation-Derived Aldehydes

Aldehyde	References
Hexanal	5, 6
2-Hydroxyhexanal	5, 6
4-Hydroxynonenal	75–85
Malondialdehyde	69–77
Oxo-valeric acid	86
Oxo-nonanoic acid	86
Acrolein	87–90

topic to date providing evidence for the involvement of oxidative processes in animal and human atherosclerotic disease.

The oxidation of LDL is a complex process during which both the protein and lipids undergo oxidative changes and form complex products. The peroxidized PUFA containing lipids decompose, generating both free and core aldehydes that covalently modify ε-amino groups of lysine residues of the protein moiety. The latter not only generates Schiff's bases, thus modifying charges on the amino acids, but also results in both intra- and intermolecular cross-links between proteolyzed apo B. Biologically relevant aldehydes formed during the oxidation of LDL are described in Table 12.1.[69–90] The take-home message from these studies was that the modification of apoprotein B might be the key determinant of lipid uptake by macrophages. However, there is also evidence in the literature for the oxidized lipid-mediated uptake of oxidized LDL.[91–94] If so, oxidized lipoproteins other than oxidized LDL as well as cells and membranes that contain similar oxidized lipids might also contribute to lipid accumulation in macrophages.

The oxidation of HDL also has been noted to affect its biological properties. For example, oxidized HDL has been noted to be a poor promoter of reverse cholesterol transport.[95] Despite enormous earlier interest on the topic, it is still not clear whether oxidation of the apoprotein alone is sufficient to influence cellular processes, although it is claimed that chlorination of tyrosines or oxidation of specific methionine residues of apoproteins could render HDL dysfunctional.

12.3 ANIMAL MODELS OF ATHEROSCLEROSIS

There are innumerable animal models of atherosclerosis.[96–101] The necessity to study the disease in a short and meaningful period of time has created numerous models that have shrunk the atherogenic process to a mere few weeks. All these models depend on the increases in plasma cholesterol levels beyond what one would see in a human being after feeding a diet rich in saturated fat and cholesterol. These models were developed with the sole purpose of studying the contribution of cholesterol carrying lipoproteins or other associated processes to the formation of macrophage-rich fatty streak lesions and have performed remarkably well to this end. These models also permit studying the contribution of individual risk factors and associated biochemical events. They were also developed for studying the molecular aspects of the development of atherosclerosis and their prevention by pharmacological agents in a reasonably short period of time. As a result, they represent "accelerated atherosclerosis" with end point of histopathologically established fatty streak lesions. What happens afterward as the disease progresses is overwhelmingly ignored. These short-term animal models of atherosclerosis have performed well to establish and test the oxidation hypothesis. The model provided evidence of lipid peroxidation and aldehyde accumulation, evidence for the presence of aldehyde–protein adducts, evidence for the presence of core-aldehydes and, above all, evidence for the attenuation of atherosclerosis by a number of antioxidants.

What is interesting in these models is that the atherogenic diet contains far less unsaturated fat than what might be required to generate quantities of aldehydes comparable to what is seen with human LDL oxidized *in vitro*. This might suggest that the degree of oxidation seen *in vitro* is unrealistic or additional changes that accompany during oxidation might contribute a lot more significantly than originally believed. Conversely, aldehydes might be constantly released by the cells from the digested lipoprotein due to lysosomal acidification to promote additional modification of extracellular lipoproteins (Fig. 12.3) and that even small amounts of aldehydes generated could be "recycled." Considering the findings that increasing the amount of PUFA or increased unsaturation, as in fish oil-derived PUFA, in the diet could actually decrease atherosclerosis in experimental animals.[102,103] one might suggest there is a threshold beyond which PUFA might have additional beneficial effects. It is also possible that in highly unsaturated PUFA, the aldehydes generated are less lipophilic and are eliminated easily as compared to lipophilic aldehydes.

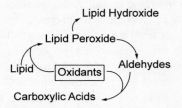

Figure 12.3 Aldehydes released from the digested lipoprotein due to lysosomal acidification.

12.3.1 Human Atherosclerosis and Animal Models

The human form of atherosclerotic disease differs considerably from the disease models in animals. As mentioned before, the animal models are predominantly set up to develop and study the very early fatty streak lesions with macrophage foam cells. Such lesions start very early in life in humans (even in infants and children) below the age of 15.[30,104–108] Maternal hypercholesterolemia seems to strongly influence the development of atherosclerotic lesions in the young.[105–107] Interestingly, while plasma LDL levels positively correlate to the severity of coronary artery disease,[31,108,109] infants and children have much lower LDL levels as compared to adults and the elderly population, suggesting that the correlation between age and plasma LDL levels could be a statistical coincidence rather than one of clinical significance except when specific subfractions with increased atherogenic potency such as oxidized low density lipoprotein (Ox-LDL) or small, dense LDL account for the increase.[110–112] Moreover, paraoxonase 1 (PON 1), a key enzyme that is implicated in the detoxification of lipid peroxides, is nearly absent in early human life.[113–117] However, the presence of fatty streak atherosclerotic lesions alone in the absence of significant other risk factors might be clinically inconsequential especially considering lower plasma lipid levels and compounding risk factors in children as compared to adults. Also, quantitatively, adults may have greater lesion surface area as compared to children, in addition to differences in the type of lesions. Acquisition of additional risk factors might significantly affect the progression or regression of the early childhood lesions.

12.3.2 Progression of Human Disease Calcification

The progression of human atherosclerotic lesions has been well established by pathologists.[118–121] They unanimously conclude that the macrophage-lipid-rich early fatty streak lesions progress to extracellular lipid-rich raised complex lesions later in adult life with or without calcification. Atherosclerotic calcification begins as early as the second decade of life, just after fatty streak formation. The lesions of younger adults have revealed small aggregates of crystalline calcium among the lipid particles of lipid cores. Calcium deposits are found more frequently and in greater amounts in elderly individuals and in more advanced lesions. In most advanced lesions, when calcification dominates, components such as lipid deposits and increased fibrous tissue may also be present. The extent of calcification also has been correlated with plaque burden.[118] It has been demonstrated that coronary calcification is detected in the vast majority of patients with a first myocardial infarction. Age-related calcium score is highly predictive for myocardial infarction event risk in prospective studies, and the association between coronary calcium and coronary heart disease (CHD) events usually remains significant after adjusting for other CHD risk factors. These study results support the concept that coronary calcium is associated with a relatively profound independent increased risk of CHD events in women and men. However, there has been a great deal of debate about the role of calcification in plaque rupture.[122–127]

Calcification has been noted as both intra- and extracellular deposits,[125] with the latter predominantly associated with extracellular lipids. Many researchers believe that coronary arterial calcification may represent an attempt to protect the weakened atherosclerotic plaque prone to rupture.[125,127] Calcified lesions and fibrotic lesions are much stiffer than cellular lesions and are unlikely to be associated with sites of plaque rupture.[125,128] In fact, a recent study showed that less than 15% of the ruptured human lesions showed evidence of associated calcification.[127] It is speculated that the plaque rupture often occurs at the interface between a calcified and noncalcified atherosclerotic areas of the lesion. Thus, calcification could be seen as a potential stabilizing force that may increase the biochemical stability of the plaque by imparting rigidity while at the same time decreasing the plaque's mechanical stability. The mechanism involved in atherosclerotic calcification or the nature of its deposition has not yet been determined. It is generally assumed that calcium phosphate precipitates in diseased coronary arteries by a mechanism similar to that found in active bone formation.[122,128–134] Some of the studies suggested the involvement of specific sulfated proteoglycans in the deposition of calcium in lesion areas.[135–138] Calcium is generally identified by van kossa stain or by computerized tomography (methods that detect calcium and not the molecule to which it is attached to). Currently, there is no hypothesis that would explain the association of calcium with lipid-rich domains.

12.3.3 Inflammation and Atherosclerosis

Yet another major difference between human and animal atherosclerosis is the contribution of inflammatory cytokines and matrix digesting enzymes (matrix metalloprotease [MMP]s) to the advancement of the vulnerable plaque.[139–145] Most of these cytokines and MMPs are induced by oxidative stress[146–171] suggesting that ongoing oxidative processes might be involved in the generation of the vulnerable plaque, inflammatory cytokines and matrix disruption, release of lytic lipids such as lyso phosphatidylcholine, aldehydes, release of reactive oxygen species, and proteolytic enzymes. It is interesting to note that while antioxidants have not been successful in preventing human cardiovascular complications, they have been reported to reduce the level of inflammatory cytokines.[169–171] Thus there is yet another paradox questioning whether the inflammation per se is responsible for cardiovascular outcome or is yet another manifestation of the disease process.

12.4 ALDEHYDE GENERATION FROM PEROXIDIZED LIPIDS

The breakdown of peroxidized lipids into aldehydes has been well documented.[69,172–174] The unsaturated aldehydes produced from these reactions have been implicated in modification of cellular proteins and other materials. The mechanism(s) by which aldehydes are generated from peroxidized lipids are still unclear. Both radical dependent and independent mechanisms have been suggested (Table 12.2).[5,6,173,174] Based on simple calculations, we predicted a glut of aldehyde products during the oxidative decomposition of LDL. Aldehydes are extremely oxidation-labile and are readily converted into carboxylic acids. Based on this, we considered the possibility that the oxidation hypothesis did not extend "far" enough to include the decomposition of peroxidized lipids.

12.4.1 The Oxidation of Aldehydes to Carboxylic Acids

While 4-hydroxynonenal (4-HNE) attracted considerable attention, the other half of the decomposition product of oxidized linoleic acid (oxo-valeric acid, oxo-nonanoic acid and other oxo-acids and core aldehydes) has attracted little attention.[175] The core aldehydes (which are esterified to lipids such as cholesterol or lysophospholipids) are seen only as markers of oxidation or as antigenic lipids or as substrates for enzymes such as platelet activating factor acetyl hydrolase (PAF-acetyl hydrolase) or phospholipases. One such product

TABLE 12.2 Potential Enzymes and Oxidants Involved in the Oxidation of Lipid Peroxide-Derived Aldehydes

Type of Enzyme and Oxidant	References
1. Increased oxygenation	2–7
2. Oxygen radicals	5,6
3. Peroxides and peracids	2–4,7
4. Peroxynitrite	5,6
5. Copper and other metal oxides	4,6
6. Xanthine oxidase, MPO, and other oxidases	2,3,4,7
7. Many flavin-containing enzymes	5,6

(oxo-nonanoic acid) that is derived from the oxidation of linoleic acid is of great biological significance as it gets readily oxidized to nonane dioic acid (azelaic acid [AZA]). AZA is a lipophilic dicarboxylic acid as opposed to short-chain dicarboxylic acids such as malonic acid, despite the reactive methylene group reactivity of the latter. Therefore, it would be expected to be formed and to accumulate in greater amounts in lipid-rich domains.

12.4.2 Proatherogenic Effects of Aldehydes

The proatherogenic effects of aldehyde products have received relatively little attention as compared to the oxidized lipids themselves. The initial burst of activities centered on the modification process per se, and the results seemed to indicate that small molecular water-soluble aldehydes such as malondialdehyde (MDA) readily diffused from the lipoprotein particle.[176] Pioneering studies by Esterbauer and coworkers brought attention to more reactive and lipophilic aldehydes such as 4-HNE into focus.[177–180] Besides being capable of protein modification and cross-linking, 4-HNE is now recognized to affect the atherogenic process by a number of different ways. Aldehydes such as 4-HNE might cause enzyme inactivation, apoptosis, modify proteins associated with cell signaling, and induce proinflammatory cytokines (Table 12.3).[177–188]

12.4.3 AZA: A Lipid Peroxidation-Derived Lipophilic Dicarboxylic Acid

AZA would have two important antiatherogenic properties. One, as other dicarboxylic acids, it would have great affinity for calcium and precipitate calcium in lipid-rich domains.[189–193] This might account for the unusually high calcification associated with lipid core domains. Second, AZA has been noted to have anti-inflammatory properties and has been noted to decrease cytokine production by macrophages and inflammatory cells.[194–196] It also has been noted to reduce oxidant production by leukocytes.[194] Thus, it has been added as an active ingredient in topical skin medications, particularly as anti-acne agent.[195–197]

The oxidation of oxo-nonanoic acid to AZA is a simple oxidation step and is inhibited by antioxidants[198] analogous to the oxidation of other aldehydes.[199–201] The oxidation of aldehydes into acids is catalyzed by both enzymatic (aldehydes oxidase, aldehyde dehydrogenase, MPO, xanthine oxidase) as well as by nonenzymatic oxidation.

12.4.4 Could Antioxidants Inhibit the Conversion of Aldehydes to Carboxylic Acids?

Thus, in human secondary prevention trials with antioxidants, there is a distinct possibility that the high levels of antioxidants could have prevented the generation of antiatherogenic and anti-inflammatory dicarboxylic acids. Such dicarboxylic acids also might be involved in calcification process, and the inhibition of their formation by antioxidants would be conducive to the formation of noncalcified vulnerable plaque (Fig. 12.4). Until the specific nature of calcium deposits are identified in human lesions, one could only speculate the role of such dicarboxylic acids in calcification. Either they can complex calcium and thus could represent calcified lesions, or they could destabilize the calcium phosphate/calcium extracellular proteoglycan complex, thus acting against calcification. It remains to be seen whether agents that would prevent the decomposition of lipid peroxides into long-chain aldehydes or enhance the conversion of aldehydes into dicarboxylic acids would have a better cardiovascular outcome (Table 12.4). The carboxylic acids, including dicarboxylic acids, are readily oxidized in the mitochondria and peroxisomes and thus could be readily eliminated. Molecules such as apo A1/HDL or enzymes such as paraoxonases or glutathione peroxidase might be inherently anti-atherogenic by not only promoting the conversion of hydroperoxides into hydroxides as well as by complexing with the aldehyde products but also by removing the hydroxides from the artery for elimination[114–117] as suggested by Fogelman and associates.

SUMMARY

Oxidative stress in general and lipid peroxidation, in particular, has been implicated in CVD. While most of the *in vitro* and animal studies support the involvement of such processes, the absence of clinically beneficial effects of antioxidants in humans has been taken as

TABLE 12.3 Proatherogenic Effects of Aldehydes

Effects	References
1. Modification of apoproteins leading to macrophage recognition and foam cells formation	69–72,80
2. Modification of apoproteins leading to immunogenicity and antibody formation	73,74
3. Induction of chemotaxis and adhesion-related proteins	75,77,85,179,171
4. Cytotoxicity and apoptosis	82,84,187
5. Initiation of oxygen radical formation	87
6. Enzyme inactivation	70,75

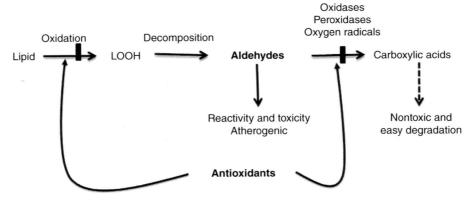

Figure 12.4 Prevention of the generation of antiatherogenic and anti-inflammatory dicarboxylic acids by antioxidants.

TABLE 12.4 Potential Carboxylic Acid Products of Lipid Peroxidation

Type of Carboxylic Acids	Compounds
Monocarboxylic acids	Hydroxy fatty acids, hydroperoxy fatty acids, epoxy fatty acids, keto acids, hexanoic acid, 2-hydroxy hexanoic acid, 4-hydroxynonenoic acid
Dicarboxylic acids	Malonic acid, azelaic acid

evidence against the oxidative theory of CVD. In this review, we point out that antioxidants could also inhibit the oxidation of toxic lipid peroxidation-derived aldehydes and could prevent the generation of potentially plaque stabilizing anti-inflammatory molecules.

ACKNOWLEDGMENT

This work was supported by NIH grants HL69038 and DK056353.

REFERENCES

1. Huo, Y., Zhao, L., Hyman, M.C., Shashkin, P., Harry, B.L., Burcin, T., Forlow, S.B., Stark, M.A., Smith, D.F., Clarke, S., Srinivasan, S., Hedrick, C.C., Pratico, D., Witztum, J.L., Nadler, J.L., Funk, C.D., Ley, K. Critical role of macrophage 12/15-lipoxygenase for atherosclerosis in apolipoprotein E-deficient mice. *Circulation* **2004**, *110*, 2024–2031.
2. Brennan, M.L., Anderson, M.M., Shih, D.M., Qu, X.D., Wang, X., Mehta, A.C., Lim, L.L., Shi, W., Hazen, S.L., Jacob, J.S., Crowley, J.R., Heinecke, J.W., Lusis, A.J. Increased atherosclerosis in myeloperoxidase-deficient mice. *J. Clin. Invest.* **2001**, *107*, 419–430.
3. McMillen, T.S., Heinecke, J.W., LeBoeuf, R.C. Expression of human myeloperoxidase by macrophages promotes atherosclerosis in mice. *Circulation* **2005**, *111*, 2798–2804.
4. Tribble, D.L., Gong, E.L., Leeuwenburgh, C., Heinecke, J.W., Carlson, E.L., Verstuyft, J.G., Epstein, C.J. Fatty streak formation in fat-fed mice expressing human copper-zinc superoxide dismutase. *Arterioscler. Thromb. Vasc. Biol.* **1997**, *17*, 1734–1740.
5. Gieseg, S., Duggan, S., Gebicki, J.M. Peroxidation of proteins before lipids in U937 cells exposed to peroxyl radicals. *Biochem. J.* **2000**, *350*, 215–218.
6. Patel, R., Diczfalusy, U., Dzeletovic, S., Wilson, M., Darley-Usmar, V. Formation of oxysterols during oxidation of low density lipoprotein by peroxynitrite, myoglobin, and copper. *J. Lipid Res.* **1996**, *37*, 2361–2371.
7. Santanam, N., Parthasarathy, S. Paradoxical actions of antioxidants in the oxidation of low density lipoprotein by peroxidases. *J. Clin. Invest.* **1995**, *95*, 2594–2600.
8. Sparrow, C.P., Olszewski, J. Cellular oxidation of low density lipoprotein is caused by thiol production in media containing transition metal ions. *J. Lipid Res.* **1993**, *34*, 219–228.
9. Parthasarathy, S. Oxidation of low-density lipoprotein by thiol compounds leads to its recognition by the acetyl LDL receptor. *Biochim. Biophys. Acta* **1987**, *917*, 337–340.
10. Lamb, D.J., Leake, D.S. Iron released from transferrin at acidic pH can catalyse the oxidation of low density lipoprotein. *FEBS Lett.* **1994**, *352*, 15–18.
11. Lamb, D.J., Hider, R.C., Leake, D.S. Hydroxypyridinones and desferrioxamine inhibit macrophage-mediated LDL oxidation by iron but not by copper. *Biochem. Soc. Trans.* **1993**, *21*, 234S.
12. Rikitake, Y., Kawashima, S., Takeshita, S., Yamashita, T., Azumi, H., Yasuhara, M., Nishi, H., Inoue, N., Yokoyama, M. Anti-oxidative properties of fluvastatin, an HMG-CoA reductase inhibitor, contribute to prevention of atherosclerosis in cholesterol-fed rabbits. *Atherosclerosis* **2001**, *154*, 87–96.

13. Stoll, L.L., McCormick, M.L., Denning, G.M., Weintraub, N.L. Antioxidant effects of statins. *Drugs Today (Barc.)* **2004**, *40*, 975–990.

14. Cominacini, L., Fratta Pasini, A., Garbin, U., Pastorino, A.M., Davoli, A., Nava, C., Campagnola, M., Rossato, P., Lo Cascio, V. Antioxidant activity of different dihydropyridines. *Biochem. Biophys. Res. Commun.* **2003**, *302*, 679–684.

15. Inouye, M., Mio, T., Sumino, K. Nilvadipine protects low-density lipoprotein cholesterol from in vivo oxidation in hypertensive patients with risk factors for atherosclerosis. *Eur. J. Clin. Pharmacol.* **2000**, *5*, 35–41.

16. Kritz, H., Oguogho, A., Aghajanian, A.A., Sinzinger, H. Semotiadil, a new calcium antagonist, is a very potent inhibitor of LDL-oxidation. *Prostaglandins Leukot. Essent. Fatty Acids* **1999**, *61*, 183–188.

17. Napoli, C., Salomone, S., Godfraind, T., Palinski, W., Capuzzi, D.M., Palumbo, G., D'Armiento, F.P., Donzelli, R., de Nigris, F., Capizzi, R.L., Mancini, M., Gonnella, J.S., Bianchi, A. 1,4-Dihydropyridine calcium channel blockers inhibit plasma and LDL oxidation and formation of oxidation-specific epitopes in the arterial wall and prolong survival in stroke-prone spontaneously hypertensive rats. *Stroke* **1999**, *30*, 1907–1915.

18. Lupo, E., Locher, R., Weisser, B., Vetter, W. In vitro antioxidant activity of calcium antagonists against LDL oxidation compared with alpha-tocopherol. *Biochem. Biophys. Res. Commun.* **1994**, *203*, 1803–1808.

19. Khan, B.V., Navalkar, S., Khan, Q.A., Rahman, S.T., Parthasarathy, S. Irbesartan, an angiotensin type 1 receptor inhibitor, regulates the vascular oxidative state in patients with coronary artery disease. *J. Am. Coll. Cardiol.* **2001**, *38*, 1662–1667.

20. Godfrey, E.G., Stewart, J., Dargie, H.J., Reid, J.L., Dominiczak, M., Hamilton, C.A., McMurray, J. Effects of ACE inhibitors on oxidation of human low density lipoprotein. *Br. J. Clin. Pharmacol.* **1994**, *37*, 63–66.

21. Ziedén, B., Wuttge, D.M., Karlberg, B.E., Olsson, A.G. Effects of in vitro addition of captopril on copper-induced low density lipoprotein oxidation. *Br. J. Clin. Pharmacol.* **1995**, *39*, 201–203.

22. Hayek, T., Attias, J., Smith, J., Breslow, J.L., Keidar, S. Antiatherosclerotic and antioxidative effects of captopril in apolipoprotein E-deficient mice. *J. Cardiovasc. Pharmacol.* **1998**, *31*, 540–544.

23. Van Antwerpen, P., Boudjeltia, K.Z., Babar, S., Legssyer, I., Moreau, P., Moguilevsky, N., Vanhaeverbeek, M., Ducobu, J., Nève, J. Thiol-containing molecules interact with the myeloperoxidase/H_2O_2/chloride system to inhibit LDL oxidation. *Biochem. Biophys. Res. Commun.* **2005**, *337*, 82–88.

24. Van Antwerpen, P., Legssyer, I., Zouaoui Boudjeltia, K., Babar, S., Moreau, P., Moguilevsky, N., Vanhaeverbeek, M., Ducobu, J., Nève, J. Captopril inhibits the oxidative modification of apolipoprotein B-100 caused by myeloperoxidase in a comparative in vitro assay of angiotensin converting enzyme inhibitors. *Eur. J. Pharmacol.* **2006**, *537*, 31–36.

25. Miura, T., Muraoka, S., Ogiso, T. Antioxidant activity of adrenergic agents derived from catechol. *Biochem. Pharmacol.* **1998**, *55*, 2001–2016.

26. Hogg, N., Struck, A., Goss, S.P., Santanam, N., Joseph, J., Parthasarathy, S., Kalyanaraman, B. Inhibition of macrophage-dependent low density lipoprotein oxidation by nitric-oxide donors. *J. Lipid Res.* **1995**, *36*, 1756–1762.

27. Hogg, N., Kalyanaraman, B., Joseph, J., Struck, A., Parthasarathy, S. Inhibition of low-density lipoprotein oxidation by nitric oxide, potential role in atherogenesis. *FEBS Lett.* **1993**, *334*, 170–174.

28. Hermann, M., Kapiotis, S., Hofbauer, R., Exner, M., Seelos, C., Held, I., Gmeiner, B. Salicylate inhibits LDL oxidation initiated by superoxide/nitric oxide radicals. *FEBS Lett.* **1999**, *445*, 212–214.

29. Whiteman, M., Kaur, H., Halliwell, B. Protection against peroxynitrite dependent tyrosine nitration and alpha 1-antiproteinase inactivation by some anti-inflammatory drugs and by the antibiotic tetracycline. *Ann. Rheum. Dis.* **1996**, *55*, 383–387.

30. Stary, H.C. Evolution and progression of atherosclerotic lesions in coronary arteries of children and young adults. *Arteriosclerosis* **1989**, *9*, 119–132.

31. Homma, S., Troxclair, D.A., Zieske, A.W., Malcom, G.T., Strong, J.P., The Pathobiological Determinants of Atherosclerosis in Youth (PDAY) Research Group. Histological topographical comparisons of atherosclerosis progression in juveniles and young adults. *Atherosclerosis* **2008**, *197*(2), 791–798.

32. McMahan, C.A., Gidding, S.S., Malcom, G.T., Schreiner, P.J., Strong, J.P., Tracy, R.E., Williams, O.D., McGill, H.C., Pathobiological Determinants of Atherosclerosis in Youth (PDAY) Research Group. Comparison of coronary heart disease risk factors in autopsied young adults from the PDAY Study with living young adults from the CARDIA study. *Cardiovasc. Pathol.* **2007**, *16*, 151–158.

33. Cromwell, W.C., Otvos, J.D. Low-density lipoprotein particle number and risk for cardiovascular disease. *Curr. Atheroscler. Rep.* **2004**, *6*, 381–387.

34. Steinberg, D., Parthasarathy, S., Crew, T.E., Khoo, J.C., Witztum, J.L. Beyond cholesterol: Modification of low-density lipoprotein that increase its atherogenecity. *N. Engl. J. Med.* **1989**, *320*, 915–924.

35. Parthasarathy, S. *Modified Lipoproteins in the Pathogenesis of Atherosclerosis*. R.G. Landes Co., Austin, TX, 1994.

36. Bemeur, C., Ste-Marie, L., Montgomery, J. Increased oxidative stress during hyperglycemic cerebral ischemia. *Neurochem. Int.* **2007**, *50*, 890–904.

37. Ceriello, A. Oxidative stress and diabetes-associated complications. *Endocr. Pract.* **2006**, *12*, S60–S62.

38. Rolo, A.P., Palmeira, C.M. Diabetes and mitochondrial function: Role of hyperglycemia and oxidative stress. *Toxicol. Appl. Pharmacol.* **2006**, *212*, 167–178.

39. Bonnefont-Rousselot, D. Glucose and reactive oxygen species. *Curr. Opin. Clin. Nutr. Metab. Care* **2002**, *5*, 561–568.
40. Hunt, J.V., Dean, R.T., Wolff, S.P. Hydroxyl radical production and autoxidative glycosylation. Glucose autoxidation as the cause of protein damage in the experimental glycation model of diabetes mellitus and ageing. *Biochem. J.* **1988**, *256*, 205–212.
41. Natarajan, R., Gerrity, R.G., Gu, J.L., Lanting, L., Thomas, L., Nadler, J.L. Role of 12-lipoxygenase and oxidant stress in hyperglycaemia-induced acceleration of atherosclerosis in a diabetic pig model. *Diabetologia* **2002**, *45*, 125–133.
42. Natarajan, R., Gu, J.L., Rossi, J., Gonzales, N., Lanting, L., Xu, L., Nadler, J. Elevated glucose and angiotensin II increase 12-lipoxygenase activity and expression in porcine aortic smooth muscle cells. *Proc. Natl. Acad. Sci. U.S.A.* **1993**, *90*, 4947–4951.
43. Masaki, H., Okano, Y., Sakurai, H. Generation of active oxygen species from advanced glycation end-products (AGE) under ultraviolet light A (UVA) irradiation. *Biochem. Biophys. Res. Commun.* **1997**, *235*, 306–310.
44. Schmidt, A.M., Hori, O., Brett, J., Yan, S.D., Wautier, J.L., Stern, D. Cellular receptors for advanced glycation end products. Implications for induction of oxidant stress and cellular dysfunction in the pathogenesis of vascular lesions. *Arterioscler. Thromb.* **1994**, *14*, 1521–1528.
45. Al-Abed, Y., Liebich, H., Voelter, W., Bucala, R. Hydroxyalkenal formation induced by advanced glycosylation of low density lipoprotein. *J. Biol. Chem.* **1996**, *271*, 2892–2896.
46. Bucala, R., Makita, Z., Koschinsky, T., Cerami, A., Vlassara, H. Lipid advanced glycosylation: Pathway for lipid oxidation in vivo. *Proc. Natl. Acad. Sci. U.S.A.* **1993**, *90*, 6434–6438.
47. Mehta, P.K., Griendling, K.K. Angiotensin II cell signaling: Physiological and pathological effects in the cardiovascular system. *Am. J. Physiol. Cell Physiol.* **2007**, *292*, C82–C97.
48. Ushio-Fukai, M., Alexander, R.W. Reactive oxygen species as mediators of angiogenesis signaling: Role of NAD(P)H oxidase. *Mol. Cell. Biochem.* **2004**, *264*, 85–97.
49. Ushio-Fukai, M. Redox signaling in angiogenesis: Role of NADPH oxidase. *Cardiovasc. Res.* **2006**, *71*, 226–235.
50. Touyz, R.M. Reactive oxygen species as mediators of calcium signaling by angiotensin II: Implications in vascular physiology and pathophysiology. *Antioxid. Redox Signal.* **2005**, *7*, 1302–1314.
51. Beckman, J.S., Beckman, T.W., Chen, J., Marshall, P.A., Freeman, B.A. Apparent hydroxyl radical production by peroxynitrite: Implications for endothelial injury from nitric oxide and superoxide. *Proc. Natl. Acad. Sci. U.S.A.* **1990**, *87*, 1620–1624.
52. Chalmers, A. Smoking and oxidative stress. *Am. J. Clin. Nutr.* **1999**, *69*, 572.
53. Cross, C.E., Van der Vliet, A., Eiserich, J.P. Cigarette smokers and oxidant stress: A continuing mystery. *Am. J. Clin. Nutr.* **1998**, *67*, 184–185.
54. Burke, A., Fitzgerald, G.A. Oxidative stress and smoking-induced vascular injury. *Prog. Cardiovasc. Dis.* **2003**, *46*, 79–90.
55. Yokode, M., Ueyama, K., Arai, N.H., Ueda, Y., Kita, T. Modification of high- and low-density lipoproteins by cigarette smoke oxidants. *Ann. N. Y. Acad. Sci.* **1996**, *15*, 245–251.
56. Yamaguchi, Y., Matsuno, S., Kagota, S., Haginaka, J., Kunitomo, M. Oxidants in cigarette smoke extract modify low-density lipoprotein in the plasma and facilitate atherogenesis in the aorta of Watanabe heritable hyperlipidemic rabbits. *Atherosclerosis* **2001**, *156*, 109–117.
57. Bloomer, R.J. Decreased blood antioxidant capacity and increased lipid peroxidation in young cigarette smokers compared to nonsmokers: Impact of dietary intake. *Nutr. J.* **2007**, *6*, 39.
58. Frei, B., Forte, T.M., Ames, B.N., Cross, C.E. Gas phase oxidants of cigarette smoke induce lipid peroxidation and changes in lipoprotein properties in human blood plasma. Protective effects of ascorbic acid. *Biochem. J.* **1991**, *277*, 133–138.
59. Bhatnagar, A. Cardiovascular pathophysiology of environmental pollutants. *Am. J. Physiol. Heart Circ. Physiol.* **2004**, *286*, H479–H485.
60. Bouloumie, A., Marumo, T., Lafontan, M., Busse, R. Leptin induces oxidative stress in human endothelial cells. *FASEB J.* **1999**, *13*, 1231–1238.
61. National Cholesterol Education Program. Third Report of the Expert Panel on Detection, Evaluation, and Treatment of High Blood Cholesterol in Adults (ATP III Final Report). National Heart, Lung, and Blood Institute, Bethesda, MD, 2002.
62. National Cholesterol Education Program. Adult Treatment Panel III, Update 2004: Implications of Recent Clinical Trials for the ATP III Guidelines. National Heart, Lung, and Blood Institute, Bethesda, MD, 2004.
63. Rosin, B.L. The progression of cardiovascular risk to cardiovascular disease. *Rev. Cardiovasc. Med.* **2007**, *8*, S3–S8.
64. Le, N.A., Walter, M.F. The role of hypertriglyceridemia in atherosclerosis. *Curr. Atheroscler. Rep.* **2007**, *9*, 110–115.
65. Steinberg, D., Lewis, A. Conner Memorial Lecture. Oxidative modification of LDL and atherogenesis. *Circulation* **1997**, *95*, 1062–1071.
66. Chisolm, G.M., Steinberg, D. The oxidative modification hypothesis of atherogenesis: An overview. *Free Radic. Biol. Med.* **2000**, *28*, 1815–1826.
67. Berliner, J. Introduction: Lipid oxidation products and atherosclerosis. *Vascul. Pharmacol.* **2002**, *38*, 187–191.
68. Matsuura, E., Kobayashi, K., Tabuchi, M., Lopez, L.R. Oxidative modification of low-density lipoprotein and immune regulation of atherosclerosis. *Prog. Lipid Res.* **2006**, *45*, 466–486.
69. Haberland, M.E., Fogelman, A.M., Edwards, P.A. Specificity of receptor-mediated recognition of malondialdehyde modified low density lipoproteins. *Proc. Natl. Acad. Sci. U.S.A.* **1982**, *79*, 1712–1716.

70. Haberland, M.E., Flessll, G.M., Scanullll, A.M., Fogelman, A.M. Malondialdehyde modification of lipoprotein (a) produces avid uptake by human monocyte-macrophages. *J. Biol. Chem.* **1992**, *267*, 4143–4151.

71. Fogelman, A.M., Shechter, I., Seager, J., Hokom, M., Child, J.S., Edwards, P.A. Malondialdehyde alteration of low density lipoproteins leads to cholesteryl ester accumulation in human monocyte-macrophages. *Proc. Natl. Acad. Sci. U.S.A.* **1980**, *77*, 2214–2218.

72. Shechter, I., Fogelman, A.M., Haberland, M.E., Seager, J., Hokom, M., Edwards, P.A. The metabolism of native and malondialdehyde altered low density lipoproteins by human monocyte-macrophages. *J. Lipid Res.* **1981**, *22*, 63–71.

73. Palinski, W., Ord, V., Plump, A.S., Breslow, J.L., Steinberg, D., Witztum, J.L. ApoE-deficient mice are a model of lipoprotein oxidation in atherogenesis: Demonstration of oxidation specific epitopes in lesions and high titers of autoantibodies to malondialdehyde-lysine in serum. *Arterioscler. Thromb.* **1994**, *14*, 605–616.

74. Palinski, W., Miller, E., Witztum, J.L. Immunization of LDL receptor-deficient rabbits with homologous malondialdehyde-modified LDL reduces atherogenesis. *Proc. Natl. Acad. Sci. U.S.A.* **1995**, *92*, 821–825.

75. Hill, G.E., Miller, J.A., Baxter, B.T., Klassen, L.W., Duryee, M.J., Tuma, D.J., Thiele, G.M. Association of malondialdehyde-acetaldehyde (MAA) adducted proteins with atherosclerotic-induced vascular inflammatory injury. *Atherosclerosis* **1998**, *141*, 107–116.

76. Kharbanda, K.K., Todero, S.L., Shubert, K.A., Sorrell, M.F., Tuma, D.J. Malondialdehyde-acetaldehyde-protein adducts increase secretion of chemokines by rat hepatic stellate cells. *Alcohol* **2001**, *25*, 123–128.

77. Wyatt, T.A., Kharbanda, K.K., Tuma, D.J., Sisson, J.H. Malondialdehyde-acetaldehyde-adducted bovine serum albumin activates protein kinase C and stimulates interleukin-8 release in bovine bronchial epithelial cells. *Alcohol* **2001**, *25*, 159–166.

78. Ramana, K.V., Bhatnagar, A., Srivastava, S., Yadav, U.C., Awasthi, S., Awasthi, Y.C., Srivastava, S.K. Mitogenic responses of vascular smooth muscle cells to lipid peroxidation-derived aldehyde 4-hydroxy-*trans*-2-nonenal (HNE), role of aldose reductase-catalyzed reduction of the HNE-glutathine conjugates in regulating cell growth. *J. Biol. Chem.* **2006**, *281*, 17652–17660.

79. Ruef, J., Rao, G.N., Li, F., Bode, C., Patterson, C., Bhatnagar, A., Runge, M.S. Induction of rat aortic smooth muscle cell growth by the lipid peroxidation product 4-hydroxy-2-nonenal. *Circulation* **1998**, *97*, 1071–1078.

80. Hoff, H.F., Cole, T.B. Macrophages uptake by low density lipoprotein modified by 4-hydroxynonenal an ultrastructure study. *Lab. Invest.* **1991**, *64*, 254–264.

81. Bounds, P.L., Winston, G.W. The reaction of xanthine oxidase with aldehydic products of lipid peroxidation. *Free Radic. Biol. Med.* **1991**, *11*, 447–453.

82. Curzio, M., Esterbauer, H., Poli, G., Biasi, F., Cecchini, G., Di Mauro, C., Cappello, N., Dianzani, M.U. Possible role of aldehydic lipid peroxidation products as chemoattractants. *Int. J. Tissue React.* **1987**, *9*, 295–306.

83. Curzio, M., Torrielli, M.V., Giroud, J.P., Esterbauer, H., Dianzani, M.U. Neutrophil chemotactic responses to aldehydes. *Res. Commun. Chem. Pathol. Pharmacol.* **1982**, *36*, 463–476.

84. Moldovan, N.I., Lupu, F., Moldovan, L., Simionescu, N. 4-Hydroxynonenal induces membrane perturbations and inhibition of basal prostacyclin production in endothelial cells, and migration of monocytes. *Cell Biol. Int.* **1994**, *18*, 985–992.

85. Minekura, H., Kumagai, T., Kawamoto, Y., Nara, F., Uchida, K. 4-Hydroxy-2-nonenal is a powerful endogenous inhibitor of endothelial response. *Biochem. Biophys. Res. Commun.* **2001**, *282*, 557–561.

86. Agarwal, A., Shiraishi, F., Visner, G.A., Nick, H.S. Linoleoyl hydroperoxide transcriptionally upregulates heme oxygenase-1 gene expression in human renal epithelial and aortic endothelial cells. *J. Am. Soc. Nephrol.* **1998**, *9*, 1990–1997.

87. Adams, J.D., Jr., Klaidman, L.K. Acrolein-induced oxygen radical formation. *Free Radic. Biol. Med.* **1993**, *15*, 187–193.

88. Lee, W.K., Ramanathan, M., Jr., Spannhake, E.W., Lane, A.P. The cigarette smoke component acrolein inhibits expression of the innate immune components IL-8 and human beta-defensin 2 by sinonasal epithelial cells. *Am. J. Rhinol.* **2007**, *21*, 658–663.

89. Valacchi, G., Pagnin, E., Phung, A., Nardini, M., Schock, B.C., Cross, C.E., van der Vliet, A. Inhibition of NFkappaB activation and IL-8 expression in human bronchial epithelial cells by acrolein. *Antioxid. Redox Signal.* **2005**, *7*, 25–31.

90. Kasahara, D.I., Poynter, M.E., Othman, Z., Hemenway, D., van der Vliet, A. Acrolein inhalation suppresses lipopolysaccharide-induced inflammatory cytokine production but does not affect acute airways neutrophilia. *J. Immunol.* **2008**, *181*, 736–745.

91. Henriksen, T., Mahoney, E.M., Steinberg, D. Enhanced macrophage degradation of biologically modified low density lipoprotein. *Arteriosclerosis* **1983**, *3*, 149–159.

92. Steinbrecher, U.P., Parthasarathy, S., Leake, D.S., Witztum, J.L., Steinberg, D. Modification of low density lipoprotein by endothelial cells involves lipid peroxidation and degradation of low density lipoprotein phospholipids. *Proc. Natl. Acad. Sci. U.S.A.* **1984**, *81*, 3883–3887.

93. Hessler, J.R., Morel, D.W., Lewis, L.J., Chisolm, G.M. Lipoprotein oxidation and lipoprotein-induced cytotoxicity. *Arteriosclerosis* **1983**, *3*, 215–222.

94. Morel, D.W., DiCorleto, P.E., Chisolm, G.M. Endothelial and smooth muscle cells alter low density lipoprotein in vitro by free radical oxidation. *Arteriosclerosis* **1984**, *4*, 357–364.

95. Bergt, C., Pennathur, S., Fu, X., Byun, J., O'Brien, K., McDonanld, T.O., Singh, P., Anantharamaiah, G.M., Chait, A., Brunzell, J., Geary, R.L., Oram, J.F., Heinecke, J.W. The myeloperoxidase product hypochlorous acid oxidizes HDL in the human artery wall and impairs

ABCA1-dependent cholesterol transport. *Proc. Natl. Acad. Sci. U.S.A.* **2004**, *101*, 13032–13037.

96. Harats, D., Shaish, A., George, J., Mulkins, M., Kurihara, H., Levkovitz, H., Sigal, E. Overexpression of 15-lipoxygenase in vascular endothelium accelerates early atherosclerosis in LDL receptor-deficient mice. *Arterioscler. Thromb. Vasc. Biol.* **2000**, *20*, 2100–2105.

97. Palinski, W., Hörkko, S., Miller, E., Steinbrecher, U.P., Powell, H.C., Curtiss, L.K., Witztum, J.L. Cloning of monoclonal autoantibodies to epitopes of oxidized lipoproteins from apolipoprotein E-deficient mice: Demonstration of epitopes of oxidized low density lipoprotein in human plasma. *J. Clin. Invest.* **1996**, *98*, 800–814.

98. Kinscherf, R., Deigner, H.P., Usinger, C., Pill, J., Wagner, M., Kamencic, H., Hou, D., Chen, M., Schmiedt, W., Schrader, M., Kovacs, G., Kato, K., Metz, J. Induction of mitochondrial manganese superoxide dismutase in macrophages by oxidized LDL: Its relevance in atherosclerosis of humans and heritable hyperlipidemic rabbits. *FASEB J.* **1997**, *11*, 1317–1328.

99. Rudel, L.L., Parks, J.S., Sawyer, J.K. Compared with dietary monounsaturated and saturated fat, polyunsaturated fat protects African green monkeys from coronary artery atherosclerosis. *Arterioscler. Thromb. Vasc. Biol.* **1995**, *15*, 2101–2110.

100. Rudel, L.L., Johnson, F.L., Sawyer, J.K., Wilson, M.S., Parks, J.S. Dietary polyunsaturated fat modifies low-density lipoproteins and reduces atherosclerosis of nonhuman primates with high and low diet responsiveness. *Am. J. Clin. Nutr.* **1995**, *62*, S463–S470.

101. Wolfe, M.S., Sawyer, J.K., Morgan, T.M., Bullock, B.C., Rudel, L.L. Dietary polyunsaturated fat decreases coronary artery atherosclerosis in a pediatric-aged population of African green monkeys. *Arterioscler. Thromb.* **1994**, *14*, 587–597.

102. Lada, A.T., Rudel, L.L. Dietary monounsaturated versus polyunsaturated fatty acids: Which is really better for protection from coronary heart disease? *Curr. Opin. Lipidol.* **2003**, *14*, 41–46.

103. Rudel, L.L., Kelley, K., Sawyer, J.K., Shah, R., Wilson, M.D. Dietary monounsaturated fatty acids promote aortic atherosclerosis in LDL receptor-null, human ApoB100-overexpressing transgenic mice. *Arterioscler. Thromb. Vasc. Biol.* **1998**, *18*, 1818–1827.

104. Zieske, A.W., Malcom, G.T., Strong, J.P. Natural history and risk factors of atherosclerosis in children and youth: The PDAY study. *Pediatr. Pathol. Mol. Med.* **2002**, *21*, 213–237.

105. Palinski, W., Napoli, C. The fetal origins of atherosclerosis: Maternal hypercholesterolemia and cholesterol-lowering or antioxidant treatment during pregnancy influence in utero programming and postnatal susceptibility to atherogenesis. *FASEB J.* **2002**, *16*, 1348–1360.

106. Napoli, C., Glass, C.K., Witztum, J.L., Deutsch, R., D'Armiento, F.P., Palinski, W. Influence of maternal hypercholesterolaemia during pregnancy on progression of early atherosclerotic lesions in childhood: Fate of Early Lesions in Children (FELIC) study. *Lancet* **1999**, *354*, 1234–1241.

107. Napoli, C., D'Armiento, F.P., Mancini, F.P., Postiglione, A., Witztum, J.L., Palumbo, G., Palinski, W. Fatty streak formation occurs in human fetal aortas and is greatly enhanced by maternal hypercholesterolemia. Intimal accumulation of low density lipoprotein and its oxidation precede monocyte recruitment into early atherosclerotic lesions. *J. Clin. Invest.* **1997**, *100*, 2680–2690.

108. McGill, H.C., Jr., Herderick, E.E., McMahan, C.A., Zieske, A.W., Malcolm, G.T., Tracy, R.E., Strong, J.P. Atherosclerosis in youth. *Minerva Pediatr.* **2002**, *54*, 437–447.

109. Ross, R. Atherosclerosis—An inflammatory disease. *N. Engl. J. Med.* **1999**, *340*, 115–126.

110. Tribble, D.L., Holl, L.G., Wood, P.D., Krauss, R.M. Variations in oxidative susceptibility among six low density lipoprotein subfractions of differing density and particle size. *Atherosclerosis* **1992**, *93*, 189–199.

111. Chait, A., Brazg, R.L., Tribble, D.L., Krauss, R.M. Susceptibility of small, dense, low-density lipoproteins to oxidative modification in subjects with the atherogenic lipoprotein phenotype pattern B. *Am. J. Med.* **1993**, *94*, 350–356.

112. Tribble, D.L., Rizzo, M., Chait, A., Lewis, D.M., Blanche, P.J., Krauss, R.M. Enhanced oxidative susceptibility and reduced antioxidant content of metabolic precursors of small, dense low-density lipoproteins. *Am. J. Med.* **2001**, *110*, 103–110.

113. Kaplan, M., Aviram, M. Oxidized low density lipoprotein: Atherogenic and proinflammatory characteristics during macrophage foam cell formation. An inhibitory role for nutritional antioxidants and serum paraoxonase. *Clin. Chem. Lab. Med.* **1999**, *37*, 777–787.

114. Navab, M., Anantharamaiah, G.M., Reddy, S.T., Fogelman, A.M. Apolipoprotein A-I mimetic peptides and their role in atherosclerosis prevention. *Nat. Clin. Pract. Cardiovasc. Med.* **2006**, *3*, 540–547.

115. Ansell, B., Navab, M., Watson, K., Fonarow, G., Fogelman, A. Anti-inflammatory properties of HDL. *Rev. Endocr. Metab. Disord.* **2004**, *5*, 351–358.

116. Van Lenten, B., Navab, M., Shih, D., Fogelman, A., Lusis, A.J. The role of high-density lipoproteins in oxidation and inflammation. *Trends Cardiovasc. Med.* **2001**, *11*, 155–161.

117. Navab, M., Berliner, J.A., Watson, A.D., Hama, S.Y., Territo, M.C., Lusis, A.J., Shih, D.M., Van Lenten, B.J., Frank, J.S., Demer, L.L., Edwards, P.A., Fogelman, A.M. The Yin and Yang of oxidation in the development of the fatty streak: A review based on the 1994 George Lyman Duff Memorial Lecture. *Arterioscler. Thromb. Vasc. Biol.* **1996**, *16*, 831–842.

118. Stary, H., Chandler, A., Glagov, S., Guyton, J., Insull, W., Jr., Rosenfeld, M., Schaffer, S.A., Schwartz, C.J., Wagner, W.D., Wissler, R.W. A definition of initial, fatty streak, and intermediate lesions of atherosclerosis: A report from the Committee on Vascular Lesions of the Council

on Arteriosclerosis, American Heart Association. *Arterioscler. Thromb.* **1994**, *14*, 840–856.

119. Eggen, D.A., Strong, J.P., McGill, H.C. Coronary calcification: Relationship to clinically significant coronary lesions and race, sex and topographic distribution. *Circulation* **1965**, *32*, 948–955.

120. Virmani, R., Ladich, E.R., Burke, A.P., Kolodgie, F.D. Histopathology of carotid atherosclerotic disease. *Neurosurgery* **2006**, *59*, S219–S227.

121. Virmani, R., Burke, A.P., Farb, A., Kolodgie, F.D. Pathology of the vulnerable plaque. *J. Am. Coll. Cardiol.* **2006**, *47*, C13–C18.

122. Johnson, R.C., Leopold, J.A., Loscalzo, J. Vascular calcification: Pathobiological mechanisms and clinical implications. *Circ. Res.* **2006**, *99*, 1044–1059.

123. Mackey, R.H., Venkitachalam, L., Sutton-Tyrrell, K. Calcifications, arterial stiffness and atherosclerosis. *Adv. Cardiol.* **2007**, *44*, 234–244.

124. Mazzini, M.J., Schulze, P.C. Proatherogenic pathways leading to vascular calcification. *Eur. J. Radiol.* **2006**, *57*, 384–389.

125. Abedin, M., Tintut, Y., Demer, L.L. Vascular calcification: Mechanisms and clinical ramifications. *Arterioscler. Thromb. Vasc. Biol.* **2004**, *24*, 1161–1170.

126. Stary, H.C. Natural history of calcium deposits in atherosclerosis progression and regression. *Z. Kardiol.* **2000**, *89*, 28–35.

127. Vengrenyuk, Y., Carlier, S., Xanthos, S., Cardoso, L., Ganatos, P., Virmani, R., Einav, S., Gilchrist, L., Weinbaum, S. A hypothesis for vulnerable plaque rupture due to stress-induced debonding around cellular microcalcifications in thin fibrous caps. *Proc. Natl. Acad. Sci. U.S.A.* **2006**, *103*, 14678–1483.

128. Beadenkopf, W.G., Assaad, D.S., Love, B.M. Calcification in the coronary arteries and its relationship to arteriosclerosis and myocardial infarction. *JAMA* **1964**, *92*, 865–871.

129. Speer, M.Y., Giachelli, C.M. Regulation of cardiovascular calcification. *Cardiovasc. Pathol.* **2004**, *13*, 63–70.

130. Doherty, T.M., Detrano, R.C. Coronary arterial calcification as an active process: A new perspective on an old problem. *Calcif. Tissue Int.* **1994**, *54*, 224–230.

131. Watson, K.E. Pathophysiology of coronary calcification. *J. Cardiovasc. Risk* **2000**, *7*, 93–97.

132. Guo, W., Morrisett, J., DeBakey, M., Lawrie, G., Hamilton, J. Quantification in situ of crystalline cholesterol and calcium phosphate hydroxyapatite in human atherosclerotic plaques by solid-state magic angle spinning NMR. *Arterioscler. Thromb. Vasc. Biol.* **2000**, *20*, 1630–1636.

133. Nadra, I., Mason, J., Philippidis, P., Florey, O., Smythe, C., McCarthy, G., Landis, R.C., Haskard, D.O. Proinflammatory activation of macrophages by basic calcium phosphate crystals via protein kinase C and MAP kinase pathways: A vicious cycle of inflammation and arterial calcification? *Circ. Res.* **2005**, *96*, 1248–1256.

134. Cozzolino, M., Dusso, A.S., Slatopolsky, E. Role of calcium-phosphate product and bone-associated proteins on vascular calcification in renal failure. *J. Am. Soc. Nephrol.* **2001**, *12*, 2511–2516.

135. Fischer, J.W., Steitz, S.A., Johnson, P.Y., Burke, A., Kolodgie, F., Virmani, R., Giachelli, C., Wight, T.N. Decorin promotes aortic smooth muscle cell calcification and colocalizes to calcified regions in human atherosclerotic lesions. *Arterioscler. Thromb. Vasc. Biol.* **2004**, *24*, 2391–2396.

136. Bini, A., Mann, K.G., Kudryk, B.J., Schoen, F.J. Noncollagenous bone matrix proteins, calcification, and thrombosis in carotid artery atherosclerosis. *Arterioscler. Thromb. Vasc. Biol.* **1999**, *19*, 1852–1861.

137. Berenson, G., Radhakrishnamurthy, B., Srinivasan, S., Vijayagopal, P., Dalferes, E. Proteoglycans and potential mechanisms related to atherosclerosis. *Ann. N. Y. Acad. Sci.* **1985**, *454*, 69–78.

138. Wight, T.N., Merrilees, M.J. Proteoglycans in atherosclerosis and restenosis: Key roles for versican. *Circ. Res.* **2004**, *94*, 1158–1167.

139. Libby, P. Changing concepts of atherogenesis. *J. Intern. Med.* **2000**, *247*, 349–358.

140. Croce, K., Libby, P. Intertwining of thrombosis and inflammation in atherosclerosis. *Curr. Opin. Hematol.* **2007**, *14*, 55–61.

141. Viles-Gonzalez, J.F., Fuster, V., Badimon, J.J. Links between inflammation and thrombogenicity in atherosclerosis. *Curr. Mol. Med.* **2006**, *6*, 489–499.

142. Yan, Z.Q., Hansson, G.K. Innate immunity, macrophage activation, and atherosclerosis. *Immunol. Rev.* **2007**, *219*, 187–203.

143. Hansson, G.K., Libby, P. The immune response in atherosclerosis: A double-edged sword. *Nat. Rev. Immunol.* **2006**, *7*, 508–519.

144. Ait-Oufella, H., Salomon, B.L., Potteaux, S., Robertson, A.K., Gourdy, P., Zoll, J., Merval, R., Esposito, B., Cohen, J.L., Fisson, S., Flavell, R.A., Hansson, G.K., Klatzmann, D., Tedgui, A., Mallat, Z. Natural regulatory T cells control the development of atherosclerosis in mice. *Nat. Med.* **2006**, *12*, 178–180.

145. Tobias, P.S., Curtiss, L.K. Toll-like receptors in atherosclerosis. *Biochem. Soc. Trans.* **2007**, *35*, 1453–1455.

146. Xing, L., Remick, D.G. Mechanisms of oxidant regulation of monocyte chemotactic protein 1 production in human whole blood and isolated mononuclear cells. *Shock* **2007**, *28*, 178–185.

147. Verhasselt, V., Goldman, M., Willems, F. Oxidative stress up-regulates IL-8 and TNF-alpha synthesis by human dendritic cells. *Eur. J. Immunol.* **1998**, *28*, 3886–3890.

148. Jaimes, E.A., Nath, K.A., Raij, L. Hydrogen peroxide downregulates IL-1-driven mesangial iNOS activity: Implications for glomerulonephritis. *Am. J. Physiol.* **1997**, *272*, F721–F728.

149. Brenneisen, P., Briviba, K., Wlaschek, M., Wenk, J., Scharffetter-Kochanek, K. Hydrogen peroxide (H_2O_2)

increases the steady-state mRNA levels of collagenase/MMP-1 in human dermal fibroblasts. *Free Radic. Biol. Med.* **1997**, *22*, 515–524.

150. Schnurr, K., Borchert, A., Kuhn, H. Inverse regulation of lipid-peroxidizing and hydroperoxyl lipid-reducing enzymes by interleukins 4 and 13. *FASEB J.* **1999**, *13*, 143–154.

151. Hsu, H.Y., Chiu, S.L., Wen, M.H., Chen, K.Y., Hua, K.F. Ligands of macrophage scavenger receptor induce cytokine expression via differential modulation of protein kinase signaling pathways. *J. Biol. Chem.* **2001**, *276*, 28719–28730.

152. Klouche, M., Gottschling, S., Gerl, V., Hell, W., Husmann, M., Dorweiler, B., Messner, M., Bhakdi, S. Atherogenic properties of enzymatically degraded LDL: Selective induction of MCP-1 and cytotoxic effects on human macrophages. *Arterioscler. Thromb. Vasc. Biol.* **1998**, *18*, 1376–1385.

153. Frostegård, J., Kjellman, B., Gidlund, M., Andersson, B., Jindal, S., Kiessling, R. Induction of heat shock protein in monocytic cells by oxidized low density lipoprotein. *Atherosclerosis* **1996**, *121*, 93–103.

154. Thai, S.F., Lewis, J.G., Williams, R.B., Johnson, S.P., Adams, D.O. Effects of oxidized LDL on mononuclear phagocytes: Inhibition of induction of four inflammatory cytokine gene RNAs, release of NO, and cytolysis of tumor cells. *J. Leukoc. Biol.* **1995**, *57*, 427–433.

155. Brand, K., Banka, C.L., Mackman, N., Terkeltaub, R.A., Fan, S.T., Curtiss, L.K. Oxidized LDL enhances lipopolysaccharide-induced tissue factor expression in human adherent monocytes. *Arterioscler. Thromb.* **1994**, *14*, 790–797.

156. Ku, G., Thomas, C.E., Akeson, A.L., Jackson, R.L. Induction of interleukin 1 beta expression from human peripheral blood monocyte-derived macrophages by 9-hydroxyoctadecadienoic acid. *J. Biol. Chem.* **1992**, *267*, 14183–14188.

157. Frostegård, J., Wu, R., Giscombe, R., Holm, G., Lefvert, A.K., Nilsson, J. Induction of T-cell activation by oxidized low density lipoprotein. *Arterioscler. Thromb.* **1992**, *12*, 461–467.

158. Lipton, B.A., Parthasarathy, S., Ord, V.A., Clinton, S.K., Libby, P., Rosenfeld, M.E. Components of the protein fraction of oxidized low density lipoprotein stimulate interleukin-1 alpha production by rabbit arterial macrophage-derived foam cells. *J. Lipid Res.* **1995**, *36*, 2232–2242.

159. Miller, Y.I., Viriyakosol, S., Worrall, D.S., Boullier, A., Butler, S., Witztum, J.L. Toll-like receptor 4-dependent and -independent cytokine secretion induced by minimally oxidized low-density lipoprotein in macrophages. *Arterioscler. Thromb. Vasc. Biol.* **2005**, *25*, 1213–1219.

160. Subbanagounder, G., Wong, J.W., Lee, H., Faull, K.F., Miller, E., Witztum, J.L., Berliner, J.A. Epoxyisoprostane and epoxycyclopentenone phospholipids regulate monocyte chemotactic protein-1 and interleukin-8 synthesis. Formation of these oxidized phospholipids in response to interleukin-1beta. *J. Biol. Chem.* **2002**, *277*, 7271–7281.

161. Xiao, D., Wang, Z., She, M. Minimally modified low-density lipoprotein induces monocyte chemotactic protein-1 expression in vivo and a novel model for monocyte adhesion to arterial intima. *Chin. Med. J.* **1999**, *112*, 438–442.

162. Cushing, S.D., Berliner, J.A., Valente, A.J., Territo, M.C., Navab, M., Parhami, F., Gerrity, R., Schwartz, C.J., Fogelman, A.M. Minimally modified low density lipoprotein induces monocyte chemotactic protein 1 in human endothelial cells and smooth muscle cells. *Proc. Natl. Acad. Sci. U.S.A.* **1990**, *87*, 5134–5138.

163. Zalba, G., Fortuño, A., Orbe, J., San José, G., Moreno, M.U., Belzunce, M., Rodríguez, J.A., Beloqui, O., Páramo, J.A., Díez, J. Phagocytic NADPH oxidase-dependent superoxide production stimulates matrix metalloproteinase-9: Implications for human atherosclerosis. *Arterioscler. Thromb. Vasc. Biol.* **2007**, *27*, 587–593.

164. Moshal, K.S., Sen, U., Tyagi, N., Henderson, B., Steed, M., Ovechkin, A.V., Tyagi, S.C. Regulation of homocysteine-induced MMP-9 by ERK1/2 pathway. *Am. J. Physiol. Cell Physiol.* **2006**, *290*, C883–C891.

165. Scholz, H., Aukrust, P., Damås, J.K., Tonstad, S., Sagen, E.L., Kolset, S.O., Hall, C., Yndestad, A., Halvorsen, B. 8-isoprostane increases scavenger receptor A and matrix metalloproteinase activity in THP-1 macrophages, resulting in long-lived foam cells. *Eur. J. Clin. Invest.* **2004**, *34*, 451–458.

166. Kameda, K., Matsunaga, T., Abe, N., Hanada, H., Ishizaka, H., Ono, H., Saitoh, M., Fukui, K., Fukuda, I., Osanai, T., Okumura, K. Correlation of oxidative stress with activity of matrix metalloproteinase in patients with coronary artery disease: Possible role for left ventricular remodeling. *Eur. Heart J.* **2003**, *24*, 2180–2185.

167. Kolev, K., Skopál, J., Simon, L., Csonka, E., Machovich, R., Nagy, Z. Matrix metalloproteinase-9 expression in post-hypoxic human brain capillary endothelial cells: H_2O_2 as a trigger and NF-kappa B as a signal transducer. *Thromb. Haemost.* **2003**, *90*, 528–537.

168. Siwik, D.A., Pagano, P.J., Colucci, W.S. Oxidative stress regulates collagen synthesis and matrix metalloproteinase activity in cardiac fibroblasts. *Am. J. Physiol. Cell Physiol.* **2001**, *280*, C53–C60.

169. Rajavashisth, T.B., Yamada, H., Mishra, N.K. Transcriptional activation of the macrophage-colony stimulating factor gene by minimally modified LDL: Involvement of nuclear factor-kappa B. *Arterioscler. Thromb. Vasc. Biol.* **1995**, *15*, 1591–1598.

170. Parhami, F., Fang, Z.T., Fogelman, A.M., Andalibi, A., Territo, M.C., Berliner, J.A. Minimally modified low density lipoprotein-induced inflammatory responses in endothelial cells are mediated by cyclic adenosine monophosphate. *J. Clin. Invest.* **1993**, *92*, 471–478.

171. Berliner, J.A., Schwartz, D.S., Territo, M.C., Andalibi, A., Almada, L., Lusis, A.J., Quismorio, D., Fang, Z.P., Fogelman, A.M. Induction of chemotactic cytokines by

minimally oxidized LDL. *Adv. Exp. Med. Biol.* **1993**, *351*, 13–18.
172. Haberland, M.E., Olch, C.L., Fogelman, A.M. Role of lysines in mediating interaction of modified low density lipoproteins with the scavenger receptor of human monocyte macrophages. *J. Biol. Chem.* **1984**, *259*, 11305–11311.
173. Wieland, E., Parthasarathy, S., Steinberg, D. Peroxidase-dependent metal-independent oxidation of low density lipoprotein in vitro: A model for in vivo oxidation? *Proc. Natl. Acad. Sci. U.S.A.* **1993**, *90*, 5929–5933.
174. Rankin, S.M., Parthasarathy, S., Steinberg, D. Evidence for a dominant role of lipoxygenase(s) in the oxidation of LDL by mouse peritoneal macrophages. *J. Lipid Res.* **1991**, *32*, 449–456.
175. Spiteller, G. Linoleic acid peroxidation—The dominant lipid peroxidation process in low density lipoprotein—and its relationship to chronic diseases. *Chem. Phys. Lipids* **1998**, *95*, 105–162.
176. Steinbrecher, U.P., Witztum, J.L., Parthasarathy, S., Steinberg, D. Decrease in reactive amino groups during oxidation or endothelial cell modification of LDL. Correlation with changes in receptor-mediated catabolism. *Arteriosclerosis* **1987**, *7*, 135–143.
177. Jürgens, G., Lang, J., Esterbauer, H. Modification of human low-density lipoprotein by the lipid peroxidation product 4-hydroxynonenal. *Biochim. Biophys. Acta* **1986**, *875*, 103–114.
178. Jessup, W., Jurgens, G., Lang, J., Esterbauer, H., Dean, R.T. Interaction of 4-hydroxynonenal-modified low-density lipoproteins with the fibroblast apolipoprotein B/E receptor. *Biochem. J.* **1986**, *234*, 245–248.
179. Esterbauer, H., Koller, E., Slee, R.G., Koster, J.F. Possible involvement of the lipid-peroxidation product 4-hydroxynonenal in the formation of fluorescent chromolipids. *Biochem. J.* **1986**, *239*, 405–409.
180. Quehenberger, O., Koller, E., Jürgens, G., Esterbauer, H. Investigation of lipid peroxidation in human low density lipoprotein. *Free Radic. Res. Commun.* **1987**, *3*, 233–242.
181. Hoff, H.F., Whitaker, T.E., O'Neil, J. Oxidation of low density lipoprotein leads to particle aggregation and altered macrophage recognition. *J. Biol. Chem.* **1992**, *267*, 602–609.
182. Nadkarni, D.V., Sayre, L.M. Structural definition of early lysine and histidine adduction chemistry of 4-hydroxynonenal. *Chem. Res. Toxicol.* **1995**, *8*, 284–291.
183. Requena, J.R., Fu, M.X., Ahmed, M.U., Jenkins, A.J., Lyons, T.J., Baynes, J.W., Thorpe, S.R. Quantification of malondialdehyde and 4-hydroxynonenal adducts to lysine residues in native and oxidized human low-density lipoprotein. *Biochem. J.* **1997**, *322*, 317–325.
184. Tsai, L., Szweda, P.A., Vinogradova, O., Szweda, L.I. Structural characterization and immunochemical detection of a fluorophore derived from 4-hydroxy-2-nonenal and lysine. *Proc. Natl. Acad. Sci. U.S.A.* **1998**, *95*, 7975–7980.
185. Xu, G., Liu, Y., Sayre, L.M. Polyclonal antibodies to a fluorescent 4-hydroxy-2-nonenal (HNE)-derived lysine-lysine cross-link: Characterization and application to HNE-treated protein and in vitro oxidized low-density lipoprotein. *Chem. Res. Toxicol.* **2000**, *13*, 406–413.
186. Marinari, U.M., Nitti, M., Pronzato, M.A., Domenicotti, C. Role of PKC-dependent pathways in HNE-induced cell protein transport and secretion. *Mol. Aspects Med.* **2003**, *24*, 205–211.
187. Awasthi, Y., Sharma, R., Cheng, J., Yang, Y., Sharma, A., Singhal, S.S., Awasthi, S. Role of 4-hydroxynonenal in stress-mediated apoptosis signaling. *Mol. Aspects Med.* **2003**, *24*, 219–230.
188. Yang, Y., Sharma, R., Sharma, A., Awasthi, S., Awasthi, Y. Lipid peroxidation and cell cycle signaling: 4-hydroxynonenal, a key molecule in stress mediated signaling. *Acta Biochim. Pol.* **2003**, *50*, 319–336.
189. Budas, G.R., Disatnik, M., Chen, C., Mochly-Rosen, D. Activation of aldehyde dehydrogenase 2 (ALDH2) confers cardioprotection in protein kinase C epsilon (PKCε) knockout mice. *J. Mol. Cell. Cardiol.* **2010**, *48*, 757–764.
190. Godfraind, T., Sturbois, X., Verbeke, N. Calcium incorporation by smooth muscle microsomes. *Biochim. Biophys. Acta* **1976**, *455*, 254–268.
191. Nishio, S., Kavanagh, J.P., Faragher, E.B., Garside, J., Blacklock, N.J. Calcium oxalate crystallisation kinetics and the effects of calcium and gamma-carboxyglutamic acid. *Br. J. Urol.* **1990**, *66*, 351–356.
192. Doyle, I.R., Ryall, R.L., Marshall, V.R. Inclusion of proteins into calcium oxalate crystals precipitated from human urine: A highly selective phenomenon. *Clin. Chem.* **1991**, *37*, 1589–1594.
193. Selvam, R., Kalaiselvi, P. Studies on calcium oxalate binding proteins: Effect of lipid peroxidation. *Nephron* **2001**, *88*, 163–167.
194. Akamatsu, H., Komura, J., Asada, Y., Miyachi, Y., Niwa, Y. Inhibitory effect of azelaic acid on neutrophil functions: A possible cause for its efficacy in treating pathogenetically unrelated diseases. *Arch. Dermatol. Res.* **1991**, *283*, 162–166.
195. Iraji, F., Sadeghinia, A., Shahmoradi, Z., Siadat, A.H., Jooya, A. Efficacy of topical azelaic acid gel in the treatment of mild-moderate acne vulgaris. *Indian J. Dermatol. Venereol. Leprol.* **2007**, *73*, 94–96.
196. Elewski, B., Thiboutot, D. A clinical overview of azelaic acid. *Cutis* **2006**, *77*, 12–16.
197. Fleischer, A.B., Jr. The evolution of azelaic acid. *Cutis* **2006**, *77*, 4–6.
198. Lin, C.C., Wu, S.J., Chang, C.H., Ng, L.T. Antioxidant activity of *Cinnamomum cassia*. *Phytother. Res.* **2003**, *7*, 726–730.
199. Bolkenius, F.N. Leukocyte-mediated inactivation of alpha 1-proteinase inhibitor is inhibited by amino analogues of alpha-tocopherol. *Biochim. Biophys. Acta* **1991**, *1095*, 23–29.

200. Nègre-Salvayre, A., Affany, A., Hariton, C., Salvayre, R. Additional antilipoperoxidant activities of alpha-tocopherol and ascorbic acid on membrane-like systems are potentiated by rutin. *Pharmacology* **1991**, *42*, 262–272.

201. Bindoli, A., Valente, M., Cavallini, L. Inhibitory action of quercetin on xanthine oxidase and xanthine dehydrogenase activity. *Pharmacol. Res. Commun.* **1985**, *17*, 831–839.

13

CYSTIC FIBROSIS

Neal S. Gould and Brian J. Day

OVERVIEW

Cystic fibrosis (CF) is an autosomal recessive genetic disorder that affects roughly 70,000 children and adults worldwide with nearly 30,000 of those in the United States.[1] CF is typically diagnosed by age 2 and the median life expectancy is limited to only 3 or 4 decades. CF is a debilitating multiorgan disease with the most serious manifestations resulting in progressive weakening of lung host defense, chronic lung infections, and loss of lung function. Other symptoms of CF include poor weight gain, malnutrition, and fibrosis of the pancreas. Many of the symptoms of CF have been known since at least the early 1900s, but they were not identified as CF. Beginning in the 1930s, the connection between CF of the pancreas and the effects on the lung and GI tract was made by Anderson.[2] Diagnosis of CF was advanced with the finding that excessive salt loss occurs in the sweat of CF patients.[3] The biochemical basis for CF was largely unknown until 1983 when Quinton reported that sweat duct cells from a CF subject lacked chloride efflux.[4] However, since the initial discovery of a defect in chloride transport, a number of other functions for the CF protein have also been described. The CF gene was identified in 1989 with the most common CF mutation having a deletion of phenylalanine at position 508 of the protein, ΔF508.[5-7]

CF is caused by mutations in the CF gene encoding for the cystic fibrosis transmembrane regulator (CFTR) protein. CFTR is an apically located ATP-binding cassette (ABC) transporter in the C-subfamily (ABCC7). CFTR was originally identified as a chloride channel due to the high salt concentration in the sweat of CF patients, but CFTR is most closely related to the MRP family of transporters best known for xenobiotic transport.[8] Over 1300 mutations have been identified in the CFTR gene that lead to CF, many of which consist of deletion, insertion, or truncation mutations.[9] CFTR mutations are commonly divided into five major groups according to their effect on CFTR function.[10] The CFTR mutation classes include Class I, defective protein synthesis; Class II, defective protein processing; Class III, defective protein regulation; Class IV, altered conductance; and Class V, reduced CFTR abundance. The most common mutation, known as the ΔF508 mutation, occurs in roughly 70% of CF patients. The ΔF508 mutation is a class II trafficking mutation, which means that a fully functional CFTR protein is synthesized but is not correctly packaged to the apical membrane, and it remains in the ER and is tagged for degradation.[11,12] Other frequent CFTR defects include truncation mutations (G542X, W1282X, S480X), substitution mutation (G551D, R347P), or splice variants, but each only makes up 1–2% or less of the mutations that cause CF.[13]

13.1 LUNG DISEASE CHARACTERISTICS IN CF

There are a number of physiological manifestations of CF, including effects on the pancreas, GI, and lung. The CF clinical phenotype is highly variable and complex due to the many genotypes and corresponding phenotypes that result in differing levels of organ involvement

Molecular Basis of Oxidative Stress: Chemistry, Mechanisms, and Disease Pathogenesis, First Edition. Edited by Frederick A. Villamena.
© 2013 John Wiley & Sons, Inc. Published 2013 by John Wiley & Sons, Inc.

TABLE 13.1 Glutathione (GSH) and Thiocyanate (SCN) are Two Major Thiols in the Lung Epithelial Lining Fluid (ELF)

Properties	Glutathione (GSH)	Thiocyanate (⁻SCN)
Molecular weight	307	58
Compound class	Peptide	Cyanate
Thiol pK_a	8.6	4
ELF levels (µM)	150–300	60–160
Synthesis	γ-GCL and GS	dietary (??)
Function	Antioxidant, cofactor	host defense, antioxidant (?)

Note: GSH and SCN share functions as being essential cofactors or substrates for enzymes and have direct and indirect antioxidant effects.

affected by the lack of functional CFTR activity.[14] The most common cause of death among CF individuals is due to progressive obstructive lung disease associated with chronic infection and inflammation; thus, we will primarily focus on the disease characteristics in the lung.

13.1.1 Lung Epithelial Lining Fluid (ELF), Host Defense, and CFTR

The lung epithelium is responsible for the composition and sustenance of the thin fluid that bathes the airway surface. The ELF contains a number of components that defends the lung from inhaled agents, including antimicrobial peptides and proteins, antioxidants, and mucus.[15] CFTR contributes some of these key components to the ELF, including GSH, a major lung antioxidant,[16] and thiocyanate (SCN), a key component of host defense with antioxidant properties (Table 13.1).[17,18] The lung also uses mucociliary clearance to remove inhaled particles and pathogens. In normal individuals, the mucociliary clearance mechanism traps bacteria and particles, preventing them from reaching the lower airways.[19] In CF individuals, the imbalance in ion transport leads to dehydration of the mucus.[20,21] This dehydration results in thick mucus, impaired mucociliary clearance, and defective host defense.[22] In CF individuals, the thick mucus layers provide a rich media for bacteria colonization that leads to chronic infections. The dehydrated and thick mucus stimulates bacteria to form biofilms that make it difficult for conventional antibiotic treatments to reach and kill the pathogens. The chronic exposure of the lung to bacteria sets up a chronic inflammatory and oxidative environment that is thought to result in much of the CF lung pathology.

13.1.2 Lung Infection and Reactive Oxygen Species (ROS) in CF

The lungs of CF individuals are uniquely susceptible to bacterial infection and over time become chronically colonized with bacteria. *Pseudomonas aeruginosa* infections are the most common in older CF subjects that acquire a mucoid phenotype with the formation of a biofilm in the lung. Infections trigger a persistent neutrophilic inflammatory response in the lung. The CF lung is hypersensitive to inflammatory stimuli with exaggerated release of proinflammatory cytokines such as interleukin (IL)-8 and decreased release of anti-inflammatory cytokines such as IL-10.[23,24] Much of the hypersensitivity to inflammatory stimuli is thought to be due to increased activity of transcriptional factors NF-kB and AP-1 and associated signaling pathways that regulate their activity.[25] These same inflammatory stimuli activate both leukocytes and epithelium to release ROS and reactive nitrogen species (RNS), including superoxide, hydrogen peroxide, hydroxyl radical, hypochlorous (HOCl) acid, nitric oxide (NO), and peroxynitrite.

Leukocytes contain NADPH oxidases (NOXs) that upon activation produce superoxide that is rapidly converted to hydrogen peroxide by either spontaneous dismutation or through reaction with superoxide dismutases. Many of the signals elicited during lung infection can activate NOXs that include toll receptor ligands, cytokines, and growth factors.[26] Infiltrating neutrophils are a hallmark of CF lung disease and they release myeloperoxidase (MPO) that generates halous acids, including HOCl acid, that is very damaging to the lung epithelium, or hypothiocyanate (HOSCN), which

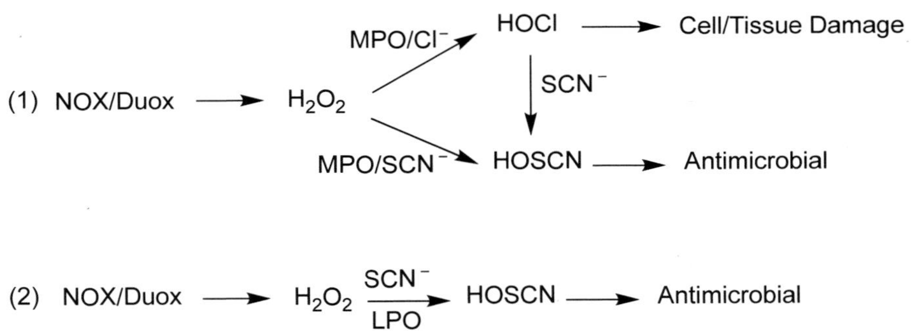

Figure 13.1 Formation of oxidants by major peroxidases in the airways. Reaction (1) illustrates the myeloperoxidase (MPO)-mediated formation of either hypochlorous (HOCl) or hypothiocyanate (HOSCN) from hydrogen peroxide (H_2O_2) based on available substrates. HOCl can also be converted to HOSCN by available thiocyanate (SCN-). Reaction (2) illustrates the less reactive pathway of hydrogen peroxide utilization by lactoperoxidase (LPO) to only form HOSCN.

is less damaging and functions as an antimicrobial (Figure 13.1). The lung epithelium can contribute to ROS production through the activation of dual oxidases (Duox) which are structurally and functionally related to NOXs and are highly expressed on the surface of lung epithelium.[27] Bacterial products are known to activate Duox formation of H_2O_2 by the lung epithelium.[28] The lung epithelium also secretes lactoperoxidase (LPO) into the airways that generates HOSCN (Fig. 13.1). One of the biological functions of NOX and Duox is to provide hydrogen peroxide, a required substrate for MPO and LPO activity.[28] Both the lung leukocytes and lung tissues express nitric oxide synthetase (NOS) that can be a source of both NO and superoxide.[29] CF patients have lower levels of exhaled NO but higher levels of NO metabolites, including nitrite, nitrate, and peroxynitrite in their lungs.[30,31] CF patients have elevations of markers of lung oxidative stress even early in life that is thought to be due to lung inflammation.[32] CFTR KO mice have increased levels of markers of mitochondrial oxidative stress even under pathogen-free conditions.[33] Under oxidative damage, free radicals can be formed from damaged tissue, with the mitochondria being a major source of ROS.[34] Furthermore, overproduction of ROS/RNS can deplete tissue and airway antioxidant defenses, essentially perpetuating the oxidative environment.

13.1.3 Inflammation in CF

Airway inflammation is a major component of CF lung disease. Signs of airway inflammation are present even in young children without evidence for bacterial colonization.[35] Airway inflammation is also enhanced as a result of recurrent infections and the persistence of the infections in CF individuals. These inflammatory signals include proinflammatory cytokines, including IL-8 and tumor necrosis factor alpha (TNFα), which have the ability to propagate the inflammatory response by activating macrophages and recruiting neutrophils, resulting in even more cytokine release in the airways.[36] Neutrophils release elastase that damages the mucociliary clearance system and stimulates mucus production, both of which are hallmarks of CF lung disease,[37] which can be reversed by antioxidants.[38] There is also a unique balance in which proinflammatory cytokine levels can be suppressed by the available of antioxidants like glutathione (GSH) in the airways. Studies have shown that thiols can negatively modulate NF-kB activity in the lung which results in less cytokine production upon stimulation.[39,40] Individuals with CF are at a unique disadvantage because GSH is low in the CF airway due to the loss of functional CFTR.[16,41]

13.1.4 Airway Antioxidants in CF

There are a number of airway antioxidants (ascorbate, GSH, uric acid, and α-tocopherol) along with surfactant proteins, albumin, and mucus that protect the airway surfaces in the lung.[42] Antioxidants in the airways are secreted by the lung epithelium to provide a protective barrier against inhaled oxidants. In the CF lung, the levels of GSH are abnormally low. This decrease is due to a multitude of factors, including decreased transport capability as well as increased oxidative stress that consumes much of the available airway GSH. Another issue in CF is the poor absorption of fat-soluble antioxidant vitamins, including α-tocopherol, vitamin D, and β-carotene,[43] as well as other elements of antioxidant systems such as ferritin[44] and selenium.[45,46] This decrease in airway antioxidants leaves the CF lung compromised and severely susceptible to oxidants from the atmospheric environment as well as those produced endogenously.

13.2 ROLE OF CFTR IN THE LUNG

CFTR is not an abundantly expressed protein in the lung. Relative protein expression of CFTR makes up only a very small fraction of the total proteins that are normally expressed in the lung epithelium.[47] Yet, the loss of lung CFTR function results in dramatic and lethal lung disease. Thus, it is important to realize that the expression level of a protein does not always dictate the importance it has on overall lung function. Despite decades of study, there are still many unknown functions and interactions that CFTR has within the lung. However, there are several well-known aspects of CFTR function in the lung that contribute to the detrimental effects of CF lung disease.

13.2.1 Chloride Transport

CFTR was first identified as a chloride transporter.[48] Early diagnostic tests were developed to identify individuals with CF by measuring elevated chloride concentrations in sweat.[49] CFTR is classically thought of as a chloride transporter, but it also has a number of other diverse transport and protein scaffold functions.[50-52] The ability to transport sodium and chloride are essential in regulating water balance and hydration of the epithelial lining fluid in the lung.[53] This imbalance in hydration leads to the classic thick mucus in the airways of CF individuals.[54] The thick mucus is also a cause of the persistent lung infections as well as many of the airway obstructions found in CF patients.[55] It is these chronic airway obstructions that are most commonly associated with mortality in CF.[56]

13.2.2 GSH Transport

With the proliferation of molecular biology techniques in science, CFTR was sequenced and found to be most closely associated with the multidrug resistance associated proteins (MRP, ABCC) family rather than classical chloride transporters.[57] The MRP family is better known for xenobiotic transport function and is the major group of proteins involved in the liver's transport of GSH into the plasma and bile.[58] Early studies found decreased levels of GSH in the airway fluids of CF patients, but whether this was due to oxidative conditions that deplete the GSH or defective transport was not known.[41] Consequently, several studies have shown CFTR mediates GSH export in cell lines expressing normal CFTR and decreased GSH export in cell with mutant forms of CFTR.[59,60] While CFTR was shown to be involved in the transport of GSH, the mechanism by which this occurs was not known but, using purified membrane vesicles expressing CFTR protein, the nucleotide-mediated transport of GSH was directly shown.[61] This gave further proof that CFTR is not just a chloride channel but can also be actively involved in the transport of other molecules. It has even been suggested that CFTR function can vary based on the type of nucleotide that is bound, with some configurations favoring transport of chloride while others facilitate the transport of GSH.[61] Currently, CFTR is the only known lung apical transporter of GSH that regulates both basal and GSH adaptive response levels.[16,18,62,63]

13.2.3 SCN Transport

SCN is another thiol-containing molecule that has also been shown to be transported by CFTR.[17,18,64] Similar to chloride and bicarbonate, SCN is a charged anion. *In vitro* studies examining anion pore properties of CFTR have shown that SCN is transported in a selective manner comparable to chloride. In fact, using biological modeling and patch clamp techniques, it has been suggested that under steady-state conditions, SCN can enter the pore and bind with more affinity to CFTR than chloride.[64] This is an interesting proposition, but in whole cell systems, chloride has been shown to still have a greater conductance through CFTR than SCN.[65] SCN can act as both an antioxidant[18] and as an alternative substrate for MPO in host defense mechanisms.[17,66] SCN is best known as a detoxification product of cyanide[67] and a specific substrate for LPO.[68] LPO generates HOSCN from H_2O_2 and SCN, which has antimicrobial properties. SCN is an important antioxidant when it comes to protecting the lung during infections and can detoxify MPO and hypochlorite-mediated cell injury.[18] In the mouse lung, SCN is roughly the same level as GSH in the normal ELF, but with a deficiency in CFTR, SCN levels fall to roughly 20% normal levels.[18] Other transporters such as pendrin have been suggested to be involved in the transport of SCN *in vitro*,[69] which may explain why CFTR KO mice still have some SCN in the airways, but it is unclear to what degree this occurs *in vivo*. Nevertheless, SCN transport relies heavily on CFTR and is a crucial aspect necessary for host defense that is defective in CF.[70]

13.3 OXIDATIVE STRESS IN THE CFTR-DEFICIENT LUNG

It is well known that there is increased oxidative stress in the lungs of CF patients[42,51,71,72] and CFTR KO mice.[33,73] The occurrence of oxidative stress in the lungs is actually due to several different factors, including increased oxidant burden from infection and host defense and decreased antioxidant transport capability into the ELF.

13.3.1 The Importance of ELF Redox Status

The ELF redox state is a highly regulated process, and even under high oxidant burdens the redox state of GSH in the ELF is not generally shifted.[74,75] While there are many redox-active molecules and proteins in the ELF, the most abundant is GSH, which makes it a good marker of overall redox status of the ELF. GSH is maintained predominately in the reduced form through the action of reductases that reduce the oxidized forms (GSSG) as well as resynthesis and transport of reduced GSH. With a shift to more oxidizing conditions, the functions of many molecules are compromised. Under oxidative conditions there can be increases in ROS/RNS, including $ONOO^-$, $O_2^{\bullet-}$, and H_2O_2. The formation of these reactive species is especially important in the airways since the lung is one of the few organs with a high oxygen concentration, which facilitates the formation of these reactive species. ROS/RNS can readily cause DNA damage and lipid peroxidation which can lead to cytotoxicity. During the repair process, increased fibrosis can occur which can limit the elasticity of the lung. Furthermore, the formation of ROS/RNS can cause damage to the pulmonary vasculature, causing an increase in vascular permeability and potentially edema. The lung uses the ELF to quickly adapt to its changing atmospheric environment by regulating the levels of antioxidants such as GSH.[76]

13.3.2 Cellular Oxidative Stress

The shift in redox status in whole organs like the lung is well known in CF, but what happens on a cellular level is commonly overlooked. It is known that CFTR can become inactivated with oxidants and that oxidative stress conditions inhibit the maturation of CFTR.[77] In CF patients, plasma lipid peroxides and oxidized proteins have been detected.[78,79] In addition, elevated urine 8-hydroxy-2-deoxyguanosine has been measured in CF subjects.[80] While these markers do not point to a specific organ, they are good general biomarkers of oxidative stress. These biomarkers suggest that there is quite a lot of lipid, protein, and DNA damage that occurs in CF. In studies examining the lungs of mice deficient in CFTR, in agreement with patient data, increased lipid peroxidation and DNA oxidation have been shown.[33] Furthermore, cells deficient in CFTR have increased cellular H_2O_2 production.[81] Mitochondria can be a major source of endogenous ROS release, especially if mitochondrial respiration is affected and the mitochondria has been shown to be affected by CFTR mutations.[33,82,83] Despite similar intracellular GSH levels in both normal and CFTR-deficient lungs, decreased mitochondrial GSH has been shown in CFTR KO mice.[33] This decrease coincides with elevated levels of mitochondrial DNA oxidation. Furthermore, in vitro studies have shown elevated ROS levels in the mitochondria of CFTR deficient cells.[33] Taken together, these data suggest that while the most drastic effect CFTR mutations have is on the extracellular environment, the intracellular environment is also affected as well. Furthermore, it is plausible that the increase in cellular ROS, particularly in the mitochondria, can affect cellular function resulting in altered respiratory function in the lung.

13.3.3 NO and CF

The dual role of NO is perplexing from the standpoint of understanding the pathophysiology of CF. Under normal circumstances, NO is an essential signaling molecule, antimicrobial, bronchodilator, and modulator of immune responses.[84] High NO concentrations can cause cytotoxic effects in the airways, suppress the effects of T_H1 $CD4^+$ cells, and react with superoxide to form $ONOO^-$ (Fig. 13.2). In fact, the role of NO in the airways may actually be quite dependant on the GSH content where NO stimulates the expression of γ-glutamylcysteine ligase, the rate-limiting enzyme in GSH synthesis.[85,86] The enzymes involved in GSH synthesis are induced by NO presumably to keep its cytotoxic effects at bay. NO and GSH can form S-nitrosoglutathione (GSNO) in both extracellular and intracellular compartments.[87] GSNO can act as a readily available source of NO as well as GSH. GSNO itself can act as a bronchodilator which can help prevent airway obstruction.[88] The paradox with NO in CF is that in CF patients, NO and GSNO have been shown to be decreased compared to controls, and inducible nitric oxide synthase (NOS2) expression has been shown to be decreased in CF as well.[89–94] This phenomenon is in stark contrast to other inflammatory diseases such as asthma where NO levels are increased.[95] For a disease that causes such a shift in redox balance and an increase in ROS, one would not think that NO would actually be decreased. Several theories have been postulated as to why NO might be decreased, one of which is that the decrease in GSH in CF leads to the preferential reaction of NO with superoxide to form $ONOO^-$ and other nitrites thereby decreasing the parent NO levels.[96] The positive signaling effects of NO are lost in CF, which potentially contributes to the increase in airway obstruction with the loss of the bronchodilation properties and the decrease in bacterial clearance without the antimicrobial properties.

13.3.4 Oxidative Stress Due to Persistent Lung Infection

A common trait among almost all individuals with CF is chronic persistent lung infections. Nearly 70% of

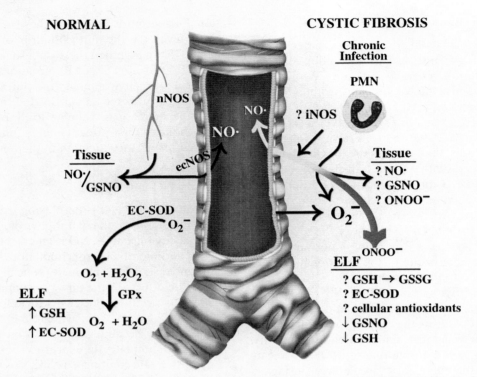

Figure 13.2 Schematic depicts reactive nitrogen and oxygen species formation in healthy and cystic fibrosis (CF) lungs and their modulation by antioxidants such as glutathione (GSH) and other antioxidants such as extracellular superoxide dismutase (EC-SOD). Under normal conditions, neuronal and endothelial forms of nitric oxide synthase (nNOS and eNOS) generate low amounts of nitric oxide (NO). Under chronic infection and inflammation seen in most CF subjects, there is increased conversion of NO to oxidative products such as peroxynitrite (ONOO$^-$) and loss of beneficial NO products such as S-nitrosoglutathione (GSNO) that has bronchodilator and anti-inflammatory effects.

people with CF have some form of lung infection. While the infectious pathogen itself can cause a variety of damage to the lung, in many instances, it is the attempts by the immune system to fight off the infection that can be just as damaging. It is thought that the dehydration of the mucus layer provides an ideal environment for pathogens to survive and evade attempts to rid the lung of these infections. The dehydrated mucus prevents either the immune cells or conventional antibiotics to eradicate the pathogens. As a result of the inability to effectively kill and clear the opportunistic pathogens, inflammatory signals are amplified. The amplification of inflammatory signals can result in increased numbers of macrophages and neutropil infiltration into the lung. Neutrophils are the main cell type that is mobilized to fight off bacterial pathogens, and they synthesize a specific enzyme called MPO that catalyzes the reaction between halides and hydrogen peroxide to form hypohalous acids, the most common of which is HOCl, a compound equivalent to household bleach solutions (Fig. 13.1).[97] HOCl is an extremely reactive and damaging oxidant, it can readily oxidize macromolecules, consume antioxidants, and at high enough concentrations can cause cytotoxicity.[98] HOCl is localized in small microenvironments in the lung, and it is easily conceivable that if HOCl concentrations reach high enough levels, lung damage can occur. The direct measurement of HOCl in the airways is not currently possible, and only estimations can be made either from stable products formed or MPO activity. In CF individuals, the MPO that is recovered in the airways is much greater than normal individuals.[99] Based on estimations of the amount of HOCl formed by MPO and the activity of MPO in the sputum of CF individuals, HOCl concentrations have been estimated to reach as high as 8 mM during infection in CF patients, more than enough to cause localized cellular toxicity. The lung is protected from excess MPO-mediated cytotoxicty with high concentrations of SCN to compete with chloride as a substrate for MPO (Fig. 13.1). When SCN is utilized by MPO, the resulting product is HOSCN, a product that is still an effective antimicrobial but much less damaging to cells than HOCl.[18,100] Additionally, HOCl can be detoxified by antioxidants like SCN and the ever-present

GSH in the airways.[101,102] This presents an interesting conundrum for CF individuals; they have persistent infections that mobilize the immune system, but their antioxidant capacity is greatly decreased in the airways. This imbalance in oxidants formed and defective capacity to maintain high levels of extracellular antioxidants can easily lead to a shift toward more oxidizing environments in the airways, which is a common finding and attractive therapeutic target in CF lung disease.

13.4 ANTIOXIDANT THERAPIES FOR CF

Several antioxidant therapies have emerged in recent years for the treatment of CF with mixed successes. While there are numerous therapies and treatments (including antibiotics, mucolytics, and gene therapy) that are either in the preclinical or early clinical trials, only those that target (or have the possibility of targeting) the antioxidant deficiencies of CF will be discussed (Table 13.2).

13.4.1 Hypertonic Saline Inhalation

The inhalation of hypertonic saline has been used as a therapy for CF patients for a number of years. Despite years of clinical use to improve lung function and mucus clearance, the exact mechanism of the beneficial effects of hypertonic saline is not known. It is thought that hypertonic saline helps to rehydrate the airways by creating an osmotic gradient and pulling more water into the airways.[103] Hyperosmotic agents have been shown to induce a wide array of different gene products which include aquaporins,[104] various transporters including CFTR,[105] membrane-associated binding proteins,[106] and cytoskeleton components.[107] It has been suggested that aquaporins are involved in the increased hydration of the epithelial lining fluid with hypertonic saline, but this has been debated.[108] However, it has been shown that hypertonic saline does increase the volume of the ELF and that it could be a result of increased CFTR protein on the plasma membrane. In addition to potentially pulling water into the airways, there has been recent research to show that GSH and SCN are also increased in the airways of mice that have inhaled hypertonic saline.[18] While it is not known whether this phenomenon exists in humans, it is another interesting aspect of hypertonic saline treatment that may contribute to its ability to improve lung function. Another interesting effect of osmotic stress is its ability to increase CFTR maturation. Osmotic agents like sorbitol and inositol have been shown to promote maturation of CFTR to the plasma membrane, which may contribute to some of its beneficial effects.[109] Hypertonic saline has other effects besides maturation based on its ability to work in CF subjects with a wide range of mutations.[110] While there are many facets of the effect of hypertonic saline on the airways in CF, it is one of the most well tolerated cost-effective treatments that increase lung function for many CF patients.[110]

13.4.2 Pharmacologic Intervention

Along the same concept of an osmotic agent causing the maturation of CFTR, some compounds have been shown to cause the maturation of CFTR to the plasma membrane, increase CFTR function, and promote the complete synthesis of CFTR protein. Pharmacologic intervention to try to restore CFTR protein function has become increasingly a popular target of researchers and pharmaceutical companies alike. Aminoglycoside antibiotics have been shown to suppress premature stop codons that result in truncated proteins.[111] Aminoglycosides have been shown to bind to a specific site on ribosomal RNA that disrupts the proofreading capability which causes misreading of the mRNA enabling premature stop codons to be misread. *In vitro* and *in vivo*, aminoglycosides have been shown to increase cAMP-mediated chloride transport and increase CFTR expression as well.[112] In a pilot study of CF patients, the aminoglycoside gentamicin was shown to stimulate nasal chloride transport.[113] In further placebo controlled double blind studies, gentamicin was shown to normalize defects in either sodium conductance, chloride conductance, or both in over 60% of the patients. While the use of gentamicin has been shown to be beneficial, there are issues with toxicity, potency, and uptake.[114] Pharmaceutical companies have begun to develop similar compounds, one of which is called Ataluren, which has also

TABLE 13.2 CF Therapies That May Target Oxidative Stress

Therapies	Effect on CFTR	Effect on Antioxidants
Hypertonic saline	↑ CFTR trafficking	↑ GSH ↑ SCN
Aminoglycosides	↑ CFTR expression and function	↑ GSH[a] ↑ SCN[a]
Flavonoid compounds	↑ CFTR function	↑ GSH[a] ↑ SCN[a]
GSH	N/A	↑ GSH
NAC	N/A	↑ GSH ↑ NAC

[a]Presumed effect on antioxidants based on increase in CFTR expression and/or function.

been shown to normalize CFTR function in nasal epithelium.[115,116]

Other targets to correct CFTR function are through CFTR modulators. These types of compounds are meant to restore function in people with reduced or defective chloride transport. Genistein, a flavonoid compound, has been shown to stimulate wild-type CFTR channels.[117] It is thought that flavonoid compounds can bind within the channel pore and increase the open time of the channel. In G551D mutations, ATP binding is severely affected, but genistein in combination with forskolin or protein kinase A, both cAMP-regulating molecules, can restore channel open time.[118] A modest repolarization of the nasal epithelium was also seen in patients with at least one copy of the G551D mutation.[119] Pharmacologic agents have also begun to be developed that are termed CFTR potentiators that may act on CFTR to keep it in the open state. Phase 2 trials of these compounds resulted in improvements in CFTR function as measured by nasal epithelium potential and sweat chloride as well as lung function measured by forced expiratory volume in 1 minute (FEV_1).[120]

While the majority of the investigation of these compounds establish changes in chloride conductance or potential, it is conceivable that GSH may also be altered in many instances. A number of flavonoids can directly stimulate GSH transport through related MRP transporters[121] and ABCG2.[122] Changes in CFTR-mediated GSH transport with these types of compounds have not been examined, but if there is an increase in CFTR protein at the plasma membrane or increases in CFTR function, there is a good chance that GSH and SCN transport may also be improved. One of the major questions facing the pharmacologic approach at restoring CFTR protein or function is how much of an increase in function is necessary to improve lung function and quality of life.

13.4.3 Oral Antioxidants

One of the easiest ways to maintain compliance with patients during a treatment regimen is through the oral administration of the therapeutic. This is not lost in the CF field, where despite the varying treatment options, many are inhaled and patients might not follow the treatment regimen completely in all cases. The alteration of diet, including taking supplements, has been established as one way to help treat some of the symptoms of CF. Malnourishment is a major concern in CF due to the decreased uptake of many essential fat-soluble vitamins and antioxidants.[123] Due to the multitude of various vitamins and dietary changes that have been tried with CF, this review will focus primarily on GSH and GSH precursors. *N*-acetylcysteine (NAC) is a precursor for GSH synthesis and has been examined with good results over short-term study. In a study of 18 CF patients that received high doses of oral NAC, blood neutrophil GSH was improved and overall neutrophil burden was improved as well.[124] Over the short time period of 4 weeks that this study covered, there was no increase in pulmonary function. However, longer treatment times are needed to properly evaluate this end point.

Oral administration alone of GSH has not been rigorously examined in CF; it has, however, been used in combination with inhaled GSH which improved both pulmonary exacerbations and function.[125] While it is not possible to discern the effects or oral versus inhaled GSH on the pulmonary function, it does highlight the ability for GSH rather than a precursor to be beneficial. There are some caveats to oral administration of any therapy and the major one in CF is absorption. The absorption of any compound into the plasma can be potentially affected by CFTR mutations, which would render the treatments ineffective if a high enough dose cannot be maintained. There is some evidence in CFTR KO mice that oral GSH absorption is impaired when CFTR function is defective.[63]

13.4.4 Inhaled Antioxidants

While not as convenient as orally administered therapies, aerosolized therapies utilizing GSH or precursors have an added benefit because they are delivered directly to the intended site of action, and the potential for malabsorption in CF patients is not a concern. Both NAC and GSH have been administered through an aerosol to treat the redox status and function of the lung in CF patients. In a small study of patients aged 16–37, aerosolized GSH was given for up to 2 weeks and was shown to increase the lung function measured by FEV_1 and forced vital capacity (FVC).[78] Lung function did seem to improve in this study, but there was little improvement on oxidative stress markers, including protein carbonyls, thiols, and lipid peroxides. Interestingly, after inhaling either 300 or 450 mg GSH three times per day, total GSH in the bronchoalveolar lavage fluid (BALF) went up, but the percentage of reduced glutathione actually went down, suggesting that the majority of the GSH that is inhaled and is in the airways is in the oxidized form. Another study also found that oxidative markers were not changed with GSH inhalation, but it did increase lymphocytes in the BALF.[126] In another placebo-controlled trial, patients were given 66 mg kg^{-1} GSH spaced over four inhalations per day for 8 weeks, and again there was improvement in peak flow measurements as well as a general feeling of lung function improvement as rated by the patients.[127] These results are quite promising for the potential to slow the loss of lung function in CF patients, but there are major caveats. Introducing significant amounts of GSSG into the airways can actually potentiate the oxidative environment in the lung, and it could cause constriction of the airways leading to even more airway obstruction. These issues would need to be addressed and solved before inhaled GSH can become an effective therapy in CF.

Besides NAC being a GSH precursor, there are properties of NAC that could directly correct the oxidative environment of the CF lung. Primarily, NAC has been shown to have antioxidant properties itself[128] and NAC being a precursor could be utilized to make GSH, which would alleviate many of the issues with increased GSSG in the airways of patients receiving inhaled GSH treatments. The majority of inhaled NAC studies are not the randomized placebo-controlled studies. However, there is still valuable information that can be deduced from these studies. Primarily over a range of 2 weeks to 3 months of being on inhaled NAC treatment, there were not any significant increases in lung function with the treatment.[129] However, there were significant improvements with respect to the viscosity of sputum and the ability to clear mucus from the airways. It has been postulated that the thiol in NAC can disassociate disulfide bonds in the mucus, thereby reducing its viscosity. The efficacy of NAC treatment for CF is ambiguous at best, and more studies are currently ongoing in the CF Clinical Trials Network. NAC does seem to have promising results as a mucolytic treatment. The mucolytic activity of NAC is still an important aspect of CF since many of the opportunistic pathogens reside within the thick mucus and cause the depletion in endogenous antioxidants. Treatments that aid in the natural clearance of bacteria in CF individuals may indirectly benefit the redox status of the airways.

SUMMARY

CF is a complex multiorgan disease. Much more is known about the causes and underlying biochemical effects of CF and CFTR mutations, but treatment options are limited and attempts at curing CF with gene therapy have been disappointing. Understanding the changes in redox status of the lung and how those redox changes occur is an important underlying aspect of CF. For years, CFTR was only thought of as a chloride channel, and that many of the symptoms of CF arose due to defective chloride transport. With the identification of the other transport capabilities of CFTR, more is being understood about the underlying antioxidant imbalances in the airways and how this can actually lead to the pathology seen in CF. While an effective treatment to cure CF is not yet available, treatments to improve the quality of life of CF individuals are being formulated and tested worldwide, several of which focus on controlling the immune response and correcting the oxidative environment in the lung.

REFERENCES

1. Strausbaugh, S.D., Davis, P.B. Cystic fibrosis: A review of epidemiology and pathobiology. *Clin. Chest Med.* **2007**, *28*(2), 279–288.
2. Anderson, D.H. Cystic fibrosis of the pancreas and its relation to celiac disease: A clinical and pathological study. *Am. J. Dis. Child.* **1938**, *56*, 344.
3. Di Sant'Agnese, P.A., Darling, R.C., Perera, G.A., Shea, E. Abnormal electrolyte composition of sweat in cystic fibrosis of the pancreas; clinical significance and relationship to the disease. *Pediatrics* **1953**, *12*(5), 549–563.
4. Quinton, P.M. Chloride impermeability in cystic fibrosis. *Nature* **1983**, *301*(5899), 421–422.
5. Rommens, J.M., Iannuzzi, M.C., Kerem, B., Drumm, M.L., Melmer, G., Dean, M., Rozmahel, R., Cole, J.L., Kennedy, D., Hidaka, N., et al. Identification of the cystic fibrosis gene: Chromosome walking and jumping. *Science* **1989**, *245*(4922), 1059–1065.
6. Kerem, B., Rommens, J.M., Buchanan, J.A., Markiewicz, D., Cox, T.K., Chakravarti, A., Buchwald, M., Tsui, L.C. Identification of the cystic fibrosis gene: Genetic analysis. *Science* **1989**, *245*(4922), 1073–1080.
7. Riordan, J.R., Rommens, J.M., Kerem, B., Alon, N., Rozmahel, R., Grzelczak, Z., Zielenski, J., Lok, S., Plavsic, N., Chou, J.L., et al. Identification of the cystic fibrosis gene: Cloning and characterization of complementary DNA. *Science* **1989**, *245*(4922), 1066–1073.
8. Higgins, C.F. ABC transporters: Physiology, structure and mechanism—An overview. *Res. Microbiol.* **2001**, *152*(3–4), 205–210.
9. Zielenski, J., Tsui, L.C. Cystic fibrosis: Genotypic and phenotypic variations. *Annu. Rev. Genet.* **1995**, *29*, 777–807.
10. Davis, P.B. Cystic fibrosis. *Pediatr. Rev.* **2001**, *22*(8), 257–264.
11. Sharma, M., Pampinella, F., Nemes, C., Benharouga, M., So, J., Du, K., Bache, K.G., Papsin, B., Zerangue, N., Stenmark, H., Lukacs, G.L. Misfolding diverts CFTR from recycling to degradation: Quality control at early endosomes. *J. Cell Biol.* **2004**, *164*(6), 923–933.
12. Du, K., Sharma, M., Lukacs, G.L. The DeltaF508 cystic fibrosis mutation impairs domain-domain interactions and arrests post-translational folding of CFTR. *Nat. Struct. Mol. Biol.* **2005**, *12*(1), 17–25.
13. Lommatzsch, S.T., Aris, R. Genetics of cystic fibrosis. *Semin. Respir. Crit. Care Med.* **2009**, *30*(5), 531–538.
14. Koch, C., Cuppens, H., Rainisio, M., Madessani, U., Harms, H., Hodson, M., Mastella, G., Navarro, J., Strandvik,

B., McKenzie, S. European Epidemiologic Registry of Cystic Fibrosis (ERCF): Comparison of major disease manifestations between patients with different classes of mutations. *Pediatr. Pulmonol.* **2001**, *31*(1), 1–12.

15. Buhl, R. Imbalance between oxidants and antioxidants in the lungs of HIV-seropositive individuals. *Chem. Biol. Interact.* **1994**, *91*(2–3), 147–158.

16. Velsor, L.W., van Heeckeren, A., Day, B.J. Antioxidant imbalance in the lungs of cystic fibrosis transmembrane conductance regulator protein mutant mice. *Am. J. Physiol. Lung Cell. Mol. Physiol.* **2001**, *281*(1), L31–L38.

17. Conner, G.E., Wijkstrom-Frei, C., Randell, S.H., Fernandez, V.E., Salathe, M. The lactoperoxidase system links anion transport to host defense in cystic fibrosis. *FEBS Lett.* **2007**, *581*(2), 271–278.

18. Gould, N.S., Gauthier, S., Kariya, C.T., Min, E., Huang, J., Day, B.J. Hypertonic saline increases lung epithelial lining fluid glutathione and thiocyanate: Two protective CFTR-dependent thiols against oxidative injury. *Respir. Res.* **2010**, *11*, 119.

19. Mall, M.A. Role of cilia, mucus, and airway surface liquid in mucociliary dysfunction: Lessons from mouse models. *J. Aerosol Med. Pulm. Drug Deliv.* **2008**, *21*(1), 13–24.

20. Knowles, M.R., Stutts, M.J., Spock, A., Fischer, N., Gatzy, J.T., Boucher, R.C. Abnormal ion permeation through cystic fibrosis respiratory epithelium. *Science* **1983**, *221*(4615), 1067–1070.

21. Matthews, L.W., Spector, S., Lemm, J., Potter, J.L. Studies on pulmonary secretions. I. The over-all chemical composition of pulmonary secretions from patients with cystic fibrosis, bronchiectasis, and laryngectomy. *Am. Rev. Respir. Dis.* **1963**, *88*, 199–204.

22. Regnis, J.A., Robinson, M., Bailey, D.L., Cook, P., Hooper, P., Chan, H.K., Gonda, I., Bautovich, G., Bye, P.T. Mucociliary clearance in patients with cystic fibrosis and in normal subjects. *Am. J. Respir. Crit. Care Med.* **1994**, *150*(1), 66–71.

23. Heeckeren, A., Walenga, R., Konstan, M.W., Bonfield, T., Davis, P.B., Ferkol, T. Excessive inflammatory response of cystic fibrosis mice to bronchopulmonary infection with *Pseudomonas aeruginosa*. *J. Clin. Invest.* **1997**, *100*(11), 2810–2815.

24. Saadane, A., Soltys, J., Berger, M. Role of IL-10 deficiency in excessive nuclear factor-kappaB activation and lung inflammation in cystic fibrosis transmembrane conductance regulator knockout mice. *J. Allergy Clin. Immunol.* **2005**, *115*(2), 405–411.

25. Rottner, M., Freyssinet, J.M., Martinez, M.C. Mechanisms of the noxious inflammatory cycle in cystic fibrosis. *Respir. Res.* **2009**, *10*, 23.

26. Lambeth, J.D., Kawahara, T., Diebold, B. Regulation of Nox and Duox enzymatic activity and expression. *Free Radic. Biol. Med.* **2007**, *43*(3), 319–331.

27. Fischer, H. Mechanisms and function of DUOX in epithelia of the lung. *Antioxid. Redox Signal.* **2009**, *11*(10), 2453–2465.

28. Gattas, M.V., Forteza, R., Fragoso, M.A., Fregien, N., Salas, P., Salathe, M., Conner, G.E. Oxidative epithelial host defense is regulated by infectious and inflammatory stimuli. *Free Radic. Biol. Med.* **2009**, *47*(10), 1450–1458.

29. Day, B.J., Patel, M., Calavetta, L., Chang, L.Y., Stamler, J.S. A mechanism of paraquat toxicity involving nitric oxide synthase. *Proc. Natl. Acad. Sci. U.S.A.* **1999**, *96*(22), 12760–12765.

30. Grasemann, H., Gaston, B., Fang, K., Paul, K., Ratjen, F. Decreased levels of nitrosothiols in the lower airways of patients with cystic fibrosis and normal pulmonary function. *J. Pediatr.* **1999**, *135*(6), 770–772.

31. Grasemann, H. Total sputum nitrate plus nitrite is raised during acute pulmonary infection in cystic fibrosis. *Am. J. Respir. Crit. Care Med.* **1999**, *159*(2), 684–685.

32. Hull, J., Vervaart, P., Grimwood, K., Phelan, P. Pulmonary oxidative stress response in young children with cystic fibrosis. *Thorax* **1997**, *52*(6), 557–560.

33. Velsor, L.W., Kariya, C., Kachadourian, R., Day, B.J. Mitochondrial oxidative stress in the lungs of cystic fibrosis transmembrane conductance regulator protein mutant mice. *Am. J. Respir. Cell Mol. Biol.* **2006**, *35*(5), 579–586.

34. Wendel, M., Heller, A.R. Mitochondrial function and dysfunction in sepsis. *Wien. Med. Wochenschr.* **2010**, *160*(5–6), 118–123.

35. Khan, T.Z., Wagener, J.S., Bost, T., Martinez, J., Accurso, F.J., Riches, D.W. Early pulmonary inflammation in infants with cystic fibrosis. *Am. J. Respir. Crit. Care Med.* **1995**, *151*(4), 1075–1082.

36. Brennan, S. Innate immune activation and cystic fibrosis. *Paediatr. Respir. Rev.* **2008**, *9*(4), 271–279.

37. Amitani, R., Wilson, R., Rutman, A., Read, R., Ward, C., Burnett, D., Stockley, R.A., Cole, P.J. Effects of human neutrophil elastase and *Pseudomonas aeruginosa* proteinases on human respiratory epithelium. *Am. J. Respir. Cell Mol. Biol.* **1991**, *4*(1), 26–32.

38. Fischer, B.M., Voynow, J.A. Neutrophil elastase induces MUC5AC gene expression in airway epithelium via a pathway involving reactive oxygen species. *Am. J. Respir. Cell Mol. Biol.* **2002**, *26*(4), 447–452.

39. Schreck, R., Rieber, P., Baeuerle, P.A. Reactive oxygen intermediates as apparently widely used messengers in the activation of the NF-kappa B transcription factor and HIV-1. *EMBO J.* **1991**, *10*(8), 2247–2258.

40. Blackwell, T.S., Blackwell, T.R., Holden, E.P., Christman, B.W., Christman, J.W. In vivo antioxidant treatment suppresses nuclear factor-kappa B activation and neutrophilic lung inflammation. *J. Immunol.* **1996**, *157*(4), 1630–1637.

41. Roum, J.H., Buhl, R., McElvaney, N.G., Borok, Z., Crystal, R.G. Systemic deficiency of glutathione in cystic fibrosis. *J. Appl. Physiol.* **1993**, *75*(6), 2419–2424.

42. van der Vliet, A., Eiserich, J.P., Marelich, G.P., Halliwell, B., Cross, C.E. Oxidative stress in cystic fibrosis: Does it occur and does it matter? *Adv. Pharmacol.* **1997**, *38*, 491–513.

43. Lagrange-Puget, M., Durieu, I., Ecochard, R., Abbas-Chorfa, F., Drai, J., Steghens, J.P., Pacheco, Y., Vital-Durand, D., Bellon, G. Longitudinal study of oxidative status in 312 cystic fibrosis patients in stable state and during bronchial exacerbation. *Pediatr. Pulmonol.* **2004**, *38*(1), 43–49.

44. Ater, J.L., Herbst, J.J., Landaw, S.A., O'Brien, R.T. Relative anemia and iron deficiency in cystic fibrosis. *Pediatrics* **1983**, *71*(5), 810–814.

45. Portal, B.C., Richard, M.J., Faure, H.S., Hadjian, A.J., Favier, A.E. Altered antioxidant status and increased lipid peroxidation in children with cystic fibrosis. *Am. J. Clin. Nutr.* **1995**, *61*(4), 843–847.

46. Ward, K.P., Arthur, J.R., Russell, G., Aggett, P.J. Blood selenium content and glutathione peroxidase activity in children with cystic fibrosis, coeliac disease, asthma, and epilepsy. *Eur. J. Pediatr.* **1984**, *142*(1), 21–24.

47. Yoshimura, K., Nakamura, H., Trapnell, B.C., Dalemans, W., Pavirani, A., Lecocq, J.P., Crystal, R.G. The cystic fibrosis gene has a "housekeeping"-type promoter and is expressed at low levels in cells of epithelial origin. *J. Biol. Chem.* **1991**, *266*(14), 9140–9144.

48. Rich, D.P., Anderson, M.P., Gregory, R.J., Cheng, S.H., Paul, S., Jefferson, D.M., McCann, J.D., Klinger, K.W., Smith, A.E., Welsh, M.J. Expression of cystic fibrosis transmembrane conductance regulator corrects defective chloride channel regulation in cystic fibrosis airway epithelial cells. *Nature* **1990**, *347*(6291), 358–363.

49. Finch, E. The sweat test in the diagnosis of fibrocystic disease of the pancreas. *J. Clin. Pathol.* **1957**, *10*(3), 270–272.

50. Kunzelmann, K., Schreiber, R. CFTR, a regulator of channels. *J. Membr. Biol.* **1999**, *168*(1), 1–8.

51. Hudson, V.M. Rethinking cystic fibrosis pathology: The critical role of abnormal reduced glutathione (GSH) transport caused by CFTR mutation. *Free Radic. Biol. Med.* **2001**, *30*(12), 1440–1461.

52. Poulsen, J.H., Fischer, H., Illek, B., Machen, T.E. Bicarbonate conductance and pH regulatory capability of cystic fibrosis transmembrane conductance regulator. *Proc. Natl. Acad. Sci. U.S.A.* **1994**, *91*(12), 5340–5344.

53. Tarran, R., Loewen, M.E., Paradiso, A.M., Olsen, J.C., Gray, M.A., Argent, B.E., Boucher, R.C., Gabriel, S.E. Regulation of murine airway surface liquid volume by CFTR and Ca2+-activated Cl-conductances. *J. Gen. Physiol.* **2002**, *120*(3), 407–418.

54. Pilewski, J.M., Frizzell, R.A. How do cystic fibrosis transmembrane conductance regulator mutations produce lung disease? *Curr. Opin. Pulm. Med.* **1995**, *1*(6), 435–443.

55. Nixon, G.M., Armstrong, D.S., Carzino, R., Carlin, J.B., Olinsky, A., Robertson, C.F., Grimwood, K., Wainwright, C. Early airway infection, inflammation, and lung function in cystic fibrosis. *Arch. Dis. Child.* **2002**, *87*(4), 306–311.

56. Moskowitz, S.M., Gibson, R.L., Effmann, E.L. Cystic fibrosis lung disease: Genetic influences, microbial interactions, and radiological assessment. *Pediatr. Radiol.* **2005**, *35*(8), 739–757.

57. Collins, F.S. Cystic fibrosis: Molecular biology and therapeutic implications. *Science* **1992**, *256*(5058), 774–779.

58. Ballatori, N., Hammond, C.L., Cunningham, J.B., Krance, S.M., Marchan, R. Molecular mechanisms of reduced glutathione transport: Role of the MRP/CFTR/ABCC and OATP/SLC21A families of membrane proteins. *Toxicol. Appl. Pharmacol.* **2005**, *204*(3), 238–255.

59. Linsdell, P., Hanrahan, J.W. Glutathione permeability of CFTR. *Am. J. Physiol.* **1998**, *275*(1 Pt 1), C323–C326.

60. Gao, L., Kim, K.J., Yankaskas, J.R., Forman, H.J. Abnormal glutathione transport in cystic fibrosis airway epithelia. *Am. J. Physiol.* **1999**, *277*(1 Pt 1), L113–L118.

61. Kogan, I., Ramjeesingh, M., Li, C., Kidd, J.F., Wang, Y., Leslie, E.M., Cole, S.P., Bear, C.E. CFTR directly mediates nucleotide-regulated glutathione flux. *EMBO J.* **2003**, *22*(9), 1981–1989.

62. Day, B.J., van Heeckeren, A.M., Min, E., Velsor, L.W. Role for cystic fibrosis transmembrane conductance regulator protein in a glutathione response to bronchopulmonary pseudomonas infection. *Infect. Immun.* **2004**, *72*(4), 2045–2051.

63. Kariya, C., Leitner, H., Min, E., van Heeckeren, C., van Heeckeren, A., Day, B.J. A role for CFTR in the elevation of glutathione levels in the lung by oral glutathione administration. *Am. J. Physiol. Lung Cell. Mol. Physiol.* **2007**, *292*(6), L1590–L1597.

64. Linsdell, P. Thiocyanate as a probe of the cystic fibrosis transmembrane conductance regulator chloride channel pore. *Can. J. Physiol. Pharmacol.* **2001**, *79*(7), 573–579.

65. Illek, B., Tam, A.W., Fischer, H., Machen, T.E. Anion selectivity of apical membrane conductance of Calu 3 human airway epithelium. *Pflugers Arch.* **1999**, *437*(6), 812–822.

66. van Dalen, C.J., Whitehouse, M.W., Winterbourn, C.C., Kettle, A.J. Thiocyanate and chloride as competing substrates for myeloperoxidase. *Biochem. J.* **1997**, *327*, 487–492.

67. Frankenberg, L., Sorbo, B. Effect of cyanide antidotes on the metabolic conversion of cyanide to thiocyanate. *Arch. Toxicol.* **1975**, *33*(2), 81–89.

68. Conner, G.E., Salathe, M., Forteza, R. Lactoperoxidase and hydrogen peroxide metabolism in the airway. *Am. J. Respir. Crit. Care Med.* **2002**, *166*(12 Pt 2), S57–S61.

69. Pedemonte, N., Caci, E., Sondo, E., Caputo, A., Rhoden, K., Pfeffer, U., Di Candia, M., Bandettini, R., Ravazzolo, R., Zegarra-Moran, O., Galietta, L.J. Thiocyanate transport in resting and IL-4-stimulated human bronchial epithelial cells: Role of pendrin and anion channels. *J. Immunol.* **2007**, *178*(8), 5144–5153.

70. Wijkstrom-Frei, C., El-Chemaly, S., Ali-Rachedi, R., Gerson, C., Cobas, M.A., Forteza, R., Salathe, M., Conner, G.E. Lactoperoxidase and human airway host defense. *Am. J. Respir. Cell Mol. Biol.* **2003**, *29*(2), 206–212.

71. Paredi, P., Kharitonov, S.A., Leak, D., Shah, P.L., Cramer, D., Hodson, M.E., Barnes, P.J. Exhaled ethane is elevated

in cystic fibrosis and correlates with carbon monoxide levels and airway obstruction. *Am. J. Respir. Crit. Care Med.* **2000**, *161*(4 Pt 1), 1247–1251.

72. Starosta, V., Rietschel, E., Paul, K., Baumann, U., Griese, M. Oxidative changes of bronchoalveolar proteins in cystic fibrosis. *Chest* **2006**, *129*(2), 431–437.

73. Boncoeur, E., Roque, T., Bonvin, E., Saint-Criq, V., Bonora, M., Clement, A., Tabary, O., Henrion-Caude, A., Jacquot, J. Cystic fibrosis transmembrane conductance regulator controls lung proteasomal degradation and nuclear factor-kappaB activity in conditions of oxidative stress. *Am. J. Pathol.* **2008**, *172*(5), 1184–1194.

74. Cantin, A.M., North, S.L., Hubbard, R.C., Crystal, R.G. Normal alveolar epithelial lining fluid contains high levels of glutathione. *J. Appl. Physiol.* **1987**, *63*(1), 152–157.

75. Gould, N.S., Min, E., Gauthier, S., Chu, H.W., Martin, R., Day, B.J. Aging adversely affects the cigarette smoke-induced glutathione adaptive response in the lung. *Am. J. Respir. Crit. Care Med.* **2010**, *182*(9), 1114–1122.

76. Gould, N.S., Day, B.J. Targeting maladaptive glutathione responses in lung disease. *Biochem. Pharmacol.* **2011**, *81*(2), 187–193.

77. Cantin, A.M., Bilodeau, G., Ouellet, C., Liao, J., Hanrahan, J.W. Oxidant stress suppresses CFTR expression. *Am. J. Physiol. Cell Physiol.* **2006**, *290*(1), C262–C270.

78. Griese, M., Ramakers, J., Krasselt, A., Starosta, V., Van Koningsbruggen, S., Fischer, R., Ratjen, F., Mullinger, B., Huber, R.M., Maier, K., Rietschel, E., Scheuch, G. Improvement of alveolar glutathione and lung function but not oxidative state in cystic fibrosis. *Am. J. Respir. Crit. Care Med.* **2004**, *169*(7), 822–828.

79. Winklhofer-Roob, B.M., Puhl, H., Khoschsorur, G., van't Hof, M.A., Esterbauer, H., Shmerling, D.H. Enhanced resistance to oxidation of low density lipoproteins and decreased lipid peroxide formation during beta-carotene supplementation in cystic fibrosis. *Free Radic. Biol. Med.* **1995**, *18*(5), 849–859.

80. Brown, R.K., McBurney, A., Lunec, J., Kelly, F.J. Oxidative damage to DNA in patients with cystic fibrosis. *Free Radic. Biol. Med.* **1995**, *18*(4), 801–806.

81. Chen, J., Kinter, M., Shank, S., Cotton, C., Kelley, T.J., Ziady, A.G. Dysfunction of Nrf-2 in CF epithelia leads to excess intracellular H2O2 and inflammatory cytokine production. *PLoS ONE* **2008**, *3*(10), e3367.

82. Antigny, F., Girardin, N., Raveau, D., Frieden, M., Becq, F., Vandebrouck, C. Dysfunction of mitochondria Ca2+ uptake in cystic fibrosis airway epithelial cells. *Mitochondrion* **2009**, *9*(4), 232–241.

83. Shapiro, B.L. Mitochondrial dysfunction, energy expenditure, and cystic fibrosis. *Lancet* **1988**, *2*(8605), 289.

84. Gaston, B., Drazen, J.M., Loscalzo, J., Stamler, J.S. The biology of nitrogen oxides in the airways. *Am. J. Respir. Crit. Care Med.* **1994**, *149*(2 Pt 1), 538–551.

85. Gegg, M.E., Beltran, B., Salas-Pino, S., Bolanos, J.P., Clark, J.B., Moncada, S., Heales, S.J. Differential effect of nitric oxide on glutathione metabolism and mitochondrial function in astrocytes and neurones: Implications for neuroprotection/neurodegeneration? *J. Neurochem.* **2003**, *86*(1), 228–237.

86. Ridnour, L.A., Sim, J.E., Choi, J., Dickinson, D.A., Forman, H.J., Ahmad, I.M., Coleman, M.C., Hunt, C.R., Goswami, P.C., Spitz, D.R. Nitric oxide-induced resistance to hydrogen peroxide stress is a glutamate cysteine ligase activity-dependent process. *Free Radic. Biol. Med.* **2005**, *38*(10), 1361–1371.

87. Singh, S.P., Wishnok, J.S., Keshive, M., Deen, W.M., Tannenbaum, S.R. The chemistry of the S-nitrosoglutathione/glutathione system. *Proc. Natl. Acad. Sci. U.S.A.* **1996**, *93*(25), 14428–14433.

88. Bannenberg, G., Xue, J., Engman, L., Cotgreave, I., Moldeus, P., Ryrfeldt, A. Characterization of bronchodilator effects and fate of S-nitrosothiols in the isolated perfused and ventilated guinea pig lung. *J. Pharmacol. Exp. Ther.* **1995**, *272*(3), 1238–1245.

89. Francoeur, C., Denis, M. Nitric oxide and interleukin-8 as inflammatory components of cystic fibrosis. *Inflammation* **1995**, *19*(5), 587–598.

90. Balfour-Lynn, I.M., Laverty, A., Dinwiddie, R. Reduced upper airway nitric oxide in cystic fibrosis. *Arch. Dis. Child.* **1996**, *75*(4), 319–322.

91. Dotsch, J., Demirakca, S., Terbrack, H.G., Huls, G., Rascher, W., Kuhl, P.G. Airway nitric oxide in asthmatic children and patients with cystic fibrosis. *Eur. Respir. J.* **1996**, *9*(12), 2537–2540.

92. Grasemann, H., Michler, E., Wallot, M., Ratjen, F. Decreased concentration of exhaled nitric oxide (NO) in patients with cystic fibrosis. *Pediatr. Pulmonol.* **1997**, *24*(3), 173–177.

93. Grasemann, H., Ioannidis, I., Tomkiewicz, R.P., de Groot, H., Rubin, B.K., Ratjen, F. Nitric oxide metabolites in cystic fibrosis lung disease. *Arch. Dis. Child.* **1998**, *78*(1), 49–53.

94. Kelley, T.J., Drumm, M.L. Inducible nitric oxide synthase expression is reduced in cystic fibrosis murine and human airway epithelial cells. *J. Clin. Invest.* **1998**, *102*(6), 1200–1207.

95. Lundberg, J.O., Nordvall, S.L., Weitzberg, E., Kollberg, H., Alving, K. Exhaled nitric oxide in paediatric asthma and cystic fibrosis. *Arch. Dis. Child.* **1996**, *75*(4), 323–326.

96. Jones, K.L., Bryan, T.W., Jinkins, P.A., Simpson, K.L., Grisham, M.B., Owens, M.W., Milligan, S.A., Markewitz, B.A., Robbins, R.A. Superoxide released from neutrophils causes a reduction in nitric oxide gas. *Am. J. Physiol.* **1998**, *275*(6 Pt 1), L1120–L1126.

97. Malle, E., Furtmuller, P.G., Sattler, W., Obinger, C. Myeloperoxidase: A target for new drug development? *Br. J. Pharmacol.* **2007**, *152*(6), 838–854.

98. Dallegri, F., Ballestrero, A., Frumento, G., Patrone, F. Role of hypochlorous acid and chloramines in the extracellular cytolysis by neutrophil polymorphonuclear leukocytes. *J. Clin. Lab. Immunol.* **1986**, *20*(1), 37–41.

99. Koller, D.Y., Urbanek, R., Gotz, M. Increased degranulation of eosinophil and neutrophil granulocytes in cystic fibrosis. *Am. J. Respir. Crit. Care Med.* **1995**, *152*(2), 629–633.

100. Xu, Y., Szep, S., Lu, Z. The antioxidant role of thiocyanate in the pathogenesis of cystic fibrosis and other inflammation-related diseases. *Proc. Natl. Acad. Sci. U.S.A.* **2009**, *106*(48), 20515–20519.

101. Winterbourn, C.C., Brennan, S.O. Characterization of the oxidation products of the reaction between reduced glutathione and hypochlorous acid. *Biochem. J.* **1997**, *326*(Pt 1), 87–92.

102. Ashby, M.T., Carlson, A.C., Scott, M.J. Redox buffering of hypochlorous acid by thiocyanate in physiologic fluids. *J. Am. Chem. Soc.* **2004**, *126*(49), 15976–15977.

103. Pettit, R.S., Johnson, C.E. Airway-rehydrating agents for the treatment of cystic fibrosis: Past, present, and future. *Ann. Pharmacother.* **2011**, *45*(1), 49–59.

104. Matsuzaki, T., Suzuki, T., Takata, K. Hypertonicity-induced expression of aquaporin 3 in MDCK cells. *Am. J. Physiol. Cell Physiol.* **2001**, *281*(1), C55–C63.

105. Ernst, S.A., Crawford, K.M., Post, M.A., Cohn, J.A. Salt stress increases abundance and glycosylation of CFTR localized at apical surfaces of salt gland secretory cells. *Am. J. Physiol.* **1994**, *267*(4 Pt 1), C990–C1001.

106. Rasmussen, M., Alexander, R.T., Darborg, B.V., Mobjerg, N., Hoffmann, E.K., Kapus, A., Pedersen, S.F. Osmotic cell shrinkage activates ezrin/radixin/moesin (ERM) proteins: Activation mechanisms and physiological implications. *Am. J. Physiol. Cell Physiol.* **2008**, *294*(1), C197–C212.

107. Ciesla, D.J., Moore, E.E., Musters, R.J., Biffl, W.L., Silliman, C.A. Hypertonic saline alteration of the PMN cytoskeleton: Implications for signal transduction and the cytotoxic response. *J. Trauma* **2001**, *50*(2), 206–212.

108. Levin, M.H., Sullivan, S., Nielson, D., Yang, B., Finkbeiner, W.E., Verkman, A.S. Hypertonic saline therapy in cystic fibrosis: Evidence against the proposed mechanism involving aquaporins. *J. Biol. Chem.* **2006**, *281*(35), 25803–25812.

109. Howard, M., Fischer, H., Roux, J., Santos, B.C., Gullans, S.R., Yancey, P.H., Welch, W.J. Mammalian osmolytes and S-nitrosoglutathione promote Delta F508 cystic fibrosis transmembrane conductance regulator (CFTR) protein maturation and function. *J. Biol. Chem.* **2003**, *278*(37), 35159–35167.

110. Donaldson, S.H., Bennett, W.D., Zeman, K.L., Knowles, M.R., Tarran, R., Boucher, R.C. Mucus clearance and lung function in cystic fibrosis with hypertonic saline. *N. Engl. J. Med.* **2006**, *354*(3), 241–250.

111. Burke, J.F., Mogg, A.E. Suppression of a nonsense mutation in mammalian cells in vivo by the aminoglycoside antibiotics G-418 and paromomycin. *Nucleic Acids Res.* **1985**, *13*(17), 6265–6272.

112. Bedwell, D.M., Kaenjak, A., Benos, D.J., Bebok, Z., Bubien, J.K., Hong, J., Tousson, A., Clancy, J.P., Sorscher, E.J. Suppression of a CFTR premature stop mutation in a bronchial epithelial cell line. *Nat. Med.* **1997**, *3*(11), 1280–1284.

113. Wilschanski, M., Famini, C., Blau, H., Rivlin, J., Augarten, A., Avital, A., Kerem, B., Kerem, E. A pilot study of the effect of gentamicin on nasal potential difference measurements in cystic fibrosis patients carrying stop mutations. *Am. J. Respir. Crit. Care Med.* **2000**, *161*(3 Pt 1), 860–865.

114. Kerem, E. Pharmacologic therapy for stop mutations: How much CFTR activity is enough? *Curr. Opin. Pulm. Med.* **2004**, *10*(6), 547–552.

115. Hamed, S.A. Drug evaluation: PTC-124—A potential treatment of cystic fibrosis and Duchenne muscular dystrophy. *IDrugs* **2006**, *9*(11), 783–789.

116. Sermet-Gaudelus, I., Boeck, K.D., Casimir, G.J., Vermeulen, F., Leal, T., Mogenet, A., Roussel, D., Fritsch, J., Hanssens, L., Hirawat, S., Miller, N.L., Constantine, S., Reha, A., Ajayi, T., Elfring, G.L., Miller, L.L. Ataluren (PTC124) induces cystic fibrosis transmembrane conductance regulator protein expression and activity in children with nonsense mutation cystic fibrosis. *Am. J. Respir. Crit. Care Med.* **2010**, *182*(10), 1262–1272.

117. Illek, B., Fischer, H., Santos, G.F., Widdicombe, J.H., Machen, T.E., Reenstra, W.W. cAMP-independent activation of CFTR Cl channels by the tyrosine kinase inhibitor genistein. *Am. J. Physiol.* **1995**, *268*(4 Pt 1), C886–C893.

118. Howell, L.D., Borchardt, R., Cohn, J.A. ATP hydrolysis by a CFTR domain: Pharmacology and effects of G551D mutation. *Biochem. Biophys. Res. Commun.* **2000**, *271*(2), 518–525.

119. Zegarra-Moran, O., Romio, L., Folli, C., Caci, E., Becq, F., Vierfond, J.M., Mettey, Y., Cabrini, G., Fanen, P., Galietta, L.J. Correction of G551D-CFTR transport defect in epithelial monolayers by genistein but not by CPX or MPB-07. *Br. J. Pharmacol.* **2002**, *137*(4), 504–512.

120. Sloane, P.A., Rowe, S.M. Cystic fibrosis transmembrane conductance regulator protein repair as a therapeutic strategy in cystic fibrosis. *Curr. Opin. Pulm. Med.* **2010**, *16*(6), 591–597.

121. Leitner, H.M., Kachadourian, R., Day, B.J. Harnessing drug resistance: Using ABC transporter proteins to target cancer cells. *Biochem. Pharmacol.* **2007**, *74*(12), 1677–1685.

122. Brechbuhl, H.M., Gould, N., Kachadourian, R., Riekhof, W.R., Voelker, D.R., Day, B.J. Glutathione transport is a unique function of the ATP-binding cassette protein ABCG2. *J. Biol. Chem.* **2010**, *285*(22), 16582–16587.

123. Kalivianakis, M., Minich, D.M., Bijleveld, C.M., van Aalderen, W.M., Stellaard, F., Laseur, M., Vonk, R.J., Verkade, H.J. Fat malabsorption in cystic fibrosis patients receiving enzyme replacement therapy is due to impaired intestinal uptake of long-chain fatty acids. *Am. J. Clin. Nutr.* **1999**, *69*(1), 127–134.

124. Ratjen, F., Wonne, R., Posselt, H.G., Stover, B., Hofmann, D., Bender, S.W. A double-blind placebo controlled trial with oral ambroxol and N-acetylcysteine for mucolytic

treatment in cystic fibrosis. *Eur. J. Pediatr.* **1985**, *144*(4), 374–378.

125. Visca, A., Bishop, C.T., Hilton, S.C., Hudson, V.M. Improvement in clinical markers in CF patients using a reduced glutathione regimen: An uncontrolled, observational study. *J. Cyst. Fibros.* **2008**, *7*(5), 433–436.

126. Hartl, D., Starosta, V., Maier, K., Beck-Speier, I., Rebhan, C., Becker, B.F., Latzin, P., Fischer, R., Ratjen, F., Huber, R.M., Rietschel, E., Krauss-Etschmann, S., Griese, M. Inhaled glutathione decreases PGE2 and increases lymphocytes in cystic fibrosis lungs. *Free Radic. Biol. Med.* **2005**, *39*(4), 463–472.

127. Bishop, C., Hudson, V.M., Hilton, S.C., Wilde, C. A pilot study of the effect of inhaled buffered reduced glutathione on the clinical status of patients with cystic fibrosis. *Chest* **2005**, *127*(1), 308–317.

128. Nagy, A.M., Vanderbist, F., Parij, N., Maes, P., Fondu, P., Neve, J. Effect of the mucoactive drug nacystelyn on the respiratory burst of human blood polymorphonuclear neutrophils. *Pulm. Pharmacol. Ther.* **1997**, *10*(5–6), 287–292.

129. Duijvestijn, Y.C., Brand, P.L. Systematic review of N-acetylcysteine in cystic fibrosis. *Acta Paediatr.* **1999**, *88*(1), 38–41.

14

BIOMARKERS OF OXIDATIVE STRESS IN NEURODEGENERATIVE DISEASES

Rukhsana Sultana, Giovanna Cenini, and D. Allan Butterfield

OVERVIEW

The involvement of oxidative stress in Alzheimer's disease (AD), Parkinson's disease (PD), and amyotrophic lateral sclerosis (ALS), suggest that free radicals play important roles in the onset and progress of neurodegenerative process. Understanding the molecular and biochemical basis of disease pathogenesis is critical for the development of potential neuroprotective therapies for neurodegenerative diseases (AD, PD, and ALS). In this chapter we discuss current knowledge of oxidative stress in relation to neurodegeneration.

14.1 INTRODUCTION

A large number of diseases have been described that involve oxidative and nitrosative stress.[1,2] Therefore, at present it is important to clarify if oxidative and nitrosative stress are strictly involved in onset and progression of diseases, and if oxidative and nitrosative stress products could be used for the identification and diagnosis of a specific pathological condition. Many products of oxidative and nitrosative stress have been proposed and studied in order to find biomarkers of disease, since a validated biomarker is especially important in the case of neurodegenerative diseases (Table 14.1). In order that an oxidative and nitrosative stress product could be used as a marker of disease, it is fundamental that it be chemically stable, accurately quantified, reflect specific oxidation pathways, and have its concentration in biological samples correlated with the severity of the disease.[3] By general definition (from NIH),[4] a biomarker is an indicator of normal processes, pathogenic processes, or pharmacological responses to a therapeutic intervention that is objectively measured. It is believed that biomarkers have great potential in predicting chances for diseases, early diagnosis, and setting standards for the development of new pharmacological treatments.

Neurodegenerative diseases are a varied group of central nervous system disorders all characterized by the progressive loss of neuronal tissues.[5] Tremendous efforts have been made in the past years to identify neuropathological, biochemical, and genetic biomarkers of neurodegenerative diseases for a diagnosis at earlier stages, which presumably would be more amenable to therapy. At the moment, the only way to do a valid neuropathological diagnosis of AD is a postmortem autopsy.[6] Having an early diagnosis of the disease might help in the early treatments of the disease or to slow down the progression of the disease (Fig. 14.1).

The brain is particularly sensitive to oxidative damage because of its high oxygen consumption, relatively low levels of antioxidant defenses, and a high content of polyunsaturated lipids that are easily oxidized.[7] Free radicals have been directly or indirectly implicated in the pathogenesis of several neurodegenerative disease associated to aging, such as AD, PD, and ALS.[8,9] Whether oxidative and nitrosative stress in these disease is casual or a secondary consequence of other processes remains

Molecular Basis of Oxidative Stress: Chemistry, Mechanisms, and Disease Pathogenesis, First Edition. Edited by Frederick A. Villamena.
© 2013 John Wiley & Sons, Inc. Published 2013 by John Wiley & Sons, Inc.

to be determined. However, monitoring the levels of indicators of such damage might be useful both to follow disease progression and to assess the efficacy of antioxidant treatments.[10] Hence, in this chapter, involvement of oxidative stress in neurodegenerative diseases such as AD, PD, and ALS is reviewed.

TABLE 14.1 Summary of Oxidative and Nitrosative Stress Markers in the Central and Peripheral Compartments in Neurodegenerative Diseases

Brain		Blood
Lipid peroxidation		
AD	HNE, MDA, Acrolein, TBARs, F2-IsoPs, F4-NP	HNE, MDA, TBARs
PD	HNE, MDA, Acrolein, IsoPs, TBARs	HNE, MDA, TBARs, F_2-IsoPs
ALS	HNE, MDA	HNE, MDA, TBARs
Protein oxidation and nitration		
AD	PC, 3NT	
PD	PC, 3NT	
ALS	PC, 3NT	
Carbohydrates oxidation		
AD	AGEs, RAGE	AGEs
PD	AGEs, RAGE	
ALS	AGEs, RAGE	AGE. RAGE
DNA/RNA oxidation		
AD	8-OHG, 8-OHdG, NPrG	8-OHG, 8-OHdG
PD	8-OHG, 8-OHdG	8-OHdG
ALS	8-OHdG	8-OHdG

AD is the most prevalent form of dementia in the elderly population. In the United States, over 5 million people suffer from AD. This disorder is a progressive disease characterized by death of neurons and synapses mainly in cerebral cortex and hippocampus regions, resulting in deterioration of cognitive functions.[11] The main neuropathological hallmarks of AD are extracellular senile plaques (SPs) and intracellular neurofibrillary tangles (NFTs). The major components of the SP are β-amyloid peptides (Aβ), while the NFT are fundamentally constituted by hyperphosphorylated-insoluble forms of the tau protein.[12]

PD is a neurodegenerative disease that affects more than 1% of all people over the age of 55. Pathological hallmarks include degeneration of dopaminergic neurons between the substantia nigra (SN) and the striatum that causes the characteristic clinical signs (slowed movements, rigidity, tremors).[13,14] Another key neuropathological mark of PD is the formation of Lewy bodies (LBs), which are cytoplasmic inclusions, composed of α-synuclein protein in the dopaminergic neurons of substantia nigra and other brain regions (cortex and magnocellular basal forebrain nuclei).[15] In a small number of families, PD is inherited in a Mendelian autosomal dominant or autosomal recessive way,[16] while AD is inherited in an autosomally dominant manner.

ALS is an age-dependent motor neuron neurodegenerative disease characterized by neuronal death of the upper and lower motor neurons, skeletal muscle atrophy,

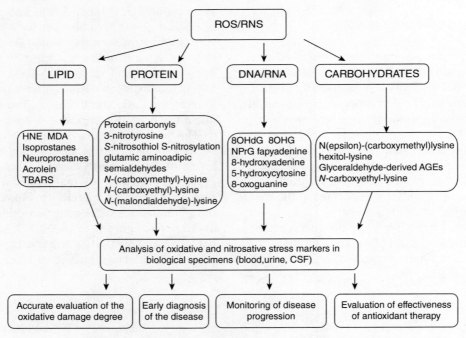

Figure 14.1 Potential use of oxidative and nitrosative markers.

paralysis, and death.[17] The primary goal for scientists with regard to the biomarkers of ALS is to show direct evidence of motor neuronal degeneration within the brain or spinal cord. Approximately 2% of all ALS and 20% of familial cases are associated with mutations in the gene for copper, zinc superoxide dismutase (SOD1).[18]

14.2 BIOMARKERS OF PROTEIN OXIDATION/NITRATION

14.2.1 Protein Carbonyls

Oxidative modification of proteins in most cases is known to affect their function. In this section of protein oxidation, we discuss protein carbonylation and protein nitration that have been used as common makers to study the effect of reactive oxygen species/reactive nitrogen species (ROS/RNS) on proteins. Protein carbonyls are formed by either the direct oxidation of certain amino acid side chains such as Lys, Arg, Pro, Thr, His, and so on, among others, by peptide backbone scission, by Michael addition reactions of His, Lys, and Cys residues with products of lipid peroxidation (e.g., 4-hydroxy-2-nonenal [HNE]), or by glycoxidation reactions with the Lys ε-amino group.[19,20] Protein carbonyls are generally detoxified by enzymes such as aldehyde dehydrogenase (ALDH) or by reduction to their corresponding alcohols by carbonyl reductase (CR).[21] The most commonly used approach for the detection of protein carbonyls is by derivatization of the carbonyl groups with hydrazine compounds such as 2,4-dinitrophenylhydrazine, followed by spectrometry, high-performance liquid chromatography (HPLC), or immunochemical detection.[22–24] In the postmortem frontal- and occipital-pole brain samples from AD, young, and age-matched controls, the levels of protein carbonyls showed an exponential increase with age, at double the rate in the frontal pole compared with the occipital pole.[23] Studies from our laboratory showed an increase of 42 and 37% of protein carbonyls in AD hippocampus and inferior parietal lobule (IPL), respectively, compared to age-matched controls.[25] Furthermore, the levels of protein carbonyls were also found to be increased in the frontal cortex of Swedish APP670/671 FAD mutation.[26] Using immunoprecipitation technique followed by Western blot we found increased oxidation of glutamine synthetase (GS), creatine kinase (CK), and beta actin in AD-affected region, and decreased activities of GS and CK were related to the increased oxidation of these proteins.[25,27]

The levels of CR were found to be increased in brain of AD and Down subjects (trisomy of chromosome 21, which harbor the gene for APP),[21] suggesting that it might be a response to the increased levels of protein carbonyls or decreased clearance of this protein which might have underwent oxidation. Reed et al. showed that CR is HNE-modified in mild cognitive impairment (MPI) brain,[28] which arguably is the earliest form of AD. Studies of animal models of amyloid beta-peptide showed increased levels of protein carbonyls, suggesting that amyloid beta-peptide plays an important role in elevating the protein carbonyls and consequently oxidative stress, cell loss, and AD pathogenesis.[29]

Our laboratory is the first to use redox proteomics techniques to identify carbonylated proteins in the IPL region of AD.[30,31] The redox proteomics approach led to the identification of a number of targets of protein carbonylation in AD brain. The identified proteins perform a wide variety of cellular functions such as energy metabolism, protein degradation, structural, neurotransmission, lipid asymmetry, pH regulation, cell cycle, tau phosphorylation, Abeta production, and mitochondrial function, all of which relate well with the histopathological, biochemical, and clinical presentation of AD.[30–33] For example, energy metabolic alterations in AD brain due to oxidative modification can be correlated well with the positron emission tomography (PET) studies that showed decreased glucose utilization in AD brain.[34] Further, the identification of an oxidatively modified brain protein does not only affect the function of this protein, but it also affects the function of other proteins that interact with it. Studies showed that sometimes a protein could perform multiple functions in a cell. For example, enolase, a protein known to be involved in the glycolytic cycle of glucose metabolism, has been reported to have a number of other nonglycolytic functions such as hypoxic-stress protein,[35] binding to polynucleotides,[36] and c-Myc binding and transcription protein,[37] and so on. Hence, oxidation of one protein could dampen a number of cellular functions in neurons and consequently be involved in AD.[38] Redox proteomics approaches also led to identification of peptidylprolyl cis/trans isomerase (Pin1) as oxidatively modified protein in AD and also in MCI. Pin1 function is critical for proper protein assembly and folding, intracellular transport, intracellular signaling, transcription, cell cycle progression, and apoptosis. Studies demonstrated that Pin1 has the ability to regulate APP processing, also phosphorylation of tau protein[39,40]; hence the oxidation of this protein could be a potential mechanism in the progression of AD. A recent study from our laboratory on APP(NLh)/APP(NLh) × PS-1(P264L)/PS-1(P264L) human double mutant knockin mice model of AD suggests that amyloid beta-peptide is involved in oxidative modification of this protein.[41] Our laboratory is further exploring the importance of Pin1 and other oxidatively modified proteins in the progression and pathogenesis of AD.

The levels of protein carbonyls were also found to be elevated in MCI brain.[42,43] Furthermore, redox proteomics studies from our laboratory in MCI brain led to the identification of a number of common targets of protein carbonylation, between AD and MCI, such as enolase, Pin1, and GS, consistent with the hypothesis that oxidative stress is critical to the pathogenesis of AD[33] and might play an important role in the progression of AD.[43,44]

Oxidative stress is elevated in PD brain, and this has been associated with mitochondrial complex I dysfunction.[45] A study using human fetal dopaminergic primary neuronal cultures overexpressing wild-type α-synuclein showed decreased mitochondrial complex I activity and increased ROS production. The increase in the ROS production is related to the metabolism of dopamine by the mitochondrial enzyme monoamine oxidase (MAO) during which molecular oxygen is converted to hydrogen peroxide (H_2O_2), an ROS. This increase in ROS is an essential component of oxidative stress. Postmortem PD brain showed increased levels of protein carbonyls in substantia nigra pars compacta compared to controls and other brain regions.[46] The increase in protein carbonyls is reported in dopaminergic neurons and has been shown to be mostly associated with high-molecular-weight proteins.[46] A recent study in PD subjects showed that increase of protein carbonyls is associated with short telomere length, suggesting that oxidative stress may be involved in the telomere abrasion in PD and consequently in the pathogenesis of PD.[47]

Both hereditary and sporadic PD demonstrate loss of dopaminergic neurons that is accompanied by oxidative stress and preceded by glutathione (GSH) depletion. GSH, the tripeptide γ-glutamyl-cysteine-glycine, is important in maintaining the proper redox balance of the cell, and in the case of neurons, it is also important in the regulation of neuronal excitability and viability. In PD brain the levels of GSH and cysteinyl-glycine (Cys-Gly) were reported to be reduced further, suggesting a role of oxidative stress in pathophysiological mechanisms of PD.[48] An *in vivo* study from our laboratory showed that gamma-glutamylcysteinyl ethyl ester (GCEE), a precursor for GSH synthesis, reduces dopamine-associated striatal neuron loss in 1-methyl-4-phenyl-1,2,3,6-tetrahydropyridine-treated mice.[49]

In the hemiparkinsonian animal model, two proteins, that is, α-enolase and β-actin, were identified as being oxidatively modified.[50] Using a redox proteomics approach, we identified carbonic anhydrase (CA-II), alpha-enolase, and lactate dehydrogenase 2 (LDH2)[51] as excessively carbonylated proteins with reduced activity in the brain stem of symptomatic mice with overexpression of an A30P mutation in α-synuclein compared to nontransgenic mice, suggesting that alteration in the cellular process due to oxidative modification of proteins might be important in the pathogenesis of PD.[51]

Oxidative stress is also implicated in the pathogenesis of ALS. Postmortem brain tissue from ALS showed increased oxidative stress.[52] The levels of protein carbonyls have been shown to be elevated in the spinal cord[53] and motor cortex[54] from familial ALS (fALS)[55] and sporadic amyotrophic lateral sclerosis (sALS)[53,56] subjects. Using redox proteomics, our laboratory found SOD1, translationally controlled tumor protein (TCTP), UCH-L1, and αB-crystallin as proteins with elevated carbonyl levels in the spinal cord of G93A-SOD1 transgenic mice compared to wild-type mice.[57] The identification of these proteins suggests the involvement of the protein carbonyl modification and thereby oxidation stress in altering the normal biological functions in the cell, which may be critical in the pathogenesis of ALS.

14.2.2 Protein Nitration

The other marker of protein oxidation, that is, protein nitration, was reported to be increased in AD brain and ventricular cerebrospinal fluid (VF),[58,59] which correlated with increased levels of nitric oxide synthase (NOS) reported in AD brain.[60,61] Furthermore, immunohistochemical studies showed the presence of nitrated tau in pre-tangles, tangles, and tau inclusions in the AD brain, suggesting nitration of tau nitration as an early event in AD pathogenesis.[62,63] Nitration of proteins led to loss of activity of glutamine synthase,[64] ubiquitin,[65] and Mn superoxide dismutase.[66,67]

Proteomics approach led to the identification of large number of proteins that are excessively nitrated in AD brain.[68,69] These proteins include alpha- and gamma-enolase, lactate dehydrogenase (LDH), neuropolypeptide h3, triose phosphate isomerase (TPI), alpha-actin, glyceraldehyde-3-phosphate dehydrogenase (GAPDH), ATP synthase alpha-chain, carbonic anhydrase-II, and voltage-dependent anion channel (VDAC).[68,69] The identified nitrated proteins are involved in regulating various cellular functions such as energy metabolism, maintenance of structure, pH regulation, and mitochondrial functions. As stated above, nitration of proteins also leads to loss of functionality,[69] and the identified nitrated proteins also correlated with AD pathology, biochemical changes, and clinical presentation. Guix et al. confirmed our finding of excess nitration of TPI in hippocampus and frontal cortex of AD subjects.[70] However, unlike other proteins, TPI activity was unaffected in AD brain. It is not clear why nitration does not affect the function of this protein, and we speculate that the structure of TPI provides protection against loss of function. Guix et al. suggested a possible link between decreased glucose metabolism, nitrosylation of TPI, and the formation of Aβ and paired helical filaments.[70] Further, Reyes et al. showed that nitrated tau protein is mostly associated with or in close proximity to amyloid plaques, implying a role of amyloid beta-peptide in inducing nitrosative stress.[71] As stated earlier in the section on protein carbonylation, the nitration of one protein could have implication in various cellular functions. For example, GAPDH is well known for its function in the glycolytic pathway of glucose metabolism; however, this protein also has other functions such as GAPDH can bind to nucleic acid and regulate transcription,[72,73] catalyze microtubule formation and polymerization,[74] bind integral membrane ion pumps associated with Ca^{2+} release,[75] and so on. Furthermore, GAPDH also interacts with a number of small molecules such as tumor necrosis factor (TNF)-alpha, GSH,[76] and so on. Hence, nitration of one protein could have detrimental effect on normal cellular functions. The nitration of proteins like GAPDH and actin also raises a question of these proteins are good to be considered as loading controls in Western blot.

Our laboratory is the first to show increased levels of 3-NT in MCI, arguably the earliest form of AD.[77] Applying proteomics, we identified increased nitration of MDH, α-enolase, multidrug resistant protein-3 (MRP3), glutathione-S-transferase Mu (GST M), glucose regulated protein precursor (GRP), aldolase, and 14-3-3 protein gamma, peroxiredoxin 6 (PR VI), DRP-2, fascin 1, and heat shock protein A8 (HSPA8) protein as specifically nitrated in MCI IPL.[78] The reported nitrated proteins are involved in the regulation of a number of important cellular functions, including energy metabolism, structural functions, cellular signaling, and antioxidant. Some of the targets of protein nitration are common between AD and MCI brain, suggesting potential involvement of these pathways in the transition of MCI to AD.[68,69,78] Further studies are required to delineate the role of nitration in the progression of the AD.

Postmortem PD brain also shows increased levels of protein nitration as indexed by increased levels of protein nitration and NOS.[79] Previous studies showed increased nitration of α-synuclein in brain of individuals with synucleinopathy, suggesting a direct link between oxidative and nitrative damage to the onset and progression of neurodegenerative synucleinopathies.[80] A recent study showed that increased nitration of alpha-synuclein could induce dopaminergic neuronal death.[81] Further, an *in vitro* study showed increased nitration of mitochondrial complex I, suggesting the involvement of nitric oxide (NO)-related events in the pathogenesis of PD.[82] Since the mitochondrial electron transport chain is critical for superoxide production, nitration of complex I might lead to increase production or leakage of superoxide consequently leading to increase production of peroxynitrate and enhanced protein nitration in PD. PD pathogenesis appears to be dependent on NO-related events; hence, compounds that prevent nitrosative damage might have therapeutic value in neurological conditions such as PD. Mythri et al. showed that curcumin protects complex I against peroxynitrite-mediated mitochondrial toxicity and oxidative stress.[83]

Spinal cords of sporadic ALS subjects showed increased levels of nitrotyrosine and NOS in the motor neurons, suggesting upregulation of protein nitration ALS.[84] So far, no proteomics studies have been performed to identify the specific target of protein nitration in ALS.

Taken together, studies conducted thus far suggest that oxidation and nitration of proteins are involved in the progression and pathology of AD, PD, and ALS. Further studies are needed to provide potential pathways involved in the progression of these diseases.

14.3 BIOMARKERS OF LIPID PEROXIDATION

Lipid peroxidation is a process resulting from damage to cellular membranes mediated by ROS that generate several relatively stable end products, including aldehydes, such as malondialdehyde (MDA), HNE, acrolein kong,[85] and isoprostanes,[86] which can be measured in plasma or tissues as markers of oxidative stress. MDA, HNE, and acrolein are able to bind DNA and proteins, in particular nucleophilic aminoacidic residues like Cys, His, and Lys generally inducing an alteration of protein conformation and function.[87] Lipid hydroperoxides and aldehydes can also be adsorbed from the diet, and then excreted in urine. For this reason, the measurements of urinary MDA and HNE can be confounded by diet and should not be used as an index of whole-body lipid peroxidation unless diet is controlled.

There are many evidences that lipid peroxidation of polyunsaturated fatty acids (PUFA) is involved in the onset and progression of many pathologies such as cardiovascular (atherosclerosis, diabetes), and neurodegenerative diseases.[88,89] For example, in the pathogenesis of AD, lipid peroxidation plays a particular role.[90,91] In fact, a number of studies demonstrated increased levels of lipid peroxidation as indicated by elevated levels of the products of lipid peroxidation such as HNE, acrolein, $F_{(2)}$-isoprostane, $F_{(4)}$-isoprostane, and neuroprostanes in AD brain.[92–94] Further, increased levels of HNE-adducted GSH were found in human postmortem brains from AD patients.[95] Normally in cells, HNE–GSH adducts are eliminated by the systems glutathione transferase (GST) and MRP-1. But in AD brain, this detoxification system was found to be a target of HNE with consequent decreased efficiency to eliminate HNE, and subsequent accumulation of HNE protein adducts in neuronal cells.[96,97] Even the proteosome, involved in the removal of damaged proteins from cells, has been demonstrated to form conjugates with HNE and neuroprostanes in both MCI and AD.[98] In addition, significant increase of free HNE in cerebrospinal fluid,[99] amygdala, hippocampus, and parahippocampal gyrus was detected in AD patients compared to control subjects.[100] Moreover, immunocytochemical studies demonstrated that HNE immunoreactivity is present in NFT, but only in some SPs in AD.[101] In blood, some studies demonstrated that HNE is significantly higher in AD compared to healthy subjects.[102,103] Proteomics studies were able to identify regionally specific HNE modification of proteins, that is, ATP synthase, GS, MnSOD, DRP-2 in AD hippocampus, and alpha-enolase, aconitase, aldolase, peroxiredoxin-6, and alpha-tubulin in AD IPL.[104] Also in hippocampus and IPL from subjects with MCI, many proteins oxidatively modified by HNE were identified by proteomics (carbonyl reductase [NADPH], alpha-enolase, lactate dehydrogenase B, phosphoglycerate kinase, heat shock protein 70, ATP synthase alpha chain, pyruvate kinase, actin, and elongation factor tau).[28] Since most of the proteins that undergo HNE modification have been shown to be dysfunctional, these results in the earliest form of AD suggest that HNE-bound proteins may play a key role in the progression and pathogenesis of AD. In addition, these results suggest that proteomics could be a valid method for identification of new markers of neurodegenerative diseases including AD.

In PD–HNE adducts of brain proteins were found.[105] Data from immunocytochemistry demonstrated that HNE levels were increased in dopaminergic cells in the substantia nigra and cerebrospinal fluid in PD.[106] Incubation of rat striatal synaptosomes with various concentrations of HNE induced a dose-dependent decrease of dopamine uptake and Na^+/K^+ ATPase activity and loss of sulfhydryl (SH) groups.[107] These data suggest that HNE is an important mediator of oxidative stress that alters dopamine uptake, and HNE may contribute to the onset and progression of PD. Although the role of oxidation in substantia nigra is well established, the significance of peripheral oxidative stress in PD is still unclear, and the results are apparently contradictory, most probably due to differences in the methods used to measure systemic oxidative stress. However, some studies reported that the concentration of HNE in the cerebrospinal fluid (CSF) and in plasma of PD patients was significantly higher than in control patients.[108] Furthermore, no significant correlation existed between the duration of symptoms and the age of the Parkinson patients with plasma or CSF concentrations of HNE.[108] There also were not significant differences between plasma and CSF concentrations of HNE between patients with untreated PD and those receiving L-DOPA therapy.[108]

HNE plays a critical role also in the motor neuron degeneration in ALS. By immunohistochemistry, HNE-protein modification was detected in ventral horn motor neurons, and immunoprecipitation analysis revealed that one of the proteins modified by HNE was the astrocytic glutamate transporter EAAT2 causing impairment of glutamate transport and excitotoxic motor neuron degeneration in ALS.[109] HNE was also found elevated in CSF of patients with sALS compared with that of the control subjects and other neurological disease.[110,111] By proteomics analysis we[112] analyzed spinal cord tissue of a model of fALS G93A-SOD1 Tg mice to study the HNE-modified proteins. Three significantly HNE-modified proteins in the spinal cord of G93A-SOD1 Tg mice in comparison to the non-Tg mice were identified: dihydropyrimidinase-related protein 2 (DRP-2; neuronal development and repair), heat-shock protein 70 (Hsp70; stress response), and possibly α-enolase (energy metabolism). Since HNE-bound proteins present structural and function alterations, the data found demonstrate that oxidative stress in the form of lipid peroxidation is implicated as a pivotal event in the motor neuron degeneration processes. Furthermore, many other studies were performed by proteomic analysis on CSF from a consistent group of ALS patients and healthy subjects, which identified some proteins with significantly different levels in ALS CSF compared to CSF from control subjects. Therefore, in ALS like in AD, proteomics analysis distinguished the disease condition from the healthy condition.[113–115] At the serum level, HNE was found significantly elevated in ALS patients compared to controls. These levels were higher even at early stages of the disease, and in the sporadic ALS serum than familial ALS, suggesting that familial and sporadic forms are qualitatively different in regard to oxidative stress.[110] This finding may reflect the presence of different mechanisms in the pathogenesis of either form of ALS. Furthermore, the level of serum HNE in ALS patients was positively correlated with the stage of disease, implicating HNE as a possible early indicator of oxidative stress and marker of the disease. The systemic presence of increased HNE in the serum of ALS patients may reflect either the presence of lipid peroxidation in the CNS with diffusion to the peripheral circulation or the activation of additional pathways originating outside the CNS, leading to the formation of HNE and its related adducts.

Another main product of lipid peroxidation, MDA, that is able to form covalent adducts with lysine residues of proteins, was found higher in plasma and serum from AD patients compared to controls subjects.[116] Other studies confirmed these results.[117–119] Further, in AD brain, the increased level of MDA is correlated to decreased levels of one of the most important antioxidant enzymes, superoxide dismutase (SOD).[120] Immunohistochemistry analysis has shown that MDA colocalized with NFT and SP.[121]

MDA was found bound to α-synuclein in the substantia nigra and frontal cortex of all dementia with LB dementias (LBDs) and PD cases examined. Furthermore, it was demonstrated that α-synuclein lipoxidation is an early event in LBDs.[122,123] The basal MDA levels were increased in PD substantia nigra and also in CSF compared with other PD brain regions and control tissue.[124,125] All these observations give support not only to the concept that lipoperoxidation precedes α-synuclein aggregation in LBDs, but also the idea that oxidative-altered proteins are present in cerebral cortex in preclinical PD. Like HNE, the MDA and thiobarbituric-acid-reactive substances (TBARS) levels also were found increased in plasma of PD subjects compared to healthy subjects.[126,127] In addition, Navarro et al. found that the levels of TBARS in frontal cortex brain mitochondria were markedly increased in PD subjects compared to controls, suggesting that PD is not only characterized by substantia nigra dysfunction, but also involves the cerebrum, leading to cognitive decline at the early stages of PD.[128] While plasma MDA levels were inversely related to the age of PD patients, there was no significant correlation between plasma MDA levels and duration of the disease.[129]

In ALS, Hall et al.[130] found an extant lipid peroxidative damage in the spinal cords of a murine model of the disease, the TgN-(SOD1-G93A)G1H mice. Lipid peroxidation was investigated in terms of changes in vitamin E and MDA levels measured by HPLC methods and by MDA–protein adduct immunoreactivity. Compared to non-Tg mice, the TgN-(SOD1-G93A)G1H mice showed an accumulation of spinal cord vitamin E and higher levels of MDA over the 30- to 120-day time span. In addition, MDA–protein adduct immunoreactivity was significantly increased in the lumbar spinal cord and in the cervical cord of the same mice. These results clearly demonstrate an early increase of lipid peroxidation in the lumbar spinal cord in the familial ALS transgenic model, which precedes the onset of clinical motor neuron disease. In another study conducted on TgN-(SOD1-G93A)G1H mice, it was observed a significant elevation in MDA in both red and white skeletal muscles was observed.[131] All these data on murine models were confirmed in spinal cord from sporadic ALS and familial ALS subjects where MDA–adduct proteins were found increased in both neurons and endothelial cells when compared to normal controls.[54] In the periphery, MDA and 2-TBARS levels have been found to be significantly higher in the plasma and serum of ALS patients than in either age-matched controls or young adults.[132–134]

Acrolein has been reported to react with DNA bases like guanine, leading to increased formation of acrolein-deoxyguanosine in AD brain.[135] In PD, α-synuclein is modified by acrolein (ACR) since histopathological observations in dopaminergic neurons from PD brains showed the colocalization of α-synuclein and acrolein.[136] Acrolein-adduct proteins, however, were not detectable in the spinal cord of sALS or fALS patients.[137]

A longitudinal study showed that levels of CFS $F_{(2)}$-IsoPs in AD patients were significantly increased during the follow-up period, and also significantly declined in patients accepting antioxidant treatment.[138] Furthermore, other studies demonstrated higher levels of the isoprostane, 8,12-iso-iPF (2alpha)-VI in CSF in AD[139,140] and MCI,[141] suggesting that this lipid peroxidation product could be another marker to identify AD at early stages. MCI brains showed increased levels of protein-bound HNE, TBARs, MDA, $F_{(2)}$-IsoPs, and $F_{(4)}$-NP.[42,142] The significance of $F_{(2)}$-IsoPs in AD and MCI plasma is still controversial, because Pratico et al. found high levels of $F_{(2)}$-IsoPs in plasma, CSF, and urine of MCI patients,[141] and the same research group showed similar results in AD patients.[143] However, in a 2007 study, plasma $F_{(2)}$-IsoPs levels were not increased in AD or MCI, and most probably, this result was affected by the high percentage of antioxidant used in MCI and AD patients studied.[92] Another research has reported that plasma and urine $F_{(2)}$-IsoPs levels did not accurately reflect CNS levels in AD patients.[144] At the present time, more work will be needed to support the validity of $F_{(2)}$-IsoPs as a plasma biomarker for AD. Like other markers of lipid peroxidation, the isofurans (IsoFs) levels are also significantly high in PD substantia nigra compared to other regions of the PD brain and compared to control and also to AD.[138] On the contrary $F_{(2)}$-isoprostanes ($F_{(2)}$-IsoPs) levels do not change in substantia nigra of PD compared to control individuals.[145] In plasma, several studies found that the $F_{(2)}$-IsoPs levels were unchanged in PD patients compared to control subjects.[92,146] But a recent study using more accurate methods demonstrated that F_2-IsoPs levels were significantly increased in plasma from PD patients.[147] Further analysis of the results revealed that most of the PD subjects analyzed had early PD, suggesting that peripheral oxidative damage is higher in the early stages of PD.

Consequently, there is much evidence for the elevation of the peripheral lipoperoxidation markers in several neurodegenerative diseases. Although it is too early for any firm conclusions to be drawn, the measurement of lipoperoxidation products, which is a simple and cheap assay to perform, can and should be incorporated into future clinical trials. Such studies would clarify and likely support the hypothesis that oxidative stress is a key component for the evolution of neurodegenerative disease, and it could be considered as a marker of these pathologies.

14.4 BIOMARKERS OF CARBOHYDRATE OXIDATION

Reducing sugars play a pivotal role in modifying proteins, forming advanced glycation end products (AGEs) in a nonenzymatic reaction named glycation. Some of the biological associations of protein glycation include some diseases such as diabetes mellitus, cardiac dysfunction, neurodegenerative disease.[148,149] For this reason, glycation has an important clinical relevance, since it could be considered a potentially useful biomarker for monitoring several diseases.

In AD, glycation is believed to play an important role in NFT formation as well in the development of SPs. Involvement of AGEs in AD was first suggested in several papers published successively during 1994–1995.[150,151] Indeed, immunohistochemical studies showed the existence of AGEs such as pyrraline and pentosidine in SPs and NFTs.[152] Tau glycation enhances the formation of paired helical filaments in AD frontal cortex, reduces its ability to bind microtubules *in vitro*, and increase the fibrillization of tau.[151] Interestingly, glycation agents such as methyglyoxal are able to activate p38 MAP kinase, which is able to phosphorilate tau,[153] an important step in the formation of NFTs.[154] Furthermore, AGE-modified tau leads to an increase in the production and secretion of amyloid beta-peptide, followed by formation of ROS.[150] Glycation by methylglyoxal promotes the formation of β-sheets, oligomers, and protofibrils.[155] Glycation likely causes increased oxidative stress, inflammation, and apoptosis. However, is not clear if glycation is an early- or a late-stage marker for AD. Some data suggest that AGE formation is a late secondary event in AD, since amyloid beta-peptide alone induces free radical generation that can promote cross-linking between peptides and sugars.[156] But glycation of AT8, a known precursor of NFTs, suggests that it could be an early event.[157] Although AGE levels increase with age, in AD, the increase is much greater (37.5% and 72.6%, respectively).[158] In addition, AGEs were found in CSF of AD patients,[159,160] suggesting that this may be explored as a biomarker for AD. Receptor for AGE (RAGE) is normally expressed in a variety of cells, including microglia, neurons, and pericytes,[161] and has been found to be a specific cell surface receptor for amyloid beta-peptide, promoting neuronal cell death and dysfunction. In addition, by immunohistochemistry, it was demonstrated that RAGE levels are increased in microglia from AD brains compared to non-AD brains, especially microglia surrounding neuritic plaques.[162] Moreover, double transgenic mice with neuronal overexpression of neuronal RAGE and mutant amyloid beta-protein precursor (mAβPP) displayed early abnormalities in spatial learning/memory, accompanied by altered activation of markers of synaptic plasticity and exaggerated neuropathological findings.[163] All these observations support the active participation of the AGE-RAGE system in AD. Researchers started to think that serum or CSF AGE could become a promising biomarker for early detection of AD. But the current results available about AGEs levels in blood from AD and non-AD patients are controversial. Many groups found that AGEs and soluble RAGE (sRAGE, a C-terminal truncated isoform of RAGE) levels were lower in blood from AD patients and from MCI compared to healthy control or other forms of dementia,[164-166] but in contrast, some other groups demonstrated that AGEs and sRAGE blood levels did not change at all or increased.[167,168] Unfortunately, the blood circulating levels of AGEs do not reflect completely what happens in the CNS. Perhaps it is just matter of methods used to measure AGEs or maybe such analyses are influenced by other external factors such as food intake.

The formation of LBs in PD is still unclear, but it seems that in addition to oxidation and phosphorylation, glycation might constitute another factor affecting the aggregation process. Glycation was first reported in the substantia nigra and locus coeruleus, showing higher immunoreactivity at the periphery of LBs in PD patients.[169] These results suggest that glycation may be involved in the chemical cross-linking and proteolytic resistance of the protein deposits. Further, a study showed that AGEs and α-synuclein are similarly distributed in very early LBs in the human brain in cases with incidental LBs disease, suggesting that most probably AGEs promote formation for LBs.[170] Although glycation was also detected in the cerebral cortex, amygdala, and substantia nigra of older control subjects, the number and levels of glycated proteins were significantly higher in PD patients.[123] sRAGE was also highly expressed in cerebral cortex of PD patients when compared to age-matched controls,[123] suggesting a role for AGEs in the disease. One important feature of PD is an acute decrease in the levels of cellular reduced glutathione (GSH) in early stages of the disease, which results in a lower activity of the glyoxalase system, an important catabolic pathway of the most important glycation agent *in vivo*, that is, methylglyoxal.[171] This would cause an increase in AGEs concentration that would increase oxidative stress, which consequently induces AGEs formation. This vicious cycle would contribute to cell damage in dopaminergic neurons and death, that is, this glycation-prone environment promotes development of PD.

Glycation was first detected in both sporadic and familial forms of ALS, in spinal cord, and brain samples.[172] Initially, it was hypothesized that glycation could be involved in the cross-linking of neurofilament protein.[173] Subsequent studies have revealed that AGEs levels in spinal cord were higher in patients carrying SOD1 mutations and in mutant SOD1 transgenic mice compared to control cases.[174] Glycation, although it is a random process, affects superoxide dismutase 1at lysine residues level, causing a decrease of its activity.[175] This could justify the observed oxidative stress in ALS. Surprisingly, levels of soluble RAGE (sRAGE) are significantly lower in the serum of ALS patients.[176] Furthermore sRAGE, lacking the transmembrane-anchoring domain, was found to ameliorate the deleterious effects of RAGE by scavenging its ligands without further activating RAGE mediated-processes.[177] Thus, sRAGE may function as an endogenous protection factor in ALS, indicating that the low sRAGE levels may pose a risk factor in the disease. Moreover, it was demonstrated that the concentration of N-ε-(carboxymethyl)lysine (CML, an AGE derived from the reaction between glyoxal and the side chain of lysine residues) was significantly increased in serum and CSF of ALS patients. This result could be a potential biomarker for diagnosis of ALS, as well as point out the relevance of glycation in ALS.[178]

14.5 BIOMARKERS OF NUCLEIC ACID OXIDATION

Among all the free radicals produced during normal metabolism and/or by exogenous sources, the hydroxyl radical (HO•) is the most toxic and most highly reactive, and it conceivably could be responsible for the most oxidative damage to biological molecules, including nucleic acids (RNA and DNA). Hydroxyl radical, produced in the vicinity of nucleic acid, can easily modify RNA and DNA because they are reactive and cannot diffuse from their site of formation. More than 20 different types of base damage by hydroxyl radicals have been identified,[179] but guanine is the most reactive of the nucleic acid bases.[180] Therefore, the oxidized bases 8-hydroxyguanosine (8-OHG) and 8-hydroxy-2′-deoxyguanosine (8-OHdG) are the most abundant among the oxidized bases, and they are used as markers of RNA and DNA oxidation, respectively.[181]

There is a considerable amount of evidence supporting early involvement of RNA oxidation in the pathological cascade of neurodegeneration, especially in AD. RNA oxidation has been observed in postmortem brains of cases with early stage AD,[182] a presymptomatic case with familial AD mutation,[183] and Down syndrome cases with early stage AD pathology.[184] One of the markers of RNA oxidation, 8-hydroxyguanosine (8OHG), is inversely correlated to Aβ deposits, NFT, and duration of dementia.[185] Increased level of 8-OHG and 1-N2-propanodeoxyguanosine (NPrG) were found not only in MCI, but also in AD brains at latest stages, suggesting that mRNA is highly sensitive to oxidative damage.[186] Ribosomal RNA also was oxidatively modified in AD brain, and oxidation of rRNA by bound redox-active iron suggested its role in impairments in protein synthesis also in MCI.[187] A decreased level of ribosomal RNA and protein synthesis rate were reported in MCI. Furthermore, approximately fivefold increased levels of oxidized RNA in CFS were reported for AD cases than controls.[188]

Elevated RNA oxidation has also been observed in both postmortem substantia nigra tissue and CSF from living PD patients. Studies about the correlation between levels of 8-OHG in the CFS and the duration of the disease suggest that RNA oxidation may occur at the early stage of the disease.[189,190]

The role of mRNA oxidation in ALS was demonstrated for the first time in the transgenic mouse model of ALS TgN-(SOD1-G93A)G1H.[191] RNA oxidation in ALS is an early event, at presymptomatic stage, before the degeneration of motor neurons. Many mRNA species that have been found oxidized in TgN-(SOD1-G93A)G1H mice are related to ALS, included SOD1, dynactin 1, vesicle-associated membrane protein 1 (VAMP), and neurofilament subunit.[191] Moreover, protein levels as a consequence of oxidized mRNA species are significantly decreased.[192]

All the data about RNA oxidation in neurodegenerative diseases suggest that oxidative modification of mRNA causes not only reduction of protein levels, but it also induces translation errors *in vivo* with alteration of protein structure and function.

A substantial body of evidence indicates that oxidative DNA damage is a feature of AD in the brain as well in peripheral tissues.[193,194] Higher concentrations of oxidized pyridines and purines were detected in lymphocytes and leukocytes of AD patients compared to controls,[195–197] and in DNA from ventricular CSF of AD patients.[198] Mecocci et al were the first to demonstrate the mitochondrial DNA oxidation in AD.[199] Later, it was reported that a DNA oxidation was not limited only to the mitochondrial compartment, but also at the nuclear level in both MCI and AD brain.[182,193] No difference in DNA oxidation in the cerebellum was observed in AD, consistent with lack of Aβ pathology and other markers of oxidative stress in this brain region.[25] Both RNA and DNA oxidation markers were found in MCI and also in AD, suggesting that nucleic acid oxidation may be an early event in the progression of AD.[200] Furthermore, the levels of base excision repair (BER) enzymes that are correlated with the number of NFTs, but not SPs, were found significantly decreased in both MCI and AD.[201]

DNA damage in PD appears to occur as the levels of 8-OHdG are increased in the substantia nigra and some other brain regions.[190,202] However, levels of another product of guanine oxidation, fapyguanine, were decrease in substantia nigra in PD.[202] The oxidative damage to the DNA occurs widely in PD brain, but the substantia nigra is particularly vulnerable. One explanation could be that one of the drugs used for PD treatment, levodopa (L-DOPA), might lead to the formation of ROS and widespread oxidative damage. In fact, it was demonstrated that levodopa induces oxidative stress and degeneration of cultured dopaminergic neurons.[203,204] Moreover, the 8-OHdG levels in serum, CFS, and urine are increased in PD patients compared to healthy people.[189,205,206] Based on this background, 8-OHdG potentially could be a good biomarker for PD.

Also familial and sporadic ALS subjects had an increased level of nuclear 8-OHdG in the motor cortex[54]; it was also 10-fold higher in the spinal cord tissue in ALS than in controls.[207] In plasma, urine, and CSF, levels of 8-OHdG are higher in ALS subjects compared to healthy people.[208–210] In addition, all these data were confirmed in a mouse model of ALS, TgN-(SOD1-G93A).[211]

Oxidative stress is involved in a number of diseases, including neurodegenerative diseases discussed above. However, so far, there are no unique set of markers that can help to differentiate these diseases or to use of the above discussed oxidative stress markers as specific biomarkers of the disease. Studies are ongoing in our laboratory to identify disease-specific biomarkers which can be used in both diagnosis or to monitor the protective efficacy of therapeutic agents. A recent comprehensive review of redox proteomics in neurodegenerative disorders from our laboratory has been accepted for publication.[212]

ACKNOWLEDGMENTS

We thank the Sanders-Brown Centre on Aging for providing AD and MCI brain samples for our studies cited in this review.

REFERENCES

1. Halliwell, B. The wanderings of a free radical. *Free Radic. Biol. Med.* **2009**, *46*(5), 531–542.
2. Gutteridge, J.M., Halliwell, B. Free radicals and antioxidants in the year 2000. A historical look to the future. *Ann. N. Y. Acad. Sci.* **2000**, *899*, 136–147.
3. Mayeux, R. Biomarkers: Potential uses and limitations. *NeuroRx* **2004**, *1*(2), 182–188.
4. Atkinson, A.J., Colburn, W.A., DeGruttola, V.G., DeMets, D.L., Downing, G.J., Hoth, D.F., Oates, J.A., Peck, C.C., Schooley, R.T., Spilker, B.A., Woodcock, J., Zeger, S.L. Biomarkers and surrogate endpoints: Preferred definitions and conceptual framework. *Clin. Pharmacol. Ther.* **2001**, *69*(3), 89–95.
5. Martin, J.B. Molecular basis of the neurodegenerative disorders. *N. Engl. J. Med.* **1999**, *340*(25), 1970–1980.
6. Love, S. Post mortem sampling of the brain and other tissues in neurodegenerative disease. *Histopathology* **2004**, *44*(4), 309–317.
7. Markesbery, W.R. Oxidative stress hypothesis in Alzheimer's disease. *Free Radic. Biol. Med.* **1997**, *23*(1), 134–147.
8. Halliwell, B. Role of free radicals in the neurodegenerative diseases: Therapeutic implications for antioxidant treatment. *Drugs Aging* **2001**, *18*(9), 685–716.
9. Sayre, L.M., Perry, G., Smith, M.A. Oxidative stress and neurotoxicity. *Chem. Res. Toxicol.* **2008**, *21*(1), 172–188.
10. Dib, M., Garrel, C., Favier, A., Robin, V., Desnuelle, C. Can malondialdehyde be used as a biological marker of progression in neurodegenerative disease? *J. Neurol.* **2002**, *249*(4), 367–374.
11. Selkoe, D.J. Alzheimer's disease: Genes, proteins, and therapy. *Physiol. Rev.* **2001**, *81*(2), 741–766.
12. Duyckaerts, C., Delatour, B., Potier, M.C. Classification and basic pathology of Alzheimer's disease. *Acta Neuropathol.* **2009**, *118*(1), 5–36.
13. Poewe, W., Mahlknecht, P. The clinical progression of Parkinson's disease. *Parkinsonism Relat. Disord.* **2009**, *15*(Suppl. 4), S28–S32.
14. Lim, S.Y., Lang, A.E. The nonmotor symptoms of Parkinson's disease—An overview. *Mov. Disord.* **2010**, *25*(Suppl. 1), S123–S130.
15. Leong, S.L., Cappai, R., Barnham, K.J., Pham, C.L. Modulation of alpha-synuclein aggregation by dopamine: A review. *Neurochem. Res.* **2009**, *34*(10), 1838–1846.
16. Gasser, T. Mendelian forms of Parkinson's disease. *Biochim. Biophys. Acta* **2009**, *1792*(7), 587–596.
17. Perry, J.J., Shin, D.S., Tainer, J.A. Amyotrophic lateral sclerosis. *Adv. Exp. Med. Biol.* **2010**, *685*, 9–20.
18. Hand, C.K., Khoris, J., Salachas, F., Gros-Louis, F., Lopes, A.A., Mayeux-Portas, V., Brewer, C.G., Brown, R.H., Jr., Meininger, V., Camu, W., Rouleau, G.A. A novel locus for familial amyotrophic lateral sclerosis, on chromosome 18q. *Am. J. Hum. Genet.* **2002**, *70*(1), 251–256.
19. Dalle-Donne, I., Aldini, G., Carini, M., Colombo, R., Rossi, R., Milzani, A. Protein carbonylation, cellular dysfunction, and disease progression. *J. Cell. Mol. Med.* **2006**, *10*(2), 389–406.
20. Butterfield, D.A., Stadtman, E.R. Protein oxidation processes in aging brain. *Adv. Cell Aging Gerontol.* **1997**, *2*, 161–191.

21. Balcz, B., Kirchner, L., Cairns, N., Fountoulakis, M., Lubec, G. Increased brain protein levels of carbonyl reductase and alcohol dehydrogenase in Down syndrome and Alzheimer's disease. *J. Neural Transm. Suppl.* **2001**, *61*, 193–201.
22. Smith, C.D., Carney, J.M., Starke-Reed, P.E., Oliver, C.N., Stadtman, E.R., Floyd, R.A., Markesbery, W.R. Excess brain protein oxidation and enzyme dysfunction in normal aging and in Alzheimer disease. *Proc. Natl. Acad. Sci. U.S.A.* **1991**, *88*(23), 10540–10543.
23. Smith, M.A., Sayre, L.M., Anderson, V.E., Harris, P.L., Beal, M.F., Kowall, N., Perry, G. Cytochemical demonstration of oxidative damage in Alzheimer disease by immunochemical enhancement of the carbonyl reaction with 2,4-dinitrophenylhydrazine. *J. Histochem. Cytochem.* **1998**, *46*(6), 731–735.
24. Sultana, R., Ravagna, A., Mohmmad-Abdul, H., Calabrese, V., Butterfield, D.A. Ferulic acid ethyl ester protects neurons against amyloid beta-peptide(1-42)-induced oxidative stress and neurotoxicity: Relationship to antioxidant activity. *J. Neurochem.* **2005**, *92*(4), 749–758.
25. Hensley, K., Hall, N., Subramaniam, R., Cole, P., Harris, M., Aksenov, M., Aksenova, M., Gabbita, S.P., Wu, J.F., Carney, J.M., et al. Brain regional correspondence between Alzheimer's disease histopathology and biomarkers of protein oxidation. *J. Neurochem.* **1995**, *65*(5), 2146–2156.
26. Bogdanovic, N., Zilmer, M., Zilmer, K., Rehema, A., Karelson, E. The Swedish APP670/671 Alzheimer's disease mutation: The first evidence for strikingly increased oxidative injury in the temporal inferior cortex. *Dement. Geriatr. Cogn. Disord.* **2001**, *12*(6), 364–370.
27. Aksenov, M.Y., Aksenova, M.V., Butterfield, D.A., Geddes, J.W., Markesbery, W.R. Protein oxidation in the brain in Alzheimer's disease. *Neuroscience* **2001**, *103*(2), 373–383.
28. Reed, T., Perluigi, M., Sultana, R., Pierce, W.M., Klein, J.B., Turner, D.M., Coccia, R., Markesbery, W.R., Butterfield, D.A. Redox proteomic identification of 4-hydroxy-2-nonenal-modified brain proteins in amnestic mild cognitive impairment: Insight into the role of lipid peroxidation in the progression and pathogenesis of Alzheimer's disease. *Neurobiol. Dis.* **2008**, *30*(1), 107–120.
29. Butterfield, D.A., Galvan, V., Lange, M.B., Tang, H., Sowell, R.A., Spilman, P., Fombonne, J., Gorostiza, O., Zhang, J., Sultana, R., Bredesen, D.E. In vivo oxidative stress in brain of Alzheimer disease transgenic mice: Requirement for methionine 35 in amyloid beta-peptide of APP. *Free Radic. Biol. Med.* **2010**, *48*(1), 136–144.
30. Castegna, A., Aksenov, M., Aksenova, M., Thongboonkerd, V., Klein, J.B., Pierce, W.M., Booze, R., Markesbery, W.R., Butterfield, D.A. Proteomic identification of oxidatively modified proteins in Alzheimer's disease brain. Part I: Creatine kinase BB, glutamine synthase, and ubiquitin carboxy-terminal hydrolase L-1. *Free Radic. Biol. Med.* **2002**, *33*(4), 562–571.
31. Castegna, A., Aksenov, M., Thongboonkerd, V., Klein, J.B., Pierce, W.M., Booze, R., Markesbery, W.R., Butterfield, D.A. Proteomic identification of oxidatively modified proteins in Alzheimer's disease brain. Part II: Dihydropyrimidinase-related protein 2, alpha-enolase and heat shock cognate 71. *J. Neurochem.* **2002**, *82*(6), 1524–1532.
32. Choi, J., Forster, M.J., McDonald, S.R., Weintraub, S.T., Carroll, C.A., Gracy, R.W. Proteomic identification of specific oxidized proteins in ApoE-knockout mice: Relevance to Alzheimer's disease. *Free Radic. Biol. Med.* **2004**, *36*(9), 1155–1162.
33. Butterfield, D.A., Reed, T., Newman, S.F., Sultana, R. Roles of amyloid beta-peptide-associated oxidative stress and brain protein modifications in the pathogenesis of Alzheimer's disease and mild cognitive impairment. *Free Radic. Biol. Med.* **2007**, *43*(5), 658–677.
34. McGeer, P.L., Kamo, H., Harrop, R., McGeer, E.G., Martin, W.R., Pate, B.D., Li, D.K. Comparison of PET, MRI, and CT with pathology in a proven case of Alzheimer's disease. *Neurology* **1986**, *36*(12), 1569–1574.
35. Aaronson, R.M., Graven, K.K., Tucci, M., McDonald, R.J., Farber, H.W. Non-neuronal enolase is an endothelial hypoxic stress protein. *J. Biol. Chem.* **1995**, *270*(46), 27752–27757.
36. al-Giery, A.G., Brewer, J.M. Characterization of the interaction of yeast enolase with polynucleotides. *Biochim. Biophys. Acta* **1992**, *1159*(2), 134–140.
37. Subramanian, A., Miller, D.M. Structural analysis of alpha-enolase. Mapping the functional domains involved in down-regulation of the c-myc protooncogene. *J. Biol. Chem.* **2000**, *275*(8), 5958–5965.
38. Butterfield, D.A., Lange, M.L. Multifunctional roles of enolase in Alzheimer's disease brain: Beyond altered glucose metabolism. *J. Neurochem.* **2009**, *111*(4), 915–933.
39. Butterfield, D.A., Abdul, H.M., Opii, W., Newman, S.F., Joshi, G., Ansari, M.A., Sultana, R. Pin1 in Alzheimer's disease. *J. Neurochem.* **2006**, *98*(6), 1697–1706.
40. Pastorino, L., Sun, A., Lu, P.J., Zhou, X.Z., Balastik, M., Finn, G., Wulf, G., Lim, J., Li, S.H., Li, X., Xia, W., Nicholson, L.K., Lu, K.P. The prolyl isomerase Pin1 regulates amyloid precursor protein processing and amyloid-beta production. *Nature* **2006**, *440*(7083), 528–534.
41. Sultana, R., Robinson, R.A., Di Domenico, F., Abdul, H.M., Clair, D.K., Markesbery, W.R., Cai, J., Pierce, W.M., Butterfield, D.A. Proteomic identification of specifically carbonylated brain proteins in APP(NLh)/APP(NLh) xPS-1(P264L)/PS-1(P264L) human double mutant knock-in mice model of Alzheimer disease as a function of age. *J. Proteomics* **2011**, *74*, 2430–2440.
42. Keller, J.N., Schmitt, F.A., Scheff, S.W., Ding, Q., Chen, Q., Butterfield, D.A., Markesbery, W.R. Evidence of increased oxidative damage in subjects with mild cognitive impairment. *Neurology* **2005**, *64*(7), 1152–1156.
43. Butterfield, D.A., Poon, H.F., St Clair, D., Keller, J.N., Pierce, W.M., Klein, J.B., Markesbery, W.R. Redox proteomics identification of oxidatively modified hippocampal proteins in mild cognitive impairment: Insights into

the development of Alzheimer's disease. *Neurobiol. Dis.* **2006**, *22*(2), 223–232.

44. Sultana, R., Perluigi, M., Newman, S.F., Pierce, W.M., Cini, C., Coccia, R., Butterfield, A. Redox proteomic analysis of carbonylated brain proteins in mild cognitive impairment and early Alzheimer's disease. *Antioxid. Redox Signal.* **2009**. doi: 10.1089/ars.2009.2810.

45. Winklhofer, K.F., Haass, C. Mitochondrial dysfunction in Parkinson's disease. *Biochim. Biophys. Acta* **2010**, *1802*(1), 29–44.

46. Floor, E., Wetzel, M.G. Increased protein oxidation in human substantia nigra pars compacta in comparison with basal ganglia and prefrontal cortex measured with an improved dinitrophenylhydrazine assay. *J. Neurochem.* **1998**, *70*(1), 268–275.

47. Watfa, G., Dragonas, C., Brosche, T., Dittrich, R., Sieber, C.C., Alecu, C., Benetos, A., Nzietchueng, R. Study of telomere length and different markers of oxidative stress in patients with Parkinson's disease. *J. Nutr. Health Aging* **2011**, *15*(4), 277–281.

48. Asanuma, M., Miyazaki, I., Diaz-Corrales, F.J., Miyoshi, K., Ogawa, N., Murata, M. Preventing effects of a novel anti-Parkinsonian agent zonisamide on dopamine quinone formation. *Neurosci. Res.* **2008**, *60*(1), 106–113.

49. Chinta, S.J., Rajagopalan, S., Butterfield, D.A., Andersen, J.K. In vitro and in vivo neuroprotection by gamma-glutamylcysteine ethyl ester against MPTP: Relevance to the role of glutathione in Parkinson's disease. *Neurosci. Lett.* **2006**, *402*(1–2), 137–141.

50. De Iuliis, A., Grigoletto, J., Recchia, A., Giusti, P., Arslan, P. A proteomic approach in the study of an animal model of Parkinson's disease. *Clin. Chim. Acta* **2005**, *357*(2), 202–209.

51. Poon, H.F., Frasier, M., Shreve, N., Calabrese, V., Wolozin, B., Butterfield, D.A. Mitochondrial associated metabolic proteins are selectively oxidized in A30P alpha-synuclein transgenic mice—A model of familial Parkinson's disease. *Neurobiol. Dis.* **2005**, *18*(3), 492–498.

52. Barber, S.C., Shaw, P.J. Oxidative stress in ALS: Key role in motor neuron injury and therapeutic target. *Free Radic. Biol. Med.* **2010**, *48*(5), 629–641.

53. Shaw, P.J., Ince, P.G., Falkous, G., Mantle, D. Oxidative damage to protein in sporadic motor neuron disease spinal cord. *Ann. Neurol.* **1995**, *38*(4), 691–695.

54. Ferrante, R.J., Browne, S.E., Shinobu, L.A., Bowling, A.C., Baik, M.J., MacGarvey, U., Kowall, N.W., Brown, R.H., Jr., Beal, M.F. Evidence of increased oxidative damage in both sporadic and familial amyotrophic lateral sclerosis. *J. Neurochem.* **1997**, *69*(5), 2064–2074.

55. Bowling, A.C., Schulz, J.B., Brown, R.H., Jr., Beal, M.F. Superoxide dismutase activity, oxidative damage, and mitochondrial energy metabolism in familial and sporadic amyotrophic lateral sclerosis. *J. Neurochem.* **1993**, *61*(6), 2322–2325.

56. Lyras, L., Evans, P.J., Shaw, P.J., Ince, P.G., Halliwell, B. Oxidative damage and motor neurone disease difficulties in the measurement of protein carbonyls in human brain tissue. *Free Radic. Res.* **1996**, *24*(5), 397–406.

57. Poon, H.F., Hensley, K., Thongboonkerd, V., Merchant, M.L., Lynn, B.C., Pierce, W.M., Klein, J.B., Calabrese, V., Butterfield, D.A. Redox proteomics analysis of oxidatively modified proteins in G93A-SOD1 transgenic mice—A model of familial amyotrophic lateral sclerosis. *Free Radic. Biol. Med.* **2005**, *39*(4), 453–462.

58. Smith, M.A., Richey Harris, P.L., Sayre, L.M., Beckman, J.S., Perry, G. Widespread peroxynitrite-mediated damage in Alzheimer's disease. *J. Neurosci.* **1997**, *17*(8), 2653–2657.

59. Williamson, K.S., Gabbita, S.P., Mou, S., West, M., Pye, Q.N., Markesbery, W.R., Cooney, R.V., Grammas, P., Reimann-Philipp, U., Floyd, R.A., Hensley, K. The nitration product 5-nitro-gamma-tocopherol is increased in the Alzheimer brain. *Nitric Oxide* **2002**, *6*(2), 221–227.

60. Fernandez-Vizarra, P., Fernandez, A.P., Castro-Blanco, S., Encinas, J.M., Serrano, J., Bentura, M.L., Munoz, P., Martinez-Murillo, R., Rodrigo, J. Expression of nitric oxide system in clinically evaluated cases of Alzheimer's disease. *Neurobiol. Dis.* **2004**, *15*(2), 287–305.

61. Hensley, K., Maidt, M.L., Yu, Z., Sang, H., Markesbery, W.R., Floyd, R.A. Electrochemical analysis of protein nitrotyrosine and dityrosine in the Alzheimer brain indicates region-specific accumulation. *J. Neurosci.* **1998**, *18*(20), 8126–8132.

62. Horiguchi, T., Uryu, K., Giasson, B.I., Ischiropoulos, H., LightFoot, R., Bellmann, C., Richter-Landsberg, C., Lee, V.M., Trojanowski, J.Q. Nitration of tau protein is linked to neurodegeneration in tauopathies. *Am. J. Pathol.* **2003**, *163*(3), 1021–1031.

63. Zhang, Y.J., Xu, Y.F., Liu, Y.H., Yin, J., Li, H.L., Wang, Q., Wang, J.Z. Peroxynitrite induces Alzheimer-like tau modifications and accumulation in rat brain and its underlying mechanisms. *FASEB J.* **2006**, *20*(9), 1431–1442.

64. Berlett, B.S., Friguet, B., Yim, M.B., Chock, P.B., Stadtman, E.R. Peroxynitrite-mediated nitration of tyrosine residues in *Escherichia coli* glutamine synthetase mimics adenylylation: Relevance to signal transduction. *Proc. Natl. Acad. Sci. U.S.A.* **1996**, *93*(5), 1776–1780.

65. Yi, D., Perkins, P.D. Identification of ubiquitin nitration and oxidation using a liquid chromatography/mass selective detector system. *J. Biomol. Tech.* **2005**, *16*(4), 364–370.

66. Anantharaman, M., Tangpong, J., Keller, J.N., Murphy, M.P., Markesbery, W.R., Kiningham, K.K., St Clair, D.K. Beta-amyloid mediated nitration of manganese superoxide dismutase: Implication for oxidative stress in a APPNLH/NLH X PS-1P264L/P264L double knock-in mouse model of Alzheimer's disease. *Am. J. Pathol.* **2006**, *168*(5), 1608–1618.

67. Yamakura, F., Taka, H., Fujimura, T., Murayama, K. Inactivation of human manganese-superoxide dismutase by peroxynitrite is caused by exclusive nitration of tyrosine

34 to 3-nitrotyrosine. *J. Biol. Chem.* **1998**, *273*(23), 14085–14089.
68. Castegna, A., Thongboonkerd, V., Klein, J.B., Lynn, B., Markesbery, W.R., Butterfield, D.A. Proteomic identification of nitrated proteins in Alzheimer's disease brain. *J. Neurochem.* **2003**, *85*(6), 1394–1401.
69. Sultana, R., Poon, H.F., Cai, J., Pierce, W.M., Merchant, M., Klein, J.B., Markesbery, W.R., Butterfield, D.A. Identification of nitrated proteins in Alzheimer's disease brain using a redox proteomics approach. *Neurobiol. Dis.* **2006**, *22*(1), 76–87.
70. Guix, F.X., Ill-Raga, G., Bravo, R., Nakaya, T., de Fabritiis, G., Coma, M., Miscione, G.P., Villa-Freixa, J., Suzuki, T., Fernandez-Busquets, X., Valverde, M.A., de Strooper, B., Munoz, F.J. Amyloid-dependent triosephosphate isomerase nitrotyrosination induces glycation and tau fibrillation. *Brain* **2009**, *132*(Pt 5), 1335–1345.
71. Reyes, J.F., Reynolds, M.R., Horowitz, P.M., Fu, Y., Guillozet-Bongaarts, A.L., Berry, R., Binder, L.I. A possible link between astrocyte activation and tau nitration in Alzheimer's disease. *Neurobiol. Dis.* **2008**, *31*(2), 198–208.
72. Baxi, M.D., Vishwanatha, J.K. Uracil DNA-glycosylase/glyceraldehyde-3-phosphate dehydrogenase is an Ap4A binding protein. *Biochemistry* **1995**, *34*(30), 9700–9707.
73. Zheng, L., Roeder, R.G., Luo, Y. S phase activation of the histone H2B promoter by OCA-S, a coactivator complex that contains GAPDH as a key component. *Cell* **2003**, *114*(2), 255–266.
74. Launay, J.F., Jellali, A., Vanier, M.T. Glyceraldehyde-3-phosphate dehydrogenase is a microtubule binding protein in a human colon tumor cell line. *Biochim. Biophys. Acta* **1989**, *996*(1–2), 103–109.
75. Patterson, R.L., van Rossum, D.B., Kaplin, A.I., Barrow, R.K., Snyder, S.H. Inositol 1,4,5-trisphosphate receptor/GAPDH complex augments Ca2+ release via locally derived NADH. *Proc. Natl. Acad. Sci. U.S.A.* **2005**, *102*(5), 1357–1359.
76. Puder, M., Soberman, R.J. Glutathione conjugates recognize the Rossmann fold of glyceraldehyde-3-phosphate dehydrogenase. *J. Biol. Chem.* **1997**, *272*(16), 10936–10940.
77. Butterfield, D.A., Reed, T.T., Perluigi, M., De Marco, C., Coccia, R., Keller, J.N., Markesbery, W.R., Sultana, R. Elevated levels of 3-nitrotyrosine in brain from subjects with amnestic mild cognitive impairment: Implications for the role of nitration in the progression of Alzheimer's disease. *Brain Res.* **2007**, *1148*, 243–248.
78. Sultana, R., Reed, T., Perluigi, M., Coccia, R., Pierce, W.M., Butterfield, D.A. Proteomic identification of nitrated brain proteins in amnestic mild cognitive impairment: A regional study. *J. Cell. Mol. Med.* **2007**, *11*(4), 839–851.
79. Jenner, P., Dexter, D.T., Sian, J., Schapira, A.H., Marsden, C.D. Oxidative stress as a cause of nigral cell death in Parkinson's disease and incidental Lewy body disease. The Royal Kings and Queens Parkinson's Disease Research Group. *Ann. Neurol.* **1992**, *32*(Suppl.), S82–S87.
80. Giasson, B.I., Duda, J.E., Murray, I.V., Chen, Q., Souza, J.M., Hurtig, H.I., Ischiropoulos, H., Trojanowski, J.Q., Lee, V.M. Oxidative damage linked to neurodegeneration by selective alpha-synuclein nitration in synucleinopathy lesions. *Science* **2000**, *290*(5493), 985–989.
81. Yu, Z., Xu, X., Xiang, Z., Zhou, J., Zhang, Z., Hu, C., He, C. Nitrated alpha-synuclein induces the loss of dopaminergic neurons in the substantia nigra of rats. *PLoS ONE* **2010**, *5*(4), e9956.
82. Chinta, S.J., Andersen, J.K. Nitrosylation and nitration of mitochondrial complex I in Parkinson's disease. *Free Radic. Res.* **2011**, *45*(1), 53–58.
83. Mythri, R.B., Jagatha, B., Pradhan, N., Andersen, J., Bharath, M.M. Mitochondrial complex I inhibition in Parkinson's disease: How can curcumin protect mitochondria? *Antioxid. Redox Signal.* **2007**, *9*(3), 399–408.
84. Abe, K., Pan, L.H., Watanabe, M., Konno, H., Kato, T., Itoyama, Y. Upregulation of protein-tyrosine nitration in the anterior horn cells of amyotrophic lateral sclerosis. *Neurol. Res.* **1997**, *19*(2), 124–128.
85. Uchida, K. 4-Hydroxy-2-nonenal: A product and mediator of oxidative stress. *Prog. Lipid Res.* **2003**, *42*(4), 318–343.
86. Cracowski, J.L., Durand, T., Bessard, G. Isoprostanes as a biomarker of lipid peroxidation in humans: Physiology, pharmacology and clinical implications. *Trends Pharmacol. Sci.* **2002**, *23*(8), 360–366.
87. Subramaniam, R., Roediger, F., Jordan, B., Mattson, M.P., Keller, J.N., Waeg, G., Butterfield, D.A. The lipid peroxidation product, 4-hydroxy-2-trans-nonenal, alters the conformation of cortical synaptosomal membrane proteins. *J. Neurochem.* **1997**, *69*(3), 1161–1169.
88. Halliwell, B. Lipid peroxidation, antioxidants and cardiovascular disease: How should we move forward? *Cardiovasc. Res.* **2000**, *47*(3), 410–418.
89. Uchida, K. Role of reactive aldehyde in cardiovascular diseases. *Free Radic. Biol. Med.* **2000**, *28*(12), 1685–1696.
90. Sayre, L.M., Smith, M.A., Perry, G. Chemistry and biochemistry of oxidative stress in neurodegenerative disease. *Curr. Med. Chem.* **2001**, *8*(7), 721–738.
91. Butterfield, D.A., Bader Lange, M.L., Sultana, R. Involvements of the lipid peroxidation product, HNE, in the pathogenesis and progression of Alzheimer's disease. *Biochim. Biophys. Acta* **2010**, *1801*(8), 924–929.
92. Irizarry, M.C., Yao, Y., Hyman, B.T., Growdon, J.H., Pratico, D. Plasma F2A isoprostane levels in Alzheimer's and Parkinson's disease. *Neurodegener Dis.* **2007**, *4*(6), 403–405.
93. Markesbery, W.R., Kryscio, R.J., Lovell, M.A., Morrow, J.D. Lipid peroxidation is an early event in the brain in amnestic mild cognitive impairment. *Ann. Neurol.* **2005**, *58*(5), 730–735.
94. Yao, Y., Zhukareva, V., Sung, S., Clark, C.M., Rokach, J., Lee, V.M., Trojanowski, J.Q., Pratico, D. Enhanced brain

levels of 8,12-iso-iPF2alpha-VI differentiate AD from frontotemporal dementia. *Neurology* **2003**, *61*(4), 475–478.

95. Volkel, W., Sicilia, T., Pahler, A., Gsell, W., Tatschner, T., Jellinger, K., Leblhuber, F., Riederer, P., Lutz, W.K., Gotz, M.E. Increased brain levels of 4-hydroxy-2-nonenal glutathione conjugates in severe Alzheimer's disease. *Neurochem. Int.* **2006**, *48*(8), 679–686.

96. Lovell, M.A., Xie, C., Markesbery, W.R. Decreased glutathione transferase activity in brain and ventricular fluid in Alzheimer's disease. *Neurology* **1998**, *51*(6), 1562–1566.

97. Sultana, R., Butterfield, D.A. Oxidatively modified GST and MRP1 in Alzheimer's disease brain: Implications for accumulation of reactive lipid peroxidation products. *Neurochem. Res.* **2004**, *29*(12), 2215–2220.

98. Cecarini, V., Ding, Q., Keller, J.N. Oxidative inactivation of the proteasome in Alzheimer's disease. *Free Radic. Res.* **2007**, *41*(6), 673–680.

99. Lovell, M.A., Ehmann, W.D., Mattson, M.P., Markesbery, W.R. Elevated 4-hydroxynonenal in ventricular fluid in Alzheimer's disease. *Neurobiol. Aging* **1997**, *18*(5), 457–461.

100. Markesbery, W.R., Lovell, M.A. Four-hydroxynonenal, a product of lipid peroxidation, is increased in the brain in Alzheimer's disease. *Neurobiol. Aging* **1998**, *19*(1), 33–36.

101. Zarkovic, K. 4-Hydroxynonenal and neurodegenerative diseases. *Mol. Aspects Med.* **2003**, *24*(4–5), 293–303.

102. McGrath, L.T., McGleenon, B.M., Brennan, S., McColl, D., Mc, I.S., Passmore, A.P. Increased oxidative stress in Alzheimer's disease as assessed with 4-hydroxynonenal but not malondialdehyde. *QJM* **2001**, *94*(9), 485–490.

103. Selley, M.L., Close, D.R., Stern, S.E. The effect of increased concentrations of homocysteine on the concentration of (E)-4-hydroxy-2-nonenal in the plasma and cerebrospinal fluid of patients with Alzheimer's disease. *Neurobiol. Aging* **2002**, *23*(3), 383–388.

104. Perluigi, M., Sultana, R., Cenini, G., Di Domenico, F., Memo, M., Pierce, W.M., Coccia, R., Butterfield, D.A. Redox proteomics identification of 4-hydroxynonenal-modified brain proteins in Alzheimer's disease: Role of lipid peroxidation in Alzheimer's disease pathogenesis. *Proteomics Clin. Appl.* **2009**, *3*(6), 682–693.

105. Castellani, R.J., Perry, G., Siedlak, S.L., Nunomura, A., Shimohama, S., Zhang, J., Montine, T., Sayre, L.M., Smith, M.A. Hydroxynonenal adducts indicate a role for lipid peroxidation in neocortical and brainstem Lewy bodies in humans. *Neurosci. Lett.* **2002**, *319*(1), 25–28.

106. Yoritaka, A., Hattori, N., Uchida, K., Tanaka, M., Stadtman, E.R., Mizuno, Y. Immunohistochemical detection of 4-hydroxynonenal protein adducts in Parkinson disease. *Proc. Natl. Acad. Sci. U.S.A.* **1996**, *93*(7), 2696–2701.

107. Morel, P., Tallineau, C., Pontcharraud, R., Piriou, A., Huguet, F. Effects of 4-hydroxynonenal, a lipid peroxidation product, on dopamine transport and Na+/K+ ATPase in rat striatal synaptosomes. *Neurochem. Int.* **1998**, *33*(6), 531–540.

108. Selley, M.L. (E)-4-hydroxy-2-nonenal may be involved in the pathogenesis of Parkinson's disease. *Free Radic. Biol. Med.* **1998**, *25*(2), 169–174.

109. Pedersen, W.A., Fu, W., Keller, J.N., Markesbery, W.R., Appel, S., Smith, R.G., Kasarskis, E., Mattson, M.P. Protein modification by the lipid peroxidation product 4-hydroxynonenal in the spinal cords of amyotrophic lateral sclerosis patients. *Ann. Neurol.* **1998**, *44*(5), 819–824.

110. Simpson, E.P., Henry, Y.K., Henkel, J.S., Smith, R.G., Appel, S.H. Increased lipid peroxidation in sera of ALS patients: A potential biomarker of disease burden. *Neurology* **2004**, *62*(10), 1758–1765.

111. Smith, R.G., Henry, Y.K., Mattson, M.P., Appel, S.H. Presence of 4-hydroxynonenal in cerebrospinal fluid of patients with sporadic amyotrophic lateral sclerosis. *Ann. Neurol.* **1998**, *44*(4), 696–699.

112. Perluigi, M., Fai Poon, H., Hensley, K., Pierce, W.M., Klein, J.B., Calabrese, V., De Marco, C., Butterfield, D.A. Proteomic analysis of 4-hydroxy-2-nonenal-modified proteins in G93A-SOD1 transgenic mice—A model of familial amyotrophic lateral sclerosis. *Free Radic. Biol. Med.* **2005**, *38*(7), 960–968.

113. Ryberg, H., An, J., Darko, S., Lustgarten, J.L., Jaffa, M., Gopalakrishnan, V., Lacomis, D., Cudkowicz, M., Bowser, R. Discovery and verification of amyotrophic lateral sclerosis biomarkers by proteomics. *Muscle Nerve* **2010**, *42*(1), 104–111.

114. Bowser, R., Lacomis, D. Applying proteomics to the diagnosis and treatment of ALS and related diseases. *Muscle Nerve* **2009**, *40*(5), 753–762.

115. Pasinetti, G.M., Ungar, L.H., Lange, D.J., Yemul, S., Deng, H., Yuan, X., Brown, R.H., Cudkowicz, M.E., Newhall, K., Peskind, E., Marcus, S., Ho, L. Identification of potential CSF biomarkers in ALS. *Neurology* **2006**, *66*(8), 1218–1222.

116. Martin-Aragon, S., Bermejo-Bescos, P., Benedi, J., Felici, E., Gil, P., Ribera, J.M., Villar, A.M. Metalloproteinase's activity and oxidative stress in mild cognitive impairment and Alzheimer's disease. *Neurochem. Res.* **2009**, *34*(2), 373–378.

117. Polidori, M.C., Mecocci, P. Plasma susceptibility to free radical-induced antioxidant consumption and lipid peroxidation is increased in very old subjects with Alzheimer disease. *J. Alzheimers Dis.* **2002**, *4*(6), 517–522.

118. Bourdel-Marchasson, I., Delmas-Beauvieux, M.C., Peuchant, E., Richard-Harston, S., Decamps, A., Reignier, B., Emeriau, J.P., Rainfray, M. Antioxidant defences and oxidative stress markers in erythrocytes and plasma from normally nourished elderly Alzheimer patients. *Age. Ageing* **2001**, *30*(3), 235–241.

119. Padurariu, M., Ciobica, A., Hritcu, L., Stoica, B., Bild, W., Stefanescu, C. Changes of some oxidative stress markers in the serum of patients with mild cognitive impairment

and Alzheimer's disease. *Neurosci. Lett.* **2010**, *469*(1), 6–10.

120. Casado, A., Encarnacion Lopez-Fernandez, M., Concepcion Casado, M., de La Torre, R. Lipid peroxidation and antioxidant enzyme activities in vascular and Alzheimer dementias. *Neurochem. Res.* **2008**, *33*(3), 450–458.

121. Dei, R., Takeda, A., Niwa, H., Li, M., Nakagomi, Y., Watanabe, M., Inagaki, T., Washimi, Y., Yasuda, Y., Horie, K., Miyata, T., Sobue, G. Lipid peroxidation and advanced glycation end products in the brain in normal aging and in Alzheimer's disease. *Acta Neuropathol.* **2002**, *104*(2), 113–122.

122. Dalfo, E., Ferrer, I. Early alpha-synuclein lipoxidation in neocortex in Lewy body diseases. *Neurobiol. Aging* **2008**, *29*(3), 408–417.

123. Dalfo, E., Portero-Otin, M., Ayala, V., Martinez, A., Pamplona, R., Ferrer, I. Evidence of oxidative stress in the neocortex in incidental Lewy body disease. *J. Neuropathol. Exp. Neurol.* **2005**, *64*(9), 816–830.

124. Dexter, D.T., Carter, C.J., Wells, F.R., Javoy-Agid, F., Agid, Y., Lees, A., Jenner, P., Marsden, C.D. Basal lipid peroxidation in substantia nigra is increased in Parkinson's disease. *J. Neurochem.* **1989**, *52*(2), 381–389.

125. Ilic, T., Jovanovic, M., Jovicic, A., Tomovic, M. Oxidative stress and Parkinson's disease. *Vojnosanit. Pregl.* **1998**, *55*(5), 463–468.

126. Chen, C.M., Liu, J.L., Wu, Y.R., Chen, Y.C., Cheng, H.S., Cheng, M.L., Chiu, D.T. Increased oxidative damage in peripheral blood correlates with severity of Parkinson's disease. *Neurobiol. Dis.* **2009**, *33*(3), 429–435.

127. Younes-Mhenni, S., Frih-Ayed, M., Kerkeni, A., Bost, M., Chazot, G. Peripheral blood markers of oxidative stress in Parkinson's disease. *Eur. Neurol.* **2007**, *58*(2), 78–83.

128. Navarro, A., Boveris, A., Bandez, M.J., Sanchez-Pino, M.J., Gomez, C., Muntane, G., Ferrer, I. Human brain cortex: Mitochondrial oxidative damage and adaptive response in Parkinson disease and in dementia with Lewy bodies. *Free Radic. Biol. Med.* **2009**, *46*(12), 1574–1580.

129. Sanyal, J., Bandyopadhyay, S.K., Banerjee, T.K., Mukherjee, S.C., Chakraborty, D.P., Ray, B.C., Rao, V.R. Plasma levels of lipid peroxides in patients with Parkinson's disease. *Eur. Rev. Med. Pharmacol. Sci.* **2009**, *13*(2), 129–132.

130. Hall, E.D., Andrus, P.K., Oostveen, J.A., Fleck, T.J., Gurney, M.E. Relationship of oxygen radical-induced lipid peroxidative damage to disease onset and progression in a transgenic model of familial ALS. *J. Neurosci. Res.* **1998**, *53*(1), 66–77.

131. Mahoney, D.J., Kaczor, J.J., Bourgeois, J., Yasuda, N., Tarnopolsky, M.A. Oxidative stress and antioxidant enzyme upregulation in SOD1-G93A mouse skeletal muscle. *Muscle Nerve* **2006**, *33*(6), 809–816.

132. Oteiza, P.I., Uchitel, O.D., Carrasquedo, F., Dubrovski, A.L., Roma, J.C., Fraga, C.G. Evaluation of antioxidants, protein, and lipid oxidation products in blood from sporadic amyotrophic lateral sclerosis patients. *Neurochem. Res.* **1997**, *22*(4), 535–539.

133. Bonnefont-Rousselot, D., Lacomblez, L., Jaudon, M., Lepage, S., Salachas, F., Bensimon, G., Bizard, C., Doppler, V., Delattre, J., Meininger, V. Blood oxidative stress in amyotrophic lateral sclerosis. *J. Neurol. Sci.* **2000**, *178*(1), 57–62.

134. Baillet, A., Chanteperdrix, V., Trocme, C., Casez, P., Garrel, C., Besson, G. The role of oxidative stress in amyotrophic lateral sclerosis and Parkinson's disease. *Neurochem. Res.* **2010**, *35*, 1530–1537.

135. Liu, X., Lovell, M.A., Lynn, B.C. Development of a method for quantification of acrolein-deoxyguanosine adducts in DNA using isotope dilution-capillary LC/MS/MS and its application to human brain tissue. *Anal. Chem.* **2005**, *77*(18), 5982–5989.

136. Shamoto-Nagai, M., Maruyama, W., Hashizume, Y., Yoshida, M., Osawa, T., Riederer, P., Naoi, M. In Parkinsonian substantia nigra, alpha-synuclein is modified by acrolein, a lipid-peroxidation product, and accumulates in the dopamine neurons with inhibition of proteasome activity. *J. Neural Transm.* **2007**, *114*(12), 1559–1567.

137. Shibata, N., Nagai, R., Miyata, S., Jono, T., Horiuchi, S., Hirano, A., Kato, S., Sasaki, S., Asayama, K., Kobayashi, M. Nonoxidative protein glycation is implicated in familial amyotrophic lateral sclerosis with superoxide dismutase-1 mutation. *Acta Neuropathol.* **2000**, *100*(3), 275–284.

138. Quinn, J.F., Montine, K.S., Moore, M., Morrow, J.D., Kaye, J.A., Montine, T.J. Suppression of longitudinal increase in CSF F2-isoprostanes in Alzheimer's disease. *J. Alzheimers Dis.* **2004**, *6*(1), 93–97.

139. Pratico, D., Sung, S. Lipid peroxidation and oxidative imbalance: Early functional events in Alzheimer's disease. *J. Alzheimers Dis.* **2004**, *6*(2), 171–175.

140. Montine, T.J., Quinn, J., Kaye, J., Morrow, J.D. F(2)-isoprostanes as biomarkers of late-onset Alzheimer's disease. *J. Mol. Neurosci.* **2007**, *33*(1), 114–119.

141. Pratico, D., Clark, C.M., Liun, F., Rokach, J., Lee, V.Y., Trojanowski, J.Q. Increase of brain oxidative stress in mild cognitive impairment: A possible predictor of Alzheimer disease. *Arch. Neurol.* **2002**, *59*(6), 972–976.

142. Butterfield, D.A., Reed, T., Perluigi, M., De Marco, C., Coccia, R., Cini, C., Sultana, R. Elevated protein-bound levels of the lipid peroxidation product, 4-hydroxy-2-nonenal, in brain from persons with mild cognitive impairment. *Neurosci. Lett.* **2006**, *397*(3), 170–173.

143. Pratico, D., Clark, C.M., Lee, V.M., Trojanowski, J.Q., Rokach, J., FitzGerald, G.A. Increased 8,12-iso-iPF2alpha-VI in Alzheimer's disease: Correlation of a noninvasive index of lipid peroxidation with disease severity. *Ann. Neurol.* **2000**, *48*(5), 809–812.

144. Montine, T.J., Quinn, J.F., Milatovic, D., Silbert, L.C., Dang, T., Sanchez, S., Terry, E., Roberts, L.J., 2nd, Kaye, J.A., Morrow, J.D. Peripheral F2-isoprostanes and F4-neuroprostanes are not increased in Alzheimer's disease. *Ann. Neurol.* **2002**, *52*(2), 175–179.

145. Fessel, J.P., Hulette, C., Powell, S., Roberts, L.J., 2nd, Zhang, J. Isofurans, but not F2-isoprostanes, are increased in the substantia nigra of patients with Parkinson's disease and with dementia with Lewy body disease. *J. Neurochem.* **2003**, *85*(3), 645–650.

146. Connolly, J., Siderowf, A., Clark, C.M., Mu, D., Pratico, D. F2 isoprostane levels in plasma and urine do not support increased lipid peroxidation in cognitively impaired Parkinson disease patients. *Cogn. Behav. Neurol.* **2008**, *21*(2), 83–86.

147. Seet, R.C., Lee, C.Y., Lim, E.C., Tan, J.J., Quek, A.M., Chong, W.L., Looi, W.F., Huang, S.H., Wang, H., Chan, Y.H., Halliwell, B. Oxidative damage in Parkinson disease: Measurement using accurate biomarkers. *Free Radic. Biol. Med.* **2010**, *48*(4), 560–566.

148. Bucala, R., Cerami, A. Advanced glycosylation: Chemistry, biology, and implications for diabetes and aging. *Adv. Pharmacol.* **1992**, *23*, 1–34.

149. Ahmed, N. Advanced glycation endproducts: Role in pathology of diabetic complications. *Diabetes Res. Clin. Pract.* **2005**, *67*(1), 3–21.

150. Yan, S.D., Yan, S.F., Chen, X., Fu, J., Chen, M., Kuppusamy, P., Smith, M.A., Perry, G., Godman, G.C., Nawroth, P., et al. Non-enzymatically glycated tau in Alzheimer's disease induces neuronal oxidant stress resulting in cytokine gene expression and release of amyloid beta-peptide. *Nat. Med.* **1995**, *1*(7), 693–699.

151. Ledesma, M.D., Bonay, P., Colaco, C., Avila, J. Analysis of microtubule-associated protein tau glycation in paired helical filaments. *J. Biol. Chem.* **1994**, *269*(34), 21614–21619.

152. Smith, M.A., Taneda, S., Richey, P.L., Miyata, S., Yan, S.D., Stern, D., Sayre, L.M., Monnier, V.M., Perry, G. Advanced Maillard reaction end products are associated with Alzheimer disease pathology. *Proc. Natl. Acad. Sci. U.S.A.* **1994**, *91*(12), 5710–5714.

153. Liu, B.F., Miyata, S., Hirota, Y., Higo, S., Miyazaki, H., Fukunaga, M., Hamada, Y., Ueyama, S., Muramoto, O., Uriuhara, A., Kasuga, M. Methylglyoxal induces apoptosis through activation of p38 mitogen-activated protein kinase in rat mesangial cells. *Kidney Int.* **2003**, *63*(3), 947–957.

154. Zhu, X., Rottkamp, C.A., Boux, H., Takeda, A., Perry, G., Smith, M.A. Activation of p38 kinase links tau phosphorylation, oxidative stress, and cell cycle-related events in Alzheimer disease. *J. Neuropathol. Exp. Neurol.* **2000**, *59*(10), 880–888.

155. Chen, K., Maley, J., Yu, P.H. Potential implications of endogenous aldehydes in beta-amyloid misfolding, oligomerization and fibrillogenesis. *J. Neurochem.* **2006**, *99*(5), 1413–1424.

156. Mattson, M.P., Carney, J.W., Butterfield, D.A. A tombstone in Alzheimer's? *Nature* **1995**, *373*(6514), 481.

157. Takeda, A., Smith, M.A., Avila, J., Nunomura, A., Siedlak, S.L., Zhu, X., Perry, G., Sayre, L.M. In Alzheimer's disease, heme oxygenase is coincident with Alz50, an epitope of tau induced by 4-hydroxy-2-nonenal modification. *J. Neurochem.* **2000**, *75*(3), 1234–1241.

158. Luth, H.J., Ogunlade, V., Kuhla, B., Kientsch-Engel, R., Stahl, P., Webster, J., Arendt, T., Munch, G. Age- and stage-dependent accumulation of advanced glycation end products in intracellular deposits in normal and Alzheimer's disease brains. *Cereb. Cortex* **2005**, *15*(2), 211–220.

159. Takeuchi, M., Sato, T., Takino, J., Kobayashi, Y., Furuno, S., Kikuchi, S., Yamagishi, S. Diagnostic utility of serum or cerebrospinal fluid levels of toxic advanced glycation end-products (TAGE) in early detection of Alzheimer's disease. *Med. Hypotheses* **2007**, *69*(6), 1358–1366.

160. Shuvaev, V.V., Laffont, I., Serot, J.M., Fujii, J., Taniguchi, N., Siest, G. Increased protein glycation in cerebrospinal fluid of Alzheimer's disease. *Neurobiol. Aging* **2001**, *22*(3), 397–402.

161. Yan, S.D., Chen, X., Fu, J., Chen, M., Zhu, H., Roher, A., Slattery, T., Zhao, L., Nagashima, M., Morser, J., Migheli, A., Nawroth, P., Stern, D., Schmidt, A.M. RAGE and amyloid-beta peptide neurotoxicity in Alzheimer's disease. *Nature* **1996**, *382*(6593), 685–691.

162. Lue, L.F., Yan, S.D., Stern, D.M., Walker, D.G. Preventing activation of receptor for advanced glycation end products in Alzheimer's disease. *Curr. Drug Targets CNS Neurol. Disord.* **2005**, *4*(3), 249–266.

163. Arancio, O., Zhang, H.P., Chen, X., Lin, C., Trinchese, F., Puzzo, D., Liu, S., Hegde, A., Yan, S.F., Stern, A., Luddy, J.S., Lue, L.F., Walker, D.G., Roher, A., Buttini, M., Mucke, L., Li, W., Schmidt, A.M., Kindy, M., Hyslop, P.A., Stern, D.M., Du Yan, S.S. RAGE potentiates Abeta-induced perturbation of neuronal function in transgenic mice. *EMBO J.* **2004**, *23*(20), 4096–4105.

164. Thome, J., Munch, G., Muller, R., Schinzel, R., Kornhuber, J., Blum-Degen, D., Sitzmann, L., Rosler, M., Heidland, A., Riederer, P. Advanced glycation end products-associated parameters in the peripheral blood of patients with Alzheimer's disease. *Life Sci.* **1996**, *59*(8), 679–685.

165. Emanuele, E., D'Angelo, A., Tomaino, C., Binetti, G., Ghidoni, R., Politi, P., Bernardi, L., Maletta, R., Bruni, A.C., Geroldi, D. Circulating levels of soluble receptor for advanced glycation end products in Alzheimer disease and vascular dementia. *Arch. Neurol.* **2005**, *62*(11), 1734–1736.

166. Staniszewska, M., Leszek, J., Malyszczak, K., Gamian, A. Are advanced glycation end-products specific biomarkers for Alzheimer's disease? *Int. J. Geriatr. Psychiatry* **2005**, *20*(9), 896–897.

167. Hernanz, A., De la Fuente, M., Navarro, M., Frank, A. Plasma aminothiol compounds, but not serum tumor necrosis factor receptor II and soluble receptor for advanced glycation end products, are related to the cognitive impairment in Alzheimer's disease and mild cognitive impairment patients. *Neuroimmunomodulation* **2007**, *14*(3–4), 163–167.

168. Riviere, S., Birlouez-Aragon, I., Vellas, B. Plasma protein glycation in Alzheimer's disease. *Glycoconj. J.* **1998**, *15*(10), 1039–1042.

169. Castellani, R., Smith, M.A., Richey, P.L., Perry, G. Glycoxidation and oxidative stress in Parkinson disease and diffuse Lewy body disease. *Brain Res.* **1996**, *737*(1–2), 195–200.
170. Munch, G., Luth, H.J., Wong, A., Arendt, T., Hirsch, E., Ravid, R., Riederer, P. Crosslinking of alpha-synuclein by advanced glycation endproducts—An early pathophysiological step in Lewy body formation? *J. Chem. Neuroanat.* **2000**, *20*(3–4), 253–257.
171. Thornalley, P.J. Glutathione-dependent detoxification of alpha-oxoaldehydes by the glyoxalase system: Involvement in disease mechanisms and antiproliferative activity of glyoxalase I inhibitors. *Chem. Biol. Interact.* **1998**, *111–112*, 137–151.
172. Kikuchi, S., Shinpo, K., Ogata, A., Tsuji, S., Takeuchi, M., Makita, Z., Tashiro, K. Detection of N epsilon-(carboxymethyl)lysine (CML) and non-CML advanced glycation end-products in the anterior horn of amyotrophic lateral sclerosis spinal cord. *Amyotroph. Lateral Scler. Other Motor Neuron Disord.* **2002**, *3*(2), 63–68.
173. Chou, S.M., Wang, H.S., Taniguchi, A., Bucala, R. Advanced glycation end products in neurofilament conglomeration of motoneurons in familial and sporadic amyotrophic lateral sclerosis. *Mol. Med.* **1998**, *4*(5), 324–332.
174. Shibata, N., Hirano, A., Hedley-Whyte, E.T., Dal Canto, M.C., Nagai, R., Uchida, K., Horiuchi, S., Kawaguchi, M., Yamamoto, T., Kobayashi, M. Selective formation of certain advanced glycation end products in spinal cord astrocytes of humans and mice with superoxide dismutase-1 mutation. *Acta Neuropathol.* **2002**, *104*(2), 171–178.
175. Arai, K., Maguchi, S., Fujii, S., Ishibashi, H., Oikawa, K., Taniguchi, N. Glycation and inactivation of human Cu-Zn-superoxide dismutase. Identification of the in vitro glycated sites. *J. Biol. Chem.* **1987**, *262*(35), 16969–16972.
176. Ilzecka, J. Serum-soluble receptor for advanced glycation end product levels in patients with amyotrophic lateral sclerosis. *Acta Neurol. Scand.* **2009**, *120*(2), 119–122.
177. Sakaguchi, T., Yan, S.F., Yan, S.D., Belov, D., Rong, L.L., Sousa, M., Andrassy, M., Marso, S.P., Duda, S., Arnold, B., Liliensiek, B., Nawroth, P.P., Stern, D.M., Schmidt, A.M., Naka, Y. Central role of RAGE-dependent neointimal expansion in arterial restenosis. *J. Clin. Invest.* **2003**, *111*(7), 959–972.
178. Kaufmann, E., Boehm, B.O., Sussmuth, S.D., Kientsch-Engel, R., Sperfeld, A., Ludolph, A.C., Tumani, H. The advanced glycation end-product N epsilon-(carboxymethyl)lysine level is elevated in cerebrospinal fluid of patients with amyotrophic lateral sclerosis. *Neurosci. Lett.* **2004**, *371*(2–3), 226–229.
179. Barciszewski, J., Barciszewska, M.Z., Siboska, G., Rattan, S.I., Clark, B.F. Some unusual nucleic acid bases are products of hydroxyl radical oxidation of DNA and RNA. *Mol. Biol. Rep.* **1999**, *26*(4), 231–238.
180. Yanagawa, H., Ogawa, Y., Ueno, M. Redox ribonucleosides. Isolation and characterization of 5-hydroxyuridine, 8-hydroxyguanosine, and 8-hydroxyadenosine from Torula yeast RNA. *J. Biol. Chem.* **1992**, *267*(19), 13320–13326.
181. Helbock, H.J., Beckman, K.B., Ames, B.N. 8-Hydroxydeoxyguanosine and 8-hydroxyguanine as biomarkers of oxidative DNA damage. *Methods Enzymol.* **1999**, *300*, 156–166.
182. Lovell, M.A., Markesbery, W.R. Oxidative damage in mild cognitive impairment and early Alzheimer's disease. *J. Neurosci. Res.* **2007**, *85*(14), 3036–3040.
183. Nunomura, A., Chiba, S., Lippa, C.F., Cras, P., Kalaria, R.N., Takeda, A., Honda, K., Smith, M.A., Perry, G. Neuronal RNA oxidation is a prominent feature of familial Alzheimer's disease. *Neurobiol. Dis.* **2004**, *17*(1), 108–113.
184. Nunomura, A., Perry, G., Pappolla, M.A., Friedland, R.P., Hirai, K., Chiba, S., Smith, M.A. Neuronal oxidative stress precedes amyloid-beta deposition in Down syndrome. *J. Neuropathol. Exp. Neurol.* **2000**, *59*(11), 1011–1017.
185. Nunomura, A., Perry, G., Aliev, G., Hirai, K., Takeda, A., Balraj, E.K., Jones, P.K., Ghanbari, H., Wataya, T., Shimohama, S., Chiba, S., Atwood, C.S., Petersen, R.B., Smith, M.A. Oxidative damage is the earliest event in Alzheimer disease. *J. Neuropathol. Exp. Neurol.* **2001**, *60*(8), 759–767.
186. Lovell, M.A., Markesbery, W.R. Oxidatively modified RNA in mild cognitive impairment. *Neurobiol. Dis.* **2008**, *29*(2), 169–175.
187. Honda, K., Smith, M.A., Zhu, X., Baus, D., Merrick, W.C., Tartakoff, A.M., Hattier, T., Harris, P.L., Siedlak, S.L., Fujioka, H., Liu, Q., Moreira, P.I., Miller, F.P., Nunomura, A., Shimohama, S., Perry, G. Ribosomal RNA in Alzheimer disease is oxidized by bound redox-active iron. *J. Biol. Chem.* **2005**, *280*(22), 20978–20986.
188. Abe, T., Tohgi, H., Isobe, C., Murata, T., Sato, C. Remarkable increase in the concentration of 8-hydroxyguanosine in cerebrospinal fluid from patients with Alzheimer's disease. *J. Neurosci. Res.* **2002**, *70*(3), 447–450.
189. Abe, T., Isobe, C., Murata, T., Sato, C., Tohgi, H. Alteration of 8-hydroxyguanosine concentrations in the cerebrospinal fluid and serum from patients with Parkinson's disease. *Neurosci. Lett.* **2003**, *336*(2), 105–108.
190. Zhang, J., Perry, G., Smith, M.A., Robertson, D., Olson, S.J., Graham, D.G., Montine, T.J. Parkinson's disease is associated with oxidative damage to cytoplasmic DNA and RNA in substantia nigra neurons. *Am. J. Pathol.* **1999**, *154*(5), 1423–1429.
191. Chang, Y., Kong, Q., Shan, X., Tian, G., Ilieva, H., Cleveland, D.W., Rothstein, J.D., Borchelt, D.R., Wong, P.C., Lin, C.L. Messenger RNA oxidation occurs early in disease pathogenesis and promotes motor neuron degeneration in ALS. *PLoS ONE* **2008**, *3*(8), e2849.
192. Kong, Q., Lin, C.L. Oxidative damage to RNA: Mechanisms, consequences, and diseases. *Cell. Mol. Life Sci.* **2010**, *67*(11), 1817–1829.

193. Wang, J., Xiong, S., Xie, C., Markesbery, W.R., Lovell, M.A. Increased oxidative damage in nuclear and mitochondrial DNA in Alzheimer's disease. *J. Neurochem.* **2005**, *93*(4), 953–962.

194. Gabbita, S.P., Lovell, M.A., Markesbery, W.R. Increased nuclear DNA oxidation in the brain in Alzheimer's disease. *J. Neurochem.* **1998**, *71*(5), 2034–2040.

195. Migliore, L., Fontana, I., Trippi, F., Colognato, R., Coppede, F., Tognoni, G., Nucciarone, B., Siciliano, G. Oxidative DNA damage in peripheral leukocytes of mild cognitive impairment and AD patients. *Neurobiol. Aging* **2005**, *26*(5), 567–573.

196. Mecocci, P., Polidori, M.C., Ingegni, T., Cherubini, A., Chionne, F., Cecchetti, R., Senin, U. Oxidative damage to DNA in lymphocytes from AD patients. *Neurology* **1998**, *51*(4), 1014–1017.

197. Kadioglu, E., Sardas, S., Aslan, S., Isik, E., Esat Karakaya, A. Detection of oxidative DNA damage in lymphocytes of patients with Alzheimer's disease. *Biomarkers* **2004**, *9*(2), 203–209.

198. Lovell, M.A., Gabbita, S.P., Markesbery, W.R. Increased DNA oxidation and decreased levels of repair products in Alzheimer's disease ventricular CSF. *J. Neurochem.* **1999**, *72*(2), 771–776.

199. Mecocci, P., MacGarvey, U., Beal, M.F. Oxidative damage to mitochondrial DNA is increased in Alzheimer's disease. *Ann. Neurol.* **1994**, *36*(5), 747–751.

200. Wang, J., Markesbery, W.R., Lovell, M.A. Increased oxidative damage in nuclear and mitochondrial DNA in mild cognitive impairment. *J. Neurochem.* **2006**, *96*(3), 825–832.

201. Weissman, L., Jo, D.G., Sorensen, M.M., de Souza-Pinto, N.C., Markesbery, W.R., Mattson, M.P., Bohr, V.A. Defective DNA base excision repair in brain from individuals with Alzheimer's disease and amnestic mild cognitive impairment. *Nucleic Acids Res.* **2007**, *35*(16), 5545–5555.

202. Alam, Z.I., Jenner, A., Daniel, S.E., Lees, A.J., Cairns, N., Marsden, C.D., Jenner, P., Halliwell, B. Oxidative DNA damage in the Parkinsonian brain: An apparent selective increase in 8-hydroxyguanine levels in substantia nigra. *J. Neurochem.* **1997**, *69*(3), 1196–1203.

203. Spencer, J.P., Jenner, A., Aruoma, O.I., Evans, P.J., Kaur, H., Dexter, D.T., Jenner, P., Lees, A.J., Marsden, D.C., Halliwell, B. Intense oxidative DNA damage promoted by L-dopa and its metabolites. Implications for neurodegenerative disease. *FEBS Lett.* **1994**, *353*(3), 246–250.

204. Walkinshaw, G., Waters, C.M. Induction of apoptosis in catecholaminergic PC12 cells by L-DOPA. Implications for the treatment of Parkinson's disease. *J. Clin. Invest.* **1995**, *95*(6), 2458–2464.

205. Kikuchi, A., Takeda, A., Onodera, H., Kimpara, T., Hisanaga, K., Sato, N., Nunomura, A., Castellani, R.J., Perry, G., Smith, M.A., Itoyama, Y. Systemic increase of oxidative nucleic acid damage in Parkinson's disease and multiple system atrophy. *Neurobiol. Dis.* **2002**, *9*(2), 244–248.

206. Sato, S., Mizuno, Y., Hattori, N. Urinary 8-hydroxydeoxyguanosine levels as a biomarker for progression of Parkinson disease. *Neurology* **2005**, *64*(6), 1081–1083.

207. Fitzmaurice, P.S., Shaw, I.C., Kleiner, H.E., Miller, R.T., Monks, T.J., Lau, S.S., Mitchell, J.D., Lynch, P.G. Evidence for DNA damage in amyotrophic lateral sclerosis. *Muscle Nerve* **1996**, *19*(6), 797–798.

208. Murata, T., Ohtsuka, C., Terayama, Y. Increased mitochondrial oxidative damage and oxidative DNA damage contributes to the neurodegenerative process in sporadic amyotrophic lateral sclerosis. *Free Radic. Res.* **2008**, *42*(3), 221–225.

209. Bogdanov, M., Brown, R.H., Matson, W., Smart, R., Hayden, D., O'Donnell, H., Flint Beal, M., Cudkowicz, M. Increased oxidative damage to DNA in ALS patients. *Free Radic. Biol. Med.* **2000**, *29*(7), 652–658.

210. Mitsumoto, H., Santella, R.M., Liu, X., Bogdanov, M., Zipprich, J., Wu, H.C., Mahata, J., Kilty, M., Bednarz, K., Bell, D., Gordon, P.H., Hornig, M., Mehrazin, M., Naini, A., Flint Beal, M., Factor-Litvak, P. Oxidative stress biomarkers in sporadic ALS. *Amyotroph. Lateral Scler.* **2008**, *9*(3), 177–183.

211. Liu, D., Wen, J., Liu, J., Li, L. The roles of free radicals in amyotrophic lateral sclerosis: Reactive oxygen species and elevated oxidation of protein, DNA, and membrane phospholipids. *FASEB J.* **1999**, *13*(15), 2318–2328.

212. Butterfield, D.A., Perluigi, M., Reed, T., Muharib, T., Hughes, C., Robinson, R., Sultana, R. Redox proteomics in selected neurodegenerative disorders: From its infancy to future applications. *Antioxid. Redox Signal.* **2012**, *17*(11), 1610–1655.

15

SYNTHETIC ANTIOXIDANTS

GRÉGORY DURAND

OVERVIEW

According to Halliwell and Gutteridge, "any substance that delays, prevents or removes oxidative damage to a target molecule" is an antioxidant.[1] Such a protection can either be achieved (a) by scavenging reactive oxygen species (ROS); (b) by inhibiting enzymatic processes that lead to the formation of ROS or by chelating trace elements involved in ROS formation, or (c) by upregulating and protecting endogenous antioxidant defenses. Aerobic organisms have progressively evolved with an antioxidant system to ensure protection against ROS, and this system can be divided in two classes. The first class is an endogenous system, which includes enzymatic defenses such as superoxide dismutase (SOD), catalase (Cat) and peroxidases (Gpx), and nonenzymatic defenses such as glutathion (GSH), lipoic acid (LA) and its reduced form (DHLA), coenzymes Q10, melatonine, uric acid, and plasma protein thiols to name a few. The second class comprises exogenous antioxidants that are obtained from the diet. To prevent oxidative damage, therapeutic strategies using natural antioxidants have been extensively developed, and promising results have been reported on animal models with vitamin E, vitamin C, and flavonoids, for instance. Unfortunately, clinical trials assessing the efficiency of antioxidant supplementation have provided inconsistent results.[2-5]

15.1 ENDOGENOUS ENZYMATIC SYSTEM OF DEFENSE

Let us start with a very brief introduction on the enzymatic system of defense. SOD and catalase are metalloproteins that efficiently catalyze the dismutation of ROS in all aerobic cells. Any reaction of the type depicted in Figure 15.1A is called a dismutation or disproportation reaction.

The dismutation reactions that SOD and catalase catalyze involve a series of one- or two-electron transfers, but they do not require any reducing equivalents and therefore do not need energy from the cell to operate. SOD enzymes, which contain either Cu–Zn, Fe, Mn, or Ni at the active site of the enzyme, catalyze the dismutation of superoxide radical anion ($O_2^{\bullet-}$) to the nonradical hydrogen peroxide (H_2O_2) and dioxygen (O_2) as depicted in Figure 15.1B.

The superoxide anion radical ($O_2^{\bullet-}$) spontaneously dismutes to O_2 and hydrogen peroxide (H_2O_2) quite rapidly (~105 $M^{-1}s^{-1}$ at pH 7), and the dismutation rate is second order with respect to initial superoxide concentration. In contrast, the reaction of superoxide with SOD is first order with respect to superoxide concentration, and SOD has the largest reaction rate of any known enzyme (~10^{-9} $M^{-1}s^{-1}$), this reaction being "diffusion limited".* Catalase enzymes contain four porphyrins heme (Fe) groups bound to the catalytic site, which allow the enzyme to dismute H_2O_2 in O_2 and H_2O as depicted in Figure 15.1C.

Another class of endogenous catalytic-H_2O_2 scavengers is the peroxidases. Peroxidases are tetrameric proteins whose each monomer contains one atom of selenium at the catalytic site. Glutathion peroxydases

---

* The reaction rate is only limited by the frequency of collision between the enzyme and superoxide. This is an approximation made when reaction rates are very high.

Molecular Basis of Oxidative Stress: Chemistry, Mechanisms, and Disease Pathogenesis, First Edition. Edited by Frederick A. Villamena.
© 2013 John Wiley & Sons, Inc. Published 2013 by John Wiley & Sons, Inc.

378 SYNTHETIC ANTIOXIDANTS

possess a selenocystein at their active site and detoxify hydrogen peroxide and lipidic hydroperoxides through reduction of the peroxide and oxidation of glutathion as substrate. A brief overview of the catalytic mechanisms of copper superoxide dismutase, manganese superoxide dismutase, and selenium glutathione peroxidase are given in Figure 15.2, Figure 15.3, and Figure 15.4, respectively.

15.2 METAL-BASED SYNTHETIC ANTIOXIDANTS

The use of native SOD and catalase as therapeutics to attenuate oxidative stress-mediated damages has been developed but has met limitations, one of them being the very large size and the fragility of such enzymes. Therefore, considering the deleterious effect of ROS and the lack of efficiency of using natural detoxification enzymes as therapeutic agents, synthetic antioxidants have emerged as a promising strategy for the treatment of diseases in which an overproduction of ROS is observed. There has been extensive research in the development of effective molecules that can provide antioxidant protection and can be used as therapeutic agents. Efforts have mainly been focused on the design of intrinsically more potent compounds compared to natural antioxidants. In this search for novel synthetic antioxidants, scientists have notably worked on the development of molecules able to mimic natural antioxidants. Among the advantages of using mimics over natural enzymes, one can cite a longer half-life, a higher cellular permeability, lower costs, as well as the lack of immunogenecity. It has been demonstrated that simple metal chelates (e.g., Mn, Fe, Cu) can react with $O_2^{\bullet-}$ and H_2O_2 but generally at low rates. Moreover, the complexes formed are relatively unstable. A particular attention was first paid to copper complexes in the development of mimics; however, since aqueous Cu (II) ions can themselves dismutate superoxide very efficiently, the study of copper mimics was complicated. Another aspect that has to be considered in the design of mimics as pharmaceutical is their toxicity. Since copper and iron ions can promote the Fenton reaction, which produces hydroxyl radicals, a metal-based drug must be very stable in its complexed state in biological milieu and should not release its redox-active ions. As Mn^{2+} is much less prone to induce hydroxyl radical production through the Fenton reaction than Cu^{2+} and Fe^{2+}, manganese has been widely favored in the design of synthetic catalytic antioxidants. The search for active and nontoxic metal containing catalytic antioxidants has resulted in the design of several series of mimics among which are the salen, the metalloporphyrin, and the mac-

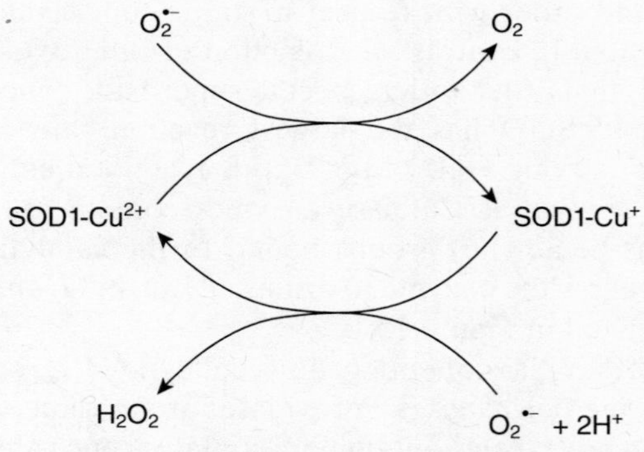

$$A + A \longrightarrow A' + A'' \quad (A)$$

$$2\,O_2^{\bullet-} \xrightarrow{2H^+} O_2 + H_2O_2 \quad (B)$$

$$2\,H_2O_2 \longrightarrow O_2 + 2\,H_2O \quad (C)$$

Figure 15.1 General equation of a dismutation reaction and two exemples of dismutation reaction catalyzed by enzymes.

Figure 15.2 Catalytic mechanisms of superoxide dismutation by copper superoxide dismutase.

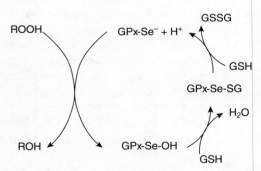

Figure 15.4 Catalytic mechanism of lipidic hydroperoxides reduction by selenium glutathion peroxidase.

$$SOD2\text{-}Mn^{3+} + O_2^{\bullet-} \rightleftharpoons [\,SOD2\text{-}Mn^{3+}-O_2^{\bullet-}\,] \longrightarrow SOD2\text{-}Mn^{2+} + O_2$$

$$SOD2\text{-}Mn^{2+} + O_2^{\bullet-} \rightleftharpoons [\,SOD2\text{-}Mn^{2+}-O_2\,]^{\bullet-} \xrightarrow{2H^+} SOD2\text{-}Mn^{3+} + H_2O_2$$

Figure 15.3 Catalytic mechanisms of superoxide dismutation by manganese superoxide dismutase.

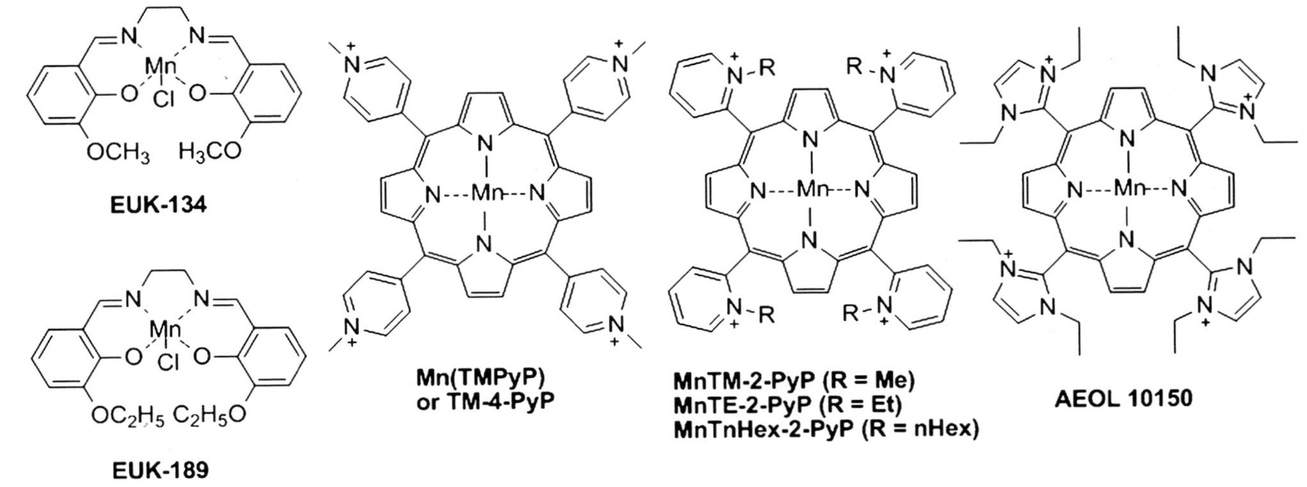

Figure 15.5 Chemical structures of salens and metalloporphyrin.

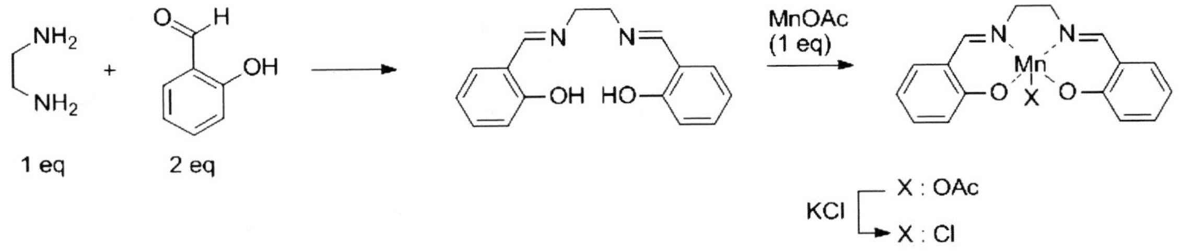

Scheme 15.1 Synthetic route of Salem EUK-8.

rocyclic series, and several good reviews have been published on this topic.[6–9]

15.2.1 Mn^{III} Complexes (Salens)

Initially, the Mn^{III} complexes (salens) were developed by Jacobsen as catalysts for chiral expoxidation, but they were also reported to be SOD and catalase mimics with a broad pharmacological efficacy.[10,11] Salens are aromatic, substituted ethylenediamine Mn^{III} complexes in which the metal center is coordinated by four axial ligands to oxygen and nitrogen atoms (Fig. 15.5). The coordination of Mn by four axial ligands results in the formation of possible several valence states that provide to salens their broad ROS-scavenging properties. In addition to their ability to scavenge $O_2^{•-}$ and H_2O_2, it has been shown that they also react with peroxynitrite (ONOO$^-$).[12]

Several salen–maganese complexes have been synthesized, with a particular attention to the EUK series. The EUK series is currently being developed by Proteome Systems (Australia). In Scheme 15.1 is represented the general synthesis of salen EUK-8 as described by Boucher.[13] This synthetic route was further used and modified in order to synthesize EUK-8 analogues.

The EUK-8 complex has been widely tested in biological models against oxidative stress-mediated toxicity. *In vitro*, EUK-8 was shown to protect organotypic hippocampal cultures from the toxicity of the β-amyloid peptide in a model of Alzheimer's disease[14] as well as hippocampal slices from hypoxia- and acidosis-induced damage.[15] It also showed potency in preventing toxin-induced cellular damage *in vivo*. Indeed, in two *in vivo* models of Parkinson's disease, a significant protection was observed either when administered by intraperitoneal injections or in the drinking water. The 3,3′-methoxy salen ring-disubstitued EUK-8 derivative, called EUK-134 (Fig. 15.5), was found more neuroprotective in a rodent stroke model.[16] The administration of EUK-134 significantly reduced brain infarct size while that of EUK-8 was substantially less effective. The enhanced activity of EUK-134 compared to EUK-8 was explained based on its greater catalase activity as they both exhibit equivalent SOD activities. Worth noting is that in certain experimental systems, administration of SOD was found ineffective or even deleterious, whereas EUK-8 was protective,[17] clearly demonstrating that catalase activity is an important element of the protective activity of salen–manganese complexes. In order to establish a structure–activity relationship, several salen-ring substituted series bearing electro-donating or electro-withdrawing substituents and exhibiting various degree of amphiphilicity were next synthesized. Doctrow et al. showed that ring substitution as well as bridge modification did not affect the SOD activity of salen–manganese complexes.[18] In contrast, the hydrogen peroxide-scavenging activities was significantly affected by the ligand structure, the catalase activity being increased by symmetrical 3- or 5-alkoxy substituents and by aromatic bridge structures. Surprisingly, the lipophilicity of the salen–maganese complexes did not appear to play a role in the protection of human dermal fibroblasts against hydrogen peroxide toxicity. Indeed, EUK-134, its ethoxy analogue EUK-189 (Fig. 15.5), and the cyclohexyl bridged EUK-159 exhibited similar protection consistent with their catalase activity. EUK-134 and EUK-189 exhibited also very similar activity in a rodent stroke model. One of the limitations of the salen–maganese complexes is their stability in biological milieu which makes difficult the determination of tissues levels and half-lives. Recently, a novel class of Mn porphyrins, the EUK-400 series, which exhibits SOD- and catalase-like activity was reported.[19] Unlike EUK-189, these new compounds are detected in the plasma after oral administration, suggesting that they are potentially orally administered drugs for oxidative stress-mediated diseases.

15.2.2 Mn^{III} (Porphyrinato) Complexes (Also Called Metalloporhyrins)

Metalloporphyrins contain either a Mn or a Fe moiety that is coordinated by four nitrogen axial ligands. In contrast to salen complexes in which the metal center is coordinated to oxygen and nitrogen atoms, the metal center of metalloporphyrins is coordinated only to nitrogen atoms (Fig. 15.5). Metalloporphyrins catalyze the dismutation of $O_2^{\bullet-}$ through the alternate reduction and oxidation of the Mn moiety similar to what occurs in SOD (Fig. 15.2 and Fig. 15.3). The alternate reduction-oxidation of Mn results in an alternate change in the valence of the complexes between Mn(III) and Mn(II).

Pasternack and Skowronek reported for the first time the superoxide scavenging activity of metalloporphyrins more than three decades ago with the tetrakis(4-N-methylpyridyl)porphin complex of Mn^{III}.[20,21] Metalloproteins also exhibit a catalase-like activity,[22] and this catalase activity is believed to be due to reversible one-electron oxidation that can undergo along the conjugated ring system. This is equivalent to what occurs for the heme prosthetic groups of endogenous catalases and peroxidases (Fig. 15.4). Groves and colleagues reported the high reactivity of metalloporphyrins (Mn(III)TMPyP) with peroxynitrite.[23] The reaction of Mn(III)TMPyP with ONOO$^-$ lead to an oxomanganese (IV) porphyrins species with a second-order rate constant of $1.8 \; 10^6 \; M^{-1}s^{-1}$.[24] The very slow first-order rate constant of reduction ($\sim0.018 \; s^{-1}$) of this oxoMn(IV) back to its Mn(III) oxidation state is much slower than the spontaneous decay of ONOO$^-$ ($0.35 \; s^{-1}$), rendering Mn(III)TMPyP unable to catalytically scavenge ONOO$^-$.[25] In contrast, when coupled with antioxidants such as ascorbate, gluthathione (GSH), or Trolox, Mn(III)TMPyP becomes a very efficient peroxynitrite reductase as shown in Figure 15.6.[26]

Moreover, aside from the SOD-like activity of Mn(III)TMPyP as demonstrated by various methods such as pulse radiolysis and stopped flow experiments, it was shown that $O_2^{\bullet-}$ can rapidly reduce oxoMn(IV) to Mn(III) oxidation state like ascorbate, gluthathione, or Trolox do. Therefore, the ability of Mn(III)TMPyP to both remove $O_2^{\bullet-}$ and ONOO$^-$ under severe conditions of oxidative stress suggests a high potency of Mn(III)TMPyP and related metalloporphyrins in preventing oxidative stress-mediated damages. The metalloporphyrins-mediated inhibition of lipid peroxidation has also been reported.[27] Their ability in inhibiting Fe^{3+}-ascorbate induced lipid peroxidation was found to be dependent on the transition metal ligated to the porphyrin, demonstrating the importance of the nature of the metal and its related redox chemistry in the antioxidant properties of the complexes.[27] As discussed earlier, the chemical properties of the ligated metal center is of importance to produce a nontoxic and catalytically active antioxidant. Although both iron and manganese have three accessible valence states, iron has a stronger preference for axial coordination than manganese, and this makes it less suitable for antioxidant properties.[28]

Manganese porphyrins were shown to be the most potent with IC$_{50}$ of 1.0 and 16 μM for MnTM-2-PyP and MnTM-4-PyP, respectively, while that of the native CuZnSOD is 15 μM. In addition, the position of the N-methylpyridyl group was also found to affect the SOD activity, with the ortho-derivative MnTM-2-PyP exhibiting a 15-fold increase in the SOD activity compared to that of the para-derivative MnTM-4-PyP (Fig. 15.5). The more positive redox potential of MnTM-2-PyP (+0.22 V) versus MnTM-4-PyP (+0.06 V) suggests an "ortho" effect, which may play a role in the SOD activity and in the inhibition of lipid peroxidation.[29] The presence of an alkylpyridinium group close to the metal

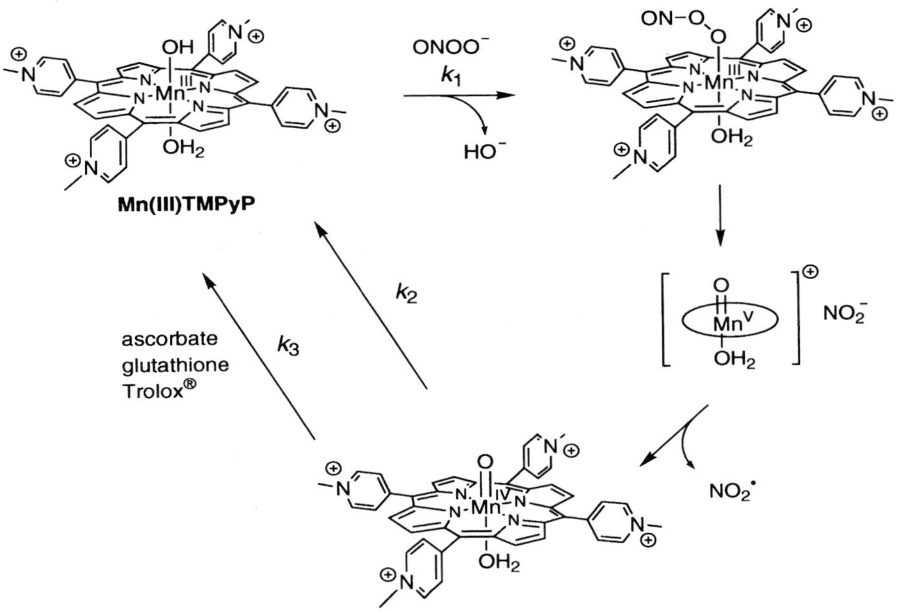

Figure 15.6 Redox antioxidant-coupled peroxynitrite reductase activity of Mn(III)TMPyP. $k_1 = 1.8 \times 10^6$ M^{-1} s^{-1}; $k_2 = 0.018$ s^{-1}; $k_3 = 5.4 \times 10^7$, 1.3×10^5, and 7.0×10^6 M^{-1} s^{-1} for ascorbate, glutathione, and Trolox, respectively. (Adapted with permission from *J. Am. Chem. Soc.* **1998**, *120*(24), 6053–6061. Copyright 1998 American Chemical Society.)

center contributes to both thermodynamic[29] (favorable redox potential) and kinetic[30] (electrostatic attraction of negatively charged species) facilitation of $O_2^{\bullet-}$ dismutation and ONOO$^-$ reduction. A more detailed study of the mechanism of inhibition of lipid peroxidation by metalloporphyrins was next examined on various lipid systems.[28] It was demonstrated that in the absence of reductants, the 5,10,15,20-tetrakis(*N*-ethylpyridinium-2-yl)porphyrin (MnTE-2-PyP) promotes lipid oxidation through an oxidation/reduction mechanism. However, in the presence of reductants at biological relevant concentration, it was shown that MnTE-2-PyP inhibits lipid oxidation most likely by redox cycling between the Mn^{3+} and Mn^{2+} redox status. It has to be underlined that the carbonate radical scavenging ability of MnTE-2-PyP was also demonstrated.

A pharmacokinetic study of MnTE-2-PyP in mouse has been conducted.[31] After a single intraperitoneal injection of MnTE-2-PyP at 10 mg/kg, the drug is rapidly absorbed into the plasma reaching its maximal concentration only 15 minutes after the injection. The maximal levels in spleen, kidney, heart, and lung were also reached rapidly within a period of 15–45 minutes, likely through a passive diffusion mechanism. However, among these organs, the levels of lung and heart were much lower than those of kidney and spleen, demonstrating both high penetration and affinity for the latter organs. In contrast, the accumulation in the liver was slower (maximal concentration after 8 hours) yet important, demonstrating a high affinity of this drug to this organ. Due to its strong binding to tissues, the elimination was slow ($t_{1/2} = 55$–135 h), therefore, confirming that there is no need for frequent dosing. After an initial high loading dose, the treatment may be followed by periodic small "maintenance" doses. The lipophilicity of metalloporphyrins has been identified as a critical parameter that dominates their efficacy in blocking the development of morphine antinociceptive tolerance through peroxynitrite-mediated pathways.[32] While MnTE-2PyP and its lipophilic derivative MnTnHex-2PyP (Fig. 15.5) exhibited similar rate constant of peroxynitrite scavenging ($\sim 10^7$ M^{-1} s^{-1}), MnTnHex-2PyP was found 30-fold more effective than the ethyl derivative. Moreover, it was also shown that it distributes in the brain at more than 10-fold higher levels than MnTE-2PyP. Very recently, Aeolus Pharmaceuticals, Inc. announced that a second study of its lead drug, AEOL 10150, has been initiated by the National Institutes of Health's and the National Institute of Allergy and Infectious Diseases to test the efficacy of AEOL 10150 as a treatment for damage to the gastrointestinal tract due to exposure to radiation. Treatment with AEOL 10150 has been shown to reduce inflammation and oxidative stress due to inhalation of a sulfur mustard analogue.[33]

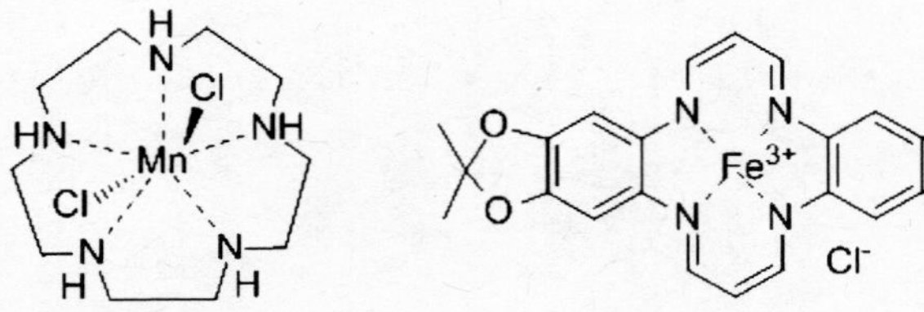

Mn([15]aneN₃)Cl₂

TAA-1/Fe

Figure 15.7 Chemical structures of the pentacoordinated maganese complex Mn([15]aneN₃)Cl₂ and the iron complex TAA-1/Fe.

15.2.3 Other Metal Complexes

There are a number of other metal complexes that exhibit antioxidant properties. The Mn^{II}(pentaazacyclo pentadecane) complexes (Fig. 15.7) have been demonstrated to possess catalytic SOD activity and *in vivo* activity against oxidative stress-mediated damages.[34,35] The parent unsubstituted compound, Mn([15]aneN₅)Cl₂ has a superoxide dismutation rate of $4.0 \ 10^7 \ M^{-1} \ s^{-1}$ at pH = 7.4. It has shown for instance high potency as antiinflammatory agent[35] or in a canine model of cardiac reperfusion injury[36] to name a few. Since in the pentaazamacrocyclic ligand-based mimetics, five coordination points hold the manganese atoms, only one-electron transfer reactions are available providing to this class of antioxidants the unique feature of being specific superoxide scavengers *in vitro*. Indeed, while the scavenging of H_2O_2 and $ONOO^-$ requires transfer of two electrons (this can be achieved by Mn(III) containing complexes), that of superoxide requires only the transfer of one electron. The dismutation of $O_2^{\bullet-}$ proceeds through alternate oxidation and reduction of the manganese atom at the center of the macrocyclic structure, resulting in an interchange valence state between Mn(II) and Mn(III). Mechanistic studies have shown the existence of two pathways for the electron transfer form Mn(II) to $O_2^{\bullet-}$[37,38]: (a) a hydrogen atom transfer from a bound water on Mn(II) to $HO_2^{\bullet}$ to yield a Mn(III) hydroxo intermediate and (b) a dissociative pathway in which superoxide anion binds to a vacant coordination site on Mn(II) followed by protonation/oxidation to yield a Mn(III) hydroperoxo species.

Since iron-based natural enzymes exist, iron (III) complexes would be attractive as catalytic antioxidants and would exhibit greater kinetic and thermodynamic stability than Mn(II) complexes for instance. However, as previously mentioned aquo iron (II and III) and iron complexes are prone to convert hydrogen peroxide to hydroxyl radicals, the so-called Fenton reaction. Therefore, the development of such iron complexes has met limitation. Nevertheless, some iron macrocyclic complexes have been developed among which is the tetraazannulene derivative (TAA1/Fe) (Fig. 15.7).[39] This complex has a good catalase-mimetic activity at physiological pH, and it protects endothelial cells against H_2O_2 toxicity. Moreover, its ligand (and not the complex) was found to strongly protect cells against iron- and/or hydrogen peroxide-mediated damages very likely by both iron chelation and hydrogen peroxide degradation.[39] Water-soluble iron porphyrins have also been developed.[40] The 5,10,15,20-tetrakis (*N*-methyl-4′-pyridyl)porphinatoiron(III) (Fe(III)TMPyP) and the 5,10,15,20-tetrakis (2,4,6-trimethyl-3,5-sulfonatophenyl) porphinatoiron(III) (Fe(III)TMPS) catalyse efficiently the decomposition of $ONOO^-$ almost exclusively to nitrate. There is strong evidence that the catalytic peroxynitrite turnover occurs through reversible oxidation of the complex to the Fe (IV).[40] A one-electron oxidation of the porphyrins to form an oxo Fe(II) intermediate and nitrogen dioxide was proposed as a common pathway for the decomposition of $ONOO^-$. However, after the formation of this oxo Fe(II) intermediate, two different processes may operate. A fast and stoechiometric pathway would dominate when the peroxynitrite concentration is comparable or lower than that of the porphyrins, whereas a 10-fold slower catalytic pathway would dominate when $ONOO^-$ is in large excess.[41] The potency of these iron (III) porphyrins was further demonstrated in *in vitro* and *in vivo* models.[42] Cytoprotection by the FeTMPS was observed at a concentration approximately 50 times lower than that of $ONOO^-$. A rodent model of acute inflammation was used to demonstrate the potency of these two catalysts *in vivo* and a dose-dependant reduction of the carrageenan-induced paw swelling was observed in treated animals, while neither the ligands themselves nor the free Fe exhibited protective effects.[42]

15.3 NONMETAL-BASED ANTIOXIDANTS

15.3.1 Ebselen

The 2-phenyl-1,2-benzisoselenazol-3(2*H*)-one, also known as ebselen or PZ51, is one of the most studied gluthation peroxidase mimics in which the selenium possesses the oxidation number +2. By various redox reactions, ebselen can be converted to several compounds, which represent the oxidation states of selenic acid (+4), selenol (−2), diselenide (−1), and selenenylsulphides (±0). As a consequence of the energetically favored five-membered ring formed with the selenium atom, ebselen possesses a high thermodynamic stability and a relatively long half-life in animals and humans,

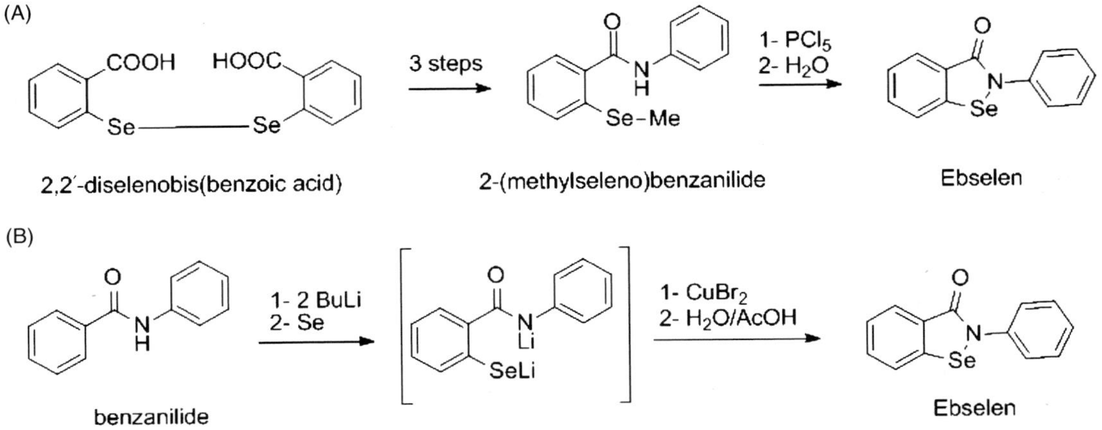

Scheme 15.2 Synthetic routes of Ebselen by (A) Weber and Renson and by (B) Engman and Hallberg.

and therefore, the selenium atom is not liberated and does not enter the selenium metabolism. Because of its unique stability and despite its pronounced redox activity, it does not exhibit toxicity unlike other selenium compounds. Because of its good lipid solubility with a log $P_{O/W}$ ~2.8,[43] it can easily diffuse into the cells. Neuroprotective effects of ebselen have been observed at ~10 μM of plasma level, and its brain level can reach ~20% of the plasma level.[44,45]

Several syntheses of ebselen have been described and the large-scale synthesis pathway that has been patented is based on those by Lesser and Weiss[46] and Weber and Renson (Scheme 15.2A).[47] The latter strategy involves a three-step conversion of 2,2′-diselenobis(benzoic acid) to 2-(methylseleno)benzanilide, which is next cyclized to give ebselen.[47] Later on, a one-pot synthetic route from benzanilide involving ortholithiation, selenium insertion, and oxidative cyclization was reported (Scheme 15.2B).[48]

Ebselen is capable of reducing hydroperoxides to their corresponding hydroxy compounds with, for instance, a second-order rate constant of reaction with hydrogen peroxide of 0.29 mM^{-1} min^{-1}.[49] Ebselen also reacts with thiols such as gluthation, N-acetyl-L-cysteine or dihydrolipoate to form selenenylsulfides, which in the presence of an excess of thiol can be converted to ebselen selenol and ebselen diselenide. The discovery of the peroxidase-like activity of ebselen was made in 1984 and since then, extensive researches have been conducted to understand the catalytic mechanism, yet it remains unclear. Good reviews on ebselen have summarized its enzyme like-catalytic activity[50,51] as well as its antiinflammatory action.[52] It has been unequivocally demonstrated that the transient formation of the ebselen selenol occurred in aqueous systems.[49] Moreover, under conditions of typical peroxidase assay, selenol was estimated to be responsible for the most part of the gluthathione-(70%) and dithiotheitol-dependant peroxidase activities of ebselen.[49] Dihydrolipoate, the reduced form of lipoic acid, has been shown to be a more efficient cofactor than glutathione for the peroxidase activity of ebselen.[53] Indeed, GSH reacts with ebselen to give a selenenylsulfide, which is subsequently converted into the diselenide form of ebselen; however, the formation of the diselenide from the GSH-selenenyl sulfide is quite slow. On the contrary, after reaction between ebselen and dihydrolipoate, the diselenide form is immediately detected, and its fast formation compared to that observed with GSH demonstrates that dihydrolipoate is a better cofactor than GSH. It is also important to stress that the peroxidase activity of ebselen is not only limited to hydrogen peroxide as organic hydroperoxides can also be reduced with rates of reaction more than 10 times higher than that of H_2O_2. The hydrophobic nature of ebselen is believed to favor the reduction of lipophilic substrates.[54] Pulse radiolysis experiments showed that ebselen reacts with trichloromethylperoxyl radical with a rate constant of 2.9 10^8 M^{-1} s^{-1} as well as with other halogenated radicals.[55] However, its reaction with peroxyl radical is not a radical scavenging reaction,[56] suggesting that ebselen does not exhibit a pronounced free radical scavenging activity. This was further confirmed by Nogochi et al. who found out that it does not inhibit AIBN initiated lipid peroxidation.[57] The quenching of singlet oxygen by ebselen was also reported but with a modest rate constant of 2.5 10^6 M^{-1} s^{-1} compared to other singlet oxygen scavenger.[58] As concluded by Schewe in his review, the radical scavenging activity of ebselen "appears to be –if any– of minor importance for the pharmacological

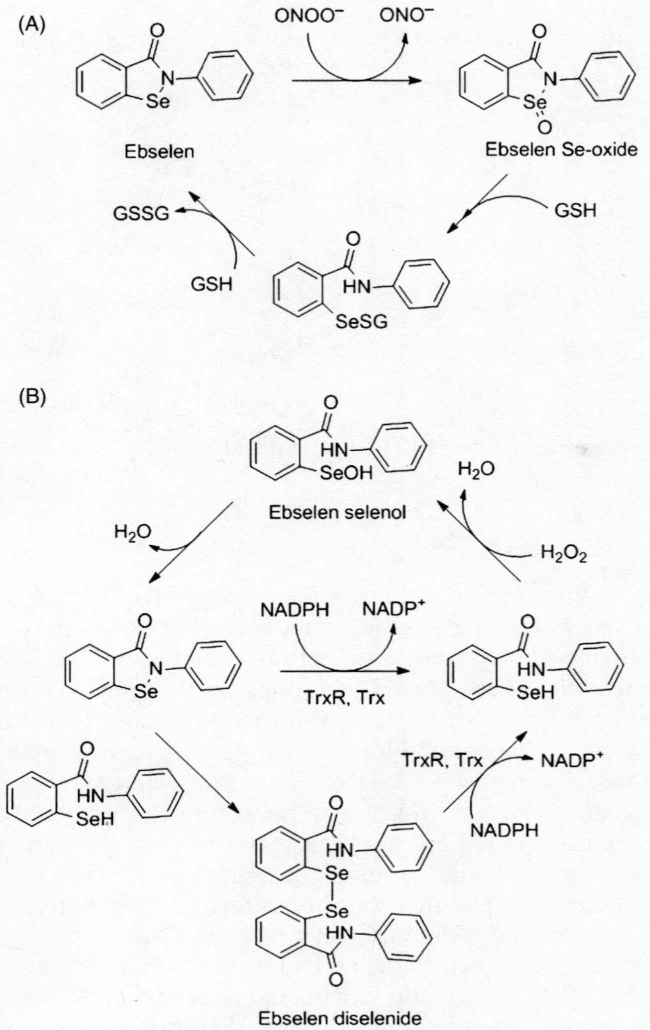

Figure 15.8 (A) Peroxynitrite-dependent oxidation of ebselen and (B) H$_2$O$_2$ reductase activity of ebselen.

a similar retention time to that of the ebselen–cysteine adduct. Therefore, the ability of ebselen to react with cellular thiol groups on proteins has made the interpretation of biological effect difficult since many enzymes possess at least a reactive thiol group in their catalytic domains. It is now admitted that the general molecular mechanism of the ebselen-mediated inhibition of several enzymes is through its reaction with essential thiol groups of these enzymes. Since the majority of these enzymes such as lipoxygenases,[62] NADPH oxidases,[63] and nitric oxide synthases[64] are implicated in inflammation, it was suggested that this may provide to ebselen its anti-inflammatory action.[52] Two mechanisms for the inhibition of lipoxygenase have been proposed, either directly by formation of an ebselen–enzyme complex or indirectly by lowering the hydroperoxide level. In the absence of glutathion, ebselen inhibits directly the pure 15-lipoxygenase with an IC$_{50}$ of 0.17 µM whereas, in the presence of glutathione, it acts indirectly by lowering hydroperoxide level with an IC$_{50}$ of 234 µM.[62] *In vitro* contradictory results were observed probably due to fact that in cells, lipoxygenases are regulated via the level of hydroperoxides. Whatever its mode of action, the ability of ebselen to inhibit lipoxygenase opens new preventive and therapeutic applications in atherosclerosis. Nitric oxide, a relevant signaling molecule, is generated in various cell types in the central nervous system. Its production by the inducible nitric oxide synthase (iNOS) is a consequence of inflammation and has been associated with brain ischemia. Ebselen was found to decrease the iNOS expression in rat hippocampal slices submitted to oxygen-glucose deprivation, a model of ischemic insults.[65] These findings were in agreement with *in vivo* studies that demonstrated that ebselen also abolished lipid peroxidation induced by a quinolinic acid-induced overstimulation of *N*-methyl-D-aspartate receptors.[66]

Because of its anti-inflammatory activity, ebselen has been used in clinical trials for the treatment of acute ischemic stroke[67] and delayed neurological deficits after aneurismal subarachnoid hemorrhage.[68] When given at a dosage of 150 mg twice a day for 2 weeks to patients with acute ischemic stroke, a significant protection was observed at 1 month. Unfortunately, although improvement was maintained at 3 months, this failed to reach statistical significance. Post hoc analysis showed that protection was effective for those having the treatment started within 24 hours after the stroke onset.[67]

Holmgren and his colleagues have shown that ebselen is an excellent substrate for mammalian thioredoxin reductase (TrxR), which is reduced by NADPH forming selenol.[69,70] They also showed that ebselen strongly enhance hydrogen peroxide reductase activity of mammalian TrxR in the presence of Trx, acting like a perox-

activity of this drug".[52] Later on, the ability of ebselen to scavenge peroxynitrite has also been demonstrated.[59,60] Ebselen rapidly reacts with peroxynitrite with a second order rate constant of ~2.0 10^6 M^{-1} s^{-1} yielding the selenoxide derivative, 2-phenyl-1,2-benzisoselenazol-3(*2H*)-one 1-oxide, as the sole selenium-containing product at 1:1 stoechiometry (Fig. 15.8A).

Daider et al. demonstrated however, that in cellular systems, ebselen is mostly present as thiol adducts and therefore loses its high reactivity toward peroxynitrite.[61] They observed that free ebselen quickly reacted with thiols in both coronary strips and in aortic microsomes to form two metabolites, one of which was identified as the ebselen–glutathione adduct, whereas the other had

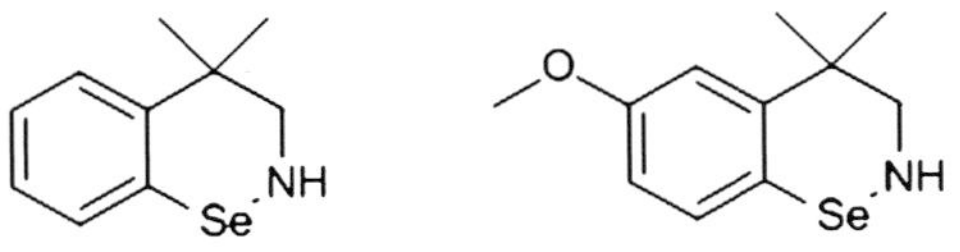

Figure 15.9 Chemical structures of the ebselen analogues BXT-51072 and BXT-51077.

iredoxin. While the reduction of ebselen by NAPDH and catalyzed by TrxR has a K_M value of 2.5 μM and a k_{cat} of 588 min^{-1}, that of ebselen diselenide, which is also a substrate of TrxR, is 100 times slower with a K_M value of 40 μM and a k_{cat} of 79 min^{-1}. Ebselen diselenide was also shown to have a higher H_2O_2 reductase activity than that of ebselen. Figure 15.8B summarizes the antioxidant mechanisms of ebselen and ebselen diselenide proposed by Holmgren and colleagues.

Several analogues of ebselen have been developed, including the BXT series (Fig. 15.9).[71] BXT-51072 was able to inhibit both H_2O_2- and TNFα-induced alterations of cytoskeleton.[72] BXT-51072 and its methoxy analogues BXT-51077 were found to be efficient inhibitors of the TNFα-induced endothelial expression of P- and E-selectin, two adhesion molecules that mediate and amplify specific tissue adhesion.[71] This clearly demonstrated the potency of these new ebselen analogues in inhibiting TNFα-induced proinflammatory responses of endothelial cells.[71,73] The BXT-series is currently being developed by Oxis International, Foster City, CA. Other organoselenides and organotellurides have also been reported to catalytically scavenge peroxides through a peroxidase-like activity.[74]

15.3.2 Edaravone

The 3-methyl-1-phenyl-2-pyrazolin-5-one, also called edaravone, Radicut, or MC-186, has been developed by Mitsubishi-Tokyo Pharmaceuticals Inc (Tokyo, Japan) and has been used in Japan to treat patients with acute brain infarction and stroke since 2001. For a recent review on the development of edaravone, see the review article by Watanabe et al.[75] Moreover, it has been shown that edaravone has also protective effects on patients with acute myocardial infarction.[76] Edaravone is a small amphipathic molecule (Molecular weight = 174.2) both soluble in water (3 g/L at 20°C) and in lipidic phase with a log P 1.28 for its neutral form.[77] Therefore, it exhibits good cell membrane permeability, and its ability to pass through the blood–brain barrier was further confirmed in dogs. The plasma concentration of edaravone in clinical cases has been estimated around 10 μM.[78] Edaravone is metabolized to its glucoronide and sulfate conjugates in the liver and is rapidly excreted in the urine.[79] Edaravone can exist in three tautomeric forms as shown in Figure 15.10.[77,80,81] While in polar media the anionic form is believed to be responsible for the scavenging activity of edaravone through a one-electron transfer mechanism,[81] in lipidic phase, the mechanism of H-atom abstraction has been hypothesized to be predominant.[77,81]

Edaravone is able to scavenge several free radicals such as hydroxyl radical, azide radical ($N_3^{\bullet}$), sulfate radical ($SO_4^{\bullet-}$), trichloroperoxyl radical ($CCl_3OO^{\bullet}$), and DPPH radical.[80,82–84] The rate constant of radical reaction with edaravone was determined by indirect method using EPR[80,85] as well as pulse radiolysis experiments.[82] A summary of the rate constants is presented in Table 15.1. Its ability to directly scavenge nitric oxide in a dose-dependent manner was also demonstrated using ESR and 2-(4-carboxyphenyl)-4,4,5,5-tetramethylimidazoline-1-oxy-3-oxide (carboxy-PTIO).[86] This indicates the additional possibility of novel neuroprotective activities against brain injury and focal cerebral ischemia through nitric oxide scavenging. Very recently, reaction of singlet oxygen ($^1O_2$) with edaravone was demonstrated in a cell-free system as well as in the presence of cells, leading to the formation of a nonidentified compound.[78] It was further demonstrated that because of its singlet oxygen scavenging, edaravone prevented $^1O_2$-induced cell death, which may contribute to its pharmacological activity. Anzai et al. demonstrated that edaravone exhibits radioprotective effects on a mice model in a dose- and injection time-dependent manner. Injection 30 minutes before the irradiation led to the greatest protection, likely by quenching short life radicals generated by the irradiation.[87]

The antioxidant action of edaravone includes the conversion of arachidonic acid to prostacyclin and the inhibition of lipoxygenase metabolism of arachidonic acid, which is mediated by hydroxyl radical trapping. Indeed, edaravone was found to directly prevent hydroxyl radical-induced injury of bovine aortic endothelial cells.[88,89] Yoshida et al. reported its ability to enhance eNOS expression; likely because of increased stability of eNOS mRNA, and its ability to reverse oxidized LDL-mediated reduction in the expression of eNOS in endothelial cells. Therefore, the preventive action of edaravone from ischemic disease consequence may be attributed to this eNOS upregulation with decreased oxidation.[90] The therapeutic effect of edaravone on alterations in endothelium-dependent relaxation as well as eNOS expression in the rabbit ear central artery after irradiation was also demonstrated. The reduced level of eNOS mRNA in irradiated vessels

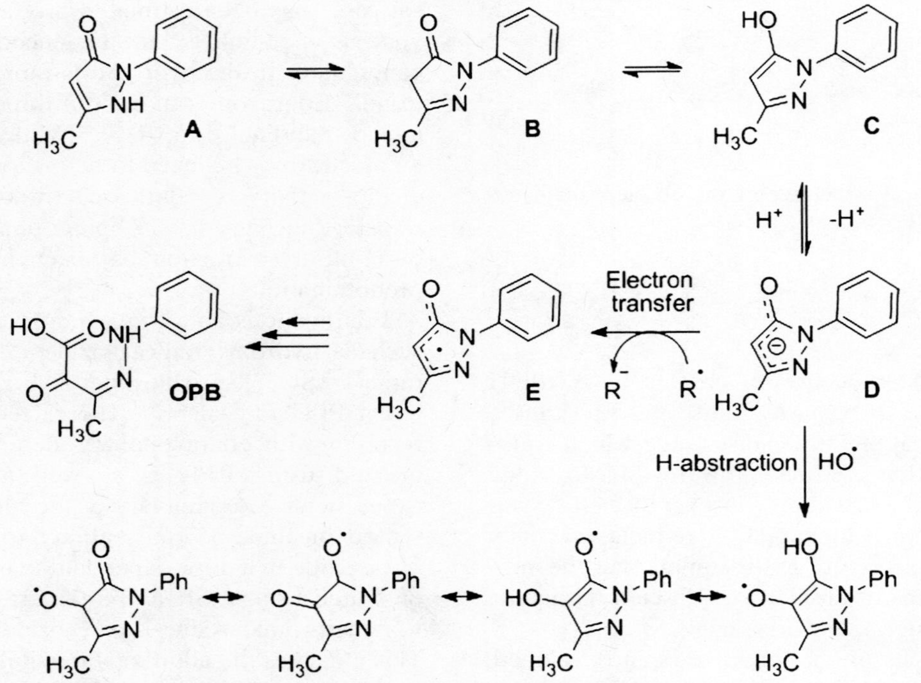

Figure 15.10 Tautomeric equilibrium of 3-methyl-1-phenyl-2-pyrazoline-5-one and its free radical scavenging mechanisms. A, amine form; B, keto form; C, enol form; D, enolate form; E, radical form; OPB, 2-oxo-3-(phenylhydrazono)1-butanoic acid; R•, peroxyl, alcoxyl or hydroxyl radical.

TABLE 15.1 Rate Constants, Oxidation Potential, and Dissociation Constant of Edaravone.

Rate constants	HO•	$3.0\ 10^{10}$ M^{-1} s$^{-1 a}$
		$8.5\ 10^{9}$ M^{-1} s$^{-1 b}$
	N$_3$•	$5.8\ 10^{9}$ M^{-1} s^{-1} (at pH 9)^b
	SO$_4$•$^-$	$6.0\ 10^{8}$ M^{-1} s^{-1} (at pH 9)^b
	CCl$_3$OO•	$5.0\ 10^{8}$ M^{-1} s$^{-1 b}$
	e$^-_{aq}$	$2.4\ 10^{9}$ M^{-1} s^{-1} (at pH 7)^b
Oxidation potential	E$_{pa}$	483 (at pH 7)^c
		480 (at pH 7.8)^c
Dissociation constant	pKa	6.9^{d,e}
		7.0^f

^{a}Determined by EPR from Abe et al.[80]
^{b}Determined by pulse radiolysis from Lin et al.[82]
^{c}vs Ag$^+$/AgCl from Nakagawa et al.[85]
^{d}From Chegaev et al.[77]
^{e}From Lin et al.[82]
^{f}From Yamamoto et al.[83]

was almost completely normalized by edaravone treatment.[91] Moreover, treatment with edaravone resulted in the prevention of spinal cord damages in rabbits increasing the number of motor neurons after ischemia. While the treatment of edaravone resulted in a significantly reduced induction of neuronal nitric oxide synthase (nNOS), those of eNOS and Cu/Zn SOD were increased. These results suggest that edaravone may be a strong candidate as therapeutic agent in the treatment of ischemic spinal cord injury.[92]

Figure 15.11 shows the putative mechanisms as proposed by Higashi et al. by which edaravone improves endothelial function in patients with cardiovascular diseases.[76] Indeed, several studies have reported that administration of edaravone before reperfusion decreases cardiac injuries in myocardial ischemia–reperfusion animal models. Intravenous infusion of edaravone at a dose of 3 mg/kg reduced the infarct size by ~50% compared to the vehicle group in rats subjected to coronary artery occlusion for 10 minutes followed by 24-hour reperfusion.[93] Reduction of myocardial damage was also demonstrated by Minhaz et al. who reported that edaravone reduced by ~50% necrotic area in isolated reperfused rat hearts after coronary artery occlusion.[94] Moreover, intraperitoneal administration of edaravone at 10 mg kg^{-1} twice a day and for 7 days attenuated pressure overloaded-induced cardiac hypertrophy by ~30% to mice subjected to thoracic aorta constriction.[95] Tsujita et al. reported for the first time in 2004 that edaravone is efficient in the treatment of myocardial reperfusion injury in humans.[96] This randomized, controlled, and open-label study in 80 patients with acute myocardial infarction demonstrated that administration

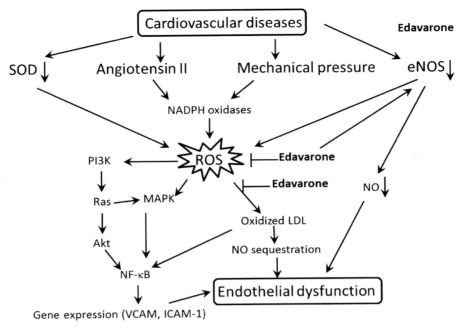

Figure 15.11 The potential mechanisms of edaravone on endothelial functions. (Adapted from reference 76.)

of edaravone at a dose of 30 mg for 10 minutes before reperfusion inhibits all the manifestations of reperfusion injury such as reperfusion arrhythmias, myocardial stunning, and lethal reperfusion injury. Indeed, serum concentration of creatine kinase-MB isoenzymes was decreased while left ventricular ejection fraction was improved.

Aside from its ability to prevent myocardial injury, edaravone exerts neuroprotective effects by inhibiting endothelial injury and by ameliorating neuronal damage in brain ischemia.[97] It provides the desirable features of NOS as it increases the beneficial eNOS whereas a decrease of the detrimental nNOS and iNOS occurs at the same time.[97] Indeed, transient ischemia in rat brain increases the expression of the three NOS isoforms at protein levels when evaluated 4 days after the ischemic insult. Treatment with edaravone immediately after the reperfusion (on day 0) downregulates the expression of nNOS and iNOS while that of eNOS is upregulated.[98] The Edaravone Acute Brain Infarction Study Group reported in a multicenter randomized clinical trial that when administered 72 hours after the onset of ischemic stroke at a dose of 30 mg twice a day for 14 days, edaravone reduced brain infarct volume and produced sustained benefits in functional outcome compared to the placebo group.[99] Recently, edaravone has been identified as a promising therapeutic agent for human motor neuron diseases including amyotrophic lateral sclerosis (ALS).[100] A small-sized open trial suggested that edaravone is safe and may delay the progression of functional motor disturbances in ALS patients.[101] Ito et al. confirmed the potency of edaravone in preventing motor neuron degeneration using a mutant SOD1 mouse model.[102] Edaravone significantly slowed down the motor function decline; however, no difference of the life span was observed between the edaravone-treated group and the control. This suggests that pathways causing motor function decline may be not independent to those causing death, generally by respiratory muscle failure.

The protective effect of edaravone against retinal damages *in vitro* and *in vivo* has been reported.[103] For instance, retinal damages induced by intravitreous injection of *N*-methyl-D-aspartate at 5 nmol were prevented by simultaneous injection of edaravone at 5 or 50 nmol, while no protection was observed by treatment with *N*-acetyl cysteine (10–1000 nmol). Moreover, when administered intravenously at 1 and 3 mg kg^{-1} immediately before NMDA injection, edaravone significantly inhibited cell death compared to the control. The protective effects of edaravone against light-induced retinal damages in mice were also demonstrated.[104] Photoreceptor damages were evaluated by measuring the outer nuclear layer thickness 5 days after light exposure and by recording the electroretinogram. Edaravone administered intraperitoneally at 3 mg kg^{-1} or intravenously at 1 mg kg^{-1} significantly inhibited the reduction of the outer nuclear layer thickness compared to the saline group.

Because of its antioxidant properties, topical administration of edaravone has been shown to prevent noise-induced hearing loss, especially if given within 21 hours of noise exposure.[105] Indeed, it is now admitted that noise exposure leads to increased levels of ROS in the cochlea, and noise-induced hearing loss have been successfully reduced by treatment with antioxidants. This suggests that edaravone may be clinically effective in the treatment of acoustic trauma. Protective effects against cisplatin-induced nephrotoxicity in rats by edaravone have also been demonstrated.[106] In addition, edaravone has been reported to counteract the development of multiple low-dose streptozotocin-induced diabetes in mice.[107]

15.3.3 Lazaroids

Lazaroids are 21-aminosteroids derived from glucocorticosteroids, but they do not exhibit glucocorticoid and mineralocorticoid activities. Belonging to this family, the tirilazad mesylate (U74006F, 21-[4-(2,6-di-1-pyrrolidinyl-4-pyrimidinyl)-1-piperazinyl]-16α-methyl-pregna-1,4,9 (11)-triene-3,20-dione) is derived from methylprednisolone in which two hydroxyl groups have been removed in 11-C and 17-C positions (Fig. 15.12). The addition at the 21-C position of a side chain comprising amino groups is responsible for the absence of any corticosteroid receptor activity.[108] Moreover, the presence of the side chain is also responsible for enhanced lipid peroxidation-inhibiting activity. Indeed, tirilazad mesylate was found to be a potent inhibitor of iron-dependant lipid peroxides[109] as well as a potent scavenger of hydroxyl, superoxide, and lipid peroxyl and phenoxy radicals.[110]

Due to its high lipophilicity, which enables a deep insertion within the lipid bilayer of cell membranes, the physical membrane stabilizing effect of tirilazad mesylate has been demonstrated.[111] It has to be noted that in solution, tirilazad mesylate is a weak radical scavenger. On the contrary, it can exert a substantial antioxidant effect in membranes and especially by a sparing and synergic effect for α-tocopherol, although it is not mediated through a reduction mechanism.[112] Moreover, U74006F has also been shown to decrease membrane fluidity, which may limit lipid peroxidation reactions. Lazaroids have been shown to locate preferentially in vascular endothelial cell membranes.[113] The ability to block arachidonic acid release from mouse tumor cells[114] and to inhibit injury-induced calcium influx may also contribute to the beneficial effects of U74006F.[115] Indeed, there is strong evidence that the beneficial effects of lazaroids are mediated through endothelium protection by preventing direct damaging effects of ischemia, inflammation, and oxidative stress.[108]

Lazaroids were initially developed as an alternative for glucocorticoids in the treatment of central nervous disorders. For instance, tirilazad reduced cerebral infarct size in rabbit. Indeed, both acute and sustained U74006F treatments with tirilazad produced a significant reduction in the severity of neuronal damage in the neocortex of rats.[116] Neurological functions were also preserved by treatment with tirilazad in a model of acute cerebral ischemia in dogs.[117] The protection efficacy of tirilazad has also been demonstrated in mouse, rat, and cat models of acute traumatic brain injury.[118] Pretreatment

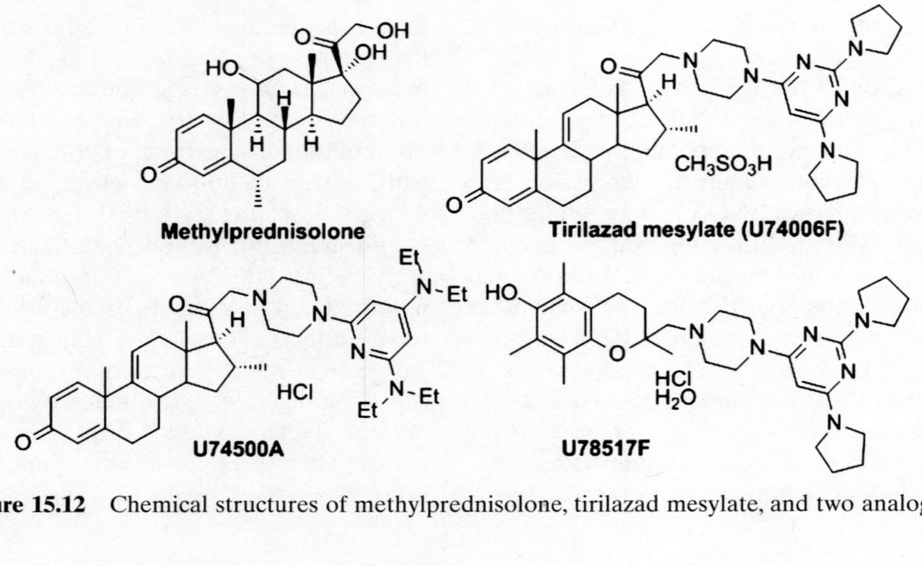

Figure 15.12 Chemical structures of methylprednisolone, tirilazad mesylate, and two analogues.

with U74006F reduced myocardial ischemia reperfusion in dogs[119]; however, when administered after the coronary artery occlusion, no significant reduction of the infarct size was observed.[120] Pretreatment with lazaroid U-74500A (Fig. 15.12), a first-generation methylprednisolone-related lazaroid, reduced significantly myocardial ischemia reperfusion in rats.[121] Administration of lazaroids in several experimental endotoxemia and other forms of sepsis showed beneficial effects as reviewed by Spapen, Zhang, and Vincent.[108] For instance, lazaroid treatment in a rat model of hepatic ischemia and reperfusion followed by endotoxin challenge resulted in a significant reduction of hepatic[122] and lung injuries.[123] U74006F was found to provide hepatic protection through stabilization of endothelial cell membranes rather than by inhibiting lipid peroxidation.[124] U74006F was also shown to prolong survival in a dog model of endotoxic shock even when administered 30 minutes after endotoxin injection. The protection of dogs was associated with a decrease of cytokines production compared to control which demonstrates the anti-inflammatory effect of U74006F.[125]

Several phase II and III clinical trials have been undertaken with tirilazad in patients with ischemic stroke.[126] Among six randomized and controlled trials assessing tirilazad in 1757 patients with presumed or confirmed acute ischemic stroke, none of them showed efficacy of tirilazad in ischemic stroke. Moreover, it was found that tirilazad worsened outcome after stroke, increasing death and disability by about 20% when given to patients with acute ischemic stroke. Although tirilazad was neuroprotective in experimental stroke,[116,127] controversial results were also reported,[128] suggesting that tirilazad could, in certain conditions, lack beneficial effect. For instance, tirilazad had no effect on infarct volume in 2 of 4 phase III trials, but it has to be noted that more than 75% of patients received treatment after 3 hours, a time when neuroprotection is less likely to be effective. One can also cite that tirilazad caused thrombophlebitis and might have induced fever, a systemic inflammatory state, which is known to be associated with poor outcome.[129]

Tirilazad was also used in clinical trials of traumatic brain injury.[118] A phase II dose–escalation study firstly demonstrated its safety in TBI patients. Two phase II multicenter clinical trials next examined its ability as therapeutic in moderately and severely injured closed TBI patients. While the trial conducted in North America has never been published, the one conducted in Europe failed to show a significant beneficial effect in moderate or severe patient categories.[130] This is in agreement with the lack of potency of tirilazad in patients with ischemic stroke as TBI is a condition known to involve cerebral ischemia. However, it was shown that moderately injured male TBI patients with traumatic subarchanoid hemorrhage (SAH) had significantly less mortality when treated with tirilazad (6%) than those of the placebo group (24%; $p < 0.026$). Tirilazad also lessened the mortality of severely injured male with traumatic SAH (34%) compared to placebo (43%; $p < 0.071$). Surprisingly, two similar trials of tirilazad in patients with SAH provided conflicting results. Patients were treated at a dose of 6 mg/kg per day for 10 days. The trial conducted in North America showed no difference between patients treated with tirilazad and those receiving placebo.[131] However, this absence of protection may be related to the fact that patients were treated with phenytoin, which is known to decrease the bioavailability of tirilazad. On the contrary, the cooperative study conducted in Europe, Australia, and New Zealand showed reduction of mortality and a better 3-month outcome in the tirilazad-treated group compared to the vehicle group, the effect being more pronounced in men.[132–134] Based on these data, tirilazad was successfully approved and marketed for use in aneurysmal SAH in several western European and Australian countries.

15.4 NITRONES

Nitrones were initially designed as spin traps for detecting transient free radicals using electron paramagnetic resonance.[135,136] As shown in Figure 15.13, free radicals react with the diamagnetic nitrone to from a paramagnetic spin adduct whose half-life is significantly longer than that of the parent radical. Among the family of nitrone, two classes have been widely employed, the cyclic nitrones derived from the 5,5-dimethyl-1-pyrroline-*N*-oxide (DMPO) and the linear ones derived from the α-phenyl-*N*-tert*-butyl nitrone (PBN) (Fig. 15.14).

Cyclic nitrones have been mainly employed as probes for the detection of free radicals because of their better

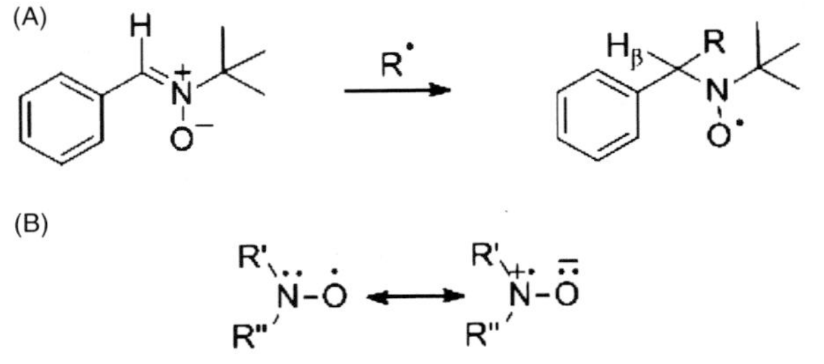

Figure 15.13 (A) The spin trapping mechanisms of PBN and (B) the two resonance structures of a nitroxide.

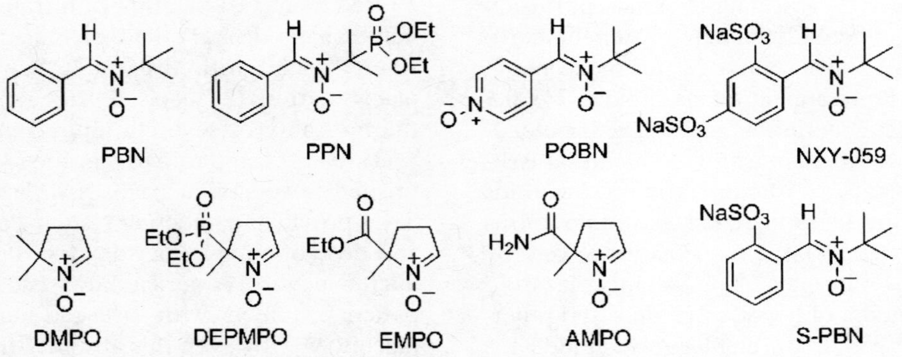

Figure 15.14 Chemical structures of simple nitrones agents.

ability to trap oxygen-centered radicals compared to linear nitrones. Therefore, several analogues of DMPO have been designed for the past two decades. One can cite the phospshorylated analogue 5-diethoxyphosphoryl-5-methyl-1pyrroline N-oxide (DEPMPO) developed by Tordo and colleagues.[137] DEPMPO leads to the formation of very stable spin-adducts. For instance it was demonstrated that its superoxide adduct is very stable and not reduced into an hydroxyl spin adduct as it is observed with DMPO. Ester analogues were also developed, such as the 5-ethoxycarbonyl-5-methyl-1pyrroline N-oxide[138] or the 2-tert-butoxycarbonyl-5-methyl-1pyrroline N-oxide (BocMPO).[139] More recently, an amido derivative (AMPO) was also reported. Using a stopped flow UV-vis kinetic method and an EPR-based competition kinetic method, the rate constant values of several cyclic nitrones were determined, the highest being that of AMPO, followed by EMPO, both DEPMPO and DMPO having the slowest reactivity.[140,141]

Although linear nitrones are considered as weaker traps compared to cyclic ones, they have been also employed in spin-trapping experiments. Indeed, PBN and its analogues efficiently trap carbone-centered radicals either in organic or aqueous phase. Moreover because of its lipophilicity, the PBN distribution within tissues and cells is much higher than that of the hydrophilic DMPO. Therefore, PBN has been successfully employed for *in vivo* experiments. In order to improve the trapping ability of PBN, several analogues were next developed, such as a phosphorylated derivative (PPN), which exhibits improved superoxide trapping capacities.[142,143]

Several methods for the synthesis of nitrones are available.[144,145] Nitrones have been used as synthetic intermediate due to their ability to react in 1,3 dipolar cycloaddition, but we will focus our attention on nitrone as spin traps and therapeutics only. In Scheme 15.3A–C are represented three general methods for the synthesis of linear nitrones. The "one-pot" synthesis (Fig. 15.3A) proceeds through the *in situ* reduction of a nitro group to its hydroxylamine, which is subsequently condensed with a benzaldehyde derivative.

This method has several advantages: (a) it is not expensive, (b) the reaction time is relatively short, and (c) a large number of substrates are stable under these conditions. In the case of more sensitive substituents to the reduction conditions, the second method as described in Scheme 15.2B is preferentially chosen. The N-tert-butylhydroxylamine is first obtained by reduction of its nitro derivative and can be purified and stored. Then, in a second step, the hydroxylamine is condensed with the benzaldehyde derivative. This allows milder conditions of coupling and an easy monitoring of the reaction. We showed in our laboratory that even bulky N-tert-butyl-hydroxylamines can be grafted onto bulky benzaldehydes. In such conditions, the use of acetic acid as a catalyst and molecular sieves significantly increase the kinetics and the yield of coupling.[146-149] Another method consists in the formation of an imine intermediate after reaction of the amino group on the carbonyl function. The imine can be oxidized with 3-chloroperbenzoic acid, for instance, followed by a thermal rearrangement of the resulting oxaziridine. Other oxidants such as H_2O_2 can also be used. However, the oxidation is sometimes difficult, requiring long reaction times and large excess of oxidant. Therefore, an alternative way is to first reduce the imine to its amino form and then oxidize it up to the nitrone.[150] The preparation of the cyclic 5,5-dimethyl-1-pyrroline-N-oxide can be achieved in a relatively simple manner (Scheme 15.3D).[145] The initial reaction is a Michael addition of the anion of 2-nitropropane onto acrolein to give 4-methyl-4-nitro-1-pentanal, and several conditions have been used. However, during the purification of the pentanal derivative by distillation under vacuum, explosion due to polymerization of unreacted acrolein may happen, and

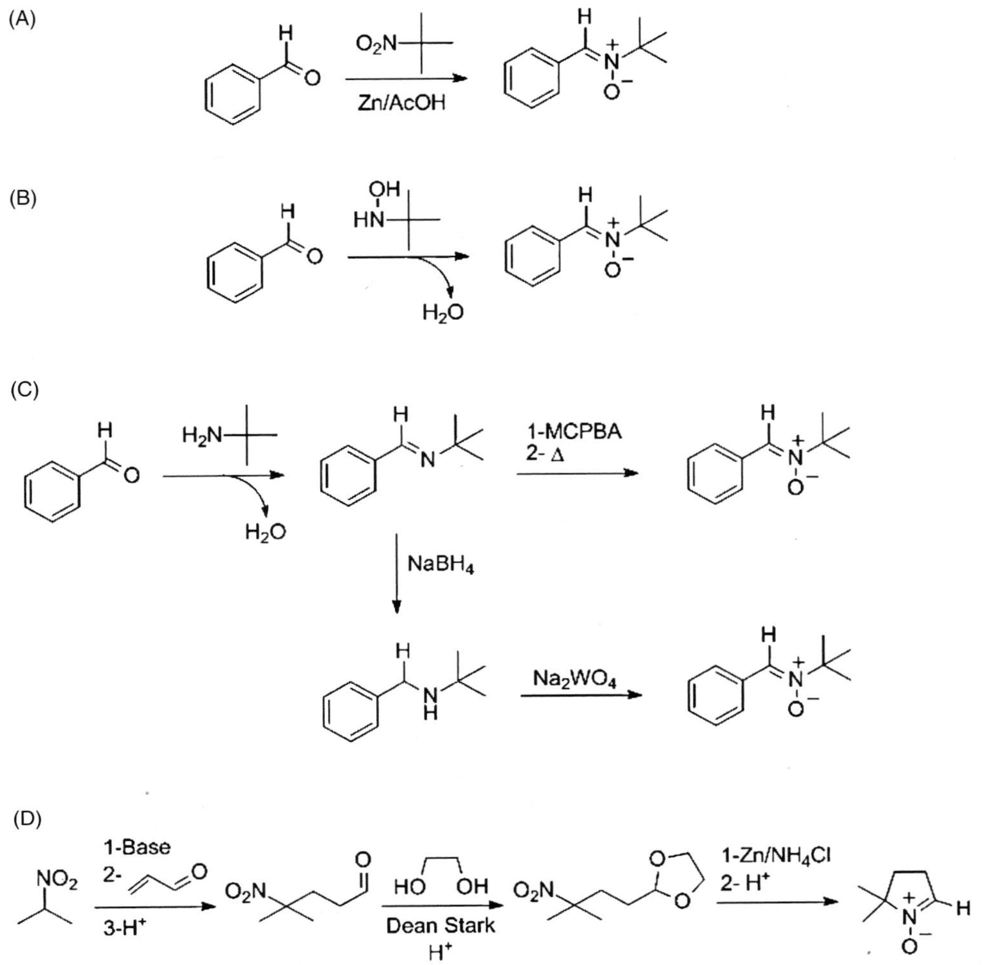

Scheme 15.3 Examples of methods for the synthesis of nitrones. (A) The "one-pot" method, (B) by condensation of N-tert-butylhydroxylamine, (C) by formation of an imine intermediate form, (D) the synthesis of DMPO.

therefore the purification requires considerable attention. Then a protection of the aldehyde function as a dioxolane group is achieved before cyclization, thereby avoiding the formation side products. Finally, reduction of the nitro group in the presence of zinc dust and ammonium chloride leads to the hydroxylamine, which, after acidic removal of the dioxolane group, reacts with the aldehyde function to form the cyclic DMPO.

Cellular toxicity of nitrone compounds such as DMPO, PBN, or their anolgues is low, rendering their use in biological systems possible without inducing any severe side effect. Janzen et al. studied the effect of intraperitoneal injection of nitrones at very high doses in rats.[151] The lethal dose of PBN was found to be approximately 100 g/100 g body weight while with DMPO at twice the lethal dose of PBN, no toxic signs were observed. A second study on the effects of nitrones on the blood chemistry in rats confirmed these findings. Neither DMPO at 232 mg/100 g body weight nor PBN at 100 g/100 g body weight were lethal.[152] Moreover, gross pathology and histopathological examinations failed to show any cellular damages with these two spin traps even at very high doses. Haseloff et al. performed cytotoxicity studies on bovine aortic endothelial cells and showed IC_{50} of 140 μM and 9 μM for DMPO and PBN, respectively, confirming *in vivo* studies.[153]

Using radiolabeled PBN, Chen et al. determined the tissue biodistribution, excretion, and metabolism of PBN in rats and showed that it is rapidly absorbed after intraperitoneal injection.[154] After 1 hour, 31.5% of the radioactivity was found in liver, 8.5% in kidney, and 3% in lung. Microdialysis technique coupled with HPLC was utilized to determine blood and brain distribution of PBN and POBN in the rats after intraperitoneal administration.[155] The results for PBN indicate a rapid localization in blood and brain, reaching a maximal concentration in less than 20 minutes. After 100 minutes, the PBN concentration in brain was twice than that of POBN, which is closely correlated to the highest lipophilicity of the former nitrone. The pharmacokinetics of nitrone spin traps was also determined using HPLC, showing that most of the spin traps were rapidly distributed in liver, heart, and blood, with a maximum concentration at 30 minutes for PBN.[156] Its plasma concentration after i.v. administration at 10 mg/kg was found to decline monoexponentially with a terminal half-life of 2.01 ± 0.35 hour.[157] The total plasma clearance value (CL_P = 12.37 ± 3.82 mL/min/kg) was close to the predicted blood clearance value (CL_B = 11.5 mL/min/kg), suggesting that PBN CL_P in rats is predominantly due to hepatic metabolism. Cellular distribution was also investigated by Cova et al. who found a better cellular penetration for PBN compared to DMPO, very likely related to its higher lipophilicity.[158] The primary metabolite of PBN, formed in the liver microsomal system, was identified as 4-hydroxyPBN, and since it circulates in body fluids and has free radical trapping ability, it was believed to provide additional effects to the pharmacological properties of PBN *in vivo*.[157,159]

15.4.1 Protective Effects of Nitrones (with Particular Attention to PBN)

15.4.1.1 Protection against Endotoxic Shock
Since the seminal work of Novelli et al.[160] nitrone spin traps, typified by α-phenyl-*N-tert*-butylnitrone (PBN), have been widely used as antioxidants in several biological models, and their use as pharmaceutical agents have been reviewed several times.[161–164] Novelli et al. first demonstrated that PBN afforded protection against death after traumatic shock, which was further confirmed by more rigorous experiments of endotoxic shock.[165–167] However, it was shown that PBN was protective in these models only if administered before the lipopolysaccharide (LPS) challenge.

15.4.1.2 Protection against Diabetes-Induced Damages
PBN was found to inhibit lipid peroxidation associated with type 1 diabetes.[168] Type 1 diabetes is characterized by a massive destruction of pancreatic β cells, which are responsible for the production of insulin. There is growing evidence implicating free radicals in the destruction of these cells. Pretreatment with PBN at 150 mg/kg i.p. reduced the severity of hyperglycemia in two distinct chemically induced model of diabetes using alloxan and streptozotocin (STZ). The protective effect of PBN in this model was demonstrated to be through inhibition of NF-κB, a transcription factor activated by both alloxan and STZ.[169]

15.4.1.3 Protection against Xenobiotic-Induced Damages
Interestingly, PBN was found effective in preventing doxorubicin-induced cardiotoxicity without altering the doxorubicin cytotoxicity. On the contrary DMPO did not provide any protection.[170] Kotamraju et al. demonstrated that PBN inhibits the doxorubicin-induced apoptotic signaling mechanism and that this antiapoptotic mechanism may be linked in part to the inhibition in formation or scavenging of hydrogen peroxide. Again DMPO was found inefficient.[171] Similarly, protection from thalidomide-induced birth defects in rabbits was observed after pretreatment with PBN, demonstrating that teratogenicity of thalidomide may involve free radical-mediated oxidative damage to embryonic cellular macromolecules.[172]

15.4.1.4 Protection against Noise-Induced Hearing Loss
There is evidence that hearing loss caused by exposure to noise involves oxidative stress. Indeed, acute acoustic trauma (AAT) induced by steady-state noise is associated with the production of ROS both with the onset of the noise trauma and some days after cessation of noise exposure as well as with a significant upregulation of iNOS activity. Therefore, several studies have been conducted to evaluate the potency of nitrones to prevent hearing loss. Fechter and colleagues showed that PBN was effective in mitigating hearing loss that occurs in rats exposed to high-level noise and to carbon dioxide[173] or acrylonitrile,[174] these two toxins strengthening the hearing loss associated with acoustic overexposure likely by increasing cochlear oxidative stress. Similar observations were made in a guinea pig model.[175] However, PBN failed to reduce auditory threshold shifts induced by noise alone. Surprisingly, recent findings by Choi et al. demonstrated that 4-hydroxyPBN decreased permanent hearing loss chinchilla exposed to high-level steady-state noise for 6 hours.[176] In animals treated with intraperitoneal injections of 4-hydroxyPBN 4 hours after the cessation of noise exposure and twice daily for another 48 hours, a dose-dependent reduction of permanent hearing threshold shifts and hair cell loss was observed. When-hydroxyPBN was combined with either *N*-acetylcysteine, or *N*-acetylcysteine plus acetyl-L-carnitine (ALCAR), the protection was even much more pronounced. Unpublished data suggest that reduction in permanent hearing loss can still be obtained by this combination of three antioxidants when given even 24 hours after the noise exposure.

15.4.1.5 Protection against Light-Induced Retinal Degeneration
Retinal degeneration is characterized by a progressive loss of photoreceptor cells leading to visual loss. Although, the primary function of photoreceptors is to absorb light, they are very sensitive to light,

and it has been demonstrated that excessive light exposure induces ROS overproduction, which can lead to retinal dysfunction and cell death. Intraperitoneal administration of PBN in rats at 50 mg kg^{-1}, 30 minutes before exposure to 2700-lux light for 24 hours resulted in a significant protection of the retina compared to control.[177] Indeed, the thickness of the outer nuclear layer (ONL), a measure of photoreceptor cell number, was significantly reduced in nontreated rats exposed to light. In contrast, when the rats were treated with PBN, the ONL was preserved. The retinal function and sensitivity of the retina was also preserved by treatment with PBN as demonstrated by electroretinograms. Tomita et al. reported later on that although PBN prevents caspase-3 gene expression in rat retinas exposed to fluorescent light, neither PBN nor light exposure had any effect on caspase-3 activity.[178] In addition, Western blot analysis showed that the c-fos protein level increased in the nuclear fraction after exposure to light and that PBN treatment decreased it. Therefore, it was concluded that inhibition of c-fos activation by PBN may be the key event in protection against light-induced retinal cell death. Recent studies in our group showed that a newly designed amphiphilic cholesterol-based α-phenyl-*N*-*tert*-butyl nitrone, called CholPBNL (Fig. 15.15), significantly protects retina against bright fluorescent light exposure when injected into the vitreous at 1 mM.[179]

Under similar conditions, PBN also exhibited protective activity at 9 mM but not at 1 mM, suggesting that this cholesterol-based nitrone may be a good candidate for the treatment of retinal diseases. PBN failed to show protective effects in two animal models of retinitis pigmentosa, P23H and S334ter. Retinitis pigmentosa (RP) is a genetic disease resulting in a progressive loss of peripheral and eventually central vision caused by mutations in rhodopsin. However, when the animals were exposed for 24 hours to 2700-lux light, PBN showed protection in the P23H mutant but not in the S334ter mutant.[180]

15.4.1.6 Protection against Fulminant Hepatitis

The effects of PBN against fulminant hepatitis with jaundice in *Long-Evans* cinnamon (LEC) rats have been reported.[181] LEC rats are mutants developing hereditary hepatitis and also exhibiting abnormal accumulation of copper in the liver. Therefore, LEC rats are considered as a model of human Wilson's disease. PBN was administered intraperitoneally every 2 days at the concentration of 128 mg/kg beginning with 13-week-old rats and continuing for 17 weeks. Treatment with PBN prevented body weight loss, reduced death rate and hepatic cell destruction. Ocular inspection also confirmed the beneficial effect of PBN on jaundice. Protective effect of PBN on the formation of oxidative damages

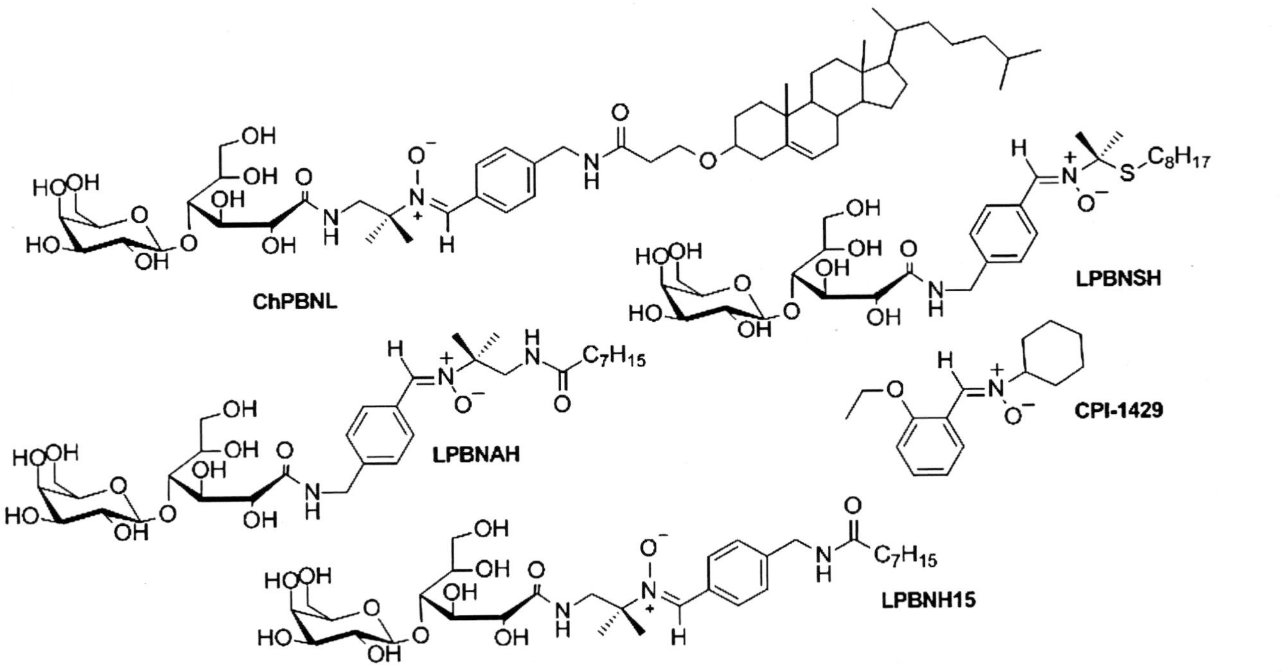

Figure 15.15 Chemical structures of the CPI-1429 nitrone and the amphiphilic derivatives ChPBNL, LPBNSH, LPBNAH, and LPBNH15.

was observed by reducing lipid peroxidation and DNA damage. Administration of a newly designed amphiphilic PBN derivative called LPBNSH[147,148] (Fig. 15.15), at extremely low concentration (0.1 mg kg^{-1} and 0.5 mg kg^{-1} every other day), resulted in a very significant protection of LEC rats, similar to that observed with PBN at 150 mg kg^{-1}.[182]

15.4.1.7 Cardioprotective Effects

The cardioprotective effects of the cyclic nitrone DMPO have also been demonstrated in isolated working rat hearts.[183] In a recent report, it was shown that the protective effects of the DMPO in myocardial ischemia reperfusion syndrome are mediated through mitochondrial respiratory chain protection against oxidative damages.[184] The amphiphilic amide nitrone referred to as LPBNAH[146–148] (Fig. 15.15) also showed protection in a rat model of reperfused isolated rats after 30 minutes of total global ischemia.[185] The addition of LPBNAH in the perfusion buffer during the first 5 minutes of postischemic reperfusion led to significant increase in functional recovery throughout the reperfusion period.

15.4.2 Antiaging Effects of Nitrones

Several studies indicated that PBN was also potent in increasing life span of mice[186,187] and rats.[188] Although the two studies with mice suffered from a lack of animal husbandry parameters and caloric intake evaluations, the data clearly showed that chronic administration of PBN prolonged significantly the mean life span. The study by Sack et al. showed that PBN administration (p.o. at 32mg/kg/day) to 24-month-old rats improved memory retention as measured using a Morris water maze, but also delayed death. The effects of PBN on cellular aging under culture conditions were also reported. Ames and collegues reported that PBN, and to a better extent, its hydrolysis product, the *N-tert*-butylhydroxylamine, both delay replicative senescence in human fibroblasts, the antioxidant activity being likely mediated through an inhibition of the mitochondrial superoxide production.[189] Von Zglinicki et al. confirmed the effect of PBN on senescence on human fibroblasts demonstrating the correlation between increased cellular longevity and decreased rates of telomere shortening.[190] A novel nitrone referred to as CPI-1429 has been shown to delay mortality as well as memory impairment in an ageing mouse model (Fig. 15.15).[161,191] The potential for CPI-1429 to ameliorate or delay progression of the learning and memory deficits associated with normal aging in C57BL/6 mice were evaluated. Separate groups of young (4–5 months) and old (23–24 months) mice were treated daily with CPI-1429 or vehicle for 2 weeks before testing and the daily treatment was maintained for a period of up to 27 weeks. While the results of the behavioral studies indicated that the performance of vehicle-treated old mice was impaired compared to that of younger mice, old mice treated with CPI-1429 (0.1 or 10 mg/kg/day) exhibited comparable rates of learning to those of young mice receiving chronic treatment with the vehicle. CPI-1429 treatment also resulted in a significant lengthening in the mortality curve of the mice.

The amphiphilic amide nitrone LPBNAH[146–148] has been shown to increase life-extension in aquatic organisms, the Bdelloid rotifer *Philodina acuticornis odiosa* Milne (Fig. 15.15).[149] Single-housed animals can be exposed to exactly the same drug regime in a controlled environment; moreover, their dates of birth and death can be precisely determined. Although this rotifer system may have limitations, such as unknown metabolic pathway, it allows a rapid screening of drugs for bioactivity, toxicity, and bioavailability. Daily treatment with PBN and LPBNAH at 5 µM from day 1 enhanced rotifer longevity, with LPBNAH being far more efficient than PBN. The life span of LPBNAH-treated rotifer group was extended to 57.2 ± 10.4 days while those of the PBN-treated and vehicle-treated rotifer group were 26.7 ± 5.2 and 22.8 ± 4.4 days, respectively. Interestingly, the total number of viable offspring per rotifer during a lifetime rose to 54 ± 1.6 animals after treatment with LPBNAH versus 18 ± 0.6 for control, but PBN treatment failed to show any effect on fecundity (16 ± 1.4). LPBNAH also increased the duration of the reproductive period per rotifer (fertile days in culture during a lifetime) to 18 ± 0.7 versus 5 ± 0.3 days for controls. Once again, the parent compound PBN has no significant effect on the duration of the reproductive period (5 ± 0.4 days). Growth effects of LPBNAH were also demonstrated in rotifer cultures. The treatment of rotifers at 5 µM per day for 15 days dramatically increased the size of these animals, which grow throughout the entire life. The average length of rotifers determined on day 15 of treatment was increased from 342 ± 8 µM for the control group to 547 ± 6 µM for the LPBNAH-treated group and PBN was without effect.

15.4.3 Neuroprotective Effects of Nitrones

Robert Floyd and John Carney were the first to discover the high potency of PBN in preventing stroke-induced damages and impairment in gerbils. The fact that PBN administered up to 1 hour after an ischemia reperfusion insult to the brain of mongolian gerbils led to protection of the animals opened a new therapeutic strategy for acute ischemic stroke in humans.[192,193] These preliminary observations were soon confirmed[194] and extended by others.[195–197] Indeed, neuroprotective activity of PBN in stroke model was demonstrated when given 30 minutes after the stroke in gerbils and even up to 3 hours in rats.[198] Intensive researches were next conducted on the neuroprotective activity of PBN and related compounds. For instance, PBN was found to protect cultured striatal neurons against NMDA-induced excitotoxicity.[199] Moreover, treatment with PBN also reduced the size of quinolinic acid-induced striatal lesion in rats, a model of the Huntington's disease.[200] PBN and 2-sulfophenyl-N-tert-butylnitrone (S-PBN) administration attenuated the damages produced by the mitochondrial toxin malonate, as well as those induced by 1-methyl-4-phenylpyridinium (MPP+) and NMDA in rats.[201] In a model of traumatic brain injury, PBN reduced the loss of ipsilateral hemispheric tissue when administered 30 minutes after a fluid percussion injury.[202] The 2-sulfophenyl-N-tert-butylnitrone led to similar protection; however, while PBN was detected in cerebral tissue, no trace of S-PBN was detected. It was concluded that the major site of free radical production in TBI is at the blood–endothelial interface.[203] Further studies on stroke demonstrated the superiority of the 2,5-disulfonate PBN (2,4-disulfophenyl-N-tert-butylnitrone), also referred to as NXY-059, compared to PBN. An extensive and general review on the neuroprotective actions of PBN, NXY-059, and S-PBN by Richard Green and colleagues sumarizes the outcomes of preclinical studies.[204] Brain damage was reduced in a rat model of transient occlusive middle cerebral arterial stroke when NXY-059 was administered up to 5 hours after the stroke, while no protection was observed with equimolar concentration of PBN.[205] Later on, Sydserff et al. reported the dose-dependant protective effects of NXY-059 in a rat model of permanent occlusive middle cerebral arterial stroke.[206] The NXY-059 plasma concentration was linearly correlated to dose, and a linear relationship between injected dose and neuroprotection was also observed. With regard to the therapeutic time window in this model, substantial neuroprotection was observed up to 4 hours after the stroke. Significant protection was also observed in a marmoset model of permanent occlusive middle cerebral arterial stroke.[207] Due to clinical relevance time point, NXY-059 was administered 4 hours postocclusion, and high dose was injected (85 mol/kg per hour) in order to ensure a plasma level of 200 μM after 24-hour infusion.[208] While the degree of motor impairment in the control animals was almost total at both 3 and 10 weeks, NXY-059 treatment resulted in a significant use of the paretic arm, and histological observations demonstrated a 28% smaller infarct in the treated animals as well. All these critical preclinical experiments conducted in several independent laboratories led to clinical trials.

15.4.4 Clinical Development of the Disulfonyl Nitrone, NXY-059

The pharmacokinetic of NXY-059 was investigated in young healthy and eldery volunteers.[209] Mean volume of distribution at steady state (V_{ss}) was lower in the elderly subjects than in the young ones (13 L vs. 16 L) with a plasma elimination half-life relatively short (2–3 hours). Renal elimination was found predominant and 80–90% of the nitrone was excreted in the urine without being metabolized irrespective of the age of the subjects. When infused i.v. at 20 mg/kg/h, it took ~300 minutes to reach a steady state plasma level, but when a loading infusion rate, that is, three times the maintenance infusion rate, was done, NXY-059 reached its steady-state level within 1 hour.[209] The tolerability of NXY-059 was next studied at various concentration in patients with acute stroke.[210] Interestingly, even at very high concentration, a loading dose of 1820 mg followed by 844 mg/h i.v., the incidence of adverse events was similar in the treated group than in the placebo one.[211] Overall, the most frequent adverse events were headache, fever, and hypertension, but in a similar manner as observed in the placebo groups, these events being generally encountered in patients with acute stroke. Although the study by Lees et al. was not designed to determine efficacy, there were more subjects among those exhibiting significant stroke severity with a good or very good outcome in the NXY-059 treated group than in the placebo.[211] There were, however, no statistically significant differences on either functional and neurological recovery.[211] Phase III clinical trials, coded SAINT for Stroke-Acute Ischemic NXY treatment, were next conducted as a series of two major studies, SAINT I[212] and SAINT II.[213] Patients with a clinical diagnosis of acute stroke were randomly assigned to receive an i.v. infusion of NXY-059 or placebo within 6 hours after the onset of stoke. The initial infusion rate was 2270 mg/hour and then reduced to 480 to 960 mg for further 71 hours with the aim of maintaining NXY-059 plasma concentration of 260 μM. In the SAINT I trial, 1699 subjects were included in the efficacy analysis, and a reduction in disability as measured by the modified Rankin score at 90 days was noted for the treated patients. However, the treatment failed to improve other outcomes such as neurological functioning. A much more thorough analysis of the data in which the outcome parameters were evaluated at 7, 30, and 90 days allowed to conclude that NXY-059 provided beneficial effects at 7 and 30 days after the stroke but not at 90 days.[214] In the SAINT II trial, the efficacy analysis was based on 3195 patients. Unfortunately, the distribution of scores on the modified Rankin scale did not differ between the NXY-059 treated group and the placebo group. Mortality was equal in the two groups, and no evidence of efficacy for any secondary end points, including scores on neurologic and activities of daily living scales, was observed. A pooled analysis of the SAINT I and II trials further concluded that NXY-059 was not effective in the combined trials.[215] Although both preclinical and clinical trials were supposed to be developed in accordance with the Stroke Therapy Academic Industry Roundtable (STAIR), the very disappointing outcome of the SAINT II trial led to the termination of NXY-059 as therapeutics for acute ischemic stroke. Critiques on how the development of NXY-059 was conducted have been raised. Savitz pointed out the need for more rigorous and strenuous testing at the preclinical stage.[216] Among Savitz's critiques about the preclinical studies, one can cite the lack of sufficient independent studies in different laboratories in order to confirm preliminary findings as well as the absence of information on NXY-059 mechanisms of action and whether it enters the brain parenchyma after embolic stroke. Clinical trial also appeared to be inadequately designed because of inappropriate treatment windows and inclusion of diverse stroke patients.

15.4.5 The Controversial Mode of Action of Nitrones

Considering the very broad activity of nitrones as discussed above and reviewed by others, there had been controversy on the mechanism of action of nitrone in biological systems. Since they were initially employed as probes for spin-trapping experiments, their biological activity had been first explained based on their radical trapping activity. However, several experimental evidences have strongly put aside the trapping activity and other mechanisms have been suggested. For instance, the rate of reaction of PBN with radical is quite slow,[135] that is, 10^5–10^7 $M^{-1}.s^{-1}$ and therefore, during spin trapping experiments in chemical milieu, PBN must be present at high concentration (10–50 mM) to trap a significant fraction of free radicals. These concentrations used are roughly 1000 times higher than those commonly employed in biological studies of protection (10–50 μM). Moreover, the concentration of nitrones in the target tissues is usually inferior at 0.5 mM, which is not sufficient to quench all the radical species. As previously discussed above, in experimental stroke model, it had been observed that PBN led to protection even if administered up to 2 hours after the start of reperfusion.[198] Preclinical and clinical studies with PBN and NXY-059 confirmed without any doubt the ability of nitrones to prevent stroke induced-damages even when administered of few hours after the onset of stroke. These findings further support the invalidation of the spin-trapping mechanism as the primary mode of action of nitrones. Another demonstration by Floyd and Carney was provided with aging gerbils that were chronically administered with PBN for 14 days.[217] Although the treatment ceased several days before the stroke, treated animals were more resistant than untreated ones. Considering the half-life of PBN ~2 hours, it is very likely that none of the nitrone is still present 3 days after cessation of the treatment, thus demonstrating that the classical mass action of spin traps is not solely responsible for the pharmacological activities of nitrones.

15.4.5.1 Antioxidant Property of PBN against Lipid Peroxydation
Its very poor antioxidant activity have been observed in simple models of lipid peroxidation demonstrating that nitrones do not act as classical chain-breaking antioxidants.[218,219] While PBN inhibited lipid peroxidation at 5 mM, antioxidants such as γ-tocopherol or BHT were a thousand-fold more efficient, exhibiting antioxidant protection at only 5 μM.[218]

15.4.5.2 Anti-Inflammatory and Anti-Apoptotic Properties of Nitrones
There is strong evidence that PBN acts to quell signal transduction processes and therefore provides potent anti-inflammatory and anti-apoptotic properties.[161,162,164,191] PBN has been shown to inhibit the activity of cytokines and transcription factors such as NF-κB, which can rapidly activate the expression of genes involved in inflammation.[220–222] The inhibition of the COX2 catalytic activity and the decrease of the steady state COX2 mRNA level was also demonstrated. Other rationales for nitrone protection in animal models showing that PBN can inhibit gene induction of heat shock proteins include c-fos and inducible nitric oxide synthase as well. One of the first and yet convincing study that illustrated the importance of neuro-inflammation in brain injury and the potency of PBN to preserve brain function by quelling exacerbated signal transduction processes is that by Floyd et al. in 2000.[223] Kainate administration to rats led to injury to the hippocampus providing a good experimental model of epilepsy. Administration of PBN prevented apoptotic neuron loss and mortality, and immunohistochemical examination showed that the Kainate-induced activation of p38 MAP kinase and NF-κB was suppressed by PBN. Since then, numerous reports have shown the ability of nitrone to suppress the proinflammatory cytokine and stressors mediated induction of genes in a wide range of biological systems. In the review by Green et al. are listed some important biochemical consequences of nitrone administration in vitro, ex vivo, and in vivo, which may be associated with neuroprotection (Table 15.2).[204]

TABLE 15.2 Some Biochemical Consequences of PBN Administration *In Vitro*, *In Vivo*, and *Ex Vivo*. (Adapted from Green et al.[204])

In vitro	1.	Inhibition of iNOS induction in HIV-1 envelope
	2.	Potentiation of H_2O_2-induced Erk and Src kinase in human
	3.	Stimulation of H_2O_2-induced activation of the prosurvival Erk signal transduction pathway
	4.	Protection of primary cerebellar neurones from glutamate toxicity
In vivo or *ex vivo*	5.	Facilitation of postischemic reperfusion following transient MCAO
	6.	Inhibition of apoptosis-associated gene expression in endotoxin-treated rats
	7.	Suppression of caspase-3 activation following global ischemia
	8.	Prevention of the decrease in stimulated (+ADP), nonstimulated (−ADP) and uncoupled rats following transient MCAO
	9.	Inhibition of endotoxin-induced induction of nitric oxide synthase
	10.	Improvement of the rate of metabolic recovery, acidosis rebound, and ATP renewal in rat brain following transient focal ischemia
	11.	Suppresion of c-fos expression in postischemic gerbil brain
	12.	Attenuation of the secondary mitochondrial dysfunction after transient focal ischemia
	13.	Reduction of the number of positive τ-oligodendrocytes after focal ischemia
	14.	Prevention of cytotoxic ischemia following malonate

15.4.5.3 Action on Membrane Enzymes

PBN was shown to act as a reversible calcium channel blocker at concentrations far lower than those required for free radical detection by EPR, and it was concluded that removal of free radicals may not contribute to the nitrone-induced vasorelaxation.[224] The potency of PBN to inhibit acetylcholinesterase activity, an enzyme that converts acetylcholine into the inactive metabolites choline and acetate, with a K_i of 0.58 mM, was also reported.[225]

15.4.5.4 Interaction with the Mitochondrial Metabolism

Hensley et al. demonstrated that PBN interacts with the mitochondrial complex I, inhibiting complex-I stimulated H_2O_2 flux and nitro blue tetrazolium reduction.[226] This site-specific interaction of PBN with mitochondrial flavin deshydrogenases, which leads to alteration of the electron transit within the enzymes, is believed to participate to the broad antioxidant and anti-inflammatory action of nitrones. Interaction of PBN with mitochondrial complex I was also demonstrated in the prevention of the doxorubicin-induced apoptosis in bovine aortic endothelial cells.[171] Indeed, it was shown that doxorubicin inactivated complex-I by a superoxide-dependant mechanism and that complex-I activity was restored by PBN, confirming the first observations by Hensley et al. Pretreatment of cells with PBN also resulted in a full inhibition of DOX-induced cytochrome c release and a complete restoration of GSH levels. Other mitochondrial-specific interactions of nitrones were next reported. An increase of endogenous mitochondrial superoxide production, following intrathecal injection of complex III inhibitor antimycin A, has been demonstrated to induce hyperalgesia in mice.[227] PBN significantly reduced hyperalgesia 30 and 60 minutes following the treatment providing transient antinociception irrespective to the administration mode of PBN, that is, intraperitoneal or intrathecal. Mitochondrial impairment has been associated with cardiac dysfunction in Chagas' disease, a parasitic disease whose about 40% of seropositive patients develop cardiomyopathy. In infected mice, oxidative stress and alteration of mitochondrial functions in the myocardium are observed as well as similar cardiac dysfunction to what is observed in human chagasic patients. Treatment of infected mice with PBN prevented mitochondrial oxidative damage and significantly improved respiratory complex activities.[228,229] The impairment of a specific site of complex III was identified as the main target in infected myocardium, leading to an increased electron leakage and O_2 production. Treatment with PBN improved the respiratory chain function by preserving electron transport chain (ETC) activity thereby limiting electron leakage and mitochondrial ROS production. The two amphiphilic amide nitrones, the LPBNAH and its reverse analogue the LPBNH15 (Fig. 15.15), have recently shown very promising results.[230] These two compounds exhibited hydroxyl radical scavenging activity and radical reducing potency in ABTS assays. Experimental and theoretical data showed that substitution of the PBN by hydrophilic and lipophilic groups alters its redox properties, with the amphiphilic amide nitrones being easier to oxidize and reduce than the parent PBN. Moreover, very high protective effects were demonstrated both in *in vitro* and *in vivo* experiments. With regard to the mitochondrial interaction, these two nitrones showed interesting properties. They both decreased electron and proton leakage as well as hydrogen peroxide formation in isolated rat brain mitochondria at nanomolar concentration. They also significantly enhanced mitochondrial membrane potential, and the dopamine-induced inhibition of complex I activity was antagonized by pretreatment with these agents. These findings strongly suggest that new nitrone analogues are more than just radical scavenging antioxidants but may act as a new class of bioenergetic agents directly interacting with mitochondrial electron and proton transport.

REFERENCES

1. Halliwell, B., Gutteridge, J.M.C. *Free Radicals in Biology and Medicine*, 4th ed. Oxford University Press, Oxford, 2007.
2. The Alpha-Tocopherol, Beta Carotene Cancer Prevention Study Group. The effect of vitamin E and beta carotene on the incidence of lung cancer and other cancers in male smokers. *N. Engl. J. Med.* **1994**, *330*, 1029–1035.
3. Ascherio, A., Rimm, E.B., Hernan, M.A., Giovannucci, E., Kawachi, I., Stampfer, M.J., Willet, W.C. Relation of consumption of vitamin E, vitamin C, and carotenoids to risk for stroke among men in the United States. *Ann. Intern. Med.* **1999**, *130*, 963–970.
4. Krirtharides, L., Stocker, R. The use of antioxidant supplements in coronary heart disease. *Atherosclerosis* **2002**, *164*, 211–219.
5. Meyers, D.G., Maloley, P.A., Weeks, D. Safety of antioxidant vitamins. *Arch. Intern. Med.* **1996**, *156*, 925–935.
6. Riley, D.P. Functional mimics of superoxide dismutase enzymes as therapeutic agents. *Chem. Rev.* **1999**, *99*, 2573–2587.
7. Day, B.J. Catalytic antioxidants: A radical approach to new therapeutics. *Drug Discov. Today* **2004**, *9*, 557–566.
8. Day, B.J. Catalase and glutathione peroxidase mimics. *Biochem. Pharmacol.* **2009**, *77*, 285–296.
9. Patel, M., Day, B.J. Metalloporphyrin class of therapeutic catalytic antioxidants. *Trends Pharmacol. Sci.* **1999**, *20*, 359–364.
10. Baudry, M., Etienne, S., Bruce, A., Palucki, M., Jacobsen, E., Malfroy, B. Salen-manganese complexes are superoxide dismutase-mimics. *Biochem. Biophys. Res. Commun.* **1993**, *192*, 964.
11. Doctrow, S.R., Huffman, K., Marcus, C.B., Malfroy, B. Salen-manganese complexes: Combined superoxide dismutase/catalase mimics with broad pharmacological efficacy. *Adv. Pharmacol.* **1997**, *38*, 247–269.
12. Sharpe, M.A., Ollosson, R., Stewart, V.C., Clark, J.B. Oxidation of nitric oxide by oxomanganese-salen complexes: A new mechanism for cellular protection by superoxide dismutase/catalase mimetics. *Biochem. J.* **2002**, *366*, 97–107.
13. Boucher, L.J. Maganese Schiff's base complexes II: Synthesis and spectroscopy of chloro-complexes of some derivatives of (salicylaldehydeethylenediimato)maganese(III). *J. Inorg. Nucl. Chem.* **1974**, *36*, 531–536.
14. Bruce, A.J., Malfroy, B., Baudry, M. beta-Amyloid toxicity in organotypic hippocampal cultures: Protection by EUK-8, a synthetic catalytic free radical scavenger. *Proc. Natl. Acad. Sci. U.S.A.* **1996**, *93*, 2312–2316.
15. Musleh, W., Bruce, A., Malfroy, B., Baudry, M. Effects of EUK-8, a synthetic catalytic superoxide scavenger, on hypoxia- and acidosis-induced damage in hippocampal slices. *Neuropharmacology* **1994**, *33*, 929–934.
16. Baker, K., Bucay-Marcus, C., Huffman, K., Kruk, H., Malfroy, B., Doctrow, S.R. Synthetic combined superoxide dismutase/catalase mimics are protective as a delayed treatment in a rat stroke model: A key role for reactive oxygen species in ischemic brain injury. *J. Pharmacol. Exp. Ther.* **1998**, *284*, 215–221.
17. Doctrow, S.R., Huffman, K., Marcus, C.B., Mulesh, W., Bruce, A., Baudry, M., Malfroy, B. Salen-manganese complexes: Combined superoxide dismutase/catalase mimics with broad pharmacological efficacy. In: *Antioxidants in Disease Mechanisms and Therapeutic Strategies*, ed. Helmut Sies. Academic Press, New York, 1997, pp. 247–270.
18. Doctrow, S.R., Huffman, K., Marcus, C.B., Tocco, G., Malfroy, E., Adinolfi, C.A., Kruk, H., Baker, K., Lazarowych, N., Mascarenhas, J., Malfroy, B. Salen-manganese complexes as catalytic scavengers of hydrogen peroxide and cytoprotective agents: Structure-activity relationship studies. *J. Med. Chem.* **2002**, *45*, 4549–4558.
19. Rosenthal, R.A., Huffman, K., Fisette, L.W., Damphouse, C.A., Callaway, W.B., Malfroy, B., Doctrow, S.R. Orally available Mn porphyrins with superoxide dismutase and catalse activity. *J. Biol. Inorg. Chem.* **2009**, *14*, 979–991.
20. Pasternack, R.F., Banth, A., Pasternack, J.M., Johnson, C.S. Catalysis of the disproportionation of superoxide by metalloporphyrins. III. *J. Inorg. Biochem.* **1981**, *15*, 261–267.
21. Pasternack, R.F., Skowronek, W.R. Catalysis of the disproportionation of superoxide by metalloporphyrins. *J. Inorg. Biochem.* **1979**, *11*, 261–267.
22. Day, B.J., Fridovich, I., Crapo, J.D. Maganic porphyrins possess catalase activity and protect endothelial cells against hydrogen peroxide-mediated injury. *Arch. Biochem. Biophys.* **1997**, *347*, 256–262.
23. Groves, J.T., Marla, S.S. Peroxynitrite-induced DNA strand scission mediated by a manganese porphyrin. *J. Am. Chem. Soc.* **1995**, *117*, 9578–9579.
24. Marla, S.S., Lee, J., Groves, J.T. Peroxynitrite rapidly permeates phospholipid membranes. *Proc. Natl. Acad. Sci. U.S.A.* **1997**, *94*, 14243–14248.
25. Groves, J.T., Lee, J., Marla, S.S. Detection and characterization of an oxomanganese(V) porphyrin complex by

rapid-mixing stopped-flow spectrophotometry. *J. Am. Chem. Soc.* **1997**, *119*, 6269–6273.
26. Lee, J., Hunt, J.A., Groves, J.T. Manganese porphyrins as redox-coupled peroxynitrite reductases. *J. Am. Chem. Soc.* **1998**, *120*, 6053–6061.
27. Day, B.J., Batini-Haberle, I., Crapo, J.D. Metalloporphyrins are potent inhibitors of lipid peroxidation. *Free Radic. Biol. Med.* **1999**, *26*, 730–736.
28. Bloodsworth, A., O'Donnell, V.B., Batinic-Haberle, I., Chumley, P.H., Hurt, J.B., Day, B.J., Crow, J.P., Freeman, B.A. Manganese-porphyrin reactions with lipids and lipoproteins. *Free Radic. Biol. Med.* **2000**, *28*, 1017–1029.
29. Batinic-Haberle, I., Benov, L., Spasojevic, I., Fridovich, I. The ortho effect makes manganese(III)meso-tetrakis(N-methylpyridinium-2-yl)porphyrin a powerful and potentially useful superoxide dismutase mimic. *J. Biol. Chem.* **1998**, *273*, 24521–24528.
30. Spasojevic, I., Batinic-Haberle, I., Rebouças, J.S., Idemori, Y.M., Fridovich, I. Electrostatic contribution in the catalysis of O_2^- dismutation by superoxide dismutase mimics. *J. Biol. Chem.* **2003**, *278*, 6831–6837.
31. Spasojevic, I., Chen, Y., Noel, T.J., Fan, P., Zhang, L., Rebouças, J.S., St. Clair, D.K., Batinic-Haberle, I. Pharmacokinetics of the potent redox-modulating manganese porphyrin, MnTE-2-PyP5+, in plasma and major organs of B6C3F1 mice. *Free Radic. Biol. Med.* **2008**, *45*, 943–949.
32. Batinic-Haberle, I., Ndengele, M.M., Cuzzocrea, S., Rebouças, J.S., Spasojevic, I., Salvemini, D. Lipophilicity is a critical parameter that dominates the efficacy of metalloporphyrins in blocking the development of morphine antinociceptive tolerance through peroxynitrite-mediated pathways. *Free Radic. Biol. Med.* **2009**, *46*, 212–219.
33. O'Neill, H.C., White, C.W., Veress, L.A., Hendry-Hofer, T.B., Loader, J.E., Min, E., Huang, J., Rancourt, R.C., Day, B.J. Treatment with the catalytic metalloporphyrin AEOL 10150 reduces inflammation and oxidative stress due to inhalation of the sulfur mustard analog 2-chloroethyl ethyl sulfide. *Free Radic. Biol. Med.* **2010**, *48*, 1188–1196.
34. Riley, D.P., Weiss, R.H. Manganese macrocyclic ligand complexes as mimics of superoxide dismutase. *J. Am. Chem. Soc.* **1994**, *116*, 387–388.
35. Weiss, R.H., Fretland, D.J., Baron, D.A., Ryan, U.S., Riley, D.P. Manganese-based superoxide dismutase mimetics inhibit neutrophil infiltration in vivo. *J. Biol. Chem.* **1996**, *271*, 26149–26156.
36. Black, S.C., Schasteen, C.S., Weiss, R.H., Riley, D.P., Driscoll, E.M., Lucchesi, B.R. Inhibition of in vivo myocardial ischemic and reperfusion injury by a synthetic manganese-based superoxide dismutase mimetic. *J. Pharmacol. Exp. Ther.* **1994**, *270*, 1208–1215.
37. Riley, D.P., Henke, S.L., Lennon, P.J., Aston, K. Computer-aided design (CAD) of synzymes: Use of molecular mechanics (MM) for the rational design of superoxide dismutase mimics. *Inorg. Chem.* **1999**, *38*, 1908–1917.
38. Riley, D.P., Lennon, P.J., Neumann, W.L., Weiss, R.H. Toward the rational design of superoxide dismutase mimics: Mechanistic studies for the elucidation of substituent effects on the catalytic activity of macrocyclic manganese(II) complexes. *J. Am. Chem. Soc.* **1997**, *119*, 6522–6528.
39. Rauen, U., Li, T., Sustmann, R., de Groot, H. Protection against iron- and hydrogen peroxide-dependent cell injuries by a novel synthetic iron catalase mimic and its precursor, the iron-free ligand. *Free Radic. Biol. Med.* **2004**, *37*, 1369–1383.
40. Stern, M.K., Jensen, M.P., Kramer, K. Peroxynitrite decomposition catalysts. *J. Am. Chem. Soc.* **1996**, *118*, 8735–8736.
41. Lee, J., Hunt, J.A., Groves, J.T. Mechanisms of iron porphyrin reactions with peroxynitrite. *J. Am. Chem. Soc.* **1998**, *120*, 7493–7501.
42. Salvemini, D., Wang, Z.-Q., Stern, M.K., Currie, M.G., Misko, T.P. Peroxynitrite decomposition catalysts: Therapeutics for peroxynitrite-mediated pathology. *Proc. Natl. Acad. Sci. U.S.A.* **1998**, *95*, 2659–2663.
43. Filipovska, A., Kelso, G.F., Brown, S.E., Beer, S.M., Smith, R.A.J., Murphy, M.P. Synthesis and characterization of a triphenylphosphonium-conjugated peroxidase mimetic. *J. Biol. Chem.* **2005**, *280*, 24113–24126.
44. Imai, H., Masayasu, H., Dewar, D., Graham, D.I., Macrae, I.M. Ebselen protects both gray and white matter in a rodent model of focal cerebral ischemia. *Stroke* **2001**, *32*, 2149–2154.
45. Ullrich, V., Weber, P., Meisch, F., von Appen, F. Ebselen-binding equilibria between plasma and target proteins. *Biochem. Pharmacol.* **1996**, *52*, 15–19.
46. Lesser, R., Weiss, R. Über selenhaltige aromatische Verbindungen (VI). *Ber. Dtsch. Chem. Ges.* **1924**, *57*, 1077.
47. Weber, R., Renson, M. *Bull. Soc. Chim. Fr.* **1976**, 1024.
48. Engman, L., Hallberg, A. Expedient synthesis of ebselen and related compounds. *J. Org. Chem.* **1989**, *54*, 2964–2966.
49. Morgenstern, R., Cotgreave, I.A., Engman, L. Determination of the relative contributions of the diselenide and selenol forms of ebselen in the mechanism of its glutathione peroxidase-like activity. *Chem. Biol. Interact.* **1992**, *84*, 77–84.
50. Sies, H. Ebselen, a selenoorganic compound as glutathione peroxidase mimic. *Free Radic. Biol. Med.* **1993**, *14*, 313–323.
51. Sies, H., Lester, P. [47] Ebselen: A glutathione peroxidase mimic. In: *Methods in Enzymology*, Vol. 234, ed. Lester Packer. Academic Press, San Diego, CA, 1994, pp. 476–482.
52. Schewe, T. Molecular actions of Ebselen—An antiinflammatory antioxidant. *Gen. Pharmacol.* **1995**, *26*, 1153–1169.
53. Haenen, G.R., De Rooij, B.M., Vermeulen, N.P., Bast, A. Mechanism of the reaction of ebselen with endogenous thiols: Dihydrolipoate is a better cofactor than glutathione

in the peroxidase activity of ebselen. *Mol. Pharmacol.* **1990**, *37*, 412–422.

54. Maiorino, M., Roveri, A., Coassin, M., Ursini, F. Kinetic mechanism and substrate specificity of glutathione peroxidase activity of ebselen (PZ51). *Biochem. Pharmacol.* **1988**, *37*, 2267–2271.

55. Schöneich, C., Narayanaswami, V., Asmus, K.-D., Sies, H. Reactivity of ebselen and related selenoorganic compounds with 1,2-dichloroethane radical cations and halogenated peroxyl radicals. *Arch. Biochem. Biophys.* **1990**, *282*, 18–25.

56. Maiorino, M., Roveri, A., Ursini, F. Antioxidant effect of ebselen (PZ 51): Peroxidase mimetic activity on phospholipid and cholesterol hydroperoxides vs free radical scavenger activity. *Arch. Biochem. Biophys.* **1992**, *295*, 404–409.

57. Noguchi, N., Yoshida, Y., Kaneda, H., Yamamoto, Y., Niki, E. Action of ebselen as an antioxidant against lipid peroxidation. *Biochem. Pharmacol.* **1992**, *44*, 39–44.

58. Scurlock, R., Rougee, M., Bensasson, R.V., Evers, M., Dereu, N. Deactivation of singlet molecular oxygen by organo-selenium compounds exhibiting glutathione peroxidase activity and by sulfur-containing homologs. *Photochem. Photobiol.* **1991**, *54*, 733–736.

59. Hiroshi, M., Kissner, R., Koppenol, W.H., Sies, H. Kinetic study of the reaction of ebselen with peroxynitrite. *FEBS Lett.* **1996**, *398*(2–3), 179–182.

60. Masumoto, H., Sies, H. The reaction of ebselen with peroxynitrite. *Chem. Res. Toxicol.* **1996**, *9*, 262–267.

61. Daiber, A., Zou, M.-H., Bachschmid, M., Ullrich, V. Ebselen as a peroxynitrite scavenger in vitro and ex vivo. *Biochem. Pharmacol.* **2000**, *59*, 153–160.

62. Schewe, C., Schewe, T., Wendel, A. Strong inhibition of mammalian lipoxygenases by the antiinflammatory seleno-organic compound ebselen in the absence of glutathione. *Biochem. Pharmacol.* **1994**, *48*, 65–74.

63. Cotgreave, I.A., Duddy, S.K., Kass, G.E.N., Thompson, D., Moldéus, P. Studies on the anti-inflammatory activity of ebselen: Ebselen interferes with granulocyte oxidative burst by dual inhibition of NADPH oxidase and protein kinase C? *Biochem. Pharmacol.* **1989**, *38*, 649–656.

64. Hattori, R., Inoue, R., Sase, K., Eizawa, H., Kosuga, K., Aoyama, T., Masayasu, H., Kawai, C., Sasayama, S., Yui, Y. Preferential inhibition of inducible nitric oxide synthase by ebselen. *Eur. J. Pharmacol.* **1994**, *267*, R1–R2.

65. Porciúncula, L.O., Rocha, J.B.T., Cimarosti, H., Vinadé, L., Ghisleni, G., Salbego, C.G., Souza, D.O. Neuroprotective effect of ebselen on rat hippocampal slices submitted to oxygen-glucose deprivation: Correlation with immunocontent of inducible nitric oxide synthase. *Neurosci. Lett.* **2003**, *346*, 101–104.

66. Rossato, J.I., Zeni, G., Mello, C.F., Rubin, M.A., Rocha, J.B.T. Ebselen blocks the quinolinic acid-induced production of thiobarbituric acid reactive species but does not prevent the behavioral alterations produced by intrastriatal quinolinic acid administration in the rat. *Neurosci. Lett.* **2002**, *318*, 137–140.

67. Yamaguchi, T., Sano, K., Takakura, K., Saito, I., Shinohara, Y., Asano, T., Yasuhara, H. Ebselen in acute ischemic stroke: A placebo-controlled, double-blind clinical trial. *Stroke* **1998**, *29*, 12–17.

68. Saito, I., Asano, T., Sano, K., Takakura, K., Abe, H., Yoshimoto, T., Kikuchi, H., Ohta, T., Ishibashi, S. Neuroprotective effect of an antioxidant, ebselen, in patients with delayed neurological deficits after aneurysmal subarachnoid hemorrhage. *Neurosurgery* **1998**, *42*, 269–277.

69. Zhao, R., Holmgren, A. A novel antioxidant mechanism of ebselen involving ebselen diselenide, a substrate of mammalian thioredoxin and thioredoxin reductase. *J. Biol. Chem.* **2002**, *277*(42), 39456–39462.

70. Zhao, R., Masayasu, H., Holmgren, A. Ebselen: A substrate for human thioredoxin reductase strongly stimulating its hydroperoxide reductase activity and a superfast thioredoxin oxidant. *Proc. Natl. Acad. Sci. U.S.A.* **2002**, *99*, 8579–8584.

71. Moutet, M., D'Alessio, P., Malette, P., Devaux, V., Chaudière, J. Glutathione peroxidase Mimmics prevent TNFα- and neutrophil-induced endothelial alterations. *Free Radic. Biol. Med.* **1998**, *25*, 270–281.

72. D'Alessio, P., Moutet, M., Marsac, C., Chaudière, J. *Pharmacological Inhibition of Endothelial Cytoskeleton Alterations Induced by Hydrogen Peroxide and TNF-α.* Marcel Dekker Inc., New York, 1997.

73. D'Alessio, P., Moutet, M., Coudrier, E., Darquenne, S., Chaudiere, J. ICAM-1 and VCAM-1 expression induced by TNF-α are inhibited by a glutathione peroxidase mimic. *Free Radic. Biol. Med.* **1998**, *24*, 979–987.

74. Nogueira, C.W., Zeni, G., Rocha, J.B.T. Organoselenium and organotellurium compounds: Toxicology and pharmacology. *Chem. Rev.* **2004**, *104*, 6255–6286.

75. Watanabe, T., Tahara, M., Todo, S. The novel antioxidant edaravone: From bench to bedside. *Cardiovasc. Ther.* **2008**, *26*, 101–114.

76. Higashi, Y., Jitsuiki, D., Chayama, K., Yoshizumi, M. Edaravone (3-methyl-1-phenyl-2-pyrazolin-5-one), a novel free radical scavenger, for treatment of cardiovascular diseases. *Recent Pat. Cardiovas. Drug Discov.* **2006**, *1*, 85–93.

77. Chegaev, K., Cena, C., Giorgis, M., Rolando, B., Tosco, P., Bertinaria, M., Fruttero, R., Carrupt, P.-A., Gasco, A. Edaravone derivatives containing NO-donor functions. *J. Med. Chem.* **2009**, *52*, 574–578.

78. Nishinaka, Y., Mori, H., Endo, N., Miyoshi, T., Yamashita, K., Adachi, S., Arai, T. Edaravone directly reacts with singlet oxygen and protects cells from attack. *Life Sci.* **2010**, *86*, 808–813.

79. Komatsu, T., Nakai, H., Takamastu, Y., Morinaka, Y., Watanabe, K., Shinoda, M., Iida, S. Pharmacokinetic studies of 3-methyl-1-phenyl-2-pyrazolin-5-one-(MCI-186): Metabolism in rats, dogs and human. *Drug Metab. Pharmacokinet.* **1996**, *11*, 451–462.

80. Abe, S., Kirima, K., Tsuchiya, K., Okamoto, M., Hasegawa, T., Houchi, H., Yoshizumi, M., Tamaki, T. The reaction

rate of edaravone (3-methyl-1-phenyl-2-pyrazolin-5-one (MCI-186)) with hydroxyl radical. *Chem. Pharm. Bull.* **2004**, *52*(2), 186–191.

81. Ono, S., Okazaki, K., Sakurai, M., Inoue, Y. Density functional study of the radical reactions of 3-methyl-1-phenyl-2-pyrazolin-5-one (MCI-186): Implication for the biological function of MCI-186 as a highly potent antioxidative radical scavenger. *J. Phys. Chem. A* **1997**, *101*, 3769–3775.

82. Lin, M., Katsumura, Y., Hata, K., Muroya, Y., Nakagawa, K. Pulse radiolysis study on free radical scavenger edaravone (3-methyl-1-phenyl-2-pyrazolin-5-one). *J. Photochem. Photobiol. B.* **2007**, *89*(1), 36–43.

83. Yamamoto, Y., Kuwahara, T., Watanabe, K., Watanabe, K. Antioxidant activity of 3-methyl-1-phenyl-2-pyrazoline-5-one. *Redox Rep.* **1996**, *2*, 333–338.

84. Wang, L.-F., Zhang, H.-Y. A theoretical investigation on DPPH radical-scavenging mechanism of edaravone. *Bioorg. Med. Chem. Lett.* **2003**, *13*, 3789–3792.

85. Nakagawa, H., Ohyama, R., Kimata, A., Suzuki, T., Miyata, N. Hydroxyl radical scavenging by edaravone derivatives: Efficient scavenging by 3-methyl-1-(pyridin-2-yl)-5-pyrazolone with an intramolecular base. *Bioorg. Med. Chem. Lett.* **2006**, *16*, 5939–5942.

86. Satoh, K., Ikeda, Y., Shioda, S., Tobe, T., Yoshikawa, T. Edaravone scavenges nitric oxide. *Redox Rep.* **2002**, *7*, 219–222.

87. Anzai, K., Furuse, M., Yoshida, A., Matsumaya, A., Moritake, T., Tsuboi, K., Ikota, N. In vivo radioprotection of mice by 3-methyl-1-phenyl-2-pyrazolin-5-one (edaravone; Radicut), a clinical drug. *J. Radiat. Res.* **2004**, *45*, 319–323.

88. Murota, S., Morita, I., Suda, N. The control of vascular endothelial cell injury. *Ann. N. Y. Acad. Sci.* **1990**, *598*, 182–187.

89. Watanabe, T., Morita, I., Nishi, H., Murota, S. Prenventive effect of MC-181 on 15-HPETE induced vascular endothelial cell injury in vivo. *Prostaglandins Leukot. Essent. Fatty Acids* **1988**, *33*, 81–87.

90. Yoshida, H., Sasaki, K., Namiki, Y., Sato, N., Tada, N. Edaravone, a novel radical scavenger, inhibits oxidative modification of low-density lipoprotein (LDL) and reverses oxidized LDL-mediated reduction in the expression of endothelial nitric oxide synthase. *Atherosclerosis* **2005**, *179*, 97–102.

91. Zhang, X.-H., Matsuda, N., Jesmin, S., Sakuraya, F., Gando, S., Kemmotsu, O., Hattori, Y. Normalization by edaravone, a free radical scavenger, of irradiation-reduced endothelial nitric oxide synthase expression. *Eur. J. Pharmacol.* **2003**, *476*, 131–137.

92. Takahashi, G., Sakurai, M., Abe, K., Itoyama, Y., Tabayashi, K. MCI-186 prevents spinal cord damage and affects enzyme levels of nitric oxide synthase and Cu/Zn superoxide dismutase after transient ischemia in rabbits. *J. Thorac. Cardiovasc. Surg.* **2003**, *126*, 1461–1466.

93. Yanagisawa, A., Miyagawa, M., Ishikawa, K., Murota, S. Cardioprotective effect of MC-186 (3-methyl-1-phenyl-2-pyrazolin-5-one) during acute ischemia-reperfusion injury in rats. *Int. J. Angiol.* **1994**, *3*, 12–15.

94. Minhaz, U., Tanaka, M., Tsukamoto, H., Watanabe, K., Koide, S., Shohtsu, A., Nakazawa, H. Effect of MCI-186 on postischemic reperfusion injury in isolated rat heart. *Free Radic. Res.* **1996**, *24*, 361–367.

95. Tsujimoto, I., Hikoso, S., Yamaguchi, O., Kashiwase, K., Nakai, A., Takeda, T., Watanabe, T., Taniike, M., Matsumura, Y., Nishida, K., Hori, M., Kogo, M., Otsu, K. The antioxidant edaravone attenuates pressure overload-induced left ventricular hypertrophy. *Hypertension* **2005**, *45*, 921–926.

96. Tsujita, K., Shimomura, H., Kawano, H., Hokamaki, J., Fukuda, M., Yamashita, T., Hida, S., Nakamura, Y., Nagayoshi, Y., Sakamoto, T., Yoshimura, M., Arai, H., Ogawa, H. Effects of edaravone on reperfusion injury in patients with acute myocardial infarction. *Am. J. Cardiol.* **2004**, *94*, 481–484.

97. Yoshida, H., Yanai, H., Namiki, Y., Fukatsu-Sasaki, K., Furutani, N., Tada, N. Neuroprotective effects of edaravone: A novel free radical scavenger in cerebrovascular injury. *CNS Drug Rev.* **2006**, *12*, 9–20.

98. Otani, H., Togashi, H., Jesmin, S., Sakuma, I., Yamaguchi, T., Matsumoto, M., Kakehata, H., Yoshioka, M. Temporal effects of edaravone, a free radical scavenger, on transient ischemia-induced neuronal dysfunction in the rat hippocampus. *Eur. J. Pharmacol.* **2005**, *512*, 129–137.

99. The Edaravone Acute Brain Infarction Study Group. Effect of a novel free radical scavenger, edaravone (MCI-186), on acute brain infarction. *Cerebrovasc. Dis.* **2003**, *15*, 222–229.

100. Takahashi, R. Edaravone in ALS. *Exp. Neurol.* **2009**, *217*, 235–236.

101. Yoshino, H., Kimura, A. Investigation of the therapeutic effects of edaravone, a free radical scavenger, on amyotrophic lateral sclerosis (Phase II study). *Amyotroph. Lateral Scler.* **2006**, *7*, 247–251.

102. Ito, H., Wate, R., Zhang, J., Ohnishi, S., Kaneko, S., Ito, H., Nakano, S., Kusaka, H. Treatment with edaravone, initiated at symptom onset, slows motor decline and decreases SOD1 deposition in ALS mice. *Exp. Neurol.* **2008**, *213*, 448–455.

103. Inokuchi, Y., Imai, S., Nakajima, Y., Shimazawa, M., Aihara, M., Araie, M., Hara, H. Edaravone, a free radical scavenger, protects against retinal damage in vitro and in vivo. *J. Pharmacol. Exp. Ther.* **2009**, *329*, 687–698.

104. Imai, S., Inokuchi, Y., Nakamura, S., Tsuruma, K., Shimazawa, M., Hara, H. Systemic administration of a free radical scavenger, edaravone, protects against light-induced photoreceptor degeneration in the mouse retina. *Eur. J. Pharmacol.* **2010**, *642*, 77–85.

105. Tanaka, K., Takemoto, T., Sugahara, K., Okuda, T., Mikuriya, T., Takeno, K., Hashimoto, M., Shimogori, H., Yamashita, H. Post-exposure administration of edaravone attenuates noise-induced hearing loss. *Eur. J. Pharmacol.* **2005**, *522*, 116–121.

106. Sueishi, K., Mishima, K., Makino, K., Itoh, Y., Tsuruya, K., Hirakata, H., Oishi, R. Protection by a radical scavenger edaravone against cisplatin-induced nephrotoxicity in rats. *Eur. J. Pharmacol.* **2002**, *451*, 203–208.

107. Fukudome, D., Matsuda, M., Kawasaki, T., Ago, Y., Matsuda, T. The radical scavenger edaravone counteracts diabetes in multiple low-dose streptozotocin-treated mice. *Eur. J. Pharmacol.* **2008**, *583*, 164–169.

108. Spapen, H., Zhang, H., Vincent, J.-L. Potential therapeutic value of lazaroids in endotoxemia and other forms of sepsis. *Shock* **1997**, *8*, 321–327.

109. Braughler, J.M., Pregenzer, J.F., Chase, R.L., Duncan, L.A., Jacobsen, E.J., McCall, J.M. Novel 21-amino steroids as potent inhibitors of iron-dependent lipid peroxidation. *J. Biol. Chem.* **1987**, *262*, 10438–10440.

110. Braughler, J.M., Pregenzer, J.F. The 21-aminosteroid inhibitors of lipid peroxidation: Reactions with lipid peroxyl and phenoxy radicals. *Free Radic. Biol. Med.* **1989**, *7*, 125–130.

111. Hinzmann, J.S., McKenna, R.L., Pierson, T.S., Han, F., Kézdy, F.J., Epps, D.E. Interaction of antioxidants with depth-dependent fluorescence quenchers and energy transfer probes in lipid bilayers. *Chem. Phys. Lipids* **1992**, *62*, 123–138.

112. Noguchi, N., Takahashi, M., Tsuchiya, J., Yamashita, H., Komuro, E., Niki, E. Action of 21-aminosteroid U74006F as an antioxidant against lipid peroxidation. *Biochem. Pharmacol.* **1998**, *55*, 785–791.

113. Audus, K.L., Guillot, F.L., Mark Braughler, J. Evidence for 21-aminosteroid association with the hydrophobic domains of brain microvessel endothelial cells. *Free Radic. Biol. Med.* **1991**, *11*, 361–371.

114. Braughler, J.M., Chase, R.L., Neff, G.L., Yonkers, P.A., Day, J.S., Hall, E.D., Sethy, V.H., Lahti, R.A. A new 21-aminosteroid antioxidant lacking glucocorticoid activity stimulates adrenocorticotropin secretion and blocks arachidonic acid release from mouse pituitary tumor (AtT-20) cells. *J. Pharmacol. Exp. Ther.* **1988**, *244*, 423–427.

115. Munns, P.L., Leach, K.L. Two novel antioxidants, U74006F and U78517F, inhibit oxidant-stimulated calcium influx. *Free Radic. Biol. Med.* **1995**, *18*, 467–478.

116. Lesiuk, H., Sutherland, G., Peeling, J., Butler, K., Saunders, J. Effect of U74006F on forebrain ischemia in rats. *Stroke* **1991**, *22*, 896–901.

117. Perkins, W.J., Milde, L.N., Milde, J.H., Michenfelder, J.D. Pretreatment with U74006F improves neurologic outcome following complete cerebral ischemia in dogs. *Stroke* **1991**, *22*, 902–909.

118. Hall, E.D., Vaishnav, R.A., Mustafa, A.G. Antioxidant therapies for traumatic brain injury. *Neurotherapeutics* **2010**, *7*, 51–61.

119. Holzgrefe, H.H., Buchanan, L.V., Gibson, J.K. Effects of U74006F, a novel inhibitor of lipid peroxidation, in stunned reperfused canine myocardium. *J. Cardiovasc. Pharmacol.* **1990**, *15*, 239–248.

120. Ovize, M., de Lorgeril, M., Ovize, A., Ciavatti, M., Delaye, J., Renaud, S. U74006F, a novel 21-aminosteroid, inhibits in vivo lipid peroxidation but fails to limit infarct size in a canine model of myocardial ischemia reperfusion. *Am. Heart J.* **1991**, *122*, 681–689.

121. Levitt, M.A., Sievers, R.E., Wolfe, C.L. Reduction of infarct size during myocardial ischemia and reperfusion by lazaroid U-74500A, a nonglucocorticoid 21-aminosteroid. *J. Cardiovasc. Pharmacol.* **1994**, *23*, 136–140.

122. Liu, P., Vonderfecht, S.L., McGuire, G.M., Fisher, M.A., Farhood, A., Jaeschke, H. The 21-aminosteroid tirilazad mesylate protects against endotoxin shock and acute liver failure in rats. *J. Pharmacol. Exp. Ther.* **1994**, *271*, 438–445.

123. McGuire, G.M., Liu, P., Jaeschke, H. Neutrophil-induced lung damage after hepatic ischemia and endotoxemia. *Free Radic. Biol. Med.* **1996**, *20*, 189–197.

124. Wang, Y., Mathews, W.R., Guido, D.M., Jaeschke, H. The 21-aminosteroid tirilazad mesylate protects against liver injury via membrane stabilization not inhibition of lipid peroxidation. *J. Pharmacol. Exp. Ther.* **1996**, *277*, 714–720.

125. Zhang, H., Spapen, H., Manikis, P., Rogiers, P., Metz, G., Buurman, W.A., Vincent, J.L. Tirilazad mesylate (U-74006F) inhibits effects of endotoxin in dogs. *Am. J. Physiol. Heart Circ. Physiol.* **1995**, *268*, 1847–1855.

126. Bath, P.M.W., Blecic, S., Bogousslavsky, J., Boysen, G., Davis, S., Diez-Tejedor, E., Ferro, J.M., Gommans, J., Hacke, W., Indredavik, B., Norrving, B., Orgogozo, J.M., Ringelstein, E.B., Sacchetti, M.L., Iddenden, R., Bath, F.J., Musch, B.C., Brosse, D.M., Naberhuis-Stehouwer, S.A. Tirilazad mesylate in acute ischemic stroke: A systematic review. *Stroke* **2000**, *31*, 2257–2265.

127. Hall, E.D., Pazara, K.E., Braughler, J.M. 21-Aminosteroid lipid peroxidation inhibitor U74006F protects against cerebral ischemia in gerbils. *Stroke* **1988**, *19*, 997–1002.

128. Beck, T., Bielenberg, G.W. Failure of the lipid peroxidation inhibitor U74006F to improve neurological outcome after transient forebrain ischemia in the rat. *Brain Res.* **1990**, *532*, 336–338.

129. Reith, J., Jorgensen, H.S., Pedersen, P.M., Nakamaya, H., Jeppesen, L.L., Olsen, T.S., Raaschou, H.O. Body temperature in acute stroke: Relation to stroke severity, infarct size, mortality, and outcome. *Lancet* **1996**, *347*, 422–425.

130. Marshall, L.F., Maas, A.I.R., Marshall, S.B., Bricolo, A., Fearnside, M., Iannotti, F., Klauber, M.R., Lagarrigue, J., Lobato, R., Persson, L., Pickard, J.D., Piek, J., Servadei, F., Wellis, G.N., Morris, G.F., Means, E.D., Musch, B. A multicenter trial on the efficacy of using tirilazad mesylate in cases of head injury. *J. Neurosurg.* **1998**, *89*, 519–525.

131. Haley, E.C., Kassell, N.F., Apperson-Hansen, C., Maile, M.H., Alves, W.M. A randomized, double-blind, vehicle-controlled trial of tirilazad mesylate in patients with aneurysmal subarachnoid hemorrhage: A cooperative study in North America. *J. Neurosurg.* **1997**, *86*, 467–474.

132. Kassell, N.F., Haley, E.C., Apperson-Hansen, C., Alves, W.M. Randomized, double-blind, vehicle-controlled trial

of tirilazad mesylate in patients with aneurysmal subarachnoid hemorrhage: A cooperative study in Europe, Australia, and New Zealand. *J. Neurosurg.* **1996**, *84*, 221–228.
133. Lanzino, G., Kassell, N.F., Dorsch, N.W.C., Pasqualin, A., Brandt, L., Schmiedek, P., Truskowski, L.L., Alves, W.M. Double-blind, randomized, vehicle-controlled study of high-dose tirilazad mesylate in women with aneurysmal subarachnoid hemorrhage. Part I. A cooperative study in Europe, Australia, New Zealand, and South Africa. *J. Neurosurg.* **1999**, *90*, 1011–1017.
134. Lanzino, G., Kassell, N.F. Double-blind, randomized, vehicle-controlled study of high-dose tirilazad mesylate in women with aneurysmal subarachnoid hemorrhage. Part II. A cooperative study in North America. *J. Neurosurg.* **1999**, *90*, 1018–1024.
135. Janzen, E.G. Spin trapping. *Acc. Chem. Res.* **1971**, *4*, 31–40.
136. Villamena, F.A., Zweier, J.L. Detection of reactive oxygen and nitrogen species by EPR spin trapping. *Antioxid. Redox Signal.* **2004**, *6*, 619–629.
137. Fréjaville, C., Karoui, H., Tuccio, B., Moigne, F.L., Culcasi, M., Pietri, S., Lauricella, R., Tordo, P. 5-(Diethoxyphosphoryl)-5-methyl-1-pyrroline N-oxide: A new efficient phosphorylated nitrone for the in vitro and in vivo spin trapping of oxygen-centered radicals. *J. Med. Chem.* **1995**, *38*, 258–265.
138. Olive, G., Mercier, A., Le Moigne, F., Rockenbauer, A., Tordo, P. 2-ethoxycarbonyl-2-methyl-3,4-dihydro-2H-pyrrole-1-oxide: Evaluation of the spin trapping properties. *Free Radic. Biol. Med.* **2000**, *28*, 403–408.
139. Zhao, H., Joseph, J., Zhang, H., Karoui, H., Kalyanaraman, B. Synthesis and biochemical applications of a solid cyclic nitrone spin trap: A relatively superior trap for detecting superoxide anions and glutathiyl radicals. *Free Radic. Biol. Med.* **2001**, *31*, 599–606.
140. Villamena, F.A., Rockenbauer, A., Gallucci, J., Velayutham, M., Hadad, C.M., Zweier, J.L. Spin trapping by 5-carbamoyl-5-methyl-1-pyrroline N-oxide (AMPO): Theoretical and experimental studies. *J. Org. Chem.* **2004**, *69*, 7994–8004.
141. Villamena, F.A., Xia, S., Merle, J.K., Lauricella, R., Tuccio, B., Hadad, C.M., Zweier, J.L. Reactivity of superoxide radical anion with cyclic nitrones: Role of intramolecular H-Bond and electrostatic effects. *J. Am. Chem. Soc* **2007**, *129*, 8177–8191.
142. Zeghdaoui, A., Tuccio, B., Finet, J.-P., Cerri, V., Tordo, P. β-Phosphorylated α-phenyl-N-tert-butylnitrone (PBN) analogs: A new series of spin traps for oxyl radicals. *J. Chem. Soc. Perkin Trans. 2* **1995**, *12*, 2087–2089.
143. Roubaud, V., Lauricella, R., Tuccio, B., Bouteiller, J.-C., Tordo, P. Decay of superoxide spin adducts of new PBN-type phosphorylated nitrones. *Res. Chem. Intermed.* **1996**, *22*, 405–416.
144. Hinton, R.D., Janzen, E.G. Synthesis and characterization of phenyl-substituted C-phenyl-N-tert-butylnitrones and some of their radical adducts. *J. Org. Chem.* **1992**, *57*, 2646–2651.
145. Rosen, G.M., Britigan, B.E., Halpern, H.J., Pou, S. *Free Radicals: Biology and Detection by Spin Trapping.* Oxford Univeristy Press, New York, 1999.
146. Durand, G., Poeggeler, B., Böker, J., Raynal, S., Polidori, A., Pappolla, M.A., Hardeland, R., Pucci, B. Fine-tuning the amphiphilicity: A crucial parameter in the design of potent alpha-phenyl-N-tert-butylnitrone analogues. *J. Med. Chem.* **2007**, *50*, 3976–3979.
147. Durand, G., Polidori, A., Ouari, O., Tordo, P., Geromel, V., Rustin, P., Pucci, B. Synthesis and preliminary biological evaluations of ionic and nonionic amphiphilic alpha-phenyl-N-tert-butylnitrone derivatives. *J. Med. Chem.* **2003**, *46*, 5230–5237.
148. Durand, G., Polidori, A., Salles, J.-P., Pucci, B. Synthesis of a new family of glycolipidic nitrones as potential antioxidant drugs for neurodegenerative disorders. *Bioorg. Med. Chem. Lett.* **2003**, *13*, 859–862.
149. Poeggeler, B., Durand, G., Polidori, A., Pappolla, M.A., Vega-Naredo, I., Coto-Montes, A., Boeker, J., Hardeland, R., Pucci, B. Mitochondrial medicine: Neuroprotection and life extension by the new amphiphilic nitrone LPBNAH acting as a highly potent antioxidant agent. *J. Neurochem.* **2005**, *95*, 962–973.
150. Bernotas, R.C., Thomas, C.E., Carr, A.A., Nieduzak, T.R., Adams, G., Ohlweiler, D.F., Hay, D.A. Synthesis and radical scavenging activity of 3,3-dialkyl-3,4-dihydroisoquinoline 2-oxides. *Bioorg. Med. Chem. Lett.* **1996**, *6*, 1105–1110.
151. Janzen, E.G., Poyer, J.L., Schaefer, C.F., Downs, P.E., BuBose, C.M. Biological spin trapping II. Toxicity of nitrones spin traps: Dose-ranging in the rat. *J. Biochem. Biophys. Methods* **1995**, *30*, 239–247.
152. Schaefer, C.F., Janzen, E.G., West, M.S., Poyer, J.L., Kosanke, S.D. Blood chemistry changes in the rat induced by high doses of nitronyl free radicals spin traps. *Free Radic. Biol. Med.* **1996**, *21*, 427–436.
153. Haseloff, R.F., Mertsch, K., Rohde, E., Baeger, I., Grigor'ev, I.A., Blasig, I.E. Cytotoxicity of spin trapping compounds. *FEBS Lett.* **1997**, *418*, 73–75.
154. Chen, G., Bray, T.M., Janzen, E.G., McCay, P.B. Excretion, metabolism and tissue distribution of a spin trapping agent, α-phenyl-N-tert-butyl-nitrone (PBN) in rats. *Free Radic. Res. Commun.* **1990**, *9*(3–6), 317–323.
155. Cheng, H.-Y., Liu, T., Feurerstein, G., Barone, F.C. Distribution of spin-trapping compounds in rat blood and brain: In vivo microdialysis determination. *Free Radic. Biol. Med.* **1993**, *14*, 243–250.
156. Liu, K.J., Kotake, Y., Lee, M., Miyake, M., Sugden, K., Yu, Z., Swartz, H.M. High-performance liquid chromatography study of the pharmacokinetics of various spin traps for application to in vivo spin trapping. *Free Radic. Biol. Med.* **1999**, *27*, 82–89.
157. Trudeau-Lame, M.E., Kalgutkar, A.S., LaFontaine, M. Pharmakokinetics and metabolism of the reactive oxygen scavenger α-phenyl-N-tert-butylnitrone in the male sprague-dawley. *Drug Metab. Dispos.* **2003**, *31*, 147–152.

158. Cova, D., De Angelis, L., Monti, E., Piccinini, F. Subcellular distribution of two spin trapping agents in rat heart: Possible explanation for their different protective effects against doxorubicin-induced cardiotoxicity. *Free Radic. Res. Commun.* **1992**, *15*, 353–360.

159. Reinke, L.A., Moore, D.R., Sang, H., Janzen, E.G., Kotake, Y. Aromatic hydroxylation in PBN spin trapping by hydroxyl radicals and cytochrome P-450. *Free Radic. Biol. Med.* **2000**, *28*, 345–350.

160. Novelli, G.P., Angiolini, P., Tani, R., Consales, G., Bordi, L. Phenyl-tert-butylnitrone is active against traumatic shock in rats. *Free Radic. Res. Commun.* **1986**, *1*, 321–327.

161. Floyd, R.A., Hensley, K., Forster, M.J., Kelleher-Andersson, J.A., Wood, P.L. Nitrones, their value as therapeutics and probes to understand aging. *Mech. Ageing Dev.* **2002**, *123*, 1021–1031.

162. Kotake, Y. Pharmacologic properties of phenyl *N-tert*-butylnitrone. *Antioxid. Redox Signal.* **1999**, *1*, 481–499.

163. Floyd, R.A. Nitrones as therapeutics in age-related diseases. *Aging Cell* **2006**, *5*, 51–57.

164. Floyd, R.A., Kopke, R.D., Choi, C.-H., Foster, S.B., Doblas, S., Towner, R.A. Nitrones as therapeutics. *Free Radic. Biol. Med.* **2008**, *45*, 1361–1374.

165. Hamburger, S.A., McCay, P.B. Endotoxin-induced mortality in rats is reduced by nitrones. *Circ. Shock* **1989**, *29*, 329–334.

166. McKechnie, K., Furman, B.L., Parrat, J.R. Modification by oxygen free radical scavengers of the metabolic and cardiovascular effects of endotoxin infusion in conscious rats. *Circ. Shock* **1986**, *19*, 429–439.

167. Progrebniak, H.W., Merino, M.J., Hahn, S.M., Mitchell, J.B., Pass, H.I. Spin trap salvage from endotoxemia: The role of cytokine down-regulation. *Surgery* **1992**, *112*, 130–139.

168. Iovino, G., Kubow, S., Marliss, E.B. Effect of α-phenyl-*N-tert*-butylnitrone on diabetes and lipid peroxidation in BB rats. *Can. J. Physiol. Pharmacol.* **1999**, *77*, 166–172.

169. Ho, E., Chen, G., Bray, T.M. Alpha-phenyl-tert-butylnitrone (PBN) inhibits NFκB activation offering protection against chemically induced diabetes. *Free Radic. Biol. Med.* **2000**, *28*, 604–614.

170. Jotti, A., Paracchini, L., Perletti, G., Piccinini, F. Cardiotoxicity induced by doxorubicin in vivo: Protective activity of the spin trap alpha-phenyl-tert-butyl nitrone. *Pharmacol. Res.* **1992**, *26*, 143–150.

171. Kotamraju, S., Konorev, E., Joseph, J., Kalyanaraman, B. Doxorubicin-induced apoptosis in endothelial cells and cardiomycetes is ameliorated by nitrone spin traps and ebselen. *J. Biol. Chem.* **2000**, *275*, 33585–33592.

172. Parman, T., Wiley, M.J., Wells, P.G. Free radical-medical oxidative DNA damage in the mechanism of thalidomide teratogenicity. *Nat. Med.* **1999**, *5*, 582–585.

173. Rao, D.B., Fechter, L.D. Protective effects of phenyl *N-tert*-butylnitrone on the potentiation of noise-induced hearing loss by carbone monoxide. *Toxicol. Appl. Pharmacol.* **2000**, *167*, 125–131.

174. Pouyatos, B., Gearhart, C.A., Fechter, L.D. Acrylonitrile potentiates hearing loss and cochlear damage induced by moderate noise exposure. *Toxicol. Appl. Pharmacol.* **2005**, *204*, 46–56.

175. Fechter, L.D., Liu, Y., Pearce, T.A. Cochlear protection from carbon monoxide exposure by free radical blockers in the guinea pig. *Toxicol. Appl. Pharmacol.* **1997**, *142*, 47–55.

176. Choi, C.-H., Chen, K., Vasquez-Weldon, A., Jackson, R.L., Floyd, R.A., Kopke, R.D. Effectiveness of 4-hydroxy phenyl N-tert-butylnitrone (4-OHPBN) alone and in combination with other antioxidant drugs in the treatment of acute acoustic trauma in chinchilla. *Free Radic. Biol. Med.* **2008**, *44*, 1772–1784.

177. Ranchon, I., Chen, S., Alvarez, K., Anderson, R.E. Systemic administration of phenyl-*N-tert*-butylnitrone protects the retina from light damage. *Invest. Ophthalmol. Vis. Sci.* **2001**, *46*, 427–434.

178. Tomita, H., Kotake, Y., Anderson, R.E. Mechanism of protection from light-induced retinal degeneration by the synthetic antioxidant phenyl-N-tert-butylnitrone. *Invest. Ophthalmol. Vis. Sci.* **2005**, *46*, 427–434.

179. Choteau, F., Durand, G., Ranchon-Cole, I., Cercy, C., Pucci, B. Cholesterol-based α-phenyl-*N-tert*-butyl nitrone derivatives as antioxidants against light-induced retinal degeneration. *Bioorg. Med. Chem. Lett.* **2010**, *20*, 7405–7409.

180. Ranchon, I., LaVail, M.M., Kotake, Y., Anderson, D.E. Free radical trap phenyl-*N-tert*-butylnitrone protects against light damage but does not rescue P23H and S334ter rhodopsin transgenic rats from inherited retinal degeneration. *J. Neurosci.* **2003**, *23*, 6050–6057.

181. Yamashita, T., Ohshima, H., Asanuma, T., Inukai, N., Miyoshi, I., Kasai, N., Kon, Y., Watanabe, T., Sato, F., Kuwabara, M. The effects of α-phenyl-*N-tert*-butyl nitrone (PBN) on copper induced rat fulminant hepatitis with jaundice. *Free Radic. Biol. Med.* **1996**, *26*, 755–761.

182. Asanuma, T., Yasui, H., Inanami, O., Waki, K., Takahashi, M., Iizuka, D., Uemura, T., Durand, G., Polidori, A., Kon, Y., Pucci, B., Kuwabara, M. A new amphiphilic derivative, N-{[4-(lactobionamido)methyl]benzylidene}-1,1-dimethyl-2-(octylsulfanyl)ethylamine N-oxide, has a protective effect against copper-induced fulminant hepatitis in *Long-Evans* cinnamon rats at an extremely low concentration compared with its original form α-phenyl-*N*-(tert-butyl) nitrone. *Chem. Biodivers.* **2007**, *4*, 2253–2267.

183. Tosaki, A., Blasig, I.E., Pali, T., Ebert, B. Heart protection and radical trapping by DMPO during reperfusion in isolated working rat hearts. *Free Radic. Biol. Med.* **1990**, *8*, 363–372.

184. Zhuo, L., Chen, Y.R., Reyes, L.A., Lee, H.L., Chen, C.L., Villamena, F.A., Zweier, J.L. The radical spin-trap 5,5-dimethyl-1-pyrroline N-oxide exerts dose-dependant protection against myocardial ischemia-reperfusion injury trough preservation of mitochondrial electron transport. *J. Pharmacol. Exp. Ther.* **2009**, *329*, 515–523.

185. Tanguy, S., Durand, G., Reboul, C., Polidori, A., Pucci, B., Dauzat, M., Obert, P. Protection against reactive oxygen species injuries in rat isolated perfused hearts: Effect of LPBNAH, a new amphiphilic spin-trap derived from PBN. *Cardiovasc. Drugs Ther.* **2006**, *20*, 147–149.

186. Edamatsu, R., Mori, A., Packer, L. The spin-trap N-tert-α-phenyl-butylnitrone prolongs the life span of the senescence accelerated mouse. *Biochem. Biophys. Res. Commun.* **1995**, *221*, 847–849.

187. Saito, K., Yoshioka, H., Cutler, R.G. A spin trap, N-tert-butyl-α-phenylnitrone extends the life span of mice. *Biosci. Biotechnol. Biochem.* **1998**, *62*, 792–794.

188. Sack, C.A., Socci, D.J., Crandall, B.M., Arendash, G.W. Antioxidant treatment with phenyl-α-tert-butyl nitrone (PBN) improves the cognitive performance. *Neurosci. Lett.* **1996**, *205*, 181–184.

189. Atamna, H., Paler-Martinez, A., Ames, B.N. N-t-butylhydroxylamine, a hydrolysis product of a-phenyl-N-t-butyl nitrone, is more potent in delaying senescence in human lung fibroblasts. *J. Biol. Chem.* **2000**, *275*, 6741–6748.

190. von Zglinicki, T., Pilger, R., Sitte, N. Accumulation of single-strand breaks is the major cause of telomere shortening in human fibroblasts. *Free Radic. Biol. Med.* **2000**, *28*, 64–74.

191. Floyd, R.A., Hensley, K., Forster, M.J., Kelleher-Andersson, J.A., Wood, P.L. Nitrones as neuroprotectants and antiaging drugs. *Ann. N. Y. Acad. Sci.* **2002**, *959*, 321–329.

192. Floyd, R.A. Role of oxygen free radicals in carcinogenesis and brain ischemia. *FASEB J.* **1990**, *4*, 2587–2597.

193. Carney, J.M., Floyd, R.A. Phenyl butyl nitrone compositions and methods for treatment of oxidative tissue damage, US Patent 5025032. Issued 18 June 1991.

194. Phillis, J.W., Clough-Helfman, C. Protection from cerebral ischemic injury in gerbils with the spin trap agent N-tert-butyl-α-phenylnitrone (PBN). *Neurosci. Lett.* **1990**, *116*, 315–319.

195. Cao, X., Phillis, J.W. α-Phenyl-tert-butyl nitrone reduces cortical infarct and edema in rats subjected to focal ischemia. *Brain Res.* **1994**, *644*, 267–272.

196. Mori, H., Arai, T., Ishii, H., Adachi, T., Endo, N., Makino, K., Mori, K. Neuroprotective effects of pterin-6-aldehyde in gerbil global brain ischemia: Comparison with those of [alpha]-phenyl-N-tert-butyl nitrone. *Neurosci. Lett.* **1998**, *241*(2–3), 99–102.

197. Yue, T.-L., Gu, J.-L., Lysko, P.G., Cheng, H.-Y., Barone, F.C., Feuerstein, G. Neuroprotective effects of phenyl-t-butyl-nitrone in gerbil global brain ischemia and in cultured rat cerebellar neurons. *Brain Res.* **1992**, *574*, 193–197.

198. Zhao, Q., Pahlmark, K., Smith, M.-I., Siesjo, B.K. Delayed treatment with the spin trap α-phenyl-N-tert-butyl nitrone (PBN) reduces infarct size following transient middle cerebral artery occlusion in rats. *Acta Physiol. Scand.* **1994**, *152*, 349–350.

199. Nakao, N., Grabson-Frodl, E.M., Widner, H., Brundin, P. Antioxidant treatment protects striatal neurons against excitotoxic insults. *Neuroscience* **1996**, *73*, 185–200.

200. Nakao, N., Brundin, P. Effects of α-phenyl-tert-butyl-nitrone on neuronal survival and motor function following intrastriatal injections of quinolinic or 3-nitropropoionic acid. *Neuroscience* **1997**, *76*, 749–761.

201. Schulz, J.B., Henshaw, D.R., Siwek, D., Jenkins, B.G., Ferrante, R.J., Cipolloni, P.B., Kowall, N.W., Rosen, B.R., Beal, M.F. Involvement of free radicals in excitotoxicity in vivo. *J. Neurochem.* **1995**, *64*, 2239–2247.

202. Marklund, N., Clausen, F., McIntosh, T.K., Hillered, L. Free radical scavenger post-treatment improves functional and morphological outcome after fluid percussion injury in the rat. *J. Neurotrauma* **2001**, *19*, 821–832.

203. Marklund, N., Lewander, T., Clausen, F., Hillered, L. Effects of the nitrone radical scavengers PBN and S-PBN on in vivo trapping of reactive oxygen species after traumatic brain injury in rats. *J. Cereb. Blood Flow Metab.* **2001**, *21*, 1259–1267.

204. Green, A.R., Ashwood, T., Odergren, T., Jackson, D.M. Nitrones as neuroprotective agents in cerebral ischemia, with particular reference to NXY-059. *Pharmacol. Ther.* **2003**, *100*, 195–214.

205. Kuroda, S., Tsuchidate, R., Smith, M.-L., Maples, K.R., Siesjo, B.K. Neuroprotective effects of a novel nitrone, NXY-059, after transient focal cerebral ischemia in the rat. *J. Cereb. Blood Flow Metab.* **1999**, *19*, 778–787.

206. Sydserff, S.G., Borelli, A.R., Green, A.R., Cross, A.J. Effect of NXY-059 on infarct volume after transient or permanent middle cerebral artery occlusion in the rat; studies on dose, plasma concentration and therapeutic time window. *Br. J. Pharmacol.* **2002**, *135*, 103–112.

207. Marshall, J.W.B., Duffin, K.J., Green, A.R., Ridley, R.M., Finklestein, S.P. NXY-059, a free radical-trapping agent, substantially lessens the functional disability resulting from cerebral ischemia in a primate species. *Stroke* **2001**, *32*, 190–198.

208. Marshall, J.W.B., Cummings, R.M., Bowes, L.J., Ridley, R.M., Green, A.R. Functional and Histological evidence for the protective effect of NXY-059 in a primate model of stroke when given 4 hours after occlusion. *Stroke* **2003**, *34*, 2228–2233.

209. Edenius, C., Strid, S., Borgå, O., Breitholtz-Emanuelsson, A., Vallén, K.L., Fransson, B. Pharmacokinetics of NXY-059, a nitrone-based free radical trapping agent, in healthy young and elderly subjects. *J. Stroke Cerebrovasc. Dis.* **2002**, *11*, 34–43.

210. Lees, K.R., Sharma, A.K., Barer, D., Ford, G.A., Kostulas, V., Cheng, Y.F., Odergren, T. Tolerability and pharmacokinetics of the nitrone NXY-059 in patients with acute stroke. *Stroke* **2001**, *32*, 675–680.

211. Lees, K.R., Barer, D., Ford, G.A., Hacke, W., Kostulas, V., Sharma, A.K., Odergren, T. Tolerability of NXY-059 at Higher target concentrations in patients with acute stroke. *Stroke* **2003**, *34*, 482–487.

212. Lees, K.R., Zivin, J.A., Ashwood, T., Davalos, A., Davis, S.M., Diener, H.-C., Grotta, J., Lyden, P., Shuaib, A., Hårdemark, H.-G., Wasiewski, W.W. NXY-059 for acute ischemic stroke. *N. Engl. J. Med.* **2006**, *354*, 588–600.

213. Shuaib, A., Lees, K.R., Lyden, P., Grotta, J., Davalos, A., Davis, S.M., Diener, H.-C., Ashwood, T., Wasiewski, W.W., Emeribe, U. NXY-059 for the treatment of acute ischemic stroke. *N. Engl. J. Med.* **2007**, *357*, 562–571.

214. Lees, K.R., Davalos, A., Davis, S.M., Diener, H.-C., Grotta, J., Lyden, P., Shuaib, A., Ashwood, T., Hårdemark, H.-G., Wasiewski, W., Emeribe, U., Zivin, J.A. Additional outcomes and subgroup analyses of NXY-059 for acute ischemic stroke in the SAINT I trial. *Stroke* **2006**, *37*, 2970–2978.

215. Diener, H.-C., Lees, K.R., Lyden, P., Grotta, J., Davalos, A., Davis, S.M., Shuaib, A., Ashwood, T., Wasiewski, W., Alderfer, V., Hårdemark, H.-G., Rodichok, L. NXY-059 for the treatment of acute stroke: Pooled analysis of the SAINT I and II trials. *Stroke* **2008**, *39*, 1751–1758.

216. Savitz, S.I. A critical appraisal of the NXY-059 neuroprotection studies for acute stroke: A need for more rigorous testing of neuroprotective agents in animal models of stroke. *Exp. Neurol.* **2007**, *205*, 20–25.

217. Floyd, R.A., Carney, J.M. *Nitrone Radical Traps Protect in Experimental Neurodegenerative Diseases*. Academin Press, London, 1996.

218. Barclay, L.R.C., Vinqvist, M.R. Do spin traps also act as classical chain-breaking antioxidants? A quantitative kinetic study of phenyl-*tert*-butylnitrone (PBN) in solution and in liposomes. *Free Radic. Biol. Med.* **2000**, *28*, 1079–1090.

219. Janzen, E.G., West, M.S., Poyer, J.L. *Comparison of Antioxidant Activity of PBN with Hindered Phenols in Initiated Rat Liver Microsomal Lipid Peroxidation*. Elsevier, San Diego, CA, 1994.

220. Kotake, Y., Sang, H., Miyajima, T., Wallis, G.L. Inhibition of NF-kB, iNOS mRNA, COX2 mRNA, and COX catalytic activity by phenyl-N-tert-butylnitrone (PBN). *Biochim. Biophys. Acta* **1998**, *1448*, 77–84.

221. Sang, H., Wallis, G.L., Stewart, C.A., Kotake, Y. Expression of cytokines and activation of transcription factors in lipopolysaccharide-administered rats and their inhibition by phenyl *N-tert*-butylnitrone (PBN). *Arch. Biochem. Biophys.* **1999**, *363*, 341–348.

222. Floyd, R.A., Hensley, K., Jaffrey, F., Maidt, L., Robinson, K., Pye, Q., Stewart, C. Increased oxidative stress brought on by pro-inflammatory cytokines in neurodegenerative processes and the protective role of nitrone-based free radicals traps. *Life Sci.* **1999**, *65*, 1893–1899.

223. Floyd, R.A., Hensley, K., Bing, G. Evidence for enhanced neuro-inflammatory processes in neurodegenerative diseases and the action of nitrones as potential therapeutics. *J. Neural Transm.* **2000**, *60*, 337–364.

224. Anderson, D.E., Yuan, X.J., Tseng, C.M., Rubin, L.J., Rosen, G.M., Tod, M.L. Nitrone spin-traps block calcium channels and induce pulmonary artery relaxation independant of free radicals. *Biochem. Biophys. Res. Commun.* **1993**, *193*, 878–885.

225. Milatovic, D., Radic, Z., Zivin, M., Dettbarn, W.D. Atypical effect of some spin trapping agents: Reversible inhibition of acetylcholinesterase. *Free Radic. Res. Commun.* **2000**, *28*, 597–603.

226. Hensley, K., Pye, Q.N., Maidt, M.L., Stewart, C.A., Robinson, K.A., Jaffrey, F., Floyd, R.A. Interaction of α-phenyl-N-tert-butyl nitrone and alternative electron acceptors with complex I indicates a substrate reduction site upstream from the rotenone binding site. *J. Neurochem.* **1998**, *71*, 2549–2557.

227. Kim, H.Y., Chung, J.M., Chung, K. Increased production of mitochondrial superoxide in the spinal cord induces pain behaviors in mice: The effect of mitochondrial electron transport complex inhibitors. *Neurosci. Lett.* **2008**, *447*, 87–91.

228. Wen, J.-J., Bhatia, V., Popov, V.L., Garg, N.J. Phenyl-{alpha}-tert-butyl nitrone reverses mitochondrial decay in acute Chagas' disease. *Am. J. Pathol.* **2006**, *169*, 1953–1964.

229. Wen, J.-J., Garg, N. Mitochondrial generation of reactive oxygen species is enhanced at the Q_0 site of the complex III in the myocardium of *Trypanosoma cruzi*-infected mice: Beneficial effects of an antioxidant. *J. Bioenerg. Biomembr.* **2008**, *40*, 587–598.

230. Durand, G., Poeggeler, B., Ortial, S., Polidori, A., Villamena, F.A., Böker, J., Hardeland, R., Pappolla, M.A., Pucci, B. Amphiphilic amide nitrones: A new class of protective agents acting as modifiers of mitochondrial metabolism. *J. Med. Chem.* **2010**, *53*, 4849–4861.

INDEX

ABCC7, 345
Acetylcholine, 194
Acetylsalicylic acid (ASA, aspirin), 248–249, 252–553, 262
Acidosis, intracellular, 317
Acrolein, 58, 61–62
Actin, 74, 78
Activator protein 1, 346
Adderall, 256
Adenine, 96
Adenosine, 316, 321
Adenosine diphosphate (ADP), 27, 114, 123–125, 316, 318
Adenosine monophosphate (AMP), 27, 316, 318
   cyclic, 243, 245, 352
Adenosine triphosphate (ATP), 27, 73, 114, 123–125, 130, 132, 190, 220, 238, 240–241, 316, 318–319, 352
   synthase, 362–364, 397,
Adenylation, 78
ADMA, *see* Asymmetric dimethylarginine
Advanced lipidation end products (ALE), 80
Aflatoxin B1, 250
Aging, 237, 252
   and DNA repair, 268–269
   and METH, 262
Airway
   inflammation, 347
   obstructions, 348
Alanine, 318
Alcohol dehydrogenase class III (ADH), 74
Aldehyde oxidase (AO), 314, 316, 318
Aldehydes
   in atherosclerosis, 329–335
   in I-R, 314
Alkoxyl radical, 51–52, 58, 62
Allopurinol, 318, 320. *See also* Inhibitors

Allysine, 79
Alzheimer's disease, 80, 194, 238–241, 250, 253–254, 263, 266, 277–278, 360
Amadori rearrangements, 80
Amino acids, free, 318–319
Aminoglycoside, 351
2-amino-3-keto butyric acid, 80
3-amino-1,2,4-triazole, 117
Amphetamine(s), 252, 255–263
   affinities to uptake transporters, 260
   bioactivation by PHSs, 252–253
   history, 255–256
   metabolism by CYPs, 257–259
   neurotoxicity, 261–263
   pharmacokinetics, 256–257
   SOD protection against toxicity, 266
   uses, 255–256
AMPO, 390. *See also* Nitrones
Amyotrophic lateral sclerosis (ALS), 239, 241, 253–4, 277–278, 360
Angiotensin II, 150, 152–156, 158
Antibiotics, 181
Antibody, 127–128
Antidepressants, 240
Antiinflammatory cytokines, 346
Antimicrobial, 346–350
Antioxidant response element (ARE), 269–270, 274–275
Antioxidants
   in cell signaling, 179, 180, 182, 185–186, 188, 190, 193–195
   enzymes, 94, 114–116, 263–268, 312–313, 318, 321
   synthetic, 377–398
Antipsychotics, 240
AP-1, *see* Activator protein 1
Apical transporter, 348

Molecular Basis of Oxidative Stress: Chemistry, Mechanisms, and Disease Pathogenesis, First Edition. Edited by Frederick A. Villamena.
© 2013 John Wiley & Sons, Inc. Published 2013 by John Wiley & Sons, Inc.

Apolipoprotein
    A1 (apo A1), 334
    B (apo B), 331
Apoptosis, 75, 216–219
    Bcl-2, 218
    Bcl-xl, 218
    BH3-only proteins, 218
    cytochrome c (Cyt c), 217–218
    death receptors, 217–218
    endoplasmic reticulum (ER) stress, 218–219
    extrinsic pathway, see Death receptors
    glutathione, 217–219
    in I-R, 313, 315, 319, 320
    intrinsic pathway, see Mitochondrial pathway
    mitochondrial pathway, 218
    multidrug-resistance proteins (MRP1), 220
    from NADPH oxidases, 145, 146, 148, 155–156, 158
    p53, 219
    and PTM, 75
    TNF-related apoptosis-inducing ligand (TRAIL), 217–218
Apurinic/apyrimidinic (AP) sites, 99, 100–101
Aquaporins, 351
Arachidonic acid (AA), 50–52, 54, 57–65, 240, 242, 246–249, 252
ARE, see Antioxidant response element
Arsenic, 94–95
Ascorbic acid, 262, 268, 347
Asparaginyl hydroxylase factor inhibiting HIF (FIH), 191–192
Aspartate, 318
Astrocytes, 239–241, 250, 254, 263, 266–268, 276–279
Asymmetric dimethylarginine (ADMA), 318
Ataluren, 351
Ataxia telangiectasia mutated (ATM), 238–239, 268
Atherosclerosis, 329–333, 384
Attention deficit–hyperactivity disorder (ADHD), 256, 260
Autophagy, 219, 319–320. See also Cell death
Azelaic acid (AZA), 334

Bacteria colonization, 346–347
BALF, see Broncho alveolar lavage fluid
Benzo[a]pyrene-7,8-dihydrodiol, 250
Beta-amyloid (AB), 241, 250
β-carotene, 347
Beta-scission reactions
    formation of fragmented lipid products, 51–52, 58–62
    role of iron and copper, 50–51, 58, 61–62
Biomarker
    for COPD, 349
    for neurodegenerative disorders, 360
Biotin, 82
Blood–brain barrier (BBB), 263
    and METH, 256
Brad 1, 189
Bradykinin, 321
Brain infarction, 385, 387
Breast cancer 1 (Brca1), 238–239, 269
Broncho alveolar lavage fluid, 352

Bronchodilation, 349
Buthionine sulfoximine, 117–118

Calcification, 331
Calcium
    in cell signaling, 186–188
    in dendritic spines, 250
    in excitotoxicity, 241
    homeostasis, 185, 188, 195
    in mitochondria, 238
Calcium-dependent proteases, 241
Calicheamicin, 100
CAMP, see Adenosine monophosphate, cyclic
Cancer, 203–235
    apoptosis, 216–219
    autophagy, 219
    catalase, 206
    cell signaling, 193–194
    cycloxygenases, 205
    DNA methylation, 216
    DNA methyl-transferases (DMTs), 216
    drug resistance, 219–220
    epigenetics, 213, 216
    glucose-6-phosphate dehydrogenase, 206
    glutathione, 205–206, 208, 217
    glutathione peroxidases (GPx), 206
    glutathione-SNO (S-nitrosoglutathione or GSNO), 208
    glutathione-S-transferases, 216, 220
    protein glutathionylation, 208
    γ-glutamylcysteine synthetase, 205
Carbohydrate oxidation biomarkers in neurodegenerative disorders, 366
Carcinogenesis, 203–235. See also Cancer
Cardiac, 311–321. See also Heart
    ischemia, 311–314, 318
    reperfusion, 311, 314, 318
    transplantation, 321
Cardiolipin, 315
Cardiomyocyte(s), 311, 314–315, 318–320
Cardioplegic,
    arrest, 312, 318
    buffer, 318
    ischemia, 312
    solution, 312
Cardiovascular disease (CVD), 330, 334–335
Catalase, 114, 143–146, 159, 238–239, 251–252, 377–378
    in brain, 262, 267
    in cancer, 206
    in I-R, 313
    and NADPH, 264
    mimetics, 380
    in neurodegenerative disease, 278
Celecoxib (Celebrex), 249, 254
Cell death, 311, 315, 319–321
    apoptosis, 319–320
    autophagy, 319–320
    necrosis, 319–320
    programmed, 315, 319
Cell signaling, 179, 180, 184–186, 193–194, 317–318

Cellular
  enzymes, 313
  organelles, 312–313
  thiolstat, 71, 84
CF, see cystic fibrosis
CFTR, see Cystic fibrosis transmembrane regulator
CGMP-dependent protein kinase (PKG)I-α, 85
Chemiluminescence, 32
Chemoprevention, 117
Chemoselective functionalization, 82
Chloride
  channel, 345, 348
  conductance, 351–352
  transport, 348
  transporter, 348, 351
ChPBNL, 393. See also Nitrones
Chromium, 94–95, 98
Chronic
  granulomatous disease (CGD), 138–141, 145–146, 154,
  infections, 346
  lung infections, 345
  obstructive pulmonary disease COPD, 80
Citric acid cycle, 180, 192
Clorgyline, 240
Cockayne syndrome B (CSB), 238–239, 269
Cofactors, 316
  BH_4, 315, 317–318, 320
Colon epithelial cells, 186
Complex I, 123, 125–129, 133, 312, 314, 316
Complex II, 126, 129, 130, 133
Complex III, 125–126, 129–133, 312, 315
Complex IV, 125–126, 132–133, 315
Conserved sequence block, 96
Copper-zinc superoxide dismutase (CuZnSOD, SOD1), see
  Superoxide dismutase
Coronary artery disease, 311, 320
Coronary heart disease (CHD), 333
COX, see Cyclooxygenase
CPI-1429, 393–394. See also Nitrones
C-Rel, see Proto-oncogene cRel
Cross-links, see Lesions
Cu,Zn superoxide dismutase (CuZnSOD, SOD1), 79, 253, 313
  in neurodegenerative diseases, 278
Cyanuric acid, 102
Cyclooxygenase-1 (COX-1), 242. See also PHS-1;
  prostaglandin H synthases
Cyclooxygenase-1b (COX-1b) (COX-3), 244
Clooygenase, 29. See also COX-1; COX-2
Cyclooxygenase-2 (COX-2), 242. See also PHS-2;
  Prostaglandin H synthases
CYP1A2, 258. See also Cytochromes P450
CYP2D6, 240, 257–259. See also Cytochromes P450
  expression in brain, 259
CYP2E1, 240. See also Cytochromes P450
Cysteinyl redox domains, 127
Cystic fibrosis (CF), 80, 345
  diagnostic tests for, 348
  gene, 345
  lung disease, 345–347, 351

Cystic fibrosis transmembrane regulator (CFTR)
  KO mice, 347–349, 352
  mutations, 349, 351
  potentiators, 352
Cytochrome b_H or heme b_H, 130–131
Cytochrome b_L or heme b_L, 130–133
Cytochrome c, 131–132
  in I-R, 314–316, 318–319. See also Nitrate/Nitrite
  reductase
Cytochromes P450 (CYPs), 240
  metabolism of amphetamines, 257–259
Cytokine, 187, 193–194
  and microglia, 241
  and NOS, 240
  as PHS activators, 245
Cytosolic phospholipase A2 (cPLA2), 240

Death receptors, 217, 219
Deoxymyoglobin, 316. See also Nitrate/Nitrite
  reductase
2-deoxypentos-4-ulose abasic site, 96
2′-deoxyribonolactone, 96, 100
Deoxyribonucleic acid (DNA), 94–103
  and histone proteins, 94
  backbone reactivity, 96
  damage, 349. See also Oxidative DNA damage
  duplex, kinetic changes to, 99
  holes, 96
  oxidation, 252–253
  repair, 268–269
2′-deoxyribose (dR), 99–102
  and hydrogen abstraction, 100
  lesions, 101
  oxidation, 100
  radical, 100
DEPMPO, 390. See also Nitrones
Detection of reactive species, 31–38
  in vitro, 32
  in vivo, 38
Dexedrine, 256
Diabetes, 194
Dicumarol, 118
Dihydroxyphenylacetic acid (DOPAC), 251–252, 255
Dilated cardiomyopathy, 313
Dimedone, 82
Dimethyl fumarate (DMF), 276–277, 279
Dioxetane, 52, 59, 61
1,2-dioxilane ring, 52
1,2-dioxolanylcarbinyl, 52, 57
Diradical, 190
Dismutation, see Superoxide radical
Disulfide, 19, 180
  formation, 19
  with nucleophiles, 19
  oxidation, 19
  reaction with thiols, 19
  reduction, 20
3,3′-dityrosine, 78
DJ-1, 269–270, 272, 277–278

DMPO, 36
    as antioxidant, 389–394
DNA, see Deoxyribonucleic acid
DNA-mediated charge transport, 96
Dopamine, 154, 156, 240, 251–252, 255, 259
    and METH neurotoxicity, 261–262
    quinones, 251, 254
Double strand breaks (DSBs), 100
DPI, 126–127
Drug resistance
    in cancer, 219
    glutathione, 219
Dual oxidases, 347. See also DUOX, DUOX1, DUOX2
DUOX, 147–148, 150, 152, 155–159
DUOX1, 147–148, 150, 152, 155, 159
DUOX2, 147–148, 150, 152, 155, 157, 159

Ebselen, 382–384
    anti-inflammatory effect of, 382, 384
    BXT-51072, 385
    BXT-51077, 385
    clinical trial (ischemic stroke), 384
    ebselen diselenide, 383–385
    ebselen selenol, 383–384
    ebselen Se-oxide, 384
    glutathione peroxydase activity of, 382
    inhibiting activity of lipoxygenases, 384
    inhibition of iNOS by, 384
    inhibition of lipid peroxidation by, 384
    peroxynitrite scavenging by, 384
    reaction with thiols, 383–384
    reduction and scavenging properties of, 383
    substrate for mammalian thioredoxin reductase, 384–385
    synthesis of, 382
Edaravone (also radicut, MC-186), 385–388
    enhancement of eNOS expression, 385, 387
    inhibition of iNOS, 385, 387
    neuroprotective effect of, 387
    pharmacokinetics of, 385
    protection against amyotrophic lateral sclerosis (ALS), 387
    protection against brain infraction, 385, 387
    protection against noise-induced hearing loss, 388
    protection against retinal degeneration, 387
    reduction of myocardial injuries, 385–387
    scavenging properties of, 385–386
Eicosanoids, 241–242
Elastase, 347
Electron leakage, 28, 312
Electron paramagnetic resonance (EPR)
    imaging, 38
    oximetry, 38
    spectroscopy, 34, 126–130, 317, 385–386, 389–390, 397
Electron transfer, 1
Electron transport chain (ETC), see Mitochondrial electron transport chain (METC)
Electronegative atoms, 1

Electronegativity, 1
Electrophile response element (EpRE), 269. See also Antioxidant response element
Electrospray ionization (ESI) mass spectrometry, 81
EMPO, 390. See also Nitrones
Endonuclease III (NTH), 102
Endoperoxides, 52–54, 57, 61–62
Endoplasmic reticulum (ER), 29, 312
    stress, 313
Endothelial, 311, 315–316, 318–320
    cell(s), 311, 316, 318–319
    dysfunction, 320
Endothelial nitric oxide synthase (eNOS, NOS3), see Nitric oxide synthase
Endothelin, 314, 321
Energy transduction, 125
ENOS/NOS3, see Nitric oxide synthase
Epithelial lining fluid, 346, 349. See also ELF
Epoxide hydrolase, 313
Epoxy
    alcohols and peroxyls, 51–53, 57
    aldehydes, 61
    fatty acids, 59, 62
    isoprostanes, 57
Erythorose abasic site, 96
Excision repair cross complementing 1 (ERCC1), 101
Excitotoxicity, 241, 253
    source of ROS, 241
4-exo-cyclization, 52
5-exo-cyclization, 52, 55, 57

ΔF508 mutation, 345
$FADH^{\bullet-}$ or $FADH^{\bullet}$ semiquinone, 129
Fat soluble antioxidant vitamins, 347
Fe(IV)=O intermediate, see Iron-oxo intermediate
Febuxostat, 320. See also Inhibitors
2Fe-2S, 3Fe-4S, 4Fe-4S, see Iron-sulfur cluster
Fe-DETC, see Iron diethyldithiocarbamate
Fe-MGD, see Iron methyldithiocarbamate
Fenton-type reaction, 125, 183–184, 378, 382
Ferritin, 116, 347
FIH, see Asparaginyl hydroxylase factor inhibiting HIF
Flavin adenine dinucleotide (FAD), 129–130, 138–140, 147, 149, 152, 186
Flavin mononucleotide (FMN), 126–128
Flavocytochrome b, 138. See also Flavocytochrome b_{558}, $gp91^{phox}$, NOX2
Flavocytochrome b_{558}, 139. See also Flavocytochrome b, $gp91^{phox}$, NOX2
Fluorescence, 32
Fluoroquinolone, 181
FMN, see Flavin mononucleotide
$FMNH^{\bullet-}$ or $FMNH^{\bullet}$ semiquinone, 126
Forced vital capacity, 352
3-formyl phosphate, 96
Forskolin, 352
Fp subcomplex, 126–127
Fragmentation

beta scission, 51–52, 58–62
formation of hydroxyalkenals, 59–61
products formed by, 58–62, 64
Free radicals
in cystic fibrosis, 347
definition of, 2
in DNA damage, 93, 95
in I-R, 311
origin of word, 2
stability, 2
π-radicals, 2
σ-radicals, 2
Fumarate and nitrate reduction (FNR), 181–182, 185
Funneling, 96
Fur proteins, 183

G551D mutation, 352
G6PD, *see* Glucose-6-phosphate dehydrogenase
Gamma irradiation, 96
Gamma-glutamylcysteine ligase, 114
G-content as predictor of damage, 95–96
Genistein, 352
Gentamicin, 351
Girard's Reagent P, 83
Glial fibrillary acidic protein (GFAP), 237, 262, 279
Glucose-6-phosphate dehydrogenase (G6PD), 263–266
in cancer, 206
deficiency, 264–265
enzymology, 264
expression in brain, 267
genetics, 263–264
protective role in blood, 265–266
protective role in brain, 267
protective role in development, 266
protein structure, 264
regulation, 264
Glutamate, 241, 318
concentrations in the brain, 241
Glutamic semialdehyde, 79
Glutamine, 318
Glutamine synthase (GS), 78
γ-glutamylcysteine
ligase, 349
synthetase in cancer, 205
Glutaredoxins (Grxs), 74–75, 208
in cancer, 208
Glutathione (GSH)
adaptive response, 348
apoptosis, 216–219
in cancer, 205–206
in cystic fibrosis, 346–347, 352
drug resistance, 219–220
glutathione-SNO (S-nitrosoglutathione or GSNO), 208
in I-R, 313
in mitochondrial dysfunction, 125, 127, 130, 133–134
multidrug-resistance proteins (MRP1), 220
in neurodegeneration, 258, 267–268, 277
and NOX, 143
and Phase 2 proteins, 114–118
in PTM, 73–74
transport, 220
Glutathione peroxidase (GPx), 115, 313
in cancer, 206
in I-R, 313
in neurodegeneration, 262, 264, 267–268, 275, 278
Glutathione reductase (GR), 115
Glutathione S-transferase (GST), 114, 118
in cancer, 216, 219–220
in I-R, 313
Glutathionylation
in cancer, 208
p53, 219
in PTM, 74
Glycoprotein, 185
Glycosylases, 94, 98, 99
DNA, 209, 211, 213
Gp91^{phox}, 137–138, 145–149. *See also* Flavocytochrome b, NOX2
GSH, *see* Glutathione
GSH/GSSG ratio, 313. *See also* Redox buffer
GSSG, *see* Oxidized glutathione
GTPase, *see* Guanosine triphosphate hydrolase
Guanidinohydantoin, 95, 98, 102, 103
Guanine, 93–103
amino acids, specific of, 94
differentation of oxidation product, 94
DNA, specific of, 93–96
hyperoxidation of, 102
mechanism of oxidation, 94–95
oligonucleotides, specific of, 94–96
oxidation and disease, 95
proteins, specific of, 94
Guanosine triphosphate hydrolase (GTPase), 137–151

Haber-Weiss reaction, 125
Heart, 311–321. *See also* Cardiac
failure, 311, 318–320
transplantation, 312
Heat shock proteins, 267
Heme a, 132
Heme a_3, 132
Heme b, 129–132
Heme in signaling, 184–186
Heme oxygenases (HO-1, HO-2), 116
in cancer, 207
in neurodegeneration, 267, 275, 278, 280
Hemoglobin, 28, 316
Hexose monophosphate pathway, 264–265
High density lipoprotein (HDL), 330, 331, 334
High performance liquid chromatography (HPLC), 33
Histochemical detection of ROS, 38
Hock cleavage, 61
Holes, 96
Homovanillic acid (HVA), 252, 255
HNE, *see* 4-hydroxy-2-nonenal
Hp, hydrophobic subcomplex of complex I, 126
Huntington's disease (HD), 241, 277–278

Hydrogen abstraction
  in formation of lipid radicals, 49–50
  by NO_2, 63–64
  in nucleotides, 100–101
  position of abstraction, 49–50
  reaction rate, 49
Hydrogen peroxide (H_2O_2), 137, 142–144, 312–316
  anions, reaction with 10
  concentrations in mitochondrial matrix, 239
  from CYPs, 240
  in cystic fibrosis, 346–347, 350
  detoxification, 11, 264, 266–268
  hydroxylation by, 11
  iron ions, reaction with 10
  from MAO, 240
  from METH, 262
  in neurodegeneration, 237–238, 251–252
  quinones, reaction with 10
  role in PTM, 71, 78
  thiols, reaction with 11
  from xanthine oxidoreductase, 240–241
5-hydroperoxyeicosatetraenoic acid (5-HPETE), 242
Hydroperoxyl radical, reactions with
  catechols, 10
  PUFA, 9
Hydroquinone and quinones
  as products, 4
  as reactants, 5
Hydroxy and hydroperoxy fatty acids, 50–51
  isoHETEs, 51–52
  isoHODEs, 52
  lipoxygenase role in formation, 50
  nitrohydroxy fatty acids, 64
5-hydroxyeicosatetraenoic acid (5-HETE), 242
4-hydroxy-2-nonenal (4-HNE), 59–61, 80, 333–334
5-hydroxytryptophan, 79
Hydroxylamine, 35, 390–391, 394
Hydroxyl radical (HO•), 142–144,
  alcohols, reactions with, 12
  aromatic hydrocarbons, reactions with, 13
  ascorbate, reactions with, 12
  carbohydrates, reactions with, 12
  carbonyls, ketones, and aldehydes, reactions with, 12
  carboxylic acids, reactions with, 12
  in cystic fibrosis, 346
  in DNA damage, 94, 96, 100, 103
  ions, reactions with, 11
  from METH, 262
  in neurodegeneration, 237–238, 251, 268
  radicals, reactions with, 11
  against synthetic antioxidants, 378–397
  thiols, reactions with, 13
  unsaturated hydrocarbons, alkenes, reactions with, 13
Hypertension, 152–156, 159
Hypertonic saline, 351
Hypochlorous acid
  amino acids, reaction with, 20–21
  anions, reaction with, 20
  in cystic fibrosis, 346–347, 350
  cytochrome c, reaction with, 21
  lipids, reaction with, 22
  metal ions, reaction with, 20
  NADPH, reaction with, 22
  from NOX, 142–144
  nucleotides, reaction with, 22
  ROS, reaction with, 20
  thiols, reaction with, 21
Hypohalous acids, 350
Hypothiocyanate, 346–347
Hypoxanthine, 240–241, 313, 316, 318
Hypoxia, 190–195, 314, 317–318

IκB, 189
Ibuprofen, 249, 254
IKK, 189
IL-8, see Interleukin-8
IL-10, see Interleukin-10
Immunocytochemical detection of ROS, 38
Immuno-spin trapping, 34
Indomethacin, 248–249, 254
Inducible nitric oxide synthase (iNOS, NOS2),
  see Nitric oxide synthase
Inflammatory stimuli, 346
Inhibitor(s), 314, 317–318, 320–321
  allopurinol, 318, 320
  febuxostat, 320
  oxypurinol, 318, 320
  uloric, 320
  zyloprim, 320
Inosine, 316, 318
Inositol, 351
Interferon-gamma (IFN-γ), 241
Interleukin-
  1 (IL-1), 241, 245
  8 (IL-8), 347
  10 (IL-10), 346
Intermembrane space (IMS), 312
Intracellular organelles, 319, 321
Ion transport, 346
Iron, role in lipid peroxidation, 50–51, 58, 61–62
Iron diethyldithiocarbamate (Fe-DETC), 37
Iron methyldithiocarbamate (Fe-MGD), 37, 317
Iron-oxo intermediate (Fe(IV)=O), 192
Iron-sulfur cluster
  in cell signaling, 180
  in mitochondrial dysfunction, 126–131
Ischemia, 241, 279
Ischemia and reperfusion (I-R), 130, 133, 241, 311–321
Ischemic preconditioning (IPC), 321
Isofuran
  biomarker of diseases, 57
  mechanism of formation, 53, 57
  oxygen dependence of formation, 57
Isolevuglandins, 54, 56
Isoprostane
  A2- D2-, E2- F2-, G2-, H2-, J2-, 52–56
  cis conformation abundance, 54

formation pathway, 52–56
isomers, 54–55
thromboxane receptor, 54
use in monitoring oxidative stress, 54, 57
vasoconstrictors, 54–56
Isopyran, 53

Jab1, 189

Kelch-like erythroid cell-derived protein with CNC homolog (ECH)-associated protein 1 (Keap1), 269–272, 276, 279
genetics 270, 281
knock-out mouse models, 280
negative regulation of Nrf2, 272
α-ketoglutarate (α-KG), 190, 192
Kinetics
activation energies, 26
bimolecular (second-order reaction), 25
reaction coordinates, 26
transition state, 26
unimolecular (first order reaction), 25
Kynurenine, 79

L-3,4-dihydroxyphenylalanine (L-DOPA), 78
Lactoperoxidase, 347
L-arginine, 28, 315–316, 318
Lazaroids (*also* methylprednisolone, U74006F), 388–389
3-methyl-1-phenyl-2-pyrazoline-5-one, 386
anti-inflammatory effect of, 388–389
clinical trial (ischemic stroke and traumatic brain injury), 389
endothelium protection by, 388
inhibition of lipid peroxidation by, 388
methylprednisolone, 388–389,
myocardial protection of, 388
neuroprotection of, 388
protection against amyotrophic lateral sclerosis (ALS), 389
protection against sepsis, 389
protection against traumatic brain injury by, 389
scavenging properties of, 388
tirilazad mesylate or U74006F, 388–389
U-74500A, 388–389
U78517F, 388
L-citrulline, 315–316
L-deprenyl (selegiline), 240, 279
L-dihydroxyphenylalanine (L-DOPA), 251–252, 255
Lesions, 93–103
alkaline treatment, 94
clustered, 101
cross-links, 95–96, 100–103
determining structure, 93
novel types, 101–103
tandem, 101
Leukocytes, 319, 346–347
Ligation-mediated polymerase chain reaction (LM-PCR), 94

Lipid
nitration, *see* Nitrated lipids
peroxides, 349
radicals, 5
Lipid peroxidation (lipoperoxidation, LPO), 49–62
in atherosclerosis, 329–334
in cancer, 209
in CF, 349
in neurodegenerative disorders, 363
Lipoic acid (LA), 377, 383
Lipopolysaccharide (LPS), 64, 240–241, 245, 250
Lipoxygenases (LOXs, 5-LOX), 50–51, 242
Low density lipoprotein (LDL), 330–333
LPBNAH, 393–394, 397. See also Nitrones
LPBNH15, 393–394, 397. See also Nitrones
LPBNSH, 393–394. See also Nitrones
Lung
disease of, 345
epithelium, 346–348
host defense, 346
infection, 346
inflammation, 347
oxidative stress, 347–350
Lymphocytes, 352
Lysine, 95, 103

Macrophage, 137, 143–158, 314, 319, 347, 350
Malabsorption of aerosolized therapies, 352
Malaria, 265–266
Malnourishment, 352
Malondialdehyde (MDA), 58, 61
in atherosclerosis, 334
formation of, 12
Manganese superoxide dismutase (MnSOD), 239. See also Superoxide dismutases
Mass spectrometry and PTM, 81
Matrix-assisted laser desorption ionization (MALDI), 81
Matrix metalloprotease (MMP), 333
Metabolite(s), 316–317, 320
Metalloporhyrins, 380–382
anti-inflammatory effect of, 381
antinociceptive properties, 381
catalase activity of, 380
inhibition of lipid peroxidation by, 380–381
lipophilicity of, 381
peroxynitrite scavenging by, 380
pharmacokinetic of, 381
scavenging properties of, 379–380
SOD activity of, 380
Metalloproteins, 180, 190, 194
Metazoans, 179, 181, 187, 189–190
Methamphetamine (METH), 252, 255. See also Amphetamine
activation of Nrf2, 279
adverse effects, 261
distribution, 257
effects of use, 260–261
pharmacokinetics, 256
pharmacology, 259

Methionine sulfoxide (MetSO), 76–77, 208
   in cancer, 208
   reductases (MSRs), 76–77, 208
2-methoxyestradiol (2-ME), 194
3,4–methylenedioxyamphetamine (MDA), 252–253, 255, 258, 262. *See also* Amphetamine
3,4–methylenedioxymethamphetamine (MDMA), 252, 255–256, 258. *See also* Amphetamine
   pharmacokinetics, 256–257
1-methyl-4-phenylpyridinium (MPP$^+$), 129
1-methyl-4-phenyl-1,2,3,6-tetrahydropyridine (MPTP), 129
   G6PD protection, 266
   and microglia, 241
   neurotoxicity, 240–241, 254, 279
   oxidation by MAO, 240
   SOD protection, 266
Michael addition, 62
Microglia, 143, 156, 239–241, 250, 253–254, 262–263, 278
Mimetics of enzymes, 378–379, 382
Mitochondria
   in cell signaling, 185, 188
   in CF, 347, 349
   dysfunction, 123–124
   in I-R, 311–321
   in neurodegeneration, 238
   mutations, 205
Mitochondrial
   disease, 129
   DNA oxidation, 349
   pathway, 217–219
   stress, 125
Mitochondrial electron transport chain (METC), 28
   in cancer, 205
   in cell signaling, 185, 193
   in I-R, 311–312, 314, 316, 318, 321
   in neurodegeneration, 238
Mitochondrial permeability transition pore (MPTP), 319
Mitogen-activated protein kinase (MAPK), 140, 151
Molybdenum enzymes, 317. *See also* Nitrate/Nitrite reductase
Monoamine oxidase, 240
   isoforms, 240
MPP$^+$, *see* 1-methyl-4-phenylpyridinium
MPTP, *see* 1-methyl-4-phenyl-1,2,3,6-tetrahydropyridine
Mucociliary clearance, 346–347
Mucus, 347–348, 350–351, 353
Multidrug resistance proteins (MRP, MRP1), 220, 345, 348, 352
Multiple sclerosis (MS), 238, 240, 253–254, 277–279
Myeloperoxidases (MPO), 10
   in atherosclerosis, 329, 333–334
   in cancer, 205
   in CF, 346–348, 350
   from NOX, 143–144
Myocardial
   apoptosis, 319
   cell death, 319
   dysfunction, 311
   infarction, 130, 311, 318–321
   injury, 317, 319–321
   ischemia, 312, 314, 317, 319, 321
   preconditioning, 321
   reperfusion, 319
   salvage, 311
Myocardium, 311–312, 314, 318–320
Myocyte(s), 314, 316–320

N-acetylcysteine, 351–353. *See also* NAC
NAD(P)H:quinone oxidoreductase 1 (NQO1), 116, 118, 251–252, 267, 270, 272
   in brain, 267
NADPH, *see* Nicotinamide adenine dinucleotide phosphate hydrogen
Naproxen, 249, 253–254
Nasal epithelium, 352
NDH, NADH dehydrogenase or Fp subcomplex, 126
Necrosis, 312, 319–320. *See also* Cell death
Nei endonuclease VIII-like 1–3 (NEIL1–3), 98–99, 102
Neocarzinostatin, 100
N-ethylmaleimide (NEM), 83
Neuronal nitric oxide synthase (nNOS, NOS1), 240, 263. *See also* Nitric oxide synthase
Neurotoxicity, 237
   and DNA repair, 268–269
   and METH, 256–259, 261–263
   and Nrf2, 277–279
   and PHS, 250–253
   reactive intermediate-mediated, 238–239
Neurotransmitters
   as substrates for, PHS 251
Neutrophil cytosolic factor
   1 (NCF1), 139. *See also* p47^{phox}
   2 (NCF2), 139. *See also* p67^{phox}
Neutrophils, 137–148, 157, 314, 319, 346–347
NF-κB, *see* Nuclear factor kappa B
N-formylkynurenine, 79
Nicotinamide adenine dinucleotide phosphate hydrogen (NADPH), 239
   and CYPs, 240
   enzymes dependent on, 252, 264–265, 267
   and NOS, 240
Nicotinamide adenine dinucleotide (NAD+), 239–240, 313–317
Nicotinamide adenine dinucleotide phosphate hydrogen (NADPH) cytochrome P450 reductase, 251–252, 259, 266
Nicotinamide adenine dinucleotide phosphate hydrogen (NADPH) oxidase (NOX), 26, 143, 152–153, 159
   in atherosclerosis, 329
   in cancer, 203–204
   in CF, 346–347
   in I-R, 314
   isoforms, 239
   in microglia, 241
   in neurodegeneration, 239, 265
Nitrate, 316, 317, 320, 347.
Nitrate/nitrite reductase(s), 316–318
   AO, 316, 318
   cyt c, 318. *See also* Cytochrome c

deoxymyoglobin, 316
hemoglobin, 316
mitochondrial enzymes, 316
molybdenum enzymes, 317
XOR, 316, 318
Nitrated lipids, 63–64
bioactivities, 64
detection in vivo, 64
mechanisms of formation, 63
Nitric oxide (•NO)
and antioxidants, 384–385
in brain, 238, 240–241
in cystic fibrosis, 347, 349–350
detection of by EPR, 36–37
in I-R, 314–321
in lipid nitration, 63
metal ions, reactions with, 15–16
in NOX, 142, 144, 152–156
oxygen, reaction with, 15
radicals, reaction with, 15
Nitric oxide synthase(s) (NOS), 28, 142, 156
in brain, 240
in cancer, 205
in cystic fibrosis, 349
eNOS/NOS3, 152–153, 315–316, 318
in I-R, 313–316, 318
iNOS in neurodegeneration, 240
iNOS/NOS2, 315
nNOS/NOS1, 315
and PTM, 73
uncoupling, 28, 318
Nitrite, 314, 316–318, 320, 347
Nitroalkanes and nitroalkenes, 63
Nitrogen dioxide (NO_2)
double bonds, reaction with, 16
in lipid nitration, 63
PUFA, reaction with, 16
radicals, reaction with, 16
thiols, reaction with, 17
Nitrones
anti-aging effects of, 394
anti-apoptotic effect of, 397
anti-inflammatory effect of, 396
antinociceptive properties of, 397
cardioprotective effects of, 394
immune spin trapping, use in, 34
inhibition of lipid peroxidation by, 396
interaction with mitochondria, 397
mode of action of, 396
neuroprotective effects of, 394–396
pharmacokinetics of, 391–392, 395–396
protection against endotoxic shock. 392
protection against fulminant hepatitis. 393
protection against noise-induced hearing loss. 392
protection against retinal degeneration. 392–393
protection against Type 1 diabetes, 392
protection against xenobiotic-induced damages. 392, 397
SAINT clinical trials, 395–396
scavenging properties of, 389–390, 396

spin trapping, use in, 34–36
synthesis of, 391
toxicity of, 391, 395
Nitronyl nitroxides, 36
Nitrotryptophan, 79
3-nitrotyrosine
in I-R, 318
in neurodegenerative disorders, 362
in post-translational protein modification, 78
N-monomethyl-L-arginine (L-NMMA), 317
N,N-diethyldithiocarbamate, 118
N-nitro-L-arginine methyl ester (L-NAME), 317
NO, see Nitric oxide
Nonheme iron, 184
Nonphagocyte NADPH Oxidase, 137, 139, 146–159
Nonsteroidal anti-inflammatory drugs (NSAIDs), 248–249, 253
Normoxia, 190–192
NOS, see Nitric oxide synthase
NOX, see Nicotinamide adenine dinucleotide phosphate hydrogen (NADPH) oxidase
NOX1, 147–158
NOX2, 137–158, 239. See also Flavocytochrome b; flavocytochrome b_{558}; gp91^{phox}
NOX3, 147–152
NOX4, 147–159
NOX5, 147–158
NOXA1, 139, 147–153, 157
NOXO1, 139, 147–151, 157
Nrf(s), see p45-nuclear factor erythroid-derived 2 related factors
Nuclear factor kappa B (Nf-κB)
in cell signaling, 194
in CF, 346–347
in neurodegeneration, 243–245, 253
in PTM, 85
Nucleic acid biomarkers in neurodegenerative disorders, 367
Nucleoside-5′-aldehyde, 96, 101
Nucleotide excision repair, 101–103
NXY-059, 390, 395–396. See also Nitrones

Octatrienyl moiety and prostane ring formation, 52
Octopamine, 240
OGG1, see Oxoguanine glycosylase 1
OhrR, 74, 184
Oligonucleotide oxidation, 93–103
2′-deoxyadenosine, 94
2′-deoxycytidine, 94
2′-deoxyguanosine, 93–94
2′-deoxythymidine, 94
4′-oxidation, 100
potentials, 93–94
Opioid agonists, 321
Organelles, 312–313, 319, 321
Osmotic
agents, 351
gradient, 351
Oxaluric acid, 102
Oxidation, 1

Oxidation potentials, *see* Redox potentials
Oxidative and nitrosative stress biomarkers, 360
Oxidative DNA damage, 209–213, 219
    apurinic/apyrimidinic (abasic; AP), 209–213
    base excision repair pathway (BER), 210–211
    2,6-diamino-4-hydroxy-5-formamidopyrimidine(FapydG), 209
    5,6-dihydroxy-5,6-dihydrothymine (thymine glycol; Tg), 209, 212–213
    DNA-glycosylases, 209, 211, 213
    double-stranded DNA breaks (DSB), 209, 211–212, 219
    measurement, 213
    mismatch repair (MMR), 212
    nonhomologous end joining (NHEJ), 212
    nucleotide excision repair (NER), 211
    oxidatively induced bistranded clustered DNA lesions (OCDLs), 211–213
    8-oxo-7,8-dihydro-2′-deoxyguanosine (8-oxodG), 209, 212–213
    single-stranded DNA break s (SSB), 209–212
    types, 209–211
    X-ray repair cross-complementing protein 1 (XRCC), 211–212
Oxidative post-translational protein modifications, 208
    p53, 219
    peroxiredoxins, 207
    protein carbonylation, 208
    protein glutathionylation, 208
    protein nitration, 208
    protein nitros(yl)ation, 208
    superoxide dismutases, 206–207, 216
    thioredoxin reducase, 207
Oxidative stress
    activation of Nrf2, 276
    in atherosclerosis, 330, 333–334
    brain susceptibility, 238
    from cell signaling, 179–195
    detection, 269
    and DNA damage, 94, 96, 98
    in I-R, 311–321
Oxidized glutathione (GSSG)
    in CF, 349
    in I-R, 313
    in mitochondrial dysfunction, 125, 127–128, 133–134
Oxidized low density lipoprotein (Ox-LDL), 330, 332
Oximetry, *see* Paramagnetic resonance oximetry
Oxoguanine glycosylase 1 (Ogg1), 238–239, 251, 268–269, 275
8′-oxo-2′-deoxyguanosine (8-oxodG), *see* 8-oxo-7,8-dihydro-2′-deoxyguanosine (8O-dG)
8-oxo-7,8-dihydro-2′-deoxyguanosine (8O-dG, 8-oxodG), 93, 95, 96–99, 251, 268, 349
    and base excision repair, 98–99
    chemistry of, 98
    and disease states, 99
    formation, 98
    history of, 97–98
    mechanism, 95
    mutagenic properties, 98

    oxidation potential of, 93–94
    structure, 97
2-oxohistidine, 79
Oxygen (triplet oxygen, dioxygen), 4
    in cell signaling, 190
    definition of, 4
    in I-R, 311–316, 318–319, 321
    photosensitization of, 13–14
    reactivity of with hydroquinones and semi-quinones, 4–5
Oxypurinol, 318, 320. *See also* inhibitors
OxyR, 182–183

P21ras
$P22^{phox}$, 137–156, 159
$P40^{phox}$, 137–156
P45-nuclear factor erythroid-derived 2 (NF-E2), 269
P45-nuclear factor erythroid-derived 2 related factor 1 (Nrf1), 269
    genetics, 270
P45-nuclear factor erythroid-derived 2 related factor 2 (Nrf2), 269–281
    activators, 276–277
    expression in brain, 277
    genetics, 270
    interactions, 272
    knockout mouse models, 280
    localization, 271
    in neurodegenerative diseases, 277–278
    and Phase 2 proteins, 116–117
    polymorphisms, 280–281
    posttranslational regulation, 271
    proteasomal degradation of, 272
    protein structure, 271
    regulators of, 272
P45-nuclear factor erythroid-derived 2 related factor 3 (Nrf3), 269
    genetics, 270
P45-nuclear factor erythroid-derived 2 related factors (Nrfs), 269–281
$P47^{phox}$, 137, 139–142, 145–151, 153–156, 158
P50, 188–189
P52, 188–189
P53, 85, 188
P65, 188–189
$P67^{phox}$, 137, 139–142, 145, 147–149, 151, 154–156
PAF acetylhydrolase, *see* Platelet-activating factor acetylhydrolase
Pancreas, 348,
Paraoxonase1 (PON1), 332
Parkinson disease (autosomal recessive, early onset) 7 (PARK7), 269. *See also* DJ-1
Parkinson's disease, 129, 194, 238, 240, 253, 277–278, 360
Pathogens, 346, 350
PBN, 36, 389–397. *See also* Nitrones
Pentadiene and pentadienyl radical
    addition of molecular oxygen to, 50
    multiple units give rise to additional products, 51
    position in polyunsaturated fatty acids, 49
    rate of hydrogen abstraction from, 49

Pentose phosphate pathway, 264–265
Peroxidase, 315, 318, 377–378
  mimetics, 380
Peroxidase activity, 315, 318
Peroxides
  diepoxides, 52, 57
  endoperoxides, 52–54, 57, 61–62
  monocyclic peroxides, 52–53
  serial cyclic peroxides, 52–53, 57
Peroxiredoxins (Prxs)
  in brain, 267
  in cancer, 207
  2-cys peroxiredoxins (2-cys Prxs), 186–187
  in I-R, 313
  in PTM, 73, 75, 84
Peroxisome proliferator-activated receptors (PPARs), 52, 58–59, 64, 242, 245, 253
Peroxisomes, 313
Peroxyl radical, 49–52, 54, 57–58, 64, 250
Peroxynitrite (OONO$^-$)
  and NOX, 142, 144–145, 156
  in brain, 240
  in cystic fibrosis, 346–347, 350
  decomposition of, 17
  formation of, 17
  in I-R, 314, 316, 318
  in lipid nitration, 63
  in PTM, 78
  reaction with CO_2, 17
  reaction with inorganic radicals, 17
  reaction with thiols, 17
PerR, 79, 182–184
Phagocyte, 137–139, 141–149, 153, 155–159
Phagosome, 138–141, 143–145
Phase 2 proteins, 113. See also Antioxidant enzymes
PHD, see Prolyl hydroxylase domain
Phosphatidylinositol
  3,4-bisphosphate, 141
  4,5-bisphosphate, 150
  3-phosphate, 141
Phosphodiesterase inhibitors, 321
Phosphodiesterases, 94
3'-phosphoglycolate aldehyde, 96
Phospholipase A2 (PLA2), 239–240, 242
Phospholipids
  carriers of polyunsaturated fatty acids, 50–51
  oxidative fragmentation, 58
  oxidative modification of, 57–61
  oxidized to PAF receptor agonists, 58–59
  reactions with aldehydes, 56, 62
2-phosphoryl-1,4-dioxobutane, 101
Photooxidants, 100
Photosensitization, 14
Phox
  and Bem (PB1) domain, 139–141, 151
  and Cdc (PC) domain, 139, 141
  homology (PX) domain, 139–141, 150
PHS-1, 242. See also Prostaglandin H synthase
PHS-2, 242. See also Prostaglandin H synthase

Piericidin A, 126–127
Pirin, 189
PK_a, 180
Platelet-activating factor acetylhydrolase (PAF-acetyl hydrolase), 59, 333
POBN, 390. See also Nitrones
Polyaromatic hydrocarbons, 4
Polymerases, 94, 99, 101
Polymorphonuclear (PMN), 319
Polypeptide, 126–129
Polyphenols, 8
Polyunsaturated fatty acid (PUFA), 329–332
  hydrogen abstraction from, 49
  mechanisms of peroxidation, 50
  peroxidation product dependence on, 50
  prediction of fragmentation products, 58
  species to give rise to prostane rings, 52
Postischemic, 311, 316–321
  blood flow, 320
  heart(s), 311, 316–321
  oxidants, 316
  reperfusion, 315–318, 321
Posttranslational protein modification, 71–92, 181. See also Oxidative post-translational protein modifications
PPARs, see Peroxisome proliferator-activated receptors
PPN, 390. See also Nitrones
Preconditioning, 321
Primer extension, 94
Pro-inflammatory cytokines, 346–347
Proliferation, 194–195
Prolyl hydroxylase domain (PHD), 190–192
Prostaglandin H synthase(s) (PHSs), 241–255
  bioactivation of amphetamines, 252
  cellular localization, 249
  cyclooxygenase activity, 247–248
  enzymology, 247
  expression in the brain, 249
  genetics of, 243
  hydroperoxidase activity, 247–248
  inactivation, 248
  inhibition, 248–249
  kinetics, 248
  in neurodegenerative diseases, 253–255
  post-transcriptional regulation, 245
  protein structure, 246
  substrates, 248, 251
  transcriptional regulation, 244–245
Prostaglandins (PGs), 241
  synthase, 242, 253
  synthesis, 242–243
  transport, 243
Prostanoid receptors, 243–244
Protein carbonylation, 80
  in cancer, 208
  in neurodegenerative disorders, 361
Protein carbonyls, 352
Protein disulfide isomerase (PDI), 73
Protein phosphorylation, 187

Protein tyrosine kinase, 78
  A (PKA), 85, 133, 352
  C (PKC), 140, 151, 155
Protein tyrosine phosphatase (PTPs), 73–74, 84, 187
Proto-oncogene cRel, 188
PrxV, 313
Pseudomonas aeruginosa, 346
PTM, see Posttranslational protein modification
2-pyrrolidone, 79

Q-cycle, 126, 130, 131, 132
Q_d, ubiquinone-binding site of complex II, 129
Q_i, ubiquinone-binding site of complex III, 131
Q_o, ubiquinone-binding site of complex III, 131–132
Q_p, ubiquinone-binding site of complex II, 129–130
Quinones, 252, 258, 267, 276
  from amphetamine metabolites, 258–259

Rac, 137, 139, 140–142, 147, 149–151, 157
Rac1, 137, 141, 149, 151, 157. See also GTPase, Rac, Rac2
Rac2, 137, 141, 142, 147, 149. See also GTPase, Rac, Rac1
Radicals, see Free radicals
Radical generation
  chemical, 30
  electrochemical, 30
  photochemical, 30
  photolysis, 29
  sonochemical, 30
Rap1A, 137, 142. See also GTPase
Rate constants, 27
Reactive nitrogen species, 14
  in CF, 346–347, 349
  in lipid nitration, 63
  in PTM, 71, 85
Reactive oxygen species (ROS), 4–14, 112, 114, 119, 237–238
  in cell signaling, 179
  in CF, 346–347
  in I-R, 311–314, 316–319, 321
  in the mitochondria, 123–125, 129–133
  from NOX, 137, 142–159
  in PTM, 71, 85
  role in DNA damage, 93–95, 101, 103
  sources in the brain, 238–263
Receptor protein tyrosine phosphatase (RPTP-α), 73–74
Receptor tyrosine kinase (RTK), 187
Redox
  buffer, 313
  chemistry, 1–2
  cycling, 31, 251–252, 258
  domains, 127, 128
  environment, 180, 186, 189, 195
  equilibrium, 186
  hemeostasis, 75
  modifications, 128
  and oxygen sensors, 179
  potentials, 24
  signaling, 71, 84–85
  status, 349, 352
Reduction, 1

Reduction potentials, see Redox potentials
RelA, 188–189
RelB, 188
Renox, 149, 154. See also NOX4
Reperfusion, 311–321
Retinoic acid, 250
Rieske iron-sulfur protein (RISP), 130–131
RIRR, 316
RNS, see Reactive nitrogen species
Rofecoxib (Vioxx), 249, 253, 254
ROS, see Reactive oxygen species
ROS-induced ROS-release (RIRR), 316
Rotenone, 126–129
Ryanodine receptors (RyRs), 188

Salen, 378–380
  AEOL 10150, 379–381
  amphiphilic antioxidants, 380, 393–394, 397
  catalase activity of, 380
  EUK-134, 379–380
  EUK-189, 379–380
  EUK-8, 379–380
  lipophilicity of, 380
  scavenging properties of, 379–380
  SOD activity of, 380
  synthesis of, 379
Salicylamine, 56
SCN, see Thiocyanate
Selenium (Se), 330, 347
S-glutathionylation, 128
SIN-1, 31
Single strand break (SSB), 99–101
Singlet oxygen ($^1O_2^*$), 13
  detection of, 35
  in DNA damage, 95
  ISC, 14
  and NOX, 144
  sensitizer, 14
  vibrational relaxation, 14
S–nitrosocysteine, 73
S–nitrosoglutathione, 73
  in CF, 349–350
S–nitrosohomocysteine, 73
S–nitrosothiols, 73–74
S-nitrosylation of protein, 82
SOD1, 253, 266. See also Superoxide dismutase
SOD2, 239, 266. See also Superoxide dismutase
SOD3, 266. See also Superoxide dismutase
SOD mimetics, see Superoxide dismutase mimetics
Sodium conductance, 351
Sorbitol, 351
SoxR, 182, 185
SoxS, 181
S-PBN, 390, 395. See also Nitrones
Spin trapping, 36, 317, 389–390, 396
  ex-vivo, 38
Spin traps, see Nitrones
Spiroiminodihydantoin, 95, 98, 102–103
Sputum, 350, 353

SQ, semiqinone, 5, 126–127
SQ$_{Nf}$, 127
SQ$_{Ns}$, 127
SQR, succinate-ubiquinone reductase or complex II, 129–130
Src homology 3 (SH3) domain, 139–141, 150–151
Stop codons, 351
Stroke, 379, 384–385, 387, 389, 394–396,
Submitochondrial particles, 123–124
Succinate dehydrogenase (SDH), 129
Sulfasalazine, 118
Sulfenic acid
    in PTM, 72–74
    in signaling, 180–182, 186, 189
Sulfhydryl oxidases, 312
Sulfhydryl radical, see Thiyl radical
Sulfiredoxin (Srx), 73
Sulfonamides, 73–74
Sulfonic acid, 73
Sulforaphane, 276–277, 279
Sulfoxide, 76
Superoxide anion ($O_2^{\cdot-}$)
    addition reaction, 7
    alkyl halides, reaction with 7
    antioxidants, reaction with 377–382, 388–390, 394, 397
    concentrations in mitochondrial matrix, 239
    from CYPs, 240
    in cystic fibrosis, 346–347, 349–350
    dismutation reaction, 5–6, 377–382
    fullerenes, reactions with, 6
    generation of, 137, 142–144
    in I-R, 312–320
    iron-sulfur [Fe-S] cluster, reaction with, 9
    metal ions, reactions with, 7
    mitochondria, 125–126, 129, 131–133
    from NADPH oxidases, 239
    in neurodegeneration, 238
    nitroxides, reactions with, 7
    from NOS, 240
    nucleophilic substitution, 7
    phenols, reactions with, 8
    proton-radical transfer, 7
    role in DNA damage, 95, 98, 103
    thiols, reactions with, 8
    tyrosyl radical, reactions with, 7
    from xanthine oxidoreductase, 240–241
Superoxide dismutase mimetics
    anti-inflammatory effect of, 382
    cardiac reperfusion injury, 382
    fullerene derivatives, 6
    metal complexes, 7
    Mn([15]aneN3)Cl2, 382
    Mn(III)TMPyP, 379–381
    MnTE-2-PyP, 379–381
    MnTM-2-PyP, 379–380
    MnTnHex-2-PyP, 379
    nitroxides, 7
    scavenging properties of, 382
    SOD activity of, 382
    TAA-1/Fe, 382
Superoxide dismutases (SOD), 114, 238, 253, 266, 275, 377–378
    in cell signaling, 180
    in CF, 346
    copper-zinc superoxide dismutase (CuZnSOD, SOD1) in cancer, 207
    in I-R, 313
    manganese superoxide dismutase (MnSOD, SOD2) in cáncer, 206–207
    mimetics, 380, 386–387
    NOX, 142–144, 152–159
    role in DNA damage, 94–95
Survival, 181–182, 187–190, 194
Sweat duct cells, 346

Taurine, 318
Tetrahydrobiopterin (BH$_4$), 28, 315–318, 320
Tetratricopeptide repeat (TPR) motif, 139–140, 149, 151
Thenoyltrifluoroacetone (TTFA), 129
Thermodynamics
    equilibrium, 23
    free energy of reaction (DG), 22–24
    half-cell reactions, 24
    potential energy, 22
    redox potentials, 24
Thiobarbituric acid reactive substances, 34, 61, 278
Thiocyanate, 346–347
    transport, 348, 350
Thiols, 8
    alcohols and ethers, reaction with, 18
    alkenes, reaction with, 18
    ascorbate, reaction with, 19
    carbohydrates, reaction with, 19
    in cell signaling, 180, 182, 186, 188
    in CF, 346–347, 352
    GS$^-$, reaction with, 19
    NO, reaction with, 19
    O$_2$, reaction with, 19
    peptides, reaction with, 18
    in PTM, 71
    PUFA, reaction with, 18
Thioredoxin (Trx)
    in brain, 267
    in cancer, 207
    in PTM, 73–75
Thrombin, 241
Thromboxanes (TXs), 241
Thyroid oxidase (ThOX), 150. See also DUOX, DUOX1, DUOX2
Tip60, 189
TNF-α, see Tumor necrosis factor factor-alpha
TNF-related apoptosis-inducing ligand (TRAIL), 217
α-tocopherol, see Vitamin E
Transcription, 179–194
Transcriptional factors, 346
Transition metals, 180, 190, 238

Transmembrane regulator in cystic fibrosis, 345–346
    in oxidative stress, 348–349
    in pharmacological intervention, 351–352
    role of in the lung, 348
Transplantation, 312, 321
Trityl radical, 2, 35
Tryptophan, 79
Tryptophan hydroxylase, 79
TTFA, *see* Thenoyltrifluoroacetone
Tumor necrosis factor factor-alpha (TNF-α), 189, 193, 241, 347
Tumors, 194
Tyramine, 240
Tyrosine, 98, 102–103
Tyrosine nitration, 128, 130
Tyrosyl radicals, 250

Ubiquinone-binding domain, 126–127
UDP-glucuronosyltransferase, 116
Uloric, 320
Unfolded protein response (UPR), 313
Urate, 313
Uric acid, 347
UV absorption of oxygenated lipids
  conjugated dienes, 50
  isoleukotrienes and leukotrienes, 51
UV-Vis spectrophotometry, 33

Vascular endothelial growth factor (VEGF), 194
Vascular smooth muscle cells (VSMCs), 186, 194
Vitamin C, *see* Ascorbic acid
Vitamin D, 347
Vitamin E, 268, 347
Vitiligo, 281
Voltage-dependent anion channel (VDAC), 185

Warburg effect, 194

Xanthine, 240, 313–318
Xanthine dehydrogenase (XDH), 240, 313, 316, 318, 320
Xanthine oxidase, 240
Xanthine oxidasereductase (*also* xanthine oxidase, XAO), 27, 240–241, 313–320, 329
  inhibitor, 318, 320
Xeroderma pigmentosum, complementation group
  C (XPC), 101
  F (XPF), 101
  G (XPG), 101

Zinc, 330
Zyloprim, 320. *See also* Inhibitors